## IMPORTANT

1. TEXTBOOKS MUST BE RETURNED BY DESIGNATED SEMESTER/TERM DEADLINES OR BE SUBJECT TO A $ PER BOOK LATE FINE.

2. LATE BOOKS/FINES ONLY ACCEPTED FOR 5 SPECIFIED DAYS FOLLOWING ESTABLISHED DEADLINES.

3. YOU WILL BE REQUIRED TO PURCHASE THIS TEXTBOOK IF YOU HAVE WRITTEN IN, UNDERLINED, HIGHLIGHTED OR IF THERE ARE ANY VISIBLE SIGNS OF WATER DAMAGE OR UNUSUAL WEAR.

CHECK OUR WEBSITE AT:
www.eiu.edu/~textbks/
NAME

PLEASE RETURN THIS TEXTBOOK TO:
TEXTBOOK RENTAL SERVICE
600 LINCOLN AVENUE
EASTERN ILLINOIS UNIVERSITY
CHARLESTON, IL 61920

# Esau's Plant Anatomy

# ESAU'S PLANT ANATOMY

## Meristems, Cells, and Tissues of the Plant Body: Their Structure, Function, and Development

### Third Edition

RAY F. EVERT

Katherine Esau Professor of Botany and Plant Pathology, Emeritus
University of Wisconsin, Madison

With the assistance of
Susan E. Eichhorn
University of Wisconsin, Madison

WILEY-
INTERSCIENCE

A John Wiley & Sons, Inc., Publication

*Library of Congress Cataloging-in-Publication Data:*

Evert, Ray Franklin.
   Esau's Plant anatomy : meristems, cells, and tissues of the plant body : their structure, function, and development /
Ray F. Evert.—3rd ed.
      p.   cm.
      Rev. ed. of: Plant anatomy / Katherine Esau. 2nd. ed. 1965.
   ISBN-13: 978-0-471-73843-5 (cloth)
   ISBN-10: 0-471-73843-3 (cloth)
   1. Plant anatomy.   2. Plant morphology.   I. Esau, Katherine, 1898- Plant anatomy.   II. Title.
   QK671.E94 2007
   571.3'2—dc22

                                                                                    2006022118

Printed in the United States of America.

10  9  8  7  6  5  4  3

# Dedicated to the late Katherine Esau, mentor and close friend

"In recognition of her distinguished service to the American community of plant biologists, and for the excellence of her pioneering research, both basic and applied, on plant structure and development, which has spanned more than six decades; for her superlative performance as an educator, in the classroom and through her books; for the encouragement and inspiration she has given a legion of young, aspiring plant biologists; for providing a special role model for women in science."

*Citation, National Medal of Science, 1989*

Katherine Esau

# Contents

# Preface

It has been over 40 years since the second edition of Esau's *Plant Anatomy* was completed. The enormous expansion of biological knowledge that has taken place during this period is unprecedented. In 1965, electron microscopy was just beginning to have an impact on plant research at the cellular level. Since then, new approaches and techniques, particularly those used in molecular-genetic research, have resulted in emphasis and direction toward the molecular realm of life. Old concepts and principles are being challenged at virtually every level, often, however, without a clear understanding of the bases upon which those concepts and principles were established.

A biologist, regardless of his or her line of specialization, cannot afford to lose sight of the whole organism, if his or her goal is an understanding of the organic world. Knowledge of the grosser aspects of structure is basic for effective research and teaching at every level of specialization. The ever-increasing trend toward a reduction of emphasis on factual information in contemporary teaching and the apparent diminution of plant anatomy and plant morphology courses at many colleges and universities make a readily accessible source of basic information on plant structure more important than ever. One consequence of these phenomena is a less precise use of terminology and an inappropriate adoption of animal terms for plant structures.

Research in plant structure has benefited greatly from the new approaches and techniques now available. Many plant anatomists are participating effectively in the interdisciplinary search for integrated concepts of growth and morphology. At the same time comparative plant anatomists continue to create new concepts on the relationships and evolution of plants and plant tissues with the aid of molecular data and cladistic analyses. The integration of ecological and systematic plant anatomy—ecophyletic anatomy—is bringing about a clearer understanding of the driving forces behind evolutionary diversifications of wood and of leaf attributes.

A thorough knowledge of the structure and development of cells and tissues is essential for a realistic interpretation of plant function, whether the function concerned is photosynthesis, the movement of water, the transport of food, or the absorption of water and minerals by roots. A full understanding of the effects of pathogenetic organisms on the plant body can only be achieved if one knows the normal structure of the plant concerned. Such horticultural practices as grafting, pruning, vegetative propagation, and the associated phenomena of callus formation, wound healing, regeneration, and development of adventitious roots and buds are more meaningful if the structural features underlying these phenomena are properly understood.

A common belief among students and many researchers alike is that we know virtually all there is to know about the anatomy of plants. Nothing could be further from the truth. Although the study of plant anatomy dates back to the last part of the 1600s, most of our knowledge of plant structure is based on temperate, often agronomic, plants. The structural features of plants growing in subtropical and tropical environments are frequently characterized as exceptions or anomalies rather than as adaptations to different environments. With the great diversity of plant species in the tropics, there is a wealth of information to be discovered on the structure and development of such plants. In addition, as noted by Dr. Esau in the preface of the first edition of *Anatomy of Seed Plants* (John Wiley & Sons, 1960) " . . . plant anatomy is interesting for its own sake. It is a gratifying experience to follow the ontogenetic and evolutionary development of structural features and gain the realization of the high degree of complexity and the remarkable orderliness in the organization of the plant."

A major goal of this book is to provide a firm foundation in the meristems, cells, and tissues of the plant body, while at the same time nothing some of the many advances being made in our understanding of their function and development through molecular research. For example, in the chapter on apical meristems, which have been the object of considerable molecular-genetic research, a historical review of the concept of apical organization is presented to provide the reader with an understanding of how that concept has evolved with the availability of more sophisticated methodology. Throughout the book, greater emphasis is made on structure-function relationships than in the previous two editions. As in the previous editions, angiosperms are emphasized, but some features of the vegetative parts of gymnosperms and seedless vascular plants are also considered.

These are exciting times for plant biologists. This is reflected, in part, in the enormity of literature output. The references cited in this book represent but a fraction of the total number of articles read in preparation of the third edition. This is particularly true of the molecular-genetic literature, which is cited most selectively. It was important not to lose focus on the anatomy. A great many of the references cited in the second edition were read anew, in part to insure continuity between the second and third editions. A large number of selected references are listed to support descriptions and interpretations and to direct the interested person toward wider reading. Undoubtedly, some pertinent papers were inadvertently overlooked. A number of review articles, books, and chapters in books with helpful reference lists are included. Additional pertinent references are listed in the addendum.

This book has been planned primarily for advanced students in various branches of plant science, for researchers (from molecular to whole plant), and for teachers of plant anatomy. At the same time, an effort has been made to attract the less-advanced student by presenting the subject in an inviting style, with numerous illustrations, and by explaining and analyzing terms and concepts as they appear in the text. It is my hope that this book will enlighten many and inspire numerous others to study plant structure and development.

R. F. E.
*Madison, Wisconsin*
*July, 2006*

# Acknowledgments

Illustrations form an important part of a book in plant anatomy. I am indebted to various persons who kindly provided illustrations of one kind or another for inclusion in the book and to others, along with publishers and scientific journals, for permission to reproduce in one form or another their published illustrations. Illustrations whose source(s) are not indicated in the figure captions are original. Numerous figures are from research articles by me or coauthored with colleagues, including my students. A great many of the illustrations are the superb work—line art and micrographs—of Dr. Esau. Some figures are expertly rendered electronic illustrations by Kandis Elliot.

Sincere thanks are extended to Laura Evert and Mary Evert for their able assistance with the process of obtaining permissions.

I am grateful to the following people, who so generously gave of their time to review parts of the manuscript: Drs. Veronica Angyalossy, Pieter Baas, Sebastian Y. Bednarek, C. E. J. Botha, Anne-Marie Catesson, Judith L. Croxdale, Nigel Chaffey, Abraham Fahn, Donna Fernandez, Peter K. Helper, Nels R. Lersten, Edward K. Merrill, Regis B. Miller, Thomas L. Rost, Alexander Schulz, L. Andrew Staehelin, Jennifer Thorsch, and Joseph E. Varner. Two of the reviewers, Judith L. Croxdale, who reviewed Chapter 9 (Epidermis), and Joseph E. Varner, who reviewed an early draft of Chapter 4 (Cell Wall), are now deceased. The reviewers offered valuable suggestions for improvement. The final responsibility for the contents of the book, including all errors and omissions, however, is mine.

Very special acknowledgment is accorded Susan E. Eichhorn. Without her assistance it would not have been possible for me to revise the second edition of *Esau's Plant Anatomy*.

# General References

ALEKSANDROV, V. G. 1966. *Anatomiia Rastenii (Anatomy of Plants),* 4th ed. Izd. Vysshaia Shkola, Moscow.

BAILEY, I. W. 1954. *Contributions to Plant Anatomy.* Chronica Botanica, Waltham, MA.

BIEBL, R., and H. GERM. 1967. *Praktikum der Pflanzenanatomie,* 2nd ed. Springer-Verlag, Vienna.

BIERHORST, D. W. 1971. *Morphology of Vascular Plants.* Macmillan, New York.

BOLD, H. C. 1973. *Morphology of Plants,* 3rd ed. Harper and Row, New York.

BOUREAU, E. 1954–1957. *Anatomie végétale: l'appareil végétatif des phanérogrames,* 3 vols. Presses Universitaires de France, Paris.

BOWES, B. G. 2000. *A Color Atlas of Plant Structure.* Iowa State University Press, Ames, IA.

BOWMAN, J., ed. 1994. *Arabidopsis: An Atlas of Morphology and Development.* Springer-Verlag, New York.

BRAUNE, W., A. LEMAN, and H. TAUBERT. 1971 (© 1970). *Pflanzenanatomisches Praktikum: zur Einführung in die Anatomie der Vegetationsorgane der höheren Pflanzen,* 2nd ed. Gustav Fischer, Stuttgart.

BUCHANAN, B. B., W. GRUISSEM, and R. L. JONES, eds. 2000. *Biochemistry and Molecular Biology of Plants.* American Society of Plant Physiologists, Rockville, MD.

CARLQUIST, S. 1961. *Comparative Plant Anatomy: A Guide to Taxonomic and Evolutionary Application of Anatomical Data in Angiosperms.* Holt, Rinehart and Winston, New York.

CARLQUIST, S. 2001. *Comparative Wood Anatomy: Systematic, Ecological, and Evolutionary Aspects of Dicotyledon Wood,* 2nd ed. Springer-Verlag, Berlin.

CHAFFEY, N. 2002. *Wood Formation in Trees: Cell and Molecular Biology Techniques.* Taylor and Francis, London.

CUTLER, D. F. 1969. *Anatomy of the Monocotyledons,* vol. IV, *Juncales.* Clarendon Press, Oxford.

CUTLER, D. F. 1978. *Applied Plant Anatomy.* Longman, London.

CUTTER, E. G. 1971. *Plant Anatomy: Experiment and Interpretation,* part 2, *Organs.* Addison-Wesley, Reading, MA.

CUTTER, E. G. 1978. *Plant Anatomy,* part 1, *Cells and Tissues,* 2nd ed. Addison-Wesley, Reading, MA.

DAVIES, P. J., ed. 2004. *Plant Hormones: Biosynthesis, Signal Transduction, Action!,* 3rd ed. Kluwer Academic, Dordrecht.

DE BARY, A. 1884. *Comparative Anatomy of the Vegetative Organs of the Phanerogams and Ferns.* Clarendon Press, Oxford.

DICKISON, W. C. 2000. *Integrative Plant Anatomy.* Harcourt/Academic Press, San Diego.

DIGGLE, P. K., and P. K. ENDRESS, eds. 1999. *Int. J. Plant Sci.* 160 (6, suppl.: *Development, Function, and Evolution of Symmetry in Plants*), S1–S166.

EAMES, A. J. 1961. *Morphology of Vascular Plants: Lower Groups.* McGraw-Hill, New York.

EAMES, A. J., and L. H. MACDANIELS. 1947. *An Introduction to Plant Anatomy,* 2nd ed. McGraw-Hill, New York.

ESAU, K. 1965. *Plant Anatomy,* 2nd ed. Wiley, New York.

ESAU, K. 1977. *Anatomy of Seed Plants,* 2nd ed. Wiley, New York.

ESCHRICH, W. 1995. *Funktionelle Pflanzenanatomie.* Springer, Berlin.

FAHN, A. 1990. *Plant Anatomy,* 4th ed. Pergamon Press, Oxford.

GIFFORD, E. M., and A. S. FOSTER. 1989. *Morphology and Evolution of Vascular Plants,* 3rd ed. Freeman, New York.

HABERLANDT, G. 1914. *Physiological Plant Anatomy.* Macmillan, London.

*Handbuch der Pflanzenanatomie (Encyclopedia of Plant Anatomy).* 1922–1943; 1951– . Gebrüder Borntraeger, Berlin.

HARTIG, R. 1891. *Lehrbuch der Anatomie und Physiologie der Pflanzen unter besonderer Berücksichtigung der Forstgewächse.* Springer, Berlin.

HAYWARD, H. E. 1938. *The Structure of Economic Plants.* Macmillan, New York.

HIGUCHI, T. 1997. *Biochemistry and Molecular Biology of Wood.* Springer, Berlin.

HOWELL, S. H. 1998. *Molecular Genetics of Plant Development.* Cambridge University Press, Cambridge.

HUBER, B. 1961. *Grundzüge der Pflanzenanatomie.* Springer-Verlag, Berlin.

IQBAL, M., ed. 1995. *The Cambial Derivatives.* Gebrüder Borntraeger, Berlin.

JANE, F. W. 1970. *The Structure of Wood,* 2nd ed. Adam and Charles Black, London.

JEFFREY, E. C. 1917. *The Anatomy of Woody Plants.* University of Chicago Press, Chicago.

JURZITZA, G. 1987. *Anatomie der Samenpflanzen.* Georg Thieme Verlag, Stuttgart.

KAUSSMANN, B. 1963. *Pflanzenanatomie: unter besonderer Berücksichtigung der Kultur- und Nutzpflanzen.* Gustav Fischer, Jena.

KAUSSMANN, B., and U. SCHIEWER. 1989. *Funktionelle Morphologie und Anatomie der Pflanzen.* Gustav Fischer, Stuttgart.

LARSON, P. R. 1994. *The Vascular Cambium. Development and Structure.* Springer-Verlag, Berlin.

MANSFIELD, W. 1916. *Histology of Medicinal Plants.* Wiley, New York.

MAUSETH, J. D. 1988. *Plant Anatomy.* Benjamin/Cummings, Menlo Park, CA.

METCALFE, C. R. 1960. *Anatomy of the Monocotyledons,* vol. I, *Gramineae.* Clarendon Press, Oxford.

METCALFE, C. R. 1971. *Anatomy of the Monocotyledons,* vol. V, *Cyperaceae.* Clarendon Press, Oxford.

METCALFE, C. R., and L. CHALK. 1950. *Anatomy of the Dicotyledons: Leaves, Stems, and Wood in Relation to Taxonomy with Notes on Economic Uses,* 2 vols. Clarendon Press, Oxford.

METCALFE, C. R., and L. CHALK, eds. 1979. *Anatomy of the Dicotyledons,* 2nd ed., vol. I. *Systematic Anatomy of Leaf and Stem, with a Brief History of the Subject.* Clarendon Press, Oxford.

METCALFE, C. R., and L. CHALK, eds. 1983. *Anatomy of the Dicotyledons,* 2nd ed., vol. II. *Wood Structure and Conclusion of the General Introduction.* Clarendon Press, Oxford.

RAUH, W. 1950. *Morphologie der Nutzpflanzen.* Quelle und Meyer, Heidelberg.

ROMBERGER, J. A. 1963. *Meristems, Growth, and Development in Woody Plants: An Analytical Review of Anatomical, Physiological, and Morphogenic Aspects.* Tech. Bull. No. 1293. USDA, Forest Service, Washington, DC.

ROMBERGER, J. A., Z. HEJNOWICZ, and J. F. HILL. 1993. *Plant Structure: Function and Development: A Treatise on Anatomy and Vegetative Development, with Special Reference to Woody Plants.* Springer-Verlag, Berlin.

RUDALL, P. 1992. *Anatomy of Flowering Plants: An Introduction to Structure and Development,* 2nd ed. Cambridge University Press, Cambridge.

SACHS, J. 1875. *Text-Book of Botany, Morphological and Physiological.* Clarendon Press, Oxford.

SINNOTT, E. W. 1960. *Plant Morphogenesis.* McGraw-Hill, New York.

SOLEREDER, H. 1908. *Systematic Anatomy of the Dicotyledons: A Handbook for Laboratories of Pure and Applied Botany,* 2 vols. Clarendon Press, Oxford.

SOLEREDER, H., and F. J. MEYER. 1928–1930, 1933. *Systematische Anatomie der Monokotyledonen,* No. 1 (*Pandales, Helobiae, Triuridales*), 1933; No. 3 (*Principes, Synanthae, Spathiflorae*), 1928; No. 4 (*Farinosae*), 1929; No. 6 (*Scitamineae, Microspermae*), 1930. Gebrüder Borntraeger, Berlin.

SRIVASTAVA, L. M. 2002. *Plant Growth and Development: Hormones and Environment.* Academic Press, Amsterdam.

STEEVES, T. A., and I. M. SUSSEX. 1989. *Patterns in Plant Development,* 2nd ed. Cambridge University Press, Cambridge.

STRASBURGER, E. 1888–1909. *Histologische Beiträge,* nos. 1–7. Gustav Fisher, Jena.

TOMLINSON, P. B. 1961. *Anatomy of the Monocotyledons,* vol. II. *Palmae.* Clarendon Press, Oxford.

TOMLINSON, P. B. 1969. *Anatomy of the Monocotyledons,* vol. III. *Commelinales—Zingiberales.* Clarendon Press, Oxford.

TROLL, W. 1954. *Praktische Einführung in die Pflanzenmorphologie,* vol. 1, *Der vegetative Aufbau.* Gustav Fischer, Jena.

TROLL, W. 1957. *Praktische Einführung in die Pflanzenmorphologie,* vol. 2, *Die blühende Pflanze.* Gustav Fischer, Jena.

WARDLAW, C. W. 1965. *Organization and Evolution in Plants.* Longmans, Green and Co., London.

# CHAPTER ONE

# Structure and Development of the Plant Body—An Overview

The complex multicellular body of a vascular plant is a result of evolutionary specialization of long duration—specialization that followed the transition of multicellular organisms from an aquatic habitat to a terrestrial one (Niklas, 1997). The requirements of the new and harsher environments led to the establishment of morphological and physiological differences among the parts of the plant body so that they became more or less strongly specialized with reference to certain functions. The recognition of these specializations by botanists became embodied in the concept of **plant organs** (Troll, 1937; Arber, 1950). At first, botanists visualized the existence of many organs, but later as the interrelationships among the plant parts came to be better understood, the number of vegetative organs was reduced to three: **stem**, **leaf**, and **root** (Eames, 1936). In this scheme, stem and leaf are commonly treated together as a morphological and functional unit, the **shoot**.

Researchers in evolution postulate that the organization of the oldest vascular plants was extremely simple, perhaps resembling that of the leafless and rootless Devonian plant *Rhynia* (Gifford and Foster, 1989; Kenrick and Crane, 1997). If the seed plants have evolved from rhyniaceous types of plants, which con-

sisted of dichotomously branched axes without appendages, the leaf, the stem, and the root would be closely interrelated through phylogenetic origin (Stewart and Rothwell, 1993; Taylor and Taylor, 1993; Raven, J. A. and Edwards, 2001). The common origin of these three organs is even more obvious in their ontogeny (development of an individual entity), for they are initiated together in the embryo as the latter develops from the unicellular zygote into a multicellular organism. At the apex of the shoot the leaf and stem increments are formed as a unit. At maturity, too, the leaf and stem imperceptibly merge with one another both externally and internally. In addition, the root and the stem constitute a continuum—a continuous structure—and have many common features in form, anatomy, function, and method of growth.

As the embryo grows and becomes a seedling, stem and root increasingly deviate from one another in their organization (Fig. 1.1). The root grows as a more or less branched cylindrical organ; the stem is composed of nodes and internodes, with leaves and branches attached at the nodes. Eventually the plant enters the reproductive stage when the shoot forms inflorescences and flowers (Fig. 1.2). The flower is sometimes called

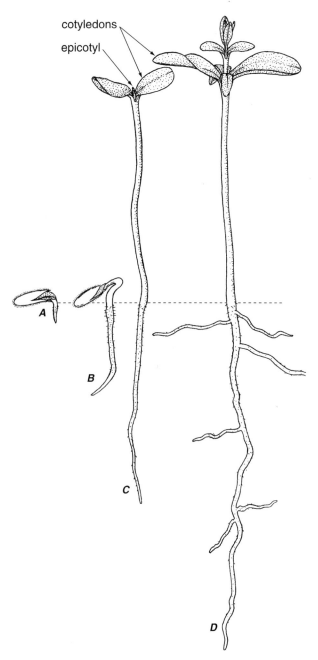

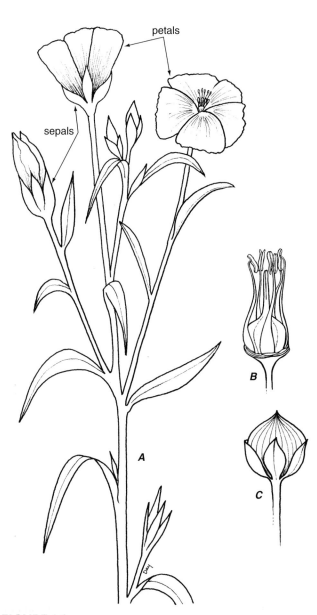

**FIGURE 1.1**

Some stages in development of the flax (*Linum usitatissimum*) seedling. **A**, germinating seed. The taproot (below interrupted line) is the first structure to penetrate the seed coat. **B**, the elongating hypocotyl (above interrupted line) has formed a hook, which subsequently will straighten out, pulling the cotyledons and shoot apex above ground. **C**, after emergence above ground, the cotyledons, which in flax persist for about 30 days, enlarge and thicken. The developing epicotyl—the stem-like axis or shoot above the cotyledons—is now apparent between the cotyledons. **D**, the developing epicotyl has given rise to several foliage leaves, and the taproot to several branch roots. (From Esau, 1977; drawn by Alva D. Grant.)

**FIGURE 1.2**

Inflorescence and flowers of flax (*Linum usitatissimum*). **A**, inflorescence, a panicle, with intact flowers showing sepals and petals. **B**, flower, from which the sepals and petals have been removed, to show the stamens and gynoecium. Flax flowers usually have five fertile stamens. The gynoecium consists of five united carpels, with five distinct styles and stigmas. **C**, mature fruit (capsule) and persistent sepals. (Drawn by Alva D. Grant.)

an organ, but the classical concept treats the flower as an assemblage of organs homologous with the shoot. This concept also implies that the floral parts—some of which are fertile (stamens and carpels) and others sterile (sepals and petals)—are homologous with the leaves. Both the leaves and the floral parts are thought to have originated from the kind of branch systems that characterized the early, leafless and rootless vascular plants (Gifford and Foster, 1989).

Despite the overlapping and intergrading of characters between plant parts, the division of the plant body into morphological categories of stem, leaf, root, and flower (where present) is commonly resorted to because it brings into focus the structural and the functional specialization of parts, the stem for support and conduction, the leaf for photosynthesis, and the root for anchorage and absorption. Such division must not be emphasized to the degree that it might obscure the essential unity of the plant body. This unity is clearly perceived if the plant is studied developmentally, an approach that reveals the gradual emergence of organs and tissues from a relatively undifferentiated body of the young embryo.

# ∎ INTERNAL ORGANIZATION OF THE PLANT BODY

The plant body consists of many different types of cell, each enclosed in its own cell wall and united with other cells by means of a cementing intercellular substance. Within this united mass certain groupings of cells are distinct from others structurally or functionally or both. These groupings are referred to as **tissues**. The structural variations of tissues are based on differences in the component cells and their type of attachment to each other. Some tissues are structurally relatively **simple** in that they consist of one cell type; others, containing more than one cell type, are **complex**.

The arrangement of tissues in the plant as a whole and in its major organs reveals a definite structural and functional organization. Tissues concerned with conduction of food and water—the **vascular tissues**—form a coherent system extending continuously through each organ and the entire plant. These tissues connect places of water intake and food synthesis with regions of growth, development, and storage. The **nonvascular tissues** are similarly continuous, and their arrangements are indicative of specific interrelations (e.g., between storage and vascular tissues) and of specialized functions (e.g., support or storage). To emphasize the organization of tissues into large entities showing topographic continuity, and revealing the basic unity of the plant body, the expression **tissue system** has been adopted (Sachs, 1875; Haberlandt, 1914; Foster, 1949).

Although the classification of cells and tissues is a somewhat arbitrary matter, for purposes of orderly description of plant structure the establishment of categories is necessary. Moreover, if the classifications issue from broad comparative studies, in which the variability and the intergrading of characters are clearly revealed and properly interpreted, they not only are descriptively useful but also reflect the natural relation of the entities classified.

### The Body of a Vascular Plant Is Composed of Three Tissue Systems

According to Sachs's (1875) convenient classification based on topographic continuity of tissues, the body of a vascular plant is composed of three tissue systems, the dermal, the vascular, and the fundamental (or ground). The **dermal tissue system** comprises the **epidermis**, that is, the primary outer protective covering of the plant body, and the **periderm**, the protective tissue that supplants the epidermis, mainly in plants that undergo a secondary increase in thickness. The **vascular tissue system** contains two kinds of conducting tissues, the **phloem** (food conduction) and the **xylem** (water conduction). The epidermis, periderm, phloem, and xylem are complex tissues.

The **fundamental tissue system** (or **ground tissue system**) includes the simple tissues that, in a sense, form the ground substance of the plant but at the same time show various degrees of specialization. **Parenchyma** is the most common of ground tissues. Parenchyma cells are characteristically living cells, capable of growth and division. Modifications of parenchyma cells are found in the various secretory structures, which may occur in the ground tissue as individual cells or as smaller or larger cell complexes. **Collenchyma** is a living thick-walled tissue closely related to parenchyma; in fact, it is commonly regarded as a form of parenchyma specialized as supporting tissue of young organs. The fundamental tissue system often contains highly specialized mechanical elements—with thick, hard, often lignified walls—combined into coherent masses as **sclerenchyma** tissue or dispersed as individual or as small groups of sclerenchyma cells.

### Structurally Stem, Leaf, and Root Differ Primarily in the Relative Distribution of the Vascular and Ground Tissues

Within the plant body the various tissues are distributed in characteristic patterns depending on plant part or plant taxon or both. Basically the patterns are alike in that the vascular tissue is embedded in ground tissue and the dermal tissue forms the outer covering. The principal differences in the structure of stem, leaf, and root lie in the relative distribution of the vascular and ground tissues (Fig. 1.3). In the stems of eudicotyledons

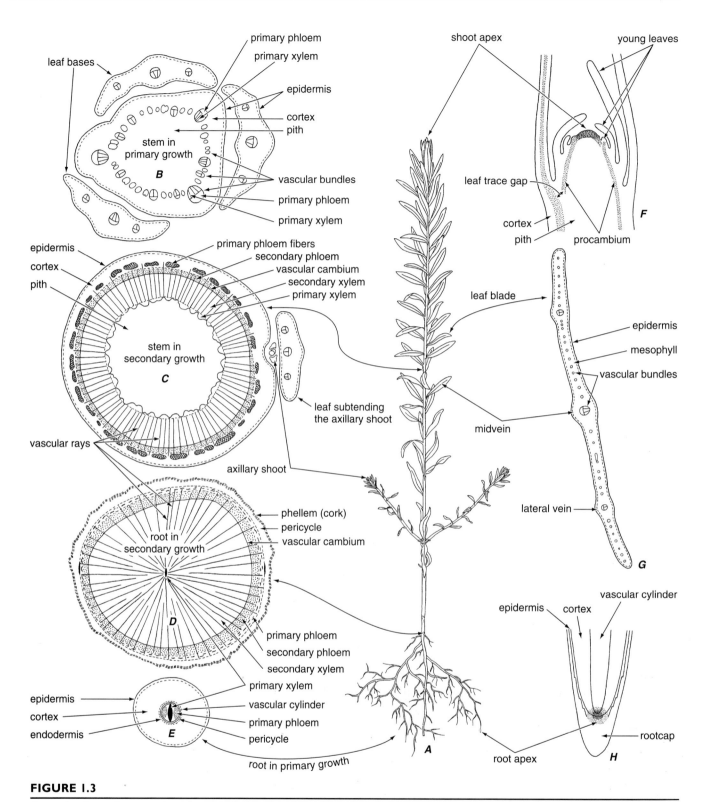

## FIGURE 1.3

Organization of a vascular plant. **A**, habit sketch of flax (*Linum usitatissimum*) in vegetative state. Transverse sections of stem at **B**, **C**, and of root at **D**, **E**. **F**, longitudinal section of terminal part of shoot with shoot apex and developing leaves. **G**, transverse section of leaf blade. **H**, longitudinal section of terminal part of root with root apex (covered by rootcap) and subjacent root regions. (A, ×²/₅; B, E, F, H, ×50; C, ×32; D, ×7; G, ×19. **A**, drawn by R. H. Miller.)

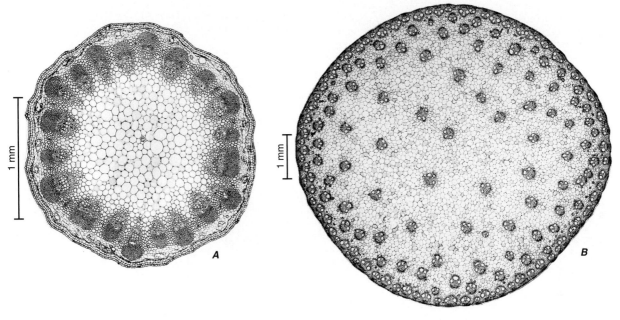

**FIGURE 1.4**

Types of stem anatomy in angiosperms. **A,** transverse section of stem of *Helianthus*, a eudicot, with discrete vascular bundles forming a single ring around a pith. **B,** transverse section of stem of *Zea*, a monocot, with the vascular bundles scattered throughout the ground tissue. The bundles are more numerous near the periphery. (From Esau, 1977.)

(eudicots), for example, the vascular tissue forms a "hollow" cylinder, with some ground tissue enclosed by the cylinder (**pith,** or **medulla**) and some located between the vascular and dermal tissues (**cortex**) (Figs. 1.3B, C and 1.4A). The primary vascular tissues may appear as a more or less continuous cylinder within the ground tissue or as a cylinder of discrete strands, or bundles, separated from one another by ground tissue. In the stems of most monocotyledons (monocots) the vascular bundles occur in more than one ring or appear scattered throughout the ground tissue (Fig. 1.4B). In the latter instance the ground tissue often cannot be distinguished as cortex and pith. In the leaf the vascular tissue forms an anastomosing system of **veins,** which thoroughly permeate the **mesophyll,** the ground tissue of the leaf that is specialized for photosynthesis (Fig. 1.3G).

The pattern formed by the vascular bundles in the stem reflects the close structural and developmental relationship between the stem and its leaves. The term "shoot" serves not only as a collective term for these two vegetative organs but also as an expression of their intimate physical and developmental association. At each node one or more vascular bundles diverge from the strands in the stem and enter the leaf or leaves attached at that node in continuity with the vasculature of the leaf (Fig. 1.5). The extensions from the vascular system in the stem toward the leaves are called **leaf traces,** and the wide gaps or regions of ground tissue in the vascular cylinder located above the level where leaf traces diverge toward the leaves are called **leaf trace gaps** (Raven et al., 2005) or **interfascicular regions** (Beck et al., 1982). A leaf trace extends from its connection with a bundle in the stem (called a **stem bundle,** or an **axial bundle**), or with another leaf trace, to the level at which it enters the leaf (Beck et al., 1982).

Compared with the stem, the internal structure of the root is usually relatively simple and closer to that of the ancestral axis (Raven and Edwards, 2001). Its relatively simple structure is due in large part to the absence of leaves and the corresponding absence of nodes and internodes. The three tissue systems in the primary stage of root growth can be readily distinguished from one another. In most roots, the vascular tissues form a solid cylinder (Fig. 1.3E), but in some they form a hollow cylinder around a pith. The vascular cylinder comprises the vascular tissues and one or more layers of nonvascular cells, the **pericycle,** which in seed plants arises from the same part of the root apex as the vascular tissues. In most seed plants branch, or lateral, roots arise in the pericycle. A morphologically differentiated

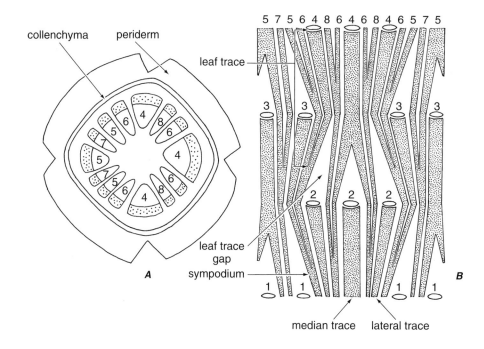

collenchyma    periderm

leaf trace

leaf trace
gap
sympodium

median trace    lateral trace

**FIGURE 1.5**

Diagrams illustrating primary vascular system in the stem of elm (*Ulmus*), a eudicot. **A**, transverse section of stem showing the discrete vascular bundles encircling the pith. **B**, longitudinal view showing the vascular cylinder as though cut through median leaf trace 5 and spread out in one plane. The transverse section (**A**) corresponds to the topmost view in **B**. The numbers in both views indicate leaf traces. Three leaf traces—a median and two lateral traces—connect the vascular system of the stem with that of the leaf. A stem bundle and its associated leaf traces are called a sympodium. (From Esau, 1977; after Smithson, 1954, with permission of the Council of the Leeds Philosophical and Literary Society.)

**endodermis** (the innermost, and compactly arranged, layer of cells of the cortex in seed plants) typically surrounds the pericycle. In the absorbing region of the root the endodermis is characterized by the presence of **Casparian strips** in its anticlinal walls (the radial and transverse walls, which are perpendicular to the surface of the root) (Fig. 1.6). In many roots the outermost layer of cortical cells is differentiated as an **exodermis**, which also exhibits Casparian strips. The Casparian strip is not merely a wall thickening but an integral band-like portion of the wall and intercellular substance that is impregnated with suberin and sometimes lignin. The presence of this hydrophobic region precludes the passage of water and solutes across the endodermis and exodermis via the anticlinal walls (Lehmann et al., 2000).

## ▌SUMMARY OF TYPES OF CELLS AND TISSUES

As implied earlier in this chapter, separation of cells and tissues into categories is, in a sense, contrary to the fact that structural features vary and intergrade with each other. Cells and tissues do, however, acquire differential properties in relation to their positions in the plant body. Some cells undergo more profound changes than others. That is, cells become specialized to varied degrees. Cells that are relatively little specialized retain living protoplasts and have the capacity to change in form and function during their lifetimes (various kinds of parenchyma cells). More highly specialized cells may develop thick, rigid cell walls, become devoid of living protoplasts, and cease to be capable of structural and functional changes (tracheary elements and various kinds of sclerenchyma cells). Between these two extremes are cells at varying levels of metabolic activity and degrees of structural and functional specialization. Classifications of cells and tissues serve to deal with the phenomena of differentiation—and the resultant diversification of plant parts—in a manner that allows making generalizations about common and divergent features among related and unrelated taxa. They make possible treating the phenomena of ontogenetic and phylogenetic specialization in a comparative and systematic way.

Table 1.1 summarizes information on the generally recognized categories of cells and tissues of seed plants without special regard to the problem of structural and functional intergrading of characteristics. The various

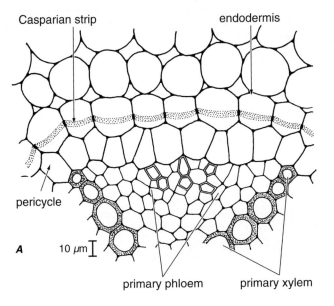

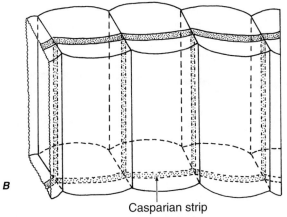

**FIGURE 1.6**

Structure of endodermis. **A**, transverse section of part of a morning glory (*Convolvulus arvensis*) root showing position of the endodermis in relation to vascular cylinder consisting of pericycle, primary xylem, and primary phloem. The endodermis is shown with transverse walls bearing Casparian strips in focus. **B**, diagram of three connected endodermal cells oriented as they are in **A**; Casparian strip occurs in transverse and radial walls (i.e., in all anticlinal walls) but is absent in tangential walls. (From Esau, 1977.)

types of cells and tissues summarized in the table are considered in detail in Chapters 7 through 15. Secretory cells—cells that produce a variety of secretions—do not form clearly delimited tissues and therefore are not included in the table. They are the topics of Chapters 16 and 17.

Secretory cells occur within other tissues as single cells or as groups or series of cells, and also in more or less definitely organized formations on the surface of the plant. The principal secretory structures on plant surfaces are glandular epidermal cells, hairs, and various glands, such as floral and extrafloral nectaries, certain hydathodes, and digestive glands. The glands are usually differentiated into secretory cells on the surfaces and nonsecretory cells support the secretory. Internal secretory structures are secretory cells, intercellular cavities or canals lined with secretory cells (resin ducts, oil ducts), and secretory cavities resulting from disintegration of secretory cells (oil cavities). Laticifers may be placed among the internal secretory structures. They are either single cells (nonarticulated laticifers) usually much branched, or series of cells united through partial dissolution of common walls (articulated laticifers). Laticifers contain a fluid called latex, which may be rich in rubber. Laticifer cells are commonly multinucleate.

# DEVELOPMENT OF THE PLANT BODY

## The Body Plan of the Plant Is Established during Embryogenesis

The highly organized body of a seed plant represents the sporophyte phase of the life cycle. It begins its existence with the product of gametic union, the unicellular **zygote**, which develops into an embryo by a process known as **embryogenesis** (Fig. 1.7). Embryogenesis establishes the body plan of the plant, consisting of two superimposed patterns: an **apical-basal pattern** along the main axis and a **radial pattern** of concentrically arranged tissue systems. Thus patterns are established in the distribution of cells, and the embryo as a whole assumes a specific, albeit relatively simple, form as contrasted with the adult sporophyte.

The initial stages of embryogenesis are essentially the same in eudicots and monocots. Formation of the embryo begins with division of the zygote within the embryo sac of the ovule. Typically the first division of the zygote is transverse and asymmetrical, with regard to the long axis of the cell, the division plane coinciding with the minimum dimension of the cell (Kaplan and Cooke, 1997). With this division the **polarity** of the embryo is established. The upper pole, consisting of a small **apical cell** (Fig. 1.7A), gives rise to most of the mature embryo. The lower pole, consisting of a larger **basal cell** (Fig. 1.7A), produces a stalk-like **suspensor** (Fig. 1.7B) that anchors the embryo at the micropyle, the opening in the ovule through which the pollen tube enters. Through a progression of divisions—in some species (e.g., *Arabidopsis*; West and Harada, 1993) quite orderly, in others (e.g., cotton and maize; Pollock and Jensen, 1964; Poethig et al., 1986) not obviously so—the embryo differentiates into a nearly spherical structure, the **embryo proper** and the suspensor. In some angiosperms polarity is already established in the egg cell and

**TABLE 1.1 ■ Tissues and Cell Types**

| Tissues | Cell Type | | Characteristics | Location | Function |
|---|---|---|---|---|---|
| Dermal | Epidermis | | Unspecialized cells; guard cells and cells forming trichomes; sclerenchyma cells | Outermost layer of cells of the primary plant body | Mechanical protection; minimizes water loss (cuticle); aeration of internal tissue via stomata |
| | Periderm | | Comprises cork tissue (phellem), cork cambium (phellogen), and phelloderm | Initial periderm generally beneath epidermis; subsequently formed periderms deeper in bark | Replaces epidermis as protective tissue in roots and stems; aeration of internal tissue via lenticels |
| Ground | Parenchyma | Parenchyma | Shape: commonly polyhedral (many-sided); variable<br>Cell wall: primary, or primary and secondary; may be lignified, suberized, or cutinized<br>Living at maturity | Throughout the plant body, as parenchyma tissue in cortex, pith, pith rays, and mesophyll; in xylem and phloem | Such metabolic processes as respiration, digestion, and photosynthesis; storage and conduction; wound healing and regeneration |
| | Collenchyma | Collenchyma | Shape: elongated<br>Cell wall: unevenly thickened, primary only—nonlignified<br>Living at maturity | On the periphery (beneath the epidermis) in young elongating stems; often as a cylinder of tissue or only in patches; in ribs along veins in some leaves | Support in primary plant body |
| | Sclerenchyma | Fiber | Shape: generally very long<br>Cell wall: primary and thick secondary—often lignified<br>Often (not always) dead at maturity | Sometimes in cortex of stems, most often associated with xylem and phloem; in leaves of monocots | Support; storage |
| | | Sclereid | Shape: variable; generally shorter than fibers<br>Cell wall: primary and thick secondary—generally lignified<br>May be living or dead at maturity | Throughout the plant body | Mechanical; protective |
| Vascular | Xylem | Tracheid | Shape: elongated and tapering<br>Cell wall: primary and secondary; lignified; contains pits but not perforations<br>Dead at maturity | Xylem | Chief water-conducting element in gymnosperms and seedless vascular plants; also found in angiosperms |
| | | Vessel element | Shape: elongated, generally not as long as tracheids; several vessel elements end-on-end constitute a vessel<br>Cell wall: primary and secondary; lignified; contains pits and perforations<br>Dead at maturity | Xylem | Chief water-conducting element in angiosperms |

**TABLE 1.1 ■** *Continued*

| Tissues | Cell Type | Characteristics | Location | Function |
|---|---|---|---|---|
| Phloem | Sieve cell | Shape: elongated and tapering<br>Cell wall: primary in most species; with sieve areas; callose often associated with wall and sieve pores<br>Living at maturity; either lacks or contains remnants of a nucleus at maturity; lacks distinction between vacuole and cytosol; contains large amounts of tubular endoplasmic reticulum; lacks proteinaceous substance known as P-protein | Phloem | Food-conducting element in gymnosperms |
| | Strasburger cell | Shape: generally elongated<br>Cell wall: primary<br>Living at maturity; associated with sieve cell, but generally not derived from same mother cell as sieve cell; has numerous plasmodesmatal connections with sieve cell | Phloem | Plays a role in the delivery of substances to the sieve cell, including informational molecules and ATP |
| | Sieve-tube element | Shape: elongated<br>Cell wall: primary, with sieve areas; sieve areas on end wall with much larger pores than those on side walls—this wall part is termed a sieve plate; callose often associated with walls and sieve pores<br>Living at maturity; either lacks a nucleus at maturity or contains only remnants of nucleus; lacks distinction between vacuole and cytosol; except for those of some monocots, contains a proteinaceous substance known as P-protein; several sieve-tube elements in a vertical series constitute a sieve tube | Phloem | Food-conducting element in angiosperms |
| | Companion cell | Shape: variable, generally elongated<br>Cell wall: primary<br>Living at maturity; closely associated with sieve-tube elements; derived from same mother cell as sieve-tube element; has numerous plasmodesmatal connections with sieve-tube element | Phloem | Plays a role in the delivery of substances to the sieve-tube element, including informational molecules and ATP |

Source: Raven et al., 2005.

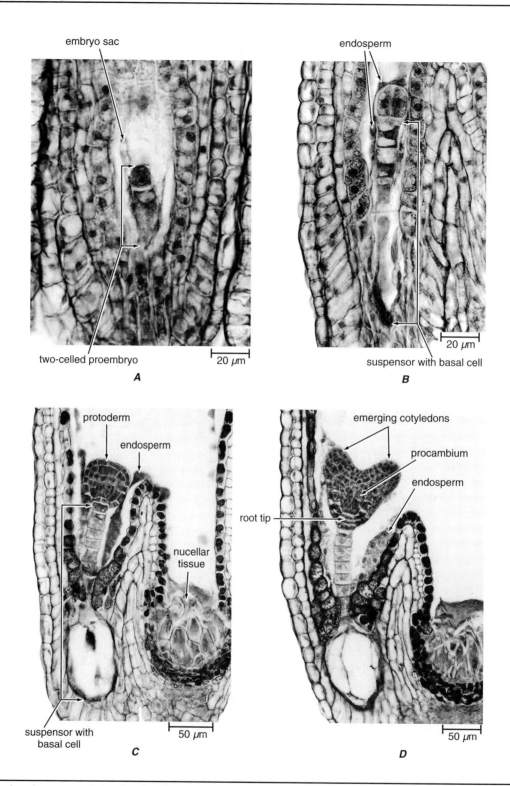

**FIGURE 1.7**

Some stages of embryogenesis in shepherd's purse (*Capsella bursa-pastoris*, Brassicaceae), a eudicot, in longitudinal sections. **A**, two-celled stage, resulting from unequal transverse division of the zygote into an upper apical cell and a lower basal cell; **B**, six-celled proembryo, consisting of a stalk-like suspensor as distinct from the two terminal cells, which develop into the embryo proper. **C**, the embryo proper is globular and has a protoderm, the primary meristem that gives rise to the epidermis. **D**, the embryo at the so-called heart stage, when the cotyledons are emerging. (Note: The basal cell of the suspensor is not the basal cell of the two-celled proembryo.)

zygote, where the nucleus and most of the cytoplasmic organelles are located in the upper (chalazal) portion of the cell, and the lower (micropylar) portion is dominated by a large vacuole.

Initially the embryo proper consists of a mass of relatively undifferentiated cells. Soon, however, cell divisions in the embryo proper and the concomitant differential growth and vacuolation of the resulting cells initiate the organization of the tissue systems (Fig. 1.7C, D). The component tissues are still meristematic, but their position and cytologic characteristics indicate a relation to mature tissues appearing in the subsequently developing seedling. The future epidermis is represented by a meristematic surface layer, the **protoderm**. Beneath it the **ground meristem** of the future cortex is distinguishable by cell vacuolation, which is more pronounced here than it is in contiguous tissues. The centrally located, less vacuolate tissue extending through the apical-basal axis is the precursor of the future primary vascular system. This meristematic tissue is the **procambium**. Longitudinal divisions and elongation of cells impart a narrow, elongated form to the procambial cells. The protoderm, ground meristem, and procambium—the so-called **primary meristems**, or **primary meristematic tissues**—extend into other regions of the embryo as embryogenesis continues.

During the early stages of embryogenesis, cell division takes place throughout the young sporophyte. As the embryo develops, however, the addition of new cells gradually becomes restricted to opposite ends of the axis, the **apical meristems** of future root and shoot (Aida and Tasaka, 2002). Meristems are embryonic tissue regions in which the addition of new cells continues while other plant parts reach maturity (Chapters 5, 6).

The mature embryo has a limited number of parts—frequently only a stem-like axis bearing one or more leaf-like appendages, the **cotyledons** (Fig. 1.8). Because of its location below the cotyledon(s), the stem-like axis is called **hypocotyl**. At its lower end (the **root pole**), the hypocotyl bears the incipient root, at its upper end (the **shoot pole**) the incipient shoot. The root may be represented by its meristem (apical meristem of the root) or by a primordial root, the **radicle**. Similarly the apical meristem of the shoot located at the shoot pole may or may not have initiated the development of a shoot. If a primordial shoot is present, it is called **plumule**.

## With Germination of the Seed, the Embryo Resumes Growth and Gradually Develops into an Adult Plant

After the seed germinates, the apical meristem of the shoot forms, in regular sequence, leaves and nodes and internodes (Figs. 1.1D and 1.3A, F). Apical meristems in

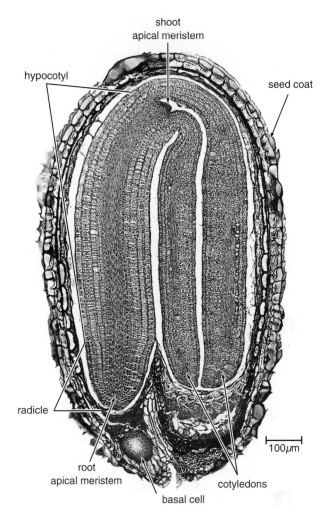

**FIGURE 1.8**

Mature shepherd's purse (*Capsella bursa-pastoris*) embryo in longitudinal section. The part of the embryo below the cotyledons is the hypocotyl. At the lower end of the hypocotyl is the embryonic root, or radicle.

the axils of leaves produce axillary shoots (exogenous origin), which in turn have other axillary shoots. As a result of such activity, the plant bears a system of branches on the main stem. If the axillary meristems remain inactive, the shoot fails to branch as, for example, in many palms. The apical meristem of the root located at the tip of the hypocotyl—or of the radicle, as the case may be—forms the primary root (first root; Groff and Kaplan, 1988). In many plants the primary root produces branch roots (secondary roots) (Figs. 1.1D and 1.3A) from new apical meristems originating from the pericycle deep in the primary root (endogenous origin). The branch roots produce further branches in turn. Thus a much branched root system results. In some plants, notably monocots, the root systems of the adult plant develop from roots arising from the stem.

The growth outlined above constitutes the vegetative stage in the life of a seed plant. At an appropriate time, determined in part by an endogenous rhythm of growth and in part by environmental factors, especially light and temperature, the vegetative apical meristem of the shoot is changed into a reproductive apical meristem, that is, in angiosperms, into a floral apical meristem, which produces a flower or an inflorescence. The vegetative stage in the life cycle of the plant is thus succeeded by the reproductive stage.

The plant organs originating from the apical meristems pass a period of expansion in length and width. The initial growth of the successively formed roots and shoots is commonly termed **primary growth**. The plant body resulting from this growth is the **primary plant body**, which consists of **primary tissues**. In most seedless vascular plants and monocots, the entire life of the sporophyte is completed in a primary plant body. The gymnosperms and most angiosperms, including some monocots, show an increase in thickness of stem and root by means of **secondary growth**.

The secondary growth may be a **cambial secondary growth** resulting from production of cells by a meristem called **cambium**. The principal cambium is the **vascular cambium**, which forms the **secondary vascular tissues** (secondary xylem and secondary phloem) and causes thereby an increase in thickness of the axis (Fig. 1.3C, D). This growth is usually accompanied by the activity of a **cork cambium**, or **phellogen**, which develops in the peripheral region of the expanding axis and gives rise to the **periderm**, a secondary protective tissue system replacing the epidermis.

The secondary growth of the axis may be diffuse in that it occurs by overall cell division and cell enlargement in ground parenchyma tissue without involving a special meristem restricted to a certain region of the axis. This kind of secondary growth has been designated **diffuse secondary growth** (Tomlinson, 1961). It is characteristic of some monocots, notably the palms, and of some plants having tuberous organs.

The tissues produced by the vascular cambium and the phellogen are more or less clearly delimited from the primary tissues and are referred to as **secondary tissues** and, in their entirely, as the **secondary plant body**. The secondary addition of vascular tissues and protective covering makes possible the development of large, much branched plant bodies, such as are characteristic of trees.

Although it is appropriate to think of a plant as becoming "adult" or "mature," in that it develops from a single cell into a complex but integrated structure capable of reproducing its own kind, an adult seed plant is a constantly changing organism. It maintains the capacity to add new increments to its body through the activity of apical meristems of shoots and roots and to increase the volume of its secondary tissues through the activity of lateral meristems. Growth and differentiation require synthesis and degradation of protoplasmic and cell wall materials and involve an exchange of organic and inorganic substances circulating by way of the conducting tissues and diffusing from cell to cell to their ultimate destinations. A variety of processes take place in specialized organs and tissue systems in providing organic substances for metabolic activities. An outstanding feature of the living state of a plant is that its perpetual changes are highly coordinated and occur in orderly sequences (Steeves and Sussex, 1989; Berleth and Sachs, 2001). Moreover, as do other living organisms, plants exhibit rhythmic phenomena, some of which clearly match environmental periodicities and indicate an ability to measure time (Simpson et al., 1999; Neff et al., 2000; Alabadi et al., 2001; Levy et al., 2002; Srivastava, 2002).

# REFERENCES

AIDA, M., and M. TASAKA. 2002. Shoot apical meristem formation in higher plant embryogenesis. In: *Meristematic Tissues in Plant Growth and Development*, pp. 58–88, M. T. McManus and B. E. Veit, eds. Sheffield Academic Press, Sheffield.

ALABADI, D., T. OYAMA, M. J. YANOVSKY, F. G. HARMON, P. MÁS, and S. A. KAY. 2001. Reciprocal regulation between *TOC1* and *LHY/CCA1* within the *Arabidopsis* circadian clock. *Science* 293, 880–883.

ARBER, A. 1950. *The Natural Philosophy of Plant Form*. Cambridge University Press, Cambridge.

BECK, C. B., R. SCHMID, and G. W. ROTHWELL. 1982. Stelar morphology and the primary vascular system of seed plants. *Bot. Rev.* 48, 692–815.

BERLETH, T., and T. SACHS. 2001. Plant morphogenesis: Long-distance coordination and local patterning. *Curr. Opin. Plant Biol.* 4, 57–62.

EAMES, A. J. 1936. *Morphology of Vascular Plants. Lower Groups*. McGraw-Hill, New York.

ESAU, K. 1977. *Anatomy of Seed Plants*, 2nd ed. Wiley, New York.

FOSTER, A. S. 1949. *Practical Plant Anatomy*, 2nd ed. Van Nostrand, New York.

GIFFORD, E. M., and A. S. FOSTER. 1989. *Morphology and Evolution of Vascular Plants*, 3rd ed. Freeman, New York.

GROFF, P. A., and D. R. KAPLAN. 1988. The relation of root systems to shoot systems in vascular plants. *Bot. Rev.* 54, 387–422.

HABERLANDT, G. 1914. *Physiological Plant Anatomy*. Macmillan, London.

KAPLAN, D. R., and T. J. COOKE. 1997. Fundamental concepts in the embryogenesis of dicotyledons: A morphological interpretation of embryo mutants. *Plant Cell* 9, 1903–1919.

KENRICK, P., and P. R. CRANE. 1997. *The Origin and Early Diversification of Land Plants: A Cladistic Study*. Smithsonian Institution Press, Washington, DC.

LEHMANN, H., R. STELZER, S. HOLZAMER, U. KUNZ, and M. GIERTH. 2000. Analytical electron microscopical investigations on the apoplastic pathways of lanthanum transport in barley roots. *Planta* 211, 816–822.

LEVY, Y. Y., S. MESNAGE, J. S. MYLNE, A. R. GENDALL, and C. DEAN. 2002. Multiple roles of *Arabidopsis VRN1* in vernalization and flowering time control. *Science* 297, 243–246.

NEFF, M. M., C. FANKHAUSER, and J. CHORY. 2000. Light: An indicator of time and place. *Genes Dev.* 14, 257–271.

NIKLAS, K. J. 1997. *The Evolutionary Biology of Plants*. University of Chicago Press, Chicago.

POETHIG, R. S., E. H. COE JR., and M. M. JOHRI. 1986. Cell lineage patterns in maize embryogenesis: A clonal analysis. *Dev. Biol.* 117, 392–404.

POLLOCK, E. G., and W. A. JENSEN. 1964. Cell development during early embryogenesis in *Capsella* and *Gossypium*. *Am. J. Bot.* 51, 915–921.

RAVEN, J. A., and D. EDWARDS. 2001. Roots: Evolutionary origins and biogeochemical significance. *J. Exp. Bot.* 52, 381–401.

RAVEN, P. H., R. F. EVERT, and S. E. EICHHORN. 2005. *Biology of Plants*, 7th ed. Freeman, New York.

SACHS, J. 1875. *Text-book of Botany, Morphological and Physiological*. Clarendon Press, Oxford.

SIMPSON, G. G., A. R. GENDALL, and C. DEAN. 1999. When to switch to flowering. *Annu. Rev. Cell Dev. Biol.* 15, 519–550.

SMITHSON, E. 1954. Development of winged cork in *Ulmus x hollandica* Mill. *Proc. Leeds Philos. Lit. Soc., Sci. Sect.*, 6, 211–220.

SRIVASTAVA, L. M. 2002. *Plant Growth and Development. Hormones and Environment*. Academic Press, Amsterdam.

STEEVES, T. A., and I. M. SUSSEX. 1989. *Patterns in Plant Development*, 2nd ed. Cambridge University Press, Cambridge.

STEWART, W. N., and G. W. ROTHWELL. 1993. *Paleobotany and the Evolution of Plants*, 2nd ed. Cambridge University Press, Cambridge.

TAYLOR, T. N., and E. L. TAYLOR. 1993. *The Biology and Evolution of Fossil Plants*. Prentice Hall, Englewood Cliffs, NJ.

TOMLINSON, P. B. 1961. *Anatomy of the Monocotyledons*. II. Palmae. Clarendon Press, Oxford.

TROLL, W. 1937. *Vergleichende Morphologie der höheren Pflanzen*, Band I, Vegetationsorgane, Teil I. Gebrüder Borntraeger, Berlin.

WEST, M. A. L., and J. J. HARADA. 1993. Embryogenesis in higher plants: An overview. *Plant Cell* 5, 1361–1369.

# CHAPTER TWO

# The Protoplast: Plasma Membrane, Nucleus, and Cytoplasmic Organelles

Cells represent the smallest structural and functional units of life (Sitte, 1992). Living organisms consist of single cells or of complexes of cells. Cells vary greatly in size, form, structure, and function. Some are measured in micrometers, others in millimeters, and still others in centimeters (fibers in certain plants). Some cells perform a number of functions; others are specialized in their activities. Despite the extraordinary diversity among cells they are remarkably similar to one another, both in their physical organization and in their biochemical properties.

The concept that the cell is the basic unit of biological structure and function is based on the **cell theory**, which was formulated in the first half of the nineteenth century by Mathias Schleiden and Theodor Schwann. In 1838, Schleiden concluded that all plant tissues are composed of cells. A year later, Schwann (1839) extended Schleiden's observation to animal tissues and proposed a cellular basis for all life. In 1858, the idea that all living organisms are composed of one or more cells took on even broader significance when Rudolf Virchow generalized that all cells arise only from preexisting cells. In its classical form, the cell theory proposed that the

bodies of all plants and animals are aggregates of individual, differentiated cells, and that the activities of the whole plant and animal might be considered the summation of the activities of the individual constituent cells, with the individual cells of prime importance.

By the latter half of the nineteenth century, an alternative to the cell theory was formulated. Known as the **organismal theory**, it maintains that the entire organism is not merely a group of independent units but rather a living unit subdivided into cells, which are connected and coordinated into a harmonious whole. An often quoted statement is that of Anton de Bary (1879), "It is the plant that forms cells, and not the cell that forms plants" (translation by Sitte, 1992). Since then substantial evidence has accumulated in favor of an organismal concept for plants (see Kaplan and Hagemann, 1991; Cooke and Lu, 1992; and Kaplan, 1992; and literature cited therein).

The organismal theory is especially applicable to plants, whose cells do not pinch apart during cell division, as do animal cells, but are partitioned initially by insertion of a cell plate (Chapter 4). The separation of plant cells is rarely complete. Contiguous plant cells remain interconnected by cytoplasmic strands known as plasmodesmata, which traverse the walls and unite the entire plant body into an organic whole. Appropriately, plants have been characterized as supracellular organisms (Lucas et al., 1993).

In its modern form the cell theory states simply that: (1) all organisms are composed of one or more cells, (2) the chemical reactions of a living organism, including its energy-related processes and its biosynthetic processes, occur within cells, (3) cells arise from other cells, and (4) cells contain the hereditary information of the organisms of which they are a part, and this information is passed on from parent to daughter cell. The cell and organismal theories are not mutually exclusive. Together, they provide a meaningful view of the structure and function at cellular and organismal levels (Sitte, 1992).

The word cell, meaning "little room," was introduced by Robert Hooke in the seventeenth century to describe the small cavities separated by cell walls in cork tissue. Later Hooke recognized that living cells in other plant tissues were filled with "juices." Eventually the contents of cells were interpreted as living matter and received the name **protoplasm**. An important step toward recognition of the complexity of protoplasm was the discovery of the nucleus by Robert Brown in 1831. This discovery was soon followed by reports of cell division. In 1846, Hugo von Mohl called attention to the distinction between protoplasmic material and cell sap, and in 1862, Albert von Kölliker used the term **cytoplasm** for the material surrounding the nucleus. The most conspicuous inclusions in the cytoplasm, the plastids, were long considered to be merely condensations of protoplasm. The concept of independent identity and continuity of these organelles was established in the nineteenth century. In 1880, Johannes Hanstein introduced the term **protoplast** to designate the unit of protoplasm inside the cell wall.

Every living cell has a means of isolating its contents from the external environment. A membrane called the **plasma membrane**, or **plasmalemma**, brings about this isolation. Plant cells have, in addition, a more or less rigid cellulosic cell wall (Chapter 4) deposited outside the plasma membrane. The plasma membrane controls the passage of materials into and out of the protoplast and so makes it possible for the cell to differ structurally and biochemically from its surroundings. Processes within a cell can release and transfer the energy necessary for growth and for the maintenance of metabolic processes. A cell is organized to retain and transfer information so that its development and that of its progeny can occur in an orderly manner. This way the integrity of the organism, of which the cells are a part, is maintained.

In the three centuries since Hooke first observed the structure of cork through his rudimentary microscope, our capacity to see the cell and its contents has increased dramatically. With improvement of the light microscope, it became possible to observe objects with a diameter of 0.2 micrometer (about 200 nanometers), an improvement on the naked eye about 500 times. With the transmission electron microscope (TEM), the limit of resolution imposed by light was greatly reduced. Because of problems with specimen preparation, contrast, and radiation damage, however, the resolution of biological objects is more like 2 nanometers. Nonetheless, this is still 100 times better than the resolution of the light microscope. The TEM has distinct disadvantages, however: the specimen to be observed must be preserved (dead) and cut into exceedingly thin, effectively two-dimensional slices. Optical microscopy using fluorescent dyes and various methods of illumination have enabled biologists to overcome these problems and to observe subcellular components in live plant cells (Fricker and Oparka, 1999; Cutler and Ehrhardt, 2000). Notable is the use of **green fluorescent protein (GFP)**, from the jelly fish *Aequorea victoria*, as a fluorescent protein tag and of confocal microscopy to visualize the fluorescent probes in intact tissues (Hepler and Gunning, 1998; Fricker and Oparka, 1999; Hawes et al., 2001). The observation of subcellular components in live plant cells is providing new and often unexpected insights into subcellular organization and dynamics.

## ▌PROKARYOTIC AND EUKARYOTIC CELLS

Based on the degree of internal organization of their cells, two fundamentally distinct groups of organisms are now recognized: prokaryotes and eukaryotes. The **prokaryotes** (*pro*, before; *karyon*, nucleus) are represented by the Archaea and Bacteria, including the cyanobacteria, and the **eukaryotes** (*eu*, true; *karyon*, nucleus) by all other living organisms (Madigan et al., 2003).

Prokaryotic cells differ most notably from eukaryotic cells in the organization of their genetic material. In prokaryotic cells, the genetic material is in the form of a large, circular molecule of deoxyribonucleic acid (DNA), with which a variety of proteins are loosely associated. This molecule, which is called the **bacterial chromosome**, is localized in a region of the cytoplasm called the **nucleoid** (Fig. 2.1). In eukaryotic cells, the nuclear DNA is linear and tightly bound to special proteins known as **histones**, forming a number of more complex chromosomes. These chromosomes are surrounded by a **nuclear envelope**, made up of two membranes, that separates them from the other cellular contents in a distinct **nucleus** (Fig. 2.2). Both prokaryotic cells and eukaryotic cells contain complexes of protein and ribonucleic acid (RNA), known as **ribosomes**, that play a crucial role in the assembly of protein molecules from their amino acid subunits.

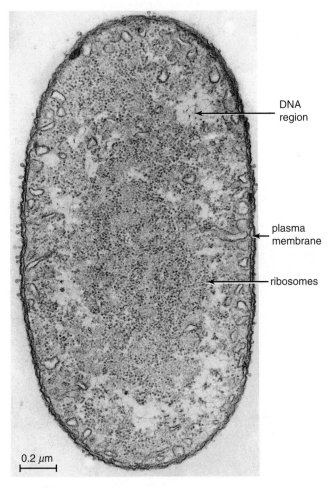

DNA region

plasma membrane

ribosomes

0.2 μm

**FIGURE 2.1**

Electron micrograph of the gram-negative bacterium, *Azotobacter vinelandii.* The granular appearance of the cytoplasm is largely due to the presence of numerous ribosomes. The clearer DNA-containing regions constitute the nucleoid. (Courtesy of Jack L. Pate.)

Eukaryotic cells are subdivided by membranes into distinct compartments that perform different functions. The cytoplasm of prokaryotic cells, by contrast, typically is not compartmentalized by membranes. Notable exceptions are the extensive system of photosynthetic membranes (thylakoids) of the cyanobacteria (Madigan et al., 2003) and the membrane-bounded entities called acidocalcisomes found in a variety of bacteria, including *Agrobacterium tumefaciens,* the plant pathogen that causes crown gall (Seufferheld et al., 2003).

The appearance of membranes under the electron microscope is remarkably similar in various organisms. When suitably preserved and stained, these membranes have a three-layered appearance, consisting of two dark layers separated by a lighter layer (Fig. 2.3). This type of membrane was named **unit membrane** by

Robertson (1962) and interpreted as a bimolecular lipid layer covered on each side with a layer of protein. Although this model of membrane structure has been superseded by the fluid mosaic model (see below), the term unit membrane remains a useful designation for a visually definable three-ply membrane.

Among the internal membranes of eukaryotic cells are those surrounding the nucleus, mitochondria, and plastids, which are characteristic components of plant cells. The cytoplasm of eukaryotic cells also contains systems of membranes (the endoplasmic reticulum and Golgi apparatus) and a complex network of nonmembranous protein filaments (actin filaments and microtubules) called the **cytoskeleton**. A cytoskeleton is absent in prokaryotic cells. Plant cells also develop multifunctional organelles, called **vacuoles**, that are bound by a membrane called the **tonoplast** (Fig. 2.2).

In addition to the plasma membrane, which controls the passage of substances into and out of the protoplast, the internal membranes control the passage of substances among compartments within the cell. This way the cell can maintain the specialized chemical environments necessary for the processes occurring in the different cytoplasmic compartments. Membranes also permit differences in electrical potential, or voltage, to become established between the cell and its environment and between adjacent compartments of the cell. Differences in the chemical concentration of various ions and molecules and the electric potential across membranes provide potential energy used to power many cellular processes.

Compartmentation of cellular contents means division of labor at the subcellular level. In a multicellular organism a division of labor occurs also at the cellular level as the cells differentiate and become more or less specialized with reference to certain functions. Functional specialization finds its expression in morphological differences among cells, a feature that accounts for the complexity of structure in a multicellular organism.

# CYTOPLASM

As mentioned previously, the term **cytoplasm** was introduced to designate the protoplasmic material surrounding the nucleus. In time, discrete entities were discovered in this material, first only those that were within the resolving power of the light microscope; later, smaller entities were discovered with the electron microscope. Thus the concept of cytoplasm has undergone an evolution; with new technologies the concept undoubtedly will continue to evolve. Most biologists today use the term cytoplasm, as originally introduced by Kölliker (1862), to designate all of the material surrounding the nucleus, and they refer to the cytoplasmic

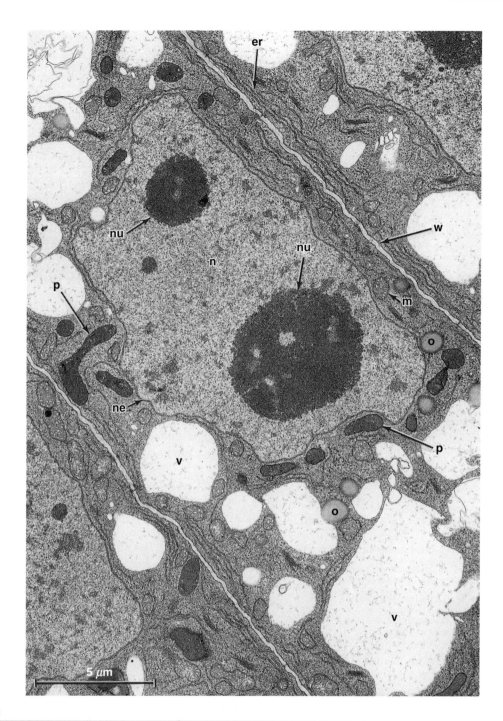

**FIGURE 2.2**

*Nicotiana tabacum* (tobacco) root tip. Longitudinal section of young cells. Details: er, endoplasmic reticulum; m, mitochondrion; n, nucleus; ne, nuclear envelope; nu, nucleolus; o, oil body; p, plastid; v, vacuole; w, cell wall. (From Esau, 1977.)

matrix, in which the nucleus, organelles, membrane systems, and nonmembranous entities are suspended, as the **cytosol**. As originally defined, however, the term cytosol was used to refer specifically "to the cytoplasm minus mitochondria and endoplasmic reticulum components" in liver cells (Lardy, 1965). **Cytoplasmic ground**

**substance** and **hyaloplasm** are terms that commonly have been used by plant cytologists to refer to the cytoplasmic matrix. Some biologists use the term cytoplasm in the sense of cytosol.

In the living plant cell the cytoplasm is always in motion; the organelles and other entities suspended in

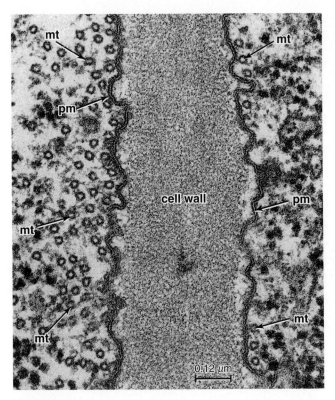

**FIGURE 2.3**

Electron micrograph showing the three-layered appearance of the plasma membranes (pm) on either side of the common wall between two cells of an *Allium cepa* leaf. Microtubules (mt) in transectional view can be seen on both sides of the wall.

**TABLE 2.1 ■ An Inventory of Plant Cell Components**

| Cell wall | Middle lamella | |
| --- | --- | --- |
| | Primary wall | |
| | Secondary wall | |
| | Plasmodesmata | |
| Protoplast | Nucleus | Nuclear envelope |
| | | Nucleoplasm |
| | | Chromatin |
| | | Nucleolus |
| | Cytoplasm | Plasma membrane |
| | | Cytosol (cytoplasmic ground substance, hyaloplasm) |
| | | Organelles bounded by two membranes: |
| | | Plastids |
| | | Mitochondria |
| | | Organelles bounded by one membrane: |
| | | Peroxisomes |
| | | Vacuoles, bounded by tonoplast |
| | | Ribosomes |
| | | Endomembrane system (major components): |
| | | Endoplasmic reticulum |
| | | Golgi apparatus |
| | | Vesicles |
| | | Cytoskeleton: |
| | | Microtubules |
| | | Actin filaments |

the cytosol can be observed being swept along in an orderly fashion in the moving currents. This movement, which is known as **cytoplasmic streaming**, or **cyclosis**, results from an interaction between bundles of actin filaments and the so-called motor protein, **myosin**, a protein molecule with an ATPase-containing "head" that is activated by actin (Baskin, 2000; Reichelt and Kendrich-Jones, 2000). Cytoplasmic streaming, a costly energy-consuming process, undoubtedly facilitates the exchange of materials within the cell (Reuzeau et al., 1997; Kost and Chua, 2002) and between the cell and its environment.

The various components of the protoplast are considered individually in the following paragraphs. Among those components are the entities called **organelles**. As with the term cytoplasm, the term organelle is used differently by different biologists. Whereas some restrict use of the term organelle to membrane-bound entities such as plastids and mitochondria, others use the term more broadly to refer also to the endoplasmic reticulum and Golgi bodies and to nonmembranous components such as microtubules and ribosomes. The term organelle is used in the restricted sense in this book (Table 2.1).

In this chapter only the plasma membrane, nucleus, and cytoplasmic organelles are considered. The remaining components of the protoplast are covered in Chapter 3.

# PLASMA MEMBRANE

Among the various membranes of the cell, the **plasma membrane** typically has the clearest dark-light-dark or unit membrane appearance in electron micrographs (Fig. 2.3; Leonard and Hodges, 1980; Robinson, 1985). The plasma membrane has several important functions: (1) it mediates the transport of substances into and out of the protoplast, (2) it coordinates the synthesis and assembly of cell wall microfibrils (cellulose), and (3) it transduces hormonal and environmental signals involved in the control of cell growth and differentiation.

The plasma membrane has the same basic structure as the internal membranes of the cell, consisting of a **lipid bilayer** in which are embedded globular proteins, many extending across the bilayer and protrude on either side (Fig. 2.4). The portion of these **transmembrane proteins** embedded in the bilayer is

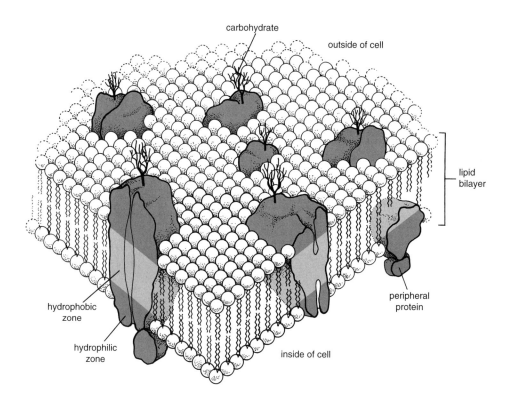

**FIGURE 2.4**

Fluid-mosaic model of membrane structure. The membrane is composed of a bilayer of lipid molecules—with their hydrophobic "tails" facing inward—and large protein molecules. Some of the proteins (transmembrane proteins) traverse the bilayer; others (peripheral proteins) are attached to the transmembrane proteins. Short carbohydrate chains are attached to most of the protruding transmembrane proteins on the outer surface of the plasma membrane. The whole structure is quite fluid; some of the transmembrane proteins float freely within the bilayer, and together with the lipid molecules move laterally within it, forming different patterns, or "mosaics," and hence the proteins can be thought of as floating in a lipid "sea." (From Raven et al., 1992.)

hydrophobic, whereas the portion or portions exposed on either side of the membrane are hydrophilic.

The inner and outer surfaces of a membrane differ considerably in chemical composition. For example, there are two major types of lipids in the plasma membrane of plant cells—**phospholipids** (the more abundant) and **sterols** (particularly stigmasterol)—and the two layers of the bilayer have different compositions of these. Moreover the transmembrane proteins have definite orientations within the bilayer, and the portions protruding on either side have different amino acid compositions and tertiary structures. Other proteins are also associated with membranes, including the **peripheral proteins**, so called because they lack discrete hydrophobic sequences and thus do not penetrate into the lipid bilayer. Transmembrane proteins and other lipid-bound proteins tightly bound to the membrane are called **integral proteins**. On the outer surface of the plasma membrane, short-chain carbohydrates (oligosaccharides) are attached to the protruding proteins, forming glycoproteins. The carbohydrates, which form a coat on the outer surface of the membranes of some

eukaryotic cells, are believed to play important roles in cell-to-cell adhesion processes and in the "recognition" of molecules (e.g., hormones, viruses, and antibiotics) that interact with the cell.

Whereas the lipid bilayer provides the basic structure and impermeable nature of cellular membranes, the proteins are responsible for most membrane functions. Most membranes are composed of 40% to 50% lipid (by weight) and 60% to 50% protein, but the amounts and types of proteins in a membrane reflect its function. Membranes involved with energy transduction, such as the internal membranes of mitochondria and chloroplasts, consist of about 75% protein. Some of the proteins are enzymes that catalyze membrane-associated reactions, whereas others are **transport proteins** involved in the transfer of specific molecules into and out of the cell or organelle. Still others act as receptors for receiving and transducing chemical signals from the cell's internal or external environment. Although some of the integral proteins appear to be anchored in place (perhaps to the cytoskeleton), the lipid bilayer is generally quite fluid. Some of the proteins float more or less freely in the bilayer, and they and the lipid molecules can move laterally within it, forming different patterns, or mosaics, that vary from time to time and place to place—hence the name **fluid-mosaic** for this model of membrane structure (Fig. 2.4; Singer and Nicolson, 1972; Jacobson et al., 1995).

Membranes contain different kinds of transport proteins (Logan et al., 1997; Chrispeels et al., 1999; Kjellbom et al., 1999; Delrot et al., 2001). Two of the types are carrier proteins and channel proteins, both of which permit the movement of a substance across a membrane only down the substance's electrochemical gradient; that is, they are passive transporters. **Carrier proteins** bind the specific solute being transported and undergo a series of conformational changes in order to transport the solute across the membrane. **Channel proteins** form water-filled pores that extend across the membrane and, when open, allow specific solutes (usually inorganic ions, e.g., $K^+$, $Na^+$, $Ca^{2+}$, $Cl^-$) to pass through them. The channels are not open continuously; instead they have "gates" that open briefly and then close again, a process referred to as **gating**.

The plasma membrane and tonoplast also contain water channel proteins called **aquaporins** that specifically facilitate the passage of water through the membranes (Schäffner, 1998; Chrispeels et al., 1999; Maeshima, 2001; Javot and Maurel, 2002). Water passes relatively freely across the lipid bilayer of biological membranes, but the aquaporins allow water to diffuse more rapidly across the plasma membrane and tonoplast. Because the vacuole and cytosol must be in constant osmotic equilibrium, rapid movement of water is essential. It has been suggested that aquaporins facilitate the rapid flow of water from the soil into root cells and to the xylem during periods of high transpiration. Aquaporins have been shown to block the influx of water into cells of the roots during periods of flooding (Tournaire-Roux et al., 2003) and to play a role in drought avoidance in rice (Lian et al., 2004). In addition evidence indicates that water movement through aquaporins increases in response to certain environmental stimuli that induce cell expansion and growth; the cyclic expression of a plasma membrane aquaporin has been implicated in the leaf unfolding mechanism in tobacco (Siefritz et al., 2004).

Carriers can be classified as uniporters and cotransporters according to how they function. **Uniporters** transport only one solute from one side of the membrane to another. With **cotransporters**, the transfer of one solute depends on the simultaneous or sequential transfer of a second solute. The second solute may be transported in the same direction, in which case the carrier protein is known as **symporter**, or in the opposite direction, as in the case of an **antiporter**.

The transport of a substance against its electrochemical gradient requires the input of energy, and is called **active transport**. In plants that energy is provided primarily by an ATP-powered **proton pump**, specifically, a membrane-bound $H^+$-ATPase (Sze et al., 1999; Palmgren, 2001). The enzyme generates a large gradient of protons ($H^+$ ions) across the membrane. This gradient provides the driving force for solute uptake by all proton-coupled cotransport systems. The tonoplast is unique among plant membranes in having two proton pumps, an $H^+$-ATPase and an $H^+$-pyrophosphatase ($H^+$-PPase) (Maeshima, 2001), although some data indicate that $H^+$-PPase may also be present in the plasma membrane of some tissues (Ratajczak et al., 1999; Maeshima, 2001).

The transport of large molecules such as most proteins and polysaccharides cannot be accommodated by the transport proteins that ferry ions and small polar molecules across the plasma membrane. These large molecules are transported by means of vesicles or sac-like structures that bud off from or fuse with the plasma membrane, a process called **vesicle-mediated transport** (Battey et al., 1999). The transport of material into the cell by vesicles that bud off of the plasma membrane is called **endocytosis** and involves portions of the plasma membrane called coated pits (Fig. 2.5; Robinson

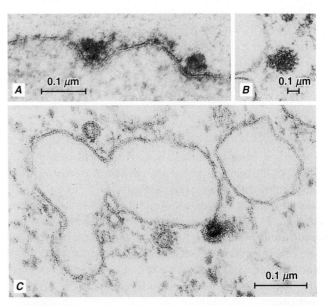

**FIGURE 2.5**

Endocytosis in maize (*Zea mays*) rootcap cells that have been exposed to a solution containing lead nitrate. **A**, granular deposits containing lead can be seen in two coated pits. **B**, a coated vesicle with lead deposits. **C**, here, one of two coated vesicles has fused with a large Golgi vesicle where it will release its contents. This coated vesicle (dark structure) still contains lead deposits, but it appears to have lost its coat, which is located just to the right of it. The coated vesicle to its left is clearly intact. (Courtesy of David G. Robinson.)

and Depta, 1988; Gaidarov et al., 1999). **Coated pits** are depressions in the plasma membrane containing specific receptors (to which the molecules to be transported into the cell must first bind) and coated on their cytoplasmic surface with **clathrin**, a protein composed of three large and three smaller polypeptide chains that together form a three-pronged structure, called a triskelion. Invaginations of the coated pits pinch off to form **coated vesicles**. Within the cell the coated vesicles shed their coats and then fuse with some other membrane-bound structures (e.g., Golgi bodies or small vacuoles). Transport by means of vesicles in the opposite direction is called **exocytosis** (Battey et al., 1999). During exocytosis, vesicles originating from within the cell fuse with the plasma membrane, expelling their contents to the outside.

Relatively large invaginations, or infoldings, of the plasma membrane are frequently encountered in tissue prepared for electron microscopy. Some form pockets between the cell wall and protoplast, and may include tubules and vesicles. Some invaginations may push the tonoplast forward and intrude into the vacuole. Others, called **multivesicular bodies**, are often detached from the plasma membrane and embedded in the cytosol or appear suspended in the vacuole. Similar formations were first observed in fungi and named lomasomes (Clowes and Juniper, 1968). Multivesicular bodies in *Nicotiana tabacum* BY-2 cells have been identified as plant prevacuolar compartments that lie on the endocytic pathway to lytic vacuoles (see below; Tse et al., 2004).

# ▌NUCLEUS

Often the most prominent structure within the protoplast of eukaryotic cells, the **nucleus** performs two important functions: (1) it controls the ongoing activities of the cell by determining which RNA and protein molecules are produced by the cell and when they are produced, and (2) it is the repository of most of the cell's genetic information, passing it on to the daughter cells in the course of cell division. The total genetic information stored in the nucleus is referred to as the **nuclear genome**.

The nucleus is bounded by a pair of membranes called the **nuclear envelope**, with a *perinuclear space* between them (Figs. 2.2 and 2.6; Dingwall and Laskey, 1992; Gerace and Foisner, 1994; Gant and Wilson, 1997; Rose et al., 2004). In various places the outer membrane of the envelope is continuous with the endoplasmic reticulum, so that the perinuclear space is continuous with the lumen of the endoplasmic reticulum. The nuclear envelope is considered a specialized, locally differentiated portion of the endoplasmic reticulum. The most distinctive feature of the nuclear envelope is

the presence of a great many cylindrical **nuclear pores**, which provide direct contact between the cytosol and the ground substance, or **nucleoplasm**, of the nucleus (Fig. 2.6). The inner and outer membranes are joined around each pore, forming the margin of its opening. Structurally complicated **nuclear pore complexes**—the largest supramolecular complexes assembled in the eukaryotic cell—span the envelope at the nuclear pores (Heese-Peck and Raikhel, 1998; Talcott and Moore, 1999; Lee, J.-Y., et al., 2000). The nuclear pore complex is roughly wheel-shaped, consisting in part of a cylindrical central channel (the hub) from which eight spokes project outwardly to an interlocking collar associated with the nuclear membrane lining the pore. The nuclear pore complexes allow relatively free passage of certain ions and small molecules through diffusion channels, which measure about 9 nanometers in diameter. The proteins and other macromolecules transported through the nuclear pore complexes greatly exceed this channel size. Their transport is mediated by a highly selective

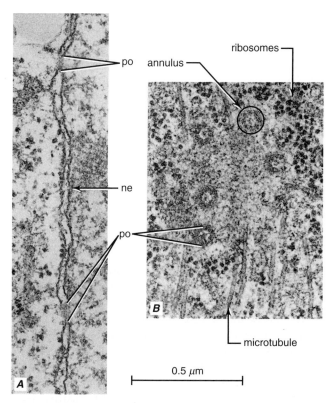

**FIGURE 2.6**

Nuclear envelope (ne) in profile (**A**) and from the surface (**B**, central part) showing pores (po). The electron-dense material in the pores in **A** is shown, in **B**, to have a form of an annulus with a central granule. The clear space between the membranes in **A** is called the perinuclear space. From a parenchyma cell in *Mimosa pudica* petiole. (From Esau, 1977.)

active (energy-dependent) transport mechanism that takes place through the central channel. The central channel has a functional diameter of up to 26 nanometers (Hicks and Raikhel, 1995; Görlich and Mattaj, 1996; Görlich, 1997).

In specially stained cells, thin threads and grains of **chromatin** can be distinguished from the nucleoplasm. Chromatin is made up of DNA combined with large amounts of proteins called *histones*. During the process of nuclear division, the chromatin becomes progressively more condensed until it takes the form of **chromosomes**. Chromosomes (chromatin) of nondividing, or *interphase*, nuclei are attached at one or more sites to the inner membrane of the nuclear envelope. Before DNA replication each chromosome is composed of a single, long DNA molecule, which carries the hereditary information. In most interphase nuclei the bulk of chromatin is diffuse and lightly staining. This uncondensed chromatin, called *euchromatin*, is genetically active and associated with high rates of RNA synthesis. The remaining, condensed chromatin, called *heterochromatin*, is genetically inactive; that is, it is not associated with RNA synthesis (Franklin and Cande, 1999). Overall, only a small percentage of the total chromosomal DNA codes for essential proteins or RNAs; apparently there is a substantial surplus of DNA in the genomes of higher organisms (Price, 1988). Nuclei may contain proteinaceous inclusions of unknown function in crystalline, fibrous, or amorphous form (Wergin et al., 1970), in addition to chromatin-containing "micropuffs" and coiled bodies composed of ribonucleoprotein (Martín et al., 1992).

Different organisms vary in the number of chromosomes present in their somatic (vegetative, or body) cells. *Haplopappus gracilis*, a desert annual, has 4 chromosomes per cell; *Arabidopsis thaliana*, 10; *Vicia faba*, broad bean, 12; *Brassica oleracea*, cabbage, 18; *Asparagus officinalis*, 20; *Triticum vulgare*, bread wheat, 42; and *Cucurbita maxima*, squash, 48. The reproductive cells, or gametes, have only half the number of chromosomes that is characteristic of the somatic cells in an organism. The number of chromosomes in the gametes is referred to as the **haploid** (single set) number and designated as *n*, and that in the somatic cells is called the **diploid** (double set) number, which is designated as 2*n*. Cells that have more than two sets of chromosomes are said to be **polyploid** (3*n*, 4*n*, 5*n*, or more).

Often the only structures discernible within a nucleus with the light microscope are spherical structures known as **nucleoli** (singular: **nucleolus**) (Fig. 2.2; Scheer et al., 1993). The nucleolus contains high concentrations of RNA and proteins, along with large loops of DNA emanating from several chromosomes. The loops of DNA, known as *nucleolar organizer regions*, contain clusters of ribosomal RNA (rRNA) genes. At these sites, newly formed rRNAs are packaged with ribosomal proteins imported from the cytosol to form ribosomal subunits (large and small). The ribosomal subunits are then transferred, via the nuclear pores, to the cytosol where they are assembled to form ribosomes. Although the nucleolus commonly is thought of as the site of ribosome manufacture, it is involved with only a part of the process. The very presence of a nucleolus is due to the accumulation of the molecules being packaged to form ribosomal subunits.

In many diploid organisms, the nucleus contains one nucleolus to each haploid set of chromosomes. The nucleoli may fuse and then appear as one large structure. The size of a nucleolus is a reflection of the level of its activity. In addition to the DNA of the nucleolar organizer region, nucleoli contain a fibrillar component consisting of rRNA already associated with protein to form fibrils, and a granular component consisting of maturing ribosomal subunits. Active nucleoli also show lightly stained regions commonly referred to as vacuoles. In living cultured cells these regions, which should not be confused with the membrane-bound vacuoles found in the cytosol, can be seen to be undergoing repeated contractions, a phenomenon that might be involved with RNA transport.

Nuclear divisions are of two kinds: **mitosis**, during which a nucleus gives rise to two daughter nuclei, each morphologically and genetically equivalent to the other and to the parent nucleus; **meiosis**, during which the parent nucleus undergoes two divisions, one of which is a reduction division. By a precise mechanism, meiosis produces four daughter nuclei, each with one-half the number of chromosomes as the parent nucleus. In plants, mitosis gives rise to somatic cells and to gametes (sperm and egg), and meiosis to meiospores. In both kinds of division (with some exceptions) the nuclear envelope breaks into fragments, which become indistinguishable from ER cisternae, and the nuclear pore complexes are disassembled. When new nuclei are assembled during telophase, ER vesicles join to form two nuclear envelopes, and new nuclear pore complexes are formed (Gerace and Foisner, 1994). The nucleoli disperse during late prophase (with some exceptions) and are newly organized during telophase.

# CELL CYCLE

Actively dividing somatic cells pass through a regular sequence of events known as the cell cycle. The cell cycle commonly is divided into interphase and mitosis (Fig. 2.7; Strange, 1992). Interphase precedes mitosis, and in most cells, mitosis is followed by **cytokinesis**, the division of the cytoplasmic portion of a cell and the separation of daughter nuclei into separate cells (Chapter 4). Hence most plant cells are uninucleate. Certain specialized cells may become multinucleate either only

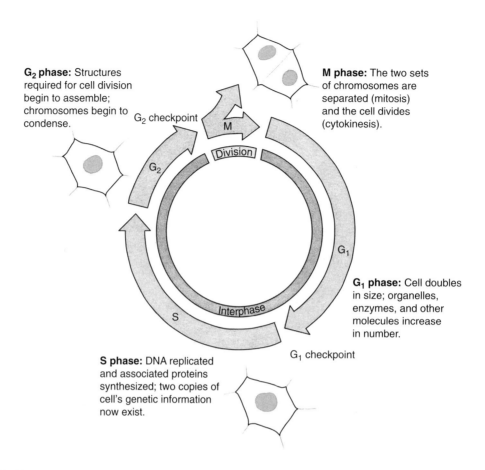

**G₂ phase:** Structures required for cell division begin to assemble; chromosomes begin to condense.

G₂ checkpoint

G₂

**M phase:** The two sets of chromosomes are separated (mitosis) and the cell divides (cytokinesis).

Division

M

Interphase

G₁

**G₁ phase:** Cell doubles in size; organelles, enzymes, and other molecules increase in number.

G₁ checkpoint

S

**S phase:** DNA replicated and associated proteins synthesized; two copies of cell's genetic information now exist.

**FIGURE 2.7**

The cell cycle. Cell division, which consists of mitosis (the division of the nucleus) and cytokinesis (the division of the cytoplasm), takes place after the completion of the three preparatory phases (G₁, S, and G₂) of interphase. Progression of the cell cycle is mainly controlled at two checkpoints, one at the end of G₁ and the other at the end of G₂. After the G₂ phase comes mitosis, which is usually followed by cytokinesis. Together, mitosis and cytokinesis constitute the M phase of the cell cycle. In cells of different species or of different tissues within the same organism, the various phases occupy different proportions of the total cycle. (From Raven et al., 2005.)

during their development (e.g., nuclear endosperm) or for life (e.g., nonarticulated laticifers). Mitosis and cytokinesis together are referred to as the **M phase** of the cell cycle.

**Interphase** can be divided into three phases, which are designated G₁, S, and G₂. The **G₁ phase** (G stands for gap) occurs after mitosis. It is a period of intense biochemical activity, during which the cell increases in size, and the various organelles, internal membranes, and other cytoplasmic components increase in number. The **S (synthesis) phase** is the period of DNA replication. At the onset of DNA replication, a diploid nucleus is said to have a 2C DNA value (C is the haploid DNA content); at completion of the S phase, the DNA value has doubled to 4C. During the S phase, many of the histones and other DNA-associated proteins are also synthesized. Following the S phase, the cell enters the **G₂** phase, which follows the S phase and precedes mitosis. The primary role of the S phase is to make sure chromosome replication is complete and to allow for repair of damaged DNA. The microtubules of the preprophase band, a ring-like band of microtubules that borders the plasma membrane and encircles the nucleus in a plane corresponding to the plane of cell division, also develop during the G₂ phase (Chapter 4; Gunning and Sammut, 1990). During mitosis the genetic material synthesized during the S phase is divided equally between two daughter nuclei, restoring the 2C DNA value.

The nature of the control or controls that regulate the cell cycle apparently is basically similar in all eukaryotic cells. In the typical cell cycle, progression is controlled at crucial transition points, called **checkpoints**—first at the G₁-S phase transition and then at the G₂-M transition (Boniotti and Griffith, 2002). The first checkpoint determines whether or not the cell enters the S phase, the second whether or not mitosis is initiated. A third checkpoint, the metaphase checkpoint, delays anaphase if some chromosomes are not properly attached to the mitotic spindle. Progression through the cycle depends on the successful formation, activation, and subsequent inactivation of cyclin-dependent protein kinases (CDKs) at the checkpoints. These kinases consist of a catalytic CDK subunit and an activating cyclin subunit (Hemerly et al., 1999; Huntley and Murray, 1999; Mironov et al., 1999; Potuschak and Doerner, 2001; Stals and Inzé, 2001). Both auxins and cytokinins have been implicated in the control of the plant cell cycle (Jacqmard et al., 1994; Ivanova and Rost, 1998; den Boer and Murray, 2000).

Cells in the $G_1$ phase have several options. In the presence of sufficient stimuli they can commit to further cell division and progress into the S phase. They may pause in their progress through the cell cycle in response to environmental factors, as during winter dormancy, and resume dividing at a later time. This specialized resting, or dormant, state is often called the **$G_0$ phase** (G-zero phase). Other fates include differentiation and **programmed cell death**, a genetically determined program that can orchestrate death of the cell (Chapter 5; Lam et al., 1999).

Some cells feature only DNA replication and gap phases without subsequent nuclear division, a process known as **endoreduplication** (Chapter 5; D'Amato, 1998; Larkins et al., 2001). The single nucleus then becomes polyploid (endopolyploidy, or endoploidy). Endoploidy may be part of the differentiation of single cells, as it is in the *Arabidopsis* trichome (Chapter 9), or that of any tissue or organ. A positive correlation exists between cell volume and the degree of polyploidy in most plant cells, indicating that polyploid nuclei might be required for the formation of large plant cells (Kondorosi et al., 2000).

# ▌PLASTIDS

Together with vacuoles and cell walls, **plastids** are characteristic components of plant cells (Bowsher and Tobin, 2001). Each plastid is surrounded by an ***envelope*** consisting of two membranes. Internally the plastid is differentiated into a more or less homogeneous matrix, the **stroma**, and a system of membranes called **thylakoids**. The principal permeability barrier between cytosol and plastid stroma is the inner membrane of the plastid envelope. The outer membrane, although a barrier to cytosolic proteins, has generally been assumed to be permeable to low molecular weight solutes (<600 Da), an assumption that may be in need of re-evaluation (Bölter and Soll, 2001). Stroma-filled tubules have been observed emanating from the surfaces of some plastids.

These so-called **stromules** can interconnect different plastids and have been shown to permit exchange of green fluorescent protein between plastids (Köhler et al., 1997; Köhler and Hanson, 2000; Arimura et al., 2001; Gray et al., 2001; Pyke and Howells, 2002; Kwok and Hanson, 2004). In a study of stromule biogenesis, increases in stromule length and frequency correlated with chromoplast differentiation; it was proposed that stromules enhance the specific metabolic activities of plastids (Waters et al., 2004).

Plastids are semiautonomous organelles widely accepted to have evolved from free-living cyanobacteria through the process of endosymbiosis (Palmer and Delwiche, 1998; Martin, 1999; McFadden, 1999; Reumann and Keegstra, 1999; Stoebe and Maier, 2002). Indeed, plastids resemble bacteria in several ways. For example, plastids, like bacteria, contain **nucleoids**, which are regions containing DNA. The DNA of the plastid, like that of the bacterium, exists in circular form (Sugiura, 1989); moreover it is not associated with histones. During the course of evolution most of the DNA of the endosymbiont (the cyanobacterium) was gradually transferred to the host nucleus; hence the genome of the modern plastid is quite small compared to the nuclear genome (Bruce, 2000; Rujan and Martin, 2001). Both plastids and bacteria contain ribosomes (70S ribosomes) that are about two-thirds as large as the ribosomes (80S ribosomes) found in the cytosol and associated with endoplasmic reticulum. (The S stands for Svedbergs, the units of the sedimentation coefficient.) In addition the process of plastid division—binary fission—is morphologically similar to bacterial cell division.

## Chloroplasts Contain Chlorophyll and Carotenoid Pigments

Mature plastids are commonly classified on the basis of the kinds of pigments they contain. **Chloroplasts** (Figs. 2.8–2.10), the sites of photosynthesis, contain **chlorophyll** and **carotenoid pigments**. The chlorophyll pigments are responsible for the green color of these plastids, which occur in green plant parts and are particularly numerous and well differentiated in leaves. In seed plants, chloroplasts are usually disk-shaped and measure between 4 and 6 micrometers in diameter. The number of chloroplasts found in a single mesophyll (middle of the leaf) cell varies widely, depending on the species and the size of the cell (Gray, 1996). A single mesophyll cell of cocoa (*Cacao theobroma*) and *Peperomia metallia* leaves may contain as few as three chloroplasts, whereas as many as 300 chloroplasts occur in a single mesophyll cell of the radish (*Raphanus sativus*) leaf. The mesophyll cells of most leaves that have been examined for plastid development contain 50 to 150 chloroplasts each. The chloroplasts are usually

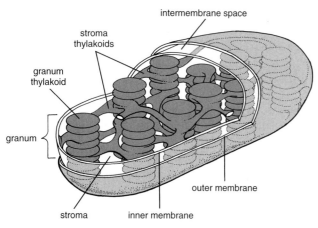

**FIGURE 2.8**

Three-dimensional structure of a chloroplast. Note that the internal membranes (thylakoids) are not connected with the plastid envelope. (From Raven et al., 1992.)

found with their broad surfaces parallel to the cell wall, preferentially on cell surfaces bordering air spaces. They can reorient in the cell under the influence of light—for example, gathering along the walls parallel with the leaf surface under low or medium light intensity, thereby optimizing light utilization for photosynthesis (Trojan and Gabryś, 1996; Williams et al., 2003). Under potentially damaging high light intensity the chloroplasts can orient themselves along walls perpendicular to the leaf surface. The blue-UV region of the spectrum is the most effective stimulus for chloroplast movement (Trojan and Gabryś, 1996; Yatsuhashi, 1996; Kagawa and Wada, 2000, 2002). In the darkness the chloroplasts are distributed either randomly around all the cell walls or their arrangement depends on local factors inside the cells (Haupt and Scheuerlein, 1990). Presumably movement of the chloroplasts involves an actin-myosin-based system.

The internal structure of the chloroplast is complex. The stroma is traversed by an elaborate system of thylakoids, consisting of **grana** (singular: **granum**)—stacks of disk-like thylakoids that resemble a stack of coins—and **stroma thylakoids** (or intergrana thylakoids) that traverse the stroma between grana and interconnect them (Figs. 2.8–2.10). The grana and stroma thylakoids and their internal compartments are believed to constitute a single, interconnected system. The thylakoids are not physically connected with the plastid envelope but are completely embedded in the stroma. Chlorophylls and carotenoid pigments—both of which are involved in light harvesting—are embedded, along with proteins, in the thylakoid membranes in discrete units of organization called photosystems. The principal function of the carotenoid pigments is that of an antioxidant,

preventing photo-oxidative damage to the chlorophyll molecules (Cunningham and Gantt, 1998; Vishnevetsky et al., 1999; Niyogi, 2000).

Chloroplasts often contain starch, phytoferritin (an iron compound) and lipid in the form of globules called **plastoglobuli** (singular: **plastoglobule**). The starch grains are temporary storage products and accumulate only when the plant is actively photosynthesizing. They may be lacking in the chloroplasts of plants kept in the dark for as little as 24 hours but often reappear after the plant has been in the light for only 3 or 4 hours.

Mature chloroplasts contain numerous copies of a circular plastid DNA molecule and the machinery for the replication, transcription, and translation of that genetic material (Gray, J. C., 1996). With the limited coding capacity (approximately 100 proteins) of the chloroplast, however, the vast majority of proteins involved with chloroplast biogenesis and function are encoded by the nuclear genome (Fulgosi and Soll, 2001). These proteins, which are synthesized on ribosomes in the cytosol, are targeted into the chloroplast as precursor proteins with the aid of an amino-terminal extension referred to as a **transit peptide**. Each protein imported into the chloroplast contains a specific transit peptide. The transit peptide both targets the protein to the chloroplast and mediates import into the stroma where it is cleaved off after import (Flügge, 1990; Smeekens et al., 1990; Theg and Scott, 1993). Transport across a thylakoid membrane is mediated by a second transit peptide unmasked when the first one is cleaved off (Cline et al., 1993; Keegstra and Cline, 1999). Evidence indicates that part of the chloroplastic protein machinery is derived from the endosymbiotic cyanobacterial ancestor of chloroplasts (Reumann and Keegstra, 1999; Bruce, 2000).

In addition to regulatory traffic from the nucleus to the chloroplast, the chloroplasts transmit signals to the nucleus to coordinate nuclear and chloroplast gene expression. Moreover plastid signals also regulate the expression of nuclear genes for nonplastid proteins and for the expression of mitochondrial genes (see references in Rodermel, 2001). Chloroplasts are not only sites of photosynthesis; they are also involved in amino acid synthesis and fatty acid synthesis and provide space for the temporary storage of starch.

## Chromoplasts Contain Only Carotenoid Pigments

**Chromoplasts** (*chroma*, color) are also pigmented plastids (Fig. 2.11). Of variable shape, they lack chlorophyll but synthesize and retain carotenoid pigments, which are often responsible for the yellow, orange, or red colors of many flowers, old leaves, some fruits, and some roots. Chromoplasts are the most heterogeneous category of plastids and are classified entirely on the structure of the carotenoid-bearing components present

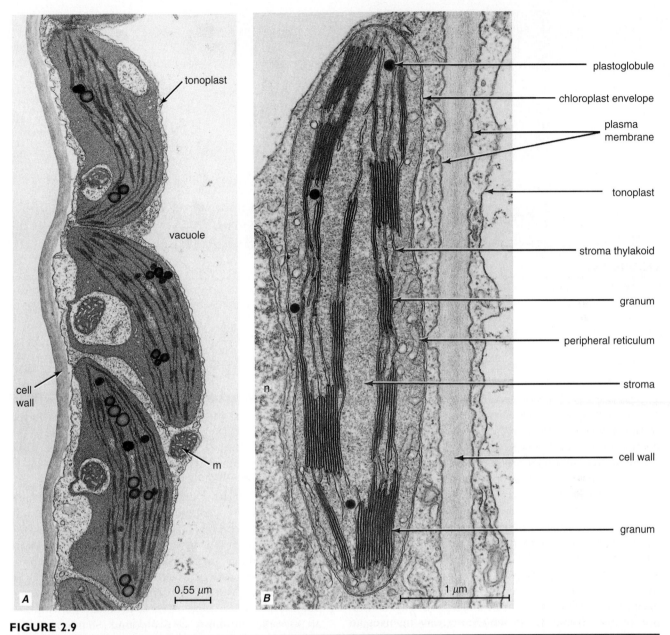

**FIGURE 2.9**

**A**, chloroplasts along the cell wall in a leaf cell of sheperd's purse (*Capsella bursa-pastoris*). Mitochondria (m) are closely associated spatially with the chloroplasts. **B**, chloroplast with grana seen in profile. From a leaf of tobacco (*Nicotiana tabacum*). (**B**, from Esau, 1977.)

in the mature plastid (Sitte et al., 1980). Most belong to one of four types: (1) **globular chromoplasts**, with many carotenoid-bearing plastoglobuli (Fig. 2.11A). Remnants of thylakoids also may be present. The plastoglobuli often are concentrated in the peripheral stroma beneath the envelope (petals of *Ranunculus repens* and yellow fruits of *Capsicum*, perianth of *Tulipa*, *Citrus* fruit); (2) **membranous chromoplasts**, which are characterized by a set of up to 20 concentric (double) carotenoid-containing membranes (Fig. 2.11B)

(*Narcissus* and *Citrus sinensis* petals); (3) **tubular chromoplasts**, in which the carotenoids are incorporated into filamentous lipoprotein "tubules" (Fig. 2.11C) (red fruits of *Capsicum*, rose hypanthium; *Tropaeolum* petals; Knoth et al., 1986); (4) **crystalline chromoplasts**, which contain crystalline inclusions of pure carotene (Fig. 2.11D) (β-carotene in *Daucus*, carrot, roots and lycopene in *Solanum lycopersicum*, tomato, fruit). Carotene crystals, commonly called **pigment bodies**, originate within thylakoids and remain

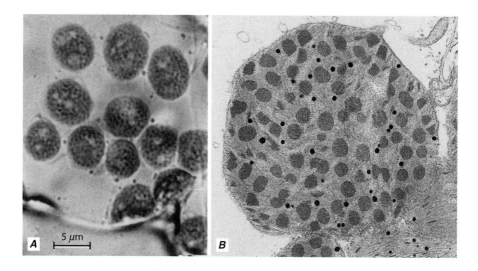

**FIGURE 2.10**

Chloroplast structure. **A**, with the light microscope grana within chloroplasts appear as dots. These chloroplasts are from a cotyledon of *Solanum lycopersicum*. **B**, an electron micrograph of a chloroplast from a bundle-sheath cell of a *Zea* leaf showing grana from the surface. (**A**, from Hagemann, 1960.)

surrounded by the plastid envelope during all stages of development. Globular chromoplasts are the most common type and are regarded as the oldest and most primitive in evolutionary terms (Camara et al., 1995).

Chromoplasts may develop from previously existing green chloroplasts by a transformation in which the chlorophyll and thylakoid membranes of the chloroplast disappear and masses of carotenoids accumulate, as occurs during the ripening of many fruits (Ziegler et al., 1983; Kuntz et al., 1989; Marano and Carrillo, 1991, 1992; Cheung et al., 1993; Ljubešić et al., 1996). Interestingly these changes apparently are accompanied by the disappearance of plastid ribosomes and rRNAs but not of the plastid DNA, which remains unchanged (Hansmann et al., 1987; Camara et al., 1989; Marano and Carrillo, 1991). With the loss of plastid ribosomes and rRNAs, protein synthesis can no longer occur in the chromoplast, indicating that it is necessary for chromoplast-specific proteins to be coded for in the nucleus and then imported into the developing chromoplast. Chromoplast development is not an irreversible phenomenon, however. For example, the chromoplasts of citrus fruit (Goldschmidt, 1988) and of the carrot root (Grönegress, 1971) are capable of reverse differentiation into chloroplasts; they lose the carotene pigment and develop a thylakoid system, chlorophyll, and photosynthetic apparatus.

The precise functions of chromoplasts are not well understood, although at times they act as attractants to insects and other animals with which they coevolved, playing an essential role in the cross-pollination of flow-

ering plants and the dispersal of fruit and seeds (Raven et al., 2005).

### Leucoplasts Are Nonpigmented Plastids

Structurally the least differentiated of mature plastids, **leucoplasts** (Fig. 2.12) generally have a uniform granular stroma, several nucleoids, and, despite reports to the contrary, typical 70S ribosomes. They lack an elaborate system of inner membranes (Carde, 1984; Miernyk, 1989). Some store starch (*amyloplasts*; Fig. 2.13), others store proteins (*proteinoplasts*), fats (*elaioplasts*), or combinations of these products. Amyloplasts are classified as simple or compound (Shannon, 1989). Simple amyloplasts, such as those of the potato tuber, contain a single starch grain, whereas compound amyloplasts contain several often tightly packed starch grains as in the endosperm of oats and rice. The starch grains of the potato tuber may become so large that the envelope is ruptured (Kirk and Tilney-Bassett, 1978). The compound amyloplasts in rootcaps play an essential role in gravity perception (Sack and Kiss, 1989; Sack, 1997).

### All Plastids Are Derived Initially from Proplastids

**Proplastids** are small, colorless plastids found in undifferentiated regions of the plant body such as root and shoot apical meristems (Mullet, 1988). Zygotes contain proplastids that are the ultimate precursors of all plastids within an adult plant. In most angiosperms the

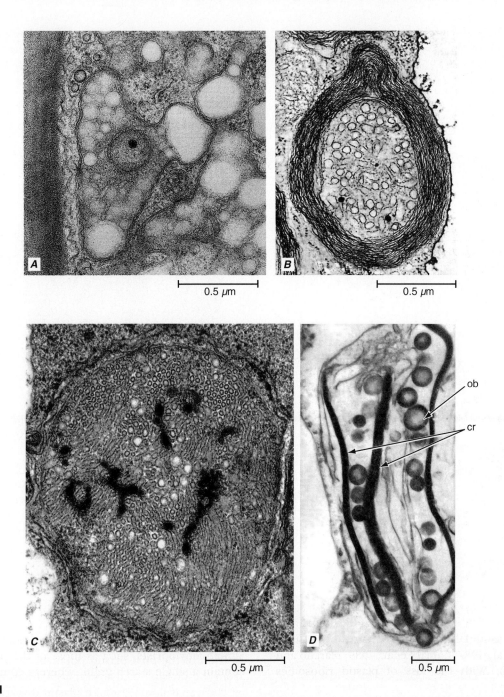

**FIGURE 2.11**

Types of chromoplasts. **A**, globular chromoplasts from *Tagetes* (marigold) petal; **B**, membranous chromoplast of *Narcissus pseudonarcissus* flower; **C**, tubular chromoplast of *Palisota barteri* fruit; **D**, crystalline chromoplast of *Solanum lycopersicum* fruit. Details: cr, crystalloids; ob, oil body. (**B**, reprinted from Hansmann et al., 1987. © 1987, with permission from Elsevier.; **C**, from Knoth et al., 1986, Fig. 7. © 1986 Springer-Verlag; **D**, from Mohr, 1979, by permission of Oxford University Press.)

proplastids of the zygote come exclusively from the cytoplasm of the egg cell (Nakamura et al., 1992). In conifers, however, the proplastids of the zygote are derived from those carried by the sperm cell. In either case the consequence is that the plastid genome of an individual plant typically is inherited from a single parent. Since all the plastids in an adult plant are derived from a single parent, all plastids (whether chloroplasts, chromoplasts, or leucoplasts) within an individual plant have identical genomes (dePamphilis and Palmer, 1989). Each proplastid contains a single circular DNA molecule.

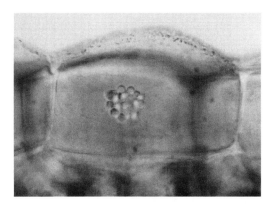

**FIGURE 2.12**

Leucoplasts clustered around the nucleus in an epidermal cell of a *Zebrina* leaf. (×620.)

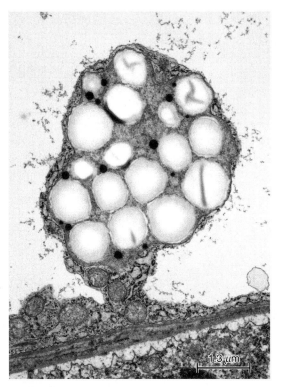

**FIGURE 2.13**

Amyloplast, a type of leucoplast, from the embryo sac of soybean (*Glycine max*). The round, clear bodies are starch grains. The smaller, dense bodies are oil bodies. Amyloplasts are involved with the synthesis and long-term storage of starch in seeds and storage organs, such as potato tubers. (Courtesy of Roland R. Dute.)

As mentioned previously, plastids reproduce by binary fission, the process of dividing into equal halves, which is characteristic of bacteria (Oross and Possingham, 1989). In meristematic cells the division of proplastids roughly keeps pace with cell division. The

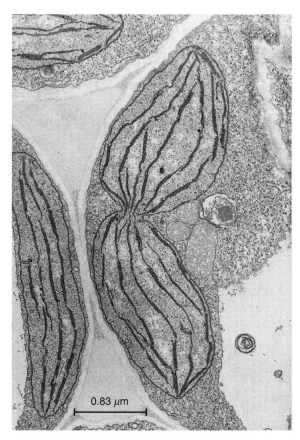

**FIGURE 2.14**

Dividing chloroplast in *Beta vulgaris* leaf. Had the division process continued, the two daughter plastids would have separated at the narrow constriction, or isthmus. Three peroxisomes can be seen to the right of the constriction.

proplastids must divide before the cells divide. The plastid population of mature cells typically exceeds that of the original proplastid population. The greater proportion of the final plastid population may be derived from the division of mature plastids during the period of cell expansion. Although plastid division apparently is controlled by the nucleus (Possingham and Lawrence, 1983), a close interaction exists between plastid DNA replication and plastid division.

Plastid division is initiated by a constriction in the middle of the plastid (Fig. 2.14). With continued narrowing of the constriction, the two daughter plastids come to be joined by a narrow isthmus, which eventually breaks. The envelope membranes of the daughter plastids then reseal. The constriction process is caused by contractile rings referred to as **plastid-dividing rings**, which are discernible with the electron microscope as electron-dense bands. There are two concentric plastid-dividing rings, an outer ring on the cytosolic face of the plastid outer membrane and an inner ring on the stromal

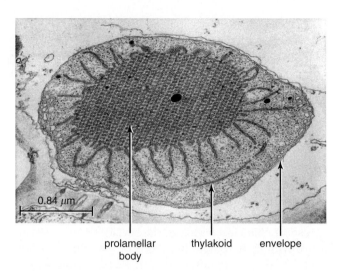

**FIGURE 2.15**

Etiolated chloroplast with a prolamellar body in a leaf cell of sugarcane *(Saccharum officinarum)*. Ribosomes are conspicuous in the plastid. (Courtesy of W. M. Laetsch.)

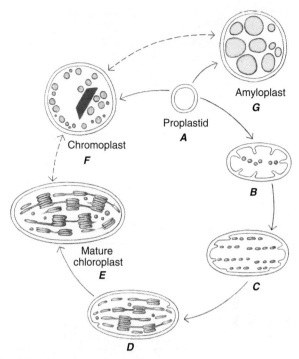

**FIGURE 2.16**

Plastid developmental cycle, beginning with the development of a chloroplast from a proplastid (**A**). Initially the proplastid contains few or no internal membranes. **B–D**, as the proplastid differentiates, flattened vesicles develop from the inner membrane of the plastid envelope and eventually align themselves into grana and stroma thylakoids. **E**, the thylakoid system of the mature chloroplast appears discontinuous with the envelope. **F, G**, proplastids may also develop into chromoplasts and leucoplasts. The leucoplast shown here is a starch-synthesizing amyloplast. Note that chromoplasts may be formed from proplastids, chloroplasts, or leucoplasts. The various kinds of plastids can change from one type to another (dashed arrows). (From Raven et al., 2005.)

face of the plastid inner membrane. Prior to the appearance of the plastid-dividing rings, two cytoskeletal-like proteins, FtsZ1 and FtsZ2—homologs of the bacterial cell division FtsZ protein—assemble into a ring at the future division site in the stroma within the plastid envelope. It has been suggested that the FtsZ ring determines the division region (Kuroiwa et al., 2002). Molecular analysis of chloroplast division indicates that the mechanism of plastid division has evolved from bacterial cell division (Osteryoung and Pyke, 1998; Osteryoung and McAndrew, 2001; Miyagishima et al., 2001).

If the development of a proplastid into a more highly differentiated form is arrested by the absence of light, it may form one or more **prolamellar bodies** (Fig. 2.15), which are quasi-crystalline bodies composed of tubular membranes (Gunning, 2001). Plastids containing prolamellar bodies are called **etioplasts** (Kirk and Tilney-Bassett, 1978). Etioplasts form in leaf cells of plants grown in the dark. During subsequent development of etioplasts into chloroplasts in the light, the membranes of the prolamellar bodies develop into thylakoids. Carotenoid synthesis has been demonstrated to be required for the formation of prolamellar bodies in etiolated seedlings of *Arabidopsis* (Park et al., 2002). In nature, the proplastids in the embryos of some seeds first develop into etioplasts; then, upon exposure to light, the etioplasts develop into chloroplasts. The various kinds of plastids are remarkable for the relative ease with which they can change from one type to another (Fig. 2.16).

# MITOCHONDRIA

**Mitochondria**, like plastids, are bounded by two membranes (Figs. 2.17 and 2.18). The inner membrane is convoluted inwardly into numerous folds known as **cristae** (singular: **crista**), which greatly increase the surface area available to enzymes and the reactions associated with them. Mitochondria are generally smaller than plastids, measuring about half a micrometer in diameter and exhibiting great variation in length and shape.

Mitochondria are the sites of respiration, a process involving the release of energy from organic molecules and its conversion to molecules of ATP (adenosine

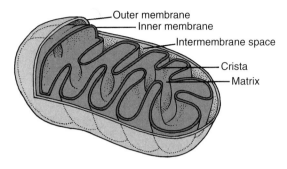

**FIGURE 2.17**

Three-dimensional structure of a mitochondrion. The inner of the two membranes bounding the mitochondrion fold inward, forming the cristae. Many of the enzymes and electron carriers involved in respiration are present in the cristae. (From Raven et al., 2005.)

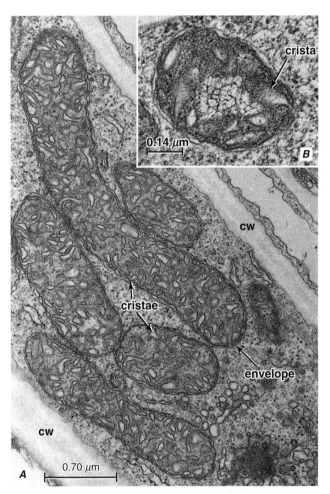

**FIGURE 2.18**

Mitochondria. **A**, in a leaf cell of tobacco *(Nicotiana tabacum)*. The envelope consists of two membranes, and the cristae are embedded in a dense stroma. **B**, mitochondrion in a leaf cell of spinach *(Spinacia oleracea)*, in a section revealing some strands of DNA in the nucleoid. Detail: cw, cell wall.

triphosphate), the principal immediate energy source for the cell (Mackenzie and McIntosh, 1999; Møller, 2001; Bowsher and Tobin, 2001). Within the innermost compartment, surrounding the cristae, is the **matrix**, a dense solution containing enzymes, coenzymes, water, phosphate, and other molecules involved with respiration. Whereas the outer membrane is fairly permeable to most small molecules, the inner one is relatively impermeable, permitting the passage of only certain molecules, such as pyruvate and ATP, while preventing the passage of others. Some enzymes of the citric-acid cycle are found in solution in the matrix. Other citric acid-cycle enzymes and the components of the electron-transport chain are built into the surfaces of the cristae. Most plant cells contain hundreds of mitochondria, the number of mitochondria per cell being related to the cell's demand for ATP.

Mitochondria are in constant motion and appear to move freely in the streaming cytoplasm from one part of the cell to another; they also fuse and divide by binary fission (Arimura et al., 2004), involving dividing rings reminiscent of the plastid-dividing rings (Osteryoung, 2000). Movement of mitochondria in cultured cells of tobacco *(Nicotiana tabacum)* has been shown to involve an actin-myosin-based system (Van Gestel et al., 2002). Mitochondria tend to congregate where energy is required. In cells in which the plasma membrane is very active in transporting materials into or out of the cell, the mitochondria often can be found arrayed along the membrane surface.

Mitochondria, like plastids, are semiautonomous organelles, containing the components necessary for the synthesis of some of their own proteins. One or more DNA-containing nucleoids and many 70S ribosomes similar to those of bacteria are found in the matrix (Fig. 2.18). The DNA is not associated with histones. Thus in plant cells genetic information is found in three different compartments: nucleus, plastid, and mitochondrion. The mitochondrial genomes of plants are much larger (200–2400 kb) than those of animals (14–42 kb), fungi (18–176 kb), and plastids (120–200 kb) (Backert et al., 1997; Giegé and Brennicke, 2001). Their structural organization is not fully understood. Linear and circular DNA molecules of variable size as well as more complex DNA molecules are consistently present (Backert et al., 1997).

Mitochondria are widely accepted to have evolved from free-living α-proteobacteria through the process of endosymbiosis (Gray, 1989). As with the chloroplast, in the course of evolution the DNA of the mitochondria was massively transferred to the nucleus (Adams et al., 2000; Gray, 2000). Evidence also indicates that some genetic information has been transferred from chloroplasts to mitochondria over long periods of evolutionary time (Nugent and Palmer, 1988; Jukes and Osawa, 1990; Nakazono and Hirai, 1993) and possibly from the nucleus to the mitochondria (Schuster and Brennicke, 1987; Marienfeld et al., 1999). Only about 30 proteins are encoded in plant mitochondrial genomes. By contrast, about 4000 proteins encoded in the nucleus are estimated to be imported from the cytosol. Nuclear encoded mitochondrial proteins contain signal peptides called *presequences* at their N-terminus to direct them into the mitochondria (Braun and Schmitz, 1999; Mackenzie and McIntosh, 1999; Giegé and Brennicke, 2001).

Genetic information found only in mitochondrial DNA may have an effect on cell development. Most notable is cytoplasmic male sterility, a maternally inherited (mitochondrial DNA is maternally inherited) trait that prevents the production of functional pollen but does not affect female fertility (Leaver and Gray, 1982). Because it prevents self-pollination, the cytoplasmic male sterility phenotype has been widely used in the commercial production of $F_1$ hybrid seed (e.g., in maize, onions, carrots, beets, and petunias).

Mitochondria have come to be regarded as key players in the regulation of programmed cell death, called *apoptosis*, in animal cells (Chapter 5; Desagher and Martinou, 2000; Ferri and Kroemer, 2001; Finkel, 2001). A primary cellular trigger for apoptosis is the release of cytochrome *c* from the mitochondrial intermembrane space. Release of the cytochrome *c* appears to be a critical event for the activation of catabolic proteases called caspases (apoptosis-specific cysteine proteases). Although mitochondria may play a role in plant programmed cell death, it is unlikely that released cytochrome *c* is involved in that process (Jones, 2000; Xu and Hanson, 2000; Young and Gallie, 2000; Yu et al., 2002; Balk et al., 2003; Yao et al., 2004).

# ▌PEROXISOMES

Unlike plastids and mitochondria, which are bounded by two membranes, **peroxisomes** (also called **microbodies**) are spherical organelles bounded by a single membrane (Figs. 2.14 and 2.19; Frederick et al., 1975; Olsen, 1998). Peroxisomes differ most notably from plastids and mitochondria, however, in their lack of DNA and ribosomes. Consequently all peroxisomal proteins are nuclear-encoded, and at least the matrix proteins are synthesized on ribosomes in the cytosol and then transported into the peroxisome. A subset of peroxisomal membrane proteins might be targeted first to the endoplasmic reticulum, and from there to the organelle by vesicle-mediated transport (Johnson and Olsen, 2001). Peroxisomes range in size from 0.5 to 1.5 µm. They lack internal membranes and have a granular interior, which sometimes contains an amorphous or crystalline body composed of protein. According to the prevailing view, peroxisomes are self-replicating organelles, new peroxisomes arising from preexisting ones by division. The existence of a vesicle-mediated pathway from the endoplasmic reticulum to the peroxisomes has led some workers to speculate that these organelles may also be generated de novo (Kunau and Erdmann, 1998; Titorenko and Rachubinski, 1998; Mullen et al., 2001), a view strongly challenged by others (Purdue and Lazarow, 2001). Biochemically peroxisomes are characterized by the presence of at least one hydrogen peroxide–producing oxidase and catalase for the removal of the hydrogen peroxide (Tolbert, 1980; Olsen, 1998). As noted by Corpas et al. (2001), an important property of peroxisomes is their "metabolic plasticity," in that their enzymatic content can vary, depending on the organism, cell type or tissue type, and environmental conditions. Peroxisomes perform a wide array of metabolic functions (Hu et al., 2002).

Two very different types of peroxisome have been studied extensively in plants (Tolbert and Essner, 1981; Trelease, 1984; Kindl, 1992). One of them occurs in green leaves, where it plays an important role in glycolic acid metabolism, which is associated with **photorespiration**, a process that consumes oxygen and releases carbon dioxide. Photorespiration involves cooperative interaction among peroxisomes, mitochondria, and chloroplasts; hence these three organelles commonly are closely associated spatially with one another (Fig. 2.19A). The biological function of photorespiration remains to be determined (Taiz and Zeiger, 2002).

The second type of peroxisome is found in the endosperm or cotyledons of germinating seeds, where it plays an essential role in the conversion of fats to carbohydrates by a series of reactions known as the glyoxylate cycle. Appropriately these peroxisomes are also called **glyoxysomes**. The two types of peroxisome are interconvertible (Kindl, 1992; Nishimura et al., 1993, 1998). For example, during the early stages of germination the cotyledons of some seeds are essentially deprived of light. As the cotyledons gradually become exposed to light, they may become green. With the depletion of fat and the appearance of chloroplasts, the glyoxysomes are converted to leaf-type peroxisomes. Glyoxysomal properties may reappear as the tissues undergo senescence.

Several studies have revealed that plant peroxisomes, like plastids and mitochondria, are motile organelles whose movement is actin dependent (Collings et al., 2002; Jedd and Chua, 2002; Mano et al., 2002; Mathur et al., 2002). The peroxisomes in leek (*Allium porrum*) and *Arabidopsis* have been shown to undergo dynamic movements along bundles of actin filaments (Collings et al., 2002; Mano et al., 2002), those in *Arabidopsis* reaching peak velocities approaching $10 \mu m \cdot s^{-1}$ (Jedd and Chua, 2002). Moreover the peroxisomes in *Arabidopsis* have been shown to be driven by myosin motors (Jedd and Chua, 2002).

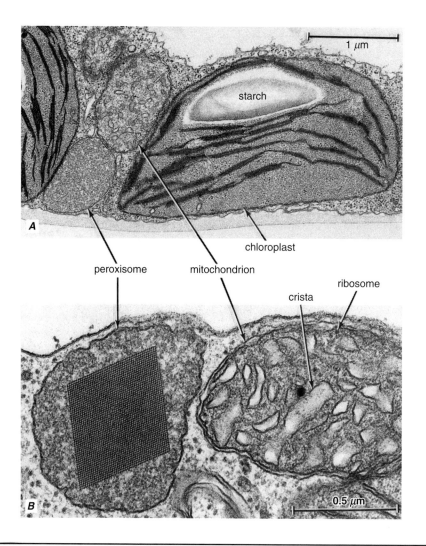

**FIGURE 2.19**

Organelles in leaf cells of sugar beet (*Beta vulgaris*, **A**) and tobacco (*Nicotiana tabacum*, **B**). The unit membranes enclosing the peroxisomes may be contrasted with the double-membraned envelopes of the other organelles. The peroxisome in **B** contains a crystal. Some ribosomes are perceptible in the chloroplast in **A** and in the mitochondrion in **B**. (From Esau, 1977.)

# ▌VACUOLES

Together with the presence of plastids and a cell wall, the vacuole is one of the three characteristics that distinguish plant cells from animal cells. As mentioned previously, vacuoles are organelles bounded by a single membrane, the **tonoplast**, or **vacuolar membrane** (Fig. 2.2). They are multifunctional organelles and are widely diverse in form, size, content, and functional dynamics (Wink, 1993; Marty, 1999). A single cell may contain more than one kind of vacuole. Some vacuoles function primarily as storage organelles, others as lytic compartments. The two types of vacuole can be characterized by the presence of specific tonoplast integral (intrinsic) proteins (TIPs): for example, whereas α-TIP is associated with the tonoplasts of protein-storage vacu-

oles, γ-TIP localizes to the tonoplasts of lytic vacuoles. Both types of TIP may colocalize to the same tonoplast of large vacuoles, apparently the result of merger of the two types of vacuole during cell enlargement (Paris et al., 1996; Miller and Anderson, 1999).

Many meristematic plant cells contain numerous small vacuoles. As the cell enlarges, the vacuoles increase in size and fuse into a single large vacuole (Fig. 2.20). Most of the increase in size of the cell in fact involves enlargement of the vacuoles. In the mature cell as much as 90% of the volume may be taken up by the vacuole, with the rest of the cytoplasm consisting of a thin peripheral layer closely pressed against the cell wall. By filling such a large proportion of the cell with "inexpensive" (in terms of energy) vacuolar contents, plants not only save "expensive" nitrogen-rich cytoplasmic

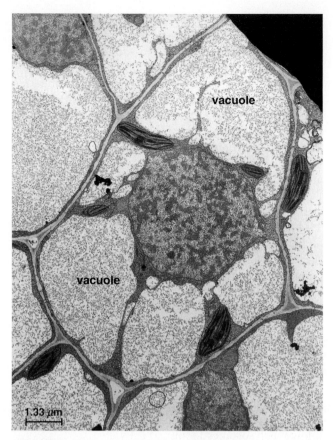

**FIGURE 2.20**

Parenchyma cell from a tobacco (*Nicotiana tabacum*) leaf, with its nucleus "suspended" in the middle of the vacuole by dense strands of cytoplasm. The dense granular substance in the nucleus is chromatin.

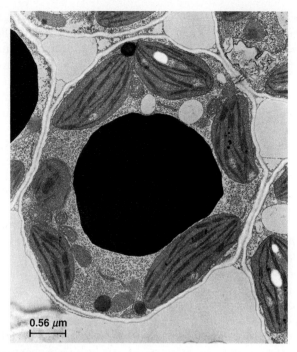

**FIGURE 2.21**

Tannin-containing vacuole in leaf cell of the sensitive plant (*Mimosa pudica*). The electron-dense tannin literally fills the central vacuole of this cell.

material but also acquire a large surface area between the thin layer of nitrogen-rich cytoplasm and the protoplast's external environment (Wiebe, 1978). Being a selectively permeable membrane, the tonoplast is involved with the regulation of osmotic phenomena associated with the vacuoles. A direct consequence of this strategy is the development of tissue rigidity, one of the principal roles of the vacuole and tonoplast.

The principal component of the non–protein-storing vacuoles is water, with other components varying according to the type of plant, organ, and cell and their developmental and physiological state (Nakamura and Matsuoka, 1993; Wink, 1993). In addition to inorganic ions such as $Ca^{2+}$, $Cl^-$, $K^+$, $Na^+$, $NO_3^-$, and $PO_4^{2-}$, such vacuoles commonly contain sugars, organic acids, and amino acids, and the aqueous solution commonly is called **cell sap**. Sometimes the concentration of a particular substance in the vacuole is sufficiently great for it to form crystals. Calcium oxalate crystals, which can assume different forms (Chapter 3), are especially

common. In most cases vacuoles do not synthesize the molecules that they accumulate but must receive them from other parts of the cytoplasm. The transport of metabolites and inorganic ions across the tonoplast is strictly controlled to ensure optimal functioning of the cell (Martinoia, 1992; Nakamura and Matsuoka, 1993; Wink, 1993).

Vacuoles are important storage compartments for various metabolites. Primary metabolites—substances that play a basic role in cell metabolism—such as sugars and organic acids are stored only temporarily in the vacuole. In photosynthesizing leaves of many species, for example, much of the sugar produced during the day is stored in the mesophyll cell vacuoles and then moved out of the vacuoles during the night for export to other parts of the plant. In CAM plants, malic acid is stored in the vacuoles during the night and released from the vacuoles and decarboxylated during the day, the $CO_2$ then becoming assimilated by the Calvin cycle in the chloroplasts (Kluge et al., 1982; Smith, 1987). In seeds, vacuoles are a primary site for the storage of proteins (Herman and Larkins, 1999).

Vacuoles also sequester toxic secondary metabolites, such as nicotine, an alkaloid, and tannins, phenolic compounds, from the rest of the cytoplasm (Fig. 2.21). Secondary metabolites play no apparent role in the plant's primary metabolism. Such substances may be sequestered permanently in the vacuoles. A great many of the secondary metabolites accumulated in the vacuoles are toxic not only to the plant itself but also to pathogens, parasites, and/or herbivores, and therefore they play an important role in plant defense. Some of the secondary metabolites stored in the vacuoles are nontoxic but are converted upon hydrolysis to such highly toxic derivatives as cyanide, mustard oils, and aglycones when the vacuoles are ruptured (Matile, 1982; Boller and Wiemken, 1986). Thus detoxification of the cytoplasm and the storage of defensive chemicals may be regarded as additional functions of vacuoles.

The vacuole is often the site of pigment deposition. The blue, violet, purple, dark red, and scarlet colors of plant cells are usually caused by a group of pigments known as the **anthocyanins**. These pigments frequently are confined to epidermal cells. Unlike most other plant pigments (e.g., chlorophylls, carotenoids), the anthocyanins are readily soluble in water and are found in solution in the vacuole. They are responsible for the red and blue colors of many fruits (grapes, plums, cherries) and vegetables (radishes, turnips, cabbages), and a host of flowers (geraniums, delphiniums, roses, petunias, peonies), and presumably serve to attract animals for pollination and seed dispersal. Anthocyanin has been implicated with the sequestration of molybdenum in vacuoles of peripheral cell layers of *Brassica* seedlings (Hale et al., 2001). In a restricted number of plant families, another class of water-soluble pigments, the nitrogen-containing **betalains**, is responsible for some of the yellow and red colors. These plants, all members of the order Chenopodiales, lack anthocyanins. The red color of beets and *Bougainvillea* flowers is due to the presence of betacyanins (red betalains). The yellow betalains are called betaxanthins (Piattelli, 1981).

Anthocyanins are also responsible for the brilliant red colors of some leaves in autumn. These pigments form in response to cold, sunny weather, when leaves stop producing chlorophyll. As the chlorophyll that is present disintegrates, the newly formed anthocyanins are unmasked. In leaves that do not form anthocyanin pigments, the breakdown of chlorophyll in autumn may unmask the more stable yellow-to-orange carotenoid pigments already present it the chloroplasts. The most spectacular autumnal coloration develops in years when cool, clear weather prevails in the fall (Kozlowski and Pallardy, 1997).

What role is played by anthocyanins found in leaves? In red-osier dogwood (*Cornus stolonifera*), anthocyanins form a pigment layer in the palisade mesophyll layer in autumn, decreasing light capture by the chloroplasts prior to leaf fall. It has been suggested that this optical masking of chlorophyll by the anthocyanins reduces the risk of photo-oxidative damage to the leaf

cells as they senesce, damage that otherwise might lower the efficiency of nutrient retrieval from the senescing leaves (Feild et al., 2001). In addition to protecting leaves from photo-oxidative damage, evidence indicates that anthocyanins protect against photoinhibition (Havaux and Kloppstech, 2001; Lee, D. W., and Gould, 2002; Steyn et al., 2002), a decline in photosynthetic efficiency resulting from excess excitation arriving at the reaction center of photosystem II. Photoinhibition is common in understory plants and occurs when they are suddenly exposed to patches of full sunlight (sunflecks) that pass through momentary openings in the upper canopy as the leaves flutter in the breeze (Pearcy, 1990).

As lytic compartments, vacuoles are involved with the breakdown of macromolecules and the recycling of components within the cell. Entire organelles, such as senescent plastids and mitochondria, may be engulfed and subsequently degraded by vacuoles containing large numbers of hydrolytic and oxidizing enzymes. The large central vacuole can sequester hydrolases, which upon breakdown of the tonoplast can result in complete autolysis of the cytoplasm, as during programmed cell death of differentiating tracheary elements (Chapter 10). Because of this digestive activity the so-called lytic vacuoles are comparable in function with the organelles known as lysosomes in animal cells.

New vacuoles have long been considered to arise from dilation of specialized regions of the smooth ER or from vesicles derived from the Golgi apparatus. Most evidence supports de novo formation of vacuoles from the ER (Robinson, 1985; Hörtensteiner et al., 1992; Herman et al., 1994).

# RIBOSOMES

**Ribosomes** are small particles, only about 17 to 23 nanometers in diameter (Fig. 2.22), consisting of protein and RNA (Davies and Larkins, 1980). Although the number of protein molecules in ribosomes greatly exceeds the number of RNA molecules, RNA constitutes about 60% of the mass of a ribosome. They are the sites at which amino acids are linked together to form proteins and are abundant in the cytoplasm of metabolically active cells (Lake, 1981). Each ribosome consists of two subunits, one small and the other large, composed of specific ribosomal RNA and protein molecules. Ribosomes occur both freely in the cytosol and attached to the endoplasmic reticulum and outer surface of the nuclear envelope. They are by far the most numerous of cellular structures and are also found in nuclei, plastids, and mitochondria. As mentioned previously, the ribosomes of plastids and mitochondria are similar in size to those of bacteria.

Ribosomes actively involved in protein synthesis occur in clusters or aggregates called **polysomes**, or

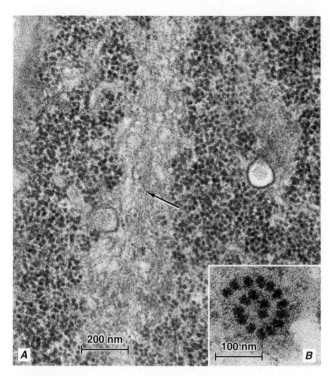

**FIGURE 2.22**

Ribosomes. **A**, in bundle-sheath cell of a maize (*Zea mays*) leaf. Arrow points to a bundle of actin filaments. **B**, a polysome (polyribosome) attached to the surface of endoplasmic reticulum in a tobacco (*Nicotiana tabacum*) leaf cell. (**B**, from Esau, 1977.)

**polyribosomes** (Fig. 2.22), united by the messenger RNA molecules carrying the genetic information from the nucleus. The amino acids from which the proteins are synthesized are brought to the polysomes by transfer RNAs located in the cytosol. The synthesis of protein, known as **translation**, consumes more energy than any other biosynthetic process. That energy is provided by the hydrolysis of guanosine triphosphate (GTP).

The synthesis of polypeptides (proteins) encoded by nuclear genes is initiated on polysomes located in the cytosol and follows one of two divergent pathways. (1) Those polysomes involved in the synthesis of polypeptides destined for the endoplasmic reticulum become associated with the endoplasmic reticulum early in the translational process. The polypeptides and their associated polysomes are directed to the endoplasmic reticulum by a targeting signal, or signal peptide, located at the amino end of each polypeptide. The polypeptides are transferred across the membrane into the lumen of the ER (or are inserted into it, in the case of integral proteins) as polypeptide synthesis continues. (2) Those polysomes involved with the synthesis of polypeptides destined for the cytosol or for import into the nucleus,

mitochondria, plastids, or peroxisomes remain free in the cytosol. The polypeptides released from the free polysomes either remain in the cytosol or are targeted to the appropriate cellular component by a targeting sequence (Holtzman, 1992). Membrane-bound and free ribosomes are both structurally and functionally identical, differing from one another only in the proteins they are making at any given time.

## REFERENCES

ADAMS, K. L., D. O. DALEY, Y.-L. QIU, J. WHELAN, and J. D. PALMER. 2000. Repeated, recent and diverse transfers of a mitochondrial gene to the nucleus in flowering plants. *Nature* 408, 354–357.

ARIMURA, S.-I., A. HIRAI, and N. TSUTSUMI. 2001. Numerous and highly developed tubular projections from plastids observed in tobacco epidermal cells. *Plant Sci.* 160, 449–454.

ARIMURA, S.-I., J. YAMAMOTO, G. P. AIDA, M. NAKAZONO, and N. TSUTSUMI. 2004. Frequent fusion and fission of plant mitochondria with unequal nucleoid distribution. *Proc. Natl. Acad. Sci. USA* 101, 7805–7808.

BACKERT, S., B. L. NIELSEN, and T. BÖRNER. 1997. The mystery of the rings: Structure and replication of mitochondrial genomes from higher plants. *Trends Plant Sci.* 2, 477–483.

BALK, J., S. K. CHEW, C. J. LEAVER, and P. F. MCCABE. 2003. The intermembrane space of plant mitochondria contains a DNase activity that may be involved in programmed cell death. *Plant J.* 34, 573–583.

BASKIN, T. I. 2000. The cytoskeleton. In: *Biochemistry and Molecular Biology of Plants*, pp. 202–258, B. B. Buchanan, W. Gruissem, and R. L. Jones, eds. American Society of Plant Physiologists, Rockville, MD.

BATTEY, N. H., N. C. JAMES, A. J. GREENLAND, and C. BROWNLEE. 1999. Exocytosis and endocytosis. *Plant Cell* 11, 643–659.

BOLLER, T., and A. WIEMKEN. 1986. Dynamics of vacuolar compartmentation. *Annu. Rev. Plant Physiol.* 37, 137–164.

BÖLTER, B., and J. SOLL. 2001. Ion channels in the outer membranes of chloroplasts and mitochondria: Open doors or regulated gates? *EMBO J.* 20, 935–940.

BONIOTTI, M. B., and M. E. GRIFFITH. 2002. "Cross-talk" between cell division cycle and development in plants. *Plant Cell* 14, 11–16.

BOWSHER, C. G., and A. K. TOBIN. 2001. Compartmentation of metabolism within mitochondria and plastids. *J. Exp. Bot.* 52, 513–527.

BRAUN, H.-P., and U. K. SCHMITZ. 1999. The protein-import apparatus of plant mitochondria. *Planta* 209, 267–274.

BRUCE, B. D. 2000. Chloroplast transit peptides: Structure, function and evolution. *Trends Cell Biol.* 10, 440–447.

CAMARA, B., J. BOUSQUET, C. CHENICLET, J.-P. CARDE, M. KUNTZ, J.-L. EVRARD, and J.-H. WEIL. 1989. Enzymology of isoprenoid biosynthesis and expression of plastid and nuclear genes during chromoplast differentiation in pepper fruits (*Capsicum annuum*). In: *Physiology, Biochemistry, and Genetics of Nongreen Plastids*, pp. 141–156, C. D. Boyer, J. C. Shannon, and R. C. Hardison, eds. American Society of Plant Physiologists, Rockville, MD.

CAMARA, B., P. HUGUENEY, F. BOUVIER, M. KUNTZ, and R. MONÉGER. 1995. Biochemistry and molecular biology of chromoplast development. *Int. Rev. Cytol.* 163, 175–247.

CARDE, J.-P. 1984. Leucoplasts: A distinct kind of organelles lacking typical 70S ribosomes and free thylakoids. *Eur. J. Cell Biol.* 34, 18–26.

CHEUNG, A. Y., T. MCNELLIS, and B. PIEKOS. 1993. Maintenance of chloroplast components during chromoplast differentiation in the tomato mutant *Green Flesh*. *Plant Physiol.* 101, 1223–1229.

CHRISPEELS, M. J., N. M. CRAWFORD, and J. I. SCHROEDER. 1999. Proteins for transport of water and mineral nutrients across the membranes of plant cells. *Plant Cell* 11, 661–676.

CLINE, K., R. HENRY, C.-J. LI, and J.-G. YUAN. 1993. Multiple pathways for protein transport into or across the thylakoid membrane. *EMBO J.* 12, 4105–4114.

CLOWES, F. A. L., and B. E. JUNIPER. 1968. *Plant Cells*. Blackwell Scientific, Oxford.

COLLINGS, D. A., J. D. I. HARPER, J. MARC, R. L. OVERALL, and R. T. MULLEN. 2002. Life in the fast lane: Actin-based motility of plant peroxisomes. *Can. J. Bot.* 80, 430–441.

COOKE, T. J., and B. LU. 1992. The independence of cell shape and overall form in multicellular algae and land plants: Cells do not act as building blocks for constructing plant organs. *Int. J. Plant Sci.* 153, S7–S27.

CORPAS, F. J., J. B. BARROSO, and L. A. DEL RÍO. 2001. Peroxisomes as a source of reactive oxygen species and nitric oxide signal molecules in plant cells. *Trends Plant Sci.* 6, 145–150.

CUNNINGHAM, F. X., JR., and E. GANTT. 1998. Genes and enzymes of carotenoid biosynthesis in plants. *Annu. Rev. Plant Physiol. Plant Mol. Biol.* 49, 557–583.

CUTLER, S., and D. EHRHARDT. 2000. Dead cells don't dance: Insights from live-cell imaging in plants. *Curr. Opin. Plant Biol.* 3, 532–537.

D'AMATO, F. 1998. Chromosome endoreduplication in plant tissue development and function. In: *Plant Cell Proliferation and Its Regulation in Growth and Development*, pp. 153–166, J. A. Bryant and D. Chiatante, eds. Wiley, Chichester.

DAVIES, E., and B. A. LARKINS. 1980. Ribosomes. In: *The Biochemistry of Plants*, vol. 1, *The Plant Cell*, pp. 413–435, N. E. Tolbert, ed. Academic Press, New York.

DE BARY, A. 1879. Besprechung. K. Prantl. Lehrbuch der Botanik für mittlere und höhere Lehranstalten. *Bot. Ztg.* 37, 221–223.

DELROT, S., R. ATANASSOVA, E. GOMÈS, and P. COUTOS-THÉVENOT. 2001. Plasma membrane transporters: A machinery for uptake of organic solutes and stress resistance. *Plant Sci.* 161, 391–404.

DEN BOER, B. G. W., and J. A. H. MURRAY. 2000. Triggering the cell cycle in plants. *Trends Cell Biol.* 10, 245–250.

DEPAMPHILIS, C. W., and J. D. PALMER. 1989. Evolution and function of plastid DNA: A review with special reference to nonphotosynthetic plants. In: *Physiology, Biochemistry, and Genetics of Nongreen Plastids*, pp. 182–202, C. D. Boyer, J. C. Shannon, and R. C. Hardison, eds. American Society of Plant Physiologists, Rockville, MD.

DESAGHER, S., and J.-C. MARTINOU. 2000. Mitochondria as the central control point of apoptosis. *Trends Cell Biol.* 10, 369–377.

DINGWALL, C., and R. LASKEY. 1992. The nuclear membrane. *Science* 258, 942–947.

ESAU, K. 1977. *Anatomy of Seed Plants*, 2nd ed. Wiley, New York.

FEILD, T. S., D. W. LEE, and N. M. HOLBROOK. 2001. Why leaves turn red in autumn. The role of anthocyanins in senescing leaves of red-osier dogwood. *Plant Physiol.* 127, 566–574.

FERRI, K. F., and G. KROEMER. 2001. Mitochondria—The suicide organelles. *BioEssays* 23, 111–115.

FINKEL, E. 2001. The mitochondrion: Is it central to apoptosis? *Science* 292, 624–626.

FLÜGGE, U.-I. 1990. Import of proteins into chloroplasts. *J. Cell Sci.* 96, 351–354.

FRANKLIN, A. E., and W. Z. CANDE. 1999. Nuclear organization and chromosome segregation. *Plant Cell* 11, 523–534.

FREDERICK, S. E., P. J. GRUBER, and E. H. NEWCOMB. 1975. Plant microbodies. *Protoplasma* 84, 1–29.

FRICKER, M. D., and K. J. OPARKA. 1999. Imaging techniques in plant transport: Meeting review. *J. Exp. Bot.* 50 (suppl. 1), 1089–1100.

FULGOSI, H., and J. SOLL. 2001. A gateway to chloroplasts—Protein translocation and beyond. *J. Plant Physiol.* 158, 273–284.

GAIDAROV, I., F. SANTINI, R. A. WARREN, and J. H. KEEN. 1999. Spatial control of coated-pit dynamics in living cells. *Nature Cell Biol.* 1, 1–7.

GANT, T. M., and K. L. WILSON. 1997. Nuclear assembly. *Annu. Rev. Cell Dev. Biol.* 13, 669–695.

GERACE, L., and R. FOISNER. 1994. Integral membrane proteins and dynamic organization of the nuclear envelope. *Trends Cell Biol.* 4, 127–131.

GIEGÉ, P., and A. BRENNICKE. 2001. From gene to protein in higher plant mitochondria. *C. R. Acad. Sci., Paris, Sci. de la Vie* 324, 209–217.

GOLDSCHMIDT, E. E. 1988. Regulatory aspects of chloro-chromoplast interconversions in senescing *Citrus* fruit peel. *Isr. J. Bot.* 37, 123–130.

GÖRLICH, D. 1997. Nuclear protein import. *Curr. Opin. Cell Biol.* 9, 412–419.

GÖRLICH, D., and I. W. MATTAJ. 1996. Nucleocytoplasmic transport. *Science* 271, 1513–1518.

GRAY, J. C. 1996. Biogenesis of chloroplasts in higher plants. In: *Membranes: Specialized Functions in Plants*, pp. 441–458, M. Smallwood, J. P. Knox, and D. J. Bowles, eds. BIOS Scientific, Oxford.

GRAY, J. C., J. A. SULLIVAN, J. M. HIBBERD, and M. R. HANSEN. 2001. Stromules: Mobile protrusions and interconnections between plastids. *Plant Biol.* 3, 223–233.

GRAY, M. W. 1989. Origin and evolution of mitochondrial DNA. Annu. Rev. *Cell Biol.* 5, 25–50.

GRAY, M. W. 2000. Mitochondrial genes on the move. *Nature* 408, 302–305.

GRÖNEGRESS, P. 1971. The greening of chromoplasts in *Daucus carota* L. *Planta* 98, 274–278.

GUNNING, B. E. S. 2001. Membrane geometry of "open" prolamellar bodies. *Protoplasma* 215, 4–15.

GUNNING, B. E. S., and M. SAMMUT. 1990. Rearrangements of microtubules involved in establishing cell division planes start immediately after DNA synthesis and are completed just before mitosis. *Plant Cell* 2, 1273–1282.

HAGEMANN, R. 1960. Die Plastidenentwicklung in Tomaten-Kotyledonen. *Biol. Zentralbl.* 79, 393–411.

HALE, K. L., S. P. McGRATH, E. LOMBI, S. M. STACK, N. TERRY, I. J. PICKERING, G. N. GEORGE, and E. A. H. PILON-SMITS. 2001. Molybdenum sequestration in *Brassica* species. A role for anthocyanins? *Plant Physiol.* 126, 1391–1402.

HANSMANN, P., R. JUNKER, H. SAUTER, and P. SITTE. 1987. Chromoplast development in daffodil coronae during anthesis. *J. Plant Physiol.* 131, 133–143.

HANSTEIN, J. 1880. *Einige Züge aus der Biologie des Protoplasmas. Botanische Abhandlungen aus dem Gebiet der Morphologie und Physiologie*, Band 4, Heft 2. Marcus, Bonn.

HAUPT, W., and R. SCHEUERLEIN. 1990. Chloroplast movement. *Plant Cell Environ.* 13, 595–614.

HAVAUX, M., and K. KLOPPSTECH. 2001. The protective functions of carotenoid and flavonoid pigments against excess visible radiation at chilling temperature investigated in *Arabidopsis npq* and *tt* mutants. *Planta* 213, 953–966.

HAWES, C., C. M. SAINT-JORE, F. BRANDIZZI, H. ZHENG, A. V. ANDREEVA, and P. BOEVINK. 2001. Cytoplasmic illuminations: in planta targeting of fluorescent proteins to cellular organelles. *Protoplasma* 215, 77–88.

HEESE-PECK, A., and N. V. RAIKHEL. 1998. The nuclear pore complex. *Plant Mol. Biol.* 38, 145–162.

HEMERLY, A. S., P. C. G. FERREIRA, M. VAN MONTAGU, and D. INZÉ. 1999. Cell cycle control and plant morphogenesis: Is there an essential link? *BioEssays* 21, 29–37.

HEPLER, P. K., and B. E. S. GUNNING. 1998. Confocal fluorescence microscopy of plant cells. *Protoplasma* 201, 121–157.

HERMAN, E. M., and B. A. LARKINS. 1999. Protein storage bodies and vacuoles. *Plant Cell* 11, 601–614.

HERMAN, E. M., X. LI, R. T. SU, P. LARSEN, H.-T. HSU, and H. SZE. 1994. Vacuolar-type H$^+$-ATPases are associated with the endoplasmic reticulum and provacuoles of root tip cells. *Plant Physiol.* 106, 1313–1324.

HICKS, G. R., and N. V. RAIKHEL. 1995. Protein import into the nucleus: An integrated view. *Annu. Rev. Cell Dev. Biol.* 11, 155–188.

HOLTZMAN, E. 1992. Intracellular targeting and sorting. *BioScience* 42, 608–620.

HÖRTENSTEINER, S., E. MARTINOIA, and N. AMRHEIN. 1992. Reappearance of hydrolytic activities and tonoplast proteins in the regenerated vacuole of evacuolated protoplasts. *Planta* 187, 113–121.

HU, J., M. AGUIRRE, C. PETO, J. ALONSO, J. ECKER, and J. CHORY. 2002. A role for peroxisomes in photomorphogenesis and development of *Arabidopsis*. *Science* 297, 405–409.

HUNTLEY, R. P., and J. A. H. MURRAY. 1999. The plant cell cycle. *Curr. Opin. Plant Biol.* 2, 440–446.

IVANOVA, M., and T. L. ROST. 1998. Cytokinins and the plant cell cycle: Problems and pitfalls of proving their function. In: *Plant Cell Proliferation and Its Regulation in Growth and Development*, pp. 45–57, J. A. Bryant and D. Chiatante, eds. Wiley, New York.

JACOBSON, K., E. D. SHEETS, and R. SIMSON. 1995. Revisiting the fluid mosaic model of membranes. *Science* 268, 1441–1442.

JACQMARD, A., C. HOUSSA, and G. BERNIER. 1994. Regulation of the cell cycle by cytokinins. In: *Cytokinins: Chemistry, Activity, and Function*, pp. 197–215, D. W. S. Mok and M. C. Mok, eds. CRC Press, Boca Raton, FL.

JAVOT, H., and C. MAUREL. 2002. The role of aquaporins in root water uptake. *Ann. Bot.* 90, 301–313.

JEDD, G., and N.-H. CHUA. 2002. Visualization of peroxisomes in living plant cells reveals acto-myosin-dependent cytoplasmic streaming and peroxisome budding. *Plant Cell Physiol.* 43, 384–392.

JOHNSON, T. L., and L. J. OLSEN. 2001. Building new models for peroxisome biogenesis. *Plant Physiol.* 127, 731–739.

JONES, A. 2000. Does the plant mitochondrion integrate cellular stress and regulate programmed cell death? *Trends Plant Sci.* 5, 225–230.

JUKES, T. H., and S. OSAWA. 1990. The genetic code in mitochondria and chloroplasts. *Experientia* 46, 1117–1126.

KAGAWA, T., and M. WADA. 2000. Blue light-induced chloroplast relocation in *Arabidopsis thaliana* as analyzed by microbeam irradiation. *Plant Cell Physiol.* 41, 84–93.

KAGAWA, T., and M. WADA. 2002. Blue light-induced chloroplast relocation. *Plant Cell Physiol.* 43, 367–371.

KAPLAN, D. R. 1992. The relationship of cells to organisms in plants: Problem and implications of an organismal perspective. *Int. J. Plant Sci.* 153, S28–S37.

KAPLAN, D. R., and W. HAGEMANN. 1991. The relationship of cell and organism in vascular plants. *BioScience* 41, 693–703.

KEEGSTRA, K., and K. CLINE. 1999. Protein import and routing systems of chloroplasts. *Plant Cell* 11, 557–570.

KINDL, H. 1992. Plant peroxisomes: Recent studies on function and biosynthesis. *Cell Biochem. Funct.* 10, 153–158.

KIRK, J. T. O., and R. A. E. TILNEY-BASSETT. 1978. *The Plastids. Their Chemistry, Structure, Growth, and Inheritance*, rev. 2nd ed. Elsevier/North-Holland Biomedical Press, Amsterdam.

KJELLBOM, P., C. LARSSON, I. JOHANSSON, M. KARLSSON, and U. JOHANSON. 1999. Aquaporins and water homeostasis in plants. *Trends Plant Sci.* 4, 308–314.

KLUGE, M., A. FISCHER, and I. C. BUCHANAN-BOLLIG. 1982. Metabolic control of CAM. In: *Crassulacean Acid Metabolism*, pp. 31–50, I. P. Ting and M. Gibbs, eds. American Society of Plant Physiologists, Rockville, MD.

KNOTH, R., P. HANSMANN, and P. SITTE. 1986. Chromoplasts of *Palisota barteri*, and the molecular structure of chromoplast tubules. *Planta* 168, 167–174.

KÖHLER, R. H., and M. R. HANSON. 2000. Plastid tubules of higher plants are tissue-specific and developmentally regulated. *J. Cell Sci.* 113, 81–89.

KÖHLER, R. H., J. CAO, W. R. ZIPFEL, W. W. WEBB, and M. R. HANSON. 1997. Exchange of protein molecules through connections between higher plant plastids. *Science* 276, 2039–2042.

KONDOROSI, E., F. ROUDIER, and E. GENDREAU. 2000. Plant cell-size control: Growing by ploidy? *Curr. Opin. Plant Biol.* 3, 488–492.

KOST, B., and N.-H. CHUA. 2002. The plant cytoskeleton: Vacuoles and cell walls make the difference. *Cell* 108, 9–12.

KOZLOWSKI, T. T., and S. G. PALLARDY. 1997. *Physiology of Woody Plants*, 2nd ed. Academic Press, San Diego.

KUNAU, W.-H., and R. ERDMANN. 1998. Peroxisome biogenesis: Back to the endoplasmic reticulum? *Curr. Biol.* 8, R299–R302.

KUNTZ, M., J.-L. EVRARD, A. d'HARLINGUE, J.-H. WEIL, and B. CAMARA. 1989. Expression of plastid and nuclear genes during chromoplast differentiation in bell pepper (*Capsicum annuum*) and sunflower (*Helianthus annuus*). *Mol. Gen. Genet.* 216, 156–163.

KUROIWA, H., T. MORI, M. TAKAHARA, S.-y. MIYAGISHIMA, and T. KUROIWA. 2002. Chloroplast division machinery as revealed by immunofluorescence and electron microscopy. *Planta* 215, 185–190.

KWOK, E. Y., and M. R. HANSON. 2004. Stromules and the dynamic nature of plastid morphology. *J. Microsc.* 214, 124–137.

LAKE, J. A. 1981. The ribosome. *Sci. Am.* 245 (August), 84–97.

LAM, E., D. PONTIER, and O. DEL POZO. 1999. Die and let live—Programmed cell death in plants. *Curr. Opin. Plant Biol.* 2, 502–507.

LARDY, H. A. 1965. On the direction of pyridine nucleotide oxidation-reduction reactions in gluconeogenesis and lipo-genesis. In: *Control of Energy Metabolism*, pp. 245–248, B. Chance, R. W. Estabrook, and J. R. Williamson, eds. Academic Press, New York.

LARKINS, B. A., B. P. DILKES, R. A. DANTE, C. M. COELHO, Y.-M. WOO, and Y. LIU. 2001. Investigating the hows and whys of DNA endoreduplication. *J. Exp. Bot.* 52, 183–192.

LEAVER, C. J., and M. W. GRAY. 1982. Mitochondrial genome organization and expression in higher plants. *Annu. Rev. Plant Physiol.* 33, 373–402.

LEE, D. W., and K. S. GOULD. 2002. Why leaves turn red. *Am. Sci.* 90, 524–531.

LEE, J.-Y., B.-C. YOO, and W. J. LUCAS. 2000. Parallels between nuclear-pore and plasmodesmal trafficking of information molecules. *Planta* 210, 177–187.

LEONARD, R. T., and T. K. HODGES. 1980. The plasma membrane. In: *The Biochemistry of Plants*, vol. 1, *The Plant Cell*, pp. 163–182, N. E. Tolbert, ed. Academic Press, New York.

LIAN, H.-L., X. YU, Q. YE, X.-S. DING, Y. KITAGAWA, S.-S. KWAK, W.-A. SU, and Z.-C. TANG. 2004. The role of aquaporin RWC3 in drought avoidance in rice. *Plant Cell Physiol.* 45, 481–489.

LJUBEŠIĆ, N., M. WRISCHER, and Z. DEVIDÉ. 1996. Chromoplast structures in *Thunbergia* flowers. *Protoplasma* 193, 174–180.

LOGAN, H., M. BASSET, A.-A. VÉRY, and H. SENTENAC. 1997. Plasma membrane transport systems in higher plants: From black boxes to molecular physiology. *Physiol. Plant.* 100, 1–15.

LUCAS, W. J., B. DING, and C. VAN DER SCHOOT. 1993. Plasmodesmata and the supracellular nature of plants. *New Phytol.* 125, 435–476.

MACKENZIE, S., and L. McINTOSH. 1999. Higher plant mitochondria. *Plant Cell* 11, 571–586.

MADIGAN, M. T., J. M. MARTINKO, and J. PARKER. 2003. *Brock Biology of Microorganisms*, 10th ed. Pearson Education, Upper Saddle River, NJ.

MAESHIMA, M. 2001. Tonoplast transporters: organization and function. *Annu. Rev. Plant Physiol. Plant Mol. Biol.* 52, 469–497.

MANO, S., C. NAKAMORI, M. HAYASHI, A. KATO, M. KONDO, and M. NISHIMURA. 2002. Distribution and characterization of peroxisomes in *Arabidopsis* by visualization with GFP: Dynamic morphology and actin-dependent movement. *Plant Cell Physiol.* 43, 331–341.

MARANO, M. R., and N. CARRILLO. 1991. Chromoplast formation during tomato fruit ripening. No evidence for plastid DNA methylation. *Plant Mol. Biol.* 16, 11–19.

MARANO, M. R., and N. CARRILLO. 1992. Constitutive transcription and stable RNA accumulation in plastids during the conversion of chloroplasts to chromoplasts in ripening tomato fruits. *Plant Physiol.* 100, 1103–1113.

MARIENFELD, J., M. UNSELD, and A. BRENNICKE. 1999. The mitochondrial genome of *Arabidopsis* is composed of both native and immigrant information. *Trends Plant Sci.* 4, 495–502.

MARTÍN, M., S. MORENO DÍAZ DE LA ESPINA, L. F. JIMÉNEZ-GARCÍA, M. E. FERNÁNDEZ-GÓMEZ, and F. J. MEDINA. 1992. Further investigations on the functional role of two nuclear bodies in onion cells. *Protoplasma* 167, 175–182.

MARTIN, W. 1999. A briefly argued case that mitochondria and plastids are descendants of endosymbionts, but that the nuclear compartment is not. *Proc. R. Soc. Lond. B* 266, 1387–1395.

MARTINOIA, E. 1992. Transport processes in vacuoles of higher plants. *Bot. Acta* 105, 232–245.

MARTY, F. 1999. Plant vacuoles. *Plant Cell* 11, 587–599.

MATHUR, J., N. MATHUR, and M. HÜLSKAMP. 2002. Simultaneous visualization of peroxisomes and cytoskeletal elements reveals actin and not microtubule-based peroxisomal motility in plants. *Plant Physiol.* 128, 1031–1045.

MATILE, P. 1982. Vacuoles come of age. *Physiol. Vég.* 20, 303–310.

MCFADDEN, G. I. 1999. Endosymbiosis and evolution of the plant cell. *Curr. Opin. Plant Biol.* 2, 513–519.

MIERNYK, J. 1989. Leucoplast isolation. In: *Physiology, Biochemistry, and Genetics of Nongreen Plastids*, pp. 15–23, C. D. Boyer, J. C. Shannon, and R. C. Hardison, eds. American Society of Plant Physiologists, Rockville, MD.

MILLER, E. A., and M. A. ANDERSON. 1999. Uncoating the mechanisms of vacuolar protein transport. *Trends Plant Sci.* 4, 46–48.

MIRONOV, V., L. DE VEYLDER, M. VAN MONTAGU, and D. INZÉ. 1999. Cyclin-dependent kinases and cell division in plants—The nexus. *Plant Cell* 11, 509–521.

MIYAGISHIMA, S.-y., M. TAKAHARA, T. MORI, H. KUROIWA, T. HIGASHIYAMA, and T. KUROIWA. 2001. Plastid division is driven by a complex mechanism that involves differential transition of the bacterial and eukaryotic division rings. *Plant Cell* 13, 2257–2268.

MOHR, W. P. 1979. Pigment bodies in fruits of crimson and high pigment lines of tomatoes. *Ann. Bot.* 44, 427–434.

MØLLER, I. M. 2001. Plant mitochondria and oxidative stress: Electron transport, NADPH turnover, and metabolism of reactive oxygen species. *Annu. Rev. Plant Physiol. Plant Mol. Biol.* 52, 561–591.

MULLEN, R. T., C. R. FLYNN, and R. N. TRELEASE. 2001. How are peroxisomes formed? The role of the endoplasmic reticulum and peroxins. *Trends Plant Sci.* 6, 256–261.

MULLET, J. E. 1988. Chloroplast development and gene expression. *Annu. Rev. Plant Physiol. Plant Mol. Biol.* 39, 475–502.

NAKAMURA, K., and K. MATSUOKA. 1993. Protein targeting to the vacuole in plant cells. *Plant Physiol.* 101, 1–5.

NAKAMURA, S., T. IKEHARA, H. UCHIDA, T. SUZUKI, T. SODMERGEN. 1992. Fluorescence microscopy of plastid nucleoids and a survey of nuclease C in higher plants with respect to mode of plastid inheritance. *Protoplasma* 169, 68–74.

NAKAZONO, M., and A. HIRAI. 1993. Identification of the entire set of transferred chloroplast DNA sequences in the mitochondrial genome of rice. *Mol. Gen. Genet.* 236, 341–346.

NISHIMURA, M., Y. TAKEUCHI, L. DE BELLIS, and I. HARA-NISHIMURA. 1993. Leaf peroxisomes are directly transformed to glyoxysomes during senescence of pumpkin cotyledons. *Protoplasma* 175, 131–137.

NISHIMURA, M., M. HAYASHI, K. TORIYAMA, A. KATO, S. MANO, K. YAMAGUCHI, M. KONDO, and H. HAYASHI. 1998. Microbody defective mutants of *Arabidopsis*. *J. Plant Res.* 111, 329–332.

NIYOGI, K. K. 2000. Safety valves for photosynthesis. *Curr. Opin. Plant Biol.* 3, 455–460.

NUGENT, J. M., and J. D. PALMER. 1988. Location, identity, amount and serial entry of chloroplast DNA sequences in crucifer mitochondrial DNAs. *Curr. Genet.* 14, 501–509.

OLSEN, L. J. 1998. The surprising complexity of peroxisome biogenesis. *Plant Mol. Biol.* 38, 163–189.

OROSS, J. W., and J. V. POSSINGHAM. 1989. Ultrastructural features of the constricted region of dividing plastids. *Protoplasma* 150, 131–138.

OSTERYOUNG, K. W. 2000. Organelle fission. Crossing the evolutionary divide. *Plant Physiol.* 123, 1213–1216.

OSTERYOUNG, K. W., and R. S. MCANDREW. 2001. The plastid division machine. *Annu. Rev. Plant Physiol. Plant Mol. Biol.* 52, 315–333.

OSTERYOUNG, K. W., and K. A. PYKE. 1998. Plastid division: Evidence for a prokaryotically derived mechanism. *Curr. Opin. Plant Biol.* 1, 475–479.

PALMER, J. D., and C. F. DELWICHE. 1998. The origin and evolution of plastids and their genomes. In: *Molecular Systematics of Plants*. II. *DNA Sequencing*, pp. 375–409, D. E. Soltis, P. S. Soltis, and J. J. Doyle, eds. Kluwer Academic, Norwell, MA.

PALMGREN, M. G. 2001. Plant plasma membrane H$^+$-ATPases: Powerhouses for nutrient uptake. *Annu. Rev. Plant Physiol. Plant Mol. Biol.* 52, 817–845.

PARIS, N., C. M. STANLEY, R. L. JONES, and J. C. ROGERS. 1996. Plant cells contain two functionally distinct vacuolar compartments. *Cell* 85, 563–572.

PARK, H., S. S. KREUNEN, A. J. CUTTRISS, D. DELLAPENNA, and B. J. POGSON. 2002. Identification of the carotenoid isomerase provides insight into carotenoid biosynthesis, prolamellar body formation, and photomorphogenesis. *Plant Cell* 14, 321–332.

PEARCY, R. W. 1990. Sunflecks and photosynthesis in plant canopies. *Annu. Rev. Plant Physiol. Plant Mol. Biol.* 41, 421–453.

PIATTELLI, M. 1981. The betalains: Structure, biosynthesis, and chemical taxonomy. In: *The Biochemistry of Plants*, vol. 7, *Secondary Plant Products*, pp. 557–575, E. E. Conn, ed. Academic Press, New York.

POSSINGHAM, J. V., and M. E. LAWRENCE. 1983. Controls to plastid division. *Int. Rev. Cytol.* 84, 1–56.

POTUSCHAK, T., and P. DOERNER. 2001. Cell cycle controls: Genome-wide analysis in *Arabidopsis. Curr. Opin. Plant Biol.* 4, 501–506.

PRICE, H. J. 1988. DNA content variation among higher plants. *Ann. Mo. Bot. Gard.* 75, 1248–1257.

PURDUE, P. E., and P. B. LAZAROW. 2001. Peroxisome biogenesis. *Annu. Rev. Cell Dev. Biol.* 17, 701–752.

PYKE, K. A., and C. A. HOWELLS. 2002. Plastid and stromule morphogenesis in tomato. *Ann. Bot.* 90, 559–566.

RATAJCZAK, R., G. HINZ, and D. G. ROBINSON. 1999. Localization of pyrophosphatase in membranes of cauliflower inflorescence cells. *Planta* 208, 205–211.

RAVEN, P. R., R. F. EVERT, and S. E. EICHHORN. 1992. *Biology of Plants*, 5th ed. Worth, New York.

RAVEN, P. R., R. F. EVERT, and S. E. EICHHORN. 2005. *Biology of Plants*, 7th ed. Freeman, New York.

REICHELT, S., and J. KENDRICK-JONES. 2000. Myosins. In: *Actin: A Dynamic Framework for Multiple Plant Cell Functions*, pp. 29–44, C. J. Staiger, F. Baluška, D. Volkmann, and P. W. Barlow, eds. Kluwer Academic, Dordrecht.

REUMANN, S., and K. KEEGSTRA. 1999. The endosymbiotic origin of the protein import machinery of chloroplastic envelope membranes. *Trends Plant Sci.* 4, 302–307.

REUZEAU, C., J. G. MCNALLY, and B. G. PICKARD. 1997. The endomembrane sheath: A key structure for understanding the plant cell? *Protoplasma* 200, 1–9.

ROBERTSON, J. D. 1962. The membrane of the living cell. *Sci. Am.* 206 (April), 64–72.

ROBINSON, D. G. 1985. Plant membranes. Endo- and plasma membranes of plant cells. Wiley, New York.

ROBINSON, D. G., and H. DEPTA. 1988. Coated vesicles. *Annu. Rev. Plant Physiol. Plant Mol. Biol.* 39, 53–99.

RODERMEL, S. 2001. Pathways of plastid-to-nucleus signaling. *Trends Plant Sci.* 6, 471–478.

ROSE, A., S. PATEL, and I. MEIER. 2004. The plant nuclear envelope. *Planta* 218, 327–336.

RUJAN, T., and W. MARTIN. 2001. How many genes in *Arabidopsis* come from cyanobacteria? An estimate from 386 protein phylogenies. *Trends Genet.* 17, 113–120.

SACK, F. D. 1997. Plastids and gravitropic sensing. *Planta* 203 (suppl. 1), S63–S68.

SACK, F. D., and J. Z. KISS. 1989. Plastids and gravity perception. In: *Physiology, Biochemistry, and Genetics of Nongreen Plastids*, pp. 171–181, C. D. Boyer, J. C. Shannon, and R. C. Hardison, eds. American Society of Plant Physiologists, Rockville, MD.

SCHÄFFNER, A. R. 1998. Aquaporin function, structure, and expression: Are there more surprises to surface in water relations? *Planta* 204, 131–139.

SCHEER, U., M. THIRY, and G. GOESSENS. 1993. Structure, function and assembly of the nucleolus. *Trends Cell Biol.* 3, 236–241.

SCHLEIDEN, M. J. 1838. Beiträge zur Phytogenesis. *Arch. Anat. Physiol. Wiss. Med. (Müller's Arch.)* 5, 137–176.

SCHUSTER, W., and A. BRENNICKE. 1987. Plastid, nuclear and reverse transcriptase sequences in the mitochondrial genome of *Oenothera*: Is genetic information transferred between organelles via RNA? *EMBO J.* 6, 2857–2863.

SCHWANN, TH. 1839. *Mikroskopische Untersuchungen über die Übereinstimmung in der Struktur und dem Wachstum der Thiere und Pflanzen.* Wilhelm Engelmann, Leipzig.

SEUFFERHELD, M., M. C. F. VIEIRA, F. A. RUIZ, C. O. RODRIGUES, S. N. J. MORENO, and R. DOCAMPO. 2003. Identification of organelles in bacteria similar to acidocalcisomes of unicellular eukaryotes. *J. Biol. Chem.* 278, 29971–29978.

SHANNON, J. C. 1989. Aqueous and nonaqueous methods for amyloplast isolation. In: *Physiology, Biochemistry, and Genetics of Nongreen Plastids*, pp. 37–48, C. D. Boyer, J. C. Shannon, and R. C. Hardison, eds. American Society of Plant Physiologists, Rockville, MD.

SIEFRITZ, F., B. OTTO, G. P. BIENERT, A. VAN DER KROL, and R. KALDENHOFF. 2004. The plasma membrane aquaporin NtAQP1 is a key component of the leaf unfolding mechanism in tobacco. *Plant J.* 37, 147–155.

SINGER, S. J., and G. L. NICOLSON. 1972. The fluid mosaic model of the structure of cell membranes. *Science* 175, 720–731.

SITTE, P. 1992. A modern concept of the "cell theory." A perspective on competing hypotheses of structure. *Int. J. Plant Sci.* 153, S1–S6.

SITTE, P., H. FALK, and B. LIEDVOGEL. 1980. Chromoplasts. In: *Pigments in Plants*, 2nd ed., pp. 117–148, F.-C. Czygan, ed. Gustav Fischer Verlag, Stuttgart.

SMEEKENS, S., P. WEISBEEK, and C. ROBINSON. 1990. Protein transport into and within chloroplasts. *Trends Biochem. Sci.* 15, 73–76.

SMITH, J. A. 1987. Vacuolar accumulation of organic acids and their anions in CAM plants. In: *Plant Vacuoles: Their Importance in Solute Compartmentation in Cells and Their Applications in Plant Biotechnology*, pp. 79–87, B. Marin, ed. Plenum Press, New York.

STALS, H., and D. INZÉ. 2001. When plant cells decide to divide. *Trends Plant Sci.* 6, 359–364.

STEYN, W. J., S. J. E. WAND, D. M. HOLCROFT, and G. JACOBS. 2002. Anthocyanins in vegetative tissues: A proposed unified function in photoprotection. *New Phytol.* 155, 349–361.

STOEBE, B., and U.-G. MAIER. 2002. One, two, three: Nature's tool box for building plastids. *Protoplasma* 219, 123–130.

STRANGE, C. 1992. Cell cycle advances. *BioScience* 42, 252–256.

SUGIURA, M. 1989. The chloroplast chromosomes in land plants. *Annu. Rev. Cell Biol.* 5, 51–70.

SZE, H., X. LI, and M. G. PALMGREN. 1999. Energization of plant cell membranes by $H^+$-pumping ATPases: Regulation and biosynthesis. *Plant Cell* 11, 677–690.

TAIZ, L., and E. ZEIGER. 2002. *Plant Physiology*, 3rd ed. Sinauer Associates, Sunderland, MA.

TALCOTT, B., and M. S. MOORE. 1999. Getting across the nuclear pore complex. *Trends Cell Biol.* 9, 312–318.

THEG, S. M., and S. V. SCOTT. 1993. Protein import into chloroplasts. *Trends Cell Biol.* 3, 186–190.

TITORENKO, V. I., and R. A. RACHUBINSKI. 1998. The endoplasmic reticulum plays an essential role in peroxisome biogenesis. *Trends Biochem. Sci.* 23, 231–233.

TOLBERT, N. E. 1980. Microbodies—Peroxisomes and glyoxysomes. In: *The Biochemistry of Plants*, vol. 1, *The Plant Cell*, pp. 359–388, N. E. Tolbert, ed. Academic Press, New York.

TOLBERT, N. E., and E. ESSNER. 1981. Microbodies: Peroxisomes and glyoxysomes. *J. Cell Biol.* 91 (suppl. 3), 271s–283s.

TOURNAIRE-ROUX, C., M. SUTKA, H. JAVOT, E. GOUT, P. GERBEAU, D.-T. LUU, R. BLIGNY, and C. MAUREL. 2003. Cytosolic pH regulates root water transport during anoxic stress through gating of aquaporins. *Nature* 425, 393–397.

TRELEASE, R. N. 1984. Biogenesis of glyoxysomes. *Annu. Rev. Plant Physiol.* 35, 321–347.

TROJAN, A., and H. GABRYŚ. 1996. Chloroplast distribution in *Arabidopsis thaliana* (L.) depends on light conditions during growth. *Plant Physiol.* 111, 419–425.

TSE, Y. C., B. MO, S. HILLMER, M. ZHAO, S. W. LO, D. G. ROBINSON, and L. JIANG. 2004. Identification of multivesicular bodies as prevacuolar compartments in *Nicotiana tabacum* BY-2 cells. *Plant Cell* 16, 672–693.

VAN GESTEL, K., R. H. KÖHLER, and J.-P. VERBELEN. 2002. Plant mitochondria move on F-actin, but their positioning in the cortical cytoplasm depends on both F-actin and microtubules. *J. Exp. Bot.* 53, 659–667.

VIRCHOW, R. 1858. *Die Cellularpathologie in ihrer Begründung auf physiologische und pathologische Gewebelehre*. A. Hirschwald, Berlin.

VISHNEVETSKY, M., M. OVADIS, and A. VAINSTEIN. 1999. Carotenoid sequestration in plants: The role of carotenoid-associated proteins. *Trends Plant Sci.* 4, 232–235.

WATERS, M. T., R. G. FRAY, and K. A. PYKE. 2004. Stromule formation is dependent upon plastid size, plastid differentiation status and the density of plastids within the cell. *Plant J.* 39, 655–667.

WERGIN, W. P., P. J. GRUBER, and E. H. NEWCOMB. 1970. Fine structural investigation of nuclear inclusions in plants. *J. Ultrastruct. Res.* 30, 533–557.

WIEBE, H. H. 1978. The significance of plant vacuoles. *BioScience* 28, 327–331.

WILLIAMS, W. E., H. L. GORTON, and S. M. WITIAK. 2003. Chloroplast movements in the field. *Plant Cell Environ.* 26, 2005–2014.

WINK, M. 1993. The plant vacuole: A multifunctional compartment. *J. Exp. Bot.* 44 (suppl.), 231–246.

XU, Y., and M. R. HANSON. 2000. Programmed cell death during pollination-induced petal senescence in *Petunia*. *Plant Physiol.* 122, 1323–1334.

YAO, N., B. J. EISFELDER, J. MARVIN, and J. T. GREENBERG. 2004. The mitochondrion—An organelle commonly involved in programmed cell death in *Arabidopsis thaliana*. *Plant J.* 40, 596–610.

YATSUHASHI, H. 1996. Photoregulation systems for light-oriented chloroplast movement *J. Plant Res.* 109, 139–146.

YOUNG, T. E., and D. R. GALLIE. 2000. Regulation of programmed cell death in maize endosperm by abscisic acid. *Plant Mol. Biol.* 42, 397–414.

YU, X.-H., T. D. PERDUE, Y. M. HEIMER, and A. M. JONES. 2002. Mitochondrial involvement in tracheary element programmed cell death. *Cell Death Differ.* 9, 189–198.

ZIEGLER, H., E. SCHÄFER, and M. M. SCHNEIDER. 1983. Some metabolic changes during chloroplast-chromoplast transition in *Capsicum annuum*. *Physiol. Vég.* 21, 485–494.

# CHAPTER THREE

# The Protoplast: Endomembrane System, Secretory Pathways, Cytoskeleton, and Stored Compounds

## ▌ENDOMEMBRANE SYSTEM

In the previous chapter various components of the protoplast were considered in isolation. With the exception of mitochondrial, plastid, and peroxisomal membranes, however, all cellular membranes—including plasma membrane, nuclear envelope, endoplasmic reticulum (ER), Golgi apparatus, tonoplast (vacuolar membrane), and various kinds of vesicles—constitute a continuous, interconnected system. This system is known as the **endomembrane system** (Fig. 3.1), whose ER is the initial source of membranes (Morré and Mollenhauer, 1974; Mollenhauer and Morré, 1980). Transition vesicles derived from the ER transport new membrane material to the Golgi apparatus, and secretory vesicles derived from the Golgi apparatus contribute to the plasma membrane. The ER and Golgi apparatus therefore constitute a functional unit, in which the Golgi apparatus serves as the main vehicle for the transformation of ER-like membranes into plasma membrane-like membranes.

Transition vesicles budding off ER membranes close to Golgi bodies are only rarely encountered because of the low volume of protein transport between the ER and the Golgi bodies in most plant cells. Transition vesicles are commonly encountered, however, in cells that produce large quantities of globulin-type storage proteins (as in legumes) or secretory proteins. In such cells proteins travel via vesicle budding with subsequent fusion from the ER through the Golgi apparatus to arrive at the storage vacuoles or at the surface of the plasma membrane (Staehelin, 1997; Vitale and Denecke, 1999).

### The Endoplasmic Reticulum Is a Continuous, Three-dimensional Membrane System That Permeates the Entire Cytosol

In profile the ER appears as two parallel membranes with a narrow space, or lumen, between them. This profile of ER should not be confused with a single unit membrane. Each of the parallel ER membranes is itself a unit membrane. The form and abundance of the ER varies greatly from cell to cell, depending on the cell type, its metabolic activity, and its stage of development. For example, cells that store or secrete large quantities of proteins have abundant **rough ER**, which consists of

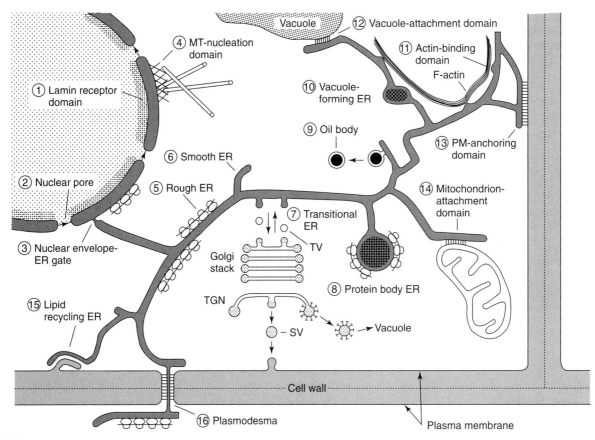

**FIGURE 3.1**

A diagrammatic representation of the endomembrane system, which includes all membranes except mitochondrial, plastid, and peroxisomal membranes. This drawing depicts 16 types of endoplasmic reticulum (ER) domains. Note the secretory pathway depicted here, involving the endoplasmic reticulum, the Golgi stack, and *trans*-Golgi network (TGN). Other details: TV, transport vesicle; SV, secretory vesicle. (From Staehelin, 1997. © Blackwell Publishing.)

flattened sacs, or *cisternae* (singular: *cisterna*), with numerous ribosomes on their outer surface. In contrast, cells that produce large quantities of lipidic compounds have extensive systems of **smooth ER**, which lacks ribosomes and is largely tubular in form. Both rough and smooth forms of ER occur within the same cell and are physically continuous. Rough and smooth ER are illustrated in Fig. 3.2A and B, respectively.

The ER is a multifunctional membrane system. Staehelin (1997) recognized 16 types of functional ER domains, or subregions, in plant cells (Fig. 3.1). Among those domains are the nuclear pores; the nuclear envelope–ER gates (connections); the transitional ER domain in the vicinity of Golgi bodies; a rough ER domain that acts as the port of entry of proteins into the secretory pathway; a smooth ER domain involved with the synthesis of lipidic molecules, including glycerolipids, isoprenoids, and flavonoids; protein body-forming and oil body-forming domains; a vacuole-forming domain; and the plasmodesmata (Fig. 3.2B), which traverse the

common walls between cells and play an important role in cell-to-cell communication (Chapter 4). This list will continue to expand as more cells are investigated by advanced techniques. In 2001 two more domains were added to Staehelin's list, a ricinosome-forming domain (Gietl and Schmid, 2001) and the "nodal ER" domain, which is unique to gravisensing columella rootcap cells (Zheng and Staehelin, 2001). Discovered in senescing endosperm of germinating castor bean (*Ricinus communis*) seeds, the **ricinosomes** bud off the ER at the beginning of programmed cell death and deliver large amounts of a papin-type cysteine endopeptidase to the cytosol in the final stages of cellular disintegration.

An extensive two-dimensional network of ER, consisting of interconnected cisternae and tubules, is located just inside the plasma membrane in the peripheral, or cortical, cytoplasm (Fig. 3.3; Hepler et al., 1990; Knebel et al., 1990; Lichtscheidl and Hepler, 1996; Ridge et al., 1999). The membranes of this **cortical ER** are continuous with those of the ER lying deeper within the

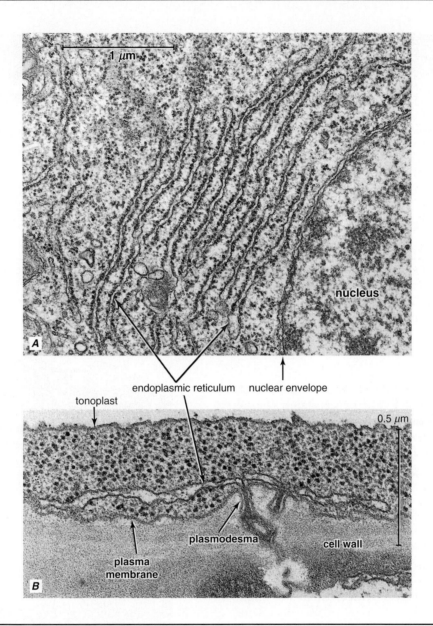

**FIGURE 3.2**

Endoplasmic reticulum (ER) seen in profile in leaf cells of tobacco (*Nicotiana tabacum*, **A**) and sugar beet (*Beta vulgaris*, **B**). The ER is associated with numerous ribosomes (rough ER) in **A**, with fewer in **B**. The largely smooth ER in **B** is connected to the electron dense cores (desmotubules) of plasmodesmata (seen only in part). Plasma membrane lines the plasmodesmatal canals. Note the three-layered appearance of tonoplast and plasma membrane in **B**. (From Esau, 1977.)

cytosol, including those in the transvacuolar strands of highly vacuolated cells. As mentioned previously, the outer nuclear membrane is also continuous with the ER. Thus the rough and smooth ER along with the nuclear envelope form a membrane continuum that encloses a single lumen and pervades the entire cytosol.

It has been suggested that the network of cortical ER serves as a structural element that stabilizes or anchors the cytoskeleton of the cell (Lichtscheidl et al., 1990). The cortical ER may function in $Ca^{2+}$ regulation; if so, it could play a profound role in a host of developmental

and physiological processes (Hepler and Wayne, 1985; Hepler et al., 1990; Lichtscheidl and Hepler, 1996).

Insights into the dynamic nature of the ER have come from studies of living cells, utilizing vital fluorescent dyes such as dihexyloxacarbocyanine iodide (DiOC) (Quader and Schnepf, 1986; Quader et al., 1989; Knebel et al., 1990), which stains endomembranes and, more recently, constructs delivering green fluorescent protein to the ER (Ridge et al., 1999). These studies have revealed that the ER membranes are in continuous motion and are constantly changing their shape and distribution

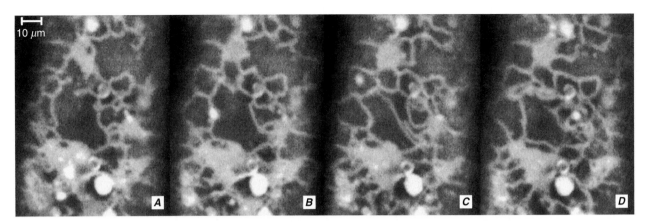

**FIGURE 3.3**

Four confocal scanning light micrographs of cortical ER membranes of tobacco BY-2 cells. The cells were grown and imaged in suspension culture in the presence of 10 μg of rhodamine 123 per ml. These micrographs, taken at 1 minute intervals, illustrate the changes that have taken place to the organization of the ER during this period of time. (From Hepler and Gunning, 1998.)

(Fig. 3.3). The ER deeper in the cell moves more actively than the cortical ER, which although constantly restructured, does not move with the rest of the ER or the organelles of the deeper streaming cytoplasm. The mobility of the cortical ER is limited by its presumed anchorage at plasmodesmata and by its adhesion to the plasma membrane (Lichtscheidl and Hepler, 1996).

### The Golgi Apparatus Is a Highly Polarized Membrane System Involved in Secretion

The term **Golgi apparatus** refers collectively to all of the Golgi body–*trans*-Golgi network complexes of a cell. Golgi bodies are also called ***dictyosomes*** or simply ***Golgi stacks***.

Each **Golgi body** consists of five to eight stacks of flattened cisternae, which often have bulbous and fenestrated margins (Fig. 3.4). The Golgi stacks are polarized structures. The opposite surfaces or poles of a stack are referred to as *cis*- and *trans*-faces. Three morphologically distinct cisternae may be recognized across the stack: *cis*-, medial-, and *trans*-cisternae, which differ from one another both structurally and biochemically (Driouich and Staehelin, 1997; Andreeva et al., 1998). The ***trans*-Golgi network (TGN)**, a tubular reticulum with clathrin-coated and noncoated budding vesicles, is associated with the *trans*-face of the Golgi stack (Fig. 3.1). Each Golgi-TGN complex is embedded in and surrounded by a ribosome-free zone called the ***Golgi matrix***.

Unlike the centralized Golgi of mammalian cells, the Golgi apparatus of plant cells consists of many separate stacks that remain functionally active during mitosis and cytokinesis (Andreeva et al., 1998; Dupree and

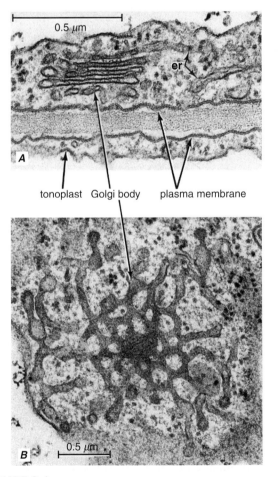

tonoplast    Golgi body    plasma membrane

**FIGURE 3.4**

Golgi bodies from a tobacco (*Nicotiana tabacum*) leaf. **A**, Golgi body in profile with the fenestrated *trans*-face toward the cell wall. **B**, Golgi body is seen from its fenestrated *trans*-face. Some of the vesicles to be pinched off are coated. Detail: er, endoplasmic reticulum. (From Esau, 1977.)

Sherrier, 1998). In living cells, stacks tagged with green fluorescent protein can be observed along bundles of actin filaments that match precisely the architecture of the ER network (Boevink et al., 1998). The stacks have been observed undergoing stop-and-go movements, oscillating rapidly between directed movement and random "wiggling." Nebenführ et al. (1999) have postulated that the stop-and-go motion of the Golgi-TGN complexes is regulated by "stop signals" produced by ER export sites and locally expanding cell wall domains to optimize ER to Golgi and Golgi to cell wall trafficking. During mitosis and cytokinesis, the Golgi stacks redistribute to specific locations as cytoplasmic streaming stops (Chapter 4; Nebenführ et al., 2000). Just prior to mitosis, the number of Golgi stacks doubles by cisternal fission, which takes place in a *cis*-to-*trans* direction (Garcia-Herdugo et al., 1988).

In most plant cells the Golgi apparatus serves two major functions: the synthesis of noncellulosic cell wall polysaccharides (hemicelluloses and pectins; Chapter 4) and protein glycosylation. Evidence obtained through the use of polyclonal antibodies indicates that different steps in polysaccharide synthesis occur in different cisternae of the Golgi body (Moore et al., 1991; Zhang and Staehelin, 1992; Driouich et al., 1993). The different polysaccharides are packaged in secretory vesicles, which migrate to and fuse with the plasma membrane (exocytosis). The vesicles then discharge their contents and the polysaccharides become part of the cell wall. In enlarging cells the vesicles contribute to growth of the plasma membrane.

The initial stage of protein glycosylation occurs in the rough ER. These glycoproteins then are transferred from the ER to the *cis*-face of the Golgi body via transition vesicles (Bednarek and Raikhel, 1992; Holtzman, 1992; Schnepf, 1993). The glycoproteins proceed stepwise across the stack to the *trans*-face and then are sorted in the TGN for delivery to the vacuole or for secretion at the cell surface. Polysaccharides destined for secretion at the cell surface are also packaged into vesicles at the TGN. A given Golgi body can process polysaccharides and glycoproteins simultaneously.

Glycoproteins and complex polysaccharides destined for secretion into the cell wall are packaged in non-coated, or smooth-surfaced, vesicles, whereas hydrolytic enzymes and storage proteins (water-soluble globulins) destined for vacuoles are packaged at the TGN into clathrin-coated vesicles and smooth, electron-dense vesicles, respectively (Herman and Larkins, 1999; Miller and Anderson, 1999; Chrispeels and Herman, 2000). The formation of Golgi-derived **dense vesicles** is not restricted to the TGN, but may also occur in the *cis*-cisternae (Hillmer et al., 2001).

Some types of storage proteins (alcohol-soluble prolamins) form aggregates and are packaged into vesicles in the ER from where they are transported directly to the protein storage vacuoles, bypassing the Golgi (Matsuoka and Bednarek, 1998; Herman and Larkins, 1999). In wheat, for example, a considerable amount of the prolamin aggregates directly into protein bodies (aleurone grains) within the rough ER, and then the protein bodies are transported intact to the vacuoles without Golgi involvement (Levanony et al., 1992). In maize, sorghum, and rice similarly formed protein bodies remain within the ER and are bounded by ER membranes (Vitale et al., 1993).

The delivery of secretory vesicles to the plasma membrane by exocytosis must be balanced by the equivalent recycling of membranes from the plasma membrane by clathrin-mediated endocytosis (Battey et al., 1999; Marty, 1999; Sanderfoot and Raikhel, 1999). Recycling is essential to support a functional endomembrane system (Battey et al., 1999).

# CYTOSKELETON

The **cytoskeleton** is a dynamic, three-dimensional network of protein filaments that extends throughout the cytosol and is intimately involved in many cellular processes, including mitosis and cytokinesis, cell expansion and differentiation, cell-to-cell communication, and the movement of organelles and other cytoplasmic components from one location to another within the cell (Seagull, 1989; Derksen et al., 1990; Goddard et al., 1994; Kost et al., 1999; Brown and Lemmon, 2001; Kost and Chua, 2002; Sheahan et al., 2004). In plant cells it consists of at least two types of protein filaments: microtubules and actin filaments. The presence of intermediate filaments, which occur in animal cells, has not been unequivocally demonstrated in plant cells. Immunofluorescence microscopy and, more recently, the use of green fluorescent protein tags to cytoskeletal proteins and confocal microscopy, have made it possible to examine the three-dimensional organization of the cytoskeleton in both fixed and living cells, and have contributed greatly to our understanding of both cytoskeletal structure and function (Lloyd, 1987; Staiger and Schliwa, 1987; Flanders et al., 1990; Marc, 1997; Collings et al., 1998; Kost et al., 1999; Kost et al., 2000).

## Microtubules Are Cylindrical Structures Composed of Tubulin Subunits

**Microtubules** are cylindrical structures about 24 nanometers in diameter and of varying lengths (Fig. 3.5). The lengths of cortical microtubules, that is, of microtubules located in the peripheral cytoplasm just inside the plasma membrane, generally correspond to the cross-sectional width of the cell facet with which they are associated (Barlow and Baluška, 2000). Each microtubule is composed of two different types of protein molecules, alpha ($\alpha$) tubulin and beta ($\beta$) tubulin. These

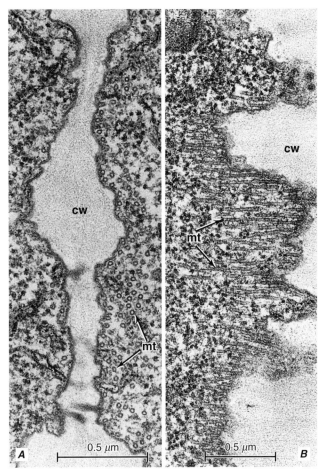

**FIGURE 3.5**

Cortical microtubules (mt) in *Allium cepa* root tip cells seen in transverse (**A**) and longitudinal (**B**) views. Other detail: cw, cell wall.

subunits come together to form soluble dimers ("two parts"), which then self-assemble into insoluble tubules. The subunits are arranged in a helix to form 13 rows, or "protofilaments," around the core of lightly contrasted material. Within each protofilament the subunits are oriented in the same direction, and all of the protofilaments are aligned in parallel with the same polarity; consequently the microtubule is a polar structure for which there can be designated plus and minus ends. The plus ends grow faster than the minus ends, and the ends of the microtubules can alternate between growing and shrinking states, a behavior called **dynamic instability** (Cassimeris et al., 1987). Indeed microtubules are dynamic structures that undergo regular sequences of breakdown, re-formation, and rearrangement into new configurations, or arrays, at specific points in the cell cycle and during differentiation (Hush et al., 1994; Vantard et al., 2000; Azimzadeh et al., 2001). The most prominent cell-cycle arrays are the interphase cortical

array, the preprophase band, the mitotic spindle, and the phragmoplast, which is located between the two newly formed daughter nuclei (Fig. 3.6; Chapter 4; Baskin and Cande, 1990; Barlow and Baluška, 2000; Kumagai and Hasezawa, 2001).

Microtubules have many functions (Wasteneys, 2004). In enlarging and differentiating cells, the cortical microtubules control the alignment of cellulose microfibrils that are being added to the wall, and the direction of cell expansion is governed, in turn, by this alignment of cellulose microfibrils in the wall (Chapter 4; Mathur and Hülskamp, 2002). In addition microtubules that make up the fibers of the mitotic spindle play a role in chromosome movement, and those forming the phragmoplast, probably with the help of kinesin-like motor proteins (Otegui et al., 2001), are involved it the formation of the cell plate (the initial partition between dividing cells).

During most of the cell cycle (interphase) microtubules radiate from all over the nuclear surface, which is the primary "nucleating site," or **microtubular organizing center (MTOC)** in the plant cell. Secondary MTOCs are located at the plasma membrane where they organize arrays of cortical microtubules, which are essential for ordered cell wall synthesis and hence for cellular morphogenesis (Wymer and Lloyd, 1996; Wymer et al., 1996). It has been suggested that the material comprising the secondary MTOCs is translocated to the cell periphery by the microtubules organized and radiating from the nuclear surface (the primary MTOC) (Baluška et al., 1997b, 1998). γ-Tubulin, which is found in all MTOCs, is believed to be essential for microtubule nucleation (Marc, 1997).

### Actin Filaments Consist of Two Linear Chains of Actin Molecules in the Form of a Helix

**Actin filaments**, also called **microfilaments** and **filamentous actin (*F actin*)**, are, like microtubules, polar structures with distinct plus and minus ends. They are composed of actin monomers that self-assemble into filaments and resemble a double-stranded helix, with an average diameter of 7 nanometers (Meagher et al., 1999; Staiger, 2000). Actin filaments occur singly and in bundles (Fig. 3.7). Actin filaments constitute a cytoskeleton system that can assemble and function independently of microtubules (e.g., actin filaments drive cytoplasmic streaming and Golgi dynamics). However, in some instances actin and microtubules can work together to perform specific functions. Some actin filaments are associated spatially with microtubules and, like microtubules, form new configurations, or arrays, at specific points in the cell cycle (Staiger and Schliwa, 1987; Lloyd, 1988; Baluška et al., 1997a; Collings et al., 1998). In cells of the transition region—a postmitotic zone interpolated between the meristem and the rapidly

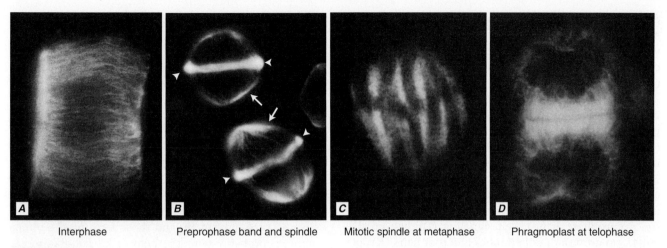

| Interphase | Preprophase band and spindle | Mitotic spindle at metaphase | Phragmoplast at telophase |

**FIGURE 3.6**

Fluorescence micrographs of microtubular arrangements, or arrays, in root tips of onion (*Allium cepa*). **A**, interphase cortical array. The microtubules lie just beneath the plasma membrane. **B**, a preprophase band of microtubules (arrowheads) encircles the nucleus at the site of the future cell plate. The prophase spindle, comprised of other microtubules (arrows), outlines the nuclear envelope (not visible). The lower cell is at a later stage than the upper one. **C**, the mitotic spindle at metaphase. **D**, during telophase new microtubules form a phragmoplast, which is involved with cell plate formation. (Reprinted with permission from Goddard et al., 1994. © American Society of Plant Biologists.)

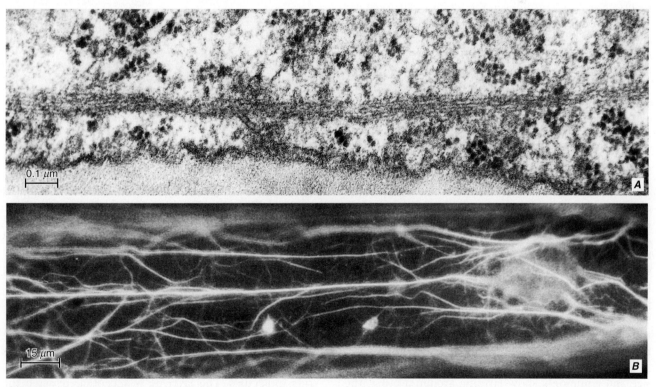

**FIGURE 3.7**

Actin filaments. **A**, a bundle of actin filaments as revealed in an electron micrograph of a leaf cell of maize (*Zea mays*). **B**, several bundles of actin filaments as revealed in a fluorescence micrograph of a stem hair of tomato (*Solanum lycopersicum*). (**B**, from Parthasarathy et al., 1985.)

elongating region—of growing maize root tips, the nuclear surface and the cortical cytoplasm associated with the two end walls have been identified as the principal organizing regions of bundles of actin filaments (Baluška et al., 1997a).

The actin cytoskeleton has been implicated in a variety of roles in plant cells, in addition to the causative role it plays—in association with myosin motor proteins (Shimmen et al., 2000)—in cytoplasmic streaming and in the movement of plastids, vesicles (Jeng and Welch, 2001), and other cytoplasmic components. Other demonstrated or proposed roles include establishing cell polarity, division plane determination (by positioning the preprophase band), cell signaling (Drøbak et al., 2004), tip growth of pollen tubes and root hairs (Kropf et al., 1998), control of plasmodesmal transport (White et al., 1994; Ding et al., 1996; Aaziz et al., 2001), and mechanosensation processes such as touch responses of leaves (Xu et al., 1996) and the grasping of support-gyrating tendrils (Engelberth et al., 1995).

# STORED COMPOUNDS

All compounds stored by plants are products of metabolism. Sometimes collectively referred to as ergastic substances, these compounds may appear, disappear, and reappear at different times in the life of a cell. Most are storage products, some are involved in plant defenses, and a few have been characterized as waste products. In most instances they form structures that are visible in light and/or electron microscopes, including starch grains, protein bodies, oil bodies, tannin-filled vacuoles, and mineral matter in the form of crystals. These substances are found in the cell wall, in the cytosol, and in organelles, including vacuoles.

### Starch Develops in the Form of Grains in Plastids

Next to cellulose, **starch** is the most abundant carbohydrate in the plant world. Moreover it is the principal storage polysaccharide in plants. During photosynthesis assimilatory starch is formed in chloroplasts (Fig. 3.8). Later it is broken down into sugars, transported to storage cells, and resynthesized as storage starch in amyloplasts (Fig. 3.9). As mentioned previously, an amyloplast may contain one (simple) or more (compound) starch grains. If several starch grains develop together, they may become enclosed in common outer layers, forming a complex starch grain (Ferri, 1974).

Starch grains, or granules, are varied in shape and size and commonly show layering around a point, the **hilum**, which may be the center of the grain or to one side (Fig. 3.9A). Fractures, often radiating from the hilum, appear during dehydration of the grains. All grains consist of two types of molecules, unbranched amylose chains and branched amylopectin molecules

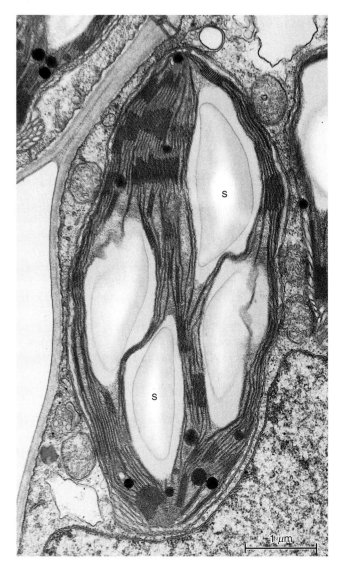

**FIGURE 3.8**

A chloroplast containing assimilatory starch (s), from a mesophyll cell of the pigweed (*Amaranthus retroflexus*) leaf. During periods of intense photosynthesis some of the carbohydrate is stored temporarily in the chloroplast as grains of assimilatory starch. At night sucrose is produced from the starch and exported from the leaf to other parts of the plant, where it is eventually used for the manufacture of other molecules needed by the plant. (From Fisher and Evert, 1982. © 1982 by The University of Chicago. All rights reserved.)

(Martin and Smith, 1995). The layering of starch grains is attributed to an alternation of these two polysaccharide molecules. The layering is accentuated when the starch grain is placed in water because of differential swelling of the two substances: amylose is soluble in water, and amylopectin is not. Amylose appears to be the predominant component of starch found in the leaves of sorghum (*Sorghum bicolor*) and maize (*Zea*

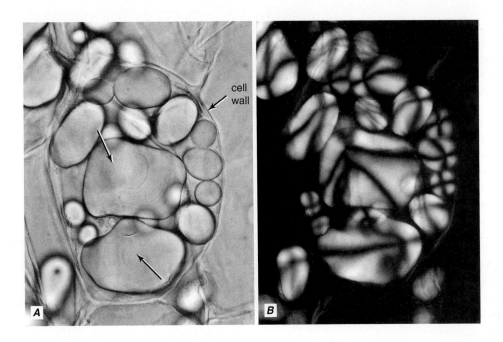

**FIGURE 3.9**

Starch grains of potato (*Solanum tuberosum*) tuber photographed with ordinary light (**A**) and with polarized light (**B**). Arrows point to the hilum of some starch grains in **A**. In **B** the starch grains show the figure of a Maltese cross. The amyloplasts in potato each contain a single starch grain. (A, B, ×620.)

*mays*), whereas the seeds contain 70% to 90% amylopectin (Vickery and Vickery, 1981). In potato tuber starch their proportion is 22% amylose and 78% amylopectin (Frey-Wyssling, 1980). Starch grains are composed of amorphous and crystalline regions, whose chains are held together by hydrogen bonds. Under polarized light, starch grains show a figure of a Maltese cross (Fig. 3.9B) (Varriano-Marston, 1983). Starch commonly stains bluish-black with iodine in potassium iodide ($I_2KI$).

Storage starch occurs widely in the plant body. It is found in parenchyma cells of the cortex, pith, and vascular tissues of roots and stems; in parenchyma cells of fleshy leaves (bulb scales), rhizomes, tubers, corms, fruits, and cotyledons; and in the endosperm of seeds. Commercial starches are obtained from various sources as, for example, the endosperm of cereals, fleshy roots of the tropical *Manihot esculenta* (tapioca starch), tubers of potato, tuberous rhizomes of *Maranta arundinacea* (arrowroot starch), and stems of *Metroxylon sagu* (sago starch).

## The Site of Protein Body Assembly Depends on Protein Composition

Storage proteins may be formed in different ways, depending in part upon whether they are composed of salt-soluble globulins or alcohol-soluble prolamins (Chrispeels, 1991; Herman and Larkins, 1999; Chrispeels

and Herman, 2000). Globulins are the major storage proteins in legumes, and prolamins in most cereals. Typically globulins aggregate in protein storage vacuoles after having been transported there from the rough ER via the Golgi apparatus. However, as indicated previously, the Golgi apparatus is not necessarily involved with prolamin transport to the vacuoles in cereals. In wheat, for example, a considerable part of the prolamins aggregate directly into **protein bodies** (aleurone grains) within the rough ER and then are transported in distinct vesicles to the vacuoles without Golgi involvement (Levanony et al., 1992). In other cereals, such as maize, sorghum, and rice, similarly formed protein bodies are not transported to vacuoles, but remain within the rough ER and bounded by ER membranes (ER domain 8, Fig. 3.1) (Vitale et al., 1993). Upon germination the stored proteins are mobilized by hydrolysis to provide energy, nitrogenous compounds, and minerals needed by the growing seedling. At the same time the protein storage vacuoles may function as lysosomal compartments, or autophagic organelles (Herman et al., 1981), taking up and digesting portions of the cytoplasm. As germination continues, the numerous small vacuoles may fuse to form one large vacuole. Although protein bodies are most abundant in seeds, they also occur in roots, stems, leaves, flowers, and fruits.

Structurally the simplest protein bodies consist of an amorphous proteinaceous matrix surrounded by a bounding membrane. Other protein bodies may contain one or more nonproteinaceous globoids (Fig. 3.10) or one or more globoids and one or more protein crystalloids, in addition to the proteinaceous matrix. Protein bodies also contain a large number of enzymes and fair amounts of phytic acid, a cation salt of myo-inositol hexaphosphoric acid, which usually is stored in the globoids. Phytic acid is an important source of phosphorous during seedling development. Some protein bodies contain calcium oxalate crystals (Apiaceae).

Proteins may occur in the form of crystalloids in the cytosol as, for example, in parenchyma cells of the potato tuber, among starch grains of *Musa*, and in the fruit parenchyma of *Capsicum*. In the potato tuber the cuboidal protein crystals typically are found in subphellogen cells. The crystals apparently are formed within vesicles from which they may or may not be released into the cytosol at maturity (Marinos, 1965; Lyshede, 1980). Proteinaceous crystalloids also occur in the nuclei. Such nuclear inclusions are widespread in occurrence among vascular plants (Wergin et al., 1970).

## Oil Bodies Bud from Smooth ER Membranes by an Oleosin-mediated Process

**Oil bodies** are more or less spherical structures that impart a granular appearance to the cytoplasm of a plant cell when viewed with the light microscope. In electron micrographs the oil bodies have an amorphous appearance (Fig. 3.10). Oil bodies are widely distributed throughout the plant body but are most abundant in fruits and seeds. Approximately 45% of the weight of sunflower, peanut, flax, and sesame seed is composed of oil (Somerville and Browse, 1991). The oil provides energy and a source of carbon to the developing seedling.

Oil bodies, also known as spherosomes or oleosomes, arise by the accumulation of *triacylglycerol molecules* at specific sites (ER domain 9, Fig. 3.1) in the interior of the ER lipid bilayer (Wanner and Thelmer, 1978; Ohlrogge and Browse, 1995). These lipid-accumulation sites are defined by the presence of 16 to 25 kDa integral membrane proteins known as *oleosins*, thumbtack-like

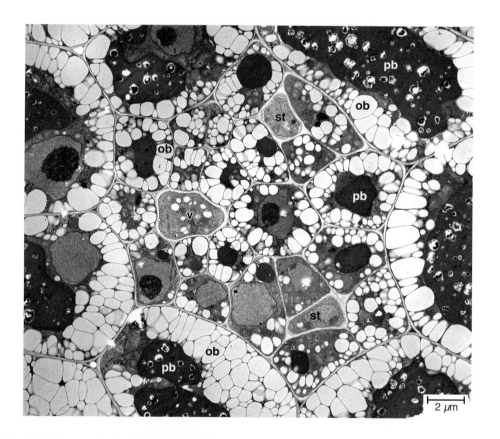

## FIGURE 3.10

Immature vascular bundle, surrounded by storage parenchyma cells, in cotyledon of *Arabidopsis thaliana* embryo. Oil bodies (ob) and globoid-containing protein bodies (pb) occupy most of the volume of the procambial cells and storage parenchyma cells. Other details: st, immature sieve tube; v, immature vessel. (From Busse and Evert, 1999. © 1999 by The University of Chicago. All rights reserved.)

molecules that cause the oil bodies to bud into the cytosol (Huang, 1996). Each oil body is bounded by a phospholipid monolayer in which the oleosins are embedded (Sommerville and Browse, 1991; Loer and Herman, 1993). The oleosins and phospholipids stabilize the oil bodies and prevent them from coalescing (Tzen and Huang, 1992; Cummins et al., 1993). Maintaining the oil bodies as small entities ensures ample surface area for the attachment of lipases and rapid mobilization of the triacylglycerols when necessary.

Storage lipids occur in all plant taxa and are probably present in every cell, at least in small amounts (Küster, 1956). Usually they are found in liquid form as oil bodies. Crystalline forms are rare. An example was reported for the endosperm of the palm *Elais*, in which the cells were filled with short needle-shaped crystals of fat (Küster, 1956). (The distinction between fats and oils is primarily physical, fats being solid at room temperature and oils liquid.) So-called essential oils are volatile oils that contribute to the essence, or odor, of plants. They are made by special cells and excreted into intercellular cavities (Chapter 17). Oils and fats may be identified by a reddish color when they are treated with Sudan III or IV.

Mention should be made of **waxes**, long-chain lipid compounds, that occur as part of the protective coating (cuticle) on the epidermis of the aerial parts of the primary plant body and on the inner surface of the primary wall of cork cells in woody roots and stems. These waxes constitute a major barrier to water loss from the surface of the plant (Chapter 9). By reducing the wetability of leaves, they also reduce the ability of fungal spores to germinate and of bacteria to grow, thereby reducing the possibility of these agents to cause disease. Most plants contain too little wax to be valuable for commercial use. Exceptions are the palm *Copernicia cerifera*, which yields the carnauba wax of commerce, and *Simmondia chinensis* (jojoba), the cotyledons of which contain a liquid wax similar in quality to the oil of the sperm whale (Rost et al., 1977; Rost and Paterson, 1978).

## Tannins Typically Occur in Vacuoles but Also Are Found in Cell Walls

**Tannins** are a heterogeneous group of polyphenolic substances, important secondary metabolites, with an astringent taste and an ability to tan leather. They usually are divided into two categories, hydrolyzable and condensed. The hydrolyzable tannins can be hydrolyzed with hot, dilute acid to form carbohydrates (mainly glucose) and phenolic acids. The condensed tannins cannot be hydrolyzed. In some of their forms, tannins are quite conspicuous in sectioned material. They appear as coarsely or finely granular material or as bodies of various sizes colored yellow, red, or brown.

No tissue appears to lack tannins completely. Tannins are abundant in leaves of many plants, in vascular tissues, in the periderm, in unripe fruits, in seed coats, and in pathologic growths like galls (Küster, 1956). Typically they occur in the vacuole (Fig. 2.21) but apparently originate in the ER (Zobel, 1985; Rao, 1988). Tannins may be present in many cells of a given tissue or isolated in specialized cells (tannin idioblasts) scattered throughout the tissue (Gonzalez, 1996; Yonemori et al., 1997). In addition they may be located in much enlarged cells called tannin sacs or in tube-like cells (Chapter 17).

Most of the vegetable extracts used for tanning come from a few eudicotyledonous plants, in particular, from the wood, bark, leaves, and/or fruit of species in the Anacardiaceae, Fabaceae, and Fagaceae (Haslam, 1981). Apparently the primary function of tannins is protective, their astringency serving as a repellent to predators and an impediment to the invasion of parasitic organisms by immobilizing extracellular enzymes. Plants that produce and secrete substantial quantities of polyphenols, including tannins, may exclude other plant species from growing under them or in their near vicinity, a phenomenon known as **allelopathy**. Tannins released from leaves decaying in water are known to be harmful against some insects (Ayres et al., 1997), including phytophagous lepidopteran larvae (Barbehenn and Martin, 1994). They apparently play an important role in habitat selection among mosquito communities from Alpine hydrosystems (Rey et al., 2000).

Phenolic compounds, mainly tannins, were synthesized in increased amounts in the leaves of beech trees (*Fagus sylvatica*) in response to environmental stress (Bussotti et al., 1998). They initially accumulated in the vacuoles, especially in those of the upper epidermis and palisade parenchyma. At a later stage the tannins appeared to be solubilized in the cytosol and retranslocated, eventually impregnating the outer epidermal cell walls. The impregnation of the walls by the tannins has been interpreted as an impermeabilization mechanism associated with a reduction in cuticular transpiration. The browning associated with growing jack pine (*Pinus banksiana*) and eucalypt (*Eucalyptus pilularis*) roots has been shown to be due to the deposition of condensed tannins in the walls of all cells external to the vascular cylinder (McKenzie and Peterson, 1995a, b). The epidermal and cortical cells in the brown "tannin zone" of the roots are dead. Condensed tannins also have been found in the *phi* thickenings of *Ceratonia siliqua* roots (Pratikakis et al., 1998). *Phi* thickenings are reticulate or band-like wall thickenings on cortical cells of certain gymnosperms (Ginkgoaceae, Araucariaceae, Taxaceae, and Cupressaceae; Gerrath et al., 2002) and a few species of angiosperms such as *Ceratonia siliqua, Pyrus malus (Malus domestica)*, and *Pelargonium hortorum* (Peterson et al., 1981).

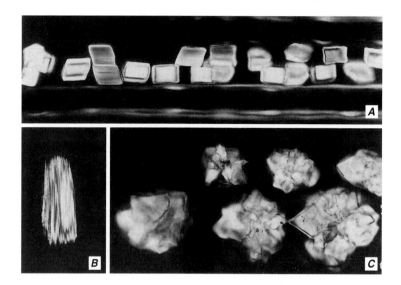

**FIGURE 3.11**

Calcium oxalate crystals seen in polarized light. **A**, prismatic crystals in phloem parenchyma of root of *Abies*. **B**, raphides in leaf of *Vitis*. **C**, druses in cortex of stem of *Tilia*. (A, ×500; B, C, ×750.)

### Crystals of Calcium Oxalate Usually Develop in Vacuoles but Also Are Found in the Cell Wall and Cuticle

Inorganic deposits in plants consist mostly of calcium salts and anhydrides of silica. Among the calcium salts, the most common is **calcium oxalate,** which occurs in the majority of plant families, notable exceptions being the Cucurbitaceae and some families of Liliales, Poales, and all Alismatidae (Prychid and Rudall, 1999). Calcium oxalate occurs as mono- and dihydrate salts in many crystalline forms. The monohydrate is the more stable and is more commonly found in plants than is the dihydrate. The most common forms of calcium oxalate crystals are (1) **prismatic crystals** (Fig. 3.11A), variously shaped prisms, usually one per cell; (2) **raphides** (Figs. 3.11B and 3.12A), needle-shaped crystals that occur in bundles; (3) **druses** (Figs. 3.11C and 3.12B), spherical aggregates of prismatic crystals; (4) **styloids,** elongated crystals with pointed or ridged ends, one or two to a cell; and (5) **crystal sand,** very small crystals, usually in masses. In some tissues calcium oxalate crystals arise in cells that resemble adjacent, crystal-free cells. In others, the crystals are formed in cells—**crystal idioblasts**—specialized to produce crystals. Crystal idioblasts contain an abundance of ER and Golgi bodies. Most crystal cells are probably alive at maturity. The location and type of calcium oxalate crystals within a given taxon may be very consistent and, hence, useful in taxonomic classification (Küster, 1956; Prychid and Rudall, 1999; Pennisi and McConnell, 2001).

Calcium oxalate crystals usually develop in vacuoles. The period of crystal cell differentiation may

**FIGURE 3.12**

Scanning electron micrographs (**A**) of raphide bundle isolated from grape (*Vitis mustangensis*) fruit and (**B**) druses from *Cercis canadensis* epidermal cells. (**A**, from Arnott and Webb, 2000. © 2000 by The University of Chicago. All rights reserved.; **B**, courtesy of Mary Alice Webb.)

correspond to that of neighboring cells, precede that of neighboring cells, or occur belatedly. The latter phenomenon is common in the nonconducting phloem in the bark of many trees and is associated with belated sclerification of many of the same cells (Chapter 14). Crystal formation commonly is preceded by the formation of some type of membrane system, or complex, that arises de novo in the vacuole and forms one or more crystal chambers (Franceschi and Horner, 1980; Arnott, 1982; Webb, 1999; Mazen et al., 2003). In raphide cells each crystal is included in an individual chamber (Fig. 3.13; Kausch and Horner, 1984; Webb et al., 1995). In addition to the crystals the vacuoles may contain mucilage (Kausch and Horner, 1983; Wang et al., 1994; Webb et al., 1995). A further stage of development may involve the deposition of a cell wall around the crystal, completely isolating the crystal from the protoplast (Ilarslan et al., 2001).

Horner and Wagner (1995) recognized two general systems of vacuolar crystal formation based in part on the presence or absence of membrane complexes in the vacuoles. System I, which is exemplified by the druses in *Capsicum* and *Vitis*, the raphides in *Psychotria*, and crystal sand in *Beta*, all eudicotyledons, is characterized by the presence of vacuolar membrane complexes and of organic paracrystalline bodies that display subunits with large periodicity. System II, which is characterized by the absence of vacuolar membrane complexes and the presence of paracrystalline bodies with closely spaced subunits, is exemplified by the raphide crystal idioblasts in *Typha, Vanilla, Yucca* (Horner and Wagner, 1995), and *Dracaena* (Pennisi et al., 2001b), all monocots.

Although uncommon in flowering plants, deposition of crystals in the cell wall and cuticle rather than in vacuoles is of frequent occurrence in conifers (Evert et al., 1970; Oladele, 1982). Among the angiosperms, calcium oxalate crystals have been reported in the cuticle of *Causarina equisetifolia* (Pant et al., 1975) and of some Aizoaceae (Öztig, 1940), between the primary epidermal cell wall and the cuticle in *Dracaena* (Pennisi et al., 2001a), and between the primary and secondary walls of the astrosclereids in *Nymphaea* and *Nuphar* (Arnott and Pautard, 1970; Kuo-Huang, 1990). In both the epidermal cells of *Dracaena sanderiana* (Pennisi et al., 2001a) and the crystal-forming sclereids of the *Nymphaea tetragona* leaf each "extracellular" crystal arises in a crystal chamber bounded by a sheath initially connected with the plasma membrane (Kuo-Huang, 1992; Kuo-Huang and Chen, 1999). After the crystals are formed in the *Nymphaea* sclereids, a thick secondary wall is deposited and the crystals are embedded between the primary and secondary cell walls.

Calcium oxalate formation has been shown to be a rapid and reversible process in *Lemna minor* (Franceschi, 1989). With an increase in the exogenous calcium concentration, crystal bundles formed in cells of the root within 30 minutes of the induction stimulus. With the source of calcium limited, the recently formed crystal bundles dissolved over a period of three hours. Obviously calcium oxalate formation is not a "dead-end process." The results of this study and of others (Kostman and Franceschi, 2000; Volk et al., 2002; Mazen et al., 2003) indicate that crystal formation is a highly controlled process and may provide a mechanism for regulating calcium levels in plant organs. The raphide idioblasts of *Pistia stratiotes* have been shown to be enriched with the calcium-binding protein calreticulum, which occurs in subdomains of the ER (Quitadamo et al., 2000; Kostman et al., 2003; Nakata et al., 2003).

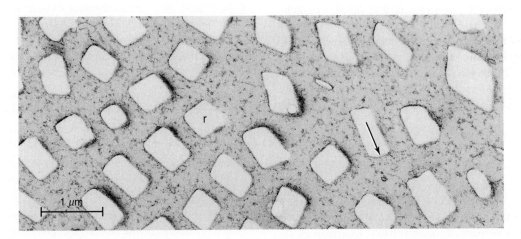

**FIGURE 3.13**

Crystal chambers in vacuole of a developing crystal cell in leaf of grape (*Vitis vulpina*) as seen with transmission electron microscope. The holes seen here were each occupied by a raphide (r). Each raphide is surrounded by a crystal chamber membrane (arrow). (From Webb et al., 1995. © Blackwell Publishing.)

It has been proposed that the calreticulum is involved with keeping cytosolic calcium activity low, while allowing for a rapid accumulation of calcium used for calcium oxalate formation (Mazen et al., 2003; Nakata et al., 2003). Other functions attributed to the calcification process include the removal of oxalate in plants unable to metabolize oxalate, protection against herbivory (Finley, 1999; Saltz and Ward, 2000; Molano-Flores, 2001), as a storage source of calcium (Ilarslan et al., 2001; Volk et al., 2002), the detoxification of heavy metals (see literature cited in Nakata, 2003), addition of mechanical strength, and addition of weight to the tissue. The weight added to tissue by calcium oxalate can be substantial. Eighty-five percent of the dry weight of some cacti reportedly consists of calcium oxalate (Cheavin, 1938).

Two types of raphide idioblasts occur in the leaves of *Colocasia esculenta* (taro; Sunell and Healey, 1985) and *Dieffenbachia maculata* (dumbcane; Sakai and Nagao, 1980): defensive and nondefensive. The defensive raphide idioblasts forcibly eject their "needles" through thin-walled papillae at the ends of the cells when the aroids (Araceae) are eaten or handled fresh. The nondefensive raphide idioblasts are not involved in the irritative property of aroids. The acidity of raphides from the edible aroids, including taro, may be due to the dual action of the sharp raphides puncturing soft skin and the presence of an irritant (a protease) in the raphides that causes swelling and soreness (Bradbury and Nixon, 1998). Paull et al. (1999) report, however, that the acridity is due entirely to an irritant (a 26 kDa protein, possibly a cysteine proteinase) found on the surface of the raphides.

**Calcium carbonate crystals** are not common in seed plants. The best known calcium carbonate formations are **cystoliths** (*kustis,* bag; *lithos,* stone), which are formed in specialized enlarged cells called **lithocysts** of the ground parenchyma and epidermis (Fig. 3.14; Chapter 9). The cystolith develops outside the plasma membrane in association with the cell wall of the lithocyst. Callose, cellulose, silica, and pectic substances also enter into the composition of cystoliths (Eschrich, 1954; Metcalfe, 1983), which are confined to a limited number (14) of plant families (Metcalfe and Chalk, 1983).

### Silica Most Commonly Is Deposited in Cell Walls

Among the seed plants the heaviest and most characteristic deposits of silica occur in the grasses (Poaceae), where they may account for 5% to 20% of the shoot's dry weight (Lewin and Reimann, 1969; Kaufman et al., 1985; Epstein, 1999). A record high silica content (41% on a dry weight basis) has been reported in the leaves of *Sasa veitchii* (Bambusoideae), which accumulate silica continuously throughout its life of about 24

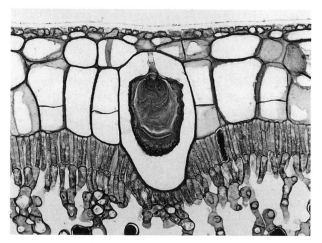

**FIGURE 3.14**

Calcium carbonate crystal. Transverse section of upper portion of rubber plant (*Ficus elastica*) leaf blade showing club-shaped cystolith in enlarged epidermal cell, the lithocyst. The cystolith consists mostly of calcium carbonate deposited on a cellulose stalk. (×155.)

months (Motomura et al., 2002). Silica deposits also occur in the roots of grasses (Sangster, 1978). In general, monocots take up and deposit more silicon than eudicots. Silicon accumulation in plants contributes to the strength of stems and provides resistance to attack by pathogenic fungi and predaceous chewing insects and other herbivores (McNaughton and Tarrants, 1983). Silica often forms bodies, termed **silica bodies** or **phytoliths,** within the lumen of the cell (Chapter 9). In the rind of *Cucurbita* fruits lignification and phytolith formation appear to be genetically linked, both being determined by a genetic locus called *hard rind (Hr)* (Piperno et al., 2002). Together with lignification of the rind, the production of phytoliths by the rind provides additional mechanical defense for the fruit.

### REFERENCES

AAZIZ, R., S. DINANT, and B. L. EPEL. 2001. Plasmodesmata and plant cytoskeleton. *Trends Plant Sci.* 6, 326–330.

ANDREEVA, A. V., M. A. KUTUZOV, D. E. EVANS, and C. R. HAWES. 1998. The structure and function of the Golgi apparatus: A hundred years of questions. *J. Exp. Bot.* 49, 1281–1291.

ARNOTT, H. J. 1982. Three systems of biomineralization in plants with comments on the associated organic matrix. In: *Biological Mineralization and Demineralization,* pp. 199–218, G. H. Nancollas, ed. Springer-Verlag, Berlin.

ARNOTT, H. J., and F. C. E. PAUTARD. 1970. Calcification in plants. In: *Biological Calcification: Cellular and Molecular Aspects,* pp.

375–446, H. Schraer, ed. Appleton-Century-Crofts, New York.

ARNOTT, H. J., and M. A. WEBB. 2000. Twinned raphides of calcium oxalate in grape (Vitis): Implications for crystal stability and function. Int. J. Plant Sci. 161, 133–142.

AYRES, M. P., T. P. CLAUSEN, S. F. MACLEAN JR., A. M. REDMAN, and P. B. REICHARDT. 1997. Diversity of structure and antiherbivore activity in condensed tannins. Ecology 78, 1696–1712.

AZIMZADEH, J., J. TRAAS, and M. PASTUGLIA. 2001. Molecular aspects of microtubule dynamics in plants. Curr. Opin. Plant Biol. 4, 513–519.

BALUŠKA, F., S. VITHA, P. W. BARLOW, and D. VOLKMANN. 1997a. Rearrangements of F-actin arrays in growing cells of intact maize root apex tissues: A major developmental switch occurs in the postmitotic transition region. Eur. J. Cell Biol. 72, 113–121.

BALUŠKA, F., D. VOLKMANN, and P. W. BARLOW. 1997b. Nuclear components with microtubular organizing properties in multicellular eukaryotes: Functional and evolutionary considerations. Int. Rev. Cytol. 175, 91–135.

BALUŠKA, F., D. VOLKMANN, and P. W. BARLOW. 1998. Tissue- and development-specific distributions of cytoskeletal elements in growing cells of the maize root apex. Plant Biosyst. 132, 251–265.

BARBEHENN, R. V., and M. M. MARTIN. 1994. Tannin sensitivity in larvae of Malacosoma disstria (Lepidoptera): Roles of the peritrophic envelope and midgut oxidation. J. Chem. Ecol. 20, 1985–2001.

BARLOW, P. W., and F. BALUŠKA. 2000. Cytoskeletal perspectives on root growth and morphogenesis. Annu. Rev. Plant Physiol. Plant Mol. Biol. 51, 289–322.

BASKIN, T. I., and W. Z. CANDE. 1990. The structure and function of the mitotic spindle in flowering plants. Annu. Rev. Plant Physiol. Plant Mol. Biol. 41, 277–315.

BATTEY, N. H., N. C. JAMES, A. J. GREENLAND, and C. BROWNLEE. 1999. Exocytosis and endocytosis. Plant Cell 11, 643–659.

BEDNAREK, S. Y., and N. V. RAIKHEL. 1992. Intracellular trafficking of secretory proteins. Plant Mol. Biol. 20, 133–150.

BOEVINK, P., K. OPARKA, S. SANTA CRUZ, B. MARTIN, A. BETTERIDGE, and C. HAWES. 1998. Stacks on tracks: The plant Golgi apparatus traffics on an actin/ER network. Plant J. 15, 441–447.

BRADBURY, J. H., and R. W. NIXON. 1998. The acidity of raphides from the edible aroids. J. Sci. Food Agric. 76, 608–616.

BROWN, R. C., and B. E. LEMMON. 2001. The cytoskeleton and spatial control of cytokinesis in the plant life cycle. Protoplasma 215, 35–49.

BUSSE, J. S., and R. F. EVERT. 1999. Pattern of differentiation of the first vascular elements in the embryo and seedling of Arabidopsis thaliana. Int. J. Plant Sci. 160, 1–13.

BUSSOTTI, F., E. GRAVANO, P. GROSSONI, and C. TANI. 1998. Occurrence of tannins in leaves of beech trees (Fagus sylvat-

ica) along an ecological gradient, detected by histochemical and ultrastructural analyses. New Phytol. 138, 469–479.

CASSIMERIS, L. U., R. A. WALKER, N. K. PRYER, and E. D. SALMON. 1987. Dynamic instability of microtubules. BioEssays 7, 149–154.

CHEAVIN, W. H. S. 1938. The crystals and cystoliths found in plant cells. Part I. Crystals. The Microscope [Brit. J. Microsc. Photomicrogr.] 2, 155–158.

CHRISPEELS, M. J. 1991. Sorting of proteins in the secretory system. Annu. Rev. Plant Physiol. Plant Mol. Biol. 42, 21–53.

CHRISPEELS, M. J., and E. H. HERMAN. 2000. Endoplasmic reticulum-derived compartments function in storage and as mediators of vacuolar remodeling via a new type of organelle, precursor protease vesicles. Plant Physiol. 123, 1227–1234.

COLLINGS, D. A., T. ASADA, N. S. ALLEN, and H. SHIBAOKA. 1998. Plasma membrane-associated actin in Bright Yellow 2 tobacco cells. Plant Physiol. 118, 917–928.

CUMMINS, I., M. J. HILLS, J. H. E. ROSS, D. H. HOBBS, M. D. WATSON, and D. J. MURPHY. 1993. Differential, temporal and spatial expression of genes involved in storage oil and oleosin accumulation in developing rapeseed embryos: Implications for the role of oleosins and the mechanisms of oil-body formation. Plant Mol. Biol. 23, 1015–1027.

DERKSEN, J., F. H. A. WILMS, and E. S. PIERSON. 1990. The plant cytoskeleton: Its significance in plant development. Acta Bot. Neerl. 39, 1–18.

DING, B., M.-O. KWON, and L. WARNBERG. 1996. Evidence that actin filaments are involved in controlling the permeability of plasmodesmata in tobacco mesophyll. Plant J. 10, 157–164.

DRIOUICH, A., and L. A. STAEHELIN. 1997. The plant Golgi apparatus: Structural organization and functional properties. In: The Golgi Apparatus, pp. 275–301, E. G. Berger and J. Roth, eds. Birkhäuser Verlag, Basel.

DRIOUICH, A., L. FAYE, and L. A. STAEHELIN. 1993. The plant Golgi apparatus: A factory for complex polysaccharides and glycoproteins. Trends Biochem. Sci. 18, 210–214.

DRØBAK, B. K., V. E. FRANKLIN-TONG, and C. J. STAIGER. 2004. The role of the actin cytoskeleton in plant cell signalling. New Phytol. 163, 13–30.

DUPREE, P., and D. J. SHERRIER. 1998. The plant Golgi apparatus. Biochim. Biophys. Acta (Mol. Cell Res.) 1404, 259–270.

ENGELBERTH, J., G. WANNER, B. GROTH, and E. W. WEILER. 1995. Functional anatomy of the mechanoreceptor cells in tendrils of Bryonia dioica Jacq. Planta 196, 539–550.

EPSTEIN, E. 1999. Silicon. Annu. Rev. Plant Physiol. Plant Mol. Biol. 50, 641–664.

ESAU, K. 1977. Anatomy of Seed Plants, 2nd ed. Wiley, New York.

ESCHRICH, W. 1954. Ein Beitrag zur Kenntnis der Kallose. Planta 44, 532–542.

EVERT, R. F., J. D. DAVIS, C. M. TUCKER, and F. J. ALFIERI. 1970. On the occurrence of nuclei in mature sieve elements. Planta 95, 281–296.

FERRI, S. 1974. Morphological and structural investigations on *Smilax aspera* leaf and storage starches. *J. Ultrastruct. Res.* 47, 420–432.

FINLEY, D. S. 1999. Patterns of calcium oxalate crystals in young tropical leaves: A possible role as an anti-herbivory defense. *Rev. Biol. Trop.* 47, 27–31.

FISHER, D. G., and R. F. EVERT. 1982. Studies on the leaf of *Amaranthus retroflexus* (Amaranthaceae): Chloroplast polymorphism. *Bot. Gaz.* 143, 146–155.

FLANDERS, D. J., D. J. RAWLINS, P. J. SHAW, and C. W. LLOYD. 1990. Re-establishment of the interphase microtubule array in vacuolated plant cells, studied by confocal microscopy and 3-D imaging. *Development* 110, 897–903.

FRANCESCHI, V. R. 1989. Calcium oxalate formation is a rapid and reversible process in *Lemna minor* L. *Protoplasma* 148, 130–137.

FRANCESCHI, V. R., and H. T. HORNER JR. 1980. Calcium oxalate crystals in plants. *Bot. Rev.* 46, 361–427.

FREY-WYSSLING, A. 1980. Why starch as our main food supply? *Ber. Dtsch. Bot. Ges.* 93, 281–287.

GARCIA-HERDUGO, G., J. A. GONZÁLEZ-REYES, F. GRACIA-NAVARRO, and P. NAVAS. 1988. Growth kinetics of the Golgi apparatus during the cell cycle in onion root meristems. *Planta* 175, 305–312.

GERRATH, J. M., L. COVINGTON, J. DOUBT, and D. W. LARSON. 2002. Occurrence of phi thickenings is correlated with gymnosperm systematics. *Can. J. Bot.* 80, 852–860.

GIETL, C., and M. SCHMID. 2001. Ricinosomes: An organelle for developmentally regulated programmed cell death in senescing plant tissues. *Naturwissenschaften* 88, 49–58.

GODDARD, R. H., S. M. WICK, C. D. SILFLOW, and D. P. SNUSTAD. 1994. Microtubule components of the plant cell cytoskeleton. *Plant Physiol.* 104, 1–6.

GONZALEZ, A. M. 1996. Nectarios extraflorales en *Turnera*, series Canaligerae y Leiocarpae. *Bonplandia* 9, 129–143.

HASLAM, E. 1981. Vegetable tannins. In: *The Biochemistry of Plants*, vol. 7, *Secondary Plant Products*, pp. 527–556, E. E. Conn, ed. Academic Press, New York.

HEPLER, P. K., and B. E. S. GUNNING. 1998. Confocal fluorescence microscopy of plant cells. *Protoplasma* 201, 121–157.

HEPLER, P. K., and R. O. WAYNE. 1985. Calcium and plant development. *Annu. Rev. Plant Physiol.* 36, 397–439.

HEPLER, P. K., B. A. PALEVITZ, S. A. LANCELLE, M. M. McCAULEY, and I. LICHTSCHEIDL. 1990. Cortical endoplasmic reticulum in plants. *J. Cell Sci.* 96, 355–373.

HERMAN, E. M., and B. A. LARKINS. 1999. Protein storage bodies and vacuoles. *Plant Cell* 11, 601–614.

HERMAN, E. M., B. BAUMGARTNER, and M. J. CHRISPEELS. 1981. Uptake and apparent digestion of cytoplasmic organelles by protein bodies (protein storage vacuoles) in mung bean *(Vigna radiata)* cotyledons. *Eur. J. Cell Biol.* 24, 226–235.

HILLMER, S., A. MOVAFEGHI, D. G. ROBINSON, and G. HINZ. 2001. Vacuolar storage proteins are sorted in the cis-cisternae of the pea cotyledon Golgi apparatus. *J. Cell Biol.* 152, 41–50.

HOLTZMAN, E. 1992. Intracellular targeting and sorting. *BioScience* 42, 608–620.

HORNER, H. T., and B. L. WAGNER. 1995. Calcium oxalate formation in higher plants. In: *Calcium Oxalate in Biological Systems*, pp. 53–72, S. R. Khan, ed. CRC Press, Boca Raton, FL.

HUANG, A. H. C. 1996. Oleosins and oil bodies in seeds and other organs. *Plant Physiol.* 110, 1055–1061.

HUSH, J. M., P. WADSWORTH, D. A. CALLAHAM, and P. K. HEPLER. 1994. Quantification of microtubule dynamics in living plant cells using fluorescence redistribution after photobleaching. *J. Cell Sci.* 107, 775–784.

ILARSLAN, H., R. G. PALMER, and H. T. HORNER. 2001. Calcium oxalate crystals in developing seeds of soybean. *Ann. Bot.* 88, 243–257.

JENG, R. L., and M. D. WELCH. 2001. Cytoskeleton: Actin and endocytosis—no longer the weakest link. *Curr. Biol.* 11, R691–R694.

KAUFMAN, P. B., P. DAYANANDAN, C. I. FRANKLIN, and Y. TAKEOKA. 1985. Structure and function of silica bodies in the epidermal system of grass shoots. *Ann. Bot.* 55, 487–507.

KAUSCH, A. P., and H. T. HORNER. 1983. The development of mucilaginous raphide crystal idioblasts in young leaves of *Typha angustifolia* L. (Typhaceae). *Am. J. Bot.* 70, 691–705.

KAUSCH, A. P., and H. T. HORNER. 1984. Differentiation of raphide crystal idioblasts in isolated root cultures of *Yucca torreyi* (Agavaceae). *Can. J. Bot.* 62, 1474–1484.

KNEBEL, W., H. QUADER, and E. SCHNEPF. 1990. Mobile and immobile endoplasmic reticulum in onion bulb epidermis cells: Short-term and long-term observations with a confocal laser scanning microscope. *Eur. J. Cell Biol.* 52, 328–340.

KOST, B., and N.-H. CHUA. 2002. The plant cytoskeleton: Vacuoles and cell walls make the difference. *Cell* 108, 9–12.

KOST, B., J. MATHUR, and N.-H. CHUA. 1999. Cytoskeleton in plant development. *Curr. Opin. Plant Biol.* 2, 462–470.

KOST, B., P. SPIELHOFER, J. MATHUR, C.-H. DONG, and N.-H. CHUA. 2000. Non-invasive F-actin visualization in living plant cells using a GFP-mouse talin fusion protein. In: *Actin: A Dynamic Framework for Multiple Plant Cell Functions*, pp. 637–659, C. J. Staiger, F. Baluška, D. Volkmann, and P. W. Barlow, eds. Kluwer, Dordrecht.

KOSTMAN, T. A., and V. R. FRANCESCHI. 2000. Cell and calcium oxalate crystal growth is coordinated to achieve high-capacity calcium regulation in plants. *Protoplasma* 214, 166–179.

KOSTMAN, T. A., V. R. FRANCESCHI, and P. A. NAKATA. 2003. Endoplasmic reticulum sub-compartments are involved in calcium sequestration within raphide crystal idioblasts of *Pistia stratiotes*. *Plant Sci.* 165, 205–212.

KROPF, D. L., S. R. BISGROVE, and W. E. HABLE. 1998. Cytoskeletal control of polar growth in plant cells. *Curr. Opin. Cell Biol.* 10, 117–122.

KUMAGAI, F., and S. HASEZAWA 2001. Dynamic organization of microtubules and microfilaments during cell cycle progression in higher plant cells. *Plant Biol.* 3, 4–16.

KUO-HUANG, L.-L. 1990. Calcium oxalate crystals in the leaves of *Nelumbo nucifera* and *Nymphaea tetragona*. *Taiwania* 35, 178–190.

KUO-HUANG, L.-L. 1992. Ultrastructural study on the development of crystal-forming sclereids of *Nymphaea tetragona*. *Taiwania* 37, 104–114.

KUO-HUANG, L.-L, and S.-J. CHEN. 1999. Subcellular localization of calcium in the crystal-forming sclereids of *Nymphaea tetragona* Georgi. *Taiwania* 44, 520–528.

KÜSTER, E. 1956. *Die Pflanzenzelle*, 3rd ed. Gustav Fischer Verlag, Jena.

LEVANONY, H., R. RUBIN, Y. ALTSCHULER, and G. GALILI. 1992. Evidence for a novel route of wheat storage proteins to vacuoles. *J. Cell Biol.* 119, 1117–1128.

LEWIN, J., and B. E. F. REIMANN. 1969. Silicon and plant growth. *Annu. Rev. Plant Physiol.* 20, 289–304.

LICHTSCHEIDL, I. K., and P. K. HEPLER. 1996. Endoplasmic reticulum in the cortex of plant cells. In: *Membranes: Specialized Functions in Plants*, pp. 383–402, M. Smallwood, J. P. Knox, and D. J. Bowles, eds. BIOS Scientific, Oxford.

LICHTSCHEIDL, I. K., S. A. LANCELLE, and P. K. HEPLER. 1990. Actin-endoplasmic reticulum complexes in *Drosera*: Their structural relationship with the plasmalemma, nucleus, and organelles in cells prepared by high pressure freezing. *Protoplasma* 155, 116–126.

LLOYD, C. W. 1987. The plant cytoskeleton: The impact of fluorescence microscopy. *Annu. Rev. Plant Physiol.* 38, 119–139.

LLOYD, C. 1988. Actin in plants. *J. Cell Sci.* 90, 185–188.

LOER, D. S., and E. M. HERMAN. 1993. Cotranslational integration of soybean (*Glycine max*) oil body membrane protein oleosin into microsomal membranes. *Plant Physiol.* 101, 993–998.

LYSHEDE, O. B. 1980. Notes on the ultrastructure of cubical protein crystals in potato tuber cells. *Bot. Tidsskr.* 74, 237–239.

MARC, J. 1997. Microtubule-organizing centres in plants. *Trends Plant Sci.* 2, 223–230.

MARINOS, N. G. 1965. Comments on the nature of a crystal-containing body in plant cells. *Protoplasma* 60, 31–33.

MARTIN, C., and A. M. SMITH. 1995. Starch biosynthesis. *Plant Cell* 7, 971–985.

MARTY, F. 1999. Plant vacuoles. *Plant Cell* 11, 587–599.

MATHUR, J., and M. HÜLSKAMP. 2002. Microtubules and microfilaments in cell morphogenesis in higher plants. *Curr. Biol.* 12, R669–R676.

MATSUOKA, K., and S. Y. BEDNAREK. 1998. Protein transport within the plant cell endomembrane system: An update. *Curr. Opin. Plant Biol.* 1, 463–469.

MAZEN, A. M. A., D. ZHANG, and V. R. FRANCESCHI. 2003. Calcium oxalate formation in *Lemna minor*: Physiological and ultrastructural aspects of high capacity calcium sequestration. *New Phytol.* 161, 435–448.

MCKENZIE, B. E., and C. A. PETERSON. 1995a. Root browning in *Pinus banksiana* Lamb. and *Eucalyptus pilularis* Sm. 1. Anatomy and permeability of the white and tannin zones. *Bot. Acta* 108, 127–137.

MCKENZIE, B. E., and C. A. PETERSON. 1995b. Root browning in *Pinus banksiana* Lamb. and *Eucalyptus pilularis* Sm. 2. Anatomy and permeability of the cork zone. *Bot. Acta* 108, 138–143.

MCNAUGHTON, S. J., and J. L. TARRANTS. 1983. Grass leaf silicification: Natural selection for an inducible defense against herbivores. *Proc. Natl. Acad. Sci. USA* 80, 790–791.

MEAGHER, R. B., E. C. MCKINNEY, and M. K. KANDASAMY. 1999. Isovariant dynamics expand and buffer the responses of complex systems: The diverse plant actin gene family. *Plant Cell* 11, 995–1006.

METCALFE, C. R. 1983. Calcareous deposits, calcified cell walls, cystoliths, and similar structures. In: *Anatomy of the Dicotyledons*, 2nd ed., vol. 2, *Wood Structure and Conclusion of the General Introduction*, pp. 94–97, C. R. Metcalfe and L. Chalk. Clarendon Press, Oxford.

METCALFE, C. R., and L. CHALK. 1983. *Anatomy of the Dicotyledons*, 2nd ed., vol. 2, *Wood Structure and Conclusion of the General Introduction*. Clarendon Press, Oxford.

MILLER, E. A., and M. A. ANDERSON. 1999. Uncoating the mechanisms of vacuolar protein transport. *Trends Plant Sci.* 4, 46–48.

MOLANO-FLORES, B. 2001. Herbivory and calcium concentrations affect calcium oxalate crystal formation in leaves of *Sida* (Malvaceae). *Ann. Bot.* 88, 387–391.

MOLLENHAUER, H. H., and D. J. MORRÉ. 1980. The Golgi apparatus. In: *The Biochemistry of Plants*, vol. 1, *The Plant Cell*, pp. 437–488, N. E. Tolbert, ed. Academic Press, New York.

MOORE, P. J., K. M. SWORDS, M. A. LYNCH, and L. A. STAEHELIN. 1991. Spatial organization of the assembly pathways of glycoproteins and complex polysaccharides in the Golgi apparatus of plants. *J. Cell Biol.* 112, 589–602.

MORRÉ, D. J., and H. H. MOLLENHAUER. 1974. The endomembrane concept: A functional integration of endoplasmic reticulum and Golgi apparatus. In: *Dynamic Aspects of Plant Ultrastructure*, pp. 84–137, A. W. Robards, ed. McGraw-Hill, (UK) Limited, London.

MOTOMURA, H., N. MITA, and M. SUZUKI. 2002. Silica accumulation in long-lived leaves of *Sasa veitchii* (Carrière) Rehder (Poaceae-Bambusoideae). *Ann. Bot.* 90, 149–152.

NAKATA, P. A. 2003. Advances in our understanding of calcium oxalate crystal formation and function in plants. *Plant Sci.* 164, 901–909.

NAKATA, P. A., T. A. KOSTMAN, and V. R. FRANCESCHI. 2003. Calreticulin is enriched in the crystal idioblasts of *Pistia stratiotes*. *Plant Physiol. Biochem.* 41, 425–430.

NEBENFÜHR, A., L. A. GALLAGHER, T. G. DUNAHAY, J. A. FROHLICK, A. M. MAZURKIEWICZ, J. B. MEEHL, and L. A. STAEHELIN. 1999.

Stop-and-go movements of plant Golgi stacks are mediated by the acto-myosin system. *Plant Physiol.* 121, 1127–1141.

NEBENFÜHR, A., J. A. FROHLICK, and L. A. STAEHELIN. 2000. Redistribution of Golgi stacks and other organelles during mitosis and cytokinesis in plant cells. *Plant Physiol.* 124, 135–151.

OHLROGGE, J., and J. BROWSE. 1995. Lipid biosynthesis. *Plant Cell* 7, 957–970.

OLADELE, F. A. 1982. Development of the crystalliferous cuticle of *Chamaecyparis lawsoniana* (A. Murr.) Parl. (Cupressaceae). *Bot. J. Linn. Soc.* 84, 273–288.

OTEGUI, M. S., D. N. MASTRONARDE, B.-H. KANG, S. Y. BEDNAREK, and L. A. STAEHELIN. 2001. Three-dimensional analysis of syncytial-type cell plates during endosperm cellularization visualized by high resolution electron tomography. *Plant Cell* 13, 2033–2051.

ÖZTIG, Ö. F. 1940. Beiträge zur Kenntnis des Baues der Blattepidermis bei den Mesembrianthemen, im besonderen den extrem xeromorphen Arten. *Flora* n.s. 34, 105–144.

PANT, D. D., D. D. NAUTIYAL, and S. SINGH. 1975. The cuticle, epidermis and stomatal ontogeny in *Casuarina equisetifolia* Forst. *Ann. Bot.* 39, 1117–1123.

PARTHASARATHY, M. V., T. D. PERDUE, A. WITZTUM, and J. ALVERNAZ. 1985. Actin network as a normal component of the cytoskeleton in many vascular plant cells. *Am. J. Bot.* 72, 1318–1323.

PAULL, R. E., C.-S. TANG, K. GROSS, and G. URUU. 1999. The nature of the taro acridity factor. *Postharvest Biol. Tech.* 16, 71–78.

PENNISI, S. V., and D. B. McCONNELL. 2001. Taxonomic relevance of calcium oxalate cuticular deposits in *Dracaena* Vand. ex L. *HortScience* 36, 1033–1036.

PENNISI, S. V., D. B. McCONNELL, L. B. GOWER, M. E. KANE, and T. LUCANSKY. 2001a. Periplasmic cuticular calcium oxalate crystal deposition in *Dracaena sanderiana*. *New Phytol.* 149, 209–218.

PENNISI, S. V., D. B. McCONNELL, L. B. GOWER, M. E. KANE, and T. LUCANSKY. 2001b. Intracellular calcium oxalate crystal structure in *Dracaena sanderiana*. *New Phytol.* 150, 111–120.

PETERSON, C. A., M. E. EMANUEL, and C. A. WEERDENBERG. 1981. The permeability of phi thickenings in apple (*Pyrus malus*) and geranium (*Pelargonium hortorum*) roots to an apoplastic fluorescent dye tracer. *Can. J. Bot.* 59, 1107–1110.

PIPERNO, D. R., I. HOLST, L. WESSEL-BEAVER, and T. C. ANDRES. 2002. Evidence for the control of phytolith formation in *Cucurbita* fruits by the hard rind (*Hr*) genetic locus: Archaeological and ecological implications. *Proc. Natl. Acad. Sci. USA* 99, 10923–10928.

PRATIKAKIS, E., S. RHIZOPOULOU, and G. K. PSARAS. 1998. A phi layer in roots of *Ceratonia siliqua* L. *Bot. Acta* 111, 93–98.

PRYCHID, C. J., and P. J. RUDALL. 1999. Calcium oxalate crystals in monocotyledons: A review of their structure and systematics. *Ann. Bot.* 84, 725–739.

QUADER, H., and E. SCHNEPF. 1986. Endoplasmic reticulum and cytoplasmic streaming: Fluorescence microscopical observations in adaxial epidermis cells of onion bulb scales. *Protoplasma* 131, 250–252.

QUADER, H., A. HOFMANN, and E. SCHNEPF. 1989. Reorganization of the endoplasmic reticulum in epidermal cells of onion bulb scales after cold stress: Involvement of cytoskeletal elements. *Planta* 177, 273–280.

QUITADAMO, I. J., T. A. KOSTMAN, M. E. SCHELLING, and V. R. FRANCESCHI. 2000. Magnetic bead purification as a rapid and efficient method for enhanced antibody specificity for plant sample immunoblotting and immunolocalization. *Plant Sci.* 153, 7–14.

RAO, K. S. 1988. Fine structural details of tannin accumulations in non-dividing cambial cells. *Ann. Bot.* 62, 575–581.

REY, D., J.-P. DAVID, D. MARTINS, M.-P. PAUTOU, A. LONG, G. MARIGO, and J.-C. MEYRAN. 2000. Role of vegetable tannins in habitat selection among mosquito communities from the Alpine hydrosystems. *C. R. Acad. Sci., Paris, Sci. de la Vie* 323, 391–398.

RIDGE, R. W., Y. UOZUMI, J. PLAZINSKI, U. A. HURLEY, and R. E. WILLIAMSON. 1999. Developmental transitions and dynamics of the cortical ER of *Arabidopsis* cells seen with green fluorescent protein. *Plant Cell Physiol.* 40, 1253–1261.

ROST, T. L., and K. E. PATERSON. 1978. Structural and histochemical characterization of the cotyledon storage organelles of jojoba (*Simmondsia chinensis*). *Protoplasma* 95, 1–10.

ROST, T. L., A. D. SIMPER, P. SCHELL, and S. ALLEN. 1977. Anatomy of jojoba (*Simmondsia chinensis*) seed and the utilization of liquid wax during germination. *Econ. Bot.* 31, 140–147.

SAKAI, W. S., and M. A. NAGAO. 1980. Raphide structure in *Dieffenbachia maculata*. *J. Am. Soc. Hortic. Sci.* 105, 124–126.

SALTZ, D., and D. WARD. 2000. Responding to a three-pronged attack: Desert lilies subject to herbivory by dorcas gazelles. *Plant Ecol.* 148, 127–138.

SANDERFOOT, A. A., and N. V. RAIKHEL. 1999. The specificity of vesicle trafficking: Coat proteins and SNAREs. *Plant Cell* 11, 629–641.

SANGSTER, A. G. 1978. Silicon in the roots of higher plants. *Am. J. Bot.* 65, 929–935.

SCHNEPF, E. 1993. Golgi apparatus and slime secretion in plants: The early implications and recent models of membrane traffic. *Protoplasma* 172, 3–11.

SEAGULL, R. W. 1989. The plant cytoskeleton. *Crit. Rev. Plant Sci.* 8, 131–167.

SHEAHAN, M. B., R. J. ROSE, and D. W. McCURDY. 2004. Organelle inheritance in plant cell division: The actin cytoskeleton is required for unbiased inheritance of chloroplasts, mitochondria and endoplasmic reticulum in dividing protoplasts. *Plant J.* 37, 379–390.

SHIMMEN, T., R. W. RIDGE, I. LAMBIRIS, J. PLAZINSKI, E. YOKOTA, and R. E. WILLIAMSON. 2000. Plant myosins. *Protoplasma* 214, 1–10.

SOMERVILLE, C., and J. BROWSE. 1991. Plant lipids: Metabolism, mutants, and membranes. *Science* 252, 80–87.

STAEHELIN, L. A. 1997. The plant ER: A dynamic organelle composed of a large number of discrete functional domains. *Plant J.* 11, 1151–1165.

STAIGER, C. J. 2000. Signaling to the actin cytoskeleton in plants. *Annu. Rev. Plant Physiol. Plant Mol. Biol.* 51, 257–288.

STAIGER, C. J., and M. SCHLIWA. 1987. Actin localization and function in higher plants. *Protoplasma* 141, 1–12.

SUNELL, L. A., and P. L. HEALEY. 1985. Distribution of calcium oxalate crystal idioblasts in leaves of taro *(Colocasia esculenta)*. *Am. J. Bot.* 72, 1854–1860.

TZEN, J. T., and A. H. HUANG. 1992. Surface structure and properties of plant seed oil bodies. *J. Cell Biol.* 117, 327–335.

VANTARD, M., R. COWLING, and C. DELICHÈRE. 2000. Cell cycle regulation of the microtubular cytoskeleton. *Plant Mol. Biol.* 43, 691–703.

VARRIANO-MARSTON, E. 1983. Polarization microscopy: Applications in cereal science. In: *New Frontiers in Food Microstructure*, pp. 71–108, D. B. Bechtel, ed. American Association of Cereal Chemists, St. Paul, MN.

VICKERY, M. L., and B. VICKERY. 1981. *Secondary Plant Metabolism.* University Park Press, Baltimore.

VITALE, A., and J. DENECKE. 1999. The endoplasmic reticulum— Gateway of the secretory pathway. *Plant Cell* 11, 615–628.

VITALE, A., A. CERIOTTI, and J. DENECKE. 1993. The role of the endoplasmic reticulum in protein synthesis, modification and intracellular transport. *J. Exp. Bot.* 44, 1417–1444.

VOLK, G. M., V. J. LYNCH-HOLM, T. A. KOSTMAN, L. J. GOSS, and V. R. FRANCESCHI. 2002. The role of druse and raphide calcium oxalate crystals in tissue calcium regulation in *Pistia stratiotes* leaves. *Plant Biol.* 4, 34–45.

WANG, Z.-Y., K. S. GOULD, and K. J. PATTERSON. 1994. Structure and development of mucilage-crystal idioblasts in the roots of five *Actinidia* species. *Int. J. Plant Sci.* 155, 342–349.

WANNER, G., and R. R. THELMER. 1978. Membranous appendices of spherosomes (oleosomes). Possible role in fat utilization in germinating oil seeds. *Planta* 140, 163–169.

WASTENEYS, G. O. 2004. Progress in understanding the role of microtubules in plant cells. *Curr. Opin. Plant Biol.* 7, 651–660.

WEBB, M. A. 1999. Cell-mediated crystallization of calcium oxalate in plants. *Plant Cell* 11, 751–761.

WEBB, M. A., J. M. CAVALETTO, N. C. CARPITA, L. E. LOPEZ, and H. J. ARNOTT. 1995. The intravacuolar organic matrix associated with calcium oxalate crystals in leaves of *Vitis. Plant J.* 7, 633–648.

WERGIN, W. P., P. J. GRUBER, and E. H. NEWCOMB. 1970. Fine structural investigation of nuclear inclusions in plants. *J. Ultrastruct. Res.* 30, 533–557.

WHITE, R. G., K. BADELT, R. L. OVERALL, and M. VESK. 1994. Actin associated with plasmodesmata. *Protoplasma* 180, 169–184.

WYMER, C., and C. LLOYD. 1996. Dynamic microtubules: Implications for cell wall patterns. *Trends Plant Sci.* 1, 222–228.

WYMER, C. L., S. A. WYMER, D. J. COSGROVE, and R. J. CYR. 1996. Plant cell growth responds to external forces and the response requires intact microtubules. *Plant Physiol.* 110, 425–430.

XU, W., P. CAMPBELL, A. K. VARGHEESE, and J. BRAAM. 1996. The *Arabidopsis XET*-related gene family: Environmental and hormonal regulation of expression. *Plant J.* 9, 879–889.

YONEMORI, K., M. OSHIDA, and A. SUGIURA. 1997. Fine structure of tannin cells in fruit and callus tissues of persimmon. *Acta Hortic.* 436, 403–413.

ZHANG, G. F., and L. A. STAEHELIN. 1992. Functional compartmentation of the Golgi apparatus of plant cells. *Plant Physiol.* 99, 1070–1083.

ZHENG, H. Q., and L. A. STAEHELIN. 2001. Nodal endoplasmic reticulum, a specialized form of endoplasmic reticulum found in gravity-sensing root tip columella cells. *Plant Physiol.* 125, 252–265.

ZOBEL, A. M. 1985. Ontogenesis of tannin coenocytes in *Sambucus racemosa* L. I. Development of the coenocytes from mononucleate tannin cells. *Ann. Bot.* 55, 765–773.

# CHAPTER FOUR

# Cell Wall

The presence of a cell wall, above all other characteristics, distinguishes plant cells from animal cells. Its presence is the basis of many of the characteristics of plants as organisms. The cell wall is rigid and therefore limits the size of the protoplast, preventing rupture of the plasma membrane when the protoplast enlarges following the uptake of water. The cell wall largely determines the size and shape of the cell, the texture of the tissue, and the final form of the plant organ. Cell types are often identified by the structure of their walls, reflecting the close relationship between cell wall structure and cell function.

Once regarded as merely an outer, inactive product of the protoplast, the cell wall is now recognized as a metabolically dynamic compartment having specific and essential functions (Bolwell, 1993; Fry, 1995; Carpita and McCann, 2000). Accordingly the primary cell wall—the wall layers formed chiefly while the cell is increasing in size—has been variously characterized as a "vital" or "indispensable organelle" (Fry, 1988; Hoson, 1991; McCann et al., 1990), a "special subcellular compartment outside the plasma membrane" (Satiat-Jeunemaitre, 1992), and a "vital extension of the cyto-

plasm" (Carpita and Gibeaut, 1993). Cell walls contain a variety of enzymes and play important roles in absorption, transport, and secretion of substances in plants. Experimental evidence indicates that molecules released from cell walls are involved in cell-to-cell signaling, influencing cellular differentiation (Fry et al., 1993; Mohnen and Hahn, 1993; Pennell, 1998; Braam, 1999; Lišková et al., 1999).

In addition the cell wall may play a role in defense against bacterial and fungal pathogens by receiving and processing information from the surface of the pathogen and transmitting this information to the plasma membrane of the host cell. Through gene-activated processes, the host cell may become resistant to attack through the production of *phytoalexins*—antibiotics that are toxic to the pathogens (Darvill and Albersheim, 1984; Bradley et al., 1992; Hammerschmidt, 1999)—or through the deposition of substances such as lignins, suberin, or callose, which may act as passive barriers to invasion (Vance et al., 1980; Perry and Evert, 1983; Pearce, 1989; Thomson et al., 1995).

Conceptually botanists long have considered the cell wall to be an integral part of the plant cell. However,

*Esau's Plant Anatomy, Third Edition,* By Ray F. Evert.
Copyright © 2006 John Wiley & Sons, Inc.

many plant cell biologists have adopted the terminology of animal cell biologists and refer to the cell wall as an "extracellular matrix," indicating that the cell wall lies outside the plant cell (Staehelin, 1991; Roberts, 1994). There are many compelling reasons not to adopt the term extracellular matrix for the plant cell wall (Robinson, 1991; Reuzeau and Pont-Lezica, 1995; Connolly and Berlyn, 1996). For example, the extracellular matrix of animal cells is completely different from the plant cell wall, the former relying on proteins for their matrix and the latter largely on polysaccharides; an animal cell is not defined by the extracellular matrix it shares with adjacent cells in the tissue, whereas the plant cell is defined by the wall made by its protoplast; animal cells are not fixed spatially but can move into a preexisting extracytoplasmic milieu, whereas plant cells cannot change their position within a common "extracellular matrix"; the presence of a cell wall is prerequisite for plant cell division; for a plant cell to grow and divide, the wall must also grow and divide (Suzuki et al., 1998). Moreover, as noted by Connolly and Berlyn (1996), the term extracellular matrix leads to confusion that is avoided by the use of well-established cell wall terms (discussed in this chapter). The term cell wall will continue to be used in this book to refer to this distinctive cellulosic component of the plant cell.

# MACROMOLECULAR COMPONENTS OF THE CELL WALL

## Cellulose Is the Principal Component of Plant Cell Walls

The principal component of plant cell walls is **cellulose**, a polysaccharide with the empirical formula $(C_6H_{10}O_5)_n$. Its molecules are linear chains of $(1{\rightarrow}4)\beta$-linked-D-glucan (repeating monomers of glucose attached end to end) (Fig. 4.1). These long, thin cellulose molecules tend to hydrogen bond together to form **microfibrils**. Considerable variation exists in the literature with respect to the diameter of microfibrils. Most values are in the range of 4 to 10 nanometers, although values as small as 1 and 2 nanometers (Preston, 1974; Ha et al., 1998; Thimm et al., 2002) and as large as 25 nanometers have been recorded (Thimm et al., 2000). The diameter of cellulose microfibrils apparently is highly dependent on the water content of the portion of wall being examined. Microfibrils of hydrated walls appear smaller than those of dehydrated walls (Thimm et al., 2000). Water, located mainly in the matrix (see below), makes up about two-thirds of the wall mass in growing tissues.

The cellulose microfibrils wind together to form fine threads that coil around one another like strands in a cable. Each "cable," or **macrofibril**, which is visible with a light microscope, measures about 0.5 micrometer

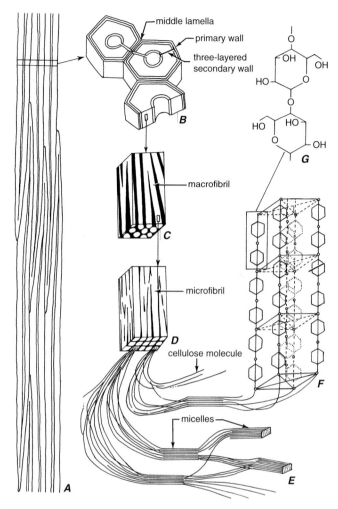

**FIGURE 4.1**

Detailed structure of cell walls. **A,** strand of fiber cells. **B,** transverse section of fiber cells showing gross layering: a layer of primary wall and three layers of secondary wall. **C,** fragment from the middle layer of the secondary wall showing microfibrils (white) of cellulose and interfibrillar spaces (black), which are filled with noncellulosic materials. **D,** fragment of a macrofibril showing microfibrils (white), which may be seen in electron micrographs. The spaces among microfibrils (black) are filled with noncellulosic materials. **E,** structure of microfibrils: chain-like molecules of cellulose, which in some parts of microfibrils are orderly arranged. These parts are the micelles. **F,** fragment of a micelle showing parts of chain-like cellulose molecules arranged in a space lattice. **G,** two glucose residues connected by an oxygen atom—a fragment of a cellulose molecule. (From Esau, 1977.)

wide and can reach four to seven micrometers in length (Fig. 4.1). Cellulose molecules wound in this fashion have a tensile strength (breaking strength) approaching that of steel (50–160 kg/mm²) (Frey-Wyssling, 1976). Cellulose microfibrils constitute 20% to 30% of the dry weight of a typical primary wall and 40% to 60% of that of the secondary wall of wood cells.

Cellulose has crystalline properties resulting from the orderly arrangement of cellulose molecules in the microfibrils (Smith, B. G., et al., 1998). Such arrangement is restricted to parts of the microfibrils that are referred to as **micelles**. Less regularly arranged glucose chains occur between and around the micelles and constitute the paracrystalline regions of the microfibril. The crystalline structure of cellulose makes the cell wall anisotropic and, consequently, doubly refractive (birefringent) when viewed with polarized light (Fig. 4.2).

### The Cellulose Microfibrils Are Embedded in a Matrix of Noncellulosic Molecules

The cellulose microfibrils of the wall are embedded in a cross-linked **matrix** of noncellulosic molecules. These molecules are the polysaccharides known as hemicelluloses and pectins, as well as the structural proteins called glycoproteins.

Principal Hemicelluoses **Hemicellulose** is a general term for a heterogeneous group of noncrystalline glycans that are tightly bound in the cell wall. The hemicelluloses vary greatly in different cell types and among taxa. Generally, one hemicellulose dominates in most cell types, with others present in smaller amounts.

**Xyloglucans** are the principal hemicelluloses of the primary cell walls of eudicots and about one-half of the monocots (Carpita and McCann, 2000), comprising about 20% to 25% of their dry weight (Kato and Matsuda, 1985). The xyloglucans consist of linear chains of (1→ 4)β-D-glucan as in cellulose, with short side chains containing xylose, galactose, and often a terminal fucose (McNeil et al., 1984; Fry, 1989; Carpita and McCann, 2000). Most of the xyloglucans apparently are tightly hydrogen-bonded with cellulose microfibrils (Moore and Staehelin, 1988; Hoson, 1991). Being tightly bound to the cellulose microfibrils, xyloglucan potentially limits the extensibility of the cell wall by tethering adjacent microfibrils and, hence, may play a significant role in regulating cell enlargement (Levy and Staehelin, 1992; Cosgrove, 1997, 1999).

Degradative by-products of xyloglucans (xyloglucan-derived oligosaccharides) exert a hormone-like anti-auxin effect on cell growth (Fry, 1989; McDougall and Fry, 1990). In addition in the seeds of some eudicots, for example, nasturtium (*Tropaeolum*), *Impatiens*, and *Annona*, xyloglucans located in thick cell walls constitute the principal storage carbohydrate (Reid, 1985). Xyloglucans apparently are lacking in the secondary cell walls (wall layers deposited inside the primary wall) of xylem elements (Fry, 1989).

Primary cell walls in which the principal hemicelluloses are xyloglucans have been designated **Type I walls** (Carpita and Gibeaut, 1993; Darley et al., 2001).

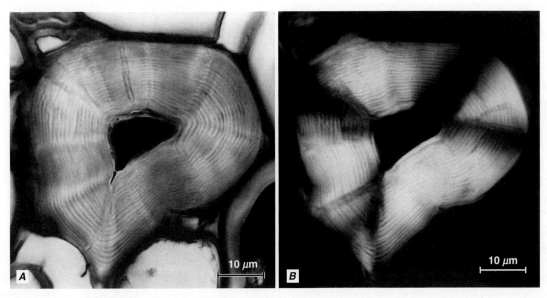

**FIGURE 4.2**

Sclereid from the root cortex of fir (*Abies*), as seen with nonpolarized (**A**) and polarized (**B**) light. Because of the crystalline nature of cellulose, the cell wall shows double refraction and appears bright in polarized light (**B**). The wall has concentric lamellation. (From Esau, 1977.)

The principal hemicelluloses in the primary cell walls of the commelinoid line of monocots (including the Poales, Zingiberales, Commelinales, and Arecales) are **glucuronoarabinoxylans**, which are characterized by a (1→4)β-D-xylose backbone. The glucuronoarabinoxylans, like the xyloglucans, can hydrogen bond to cellulose and to each other. The primary walls of the Poales are distinguished from those of other commelinoid monocots by the presence of **mixed-linked** (1→3), (1→4)β-D-glucans (Carpita, 1996; Buckeridge et al., 1999; Smith, B. G., and Harris, 1999; Trethewey and Harris, 2002). The commelinoid cell walls have been designated **Type II walls** (Carpita and Gibeaut, 1993; Darley et al., 2001).

**Xylans** are the major noncellulosic polysaccharides of the secondary walls of all angiosperms (Bacic et al., 1988; Awano et al., 2000; Awano et al., 2002). **Glucomannans** form the major hemicelluloses of the secondary cell walls of gymnosperms (Brett and Waldron, 1990).

Pectins The **pectins** are probably the most chemically diverse of the noncellulosic polysaccharides (Bacic et al., 1988; Levy and Staehelin, 1992; Willats et al., 2001). They are characteristic of the primary walls of eudicots and to a lesser extent of monocots. Pectins may account for 30% to 50% of the dry weight of the primary walls of eudicots compared with only 2% to 3% that of monocots (Goldberg et al., 1989; Hayashi, 1991). The grasses often contain only trace amounts of pectin (Fry, 1988). Pectins may be lacking from secondary walls entirely.

Two fundamental constituents of pectins are **polygalacturonic acid** and **rhamnogalacturonan**. They and the other pectin constituents form a gel in which the cellulose-hemicellulose network is embedded (Roberts, 1990; Carpita and Gibeaut, 1993).

Pectins are highly hydrophilic and are best known for their ability to form gels. The water that is introduced into the wall by pectins imparts plastic properties to the wall and modulates the ability of the wall to be stretched (Goldberg et al., 1989). The cell walls of meristems are especially low in $Ca^{2+}$, but the amount increases markedly in the walls of meristem derivatives as they elongate and differentiate. Extensive $Ca^{2+}$ cross-linking of pectins occurs after cell elongation is completed, precluding further stretching. Evidence also exists for boron cross-linking of pectin (Blevins and Lukaszewski, 1998; Matoh and Kobayashi, 1998; Ishii et al., 1999).

It appears that the porosity of the cell wall is determined largely by the organization of the pectins rather than by the cellulose or hemicelluloses (Baron-Epel et al., 1988). The diameters of pores range from about 4.0 to 6.8 nanometers (Carpita et al., 1979; Carpita, 1982; Baron-Epel et al., 1988), which would allow the passage of substances such as salts, sugars, amino acids, and phytohormones. Molecules with diameters larger than those of the pores would be impeded in their ability to penetrate such walls. The wall is an effective physical barrier against most potentially pathogenic organisms. Its pores are too small to permit even viruses to penetrate it to the protoplast (Brett and Waldron, 1990). Pectic breakdown fragments may play a role as signaling molecules (Aldington and Fry, 1993; Fry et al., 1993).

Proteins In addition to the polysaccharides described above, the cell wall matrix may contain structural proteins (glycoproteins). Structural proteins make up about 10% of the dry weight of many primary walls. Among the major classes of structural proteins are the **hydroxyproline-rich proteins (HRGPs)**, the **proline-rich proteins (PRPs)**, and the **glycine-rich proteins (GRPs)**. Structural proteins are highly specific for certain cell types and tissues (Ye and Varner, 1991; Keller, 1993; Cassab, 1998). Relatively little is known about their biological function.

The best characterized structural proteins are the **extensins**, a family of HRGPs, so named because they were originally assumed to be involved with the cell wall's extensibility, an idea that now has been abandoned. It appears that extensin may play a structural role in development. For example, an extensin has been found in the palisade epidermal cells and hourglass cells that make up the outer two cell layers of the soybean seed coat (Cassab and Varner, 1987). Having relatively thick secondary walls, these cells provide mechanical protection for the enveloped embryo. A gene-encoding tobacco extensin was specifically expressed in one or two layers of cells located at the tips of emerging lateral roots. It has been suggested that the deposition of extensin might strengthen the walls and aid in the mechanical penetration of the cortex and epidermis (Keller and Lamb, 1989).

All three of the structural cell wall proteins—HRGPs, PRPs, and GRPs—have been found in the vascular tissues of stems (Showalter, 1993; Cassab, 1998). HRGPs are mainly associated with phloem, cambium, and sclerenchyma, whereas PRPs and GRPs are most often localized to xylem. GRPs have been localized in the modified primary walls of early tracheary elements (protoxylem elements) (Ryser et al., 1997) (Chapter 10). Once believed to be associated with lignification in the xylem, the deposition of GRPs and lignification have been shown to be independent processes. In bean hypocotyls the GRP apparently is not produced by the tracheary elements but by xylem parenchyma cells, which export the protein to the primary walls of the protoxylem elements (Ryser and Keller, 1992). PRPs have been implicated with lignification. Some PRPs constitute part of the nodule cell walls of legume roots and may play a role in nodule formation (Showalter, 1993).

Unlike the proteins listed above, **arabinogalactan proteins (AGPs)**, which are widely distributed in the plant kingdom, do not have an apparent structural function. AGPs are soluble and diffusible and occur in the plasma membrane, cell wall, and intercellular spaces (Serpe and Nothnagel, 1999); consequently they are good candidates to act as messengers in cell–cell interaction during differentiation. AGPs have been found to be important for somatic embryogenesis of carrot (*Daucus carota*) (Kreuger and van Holst, 1993), to play a role in the control of root epidermal cell expansion in *Arabidopsis thaliana* (Ding, L., and Zhu, 1997), and to be involved in tip growth of lily (*Lilium longiflorum*) pollen tubes (Jauh and Lord, 1996). AGPs apparently play multiple roles in plant development (Majewska-Sawka and Nothnagel, 2000). Another class of cell wall protein, named expansin (Li et al., 2002), has been shown to function as a wall-loosening agent to promote cell expansion (see below).

Numerous enzymes have been reported in primary cell walls, including peroxidases, laccases, phosphatases, invertases, cellulases, pectinases, pectin methylesterases, malate dehydrogenase, chitinases, and (1→3)β-glucanases (Fry, 1988; Varner and Lin, 1989). Some cell wall enzymes, such as the chitinases, (1→3)β-glucanases, and peroxidases, may be involved in the defense mechanisms of plants. Peroxidases and laccases may also catalyze lignification (Czaninski et al., 1993; O'Malley et al., 1993; Østergaard et al., 2000). Cellulase and pectinase play major roles in cell wall degradation, notably during leaf abscission and formation of the perforation plate in developing vessel elements.

Most information on cell wall proteins comes from studies on the primary cell walls of eudicots. Little is known about the cell wall proteins in monocots and gymnosperms, although extensins (HRGPs) and PRPs apparently exist in both groups, and GRPs in monocots (Levy and Staehelin, 1992; Keller, 1993; Showalter, 1993). Far less is known about the proteins in secondary cell walls. An extensin-like protein has been localized in the secondary cell walls of mature loblolly pine (*Pinus taeda*) wood (Bao et al., 1992).

### Callose Is a Widely Distributed Cell Wall Polysaccharide

**Callose**, a linear (1→3)β-D-glucan, is deposited between the plasma membrane and the existing cellulosic cell wall (Stone and Clarke, 1992; Kauss, 1996). It is probably best known in the sieve elements of angiosperm phloem, where it is associated with developing sieve pores (Fig. 4.3) and commonly found lining fully developed pores (Chapter 13; Evert, 1990). Callose is deposited very rapidly in response to mechanical wounding and environmental- or pathogen-induced stress, sealing

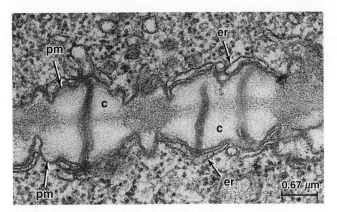

**FIGURE 4.3**

Callose (c) at developing sieve pores in wall between immature sieve elements of a *Cucurbita* root tip. A single plasmodesma traverses each pore site. A plasma membrane (pm) covers the wall including the callose at the pore sites. Endoplasmic reticulum (er) cisternae cover the pore sites.

the plasmodesmata between contiguous cells (Radford et al., 1998) or forming cell wall appositions ("papillae") opposite sites of attempted invasion of host cells by fungi (Perry and Evert, 1983). The callose associated with fully developed sieve pores may also be "wound" callose.

In addition to its association with developing sieve pores, callose also occurs in the normal course of development in pollen tubes (Ferguson et al., 1998), in cotton fibers during the early stages of secondary wall synthesis (Maltby et al., 1979), and temporarily during microsporogenesis and megasporogenesis (Rodkiewicz, 1970; Horner and Rogers, 1974). Callose is also transiently associated with the cell plates of dividing cells (Samuels et al., 1995). Callose is characterized histologically by its staining reactions with either resorcin blue as diachrome or alkaline aniline as flurochrome (Eschrich and Currier, 1964; Kauss, 1989). In transmission electron microscope sections, callose can be immunogold-labeled by using specific antibody probes (Benhamou, 1992; Dahiya and Brewin, 2000).

### Lignins Are Phenolic Polymers Deposited Mainly in Cell Walls of Supporting and Conducting Tissues

**Lignins** are phenolic polymers formed from the polymerization of three main monomeric units, the monolignols *p*-**coumaryl**, **coniferyl**, and **sinapyl alcohols** (Ros Barceló, 1997; Whetten et al., 1998; Hatfield and Vermerris, 2001). Generally, lignins are classified as *guaiacyl* (formed predominately from coniferyl alcohol), *guaiacyl-syringyl* (copolymers of coniferyl and sinapyl alcohols), or *guaiacyl-syringyl-p-hydroxyphenyl lignins* (formed from all three monomers), according to whether they are from gymnosperms, woody angiosperms, or grasses, respectively. Caution must be taken, however, with such broad generalization. The structures of "gymnosperm lignin" and "angiosperm lignin" hold at best only for lignins from the secondary xylem of the corresponding woods (Monties, 1989). Great variation exists in the monomeric composition of lignins of different species, organs, tissues, and even cell wall fractions (Wu, 1993; Terashima et al., 1998; Whetten et al., 1998; Sederoff et al., 1999; Grünwald et al., 2002). All lignins contain some $p$-hydroxyphenyl lignins, although they are generally ignored. In addition both guaiacyl and guaiacyl-syringyl lignins are found in gymnosperms and angiosperms (Lewis and Yamamoto, 1990).

Typically lignification begins in the intercellular substance at the corners of the cells, and extends to the intercorner middle lamellae; it then spreads to the first-formed (primary) wall layers and finally to those (the secondary wall layers) formed last (Terashima et al., 1993; Higuchi, 1997; Terashima, 2000; Grünwald et al., 2002). Other patterns of lignification have been reported (Calvin, 1967; Vallet et al., 1996; Engels and Jung, 1998). Apparently the lignin is covalently linked to wall polysaccharides (Iiyama et al., 1994). Results of experiments in which the enzyme cinnamoyl GA reductase was genetically down-regulated from the monoliguol biosynthetic pathway during secondary wall formation in the fibers of tobacco and *Arabidopsis thaliana* suggest that the mode of lignin polymerization may play a significant role in determining the three-dimensional organization of the polysaccharide matrix (Ruel et al., 2002).

Lignin is not restricted to the primary walls of cells that deposit secondary walls. For example, lignin has been reported to occur throughout the primary walls of rapidly expanding maize coleoptile parenchyma cells (Müsel et al., 1997). In addition lignin commonly is deposited in the primary walls of parenchymatous elements in response to wounding or to attack by parasites or pathogens (Walter, 1992).

Lignification is an irreversible process and typically is preceded by the deposition of cellulose and noncellulosic matrix components (hemicelluloses, pectins, and structural proteins) (Terashima et al., 1993; Hafrén et al., 1999; Lewis, 1999; Grünwald et al., 2002). Lignin is a hydrophobic filler that replaces the wall's water (Fig. 4.4). In the intercellular substance, lignin functions as a binding agent imparting compressive strength and bending stiffness to the woody stem. Lignin has no effect on the wall's tensile strength (Grisebach, 1981).

By waterproofing the walls of the xylem, lignin limits lateral diffusion, thus facilitating longitudinal transport of water in conducting xylem. It has been suggested that this may have been a main function of lignin during

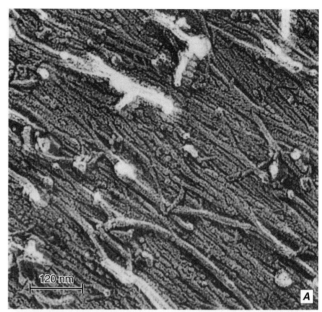

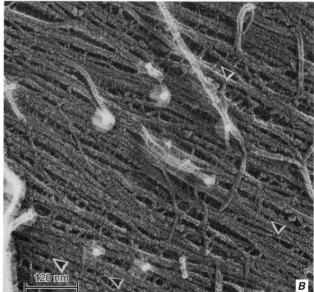

**FIGURE 4.4**

Structure of the $S_2$ wall layer of mature *Pinus thunbergii* tracheids before (**A**) and after (**B**) delignification. Note the closed structure of the fully lignified $S_2$ wall in **A**. After delignification (**B**), cross-links (arrowheads) are visible between microfibrils; in addition pores (gaps) can be seen in the wall. A rapid-freezing deep-etching technique combined with transmission electron microscopy was used to obtain these images. (From Hafrén et al., 1999, by permission of Oxford University Press.)

plant evolution (Monties, 1989). The mechanical rigidity of lignin strengthens the xylem, enabling the tracheary elements to endure the negative pressure generated from transpiration without collapse of the tissue. Lignified cell walls are resistant to microbial attack (Vance et al., 1980; Nicholson and Hammerschmidt, 1992). Lignin may have first functioned as an antimicrobial agent and only later assumed a role in water transport and mechanical support in the evolution of land plants (Sederoff and Chang, 1991).

Two tests, the Wiesner and Mäule tests, are commonly used for qualitative determination of lignin (Vance et al., 1980; Chen, 1991; Pomar et al., 2002). The Wiesner test is applicable to all lignins. In this test the cell walls of tissues containing lignin give a bright purple-red color when treated with phloroglucinol in concentrated hydrochloric acid. Predominantly syringyl lignins give only a weak reaction. The Mäule test is specific for syringyl groups. In this test the cell walls of tissues containing syringyl lignin give a deep rose-red color when treated successively with aqueous permanganate, hydrochloric acid, and ammonia. Polyclonal lignin antibodies also are available for immunogold labeling of different lignin types (Ruel et al., 1994; Grünwald et al., 2002).

### Cutin and Suberin Are Insoluble Lipid Polymers Found Most Commonly in the Protective Surface Tissues of the Plant

The major function of cutin and suberin is to form a matrix in which **waxes**—long-chain lipid compounds—are embedded (Post-Beittenmiller, 1996). Together the cutin-wax or suberin-wax combinations form barrier layers that help prevent the loss of water and other molecules from the aerial parts of the plant (Kolattukudy, 1980).

**Cutin**, together with its embedded waxes, forms the **cuticle**, which covers the surface of the epidermis of all aerial parts. The cuticle consists of several layers having varying amounts of cutin, waxes, and cellulose (Chapter 9).

**Suberin** is a major component of the cell walls of the secondary protective tissue, cork, or phellem (Chapter 15), of the endodermal and exodermal cells of roots, and of the bundle sheath cells enclosing the leaf veins of many of the Cyperaceae, Juncaceae, and Poaceae. In addition to reducing the loss of water from the aerial parts of the plant, suberin restricts apoplastic (via the cell wall) movement of water and solutes and forms a barrier to microbial penetration. Cell wall suberin is characterized by the presence of two domains, a polyphenolic domain and a polyaliphatic domain (Bernards and Lewis, 1998; Bernards, 2002). The polyphenolic domain is incorporated *within* the primary wall and is covalently linked to the polyaliphatic domain, which is deposited on the inner surface of the primary wall, that is, between the primary wall and the plasma membrane. As seen with the electron microscope, the polyaliphatic domain has a lamellar, or layered, appearance, with light bands alternating

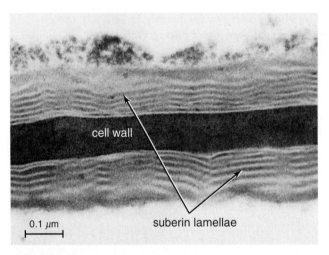

**FIGURE 4.5**

Electron micrograph showing suberin lamellae in the walls between two cork cells of a potato tuber (*Solanum tuberosum*). Note the alternating light and dark bands. Cork cells form the outermost layer of a protective covering in plant parts such as potato tubers and woody stems and roots. (From Thomson et al., 1995.)

with dark bands (Fig. 4.5). It has been proposed that the light bands comprise predominantly aliphatic zones and the darker bands phenolic rich zones, and that very long chain fatty acids and waxes likely span successive lamellar bands or intercalate within the polyester network of the polyaliphatic domain (Bernards, 2002). In times past, the light bands have been considered to be composed largely of waxes and the dark bands of suberin (Kolattukudy and Soliday, 1985). Some cuticles also have a lamellar appearance.

## ▌CELL WALL LAYERS

Plant cell walls vary greatly in thickness, depending partly on the role the cells play and partly on the age of the individual cell. Generally, young cells have thinner walls than fully developed ones, but in some cells the wall does not thicken much after the cell ceases to grow. Each protoplast forms its wall from the outside inward so that the youngest layer of a given wall is in the innermost position, next to the protoplast. The cellulosic layers formed first make up the **primary wall**. The region of union of the primary walls of adjacent cells is called the **middle lamella**, or **intercellular substance**. Many cells deposit additional wall layers; these form the **secondary wall**. Being deposited after the primary wall, the secondary wall is laid down by the protoplast of the cell on the inner surface of the primary wall (Fig. 4.1).

## The Middle Lamella Frequently Is Difficult to Distinguish from the Primary Wall

It is especially difficult to distinguish the middle lamella from the primary wall in cells that develop thick secondary walls. In cells in which the distinction between the middle lamella and primary walls is obscured, the two adjacent primary walls and the middle lamella, and perhaps the first layer of the secondary wall, may be called **compound middle lamella**. Hence the term compound middle lamella would mean sometimes a three-ply, and other times a five-ply, structure (Kerr and Bailey, 1934).

Electron microscopy rarely reveals the middle lamella as a well-delimited layer except along the corners of cells where the intercellular material is most abundant. Recognition of the middle lamella is based chiefly on microchemical tests and maceration techniques. The middle lamella is largely pectic in nature but often becomes lignified in cells with secondary walls.

## The Primary Wall Is Deposited While the Cell Is Increasing in Size

The primary wall, composed of the first-formed wall layers, is deposited before and during the growth of the cell. Actively dividing cells commonly have only primary walls, as do most mature cells involved with such metabolic processes as photosynthesis, secretion, and storage. Such cells have relatively thin primary walls, and the primary wall is usually thin in cells having secondary walls. The primary wall may attain considerable thickness, however, as in the collenchyma of stems and leaves and the endosperm of some seeds, although these wall thickenings are considered secondary by some (Frey-Wyssling, 1976). Thick primary walls may show layering, or a polylamellate texture, caused by variations in the orientation of the cellulose microfibrils, from one layer to the next (see below). Regardless of the thickness of the wall, living cells with only primary walls may remove the thickening previously acquired, lose their specialized cellular form, divide, and differentiate into new cell types. For this reason it is principally cells with only primary walls that are involved in wound healing and regeneration in plants.

Current models of the architecture of the primary (growing) cell wall envision a network of cellulose microfibrils intertwined with hemicelluloses, such as xyloglucan, and embedded in a gel of pectins. In one model (Fig. 4.6A) the hemicelluloses coat the surface of the cellulose, to which they are noncovalently bonded, and form cross-links, or tethers, that bind the cellulose microfibrils together. It has been estimated that such a cellulose-xyloglucan network may contribute as much as 70% of the total strength to normal primary walls (Shedletzky et al., 1992). Evidence for cellulose-hemicellulose cross-links has been provided with the electron microscope (Fig. 4.7; McCann et al., 1990; Hafrén et al., 1999; Fujino et al., 2000). Pauly et al. (1999) found three district xyloglucan fractions in the stem cell walls of *Pisum sativum*. Approximately 8% of the dry weight of the walls consisted of xyloglucan that can be solubilized by treatment with a xyloglucan-specific endoglucanase. This material corresponds to the xyloglucan domain proposed to form the cross-links between cellulose microfibrils. Briefly, a second domain (10% of the wall dry weight) consisted of xyloglucan proposed to be closely associated with the surface of the cellulose microfibrils, and a third (3% of the wall dry weight) of xyloglucan proposed to be entrapped within or between cellulose microfibrils.

In an alternative model of the primary wall (Fig. 4.6B), there are no direct microfibril-microfibril links. Instead, hemicelluloses tightly bound to the microfibrils are sheathed in a layer of less tightly bound hemicelluloses, which in turn are embedded in the pectin matrix, filling the spaces between microfibrils (Talbott and Ray, 1992). In this model, wall strength may depend in part on the presence of many noncovalent interactions between laterally aligned matrix molecules (Cosgrove, 1999). It is pertinent to note that a study utilizing solid state $^{13}C$ nuclear magnetic resonance spectroscopy found little evidence for substantial interaction between cellulose and hemicelluloses in the primary cell walls of three monocots (Italian ryegrass, pineapple, and onion) and one eudicot (cabbage) (Smith, B. G., et al., 1998). The authors suggested, however, that a relatively small number of hemicellulose molecules would be sufficient to cross-link the cellulose microfibrils. A similar finding was reported for the primary cell walls of *Arabidopsis thaliana* (Newman et al., 1996). The mean orientation of the cellulose microfibrils apparently is a key factor in determining the mechanical properties of the cell wall (Kerstens et al., 2001).

## The Secondary Wall Is Deposited inside the Primary Wall Largely, If Not Entirely, after the Primary Wall Has Stopped Increasing in Surface Area

Although the secondary wall commonly is thought of as being deposited after the increase in surface area of the primary wall has ceased, evidence has long existed that the initial layer of secondary wall becomes slightly extended because its deposition begins somewhat before increase in surface of the wall ceases (Roelofsen, 1959). That secondary wall deposition begins before cell expansion ceases has been reported for both conifer tracheids (Abe et al., 1997) and for the fibers in bamboo culms (MacAdam and Nelson, 2002; Gritsch and Murphy, 2005).

Secondary walls are particularly important in specialized cells that have strengthening function and in those involved in the conduction of water; in these cells the

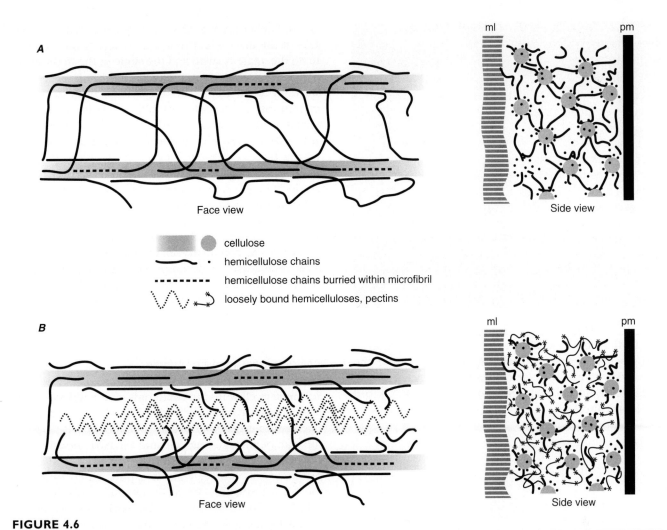

**FIGURE 4.6**

Two models of the growing (primary) cell wall. In the model depicted in **A**, the mechanical strength of the wall is attributed to tethering of cellulose microfibrils by xyloglucans that are bonded noncovalently to the microfibril surface and entrapped within the microfibril. Pectins (not shown) form a co-extensive matrix in which the cellulose-xyloglucan network is embedded. The alternative model depicted in **B** differs from that depicted in **A** primarily by the lack of polymers that directly cross-link the microfibrils. Instead, the tightly bound hemicelluloses, such as xyloglucan, are viewed as sheathed in a layer of less tightly bound polysaccharides. The latter in turn are embedded in the pectin matrix that fills the spaces between the microfibrils. Details: ml, middle lamella; pm, plasma membrane. (Adapted from Cosgrove, 1999. Reprinted, with permission, from the *Annual Review of Plant Physiology and Plant Molecular Biology*, vol. 50, © 1999 by Annual Reviews. www.annualreviews.org)

protoplast often dies after the secondary wall has been deposited. Cellulose is more abundant in secondary walls than in primary walls, and pectic substances are lacking; the secondary wall is therefore rigid and not readily stretched. Structural proteins and enzymes, which are relatively abundant in primary cell walls, apparently are either absent or present in small amounts in secondary walls. As mentioned previously, an extensin-like protein has been localized in the secondary walls of mature loblolly pine wood (Bao et al., 1992). Lignin is common in the secondary walls of cells found in wood.

In thick-walled wood cells, three distinct layers—designated $S_1$, $S_2$, and $S_3$, for outer, middle, and inner layer, respectively—can frequently be distinguished in the secondary wall. The $S_2$ layer is the thickest. The $S_3$ layer may be very thin or lacking entirely. Some wood anatomists consider the $S_3$ layer to be sufficiently distinct from the $S_1$ and $S_2$ layers to be called *tertiary wall*.

The separation of the secondary wall into the three S layers results mainly from different orientations of microfibrils in the three layers (Frey-Wyssling, 1976). Typically the microfibrils are helically oriented in the various layers (Fig. 4.8). In $S_1$, the fibrils run along

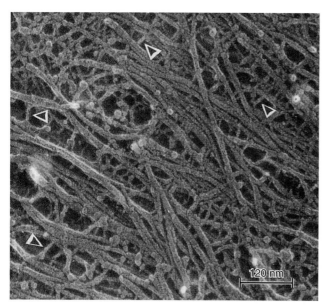

**FIGURE 4.7**

Tangential view of a newly synthesized primary wall in the vascular cambium of *Pinus thunbergii*. Note the numerous cross-links (arrowheads) between microfibrils. (From Hafrén et al., 1999, by permission of Oxford University Press.)

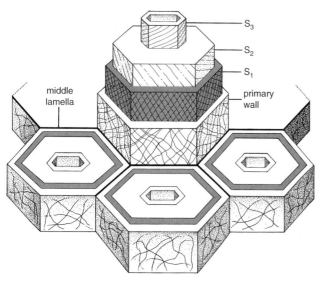

**FIGURE 4.8**

The layers of secondary cell walls. Diagram showing the organization of the cellulose microfibrils and the three layers ($S_1$, $S_2$, $S_3$) of the secondary wall. The different orientations of the three layers strengthen the secondary wall. (From Raven et al., 2005.)

crossed helices, which make a large angle with the long axis of the cell so that this layer is highly birefringent. In $S_2$, the angle is small and the slope of the helix steep; hence the cellulose microfibrils in this layer do not show up with the polarizing microscope. In $S_3$, the microfibrils are deposited as in $S_1$, at a large angle to the long axis of the cell. In at least some wood fibers the $S_1$ and $S_2$ layers are interconnected by a transition zone with a helicoidal texture (Vian et al., 1986; Reis and Vian, 2004). The primary wall differs from the secondary in having a rather random arrangement of microfibrils. In fibers and tracheids of most woody species the inner surface of the $S_3$ layer is covered with a noncellulosic film, often bearing lumps called **warts**. Once thought to be composed of cytoplasmic debris left over from remnants of a decomposed protoplast, the warts are now considered as outgrowths of the cell wall formed largely from lignin precursors (Frey-Wyssling, 1976; Castro, 1991).

# PITS AND PRIMARY PIT-FIELDS

Secondary cell walls are commonly characterized by the presence of cavities called **pits** (Figs. 4.9B–D and 4.10B, C). A pit in a cell wall usually occurs opposite a pit in the wall of an adjoining cell, and the two opposing pits constitute a **pit-pair**. The middle lamella and the two primary walls between the two pit cavities are called the **pit membrane**. Pits arise during ontogeny of the cell and result from differential deposition of secondary wall material; none is deposited over the pit membrane so that the pits are actual discontinuities in the secondary wall.

Whereas secondary walls have pits, primary walls have **primary pits**, which are thin areas, not interruptions, in the primary wall (Figs. 4.9A and 4.10A). In this book the term **primary pit-field** is used to describe both a solitary primary pit and a cluster of primary pits. During the deposition of the secondary wall, the pits are formed over the primary pit-fields. Several pits may arise over one primary pit-field.

Plasmodesmata (see below) are commonly aggregated in the primary pit-fields (Fig. 4.9A). When a secondary wall develops, the plasmodesmata remain in the pit membrane as connections between the protoplasts of the adjoining cells. Plasmodesmata are not restricted to primary pit-fields. A scattering of plasmodesmata through a wall of uniform thickness is of common occurrence. Moreover in many instances the primary wall is thickened specifically where plasmodesmata occur.

Pits vary in size and detailed structure (Chapters 8 and 10), but two principal types are recognized in cells with secondary walls: **simple pits** and **bordered pits** (Fig. 4.9C, D). The basic difference between the two

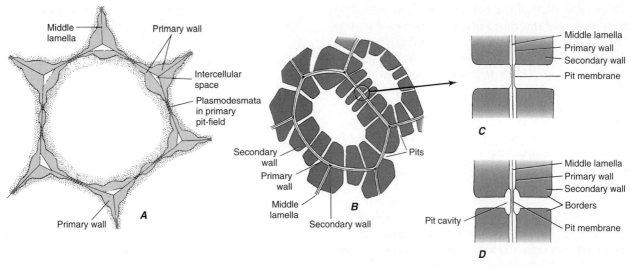

**FIGURE 4.9**

Primary pit-fields, pits, and plasmodesmata. **A**, parenchyma cell with primary walls and primary pit-fields, thin areas in the walls. As shown here, plasmodesmata commonly traverse the wall at the primary pit-fields. **B**, cells with secondary walls and numerous simple pits. **C**, a simple pit-pair. **D**, a bordered pit-pair. (From Raven et al., 2005.)

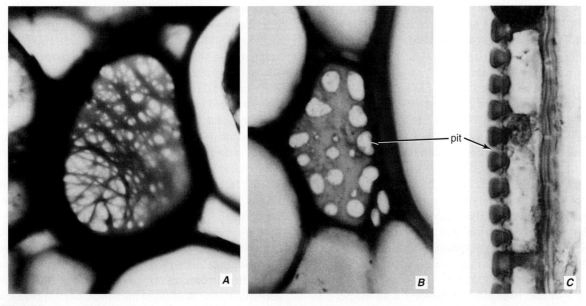

**FIGURE 4.10**

Primary pit-fields and pits. Parenchyma cells from root cortex of *Abies* (**A**), xylem of *Nicotiana* (**B**), and *Vitis* (**C**). **A**, surface view of reticulum of cellulose; unstained meshes are thin places penetrated by plasmodesmata (not visible). **B**, pits in surface view and, **C**, in section. In **C**, pitted wall between parenchyma cell and vessel. (A, ×930; B, ×1100; C, ×1215.)

kinds of pit is that, in the bordered pit, the secondary wall arches over the pit cavity and narrows down its opening to the lumen of the cell. The overarching secondary wall constitutes the **border**. In simple pits, no such overarching occurs. In bordered pits, the part of the cavity enclosed by the border is called the **pit chamber**, and the opening in the border is the **aperture**.

A combination of simple pits is termed a ***simple pit-pair***, and of two opposing bordered pits a ***bordered pit-pair***. Combinations of simple pits and bordered pits, called ***half-bordered pit-pairs***, are found in the xylem. A pit may have no complementary structure, for example, as when it occurs opposite an intercellular space. Such pits are called ***blind pits***. In addition two or more pits may oppose a single pit in an adjoining cell, a combination that has been named ***unilaterally compound pitting***.

Simple pits are found in certain parenchyma cells, in extraxylary fibers, and in sclereids (Chapter 8). In a simple pit, the cavity may be uniform in width, or it may be slightly wider or slightly narrower toward the lumen of the cell. If it narrows down toward the lumen, the simple pit intergrades with the bordered pit in its structure. Depending on the thickness of the secondary wall, the simple pit may be shallow or it may form a canal extending from the cell lumen toward the pit membrane. Pits may coalesce as the wall increases in thickness and form ***branched*** or ***ramiform*** (from the Latin *ramus*, branch) ***pits*** (Chapter 8).

Both simple and bordered pits occur in the secondary walls of tracheary elements (Chapters 10 and 11). In conifer tracheids the bordered pit-pairs have an especially elaborate structure (Chapter 10).

If the secondary wall is very thick, the pit border is correspondingly thick. The chamber of such a pit is rather small and is connected with the cell lumen through a narrow passage in the border, the **pit canal**. The canal has an **outer aperture** opening into the pit chamber and an **inner aperture** facing the lumen of the cell. In certain pits the pit canal resembles a compressed funnel, and its two apertures differ in size and form (Fig. 4.11). The outer aperture is small and circular, the inner extended and slitlike. In a pit-pair the inner apertures of the two pits are crossed with respect to each other (Chapter 10). This arrangement is related to the helical deposition of the microfibrils in the secondary wall.

# ORIGIN OF CELL WALL DURING CELL DIVISION

## Cytokinesis Occurs by the Formation of a Phragmoplast and Cell Plate

During vegetative growth, cell division (**cytokinesis**) usually follows nuclear division (**karyokinesis**, or

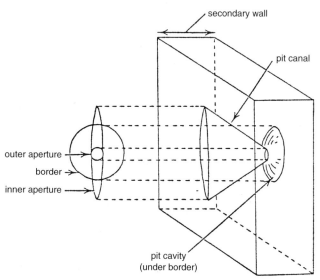

**FIGURE 4.11**

Diagram of a bordered pit with extended inner aperture and reduced border. (From Esau, 1977; after Record, 1934.)

**mitosis**). The mother cell divides, resulting in the formation of two daughter cells. Cytokinesis is initiated in late anaphase with the formation of the **phragmoplast**, an initially barrel-shaped system of microtubules—remnants of the mitotic spindle—that appears between the two sets of daughter chromosomes (Fig. 4.12A). The phragmoplast, like the mitotic spindle that preceded it, is composed of two opposing and overlapping sets of microtubules that form on either side of the division plane (not depicted in Fig. 4.12A). Actin filaments are also a prominent component of the phragmoplast, and they likewise are aligned perpendicular to the division plane. In contrast to the microtubules, the actin filaments, although organized into two opposing sets, do not overlap.

The phragmoplast serves as the framework for the assembly of the **cell plate**, the initial partition between the daughter cells (Fig. 4.13). The cell plate is formed from the fusion of Golgi-derived vesicles apparently directed to the division plane by the phragmoplast microtubules, possibly with the help of motor proteins. The role of the actin filaments is less clear. When the cell plate is initiated, the phragmoplast does not extend to the walls of the dividing cell. During cell plate expansion, the phragmoplast microtubules depolymerized in the center and are successively repolymerized at the margins of the cell plate. The cell plate—preceded by the phragmoplast (Fig. 4.12B, C)—grows outward (**centrifugally**) until it reaches the walls of the dividing cell, completing the separation of the two daughter cells. It is pertinent to note that, in addition to microtubules and actin fila-

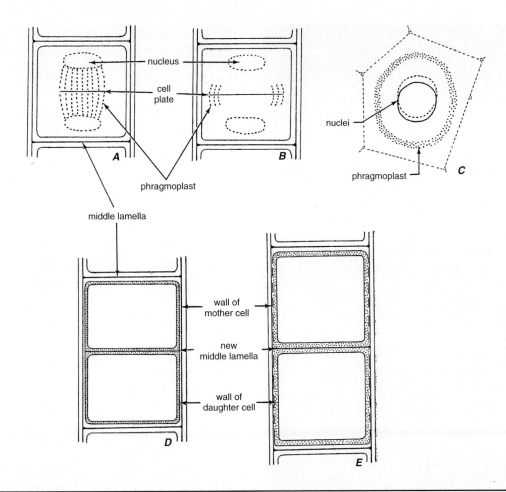

**FIGURE 4.12**

Formation of wall during cell division. **A**, formation of cell plate in the equatorial plane of phragmoplast at telophase. **B**, **C**, phragmoplast now appears along the margin of the circular cell plate (in side view in **B**; in surface view in **C**). **D**, cell division is completed and each sister cell has formed its own primary wall (stipples). **E**, sister cells have enlarged, their primary walls have thickened, and the mother cell wall has been torn along the vertical sides of the cells. (From Esau, 1977.)

ments, some researchers include the Golgi-derived vesicles and endoplasmic reticulum early associated with the developing cell plate as part of the phragmoplast (Staehelin and Hepler, 1996; Smith, L. G., 1999).

The actual process of cell-plate formation is rather complex and consists of several stages (Fig. 4.14; Samuels et al., 1995; Staehelin and Hepler, 1996; Nebenführ et al., 2000; Verma, 2001): (1) the arrival of Golgi-derived vesicles in the division plane; (2) the formation of 20-nanometer tubes (**fusion tubes**) that grow out of the vesicles and fuse with others, giving rise to a continuous, interwoven, **tubulo-vesicular network**, with a dense fibrous coat; (3) transformation of the tubulo-vesicular network into a **tubular network** and then into a **fenestrated plate-like structure**, during which the dense membrane coat and the associated phragmoplast microtubules are disassembled; (4) the formation of numerous finger-like projections at the margins of the cell plate that fuse with the plasma membrane of the mother cell wall; (5) maturation of the cell plate into a new cell wall. The latter stage includes closing of the fenestrae. At this time segments of tubular endoplasmic reticulum are entrapped within the developing wall and plasmodesmata are formed. Soon after the phragmoplast cytoskeleton disappears and the new cell plate begins to mature in root cells of cress (*Lepidium sativum*) and maize, myosin begins to localize in the newly formed plasmodesmata (Reichelt et al., 1999; Baluška et al., 2000). At the same time bundles of actin filaments appear to become attached to the plasmodesmata.

A number of proteins have been implicated in cell-plate formation (Heese et al., 1998; Smith, L. G., 1999; Harper et al., 2000; Lee, Y.-R. J., and Liu, 2000; Otegui

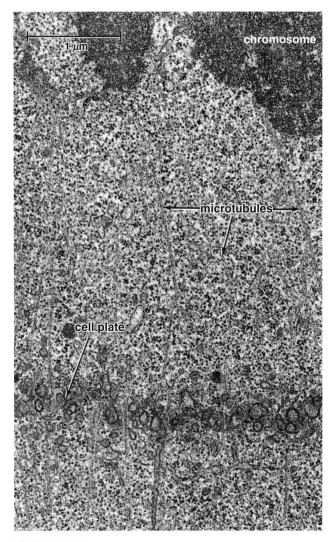

**FIGURE 4.13**

Details of early cytokinesis in a tobacco (*Nicotiana tabacum*) mesophyll cell. Cell plate is still composed of individual vesicles. Phragmoplast microtubules occur on both sides of the cell plate, some traversing the plate. Some chromosome material of one of the two future daughter nuclei is shown. (From Esau, 1977.)

and Staehelin, 2000; Assaad, 2001). For example, **phragmoplastin**, a dynamin-like protein, fused with green fluorescence protein, has been localized to the developing cell plate in tobacco BY-2 cells (Gu and Verma, 1997). Phragmoplastin may be involved with the formation of the fusion tubes and vesicle fusion at the cell plate. Overexpression of phragmoplastin in transgenic tobacco seedlings resulted in the accumulation of callose at the cell plate and arrested plant growth (Geisler-Lee et al., 2002). More direct functional evidence has been obtained for involvement of the *Arabidopsis thaliana* KNOLLE protein in cell-plate formation (Lukowitz et al.,

1996). A syntaxin-related protein, KNOLLE apparently serves as a docking receptor for vesicles that are transported by the phragmoplast. In the absence of KNOLLE proteins the vesicles fail to fuse (Lauber et al., 1997).

### Initially Callose Is the Principal Cell Wall Polysaccharide Present in the Developing Cell Plate

Callose begins to accumulate in the lumen of the developing cell plate during the tubulo-vesicular stage but is most abundant during the conversion of the tubular network into the fenestrated sheet. It has been suggested that the callose may exert a spreading force on the membranes, facilitating their conversion into a plate-like structure (Samuels et al., 1995).

The pattern of deposition of cellulose and matrix components and their eventual replacement of callose in the developing cell plate is not at all clear. In tobacco BY-2 cells, xyloglucans and pectins were localized beginning with the tubulo-vesicular stage, but their concentrations only seemed to increase substantially after completion of the cell plate (Samuels et al., 1995). Cellulose began to be synthesized in significant amounts when the cell plate reached the fenestrated sheet stage. In root meristem cells of *Phaseolus vulgaris*, by contrast, cellulose, hemicelluloses, and pectins reportedly were deposited simultaneously along the plate (Matar and Catesson, 1988).

According to a long held view (Priestly and Scott, 1939), fusion of the cell plate—which was regarded as a new middle lamella—with the mother cell walls occurs when the primary wall is broken opposite the cell plate during expansion of the daughter protoplasts. The new, centrifugally developing middle lamella then makes contact with the mother cell middle lamellae outside the stretched and broken mother cell walls. More recently it has been shown that the cell plate is not the middle lamella per se, and that the pectic middle lamella does not begin to develop until after the cell plate contacts the mother cell walls. The middle lamella then extends **centripetally** (from the outside toward the inside) within the cell plate from the junction with the mother cell walls (Matar and Catesson, 1988). This is preceded by formation of a buttress-like zone at that junction. The buttress-like zone is the starting point for a sequence of changes in fibrillar architecture leading to the fusion and continuity of the fibrillar skeleton of the two walls. The middle lamella of the mother cell wall then produces a wedge-shaped protuberance that penetrates into the buttress and progresses toward the cell plate.

### The Preprophase Band Predicts the Plane of the Future Cell Plate

Before the cell divides, the nucleus assumes a proper position for the event. If the cell about to divide is highly vacuolated, a layer of cytoplasm, the **phragmosome**,

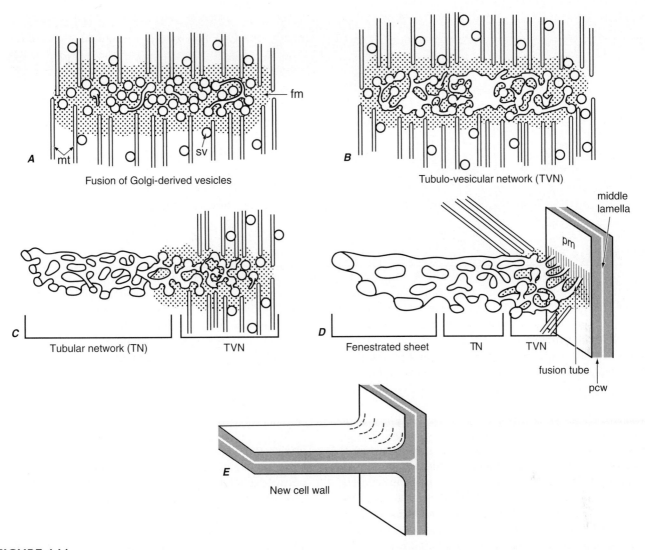

**FIGURE 4.14**

Stages of cell plate development. **A**, the fusion of Golgi-derived secretory vesicles (sv) at the equatorial zone, among phragmoplast microtubules (mt) and a cytoplasmic fuzzy matrix (fm). **B**, fused Golgi-derived vesicles give rise to tubulo-vesicular network covered by a "fuzzy coat." **C**, a tubular network (TN) forms as the lumen of the tubulo-vesicular network (TVN) becomes filled with cell wall polysaccharides, especially callose. Fuzzy matrix surrounding the network and microtubules disappears, further distinguishing this stage from the tubulo-vesicular network. **D**, the tubule areas expand, forming an almost continuous sheet. Numerous finger-like projections extend from the margins of the cell plate and fuse with the plasma membrane (pm) of the parent cell wall (pcw) at the site previously occupied by the preprophase band. **E**, maturation of the cell plate into a new cell wall. (After Samuels et al., 1995. Reproduced from *The Journal of Cell Biology* 1995, vol. 130, 1345–1357, by copyright permission of the Rockefeller University Press.)

spreads across the future division plane and the nucleus becomes located in this layer (Fig. 4.15; Sinnott and Bloch, 1941; Gunning, 1982; Venverloo and Libbenga, 1987). The phragmosome contains both microtubules and actin filaments (Goosen-de Roo et al., 1984), both of which apparently are involved with its development. In addition most vegetative cells display a **preprophase band**, a cortical belt of microtubules and actin fila-

ments, that predicts the plane of the future cell plate (Gunning, 1982; Gunning and Wick, 1985; Vos et al., 2004). Examination of dividing root-tip cells of *Pinus brutia* by confocal-laser-scanning microscopy, using immunolocalization techniques to identify microtubules and endoplasmic reticulum (ER), revealed that tubules of ER formed a dense ring-like formation at the site of the preprophase band (Zachariadis et al., 2001).

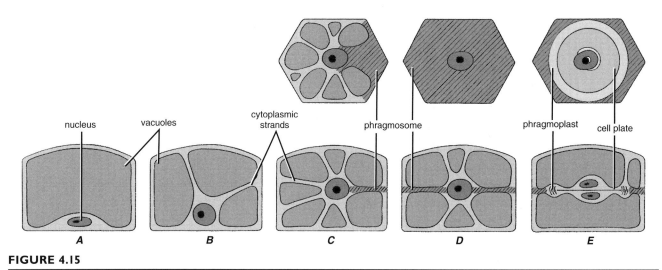

**FIGURE 4.15**

Cell division in a highly vacuolated cell. **A**, initially, the nucleus lies along one wall of the cell, which contains a large central vacuole. **B**, strands of cytoplasm penetrate the vacuole, providing a pathway for the nucleus to migrate to the center of the cell. **C**, the nucleus has reached the center of the cell and is suspended there by numerous cytoplasmic strands. Some of the strands have begun to merge to form the phragmosome through which cell division will take place. **D**, the phragmosome, which forms a layer that bisects the cell, is fully formed. **E**, when mitosis is completed, the cell will divide in the plane occupied by the phragmosome. (Courtesy of W. H. Freeman; after Venverloo and Libbenga, 1987. © 1987, with permission from Elsevier.)

Development of the "ER preprophase band" closely resembled that of the "microtubule preprophase band." The preprophase band disappears after initiation of the mitotic spindle and breakdown of the nuclear envelope (Dixit and Cyr, 2002), long before the initiation of the cell plate, yet the developing cell plate fuses with the parent wall precisely at the zone demarcated earlier by the band. Actin filaments have been found bridging the gap between the leading edge of the phragmoplast-cell plate and a cortical actin network in the vicinity of this zone (Lloyd and Traas, 1988; Schmit and Lambert, 1988; Goodbody and Lloyd, 1990). These filaments probably aid in guiding cell plate growth, utilizing an acto-myosin-based mechanism (Molchan et al., 2002). In some vacuolated cells the mitotic spindle and phragmoplast are laterally displaced, and the growing cell plate anchors on one side of the cell at an early stage of development, a mode of cytokinesis termed "polarized cytokinesis" by Cutler and Ehrhardt (2002).

## GROWTH OF THE CELL WALL

After completion of the cell plate, additional wall material is deposited on either side of it, resulting in an increase in thickness of the new partition. New wall material is deposited around each of the daughter protoplasts in a mosaic fashion, so that the new walls of meristematic cells are characterized by a heterogeneous distribution of polysaccharides (Matar and Catesson, 1988).

Matrix materials, including glycoproteins, are delivered to the wall in Golgi vesicles. Cellulose microfibrils, by contrast, are synthesized by **cellulose synthase complexes** that appear as rings, or **rosettes**, of six hexagonally arranged particles that span the plasma membrane (Fig. 4.16; Delmer and Stone, 1988; Hotchkiss, 1989; Fujino and Itoh, 1998; Delmer, 1999; Hafrén et al., 1999; Kimura et al., 1999; Taylor et al., 2000). Each rosette synthesizes cellulose from the glucose derivative UDP-glucose (uridine diphosphate glucose). Two enzymes required for cellulose synthesis in *Arabidopsis* have been identified through the analysis of mutants, CesA glycosyltransferaces and KOR membrane-associated endo-1,4-β-glucanases (Williamson et al., 2002). CesA proteins are components of the cellulose synthase complex, which likely contains 18 to 36 such proteins. At least three CesA proteins are required for cellulose synthesis in the secondary cell wall of developing *Arabidopsis* xylem vessels (Taylor et al., 2000, 2003). Moreover, all three CesAs are required for proper localization of these proteins to the regions of the plasma membrane associated with cell wall thickening (Gardiner et al., 2003b). The cortical microtubules assemble at the site of secondary cell wall formation before the latter begins and are required continually to maintain normal CesA protein localization (Gardiner et al., 2003b).

During cellulose synthesis the rosettes, which move in the plane of the membrane, exude the microfibrils on the outer surface of the membrane. The rosettes, which are formed by the endoplasmic reticulum, are inserted

**FIGURE 4.16**

Freeze fracture replicas of rosettes associated with cellulose microfibril biogenesis in a differentiating tracheary element of *Zinnia elegans*. The rosettes shown here exist in the leaflet of the plasma membrane bilayer that is nearest the cytoplasm (the PF face). Several rosettes are shown (surrounded by circles) in the main micrograph. The inset shows one rosette at higher magnification and after high resolution rotary shadowing at ultracold temperature with a minimum amount of platinum/carbon. (Courtesy of Mark J. Grimson and Candace H. Haigler.)

into the plasma membrane via Golgi vesicles (Haigler and Brown, 1986) and apparently are pushed forward by the synthesis (polymerization) and crystallization forces of the cellulose microfibrils (Delmer and Amor, 1995).

The orientation of the cellulose microfibrils in elongating plant cells and under the secondary wall thickenings of xylem vessels normally parallels that of the underlying cortical microtubules. This observation led to the widely accepted hypothesis—dubbed the ***alignment hypothesis*** by Baskin (2001)—that the orientation of the nascent cellulose microfibrils is determined by the underlying cortical microtubules (Abe et al., 1995a, b; Wymer and Lloyd, 1996; Fisher, D. D., and Cyr, 1998), which channel the rosettes through the plane of the plasma membrane (Herth, 1980; Giddings and Staehelin, 1988). The alignment hypothesis seems inadequate, however, to explain the deposition of walls in nonelongating plant cells in which the cortical microtubules do not parallel the nascent microfibrils (reviewed in Emons et al., 1992 and Baskin, 2001). In addition, studies using drugs and a temperature sensitive mutant (*mor1-1*) of *Arabidopsis* showed that the disordering or complete loss of cortical microtubules did little to alter the parallel arrangement of cellulose microfibrils in expanding root cells (Himmelspach et al., 2003; Sugimoto et al., 2003).

Other hypotheses have been proposed to explain the mechanism of cellulose microfibril deposition. One of these is the ***liquid crystalline self-assembly hypoth-***

***esis***. Noting the similarity of helicoidal cell walls (see below), the cellulose microfibrils of which do not match the cortical microtubules, and cholesteric liquid crystals, Bouligand (1976) proposed that helicoidal wall structure could arise from a liquid crystalline self-organizing principle. (See critique of this hypothesis by Emons and Mulder, 2000.)

Baskin (2001) has proposed a ***template incorporation mechanism***, in which the nascent microfibrils can be oriented by microtubules or become incorporated into the cell wall by binding to a scaffold built and oriented around either already incorporated microfibrils or membrane proteins or both. In this model the cortical microtubules serve to bind and orient components of the scaffold at the plasma membrane. Microtubules are not required for cellulose synthesis or for the formation of cellulose microfibrils.

A ***geometrical model*** for cellulose microfibril deposition has been formulated from extensive observations on the helicoidal (secondary) cell wall structure of *Equisetum hyemale* root hairs (Emons, 1994; Emons and Mulder, 1997, 1998, 2000, 2001). The model, which is purely mathematical, quantitatively relates the deposition angle of cellulose microfibrils (with respect to the cell axis) to (1) the density of active synthases in the plasma membrane, (2) the distance between individual microfibrils within a lamella, and (3) the geometry of the cell. The crucial factor in the model is the coupling of the rosette trajectories, and hence the orientation of the microfibrils being deposited, to the local number, or density, of active rosettes. This provides the cell with a route to manipulate the structure of the cell wall by creating controlled local variations of the number of active rosettes (Emons and Mulder, 2000; Mulder and Emons, 2001; Mulder et al., 2004). A feedback mechanism would preclude the density of rosettes to rise beyond a maximum dictated by the geometry of the cell.

Electron microscopy has shown that cortical microtubules are linked to the inner leaflet of the plasma membrane by protein cross-bridges (Gunning and Hardham, 1982; Vesk et al., 1996). Studies on tobacco (Marc et al., 1996; Gardiner et al., 2001; Dhonukshe et al., 2003) and *Arabidopsis* (Gardiner et al., 2003a) plasma membranes indicate that this protein is a 90-kDa protein, phospholipaseD (PLD). It has been suggested that production of the signaling molecule phosphatidic acid (PA) by PLD may be required for normal microtubule organization and hence normal growth in *Arabidopsis* (Gardiner et al., 2003a).

### The Orientation of Cellulose Microfibrils within the Primary Wall Influences the Direction of Cell Expansion

In cells that enlarge more or less uniformly in all directions, the microfibrils are laid down in a random array (multidirectionally), forming an irregular network. Such cells are found in the pith of stems, in storage tissues, and in tissue cultures. In contrast, in many elongating cells, the microfibrils of the lateral, or side, walls are deposited at approximately right angles (transverse) to the axis of elongation. As the wall increases in surface area the orientation of the outer microfibrils becomes more nearly longitudinal, or parallel, to the long axis of the cell, as if passively reoriented by cell expansion (the **multinet growth hypothesis**) (Roelofsen, 1959; Preston, 1982). Longitudinal alignment of microfibrils encourages expansion primarily in a lateral direction (Abe et al., 1995b).

The structure of primary cell walls is not always so simple as that of cells encompassed by the multinet growth hypothesis. In many cells the orientation of cellulose microfibril deposition changes rhythmically, resulting in a wall with **helicoidal structure**, which consists of cellulose microfibrils arranged in one microfibril-thick lamellae. The cellulose microfibrils within each lamella lie more or less parallel to each other in one plane and form helices around the cell. Between successive lamellae the angle of inclination is rotated with respect to that of the previous lamella (Satiat-Jeunemaitre et al., 1992; Vian et al., 1993; Wolters-Arts et al., 1993; Emons, 1994; Wymer and Lloyd, 1996).

Also described as **polylamellate**, helicoidal-like wall texture is found in various primary and secondary walls.

The most conspicuous helicoidal type of wall texture is found in secondary walls (Figs. 4.2 and 4.17; Roland et al., 1989; Emons and Mulder, 1998; Reis and Vian, 2004). Polylamellate or helicoidal primary cell walls have been recorded for parenchyma (Deshpande, 1976b; Satiat-Jeunemaitre et al., 1992), collenchyma (Chafe, 1970; Deshpande, 1976a; Vian et al., 1993), and epidermal (Chafe and Wardrop, 1972; Satiat-Jeunemaitre et al., 1992) cells, as well as the nacreous wall layer of sieve tubes (Deshpande, 1976c). The primary walls of collenchyma cells usually are described as having a ***crossed polylamellate structure***, in which lamellae having a generally transverse orientation of microfibrils alternate with lamellae having a generally longitudinal, or vertical, orientation. It is likely that these orientations represent helices of shallow and steep pitch, respectively (Chafe and Wardrop, 1972). During cell expansion or elongation the helicoidal organization of the primary wall may become completely dispersed, changing progressively from the helicoidal to a random pattern. When extension of collenchyma cells ceases, the heli-

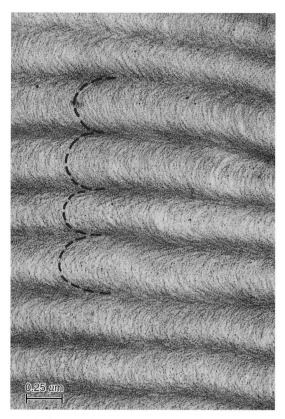

**FIGURE 4.17**

Helicoidal pattern in the secondary wall of a stone cell of pear (*Pyrus malus*). On oblique sections the lamellae appear as regular layers of arcs. Darker bands are regions in which microfibrils are oriented parallel to the section surface. (From Roland et al., 1987.)

coidal pattern of deposition typically continues to thicken the wall (Vian et al., 1993).

The presence of helicoidal cell walls, with their successive layers of cellulose microfibrils at varying angles, makes it difficult to perceive microtubules undergoing such rapid reorientations. That the angle of the microtubule array does shift to match the angle of each new layer of microfibrils has been demonstrated for the tracheids of the conifer *Abies sachalinensis* (Japanese black pine) (Abe et al., 1995a, b) and the fibers in the secondary xylem of the angiosperm *Aesculus hippocastanum* (horse chestnut) (Chaffey et al., 1999). Moreover, during a study on microinjected epidermal cells of *Pisum sativum*, the change in alignment from transverse to longitudinal of some of the rhodamine-labeled microtubules was found to be as rapid as 40 minutes, a reflection of the dynamic properties of the microtubules (Yuan et al., 1994). In addition the half-time of turnover of cortical microtubules in *Tradescantia* stamen hair cells, as determined by the FRAP (fluorescence redistribution after photobleaching) technique, was only about 60 seconds (Hush et al., 1994).

### When Considering the Mechanism of Wall Growth, It Is Necessary to Distinguish between Growth in Surface (Wall Expansion) and Growth in Thickness

Growth in thickness is particularly obvious in secondary walls, but it is common also in primary walls. As primary walls of growing cells expand, they typically maintain their thickness. According to the classical concept, increase in thickness of the wall occurs by two methods of deposition of wall material, apposition and intussusception. In **apposition**, the building units are placed one on top of the other; in **intussusception**, the units of new materials are inserted into the existing structure. Intussusception is probably the rule when lignin or cutin are incorporated into the wall. Both xylan and lignin have been shown to penetrate the differentiating secondary walls of *Fagus crentata* fibers simultaneously, accumulating on or around the recently deposited microfibrils (Awano et al., 2002). With regard to cellulose microfibrils, intussusception would result in an interweaving of the fibrils. In some walls microfibrils appear to be interwoven, but this is probably due to compression of the lamellae during cellulose deposition.

## ❚ EXPANSION OF THE PRIMARY CELL WALL

Expansion, or extension, of the cell wall is a complex process that requires respiration, polysaccharide and protein synthesis, wall stress relaxation (loosening of wall structure), and turgor pressure (McQueen-Mason, 1995; Cosgrove, 1997, 1998, 1999; Darley et al., 2001). Wall stress relaxation is crucial because it is the means

by which the cell lowers its water potential, leading to the uptake of water by the protoplast and turgor-driven extension of the wall. The rate at which an individual cell will expand is controlled by (1) the amount of turgor pressure inside the cell pushing against the cell wall and (2) the extensibility of the wall. **Extensibility**, a physical property of the wall, refers to the ability of the wall to expand or extend permanently when a force is applied to it.* The walls of growing cells exhibit a steady, long-term extension referred to as **creep** (Shieh and Cosgrove, 1998).

During growth the primary wall must yield enough to allow an appropriate extent of expansion, while at the same time remaining strong enough to constrain the protoplast. A number of factors are capable of influencing the extensibility of the wall. Among those factors are the plant hormones (Shibaoka, 1991; Zandomeni and Schopfer, 1993). Although hormones can affect the extensibility of the cell wall, they have little if any direct influence on the turgor pressure. Auxin and gibberellins increase the extensibility of cell walls, whereas abscisic acid and ethylene decrease their extensibility. Some hormones influence the arrangement of the cortical microtubules. Gibberellins, for example, promote a transverse arrangement, resulting in greater elongation.

The mechanisms by which hormones alter the extensibility of cell walls are not well understood. The most coherent explanation for the effect of a plant hormone on cell wall extensibility is the **acid-growth hypothesis** (Brett and Waldron, 1990; Kutschera, 1991), whereby auxin activates a proton-pumping ATPase in the plasma membrane. Protons are pumped from the cytosol into the cell wall. The resulting drop in pH is thought to cause a loosening of the cell wall structure, permitting turgor-driven extension of the wall polymer network. An alternative hypothesis is that auxin activates the expression of specific genes that influence the delivery of new wall materials in such a way as to affect cell wall extensibility (Takahashi et al., 1995; Abel and Theologis, 1996). There is little experimental evidence in support of this second hypothesis. By contrast, there is no doubt that growing cell walls enlarge more rapidly at acidic pH (below 5.5) than at neutral pH.

A novel class of wall proteins called **expansins** have been found to be the principal protein mediators of acid growth (Cosgrove, 1998, 1999, 2000, 2001; Shieh and Cosgrove, 1998; Li et al., 2002). Expansins apparently cause wall creep by loosening noncovalent associations between wall polysaccharides. Given the first model of

---

* Heyn (1931, 1940) defined the term "extensibility" simply as the ability of the wall to undergo changes in length, and distinguished between plastic extensibility and elastic extensibility. Plastic extensibility (plasticity) is the ability of the wall to extend irreversibly; elastic extensibility (elasticity) denotes the capacity for reversible enlargement (Kutschera, 1996).

primary wall architecture described above, a logical target would be the interface between cellulose and one or more hemicelluloses. Beyond its role of cell wall loosening in growing tissues, expansin has been implicated in leaf initiation (Fleming et al., 1997, 1999; Reinhardt et al., 1998), leaf abscission (Cho and Cosgrove, 2000), fruit ripening (Rose and Bennett, 1999; Catalá et al., 2000; Rose et al., 2000; Brummell and Harpster, 2001), and growth of both pollen tubes (Cosgrove et al., 1997; Cosgrove, 1998) and cotton fibers (Shimizu et al., 1997).

Cosgrove (1999) has proposed distinguishing between primary and secondary wall-loosening agents. He defines ***primary wall-loosening agents*** as those substances and processes that are competent and sufficient to induce extension of walls in vitro. Expansins are prime examples. ***Secondary wall-loosening agents***, which do not possess such activity, are defined as those substances and processes that modify wall structure to enhance the action of primary agents. Plant endoglucanases, xyloglucan endotransglycoylases (XETs), and pectinases, as well as the secretion of specific wall polymers and the production of hydroxyl radicals, might function as secondary wall-loosening agents. XETs are of special interest because they can cut and rejoin xyloglucan chains, allowing the cell wall to expand without undermining its structure (Campbell and Braam, 1999; Bourquin et al., 2002).

## CESSATION OF WALL EXPANSION

Cessation of growth during cell maturation is generally irreversible and is attributable to a loss of wall extensibility (plasticity). Growth cessation is not due to a decline in turgor pressure but rather to a mechanical stiffening, or rigidication, of the cell wall (Kutschera, 1996). Several factors may contribute to the physical changes accompanying wall maturation. These include (1) a reduction in wall-loosening processes, (2) an increase in cross-linking of cell wall components, and (3) a change in wall composition, resulting in a more rigid structure or one less susceptible to wall loosening (Cosgrove, 1997).

Cell walls lose their capacity for acid-induced extension as they approach maturity (Van Volkenburgh et al., 1985; Cosgrove, 1989), a condition that cannot be restored by application of exogenous expansin (McQueen-Mason, 1995). Thus cessation of wall expansion is associated with both the loss of expansin expression and wall stiffening. A number of modifications may contribute to wall stiffening, including formation of tighter complexes between the hemicelluloses and cellulose, de-esterification of pectins, more extensive $Ca^{2+}$ cross-linkings of pectins, cross-linking of extensions, and lignification.

## INTERCELLULAR SPACES

A large volume of the plant body is occupied by a system of **intercellular spaces**, air spaces that are essential for the aeration of the internal tissues. Although the intercellular spaces are most characteristic of mature tissues, they also extend into meristematic tissues where the dividing cells are intensively respiring. Examples of tissues having large and well-interconnected intercellular spaces are found in foliage leaves and submerged organs of water plants (Chapter 7).

The most common intercellular spaces develop by separation of contiguous primary walls through the middle lamella (Fig. 4.18). The process typically starts at the junction of three or more cells and spreads to other wall parts. This type of intercellular space is called **schizogenous**, that is, arising by splitting, although it commonly is thought of as being initiated by enzymic removal of pectins. The middle lamella may or may not be directly involved in the initiation of the intercellular space. Wall separation may be preceded by the accumulation and subsequent degradation of an electron dense intra-wall material (Kollöffel and Linssen, 1984; Jeffree et al., 1986) or by a special "splitting layer," which is distinct from the pectic middle lamella (Roland, 1978). Cleavage of the splitting layer results in separation of the contiguous walls. Formation of the large schizogenous intercellular spaces in the mesophyll of leaves is directly related to cell morphogenesis. Local differences in wall extensibility, resulting from differential wall thickening, lead to the production of lobed cells and at the same time generate mechanical tensions that initiate the intercellular spaces (Jung and Wernicke, 1990; Apostolakos et al., 1991; Panteris et al., 1993).

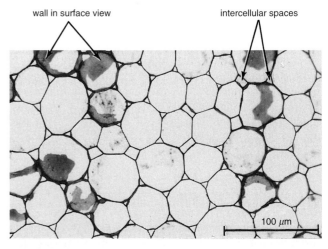

wall in surface view    intercellular spaces

100 µm

**FIGURE 4.18**

A thin-walled type of parenchyma, with regularly shaped cells and schizogenous intercellular spaces, from petiole of celery (*Apium*). (From Esau, 1977.)

Copious pectin formation may result in a pectic sol, which may partially or completely fill the smaller intercellular spaces. Some totally unexpected substances have been found in intercellular spaces, such as the threonine-rich hydroxyproline-rich glycoprotein found in filled intercellular spaces of maize root tips (Roberts, 1990). Several kinds of intercellular pectic protuberances may develop with the formation of intercellular spaces during tissue expansion (Potgieter and Van Wyk, 1992).

Some intercellular spaces result from breakdown of entire cells and are called **lysigenous** (arising by dissolution). Some roots have extensive lysigenous intercellular spaces. Intercellular spaces may also result from tearing or breaking of cells. Such spaces are called **rhexigenous**. Examples of rhexigenous intercellular spaces are the protoxylem lacunae that result from tearing of the first-formed primary xylem elements (protoxylem elements) during elongation of the plant part and the relatively large intercellular spaces found in the bark of some trees that arise during dilatation growth. Schizogeny, lysigeny, and/or rhexigeny may be combined in the formation of spaces.

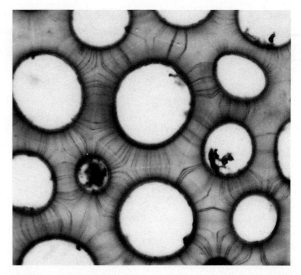

**FIGURE 4.19**

Light micrograph of plasmodesmata in the thick primary walls of persimmon (*Diospyros*) endosperm, the nutritive tissues within the seed. The plasmodesmata appear as fine lines extending from cell to cell across the walls. (×620.)

# ❚ PLASMODESMATA

As mentioned previously, the protoplasts of adjacent plant cells are interconnected by narrow strands of cytoplasm called **plasmodesmata** (singular: **plasmodesma**), which provide potential pathways for the passage of substances from cell to cell (van Bel and van Kesteren, 1999; Haywood et al., 2002). Although such structures have long been visible with the light microscope (Fig. 4.19)—they were first described by Tangl in 1879—it was not until they could be observed with an electron microscope that their nature as cytoplasmic strands was confirmed.

Plasmodesmata are the structural and functional analogs of the *gap junctions* found between animal cells (Robards and Lucas, 1990). At the gap junctions the plasma membranes of adjacent cells are associated in placques having narrow channels called "connexons," through which the protoplasts of the two cells communicate. The presence of a cell wall precludes direct contact between the plasma membranes of adjacent plant cells; consequently the plant body essentially is partitioned into two compartments, the symplast (or symplasm) and the apoplast (or apoplasm) (Münch, 1930). The **symplast** is composed of the plasma membrane-bound protoplasts and their interconnections, the plasmodesmata; the **apoplast** consists of the cell wall continuum and intercellular spaces. Accordingly the movement of substances from cell to cell by means of plasmodesmata is called **symplastic transport** (symplasmic transport), and the movement of substances in the cell wall continuum is called **apoplastic transport** (apoplasmic transport).

### Plasmodesmata May Be Classified as Primary or Secondary According to Their Origin

Many plasmodesmata are formed during cytokinesis as strands of tubular endoplasmic reticulum become entrapped within the developing cell plate (Fig. 4.20). The plasmodesmata formed during cytokinesis are called **primary plasmodesmata**. Plasmodesmata can also be formed de novo across existing cell walls. These postcytokinetically formed plasmodesmata are referred to as **secondary plasmodesmata**, and their formation is essential to establish communication between ontogenetically unrelated cells (Ding, B., and Lucas, 1996).

The formation of secondary plasmodesmata occurs naturally between neighboring cells not derived from the same cell lineage or precursor cell. According to one proposed mechanism, secondary plasmodesmata development involves the activity of localized wall degrading enzymes—pectinases, hemicellulases, and possibly cellulases—that allow cytoplasmic strands to penetrate the otherwise intact wall. The control of these enzymes presumably is mediated by the plasma membrane (Jones, 1976). Studies on cultures of regenerating protoplasts (Monzer, 1991; Ehlers and Kollmann, 1996) and graft interfaces (Kollmann and Glockmann, 1991) indicate,

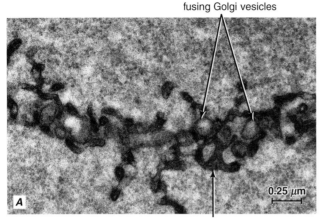

fusing Golgi vesicles

0.25 μm

A

endoplasmic reticulum

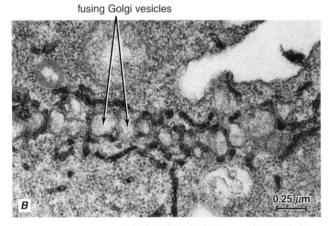

fusing Golgi vesicles

0.25 μm

B

endoplasmic reticulum          desmotubule

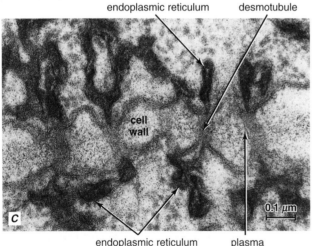

cell wall

0.1 μm

C

endoplasmic reticulum          plasma membrane

**FIGURE 4.20**

Progressive stages of cell plate formation in root cells of lettuce (*Lactuca sativa*), showing association of the endoplasmic reticulum with the developing cell plate and the origin of plasmodesmata. **A**, a relatively early stage of cell plate formation, with numerous small, fusing Golgi vesicles and loosely arranged elements of tubular (smooth) endoplasmic reticulum. **B**, an advanced stage of cell plate formation, revealing a persistent close relationship between the endoplasmic reticulum and fusing vesicles. Strands of tubular endoplasmic reticulum become trapped during cell plate consolidation. **C**, mature plasmodesmata, which consist of a plasma membrane-lined channel and a tubule, the desmotubule, of the endoplasmic reticulum. (From Hepler, 1982.)

associated endoplasmic reticulum of the two cells fuse, forming a continuous plasmodesma. A failure in coordination between neighboring cells may result in formation of a half-plasmodesma. Secondary plasmodesmata typically are branched, and many are characterized by the presence of a **median cavity** in the region of the middle lamella (Fig. 4.21).

Primary plasmodesmata may also undergo branching. A mechanism by which such branching may occur has been demonstrated by Ehlers and Kollmann (1996) (Fig. 4.22). Briefly, during normal wall thickening the primary plasmodesma, with its included tubule of endoplasmic reticulum, must elongate, requiring the addition of new parts to the original plasmodesmatal structure of the cell plate. If the original unbranched tubule of endoplasmic reticulum is connected to branched endoplasmic reticulum of the cytoplasm, enclosure of the branched endoplasmic reticulum by new wall material will lead to the formation of branched plasmodesmata.

Primary plasmodesmata may also be modified into what appear as highly branched plasmodesmata through the lateral fusion of neighboring plasmodesmata in the region of the middle lamella. Notable examples of such plasmodesmata are found in developing leaves (Ding, B., et al., 1992a, 1993; Itaya et al., 1998; Oparka et al., 1999; Pickard and Beachy, 1999). Apparently some such "branched plasmodesmata" are further modified by the de novo formation of additional endoplasmic reticulum-containing strands across the cell wall. It has been suggested that plasmodesmatal aggregates such as these be referred to as "*complex secondary plasmodesmata*" (Ding, B., 1998; Ding, B., et al., 1999).

Both primary and secondary plasmodesmata may be unbranched or branched, and at times it is difficult to determine their origin, that is, whether they are primary

however, that continuous secondary plasmodesmata arise from the fusion of opposite secondary half-plasmodesmata formed simultaneously by neighboring cells. At such sites a segment of endoplasmic reticulum becomes attached to the plasma membrane on both sides of the extremely narrow cell wall. With the removal of wall material at the site, the plasma membrane and

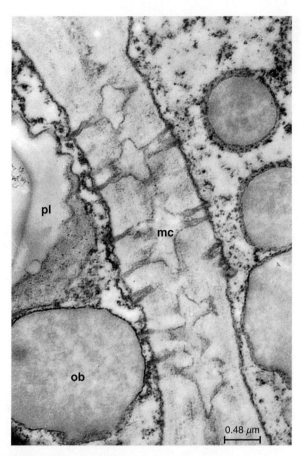

**FIGURE 4.21**

Branched plasmodesmata in radial walls of ray parenchyma cells in secondary phloem of white pine (*Pinus strobus*). Note the median cavities (mc) in the region of the middle lamella. Other details: ob, oil body; pl, plastid. (From Murmanis and Evert, 1967, Fig. 10. © 1967, Springer-Verlag.)

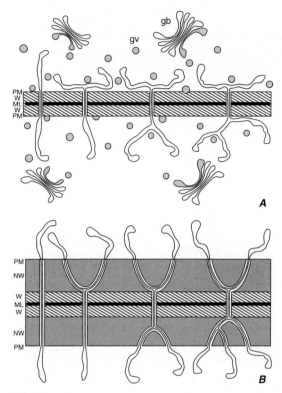

**FIGURE 4.22**

Branching of primary plasmodesmata. Initially unbranched (**A**), branches develop from the enclosure of branched ER tubules within newly deposited wall material (**B**). Details: gb, Golgi body; gv, Golgi vesicle; ML, middle lamella; NW, subsequently formed new wall layers; PM, plasma membrane; W, first-formed wall layers. (From Ehlers and Kollmann, 1996, Fig. 35a,b. © 1996, Springer-Verlag.)

or secondary. Under such circumstances they may simply be designated as "branched" or "unbranched" (or "single"). A detailed review on the structure, origin, and functioning of primary and secondary plasmodesmata is provided by Ehlers and Kollmann (2001).

### Plasmodesmata Contain Two Types of Membranes: Plasma Membrane and Desmotubule

A plasmodesma is a plasma membrane-lined channel typically traversed by a tubular strand of tightly constricted endoplasmic reticulum called a **desmotubule** (Figs. 4.23 and 4.24). In most plasmodesmata the desmotubule does not resemble the adjoining endoplasmic reticulum. It is much smaller in diameter and contains a central, rod-like structure. Considerable controversy has centered on the interpretation of the central rod (Esau and Thorsch, 1985). Most investigators believe it represents the merger of the inner leaflets, or inner por-

tions, of the bilayers of the endoplasmic reticulum forming the desmotubule. If this interpretation is correct, the desmotubule lacks a lumen, or opening, and the major pathway through which substances move from cell to cell via plasmodesmata is the region between the desmotubule and the plasma membrane. This region, called the **cytoplasmic sleeve**, is subdivided into 2.5-nanometer diameter microchannels by globular particles embedded in both the plasma membrane and desmotubule and interconnected by spoke-like extensions (Tilney et al., 1990; Ding, B., et al., 1992b; Botha et al., 1993). Some plasmodesmata appear conspicuously narrower at their ends, or orifices, forming so-called neck constrictions. Neck constrictions can result, however, from the deposition of wound callose induced during tissue manipulation or fixation (Radford et al., 1998). Most information on the architecture of plasmodesmata comes from studies of unbranched primary plasmodesmata. Little is known about the substructure of secondary plasmodesmata.

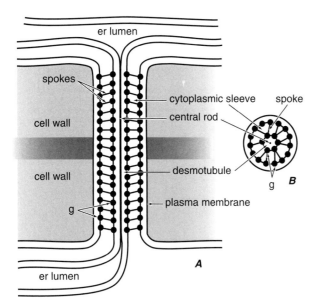

**FIGURE 4.23**

Diagrammatic representation of a primary plasmodesma in longitudinal (**A**) and transverse (**B**) views. Globular integral membrane proteins (g) are located in the inner and outer leaflets of the plasma membrane and desmotubule, respectively, and are interconnected by spoke-like extensions. Note that the cytoplasmic sleeve is divided into a number of microchannels.

Desmotubules do not always appear completely constricted. In some plasmodesmata, such as those between mesophyll cells and between mesophyll cells and bundle sheath cells in maize (Evert et al., 1977) and sugarcane (Robinson-Beers and Evert, 1991) leaves, the desmotubules appear constricted only at the neck constrictions; between the neck constrictions they appear as open tubules (Fig. 4.25). The desmotubules in the trichome cell plasmodesmata of the *Nicotiana clevelandii* leaf appear open for their entire length (Waigmann et al., 1997).

Although an open desmotubule has at times been proposed as a pathway for transport (Gamalei et al., 1994), there is no direct evidence in support of this. By contrast, it has been shown that osmotic treatment that enhances intercellular transport of sucrose via plasmodesmata in pea root tips results from widening of the cytoplasmic sleeves, not from changes in the diameter of the desmotubules (Schulz, A., 1995). In addition, in *Nicotiana tabacum* leaves, green fluorescent protein targeted to the endoplasmic reticulum was confined to single cells, indicating that the desmotubule is not a functional pathway for green fluorescent protein trafficking through either single or branched plasmodesmata (Oparka et al., 1999). Transport of lipid molecules can occur, however, via the lipid bilayers of the desmotubule (Grabski et al., 1993).

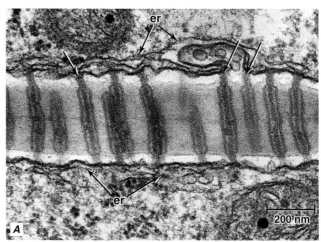

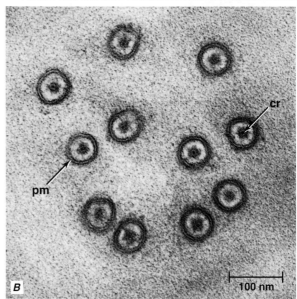

**FIGURE 4.24**

Plasmodesmata in cell walls of sugarcane (*Saccharum*) leaf in longitudinal (**A**) and transverse (**B**) views. Note connections (arrows) between endoplasmic reticulum (er) and desmotubules and the subtle neck constrictions in **A**. In **B**, the inner leaflet of the desmotubule appears as a centrally located dot (cr, the central rod). The cytoplasmic sleeve appears somewhat mottled, in part, because of the presence of electron-opaque, spoke-like structures that extend from the outer leaflet of the desmotubule toward the inner leaflet of the plasma membrane (pm) and alternate with electron-lucent regions. (From Robinson-Beers and Evert, 1991, Figs. 14 and 15. © 1991, Springer-Verlag.)

### Plasmodesmata Enable Cells to Communicate

The successful existence of multicellular organisms depends on the ability of individual cells to communicate with one another. Although cellular differentiation depends on the control of gene expression, the fate of

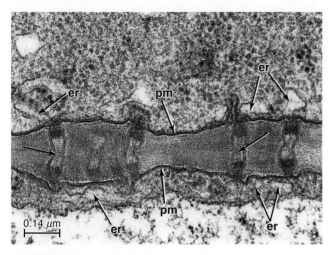

**FIGURE 4.25**

Plasmodesmata in the common wall between two mesophyll cells of a maize (*Zea mays*) leaf. Note the open appearance of the desmotubules (arrows). Details: er, endoplasmic reticulum; pm, plasma membrane. (From Evert et al., 1977, Fig. 8. © 1977, Springer-Verlag.)

a plant cell—that is, what kind of cell it will become—is determined by its final position in the developing organ rather than on lineage relationships. Therefore one aspect of plant cell interaction is the communication, or signaling, of positional information from one cell to another.

Early evidence for transport between cells via plasmodesmata came from studies utilizing fluorescent dyes (Goodwin, 1983; Erwee and Goodwin, 1985; Tucker and Spanswick, 1985; Terry and Robards, 1987; Tucker et al., 1989) and electrical currents (Spanswick, 1976; Drake, 1979; Overall and Gunning, 1982). The passage of pulses of electric current from one cell to another can be monitored by means of receiver electrodes placed in neighboring cells. The magnitude of the electrical force varies with the frequency, or density, of plasmodesmata and with the number of cells and length of cells between the injection and receiver electrodes, indicating that plasmodesmata can serve as a path for electrical signaling between plant cells.

In the dye-coupling experiments the dyes, which do not easily cross the plasma membrane, can be observed moving from the injected cells into the neighboring cells and beyond. Results of such studies early established the size of molecules free to flow by passive diffusion between such cells at approximately 1 kDa (1000 daltons; a dalton is the weight of a single hydrogen atom). Such **size exclusion limits** allow sugars, amino acids, phytohormones, and nutrients to move freely across the plasmodesmata. More recent studies indicate that plasmodesmata in different cell types can have dif-

ferent basal size exclusion limits. For example, fluorescent dextrans of approximately 7 kDa can diffuse between leaf trichome cells of *Nicotiana clevelandii* (Waigmann and Zambryski, 1995), and dextrans of at least 10 kDa move through plasmodesmata connecting the sieve elements and companion cells in the stem phloem of *Vicia faba* (Kempers and van Bel, 1997). Moreover plasmodesmata may fluctuate their size exclusion limits as a function of growth conditions (Crawford and Zambryski, 2001).

It now is known that plasmodesmata also have the capacity to mediate the cell-to-cell transport of macromolecules, including proteins and nucleic acids (Lucas et al., 1993; Mezitt and Lucas, 1996; Ding, 1997; Lucas, 1999; Haywood et al., 2002). Based on these findings, Lucas and co-workers (Ding et al., 1993; Lucas et al., 1993) have hypothesized that plants function as **supracellular organisms**, rather than multicellular organisms. Accordingly the dynamics of the plant body, including cell differentiation, tissue formation, organogenesis, and specialized physiological functions, are subject to plasmodesmatal regulation. The plasmodesmata presumably accomplish such regulatory roles by trafficking informational molecules that "orchestrate" both metabolic activity and gene expression.

Initial insight into the dynamic function of plasmodesmata came from studies on plant viruses, which have long been known to move relatively short distances from cell to cell via plasmodesmata (Fig. 4.26; Wolf et al., 1989; Robards and Lucas, 1990; Citovsky, 1993; Leisner and Turgeon, 1993). Such studies revealed that plant viruses encode for nonstructural proteins, called **movement proteins**, that function in the cell-to-cell spread of infectious material. When expressed in transgenic plants, the movement proteins are targeted to plasmodesmata, bringing about an increase in their size exclusion limit. Considerable evidence has accumulated implicating both the endoplasmic reticulum and cytoskeletal elements (microtubules and actin filaments) in the targeting of movement proteins, and presumably viral nucleic acid-protein complexes, to the plasmodesmata (Reichel et al., 1999).

Evidence for the involvement of plasmodesmata in the trafficking of endogenous proteins between plant cells comes from studies on phloem transport. More than 200 proteins, ranging in molecular weight from 10 to 200 kDa have been found in phloem exudate (sieve-tube sap) (Fisher et al., 1992; Nakamura et al., 1993; Sakuth et al., 1993; Ishiwatari et al., 1995; Schobert et al., 1995). Inasmuch as mature sieve elements lack nuclei and ribosomes (Evert, 1990), most, if not all, of the proteins are likely synthesized in the companion cells and then transported into their associated sieve elements via the pore-plasmodesmata connections in their common walls (Chapter 13). Proteins found in sieve-tube sap have been shown to have the capacity to increase the size exclusion

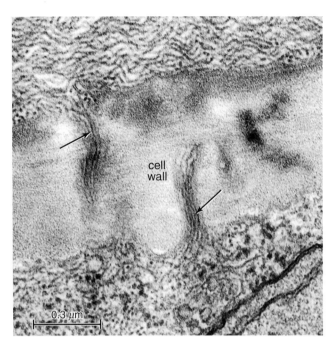

**FIGURE 4.26**

Beet yellows virus particles (arrows) in plasmodesmata, moving from a sieve-tube element of the phloem (above) to its sister cell, a companion cell (below) in sugar beet (*Beta vulgaris*).

limit of mesophyll plasmodesmata and to traffic from cell to cell (Balachandran et al., 1997; Ishiwatari et al., 1998). In pumpkin (*Cucurbita maxima*) all of the sieve-tube sap proteins, regardless of size, appear to induce a similar increase in plasmodesmatal size inclusion limit, approximately 25 kDa (Balachandran et al., 1997). As some of these proteins range up to 200 kDa in size, it is likely that unfolding of the larger proteins is required for their transport through the plasmodesmata. Chaperones have been found in the sieve-tube sap of several plant species (Schobert et al., 1995, 1998), and they are believed to mediate the unfolding/refolding process of protein transport between companion cells and sieve elements (Crawford and Zambryski, 1999; Lee et al., 2000).

The concept that plasmodesmata play a role in development is supported by molecular, genetic, and micro-injection studies of the plant transcription factor called KNOTTED1 (KN1). In maize, KN1 is involved in maintaining the shoot apical meristem in an undifferentiated state (Sinha et al., 1993). During development the RNA encoding KN1 is found in all cell layers of the meristem except the outermost (L1) layer. The KN1 protein is present, however, in all layers, including the L1, indicating that the KN1 protein produced in the inner layers traffics into the L1 layer (Jackson et al., 1994). Microin-

jection of KN1 protein into mesophyll cells of either maize or tobacco show that the KN1 protein can indeed traffic from cell to cell and increase the plasmodesmatal size exclusion limit from 1 kDa to greater than 40 kDa (Lucas et al., 1995). Two more recent studies, one using grafting experiments to investigate the autonomy of a dominant leaf mutant, called *Mouse ears (Me)*, of tomato (Kim, M., et al., 2001) and a second on the role of the *SHORT-ROOT (SHR)* gene in patterning of the *Arabidopsis* root (Nakajima et al., 2001), provide further evidence for the role of plasmodesmata in development.

At present little information is available concerning the mechanism by which the plasmodesmata dilate or undergo transient changes in conductivity (**gating**) (Schulz, A., 1999). Several factors have been shown to affect the size exclusion limit, including changes in cytoplasmic levels of $Ca^{2+}$ (Holdaway-Clarke et al., 2000) and ATPase (Cleland et al., 1994). Both actin and myosin have been localized at or in plasmodesmata (White et al., 1994; Radford and White, 1998; Overall et al., 2000; Baluška et al., 2001), and both have been implicated in the control of plasmodesmatal permeability. With regard to actin, experimental evidence has been presented for the involvement of actin filaments in the control of plasmodesmatal permeability in tobacco mesophyll (Ding et al., 1996). Clearly, a close relationship exists between plasmodesmata and the cytoskeleton (Aaziz et al., 2001).

Regulation of plasmodesmatal permeability is thought to occur at the neck region (White et al., 1994; Blackman et al., 1999). Some investigators have proposed that sphincter-like structures at the neck region regulate plasmodesmatal transport in some species (Olesen, 1979; Olesen and Robards, 1990; Badelt et al., 1994; Overall and Blackman, 1996). Functional parallels have been noted between plasmodesmatal transport of macromolecules and the transport of proteins and nucleic acids via the nuclear pore complexes permeating the nuclear envelope (Lee et al., 2000).

### The Symplast Undergoes Reorganization throughout the Course of Plant Growth and Development

Studies of plant embryos indicate that initially all of the cells of the young plant body are interconnected by plasmodesmata and integrated into a single symplast (Schulz and Jensen, 1968; Mansfield and Briarty, 1991; Kim et al., 2002). As the plant resumes growth and develops further, individual cells or groups of cells become more or less isolated symplastically so that the plant becomes divided into a mosaic of **symplastic domains** (Erwee and Goodwin, 1985). The establishment of symplastic domains is generally considered essential for subsets of cells to pursue specific developmental pathways and to function as distinct compartments within the plant body (Fisher and Oparka, 1996;

McLean et al., 1997; Kragler et al., 1998; Nelson and van Bel, 1998; Ding et al., 1999).

Communication and transport between domains are closely linked to the frequency, distribution, and function of plasmodesmata. Although plasmodesmatal frequency has often been used as an indicator of symplastic continuity at different interfaces, the use of such data is speculative because it is presumed that all of the plasmodesmata are capable of intercellular transport. Changes in symplastic continuity can occur by the isolation of cells initially connected symplastically with other cells, as in the case of guard cells (Palevitz and Hepler, 1985) and root hairs (Duckett et al., 1994), or by the establishment of new, secondary plasmodesmata, as during leaf maturation (Turgeon, 1996; Volk et al., 1996) and the union of the vascular tissues of lateral and parent roots (Oparka et al., 1995).

# ▊ REFERENCES

AAZIZ, R., S. DINANT, and B. L. EPEL. 2001. Plasmodesmata and plant cytoskeleton. *Trends Plant Sci.* 6, 326–330.

ABE, H., R. FUNADA, H. IMAIZUMI, J. OHTANI, and K. FUKAZAWA. 1995a. Dynamic changes in the arrangement of cortical microtubules in conifer tracheids during differentiation. *Planta* 197, 418–421.

ABE, H., R. FUNADA, J. OHTANI, and K. FUKAZAWA. 1995b. Changes in the arrangement of microtubules and microfibrils in differentiating conifer tracheids during the expansion of cells. *Ann. Bot.* 75, 305–310.

ABE, H., R. FUNADA, J. OHTANI, and K. FUKAZAWA. 1997. Changes in the arrangement of cellulose microfibrils associated with the cessation of cell expansion in tracheids. *Trees* 11, 328–332.

ABEL, S., and A. THEOLOGIS. 1996. Early genes and auxin action. *Plant Physiol.* 111, 9–17.

ALDINGTON, S., and S. C. FRY. 1993. Oligosaccharins. *Adv. Bot. Res.* 19, 1–101.

APOSTOLAKOS, P., B. GALATIS, and E. PANTERIS. 1991. Microtubules in cell morphogenesis and intercellular space formation in *Zea mays* leaf mesophyll and *Pilea cadierei* epithem. *J. Plant Physiol.* 137, 591–601.

ASSAAD, F. F. 2001. Plant cytokinesis. Exploring the links. *Plant Physiol.* 126, 509–516.

AWANO, T., K. TAKABE, M. FUJITA, and G. DANIEL. 2000. Deposition of glucuronoxylans on the secondary cell wall of Japanese beech as observed by immuno-scanning electron microscopy. *Protoplasma* 212, 72–79.

AWANO, T., K. TAKABE, and M. FUJITA. 2002. Xylan deposition on secondary wall of *Fagus crenata* fiber. *Protoplasma* 219, 106–115.

BACIC, A., P. J. HARRIS, and B. A. STONE. 1988. Structure and function of plant cell walls. In: *The Biochemistry of Plants*, vol.

14, *Carbohydrates*, pp. 297–371, J. Preiss, ed. Academic Press, New York.

BADELT, K., R. G. WHITE, R. L. OVERALL, and M. VESK. 1994. Ultrastructural specializations of the cell wall sleeve around plasmodesmata. *Am. J. Bot.* 81, 1422–1427.

BALACHANDRAN, S., Y. XIANG, C. SCHOBERT, G. A. THOMPSON, and W. J. LUCAS. 1997. Phloem sap proteins from *Cucurbita maxima* and *Ricinus communis* have the capacity to traffic cell to cell through plasmodesmata. *Proc. Natl. Acad. Sci. USA* 94, 14150–14155.

BALUŠKA, F., P. W. BARLOW, and D. VOLKMANN. 2000. Actin and myosin in developing root apex cells. In: *Actin: A Dynamic Framework for Multiple Plant Cell Functions*, pp. 457–476, C. J. Staiger, F. Baluška, D. Volkmann, and P. W. Barlow, eds. Kluwer Academic, Dordrecht.

BALUŠKA, F., F. CVRČKOVÁ, J. KENDRICK-JONES, and D. VOLKMANN. 2001. Sink plasmodesmata as gateways for phloem unloading. Myosin VIII and calreticulin as molecular determinants of sink strength? *Plant Physiol.* 126, 39–46.

BAO, W., D. M. O'MALLEY, and R. R. SEDEROFF. 1992. Wood contains a cell-wall structural protein. *Proc. Natl. Acad. Sci. USA* 89, 6604–6608.

BARON-EPEL, O., P. K. GHARYAL, and M. SCHINDLER. 1988. Pectins as mediators of wall porosity in soybean cells. *Planta* 175, 389–395.

BASKIN, T. I. 2001. On the alignment of cellulose microfibrils by cortical microtubules: A review and a model. *Protoplasma* 215, 150–171.

BENHAMOU, N. 1992. Ultrastructural detection of β-1,3-glucans in tobacco root tissues infected by *Phytophthora parasitica* var. *nicotianae* using a gold-complexed tobacco β-1,3-glucanase. *Physiol. Mol. Plant Pathol.* 41, 351–370.

BERNARDS, M. A. 2002. Demystifying suberin. *Can. J. Bot.* 80, 227–240.

BERNARDS, M. A., and N. G. LEWIS. 1998. The macromolecular aromatic domain in suberized tissue: A hanging paradigm. *Phytochemistry* 47, 915–933.

BLACKMAN, L. M., J. D. I. HARPER, and R. L. OVERALL. 1999. Localization of a centrin-like protein to higher plant plasmodesmata. *Eur. J. Cell Biol.* 78, 297–304.

BLEVINS, D. G., and K. M. LUKASZEWSKI. 1998. Boron in plant structure and function. *Annu. Rev. Plant Physiol. Plant Mol. Biol.* 49, 481–500.

BOLWELL, G. P. 1993. Dynamic aspects of the plant extracellular matrix. *Int. Rev. Cytol.* 146, 261–324.

BOTHA, C. E. J., B. J. HARTLEY, and R. H. M. CROSS. 1993. The ultrastructure and computer-enhanced digital image analysis of plasmodesmata at the Kranz mesophyll-bundle sheath interface of *Themeda triandra* var. *imberbis* (Retz) A. Camus in conventionally-fixed leaf blades. *Ann. Bot.* 72, 255–261.

BOULIGAND, Y. 1976. Les analogues biologiques des cristaux liquides. *La Recherche* 7, 474–476.

BOURQUIN, V., N. NISHIKUBO, H. ABE, H. BRUMER, S. DENMAN, M. EKLUND, M. CHRISTIERNIN, T. T. TEERI, B. SUNDBERG, and E. J. MELLEROWICZ. 2002. Xyloglucan endotransglycosylases have a function during the formation of secondary cell walls of vascular tissues. *Plant Cell* 14, 3073–3088.

BRAAM, J. 1999. If walls could talk. *Curr. Opin. Plant Biol.* 2, 521–524.

BRADLEY, D. J., P. KJELLBOM, and C. J. LAMB. 1992. Elicitor- and wound-induced oxidative cross-linking of a proline-rich plant cell wall protein: A novel, rapid defense response. *Cell* 70, 21–30.

BRETT, C., and K. WALDRON. 1990. *Physiology and Biochemistry of Plant Cell Walls.* Unwin Hyman, London.

BRUMMELL, D. A., and M. H. HARPSTER. 2001. Cell wall metabolism in fruit softening and quality and its manipulation in transgenic plants. *Plant Mol. Biol.* 47, 311–340.

BUCKERIDGE, M. S., C. E. VERGARA, and N. C. CARPITA. 1999. The mechanism of synthesis of a mixed-linkage (1→3), (1→4)β-D-glucan in maize. Evidence for multiple sites of glucosyl transfer in the synthase complex. *Plant Physiol.* 120, 1105–1116.

CALVIN, C. L. 1967. The vascular tissues and development of sclerenchyma in the stem of the mistletoe, *Phoradendron flavescens. Bot. Gaz.* 128, 35–59.

CAMPBELL, P., and J. BRAAM. 1999. Xyloglucan endotransglycosylases: Diversity of genes, enzymes and potential wall-modifying functions. *Trends Plant Sci.* 4, 361–366.

CARPITA, N. C. 1982. Limiting diameters of pores and the surface structure of plant cell walls. *Science* 218, 813–814.

CARPITA, N. C. 1996. Structure and biogenesis of the cell walls of grasses. *Annu. Rev. Plant Physiol. Plant Mol. Biol.* 47, 445–476.

CARPITA, N. C., and D. M. GIBEAUT. 1993. Structural models of primary cell walls in flowering plants: Consistency of molecular structure with the physical properties of the walls during growth. *Plant J.* 3, 1–30.

CARPITA, N., and M. MCCANN. 2000. The cell wall. In: *Biochemistry and Molecular Biology of Plants*, pp. 52–108, B. Buchanan, W. Gruissem, and R. Jones, eds. American Society of Plant Physiologists, Rockville, MD.

CARPITA, N., D. SABULARSE, D. MONTEZINOS, and D. P. DELMER. 1979. Determination of the pore size of cell walls of living plant cells. *Science* 205, 1144–1147.

CASSAB, G. I. 1998. Plant cell wall proteins. *Annu. Rev. Plant Physiol. Plant Mol. Biol.* 49, 281–309.

CASSAB, G. I., and J. E. VARNER. 1987. Immunocytolocalization of extensin in developing soybean seed coats by immunogold-silver staining and by tissue printing on nitrocellulose paper. *J. Cell Biol.* 105, 2581–2588.

CASTRO, M. A. 1991. Ultrastructure of vestures on the vessel wall in some species of *Prosopis* (Leguminosae-Mimosoideae). *IAWA Bull.* n.s. 12, 425–430.

CATALÁ, C., J. K. C. ROSE, and A. B. BENNETT. 2000. Auxin-regulated genes encoding cell wall-modifying proteins are expressed during early tomato fruit growth. *Plant Physiol.* 122, 527–534.

CHAFE, S. C. 1970. The fine structure of the collenchyma cell wall. *Planta* 90, 12–21.

CHAFE, S. C., and A. B. WARDROP. 1972. Fine structural observations on the epidermis. I. The epidermal cell wall. *Planta* 107, 269–278.

CHAFFEY, N., J. BARNETT, and P. BARLOW. 1999. A cytoskeletal basis for wood formation in angiosperm trees: The involvement of cortical microtubules. *Planta* 208, 19–30.

CHEN, C.-L. 1991. Lignins: Occurrence in woody tissues, isolation, reactions, and structure. In: *Wood Structure and Composition*, pp. 183–261, M. Lewin and I. S. Goldstein, eds. Dekker, New York.

CHO, H.-T., and D. J. COSGROVE. 2000. Altered expression of expansin modulates leaf growth and pedicel abscission in *Arabidopsis thaliana. Proc. Natl. Acad. Sci. USA* 97, 9783–9788.

CITOVSKY, V. 1993. Probing plasmodesmal transport with plant viruses. *Plant Physiol.* 102, 1071–1076.

CLELAND, R. E., T. FUJIWARA, and W. J. LUCAS. 1994. Plasmodesmal-mediated cell-to-cell transport in wheat roots is modulated by anaerobic stress. *Protoplasma* 178, 81–85.

CONNOLLY, J. H., and G. BERLYN. 1996. The plant extracellular matrix. *Can. J. Bot.* 74, 1545–1546.

COSGROVE, D. J. 1989. Characterization of long-term extension of isolated cell walls from growing cucumber hypocotyls. *Planta* 177, 121–130.

COSGROVE, D. J. 1997. Assembly and enlargement of the primary cell wall in plants. *Annu. Rev. Cell Dev. Biol.* 13, 171–201.

COSGROVE, D. J. 1998. Cell wall loosening by expansins. *Plant Physiol.* 118, 333–339.

COSGROVE, D. J. 1999. Enzymes and other agents that enhance cell wall extensibility. *Annu. Rev. Plant Physiol. Plant Mol. Biol.* 50, 391–417.

COSGROVE, D. J. 2000. New genes and new biological roles for expansins. *Curr. Opin. Plant Biol.* 3, 73–78.

COSGROVE, D. J. 2001. Wall structure and wall loosening. A look backwards and forwards. *Plant Physiol.* 125, 131–134.

COSGROVE, D. J., P. BEDINGER, and D. M. DURACHKO. 1997. Group I allergens of grass pollen as cell wall-loosening agents. *Proc. Natl. Acad. Sci. USA* 94, 6559–6564.

CRAWFORD, K. M., and P. C. ZAMBRYSKI. 1999. Phloem transport: Are you chaperoned? *Curr. Biol.* 9, R281–R285.

CRAWFORD, K. M., and P. C. ZAMBRYSKI. 2001. Non-targeted and targeted protein movement through plasmodesmata in leaves in different developmental and physiological states. *Plant Physiol.* 125, 1802–1812.

CUTLER, S. R., and D. W. EHRHARDT. 2002. Polarized cytokinesis in vacuolate cells of *Arabidopsis. Proc. Natl. Acad. Sci. USA* 99, 2812–2817.

CZANINSKI, Y., R. M. SACHOT, and A. M. CATESSON. 1993. Cytochemical localization of hydrogen peroxide in lignifying cell walls. *Ann. Bot.* 72, 547–550,

DAHIYA, P., and N. J. BREWIN. 2000. Immunogold localization of callose and other cell wall components in pea nodule transfer cells. *Protoplasma* 214, 210–218.

DARLEY, C. P., A. M. FORRESTER, and S. J. McQUEEN-MASON. 2001. The molecular basis of plant cell wall extension. *Plant Mol. Biol.* 47, 179–195.

DARVILL, A. G., and P. ALBERSHEIM. 1984. Phytoalexins and their elicitors—A defense against microbial infection in plants. *Annu. Rev. Plant Physiol.* 35, 243–275.

DELMER, D. P. 1999. Cellulose biosynthesis: Exciting times for a difficult field of study. *Annu. Rev. Plant Physiol. Plant Mol. Biol.* 50, 245–276.

DELMER, D. P., and Y. AMOR. 1995. Cellulose biosynthesis. *Plant Cell* 7, 987–1000.

DELMER, D. P., and B. A. STONE. 1988. Biosynthesis of plant cell walls. In: *The Biochemistry of Plants*, vol. 14, *Carbohydrates*, pp. 373–420, J. Preiss, ed. Academic Press, New York.

DESHPANDE, B. P. 1976a. Observations on the fine structure of plant cell walls. I. Use of permanganate staining. *Ann. Bot.* 40, 433–437.

DESHPANDE, B. P. 1976b. Observations on the fine structure of plant cell walls. II. The microfibrillar framework of the parenchymatous cell wall in *Cucurbita*. *Ann. Bot.* 40, 439–442.

DESHPANDE, B. P. 1976c. Observations on the fine structure of plant cell walls. III. The sieve tube wall of *Cucurbita*. *Ann. Bot.* 40, 443–446.

DHONUKSHE, P., A. M. LAXALT, J. GOEDHART, T. W. J. GADELLA, and T. MUNNIK. 2003. Phospholipase D activation correlates with microtubule reorganization in living plant cells. *Plant Cell* 15, 2666–2679.

DING, B. 1997. Cell-to-cell transport of macromolecules through plasmodesmata: A novel signalling pathway in plants. *Trends Cell Biol.* 7, 5–9.

DING, B. 1998. Intercellular protein trafficking through plasmodesmata. *Plant Mol. Biol.* 38, 279–310.

DING, B., and W. J. LUCAS. 1996. Secondary plasmodesmata: Biogenesis, special functions and evolution. In: *Membranes: Specialized Functions in Plants*, pp. 489–506, M. Smallwood, J. P. Knox, and D. J. Bowles, eds. BIOS Scientific, Oxford.

DING, B., J. S. HAUDENSHIELD, R. J. HULL, S. WOLF, R. N. BEACHY, and W. J. LUCAS. 1992a. Secondary plasmodesmata are specific sites of localization of the tobacco mosaic virus movement protein in transgenic tobacco plants. *Plant Cell* 4, 915–928.

DING, B., R. TURGEON, and M. V. PARTHASARATHY. 1992b. Substructure of freeze-substituted plasmodesmata. *Protoplasma* 169, 28–41.

DING, B., J. S. HAUDENSHIELD, L. WILLMITZER, and W. J. LUCAS. 1993. Correlation between arrested secondary plasmodesmal development and onset of accelerated leaf senescence in yeast acid invertase transgenic tobacco plants. *Plant J.* 4, 179–189.

DING, B., M.-O. KWON, and L. WARNBERG. 1996. Evidence that actin filaments are involved in controlling the permeability of plasmodesmata in tobacco mesophyll. *Plant J.* 10, 157–164.

DING, B., A. ITAYA, and Y.-M. WOO. 1999. Plasmodesmata and cell-to-cell communication in plants. *Int. Rev. Cytol.* 190, 251–316.

DING, L., and J.-K. ZHU. 1997. A role for arabinogalactan-proteins in root epidermal cell expansion. *Planta* 203, 289–294.

DIXIT, R., and R. J. CYR. 2002. Spatio-temporal relationship between nuclear-envelope breakdown and preprophase band disappearance in cultured tobacco cells. *Protoplasma* 219, 116–121.

DRAKE, G. 1979. Electrical coupling, potentials, and resistances in oat coleoptiles: Effects of azide and cyanide. *J. Exp. Bot.* 30, 719–725.

DUCKETT, C. M., K. J. OPARKA, D. A. M. PRIOR, L. DOLAN, and K. ROBERTS. 1994. Dye-coupling in the root epidermis of *Arabidopsis* is progressively reduced during development. *Development* 120, 3247–3255.

EHLERS, K., and R. KOLLMANN. 1996. Formation of branched plasmodesmata in regenerating *Solanum nigrum*-protoplasts. *Planta* 199, 126–138.

EHLERS, K., and R. KOLLMANN. 2001. Primary and secondary plasmodesmata: Structure, origin, and functioning. *Protoplasma* 216, 1–30.

EMONS, A. M. C. 1994. Winding threads around plant cells: A geometrical model for microfibril deposition. *Plant Cell Environ.* 17, 3–14.

EMONS, A. M. C., and B. M. MULDER. 1997. Plant cell wall architecture. *Comm. Modern Biol. Part C. Comm. Theor. Biol.* 4, 115–131.

EMONS, A. M. C. and B. M. MULDER. 1998. The making of the architecture of the plant cell wall: How cells exploit geometry. *Proc. Natl. Acad. Sci. USA* 95, 7215–7219.

EMONS, A. M. C., and B. M. MULDER. 2000. How the deposition of cellulose microfibrils builds cell wall architecture. *Trends Plant Sci.* 5, 35–40.

EMONS, A. M. C., and B. M. MULDER. 2001. Microfibrils build architecture. A geometrical model. In: *Molecular Breeding of Woody Plants*, pp. 111–119, N. Morohoshi and A. Komamine, eds. Elsevier Science B. V., Amsterdam.

EMONS, A. M. C., J. DERKSEN, and M. M. A. SASSEN. 1992. Do microtubules orient plant cell wall microfibrils? *Physiol. Plant.* 84, 486–493.

ENGELS, F. M., and H. G. JUNG. 1998. Alfalfa stem tissues: Cell-wall development and lignification. *Ann. Bot.* 82, 561–568.

ERWEE, M. G., and P. B. GOODWIN. 1985. Symplast domains in extrastelar tissues of *Egeria densa* Planch. *Planta* 163, 9–19.

ESAU, K. 1997. *Anatomy of Seed Plants*, 2nd ed. Wiley, New York.

ESAU, K., and J. THORSCH. 1985. Sieve plate pores and plasmodesmata, the communication channels of the symplast: Ultrastructural aspects and developmental relations. *Am. J. Bot.* 72, 1641–1653.

ESCHRICH, W., and H. B. CURRIER. 1964. Identification of callose by its diachrome and fluorochrome reactions. *Stain Technol.* 39, 303–307.

EVERT, R. F. 1990. Dicotyledons. In: *Sieve Elements: Comparative Structure, Induction and Development*, pp. 103–137, H.-D. Behnke and R. D. Sjolund, eds. Springer-Verlag, Berlin.

EVERT, R. F., W. ESCHRICH, and W. HEYSER. 1977. Distribution and structure of the plasmodesmata in mesophyll and bundle-sheath cells of *Zea mays* L. *Planta* 136, 77–89.

FERGUSON, C., T. T. TEERI, M. SIIKA-AHO, S. M. READ, and A. BACIC. 1998. Location of cellulose and callose in pollen tubes and grains of *Nicotiana tabacum*. *Planta* 206, 452–460.

FISHER, D. B., and K. J. OPARKA. 1996. Post-phloem transport: Principles and problems. *J. Exp. Bot.* 47 (spec. iss.), 1141–1154.

FISHER, D. B., Y. WU, and M. S. B. KU. 1992. Turnover of soluble proteins in the wheat sieve tube. *Plant Physiol.* 100, 1433–1441.

FISHER, D. D., and R. J. CYR. 1998. Extending the microtubule/microfibril paradigm. Cellulose synthesis is required for normal cortical microtubule alignment in elongating cells. *Plant Physiol.* 116, 1043–1051.

FLEMING, A. J., S. MCQUEEN-MASON, T. MANDEL, and C. KUHLEMEIER. 1997. Induction of leaf primordia by the cell wall protein expansin. *Science* 276, 1415–1418.

FLEMING, A. J., D. CADERAS, E. WEHRLI, S. MCQUEEN-MASON, and C. KUHLEMEIER. 1999. Analysis of expansin-induced morphogenesis of the apical meristem of tomato. *Planta* 208, 166–174.

FREY-WYSSLING, A. 1976. The plant cell wall. In *Handbuch der Pflanzenanatomie*, Band 3, Teil 4. Abt. *Cytologie*, 3rd rev. ed. Gebrüder Borntraeger, Berlin.

FRY, S. C. 1988. *The Growing Plant Cell Wall: Chemical and Metabolic Analysis*. Longman Scientific Burnt Mill, Harlow, Essex.

FRY, S. C. 1989. The structure and functions of xyloglucan. *J. Exp. Bot.* 40, 1–11.

FRY, S. C. 1995. Polysaccharide-modifying enzymes in the plant cell wall. *Annu. Rev. Plant Physiol. Plant Mol. Biol.* 46, 497–520.

FRY, S. C., S. ALDINGTON, P. R. HETHERINGTON, and J. AITKEN. 1993. Oligosaccharides as signals and substrates in the plant cell wall. *Plant Physiol.* 103, 1–5.

FUJINO, T., and T. ITOH. 1998. Changes in the three dimensional architecture of the cell wall during lignification of xylem cells in *Eucalyptus tereticornis*. *Holzforschung* 52, 111–116.

FUJINO, T., Y. SONE, Y. MITSUISHI, and T. ITOH. 2000. Characterization of cross-links between cellulose microfibrils, and their occurrence during elongation growth in pea epicotyl. *Plant Cell Physiol.* 41, 486–494.

GAMALEI, Y. V., A. J. E. VAN BEL, M. V. PAKHOMOVA, and A. V. SJUTKINA. 1994. Effects of temperature on the conformation of the endoplasmic reticulum and on starch accumulation in leaves with the symplasmic minor-vein configuration. *Planta* 194, 443–453.

GARDINER, J. C., J. D. I. HARPER, N. D. WEERAKOON, D. A. COLLINGS, S. RITCHIE, S. GILORY, R. J. CYR, and J. MARC. 2001. A 90-kD phospholipase D from tobacco binds to microtubules and the plasma membrane. *Plant Cell* 13, 2143–2158.

GARDINER, J., D. A. COLLINGS, J. D. I. HARPER, and J. MARC. 2003a. The effects of the phospholipase D-antagonist 1-butanol on seedling development and microtubule organisation in *Arabidopsis*. *Plant Cell Physiol.* 44, 687–696.

GARDINER, J. C., N. G. TAYLOR, and S. R. TURNER. 2003b. Control of cellulose synthase complex localization in developing xylem. *Plant Cell* 15, 1740–1748.

GEISLER-LEE, C. J., Z. HONG, and D. P. S. VERMA. 2002. Overexpression of the cell plate-associated dynamin-like GTPase, phragmoplastin, results in the accumulation of callose at the cell plate and arrest of plant growth. *Plant Sci.* 163, 33–42.

GIDDINGS, T. H., JR., and L. A. STAEHELIN. 1988. Spatial relationship between microtubules and plasma-membrane rosettes during the deposition of primary wall microfibrils in *Closterium* sp. *Planta* 173, 22–30.

GOLDBERG, R., P. DEVILLERS, R. PRAT, C. MORVAN, V. MICHON, and C. HERVÉ DU PENHOAT. 1989. Control of cell wall plasticity. In: *Plant Cell Wall Polymers. Biogenesis and Biodegradation*, pp. 312–323, N. G. Lewis and M. C. Paice, eds. American Chemical Society, Washington, DC.

GOODBODY, K. C., and C. W. LLOYD. 1990. Actin filaments line up across *Tradescantia* epidermal cells, anticipating wound-induced division planes. *Protoplasma* 157, 92–101.

GOODWIN, P. B. 1983. Molecular size limit for movement in the symplast of the *Elodea* leaf. *Planta* 157, 124–130.

GOOSEN-DE ROO, L., R. BAKHUIZEN, P. C. VAN SPRONSEN, and K. R. LIBBENGA. 1984. The presence of extended phragmosomes containing cytoskeletal elements in fusiform cambial cells of *Fraxinus excelsior* L. *Protoplasma* 122, 145–152.

GRABSKI, S., A. W. DE FEIJTER, and M. SCHINDLER. 1993. Endoplasmic reticulum forms a dynamic continuum for lipid diffusion between contiguous soybean root cells. *Plant Cell* 5, 25–38.

GRISEBACH, H. 1981. Lignins. In: *The Biochemistry of Plants*, vol. 7, *Secondary Plant Products*, pp. 457–478, E. E. Conn, ed. Academic Press, New York.

GRITSCH, C. S., and R. J. MURPHY. 2005. Ultrastructure of fibre and parenchyma cell walls during early stages of culm development in *Dendrocalamus asper*. *Ann. Bot.* 95, 619–629.

GRÜNWALD, C., K. RUEL, Y. S. KIM, and U. SCHMITT. 2002. On the cytochemistry of cell wall formation in poplar trees. *Plant Biol.* 4, 13–21.

GU, X., and D. P. S. VERMA. 1997. Dynamics of phragmoplastin in living cells during cell plate formation and uncoupling of cell elongation from the plane of cell division. *Plant Cell* 9, 157–169.

GUNNING, B. E. S. 1982. The cytokinetic apparatus: Its development and spatial regulation. In: *The Cytoskeleton in Plant Growth and Development*, pp. 229–292, C. W. Lloyd, ed. Academic Press, London.

GUNNING, B. E. S., and A. R. HARDHAM. 1982. Microtubules. *Annu. Rev. Plant Physiol.* 33, 651–698.

GUNNING, B. E. S., and S. M. WICK. 1985. Preprophase bands, phragmoplasts, and spatial control of cytokinesis. *J. Cell Sci.* suppl. 2, 157–179.

HA, M.-A., D. C. APPERLEY, B. W. EVANS, I. M. HUXHAM, W. G. JARDINE, R. J. VIËTOR, D. REIS, B. VIAN, and M. C. JARVIS. 1998. Fine structure in cellulose microfibrils: NMR evidence from onion and quince. *Plant J.* 16, 183–190.

HAFRÉN, J., T. FUJINO, and T. ITOH. 1999. Changes in cell wall architecture of differentiating tracheids of *Pinus thunbergii* during lignification. *Plant Cell Physiol.* 40, 532–541.

HAIGLER, C. H., and R. M. BROWN JR. 1986. Transport of rosettes from the Golgi apparatus to the plasma membrane in isolated mesophyll cells of *Zinnia elegans* during differentiation to tracheary elements in suspension culture. *Protoplasma* 134, 111–120.

HAMMERSCHMIDT, R. 1999. Phytoalexins: What have we learned after 60 years? *Annu. Rev. Phytopathol.* 37, 285–306.

HARPER, J. D. I., L. C. FOWKE, S. GILMER, R. L. OVERALL, and J. MARC. 2000. A centrin homologue is localized across the developing cell plate in gymnosperms and angiosperms. *Protoplasma* 211, 207–216.

HATFIELD, R., and W. VERMERRIS. 2001. Lignin formation in plants. The dilemma of linkage specificity. *Plant Physiol.* 126, 1351–1557.

HAYASHI, T. 1991. Biochemistry of xyloglucans in regulating cell elongation and expansion. In: *The Cytoskeletal Basis of Plant Growth and Form*, pp. 131–144, C. W. Lloyd, ed. Academic Press, San Diego.

HAYWOOD, V., F. KRAGLER, and W. J. LUCAS. 2002. Plamodesmata: Pathways for protein and ribonucleoprotein signaling. *Plant Cell* 14 (suppl.), S303–S325.

HEESE, M., U. MAYER, and G. JÜRGENS. 1998. Cytokinesis in flowering plants: Cellular process and developmental integration. *Curr. Opin. Plant Biol.* 1, 486–491.

HEPLER, P. K. 1982. Endoplasmic reticulum in the formation of the cell plate and plasmodesmata. *Protoplasma* 111, 121–133.

HERTH, W. 1980. Calcofluor white and Congo red inhibit chitin microfibril assembly of *Poterioochromonas*: Evidence for a gap between polymerization and microfibril formation. *J. Cell Biol.* 87, 442–450.

HEYN, A. N. J. 1931. Der Mechanismus der Zellstreckung. *Rec. Trav. Bot. Neerl.* 28, 113–244.

HEYN, A. N. J. 1940. The physiology of cell elongation. *Bot. Rev.* 6, 515–574.

HIGUCHI, T. 1997. *Biochemistry and Molecular Biology of Wood*. Springer-Verlag, Berlin.

HIMMELSPACH, R., R. E. WILLIAMSON, and G. O. WASTENEYS. 2003. Cellulose microfibril alignment recovers from DCB-induced disruption despite microtubule disorganization. *Plant J.* 36, 565–575.

HOLLAWAY-CLARKE, T. L., N. A. WALKER, P. K. HEPLER, and R. L. OVERALL. 2000. Physiological elevations in cytoplasmic free calcium by cold or ion injection result in transient closure of higher plant plasmodesmata. *Planta* 210, 329–335.

HORNER, H. T., and M. A. ROGERS. 1974. A comparative light and electron microscopic study of microsporogenesis in male-fertile and cytoplasmic male-sterile pepper (*Capsicum annuum*). *Can. J. Bot.* 52, 435–441.

HOSON, T. 1991. Structure and function of plant cell walls: Immunological approaches. *Int. Rev. Cytol.* 130, 233–268.

HOTCHKISS, A. T., JR. 1989. Cellulose biosynthesis. The terminal complex hypothesis and its relationship to other contemporary research topics. In: *Plant Cell Wall Polymers: Biogenesis and Biodegradation*, pp. 232–247, N. G. Lewis and M. G. Paice, eds. American Chemical Society, Washington, DC.

HUSH, J. M., P. WADSWORTH, D. A. CALLAHAM, and P. K. HEPLER. 1994. Quantification of microtubule dynamics in living plant cells using fluorescence redistribution after photobleaching. *J. Cell Sci.* 107, 775–784.

IIYAMA, K., T. B.-T. LAM, and B. A. STONE. 1994. Covalent cross-links in the cell wall. *Plant Physiol.* 104, 315–320.

ISHII, T., T. MATSUNAGA, P. PELLERIN, M. A. O'NEILL, A. DARVILL, and P. ALBERSHEIM. 1999. The plant cell wall polysaccharide rhamnogalacturonan II self-assembles into a covalently cross-linked dimer. *J. Biol. Chem.* 274, 13098–13104.

ISHIWATARI, Y., C. HONDA, I. KAWASHIMA, S-I. NAKAMURA, H. HIRANO, S. MORI, T. FUJIWARA, H. HAYASHI, and M. CHINO. 1995. Thioredoxin h is one of the major proteins in rice phloem sap. *Planta* 195, 456–463.

ISHIWATARI, Y., T. FUJIWARA, K. C. MCFARLAND, K. NEMOTO, H. HAYASHI, M. CHINO, and W. J. LUCAS. 1998. Rice phloem thioredoxin h has the capacity to mediate its own cell-to-cell transport through plasmodesmata. *Planta* 205, 12–22.

ITAYA, A., Y.-M. WOO, C. MASUTA, Y. BAO, R. S. NELSON, and B. DING. 1998. Developmental regulation of intercellular protein trafficking through plasmodesmata in tobacco leaf epidermis. *Plant Physiol.* 118, 373–385.

JACKSON, D., B. VEIT, and S. HAKE. 1994. Expression of maize *KNOTTED1* related homeobox genes in the shoot apical meristem predicts patterns of morphogenesis in the vegetative shoot. *Development* 120, 405–413.

JAUH, G. Y. and E. M. LORD. 1996. Localization of pectins and arabinogalactan-proteins in lily (*Lilium longiflorum* L.) pollen tube and style, and their possible roles in pollination. *Planta* 199, 251–261.

JEFFREE, C. E., J. E. DALE, and S. C. FRY. 1986. The genesis of intercellular spaces in developing leaves of *Phaseolus vulgaris* L. *Protoplasma* 132, 90–98.

Jones, M. G. K. 1976. The origin and development of plasmo-desmata. In: *Intercellular Communication in Plants: Studies on Plasmodesmata*, pp. 81–105, B. E. S. Gunning and A. W. Robards, eds. Springer-Verlag, Berlin.

Jung, G., and W. Wernicke. 1990. Cell shaping and microtubules in developing mesophyll of wheat (*Triticum aestivum* L.). *Protoplasma* 153, 141–148.

Kato, Y., and K. Matsuda. 1985. Xyloglucan in cell walls of sus-pension-cultured rice cells. *Plant Cell Physiol.* 26, 437–445.

Kauss, H. 1989. Fluorometric measurement of callose and other 1,3-β-glucans. In: *Plant Fibers*, pp. 127–137, H. F. Linskens and J. F. Jackson, eds. Springer-Verlag, Berlin.

Kauss, H. 1996. Callose synthesis. In: *Membranes: Specialized Functions in Plants*, pp. 77–92, M. Smallwood, J. P. Knox, and D. J. Bowles, eds. BIOS Scientific, Oxford.

Keller, B. 1993. Structural cell wall proteins. *Plant Physiol.* 101, 1127–1130.

Keller, B., and C. J. Lamb. 1989. Specific expression of a novel cell wall hydroxyproline-rich glycoprotein gene in lateral root initiation. *Genes Dev.* 3, 1639–1646.

Kempers, R., and A. J. E. van Bel. 1997. Symplasmic connections between sieve element and companion cell in stem phloem of *Vicia faba* L. have a molecular exclusion limit of at least 10 kDa. *Planta* 201, 195–201.

Kerr, T., and I. W. Bailey. 1934. The cambium and its derivative tissues. X. Structure, optical properties and chemical com-position of the so-called middle lamella. *J. Arnold Arb.* 15, 327–349.

Kerstens, S., W. F. Decraemer, and J.-P. Verbelen. 2001. Cell walls at the plant surface behave mechanically like fiber-rein-forced composite materials. *Plant Physiol.* 127, 381–385.

Kim, I., F. D. Hempel, K. Sha, J. Pfluger, and P. C. Zambryski. 2002. Identification of a developmental transition in plasmo-desmatal function during embryogenesis in *Arabidopsis thali-ana*. *Development* 129, 1261–1272.

Kim, M., W. Canio, S. Kessler, and N. Sinha. 2001. Develop-mental changes due to long-distance movement of a homeo-box fusion transcript in tomato. *Science* 293, 287–289.

Kimura, S., W. Laosinchai, T. Itoh, X. Cui, C. R. Linder, and R. M. Brown Jr. 1999. Immunogold labeling of rosette ter-minal cellulose-synthesizing complexes in the vascular plant *Vigna angularis*. *Plant Cell* 11, 2075–2085.

Kolattukudy, P. E. 1980. Biopolyester membranes of plants: Cutin and suberin. *Science* 208, 990–1000.

Kolattukudy, P. E., and C. L. Soliday. 1985. Effects of stress on the defensive barriers of plants. In: *Cellular and Molecular Biology of Plant Stress*, pp. 381–400, J. L. Key and T. Kosuge, eds. Alan R. Liss, New York.

Kollmann, R., and C. Glockmann. 1991. Studies on graft unions. III. On the mechanism of secondary formation of plasmodesmata at the graft interface. *Protoplasma* 165, 71–85.

Kollöffel, C., and P. W. T. Linssen. 1984. The formation of intercellular spaces in the cotyledons of developing and ger-minating pea seeds. *Protoplasma* 120, 12–19.

Kragler, F., W. J. Lucas, and J. Monzer. 1998. Plasmodesmata: Dynamics, domains and patterning. *Ann. Bot.* 81, 1–10.

Kreuger, M., and G.-J. van Holst. 1993. Arabinogalactan pro-teins are essential in somatic embryogenesis of *Daucus carota* L. *Planta* 189, 243–248.

Kutschera, U. 1991. Regulation of cell expansion. In: *The Cyto-skeletal Basis of Plant Growth and Form*, pp. 149–158, C. W. Lloyd, ed. Academic Press, London.

Kutschera, U. 1996. Cessation of cell elongation in rye coleop-tiles is accompanied by a loss of cell-wall plasticity. *J. Exp. Bot.* 47, 1387–1394.

Lauber, M. H., I. Waizenegger, T. Steinmann, H. Schwarz, U. Mayer, I. Hwang, W. Lukowitz, and G. Jürgens. 1997. The *Arabidopsis* KNOLLE protein in a cytokinesis-specific syn-taxin. *J. Cell Biol.* 139, 1485–1493.

Lee, J.-Y., B.-C. Yoo, and W. J. Lucas. 2000. Parallels between nuclear-pore and plasmodesmal trafficking of information molecules. *Planta* 210, 177–187.

Lee, Y.-R. J., and B. Liu. 2000. Identification of a phragmoplast-associated kinesin-related protein in higher plants. *Curr. Biol.* 10, 797–800.

Leisner, S. M., and R. Turgeon. 1993. Movement of virus and photoassimilate in the phloem: A comparative analysis. *BioEs-says* 15, 741–748.

Levy, S., and L. A. Staehelin. 1992. Synthesis, assembly and func-tion of plant cell wall macromolecules. *Curr. Opin. Cell Biol.* 4, 856–862.

Lewis, N. G. 1999. A 20th century roller coaster ride: A short account of lignification. *Curr. Opin. Plant Biol.* 2, 153–162.

Lewis, N. G., and E. Yamamoto. 1990. Lignin: Occurrence, bio-genesis and biodegradation. *Annu. Rev. Plant Physiol. Plant Mol. Biol.* 41, 455–496.

Li, Y., C. P. Darley, V. Ongaro, A. Fleming, O. Schipper, S. L. Baldauf, and S. J. McQueen-Mason. 2002. Plant expansins are a complex multigene family with an ancient evolutionary origin. *Plant Physiol.* 128, 854–864.

Lišková, D., D. Kákoniová, M. Kubačková, K. Sadloňová-Kollárová, P. Capek, L. Bilisics, J. Vojtaššák, and L. Slováková. 1999. Biologically active oligosaccharides. In: *Advances in Regulation of Plant Growth and Development*, pp. 119–130, M. Strnad, P. Peč, and E. Beck, eds. Peres Publishers, Prague.

Lloyd, C. W., and J. A. Traas. 1988. The role of F-actin in determining the division plane of carrot suspension cells. Drug studies. *Development* 102, 211–221.

Lucas, W. J. 1999. Plasmodesmata and the cell-to-cell transport of proteins and nucleoprotein complexes. *J. Exp. Bot.* 50, 979–987.

LUCAS, W. J., B. DING, and C. VAN DER SCHOOT. 1993. Plasmodesmata and the supracellular nature of plants. *New Phytol.* 125, 435–476.

LUCAS, W. J., S. BOUCHÉ-PILLON, D. P. JACKSON, L. NGUYEN, L. BAKER, B. DING, and S. HAKE. 1995. Selective trafficking of KNOTTED1 homeodomain protein and its mRNA through plasmodesmata. *Science* 270, 1980–1983.

LUKOWITZ, W., U. MAYER, and G. JÜRGENS. 1996. Cytokinesis in the *Arabidopsis* embryo involves the syntaxin-related *KNOLLE* gene product. *Cell* 84, 61–71.

MACADAM, J. W., and C. J. NELSON. 2002. Secondary cell wall deposition causes radial growth of fibre cells in the maturation zone of elongating tall fescue leaf blades. *Ann. Bot.* 89, 89–96.

MAJEWSKA-SAWKA, A., and E. A. NOTHNAGEL. 2000. The multiple roles of arabinogalactan proteins in plant development. *Plant Physiol.* 122, 3–9.

MALTBY, D., N. C. CARPITA, D. MONTEZINOS, C. KULOW, and D. P. DELMER. 1979. β-1,3-glucan in developing cotton fibers. Structure, localization, and relationship of synthesis to that of secondary wall cellulose. *Plant Physiol.* 63, 1158–1164.

MANSFIELD, S. G., and L. G. BRIARTY. 1991. Early embryogenesis in *Arabidopsis thaliana*. II. The developing embryo. *Can. J. Bot.* 69, 461–476.

MARC, J., D. E. SHARKEY, N. A. DURSO, M. ZHANG, and R. J. CYR. 1996. Isolation of a 90-kD microtubule-associated protein from tobacco membranes. *Plant Cell* 8, 2127–2138.

MATAR, D., and A. M. CATESSON. 1988. Cell plate development and delayed formation of the pectic middle lamella in root meristems. *Protoplasma* 146, 10–17.

MATOH, T., and M. KOBAYASHI. 1998. Boron and calcium, essential inorganic constituents of pectic polysaccharides in higher plant cell walls. *J. Plant Res.* 111, 179–190.

MCCANN, M. C., B. WELLS, and K. ROBERTS. 1990. Direct visualization of cross-links in the primary plant cell wall. *J. Cell Sci.* 96, 323–334.

MCDOUGALL, G. J., and S. C. FRY. 1990. Xyloglucan oligosaccharides promote growth and activate cellulase: Evidence for a role of cellulase in cell expansion. *Plant Physiol.* 93, 1042–1048.

MCLEAN, B. G., F. D. HEMPEL, and P. C. ZAMBRYSKI. 1997. Plant intercellular communication via plasmodesmata. *Plant Cell* 9, 1043–1054.

MCNEIL, M., A. G. DARVILL, S. C. FRY, and P. ALBERSHEIM. 1984. Structure and function of the primary cell walls of plants. *Annu. Rev. Biochem.* 5, 625–663.

MCQUEEN-MASON, S. J. 1995. Expansins and cell wall expansion. *J. Exp. Bot.* 46, 1639–1650.

MEZITT, L. A., and W. J. LUCAS. 1996. Plasmodesmal cell-to-cell transport of proteins and nucleic acids. *Plant Mol. Biol.* 32, 251–273.

MOHNEN, D., and M. G. HAHN. 1993. Cell wall carbohydrates as signals in plants. *Semin. Cell Biol.* 4, 93–102.

MOLCHAN, T. M., A. H. VALSTER, and P. K. HEPLER. 2002. Actomyosin promotes cell plate alignment and late lateral expansion in *Tradescantia* stamen hair cells. *Planta* 214, 683–693.

MONTIES, B. 1989. Lignins. In: *Methods in Plant Biochemistry*, vol. 1, *Plant Phenolics*, pp. 113–157, J. B. Harborne, ed. Academic Press, London.

MONZER, J. 1991. Ultrastructure of secondary plasmodesmata formation in regenerating *Solanum nigrum*-protoplast cultures. *Protoplasma* 165, 86–95.

MOORE, P. J., and L. A. STAEHELIN. 1988. Immunogold localization of the cell-wall-matrix polysaccharides rhamnogalacturonan I and xyloglucan during cell expansion and cytokinesis in *Trifolium pratense* L.; implication for secretory pathways. *Planta* 174, 433–445.

MOREJOHN, L. C. 1991. The molecular pharmacology of plant tubulin and microtubules. In: *The Cytoskeletal Basis of Plant Growth and Form*, pp. 29–43, C. W. Lloyd, ed. Academic Press, London.

MULDER, B. M., and A. M. C. EMONS. 2001. A dynamical model for plant cell wall architecture formation. *J. Math. Biol.* 42, 261–289.

MULDER, B., J. SCHEL, and A. M. EMONS. 2004. How the geometrical model for plant cell wall formation enables the production of a random texture. *Cellulose* 11, 395–401.

MÜNCH, E. 1930. *Die Stoffbewegungen in der Pflanze*. Gustav Fischer, Jena.

MURMANIS, L., and R. F. EVERT. 1967. Parenchyma cells of secondary phloem in *Pinus strobus*. *Planta* 73, 301–318.

MÜSEL, G., T. SCHINDLER, R. BERGFELD, K. RUEL, G. JACQUET, C. LAPIERRE, V. SPETH, and P. SCHOPFER. 1997. Structure and distribution of lignin in primary and secondary walls of maize coleoptiles analyzed by chemical and immunological probes. *Planta* 201, 146–159.

NAKAJIMA, K., G. SENA, T. NAWY, and P. N. BENFEY. 2001. Intercellular movement of the putative transcription factor SHR in root patterning. *Nature* 413, 307–311.

NAKAMURA, S.-I., H. HAYASHI, S. MORI, and M. CHINO. 1993. Protein phosphorylation in the sieve tubes of rice plants. *Plant Cell Physiol.* 34, 927–933.

NEBENFÜHR, A., J. A. FROHLICK, and L. A. STAEHELIN. 2000. Redistribution of Golgi stacks and other organelles during mitosis and cytokinesis in plant cells. *Plant Physiol.* 124, 135–151.

NELSON, R. S., and A. J. E. VAN BEL. 1998. The mystery of virus trafficking into, through and out of vascular tissue. *Prog. Bot.* 59, 476–533.

NEWMAN, R. H., L. M. DAVIES, and P. J. HARRIS. 1996. Solid-state $^{13}$C nuclear magnetic resonance characterization of cellulose in the cell walls of *Arabidopsis thaliana* leaves. *Plant Physiol.* 111, 475–485.

NICHOLSON, R. L., and R. HAMMERSCHMIDT. 1992. Phenolic compounds and their role in disease resistance. *Annu. Rev. Phytopathol.* 30, 369–389.

OLESEN, P. 1979. The neck constriction in plasmodesmata. Evidence for a peripheral sphincter-like structure revealed by fixation with tannic acid. *Planta* 144, 349–358.

OLESEN, P., and A. W. ROBARDS. 1990. The neck region of plasmodesmata: general architecture and some functional aspects. In: *Parallels in Cell to Cell Junctions in Plants and Animals*, pp. 145–170, A. W. Robards, H. Jongsma, W. J. Lucas, J. Pitts, and D. Spray, eds. Springer-Verlag, Berlin.

O'MALLEY, D. M., R. WHETTEN, W. BAO, C.-L. CHEN, and R. R. SEDEROFF. 1993. The role of laccase in lignification. *Plant J.* 4, 751–757.

OPARKA, K. J., D. A. M. PRIOR, and K. M. WRIGHT. 1995. Symplastic communication between primary and developing lateral roots of *Arabidopsis thaliana*. *J. Exp. Bot.* 46, 187–197.

OPARKA, K. J., A. G. ROBERTS, P. BOEVINK, S. SANTA CRUZ, I. ROBERTS, K. S. PRADEL, A. IMLAU, G. KOTLIZKY, N. SAUER, and B. EPEL. 1999. Simple, but not branched, plasmodesmata allow the nonspecific trafficking of proteins in developing tobacco leaves. *Cell* 97, 743–754.

ØSTERGAARD, L., K. TEILUM, O. MIRZA, O. MATTSSON, M. PETERSEN, K. G. WELINDER, J. MUNDY, M. GAJHEDE, and A. HENRIKSEN. 2000. *Arabidopsis* ATP A2 peroxidase. Expression and high-resolution structure of a plant peroxidase with implications for lignification. *Plant Mol. Biol.* 44, 231–243.

OTEGUI, M., and L. A. STAEHELIN. 2000. Cytokinesis in flowering plants: More than one way to divide a cell. *Curr. Opin. Plant Biol.* 3, 493–502.

OVERALL, R. L., and L. M. BLACKMAN. 1996. A model of the macromolecular structure of plasmodesmata. *Trends Plant Sci.* 1, 307–311.

OVERALL, R. L., and B. E. S. GUNNING. 1982. Intercellular communication in *Azolla* roots. II. Electrical coupling. *Protoplasma* 111, 151–160.

OVERALL, R. L., R. G. WHITE, L. M. BLACKMAN, and J. E. RADFORD. 2000. Actin and myosin in plasmodesmata. In: *Actin: A Dynamic Framework for Multiple Plant Cell Function*, pp. 497–515, C. J. Staiger, F. Baluška, D. Volkmann, and P. W. Barlow, eds. Kluwer/Academic Press, Dordrecht.

PALEVITZ, B. A., and P. K. HEPLER. 1985. Changes in dye coupling of stomatal cells of *Allium* and *Commelina* demonstrated by microinjection of Lucifer yellow. *Planta* 164, 473–479.

PANTERIS, E., P. APOSTOLAKOS, and B. GALATIS. 1993. Microtubule organization, mesophyll cell morphogenesis, and intercellular space formation in *Adiantum capillus veneris* leaflets. *Protoplasma* 172, 97–110.

PAULY, M., P. ALBERSHEIM, A. DARVILL, and W. S. YORK. 1999. Molecular domains of the cellulose/xyloglucan network in the cell walls of higher plants. *Plant J.* 20, 629–639.

PEARCE, R. B. 1989. Cell wall alterations and antimicrobial defense in perennial plants. In: *Plant Cell Wall Polymers. Biogenesis and Biodegradation*, pp. 346–360, N. G. Lewis and M. G. Paice, eds. American Chemical Society, Washington, DC.

PENNELL, R. 1998. Cell walls: Structures and signals. *Curr. Opin. Plant Biol.* 1, 504–510.

PERRY, J. W., and R. F. EVERT. 1983. Histopathology of *Verticillium dahliae* within mature roots of Russet Burbank potatoes. *Can. J. Bot.* 61, 3405–3421.

PICKARD, B. G., and R. N. BEACHY. 1999. Intercellular connections are developmentally controlled to help move molecules through the plant. *Cell* 98, 5–8.

POMAR, F., F. MERINO, and A. ROS BARCELÓ. 2002. O-4-linked coniferyl and sinapyl aldehydes in lignifying cell walls are the main targets of the Wiesner (phloroglucinol-HCl) reaction. *Protoplasma* 220, 17–28.

POST-BEITTENMILLER, D. 1996. Biochemistry and molecular biology of wax production in plants. *Annu. Rev. Plant Physiol. Plant Mol. Biol.* 47, 405–430.

POTGIETER, M. J., and A. E. VAN WYK. 1992. Intercellular pectic protuberances in plants: Their structure and taxonomic significance. *Bot. Bull. Acad. Sin.* 33, 295–316.

PRESTON, R. D. 1974. Plant cell walls. In: *Dynamic Aspects of Plant Ultrastructure*, pp. 256–309, A. W. Robards, ed. McGraw-Hill, London.

PRESTON, R. D. 1982. The case for multinet growth in growing walls of plant cells. *Planta* 155, 356–363.

PRIESTLEY, J. H., and L. I. SCOTT. 1939. The formation of a new cell wall at cell division. *Proc. Leeds Philos. Lit. Soc., Sci. Sect.*, 3, 532–545.

RADFORD, J. E., and R. G. WHITE. 1998. Localization of a myosin-like protein to plasmodesmata. *Plant J.* 14, 743–750.

RADFORD, J. E., M. VESK, and R. L. OVERALL. 1998. Callose deposition at plasmodesmata. *Protoplasma* 201, 30–37.

RAVEN, P. H., R. F. EVERT, and S. E. EICHHORN. 2005. *Biology of Plants*, 7th ed. Freeman, New York.

RECORD, S. J. 1934. *Identification of the Timbers of Temperate North America*. Wiley, New York.

REICHEL, C., P. MÁS, and R. N. BEACHY. 1999. The role of the ER and cytoskeleton in plant viral trafficking. *Trends Plant Sci.* 4, 458–462.

REICHELT, S., A. E. KNIGHT, T. P. HODGE, F. BALUŠKA, J. SAMAJ, D. VOLKMANN, and J. KENDRICK-JONES. 1999. Characterization of the unconventional myosin VIII in plant cells and its localization at the post-cytokinetic cell wall. *Plant J.* 19, 555–567.

REID, J. S. G. 1985. Cell wall storage carbohydrates in seeds—Biochemistry of the seed "gums" and "hemicelluloses." *Adv. Bot. Res.* 11, 125–155.

REINHARDT, D., F. WITTWER, T. MANDEL, and C. KUHLEMEIER. 1998. Localized upregulation of a new expansin gene predicts the site of leaf formation in the tomato meristem. *Plant Cell* 10, 1427–1437.

REIS, D., and B. VIAN. 2004. Helicoidal pattern in secondary cell walls and possible role of xylans in their construction. *C. R. Biologies* 327, 785–790.

REUZEAU, C., and R. F. PONT-LEZICA. 1995. Comparing plant and animal extracellular matrix-cytoskeleton connections—Are they alike? *Protoplasma* 186, 113–121.

ROBARDS, A. W., and W. J. LUCAS. 1990. Plasmodesmata. *Annu. Rev. Plant Physiol. Plant Mol. Biol.* 41, 369–419.

ROBERTS, K. 1990. Structures at the plant cell surface. *Curr. Opin. Cell Biol.* 2, 920–928.

ROBERTS, K. 1994. The plant extracellular matrix: In a new expansive mood. *Curr. Opin. Cell Biol.* 6, 688–694.

ROBINSON, D. G. 1991. What is a plant cell? The last word. *Plant Cell* 3, 1145–1146.

ROBINSON-BEERS, K., and R. F. EVERT. 1991. Fine structure of plasmodesmata in mature leaves of sugarcane. *Planta* 184, 307–318.

RODKIEWICZ, B. 1970. Callose in cell walls during megasporogenesis in angiosperms. *Planta* 93, 39–47.

ROELOFSEN, P. A. 1959. The plant cell-wall. *Handbuch der Pflanzenanatomie*, Band III, Teil 4, *Cytologie*. Gebrüder Borntraeger, Berlin-Nikolassee.

ROLAND, J. C. 1978. Cell wall differentiation and stages involved with intercellular gas space opening. *J. Cell Sci.* 32, 325–336.

ROLAND, J. C., D. REIS, B. VIAN, B. SATIAT-JEUNEMAITRE, and M. MOSINIAK. 1987. Morphogenesis of plant cell walls at the supramolecular level: Internal geometry and versatility of helicoidal expression. *Protoplasma* 140, 75–91.

ROLAND, J.-C., D. REIS, B. VIAN, and S. ROY. 1989. The helicoidal plant cell wall as a performing cellulose-based composite. *Biol. Cell* 67, 209–220.

ROS BARCELÓ, A. 1997. Lignification in plant cell walls. *Int. Rev. Cytol.* 176, 87–132.

ROSE, J. K. C., and A. B. BENNETT. 1999. Cooperative disassembly of the cellulose-xyloglucan network of plant cell walls: Parallels between cell expansion and fruit ripening. *Trends Plant Sci.* 4, 176–183.

ROSE, J. K. C., D. J. COSGROVE, P. ALBERSHEIM, A. G. DARVILL, and A. B. BENNETT. 2000. Detection of expansin proteins and activity during tomato fruit ontogeny. *Plant Physiol.* 123, 1583–1592.

RUEL, K., O. FAIX, and J.-P. JOSELEAU. 1994. New immunogold probes for studying the distribution of the different lignin types during plant cell wall biogenesis. *J. Trace Microprobe Tech.* 12, 247–265.

RUEL, K., M.-D. MONTIEL, T. GOUJON, L. JOUANIN, V. BURLAT, and J.-P. JOSELEAU. 2002. Interrelation between lignin deposition and polysaccharide matrices during the assembly of plant cell walls. *Plant Biol.* 4, 2–8.

RYSER, U., and B. KELLER. 1992. Ultrastructural localization of a bean glycine-rich protein in unlignified primary walls of protoxylem cells. *Plant Cell* 4, 773–783.

RYSER, U., M. SCHORDERET, G.-F. ZHAO, D. STUDER, K. RUEL, G. HAUF, and B. KELLER. 1997. Structural cell-wall proteins in protoxylem development: evidence for a repair process mediated by a glycine-rich protein. *Plant J.* 12, 97–111.

SAKUTH, T., C. SCHOBERT, A. PECSVARADI, A. EICHHOLZ, E. KOMOR, and G. ORLICH. 1993. Specific proteins in the sieve-tube exudate of *Ricinus communis* L. seedlings: Separation, characterization and *in vivo* labelling. *Planta* 191, 207–213.

SAMUELS, A. L., T. H. GIDDINGS Jr., and L. A. STAEHELIN. 1995. Cytokinesis in tobacco BY-2 and root tip cells: A new model of cell plate formation in higher plants. *J. Cell Biol.* 130, 1345–1357.

SATIAT-JEUNEMAITRE, B. 1992. Spatial and temporal regulations in helicoidal extracellular matrices: Comparison between plant and animal systems. *Tissue Cell* 24, 315–334.

SATIAT-JEUNEMAITRE, B., B. MARTIN, and C. HAWES. 1992. Plant cell wall architecture is revealed by rapid-freezing and deep-etching. *Protoplasma* 167, 33–42.

SCHMIT, A.-C., and A.-M. LAMBERT. 1988. Plant actin filament and microtubule interactions during anaphase-telophase transition: Effects of antagonist drugs. *Biol. Cell* 64, 309–319.

SCHOBERT, C., P. GROßMANN, M. GOTTSCHALK, E. KOMOR, A. PECSVARADI, and U. ZUR NIEDEN. 1995. Sieve-tube exudate from *Ricinus communis* L. seedlings contains ubiquitin and chaperones. *Planta* 196, 205–210.

SCHOBERT, C., L. BAKER, J. SZEDERKÉNYI, P. GROßMANN, E. KOMOR, H. HAYASHI, M. CHINO, and W. J. LUCAS. 1998. Identification of immunologically related proteins in sieve-tube exudate collected from monocotyledonous and dicotyledonous plants. *Planta* 206, 245–252.

SCHULZ, A. 1995. Plasmodesmal widening accompanies the short-term increase in symplasmic phloem loading in pea root tips under osmotic stress. *Protoplasma* 188, 22–37.

SCHULZ, A. 1999. Physiological control of plasmodesmal gating. In: *Plasmodesmata: Structure, Function, Role in Cell Communication*, pp. 173–204, A. J. E. van Bel and W. J. P. van Kestern, eds. Springer, Berlin.

SCHULZ, R., and W. A. JENSEN. 1968. *Capsella* embryogenesis: The egg, zygote, and young embryo. *Am. J. Bot.* 55, 807–819.

SEDEROFF, R., and H.-M. CHANG. 1991. Lignin biosynthesis. In: *Wood Structure and Composition*, pp. 263–285, M. Lewin and I. S. Goldstein, eds. Dekker, New York.

SEDEROFF, R. R., J. J. MACKAY, J. RALPH, and R. D. HATFIELD. 1999. Unexpected variation in lignin. *Curr. Opin. Plant Biol.* 2, 145–152.

SERPE, M. D., and E. A. NOTHNAGEL. 1999. Arabinogalactan-proteins in the multiple domains of the plant cell surface. *Adv. Bot. Res.* 30, 207–289.

SHEDLETZKY, E., M. SHUMEL, T. TRAININ, S. KALMAN, and D. DELMER. 1992. Cell wall structure in cells adapted to growth on the cellulose-synthesis inhibitor 2,6-dichlorobenzonitrile. A comparison between two dicotyledonous plants and a graminaceous monocot. *Plant Physiol.* 100, 120–130.

SHIBAOKA, H. 1991. Microtubules and the regulation of cell morphogenesis by plant hormones. In: *The Cytoskeletal Basis of*

*Plant Growth and Form*, pp. 159–168, C. W. Lloyd, ed. Academic Press, London.

SHIEH, M. W., and D. J. COSGROVE. 1998. Expansins. *J. Plant Res.* 111, 149–157.

SHIMIZU, Y., S. AOTSUKA, O. HASEGAWA, T. KAWADA, T. SAKUNO, F. SAKAI, and T. HAYASHI. 1997. Changes in levels of mRNAs for cell wall-related enzymes in growing cotton fiber cells. *Plant Cell Physiol.* 38, 375–378.

SHOWALTER, A. M. 1993. Structure and function of plant cell wall proteins. *Plant Cell* 5, 9–23.

SINHA, N. R., R. E. WILLIAMS, and S. HAKE. 1993. Overexpression of the maize homeobox gene, *KNOTTED-1*, causes a switch from determinate to indeterminate cell fates. *Genes Dev.* 7, 787–795.

SINNOTT, E. W., and R. BLOCH. 1941. Division in vacuolate plant cells. *Am. J. Bot.* 28, 225–232.

SMITH, B. G., and P. J. HARRIS. 1999. The polysaccharide composition of Poales cell walls: Poaceae cell walls are not unique. *Biochem. System. Ecol.* 27, 33–53.

SMITH, B. G., P. J. HARRIS, L. D. MELTON, and R. H. NEWMAN. 1998. Crystalline cellulose in hydrated primary cell walls of three monocotyledons and one dicotyledon. *Plant Cell Physiol.* 39, 711–720.

SMITH, L. G. 1999. Divide and conquer: Cytokinesis in plant cells. *Curr. Opin. Plant Biol.* 2, 447–453.

SPANSWICK, R. M. 1976. Symplasmic transport in tissues. In: *Encyclopedia of Plant Physiology*, n.s., vol. 2, *Transport in Plants II, Part B, Tissues and Organs*, pp. 35–53, U. Lüttge and M. G. Pitman, eds. Springer-Verlag, Berlin.

STAEHELIN, A. 1991. What is a plant cell? A response. *Plant Cell* 3, 553.

STAEHELIN, L. A., and P. K. HEPLER. 1996. Cytokinesis in higher plants. *Cell* 84, 821–824.

STONE, B. A., and A. E. CLARKE. 1992. *Chemistry and Biology of (1→3)-β-glucans*. La Trobe University Press, Bundoora, Victoria, Australia.

SUGIMOTO, K., R. HIMMELSPACH, R. E. WILLIAMSON, and G. O. WASTENEYS. 2003. Mutation or drug-dependent microtubule disruption causes radial swelling without altering parallel cellulose microfibril deposition in *Arabidopsis* root cells. *Plant Cell* 15, 1414–1429.

SUZUKI, K., T. ITOH, and H. SASAMOTO. 1998. Cell wall architecture prerequisite for the cell division in the protoplasts of white poplar, *Populus alba* L. *Plant Cell Physiol.* 39, 632–638.

TAKAHASHI, Y., S. ISHIDA, and T. NAGATA. 1995. Auxin-regulated genes. *Plant Cell Physiol.* 36, 383–390.

TALBOTT, L. D., and P. M. RAY. 1992. Molecular size and separability features of pea cell wall polysaccharides. Implications for models of primary wall structure. *Plant Physiol.* 98, 357–368.

TANGL, E. 1879. Ueber offene Communicationen zwischen den Zellen des Endospersms einiger Samen. *Jahrb. Wiss. Bot.* 12, 170–190.

TAYLOR, N. G., S. LAURIE, and S. R. TURNER. 2000. Multiple cellulose synthase catalytic subunits are required for cellulose synthesis in *Arabidopsis*. *Plant Cell* 12, 2529–2539.

TAYLOR, N. G., R. M. HOWELLS, A. K. HUTTLY, K. VICKERS, and S. R. TURNER. 2003. Interactions among three distinct CesA proteins essential for cellulose synthesis. *Proc. Natl. Acad. Sci. USA* 100, 1450–1455.

TERASHIMA, N. 2000. Formation and ultrastructure of lignified plant cell walls. In: *New Horizons in Wood Anatomy*, pp. 169–180, Y. S. Kim, ed. Chonnam National University Press, Kwangju, S. Korea.

TERASHIMA, N., K. FUKUSHIMA, L.-F. HE, and K. TAKABE. 1993. Comprehensive model of the lignified plant cell wall. In: *Forage Cell Wall Structure and Digestibility*, pp. 247–270, H. G. Jung, D. R. Buxton, R. D. Hatfield, and J. Ralph, eds. American Society of Agronomy, Madison, WI.

TERASHIMA, N., J. NAKASHIMA, and K. TAKABE. 1998. Proposed structure for protolignin in plant cell walls. In: *Lignin and Lignan Biosynthesis*, pp. 180–193, N. G. Lewis and S. Sarkanen, eds. American Chemical Society, Washington, DC.

TERRY, B. R., and A. W. ROBARDS. 1987. Hydrodynamic radius alone governs the mobility of molecules through plasmodesmata. *Planta* 171, 145–157.

THIMM, J. C., D. J. BURRITT, W. A. DUCKER, and L. D. MELTON. 2000. Celery (*Apium graveolens* L.) parenchyma cell walls examined by atomic force microscopy: Effect of dehydration on cellulose microfibrils. *Planta* 212, 25–32.

THIMM, J. C., D. J. BURRITT, I. M. SIMS, R. H. NEWMAN, W. A. DUCKER, and L. D. MELTON. 2002. Celery (*Apium graveolens*) parenchyma cell walls: Cell walls with minimal xyloglucan. *Physiol. Plant.* 116, 164–171.

THOMSON, N., R. F. EVERT, and A. KELMAN. 1995. Wound healing in whole potato tubers: A cytochemical, fluorescence, and ultrastructural analysis of cut and bruised wounds. *Can. J. Bot.* 73, 1436–1450.

TILNEY, L. G., T. J. COOKE, P. S. CONNELLY, and M. S. TILNEY. 1990. The distribution of plasmodesmata and its relationship to morphogenesis in fern gametophytes. *Development* 110, 1209–1221.

TRETHEWEY, J. A. K., and P. J. HARRIS. 2002. Location of (1→3)-- and (1→3), (1→4)-β-D-glucans in vegetative cell walls of barley (*Hordeum vulgare*) using immunogold labelling. *New Phytol.* 154, 347–358.

TUCKER, E. B., and R. M. SPANSWICK. 1985. Translocation in the staminal hairs of *Setcreasea purpurea*. II. Kinetics of intercellular transport. *Protoplasma* 128, 167–172.

TUCKER, J. E., D. MAUZERALL, and E. B. TUCKER. 1989. Symplastic transport of carboxyfluorescein in staminal hairs of *Setcreasea purpurea* is diffusive and includes loss to the vacuole. *Plant Physiol.* 90, 1143–1147.

TURGEON, R. 1996. Phloem loading and plasmodesmata. *Trends Plant Sci.* 1, 418–423.

VALLET, C., B. CHABBERT, Y. CZANINSKI, and B. MONTIES. 1996. Histochemistry of lignin deposition during sclerenchyma differentiation in alfalfa stems. *Ann. Bot.* 78, 625–632.

VAN BEL, A. J. E., and W. J. P. VAN KESTEREN, eds. 1999. *Plasmodesmata: Structure, Function, Role in Cell Communication.* Springer-Verlag, Berlin.

VANCE, C. P., T. K. KIRK, and R. T. SHERWOOD. 1980. Lignification as a mechanism of disease resistance. *Annu. Rev. Phytopathol.* 18, 259–288.

VAN VOLKENBURGH, E., M. G. SCHMIDT, and R. E. CLELAND. 1985. Loss of capacity for acid-induced wall loosening as the principal cause of the cessation of cell enlargement in light-grown bean leaves. *Planta* 163, 500–505.

VARNER, J. E., and L.-S. LIN. 1989. Plant cell wall architecture. *Cell* 56, 231–239.

VENVERLOO, C. J., and K. R. LIBBENGA. 1987. Regulation of the plane of cell division in vacuolated cells. I. The function of nuclear positioning and phragmosome formation. *J. Plant Physiol.* 131, 267–284.

VERMA, D. P. S. 2001. Cytokinesis and building of the cell plate in plants. *Annu. Rev. Plant Physiol. Plant Mol. Biol.* 52, 751–784.

VESK, P. A., M. VESK, and B. E. S. GUNNING. 1996. Field emission scanning electron microscopy of microtubule arrays in higher plant cells. *Protoplasma* 195, 168–182.

VIAN, B., D. REIS, M. MOSINIAK, and J. C. ROLAND. 1986. The glucuronoxylans and the helicoidal shift in cellulose microfibrils in linden wood: Cytochemistry *in muro* and on isolated molecules. *Protoplasma* 131, 185–199.

VIAN, B., J.-C. ROLAND, and D. REIS. 1993. Primary cell wall texture and its relation to surface expansion. *Int. J. Plant Sci.* 154, 1–9.

VOLK, G. M., R. TURGEON, and D. U. BEEBE. 1996. Secondary plasmodesmata formation in the minor-vein phloem of *Cucumis melo* L. and *Cucurbita pepo* L. *Planta* 199, 425–432.

VOS, J. W., M. DOGTEROM, and A. M. C. EMONS. 2004. Microtubules become more dynamic but not shorter during preprophase band formation: A possible "search-and-capture" mechanism for microtubule translocation. *Cell Motil. Cytoskel.* 57, 246–258.

WAIGMANN, E., and P. C. ZAMBRYSKI. 1995. Tobacco mosaic virus movement protein-mediated protein transport between trichome cells. *Plant Cell* 7, 2069–2079.

WAIGMANN, E., A. TURNER, J. PEART, K. ROBERTS, and P. ZAMBRYSKI. 1997. Ultrastructural analysis of leaf trichome plasmodesmata reveals major differences from mesophyll plasmodesmata. *Planta* 203, 75–84.

WALTER, M. H. 1992. Regulation of lignification in defense. In: *Genes Involved in Plant Defense*, pp. 327–352, T. Boller and F. Meins, eds. Springer-Verlag, Vienna.

WHETTEN, R. W., J. J. MACKAY, and R. R. SEDEROFF. 1998. Recent advances in understanding lignin biosynthesis. *Annu. Rev. Plant Physiol. Plant Mol. Biol.* 49, 585–609.

WHITE, R. G., K. BADELT, R. L. OVERALL, and M. VEST. 1994. Actin associated with plasmodesmata. *Protoplasma* 180, 169–184.

WILLATS, W. G. T., L. MCCARTNEY, W. MACKIE, and J. P. KNOX. 2001. Pectin: Cell biology and prospects for functional analysis. *Plant Mol. Biol.* 47, 9–27.

WILLIAMSON, R. E., J. E. BURN, and C. H. HOCART. 2002. Towards the mechanism of cellulose synthesis. *Trends Plant Sci.* 7, 461–467.

WOLF, S., C. M. DEOM, R. N. BEACHY, and W. J. LUCAS. 1989. Movement protein of tobacco mosaic virus modifies plasmodesmatal size exclusion limit. *Science* 246, 377–379.

WOLTERS-ARTS, A. M. C., T. VAN AMSTEL, and J. DERKSEN. 1993. Tracing cellulose microfibril orientation in inner primary cell walls. *Protoplasma* 175, 102–111.

WU, J. 1993. Variation in the distribution of guaiacyl and syringyl lignin in the cell walls of hardwoods. *Mem. Fac. Agric. Hokkaido Univ.* 18, 219–268.

WYMER, C., and C. LLOYD. 1996. Dynamic microtubules: Implications for cell wall patterns. *Trends Plant Sci.* 1, 222–228.

YE, Z.-H., and J. E. VARNER. 1991. Tissue-specific expression of cell wall proteins in developing soybean tissues. *Plant Cell* 3, 23–37.

YUAN, M., P. J. SHAW, R. M. WARN, and C. W. LLOYD. 1994. Dynamic reorientation of cortical microtubules, from transverse to longitudinal, in living plant cells. *Proc. Natl. Acad. Sci. USA* 91, 6050–6053.

ZACHARIADIS, M., H. QUADER, B. GALATIS, and P. APOSTOLAKOS. 2001. Endoplasmic reticulum preprophase band in dividing root-tip cells of *Pinus brutia*. *Planta* 213, 824–827.

ZANDOMENI, K., and P. SCHOPFER. 1993. Reorientation of microtubules at the outer epidermal wall of maize coleoptiles by phytochrome, blue-light receptor, and auxin. *Protoplasma* 173, 103–112.

# CHAPTER FIVE

# Meristems and Differentiation

## ▌MERISTEMS

Beginning with the division of the zygote, the vascular plant produces new cells and develops new organs, generally until it dies. In the early embryonic development, reproduction of cells occurs throughout the organism, but as the embryo becomes an independent plant, the addition of new cells is gradually restricted to certain of its regions. The now localized growing tissues, which remain embryonic in character, are maintained throughout the life of the plant so that its body is a composite of adult and juvenile tissues. These embryonic tissue regions, primarily concerned with formation of new cells, are the **meristems** (Fig. 5.1; McManus and Veit, 2002).

The restriction of cell reproduction to certain parts of the plant is a result of evolutionary specialization. In the most primitive (least specialized) plants all cells are essentially alike and all take part in such vital activities as cell multiplication, photosynthesis, secretion, storage, and transport. With progressive specialization of cells and tissues, the function of cell reproduction became largely confined to the meristems and their immediate derivatives.

The term meristem (from the Greek *merismos*, division) emphasizes cell division activity as a characteristic of a meristematic tissue. Living tissues other than the meristems may be induced to produce new cells, but the meristems maintain such activity indefinitely, for they not only add cells to the plant body but also perpetuate themselves; that is, some of the products of cell division in the meristems do not develop into adult cells but remain meristematic. Those cells that maintain the meristem as a continuing source of new cells are referred to as the **initiating cells**, or **meristematic initials**, or simply **initials**. Their products, which after a variable number of cell divisions give rise to the **body cells**, are the **derivatives** of the initials. One can also say that a dividing cell is a precursor of derivatives. A given initial in a meristem is a precursor of two derivatives, one of which is a new initial and the other a precursor of body cells.

The concept of initials and derivatives should include the qualification that the initials are not inherently different from their derivatives. The concept of initials and derivatives is considered from various aspects in connection with the descriptions of the vascular cambium (Chapter 12) and the apical meristems of shoot and root

*Esau's Plant Anatomy, Third Edition,* By Ray F. Evert.
Copyright © 2006 John Wiley & Sons, Inc.

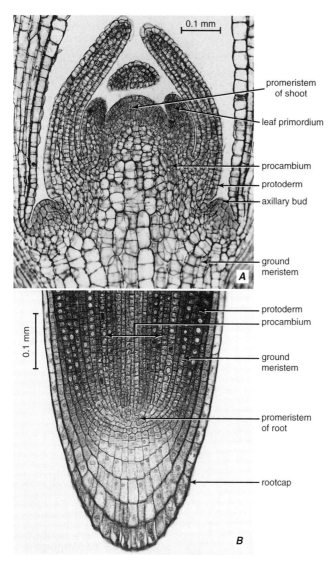

0.1 mm

promeristem of shoot

leaf primordium

procambium

protoderm

axillary bud

ground meristem

*A*

protoderm

procambium

ground meristem

promeristem of root

rootcap

*B*

0.1 mm

**FIGURE 5.1**

Shoot tip (A) and root tip (B) of seedling of flax (*Linum usitatissimum*) in longitudinal sections. Both illustrate apical meristems and derivative primary meristematic tissues, protoderm, ground meristem, and procambium, the least determined part of which is called promeristem. **A**, primordia of leaves and axillary buds are present. **B**, the rootcap covers the apical meristem. Note the files, or lineages, of cells behind root apical meristem. (**A**, from Sass, 1958. © Blackwell Publishing; **B**, from Esau, 1977.)

(Chapter 6). Suffice to point out here that, according to a common view, certain cells in the meristems act as initials mainly because they occupy the proper position for such activity.

The retention of meristems and the ability to produce new organs distinguish the higher plant from the higher animal. In the latter a fixed number of organs is produced during embryonic development, although the

adult tissues and organs in the fully grown animal are maintained throughout life by populations of cells that reside within the tissue or organ. These cells, called **stem cells** (Weissman, 2000; Fuchs, E., and Segre, 2000; Weissman et al., 2001; Lanza et al., 2004a, b), have been compared to the initials of plant meristems, and the term stem cell, for initial, has been adopted by a number of plant biologists. The term stem cell has not been adopted in this book.

Although animal stem cells may be analogous to the initial cells of plant meristems, they are not equivalent to them. Initials are **totipotent** (from the Latin *totus*, entire); that is, they have the capacity to produce the entire spectrum of cell types, even to develop into complete plants. In most animals only the zygote, or fertilized egg, is truly totipotent. Embryonic stem cells are very nearly totipotent, but early in ontogeny (soon after the blastocyst stage) they give rise to adult stem cells, which are **pluripotent** and restricted in the number of cell types they can produce—typically to those of the adult tissues and organs in which they reside (Lanza et al., 2004a, b). Some adult stem cells reportedly are able to migrate (a trait not shared by initials) from their original niches and assume the morphologies and functions typical of their new environment (Blau et al., 2001).

Many living cells in a mature plant part remain developmentally totipotent. Thus the development and organization of a plant may be characterized as having **plasticity** (Pigliucci, 1998), a property that is interpreted as an evolutionary response to the sessile form of life. Being stationary, the plant cannot escape from the environment in which it is growing and must adjust itself to adverse environmental conditions and predation by not undergoing irreversible changes (Trewavas, 1980).

## Classification of Meristems

A Common Classification of Meristems Is Based on Their Position in the Plant Body There are **apical meristems**, that is, meristems located at the apices of main and lateral shoots and roots (Fig. 5.1), and **lateral meristems**, that is, meristems arranged parallel with the sides of the axis, usually that of stem and root. The vascular and cork cambia are lateral meristems.

The third term based on the position of a meristem is **intercalary meristem**. This term refers to meristematic tissue derived from the apical meristem and continuing meristematic activity some distance from that meristem. The word intercalary implies that the meristem is inserted (intercalated) between tissues that are no longer meristematic. The best known examples of intercalary meristems are those in internodes and leaf sheaths of monocots, particularly grasses (Fig. 5.2). These kinds of growth regions contain differentiated conducting tissue elements and eventually are transformed into mature tissues, although their parenchyma

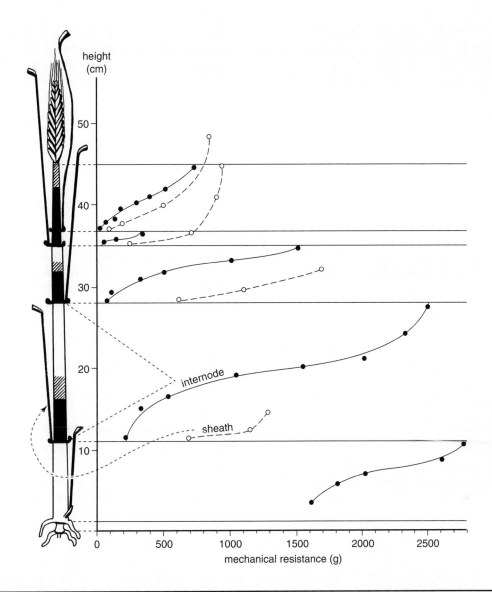

**FIGURE 5.2**

Distribution of growth regions in a culm of rye plant. The plant has five internodes and a spike. The leaf sheaths are represented as extending upward from each node and terminating where leaf blades (shown only in part) diverge from them. The youngest tissue in internodes (intercalary meristems) is represented in black, somewhat older tissues is hatched, and the most mature is left white. Curves to the right indicate mechanical resistance of internodal tissues (solid lines) and of sheaths (broken lines) at the various levels of shoot. Resistance was equated with the pressure, expressed in grams, necessary to make a transverse cut through the internode or sheath. (After Prat, 1935. © Masson, Paris.)

cells long retain the capacity to resume growth (Chapter 7). As meristems, the intercalary meristems are not of the same rank as the apical and lateral meristems because they do not contain cells that can be called initials.

In descriptions of primary differentiation in shoot and root tips, the initiating cells and their most recent derivatives are frequently distinguished from the partly differentiated but still meristematic subjacent tissues under the name of **promeristem** (or **protomeristem**; Jackson, 1953). The subjacent meristematic tissues are classified according to the tissue systems that are derived

from them, namely into **protoderm**, which differentiates into the epidermis; **procambium** (also called **provascular tissue**[*]), which gives rise to the primary vascular tissues; and the **ground meristem**, precursor of the fundamental or ground tissue system. If the term

---

[*] Some workers distinguish between provascular cells and procambial cells, provascular cells being regarded as uncommitted but vascular-competent cells and procambial cells as cells that have already progressed toward differentiation into xylem and phloem (Clay and Nelson, 2002).

meristem is used broadly, protoderm, procambium, and ground meristem are referred to as the **primary meristems** (Haberlandt, 1914). In a more restricted sense of meristem (combination of initials and immediate derivatives), these three tissues constitute partly determined primary meristematic tissues.

The terms protoderm, procambium, and ground meristem serve well for describing the pattern of tissue differentiation in plant organs, and they are correlated with the equally simple and convenient classification of mature tissues into the three tissue systems, epidermal, vascular, and fundamental, reviewed in the first chapter. It seems to be immaterial whether the protoderm, procambium, and ground meristem are called meristems or meristematic tissues as long as it is understood that the future development of these tissues is at least partly determined.

*Meristems Are Also Classified According to the Nature of Cells That Give Origin to Their Initial Cells* If the initials are direct descendants of embryonic cells that never ceased to be concerned with meristematic activity, the resulting meristem is called **primary**. If, however, the initials originate from cells that had differentiated, then resumed meristematic activity, the resulting meristem is called **secondary**.

The cork cambium (phellogen) is a good example of a secondary meristem, for it arises from the epidermis or various parenchymatous tissues in the cortex and deeper layers of the bark. The vascular cambium has a more varied origin related to the organization of the primary vascular system. This system differentiates from the procambium, which is ultimately derived from an apical meristem. Commonly the procambium and the primary vascular tissues originating from it occur in bundles (fascicles) more or less already separated from each other by interfascicular parenchyma (Fig. 5.3A). At the end of primary growth, the remainder of the procambium between the primary xylem and the primary phloem becomes the fascicular part of the cambium. This cambium is complemented by interfascicular cambium arising in the interfascicular parenchyma (Fig. 5.3B). Thus a continuous cylinder of cambium (a ring in transections) is formed, partly fascicular and partly interfascicular in origin. According to the definition of primary and secondary meristems, the fascicular cambium is a primary meristem—a derivative of the apical meristem through the procambium—whereas the interfascicular cambium is a secondary meristem—a derivative of parenchyma tissue that secondarily resumed meristematic activity.

In many woody plants the parts of the cambium originating in the two sites become indistinguishable in later secondary growth. Moreover, in view of the evidence obtained from tissue culture work that living plant cells long retain their potentiality for growth

(Street, 1977), the appearance of cambial divisions in the interfascicular parenchyma does not indicate a major change in the character of the cells involved. Thus the grouping of the vascular cambium partly with the primary and partly with the secondary meristems is purely theoretical. This conclusion, however, does not detract from the value of the classification of mature tissues into primary and secondary as reviewed in Chapter 1.

### Characteristics of Meristematic Cells

Meristematic cells are fundamentally similar to young parenchyma cells. During cell division the cells at the shoot apices are relatively thin-walled, rather poor in storage materials, and their plastids are in proplastid stages. The endoplasmic reticulum is small in amount and mitochondria have few cristae. Golgi bodies and microtubules are present as is characteristic of cells with growing cell walls. The vacuoles are small and dispersed.

Deeper layers of apical meristems may be more highly vacuolated and contain starch (Steeves et al., 1969). In some taxa, notably ferns, conifers, and *Ginkgo*, conspicuously vacuolated cells occur at the highest position in the apical dome (Chapter 6). Before seed germination, the meristems of embryos contain storage materials.

During periods of cell division, the vascular cambium cells are highly vacuolated, with one or two large vacuoles limiting the dense cytoplasm to a thin parietal layer (Chapter 12), which contains rough endoplasmic reticulum and other cell components. During dormancy the vacuolar system assumes the form of numerous interconnected vacuoles. Winter vacuoles sometimes contain polyphenols and protein bodies. At this time the endoplasmic reticulum is smooth and the ribosomes are free in the cytosol.

Meristematic cells are often described as having large nuclei. But the ratio of cell size to nuclear size—the cytonuclear ratio—varies considerably (Trombetta, 1942). Generally, larger meristematic cells have smaller nuclei in proportion to cell size.

The nuclei show characteristic structural variations during changes in mitotic activity (Cottignies, 1977). In the dormant cambium, for example, when the nucleus is blocked in the $G_1$ phase of the mitotic cycle, RNA synthesis is absent and the nucleolus is small, compact, and largely fibrillar in texture. When the cell is active and RNA synthesis is in progress, the nucleolus is large, and has prominent vacuoles and an extensive granular zone, which is intermingled with the fibrillar zone.

The preceding review indicates that meristematic cells vary in size, shape, and cytoplasmic features. In recognition of this variability, the term **eumeristem**

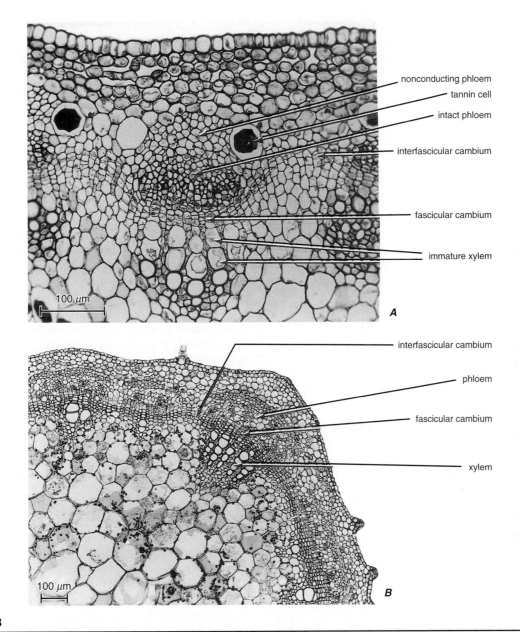

**FIGURE 5.3**

Transverse sections of stems showing an earlier (**A**) and a later (**B**) stage in activity of fascicular and interfascicular cambia. **A**, *Lotus corniculatus*. **B**, *Medicago sativa*, alfalfa. (**A**, courtesy of J. E. Sass; **B**, from Sass, 1958. © Blackwell Publishing.)

(true meristem) has been suggestsed for designating a meristem composed of small, approximately isodiametric cells with thin walls and rich in cytoplasm (Kaplan, R., 1937). For descriptive purposes this term is often convenient to use, but it should not imply that some cells are more typically meristematic than others.

## Growth Patterns in Meristems

Meristems and meristematic tissues show varied arrangements of cells resulting from differences in patterns of cell division and enlargement. Apical meristems having only one initiating cell occupying the very apex (*Equisetum* and many ferns) have an orderly distribution of the recently formed, still meristematic cells (Chapter 6). In seed plants the pattern of cell division appears less precise. It is not random, however, for an apical meristem grows as an organized whole and the divisions and enlargement of individual cells are related to the internal distribution of growth and to the external form of the apex. These correlative influences bring about a differentiation of distinctive zones in the meristems. In some parts of the meristem, the cells may divide sluggishly and attain considerable dimensions; in others, they may divide frequently and remain small. Some cell complexes divide in various planes (volume growth), and others only by walls at right angles to the surface of the meristem (**anticlinal divisions**, surface growth).

The lateral meristems are particularly distinguished by divisions parallel with the nearest surface of the organ (Fig. 5.4A; **periclinal divisions**), which result in establishment of rows of cells parallel with the radii of the axes (radial seriation or alignment) and an increase in thickness of the organ. In cylindrical bodies, such as stems and roots, the term **tangential** division (or tangential longitudinal) is commonly used instead of periclinal division. If the anticlinal division occurs parallel with the radius of the cylinder, it is referred to as a **radial** anticlinal (or radial longitudinal) division (Fig. 5.4B). If the new wall is laid down at right angles to the

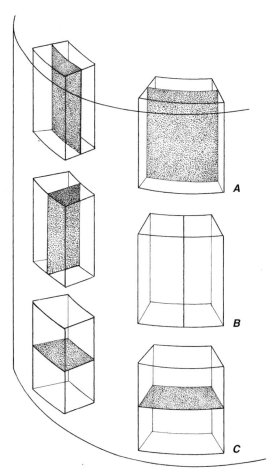

**FIGURE 5.4**

Diagrams illustrating planes of division in a cylindrical plant structure. **A,** periclinal (parallel with the surface). **B,** radial anticlinal (parallel with the radius). **C,** transverse (anticlinal division at right angles to the long axis).

longitudinal axis of the cylinder, the anticlinal division is **transverse** (Fig. 5.4C).

Organs arising at the same apical meristem may assume varied forms because the still-meristematic derivatives of the apical meristem (primary meristems) often exhibit distinct patterns of growth. Some of these patterns are so characteristic that the meristematic tissues showing them have received special names. These are mass meristem (or block meristem), rib meristem (or file meristem), and plate meristem (Schüepp, 1926). The **mass meristem** grows by divisions in all planes and produces bodies that are isodiametric or spheroidal or have no definite shape. Such growth occurs, for example, during the formation of spores, sperms (in seedless vascular plants), and endosperm. Divisions in various planes are associated with the spheroidal shape of many angiosperm embryos at a certain stage of development. The **rib meristem** gives rise to a complex of parallel longitudinal files ("ribs") of cells by divisions at right angles to the longitudinal axis of the cell row. This pattern of growth occurs in elongating cylindrical plant parts as illustrated by the cortex of roots and pith and cortex of stems (Fig. 5.1). The **plate meristem** shows chiefly anticlinal divisions so that the number of layers originally established in the young organ does not increase any further, and a plate-like structure is produced. The flat blades of angiosperm leaves exemplify the result of growth by a plate meristem (Fig. 5.5). The plate meristem and the rib meristem are growth forms that occur mainly in the ground meristem. They are associated with the two basic forms of the plant body, the thin spreading lamina (blade) of the leaf-like organs and the elongated cylindrical plant parts such as root, stem, and petiole.

## Meristematic Activity and Plant Growth

Meristems and their meristematic derivatives are concerned with growth in the broad sense of the term, meaning irreversible increase in size, including volume and surface. In multicellular plants growth is based on two processes, cell division and cell enlargement. The recent derivatives of meristematic initials produce other derivatives by cell division, and the successive generations of derivatives increase in size. Cell enlargement becomes dominant over cell division and, in time, replaces it entirely. When the derivatives farthest from the meristem initials cease to divide and to expand, they acquire the characteristic specific for the tissues in which they are located; that is, the cells differentiate and eventually mature.

Although cell division as such does not contribute to the volume of the growing entity (Green, 1969, 1976), addition of cells is a primary requirement for growth of a multicellular organism. Cell division and cell expansion are different stages of the growth process, in which cell enlargement determines the final size of the plant

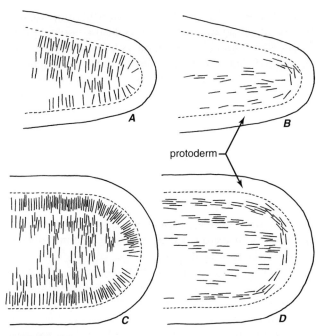

protoderm

**FIGURE 5.5**

Diagrams illustrating meristematic activity at margins of *Lupinus* leaflets of two sizes: **A**, **B**, 600 μm long; **C**, **D**, 8500 μm long. The short lines indicate equatorial planes or cell plates of dividing cells. The division figures were recorded in serial sections and assembled, with anticlinal divisions in **A** and **C** and periclinal divisions in **B** and **D**. Periclinal divisions at margin establish the number of layers in the leaf. Anticlinal divisions extend the layers (plate-meristem activity). (From Esau, 1977; redrawn from Fuchs, 1968.)

and its parts. As an integral step in cell **ontogeny** (development of an individual entity), the enlargement of a cell serves as the transition between the stages of division and maturation (Hanson and Trewavas, 1982).

Cell division rarely occurs without being accompanied by cell enlargement, at least to the extent that the original cell size is maintained in the mass of dividing cells. Some angiosperm embryos are an exception in this regard. During the first two or three cycles of cell division in the zygote, developing into such an embryo, little or no cell expansion is detectable (Dyer, 1976). Cell division without an increase in size of the entity concerned occurs also in the formation of the male gametophyte within the microspore (pollen grain). Likewise cell division is not associated with cell enlargement when a multinucleate endosperm is transformed into a cellular tissue. Commonly, however, growth may be approximately divided into two stages: growth by cell division and limited cell enlargement, and growth without cell division but with much cell enlargement.

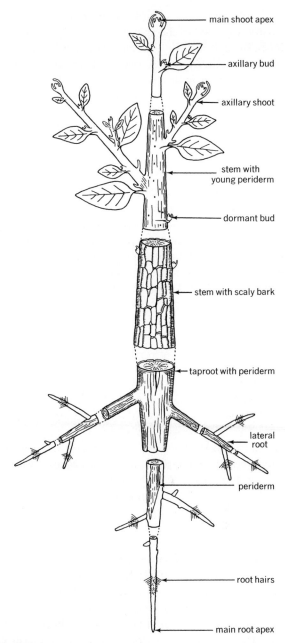

main shoot apex

axillary bud

axillary shoot

stem with young periderm

dormant bud

stem with scaly bark

taproot with periderm

lateral root

periderm

root hairs

main root apex

**FIGURE 5.6**

Diagram of a woody angiospermous plant illustrating branching of shoot and root, increase in thickness of stem and root by secondary growth, and development of periderm and bark on the thickened axis. The apices of main and lateral (axillary) shoots bear leaf primordia of various sizes. Root hairs occur some distance from the apices of the main (taproot) and lateral roots. (From Esau, 1977; adapted from Rauh, 1950.)

Since apical meristems occur at the apices of all shoots and roots, main and lateral, their number in a given plant is relatively large. Vascular plants that undergo secondary increase in thickness of stem and root (Fig. 5.6) possess additionally extensive meristems, the vascular and cork cambia. The primary growth of the plant, initiated in the apical meristems, expands the plant body, determines its height, and through increase in its surface, the area of contact with air and soil. Eventually primary growth also gives rise to the reproductive organs. The secondary growth resulting from the activity of the cambia increases the volume of conducting tissues and forms supportive and protective tissues.

Usually not all of the apical meristems present on a plant are active simultaneously. Suppression of lateral bud growth while the terminal shoot is actively growing (**apical dominance**; Cline, 1997, 2000; Napoli et al., 1999; Shimizu-Sato and Mori, 2001) is a common phenomenon. Then the activity of cambia varies in intensity, and both apical and lateral meristems may show seasonal fluctuations in meristematic activity, with a slowing down or a complete cessation of cell division during winter in temperate zones.

# ▌ DIFFERENTIATION

## Terms and Concepts

The development of a plant consists of the closely interrelated phenomena of growth, differentiation, and morphogenesis. **Differentiation** refers to the succession of changes in form, structure, and function of progenies of meristematic derivatives, and their organization into tissues and organs. One may speak of differentiation of a single cell, a tissue (**histogenesis**), an organ (**organogenesis**), and the plant as a whole. Differentiation also refers to processes by which a fertilized egg cell produces a progeny of cells showing increasing heterogeneity, specialization, and patterned organization. The term is imprecise, especially if it is used to contrast differentiated from undifferentiated cells. A meristematic cell and an egg cell are cytologically complex and are themselves products of differentiation. Describing such cells as undifferentiated is simply a convention (Harris, 1974).

The degree of differentiation and the attendant **specialization** (structural adaptation to a particular function) vary greatly (Fig. 5.7). Some cells diverge relatively little from their meristematic precursors and retain the power to divide (various parenchyma cells). Others are more profoundly modified and lose most, or all, of their meristematic potentialities (sieve elements, laticifers, tracheary elements, various sclereids). Thus cells differentiating in a multicellular body become different from their meristematic precursors, as well as from cells in other tissues of the same plant.

The variously differentiated cells may be called **mature** in the sense that they have reached the kind of specialization and physiological stability that normally characterizes them as components of certain tissues of an adult plant. This definition of maturity must include the qualification that mature cells with complete protoplasts may resume meristematic activity when properly stimulated. The stimulation can be induced by accidental wounding, invasion by parasites, and infection with agents of diseases (Beers and McDowell, 2001). Normal internal stress causing a selective breakdown of tissues may elicit growth reactions resembling "repair reactions." Such reactions are common in barks during the secondary expansion of the axis in width and in abscission regions where leaves or other organs are normally severed from the plant. Stimulation of cells to resume growth by isolation from the plant followed by culture in vitro has been a particularly useful experimental approach to the study of meristematic potentialities of mature cells (Street, 1977).

In studies on resumption of meristematic activity by nonmeristematic cells the terms **dedifferentiation**—loss of previously acquired characteristics—and **redifferentiation**— acquisition of new characteristics—are often used. The entire process is referred to as **transdifferentiation**. Like differentiation, dedifferentiation is not a precise term. Dedifferentiating cells do not revert to the status of a fertilized egg, or even to that of an embryonic cell, but they may lose some of the specialized features and increase the amount of subcellular components involved in DNA and protein synthesis.

Discussions of differentiation may include a reference to determination (McDaniel, 1984a, b; Lyndon, 1998), a phenomenon that may be regarded as one of the aspects of differentiation. **Determination** means progressive commitment to a specific course of development that brings about a weakening or loss of capacity to resume growth. Some cells are determined earlier and more completely than others and some maintain their totipotency after differentiation. Differentiation is associated with growth and both phenomena occur at all morphological levels from subcellular structures to the entire plant. Growth without differentiation is encountered in abnormal structures such as tumors. Callus tissue also may be induced to grow essentially without differentiation.

The term competence appears frequently in discussions of differentiation. As defined by McDaniel (1984a, b), **competence** refers to the ability of a cell to develop in response to a specific signal, such as light. It implies that the competent cell is able to recognize the signal and translate it into a response.

During its development, a plant assumes a specific form. Thus the plant undergoes **morphogenesis** (from Greek words for shape and origin). The term commonly is used with reference to both external form and internal organization, and just like differentiation, morphogenesis is revealed at all levels of organization, from cell components to whole plants. D. R. Kaplan and W. Hagemann (1991) emphasize, however, that although

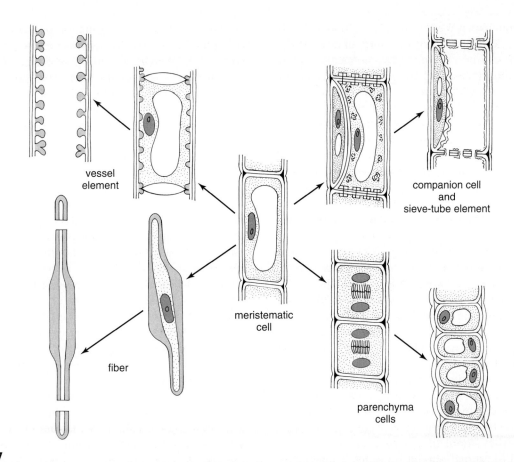

**FIGURE 5.7**

Diagram illustrating some of the cell types that may originate from a meristematic cell of the procambium or vascular cambium. The meristematic cell depicted here (at the center), with a single large vacuole, is typical of the meristematic cells of the vascular cambium. Procambial cells typically contain several small vacuoles. The meristematic cells or precursors of all these cells had identical genomes. The different cell types become distinct from one another because they express sets of genes not expressed by other cell types. Of the four cell types depicted here, the parenchyma cells are the least specialized. Both the mature vessel element, which is specialized for the conduction of water, and the mature fiber, which is specialized for support, lack a protoplast. The mature sieve-tube element, which is specialized for the transport of sugars and other assimilates, retains a living protoplast but lacks a nucleus and vacuole. It depends on its sister cell, the companion cell, for life support. (From Raven et al., 2005.)

some aspects of plant anatomy and morphology are correlated, cell and tissue differentiation follows organogenesis or morphogenesis. Noting a tendency among some plant biologists to confuse anatomical characters for morphological features in interpreting plant developmental mechanisms, D. R. Kaplan (2001) further stresses "whereas the anatomy may be determined by the morphology . . . , the anatomy does not determine the morphology."

## Senescence (Programmed Cell Death)

The natural termination of life of a plant as a result of senescence may be regarded as a normal stage in plant development, a sequel to the events of differentiation and maturation (Leopold, 1978; Noodén and Leopold,

1988; Greenberg, 1996). The term **senescence** specifically refers to the series of changes in a living organism that lead to its death (Noodén and Thompson, 1985; Greenberg, 1996; Pennell and Lamb, 1997). Senescence may affect the whole organism or some of its organs, tissues, or cells. Annual plants that bloom only once in their lifetime (**monocarpy**: fruiting once only) senesce within one season. In deciduous trees the leaves commonly senesce at the end of seasonal growth. Fruits ripen and senesce in a few weeks, isolated leaves and flowers in a few days. Senescing individual cells are rootcap cells, which are continuously shed from a growing root. Since senescence occurs in orderly sequences in the life of the plant and is an active degenerative process, it is considered to be genetically controlled, or programmed—a process of **programmed**

**cell death** (Buchanan-Wollaston, 1997; Noodén et al., 1997; Dangl et al., 2000; Kuriyama and Fukuda, 2002).

Senescence can be controlled by chemicals, including growth substances, and by environmental conditions (Dangl et al., 2000). Treatment of soybean leaves with auxin and cytokinins, for example, prevented senescence normally induced by seed development (Thimann, 1978). The treated leaves maintained their photosynthetic activity and nitrogen assimilation instead of releasing the reserves to the reproductive structures and becoming senescent. In contrast, senescence can be induced by ethylene (Grbić and Bleecker, 1995), which stimulates the expression of an array of senescence-associated genes (SAGs; Lohman et al., 1994).

Although the term senescence is derived from the Latin *senesco*, to grow old, it is not considered to be synonymous with the word aging (Leopold, 1978; Noodén and Thompson, 1985; Noodén, 1988). Like senescence, aging is an integral part of a life cycle of an individual organism and is not easily distinguishable from senescence. **Aging** is defined as an accumulation of changes that lower the vitality of a living entity without being lethal themselves. Aging, however, may lead into senescence. The ambiguity of the term aging has been enhanced by its use in experimental work to denote the practice of culturing slices of storage tissue in conditions that stimulate increased metabolic activity. This kind of "aging" should be called rejuvenation (Beevers, 1976).

Common changes in senescing cells of leaves are a decline of chlorophyll content, increase in amount of red (anthocyanin) and yellow (carotenoids) pigments, proteolysis and decrease in content of nucleic acids, and an increase in leakiness of cells (Leopold, 1978; Huang et al., 1997; Fink, 1999; Jing et al., 2003). The leakiness is associated with disorganization of membrane lipids (Simon, 1977; Thompson et al., 1997). In naturally senescing wheat leaves, chloroplasts accumulate lipids in the form of plastoglobuli, grana and intergrana lamellae become distended and break up into vesicles, the stroma disintegrates, and finally the plastid envelope breaks and releases the organelle contents (Hurkman, 1979). During senescence much of the cell's biochemical processes are directed toward salvaging and redistributing metabolites and structural materials, especially reserves of nitrogen and phosphorous. Peroxisomes are converted to glyoxysomes, which convert lipids to sugars. In green cells most of the protein is represented by Rubisco, which is located in the stromal phase of the chloroplast. Thus far over 100 so-called senescence-associated genes—the expression levels of which are up-regulated during leaf senescence—have been identified in diverse plant species (see literature cited in Jing et al., 2003).

Other examples of programmed cell death in plants include the maturation of tracheary elements (Chapter 10; Fukuda, 1997); the formation of aerenchyma tissue (Chapter 7) in roots in response to oxygen deficiency (hypoxia) brought about by flooding (Drew et al., 2000); destruction of the suspensor during embryogeny (Wredle et al., 2001); death of three of the four megaspores during megagametogenesis; death of cereal aleurone cells following their production of large amounts of $\alpha$-amylases, which are required for the breakdown and mobilization of starch to provide an energy source for the developing seedling (Fath et al., 2000; Richards et al., 2001); and the developmental remodeling of leaf shape (Gunawardena et al., 2004). Programmed cell death also plays an important role in resistance against pathogens (Mittler et al., 1997). The rapid cell death—known as the **hypersensitive response** or **HR**—that occurs in response to pathogen attack is closely related to active resistance (Greenberg, 1997; Pontier et al., 1998; Lam et al., 2001; Loake, 2001). The exact process by which the HR resists pathogens is still problematic. It has been suggested that the HR directly kills the pathogen and/or limits its growth by interfering with its acquisition of nutrients (Heath, 2000).

Programmed cell death in plants is triggered by hormonal signals, involves changes in cytosolic $Ca^{2+}$ concentrations (He et al., 1996; Huang et al., 1997), and activation of hydrolytic enzymes sequestered in the vacuole. With collapse of the vacuole, the enzymes are released, allowing them to attack the nucleus and cytoplasmic components of the protoplast. Ethylene induces programmed cell death and aerenchyma formation in roots following hypoxia and, as mentioned previously, promotes leaf senescence (He et al., 1996; Drew et al., 2000). When added to tobacco TBY-2 cells that had just completed the S-phase, ethylene resulted in a substantial peak of mortality at the $G_2/M$ checkpoint of the cell cycle, lending support to the hypothesis that programmed cell death can be tightly linked to cell cycle checkpoints (Herbert et al., 2001). Programmed cell death in aleurone cells is initiated by gibberellic acid (Fath et al., 2000), and brassinosteroids have been shown to induce programmed cell death in tracheary elements (Yamamoto et al., 2001).

The terms programmed cell death and apoptosis are often used interchangeably. **Apoptosis**, however, was originally used to refer to particular features of programmed cell death in animal cells (Chapter 2; Kerr et al., 1972; Kerr and Harmon, 1991). Those features include nuclear shrinkage, chromosome condensation, DNA fragmentation, cellular shrinkage, membrane blebbing, and the formation of membrane-bound "apoptotic bodies" that are engulfed and degraded by adjacent cells. Thus far, none of the programmed cell deaths recorded for plant cells possess all of the features characteristic of apoptosis (Lee and Chen, 2002; Watanabe et al., 2002; and literature cited therein).

## Cellular Changes in Differentiation

During differentiation, histologic diversity results from changes in the characteristics of individual cells and from alterations in intercellular relationships. The common features of more or less differentiated cells, including the structure and function of individual cell components, are described in Chapters 2 and 3. The changes in cell wall structure during cell differentiation are considered in Chapter 4. Differential increase in thickness of primary and secondary cell walls, changes in cell wall texture and chemistry, and development of special sculptural patterns introduce differences among cells.

A Cytologic Phenomenon Commonly Observed in Differentiating Cells of Angiosperms Is Endopolyploidy **Endopolyploidy** refers to a condition that arises from a DNA replication cycle within the nuclear envelope and without spindle formation. Hence the newly formed strands of DNA remain in the same nucleus, and the nucleus becomes polyploid. This type of DNA replication cycle is called an **endocycle** (Nagl, 1978, 1981). In some endocycles structural changes resembling those seen in mitosis occur, and the replicated strands of DNA become separate chromosomes (an endomitotic cycle). The most common endocycle in plants is the **endoreduplication**, or **endoreplication**, **cycle**, in which no mitosis-like structural changes take place (D'Amato, 1998; Traas et al., 1998; Joubès and Chevalier, 2000; Edgar and Orr-Weaver, 2001). During endoreduplication, polytene chromosomes are formed. Such chromosomes contain numerous strands of DNA attached side by side in the form of a cable. **Polyteny** thus results from the replication of DNA without separation of the sister chromosomes and hence without change in the chromosome number.

Endocycles are sometimes interpreted as exceptional phenomena without functional significance. According to another view, growth involving endocycles has important advantages because it provides a mechanism to increase the level of gene expression (Nagl, 1981; Larkins et al., 2001). In addition, during an endocycle, RNA synthesis is not interrupted as it is during the mitotic cycle. Hence the cell has continuously high rates of RNA and protein syntheses, activities that promote rapid growth and an early entry into a functional state. In contrast, growing tissues, in which mitotic activity is taking place, show a delay in assuming mature physiological activity. While, for example, the embryo proper of *Phaseolus* still shows meristematic activity, the suspensor containing polytene chromosomes is a metabolically highly active structure providing nutrients for the growing embryo.

There is increasing evidence that a positive correlation exists between ploidy level and cell size (Kondorosi et al., 2000; Kudo and Kimura, 2002; Sugimoto-Shirasu et al., 2002). Endoreduplication therefore may be an important strategy of cell growth (Edgar and Orr-Weaver, 2001). It may also be required for differentiation of specific cell types. In *Arabidopsis*, for example, trichome initiation is closely linked with the onset of endoreduplication (Chapter 9; Hülskamp et al., 1994).

As noted by Nagl (1978), endocycles can be understood as an evolutionary strategy. Phyla with species having rather low basic nuclear DNA contents throughout always display endopolyploidy, whereas those with species having mainly high DNA values lack it. As proposed by Mizukami (2001), "presumably, endoreduplication has evolved as a developmental means of providing differential gene expression in species with a small genome."

One of the Early Visible Changes in Differentiating Tissues Is the Unequal Increase in Cell Size Some cells continue to divide with small increase in size, whereas others cease dividing and enlarge considerably (Fig. 5.8). Examples of differential growth in cell size are found in the elongation of procambial cells and lack of similar elongation in the adjacent parenchyma cells of pith and cortex; elongation of protophloem (first-formed) sieve elements in roots, in contrast to the adjacent pericyclic cells that continue to divide transversely (Fig. 5.8A); widening of vessel elements in contrast to the surrounding cells that remain narrow (Fig. 5.8E). Size differences between two adjacent cells may result also from asymmetric, or unequal, cell divisions, which generate cells with different fates (Gallagher and Smith, 1997; Scheres and Benfey, 1999). For example, during pollen formation, an asymmetric cell division produces a larger tube, or vegetative, cell and a smaller generative cell (Twell et al., 1998). In some plants, root hairs develop from cells that are the smaller of two sister cells formed by the asymmetric division of a protodermal cell (Fig. 5.8B, C; Chapter 9). Unequal divisions also occur in the formation of stomata (Chapter 9; Larkin et al., 1997; Gallagher and Smith, 2000).

The increase in cell size may be relatively uniform along all diameters, but frequently the cell enlarges more in one direction than in others. Such cells may become strikingly different in shape from their meristematic precursors (long primary phloem fibers, branched sclereids). Many, however, become modified in a more moderate manner by increasing the number of cell facets but retaining a polyhedral shape (Hulbary, 1944).

The predominant cell arrangement in a tissue may be determined by the growth form of its meristem (rib meristem, plate meristem). The relative position of walls in contiguous cell rows also gives a distinctive appearance to a tissue (Sinnott, 1960). The walls at right angles to the cell row may alternate with one another in adjacent rows or may occur in the same plane.

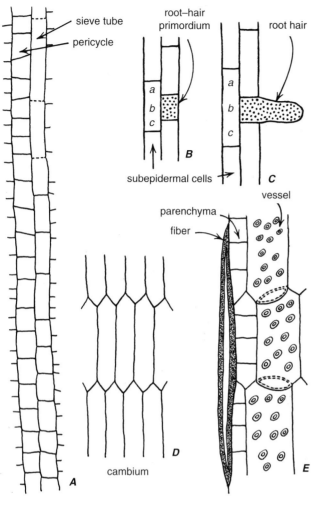

**FIGURE 5.8**

Intercellular adjustments during tissue differentiation. **A,** series of cells from the root tip of tobacco. Parenchyma cells continued dividing; phloem cells ceased dividing and began elongating. **B, C,** development of root hair from the smaller of two sister cells resulting from transverse division of protodermal cells. **C,** the root-hair cell is extended at a right angle to the root but not in the direction of the root's elongation. Apparently in the cell adjacent to the root hair, wall parts *a* and *c* continued to elongate, whereas part *b* ceased to elongate after the root-hair primordium was formed. **D, E,** the cambium and xylem that can develop from such cambium, both in tangential sections. **E,** shows the results of following developmental changes in the cambial derivatives: parenchyma cells were formed by transverse divisions of such derivatives, vessel elements expanded laterally, and the fiber elongated by apical intrusive growth.

**Intercellular Adjustment in Differentiating Tissue Involves Coordinated and Intrusive Growth** Enlargement and change in shape of cells in a differentiating tissue are accompanied by more or less profound changes in the spatial relations between cells. A familiar phenomenon is the appearance of intercellular spaces along the line of union of three or more cells (Chapter 4). In some tissues, the formation of intercellular spaces does not change the general arrangement of cells; in others, it profoundly modifies the appearance of the tissue (Hulbary, 1944). The role of the cytoskeleton, particularly of microtubules and orientation of cellulose microfibrils in the shaping of cells, is considered in Chapter 4.

With regard to the growth of cell walls during tissue differentiation, two kinds of intercellular adjustments are visualized: (1) contiguous growing wall layers belonging to two neighboring cells do not separate but expand jointly; (2) contiguous wall layers do separate and the growing cells intrude into the resulting spaces. The first method of growth, originally named symplastic growth (Priestley, 1930), is common in organs expanding during their primary growth. Whether all cells in a complex are dividing, or whether some have ceased to divide and are increasing in length and in width, the walls of contiguous cells appear to grow in unison without separation or buckling. In this **coordinated growth** it is possible that part of the common wall between two cells is expanding and another is not.

The intercellular adjustment that involves an intrusion of cells among others is called **intrusive growth** (Sinnott and Bloch, 1939) or *interpositional growth* (Schoch-Bodmer, 1945). The evidence for such growth is based on light-microscope observations (Bailey, 1944; Bannan, 1956; Bannan and Whalley, 1950; Schoch-Bodmer and Huber, 1951, 1952). It is common in elongating cambial initials, primary and secondary fibers in vascular tissues, tracheids, laticifers, and some sclereids. Intrusive growth may be exceptionally intensive, as in certain woody Liliaceae in which the secondary tracheids become 15 to 40 times longer than their meristematic counterparts (Cheadle, 1937). The elongating cells grow at their apices (**apical intrusive growth**), usually at both ends. The localization of specific expansion gene expression to the ends of differentiating *Zinnia* xylem cells indicates that expansins may be involved in the elongation by intrusive growth of the primary cell walls of such cells (Im et al., 2000). The intercellular material through which the elongating cells intrude is probably hydrolyzed in front of the advancing tip, and the primary walls of adjacent cells become separated from each other in the same manner as they do during the formation of intercellular spaces (Chapter 4).

If plasmodesmata are present, they are probably ruptured by the intrusion of the growing cell. The reported separation of members of pairs of primary pit-fields

(Neeff, 1914) indicates occurrence of such rupture. Pit-pairs later appear between pairs of cells that come in contact through intrusive growth (Bannan, 1950; Bannan and Whalley, 1950). Such pit-pairs are characterized by the presence of secondary plasmodesmata (Chapter 4). Intrusive growth is associated also with the lateral expansion of cells that attain considerable width as do, for example, the vessel elements in the xylem (Chapter 10).

Early botanists presumed that gliding (or sliding) growth occurred during differential elongation or lateral expansion of cells intruding among other cells. In gliding growth a section of wall of a growing cell was supposed to separate from and glide over the wall of the adjacent cell (Krabbe, 1886; Neeff, 1914). This concept was replaced by that of intrusive growth. Whether such localized cell extension involves some gliding of the new wall part over the old with which a new contact is made (Bannan, 1951), or whether the new wall is apposed along the free surfaces of the cells that were separated (interposition; Schoch-Bodmer, 1945), remains uncertain.

## CAUSAL FACTORS IN DIFFERENTIATION

Studies on differentiation and morphogenesis include observations on plants developing normally and those the development of which is subjected to experimental manipulations. Examples of experimental treatments are use of growth-regulating chemicals, surgical procedures, exposure to radiation, confinement to controlled temperatures and light, interference with normal gravity effects, and growth under selected day length conditions. Considerable evidence has been gathered on effects of mechanical perturbations on plants. Slight rubbing or bending of stems causes marked retardation of growth in length and an increase in radial growth of all species tested. The response to mechanical interference is termed **thigmomorphogenesis** (Jaffe, 1980; Giridhar and Jaffe, 1988), *thigm* meaning touch in Greek. In nature, wind is apparently the environmental factor most responsible for thigmomorphogenesis. The molecular genetic approach to the study of differentiation and morphogenesis, involving the identification of mutations that perturb processes of interest, has contributed greatly to our understanding of factors regulating various aspects of plant development (Žárský and Cvrčková, 1999).

### Tissue Culture Techniques Have Been Useful for the Determination of Requirements for Growth and Differentiation

Studies on intact plants and on those treated experimentally clearly showed that the organized patterned development of a higher plant depends on internal controlling mechanisms the action of which is modified to a small or larger degree by environmental factors (Steward et al., 1981).

The demands of patterned differentiation in a plant place a restraint on the meristematic potentialities of cells. When this restraint is interrupted by excision from the organized body of the plant, living cells are able to resume growth. In research on tissue culture in vitro (outside a living organism) the release of cells from controlling mechanisms in the intact plant is utilized to explore conditions that favor meristematic activity or, to the contrary, induce differentiation and morphogenesis (Gautheret, 1977; Street, 1977; Williams and Maheswaran, 1986; Vasil, 1991). Since the capacity of a cell to respond with growth to stimuli provided in a tissue culture is not necessarily predictable (Halperin, 1969), much of the tissue culture research has dealt with testing of explants from different taxa and different parts of a plant for their meristematic potentialities. The other aim of this research has been the study of effects on the explant of the various constituents of the culture medium, particularly of the growth-regulating substances. Originally plant tissue culture served mainly in specialized botanical research. Later the technique came into wide use for propagation of economically important plants, obtaining disease-free plants, and culturing cells and tissues as sources of medicinal and other kinds of plant constituents (Murashige, 1979; Withers and Anderson, 1986; Jain et al., 1995; Ma et al., 2003). The study of isolated *Zinnia* mesophyll cells in culture has provided valuable information on cell differentiation and programmed cell death in plants (Chapter 10).

In early work in tissue culture, secondary phloem tissue from carrot root was popular experimental material (Fig. 5.9; Steward et al., 1964). Cultured in dissociated state in a liquid medium containing coconut endosperm, the explants first developed into randomly proliferating callus tissue, then produced a more orderly type of growth: nodules with centrally located xylem and phloem outside the xylem (Esau, 1965, p. 97). The nodule eventually gave rise to roots, then to shoots opposite the roots. The resulting plantlets assumed the form of young carrot plants, which when transferred to soil formed typical tap roots and flowered. Isolated carrot cells can undergo morphogenesis in other ways than through callus (Jones, 1974). Often small, sparsely vacuolated cells become detached from primary explants and assume the form of **embryoids**, plantlets that resemble zygotic embryos in their development into plants. The process of initiation and development of embryoids from plant somatic cells is termed **somatic embryogenesis** (Griga, 1999).

Refinements of techniques made possible the isolation of protoplasts by enzymic removal of the cell wall from single cells. Such protoplasts make the plasma

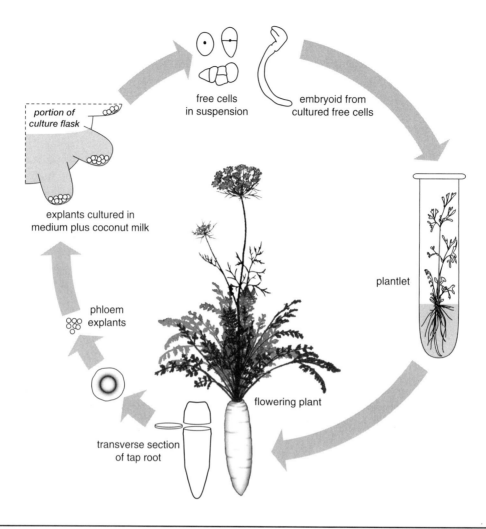

**FIGURE 5.9**

Development of carrot plants from cells in tissue culture. Cultured cells are obtained from phloem of the carrot tap root. (Adapted, with permission, from F. C. Steward, M. O. Mapes, A. E. Kent, and R. D. Holsten. 1964. Growth and development of cultured plant cells. *Science* 143, 20-27. © 1964 AAAS.)

membrane accessible for a wide range of experiments. Isolated protoplasts may be induced to fuse and produce somatic hybrids, a technique particularly valuable with plants for which plant breeding methods have limited success, for example, the potato plant (Shepard et al., 1980). Isolated protoplasts eventually regenerate the cell wall and may undergo division and produce whole plants (Power and Cocking, 1971; Lörz et al., 1979). Today genetic engineering—the application of recombinant DNA technology—allows individual genes to be inserted into plant cells, with or without the wall removed, in a way that is both precise and simple (Slater et al., 2003; Peña, 2004; Poupin and Arce-Johnson, 2005; Vasil, 2005). In addition the species involved in the gene transfer do not have to be capable of hybridizing with one another.

Much research in cell culture has been based on the use of anthers and pollen grains (Raghavan, 1976, 1986;

Bárány et al., 2005; Chanana et al., 2005; Maraschin et al., 2005). Under appropriate culture conditions, pollen grains enclosed in anthers can give rise to embryoids, which are liberated when the anthers open. For isolated pollen cultures, anthers or whole flower buds are ground in a liquid medium, the suspension is filtered to isolate the pollen, which is then cultured in suspension or on agar.

In successful culture the pollen grain is diverted from the normal gametophytic development into vegetative sporophytic development leading to embryoid formation, directly or through callus growth (Geier and Kohlenbach, 1973). This process is called **androgenesis**. The common pathway is through the vegetative cell (Sunderland and Dunwell, 1977; Chapter 9 in Street, 1977).

Since a pollen grain possesses only one genome, haploid plants are obtained. These have many uses in plant breeding and have been especially important in mutation research. Induced mutations are immediately expressed in the haploid phenotype, whereas in a diploid higher plant the normally recessive mutations are revealed only in the progeny of a mutagenized plant.

## The Analysis of Genetic Mosaics Can Reveal Patterns of Cell Division and Cell Fate in Developing Plants

The term **genetic mosaic** refers to a plant in which cells of different genotypes occur. In flowering plants genetic mosaics, referred to as **chimeras**, occur in the shoot apical meristem (Fig. 5.10) (Tilney-Bassett, 1986; Poethig, 1987; Szymkowiak and Sussex, 1996; Marcotrigiano, 1997, 2001). In some apical meristems, entire parallel layers of cells are found that differ genetically from one another. Such chimeras are called **periclinal chimeras**. In others, only part of a layer (or layers) is genetically different (**mericlinal chimeras**); in still others, a sharply defined boundary of genetically dissimilar cells extends through all layers (**sectorial chimeras**). The differences serve as markers that may be followed through continuous cell lineages to similar differences in the cell layers of the apical meristem. Some chimeras have combinations of layers with diploid and polyploid nuclei (**cytochimeras**). Polyploidization

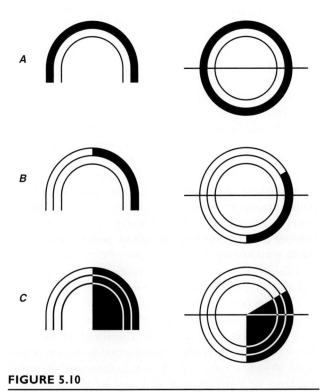

**FIGURE 5.10**

Chimerical shoot apices. **A**, periclinal; **B**, mericlinal; **C**, sectorial. On the right, transverse sections are shown; on the left, longitudinal sections in the plane indicated by the line.

of nuclei may be induced by treating shoot apices with colchicine (Fig. 5.11). As a result one or another layer in the apical meristem becomes populated with polyploid nuclei, and the change is propagated through the progeny of the layer in the differentiating plant body (Dermen, 1953). Periclinal chimeras are also available among mutants with defective, colorless plastids. As in the nuclear chimeras the deviating feature—the defective plastids in this instance—can be traced by lineage between the apical meristem and the mature tissues (Stewart et al., 1974). Anthocyanin pigmentation is another common marker.

Another type of genetic mosaic is one in which clones of genetically different cells are scattered about the plant body (Fig. 5.12). Simply referred to as genetic mosaics, such clones may be generated experimentally with ionizing radiation. The resulting chromosomal rearrangements allow the phenotypic expression of recessive cell-autonomous mutations. The cell lineages, or clones, derived from such cells are permanently marked and analysis of the clones can be used to generate fate maps of cells from any region of the plant body. As noted by Poethig (1987), who has studied leaf development by **clonal analysis** (Poethig, 1984a; Poethig and Sussex, 1985; Poethig et al., 1986), this technique is not, however, a substitute for histological studies on plant development. To interpret clonal patterns accurately, "it is essential to have a clear understanding of the developmental histology and morphology of the system in question" (Poethig, 1987).

## Gene Technologies Have Dramatically Increased Our Understanding of Plant Development

Ultimately it is the genes that determine the characteristics of the plant. Advances in DNA-sequencing technology has made it possible to sequence entire genomes and has given rise to the new science of genomics. **Genomics** encompasses the study of the content, organization, and function of entire genomes (Grotewold, 2004). The first plant genome sequenced in its entirety is that of *Arabidopsis thaliana* (*Arabidopsis* Genome Initiative, 2000). More recently, a highly accurate sequence of the rice (*Oryza sativa*) genome has been completed (International Rice Genome Sequencing Project, 2005).

A broad aim of genomics is to identify genes, to determine which genes are expressed (and under what conditions), and to determine their function or that of their protein products. How does one go about determining the function of a gene? A very successful procedure has been to uncover mutations that have a visible, or phenotypic, effect on plant development. Large populations of mutagen-treated *Arabidopsis* plants have been screened for such mutations. Collections of mutants, in which genes have been inactivated by insertion of a

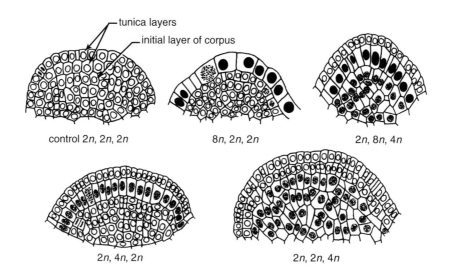

**FIGURE 5.11**

Shoot apices of *Datura* from a diploid plant (upper left) and from several periclinal cytochimeras. Chromosomal combinations are indicated by values given below each drawing. The first figure of each group of three refers to the first tunica layer; the second, to the second tunica layer; the third, to the initial layer of the corpus. The three layers commonly are designated L1, L2, and L3. Octoploid cells are the largest, and their nuclei are shown in black for emphasis; tetraploid cells are somewhat smaller, and their nuclei are stippled; diploid cells are the smallest, and their nuclei are shown by circles. Chromosomal characteristics of tunica layers are perpetuated only in these layers and their derivatives (anticlinal divisions in tunica); those of the initial layer of the corpus are immediately transmitted to the subjacent layers (divisions in various planes). (After Satina et al., 1940.)

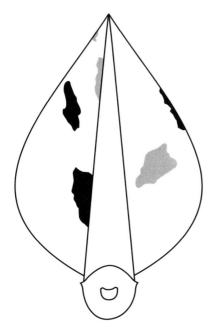

**FIGURE 5.12**

Clones in the subepidermal layer of a tobacco (*Nicotiana tabacum*) leaf irradiated prior to the initiation of the blade (axis = 100 μm). At this stage, clones are usually confined to either the upper (black) or lower (gray) subepidermal layer. None of the clones extend from the margin to the midrib. (Redrawn from Poethig, 1984b. In: *Pattern Formation*. Macmillan. © 1984. Reproduced with permission of McGraw-Hill Companies.)

large piece of DNA such as the T-DNA of *Agrobacterium tumefaciens* also are being developed for a large number of genes (Bevan, 2002). These so-called **knockout mutants**, each with a different gene inactivated, are then screened for changes in their phenotype or how they function in a defined environment. Any of the changes identified are traced back to the specific sequence mutated. With regard to *Arabidopsis*, genes have been identified that are responsible for major events in embryogenesis (Laux and Jürgens, 1994), for the formation and maintenance of the shoot apical meristem (Bowman and Eshed, 2000; Doerner, 2000b; Haecker and Laux, 2001), and for the control of flower formation and floral organ development (Theißen and Saedler, 1999).

A series of processes intervene between the primary action of genes and their final expression. The explanation of the initial effect of genes on differentiation is sought at the molecular level. It is discussed in terms of gene activation and repression, transcription (the synthesis of messenger RNA from a portion of one strand of the double-stranded DNA helix), and translation (the synthesis of a polypeptide directed by the nucleotide sequence of mRNA). The difference between the many cell types of a multicellular organism is the result of **selective gene expression**—that is, only certain genes are expressed and transcribed into mRNA. As a result the proteins that mediate cell differentiation are synthesized selectively. In a given cell some genes are expressed continuously, others only as their products are needed, and still others not at all. The mechanisms that control gene expression—that turn genes "on" and "off"—are collectively called **gene regulation**.

## Polarity Is a Key Component of Biological Pattern Formation and Is Related to the Phenomenon of Gradients

Polarity refers to the orientation of activities in space. It is an essential component of biological pattern formation (Sachs, 1991). Polarity manifests itself early in the life of the plant. It is displayed in the egg cell, in which the nucleus is located in the chalazal end and a large vacuole in the micropylar end, and in the bipolar development of the embryo from the zygote. It is later expressed in the organization of the plant into root and shoot and is also evident in various phenomena at the cellular level (Grebe et al., 2001). Transplantation studies (Gulline, 1960) and tissue culture studies (Wetmore and Sorokin, 1955) indicate that polarity is exhibited not only by the plant as a whole but also by its parts, even if these are isolated from the plant. Polarity in stems is a familiar phenomenon. In plants that are propagated by stem cuttings, for example, roots will form at the lower end of the stem, leaves and buds at the upper end. Moreover polarity cannot be changed by reversing a stem cutting by planting it upside down.

The stability of polarity has been clearly demonstrated in an experiment with centrifugation of differentiating fern spores (Bassel and Miller, 1982). The normal first division of the spore in *Onoclea sensibilis* is preceded by migration of the nucleus from the center of the ellipsoidal spore to one end of it. A highly asymmetric division follows. The large cell forms a protonema, and the small cell develops into a rhizoid. Centrifugation of the spore does not change this pattern of division, even though the contents of the spore become displaced and stratified. Only when the centrifugation occurs immediately before or during mitosis or cytokinesis is the asymmetric division blocked.

Polar behavior of individual cells in the intact plant is illustrated by unequal divisions that result in physiologically, and often also morphologically, different daughter cells (Gallagher and Smith, 1997). In the epidermis of some roots, unequal divisions occur after which the smaller of the two products of division forms a root hair. Before the division most of the organelle-rich cytoplasm accumulates at either the proximal end (the end toward the root apex) or distal end of the cell. The nucleus migrates in the same direction and then divides. Cell-plate formation separates the small future root-hair cell from the longer epidermal cell that produces no root hair (Sinnott, 1960). Biochemical differences

between the two kinds of cell also become evident (Avers and Grimm, 1959). The common presumption is that unequal divisions depend on polarization of cytoplasm, for there is no evidence of unequal distribution of chromosomal material (Stebbins and Jain, 1960).

Polarity is related to the phenomenon of gradients, since the differences between the two poles of the plant axis appear in graded series. There are physiological gradients, for example, those expressed in the rates of metabolic processes, in concentration of auxins, and in concentration of sugar in the conducting system; there are also gradients in anatomical differentiation and in the development of the external features (Prat, 1948, 1951). The plant axis shows transitional anatomical and histological characteristics in the transition between root and stem. The differentiation of derivatives of meristems, in general, occurs in graded series, and adjacent but different tissues may show different gradients. Externally, graduated development is seen in the change in the form of the successive leaves along the axis, from the usually smaller and simpler juvenile form to the large and more elaborate adult form (Fig. 5.13). Subsequently, after the reproductive state is induced, smaller leaves again are gradually produced, the series becoming completed with inflorescence bracts that support subdivisions of the inflorescence or the individual flowers.

## Plant Cells Differentiate According to Position

Although cellular differentiation depends on the control of gene expression, the fate of a plant cell—that is, what kind of cell it will become—is determined by its final position in the developing organ. Even though distinct cell lineages may be established, such as those in a root, position not lineage determines cell fate. The concept that the function of a cell in a multicellular organism is early determined by its position in that organism goes back to the second half of the nineteenth century (Vöchting, 1878, p. 241). It was not until the early 1970s, however, that occasional cell displacements, noted in chimeras, provided evidence that cell fate in stem and leaf was determined by position rather than by lineage, even at late stages of development (Stewart and Burk, 1970). Since then conclusive evidence has accumulated through analysis of genetic mosaics that the position of a cell, not its clonal origin, determines its fate (Irish, 1991; Szymkowiak and Sussex, 1996; Kidner et al., 2000). If an undifferentiated cell is displaced from its original position, it will differentiate into a cell type appropriate to its new position, without any effect on the organization of the plant (Tilney-Bassett, 1986). Laser ablation experiments on *Arabidopsis* root tips (van den Berg et al., 1995) also have shown that ablated cells can be replaced by cells from other lineages, the "replacement cells" differentiating according to their new position.

**FIGURE 5.13**

First 10 leaves from the main shoot of a potato plant (*Solanum tuberosum*). The leaves undergo a transition from simple to pinnately compound. (×0.1. From McCauley and Evert, 1988.)

Inasmuch as the fate of plant cells is dependent on their position within the plant, it is obvious that the cells must be able to communicate with each other, that is, to exchange positional information. Positional information has been shown to play a role in the differentiation of photosynthetic cell types in the maize leaf (Langdale et al., 1989), the spacing of trichomes in the leaf epidermis of *Arabidopsis* (Larkin et al., 1996), and in maintaining a balance of cell types in the shoot and root apical meristems of *Arabidopsis* (Scheres and Wolkenfelt, 1998; Fletcher and Meyerowitz, 2000; Irish and Jenik, 2001). The mechanistic basis of cell-to-cell signaling in plants remains to be elucidated. Some signaling processes in plants appear to be mediated by transmembrane receptor-like kinases (Irish and Jenik, 2001); others employ plasmodesmata (Chapter 4; Zambryski and Crawford, 2000).

# PLANT HORMONES

**Plant hormones**, or **phytohormones**, are chemical signals that play a major role in regulating growth and development, and hence are briefly considered here (Davies, P. J., 2004; Taiz and Zeiger, 2002; Crozier et al., 2000; Weyers and Paterson, 2001). The term hormone (from the Greek *horman*, meaning to set in motion) was adopted from animal physiology. The basic feature of animal hormones—that they are active at a distance from where they are synthesized—does not apply equally to plant hormones. Whereas some plant hormones are produced in one tissue and transported to another tissue, where they produce specific physiological responses, others act within the same tissue where they are produced. In both cases they help coordinate growth and development by acting as chemical messengers, or signals, between cells.

Plant hormones have multiple activities. Some, rather than acting as stimulators, have inhibitory influences. The response to a particular hormone depends not only on its chemical structure but on how it is "read" by its target tissue. A given hormone can elicit different responses in different tissues or at different stages of development of the same tissue. Some plant hormones are able to influence the biosynthesis of another or to interfere in the signal transduction of another. Tissues may require different amounts of hormones. Such differences are referred to as differences in **sensitivity**. Thus plant systems may vary the intensity of hormone signals by alternating hormone concentrations or by changing the sensitivity to hormones that are already present.

Traditionally five classes of plant hormones have received the most attention: auxins, cytokinins, ethylene, abscisic acid, and gibberellins (Kende and Zeevaart, 1997). It has become increasingly clear, however, that additional chemical signals are used by plants (Creelman and Mullet, 1997), including the *brassinosteroids*—a group of naturally occurring polyhydroxy steroids—that have been identified in many plants and appear to be required for normal growth of most plant tissues; *salicylic acid*—a phenolic compound with a structure similar to aspirin—the production of which is associated with disease resistance and has been linked to the hypersensitive response; the *jasmonates*—a class of compounds known as oxylipins—that play a role in the regulation of seed germination, root growth, storage-protein accumulation, and synthesis of defense proteins; *systemin*—an 18-amino acid polypeptide—that is secreted by wounded cells and then transported via the phloem to the upper, unwounded leaves to activate chemical defenses against herbivores, a phenomenon called *systemic acquired resistance* (Hammond-Kosack and Jones, 2000); the *polyamines*—low molecular weight, strongly basic molecules—that are essential for growth and development and affect the processes of mitosis and meiosis; and the gas *nitric oxide (NO)* that has been found to serve as a signal in hormonal and defense responses. NO has been reported to repress the floral transition in *Arabidopsis* (He, Y., et al., 2004). In their multiple activities the hormones interact with one another; actually the interactions and balance among the growth substances, rather than the activity of a single substance, regulate normal growth and development.

In the following paragraphs some salient features of each of the traditional groups of plant hormones are considered.

## Auxins

The principal naturally occurring auxin is **indole-3-acetic acid (IAA)**. IAA is synthesized primarily in leaf primordia and young leaves, and has been implicated in many aspects of plant development, including the overall polarity of the plant root-shoot axis, which is established during embryogenesis. This structural polarity is traceable to the polar, or unidirectional, transport of IAA in the plant. Polar auxin transport proceeds in a cell-to-cell fashion through the action of specific membrane-bound influx (AUX1) and efflux (PIN) carriers (Steinmann et al., 1999; Friml et al., 2002; Marchant et al., 2002; Friml, 2003; Volger and Kuhlemeier, 2003). The steady stream of auxin from the leaves and downward in the stem leads to the flow of auxin along narrow files of cells and results in the formation of continuous strands of vascular tissues (Aloni, 1995; Berleth and Mattsson, 2000; Berleth et al., 2000)—*the canalization hypothesis* of Sachs (1981).

In both shoots and roots the **polar transport** is always **basipetal**—from the shoot tip and leaves downward in the stem, and from the root tip toward the base of the root (the root-shoot junction). The velocity of polar auxin transport—5 to 20 centimeters per hour—is faster than the rate of passive diffusion. In addition to the polar transport of auxin it has recently been recognized that most of the IAA synthesized in mature leaves apparently is transported throughout the plant over long distances **nonpolarly** via the phloem at rates considerably higher than those of polar transport. Relatively high concentrations of free IAA have been detected in the phloem (sieve tube) sap of *Ricinus communis*, indicating that auxin may be transported over long distances in the phloem (Baker, 2000). Additional research indicates that in *Arabidopsis* the auxin influx carrier AUX1 is involved with phloem loading in the leaf and with phloem unloading in the root (Swarup et al., 2001; Marchant et al., 2002), lending further support for auxin transport in the phloem. In plants capable of secondary growth, auxin transport also occurs in the region of the vascular cambium (Sundberg et al., 2000).

In an elegant study Aloni and co-workers (2003), utilizing a combination of molecular and localization procedures, demonstrated the pattern of free auxin (IAA) production in developing *Arabidopsis* leaves (Fig. 5.14). The stipules are the first major sites of high free auxin production. In the developing blades the hydathodes are the primary sites of high free auxin production, first those in the blade tips and then, progressing basipetally, those along the margins. Trichomes and mesophyll cells are secondary sites of free auxin production. During blade development the sites and concentrations of free auxin production shift from the elongating tip basipetally along the expanding margins and finally to the central regions of the blade. The orderly shifts in the sites and concentrations of free auxin presumably control venation pattern formation and vascular differentiation in the leaf, the intense production of auxin in the hydathodes inducing differentiation of the midvein and the secondary bundles, and the low free auxin production in the lamina—particularly in association with the trichomes—inducing differentiation of the tertiary and quaternary veins and vein endings. The results of this study are in accord with the *leaf-venation hypothesis* proposed by Aloni (2001) to explain the hormonal control of vascular differentiation in leaves of eudicots.

A gene, named *VASCULAR HIGHWAY1 (VH1)*, the expression of which is provascular/procambial cell-specific, has been identified in developing leaves of *Arabidopsis* (Clay and Nelson, 2002). The expression pattern of *VH1* corresponds to the pattern of vein formation in developing leaves and, as noted by Clay and Nelson (2002), is consistent with the canalization hypothesis of patterned vascular differentiation based on the production and distribution of auxin (Sachs, 1981).

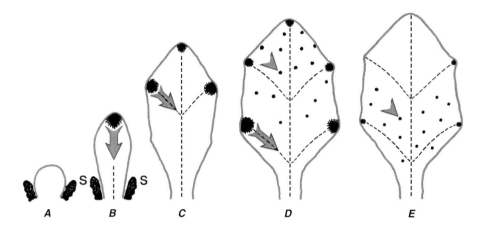

**FIGURE 5.14**

Gradual changes in sites (indicated by spots) and concentrations (indicated by spot sizes) of free–IAA production during leaf primordium development in *Arabidopsis*. Initial production of IAA occurs in the stipules (s) (**A**). Arrows show the direction of the basipetal polar IAA movement, in the blade, descending from the differentiating hydathodes (**B-D**); arrowheads show the location of secondary free–auxin production sites in the blade (**D, E**). Experimental evidence indicates that although the midvein develops acropetally (**B**), it is induced by the basipetal polar IAA flow. (From Aloni, 2004, Fig. 1. © 2004, with kind permission of Springer Science and Business Media.)

Experimental evidence has also been provided for a role of polar auxin transport in vascular patterning in the rice *(Oryza sativa)* leaf (Scarpella et al., 2002). It has been proposed that the *Oshox1* gene of rice, which is expressed in procambial cells (Scarpella et al., 2000), promotes fate commitment of procambial cells by increasing their auxin conductivity properties (Scarpella et al., 2002).

Auxins provide signals that coordinate a multiplicity of developmental processes at many levels throughout the plant body (Berleth and Sachs, 2001). Auxin has been implicated in regulating the pattern of cell division, elongation, and differentiation (Chen, 2001; Ljung et al., 2001; Friml, 2003). In *Arabidopsis* leaves high levels of IAA are strongly correlated with high rates of cell division. Dividing mesophyll tissue contained tenfold higher IAA levels than tissue growing solely by elongation. Although the youngest leaves exhibited the highest capacity to synthesize IAA, all other parts of the young *Arabidopsis* plant (including cotyledons, expanding leaves, and roots) showed a capacity to synthesize IAA de novo (Ljung et al., 2001). The gradient of auxin caused by its polar movement provides important developmental signals during embryogenesis (Hobbie et al., 2000; Berleth, 2001; Hamann, 2001), leaf vascular patterning (Mattsson et al., 1999; Aloni et al., 2003), and lateral organ formation in shoot and root (Reinhardt et al., 2000; Casimiro et al., 2001; Paquette and Benfey, 2001; Scarpella et al., 2002; Bhalerao et al., 2002). Auxin has been implicated in both gravitropism and phototropism (Marchant et al., 1999; Rashotte et al., 2001; Muday, 2001; Parry et al., 2001) and in organization and main-

tenance of both shoot and root apical meristems (Sachs, 1993; Sabatini et al., 1999; Doerner, 2000a, b; Kerk et al., 2000). Together with ethylene, auxin plays an important role in root-hair development in *Arabidopsis* (Rahman et al., 2002). Some other activities of auxin are the inhibition of axillary bud development as a part of apical dominance phenomenon and retardation of abscission.

### Cytokinins

**Cytokinins** are so named because in conjunction with auxin, they promote cell division. Roots are rich sources of cytokinins, which are transported in the xylem from the root to the shoot (Letham, 1994). Among the regulatory functions proposed for the root-synthesized cytokinins has been that of triggering the release of lateral buds from dormancy, acting counter to auxin, which inhibits lateral bud growth. The results of experiments with transgenic plants, in which systemic and local cytokinin syntheses are controllable (reviewed in Schmülling, 2002), indicate, however, that locally synthesized cytokinin, not root-derived cytokinin, is needed to release buds from dormancy. A more likely role for root-derived cytokinins appears to be one of carrying information about the nutritional status, particularly nitrogen nutrition, in root and shoot (Sakakibara et al., 1998; Yong et al., 2000). Cytokinin produced in the rootcap has been implicated in the early response of *Arabidopsis* roots to gravity (Aloni et al., 2004). Cytokinins play on important role in formation of the provascular tissue during embryogenesis (Mähönen et al.,

2000) and in the control of meristem activity and organ growth during postembryonic development (Coenen and Lomax, 1997). Whereas they are promoters of shoot development, cytokinins are inhibitors of root development (Werner et al., 2001).

## Ethylene

**Ethylene,** a simple hydrocarbon ($H_2C=CH_2$), can be produced by virtually all parts of seed plants (Mattoo and Suttle, 1991). Being a gas, it moves by diffusion from its site of synthesis. The rate of ethylene production varies among plant tissues and depends on the stage of development. Shoot apices of seedlings are important sites of ethylene production, as are the nodal regions of stems, which produce considerably more ethylene than the internodes (on an equal tissue weight basis).

Ethylene production increases during leaf abscission and during the ripening of some fruits. During the ripening of avocados, tomatoes, and pome fruits, such as apples and pears, there is a large increase in cellular respiration. This phase is known as the **climacteric**, and the fruits are called climacteric fruits. In climacteric fruits increased ethylene synthesis precedes and is responsible for many of the ripening processes. An increase in ethylene production also occurs in most tissues in response to both biotic (disease, insect damage) and abiotic (flooding, temperature, drought) stress (Lynch and Brown, 1997). As mentioned previously, lysigenous aerenchyma formation is an ethylene-mediated response to flooding (Grichko and Glick, 2001).

Ethylene often has effects opposite those of auxin. Whereas auxin prevents leaf abscission, ethylene promotes leaf abscission. Ethylene production in the abscission zone is regulated by auxin. In most plant species ethylene has an inhibitory effect on cell elongation (Abeles et al., 1992); auxin promotes cell elongation. In some semiaquatic species (*Ranunculus sceleratus, Callitriche platycarpa, Nymphoides peltata,* deepwater rice), however, ethylene elicits rapid stem growth.

## Abscisic Acid

**Abscisic Acid (ABA)** is a misnomer for this compound. It was so named because originally it was thought to be involved with abscission, a process now known to be triggered by ethylene. ABA is synthesized in almost all cells containing amyloplasts or chloroplasts; hence it can be detected in living tissue from root tip to shoot tip (Milborrow, 1984). ABA is transported in both the xylem and the phloem, although it is generally much more abundant in the phloem. ABA synthesized in the roots in response to water stress is transported upward in the xylem to the leaves where it can induce closing of stomata (Chapter 9; Davies and Zhang, 1991).

ABA levels increase during early seed development in many plant species, stimulating the production of seed storage proteins (Koornneef et al., 1989) and preventing premature germination. The breaking of dormancy in many seeds is correlated with declining ABA levels in the seed.

## Gibberellins

**Gibberellins (GAs)** are tetracyclic diterpenoids. Over 125 GAs have been identified, but only a few are known to be biologically active. Developing seeds and fruits exhibit the highest levels of GAs. Young, actively growing buds, leaves, and the upper internodes of pea seedlings have been identified as sites of GA synthesis (Coolbaugh, 1985; Sherriff et al., 1994). GAs synthesized in the shoot can be transported throughout the plant via the phloem.

GAs have dramatic effects on stem and leaf elongation by stimulating both cell division and cell elongation. Their role in stem growth is most clearly demonstrated when they are applied to many dwarf plants. Under GA treatment such plants become indistinguishable from normal nonmutant plants, indicating that the mutants are unable to synthesize GA and that growth requires GA. The *Arabidopsis gal-3* dwarf mutant (Zeevaart and Talon, 1992) illustrates multiple effects of GA deficiency. In addition to being dwarfed, the plants are bushier and have darker leaves. Also *gal-3* flowering is delayed, *gal-3* flowers are male sterile, and *gal-3* seeds fail to germinate. All of the wild-type characters are restored to the mutant upon the addition of GA. Studies on tobacco and pea indicate that IAA from the apical bud is required for normal $GA_1$ biosynthesis in stems (Ross et al., 2002). $GA_1$ may be the only GA controlling stem elongation. $GA_1$-induced elongation is usually accompanied by an increase in IAA content.

GAs control a wide array of plant developmental processes (Richards et al., 2001). They are important for normal root elongation in pea (Yaxley et al., 2001), for seed development, and for pollen tube growth in *Arabidopsis* (Singh et al., 2002), and they are essential for seed germination in a number of plant species (Yamaguchi and Kamiya, 2002). In many species of seed plants, GAs can substitute for the dormancy-breaking cold or light required for seeds to germinate. As mentioned previously, in cereal grains GAs regulate the production and secretion of the enzyme ($\alpha$-amylase), leading to hydrolysis of starch stored in the endosperm. GAs may also serve as long-day flowering signals (King et al., 2001). Application of GAs to long-day plants and biennials may cause bolting and flowering without appropriate cold or long-day exposure.

## REFERENCES

ABELES, F. B., P. W. MORGAN, and M. E. SALTVEIT Jr. 1992. *Ethylene in Plant Biology*, 2nd ed. Academic Press, San Diego.

ALONI, R. 1995. The induction of vascular tissues by auxin and cytokinin. In: *Plant Hormones: Physiology, Biochemistry and Molecular Biology*, 2nd ed., pp. 531–546, P. J. Davies, ed. Kluwer Academic, Dordrecht.

ALONI, R. 2001. Foliar and axial aspects of vascular differentiation: Hypotheses and evidence. *J. Plant Growth Regul.* 20, 22–34.

ALONI, R. 2004. The induction of vascular tissue by auxin. In: *Plant Hormones—Biosynthesis, Signal Transduction, Action!*, 3rd ed., pp. 471–492, P. J. Davies, ed. Kluwer Academic, Dordrecht.

ALONI, R., K. SCHWALM, M. LANGHANS, and C. I. ULLRICH. 2003. Gradual shifts in sites of free-auxin production during leaf-primordium development and their role in vascular differentiation and leaf morphogenesis in *Arabidopsis*. *Planta* 216, 841–853.

ALONI, R., M. LANGHANS, E. ALONI, and C. I. ULLRICH. 2004. Role of cytokinin in the regulation of root gravitropism. *Planta* 220, 177–182.

*ARABIDOPSIS* GENOME INITIATIVE, THE. 2000. Analysis of the genome sequence of the flowering plant *Arabidopsis thaliana*. *Nature* 408, 796–815.

AVERS, C. J., and R. B. GRIMM. 1959. Comparative enzyme differentiations in grass roots. II. Peroxidase. *J. Exp. Bot.* 10, 341–344.

BAILEY, I. W. 1944. The development of vessels in angiosperms and its significance in morphological research. *Am. J. Bot.* 31, 421–428.

BAKER, D. A. 2000. Vascular transport of auxins and cytokinins in *Ricinus*. *Plant Growth Regul.* 32, 157–160.

BANNAN, M. W. 1950. The frequency of anticlinal divisions in fusiform cambial cells of *Chamaecyparis*. *Am. J. Bot.* 37, 511–519.

BANNAN, M. W. 1951. The reduction of fusiform cambial cells in *Chamaecyparis* and *Thuja*. *Can. J. Bot.* 29, 57–67.

BANNAN, M. W. 1956. Some aspects of the elongation of fusiform cambial cells in *Thuja occidentalis* L. *Can. J. Bot.* 34, 175–196.

BANNAN, M. W., and B. E. WHALLEY. 1950. The elongation of fusiform cambial cells in *Chamaecyparis*. *Can. J. Res., Sect. C* 28, 341–355.

BÁRÁNY, I., P. GONZÁLEZ-MELENDI, B. FADÓN, J. MITYKÓ, M. C. RISUEÑO, and P. S. TESTILLANO. 2005. Microspore-derived embryogenesis in pepper (*Capsicum annuum* L.): Subcellular rearrangements through development. *Biol. Cell* 97, 709–722.

BASSEL, A. R., and J. H. MILLER. 1982. The effects of centrifugation on asymmetric cell division and differentiation of fern spores. *Ann. Bot.* 50, 185–198.

BEERS, E. P., and J. M. MCDOWELL. 2001. Regulation and execution of programmed cell death in response to pathogens, stress and developmental cues. *Curr. Opin. Plant Biol.* 4, 561–567.

BEEVERS, L. 1976. Senescence. In: *Plant Biochemistry*, 3rd ed., pp. 771–794, J. Bonner and J. E. Varner, eds. Academic Press, New York.

BERLETH, T. 2001. Top-down and inside-out: Directionality of signaling in vascular and embryo development. *J. Plant Growth Regul.* 20, 14–21.

BERLETH, T., and J. MATTSSON. 2000. Vascular development: Tracing signals along veins. *Curr. Opin. Plant Biol.* 3, 406–411.

BERLETH, T., and T. SACHS. 2001. Plant morphogenesis: Long-distance coordination and local patterning. *Curr. Opin. Plant Biol.* 4, 57–62.

BERLETH, T., J. MATTSSON, and C. S. HARDTKE. 2000. Vascular continuity and auxin signals. *Trends Plant Sci.* 5, 387–393.

BEVAN, M. 2002. Genomics and plant cells: Application of genomics strategies to *Arabidopsis* cell biology. *Philos. Trans. R. Soc. Lond. B* 357, 731–736.

BHALERAO, R. P., J. EKLÖF, K. LJUNG, A. MARCHANT, M. BENNETT, and G. SANDBERG. 2002. Shoot-derived auxin is essential for early lateral root emergence in *Arabidopsis* seedlings. *Plant J.* 29, 325–332.

BLAU, H. M., T. R. BRAZELTON, and J. M. WEIMANN. 2001. The evolving concept of a stem cell: Entity or function? *Cell* 105, 829–841.

BOWMAN, J. L., and Y. ESHED. 2000. Formation and maintenance of the shoot apical meristem. *Trends Plant Sci.* 5, 110–115.

BUCHANAN-WOLLASTON, V. 1997. The molecular biology of leaf senescence. *J. Exp. Bot.* 48, 181–199.

CASIMIRO, I., A. MARCHANT, R. P. BHALERAO, T. BEECKMAN, S. DHOOGE, R. SWARUP, N. GRAHAM, D. INZÉ, G. SANDBERG, P. J. CASERO, and M. BENNETT. 2001. Auxin transport promotes *Arabidopsis* lateral root initiation. *Plant Cell* 13, 843–852.

CHANANA, N. P., V. DHAWAN, and S. S. BHOJWANI. 2005. Morphogenesis in isolated microspore cultures of *Brassica juncea*. *Plant Cell Tissue Org. Cult.* 83, 169–177.

CHEADLE, V. I. 1937. Secondary growth by means of a thickening ring in certain monocotyledons. *Bot. Gaz.* 98, 535–555.

CHEN, J.-G. 2001. Dual auxin signaling pathways control cell elongation and division. *J. Plant Growth Regul.* 20, 255–264.

CLAY, N. K., and T. NELSON. 2002. VH1, a provascular cell-specific receptor kinase that influences leaf cell patterns in *Arabidopsis*. *Plant Cell* 14, 2707–2722.

CLINE, M. G. 1997. Concepts and terminology of apical dominance. *Am. J. Bot.* 84, 1064–1069.

CLINE, M. G. 2000. Execution of the auxin replacement apical dominance experiment in temperate woody species. *Am. J. Bot.* 87, 182–190.

COENEN, C., and T. L. LOMAX. 1997. Auxin-cytokinin interactions in higher plants: Old problems and new tools. *Trends Plant Sci.* 2, 351–356.

COOLBAUGH, R. C. 1985. Sites of gibberellin biosynthesis in pea seedlings. *Plant Physiol.* 78, 655–657.

COTTIGNIES, A. 1977. Le nucléole dans le point végétatif dormant et non dormant du *Fraxinus excelsior L. Z. Pflanzenphysiol.* 83, 189–200.

CREELMAN, R. A., and J. E. MULLET. 1997. Oligosaccharins, brassinolides, and jasmonates: Nontraditional regulators of plant growth, development, and gene expression. *Plant Cell* 9, 1211–1223.

CROZIER, A., Y. KAMIYA, G. BISHOP, and T. YOKOTA. 2000. Biosynthesis of hormones and elicitor molecules. In: *Biochemistry and Molecular Biology of Plants*, pp. 850–929, B. B. Buchanan, W. Gruissem, and R. L. Jones, eds. American Society of Plant Physiologists, Rockville, MD.

D'AMATO, F. 1998. Chromosome endoreduplication in plant tissue development and function. In: *Plant Cell Proliferation and Its Regulation in Growth and Development*, pp. 153–166, J. A. Bryant and D. Chiatante, eds. Wiley, New York.

DANGL, J. L., R. A. DIETRICH, and H. THOMAS. 2000. Senescence and programmed cell death. In: *Biochemistry and Molecular Biology of Plants*, pp. 1044–1100, B. B. Buchanan, W. Gruissem, and R. L. Jones, eds. American Society of Plant Physiologists, Rockville, MD.

DAVIES, P. J., ed. 2004. *Plant Hormones—Biosynthesis, Signal Transduction, Action!*, 3rd ed. Kluwer Academic, Dordrecht.

DAVIES, W. J., and J. ZHANG. 1991. Root signals and the regulation of growth and development of plants in drying soil. *Annu. Rev. Plant Physiol. Plant Mol. Biol.* 42, 55–76.

DERMEN, H. 1953. Periclinal cytochimeras and origin of tissues in stem and leaf of peach. *Am. J. Bot.* 40, 154–168.

DOERNER, P. 2000a. Root patterning: Does auxin provide positional cues? *Curr. Biol.* 10, R201–R203.

DOERNER, P. 2000b. Plant stem cells: The only constant thing is change. *Curr. Biol.* 10, R826–R829.

DREW, M. C., C.-J. HE, and P. W. MORGAN. 2000. Programmed cell death and aerenchyma formation in roots. *Trends Plant Sci.* 5, 123–127.

DYER, A. F. 1976. Modifications and errors of mitotic cell division in relation to differentiation. In: *Cell Division in Higher Plants*, pp. 199–249, M. M. Yeoman, ed. Academic Press, London.

EDGAR, B. A., and T. L. ORR-WEAVER. 2001. Endoreplication cell cycles: More for less. *Cell* 105, 297–306.

ESAU, K. 1965. *Vascular Differentiation in Plants.* Holt, Reinhart and Winston, New York.

ESAU, K. 1977. *Anatomy of Seed Plants*, 2nd ed. Wiley, New York.

FATH, A., P. BETHKE, J. LONSDALE, R. MEZA-ROMERO, and R. JONES. 2000. Programmed cell death in cereal aleurone. *Plant Mol. Biol.* 44, 255–266.

FINK, S. 1999. *Pathological and Regenerative Plant Anatomy. Encyclopedia of Plant Anatomy*, Band 14, Teil 6. Gebrüder Borntraeger, Berlin.

FLETCHER, J. C., and E. M. MEYEROWITZ. 2000. Cell signalling within the shoot meristem. *Curr. Opin Plant Biol.* 3, 23–30.

FRIML, J. 2003. Auxin transport—Shaping the plant. *Curr. Opin. Plant Biol.* 6, 7–12.

FRIML, J., E. BENKOVÁ, I. BLILOU, J. WISNIEWSKA, T. HAMANN, K. LJUNG, S. WOODY, G. SANDBERG, B. SCHERES, G. JÜRGENS, and K. PALME. 2002. AtPIN4 mediates sink-driven auxin gradients and root patterning in *Arabidopsis. Cell* 108, 661–673.

FUCHS, E., and J. A. SEGRE. 2000. Stem cells: A new lease on life. *Cell* 100, 143–155.

FUCHS, M. C. 1968. Localisation des divisions dos le méristème des feuilles des *Lupinus albus* L., *Tropaeolum peregrinum* L., *Limonium sinyatum* (L.) Miller et *Nemophila maculata* Benth. *C. R. Acad. Sci., Paris,* Sér. D 267, 722–725.

FUKUDA, H. 1997. Programmed cell death during vascular system formation. *Cell Death Differ.* 4, 684–688.

GALLAGHER, K., and L. G. SMITH. 1997. Asymmetric cell division and cell fate in plants. *Curr. Opin. Cell Biol.* 9, 842–848.

GALLAGHER, K., and L. G. SMITH. 2000. Roles of polarity and nuclear determinants in specifying daughter cell fates after an asymmetric cell division in the maize leaf. *Curr. Biol.* 10, 1229–1232.

GAUTHERET, R. J. 1977. *La Culture des tissus et des cellules des végétaux: Résultats généraux et réalisations pratiques.* Masson, Paris.

GEIER, T., and H. W. KOHLENBACH. 1973. Entwicklung von Embryonen und embryogenem Kallus aus Pollenkörnern von *Datura meteloides* und *Datura innoxia. Protoplasma* 78, 381–396.

GIRIDHAR, G., and M. J. JAFFE. 1988. Thigmomorphogenesis: XXIII. Promotion of foliar senescence by mechanical perturbation of *Avena sativa* and four other species. *Physiol Plant.* 74, 473–480.

GRBIĆ, V., and A. B. BLEECKER. 1995. Ethylene regulates the timing of leaf senescence in *Arabidopsis. Plant J.* 8, 595–602.

GREBE, M., J. Xu, and B. SCHERES. 2001. Cell axiality and polarity in plants—Adding pieces to the puzzle. *Curr. Opin. Plant Biol.* 4, 520–526.

GREEN, P. B. 1969. Cell morphogenesis. *Annu. Rev. Plant Physiol.* 20, 365–394.

GREEN, P. B. 1976. Growth and cell pattern formation on an axis: Critique of concepts, terminology, and modes of study. *Bot. Gaz.* 137, 187–202.

GREENBERG, J. T. 1996. Programmed cell death: A way of life for plants. *Proc. Natl. Acad. Sci. USA* 93, 12094–12097.

GREENBERG, J. T. 1997. Programmed cell death in plant-pathogen interactions. *Annu. Rev. Plant Physiol. Plant Mol. Biol.* 48, 525–545.

GRICHKO, V. P., and B. R. GLICK. 2001. Ethylene and flooding stress in plants. *Plant Physiol. Biochem.* 39, 1–9.

GRIGA, M. 1999. Somatic embryogenesis in grain legumes. In: *Advances in Regulation of Plant Growth and Development*, pp. 233–249, M. Strnad, P. Peč, and E. Beck, eds. Peres Publishers, Prague.

GROTEWOLD, E., ed. 2004. *Plant Functional Genomics*. Humana Press Inc., Totowa, NJ.

GULLINE, H. F. 1960. Experimental morphogenesis in adventitious buds of flax. *Aust. J. Bot.* 8, 1–10.

GUNAWARDENA, A. H. L. A. N., J. S. GREENWOOD, and N. G. DENGLER. 2004. Programmed cell death remodels lace plant leaf shape during development. *Plant Cell* 16, 60–73.

HABERLANDT, G. 1914. *Physiological Plant Anatomy*. Macmillan, London.

HAECKER, A., and T. LAUX. 2001. Cell-cell signaling in the shoot meristem. *Curr. Opin. Plant Biol.* 4, 441–446.

HALPERIN, W. 1969. Morphogenesis in cell cultures. *Annu. Rev. Plant Physiol.* 20, 395–418.

HAMANN, T. 2001. The role of auxin in apical-basal pattern formation during *Arabidopsis* embryogenesis. *J. Plant Growth Regul.* 20, 292–299.

HAMMOND-KOSACK, K., and J. D. G. JONES. 2000. Responses to plant pathogens. In: *Biochemistry and Molecular Biology of Plants*, pp. 1102–1156, B. B. Buchanan, W. Gruissem, and R. L. Jones, eds. American Society of Plant Physiologists, Rockville, MD.

HANSON, J. B., and A. J. TREWAVAS. 1982. Regulation of plant cell growth: The changing perspective. *New Phytol.* 90, 1–18.

HARRIS, H. 1974. *Nucleus and Cytoplasm*, 3rd. ed. Clarendon Press, Oxford.

HE, C.-J., P. W. MORGAN, and M. C. DREW. 1996. Transduction of an ethylene signal is required for cell death and lysis in the root cortex of maize during aerenchyma formation induced by hypoxia. *Plant Physiol.* 112, 463–472.

HE, Y., R.-H. TANG, Y. HAO, R. D. STEVENS, C. W. COOK, S. M. AHN, L. JING, Z. YANG, L. CHEN, F. GUO, F. FIORANI, R. B. JACKSON, N. M. CRAWFORD, and Z.-M. PEI. 2004. Nitric oxide represses the *Arabidopsis* floral transition. *Science* 305, 1968–1971.

HEATH, M. C. 2000. Hypersensitive response-related death. *Plant Mol. Biol.* 44, 321–334.

HERBERT, R. J., B. VILHAR, C. EVETT, C. B. ORCHARD, H. J. ROGERS, M. S. DAVIES, and D. FRANCIS. 2001. Ethylene induces cell death at particular phases of the cell cycle in the tobacco TBY-2 cell line. *J. Exp. Bot.* 52, 1615–1623.

HOBBIE, L., M. MCGOVERN, L. R. HURWITZ, A. PIERRO, N. Y. LIU, A. BANDYOPADHYAY, and M. ESTELLE. 2000. The *axr6* mutants of *Arabidopsis thaliana* define a gene involved in auxin response and early development. *Development* 127, 23–32.

HUANG, F.-Y., S. PHILOSOPH-HADAS, S. MEIR, D. A. CALLAHAM, R. SABATO, A. ZELCER, and P. K. HEPLER. 1997. Increases in cytosolic $Ca^{2+}$ in parsley mesophyll cells correlate with leaf senescence. *Plant Physiol.* 115, 51–60.

HULBARY, R. L. 1944. The influence of air spaces on the three-dimensional shapes of cells in *Elodea* stems, and a comparison with pith cells of *Ailanthus*. *Am. J. Bot.* 31, 561–580.

HÜLSKAMP, M., S. MISÉRA, and G. JÜRGENS. 1994. Genetic dissection of trichome cell development in *Arabidopsis*. *Cell* 76, 555–566.

HURKMAN, W. J. 1979. Ultrastructural changes of chloroplasts in attached and detached, aging primary wheat leaves. *Am. J. Bot.* 66, 64–70.

IM, K.-H., D. J. COSGROVE, and A. M. JONES. 2000. Subcellular localization of expansin mRNA in xylem cells. *Plant Physiol.* 123, 463–470.

INTERNATIONAL RICE GENOME SEQUENCING PROJECT. 2005. The map-based sequence of the rice genome. *Nature* 436, 793–800.

IRISH, V. F. 1991. Cell lineage in plant development. *Curr. Opin. Cell Biol.* 3, 983–987.

IRISH, V. F., and P. D. JENIK. 2001. Cell lineage, cell signaling and the control of plant morphogenesis. *Curr. Opin. Gen. Dev.* 11, 424–430.

JACKSON, B. D. 1953. *A Glossary of Botanic Terms, with Their Derivation and Accent*, rev. and enl. 4th ed., J. B. Lippincott, Philadelphia.

JAFFE, M. J. 1980. Morphogenetic responses of plants to mechanical stimuli or stress. *BioScience* 30, 239–243.

JAIN, S. M., P. K. GUPTA, and R. J. NEWTON, eds. 1995. *Somatic Embryogenesis in Woody Plants*, vols. 1–6. Kluwer Academic, Dordrecht.

JING, H.-C., J. HILLE, and P. P. DIJKWEL. 2003. Ageing in plants: Conserved strategies and novel pathways. *Plant Biol.* 5, 455–464.

JONES, L. H. 1974. Factors influencing embryogenesis in carrot cultures (*Daucus carota* L.) *Ann. Bot.* 38, 1077–1088.

JOUBÈS, J., and C. CHEVALIER. 2000. Endoreduplication in higher plants. *Plant Mol. Biol.* 43, 735–745.

KAPLAN, D. R. 2001. Fundamental concepts of leaf morphology and morphogenesis: A contribution to the interpretation of molecular genetic mutants. *Int. J. Plant Sci.* 162, 465–474.

KAPLAN, D. R., and W. HAGEMANN. 1991. The relationship of cell and organism in vascular plants. *BioScience* 41, 693–703.

KAPLAN, R. 1937. Über die Bildung der Stele aus dem Urmeristem von Pteridophyten und Spermatophyten. *Planta* 27, 224–268.

KENDE, H., and J. A. D. ZEEVAART. 1997. The five "classical" plant hormones. *Plant Cell* 9, 1197–1210.

KERK, N. M., K. JIANG, and L. J. FELDMAN. 2000. Auxin metabolism in the root apical meristem. *Plant Physiol.* 122, 925–932.

KERR, J. F. R., and B. V. HARMON. 1991. Definition and incidence of apoptosis: A historical perspective. In: *Apoptosis: The Molecular Basis of Cell Death*, pp. 5–29, L. D. Tomei and F. O. Cope, eds. Cold Spring Harbor Laboratory Press, Cold Spring Harbor, NY.

KERR, J. F. R., A. H. WYLLIE, and A. R. CURRIE. 1972. Apoptosis: A basic biological phenomenon with wide-ranging implications in tissue kinetics. *Brit. J. Cancer* 26, 239–257.

KIDNER, C., V. SUNDARESAN, K. ROBERTS, and L. DOLAN 2000. Clonal analysis of the *Arabidopsis* root confirms that position, not lineage, determines cell fate. *Planta* 211, 191–199.

KING, R. W., T. MORITZ, L. T. EVANS, O. JUNTTILA, and A. J. HERLT. 2001. Long-day induction of flowering in *Lolium temulentum* involves sequential increases in specific gibberellins at the shoot apex. *Plant Physiol.* 127, 624–632.

KONDOROSI, E., F. ROUDIER, and E. GENDREAU. 2000. Plant cell-size control: Growing by ploidy? *Curr. Opin. Plant Biol.* 3, 488–492.

KOORNNEEF, M., C. J. HANHART, H. W. M. HILHORST, and C. M. KARSSEN. 1989. *In vivo* inhibition of seed development and reserve protein accumulation in recombinants of abscisic acid biosynthesis and responsiveness mutants in *Arabidopsis thaliana. Plant Physiol.* 90, 463–469.

KRABBE, G. 1886. *Das gleitende Wachsthum bei der Gewebebildung der Gefässpflanzen.* Gebrüder Borntraeger, Berlin.

KUDO, N., and Y. KIMURA. 2002. Nuclear DNA endoreduplication during petal development in cabbage: Relationship between ploidy levels and cell size. *J. Exp. Bot.* 53, 1017–1023.

KURIYAMA, H., and H. FUKUDA. 2002. Developmental programmed cell death in plants. *Curr. Opin. Plant Biol.* 5, 568–573.

LAM, E., N. KATO, and M. LAWTON. 2001. Programmed cell death, mitochondria and the plant hypersensitive response. *Nature* 411, 848–853.

LANGDALE, J. A., B. LANE, M. FREELING, and T. NELSON. 1989. Cell lineage analysis of maize bundle sheath and mesophyll cells. *Dev. Biol.* 133, 128–139.

LANZA, R., J. GEARHART, B. HOGAN, D. MELTON, R. PEDERSEN, J. THOMSON, and M. WEST, eds. 2004a. *Handbook of Stem Cells,* vol. 1, *Embryonic.* Elsevier Academic Press, Amsterdam.

LANZA, R., H. BLAU, D. MELTON, M. MOORE, E. D. THOMAS (Hon.), C. VERFAILLE, I. WEISSMAN, and M. WEST, eds. 2004b. *Handbook of Stem Cells,* vol. 2, *Adult and Fetal.* Elsevier Academic Press, Amsterdam.

LARKIN, J. C., N. YOUNG, M. PRIGGE, and M. D. MARKS. 1996. The control of trichome spacing and number in *Arabidopsis. Development* 122, 997–1005.

LARKIN, J. C., M. D. MARKS, J. NADEAU, and F. SACK. 1997. Epidermal cell fate and patterning in leaves. *Plant Cell* 9, 1109–1120.

LARKINS, B. A., B. P. DILKES, R. A. DANTE, C. M. COELHO, Y.-M. WOO, and Y. LIU. 2001. Investigating the hows and whys of DNA endoreduplication. *J. Exp. Bot.* 52, 183–192.

LAUX, T., and G. JÜRGENS. 1994. Establishing the body plan of the *Arabidopsis* embryo. *Acta Bot. Neerl.* 43, 247–260.

LEE, R.-H., and S.-C. G. CHEN. 2002. Programmed cell death during rice leaf senescence is nonapoptotic. *New Phytol.* 155, 25–32.

LEOPOLD, A. C. 1978. The biological significance of death in plants. In: *The Biology of Aging,* pp. 101–114, J. A. Behnke, C. E. Finch, and G. B. Moment, eds. Plenum, New York.

LETHAM, D. S. 1994. Cytokinins as phytohormones—Sites of biosynthesis, translocation, and function of translocated cytokinin. In: *Cytokinins: Chemistry, Activity, and Function,* pp. 57–80, D. W. S. Mok and M. C. Mok, eds. CRC Press, Boca Raton, FL.

LJUNG, K., R. P. BHALERAO, and G. SANDBERG. 2001. Sites and homeostatic control of auxin biosynthesis in *Arabidopsis* during vegetative growth. *Plant J.* 28, 465–474.

LOAKE, G. 2001. Plant cell death: Unmasking the gatekeepers. *Curr. Biol.* 11, R1028–R1031.

LOHMAN, K. N., S. GAN, M. C. JOHN, and R. M. AMASINO. 1994. Molecular analysis of natural leaf senescence in *Arabidopsis thaliana. Physiol. Plant.* 92, 322–328.

LÖRZ, H., W. WERNICKE, and I. POTRYKUS. 1979. Culture and plant regeneration of *Hyoscyamus* protoplasts. *Planta Med.* 36, 21–29.

LYNCH, J., and K. M. BROWN. 1997. Ethylene and plant responses to nutritional stress. *Physiol. Plant.* 100, 613–619.

LYNDON, R. F. 1998. *The Shoot Apical Meristem. Its Growth and Development.* Cambridge University Press, Cambridge.

MA, J., K.-C. PASCAL, M. W. DRAKE, and P. CHRISTOU. 2003. The production of recombinant pharmaceutical proteins in plants. *Nat. Rev.* 4, 794–805.

MÄHÖNEN, A. P., M. BONKE, L. KAUPPINEN, M. RIIKONEN, P. N. BENFEY, and Y. HELARIUTTA. 2000. A novel two-component hybrid molecule regulates vascular morphogenesis of the *Arabidopsis* root. *Genes Dev.* 14, 2938–2943.

MARASCHIN, S. F., W. DE PRIESTER, H. P. SPAINK, and M. WANG. 2005. Androgenic switch: an example of plant embryogenesis from the male gametophyte perspective. *J. Exp. Bot.* 56, 1711–1726.

MARCHANT, A., J. KARGUL, S. T. MAY, P. MULLER, A. DELBARRE, C. PERROT-RECHENMANN, and M. J. BENNETT. 1999. AUX1 regulates root gravitropism in *Arabidopsis* by facilitating auxin uptake within root apical tissues. *EMBO J.* 18, 2066–2073.

MARCHANT, A., R. BHALERAO, I. CASIMIRO, J. EKLÖF, P. J. CASERO, M. BENNETT, and G. SANDBERG. 2002. AUX1 promotes lateral root formation by facilitating indole-3-acetic acid distribution between sink and source tissues in the *Arabidopsis* seedling. *Plant Cell* 14, 589–597.

MARCOTRIGIANO, M. 1997. Chimeras and variegation: Patterns of deceit. *HortScience* 32, 773–784.

MARCOTRIGIANO, M. 2001. Genetic mosaics and the analysis of leaf development. *Int. J. Plant Sci.* 162, 513–525.

MATTOO, A. K., and J. C. SUTTLE, eds. 1991. *The Plant Hormone Ethylene.* CRC Press, Boca Raton, FL.

MATTSSON, J., Z. R. SUNG, and T. BERLETH. 1999. Responses of plant vascular systems to auxin transport inhibition. *Development* 126, 2979–2991.

MCCAULEY, M. M., and R. F. EVERT. 1988. Morphology and vasculature of the leaf of potato (*Solanum tuberosum*). *Am. J. Bot.* 75, 377–390.

MCDANIEL, C. N. 1984a. Competence, determination, and induction in plant development. In: *Pattern Formation. A Primer in*

*Developmental Biology*, pp. 393–412, G. M. Malacinski, ed. and S. V. Bryant, consulting ed. Macmillan, New York.

MCDANIEL, C. N. 1984b. Shoot meristem development. In: *Positional Controls in Plant Development*, pp. 319–347, P. W. Barlow and D. J. Carr, eds. Cambridge University Press, Cambridge.

MCMANUS, M. T., and B. E. VEIT, eds. 2002. *Meristematic Tissues in Plant Growth and Development*. Sheffield Academic Press, Sheffield, UK.

MILBORROW, B. V. 1984. Inhibitors. In: *Advanced Plant Physiology*, pp. 76–110, M. B. Wilkins, ed. Longman Scientific & Technical, Essex, England.

MITTLER, R., O. DEL POZO, L. MEISEL, and E. LAM. 1997. Pathogen-induced programmed cell death in plants, a possible defense mechanism. *Dev. Genet.* 21, 279–289.

MIZUKAMI, Y. 2001. A matter of size: Developmental control of organ size in plants. *Curr. Opin. Plant Biol.* 4, 533–539.

MUDAY, G. K. 2001. Auxins and tropisms. *J. Plant Growth Regul.* 20, 226–243.

MURASHIGE, T. 1979. Plant tissue culture and its importance to agriculture. In: *Practical Tissue Culture Applications*, pp. 27–44, K. Maramorosch and H. Hirumi, eds. Academic Press, New York.

NAGL, W. 1978. *Endopolyploidy and Polyteny in Differentiation and Evolution*. North-Holland, Amsterdam.

NAGL, W. 1981. Polytene chromosomes in plants. *Int. Rev. Cytol.* 73, 21–53.

NAPOLI, C. A., C. A. BEVERIDGE, and K. C. SNOWDEN. 1999. Reevaluating concepts of apical dominance and the control of axillary bud outgrowth. *Curr. Topics Dev. Biol.* 44, 127–169.

NEEFF, F. 1914. Über Zellumlagerung. Ein Beitrag zur experimentellen Anatomie. *Z. Bot.* 6, 465–547.

NOODÉN, L. D. 1988. The phenomena of senescence and aging. In: *Senescence and Aging in Plants*, pp. 1–50, L. D. Noodén and A. C. Leopold, eds. Academic Press, San Diego.

NOODÉN, L. D., and A. C. LEOPOLD, eds. 1988. *Senescence and Aging in Plants*. Academic Press, San Diego.

NOODÉN, L. D., and J. E. THOMPSON. 1985. Aging and senescence in plants. In: *Handbook of the Biology of Aging*, 2nd ed., pp. 105–127, C. E. Finch and E. L. Schneider, eds. Van Nostrand Reinhold, New York.

NOODÉN, L. D., J. J. GUIAMÉT, and I. JOHN. 1997. Senescence mechanisms. *Physiol. Plant.* 101, 746–753.

PAQUETTE, A. J., and P. N. BENFEY. 2001. Axis formation and polarity in plants. *Curr. Opin. Gen. Dev* 11, 405–409.

PARRY, G., A. DELBARRE, A. MARCHANT, R. SWARUP, R. NAPIER, C. PERROT-RECHENMANN, and M. J. BENNETT. 2001. Novel auxin transport inhibitors phenocopy the auxin influx carrier mutation *aux1*. *Plant J.* 25, 399–406.

PEÑA, L., ed. 2004. *Transgenic Plants*. Humana Press, Inc., Totowa, NJ.

PENNELL, R. I., and C. LAMB. 1997. Programmed cell death in plants. *Plant Cell* 9, 1157–1168.

PIGLIUCCI, M. 1998. Developmental phenotypic plasticity: Where internal programming meets the external environment. *Curr. Opin. Plant Biol.* 1, 87–91.

POETHIG, R. S. 1984a. Cellular parameters of leaf morphogenesis in maize and tobacco. In: *Contemporary Problems in Plant Anatomy*, pp. 235–259, R. A. White and W. C. Dickison, eds. Academic Press, New York.

POETHIG, R. S. 1984b. Patterns and problems in angiosperm leaf morphogenesis. In: *Pattern Formation. A Primer in Developmental Biology*, pp. 413–432, G. M. Malacinski, ed. and S. V. Bryant, consulting ed. Macmillan, New York.

POETHIG, R. S. 1987. Clonal analysis of cell lineage patterns in plant development. *Am. J. Bot.* 74, 581–594.

POETHIG, R. S., and I. M. SUSSEX. 1985. The cellular parameters of leaf development in tobacco: A clonal analysis. *Planta* 165, 170–184.

POETHIG, R. S., E. H. COE JR., and M. M. JOHRI. 1986. Cell lineage patterns in maize embryogenesis: A clonal analysis. *Dev. Biol.* 117, 392–404.

PONTIER, D., C. BALAGUÉ, and D. ROBY. 1998. The hypersensitive response. A programmed cell death associated with plant resistance. *C.R. Acad. Sci., Paris, Sci. de la Vie* 321, 721–734.

POUPIN, M. J., and P. ARCE-JOHNSON. 2005. Transgenic trees for a new era. *In Vitro Cell. Dev. Biol.—Plant* 41, 91–101.

POWER, J. B., and E. C. COCKING. 1971. Fusion of plant protoplasts. *Sci. Prog. Oxf.* 59, 181–198.

PRAT, H. 1935. Recherches sur la structure et le mode de croissance de chaumes. *Ann. Sci. Nat. Bot.*, Sér. 10, 17, 81–145.

PRAT, H. 1948. Histo-physiological gradients and plant organogenesis. *Bot. Rev.* 14, 603–643.

PRAT, H. 1951. Histo-physiological gradients and plant organogenesis. (Part II). *Bot. Rev.* 17, 693–746.

PRIESTLEY, J. H. 1930. Studies in the physiology of cambial activity. II. The concept of sliding growth. *New Phytol.* 29, 96–140.

RAGHAVAN, V. 1976. *Experimental Embryogenesis in Vascular Plants*. Academic Press, London.

RAGHAVAN, V. 1986. *Embryogenesis in Angiosperms. A Developmental and Experimental Study*. Cambridge University Press, Cambridge.

RAHMAN, A., S. HOSOKAWA, Y. OONO, T. AMAKAWA, N. GOTO, and S. TSURUMI. 2002. Auxin and ethylene response interactions during *Arabidopsis* root hair development dissected by auxin influx modulators. *Plant Physiol.* 130, 1908–1917.

RASHOTTE, A. M., A. DELONG, and G. K. MUDAY. 2001. Genetic and chemical reductions in protein phosphatase activity alter auxin transport, gravity response, and lateral root growth. *Plant Cell* 13, 1683–1697.

RAUH, W. 1950. *Morphologie der Nutzpflanzen*. Quelle & Meyer, Heidelberg.

RAVEN, P. H., R. F. EVERT, and S. E. EICHHORN. 2005. *Biology of Plants*, 7th ed. Freeman, New York.

REINHARDT, D., T. MANDEL, and C. KUHLEMEIER. 2000. Auxin regulates the initiation and radial position of plant lateral organs. *Plant Cell* 12, 507–518.

RICHARDS, D. E., K. E. KING, T. AIT-ALI, and N. P. HARBERD. 2001. How gibberellin regulates plant growth and development: A molecular genetic analysis of gibberellin signaling. *Annu. Rev. Plant Physiol. Plant Mol. Biol.* 52, 67–88.

ROSS, J. J., D. P. O'NEILL, C. M. WOLBANG, G. M. SYMONS, and J. B. REID. 2002. Auxin-gibberellin interactions and their role in plant growth. *J. Plant Growth Regul.* 20, 346–353.

SABATINI, S., D. BEIS, H. WOLKENFELT, J. MURFETT, T. GUILFOYLE, J. MALAMY, P. BENFEY, O. LEYSER, N. BECHTOLD, P. WEISBEEK, and B. SCHERES. 1999. An auxin-dependent distal organizer of pattern and polarity in the *Arabidopsis* root. *Cell* 99, 463–472.

SACHS, T. 1981. The control of the patterned differentiation of vascular tissues. *Adv. Bot. Res.* 9, 152–262.

SACHS, T. 1991. Cell polarity and tissue patterning in plants. *Development* suppl. 1, 83–93.

SACHS, T. 1993. The role of auxin in the polar organisation of apical meristems. *Aust. J. Plant Physiol.* 20, 541–553.

SAKAKIBARA, H., M. SUZUKI, K. TAKEI, A. DEJI, M. TANIGUCHI, and T. SUGIYAMA. 1998. A response-regulator homologue possibly involved in nitrogen signal transduction mediated by cytokinin in maize. *Plant J.* 14, 337–344.

SASS, J. E. 1958. *Botanical Microtechnique*, 3rd ed. Iowa State College Press, Ames, IA.

SATINA, S., A. F. BLAKESLEE, and A. G. AVERY. 1940. Demonstration of the three germ layers in the shoot apex of *Datura* by means of induced polyploidy in periclinal chimeras. *Am. J. Bot.* 27, 895–905.

SCARPELLA, E., S. RUEB, K. J. M. BOOT, J. H. C. HOGE, and A. H. MEIJER. 2000. A role for the rice homeobox gene *Oshox1* in provascular cell fate commitment. *Development* 127, 3655–3669.

SCARPELLA, E., K. J. M. BOOT, S. RUEB, and A. H. MEIJER. 2002. The procambium specification gene *Oshox1* promotes polar auxin transport capacity and reduces its sensitivity toward inhibition. *Plant Physiol.* 130, 1349–1360.

SCHERES, B., and P. N. BENFEY. 1999. Asymmetric cell division in plants. *Annu. Rev. Plant Physiol. Plant Mol. Biol.* 50, 505–537.

SCHERES, B., and H. WOLKENFELT. 1998. The *Arabidopsis* root as a model to study plant development. *Plant Physiol. Biochem.* 36, 21–32.

SCHMÜLLING, T. 2002. New insights into the functions of cytokinins in plant development. *J. Plant Growth Regul.* 21, 40–49.

SCHOCH-BODMER, H. 1945. Interpositionswachstum, symplastisches und gleitendes Wachstum. *Ber. Schweiz. Bot. Ges.* 55, 313–319.

SCHOCH-BODMER, H., and P. HUBER. 1951. Das Spitzenwachstum der Bastfasern bei *Linum usitatissimum* und *Linum perenne*. *Ber. Schweiz. Bot. Ges.* 61, 377–404.

SCHOCH-BODMER, H., and P. HUBER. 1952. Local apical growth and forking in secondary fibres. *Proc. Leeds Philos. Lit. Soc., Sci. Sect.*, 6, 25–32.

SCHÜEPP, O. 1926. *Meristeme. Handbuch der Pflanzenanatomie*, Band 4, Lief 16. Gebrüder Borntraeger, Berlin.

SHEPARD, J. F., D. BIDNEY, and E. SHAHIN. 1980. Potato protoplasts in crop improvement. *Science* 208, 17–24.

SHERRIFF, L. J., M. J. MCKAY, J. J. ROSS, J. B. REID, and C. L. WILLIS. 1994. Decapitation reduces the metabolism of gibberellin A20 to A1 in *Pisum sativum* L., decreasing the Le/le difference. *Plant Physiol.* 104, 277–280.

SHIMIZU-SATO, S., and H. MORI. 2001. Control of outgrowth and dormancy in axillary buds. *Plant Physiol.* 127, 1405–1413.

SIMON, E. W. 1977. Membranes in ripening and senescence. *Ann. Appl. Biol.* 85, 417–421.

SINGH, D. P., A. M. JERMAKOW, and S. M. SWAIN. 2002. Gibberellins are required for seed development and pollen tube growth in *Arabidopsis*. *Plant Cell* 14, 3133–3147.

SINNOTT, E. W. 1960. *Plant Morphogenesis*. McGraw-Hill, New York.

SINNOTT, E. W., and R. BLOCH. 1939. Changes in intercellular relationships during the growth and differentiation of living plant tissues. *Am. J. Bot.* 26, 625–634.

SLATER, A., N. W. SCOTT, and M. R. FOWLER. 2003. *Plant Biotechnology—The Genetic Manipulation of Plants*. Oxford University Press, Oxford.

STEBBINS, G. L., and S. K. JAIN. 1960. Developmental studies of cell differentiation in the epidermis of monocotyledons. I. *Allium, Rhoeo,* and *Commelina. Dev. Biol.* 2, 409–426.

STEEVES, T. A., M. A. HICKS, J. M. NAYLOR, and P. RENNIE. 1969. Analytical studies of the shoot apex of *Helianthus annuus. Can. J. Bot.* 47, 1367–1375.

STEINMANN, T., N. GELDNER, M. GREBE, S. MANGOLD, C. L. JACKSON, S. PARIS, L. GÄLWEILER, K. PALME, and G. JÜRGENS. 1999. Coordinated polar localization of auxin efflux carrier PIN1 by GNOM ARF GEF. *Science* 286, 316–318.

STEWARD, F. C., M. O. MAPES, A. E. KENT, and R. D. HOLSTEN. 1964. Growth and development of cultured plant cells. *Science* 143, 20–27.

STEWARD, F. C., U. MORENO, and W. M. ROCA. 1981. Growth, form and composition of potato plants as affected by environment. *Ann. Bot.* 48 (suppl. 2), 1–45.

STEWART, R. N., and L. G. BURK. 1970. Independence of tissues derived from apical layers in ontogeny of the tobacco leaf and ovary. *Am. J. Bot.* 57, 1010–1016.

STEWART, R. N., P. SEMENIUK, and H. DERMEN. 1974. Competition and accommodation between apical layers and their derivatives in the ontogeny of chimeral shoots of *Pelargonium* x *Hortorum. Am. J. Bot.* 61, 54–67.

STREET, H. E., ed. 1977. *Plant Tissue and Cell Culture*, 2nd ed. Blackwell, Oxford.

SUGIMOTO-SHIRASU, K., N. J. STACEY, J. CORSAR, K. ROBERTS, and M. C. MCCANN. 2002. DNA topoisomerase VI is essential for endoreduplication in *Arabidopsis. Curr. Biol.* 12, 1782–1786.

SUNDBERG, B., C. UGGLA, and H. TUOMINEN. 2000. Cambial growth and auxin gradients. In: *Cell and Molecular Biology of Wood Formation*, pp. 169–188, R. A. Savidge, J. R. Barnett, and R. Napier, eds. BIOS Scientific, Oxford.

SUNDERLAND, N., and J. M. DUNWELL. 1977. Anther and pollen culture. In: *Plant Tissue and Cell Culture*, 2nd. ed., pp. 223–265, H. E. Street, ed. Blackwell, Oxford.

SWARUP, R., J. FRIML, A. MARCHANT, K. LJUNG, G. SANDBERG, K. PALME, and M. BENNETT. 2001. Localization of the auxin permease AUX1 suggests two functionally distinct hormone transport pathways operate in the *Arabidopsis* root apex. *Genes Dev.* 15, 2648–2653.

SZYMKOWIAK, E. J., and I. M. SUSSEX. 1996. What chimeras can tell us about plant development. *Annu. Rev. Plant Physiol. Plant Mol. Biol.* 47, 351–376.

TAIZ, L., and E. ZEIGER. 2002. *Plant Physiology*, 3rd ed. Sinauer Associates Inc., Sunderland, MA.

THEIßEN, G., and H. SAEDLER. 1999. The golden decade of molecular floral development (1990–1999): A cheerful obituary. *Dev. Genet.* 25, 181–193.

THIMANN, K. V. 1978. Senescence. *Bot. Mag., Tokyo,* spec. iss. 1, 19–43.

THOMPSON, J. E., C. D. FROESE, Y. HONG, K. A. HUDAK, and M. D. SMITH. 1997. Membrane deterioration during senescence. *Can. J. Bot.* 75, 867–879.

TILNEY-BASSETT, R. A. E. 1986. *Plant Chimeras.* Edward Arnold, London.

TRAAS, J., M. HÜLSKAMP, E. GENDREAU, and H. HÖFTE. 1998. Endoreplication and development: rule without dividing? *Curr. Opin. Plant Biol.* 1, 498–503.

TREWAVAS, A. 1980. Possible control points in plant development. In: *The Molecular Biology of Plant Development*. Botanical Monographs, vol. 18, pp. 7–27, H. Smith and D. Grierson, eds. University of California Press, Berkeley.

TROMBETTA, V. V. 1942. The cytonuclear ratio. *Bot. Rev.* 8, 317–336.

TWELL, D., S. K. PARK, and E. LALANNE. 1998. Asymmetric division and cell-fate determination in developing pollen. *Trends Plant Sci.* 3, 305–310.

VAN DEN BERG, C., V. WILLEMSEN, W. HAGE, P. WEISBEEK, and B. SCHERES. 1995. Cell fate in the *Arabidopsis* root meristem determined by directional signalling. *Nature* 378, 62–65.

VASIL, I. 1991. Plant tissue culture and molecular biology as tools in understanding plant development and plant improvement. *Curr. Opin. Biotech.* 2, 158–163.

VASIL, I. K. 2005. The story of transgenic cereals: the challenge, the debate, and the solution—A historical perspective. *In Vitro Cell. Dev. Biol.—Plant* 41, 577–583.

VÖCHTING, H. 1878. *Über Organbildung im Pflanzenreich: Physiologische Untersuchungen über Wachsthumsursachen und Lebenseinheiten.* Max Cohen, Bonn.

VOGLER, H., and C. KUHLEMEIER. 2003. Simple hormones but complex signaling. *Curr. Opin. Plant Biol.* 6, 51–56.

WATANABE, M., D. SETOGUCHI, K. UEHARA, W. OHTSUKA, and Y. WATANABE. 2002. Apoptosis-like cell death of *Brassica napus* leaf protoplasts. *New Phytol.* 156, 417–426.

WEISSMAN, I. L. 2000. Stem cells: Units of development, units of regeneration, and units in evolution. *Cell* 100, 157–168.

WEISSMAN, I. L., D. J. ANDERSON, and F. GAGE. 2001. Stem and progenitor cells: Origins, phenotypes, lineage commitments, and transdifferentiations. *Annu. Rev. Cell Dev. Biol.* 17, 387–403.

WERNER, T., V. MOTYKA, M. STRNAD, and T. SCHMÜLLING. 2001. Regulation of plant growth by cytokinin. *Proc. Natl. Acad. Sci. USA* 98, 10487–10492.

WETMORE, R. H., and S. SOROKIN. 1955. On the differentiation of xylem. *J. Arnold Arbor.* 36, 305–317.

WEYERS, J. D. B., and N. W. PATERSON. 2001. Plant hormones and the control of physiological processes. *New Phytol.* 152, 375–407.

WILLIAMS, E. G., and G. MAHESWARAN. 1986. Somatic embryogenesis: Factors influencing coordinated behaviour of cells as an embryogenic group. *Ann. Bot.* 57, 443–462.

WITHERS, L., and P. G. ANDERSON, eds. 1986. *Plant Tissue Culture and Its Agricultural Applications.* Butterworths, London.

WREDLE, U., B. WALLES, and I. HAKMAN. 2001. DNA fragmentation and nuclear degradation during programmed cell death in the suspensor and endosperm of *Vicia faba. Int. J. Plant Sci.* 162, 1053–1063.

YAMAGUCHI, S., and Y. KAMIYA. 2002. Gibberellins and light-stimulated seed germination. *J. Plant Growth Regul.* 20, 369–376.

YAMAMOTO, R., S. FUJIOKA, T. DEMURA, S. TAKATSUTO, S. YOSHIDA, and H. FUKUDA. 2001. Brassinosteroid levels increase drastically prior to morphogenesis of tracheary elements. *Plant Physiol.* 125, 556–563.

YAXLEY, J. R., J. J. ROSS, L. J. SHERRIFF, and J. B. REID. 2001. Gibberellin biosynthesis mutations and root development in pea. *Plant Physiol.* 125, 627–633.

YONG, J. W. H., S. C. WONG, D. S. LETHAM, C. H. HOCART, and G. D. FARQUHAR. 2000. Effects of elevated [$CO_2$] and nitrogen nutrition on cytokinins in the xylem sap and leaves of cotton. *Plant Physiol.* 124, 767–779.

ZAMBRYSKI, P., and K. CRAWFORD. 2000. Plasmodesmata: Gatekeepers for cell-to-cell transport of developmental signals in plants. *Annu. Rev. Cell Dev. Biol.* 16, 393–421.

ŽÁRSKÝ, V., and F. CVRČKOVÁ. 1999. Rab and Rho GTPases in yeast and plant cell growth and morphogenesis. In: *Advances in Regulation of Plant Growth and Development*, pp. 49–57, M. Strnad, P. Peč, and E. Beck, eds. Peres Publishers, Prague.

ZEEVAART, J. A. D., and M. TALON. 1992. Gibberellin mutants in *Arabidopsis thaliana*. In: *Progress in Plant Growth Regulation*, pp. 34–42, C. M. Karssen, L. C. van Loon, and D. Vreugdenhil, eds. Kluwer Academic, Dordrecht.

# CHAPTER SIX

# Apical Meristems

The term **apical meristem** refers to a group of meristematic cells at the apex of shoot and root that by cell division lay the foundation of the primary plant body. As indicated in Chapter 5, meristems are composed of initials, which perpetuate the meristems, and their derivatives. Also the derivatives usually divide and produce one or more generations of cells before the cytologic changes, denoting differentiation of specific cells and tissues, occur near the tip of the shoot or root. The divisions continue at all levels where such changes are already discernible. Therefore growth, in the sense of cell division, is not limited to the very tip of shoot or root but extends to levels considerably removed from the region usually called the apical meristem. In fact the divisions some distance from the apex are more abundant than at the apex (Buvat, 1952). In the shoot a more intensive meristematic activity is observed at levels where new leaves are initiated than at the tip, and during the elongation of the stem, cell division extends several internodes below the apical meristem (Sachs, 1965). The change from apical meristem to adult primary tissues is gradual and involves the intergrading of the phenomena of cell division, cell enlargement, and cell differentiation, so one cannot restrict the term meristem to the apex of shoot and root. The parts of shoot and root where future tissues and organs are already partly determined but where cell division and cell enlargement are still in progress are also meristematic.

The profuse and inconsistent terminology in the voluminous literature on apical meristems (Wardlaw, 1957; Clowes, 1961; Cutter, 1965; Gifford and Corson, 1971; Medford, 1992; Lyndon, 1998) reflects the complexity of the subject matter. Most commonly the term apical meristem is used in a wider sense than merely the initials and their immediate derivatives; it also includes variable lengths of shoot and root proximal to the apex. **Shoot apex** and **root apex** often are employed as synonyms of apical meristem, although a distinction is sometimes made between the shoot apical meristem and the shoot apex: the apical meristem denotes only the part of the shoot lying distal to the youngest leaf primordium, whereas the shoot apex includes the apical meristem together with the subapical region bearing young leaf primordia (Cutter, 1965). When determinations of the dimensions of apices of shoots are made,

---

*Esau's Plant Anatomy, Third Edition*, By Ray F. Evert.

only the part above the youngest leaf primordium, or the youngest node, is measured.

When it is important to differentiate the least determined part of the apical meristem the term **promeristem** or **protomeristem** (Jackson, 1953) is used: it refers to the initials and their most recent derivatives, which exhibit no evidence of tissue differentiation and are presumed to be in the same physiological state as the initials (Sussex and Steeves, 1967; Steeves and Sussex, 1989). Johnson and Tolbert's (1960) **metameristem** refers to the same group of cells as the promeristem (protomeristem). They define it specifically as "the central part of the shoot apex which maintains itself, contributes to the growth and organization of the apex, but exhibits little or no evidence of tissue separation." Thus this least determined part of the apical meristem corresponds to the general area termed the central zone in shoot apices (see below). The promeristem of Clowes (1961), on the other hand, includes only the initials.

## ▌EVOLUTION OF THE CONCEPT OF APICAL ORGANIZATION

### Apical Meristems Originally Were Envisioned as Having a Single Initial Cell

Following Wolff's (1759) recognition of the shoot apex as an undeveloped region from which growth of the plant proceeded and the discovery of a single, morphologically distinct initial cell at the apex of seedless vascular plants, the idea developed that such cells exist in seed plants as well. The **apical cell** (Fig. 6.1) was interpreted as a consistent structural and functional unit of apical meristems governing the whole process of growth (the **apical-cell theory**). Subsequent researchers refuted the assumption of a universal occurrence of single apical cells and replaced it by a concept of independent origin of different parts of the plant body.

### The Apical-Cell Theory Was Superseded by the Histogen Theory

The **histogen theory** was developed by Hanstein (1868, 1870) on the basis of extensive studies of angiosperm shoot apices and embryos. According to this theory the main body of the plant arises not from superficial cells but from a massive meristem of considerable depth comprising three parts, the **histogens**, which may be distinguished by their origin and course of development. The outermost part, the **dermatogen** (from the Greek words meaning skin and to bring forth), is the precursor of the epidermis; the second, the **periblem** (from the Greek, clothing), gives rise to the cortex; the third, the **plerome** (from the Greek, that which fills), constitutes the inner mass of the axis. The dermatogen, each layer

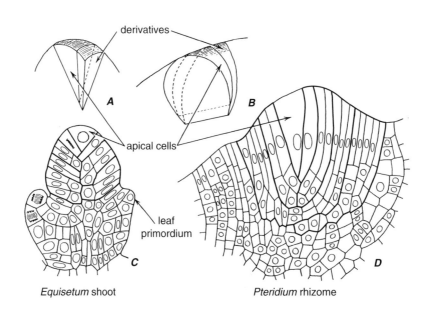

**FIGURE 6.1**

Shoot apices with apical cells. **A**, **B**, two forms of the apical cell, pyramidal, or four-sided, (**A**) and lenticular, or three-sided, (**B**). Cells are cut off from three faces of the pyramidal cell, from two in the lenticular. In both **A** and **B** a derivative cell is shown attached to the right side of the apical cell. **C**, **D**, apical cells in longitudinal sections of shoot (**C**) and rhizome (**D**). In **C**, apical cells in leaf primordia; one of these (left) is dividing. Subdivided derivatives—merophytes—of the apical cell are indicated by slightly thicker walls. (C, D, ×230. A, B, adapted from Schüepp, 1926. www.schweizerbart.de)

of the periblem, and the plerome begin with one or several initials distributed in superposed tiers in the most distal part of the apical meristem.

Hanstein's "dermatogen" is not equivalent to Haberlandt's (1914) "protoderm." Haberlandt's protoderm refers to the outermost layer of the apical meristem regardless of whether this layer arises from independent initials or not, and regardless of whether it gives rise to the epidermis or to some subepidermal tissue as well. In many apices the epidermis does originate from an independent layer in the apical meristem; in such apices the protoderm and dermatogen may coincide. The periblem and plerome in the sense of Hanstein are discernible in many roots but are seldom delimited in shoots. Thus the subdivision into dermatogen, periblem, and plerome has no universal application. But the fatal flaw of Hanstein's histogen theory is its presumption that the destinies of the different regions of the plant body are determined by the discrete origin of these regions in the apical meristem.

## The Tunica-Corpus Concept of Apical Organization Applies Largely to Angiosperms

The apical-cell and histogen theories were developed with reference to both root and shoot apices. The third theory of apical structure, the **tunica-corpus theory** of A. Schmidt (1924), was an outcome of observations on angiosperm shoot apices. It states that the initial region of the apical meristem consists of (1) the **tunica**, one or more peripheral layers of cells that divide in planes perpendicular to the surface of the meristem (anticlinal divisions), and (2) the **corpus**, a body of cells several layers deep in which the cells divide in various planes (Fig. 6.2). Thus, whereas the corpus adds bulk to the apical meristem by increase in volume, the one or more layers of tunica maintain their continuity over the enlarging mass by surface growth. Each layer of tunica arises from a small group of separate initials, and the corpus has its own initials located beneath those of the tunica. In other words, the number of tiers of initials is equal to the number of tunica layers plus one, the tier of corpus initials. In contrast to the histogen theory, the tunica-corpus theory does not imply any relation between the configuration of cells at the apex and histogenesis below the apex. Although the epidermis usually arises from the outermost layer of tunica, which thus coincides with Hanstein's dermatogen, the underlying tissues may have their origin in the tunica or corpus or both, depending on the plant species and the number of tunica layers.

As more plants came to be examined, the tunica-corpus concept underwent some modifications, especially with regard to the strictness of the definition of the tunica. According to one view, tunica should include only those layers that never show any periclinal divisions in the median position, that is, above the level of origin of leaf primordia (Jentsch, 1957). If the apex contains additional parallel layers that periodically divide periclinally, these layers are assigned to the corpus and the latter is characterized as being stratified (Sussex, 1955; Tolbert and Johnson, 1966). Other workers treat the tunica more loosely and describe it as fluctuating in number of layers: one or more of the inner layers of the tunica may divide periclinally and thus become part of the corpus (Clowes, 1961). Because of the differing usage of the term tunica, its usefulness in accurately describing growth relations at the shoot apex was questioned by Popham (1951), who proposed the term **mantle** to include "all layers at the summit of the apex in which anticlinal divisions are sufficiently frequent to result in the perpetuation of definite cell layers"; the mantle overarches a body of cells called the **core**. The term corpus was avoided.

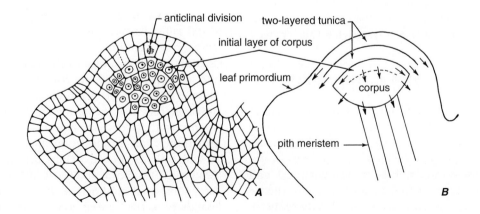

**FIGURE 6.2**

Shoot apex of *Pisum* (pea). Cellular details in **A**, interpretative diagram in **B**. The pith meristem does not show typical rib-meristem form of growth. (From Esau, 1977.)

### The Shoot Apices of Most Gymnosperms and Angiosperms Show a Cytohistological Zonation

The tunica-corpus concept, which was developed with reference to angiosperm shoot apices, proved to be largely unsuitable for the characterization of the apical meristem of gymnosperms (Foster, 1938, 1941; Johnson, 1951; Gifford and Corson, 1971; Cecich, 1980). With few exceptions (*Gnetum*, *Ephedra*, and several species of conifers) the gymnosperms do not show a tunica-corpus organization in the shoot apex; that is, they do not have stable surface layers dividing only anticlinally. The outermost layer of the apical meristem undergoes periclinal and anticlinal divisions and contributes cells to the peripheral and interior tissues of the shoot. The surface cells located in a median position in the apical meristem are interpreted as initials. Studies of gymnosperm apices led to the recognition of a zonation—termed a **cytohistological zonation**—based not only on planes of division but also on cytologic and histologic differentiation and the degree of meristematic activity of the component cell complexes (Fig. 6.3). Similar zonations, superimposed on a tunica-corpus organization, have since been observed in most angiosperms (Clowes, 1961; Gifford and Corson, 1971).

The cytologic zones that may be recognized in shoot apical meristems vary in degree of differentiation and in details of grouping of cells. The zonation may be succinctly characterized by dividing the apical meristem into a **central zone** and two zones derived from it. One of these, the **rib zone**, or **rib (pith) meristem**, appears directly below the central zone and is centrally located in the apex. It usually becomes the pith after additional meristematic activity has occurred. The other, the **peripheral zone**, or **peripheral meristem**, encircles the other zones. The peripheral zone typically is the most meristematic of all three zones and has the densest protoplasts and the smallest cell dimensions. It may be described as a **eumeristem**. Leaf primordia and the procambium arise here, as well as the cortical ground tissue. In species with a tunica-corpus organization, the central zone corresponds to the corpus and the portion(s) of the tunica layer(s) overlying the corpus.

## INQUIRIES INTO THE IDENTITY OF APICAL INITIALS

The next development in the interpretation of the shoot apical meristem resulted from the efforts of French cytologists (Buvat, 1955a; Nougarède, 1967). Meristematic activity drew the chief attention of this work. Counts of mitoses, and cytological, histochemical, and ultrastructural studies served to formulate the theory that after the apical structure is organized in the embryo, the central zone of cells becomes the waiting meristem (Fig. 6.4; **méristème d'attente**). The méristème d'attente stays in a quiescent state until the reproductive stage is

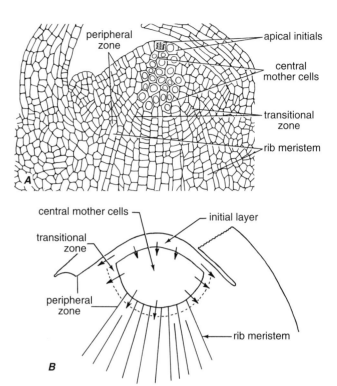

**FIGURE 6.3**

Shoot tip of *Pinus strobus* in longitudinal view. Cellular details in **A**, interpretative diagram in **B**. Apical initials contribute cells to the surface layer by anticlinal divisions and to the central mother-cell zone by periclinal divisions. The mother-cell zone (cells with nuclei) contributes cells to the transitional zone composed of actively dividing cells arranged in a series radiating from the mother-cell zone. Products of these divisions form the rib meristem and the subsurface layers of the peripheral zone. (A, ×139. **A**, as drawn from a slide by A. R. Spurr; **B**, from Esau, 1977.)

reached and meristematic activity is resumed in the distal cells. During the vegetative stage, meristematic activity is centered in the initiating ring (**anneau initial**), corresponding to the peripheral zone, and in the medullary (pith) meristem (**méristème medullaire**). The concept of the inactive central zone in the apical meristem was extended from the shoots of angiosperms to those of gymnosperms (Camefort, 1956; who called the central zone "zone apicale") and the seedless vascular plants (Buvat, 1955b), and to roots (Buvat and Genevès, 1951; Buvat and Liard, 1953). The concept was later somewhat modified in that variations in degree of inactivity of the central zone in relation to the size of the apex and its stage of development came to be recognized (Catesson, 1953; Lance, 1957; Loiseau, 1959). With regard to root apices, the occurrence of an inactive

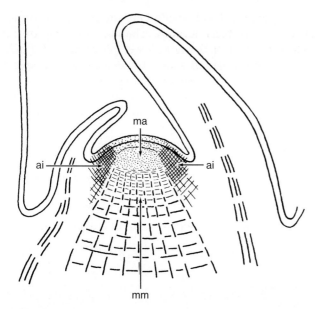

**FIGURE 6.4**

Diagram of a shoot apex of *Cheiranthus cheiri* interpreted according to the méristème d'attente concept. Details: ai, anneau initial; ma, méristème d'attente; mm, méristème medullaire. (After Buvat, 1955a. © Masson, Paris.)

**TABLE 6.1 ■ Cell Doubling Time (Mean Cell Generation Time) at the Summit and on the Flanks of Vegetative Shoot Apical Meristems of Angiosperms**

| Species | Cell Doubling Time (h) | |
| | Summit[c] | Flanks[d] |
| --- | --- | --- |
| *Trifolium repens* | 108 | 69 |
| *Pisum* (probably *P. sativum*) | 69 | 28 |
| *Pisum* (main apex) | 49 | 31 |
| *Pisum* (axillary bud, initiated) | 127 | 65 |
| *Pisum* (axillary bud, released) | 40 | 33 |
| *Oryza* (rice) | 86 | 11 |
| *Rudbeckia bicolor* | >40 | 30 |
| *Solanum* (potato) | 117 | 74 |
| *Datura stramonium* | 76 | 36 |
| *Coleus blumei* | 250 | 130 |
| *Sinapis alba* | 288 | 157 |
| *Chrysanthemum*[a] | 144 | 50 |
| *Chrysanthemum*[b] | 102 | 32 |
| *Chrysanthemum* | 139 | 48 |
| *Chrysanthemum segetum* | 140 | 54 |
| *Helianthus annuus* | 83 | 37 |

Source: From Lyndon, 1998.
[a] photon flux = 70 µmol/m².
[b] photon flux = 200 µmol/m².
[c] Or central zone.
[d] Or peripheral zone.

center in the meristem found confirmation in many studies, resulting in the development of the concept of **quiescent center** by Clowes (1961).

The revision of the concept of apical initials by the French workers served as a considerable stimulant for further research on apical meristems (Cutter, 1965; Nougarède, 1967; Gifford and Corson, 1971). Counts of mitoses in different regions of the shoot apex, feeding root tips with radiolabeled compounds to detect the location and synthesis of DNA, RNA, and protein, histochemical tests, experimental manipulations, and tracing of cell patterns in fixed and living shoot apices provided data that, in essence, corroborated the postulate of the relative infrequency of mitotic activity in the central zone (Table 6.1) (Davis et al., 1979; Lyndon, 1976, 1998).

Recognition of the relative infrequency of mitotic activity in the central zone has not led to an abandonment of the concept that the most distal cells are the true initials and the ultimate source of all body cells in the shoot. Considering the geometry of the apex, one can deduce a priori that in view of the exponential growth of the derivatives of the apical meristem, a few divisions in the distalmost cells would result in the propagation of any distinctive genome characteristic of these cells through large populations of cells. As noted earlier, the tunica-corpus theory postulated the presence of a small group of initials in each layer of the apical meristem. Clonal analyses are often cited as pro-

viding evidence for one to three initials in each layer (Stewart and Dermen, 1970, 1979; Zagórska-Marek and Turzańska, 2000; Korn, 2001).

The relation between the initials and the immediate derivatives in the apical meristem is flexible. A cell functions as an initial not because of any inherent properties but because of its position. (See the similar concept of initials in the vascular cambium, Chapter 12.) At the time of division of an initial it is impossible to predict which of the two daughter cells will "inherit" the initial function and which will become the derivative. It is also known that a given initial may be replaced by a cell that through prior history would be classified as a derivative of an initial (Soma and Ball, 1964; Ball, 1972; Ruth et al., 1985; Hara, 1995; Zagórska-Marek and Turzańska, 2000).

Inasmuch as no cells are permanent initials, Newman (1965) maintained that in order to understand structure and functioning of a meristem, a distinction must be made between the "continuing meristematic residue"—that is, the source of cellular structure that functions as initials—and the "general meristem," a region of elaboration. Emergence of new cells from the continuing meristematic residue is a very slow, continuous process of long duration, whereas the passage of cells in the general meristem is a very rapid, continuous process of only short duration. This concept is used in Newman's classification of apical meristems designed for all groups of vascular plants: (1) **monopodial**, as in ferns—the residue is in the superficial layer and any kind of division contributes to growth in length and breadth; (2) **simplex**, as in gymnosperms—the residue is in a single, superficial layer and both anticlinal and periclinal divisions are needed for bulk growth; (3) **duplex**, as in angiosperms—the residue occurs in at least two surface layers with two contrasting modes of growth, anticlinal divisions near the surface and divisions in at least two planes deeper in the apical meristem.

# VEGETATIVE SHOOT APEX

The vegetative shoot apex is a dynamic structure that in addition to adding cells to the primary plant body, repetitively produces units, or modules, called **phytomeres** (Fig. 6.5). Each phytomere consists of a node,

with its attached leaf, a subjacent internode, and a bud at the base of the internode. The bud is located in the axil of the leaf of the next lower phytomere and may develop into a lateral shoot. In seed plants the apical meristem of the first shoot is organized in the embryo before or after the appearance of the cotyledon or cotyledons (Saint-Côme, 1966; Nougarède, 1967; Gregory and Romberger, 1972).

Vegetative shoot apices vary in shape, size, cytologic zonation, and meristematic activity (Fig. 6.6). The shoot apices of conifers are commonly relatively narrow and conical in form; in *Ginkgo* and in the cycads they are rather broad and flat. The apical meristem of some monocots (grasses, *Elodea*) and eudicots (*Hippuris*) is

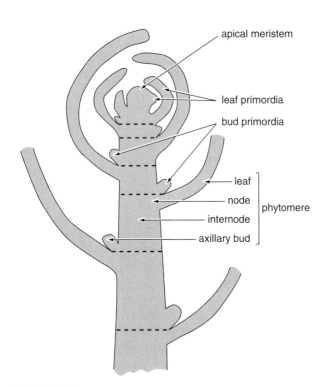

**FIGURE 6.5**

Diagram of a longitudinal section of a eudicot shoot tip. Activity of the apical meristem, which repetitively produces leaf and bud primordia, results in a succession of repeated units called phytomeres. Each phytomere consists of a node, its attached leaf, the internode below that leaf, and the bud at the base of the internode. The boundaries of the phytomeres are indicated by the dashed lines. Note that the internodes are increasing in length the farther they are from the apical meristem. Internodal elongation accounts for most of the increase in length of the stem.

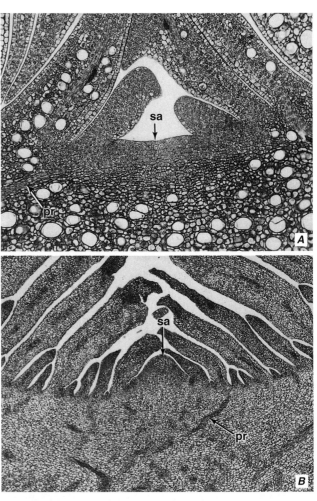

**FIGURE 6.6**

Distinctive forms of shoot apices (sa): flat or slightly concave in *Drimys* (**A**) and conical but inserted on broad base bearing leaf primordia in *Washingtonia* palm (**B**). Longitudinal sections. The large cavities in the *Drimys* shoot apex are oil cells. Other detail: pr, procambium. (A, ×90; B, ×19. **A**, slide by Ernest M. Gifford; **B**, from Ball, 1941.)

narrow and elongated, with the distal portion much elevated above the youngest node. In many eudicots, the distal portion barely rises above the leaf primordia, or even appears sunken (Gifford, 1950). In some plants the axis increases in width close to the apex and the peripheral region bearing the leaf primordia becomes elevated above the apical meristem leaving the latter in a pit-like depression (Ball, 1941; rosette type of eudicots, Rauh and Rappert, 1954). Examples of widths of apices measured in micrometers at insertion of the youngest leaf primordia are: 280, *Equisetum hiemale*; 1000, *Dryopteris dilatata*; 2000 to 3300, *Cycas revoluta*; 280, *Pinus mugo*; 140, *Taxus baccata*; 400, *Ginkgo biloba*; 288, *Washingtonia filifera*; 130, *Zea mays*; 500, *Nuphar lutea* (Clowes, 1961). At germination the shoot apical meristem of the *Arabidopsis thaliana* (Wassilewskija ecotype) embryo measures approximately 35 by 55 micrometers (Medford et al., 1992). The shape and size of the apex change during the development of a plant from embryo to reproduction, between initiation of successive leaves, and in relation to seasonal changes. An example of change in width during growth is available for *Phoenix canariensis* (Ball, 1941). The diameter in micrometers was found to be 80 in the embryo, 140 in the seedling, and 528 in the adult plant.

In the following paragraphs further aspects on the structure and function of the shoot apical meristems of each of the major groups of vascular plants are considered. We begin with the seedless vascular plants.

## The Presence of an Apical Cell Is Characteristic of Shoot Apices in Seedless Vascular Plants

In most seedless vascular plants—the leptosporangiate (more specialized) ferns, *Osmunda*—growth at the shoot apex proceeds from a superficial layer of large, highly vacuolated cells, with a more or less distinct apical cell, *the* initial cell, in the center. In some seedless vascular plants (*Equisetum, Psilotum*, species of *Selaginella*), the apical cell is enlarged and fairly conspicuous; in still others (eusporangiate ferns, *Lycopodium, Isoetes*), distinctive apical cells are wanting, and the situation is less clear (Guttenberg, 1966). Both single apical cells and groups of apical initials have been reported in the same species of *Lycopodium* (Schüepp, 1926; Härtel, 1938) and of some eusporangiate ferns (Campbell, 1911; Bower, 1923; Bhambie and Puri, 1985). It is probable, however, that a single apical cell is present in the shoot apices of nearly all seedless vascular plants (Bierhorst, 1977; White, R. A., and Turner, 1995).

Most commonly the apical cell is pyramidal (tetrahedral) in shape (Fig. 6.1A, C). The base of the pyramid is turned toward the free surface, and the other three sides downward. In apices with a tetrahedral apical cell, the derivative cells form an orderly pattern, which is initiated by the orderliness of divisions of the apical cells:

the successive divisions follow one another in acropetal sequence along a helix. The term **merophyte** is used to refer to the immediate unicellular derivatives of an apical cell and also to the multicellular structural units derived from them (Gifford, 1983). Tetrahedral apical cells are found in *Equisetum* and most leptosporangiate ferns.

Apical cells may be three-sided, with two sides along which new cells are cut off. Such apical cells are characteristic of bilaterally symmetrical shoots, as in the water ferns *Salvinia, Marsilea*, and *Azolla* (Guttenberg, 1966; Croxdale, 1978, 1979; Schmidt, K. D., 1978; Lemon and Posluszny, 1997). The flattened rhizome apex of *Pteridium* also bears a three-sided apical cell (Fig. 6.1B, D; Gottlieb and Steeves, 1961).

Some workers describe the shoot apices of ferns on a zonation basis (McAlpin and White, 1974; White, R. A., and Turner, 1995). According to this concept the promeristem is composed of two zones or layers of meristematic cells, a surface layer and a subsurface layer. Subjacent to the promeristem are distinct meristematic zones "transitional to the developing tissues of the cortex, stele, and pith" (White, R. A., and Turner, 1995). A second concept, developed in relation to the shoot apices of *Matteuccia struthiopteris* and *Osmunda cinnamomea*, considers the promeristem to consist of only the surface layer, which possesses a single apical cell (Ma and Steeves, 1994, 1995). Immediately below the surface layer is prestelar tissue consisting of provascular tissue (defined as tissue in the initial stage of vascularization and in which procambium is subsequently formed) and pith mother cells, which represent the initial differentiation of the pith.

Although the apical cell at the tips of shoot and root apical meristems of seedless vascular plants was considered by early plant morphologists to be the ultimate source of all cells in shoot and root, with the advent of the méristème d'attente concept, a formative role for the apical cell began to be questioned. Some workers concluded that the apical cell is active mitotically only in very young plants and then becomes mitotically inactive and comprises a "quiescent center" comparable to the multicellular quiescent center in angiosperm roots. The apical cells of certain ferns were reported to be highly polyploid as a result of endoreduplication (Chapter 5), a condition that would support the contention that the apical cells are mitotically inactive (D'Amato, 1975). Subsequent studies involving the determination of the mitotic index, the duration of the cell cycle and of mitosis, and measurement of DNA content in shoot and root apices of certain ferns clearly indicate, however, that the apical cell remains mitotically active during active shoot and root growth (Gifford et al., 1979; Kurth, 1981). No evidence of endoreduplication was found in the apical meristem during development. These studies, in addition to "rediscovery" of the merophyte as a single derivative of the apical cell (Bierhorst, 1977), reaffirmed the classical concept for the role of the apical cell.

### The Zonation Found in the *Ginkgo* Apex Has Served as a Basis for the Interpretation of Shoot Apices in Other Gymnosperms

The presence of cytologic zones in the apical meristem was first recognized by Foster (1938) in the shoot apex of *Ginkgo biloba* (Fig. 6.7). In *Ginkgo* all cells of the apex are derived from a group of surface initials termed the **apical initial group**. The subjacent group of cells originating from the surface initials constitutes the **central mother cell zone**. This entire assemblage of cells, including the lateral derivatives of the apical initial group, is conspicuously vacuolated, a feature associated with a relatively low rate of mitotic activity. Moreover the central mother cell zone cells often have thickened and distinctly pitted walls. The apical surface initials and central mother cells constitute the promeristem. Surrounding the central mother cell zone is the **peripheral zone (peripheral meristem)** and beneath it the **rib**, or **pith**, **meristem**. The peripheral zone originates

in part from the lateral derivatives of the apical initials and in part from the central mother cells. The derivatives produced at the base of the mother cell zone become pith cells as they pass through the rib meristem form of growth. During active growth a cup-shaped region of orderly dividing cells, the **transitional zone**, delimits the mother cell zone and may extend to the surface of the apical zone.

The details of the structural pattern just reviewed vary in the different groups of gymnosperms. The cycads have very wide apices with a large number of surface cells contributing derivatives to the deeper layers by periclinal divisions. Foster (1941, 1943) interpreted this extended surface and its immediate derivatives as the initiation zone; others would confine the initials to a relatively small number of surface cells (Clowes, 1961; Guttenberg, 1961). The periclinal derivatives of the surface layer converge toward the mother cell zone, a pattern apparently characteristic of cycads. In other seed plants the cell layers typically diverge from the point of initiation. The convergent pattern results from numerous anticlinal divisions in the surface cells and their recent derivatives—

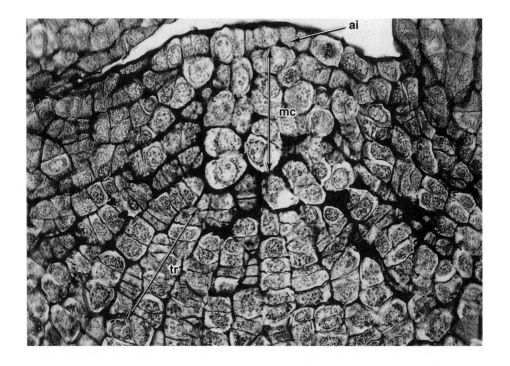

**FIGURE 6.7**

Longitudinal section of the shoot apex of *Ginkgo biloba*. Apical initial group (ai) contributes to surface layer by anticlinal division. It also adds cells by periclinal divisions to the central mother-cell group (mc). Growth in volume is by cell enlargement, and occasional divisions in various planes characterize the central mother-cell zone. Outermost products of divisions in this zone become displaced toward the transitional zone (tr) where they divide by walls periclinal with reference to the mother-cell zone. Derivatives of these divisions form peripheral subsurface layers and prospective pith, the rib-meristem zone. (×430. From Foster, 1938.)

evidence of surface growth through a tissue of some depth. This growth appears to be associated with the large width of the apex. The mother cell group is relatively indistinct in cycads. The extensive peripheral zone arises from the immediate derivatives of the surface initials and from the mother cells. The rib meristem is more or less pronounced beneath the mother cell zone.

Most conifers have periclinally dividing apical initials in the surface layer. A contrasting organization, with a cell layer dividing almost exclusively or predominantly by anticlinal walls, has been described in *Araucaria*, *Cupressus*, *Thujopsis* (Guttenberg, 1961), *Agathis* (Jackman, 1960), and *Juniperus* (Ruth et al., 1985). In these plants the apices have been interpreted as having a tunica-corpus organization. The mother cell group may be well differentiated in conifers and a transitional zone may be present. In conifers with narrow apices, mother cells are few and may or may not be enlarged and vacuolated. In such apices a small mother cell group, three or four cells in depth, is abruptly succeeded below by highly vacuolated pith cells without the interposition of a rib meristem, and the peripheral zone is also only a few cells wide.

Coniferous shoot apices have been studied with regard to seasonal variations in structure. In some species (*Pinus lambertiana* and *P. ponderosa*, Sacher, 1954; *Abies concolor*, Parke, 1959; *Cephalotaxus drupacea*, Singh, 1961) the basic zonation does not change, but the height of the apical dome above the youngest node is greater during growth than during rest, or dormancy (Fig. 6.8). Because of this difference the zones are differentially distributed in the two kinds of apices with regard to the youngest node: the rib meristem occurs below this node in resting apices and partly above it in active apices. This observation calls attention to a terminological problem. If the apical meristem is defined strictly, as the part of the apex above the youngest node, it must be interpreted as varying in its composition during different growth phases (Parke, 1959). A loss of zonation and the appearance of a tunica-corpus-like structure has been reported for the dormant meristems in *Tsuga heterophylla* (Owens and Molder, 1973) and *Picea mariana* (Riding, 1976).

The Gnetophyta commonly have a definite separation into a surface layer and an inner core derived from its own initials. Therefore the shoot apices of *Ephedra* and *Gnetum* have been described as having a tunica-corpus pattern of growth (Johnson, 1951; Seeliger, 1954). The tunica is uniseriate, and the corpus is comparable to the central mother cell zone in its morphology and manner of division. The shoot apex of *Welwitschia* typically produces only one pair of foliage leaves and does not possess distinct zonation. Periclinal divisions have been observed in the surface layer (Rodin, 1953).

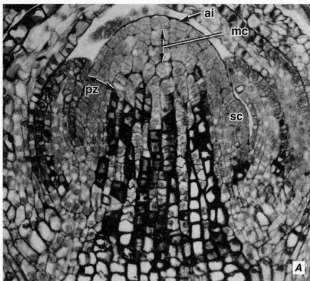

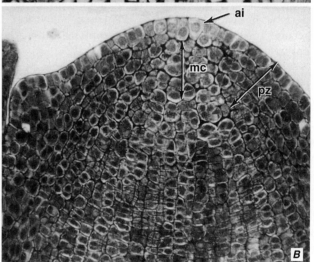

**FIGURE 6.8**

Longitudinal sections of shoot tips of *Abies* during the first phase of seasonal growth (**A**) and during the winter rest phase (**B**). In **A**, scale primordia (sc) are being initiated, and the tannin content in the pith distinguishes this region from the apex and the peripheral zone (pz). Results of recent divisions are evident in the apex. **B**, zonation less distinctive than in **A**. Other details: ai, apical initial group; mc, mother cells. (A, ×270; B, ×350. **B**, from Parke, 1959.)

## The Presence of a Zonation Superimposed on a Tunica-Corpus Configuration Is Characteristic of Angiosperm Shoot Apices

As noted previously, the corpus and each layer of tunica are visualized as having their own initials. In the tunica the initials are disposed in the median axial position. By anticlinal divisions these cells form progenies of new cells, some of which remain at the summit as initials;

others function as derivatives that, by subsequent divisions, contribute cells to the peripheral part of the shoot. The initials of the corpus appear beneath those of the tunica. By periclinal divisions these initials give derivatives to the corpus below, the cells of which divide in various planes. Cells produced by divisions in the corpus are added to the center of the axis, that is, to the rib meristem, and also to the peripheral meristem. Together the corpus and the tunica layer(s) overlying the corpus constitute the central zone or promeristem of the meristem.

The initials of the corpus may form a well-defined layer, in contrast to the less orderly arranged cells in the mass of the corpus. When this pattern is present, the delimitation between the tunica and the corpus may be difficult to determine. However, if the apices are collected at different stages of development, the uppermost layer of the corpus will be found undergoing periodic periclinal divisions. After such a division a second orderly layer appears temporarily in the corpus.

The number of tunica layers varies in angiosperms (Gifford and Corson, 1971). More than half of the species studied among eudicots have a two-layered tunica (Fig.

6.9). The reports of higher numbers, four and five or more (Hara, 1962), are subject to the qualification that some workers include the innermost parallel layer or layers in the tunica, others in the corpus. One and two are the common numbers of tunica layers in the monocots. Two tunica layers are common in festucoid grasses and a single layer in panicoid grasses (Fig. 6.10) (Brown et al., 1957). An absence of tunica-corpus organization, with the outermost layer dividing periclinally, has also been observed (*Saccharum*, Thielke, 1962). The number of parallel layers in the shoot apex may vary during the ontogeny of the plant (Mia, 1960; Gifford and Tepper, 1962) and under the influence of seasonal growth changes (Hara, 1962). There may also be periodic changes in stratification in relation to the initiation of leaves (Sussex, 1955).

The view that the layers in the apical meristem with a tunica-corpus organization are clonally distinct cell layers is supported by observations on periclinal cytochimeras (Chapter 5). Most of the plants studied by reference to cytochimeras are eudicots with a two-layered tunica. In these plants, periclinal cytochimeras have clearly revealed the existence of three independent layers (two layers of tunica and one of corpus initials)

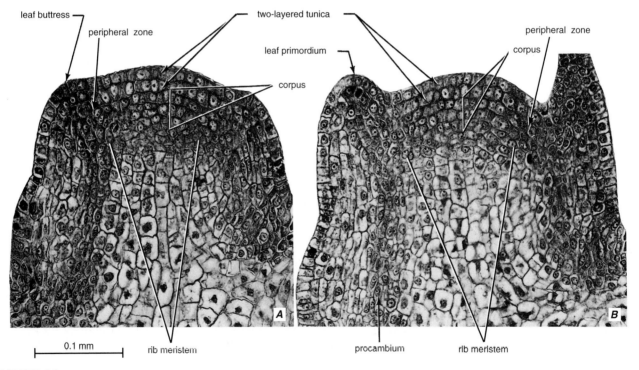

**FIGURE 6.9**

Longitudinal sections of shoot apices of potato (*Solanum tuberosum*) showing tunica-corpus organization of the apical meristem and two stages in the initiation of a leaf primordium; leaf buttress stage in **A**, and beginning of upward growth in **B**. A procambial strand, which will differentiate upward into the developing leaf, can be seen beneath the leaf buttress. (From Sussex, 1955.)

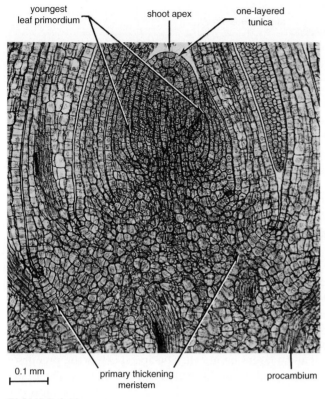

youngest
leaf primordium

shoot apex

one-layered
tunica

0.1 mm

primary thickening
meristem

procambium

**FIGURE 6.10**

Longitudinal section of a shoot apex of maize (*Zea mays*), a panicoid grass, with a one-layered tunica. Parts of each leaf occur on both sides of the axis because the leaves encircle the stem in their growth. (From Esau, 1977.)

in the apical meristem (Fig. 5.11; Satina et al., 1940). These three cell layers commonly are designated L1, L2, and L3, with the outermost layer being designated L1 and the innermost L3. Some workers erroneously designate the entire corpus L3, rather than just the initial layer of the corpus (e.g., Bowman and Eshed, 2000; Vernoux et al., 2000a; Clark, 2001).

The stage of plant development at which zonation is established in the vegetative shoot apex can vary among species. In the Cactaceae, for example, zonation is already established at the time of germination in certain species, whereas in others only a tunica-corpus organization is present at that time (Mauseth, 1978). In some cactus species zonation is not completed until more than 30 leaves have been produced. Similarly, in the shoot apex of *Coleus*, zonation is not completed until five pairs of leaves have been initiated (Saint-Côme, 1966). Thus, although zonation is a characteristic feature of these meristems, it is not essential for leaf production or for normal functioning of the meristem, in general. Sekhar and Sawhney (1985) were unable to recognize a zonation pattern in the shoot apices of tomato (*Solanum lycopersicum*).

As indicated in Chapter 5, a number of plant biologists have adopted the term stem cell to refer to the initials and/or their recent derivatives in the apical meristem. Some workers confusingly use both terms in their descriptions of the shoot apical meristem. Following are some examples. "The stem cells are not permanent initial cells . . ." (Fletcher, 2004). "It is now generally accepted that the central zone acts as a population of stem cells . . . generating the initials for the other two zones whilst maintaining itself" (Vernoux et al., 2000a). "The central zone acts as a reservoir of stem cells, which replenish both the peripheral and rib zones, as well as maintaining the integrity of the central zone. It should be noted that these cells do not act as permanent initials, but rather their behavior is governed in a position-dependent manner" (Bowman and Eshed, 2000). "It is now widely assumed that central cells function as stem cells and serve as initials or source cells for the two other zones of the shoot apical meristem" (Laufs et al., 1998a). While characterizing the shoot apical meristem as "a group of stem cells," another worker (Meyerowitz, 1997) designated the central zone the "zone of initials."

Some workers have noted the ambiguity of the term stem cell with reference to plants and, for the most part, have avoided its use in their descriptions of the shoot apical meristem (Evans, M. M. S., and Barton, 1997). In order to avoid any confusion inherent in use of the term stem cell in plant biology, Barton (1998) adopted the term promeristem, which she noted conceptually consists of the apical initials and their recent derivatives, "to refer to the hypothetical population of cells that have not yet been specified as leaf or stem. . . ." This is entirely appropriate because the terms promeristem and central zone are essentially synonyms. As noted previously, the term stem cell has not been adopted in this book.

## THE VEGETATIVE SHOOT APEX OF *ARABIDOPSIS THALIANA*

The vegetative shoot apex of *Arabidopsis* has a two-layered tunica overlying a shallow corpus (Vaughn, 1955; Medford et al., 1992). Superimposed on the tunica-corpus organization are the three zones characteristic of angiosperm shoot apices: a central zone about five cells deep and three to four cells wide, as seen in median longitudinal sections; a peripheral zone of deeply staining cells; and a rib meristem. A morphometric study on the *Arabidopsis* shoot apex has shown the mitotic index (the percentage of nuclei in division at a given time) of the peripheral zone to be approximately 50% higher than that of the central zone (Laufs et al., 1998b). Invaluable information on the function of the shoot apical meristem has come from genetic and molecular

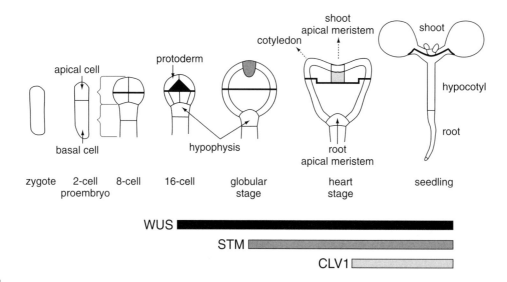

**FIGURE 6.11**

Formation of the shoot apical meristem (SAM) during *Arabidopsis* embryogenesis. The first indication of SAM development is the initiation of *WUS* expression at the 16-cell stage, long before the SAM is discernible. Subsequently *STM* and *CLV1* expression is initiated. The onset of *STM* expression is independent of *WUS* activity, and the initiation of *CLV1* expression is independent of *STM*. Bars denote the stages at which the mRNA for each of the genes is detected. Note that division of the zygote gives rise to a small apical cell and a larger basal cell. The apical cell is the precursor of the embryo proper. Vertical and transverse divisions of the apical cell result in an 8-celled proembryo. The upper four cells are the source of the apical meristem and cotyledons, the lower four of the hypocotyl. The uppermost cell of the filamentous suspensor divides transversely and the upper cell becomes the hypophysis. The hypophysis gives rise to the central cells of the root apical meristem and columella rootcap. The rest of the root meristem and lateral rootcap are derived from the embryo proper. (See Fig. 1.7.) (After Lenhard and Laux, 1999. © 1999, with permission from Elsevier.)

studies of *Arabidopsis thaliana*. Results of only a few such studies will be considered here.

The primary shoot apical meristem in *Arabidopsis* becomes apparent relatively late in embryogenesis, after the cotyledons are initiated (Fig. 6.11; Barton and Poethig, 1993). (See Kaplan and Cooke, 1997, for a discussion on the origin of the shoot apical meristem and cotyledons during angiosperm embryogenesis.) Establishment of the shoot apical meristem requires the activity of the *SHOOTMERISTEMLESS* (*STM*) gene, which is first expressed in one or two cells of the late globular stage embryo (Long et al., 1996; Long and Barton, 1998). Severe loss-of-function *stm* mutations result in seedlings having normal roots, hypocotyls, and cotyledons, but lacking shoot apical meristems (Barton and Poethig, 1993). STM mRNA is present in the central and peripheral zones of all vegetative shoot apices but is absent in developing leaf primordia (Long et al., 1996).

Whereas the *STM* gene is required for establishment of the shoot apical meristem, the *WUSCHEL* (*WUS*) gene, in addition to the *STM* gene, is necessary for the maintenance of initial function. In *wus* mutants, the initials undergo differentiation (Laux et al., 1996). *WUS* expression begins at the 16-cell stage of embryo development, in advance of *STM* expression, and long before

the meristem is evident (Fig. 6.11). In the fully developed meristem, *WUS* expression is restricted to a small group of central zone cells beneath the L3 layer (the initial layer of the corpus) and persists throughout shoot development (Mayer et al., 1998; Vernoux et al., 2000a). Thus *WUS* is not expressed within the initials, indicating that signaling must occur between the two groups of cells (Gallois et al., 2002).

In addition to meristem-promoting genes, such as *STM* and *WUS*, other genes regulate meristem size by repressing initial activity (Fig. 6.12). These are the *CLAVATA* (*CLV*) genes (*CLV1*, *CLV2*, *CLV3*), mutations in which cause an accumulation of undifferentiated cells in the central zone, bringing about an increase in size of the meristem (Clark et al., 1993, 1995; Kayes and Clark, 1998; Fletcher, 2002). The accumulation of cells apparently is due to a failure to promote cells in the peripheral zone toward differentiation. *CLV3* expression is primarily restricted to the L1 and L2 layers and to a few L3 cells of the central zone, and probably marks the initials in these layers; the *CLV1*-expressing cells underlie the L1 and L2 layers (Fletcher et al., 1999). *WUS* is expressed in the deepest region of the meristem. It has been proposed that the *WUS*-expressing cells act as an "organizing center," which confers initial cell

**FIGURE 6.12**

Diagram of the central zone of the *Arabidopsis* shoot apical meristem showing the approximate overlapping expression domains, *CLV3, CLV1,* and *WUS. CLV3* expression is primarily restricted to the L1 and L2 layers and to a few L3 cells, *CLV1* expression to cells underlying the L1 and L2 layers. *WUS* is expressed by cells in the deepest regions of the meristem. (After Fletcher, 2004. © 2004, with permission from Elsevier.)

identity to the overlying neighboring cells, while signals from the *CLV1/CLV3* regions act negatively to dampen such activity (Meyerowitz, 1997; Mayer et al., 1998; Fletcher et al., 1999). More specifically, it has been proposed that CLV3 protein secreted by the initials in the apex moves through the apoplast and binds to a CLV1/CLV2 receptor complex at the plasma membrane of the underlying cells (Rojo et al., 2002). Signaling by CLV3 through the CLV1/CLV2 receptor complex causes the down-regulation of *WUS*, maintaining the appropriate amount of initial activity throughout development. Thus, through this feedback loop, a balance is maintained between the proliferation of promeristem cells and the loss of promeristem cells through differentiation and the initiation of lateral organs in the peripheral zone (Schoof et al., 2000; Simon, 2001; Fletcher, 2004).

# ORIGIN OF LEAVES

The shoot apex produces lateral organs, and therefore the structure and activity of the shoot apical meristem must be considered in relation to the origin of lateral organs, especially the leaves, which are initiated in the peripheral zone of the shoot apex. In this chapter only those features of leaf origin are considered that are related to the structure and activity of the apical meristem.

Down-regulation of the *KNOTTED1* class of homeobox-containing plant genes—originally identified in maize—provides an early molecular marker of leaf initiation in the shoot apical meristem (Smith et al., 1992; Brutnell and Langdale, 1998; van Lijsebettens and Clarke, 1998; Sinha, 1999). The *KN1* gene in maize is specifically down regulated at the site of leaf primordium initiation. Down-regulation of the *KNOTTED1* class genes *KNAT1* and *STM1* (Long and Barton, 2000) in *Arabidopsis* also marks the sites of primordia initiation. The *HBK1* gene, which is found in the shoot apical meristem

of the conifer *Picea abies*, may play a role similar to the *KNOTTED* genes in angiosperms (Sundås-Larsson et al., 1998).

## Throughout the Vegetative Period the Shoot Apical Meristem Produces Leaves in a Regular Order

The order, or arrangement, of leaves on a stem is called **phyllotaxis** (or **phyllotaxy**; from the Greek *phyllon,* leaf, and *taxis,* arrangement; Schwabe, 1984; Jean, 1994). The most common phyllotaxis is the **spiral**, with one leaf at each node and the leaves forming a helical pattern around the stem and with an angle of divergence between successive leaves of 137.5° (*Quercus, Croton, Morus alba, Hectorella caespitosa*). In other plants with a single leaf at each node, the leaves are disposed 180° apart in two opposite ranks, as in the grasses. This type of phyllotaxis is called **distichous**. In some plants the leaves are disposed 90° from each other in pairs at each node and the phyllotaxis is said to be **opposite** (*Acer, Lonicera*). If each successive pair is at right angles to the previous pair the arrangement is termed **decussate** (Labiatae, including *Coleus*). Plants with three or more leaves at each node (*Nerium oleander, Veronicastrum virginicum*) are said to have **whorled phyllotaxis**.

The first histological events commonly associated with leaf initiation are changes in the rates and planes of cell division in the peripheral zone of the apical meristem, resulting in the formation of a protrusion (called a **leaf buttress**) on a side of the axis (Fig. 6.9). In shoots with a helical leaf arrangement the divisions alternate in different sectors around the circumference of the apical meristem, and the resulting periodic enlargement of the apex, as seen from above, is asymmetric. In shoots with decussate leaf arrangement, the enlargement is symmetrical because the intensified meristematic activity occurs simultaneously on opposite sides (Fig. 6.13). Thus the initiation of leaves causes periodic changes in the size and form of the shoot apex. The period, or interval, between the initiation of two successive leaf primordia (or pairs or whorls of primordia with an opposite or whorled leaf arrangement) is designated **plastochron**. The changes in the morphology of the shoot apex occurring during one plastochron may be referred to as **plastochronic changes**.

The term plastochron was originally formulated in a rather general sense for a time interval between two successive similar events occurring in a series of similar periodically repeated events (Askenasy, 1880). In this sense the term may be applied to the time interval between a variety of corresponding stages in the development of successive leaves, for example, the initiation of divisions in the sites of origin of primordia, the beginning of upward growth of the primordium from the buttress, and the initiation of the lamina. Plastochron

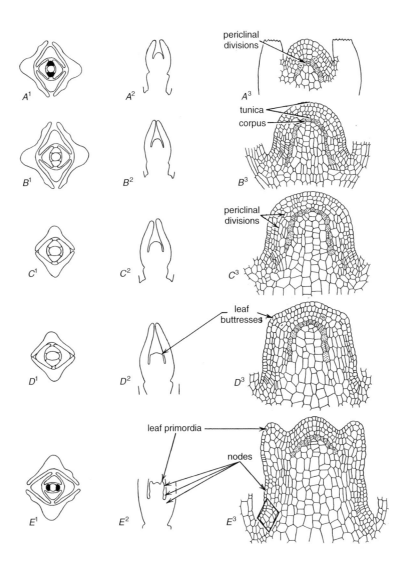

**FIGURE 6.13**

Leaf initiation in shoot tip of *Hypericum uralum*, with a decussate (opposite, with the alternate leaf pairs at right angle to each other) leaf arrangement. Before the initiation of a new leaf primordium the apical meristem appears as a small rounded mound (A). It gradually widens (B, C). Then leaf buttresses are initiated on its sides (D). While the new leaf primordia grow upward from the buttresses, the apical meristem again assumes the appearance of a small mound (E). Early stage of leaf pair shown in black in A¹, and ending shortly after emergence of leaf pair shown in black in E¹. Sections are transverse in A¹–E¹, longitudinal in A²–E² and A³–E³. A³–E³, stippling indicates outer-boundary cells (initial layer) of corpus and their immediate derivatives. E³, four-sided figure, presumptive place of origin of axillary bud. (Adapted from Zimmermann, 1928.)

may be used also with reference to the development of internodes and of axillary buds, to stages of vascularization of the shoot, and to the development of floral parts.

The length of the plastochron is generally measured as the reciprocal of the rate of primordium initiation. Successive plastochrons may be of equal duration, at least during part of vegetative growth of genetically uniform material growing in a controlled environment (Stein and Stein, 1960). The stage of development of the plant and environmental conditions are known to affect the length of the plastochrons. In *Zea mays*, for example, the successive plastochrons in the embryo lengthen from 3.5 to 13.5 days, whereas those in the seedling shorten from 3.6 to 0.5 days (Abbe and Phinney, 1951; Abbe and Stein, 1954). In *Lonicera nitida* the duration of plastochrons varied from 1.5 to 5.5 days, apparently in relation to changing temperature (Edgar, 1961). Temperature also affected the rate of primordium initiation in *Glycine max* (Snyder and Bunce, 1983) and *Cucumis sativus* (Markovskaya et al., 1991). The rate of production of leaves is also affected by light (Mohr and Pinnig, 1962; Snyder and Bunce, 1983; Nougarède et al., 1990; Schultz, 1993). The *PLASTOCHRON1* (*PLA1*) gene in rice has been implicated in regulation of the duration of the vegetative phase by controlling the rate of leaf production in the meristem (Itoh et al., 1998).

A commonly used measurement of developmental time, in relation to the shoot apex, is the **phyllochron**, which refers specifically to the interval between the visible appearance or emergence of successive leaves in the intact plant, the inverse of which is the rate of leaf appearance (Lyndon, 1998). The durations of the plastochron and phyllochron do not necessarily correspond. The rate of primordium initiation and that of leaf emergence are similar only when the period between the two events is constant, which often is not the case. In *Cyclamen persicum*, for example, the rate of primordium initiation exceeds that of leaf emergence early in the growing season and primordia accumulate in the shoot apex; this trend is reversed late in the growing season (Sundberg, 1982). In *Triticum aestivum* and *Hordeum vulgare* the earlier leaves were found to emerge faster than the later ones, whereas in *Brassica napus* the opposite pattern was observed (Miralles et al., 2001). Possibly the greatest difference between the plastochron and phyllochron durations is exhibited by the conifers. In *Picea sitchensis*, for example, hundreds of needle primordia accumulate during bud-set in the autumn (Cannell and Cahalan, 1979). During bud-break, in the spring, the opposite happens as the leaves rapidly enlarge.

If the shoot apex undergoes plastochronic changes in size, both its volume and surface area change. To designate these changes, the expressions **maximal-area phase** and **minimal-area phase** were introduced (Schmidt, A., 1924). For a shoot with decussate phyllotaxy, the apex reaches its maximal-area phase just before the pair of leaf primordia emerge (Fig. 6.13B). As the leaf primordia become elevated, the apical meristem is decreased in width (Fig. 6.13E). The apex enters the minimal-area phase of plastochronic growth. Before a pair of new primordia is formed, the apex returns to the maximal phase. The extension now occurs perpendicular to the longest diameter of the preceding maximal phase, but the enlargement of the apical meristem is evident also between the members of the pair of leaves the growth of which had previously caused the reduction in the size of the apex.

The relation between the growing leaf primordium and the apical meristem varies greatly in different species. Figure 6.14 illustrates one extreme in which the apical meristem almost vanishes between the enlarging leaf primordia (Fig. 6.14D). In another species, the apical meristem is affected much less (Fig. 6.9), and in species in which the apical meristem is elevated considerably above the organogenic region, the apex does not undergo plastochronic changes in size (Fig. 6.10).

## The Initiation of a Leaf Primordium Is Associated with an Increase in the Frequency of Periclinal Divisions at the Initiation Site

In the eudicots and monocots with a two-layered tunica, the first periclinal divisions occur most frequently in the L2 layer, followed by similar divisions in the L3 layer and by anticlinal divisions in the L1 layer (Guttenberg, 1960; Steward and Dermen, 1979). In some monocots, leaf primordia are initiated by periclinal divisions in the L1 layer. In both *Triticum aestivum* (Evans, L. S., and Berg, 1972) and *Zea mays* (Sharman, 1942; Scanlon and Freeling, 1998), the first periclinal divisions occur in the L1 layer, followed by similar divisions in the L2 layer on one side of the meristem. Periclinal divisions then spread laterally in both layers, forming a ring that encircles the meristem. Since the initiation of leaves in angiosperms follows a relatively consistent pattern, whereas the depth of the tunica is variable, the tunica and the corpus are variously concerned with leaf formation, depending on their quantitative relationship in a given apex. Thus leaf primordia are initiated by groups of cells that span two or more cell layers in the meristem. The total number of cells involved has been estimated at about 100 in cotton (Dolan and Poethig, 1991, 1998), tobacco (Poethig and Sussex, 1985a, b), and *Impatiens* (Battey and Lyndon, 1988), 100 to 250 in maize (Poethig, 1984; McDaniel and Poethig, 1988), and 30 in *Arabidopsis* (Hall and Langdale, 1996). These cells—the immediate precursors of the leaf primordia—are referred to by some workers as **founder cells** (sometimes referred to as "anlagen," which means primordia).

Either concomitant with or preceding the periclinal divisions associated with initiation of a leaf primordium, one or more procambial strands (leaf traces), which will differentiate upward into the developing leaf, may already be present at the base of the leaf site (Fig. 6.9). Precocious procambial strands have been observed in both eudicots (*Garrya elliptica*, Reeve, 1942; *Linum usitatissimum*, Girolami, 1953, 1954; *Xanthium chinense*, McGahan, 1955; *Acer pseudoplatanus*, White, D. J. B., 1955; *Xanthium pennsylvanicum*, Millington and Fisk, 1956; *Michelia fuscata*, Tucker, 1962; *Populus deltoides*, Larson, 1975; *Arabidopsis thaliana*, Lynn et al., 1999) and monocots (*Alstroemeria*, Priestly et al., 1935; *Andropogon gerardii*, Maze, 1977). In *Arabidopsis*, the precocious leaf trace was detected as a high-density region of *PINHEAD* (*PNH*) expression (Lynn et al., 1999). Expression of the *PNH* preceded down-regulation of *STM* at the leaf site and hence may be regarded as an earlier marker of leaf formation than the loss of *STM* expression.

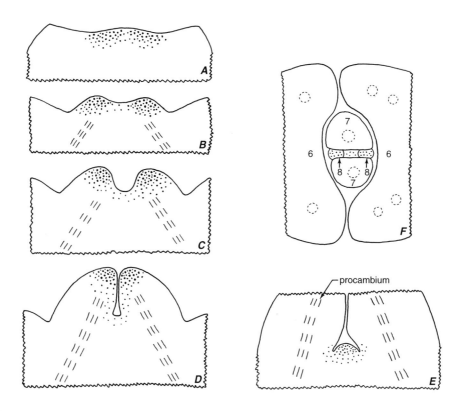

**FIGURE 6.14**

Outlines of developing leaf primordia of *Kalanchoë* from longitudinal (**A–E**) and transverse (**F**) sections of shoots sampled during the initiation and development of the eighth leaf pair. **A**, after plastochron 7; apex in maximal phase. **B**, early plastochron 8; leaf pair 8 has been initiated. **C**, leaves of pair 8 somewhat elongated. **D**, midphase of plastochron 8; apex in minimal phase. **E**, early plastochron 9; the primordia of pair 9 alternate with those of pair 8 and therefore do not appear in the plane of figure **E**; the enlarging apex between the two primordia 8 is visible. **F**, early plastochron 8, phase similar to that in **B**. (From Esau, 1977; after photomicrographs in Stein and Stein, 1960.)

In the gymnosperms the leaves also arise in the peripheral zone. According to Owens (1968) the first indication of leaf initiation in Douglas fir (*Pseudotsuga menziesii*) is the differentiation of a procambial strand in the peripheral zone "to supply the presumptive primordium." Precocious procambial strands have been observed in other gymnosperms (*Sequoia sempervirens,* Crafts, 1943; *Ginkgo biloba,* Gunckel and Wet-more, 1946; *Pseudotsuga taxifolia,* Sterling, 1947). The divisions associated with initiation of the leaf primordia in gymnosperms commonly occur in the second or third layer from the surface. The surface layer may contribute cells to the internal tissue of the primordium by periclinal and other divisions (Guttenberg, 1961; Owens, 1968). In the seedless vascular plants the leaves arise from either single superficial cells or groups of cells, one of which enlarges and becomes the conspicuous apical cell of the primordium (White and Turner, 1995).

It is pertinent to note that although changes in the rates and planes of cell division long have been associated with the initiation of leaf primordia, evidence indicates that new primordia can be initiated in the absence of cell division (Foard, 1971). In addition it has been demonstrated that already existing leaves with down-regulated cell cycle activity (Hemerly et al., 1995) and ones with a mutation that interferes with correct cell plate orientation (Smith et al., 1996) can develop almost normal shapes. These observations support the concept that during plant development shape is acquired independently from the patterns of cell division (Kaplan and Hagemann, 1991). Apparently it is the regulation of cell expansion rather than the pattern of cell division that is responsible for primordium initiation and the final shape and size of the plant and its organs (Reinhardt et al., 1998).

Initiation of a leaf primordium is accompanied by changes in the orientation and patterning of the cellulose microfibrils in the outer epidermal cell walls, as the epidermis shifts cellulose reinforcement to accommodate formation of the new organ (Green and Brooks, 1978; Green, 1985, 1989; Selker et al., 1992; Lyndon, 1994). The orientation of the microfibrils can be visualized with polarized light in thin sections made parallel to the apex surface (Green, 1980). In *Graptopetalum* the newly oriented microfibrils are arranged in circular arrays, marking the sites at which the new pair of leaf primordia will emerge (Green and Brooks, 1978). Other vegetative apices have also been examined to follow the changes in microfibril orientation that accompany leaf initiation, including *Vinca* (Green, 1985; Sakaguchi et al., 1988; Jesuthasan and Green, 1989) and *Kalanchoë* (Nelson, 1990), both of which exhibit decussate phyllotaxis, and *Ribes* (Green, 1985) and *Anacharis* (Green, 1986) with spiral phyllotaxis. Regardless of the type of phyllotaxis, the leaves arise from specific fields of cellulose reinforcement on the surface of the shoot apex (Green, 1986).

### Leaf Primordia Arise at Sites That Are Correlated with the Phyllotaxis of the Shoot

The mechanisms underlying the orderly initiation of leaves around the circumference of the apical meristem have been of interest to botanists for a long time. An early view—based on the results of surgical manipulations—was that a new leaf primordium arises in "the first available space"; that is, a new primordium arises when sufficient width and distance from the summit of the apex are attained (Snow and Snow, 1932). While confirming these earlier observations, Wardlaw (1949) proposed "the physiological field theory." As each new leaf is initiated, it is surrounded by a physiological field within which the initiation of new primordia are inhibited. Not until the position for the next leaf primordium comes to lie outside the existing fields can a new primordium be initiated. More recently it has been suggested that "biophysical forces" in the growing apex determine leaf initiation sites (Green, 1986). In this hypothesis, a leaf primordium is initiated when a region of the tunica surface bulges or buckles, a condition brought about in part by a localized reduction in the surface layer's ability to resist pressure from the tissues below (Jesuthasan and Green, 1989; Green, 1999). It has been suggested that local stress variations created by the buckling trigger the periclinal divisions commonly associated with lateral organ formation (Green and Selker, 1991; Dumais and Steele, 2000).

Support for the biophysical forces hypothesis comes in part from studies in which the localized application of expansin to the shoot apical meristem of tomato induced the formation of leaf-like outgrowths (Fleming et al., 1997, 1999). Apparently the expansin promoted cell wall extensibility in the outer cell layer of the tunica resulting in outward bulging of the tissue. In situ hybridization analyses have shown that expansin genes are specifically expressed at the site of primordium initiation in both tomato (Fleming et al., 1997; Reinhardt et al., 1998; Pien et al., 2001) and rice (Cho and Kende, 1997). Moreover expansin expressed in transformed

plants induced primordia capable of developing into normal leaves (Pien et al., 2001). These studies further support the view that the primary event in morphogenesis is the expansion of tissue, which then is subdivided into smaller units by cell division (Reinhardt et al., 1998; Fleming et al., 1999).

Several studies have implicated auxin in the regulation of phyllotaxis (Cleland, 2001). In one such study, when vegetative tomato shoot apices were cultured on a synthetic medium containing a specific inhibitor of auxin transport, leaf production was completely suppressed, resulting in the formation of pin-like naked stems with otherwise normal meristems at the tips (Reinhardt et al., 2000). Microapplication of IAA to the surface of such apices restored leaf formation. Exogenous IAA also induced flower formation on *Arabidopsis pin-formed1* (*pin1*) inflorescence apices. Flower formation is blocked in *pin1* apices because of a mutation in a putative auxin transport protein. *PIN1* itself is upregulated in developing leaf primordia (Vernoux et al., 2000b), indicating that a sufficient amount of auxin must accumulate to initiate cell expansion and leaf primordium formation. For auxin to accumulate at that site, it must be transported there from preexisting leaf primordia and developing leaves, the sources of the auxin. A model has been proposed in which auxin efflux carriers control the delivery of auxin to the shoot apical meristem, while influx and efflux carriers regulate its distribution within the meristem (Stieger et al., 2002). Whereas the efflux carrier plays a role in the redistribution of auxin within the meristem, the influx carrier presumably is required for correct leaf positioning, or phyllotaxy.

The arrangement of leaves is correlated with the architecture of the vascular system in the stem so that the spatial relation of the leaves to one another is part of an overall pattern in shoot organization (Esau, 1965; Larson, 1975; Kirchoff, 1984; Jean, 1989). The developmental relationship between the leaves and the leaf traces in the stem suggests that the procambial strands (leaf traces) associated with the prospective primordium sites provide a transport pathway for auxin or some other substance that promotes primordium initiation (the "procambial strand hypothesis," Larson, 1983). Obviously, multiple factors and events are involved with the orderly initiation of leaves, and they are not necessarily limited to the apical region.

## ORIGIN OF BRANCHES

In seedless vascular plants, such as *Psilotum, Lycopodium, Selaginella,* and certain ferns, branching occurs at the apex, without reference to the leaves (Gifford and Foster, 1989). The original apical meristem undergoes a median division into two equal parts, each of which forms a shoot. This type or process of branching is described as **dichotomous**. When a branch arises laterally at the apex, the branching is termed **monopodial**. Monopodial branching is the prevalent type of branching in seed plants. The branches commonly originate as buds in the axils of leaves, and in their nascent state are referred to as **axillary buds**. Judged from most investigations the term axillary is somewhat inaccurate because the buds generally arise on the stem (Figs. 6.13E and 6.15) but become displaced closer to the leaf base, or even onto the leaf itself, by subsequent growth readjustments. Such relationships have been observed in ferns (Wardlaw, 1943), eudicots (Koch, 1893; Garrison, 1949a, 1955; Gifford, 1951), and Poaceae (Evans and Grover, 1940; Sharman, 1942, 1945; McDaniel and Poethig, 1988). In the grasses, the lack of developmental relation between the bud and the subtending (axillant) leaf is particularly clear. The bud originates close to the leaf located above it (Fig. 6.16). Later the bud becomes separated from the leaf by the interpolation of an internode between it and the leaf. A rather similar origin of lateral buds has been observed in other monocots (*Tradescantia*, Guttenberg, 1960; *Musa*, Barker and Steward, 1962). In the conifers, bud development resembles that in eudicots.

### In Most Seed Plants Axillary Meristems Originate from Detached Meristems

Axillary buds arise at variable plastochronic distances from the apical meristem, most frequently in the axil of the second or third leaf from the apex; hence they are commonly initiated somewhat later than the leaves subtending them. In some seed plants the buds are initiated in the apical meristem itself immediately following inception of the axillant leaf, so that the bud is formed in continuity with the apical meristem (Garrison, 1955; Cutter, 1964). In most seed plants, however, the axillary buds are initiated at a later time in meristematic tissue derived from the apical meristem but separated from it by vacuolated cells (Garrison, 1949a, b; Gifford, 1951; Sussex, 1955; Bieniek and Millington, 1967; Shah and Unnikrishnan, 1969, 1971; Remphrey and Steeves, 1984; Tian and Marcotrigiano, 1994). These pockets of meristematic cells, which remain spatially associated with the leaf axil, are called **detached meristems**. Less commonly, buds have been reported to develop from somewhat differentiated, vacuolated cells that dedifferentiate and renew meristematic activity (Koch, 1893; Majumdar and Datta, 1946). In a few cases the axillary meristems appear to arise from the adaxial (upper) surface of the leaf primordia; that is, they apparently are foliar in origin (*Heracleum, Leonurus*, Majumdar, 1942; Majumdar and Datta, 1946; *Arabidopsis*, Furner and Pumfrey, 1992; Irish and Sussex, 1992; Talbert et al., 1995; Evans and Barton, 1997; Long and Barton, 2000).

Even though different populations of meristematic cells may give rise to the axillary buds and to their axillant leaves, experimental evidence indicates that an axillary bud is determined by its leaf (Snow and Snow, 1942). If, for instance, a leaf primordium is surgically

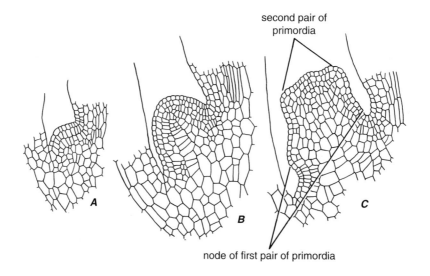

**FIGURE 6.15**

Origin of axillary bud in *Hypericum uralum*. It is formed by derivatives of three outer layers of tunica of the main shoot. Two outer layers divide anticlinally and maintain their individuality as two outer layers of tunica of the bud (A–C). The third layer divides periclinally and otherwise and gives rise to third and fourth layers of tunica and to the corpus of bud. Third tunica layer is evident in bud in **C**; the fourth appears later. **C**, second pair of leaf primordia is being initiated. The first pair is orientated in a plane perpendicular to surface of drawing. (Adapted from Zimmermann, 1928.)

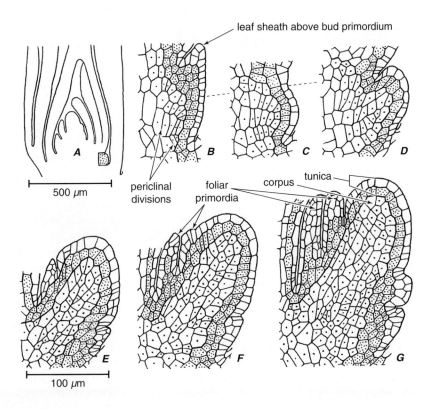

leaf sheath above bud primordium

periclinal divisions

foliar primordia

corpus

tunica

500 μm

100 μm

**FIGURE 6.16**

Development of lateral bud in *Agropyron repens* (quackgrass). Median longitudinal sections. **A,** low-power view of the shoot tip with several leaf primordia. The stippled part indicates the position of bud. It is formed by derivatives of the two-layered tunica and the corpus. **B-G,** derivatives of the second layer of tunica are stippled, and those of the corpus are indicated by a single dot in each cell. The bud is initiated by periclinal divisions in corpus derivatives (**B, C**). Anticlinal divisions occur in tunica derivatives. Bud emerges above the surface of the stem (**D**). By rib-meristem growth the corpus derivatives elongate the core of the bud (**E-G**). They also organize its corpus. Tunica derivatives remain in a biseriate arrangement at the apex of bud and form its two-layered tunica (**E, G**). Leaf primordia arise on bud (**E-G**). (Adapted from Sharman, 1945. © 1945 by The University of Chicago. All rights reserved.)

removed before its bud is initiated, the bud will fail to develop. On the other hand, if only an extremely small portion of the leaf base remains, it is often adequate to promote bud formation (Snow and Snow, 1932). Further evidence for an inductive relationship between leaf and axillary bud development is provided by the *Arabidopsis phabulosa-1d* (*phb-1d*) mutation. In *Arabidopsis,* the axillary meristems normally develop in close association with the adaxial surface of the leaf base. In *phb-1d*, abaxial (lower) leaf fate is transformed into adaxial fate, resulting in the formation of ectopic axillary meristems on the undersides of leaves (McConnell and Barton, 1998). Apparently the adaxial basal leaf fate plays an important role in promoting axillary bud development.

Buds do not always develop in the axil of every foliage leaf (Cutter, 1964; Cannell and Bowler, 1978; Wildeman and Steeves, 1982); in rare instances buds are absent altogether (Champagnat et al., 1963; Rees, 1964). In *Stellaria media* the first pair of foliage leaves often lack axillary buds and in later-formed leaf pairs generally only one leaf of a pair has a bud (Tepper, 1992). Application of benzyladenine to the shoot tips of five- to seven-day old *Stellaria* seedlings promoted axillary bud development in the normally empty axils, implicating cytokinins in the initiation of axillary buds during normal plant development (Tepper, 1992).

Whereas the axils of some leaves have one bud and others none, it is not uncommon for multiple buds (accessory buds, in addition to the axillary bud) to exist in association with a single leaf in certain species (Wardlaw, 1968). In some species, the origin of the first accessory-bud meristem is from the axillary bud and the second accessory bud originates from the first one (Shah and Unnikrishnan, 1969, 1971). In others, both the axillary and accessory buds originate from the same group of meristematic cells, which are of apical meristem origin (Garrison, 1955).

When a bud is formed, periclinal and anticlinal divisions occur in a variable number of cell layers in the leaf axil, the bud meristem is elevated above the surface, and the apical meristem of the bud is organized (Figs. 6.15, 6.16, and 6.17B). In many plants orderly divisions occur along the basal and lateral limits of the incipient bud and form a zone of parallel curving layers (Figs. 6.16C and 6.17A) referred to as the **shell zone** because of its shell-like shape (Schmidt, A., 1924; Shah and Patel, 1972). In some plants the shell zone appears later, after the bud has undergone some development. Whereas some investigators consider the shell zone to be an integral part of the developing bud, others do not (Remphrey and Steeves, 1984). The shell zone disappears at various stages of bud development in different species. In many species the incipient bud is connected to the vascular system of the main axis by two strands of procambial cells, the **bud traces**, providing potential conducting channels to the bud early in its development (Garrison, 1949a, b, 1955; Shah and Unnikrishnan, 1969; Larson and Pizzolato, 1977; Remphrey and Steeves, 1984). If the axillary bud is not dormant, its upward growth is followed by the initiation of leaf primordia, beginning with the prophylls.

### Shoots May Develop from Adventitious Buds

**Adventitious buds** arise with no direct relation to the apical meristem. Adventitious buds may develop on roots, stems, hypocotyls, and leaves. They originate in callus tissue of cuttings or near wounds, in the vascular cambium or on the periphery of the vascular cylinder. The epidermis may produce adventitious buds. Depending on the depth of the initiating tissue, the buds are described as having an **exogenous** origin (from relatively superficial tissues) or an **endogenous** origin (from tissues deep within the parent axis) (Priestley and Swingle, 1929). If the adventitious buds arise in mature tissues, their initiation involves the phenomenon of dedifferentiation.

# ROOT APEX

In contrast to the apical meristem of the shoot, that of the root produces cells not only toward the axis but also away from it, for it initiates the rootcap. Because of the presence of the rootcap the distal part of the apical meristem of the root is not terminal but subterminal in position, in the sense that it is located beneath the rootcap. The root apex further differs from the shoot meristem in that it forms no lateral appendages comparable to the leaves and no branches. The root branches are usually initiated beyond the region of most active growth and arise endogenously. Because of the absence of leaves the root apex shows no periodic changes in shape and structure such as commonly occur in shoot

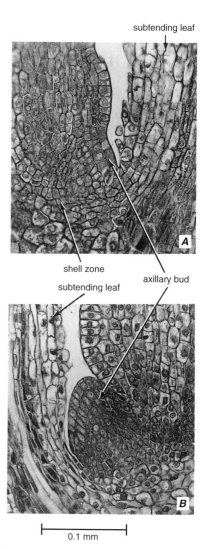

subtending leaf

shell zone

subtending leaf

axillary bud

0.1 mm

**FIGURE 6.17**

Origin of axillary buds in potato (*Solanum tuberosum*). Longitudinal sections of nodes showing an earlier (**A**) and a later (**B**) stage of bud development. (From Sussex, 1955.)

apices in relation to leaf initiation. The root also produces no nodes and internodes and therefore grows more uniformly in length than the shoot, in which the internodes elongate much more than the nodes. The rib-meristem type of growth is characteristic of the elongating root cortex.

The distal part of the apical meristem of the root, like that of the shoot, may be termed promeristem and, as such, contrasted with the subjacent primary meristematic tissues. The young root axis is more or less clearly separated into the future central cylinder and cortex. In their meristematic state the tissues of these two regions consist of procambium and ground meristem, respectively. The term procambium may be applied to the entire central cylinder of the root if this cylinder eventually differentiates into a solid vascular core. Many roots, however, have a pith in the center. This region is often interpreted as potential vascular tissue that, in the course of evolution, ceased to differentiate as such. In this context the pith is regarded as part of the vascular cylinder originating from the procambium. The contrary view is that the pith in the root is ground tissue similar to that of the pith in stems and differentiating from a ground meristem. The term protoderm, if used to designate the surface layer regardless of its developmental relation to other tissues, may be applied to the outer layer of the young root (Chapter 9). Usually the root protoderm does not arise from a separate layer of the promeristem. It has a common origin with either the cortex or the rootcap.

## Apical Organization in Roots May Be either Open or Closed

The architecture, or cellular configuration, of the apical meristems of roots has been studied most often for the purpose of revealing the origin of the tissue systems, and has served for the establishment of the so-called types (Schüepp, 1926; Popham, 1966) and for discussions of the trends in the evolution of root apical organization (Voronine, 1956; Voronkina, 1975). By analyzing the patterns of cells in an apical meristem, it is possible to trace out planes of cell division and the direction of growth. In one type of analysis the differentiating tissues are followed to the apex of the root in order to determine whether there are specific cells that appear to be the source of one or more of the discrete tissues. Thus the implication is made that a spatial correlation of tissues with certain cells or groups of cells at the apex indicates an ontogenetic relation between the two, in other words, that the apical cells function as initials.

The analysis of origin of root tissues in terms of distinct initials at the apex corresponds to the approach used by Hanstein (1868, 1870) when he formulated the histogen theory. As discussed earlier in this chapter, Hanstein considered the body of the plant to arise from a massive meristem comprising three precursors of tissue regions, the histogens, each beginning with one to several initials at the apex arranged in superposed tiers. The histogens are the dermatogen (precursor of epidermis), the plerome (precursor of the central vascular cylinder), and the periblem (precursor of the cortex). Although the subdivision into the three histogens does not have universal application—it is seldom discernible in shoots, and many roots lack a dermatogen in the sense of Hanstein (1870), that is, an independent layer that gives rise to the epidermis—it has often been used for descriptions of tissue regions in the root.

Figure 6.18 depicts the principal patterns of spatial relation between tissue regions and cells at the apex of the root. In the majority of ferns and *Equisetum*, all tissues arc derived from a single apical cell (Fig. 6.18A, B; Gifford, 1983, 1993). These plants usually have the same structure in both the root and the shoot. In some gymnosperms and angiosperms, all tissue regions of the root or all except the central cylinder appear to arise from a common meristematic group of cells (Fig. 6.18C, D); in others, one or more of these regions can be traced to separate initials (Fig. 6.18E–H). The two kinds of organization are classified as the **open** and the **closed**, respectively (Guttenberg, 1960). The distinction between open and closed meristems is not always clear-cut (Seago and Heimsch, 1969; Clowes, 1994). Both types of meristem have been reported to originate from the closed pattern in the embryonic root or the primordium of a lateral or adventitious root. During later elongation of the root the closed pattern may be retained or replaced by an open one (Guttenberg, 1960; Seago and Heimsch, 1969; Byrne and Heimsch, 1970; Armstrong and Heimsch, 1976; Vallade et al., 1983; Verdaguer and Molinas, 1999; Baum et al., 2002; Chapman et al., 2003). In pea (*Pisum sativum*) both embryonic and adult roots have open meristems (Clowes, 1978b).

In most ferns the apical cell is tetrahedral (Gifford, 1983, 1991). It cuts off segments, or merophytes, on the three lateral (proximal) faces and thus produces the tissues of the main body of the root (Fig. 6.18A, B). The rootcap has its origin either from the fourth (distal) face of the apical cell (*Marsilea*, Vallade et al., 1983; *Asplenium*, Gifford, 1991) or from a separate meristem formed early in root development (*Azolla*, Nitayangkura et al., 1980). The tetrahedral apical cell of the *Equisetum* root contributes both to the main body of the root and to the rootcap, but early root development in *Equisetum* is markedly different from that of most ferns (Gifford, 1993). Inasmuch as the rootcap in *Azolla* is discrete from the rest of the root, the apical meristem of the *Azolla* root is classified as closed. Conversely, the root apices of *Equisetum* and ferns with apical cells that cut off cells from all four faces are classified as open (Clowes, 1984).

Another approach to an analysis of the relationship between cell patterns and growth in root tips is that represented by the body-cap (Körper-Kappe) concept of Schüepp (1917), which emphasizes the planes of those divisions that are responsible for the increase in number of vertical cell files in the meristematic region of the root. Many of the files divide in two, and where they do so, a cell divides transversely; then one of the two new cells divides longitudinally and each daughter cell of this division becomes the source of a new file. The combination of the transverse and the longitudinal division results in an approximately T- (or Y-) shaped wall pattern, and therefore such divisions have been named T divisions. The direction of the top stroke (horizontal bar) of the T varies in different root parts. In the cap it

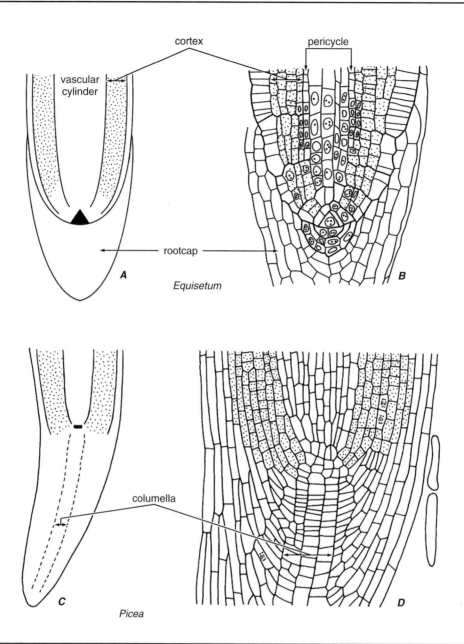

**FIGURE 6.18**

Apical meristem and derivative regions in roots. **A**, **B**, horsetail *(Equisetum)*. A single apical cell (black triangle) is the source of all parts of the root and rootcap. Heavier lines in **B** outline merophyte boundaries. The innermost boundaries of older merophytes are difficult to determine. **C**, **D**, spruce *(Picea)*. All regions of root arise from one group of initials. The rootcap has a central columella of transversely dividing cells. The columella also gives off derivatives laterally. **E**, **F**, radish *(Raphanus)*. Three layers of initials. The epidermis has common origin with the rootcap and becomes delimited on the sides of the root by periclinal walls (arrows in **F**). **G**, **H**, grass *(Stipa)*. Three layers of initials, those of rootcap forming a calyptrogen. The epidermis has common origin with the cortex. (**B**, after Gifford, 1993; **C–H**, from Esau, 1977.)

is directed toward the base of the root, in the body toward the apex (Fig. 6.19). Whereas a clear boundary exists between the body and the cap in some roots (those with separate rootcap initials), in others the boundary is not sharply delimited (e.g., in *Fagus sylvatica* where the transition between body and cap is very gradual; Clowes, 1950).

The two types of apical organization in angiosperms, the closed and the open, require separate consideration. The closed pattern is often characterized by the presence of three tiers or layers of initials (Fig. 6.20). One tier appears at the apex of the central cylinder, the second terminates the cortex, and the third gives rise to the rootcap. The three-tiered meristems may be

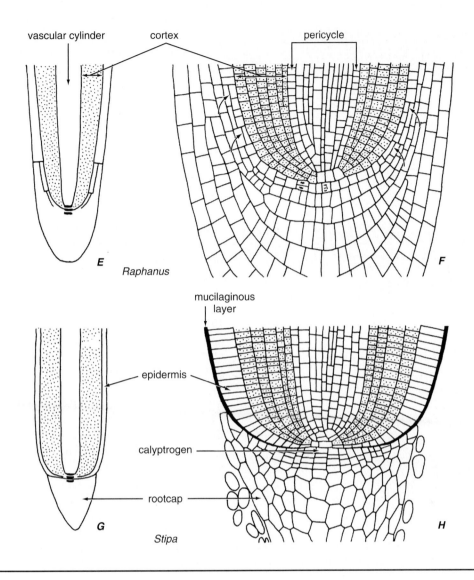

vascular cylinder    cortex    pericycle

E    *Raphanus*    F

mucilaginous layer

epidermis

calyptrogen

rootcap

G    *Stipa*    H

**FIGURE 6.18**

(*Continued*)

grouped according to the origin of the epidermis (rhizodermis of some authors; Chapter 9; Clowes, 1994). In one group, the epidermis has common origin with the rootcap and becomes distinct as such after a series of T divisions along the periphery of the root (Figs. 6.18E, F, 6.19C, and 6.21A). In the second, the epidermis and cortex have common initials (Figs. 6.18G, H, and 6.21B), whereas the rootcap arises from its own initials that constitute the rootcap meristem, or **calyptrogen** (from the Greek *calyptra*, veil, and *tenos*, offspring; Janczewski, 1874). If the rootcap and the epidermis have common origin, the cell layer concerned is called **dermatocalyptrogen** (Guttenberg, 1960). As succinctly put by Clowes (1994), "Where there is a discrete region of meristem producing either cap cells alone or cap and epidermal cells the meristem is called closed."

Roots with a dermatocalyptrogen are common in eudicots (representatives of Rosaceae, Solanaceae,

Brassicaceae, Scrophulariaceae, and Asteraceae; Schüepp, 1926). Roots with a calyptrogen are characteristic of monocots (Poaceae, Zingiberaceae, some Palmae; Guttenberg, 1960; Hagemann, 1957; Pillai et al., 1961). Sometimes the epidermis appears to terminate in the distal zone with its own initials (Shimabuku, 1960). In some aquatic monocots (*Hydrocharis* and *Stratiotes* in the Hydrocharitaceae, *Pistia* in the Avaceae, *Lemna* in the Lemnaceae) the epidermis is regularly independent from the cortex and rootcap (Clowes, 1990, 1994).

An analysis of root meristems on the basis of the body-cap concept reveals the difference in origin of the epidermis. In the root with a calyptrogen the cap includes only the rootcap (Fig. 6.19A), in one with a dermatocalyptrogen the cap extends into the epidermis (Fig. 6.19C). The body-cap configurations show other variations that elucidate patterns of growth of roots. In some roots the central core of the rootcap is distinct

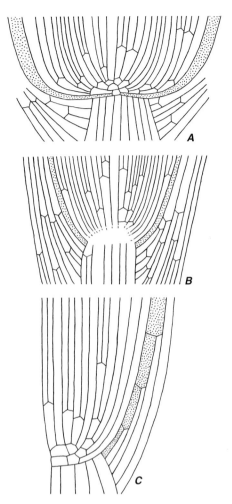

**FIGURE 6.19**

Interpretations of root apices of *Zea* (**A**), *Allium* (**B**), and *Nicotiana* (**C**) in terms of body-cap concept. Body, upper stroke of T points toward apex; cap, upper stroke of T points toward base of root. Protoderm is stippled. It is part of the body in **A**, and probably in **B**, and part of the cap in C. All three rootcaps have a distinct core, or columella.

from the peripheral part in having few or no longitudinal divisions. If conspicuous enough, such a core is referred to as **columella** (Fig. 6.19) (Clowes, 1961). The few T divisions that occur in the columella may be oriented according to the body pattern; then only the peripheral parts of the rootcap show the cap patterns. In the *Arabidopsis* root, which has a dermatocalyptrogen layer, the columella portion of the rootcap arises from so-called columella initials and the peripheral portion of the rootcap and the protoderm from so-called rootcap/protoderm initials, which form a collar around the columella initials (Baum and Rost, 1996; Wenzel and Rost, 2001). Cell divisions in the columella and rootcap/protoderm initials and their daughter cells occur in a highly coordinated pattern (Wenzel and Rost, 2001).

Apices with an open type of organization are difficult to analyze (Figs. 6.18C, D, 6.19B, and 6.22). One common interpretation is that such roots have a ***transversal***, or ***transverse, meristem*** without any boundaries with reference to the derivative regions of the root (Popham, 1955). The other view is that the central cylinder has its own initials. In some such roots, the central cylinder appears to abut the central files of cap cells, whereas in others of the same species, one or more tiers of cortical cells occurs between the "stelar pole" and the discernible central files of the cap (Clowes, 1994). Clowes (1981) attributes this difference in cell pattern to an instability in the boundary between the cap and the rest of the root resulting from the cells in this region being only transiently quiescent. Analysis of body-cap configurations indicates that the open meristems of monocots show closer affinity between epidermis and cortex, and those of eudicots between epidermis and cap (Clowes, 1994). The latter condition is exemplified by the open apical meristem of the *Trifolium repens* root (Wenzel et al., 2001).

Groot et al. (2004) distinguish between two types of open root apical meristems in eudicots, basic-open and intermediate-open. In the basic-open meristem, the cell files terminate apically in a relatively large region of initials, and the fate of the initials' derivatives are not immediately evident. In the intermediate-open meristem, the initial region is much shorter than in the basic-open type, so the fate of a derivative is usually evident immediately after it has been cut off its initial cell. The cell files in intermediate-open meristems appear to converge on the initial region, but the initials are shared between the rootcap and both the cortex and the vascular cylinder. Mapping root apical meristem organization on a phylogenetic tree, Groot et al. (2004) determined that the intermediate-open meristem is ancestral and the basic-open and closed types are derived.

The root apical meristems of gymnosperms, with their open organization (Fig. 6.18C, D), lack an epidermis per se (Guttenberg, 1961; Clowes, 1994). That is because no individualized progenitor of an epidermis (dermatogen or protoderm) exists in the meristem. Instead, what serves as an epidermis is whatever tissue is exposed by sloughing of the outer cells of the cortex/rootcap complex, as in *Pseudotsuga* (Allen, 1947; Vallade et al., 1983), *Abies* (Wilcox, 1954), *Ephedra* (Peterson and Vermeer, 1980), and *Pinus* (Clowes, 1994).

Apical meristems with no separate initials for the root regions have been described in eudicots (representatives of Musaceae, Palmae; Pillai and Pillai, 1961a, b), and some gymnosperms (Guttenberg, 1961; Wilcox, 1954). In some gymnosperm roots only the central cylinder appears to have a separate initial layer (Vallade et al., 1983).

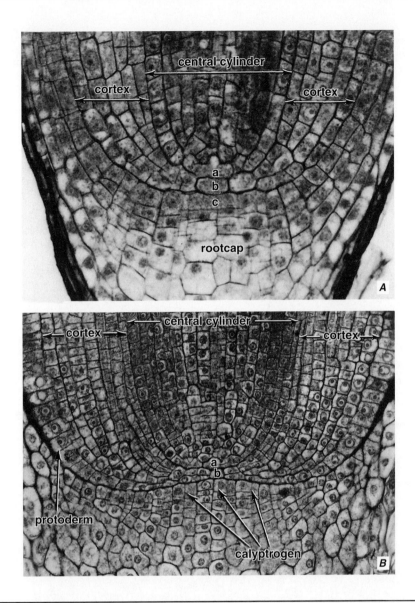

**FIGURE 6.20**

Longitudinal sections of root apical meristems of *Nicotiana tabacum* (**A**) and *Zea mays* (**B**). These apices have a closed organization with three distinct tiers or layers of initials, designated a, b, and c in **A**. In *Nicotiana* (**A**), the epidermis and rootcap have common initials (c); a gives rise to the central cylinder and b to the cortex. In *Zea* (**B**), the epidermis and cortex have common initials (b) and the rootcap arises from a calyptrogen; a designates the initial layer of the central cylinder. (See Fig. 6.21.) (A, ×455; B, ×280. **B**, slide by Ernest M. Gifford.)

### The Quiescent Center Is Not Completely Devoid of Divisions under Normal Conditions

By means of analyses of the organization of root tips in terms of the histogen and the body-cap concepts, information is obtained about growth that already has taken place and has produced the pattern now discernible. The discovery by Clowes (1954, 1956) of a quiescent center in the root apex brought about a fundamental change of view about the behavior of root meristems. Extensive research on normally developing roots and on those treated surgically, also on irradiated roots and on roots

that were fed labeled compounds involved with DNA synthesis, has shown that, as a general phenomenon, the initials that are responsible for the original cell pattern—the minimal constructional center of Clowes (1954)—largely cease to be mitotically active during the later growth of the root (Fig. 6.23) (Clowes, 1961, 1967, 1969). They are supplanted in this activity by cells on the margin of the relatively inactive region, or quiescent center.

A quiescent center arises twice in primary roots, first during embryogeny and then again during the early stages of germination of the seed. At the time of emergence from the seed the root is without a quiescent

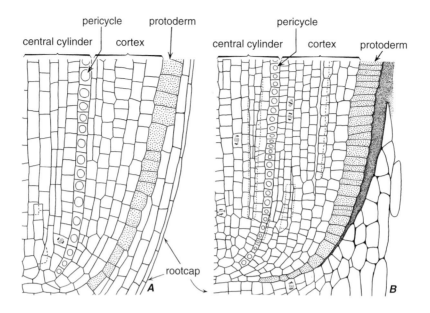

**FIGURE 6.21**

Longitudinal sections of root tips of *Nicotiana tabacum* (**A**) and *Zea mays* (**B**), illustrating two contrasting methods of origin of epidermis. **A**, epidermis separates from the rootcap by periclinal divisions. **B**, epidermis arose from same initials as the cortex through periclinal division in a recent derivative of a cortical initial. The densely stippled area in **B** indicates the gelatinized wall between the rootcap and protoderm. (A, ×285; B, ×210.)

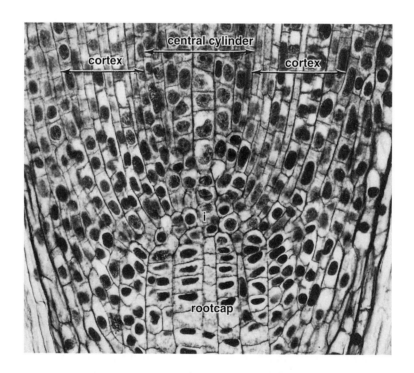

**FIGURE 6.22**

Longitudinal section of apical meristem of *Allium sativum* root. This apex has an open organization; the tissue regions merge in a common initial (i) group. (×600. From Mann, 1952. *Hilgardia* 21 (8), 195–251. © 1952 Regents, University of California.)

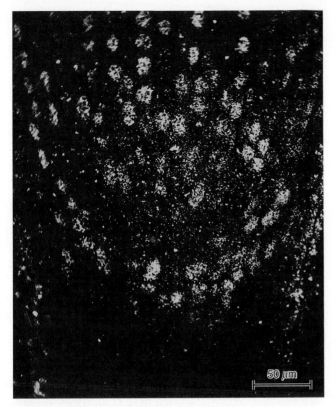

**FIGURE 6.23**

Quiescent center. Autoradiograph of an *Allium sativum* root tip, as seen in longitudinal section, fed with tritiated thymidine for 48 hours. In the rapidly dividing cells around the quiescent center, the radioactive material was quickly incorporated into the nuclear DNA. (From Thompson and Clowes, 1968, by permission of Oxford University Press.)

**TABLE 6.2 ■ Average Duration of Mitotic Cycle in Hours Calculated from Metaphase Accumulations in Dividing Nuclei in Root Meristems Treated with Inhibitors Blocking Mitosis**

| Species | Quiescent Center | Rootcap Initials | Central Cylinder | |
| --- | --- | --- | --- | --- |
| | | | Just above[a] QC[b] | 200–250 μm above[a] QC[b] |
| *Zea mays* | 174 | 12 | 28 | 29 |
| *Vicia faba* | 292 | 44 | 37 | 26 |
| *Sinapis alba* | 520 | 35 | 32 | 25 |
| *Allium sativum* | 173 | 33 | 35 | 26 |

Source: From Esau, 1977; adapted from Clowes, 1969.
[a] Toward the base of root.
[b] Quiescent center.

center (Jones, 1977; Clowes, 1978a, b; Feldman, 1984). In lateral root primordia of *Zea mays* a quiescent center also appears twice, first while the primordium is still embedded in the cortex and then either before or after emergence from the parent root (Clowes, 1978a).

The quiescent center excludes the initials of the rootcap, is hemispherical or discoid in shape, and in some species studied contains as few as four cells (*Petunia hybrida*, Vallade et al., 1978; *Arabidopsis thaliana*, Benfey and Scheres, 2000) and in others over a thousand (*Zea mays*, Feldman and Torrey, 1976). The quiescent center is variable in volume apparently in relation to root size, for it is smaller or entirely absent in thin roots (Clowes, 1984). In the root system of *Euphorbia esula* the vigorous perennial long roots have distinctive quiescent centers, whereas the determinate laterals (short roots) lack such centers throughout their brief development (Raju et al., 1964, 1976). Seedless vascular plants with tetrahedral apical cells lack quies-

cent centers (Gunning et al., 1978; Kurth, 1981; Gifford and Kurth, 1982; Gifford, 1991).

The relatively inactive state of the quiescent center cells does not mean that they have become permanently nonfunctional. Quiescent center cells do divide occasionally and serve to renew the more actively dividing regions around them, the cells of which are unstable and displaced from time to time (Barlow, 1976; Kidner et al., 2000). The quiescent center in the long roots of *Euphorbia esula* apparently undergo a seasonal fluctuation in cell production (Raju et al., 1976). At the height of the growing season, they exhibit a well-developed quiescent center, but during reactivation of growth early in the growing season, a quiescent center is not discernible. In roots injured experimentally by radiation or surgical treatments the quiescent center is able to repopulate the meristem (Clowes, 1976). It also resumes division during recovery from a period of dormancy induced by cold (Clowes and Stewart, 1967; Barlow and Rathfelder, 1985). When the rootcap is removed, the cells of the quiescent center begin to grow and undergo a controlled sequence of divisions that regenerate the rootcap (Barlow, 1973; Barlow and Hines, 1982).

By labeling nuclei with tritiated thymidine and by blocking the cell cycle at metaphase with inhibitors, one can obtain quantitative data on the duration of the mitotic cycle in different regions of the root meristem (Clowes, 1969). These data indicate that the cells of the quiescent center divide approximately 10 times slower than the adjacent cells (Table 6.2). Pulse labeling with thymidine has shown moreover that the differences in the duration of mitotic cycles are largely caused by differences in the duration of $G_1$, the phase between the end of mitosis and the beginning of DNA synthesis.

The paucity of mitotic activity in the quiescent center led Clowes (1954, 1961) to suggest that the initials of the root apex are located just outside the quiescent center along its margin, and he designated this group of cells the promeristem of the root. Barlow (1978) and Steeves and Sussex (1989) noted that it is more realistic, however, to consider the slowly dividing cells of the quiescent center—cells that could act as the ultimate source of cells for the whole root—as the true initials, and more actively dividing cells immediately surrounding them as their derivatives, a view advanced earlier by Guttenberg (1964). Adopting that view, the quiescent center of the root is strikingly similar to the central zone, or promeristem, of the shoot, and it may be regarded as the promeristem of the root. Some workers consider the root promeristem as comprising the quiescent center and its immediate, actively dividing derivatives (Kuras, 1978; Vallade et al., 1983). In the seedless vascular plants the promeristem would consist of the apical cell only. Today there is no uniformity in the use of terms to describe the slowly dividing region and its actively dividing derivatives in the roots of seed plants. More often the actively dividing cells bordering the quiescent center are referred to as initials and the cells of the quiescent center simply as quiescent-center cells.

Many views have been expressed on possible causes of the appearance of a quiescent center in a growing root. According to a proposal based on the analyses of growth patterns in root tips, quiescence in the particular location of the root meristem results from antagonistic directions of cell growth in various parts of the meristem (Clowes, 1972, 1984; Barlow, 1973), the rootcap or rootcap meristem being particularly important in the suppression of growth. During embryogenesis the appearance of the quiescent center coincides with the appearance of the rootcap meristem (Clowes, 1978a, b). Moreover, as mentioned previously, if the cap is damaged or removed, the quiescent center activates and gives rise to a new rootcap meristem, which in turn produces a new cap; quiescence then resumes. This behavior prompted Barlow and Adam (1989) to suggest that activation of the quiescent center, after damage or removal of the cap, results from an interruption or modification of signaling—possibly by hormones—between the rootcap or its initials and the quiescent center. A likely candidate for that hormone is auxin, which has been implicated in formation of the root pole during embryogenesis and maintenance of tissue organization in the seedling root in *Arabidopsis* (Sabatini et al., 1999; Costa and Dolan, 2000). It has been hypothesized that the origin and maintenance of the quiescent center in maize root tips are a consequence of the polar auxin supply, and that the rootcap initials play an important role in regulating polar auxin movements toward the root tip (Kerk and Feldman, 1994). High levels of auxin bring about elevated levels of ascorbic acid oxidase (AAO) and the resultant depletion of ascorbic acid within the quiescent center. Inasmuch as ascorbic acid

is necessary for the $G_1$ to S transition in the cell cycle in root tips (Liso et al., 1984, 1988), Kerk and Feldman (1995) have proposed that the depletion of ascorbic acid in root tips may be responsible for the formation and maintenance of the quiescent center. More recently Kerk et al. (2000) reported that AAO also oxidatively decarboxylates auxin in maize root tips, thereby providing another mechanism for regulating auxin levels within the quiescent center and other root tissues. An intact rootcap must be present for this metabolic process to occur.

# THE ROOT APEX OF *ARABIDOPSIS THALIANA*

The apical meristem of the *Arabidopsis* root has a closed organization with three layers of initials (Fig. 6.24). The lower layer, a dermatocalyptrogen, consists of columella rootcap initials and the initials of the lateral rootcap cells and epidermis. The middle layer consists of the initials of the cortex (from which the parenchymatous and endodermal cortical cells are derived), and the upper layer of initials of the vascular cylinder (pericycle and vascular tissue), sometimes erroneously referred to as a vascular bundle (van den Berg et al., 1998; Burgeff et al., 2002). At the center of the middle layer is a set of four cells that rarely divide during early root development. Several terms have been used to describe these centrally located cells, including "central cells" (Costa and Dolan, 2000; Kidner et al., 2000), "quiescent-center cells" (Dolan et al., 1993; van den Berg et al., 1998; Scheres and Heidstra, 2000), "central ground meristem initial cells" (Baum and Rost, 1997), "central cortex initials" (Zhu et al., 1998a), and "central initials" (Baum et al., 2002).

The embryonic origin of the primary root of *Arabidopsis* is well documented (Scheres et al., 1994). Briefly, embryogenesis begins with an asymmetric transverse division of the zygote, giving rise to a small apical cell and a larger basal cell. The apical cell gives rise to the embryo proper and the basal cell to a stalk-like suspensor, the uppermost cell of which is called the **hypophysis**, or **hypophyseal cell** (Fig. 6.11). At the early heart stage (triangular stage) of embryogenesis the hypophysis divides to form a lens-shaped cell, which is the progenitor of the four central cells. The lower hypophyseal-cell derivative gives rise to the initials for the central portion (columella) of the rootcap. All other initials of the meristem are derived from the embryo proper and are recognizable at the late heart stage.

Laser ablation experiments have clearly demonstrated that positional information rather than cell lineage relationships plays the most important role in the determination of cell fate in the *Arabidopsis* root (van den Berg et al., 1995, 1997a; Scheres and Wolkenfelt, 1998). In these experiments specific cells were ablated with a

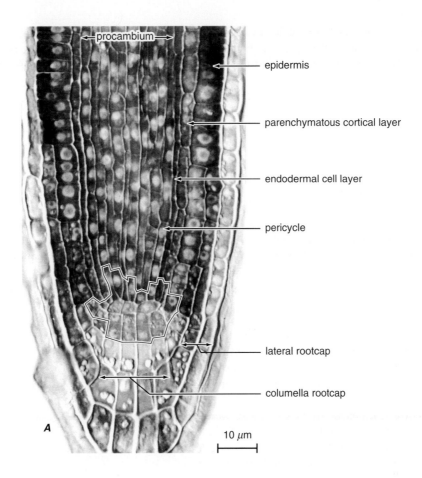

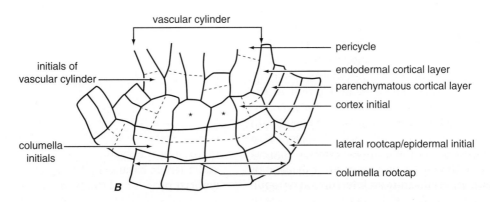

**FIGURE 6.24**

**A**, median longitudinal section of an *Arabidopsis* root tip. **B**, drawing of the promeristem showing the relationships of the layers of initials to the tissue regions of the root. The upper layer consists of the initials of the vascular cylinder, the middle layer of the central cells (asterisks) and initials of the cortex, and the lower layer of columella rootcap initials and initials of the lateral rootcap cells and epidermis. The dashed lines indicate the planes of cell division in the cortical initials and in the lateral rootcap/epidermal initials. (Reprinted with permission; **A**, from, **B**, after Schiefelbein et al., 1997. © American Society of Plant Biologists.)

laser and the effect on neighboring cells observed. For example, when all four quiescent-center cells were ablated, they were replaced by the initials of the vascular cylinder. Ablated cortical initials were replaced by pericycle cells, which then switched fate and behaved as cortical initials.

Ablation of a single quiescent-center cell resulted in cessation of cell division and the progression of differentiation in the columella and cortical cell initials with which it was in contact. These results indicated that a major role of the quiescent-center cells is to inhibit differentiation of the contacting initial cells through signals that act at the single-cell range (van den Berg et al., 1997b; Scheres and Wolkenfelt, 1998; van den Berg et al., 1998). Ablation of a single daughter cell of a cortical initial had no effect on subsequent divisions occurring in that initial, which was in contact with other cortical daughter cells of neighboring cortical initials. When all cortical daughter cells bordering a cortical initial were ablated, however, that initial was unable to generate files of parenchymatous and endodermal cortical cells. Apparently the cortical initials—perhaps all initials—depend on positional information from more mature daughter cells within the same cell layer. In other words, the initials of the root apical meristem apparently lack intrinsic pattern-generating information (van den Berg et al., 1995, 1997b). This is contrary to the traditional view of meristems as autonomous pattern-generating machines.

During growth of the *Arabidopsis* primary root the once quiescent central cells become mitotically active, and they and the initial cells become disorganized and vacuolated, as the apex is transformed from a closed to open organization (Baum et al., 2002). As noted by Baum et al. (2002), these changes, in addition to the accompanying decrease in the number of plasmodesmata (Zhu et al., 1998a), are phenomena associated with root determinacy, the final developmental stage of root growth. The presence of determinate primary roots is not unique to *Arabidopsis*. Determinate root growth associated with the transformation of the apical meristem from a closed organization to an open one apparently is a common phenomenon (Chapman et al., 2003).

## ▌GROWTH OF THE ROOT TIP

The region of actively dividing cells—the apical meristem—extends for a considerable distance basipetally from the apex, that is, toward the older part of the root. At one level of organization both the rootcap and the root proper may be envisioned as consisting of files of cells that emanate from the promeristem. Relatively close to the promeristem some files divide longitudinally—either radially or periclinally—by T divisions to

generate new files of cells. Such divisions are called **formative divisions** because they are important in determining pattern formation (Gunning et al., 1978). The radial divisions increase the number of cells in an individual cell layer, whereas the periclinal divisions increase the number of layers and, hence, the diameter of the root. By dividing transversely, the numbers of cells in each file is increased. The transverse divisions, called **proliferative divisions**, determine the extent of the meristem. In some roots, groups of cells of common ancestry, called **cell packets**, have been recognized in the files (Fig. 6.25; Barlow, 1983, 1987). The packets are each derived from a single mother cell and are useful in the study of cell division in the root.

Although the traditional model of root structure divides the root tip into three more or less distinct regions—cell division (the meristem), elongation, and maturation (Ivanov, 1983)—at the same level of the root, these processes overlap not only in different tissue

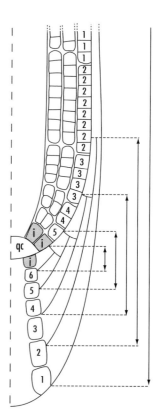

**FIGURE 6.25**

Patterns of growth in the root apex of *Petunia hybrida*. The numbers indicate the sequence of formation of cells via transverse divisions (proliferative divisions) in columella of the rootcap and in the cortex, where cells of common ancestry occur in packets. The arrows indicate the growth of the lateral rootcap-epidermis complex. Details: qc, quiescent center; i, initial. (From Vallade et al., 1983.)

regions but also in the different cell files of the same tissue region, and even in individual cells. Typically the meristematic cortex vacuolates and develops intercellular spaces close to the apex, where the central-cylinder meristem (procambium) still appears dense. In the central cylinder, the precursors of the innermost xylem vessels (metaxylem vessels) cease dividing, enlarge, and vacuolate considerably in advance of the other precursors, and the first sieve tubes commonly mature in the part of the root where cell division is still in progress. In individual cells, division, elongation, and vacuolation are combined.

As indicated, the level at which transverse divisions cease along the axis of the root differs among tissues. In the barley (*Hordeum vulgare*) root tip, for example, cells of the central metaxylem stopped dividing at a distance of 300 to 350 micrometers from the initials and those of the epidermis at 600 to 750 micrometers. The pericycle exhibited the longest duration of cell division, dividing up to a distance of 1000 to 1150 micrometers, the longest opposite the xylem poles (Luxová, 1975). In the *Vicia faba* root the pericycle also divided the longest but the cells of the protophloem (first-formed phloem) were the first to cease dividing. Mature protophloem sieve tubes were found at a distance of 600 to 700 micrometers from the apex (Luxová and Murín, 1973).

In the *Pisum sativum* root the distribution patterns of cell division were found to correspond to the tissue differentiation patterns in corresponding cylinders and vascular tissue sectors (Fig. 6.26; Rost et al., 1988). At approximately the 350 to 500 micrometers level from the root proper/cap junction, the tracheary elements of the xylem and the parenchyma cells of the pith and middle cortex had stopped dividing. At this level cell division was essentially restricted to two cylinders, an "outer cortical cylinder" (composed of the inner rootcap, the epidermis, and the outer cortex) and an "inner cortex cylinder" (composed of the inner cortex, the pericycle, and the vascular tissue). With maturation of the protophloem, all cells in the phloem sector of the "inner cortical cylinder," including the one layer of pericycle there, the endodermis, and the phloem parenchyma had stopped dividing. In the xylem sectors, the 3 to 4-layered pericycle continued to divide until about the 10 millimeter level, following maturation of the protoxylem tracheary elements. Inasmuch as proliferative divisions in the various tissues or cell files do not stop at exactly the same distance from the root apex, the basal boundary of the meristem, or region of cell division, is not clearly defined (Webster and MacLeod, 1980). Rost and Baum (1988) used the term "relative meristem height" for this diffuse boundary in *Pisum sativum*.

Studies using the kinematic method—by which local rates of cell division and cell expansion can be measured simultaneously (Baskin, 2000)—have clearly established that although different tissues or cell files stop dividing at different distances from the apex, while they divide, cells in all tissues divide at about the same rate. In contrast to the constancy of the cell division rate, the number of dividing cells in the meristem varies widely, indicating that the root must control exit from the cell cycle at the base of the meristem (Baskin, 2000). In addition it is now clearly established that cell division continues well into the region where cell length increases rapidly (Ivanov and Dubrovsky, 1997; Sacks et al., 1997; Beemster and Baskin, 1998). Thus a **transition zone** (Baluška et al., 1996) apparently exists in the basal part of the meristem and the region where cells expand rapidly or, to be more exacting, "where cells are undergoing their final division as well as expanding rapidly" (Beemster and Baskin, 1998). It has been hypothesized that the division and elongation regions are coupled and may actually constitute one developmental zone (Scheres and Heidstra, 2000).

The control of cell division and the coordination of development between tissues and cell files in the root, as elsewhere in the plant body, require cell-to-cell communication, and very likely involve directional movement of position-dependent signals, such as transcription factors or hormones (Barlow, 1984; Lucas, 1995; van den Berg et al., 1995; Zhu et al., 1998a). A potential pathway for the movement of such putative positional signals are the plasmodesmata, which link the cells symplastically. In the *Arabidopsis* root the initial cells, although uniformly interconnected, had fewer plasmodesmata in their common walls than in the walls between them and their derivatives (Zhu et al., 1998a, b). The frequency of plasmodesmata was greatest across the transverse walls of the cell files (primary plasmodesmata). The longitudinal walls between cell files and the common walls between neighboring tissues were traversed by secondary plasmodesmata. Not surprisingly, small symplastically mobile fluorescent dyes were found to diffuse preferentially across transverse walls of ground meristem cells and their progeny cortical cells (Zhu et al., 1998a).

With increasing age of the *Arabidopsis* root, the frequency of all plasmodesmata decreases (Zhu et al., 1998b), a phenomenon associated with the programmed cell death of the outer rootcap cells (Zhu and Rost, 2000). Earlier Gunning (1978) suggested that the limited life span of the determinate root of *Azolla pinnata* is due to a programmed senescence associated with the progressive diminution in the frequency of plasmodesmata between the apical cell and its lateral derivatives. The reduction in plasmodesmatal frequency begins at about the thirty-fifth cell division and ultimately results in the symplastic isolation of the apical cell, which no longer divides.

The tip of the root does not grow continuously at the same rate, especially in perennial plants (Kozlowski and

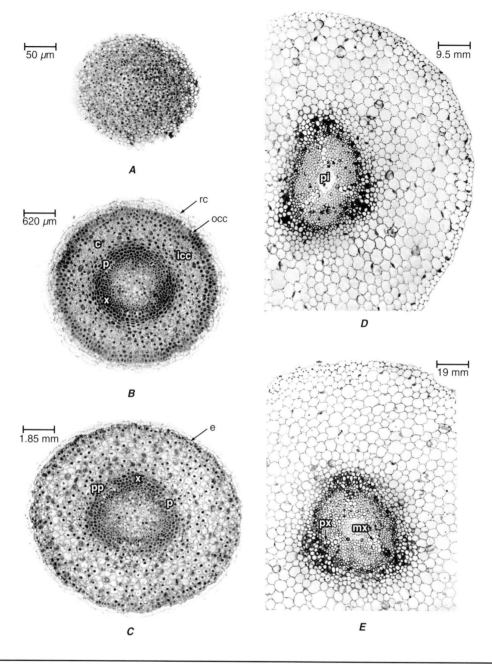

**FIGURE 6.26**

Transverse sections of the *Pisum sativum* root showing the developmental cylinder and sector changes at different levels. Details: c, middle cortex; e, epidermis; icc, inner cortex/pericycle/vascular tissue cylinder; mx, mature metaxylem; occ, rootcap/epidermis/outer cortex cylinder; p, phloem sector; pi, pith; pp, mature protophloem; px, mature protoxylem; rc, rootcap; x, xylem sector. (From Rost et al., 1988.)

Pallardy, 1997). In noble fir (*Abies procera*), for example, the roots show periodic deceleration of growth and have periods of dormancy (Wilcox, 1954). Dormancy is preceded by lignification of cell walls and by a deposition of suberin—a double process called **metacutization**—in the cortex and rootcap throughout a layer of cells that is continuous with the endodermis and completely covers the apical meristem. The latter thus becomes sealed off by a protective layer on all sides except toward the base of the root. Externally such root tips appear brown. When growth is resumed, the brown covering is broken and the root tip pushes beyond it. Studies on excised roots indicate that roots may have a growth rhythm not dependent on seasonal changes but determined by internal factors (Street and Roberts, 1952).

# REFERENCES

ABBE, E. C., and B. O. PHINNEY. 1951. The growth of the shoot apex in maize: External features. *Am. J. Bot.* 38, 737–743.

ABBE, E. C., and O. L. STEIN. 1954. The growth of the shoot apex in maize: Embryogeny. *Am. J. Bot.* 41, 285–293.

ALLEN, G. S. 1947. Embryogeny and the development of the apical meristems of *Pseudotsuga*. III. Development of the apical meristems. *Am. J. Bot.* 34, 204–211.

ARMSTRONG, J. E., and C. HEIMSCH. 1976. Ontogenetic reorganization of the root meristem in the Compositae. *Am. J. Bot.* 63, 212–219.

ASKENASY, E. 1880. Ueber eine neue Methode, um die Vertheilung der Wachsthumsintensität in wachsenden Theilen zu bestimmen. *Verhandlungen des Naturhistorisch-medicinischen Vereins zu Heidelberg,* n.f. 2, 70–153.

BALL, E. 1941. The development of the shoot apex and of the primary thickening meristem in *Phoenix canariensis* Chaub., with comparisons to *Washingtonia filifera* Wats. and *Trachycarpus excelsa* Wendl. *Am. J. Bot.* 28, 820–832.

BALL, E. 1972. The surface "histogen" of living shoot apices. In: *The Dynamics of Meristem Cell Populations,* pp. 75–97, M. W. Miller and C. C. Kuehnert, eds. Plenum Press, New York.

BALUŠKA, F., D. VOLKMANN, and P. W. BARLOW. 1996. Specialized zones of development in roots. View from the cellular level. *Plant Physiol.* 112, 3–4.

BARKER, W. G., and F. C. STEWARD. 1962. Growth and development of the banana plant. I. The growing regions of the vegetative shoot. *Ann. Bot.* 26, 389–411.

BARLOW, P. W. 1973. Mitotic cycles in root meristems. In: *The Cell Cycle in Development and Differentiation,* pp. 133–165, M. Balls and F. S. Billett, eds. Cambridge University Press, Cambridge.

BARLOW, P. W. 1976. Towards an understanding of the behaviour of root meristems. *J. Theoret. Biol.* 57, 433–451.

BARLOW, P. W. 1978. RNA metabolism in the quiescent centre and neighbouring cells in the root meristem of *Zea mays.* *Z. Pflanzenphysiol.* 86, 147–157.

BARLOW, P. W. 1983. Cell packets and cell kinetics in the root meristem of *Zea mays.* In: *Wurzelökologie und ihre Nutzanwendung* (Root ecology and its practical application), pp. 711–720, W. Böhm, L. Kutschera, and E. Lichtenegger, eds. Bundesanstalt für Alpenländische Landwirtschaft Gumpenstein, Irdning, Austria.

BARLOW, P. W. 1984. Positional controls in root development. In: *Positional Controls in Plant Development,* pp. 281–318, P. W. Barlow and D. J. Carr, eds. Cambridge University Press, Cambridge.

BARLOW, P. W. 1987. Cellular packets, cell division and morphogenesis in the primary root meristem of *Zea mays* L. *New Phytol.* 105, 27–56.

BARLOW, P. W., and J. S. ADAM. 1989. The response of the primary root meristem of *Zea mays* L. to various periods of cold. *J. Exp. Bot.* 40, 81–88.

BARLOW, P. W., and E. R. HINES. 1982. Regeneration of the rootcap of *Zea mays* L. and *Pisum sativum* L.: A study with the scanning electron microscope. *Ann. Bot.* 49, 521–529.

BARLOW, P. W., and E. L. RATHFELDER. 1985. Cell division and regeneration in primary root meristems of *Zea mays* recovering from cold treatment. *Environ. Exp. Bot.* 25, 303–314.

BARTON, M. K. 1998. Cell type specification and self renewal in the vegetative shoot apical meristem. *Curr. Opin. Plant Biol.* 1, 37–42.

BARTON, M. K., and R. S. POETHIG. 1993. Formation of the shoot apical meristem in *Arabidopsis thaliana:* Analysis of development in the wild type and in the shoot meristemless mutant. *Development* 119, 823–831.

BASKIN, T. I. 2000. On the constancy of cell division rate in the root meristem. *Plant Mol. Biol.* 43, 545–554.

BATTEY, N. H., and R. F. LYNDON. 1988. Determination and differentiation of leaf and petal primordia in *Impatiens balsamina* L. *Ann. Bot.* 61, 9–16.

BAUM, S. F., and T. L. ROST. 1996. Root apical organization in *Arabidopsis thaliana*. 1. Root cap and protoderm. *Protoplasma.* 192, 178–188.

BAUM, S. F., and T. L. ROST. 1997. The cellular organization of the root apex and its dynamic behavior during root growth. In: *Radical Biology: Advances and Perspectives on the Function of Plant Roots,* pp. 15–22, H. E. Flores, J. P. Lynch, and D. Eissenstat, eds. American Society of Plant Physiologists, Rockville, MD.

BAUM, S. F., J. G. DUBROVSKY, and T. L. ROST. 2002. Apical organization and maturation of the cortex and vascular cylinder in *Arabidopsis thaliana* (Brassicaceae) roots. *Am. J. Bot.* 89, 908–920.

BEEMSTER, G. T. S., and T. I. BASKIN. 1998. Analysis of cell division and elongation underlying the developmental acceleration of root growth in *Arabidopsis thaliana.* *Plant Physiol.* 116, 1515–1526.

BENFEY, P. N., and B. SCHERES. 2000. Root development. *Curr. Biol.* 10, R813–R815.

BHAMBIE, S., and V. PURI. 1985. Shoot and root apical meristems in pteridophytes. In: *Trends in Plant Research,* pp. 55–81, C. M. Govil, Y. S. Murty, V. Puri, and V. Kumar, eds. Bishen Singh Mahendra Pal Singh, Dehra Dun, India.

BIENIEK, M. E., and W. F. MILLINGTON. 1967. Differentiation of lateral shoots as thorns in *Ulex europaeus.* *Am. J. Bot.* 54, 61–70.

BIERHORST, D. W. 1977. On the stem apex, leaf initiation and early leaf ontogeny in Filicalean ferns. *Am. J. Bot.* 64, 125–152.

BOWER, F. O. 1923. *The Ferns,* vol. I, *Analytical Examination of the Criteria of Comparison.* Cambridge University Press, Cambridge.

BOWMAN, J. L., and Y. ESHED. 2000. Formation and maintenance of the shoot apical meristem. *Trends Plant Sci.* 5, 110–115.

BROWN, W. V., C. HEIMSCH, and H. P. EMERY. 1957. The organization of the grass shoot apex and systematics. *Am. J. Bot.* 44, 590–595.

BRUTNELL, T. P., and J. A. LANGDALE. 1998. Signals in leaf development. *Adv. Bot. Res.* 28, 161–195.

BURGEFF, C., S. J. LILJEGREN, R. TAPIA-LÓPEZ, M. F. YANOFSKY, and E. R. ALVAREZ-BUYLLA. 2002. MADS-box gene expression in lateral primordia, meristems and differentiated tissues of *Arabidopsis thaliana* roots. *Planta* 214, 365–372.

BUVAT, R. 1952. Structure, évolution et fonctionnement du méristème apical de quelques Dicotylédones. *Ann. Sci. Nat. Bot. Biol. Vég., Sér. 11*, 13, 199–300.

BUVAT, R. 1955a. Le méristème apical de la tige. *L'Année Biologique* 31, 595–656.

BUVAT, R. 1955b. Sur la structure et le fonctionnement du point végétatif de *Selaginella caulescens* Spring var. *amoena. C.R. Séances Acad. Sci.* 241, 1833–1836.

BUVAT, R., and L. GENEVÈS. 1951. Sur l'inexistence des initiales axiales dans la racine d'*Allium cepa* L. (Liliacées). *C.R. Séances Acad. Sci.* 232, 1579–1581.

BUVAT, R., and O. LIARD. 1953. Nouvelle constatation de l'inertie des soi-disant initiales axiales dans le méristème radiculaire de *Triticum vulgare. C.R. Séances Acad. Sci.* 236, 1193–1195.

BYRNE, J. M., and C. HEIMSCH. 1970. The root apex of *Malva sylvestris.* I. Structural development *Am. J. Bot.* 57, 1170–1178.

CAMEFORT, H. 1956. Étude de la structure du point végétatif et des variations phyllotaxiques chez quelques gymnospermes. *Ann. Sci. Nat. Bot. Biol. Vég., Sér. 11*, 17, 1–185.

CAMPBELL, D. H. 1911. *The Eusporagiatae.* Publ. no. 140. Carnegie Institution of Washington, Washington, DC.

CANNELL, M. G. R., and K. C. BOWLER. 1978. Spatial arrangement of lateral buds at the time that they form on leaders of *Picea* and *Larix. Can. J. For. Res.* 8, 129–137.

CANNELL, M. G. R., and C. M. CAHALAN. 1979. Shoot apical meristems of *Picea sitchensis* seedlings accelerate in growth following bud-set. *Ann. Bot.* 44, 209–214.

CATESSON, A. M. 1953. Structure, évolution et fonctionnement du point végétatif d'une Monocotylédone: *Luzula pedemontana* Boiss. et Reut. (Joncacées). *Ann. Sci. Nat. Bot. Biol. Vég., Sér. 11*, 14, 253–291.

CECICH, R. A. 1980. The apical meristem. In: *Control of Shoot Growth in Trees*, pp. 1–11, C. H. A. Little, ed. Maritimes Forest Research Centre, Fredericton, N.B., Canada.

CHAMPAGNAT, M., C. CULEM, and J. QUIQUEMPOIS. 1963. Aisselles vides et bourgeonnemt axillaire épidermique chez *Linum usitatissimum* L. *Mém. Soc. Bot. Fr., March*, 122–138.

CHAPMAN, K., E. P. GROOT, S. A. NICHOL, and T. L. ROST. 2003. Primary root growth and the pattern of root apical meristem organization are coupled. *J. Plant Growth Regul.* 21, 287–295.

CHO, H. T., and H. KENDE. 1997. Expression of expansin genes is correlated with growth in deepwater rice. *Plant Cell* 9, 1661–1671.

CLARK, S. E. 2001. Meristems: Start your signaling. *Curr. Opin. Plant Biol.* 4, 28–32.

CLARK, S. E., M. P. RUNNING, and E. M. Meyerowitz, 1993. *CLAVATA1*, a regulator of meristem and flower development in *Arabidopsis. Development* 119, 397–418.

CLARK, S. E., M. P. RUNNING, and E. M. MEYEROWITZ. 1995. *CLAVATA3* is a specific regulator of shoot and floral meristem development affecting the same processes as *CLAVATA1. Development* 121, 2057–2067.

CLELAND, R. E. 2001. Unlocking the mysteries of leaf primordia formation. *Proc. Natl. Acad. Sci. USA* 98, 10981–10982.

CLOWES, F. A. L. 1950. Root apical meristems of *Fagus sylvatica. New Phytol.* 49, 248–268.

CLOWES, F. A. L. 1954. The promeristem and the minimal constructional centre in grass root apices. *New Phytol.* 53, 108–116.

CLOWES, F. A. L. 1956. Nucleic acids in root apical meristems of *Zea. New Phytol.* 55, 29–35.

CLOWES, F. A. L. 1961. *Apical Meristems.* Botanical monographs, vol. 2. Blackwell Scientific, Oxford.

CLOWES, F. A. L. 1967. The functioning of meristems. *Sci. Prog. Oxf.* 55, 529–542.

CLOWES, F. A. L. 1969. Anatomical aspects of structure and development. In: *Root Growth*, pp. 3–19, W. J. Whittingham, ed. Butterworths, London.

CLOWES, F. A. L. 1972. The control of cell proliferation within root meristems. In: *The Dynamics of Meristem Cell Populations*, pp. 133–147, M. W. Miller and C. C. Kuehnert, eds. Plenum Press, New York.

CLOWES, F. A. L. 1976. The root apex. In: *Cell Division in Higher Plants*, pp. 254–284, M. M. Yeoman, ed. Academic Press, New York.

CLOWES, F. A. L. 1978a. Origin of the quiescent centre in *Zea mays. New Phytol.* 80, 409–419.

CLOWES, F. A. L. 1978b. Origin of quiescence at the root pole of pea embryos. *Ann. Bot.* 42, 1237–1239.

CLOWES, F. A. L. 1981. The difference between open and closed meristems. *Ann. Bot.* 48, 761–767.

CLOWES, F. A. L. 1984. Size and activity of quiescent centres of roots. *New Phytol.* 96, 13–21.

CLOWES, F. A. L. 1990. The discrete root epidermis of floating plants. *New Phytol.* 115, 11–15.

CLOWES, F. A. L. 1994. Origin of the epidermis in root meristems. *New Phytol.* 127, 335–347.

CLOWES, F. A. L., and H. E. STEWART. 1967. Recovery from dormancy in roots. *New Phytol.* 66, 115–123.

COSTA, S., and L. DOLAN. 2000. Development of the root pole and cell patterning in *Arabidopsis* roots. *Curr. Opin. Gen. Dev.* 10, 405–409.

CRAFTS, A. S. 1943. Vascular differentiation in the shoot apex of *Sequoia sempervirens*. *Am. J. Bot.* 30, 110–121.

CROXDALE, J. G. 1978. *Salvinia* leaves. I. Origin and early differentiation of floating and submerged leaves. *Can. J. Bot.* 56, 1982–1991.

CROXDALE, J. G. 1979. *Salvinia* leaves. II. Morphogenesis of the floating leaf. *Can. J. Bot.* 57, 1951–1959.

CUTTER, E. G. 1964. Observations on leaf and bud formation in *Hydrocharis morsus-ranae*. *Am. J. Bot.* 51, 318–324.

CUTTER, E. G. 1965. Recent experimental studies of the shoot apex and shoot morphogenesis. *Bot. Rev.* 31, 7–113.

D'AMATO, F. 1975. Recent findings on the organization of apical meristems with single apical cells. *G. Bot. Ital.* 109, 321–334.

DAVIS, E. L., P. RENNIE, and T. A. STEEVES. 1979. Further analytical and experimental studies on the shoot apex of *Helianthus annuus*: Variable activity in the central zone. *Can. J. Bot.* 57, 971–980.

DOLAN, L., and R. S. POETHIG. 1991. Genetic analysis of leaf development in cotton. *Development* suppl. 1, 39–46.

DOLAN, L., and R. S. POETHIG. 1998. Clonal analysis of leaf development in cotton. *Am. J. Bot.* 85, 315–321.

DOLAN, L., K. JANMAAT, V. WILLEMSEN, P. LINSTEAD, S. POETHIG, K. ROBERTS, and B. SCHERES. 1993. Cellular organization of the *Arabidopsis thaliana* root. *Development* 119, 71–84.

DUMAIS, J., and C. R. STEELE. 2000. New evidence for the role of mechanical forces in the shoot apical meristem. *J. Plant Growth Regul.* 19, 7–18.

EDGAR, E. 1961. *Fluctuations in Mitotic Index in the Shoot Apex of Lonicera nitida*. Publ. no. 1. University of Canterbury, Christchurch, NZ.

ESAU, K. 1965. *Vascular Differentiation in Plants*. Holt, Rinehart and Winston, New York.

ESAU, K. 1977. *Anatomy of Seed Plants*, 2nd ed. Wiley, New York.

EVANS, L. S., and A. R. BERG. 1972. Early histogenesis and semiquantitative histochemistry of leaf initiation in *Triticum aestivum*. *Am. J. Bot.* 59, 973–980.

EVANS, M. M. S., and M. K. BARTON. 1997. Genetics of angiosperm shoot apical meristem development. *Annu. Rev. Plant Physiol. Plant Mol. Biol.* 48, 673–701.

EVANS, M. W., and F. O. GROVER. 1940. Developmental morphology of the growing point of the shoot and the inflorescence in grasses. *J. Agric. Res.* 61, 481–520.

FELDMAN, L. J. 1984. The development and dynamics of the root apical meristem. *Am. J. Bot.* 71, 1308–1314.

FELDMAN, L. J., and J. G. TORREY. 1976. The isolation and culture in vitro of the quiescent center of *Zea mays*. *Am. J. Bot.* 63, 345–355.

FLEMING, A. J., S. McQUEEN-MASON, T. MANDEL, and C. KUHLEMEIER. 1997. Induction of leaf primordia by the cell wall protein expansin. *Science* 276, 1415–1418.

FLEMING, A. J., D. CADERAS, E. WEHRLI, S. McQUEEN-MASON, and C. KUHLEMEIER. 1999. Analysis of expansin-induced morphogenesis on the apical meristem of tomato. *Planta* 208, 166–174.

FLETCHER, J. C. 2002. The vegetative meristem. In: *Meristematic Tissues in Plant Growth and Development*, pp. 16–57, M. T. McManus and B. E. Veit, eds. Sheffield Academic Press, Sheffield.

FLETCHER, J. C. 2004. Stem cell maintenance in higher plants. In: *Handbook of Stem Cells*, vol. 2., *Adult and Fetal*, pp. 631–641, R. Lanza, H. Blau, D. Melton, M. Moore, E. D. Thomas (Hon.), C. Verfaille, I. Weissman, and M. West, eds. Elsevier Academic Press, Amsterdam.

FLETCHER, J. C., U. BRAND, M. P. RUNNING, R. SIMON, and E. M. MEYEROWITZ. 1999. Signaling of cell fate decisions by *CLAVATA3* in *Arabidopsis* shoot meristems. *Science* 283, 1911–1914.

FOARD, D. E. 1971. The initial protrusion of a leaf primordium can form without concurrent periclinal cell divisions. *Can. J. Bot.* 49, 1601–1603.

FOSTER, A. S. 1938. Structure and growth of the shoot apex of *Ginkgo biloba*. *Bull. Torrey Bot. Club* 65, 531–556.

FOSTER, A. S. 1941. Comparative studies on the shoot apex in seed plants. *Bull. Torrey Bot. Club* 68, 339–350.

FOSTER, A. S. 1943. Zonal structure and growth of the shoot apex in *Microcycas calocoma* (Miq.) A. DC. *Am. J. Bot.* 30, 56–73.

FURNER, I. J., and J. E. PUMFREY. 1992. Cell fate in the shoot apical meristem of *Arabidopsis thaliana*. *Development* 115, 755–764.

GALLOIS, J.-L., C. WOODWARD, G. V. REDDY, and R. SABLOWSKI. 2002. Combined *SHOOT MERISTEMLESS* and *WUSCHEL* trigger ectopic organogenesis in *Arabidopsis*. *Development* 129, 3207–3217.

GARRISON, R. 1949a. Origin and development of axillary buds: *Syringa vulgaris* L. *Am. J. Bot.* 36, 205–213.

GARRISON, R. 1949b. Origin and development of axillary buds: *Betula papyrifera* Marsh. and *Euptelea polyandra* Sieb. et Zucc. *Am. J. Bot.* 36, 379–389.

GARRISON, R. 1955. Studies in the development of axillary buds. *Am. J. Bot.* 42, 257–266.

GIFFORD, E. M., Jr. 1950. The structure and development of the shoot apex in certain woody Ranales. *Am. J. Bot.* 37, 595–611.

GIFFORD, E. M., Jr. 1951. Ontogeny of the vegetative axillary bud in *Drimys winteri* var. *chilensis*. *Am. J. Bot.* 38, 234–243.

GIFFORD, E. M., Jr. 1983. Concept of apical cells in bryophytes and pteridophytes. *Annu. Rev. Plant Physiol.* 34, 419–440.

GIFFORD, E. M. 1991. The root apical meristem of *Asplenium bulbiferum*: Structure and development. *Am. J. Bot.* 78, 370–376.

GIFFORD, E. M. 1993. The root apical meristem of *Equisetum diffusum*: Structure and development. *Am. J. Bot.* 80, 468–473.

GIFFORD, E. M., Jr., and G. E. CORSON Jr. 1971. The shoot apex in seed plants. *Bot. Rev.* 37, 143–229.

GIFFORD, E. M., and A. S. FOSTER. 1989. *Morphology and Evolution of Vascular Plants*, 3rd ed. Freeman, New York.

GIFFORD, E. M., Jr., and E. KURTH. 1982. Quantitative studies on the root apical meristem of *Equisetum scirpoides. Am. J. Bot.* 69, 464–473.

GIFFORD, E. M., Jr., and H. B. TEPPER. 1962. Ontogenetic and histochemical changes in the vegetative shoot tip of *Chenopodium album. Am. J. Bot.* 49, 902–911.

GIFFORD, E. M., Jr., V. S. POLITO, and S. NITAYANGKURA. 1979. The apical cell in shoot and roots of certain ferns: A re-evaluation of its functional role in histogenesis. *Plant Sci. Lett.* 15, 305–311.

GIROLAMI, G. 1953. Relation between phyllotaxis and primary vascular organization in *Linum. Am. J. Bot.* 40, 618–625.

GIROLAMI, G. 1954. Leaf histogenesis in *Linum usitatissimum. Am. J. Bot.* 41, 264–273.

GOTTLIEB, J. E., and T. A. STEEVES. 1961. Development of the bracken fern, *Pteridium aquilinum* (L.) Kuhn. III. Ontogenetic changes in the shoot apex and in the pattern of differentiation. *Phytomorphology* 11, 230–242.

GREEN, P. B. 1980. Organogenesis—A biophysical view. *Annu. Rev. Plant Physiol.* 31, 51–82.

GREEN, P. B. 1985. Surface of the shoot apex: A reinforcement-field theory for phyllotaxis. *J. Cell Sci.* suppl. 2, 181–201.

GREEN, P. B. 1986. Plasticity in shoot development: A biophysical view. In: *Plasticity in Plants*, pp. 211–232, D. H. Jennings and A. J. Trewavas, eds. Company of Biologists Ltd., Cambridge.

GREEN, P. B. 1989. Shoot morphogenesis, vegetative through floral, from a biophysical perspective. In: *Plant Reproduction: from Floral Induction to Pollination*, pp. 58–75, E. Lord and G. Bernier, eds. American Society of Plant Physiologists, Rockville, MD.

GREEN, P. B. 1999. Expression of pattern in plants: Combining molecular and calculus-based biophysical paradigms. *Am. J. Bot.* 86, 1059–1076.

GREEN, P. B., and K. E. BROOKS. 1978. Stem formation from a succulent leaf: Its bearing on theories of axiation. *Am. J. Bot.* 65, 13–26.

GREEN, P. B., and J. M. L. SELKER. 1991. Mutual alignments of cell walls, cellulose, and cytoskeletons: Their role in meristems. In: *The Cytoskeletal Basis of Plant Growth and Form*, pp. 303–322, C. W. Lloyd, ed. Academic Press, New York.

GREGORY, R. A., and J. A. ROMBERGER. 1972. The shoot apical ontogeny of the *Picea abies* seedling. I. Anatomy, apical dome diameter, and plastochron duration. *Am. J. Bot.* 59, 587–597.

GROOT, E. P., J. A. DOYLE, S. A. NICHOL, and T. L. ROST. 2004. Phylogenetic distribution and evolution of root apical meristem organization in dicotyledonous angiosperms. *Int. J. Plant Sci.* 165, 97–105.

GUNCKEL, J. E., and R. H. WETMORE. 1946. Studies of development in long shoots and short shoots of *Ginkgo biloba* L. I. The origin and pattern of development of the cortex, pith and procambium. *Am. J. Bot.* 33, 285–295.

GUNNING, B. E. S. 1978. Age-related and origin-related control of the numbers of plasmodesmata in cell walls of developing *Azolla* roots. *Planta* 143, 181–190.

GUNNING, B. E. S., J. E. HUGHES, and A. R. HARDHAM. 1978. Formative and proliferative cell divisions, cell differentiation, and developmental changes in the meristem of *Azolla* roots. *Planta* 143, 121–144.

GUTTENBERG, H. VON. 1960. *Grundzüge der Histogenese höherer Pflanzen. I. Die Angiospermen. Handbuch der Pflanzenanatomie*, Band 8, Teil 3. Gebrüder Borntraeger, Berlin.

GUTTENBERG, H. VON. 1961. *Grundzüge der Histogenese höherer Pflanzen. II. Die Gymnospermen. Handbuch der Pflanzenanatomie*, Band 8, Teil 4. Gebrüder Borntraeger, Berlin.

GUTTENBERG, H. VON. 1964. Die Entwicklung der Wurzel. *Phytomorphology* 14, 265–287.

GUTTENBERG, H. VON. 1966. *Histogenese der Pteridophyten*, 2nd ed. *Handbuch der Pflanzenanatomie*, Band 7, Teil 2. Gebrüder Borntraeger, Berlin.

HABERLANDT, G. 1914. *Physiological Plant Anatomy*. Macmillan, London.

HAGEMANN, R. 1957. Anatomische Untersuchungen an Gerstenwurzeln. *Kulturpflanze* 5, 75–107.

HALL, L. N., and J. A. LANGDALE. 1996. Molecular genetics of cellular differentiation in leaves. *New Phytol.* 132, 533–553.

HANSTEIN, J. 1868. Die Scheitelzellgruppe im Vegetationspunkt der Phanerogamen. In: Festschr. Friedrich Wilhelms Universität Bonn. *Niederrhein. Ges. Natur und Heilkunde*, pp. 109–134. Marcus, Bonn.

HANSTEIN, J. 1870. Die Entwicklung der keimes der Monokotylen und Dikotylen. In: *Botanische Abhandlungen aus dem Gebiet der Morphologie und Physiologie*, vol. 1, pt. 1, J. Hanstein, ed. Marcus, Bonn.

HARA, N. 1962. Structure and seasonal activity of the vegetative shoot apex of *Daphne pseudomezereum. Bot. Gaz.* 124, 30–42.

HARA, N. 1995. Developmental anatomy of the three-dimensional structure of the vegetative shoot apex. *J. Plant Res.* 108, 115–125.

HÄRTEL, K. 1938. Studien an Vegetationspunkten einheimischer Lycopodien. *Beit. Biol. Pflanz.* 25, 125–168.

HEMERLY, A., J. DE ALMEIDA ENGLER, C. BERGOUNIOUX, M. VAN MONTAGU, G. ENGLER, D. INZÉ, and P. FERREIRA. 1995. Dominant negative mutants of the Cdc2 kinase uncouple cell division from iterative plant development. *EMBO J.* 14, 3925–3936.

IRISH, V. F., and I. M. SUSSEX. 1992. A fate map of the *Arabidopsis* embryonic shoot apical meristem. *Development* 115, 745–753.

ITOH, J.-I., A. HASEGAWA, H. KITANO, and Y. NAGATO. 1998. A recessive heterochronic mutation, *plastochron1*, shortens the plastochron and elongates the vegetative phase in rice. *Plant Cell* 10, 1511–1521.

IVANOV, V. B. 1983. Growth and reproduction of cells in roots. In: *Progress in Science Series. Plant Physiology*, vol. 1, pp. 1–40. Amerind Publishing, New Delhi.

IVANOV, V. B., and J. G. DUBROVSKY. 1997. Estimation of the cell-cycle duration in the root apical meristem: A model of linkage between cell-cycle duration, rate of cell production, and rate of root growth. *Int. J. Plant Sci.* 158, 757–763.

JACKMAN, V. H. 1960. The shoot apices of some New Zealand gymnosperms. *Phytomorphology* 10, 145–157.

JACKSON, B. D. 1953. *A Glossary of Botanic Terms with Their Derivation and Accent.*, 4th ed., rev. and enl. J. B. Lippincott, Philadelphia.

JANCZEWSKI, E. VON. 1874. Das Spitzenwachsthum der Phanerogamenwurzeln. *Bot. Ztg.* 32, 113–116.

JEAN, R. V. 1989. Phyllotaxis: A reappraisal. *Can. J. Bot.* 67, 3103–3107.

JEAN, R. V. 1994. *Phyllotaxis: A Systemic Study of Plant Pattern Morphogenesis.* Cambridge University Press, Cambridge.

JENTSCH, R. 1957. Untersuchungen an den Sprossvegetationspunkten einiger Saxifragaceen. *Flora* 144, 251–289.

JESUTHASAN, S., and P. B. GREEN. 1989. On the mechanism of decussate phyllotaxis: Biophysical studies on the tunica layer of *Vinca major*. *Am. J. Bot.* 76, 1152–1166.

JOHNSON, M. A. 1951. The shoot apex in gymnosperms. *Phytomorphology* 1, 188–204.

JOHNSON, M. A., and R. J. TOLBERT. 1960. The shoot apex in *Bombax*. *Bull. Torrey Bot. Club* 87, 173–186.

JONES, P. A. 1977. Development of the quiescent center in maturing embryonic radicles of pea (*Pisum sativum* L. cv. Alaska). *Planta* 135, 233–240.

KAPLAN, D. R., and T. J. COOKE. 1997. Fundamental concepts in the embryogenesis of dicotyledons: A morphological interpretation of embryo mutants. *Plant Cell* 9, 1903–1919.

KAPLAN, D. R., and W. HAGEMANN. 1991. The relationship of cell and organism in vascular plants. *BioScience* 41, 693–703.

KAYES, J. M., and S. E. CLARK. 1998. *CLAVATA2*, a regulator of meristem and organ development in *Arabidopsis*. *Development* 125, 3843–3851.

KERK, N., and L. FELDMAN. 1994. The quiescent center in roots of maize: Initiation, maintenance and role in organization of the root apical meristem. *Protoplasma* 183, 100–106.

KERK, N. M., and L. J. FELDMAN. 1995. A biochemical model for the initiation and maintenance of the quiescent center: Implications for organization of root meristems. *Development* 121, 2825–2833.

KERK, N. M., K. JIANG, and L. J. FELDMAN. 2000. Auxin metabolism in the root apical meristem. *Plant Physiol.* 122, 925–932.

KIDNER, C., V. SUNDARESAN, K. ROBERTS, and L. DOLAN. 2000. Clonal analysis of the *Arabidopsis* root confirms that position, not lineage, determines cell fate. *Planta* 211, 191–199.

KIRCHOFF, B. K. 1984. On the relationship between phyllotaxy and vasculature: A synthesis. *Bot. J. Linn. Soc.* 89, 37–51.

KOCH, L. 1893. Die vegetative Verzweigung der höheren Gewächse. *Jahrb. Wiss. Bot.* 25, 380–488.

KORN, R. W. 2001. Analysis of shoot apical organization in six species of the Cupressaceae based on chimeric behavior. *Am. J. Bot.* 88, 1945–1952.

KOZLOWSKI, T. T., and S. G. PALLARDY. 1997. *Physiology of Woody Plants*, 2nd ed. Academic Press, San Diego.

KURAS, M. 1978. Activation of embryo during rape (*Brassica napus* L.) seed germination. I. Structure of embryo and organization of root apical meristem. *Acta Soc. Bot. Pol.* 47, 65–82.

KURTH, E. 1981. Mitotic activity in the root apex of the water fern *Marsilea vestita* Hook. and Grev. *Am. J. Bot.* 68, 881–896.

LANCE, A. 1957. Recherches cytologiques sur l'évolution de quelques méristème apicaux et sur ses variations provoquées par traitements photopériodiques. *Ann. Sci. Nat. Bot. Biol. Vég.*, Sér. 11, 18, 91–421.

LARSON, P. R. 1975. Development and organization of the primary vascular system in *Populus deltoides* according to phyllotaxy. *Am. J. Bot.* 62, 1084–1099.

LARSON, P. R. 1983. Primary vascularization and siting of primordia. In: *The Growth and Functioning of Leaves*, pp. 25–51, J. E. Dale and F. L. Milthorpe, eds. Cambridge University Press, Cambridge.

LARSON, P. R., and T. D. PIZZOLATO. 1977. Axillary bud development in *Populus deltoides*. I. Origin and early ontogeny. *Am. J. Bot.* 64, 835–848.

LAUFS, P., C. JONAK, and J. TRAAS. 1998a. Cells and domains: Two views of the shoot meristem in *Arabidopsis*. *Plant Physiol. Biochem.* 36, 33–45.

LAUFS, P., O. GRANDJEAN, C. JONAK, K. KIÊU, and J. TRAAS. 1998b. Cellular parameters of the shoot apical meristem in *Arabidopsis*. *Plant Cell* 10, 1375–1389.

LAUX, T., K. F. X. MAYER, J. BERGER, and G. JÜRGENS. 1996. The *WUSCHEL* gene is required for shoot and floral meristem integrity in *Arabidopsis*. *Development* 122, 87–96.

LEMON, G. D., and U. POSLUSZNY. 1997. Shoot morphology and organogenesis of the aquatic floating fern *Salvinia molesta* D. S. Mitchell, examined with the aid of laser scanning confocal microscopy. *Int. J. Plant Sci.* 158, 693–703.

LENHARD, M., and T. LAUX. 1999. Shoot meristem formation and maintenance. *Curr. Opin. Plant Biol.* 2, 44–50.

LISO, R., G. CALABRESE, M. B. BITONTI, and O. ARRIGONI. 1984. Relationship between ascorbic acid and cell division. *Exp. Cell Res.* 150, 314–320.

LISO, R., A. M. INNOCENTI, M. B. BITONTI, and O. ARRIGONI. 1988. Ascorbic acid-induced progression of quiescent centre cells from G$_1$ to S phase. *New Phytol.* 110, 469–471.

LOISEAU, J. E. 1959. Observation and expérimentation sur la phyllotaxie et le fonctionnement du sommet végétatif chez quelques Balsaminacées. *Ann. Sci. Nat. Bot. Biol. Vég., Sér. 11*, 20, 1–24.

LONG, J. A., and M. K. BARTON. 1998. The development of apical embryonic pattern in *Arabidopsis*. *Development* 125, 3027–3035.

LONG, J., and M. K. BARTON. 2000. Initiation of axillary and floral meristems in *Arabidopsis*. *Dev. Biol.* 218, 341–353.

LONG, J. A., E. I. MOAN, J. I. MEDFORD, and M. K. BARTON. 1996. A member of the KNOTTED class of homeodomain proteins encoded by the *STM* gene of *Arabidopsis*. *Nature* 379, 66–69.

LUCAS, W. J. 1995. Plasmodessmata: Intercellular channels for macromolecular transport in plants. *Curr. Opin Cell Biol.* 7, 673–680.

LUXOVÁ, M. 1975. Some aspects of the differentiation of primary root tissues. In: *The Development and Function of Roots*, pp. 73–90, J. G. Torrey and D. T. Clarkson, eds. Academic Press, London.

LUXOVÁ, M., and A. MURÍN. 1973. The extent and differences in mitotic activity of the root tip of *Vicia faba*. L. *Biol. Plant* 15, 37–43.

LYNDON, R. F. 1976. The shoot apex. In: *Cell Division in Higher Plants*, pp. 285–314, M. M. Yeoman, ed. Academic Press, New York.

LYNDON, R. F. 1994. Control of organogenesis at the shoot apex. *New Phytol.* 128, 1–18.

LYNDON, R. F. 1998. *The Shoot Apical Meristem. Its Growth and Development*. Cambridge University Press, Cambridge.

LYNN, K., A. FERNANDEZ, M. AIDA, J. SEDBROOK, M. TASAKA, P. MASSON, and M. K. BARTON. 1999. The *PINHEAD/ZWILLE* gene acts pleiotropically in *Arabidopsis* development and has overlapping functions with the *ARGONAUTE1* gene. *Development* 126, 469–481.

MA, Y., and T. A. STEEVES. 1994. Vascular differentiation in the shoot apex of *Matteuccia struthiopteris*. *Ann. Bot.* 74, 573–585.

MA, Y., and T. A. STEEVES. 1995. Characterization of stelar initiation in shoot apices of ferns. *Ann. Bot.* 75, 105–117.

MAJUMDAR, G. P. 1942. The organization of the shoot in *Heracleum* in the light of development. *Ann. Bot. n.s.* 6, 49–81.

MAJUMDAR, G. P., and A. DATTA. 1946. Developmental studies. I. Origin and development of axillary buds with special reference to two dicotyledons. *Proc. Indian Acad. Sci.* 23B, 249–259.

MANN, L. K. 1952. Anatomy of the garlic bulb and factors affecting bud development. *Hilgardia* 21, 195–251.

MARKOVSKAYA, E. F., N. V. VASILEVSKAYA, and M. I. SYSOEVA. 1991. Change of the temperature dependence of apical meristem differentiation in ontogenesis of the indeterminate species. *Sov. J. Dev. Biol.* 22, 394–397.

MAUSETH, J. D. 1978. An investigation of the morphogenetic mechanisms which control the development of zonation in seedling shoot apical meristems. *Am. J. Bot.* 65, 158–167.

MAYER, K. F. X., H. SCHOOF, A. HAECKER, M. LENHARD, G. JÜRGENS, and T. LAUX. 1998. Role of *WUSCHEL* in regulating stem cell fate in the *Arabidopsis* shoot meristem. *Cell* 95, 805–815.

MAZE, J. 1977. The vascular system of the inflorescence axis of *Andropogon gerardii* (Gramineae) and its bearing on concepts of monocotyledon vascular tissue. *Am. J. Bot.* 64, 504–515.

MCALPIN, B. W., and R. A. WHITE. 1974. Shoot organization in the Filicales: The promeristem. *Am. J. Bot.* 61, 562–579.

MCCONNELL, J. R., and M. K. BARTON. 1998. Leaf polarity and meristem formation in *Arabidopsis*. *Development* 125, 2935–2942.

MCDANIEL, C. N., and R. S. POETHIG. 1988. Cell-lineage patterns in the shoot apical meristem of the germinating maize embryo. *Planta* 175, 13–22.

MCGAHAN, M. W. 1955. Vascular differentiation in the vegetative shoot of *Xanthium chinense*. *Am. J. Bot.* 42, 132–140.

MEDFORD, J. I. 1992. Vegetative apical meristems. *Plant Cell* 4, 1029–1039.

MEDFORD, J. I., F. J. BEHRINGER, J. D. CALLOS, and K. A. FELDMANN. 1992. Normal and abnormal development in the *Arabidopsis* vegetative shoot apex. *Plant Cell* 4, 631–643.

MEYEROWITZ, E. M. 1997. Genetic control of cell division patterns in developing plants. *Cell* 88, 299–308.

MIA, A. J. 1960. Structure of the shoot apex of *Rauwolfia vomitoria*. *Bot. Gaz.* 122, 121–124.

MILLINGTON, W. F., and E. L. FISK. 1956. Shoot development in *Xanthium pensylvanicum*. I. The vegetative plant. *Am. J. Bot.* 43, 655–665.

MIRALLES, D. J., B. C. FERRO, and G. A. SLAFER. 2001. Developmental responses to sowing date in wheat, barley and rapeseed. *Field Crops Res.* 71, 211–223.

MOHR, H., and E. PINNIG. 1962. Der Einfluss des Lichtes auf die Bildung von Blattprimordien am Vegetationskegel der Keimlinge von *Sinapis alba* L. *Planta* 58, 569–579.

NELSON, A. J. 1990. Net alignment of cellulose in the periclinal walls of the shoot apex surface cells of *Kalanchoë blossfeldiana*. I. Transition from vegetative to reproductive morphogenesis. *Can. J. Bot.* 68, 2668–2677.

NEWMAN, I. V. 1965. Patterns in the meristems of vascular plants. III. Pursuing the patterns in the apical meristems where no cell is a permanent cell. *J. Linn. Soc. Lond. Bot.* 59, 185–214.

NITAYANGKURA, S., E. M. GIFFORD Jr., and T. L. ROST. 1980. Mitotic activity in the root apical meristem of *Azolla filiculoides* Lam., with special reference to the apical cell. *Am. J. Bot.* 67, 1484–1492.

NOUGARÈDE, A. 1967. Experimental cytology of the shoot apical cells during vegetative growth and flowering. *Int. Rev. Cytol.* 21, 203–351.

NOUGARÈDE, A., M. N. DIMICHELE, P. RONDET, and R. SAINT-CÔME. 1990. Plastochrone cycle cellulaire et teneurs en ADN nucléaire du méristème caulinaire de plants de *Chrysanthemum segetum* soumis à deux conditions lumineuses différentes, sous une photopériode de 16 heures. *Can. J. Bot.* 68, 2389–2396.

OWENS, J. N. 1968. Initiation and development of leaves in Douglas fir. *Can. J. Bot.* 46, 271–283.

OWENS, J. N., and M. MOLDER. 1973. Bud development in western hemlock. I. Annual growth cycle of vegetative buds. *Can. J. Bot.* 51, 2223–2231.

PARKE, R. V. 1959. Growth periodicity and the shoot tip of *Abies concolor*. *Am. J. Bot.* 46, 110–118.

PETERSON, R., and J. VERMEER. 1980. Root apex structure in *Ephedra monosperma* and *Ephedra chilensis* (Ephedraceae). *Am. J. Bot.* 67, 815–823.

PIEN, S., J. WYRZYKOWSKA, S. MCQUEEN-MASON, C. SMART, and A. FLEMING. 2001. Local expression of expansin induces the entire process of leaf development and modifies leaf shape. *Proc. Natl. Acad. Sci. USA* 98, 11812–11817.

PILLAI, S. K., and A. PILLAI. 1961a. Root apical organization in monocotyledons—Musaceae. *Indian Bot. Soc. J.* 40, 444–455.

PILLAI, S. K., and A. PILLAI. 1961b. Root apical organization in monocotyledons—Palmae. *Proc. Indian Acad. Sci., Sect. B*, 54, 218–233.

PILLAI, S. K., A. PILLAI, and S. SACHDEVA. 1961. Root apical organization in monocotyledons—Zingiberaceae. *Proc. Indian Acad. Sci., Sect. B*, 53, 240–256.

POETHIG, R. S. 1984. Patterns and problems in angiosperm leaf morphogenesis. In: *Pattern Formation. A Primer in Developmental Biology*, pp. 413–432, G. M. Malacinski ed. Macmillan, New York.

POETHIG, R. S., and I. M. SUSSEX. 1985a. The developmental morphology and growth dynamics of the tobacco leaf. *Planta* 165, 158–169.

POETHIG, R. S., and I. M. SUSSEX. 1985b. The cellular parameters of leaf development in tobacco: A clonal analysis. *Planta* 165, 170–184.

POPHAM, R. A. 1951. Principal types of vegetative shoot apex organization in vascular plants. *Ohio J. Sci.* 51, 249–270.

POPHAM, R. A. 1955. Zonation of primary and lateral root apices of *Pisum sativum*. *Am. J. Bot.* 42, 267–273.

POPHAM, R. A. 1966. *Laboratory Manual for Plant Anatomy*. Mosby, St. Louis.

PRIESTLEY, J. H., and C. F. SWINGLE. 1929. Vegetative propagation from the standpoint of plant anatomy. *USDA Tech. Bull.* no. 151.

PRIESTLY, J. H., L. I. SCOTT, and E. C. GILLETT. 1935. The development of the shoot in *Alstroemeria* and the unit of shoot growth in monocotyledons. *Ann. Bot.* 49, 161–179.

RAJU, M. V. S., T. A. STEEVES, and J. M. NAYLOR. 1964. Developmental studies of *Euphorbia esula* L.: Apices of long and short roots. *Can. J. Bot.* 42, 1615–1628.

RAJU, M. V. S., T. A. STEEVES, and J. MAZE. 1976. Developmental studies on *Euphorbia esula*: Seasonal variations in the apices of long roots. *Can. J. Bot.* 4, 605–610.

RAUH, W., and F. RAPPERT. 1954. Über das Vorkommen und die Histogenese von Scheitelgruben bei krautigen Dikotylen, mit besonderer Berücksichtigung der Ganz- und Halbrosettenpflanzen. *Planta* 43, 325–360.

REES, A. R. 1964. The apical organization and phyllotaxis of the oil palm. *Ann. Bot.* 28, 57–69.

REEVE, R. M. 1942. Structure and growth of the vegetative shoot apex of *Garrya elliptica* Dougl. *Am. J. Bot.* 29, 697–711.

REINHARDT, D., F. WITTWER, T. MANDEL, and C. KUHLEMEIER. 1998. Localized upregulation of a new expansin gene predicts the site of leaf formation in the tomato meristem. *Plant Cell* 10, 1427–1437.

REINHARDT, D., T. MANDEL, and C. KUHLEMEIER. 2000. Auxin regulates the initiation and radial position of plant lateral organs. *Plant Cell* 12, 507–518.

REMPHREY, W. R., and T. A. STEEVES. 1984. Shoot ontogeny in *Arctostaphylos uva-ursi* (bearberry): Origin and early development of lateral vegetative and floral buds. *Can. J. Bot.* 62, 1933–1939.

RIDING, R. T. 1976. The shoot apex of trees of *Picea mariana* of differing rooting potential. *Can. J. Bot.* 54, 2672–2678.

RODIN, R. J. 1953. Seedling morphology of *Welwitschia*. *Am. J. Bot.* 40, 371–378.

ROJO, E., V. K. SHARMA, V. KOVALEVA, N. V. RAIKHEL, and J. C. FLETCHER. 2002. CLV3 is localized to the extracellular space, where it activates the *Arabidopsis* CLAVATA stem cell signaling pathway. *Plant Cell* 14, 969–977.

ROST, T. L., and S. BAUM. 1988. On the correlation of primary root length, meristem size and protoxylem tracheary element position in pea seedlings. *Am. J. Bot.* 75, 414–424.

ROST, T. L., T. J. JONES, and R. H. FALK. 1988. Distribution and relationship of cell division and maturation events in *Pisum sativum* (Fabaceae) seedling roots. *Am. J. Bot.* 75, 1571–1583.

RUTH, J., E. J. KLEKOWSKI JR., and O. L. STEIN. 1985. Impermanent initials of the shoot apex and diplontic selection in a juniper chimera. *Am. J. Bot.* 72, 1127–1135.

SABATINI, S., D. BEIS, H. WOLKENFELT, J. MURFETT, T. GUILFOYLE, J. MALAMY, P. BENFEY, O. LEYSER, N. BECHTOLD, P. WEISBEEK, and B. SCHERES. 1999. An auxin-dependent distal organizer of pattern and polarity in the *Arabidopsis* root. *Cell* 99, 463–472.

SACHER, J. A. 1954. Structure and seasonal activity of the shoot apices of *Pinus lambertiana* and *Pinus ponderosa*. *Am. J. Bot.* 41, 749–759.

SACHS, R. M. 1965. Stem elongation. *Annu. Rev. Plant Physiol.* 16, 73–96.

SACKS, M. M., W. K. SILK, and P. BURMAN. 1997. Effect of water stress on cortical cell division rates within the apical meristem of primary roots of maize. *Plant Physiol.* 114, 519–527.

SAINT-CÔME, R. 1966. Applications des techniques histoautoradiographiques et des méthodes statistiques à l'étude du fonctionnement apical chez le *Coleus blumei* Benth. *Rev. Gén. Bot.* 73, 241–324.

SAKAGUCHI, S., T. HOGETSU, and N. HARA. 1988. Arrangement of cortical microtubules at the surface of the shoot apex in *Vinca major* L.: Observations by immunofluorescence microscopy *Bot. Mag., Tokyo* 101, 497–507.

SATINA, S., A. F. BLAKESLEE, and A. G. AVERY. 1940. Demonstration of the three germ layers in the shoot apex of *Datura* by means of induced polyploidy in periclinal chimeras. *Am. J. Bot.* 27, 895–905.

SCANLON, M. J., and M. FREELING. 1998. The narrow sheath leaf domain deletion: A genetic tool used to reveal developmental homologies among modified maize organs. *Plant J.* 13, 547–561.

SCHERES, B., and R. HEIDSTRA. 2000. Digging out roots: Pattern formation, cell division, and morphogenesis in plants. *Curr. Topics Dev. Biol.* 45, 207–247.

SCHERES, B., and H. WOLKENFELT. 1998. The *Arabidopsis* root as a model to study plant development. *Plant Physiol. Biochem.* 36, 21–32.

SCHERES, B., H. WOLKENFELT, V. WILLEMSEN, M. TERLOUW, E. LAWSON, C. DEAN, and P. WEISBEEK. 1994. Embryonic origin of the *Arabidopsis* primary root and root meristem initials. *Development* 120, 2475–2487.

SCHIEFELBEIN, J. W., J. D. MASUCCI, and H. WANG. 1997. Building a root: The control of patterning and morphogenesis during root development. *Plant Cell* 9, 1089–1098.

SCHMIDT, A. 1924. Histologische Studien an phanerogamen Vegetationspunkten. *Bot. Arch.* 8, 345–404.

SCHMIDT, K. D. 1978. Ein Beitrag zum Verständis von Morphologie und Anatomie der Marsileaceae. *Beitr. Biol. Pflanz.* 54, 41–91.

SCHOOF, H., M. LENHARD, A. HAECKER, K. F. X. MAYER, G. JÜRGENS, and T. LAUX. 2000. The stem cell population of *Arabidopsis* shoot meristems is maintained by a regulatory loop between the *CLAVATA* and *WUSCHEL* genes. *Cell* 100, 635–644.

SCHÜEPP, O. 1917. Untersuchungen über Wachstum und Formwechsel von Vegetationspunkten. *Jahrb. Wiss. Bot.* 57, 17–79.

SCHÜEPP, O. 1926. *Meristeme. Handbuch der Pflanzenanatomie*, Band 4, Lief 16. Gebrüder Borntraeger, Berlin.

SCHULTZ, H. R. 1993. Photosynthesis of sun and shade leaves of field-grown grapevine (*Vitis vinifera* L.) in relation to leaf age. Suitability of the plastochron concept for the expression of physiological age. *Vitis* 32, 197–205.

SCHWABE, W. W. 1984. Phyllotaxis. In: *Positional Controls in Plant Development*, pp. 403–440, P. W. Barlow and D. J. Carr, eds. Cambridge University Press, Cambridge.

SEAGO, J. L., and C. HEIMSCH. 1969. Apical organization in roots of the Convolvulaceae. *Am. J. Bot.* 56, 131–138.

SEELIGER, I. 1954. Studien am Sprossbegetationskegel von *Ephedra fragilis* var. *campylopoda* (C. A. Mey.) Stapf. *Flora* 141, 114–162.

SEKHAR, K. N. C., and V. K. SAWHNEY. 1985. Ultrastructure of the shoot apex of tomato (*Lycopersicon esculentum*). *Am. J. Bot.* 72, 1813–1822.

SELKER, J. M. L., G. L. STEUCEK, and P. B. GREEN. 1992. Biophysical mechanisms for morphogenetic progressions at the shoot apex. *Dev. Biol.* 153, 29–43.

SHAH, J. J., and J. D. PATEL. 1972. The shell zone: Its differentiation and probable function in some dicotyledons. *Am. J. Bot.* 59, 683–690.

SHAH, J. J., and K. UNNIKRISHNAN. 1969. Ontogeny of axillary and accessory buds in *Clerodendrum phlomidis* L. *Ann. Bot.* 33, 389–398.

SHAH, J. J., and K. UNNIKRISHNAN. 1971. Ontogeny of axillary and accessory buds in *Duranta repens* L. *Bot. Gaz.* 132, 81–91.

SHARMAN, B. C. 1942. Developmental anatomy of the shoot of *Zea mays* L. *Ann. Bot.* n.s. 6, 245–282.

SHARMAN, B. C. 1945. Leaf and bud initiation in the Graminae. *Bot. Gaz.* 106, 269–289.

SHIMABUKU, K. 1960. Observation on the apical meristem of rice roots. *Bot. Mag., Tokyo* 73, 22–28.

SIMON, R. 2001. Function of plant shoot meristems. *Semin. Cell Dev. Biol.* 12, 357–362.

SINGH, H. 1961. Seasonal variations in the shoot apex of *Cephalotaxus drupacea* Sieb. et Zucc. *Phytomorphology* 11, 146–153.

SINHA, N. 1999. Leaf development in angiosperms. *Annu. Rev. Plant Physiol. Plant Mol. Biol.* 50, 419–446.

SMITH, L. G., B. GREENE, B. VEIT, and S. HAKE. 1992. A dominant mutation in the maize homeobox gene, *Knotted-1*, causes its ectopic expression in leaf cells with altered fates. *Development* 116, 21–30.

SMITH, L. G., S. HAKE, and A. W. SYLVESTER. 1996. The *tangled-1* mutation alters cell division orientations throughout maize leaf development without altering leaf shape. *Development* 122, 481–489.

SNOW, M., and R. SNOW. 1932. Experiments on phyllotaxis. I. The effect of isolating a primordium. *Philos. Trans. R. Soc. Lond. B* 221, 1–43.

SNOW, M., and R. SNOW. 1942. The determination of axillary buds. *New Phytol.* 41, 13–22.

SNYDER, F. W., and J. A. BUNCE. 1983. Use of the plastochron index to evaluate effects of light, temperature and nitrogen on growth of soya bean (*Glycine max* L. Merr). *Ann. Bot.* 52, 895–903.

SOMA, K., and E. BALL. 1964. Studies of the surface growth of the shoot apex of *Lupinus albus*. *Brookhaven Symp. Biol.* 16, 13–45.

STEEVES, T. A., and I. M. SUSSEX. 1989. *Patterns in Plant Development*, 2nd ed. Cambridge University Press, Cambridge.

STEIN, D. B., and O. L. STEIN. 1960. The growth of the stem tip of *Kalanchoë* cv. "Brilliant Star." *Am. J. Bot.* 47, 132–140.

STERLING, C. 1947. Organization of the shoot of *Pseudotsuga taxifolia* (Lamb.) Britt. II. Vascularization. *Am. J. Bot.* 34, 272–280.

STEWART, R. N., and H. DERMEN. 1970. Determination of number and mitotic activity of shoot apical initial cells by analysis of mericlinal chimeras. *Am. J. Bot.* 57, 816–826.

STEWART, R. N., and H. DERMEN. 1979. Ontogeny in monocotyledons as revealed by studies of the developmental anatomy of periclinal chloroplast chimeras. *Am. J. Bot.* 66, 47–58.

STIEGER, P. A., D. REINHARDT, and C. KUHLEMEIER. 2002. The auxin influx carrier is essential for correct leaf positioning. *Plant J.* 32, 509–517.

STREET, H. E., and E. H. ROBERTS. 1952. Factors controlling meristematic activity in excised roots. I. Experiments showing the operation of internal factors. *Physiol. Plant.* 5, 498–509.

SUNDÅS-LARSSON, A., M. SVENSON, H. LIAO, and P. ENGSTRÖM. 1998. A homeobox gene with potential developmental control function in the meristem of the conifer *Picea abies*. *Proc. Natl. Acad. Sci. USA* 95, 15118–15122.

SUNDBERG, M. D. 1982. Leaf initiation in *Cyclamen persicum* (Primulaceae). *Can. J. Bot.* 60, 2231–2234.

SUSSEX, I. M. 1955. Morphogenesis in *Solanum tuberosum* L.: Apical structure and developmental pattern of the juvenile shoot. *Phytomorphology* 5, 253–273.

SUSSEX, I. M., and T. A. STEEVES. 1967. Apical initials and the concept of promeristem. *Phytomorphology* 17, 387–391.

TALBERT, P. B., H. T. ADLER, D. W. PARKS, and L. COMAI. 1995. The *REVOLUTA* gene is necessary for apical meristem development and for limiting cell divisions in the leaves and stems of *Arabidopsis thaliana*. *Development* 121, 2723–2735.

TEPPER, H. B. 1992. Benzyladenine promotes shoot initiation in empty leaf axils of *Stellaria media* L. *J. Plant Physiol.* 140, 241–243.

THIELKE, C. 1962. Histologische Untersuchungen am Sprosscheitel von *Saccharum*. II. Mitteilung. Die Sprossscheitel von *Saccharum sinense*. *Planta* 58, 175–192.

THOMPSON, J., and F. A. L. CLOWES. 1968. The quiescent centre and rates of mitosis in the root meristem of *Allium sativum*. *Ann. Bot.* 32, 1–13.

TIAN, H.-C., and M. MARCOTRIGIANO. 1994. Cell-layer interactions influence the number and position of lateral shoot meristems in *Nicotiana*. *Dev. Biol.* 162, 579–589.

TOLBERT, R. J., and M. A. JOHNSON. 1966. A survey of the vegetative apices in the family Malvaceae. *Am. J. Bot.* 53, 961–970.

TUCKER, S. C. 1962. Ontogeny and phyllotaxis of the terminal vegetative shoots of *Michelia fuscata*. *Am. J. Bot.* 49, 722–737.

VALLADE, J., J. ALABOUVETTE, and F. BUGNON. 1978. Apports de l'ontogenèse à l'interprétation structurale et fonctionnelle du méristème racinaire du *Petunia hybrida*. *Rev. Cytol. Biol. Vég. Bot.* 1, 23–47.

VALLADE, J., F. BUGNON, G. GAMBADE, and J. ALABOUVETTE. 1983. L'activité édificatrice du proméristème racinaire: Essai d'interprétation morphogénétique. *Bull. Sci. Bourg.* 36, 57–76.

VAN DEN BERG, C., V. WILLEMSEN, W. HAGE, P. WEISBEEK, and B. SCHERES. 1995. Cell fate in the *Arabidopsis* root meristem determined by directional signaling. *Nature* 378, 62–65.

VAN DEN BERG, C., W. HAGE, V. WILLEMSEN, N. VAN DER WERFF, H. WOLKENFELT, H. McKHANN, P. WEISBEEK, and B. SCHERES. 1997a. The acquisition of cell fate in the *Arabidopsis thaliana* root meristem. In: *Biology of Root Formation and Development*, pp. 21–29, A. Altman and Y. Waisel, eds. Plenum Press, New York.

VAN DEN BERG, C., V. WILLEMSEN, G. HENDRIKS, P. WEISBEEK, and B. SCHERES. 1997b. Short-range control of cell differentiation in the *Arabidopsis* root meristem. *Nature* 39, 287–289.

VAN DEN BERG, C., P. WEISBEEK, and B. SCHERES. 1998. Cell fate and cell differentiation status in the *Arabidopsis* root. *Planta* 205, 483–491.

VAN LIJSEBETTENS, M., and J. CLARKE. 1998. Leaf development in *Arabidopsis*. *Plant Physiol. Biochem.* 36, 47–60.

VAUGHN, J. G. 1955. The morphology and growth of the vegetative and reproductive apices of *Arabidopsis thaliana* (L.) Heynh., *Capsella bursa-pastoris* (L.) Medic. and *Anagallis arvensis* L. *J. Linn. Soc. Lond. Bot.* 55, 279–301.

VERDAGUER, D., and M. MOLINAS. 1999. Developmental anatomy and apical organization of the primary root of cork oak (*Quercus suber* L.). *Int. J. Plant Sci.* 160, 471–481.

VERNOUX, T., D. AUTRAN, and J. TRAAS. 2000a. Developmental control of cell division patterns in the shoot apex. *Plant Mol. Biol.* 43, 569–581.

VERNOUX, T., J. KRONENBERGER, O. GRANDJEAN, P. LAUFS, and J. TRAAS. 2000b. *PIN-FORMED 1* regulates cell fate at the periphery of the shoot apical meristem. *Development* 127, 5157–5165.

VORONINE, N. S. 1956. Ob evoliûtsii korneĭ rasteniĭ (De l'évolution des racines des plantes). *Biul. Moskov. Obshch. Isp. Priody, Otd. Biol.* 61, 47–58.

VORONKINA, N. V. 1975. Histogenesis in root apices of angiospermous plants and possible ways of its evolution. *Bot. Zh.* 60, 170–187.

WARDLAW, C. W. 1943. Experimental and analytical studies of pteridophytes. II. Experimental observations on the development of buds in *Onoclea sensibilis* and in species of *Dryopteris*. *Ann. Bot.* n.s. 7, 357–377.

WARDLAW, C. W. 1949. Experiments on organogenesis in ferns. *Growth* (suppl.) 13, 93–131.

WARDLAW, C. W. 1957. The reactivity of the apical meristem as ascertained by cytological and other techniques. *New Phytol.* 56, 221–229.

WARDLAW, C. W. 1968. *Morphogenesis in Plants: A Contemporary Study.* Methuen, London.

WEBSTER, P. L., and R. D. MACLEOD. 1980. Characteristics of root apical meristem cell population kinetics: A review of analyses and concepts. *Environ. Exp. Bot.* 20, 335–358.

WENZEL, C. L., and T. L. ROST. 2001. Cell division patterns of the protoderm and root cap in the "closed" root apical meristem of *Arabidopsis thaliana. Protoplasma* 218, 203–213.

WENZEL, C. L., K. L. TONG, and T. L. ROST. 2001. Modular construction of the protoderm and peripheral root cap in the "open" root apical meristem of *Trifolium repens* cv. Ladino. *Protoplasma* 218, 214–224.

WHITE, D. J. B. 1955. The architecture of the stem apex and the origin and development of the axillary buds in seedlings of *Acer pseudoplatanus* L. *Ann. Bot.* n.s. 19, 437–449.

WHITE, R. A., and M. D. TURNER. 1995. Anatomy and development of the fern sporophyte. *Bot. Rev.* 61, 281–305.

WILCOX, H. 1954. Primary organization of active and dormant roots of noble fir, *Abies procera. Am. J. Bot.* 41, 812–821.

WILDEMAN, A. G., and T. A. STEEVES. 1982. The morphology and growth cycle of *Anemone patens. Can. J. Bot.* 60, 1126–1137.

WOLFF, C. F. 1759. *Theoria Generationis.* Wilhelm Engelmann, Leipzig.

ZAGÓRSKA-MAREK, B., and M. TURZAŃSKA. 2000. Clonal analysis provides evidence for transient initial cells in shoot apical meristems of seed plants. *J. Plant Growth Regul.* 19, 55–64.

ZHU, T., W. J. LUCAS, and T. L. ROST. 1998a. Directional cell-to-cell communication in the *Arabidopsis* root apical meristem. I. An ultrastructural and functional analysis. *Protoplasma* 203, 35–47.

ZHU, T., R. L. O'QUINN, W. J. LUCAS, and T. L. ROST. 1998b. Directional cell-to-cell communication in the *Arabidopsis* root apical meristem. II. Dynamics of plasmodesmatal formation. *Protoplasma* 204, 84–93.

ZHU, T., and T. L. ROST. 2000. Directional cell-to-cell communication in the *Arabidopsis* root apical meristem. III. Plasmodesmata turnover and apoptosis in meristem and root cap cells during four weeks after germination. *Protoplasma* 213, 99–107.

ZIMMERMANN, W. 1928. Histologische Studien am Vegetationspunkt von *Hypericum uralum. Jahrb. Wiss. Bot.* 68, 289–344.

# CHAPTER SEVEN

# Parenchyma and Collenchyma

## ▮ PARENCHYMA

The term **parenchyma** refers to a tissue composed of living cells variable in their morphology and physiology, but generally having thin walls and a polyhedral shape (Fig. 7.1), and concerned with vegetative activities of the plant. The individual cells of such a tissue are **parenchyma cells**. The word parenchyma is derived from the Greek *para*, beside, and *en-chein*, to pour, a combination of words that expresses the ancient concept of parenchyma as a semiliquid substance "poured beside" other tissues that are formed earlier and are more solid.

Parenchyma is often spoken of as *the* fundamental or ground tissue. It fits this definition from morphological as well as physiological aspects. In the plant body as a whole or in its organs parenchyma appears as a ground substance in which other tissues, notably the vascular, are embedded. It is the foundation of the plant in the sense that the reproductive cells (spores and gametes) are parenchymatous in nature. Inasmuch as the presumed ancestors of plants consisted entirely of parenchymatous cells

(Graham, 1993), parenchyma may be considered the phylogenetic precursor of all other tissues.

This tissue is the principal seat of such essential activities as photosynthesis, assimilation, respiration, storage, secretion, and excretion—in short, activities depending on the presence of complete living protoplasts. Parenchyma cells that occur in the xylem and phloem play an important role in the movement of water and the transport of food substances.

Developmentally, parenchyma cells are relatively undifferentiated. They are unspecialized morphologically and physiologically, compared with such cells as sieve elements, tracheids, and fibers because, in contrast to these three examples of cell categories, parenchyma cells may change functions or combine several different ones. However, parenchyma cells may also be distinctly specialized, for example, with reference to photosynthesis, storage of specific substances, or deposition of materials that are in excess in the plant body. Whether they are specialized or not, parenchyma cells are highly complex physiologically because they possess living protoplasts.

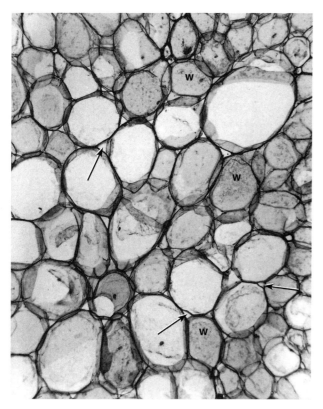

**FIGURE 7.1**

Parenchyma from tomato (*Solanum lycopersicum*) stem. Details: arrows point to intercellular spaces; w, walls in surface view. (×49.)

Characteristically living at maturity, parenchyma cells are able to resume meristematic activity: to dedifferentiate, divide, and redifferentiate. Because of this ability, parenchyma cells with only primary walls play an important role in wound healing, regeneration, the formation of adventitious roots and shoots, and the union of grafts. Furthermore single parenchyma cells, having all the genes present in the fertilized egg, or zygote, have the ability to become embryonic cells and then, given proper conditions for growth and development, to develop into an entire plant. Such cells are said to be **totipotent** (Chapter 5). The goal of workers involved with plant multiplication, utilizing tissue culture methods, or micropropagation, is to induce individual cells to express their totipotency (Bengochea and Dodds, 1986).

### Parenchyma Cells May Occur in Continuous Masses as Parenchyma Tissue or Be Associated with Other Cell Types in Morphologically Heterogeneous Tissues

Examples of plant parts consisting largely or entirely of parenchyma cells are the pith and cortex of stems and roots, the photosynthetic tissue (mesophyll) of leaves (see Fig. 7.3A), the flesh of succulent fruits, and the endosperm of seeds. As components of heterogeneous, or complex, tissues, parenchyma cells form the vascular rays and the vertical files of living cells in the xylem (Chapters 10 and 11) and the phloem (Chapters 13 and 14). Sometimes an essentially parenchymatous tissue contains parenchymatous or nonparenchymatous cells or groups of cells, morphologically or physiologically distinct from the main mass of cells in the tissue. Sclereids, for example, may be found in the leaf mesophyll and in the pith and cortical parenchyma (Chapter 8). Laticifers occur in various parenchymatous regions of the plant containing latex (Chapter 17). Sieve tubes traverse the cortical parenchyma of certain plants (Chapter 13).

The parenchyma tissue of the primary plant body, that is, the parenchyma of the cortex and pith, of the mesophyll of leaves, and of flower parts, differentiates from the ground meristem. The parenchyma cells associated with the primary and secondary vascular tissues are formed by the procambium and the vascular cambium, respectively. Parenchyma may also arise from the phellogen in the form of phelloderm, and it may be increased in amount by diffuse secondary growth.

The variable structure of parenchyma tissue (Fig. 7.2) and the distribution of parenchyma cells in the plant body clearly illustrate the problems involved in the proper definition and classification of a tissue. On the one hand, parenchyma may fit the most restricted definition of a tissue as a group of cells having a common origin, essentially the same structure, and the same function. On the other hand the homogeneity of a parenchyma tissue may be broken by the presence of varying numbers of nonparenchymatous cells, or parenchyma cells may occur as one of many cell categories in a complex tissue.

Thus the spatial delimitation of the parenchyma as a tissue is not precise in the plant body. Furthermore parenchyma cells may intergrade with cells that are distinctly nonparenchymatous. Parenchyma cells may be more or less elongated and have thick walls, a combination of characters suggesting specialization with regard to support. A certain category of parenchyma cells is so distinctly differentiated as a supporting tissue that it is designated by the special name of collenchyma, considered later in this chapter. Parenchyma cells may develop relatively thick lignified walls and assume some of the characteristics of sclerenchyma cells (Chapter 8). Tannin may be found in ordinary parenchyma cells and also in cells basically parenchymatous but of such distinct form (vesicles, sacs, or tubes) that they are designated as idioblasts. Similarly certain secretory cells differ from other parenchyma cells mainly in their function; others are so much modified that they are commonly treated as a special category of elements (laticifers; Chapter 17).

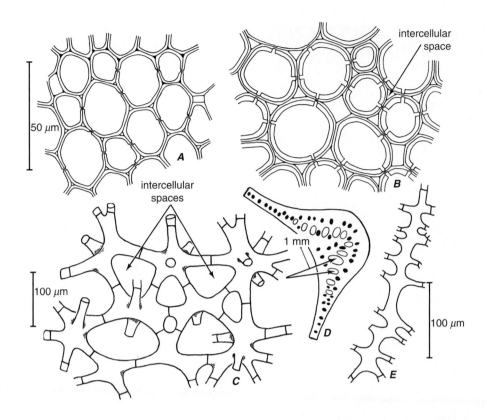

**FIGURE 7.2**

Shape and wall structure of parenchyma cells. (Cell contents are omitted.) **A**, **B**, parenchyma from the stem pith of birch (*Betula*). In younger stem (**A**), the cells have only primary walls; in older stem (**B**), secondary walls also occur. **C**, **D**, parenchyma of the aerenchyma type (**C**), which occurs in lacunae of petioles and midribs (**D**) of *Canna* leaves. The cells have many "arms." **E**, long "armed" cell from the mesophyll of a disc flower of *Gaillardia*. (From Esau, 1977.)

This chapter is restricted to a consideration of parenchyma concerned with the most ordinary vegetative activities of plants, excluding the meristematic. The parenchyma cells of the xylem and phloem are described in chapters dealing with these two tissues, and the general characteristics of the protoplasts of parenchyma cells are discussed in Chapters 2 and 3. Much of the discussion in Chapters 2 and 3 is relevant to the next topic.

### The Contents of Parenchyma Cells Are a Reflection of the Activities of the Cells

Parenchyma tissue specialized for photosynthesis contains numerous chloroplasts and is called **chlorenchyma**. The greatest expression of chlorenchyma is represented by the mesophyll of leaves (Fig. 7.3A), but chloroplasts may be abundant also in the cortex of a stem (Fig. 7.3B). Chloroplasts may occur in deeper stem tissues, including the secondary xylem and even the pith. Typically photosynthesizing cells are conspicuously vacuolated, and the tissue is permeated by an extensive system of intercellular spaces. In contrast, secretory types of parenchyma cells have dense protoplasts, especially rich in ribosomes, and have either numerous Golgi bodies or a massively developed endoplasmic reticulum, depending on the type of secretory product formed (Chapter 16).

Parenchyma cells may assume distinctive characteristics by accumulating specific kinds of substances. In starch-storing cells such as those of the potato tuber (Fig. 3.9), the endosperm of cereals, and the cotyledons of many embryos, the abundant starch-containing amyloplasts may virtually obscure all other cytoplasmic components. In many seeds the storage parenchyma cells are characterized by an abundance of protein and/or oil bodies (Fig. 3.10). Parenchyma cells of flowers and fruits often contain chromoplasts (Fig. 2.11). In various parts of the plant, parenchyma cells may become conspicuous by accumulating anthocyanins or tannins in their vacuoles (Fig. 2.21) or by depositing crystals of one or another form (Figs. 3.11–3.14).

Water is abundant in all active vacuolated parenchyma cells so that the parenchyma plays a major role as a water reservoir. In a study of a species of bamboo the variations in moisture content of the different parts

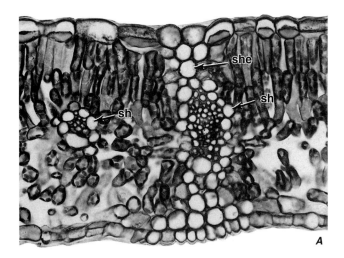

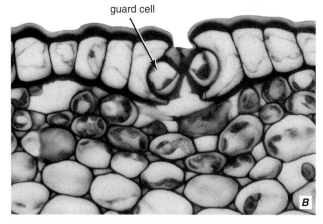

**FIGURE 7.3**

**A,** transverse section of pear (*Pyrus*) leaf. The two vascular bundles (veins) seen here are embedded in mesophyll. With its numerous chloroplasts, the mesophyll of leaves is the principal photosynthetic tissue of the plant. The veins are separated from the mesophyll by parenchymatous bundle sheaths (sh). Bundle-sheath extensions (one at she) connect the sheath of the larger vein with both epidermal layers. **B,** transverse section of *Asparagus* stem showing epidermis and some cortex. Chlorenchyma beneath epidermis and substomatal chamber beneath guard cells. (A, ×280; B, ×760.)

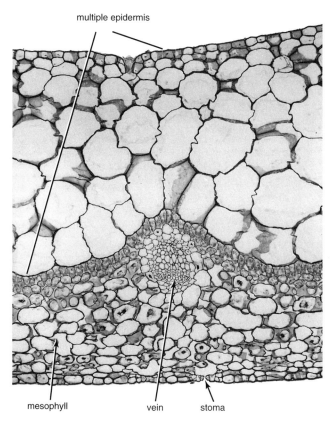

**FIGURE 7.4**

Transverse section of leaf blade of *Peperomia*. The very thick multiple epidermis visible on its upper surface presumably functions as water-storage tissue. (×110.)

of culms were found to be clearly associated with the proportions of parenchyma cells in the tissue system (Liese and Grover, 1961).

Parenchyma may be rather specialized as a water-storage tissue. Many succulent plants, such as the Cactaceae, *Aloe, Agave, Sansevieria* (Koller and Rost, 1988a, b), *Mesembryanthemum*, and *Peperomia* (Fig. 7.4), contain in their photosynthetic organs chlorophyll-free parenchyma cells full of water. This water tissue consists of living cells of particularly large size and usually with thin walls. The cells are often in rows and

may be elongated like palisade cells. Each has a thin layer of relatively dense parietal cytoplasm, a nucleus, and a large vacuole with watery or somewhat mucilaginous contents. The mucilages seem to increase the capacity of the cells to absorb and to retain water and may occur in the protoplasts and in the walls.

In the underground storage organs there is usually no separate water-storing tissue, but the cells containing starch and other food materials have a high water content. Potato tubers may start shoot growth in air and provide the growing parts with moisture for the initial growth (Netolitzky, 1935). A high water content is characteristic not only of underground storage organs, such as tubers and bulbs, but also of buds and of fleshy enlargements on aerial stems. In all these structures the storage of water is combined with that of the storage of food reserves.

### The Cell Walls of Parenchyma Cells May Be Thick or Thin

Parenchyma cells, including chlorenchyma and most storage cells, typically have thin, nonlignified primary walls (Figs. 7.1 and 7.2). Plasmodesmata are common in such walls, sometimes aggregated in primary pit-fields or in thickened wall portions, sometimes distributed throughout walls of uniform thickness. Some storage parenchyma develop remarkably thick walls (Bailey, 1938). As mentioned previously, xyloglucans located in such walls constitute the principal storage carbohydrate (Chapter 4). Thick walls occur, for example, in the endosperm of the date palm (*Phoenix dactylifera*), persimmon (*Diospyros*; Fig. 4.19), *Asparagus*, and *Coffea arabica*. They become thinner during germination. Relatively thick and often lignified secondary walls also occur in parenchyma cells of the wood (secondary xylem) and the pith, making it difficult to distinguish between such sclerified parenchyma cells and typical sclerenchyma cells.

The mechanical strength of typical parenchyma tissue is derived largely from the hydraulic property of its cells (Romberger et al., 1993). Consisting of cells with thin, nonlignified primary walls, parenchyma is rigid only when its cells are near or at full turgor. As noted by Niklas (1992), the degree to which parenchyma is utilized for mechanical support also depends on how closely its cells are packed together. In that regard aerenchyma tissue, with its large volume of intercellular space, might be expected to provide little mechanical support to the organs. It has been suggested, however, that aerenchyma with a honeycomb-like system of intercellular spaces is structurally efficient, providing the necessary strength with the smallest amount of tissue (Williams and Barber, 1961).

## Some Parenchyma Cells—Transfer Cells—Contain Wall Ingrowths

**Transfer cells** are specialized parenchyma cells containing cell wall ingrowths, which often greatly increase the surface area of the plasma membrane (Fig. 7.5). The ingrowths develop relatively late in cell maturation and are deposited on the original primary wall; hence they may be considered a specialized form of secondary wall (Pate and Gunning, 1972). Transfer cells play an important role in the transfer of solutes over short distances (Gunning, 1977). Their presence is generally correlated with the existence of intensive solute fluxes—in either inward (uptake) or outward (secretion) directions—across the plasma membrane. The wall ingrowths form just as intensive transport begins and are best developed on the cell surfaces presumably most actively involved with solute transport (Gunning and Pate, 1969). The plasma membrane closely follows the contours of the wall ingrowths however tortuous they may be, forming a so-called wall-membrane apparatus, which is bordered by numerous mitochondria and conspicuous endoplasmic reticulum.

High densities of a plasma membrane $H^+$-ATPase and sucrose transport proteins have been co-localized to the

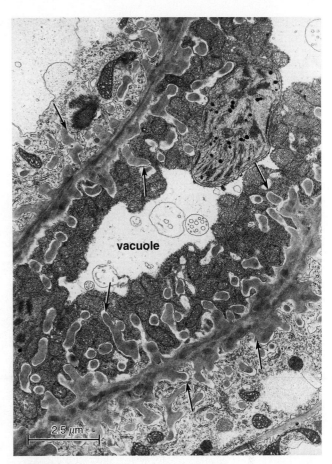

**FIGURE 7.5**

Longitudinal section of a portion of the phloem from a small vein of a *Sonchus deraceus* (sow thistle) leaf. The cell, with the dense cytoplasm, in the center of this electron micrograph is a companion cell. Phloem parenchyma cells occur on both sides of the companion cell. All three cells contain wall ingrowths (arrows); all three cells are transfer cells.

wall ingrowths of the seed coat and cotyledon transfer cells at the maternal/filial interface in developing *Vicia faba* seed (Harrington et al., 1997a, b), indicating that these transfer cells are the sites of membrane transport of sucrose to and from the seed apoplast. The transport of sucrose across the membrane involves a proton/sucrose cotransport mechanism (McDonald et al., 1996a, b).

Morphologically two categories of wall ingrowths can be recognized for most transfer cells: reticulate and flange (Fig. 7.6; Talbot et al., 2002). **Reticulate**-type wall ingrowths originate as small, randomly distributed papillae from the underlying wall. The papillae then branch and fuse laterally to form a complex labyrinth of variable morphology. **Flange**-type ingrowths arise as curvilinear, rib-shaped projections that are in contact with the underlying wall along their length. The

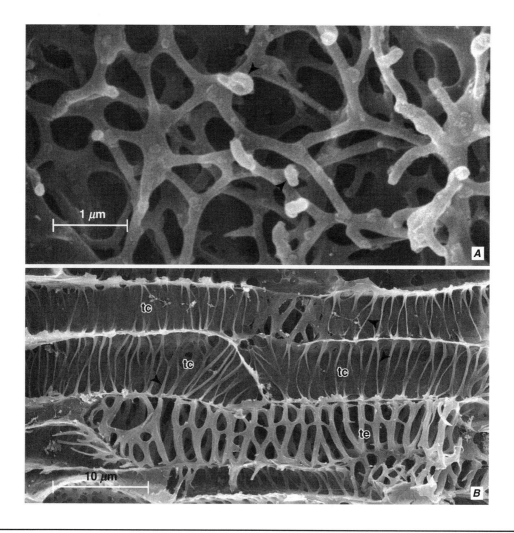

**FIGURE 7.6**

**A**, reticulate wall ingrowths in xylem parenchyma transfer cells in root nodule of *Vicia faba*. Arrowheads point to new wall ingrowths deposited on the most recently formed layer of wall ingrowth. **B**, flange wall ingrowth in xylem parenchyma transfer cells (tc) of longitudinally fractured vegetative nodes of *Triticum aestivum*. Flange ingrowths are roughly parallel, long bar-like thickenings (arrowheads) that are similar but much thinner than wall thickenings of the adjacent tracheary element (te). (From Talbot et al., 2002.)

projections become variously elaborated in different transfer cell types. Some transfer cells exhibit both reticulate and flange-like wall ingrowths; some others have wall ingrowths that fit neither category.

Transfer cells occur in a wide range of locations in the plant body: in the xylem and phloem of small, or minor, veins in cotyledons and foliage leaves of many herbaceous eudicots (Pate and Gunning, 1969; van Bel et al., 1993); in association with the xylem and phloem of leaf traces at the nodes of both eudicots and monocots (Gunning et al., 1970); in various tissues of reproductive structures (placentae, embryo sacs, aleurone cells, endosperm; Rost and Lersten, 1970; Pate and Gunning, 1972; Wang and Xi, 1992; Diane et al., 2002; Gómez et al., 2002); in root nodules (Joshi et al., 1993); and in various glandular structures (nectaries, salt glands, the glands of

carnivorous plants; Pate and Gunning, 1972; Ponzi and Pizzolongo, 1992). Each of these locations is a potential site of intensive short-distance solute transfer. Transfer cells can also be induced to form by external stimuli, such as by nematode infection (Sharma and Tiagi, 1989; Dorhout et al., 1993), in a plant that does not normally develop such cells.

Presumably the greater the surface area of the wall-membrane apparatus, the greater the total flux possible across it. In one study designed to test this hypothesis (Wimmers and Turgeon, 1991), the size and number of wall ingrowths in minor-vein phloem transfer cells of *Pisum sativum* leaves were increased significantly by growing the plants under a relatively high photon flux density. Remarkably, a resultant 47% increase in surface area of the plasma membrane of high-light leaves over low-light leaves was paralleled by a 47% increase in the flux of exogenous sucrose into the transfer cells and their associated sieve elements.

Wall ingrowths are not a prerequisite for solute transport across the plasma membrane. Cells without such modification may be similarly concerned with the transfer of substances between cells.

## Parenchyma Cells Vary Greatly in Shape and Arrangement

Parenchyma cells are commonly described as having a **polyhedral shape**, that is, as having many sides, or facets, but they vary greatly in shape even in the same plant (Figs. 7.2 and 7.7). Typically the ground tissue parenchyma consists of cells that are not much longer than wide and may be nearly **isodiametric**. In contrast, parenchyma cells may be more or less elongated or variously lobed or branched. In relatively homogeneous parenchyma the number of facets tends to approach 14. A geometrically perfect 14-sided figure is a polyhedron with 8 hexagonal and 6 quadrilateral facets (Fig. 7.7A) (orthic tetrakaidecahedron). Plant cells rarely approach this ideal form (Fig. 7.7B; Matzke, 1940) and show variable numbers of facets even in such homogeneous parenchyma as is often found in the pith

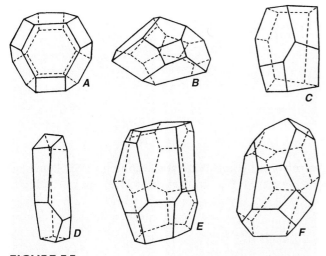

**FIGURE 7.7**

The shape of parenchyma cells. **A**, diagram of the orthic tetrakaidecahedron, a 14-sided polyhedron. **B**, diagram of a cell from the pith of *Ailanthus*. It has 1 heptagonal, 4 hexagonal, 5 pentagonal, and 4 quadrilateral faces, a total of 14 faces. An example of a cell approximating an orthic tetrakaihedron. **C–F**, diagrams of pith cells of *Eupatorium*. The numbers of facets are 10 (**C**), 9 (**D**), 16 (**E**), and 20 (**F**). (From Esau, 1977; **A, B**, after Matzke, 1940; **C–F**, after Marvin, 1944.)

of stems (Fig. 7.7C–F). The occurrence of smaller and larger cells in the same tissue, the development of intercellular spaces, and the change of cells from nearly isodiametric to some other shape are factors that determine the number of facets per cell (Matzke and Duffy, 1956). Small cells have fewer than 14 facets and large cells more than 14. The presence of intercellular spaces, particularly of large ones, reduces the number of contacts (Hulbary, 1944).

Pressure and surface tension have long been considered as factors influencing the shapes and sizes of cells. During differentiation of the "armed," or "stellate," parenchyma cells of the mesophyll of *Canna* leaves and the pith of *Juncus*, lateral tension appears to be one of the factors determining the final shape (Maas Geesteranus, 1941). The arms evidently elongate through their entire length. Korn (1980) has suggested that the shapes and sizes of cells are the result of three cellular processes: (1) the rate of wall expansion, (2) the duration of the cell cycle, and (3) the placement of the cell plate usually near the center of the longest wall, thus avoiding the intersection of existing partitions between adjacent cells and providing for nearly equal cell division. Subcellular factors influencing cell expansion and the formation of intercellular spaces are discussed in Chapter 4.

The arrangement of cells varies in different kinds of parenchyma. Storage parenchyma of fleshy roots and stems has abundant intercellular spaces, but the endosperm of seeds is usually a compact tissue, with small intercellular spaces at most. The extensive development of intercellular spaces in the leaf mesophyll, and in chlorenchyma in general, obviously is associated with the gaseous exchange in a photosynthetic tissue. Throughout the plant body, however, the ground tissue typically is permeated by a less conspicuous labyrinth of intercellular spaces, which also is essential for the diffusion-dependent flow of gases (Prat et al., 1997). In herbaceous species the labyrinth of intercellular spaces may extend from the substomatal chambers of the leaves to within a very short distance of the rootcap, via the cortical parenchyma of stem and root (Armstrong, W., 1979).

The intercellular spaces in the various tissues just described are commonly of schizogenous origin (Chapter 4). Such spaces can become very large if the cells separate along a considerable area of their contact with other cells. The separation is combined with an expansion of the tissue as a whole. In the growing tissue the cells maintain their limited connection with one another by differential growth and assume a lobed or "armed" form (Fig. 7.2C, E; Kaul, 1971). In some species the cells not only grow but also divide next to the intercellular spaces. In these divisions the new walls are formed perpendicular to the walls outlining the spaces (Hulbary, 1944).

### Some Parenchyma Tissue—Aerenchyma—Contains Particularly Large Intercellular Spaces

Air spaces are particularly well developed in angiosperms growing in aquatic and semi-aquatic habitats or waterlogged soils (Armstrong, W., 1979; Kozlowski, 1984; Bacanamwo and Purcell, 1999; Drew et al., 2000). Because of the prominence of intercellular spaces, the tissue is called **aerenchyma**, a term originally used for a phellogen-derived, nonsuberized cork (phellem) tissue containing numerous air chambers (Schenck, 1889). Aerenchyma development in the roots of some species occurs entirely by the enlargement of schizogenous intercellular spaces; in others, aerenchyma development involves various degrees of lysigeny (Smirnoff and Crawford, 1983; Justin and Armstrong, 1987; Armstrong and Armstrong, 1994). Interestingly, regardless of the degree of lysigeny, the cortical cells surrounding lateral roots always remain intact, indicating that aerenchyma formation is a controlled process. Ethylene has been implicated in the lysigenous development of aerenchyma in the roots of waterlogged plants (Kawase, 1981; Kozlowski, 1984; Justin and Armstrong, 1991; Drew, 1992). As mentioned previously (Chapter 5), the deficiency of oxygen in such plants triggers the production of ethylene, which in turn induces programmed cell death and aerenchyma development. Aerenchyma formation occurs naturally (constitutively) in the roots of some species, that is, apparently without any requirement for an external stimulus. Most notable among these are the roots of rice (*Oryza sativa*) (Fig. 7.8; Webb and Jackson, 1986).

The aerenchyma found in leaves and stems of aquatic plants generally differs structurally from that found in the roots (Armstrong, W., 1979). The tissue occurs as large longitudinal air spaces, or lacunae, sometimes containing stellate cells and often intersected at regular intervals by thin, transversely oriented plates of cells, called *diaphragms*, typically with intercellular spaces (Fig. 7.9; Kaul, 1971, 1973, 1974; Matsukura et al., 2000). In the shoots of some species all diaphragms are alike; in others, two or three types of diaphragm are produced. In the leaves of *Typha latifolia*, for example, diaphragms consisting entirely of stellate cells alternate with ones that are vascularized (Kaul, 1974). Despite suggestions that aerenchymatous tissue is often water- or fluid-filled (Canny, 1995), there is substantial evidence that the lacunae are usually gas-filled (Constable et al., 1992; Drew, 1997). The presence of aerenchyma, which is continuous from shoots to roots, enhances the diffusion of air from the leaves to the roots and enables wetland and waterlogged plants to maintain levels of oxygen sufficient to support respiration. Oxygen in excess of that consumed by respiring cells often diffuses from the roots into the soil atmosphere (Hook et al., 1971). This benefits the plant by creating a locally

**FIGURE 7.8**

Scanning electron micrograph of rice (*Oryza sativa*) root, in transverse section, showing aerenchyma tissue. (×80. Courtesy of P. Dayanandan.)

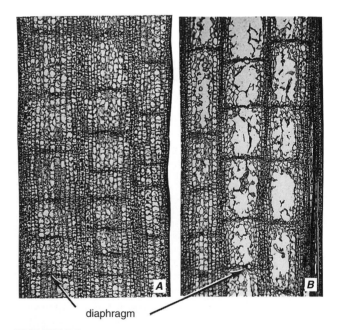

diaphragm

**FIGURE 7.9**

**A, B,** two stages in aerenchyma formation in midribs of leaf sheath of rice (*Oryza sativa*). Diaphragms remain intact between lacunae. (Both, ×190. From Kaufman, 1959.)

aerobic rhizosphere in an otherwise anaerobic soil (Topa and McLeod, 1986).

Other developmental phenomena associated with flooding are the development of adventitious roots (Visser et al., 1996; Shiba and Daimon, 2003) and the formation of lenticels at the base of the stem and on older roots (Hook, 1984). In some woody species, an aerenchymatous phellem may provide an alternative pathway for gaseous exchange between the roots and the shoots following destruction of the cortical aerenchyma with secondary growth (Stevens et al., 2002).

# ▌COLLENCHYMA

**Collenchyma** is a living tissue composed of more or less elongated cells with thickened primary walls (Fig. 7.10). It is a simple tissue, for it consists of a single cell type, the **collenchyma cell**. Collenchyma cells and parenchyma cells are similar to one another both physiologically and structurally. Both have complete protoplasts capable of resuming meristematic activity, and their cell walls are typically primary and nonlignified.

The difference between the two lies chiefly in the thicker walls of collenchyma cells; in addition the more highly specialized collenchyma cells are longer than most kinds of parenchyma cells. Where collenchyma cells and parenchyma cells lie next to each other, they intergrade both in wall thickness and form. The walls of parenchyma cells abutting collenchyma may be thickened—"collenchymatously thickened"—like those of the collenchyma cells. Both cell types contain chloroplasts (Maksymowych et al., 1993). Chloroplasts are most numerous in collenchyma cells that approach parenchyma cells in form. Long, narrow collenchyma cells contain only a few small chloroplasts or none at all. Because of the similarities between the two tissues and the structural and functional variability of parenchyma, collenchyma commonly is considered as a thick-walled kind of parenchyma structurally specialized as a supporting tissue. The terms parenchyma and collenchyma are also related, but in the latter the first part of the word, derived from the Greek *colla,* glue, refers to the thick glistening wall characteristic of collenchyma.

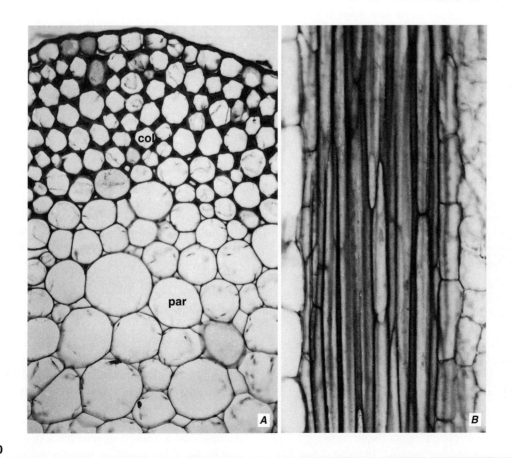

**FIGURE 7.10**

Collenchyma (col) of sugar beet (*Beta*) petiole in transverse section (**A**) and of grapevine (*Vitis*) stem in longitudinal section (**B**). Other detail: par, parenchyma. (×285.)

Collenchyma differs from the other representative supporting tissue, sclerenchyma (Chapter 8), in wall structure and condition of the protoplast. Collenchyma has relatively soft, pliable, nonlignified primary walls, whereas sclerenchyma has hard, more or less rigid, secondary walls, which commonly are lignified. Collenchyma cells retain active protoplasts capable of removing the wall thickenings when the cells are induced to resume meristematic activity, as in formation of a cork cambium (Chapter 15) or in response to wounding. Sclerenchyma walls are more permanent than those of collenchyma. They are not readily removed even if the protoplast is retained in the cell. Many sclerenchyma cells lack protoplasts at maturity. In some collenchyma cells, the products of transverse division remain together, enclosed by the common mother-cell wall (Majumdar, 1941; Majumdar and Preston, 1941). Such cell complexes resemble septate fibers.

### The Structure of the Cell Walls of Collenchyma Is the Most Distinctive Characteristic of This Tissue

The walls of collenchyma cells are thick and glistening in fresh sections (Fig. 7.11), and often the thickening is unevenly distributed. They contain, in addition to cellulose, large amounts of pectins and hemicelluloses and no lignin (Roelofsen, 1959; Jarvis and Apperley, 1990).

**FIGURE 7.11**

Transverse section of collenchyma tissue from a petiole in rhubarb (*Rheum rhabarbarum*). In fresh tissue like this, the unevenly thickened collenchyma cell walls have a glistening appearance. (×400.)

In some species, collenchyma walls have an alternation of layers rich in cellulose and poor in pectins with layers that are rich in pectins and poor in cellulose (Beer and Setterfield, 1958; Preston, 1974; Dayanandan et al., 1976). Since the pectins are hydrophilic, collenchyma walls are rich in water (Jarvis and Apperley, 1990). This feature can be demonstrated by treating fresh sections of collenchyma with alcohol. The dehydrating action of alcohol causes a noticeable shrinkage of collenchyma walls. Ultrastructurally collenchyma walls of various types have been described as having a crossed polylamellate (Wardrop, 1969; Chafe, 1970; Deshpande, 1976; Lloyd, 1984) or helicoidal structure (Chapter 4; Vian et al., 1993). Primary pit-fields are often present in collenchyma walls, especially in those that are rather uniform in thickness (Duchaigne, 1955).

The distribution of wall thickening in collenchyma shows several patterns (Fig. 7.12; Chafe, 1970). If the wall is unevenly thickened, it attains its greatest thickness either in the corners of the cell or on two opposite walls, the inner and the outer tangential walls (walls parallel with the surface of the plant part). Collenchyma with the thickenings on the tangential walls is called **lamellar**, or **plate**, **collenchyma** (Fig. 7.12A). Lamellar collenchyma is especially well developed in the stem cortex of *Sambucus nigra*. They may also be found in the stem cortex of *Sanguisorba*, *Rheum*, and *Eupatorium* and the petiole of *Cochlearia armoracia*. Collenchyma with wall thickenings localized in the corners commonly is called **angular collenchyma** (Fig. 7.12B). Examples of angular collenchyma are found in the stems of *Atropa belladonna* and *Solanum tuberosum* and petioles of *Begonia*, *Beta*, *Coleus*, *Cucurbita*, *Morus*, *Ricinus*, and *Vitis*.

Collenchyma may or may not contain intercellular spaces. If spaces are present in the angular type of collenchyma, the thickened walls occur next to the intercellular spaces. Collenchyma with such distribution of wall thickening is sometimes classified as a special type, the **lacunar**, or **lacunate**, **collenchyma** (Fig. 7.12C). When collenchyma develops no intercellular spaces, the corners where several cells meet show a thickened middle lamella. Such thickening is sometimes exaggerated by an accumulation of intercellular material in the potential intercellular spaces. The rate of this accumulation apparently varies, for intercellular spaces may arise in early stages of development, only to be closed later by pectic substances. Where the intercellular spaces are large, the pectic substances fail to fill them and form crests or wartlike accumulations protruding into the intercellular spaces (Duchaigne, 1955; Carlquist, 1956). The presence of intercellular spaces is not universally accepted as a valid criterion for a distinct type of collenchyma. What could be interpreted as lacunar collenchyma can be found in the stem cortex of *Brunellia* and *Salvia*, and of various Asteraceae and Malvaceae.

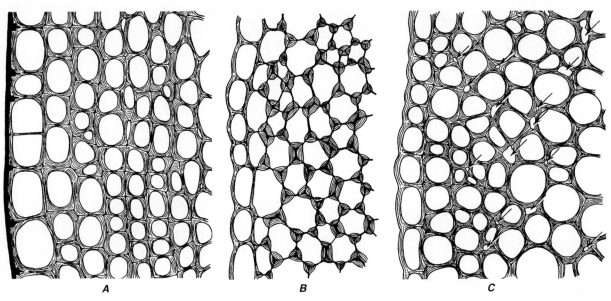

**FIGURE 7.12**

Collenchyma in stems (transverse sections). In all drawings the epidermal layer is to the left. **A**, *Sambucus*; thickenings mainly on tangential walls (lamellar collenchyma). **B**, *Cucurbita*; thickenings in the angles (angular collenchyma). **C**, *Lactuca*; numerous intercellular spaces (indicated by arrows) and the most prominent thickenings located next to these spaces (lacunar collenchyma). Thick cuticle (shown in black) in **A**. (All, ×320.)

A fourth type of collenchyma, **annular collenchyma**, is recognized by some plant anatomists (Metcalfe, 1979). Such collenchyma is characterized by cell walls that are more uniformly thickened and lumina that are more of less circular in outline, as seen in transverse sections. The distinction between annular and angular collenchyma is not clear-cut because the degree of restriction of wall thickenings to the corners of the angular collenchyma varies in relation to the amount of wall thickening present on other parts of the wall. If the general wall thickening becomes massive, the thickening in the corners is obscured and the lumen assumes a circular outline, instead of an angular one (Duchaigne, 1955; Vian et al., 1993).

Collenchyma walls are generally regarded as exemplifying thick primary walls, the thickening being deposited while the cell is growing. In other words, the cell wall increases simultaneously in surface area and in thickness. How much, if any, of the thickening is deposited after cells have stopped growing is generally impossible to determine, so it is generally impossible to delimit primary from secondary wall layers in such cells.

Collenchyma walls may become modified in older plant parts. In woody species with secondary growth, collenchyma follows, at least for a time, the increase in circumference of the axis by active growth with retention of the original characteristics. In some plants (*Tilia, Acer, Aesculus*) collenchyma cells enlarge and their walls become thinner (de Bary, 1884). Apparently it is

not known whether this reduction in wall thickness results from a removal of wall material or from stretching and dehydration. Collenchyma may change into sclerenchyma by a deposition of lignified secondary walls bearing simple pits (Duchaigne, 1955; Wardrop, 1969; Calvin and Null, 1977).

### Collenchyma Characteristically Occurs in a Peripheral Position

Collenchyma is the typical supporting tissue, first, of growing organs and, second, of those mature herbaceous organs that are only slightly modified by secondary growth or lack such growth entirely. It is the first supporting tissue in stems, leaves, and floral parts, and it is the main supporting tissue in many mature eudicot leaves and some green stems. Roots rarely have collenchyma but collenchyma may occur in the cortex (Guttenberg, 1940), particularly if the root is exposed to light (Van Fleet, 1950). Collenchyma is absent in stems and leaves of many of the monocots that early develop sclerenchyma (Falkenberg, 1876; Giltay, 1882). Collenchymatous tissue typically replaces sclerenchyma at the junction of the blade and sheath (the blade joint) and in the pulvinus of grass leaves (Percival, 1921; Esau, 1965; Dayanandan et al., 1977; Paiva and Machado, 2003). Massive collenchymatous bundle caps differentiate in connection with the leaf sheath bundles.

The peripheral position of collenchyma is highly characteristic (Fig. 7.13). It may be present immediately

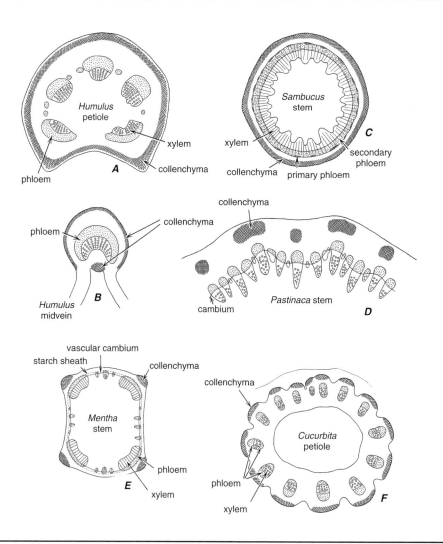

**FIGURE 7.13**

Distribution of collenchyma (crosshatched) and vascular tissues in various plant parts. Transverse sections. (A, B, ×19; C–F, ×9.5.)

beneath the epidermis or it may be separated from the epidermis by one or more layers of parenchyma. It has its origin in the ground meristem. If collenchyma is located next to the epidermis, the inner tangential walls of the epidermis may be thickened like the walls of collenchyma. Sometimes the entire epidermal cells are collenchymatous. In stems, collenchyma frequently forms a continuous layer around the circumference of the axis (Fig. 7.13C). Sometimes it occurs in strands, often within externally visible ridges (ribs) found in many herbaceous stems and in those of woody stems that have not yet undergone secondary growth (Fig. 7.13D, E). The distribution of collenchyma in the petioles shows patterns similar to those encountered in the stems (Fig. 7.13A, F). In the leaf blade, collenchyma occurs in the ribs accompanying the larger vascular bundles (major veins), sometimes on both sides of the rib (Fig. 7.13B) and sometimes on one side only, usually

the lower. Collenchyma also differentiates along the margins of the leaf blade.

In many plants the parenchyma that occurs in the outermost (phloem-side) or innermost (xylem-side) part of a vascular bundle, or completely surrounds a bundle, consists of long cells with thick primary walls. The wall thickening may resemble that of collenchyma, especially the annular type (Esau, 1936; Dayanandan et al., 1976). This tissue is often called collenchyma, but because of its association with the vascular tissues, it has a history of development somewhat different from that of the independent collenchyma, which originates in the ground meristem. It is preferable therefore to refer to such elongated, vascular bundle-associated cells with thick primary cell walls as collenchymatous parenchyma cells or collenchymatously thickened parenchyma, if their resemblance to collenchyma cells must be stressed. This designation may be applied to parenchyma resembling collenchyma in any location in the plant.

## Collenchyma Appears to Be Particularly Well Adapted for Support of Growing Leaves and Stems

Collenchyma walls begin to thicken early during the development of the shoot and, because the cells are capable of increasing simultaneously the surface and thickness of their walls, they can develop and maintain thick walls while the organ is still elongating. In addition, because the wall thickenings are plastic and capable of extension, they do not hinder the elongation of stem and leaf. In celery petioles the collenchyma cells lengthened by a factor of about 30 while the walls strongly increased simultaneously in thickness and surface area (Frey-Wyssling and Mühlethaler, 1965). In a more advanced state of development collenchyma continues to be a supporting tissue in plant parts (many leaves, herbaceous stems) that do not develop much sclerenchyma. With regard to the supporting role of collenchyma, it is of interest that in developing plant parts subjected to mechanical stresses (by exposure to wind, attachment of weights to inclined shoots), the wall thickening in collenchyma begins earlier than in plants not subjected to such stresses (Venning, 1949; Razdorskii, 1955; Walker, 1960). In addition stressed shoots may exhibit a significantly greater proportion of collenchyma (Patterson, 1992). Such stresses do not affect the type of collenchyma formed. In addition to its role as a supporting tissue, collenchyma has also been implicated in the resistance of oaks to mistletoe colonization (Hariri et al., 1992) and to stem feeding by insects (Oghiakhe et al., 1993).

A comparison of collenchyma with fibers is particularly interesting. In one study, the collenchyma strands elongated from 2% to 2.5% before breaking, whereas the fiber strands extended less than 1.5% before breaking (Ambronn, 1881). The collenchyma strands were capable of supporting 10 to 12 kg per mm², and the fiber strands 15 to 20 kg per mm². The fiber strands regained their original length even after they had been subjected to a tension of 15 to 20 kg per mm², whereas the collenchyma strands remained permanently extended after they had been made to support only 1.5 to 2 kg per mm². In other words, the tensile strength of collenchyma compares favorably with that of fibers, but collenchyma is plastic and sclerenchyma elastic. If fibers were to develop in growing plant parts, they would hinder elongation because of their tendency to regain their original length when stretched. Collenchyma, on the other hand, would remain extended under the same conditions. The importance of the plasticity of collenchyma cell walls is further emphasized by the observation that much internodal elongation occurs after the collenchyma cells have thickened their walls. In the celery petiole, wall thickening continues for some time after cessation of growth (Vian et al., 1993).

Mature collenchyma is a strong, flexible tissue consisting of long overlapping cells (in the center of the strands some may reach 2 millimeters in length; Duchaigne, 1955) with thick, nonlignified walls. In old plant parts collenchyma may harden and become less plastic than in younger parts, or as mentioned previously, it may change into sclerenchyma by the deposition of lignified secondary walls. The loss of the capacity for extension growth by mature celery collenchyma has been attributed to both the net longitudinal orientation of its microfibrils and the relative lack of methylated pectins (Fenwick et al., 1997). Cross-linking of pectins and hemicelluloses may also serve to rigidify mature collenchyma cell walls (Liu et al., 1999). If collenchyma does not undergo these changes, its role as a supporting tissue may become less important because of the development of sclerenchyma in the deeper parts of the stem or petiole. Moreover, in stems with secondary growth, the xylem becomes the chief supporting tissue because of the predominance of cells with lignified secondary walls and the abundance of long, overlapping cells in that tissue.

## ▌ REFERENCES ▐

AMBRONN, H. 1881. Über die Entwickelungsgeschichte und die mechanischen Eigenschaftern des Collenchyms. Ein Beitrag zur Kenntnis des mechanischen Gewebesystems. *Jahrb. Wiss. Bot.* 12, 473–541.

ARMSTRONG, J., and W. ARMSTRONG. 1994. Chlorophyll development in mature lysigenous and schizogenous root aerenchymas provides evidence of continuing cortical cell viability. *New Phytol.* 126, 493–497.

ARMSTRONG, W. 1979. Aeration in higher plants. *Adv. Bot. Res.* 7, 225–332.

BACANAMWO, M., and L. C. PURCELL. 1999. Soybean root morphological and anatomical traits associated with acclimation to flooding. *Crop Sci.* 39, 143–149.

BAILEY, I. W. 1938. Cell wall structure of higher plants. *Ind. Eng. Chem.* 30, 40–47.

BEER, M., and G. SETTERFIELD. 1958. Fine structure in thickened primary walls of collenchyma cells of celery petioles. *Am. J. Bot.* 45, 571–580.

BENGOCHEA, T., and J. H. DODDS. 1986. *Plant Protoplasts. A Biotechnological Tool for Plant Improvement.* Chapman and Hall, London.

CALVIN, C. L., and R. L. NULL. 1977. On the development of collenchyma in carrot. *Phytomorphology* 27, 323–331.

CANNY, M. J. 1995. Apoplastic water and solute movement: New rules for an old space. *Annu. Rev. Plant Physiol. Plant Mol. Biol.* 46, 215–236.

CARLQUIST, S. 1956. On the occurrence of intercellular pectic warts in Compositae. *Am. J. Bot.* 43, 425–429.

CHAFE, S. C. 1970. The fine structure of the collenchyma cell wall. *Planta* 90, 12–21.

CONSTABLE, J. V. H., J. B. GRACE, and D. J. LONGSTRETH. 1992. High carbon dioxide concentrations in aerenchyma of *Typha latifolia*. *Am. J. Bot.* 79, 415–418.

DAYANANDAN, P., F. V. HEBARD, and P. B. KAUFMAN. 1976. Cell elongation in the grass pulvinus in response to geotropic stimulation and auxin application. *Planta* 131, 245–252.

DAYANANDAN, P., F. V. HEBARD, V. D. BALDWIN, and P. B. KAUFMAN. 1977. Structure of gravity-sensitive sheath and internodal pulvini in grass shoots. *Am. J. Bot.* 64, 1189–1199.

DE BARY, A. 1884. *Comparative Anatomy of the Vegetative Organs of the Phanerogams and Ferns.* Clarendon Press, Oxford.

DESHPANDE, B. P. 1976. Observations on the fine structure of plant cell walls. I. Use of permanganate staining. *Ann. Bot.* 40, 433–437.

DIANE, N., H. H. HILGER, and M. GOTTSCHLING. 2002. Transfer cells in the seeds of Boraginales. *Bot. J. Linn. Soc.* 140, 155–164.

DORHOUT, R., F. J. GOMMERS, and C. KOLLÖFFEL. 1993. Phloem transport of carboxyfluorescein through tomato roots infected with *Meloidogyne incognita*. *Physiol. Mol. Plant Pathol.* 43, 1–10.

DREW, M. C. 1992. Soil aeration and plant root metabolism. *Soil Sci.* 154, 259–268.

DREW, M. C. 1997. Oxygen deficiency and root metabolism: Injury and acclimation under hypoxia and anoxia. *Annu. Rev. Plant Physiol. Plant Mol. Biol.* 48, 223–250.

DREW, M. C., C.-J. HE, and P. W. MORGAN. 2000. Programmed cell death and aerenchyma formation in roots. *Trends Plant Sci.* 5, 123–127.

DUCHAIGNE, A. 1955. Les divers types de collenchymes chez les Dicotylédones: Leur ontogénie et leur lignification. *Ann. Sci. Nat. Bot. Biol Vég., Sér. II,* 16, 455–479.

ESAU, K. 1936. Ontogeny and structure of collenchyma and of vascular tissues in celery petioles. *Hilgardia* 10, 431–476.

ESAU, K. 1965. *Vascular Differentiation in Plants.* Holt, Reinhart and Winston, New York.

ESAU, K. 1977. *Anatomy of Seed Plants,* 2nd ed. Wiley, New York.

FALKENBERG, P. 1876. *Vergleichende Untersuchungen über den Bau der Vegetationsorgane der Monocotyledonen.* Ferdinand Enke, Stuttgart.

FENWICK, K. M., M. C. JARVIS, and D. C. APPERLEY. 1997. Estimation of polymer rigidity in cell walls of growing and nongrowing celery collenchyma by solid-state nuclear magnetic resonance in vivo. *Plant Physiol.* 115, 587–592.

FREY-WYSSLING, A., and K. MÜHLETHALER. 1965. Ultrastructural plant cytology, with an introduction to molecular biology. Elsevier, Amsterdam.

GILTAY, E. 1882. Sur le collenchyme. *Arch. Néerl. Sci. Exact. Nat.* 17, 432–459.

GÓMEZ, E., J. ROYO, Y. GUO, R. THOMPSON, and G. HUEROS. 2002. Establishment of cereal endosperm expression domains: Identification and properties of a maize transfer cell-specific transcription factor, ZmMRP-1. *Plant Cell* 14, 599–610.

GRAHAM, L. E. 1993. *Origin of Land Plants.* Wiley, New York.

GUNNING, B. E. S. 1977. Transfer cells and their roles in transport of solutes in plants. *Sci. Prog. Oxf.* 64, 539–568.

GUNNING, B. E. S., and J. S. PATE. 1969. "Transfer cells." Plant cells with wall ingrowths, specialized in relation to short distance transport of solutes—Their occurrence, structure, and development. *Protoplasma* 68, 107–133.

GUNNING, B. E. S., J. S. PATE, and L. W. GREEN. 1970. Transfer cells in the vascular system of stems: Taxonomy, association with nodes, and structure. *Protoplasma* 71, 147–171.

GUTTENBERG, H. VON. 1940. *Der primäre Bau der Angiospermenwurzel. Handbuch der Pflanzenanatomie,* Band 8, Lief 39. Gebrüder Borntraeger, Berlin.

HARIRI, E. B., B. JEUNE, S. BAUDINO, K. URECH, and G. SALLÉ. 1992. Élaboration d'un coefficient de résistance au gui chez le chêne. *Can. J. Bot.* 70, 1239–1246.

HARRINGTON, G. N., V. R. FRANCESCHI, C. E. OFFLER, J. W. PATRICK, M. TEGEDER, W. B. FROMMER, J. F. HARPER, and W. D. HITZ. 1997a. Cell specific expression of three genes involved in plasma membrane sucrose transport in developing *Vicia faba* seed. *Protoplasma* 197, 160–173.

HARRINGTON, G. N., Y. NUSSBAUMER, X.-D. WANG, M. TEGEDER, V. R. FRANCESCHI, W. B. FROMMER, J. W. PATRICK, and C. E. OFFLER. 1997b. Spatial and temporal expression of sucrose transport-related genes in developing cotyledons of *Vicia faba* L. *Protoplasma* 200, 35–50.

HOOK, D. D. 1984. Adaptations to flooding with fresh water. In: *Flooding and Plant Growth,* pp. 265–294, T. T. Kozlowski, ed. Academic Press, Orlando, FL.

HOOK, D. D., C. L. BROWN, and P. P. KORMANIK. 1971. Inductive flood tolerance in swamp tupelo [*Nyssa sylvatica* var. *biflora* (Walt.) Sarg.]. *J. Exp. Bot.* 22, 78–89.

HULBARY, R. L. 1944. The influence of air spaces on the three-dimensional shapes of cells in *Elodea* stems, and a comparison with pith cells of *Ailanthus*. *Am. J. Bot.* 31, 561–580.

JARVIS, M. C., and D. C. APPERLEY. 1990. Direct observation of cell wall structure in living plant tissues by solid-state $^{13}C$ NMR spectroscopy. *Plant Physiol.* 92, 61–65.

JOSHI, P. A., G. CAETANO-ANOLLÉS, E. T. GRAHAM, and P. M. GRESSHOFF. 1993. Ultrastructure of transfer cells in spontaneous nodules of alfalfa (*Medicago sativa*). *Protoplasma* 172, 64–76.

JUSTIN, S. H. F. W., and W. ARMSTRONG. 1987. The anatomical characteristics of roots and plant response to soil flooding. *New Phytol.* 106, 465–495.

JUSTIN, S. H. F. W., and W. ARMSTRONG. 1991. Evidence for the involvement of ethene in aerenchyma formation in adventitious roots of rice (*Oryza sativa* L.). *New Phytol.* 118, 49–62.

KAUFMAN, P. B. 1959. Development of the shoot of *Oryza sativa* L.—II. Leaf histogenesis. *Phytomorphology* 9, 297–311.

KAUL, R. B. 1971. Diaphragms and aerenchyma in *Scirpus validus*. *Am. J. Bot.* 58, 808–816.

KAUL, R. B. 1973. Development of foliar diaphragms in *Sparganium eurycarpum*. *Am. J. Bot.* 60, 944–949.

KAUL, R. B. 1974. Ontogeny of foliar diaphragms in *Typha latifolia*. *Am. J. Bot.* 61, 318–323.

KAWASE, M. 1981. Effect of ethylene on aerenchyma development. *Am. J. Bot.* 68, 651–658.

KOLLER, A. L., and T. L. ROST. 1988a. Leaf anatomy in *Sansevieria* (Agavaceae). *Am. J. Bot.* 75, 615–633.

KOLLER, A. L., and T. L. ROST. 1988b. Structural analysis of water-storage tissue in leaves of *Sansevieria* (Agavaceae). *Bot. Gaz.* 149, 260–274.

KORN, R. W. 1980. The changing shape of plant cells: Transformations during cell proliferation. *Ann. Bot.* n.s. 46, 649–666.

KOZLOWSKI, T. T. 1984. Plant responses to flooding of soil. *BioScience* 34, 162–167.

LIESE, W., and P. N. GROVER. 1961. Untersuchungen über dem Wassergehalt von indischen Bambushalmen. *Ber. Dtsch. Bot. Ges.* 74, 105–117.

LIU, L., K.-E. L. ERIKSSON, and J. F. D. DEAN. 1999. Localization of hydrogen peroxide production in *Zinnia elegans* L. stems. *Phytochemistry* 52, 545–554.

LLOYD, C. W. 1984. Toward a dynamic helical model for the influence of microtubules on wall patterns in plants. *Int. Rev. Cytol.* 86, 1–51.

MAAS GEESTERANUS, R. A. 1941. On the development of the stellate form of the pith cells of *Juncus* species. *Proc. Sect. Sci. K. Ned. Akad. Wet.* 44, 489–501; 648–653.

MAJUMDAR, G. P. 1941. The collenchyma of *Heracleum Sphondylium* L. *Proc. Leeds Philos. Lit. Soc., Sci. Sect.* 4, 25–41.

MAJUMDAR, G. P., and R. D. PRESTON. 1941. The fine structure of collenchyma cells in *Heracleum sphondylium* L. *Proc. R. Soc. Lond. B.* 130, 201–217.

MAKSYMOWYCH, R., N. DOLLAHON, L. P. diCOLA, and J. A. J. ORKWISZEWSKI. 1993. Chloroplasts in tissues of some herbaceous stems. *Acta Soc. Bot. Pol.* 62. 123–126.

MARVIN, J. W. 1944. Cell shape and cell volume relations in the pith of *Eupatorium perfoliatum* L. *Am. J. Bot.* 31, 208–218.

MATSUKURA C., M. KAWAI, K. TOYOFUKU, R. A. BARRERO, H. UCHIMIYA, and J. YAMAGUCHI. 2000. Transverse vein differentiation associated with gas space formation—The middle cell layer in leaf sheath development of rice. *Ann. Bot.* 85, 19–27.

MATZKE, E. B. 1940. What shape is a cell? *Teach. Biol.* 10, 34–40.

MATZKE, E. B., and R. M. DUFFY. 1956. Progressive three-dimensional shape changes of dividing cells within the apical meristem of *Anacharis densa*. *Am. J. Bot.* 43, 205–225.

MCDONALD, R., S. FIEUW, and J. W. PATRICK. 1996a. Sugar uptake by the dermal transfer cells of developing cotyledons of *Vicia faba* L. Experimental systems and general transport properties. *Planta* 198, 54–65.

MCDONALD, R., S. FIEUW, and J. W. PATRICK. 1996b. Sugar uptake by the dermal transfer cells of developing cotyledons of *Vicia faba* L. Mechanism of energy coupling. *Planta* 198, 502–509.

METCALFE, C. R. 1979. Some basic types of cells and tissues. In: *Anatomy of the Dicotyledons*, 2nd ed., vol. I, *Systematic Anatomy of Leaf and Stem, with a Brief History of the Subject*, pp. 54–62, C. R. Metcalfe and L. Chalk, eds. Clarendon Press, Oxford.

NETOLITZKY, F. 1935. *Das Trophische Parenchym. C. Speichergewebe.* Handbuch der Pflanzenanatomie, Band 4, Lief 31. Gebrüder Borntraeger, Berlin.

NIKLAS, K. J. 1992. *Plant Biomechanics: An Engineering Approach to Plant Form and Function*. University of Chicago Press, Chicago.

OGHIAKHE, S., L. E. N. JACKAI, C. J. HODGSON, and Q. N. NG. 1993. Anatomical and biochemical parameters of resistance of the wild cowpea, *Vigna vexillata* Benth. (Acc. TVNu 72) to *Maruca testulalis* Geyer (Lepidoptera: Pyralidae). *Insect Sci. Appl.* 14, 315–323.

PAIVA, E. A. S., and S. R. MACHADO. 2003. Collenchyma in *Panicum maximum* (Poaceae): Localisation and possible role. *Aust. J. Bot.* 51, 69–73.

PATE, J. S., and B. E. S. GUNNING. 1969. Vascular transfer cells in angiosperm leaves. A taxonomic and morphological survey. *Protoplasma* 68, 135–156.

PATE, J. S., and B. E. S. GUNNING. 1972. Transfer cells. *Annu. Rev. Plant Physiol.* 23, 173–196.

PATTERSON, M. R. 1992. Role of mechanical loading in growth of sunflower (*Helianthus annuus*) seedlings. *J. Exp. Bot.* 43, 933–939.

PERCIVAL, J. 1921. *The Wheat Plant*. Dutton, New York.

PONZI, R., and P. PIZZOLONGO. 1992. Structure and function of *Rhinanthus minor* L. trichome hydathode. *Phytomorphology* 42, 1–6.

PRAT, R., J. P. ANDRÉ, S. MUTAFTSCHIEV, and A.-M. CATESSON. 1997. Three-dimensional study of the intercellular gas space in *Vigna radiata* hypocotyl. *Protoplasma* 196, 69–77.

PRESTON, R. D. 1974. *The Physical Biology of Plant Cell Walls*. Chapman & Hall, London.

RAZDORSKII, V. F. 1955. *Arkhitektonika rastenii (Architectonics of Plants)*. Sovetskaia Nauka, Moskva.

ROELOFSEN, P. A. 1959. *The Plant Cell Wall. Handbuch der Pflanzenanatomie*, Band 3, Teil 4, *Cytologie*. Gebrüder Borntraeger, Berlin-Nikolassee.

ROMBERGER, J. A., Z. HEJNOWICZ, and J. F. HILL. 1993. *Plant Structure: Function and Development. A Treatise on Anatomy and Vegetative Development, with Special Reference to Woody Plants*. Springer-Verlag, Berlin.

ROST, T. L., and N. R. LERSTEN. 1970. Transfer aleurone cells in *Setaria lutescens* (Gramineae). *Protoplasma* 71, 403–408.

SCHENCK, H. 1889. Über das Aëenchym, ein dem Kork homologes Gewebe bei Sumpflanzen. *Jahrb. Wiss. Bot.* 20, 526–574.

SHARMA, R. K., and B. TIAGI. 1989. Giant cell formation in pea roots incited by *Meloidogyne incognita* infection. *J. Phytol. Res.* 2, 185–191.

SHIBA, H., and H. DAIMON. 2003. Histological observation of secondary aerenchyma formed immediately after flooding in *Sesbania cannabina* and *S. rostrata*. *Plant Soil* 255, 209–215.

SMIRNOFF, N., and R. M. M. CRAWFORD. 1983. Variation in the structure and response to flooding of root aerenchyma in some wetland plants. *Ann. Bot.* 51, 237–249.

STEVENS, K. J., R. L. PETERSON, and R. J. READER. 2002. The aerenchymatous phellem of *Lythrum salicaria* (L.): A pathway for gas transport and its role in flood tolerance. *Ann. Bot.* 89, 621–625.

TALBOT, M. J., C. E. OFFLER, and D. W. MCCURDY. 2002. Transfer cell wall architecture: A contribution towards understanding localized wall deposition. *Protoplasma* 219, 197–209.

TOPA, M. A., and K. W. MCLEOD. 1986. Aerenchyma and lenticel formation in pine seedlings: A possible avoidance mechanism to anaerobic growth conditions. *Physiol. Plant.* 68, 540–550.

VAN BEL, A. J. E., A. AMMERLAAN, and A. A. VAN DIJK. 1993. A three-step screening procedure to identify the mode of phloem loading in intact leaves: Evidence for symplasmic and apoplasmic phloem loading associated with the type of companion cell. *Planta* 192, 31–39.

VAN FLEET, D. S. 1950. A comparison of histochemical and anatomical characteristics of the hypodermis with the endodermis in vascular plants. *Am. J. Bot.* 37, 721–725.

VENNING, F. D. 1949. Stimulation by wind motion of collenchyma formation in celery petioles. *Bot. Gaz.* 110, 511–514.

VIAN, B., J.-C. ROLAND, and D. REIS. 1993. Primary cell wall texture and its relation to surface expansion. *Int. J. Plant Sci.* 154, 1–9.

VISSER, E. J. W., C. W. P. M. BLOM, and L. A. C. J. VOESENEK. 1996. Flooding-induced adventitious rooting in *Rumex*: Morphology and development in an ecological perspective. *Acta Bot. Neerl.* 45, 17–28.

WALKER, W. S. 1960. The effects of mechanical stimulation and etiolation on the collenchyma of *Datura stramonium*. *Am. J. Bot.* 47, 717–724.

WANG, C.-G., and X.-Y. XI. 1992. Structure of embryo sac before and after fertilization and distribution of transfer cells in ovules of green gram. *Acta Bot. Sin.* 34, 496–501.

WARDROP, A. B. 1969. The structure of the cell wall in lignified collenchyma of *Eryngium* sp. (Umbelliferae). *Aust. J. Bot.* 17, 229–240.

WEBB, J., and M. B. JACKSON. 1986. A transmission and cryo-scanning electron microscopy study of the formation of aerenchyma (cortical gas-filled space) in adventitious roots of rice (*Oryza sativa*). *J. Exp. Bot.* 37, 832–841.

WILLIAMS, W. T., and D. A. BARBER. 1961. The functional significance of aerenchyma in plants. In: *Mechanisms in Biological Competition. Symp. Soc. Exp. Biol.* 15, 132–144.

WIMMERS, L. E., and R. TURGEON. 1991. Transfer cells and solute uptake in minor veins of *Pisum sativum* leaves. *Planta* 186, 2–12.

# CHAPTER EIGHT

# Sclerenchyma

The term **sclerenchyma** refers to a tissue composed of cells with secondary walls, often lignified, whose principal function is mechanical or support. These cells are supposed to enable plant organs to withstand various strains, such as may result from stretching, bending, weight, and pressure without undue damage to the thin-walled softer cells. The word is derived from the Greek *skleros*, meaning "hard" and *enchyma*, an infusion; it emphasizes the hardness of sclerenchyma walls. The individual cells of sclerenchyma are termed **sclerenchyma cells**. In addition to comprising sclerenchyma tissue, sclerenchyma cells like parenchyma cells may occur singly or in groups in other tissues. In the previous chapter (Chapter 7), it was noted that both parenchyma cells and collenchyma cells may become *sclerified*. Especially notable in that regard are the parenchyma cells of the secondary xylem, the water-conducting cells (tracheary elements) of which also have secondary walls. Thus secondary walls are not unique to sclerenchyma cells, and therefore the delimitation between typical sclerenchyma cells and sclerified parenchyma or collenchyma cells, on the one hand, and tracheary elements, on the other, is not sharp. Sclerenchyma cells may or may not retain their protoplasts at maturity. This variability adds to the difficulty of distinguishing between sclerenchyma cells and sclerified parenchyma cells.

Sclerenchyma cells are usually divided into two categories, fibers and sclereids. **Fibers** are described as long cells, and **sclereids** as relatively short cells. Sclereids, however, may grade from short to conspicuously elongated, not only in different plants but also in the same individual. The fibers, similarly, may be shorter or longer. Sclereids are generally thought of as having more conspicuous pitting in their walls than fibers, but this difference is not constant. Sometimes the origin of the two categories of cells is considered to be the distinguishing characteristic: sclereids often are said to arise through secondary sclerosis of parenchyma cells, fibers from meristematic cells that are early determined as fibers. This criterion does not entirely hold, however. Some sclereids differentiate from cells early individualized as sclereids (*Camellia*, Foster, 1944; *Monstera*, Bloch, 1946), and in certain plants parenchyma cells of the phloem differentiate into fiber-like cells only in that part of the tissue no longer concerned with conduction

(Chapter 14; Esau, 1969; Kuo-Huang, 1990). When it is difficult to classify a cell in terms of a fiber or a sclereid, the term **fiber-sclereid** may be used.

# FIBERS

Fibers typically are long, spindle-shaped cells, with more or less thick secondary walls, and they usually occur in strands (Fig. 8.1). Such strands constitute the "fibers" of commerce. The process of *retting* (technical form of the word rotting) used in the extraction of fibers from the plant results in a separation of the fiber bundles from the associated nonfibrous cells. Within a strand, the fibers overlap, a feature that imparts strength to the fiber bundles. In contrast to the thickened primary walls of collenchyma cells, the fiber walls are not highly hydrated. They are therefore harder than collenchyma walls and are elastic rather than plastic. Fibers serve as supporting elements in plant parts that are no longer elongating. The degree of lignification varies, and typi-

cally the simple or slightly bordered pits are relatively scarce and slit-like. Many fibers retain their protoplasts at maturity.

## Fibers Are Widely Distributed in the Plant Body

Fibers occur in separate strands or cylinders in the cortex and the phloem, as sheaths or bundle caps associated with the vascular bundles, or in groups or scattered in the xylem and the phloem. In the stems of monocots and eudicots the fibers are arranged in several characteristic patterns (Schwendener, 1874; de Bary, 1884; Haberlandt, 1914; Tobler, 1957). In many Poaceae the fibers form a system having the shape of a ribbed hollow cylinder, with the ribs connected to the epidermis (Fig. 8.2A). In *Zea, Saccharum, Andropogon, Sorghum* (Fig. 8.2B), and other related genera the vascular bundles have prominent sheaths of fibers, and the peripheral bundles may be irregularly fused with each other or united by sclerified parenchyma into a sclerenchymatous cylinder. The hypodermal parenchyma may be

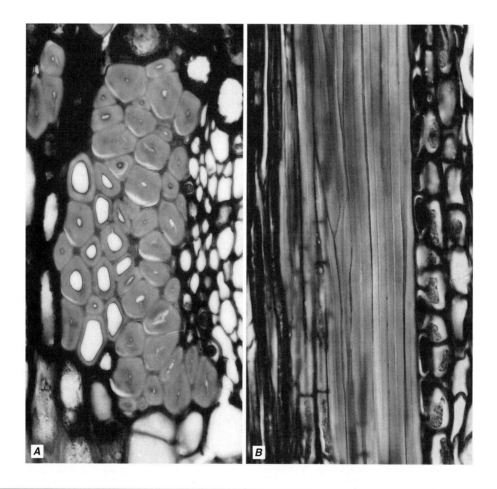

**FIGURE 8.1**

Primary phloem fibers from the stem of basswood (*Tilia americana*), seen here in both (**A**) transverse and (**B**) longitudinal views. The secondary walls of these long, thick-walled fibers contain relatively inconspicuous pits. Only a portion of the fibers can be seen in (**B**). (A, ×620; B, ×375.)

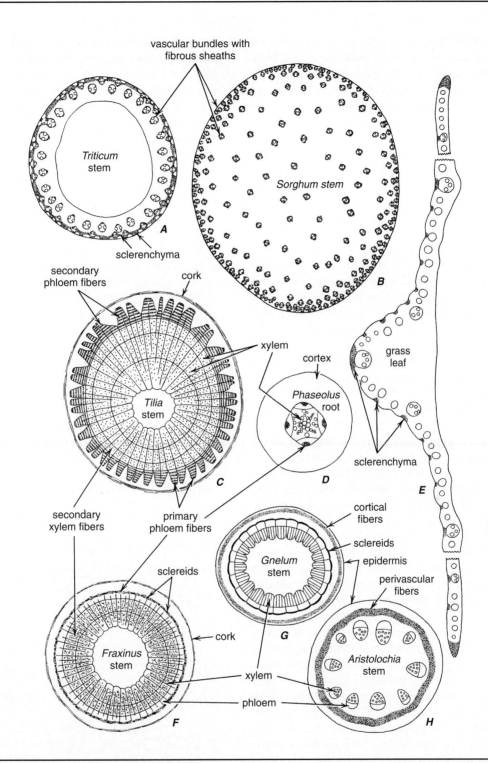

**FIGURE 8.2**

Transverse sections of various plant organs showing distribution of sclerenchyma (stippled), mainly fibers, and of vascular tissues. **A**, *Triticum* stem, sclerenchyma ensheathes vascular bundles and forms layers in peripheral part of stem. **B**, *Sorghum* stem, sclerenchyma in fibrous sheaths about vascular bundles. **C**, *Tilia* stem, fibers in primary and secondary phloem and in secondary xylem. **D**, *Phaseolus* root, fibers in primary phloem. **E**, grass leaf, sclerenchyma in strands beneath abaxial epidermis and along margins of blade. **F**, *Fraxinus* stem, fibers in primary phloem and secondary xylem; phloem fibers alternate with sclereids. **G**, *Gnetum gnemon* stem fibers in cortex and sclereids in perivascular position. **H**, *Aristolochia* stem, cylinder of fibers inside starch sheath in perivascular position. (A, G, ×14; B, C, F, ×7; D, ×9.5; E, ×29.5; H, ×13.)

strongly sclerified (Magee, 1948). A hypodermis containing long fibers, some over 1 mm long, has been recorded in *Zea mays* (Murdy, 1960). (A hypodermis is comprised of one or more layers of cells located beneath the epidermis and distinct from other neighboring cells of the ground tissue.) In the palms the central cylinder is demarcated by a sclerotic zone that may be several inches wide (Tomlinson, 1961). It consists of vascular bundles with massive, radially extended fibrous sheaths. The associated ground parenchyma also becomes sclerotic. In addition fiber strands occur in the cortex and a few in the central cylinder. Other patterns may be found in the monocots, and patterns may vary at different levels of the stem in the same plant (Murdy, 1960). Fibers may be prominent in the leaves of monocots (Fig. 8.2E). Here they form sheaths enclosing the vascular bundles, or strands, extending between the epidermis and the vascular bundles, or subepidermal strands not associated with the vascular bundles.

In stems of angiosperms, fibers frequently occur in the outermost part of the primary phloem, forming more or less extensive anastomosing strands or tangential plates (Fig. 8.2C, F). In some plants no other than the peripheral fibers (primary phloem fibers) occur in the phloem (*Alnus, Betula, Linum, Nerium*). Others develop fibers in the secondary phloem also, few (*Nicotiana, Catalpa, Boehmeria*) or many (*Clematis, Juglans, Magnolia, Quercus, Robinia, Tilia, Vitis*). Some eudicots have complete cylinders of fibers, either close to the vascular tissues (*Geranium, Pelargonium, Lonicera*, some Saxifragaceae, Caryophyllaceae, Berberidaceae, Primulaceae) or at a distance from them, but still located to the inside of the innermost layer of the cortex (Fig. 8.2H; *Aristolochia, Cucurbita*). In eudicot stems without secondary growth the isolated vascular bundles may be accompanied by fiber strands on both their inner and outer sides (*Polygonum, Rheum, Senecio*). Plants having phloem internal to the xylem may have fibers associated with this phloem (*Nicotiana*). Finally, a highly characteristic location for fibers in the angiosperms is the primary and the secondary xylem where they have varied arrangements (Chapter 11). Roots show a distribution of fibers similar to that of the stems and may have fibers in the primary (Fig. 8.2D) and in the secondary body. Conifers usually have no fibers in the primary phloem but may have them in the secondary phloem (*Sequoia, Taxus, Thuja*). Cortical fibers are sometimes present in stems (Fig. 8.2G).

### Fibers May Be Divided into Two Large Groups, Xylary and Extraxylary

**Xylary fibers** are fibers of the xylem, and **extraxylary fibers** are fibers located outside the xylem. Among the extraxylary fibers are the **phloem fibers**. Phloem fibers occur in many stems. The flax (*Linum usitatissimum*)

stem has only one band of fibers, several layers in depth, located on the outer periphery of the vascular cylinder (Fig. 8.3). These fibers originate in the earliest part of the primary phloem (the protophloem) but mature as fibers after this part of the phloem ceases to function in conduction (Fig. 8.4). Flax fibers are, therefore, *primary phloem fibers*, or *protophloem fibers*. The stems of *Sambucus* (elderberry), *Tilia* (basswood), *Liriodendron* (tulip tree), *Vitis* (grapevine), *Robinia pseudoacacia* (black locust), and many others, have both primary phloem fibers and *secondary phloem fibers*, which are located within the secondary phloem (Fig. 8.2C).

Two other groups of extraxylary fibers encountered in the stems of eudicots are the cortical fibers and the perivascular fibers. **Cortical fibers**, as the name implies, originate in the cortex (Fig. 8.2G). **Perivascular fibers** are located on the periphery of the vascular cylinder inside the innermost cortical layer (Fig. 8.2H; *Aristolochia* and *Cucurbita*). They do not originate as part of the phloem tissue but outside it. Perivascular fibers are commonly referred to as pericyclic fibers. However, the designation pericyclic is often used with reference to the primary phloem fibers as well (Esau, 1979). (See Blyth, 1958, for an evaluation of the term pericycle.) Extraxylary fibers also include the fibers of the

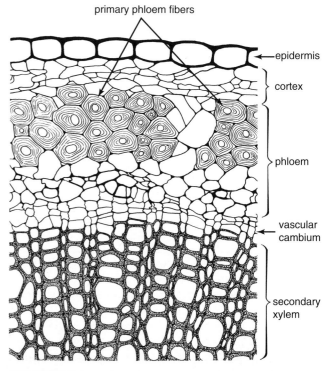

primary phloem fibers

epidermis

cortex

phloem

vascular cambium

secondary xylem

**FIGURE 8.3**

Transverse section of stem of *Linum usitatissimum* showing position of primary phloem fibers. (×320.)

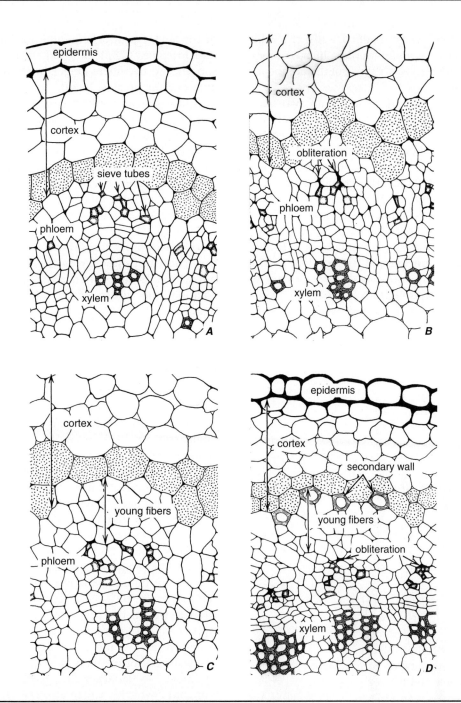

**FIGURE 8.4**

Development of primary phloem fibers in *Linum perenne* L. **A**, first primary sieve tubes are mature. **B**, **C**, new sieve tubes differentiate while older ones are obliterated. **D**, cells remaining after obliteration of sieve tubes begin to develop secondary walls characteristic of flax fibers. (A-C, ×745; D, ×395.)

monocots, whether or not associated with the vascular bundles.

The cell walls of the extraxylary fibers are often very thick. In the phloem fibers of flax, the secondary wall may amount to 90% of the cross-sectional area of the cell (Fig. 8.3). The secondary walls of these extraxylary fibers have a distinct polylamellate structure, the individual lamellae varying in thickness from 0.1 to 0.2 μm.

Not all extraxylary fibers have such wall structure. In mature bamboo culms, some fiber walls show a high degree of polylamellation, whereas others have no clearly visible lamellae (Murphy and Alvin, 1992). In addition the secondary walls of the secondary phloem fibers of most woody angiosperms and conifers consist of only two layers, a thin outer (S$_1$) layer and a thick inner (S$_2$) layer (Holdheide, 1951; Nanko et al., 1977).

Some extraxylary fibers have lignified walls; the walls of others contain little or no lignin (flax, hemp, ramie). Some extraxylary fibers, notably those of the monocots, are strongly lignified.

Wood fibers are commonly divided into two main groups, the **libriform fibers** and the **fiber-tracheids** (Fig. 8.5B, C), both of which typically have lignified cell walls. The libriform fibers resemble phloem fibers. Libriform is derived from the Latin *liber,* meaning "inner bark," that is, phloem. Although the distinction between the two groups of wood fibers has long been based primarily on the presence of simple pits in libriform fibers and of bordered pits in fiber-tracheids (IAWA Committee on Nomenclature, 1964), truly simple pits in fiber walls are extremely uncommon (Baas, 1986). The extremes of the two types of wood fiber are easy to distinguish, but imperceptible gradations occur between them. Fiber-tracheids also intergrade with tracheids, which have distinctly bordered pits (Fig. 8.5A). Commonly the thickness of the wall increases in the sequence

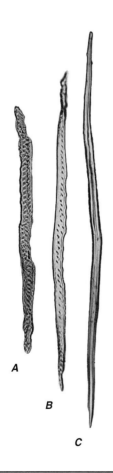

**FIGURE 8.5**

**A**, tracheid, **B**, fiber-tracheid, and **C**, libriform fiber in the secondary xylem, or wood, of a red oak tree (*Quercus rubra*). The spotted appearance of these cells is due to pits in the walls; pits are not discernible in **C**. (All, ×172.)

of tracheid, fiber-tracheid, and libriform fiber. In addition, in a given sample of wood, the tracheids are usually shorter and the fibers longer, with the libriform fibers attaining the greatest length.

Although commonly regarded as dead cells at maturity, living protoplasts are retained in the libriform fibers and fiber-tracheids in many woody plants (Fahn and Leshem, 1963; Wolkinger, 1971; Dumbroff and Elmore, 1977). (Fibers with living protoplasts have been found in bamboo culms over nine years of age; Murphy and Alvin, 1997.) These fibers often contain numerous starch grains; hence, in addition to support, they function in the storage of carbohydrates. The secondary walls of wood fibers differ from those of phloem fibers in that they consist of three layers designated $S_1$, $S_2$, and $S_3$ for outer, middle, and inner, respectively (Chapter 4). In addition the walls of wood fibers typically are lignified.

### Both Xylary and Extraxylary Fibers May Be Septate or Gelatinous

The phloem and/or xylem fibers of some eudicots undergo regular mitotic divisions after the secondary wall is deposited and are partitioned into two or more compartments by cross-walls, or **septae** (Fig. 8.6A) (Parameswaran and Liese, 1969; Chalk, 1983; Ohtani, 1987). Such fibers, called **septate fibers**, also occur in some monocots where they are nonvascular in origin (in Palmae and Bambuscoideae; Tomlinson, 1961; Parameswaran and Liese, 1977; Gritsch and Murphy, 2005). (Sclereids may also become partitioned by septae; Fig. 8.6B; Bailey, 1961.) The septae consist of a middle lamella and two primary walls and, apparently, they may or may not be lignified. The septae are in contact but not fused with the secondary wall and are separated by the latter from the original primary wall of the fiber. Apparently the primary walls of the septae continue over part or all of the inner surface of the fiber secondary wall (Butterfield and Meylan, 1976; Ohtani, 1987). Additional secondary wall may develop after the division and cover the septae also (Fig. 8.6B). In bamboos the septate fibers are characterized by thick polylamellate secondary walls. In addition to a middle lamella and primary wall layers, the septae of these fibers have secondary wall lamellae that continue on the longitudinal walls of the fibers (Parameswaran and Liese, 1977). Plasmodesmata interconnect the protoplasts via the septae of the fibers, which are living at maturity. Starch commonly is found in septate fibers, indicating a storage function, in addition to a supporting role, for these cells. Some septate fibers also contain crystals of calcium oxalate (Purkayastha, 1958; Chalk, 1983).

Another type of fiber that is neither strictly xylary nor extraxylary is the **gelatinous fiber**. Gelatinous fibers are identified by the presence of a so-called gelati-

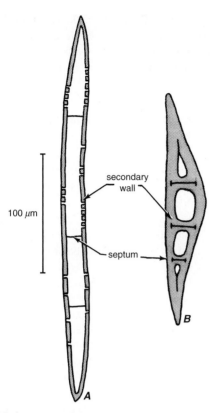

FIGURE 8.6

A, septate fiber from phloem of grapevine (*Vitis*) stem. The septae are in contact with the pitted secondary wall. B, septate sclereid from the phloem of *Pereskia* (Cactaceae) in which the septae are covered with secondary wall material. (From Esau, 1977; B, after Bailey, 1961.)

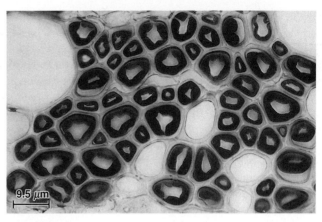

FIGURE 8.7

Gelatinous fibers as seen in transverse section in the wood of *Fagus* sp. In most of these fibers, the darkly stained gelatinous layer has pulled away from the rest of the wall. (Courtesy of Susanna M. Jutte.)

nous layer (G-layer), an innermost secondary wall layer that can be distinguished from the outer secondary wall layer(s) by its high cellulose content and lack of lignin (Fig. 8.7). The cellulose microfibrils of the G-layer are oriented parallel to the long axis of the cell, and hence this layer is isotropic or slightly birefringent when viewed in transverse section under polarized light (Wardrop, 1964). Being hygroscopic, the G-layer has the capacity to absorb much water. Upon swelling, the G-layer may occlude the lumen of the cell; upon drying, it commonly pulls away from the rest of the wall. Gelatinous fibers have been found in the xylem and phloem of roots, stems, and leaves of eudicots (Patel, 1964; Fisher and Stevenson, 1981; Sperry, 1982) and in nonvascular tissue of monocot leaves (Staff, 1974). They have been most extensively studied in tension wood (Chapter 11). Also referred to by the more general term **reaction fiber**, it is assumed that gelatinous fibers contract during development, generating contractile force sufficient eventually to bend a leaning or crooked stem toward a more normal position (Fisher and Stevenson,

1981). Gelatinous fibers in leaves may assist in the maintenance of leaf orientation with respect to gravity and in the display of leaflets to the sun (Sperry, 1982).

### Commercial Fibers Are Separated into Soft Fibers and Hard Fibers

The phloem fibers of eudicots represent the bast fibers of commerce (Harris, M., 1954; Needles, 1981). These fibers are classified as soft fibers because, whether or not lignified, they are relatively soft and flexible. Some of the well-known sources and usages of bast fibers are hemp (*Cannabis sativa*), cordage; jute (*Corchorus capsularis*), cordage, coarse textiles; flax (*Linum usitatissimum*), textiles (e.g., linen), thread; and ramie (*Boehmeria nivea*), textiles. Phloem fibers of some eudicots are used for making paper (Carpenter, 1963).

The fibers of monocots—usually called leaf fibers because they are obtained from leaves—are classified as hard fibers. They have strongly lignified walls and are hard and stiff. Examples of sources and uses of leaf fibers are abaca, or Manila hemp (*Musa textilis*), cordage; bowstring hemp (*Sansevieria*, entire genus), cordage; henequen and sisal (*Agave* species), cordage, coarse textiles; New Zealand hemp (*Phormium tenax*), cordage; and pineapple fiber (*Ananas comosus*), textiles. Leaf fibers of monocots (together with the xylem) serve as raw material for making paper (Carpenter, 1963): maize (*Zea mays*), sugar cane (*Saccharum officinarum*), esparto grass (*Stipa tenacissima*), and others.

The length of individual fiber cells varies considerably in different species. Examples of ranges of lengths in millimeters may be cited from M. Harris's (1954) handbook. Bast fibers: jute, 0.8–6.0; hemp, 5–55; flax,

9–70; ramie, 50–250. Leaf fibers: sisal, 0.8–8.0; bowstring hemp, 1–7; abaca, 2–12; New Zealand hemp, 2–15.

In commerce, the term fiber is often applied to materials that include, in the botanical sense, other types of cells besides fibers and also to structures that are not fibers at all. In fact the fibers obtained from the leaves of monocots represent vascular bundles together with associated fibers. Cotton fibers are epidermal hairs of seeds of *Gossypium* (Chapter 9); raffia is composed of leaf segments of *Raphia* palm; rattan is made from stems of *Calamus* palm.

# SCLEREIDS

Sclereids typically are short cells with thick secondary walls, strongly lignified, and provided with numerous simple pits. Some sclereids have relatively thin secondary walls, however, and may be difficult to distinguish from sclerified parenchyma cells. The thick-walled forms, on the other hand, may strongly contrast with parenchyma cells: their walls may be so massive as almost to occlude the lumina, and their prominent pits often are ramiform (Fig. 8.8). The secondary wall typically appears multilayered, reflecting its helicoidal construction (Roland et al., 1987, 1989). Crystals are embedded in the secondary walls of certain species (Fig. 8.9) (Kuo-Huang, 1990). Many sclereids retain living protoplasts at maturity.

### Based on Shape and Size, Sclereids May Be Classified into a Number of Types

The most commonly recognized categories of sclereids are (1) **brachysclereids**, or **stone cells**, roughly isodiametric or somewhat elongated cells, widely distributed in cortex, phloem, and pith of stems, and in the flesh of fruit (Figs. 8.8 and 8.10A–D); (2) **macrosclereids**, elongated and columnar (rod-like) cells, exemplified by sclereids forming the palisade-like epidermal layer of leguminous seed coats (see Fig. 8.14); (3) **osteosclereids**, bone cells, also columnar but with enlarged ends as in the subepidermal layer of some seed coats (see Fig. 8.14E); and (4) **astrosclereids**, star-cells, with lobes or arms diverging from a central body (Fig. 8.10L), often found in the leaves of eudicots. Other less commonly recognized types include **trichosclereids**, thin-walled sclereids resembling hairs, with branches projecting

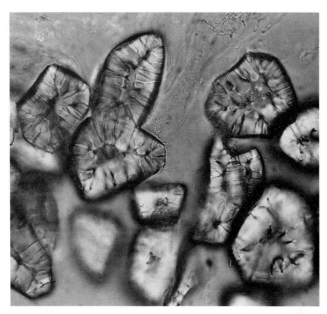

**FIGURE 8.8**

Sclereids (stone cells) from fresh tissue of pear (*Pyrus communis*) fruit. The secondary walls contain conspicuous simple pits with many branches, known as ramiform pits. During formation of the clusters of stone cells in the flesh of the pear fruit, cell divisions occur concentrically around some of the sclereids formed earlier. The newly formed cells differentiate as stone cells, adding to the cluster. (×400.)

**FIGURE 8.9**

Branched sclereid from a leaf of the water lily (*Nymphaea odorata*) as seen in polarized light. Numerous small angular crystals are embedded in the wall of this sclereid. (×230.)

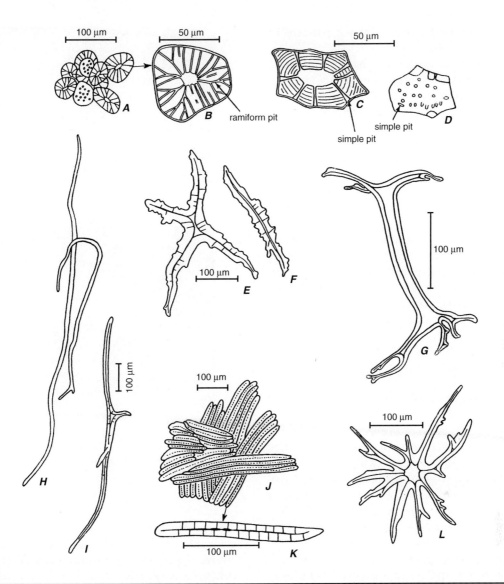

**FIGURE 8.10**

Sclereids. **A**, **B**, stone cells from fruit flesh of pear (*Pyrus*). **C**, **D**, sclereids from stem cortex of wax plant (*Hoya*), in sectional (**C**) and surface (**D**) views. **E**, **F**, sclereids from petiole of *Camellia*. **G**, columnar sclereid with ramified ends from palisade mesophyll of *Hakea*. **H**, **I**, filiform sclereids from leaf mesophyll of olive (*Olea*). **J**, **K**, sclereids from endocarp of fruit of apple (*Malus*). **L**, astrosclereid from stem cortex of *Trochodendron*. (From Esau, 1977.)

into intercellular spaces, and **filiform sclereids**, long slender cells resembling fibers (Fig. 8.10H, I; see also Fig. 8.13). Astrosclereids and trichosclereids are structurally similar, and trichosclereids intergrade with filiform sclereids. Osteosclereids may be branched at their ends (as in Fig. 8.10G), and consequently they resemble trichosclereids. This classification is rather arbitrary and does not cover all the forms of sclereids known (Bailey, 1961; Rao, T. A., 1991). Moreover it is of limited utility because, as indicated, the various forms frequently intergrade.

### Sclereids Like Fibers Are Widely Distributed in the Plant Body

The distribution of sclereids among other cells is of special interest with regard to problems of cell differentiation in plants. They may occur in more or less extensive layers or clusters, but frequently they appear isolated among other types of cells from which they may differ sharply by their thick walls and often bizarre shapes. As isolated cells they are classified as **idioblasts** (Foster, 1956). The differentiation of idioblasts poses many still unresolved questions regarding the causal relationships in the development of tissue patterns in plants.

Sclereids occur in the epidermis, the ground tissue, and the vascular tissues. In the following paragraphs, sclereids are described by examples from different parts of the plant body, excluding those sclereids that occur in the vascular tissues.

Sclereids in Stems A continuous cylinder of sclereids occurs on the periphery of the vascular region in the stem of *Hoya carnosa* and groups of sclereids in the pith of stems of *Hoya* and *Podocarpus*. These sclereids have moderately thick walls and numerous pits (Fig. 8.10C, D). In shape and size they resemble the adjacent parenchyma cells. This resemblance is often taken as an indication that such sclereids are by origin sclerified parenchyma cells. Their sclerification, however, has advanced so far that they may be grouped with the sclereids rather than parenchyma cells. This simple type of sclereid exemplifies a stone cell, or brachysclereid. A much branched astrosclereid is found in the cortex of *Trochodendron* stem (Fig. 8.10L). Somewhat less profusely branched sclereids occur in the cortex of the Douglas fir (*Pseudotsuga taxifolia*).

Sclereids in Leaves Leaves are an especially rich source of sclereids with regard to variety of form, although they are rare in the leaves of monocots (Rao, T. A., and Das, 1979). In the mesophyll, two main distributional patterns of sclereids are recognized: the **terminal**, with sclereids confined to the ends of the small veins (Fig. 8.11; *Arthrocnemum, Boronia, Hakea, Mouriria*), and the **diffuse**, with solitary sclereids or groups of sclereids dispersed throughout the tissue without any spatial relationship to the vein endings (*Olea, Osmanthus, Pseudotsuga, Trochodendron*) (Foster, 1956; Rao, T. A., 1991). In some protective foliar structures, like the clove scales of garlic (*Allium sativum*), the sclereids form part of the entire epidermis (Fig. 8.12).

Sclereids with definite branches or only with spicules (short, conical, or irregular projections) occur in the ground tissue of *Camellia* petiole (Fig. 8.10E, F) and in the mesophyll of *Trochodendron* leaf. The mesophyll of *Osmanthus* and *Hakea* contains columnar sclereids, ramified at each end, that is, osteosclereids (Fig. 8.10G). In the leaves of *Hakea suaveolens*, the terminal sclereids apparently play dual roles of support and water conduction. When a detached shoot was allowed to absorb through the cut end a solution of the fluorochrome berberine sulfate, the pattern of fluorescence observed in the leaves indicated that the berberine solution had moved from the enlarged tracheids (tracheoids) of the vein endings to the walls of the upper epidermal cells via the weakly lignified walls of the sclereids (Heide-Jørgensen, 1990). From the epidermis the solution moved downward into the walls of the palisade paren-

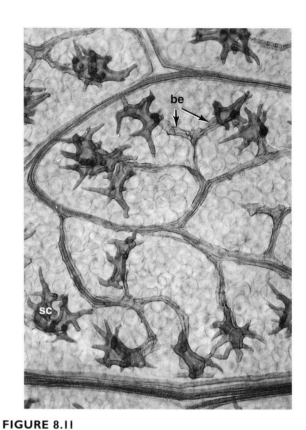

**FIGURE 8.11**

Cleared *Boronia* leaf. Sclereids (sc) at bundle ends (be). (×93. From Foster, 1955.)

chyma. Apparently the sclereids serve as vein extensions that conduct water to the epidermis and provide a rapid supply of water to the palisade cells. *Monstera deliciosa*, *Nymphaea* (water lily), and *Nuphar* (yellow pond lily) have typical trichosclereids with branches extending into large intercellular spaces, or air chambers, characteristic of the leaves of these species. Small prismatic sclereids are embedded within the *Nymphaea* sclereid walls (Fig. 8.9; Kuo-Huang, 1992). Branched sclereids may be found in leaves of conifers such as *Pseudotsuga taxifolia*.

The filiform sclereids of the olive (*Olea europaea*) leaf originate in both palisade and spongy parenchyma, average one millimeter in length, and permeate the mesophyll in the form of a dense network or mat (Fig. 8.13). Part of the network consists of T-shaped sclereids, the basal parts of which extend from the upper epidermis and palisade parenchyma into the underlying spongy parenchyma. The rest of the network consists of branched "polymorphic" sclereids that traverse the mesophyll layers, in what has been described as a chaotic pattern (Karabourniotis et al., 1994). It has been demonstrated that the T-shaped sclereids are capable of transmitting light from the upper epidermis to the spongy parenchyma, indicating that they may act

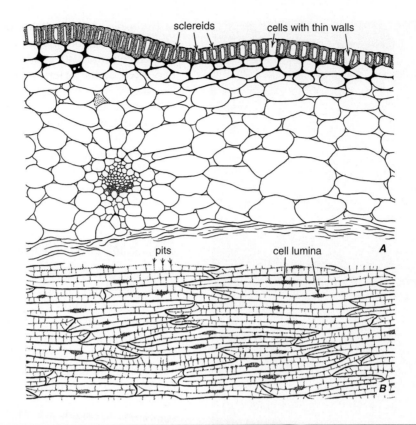

**FIGURE 8.12**

Epidermal sclereids in a protective bulb scale of *Allium sativum* (garlic). **A**, section of scale, with sclereid walls stippled. **B**, surface view of scale showing the solid layer of epidermal sclereids overlapping each other. (Both, ×99. From Esau, 1977; after Mann, 1952. *Hilgardia* 21 (8), 195–251. © 1952 Regents, University of California.)

like synthetic optical fibers and may help to improve the light microenvironment within the mesophyll of this thick and compact sclerophyllous leaf (Karabourniotis et al., 1994). The osteosclereids in the leaves of the evergreen sclerophyll *Phillyrea latifolia* apparently play a similar optical role of guiding light within the mesophyll (Karabourniotis, 1998).

Sclereids in Fruits Sclereids occur in various locations in fruits. In pear (*Pyrus*) and quince (*Cydonia*), single or clustered stone cells, or brachysclereids, are scattered in the fleshy parts of the fruit (Figs. 8.8 and 8.10A, B). The clusters of sclereids give pears their characteristic gritty texture. During formation of the clusters, cell divisions occur concentrically around some of the sclereids formed earlier (Staritsky, 1970). The radiating pattern of parenchyma cells around the mature cluster of sclereids is related to this mode of development. The sclereids of pear and quince often show ramiform pits resulting from a fusion of one or more cavities during the increase in thickness of the wall.

The apple (*Malus*) furnishes another example of sclereids in fruits. The cartilaginous endocarp enclosing

the seeds consists of obliquely oriented layers of elongated sclereids (Fig. 8.10J, K). Sclereids also compose the hard shells of nutlike fruits and the stony endocarp of stone fruits (drupes). In the drupe of *Ozoroa paniculosa* (Anacardiaceae), the resin tree, which is widely distributed in the savanna regions of southern Africa, the endocarp consists of consecutive layers of macrosclereids, osteosclereids, brachysclereids, and crystalliferous sclereids (Von Teichman and Van Wyk, 1993).

Sclereids in Seeds The hardening of seed coats during ripening of the seeds often results from a development of secondary walls in the epidermis and in the layer or layers beneath the epidermis. The leguminous seeds furnish a good example of such sclerification. In seeds of bean (*Phaseolus*), pea (*Pisum*), and soybean (*Glycine*), columnar macrosclereids comprise the epidermis and prismatic sclereids or bone-shaped osteosclereids occur beneath the epidermis (Fig. 8.14). During development of the pea seed coat the protodermal cells, from which the macrosclereids are derived, undergo extensive anticlinal division followed by cell elongation and then secondary wall formation (Harris, 1983). The precursors

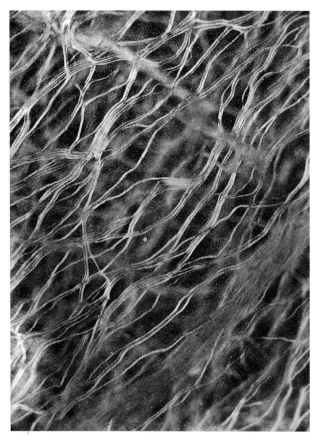

**FIGURE 8.13**

Filiform sclereids of *Olea* (olive), doubly refractive in polarized light, as seen in cleared leaf. (×57.)

of the osteosclereids divide both anticlinally and periclinally but do not begin to differentiate into bone-shaped cells until after thick secondary walls have been deposited by the macrosclereids (Harris, W. M., 1984). Secondary wall formation occurs first in a median portion of the developing osteosclereid, preventing further expansion there, while the thin primary walls at the ends of the cell continue to expand. Apparently neither the macrosclereids nor the osteosclereids of the pea seed coat are lignified. Pitting in their walls is inconspicuous. The seed coat of the coconut (*Cocos nucifera*) contains sclereids with numerous ramiform pits.

# ORIGIN AND DEVELOPMENT OF FIBERS AND SCLEREIDS

As indicated by their wide distribution in the plant body, fibers arise from various meristems: those of the xylem and phloem from the procambium and vascular cambium; most extraxylary fibers other than phloem fibers from the ground meristem; and the fibers of some Poaceae and Cyperaceae from the protoderm. Sclereids also arise from different meristems: those of the vascular tissues from derivatives of procambial and cambial cells; stone cells embedded in cork tissue from the cork cambium, or phellogen; the macrosclereids of seed coats from the protoderm; and many other sclereids from the ground meristem.

The development of the usually long fibers and of branched and long sclereids involves remarkable intercellular adjustments. Of particular interest is the

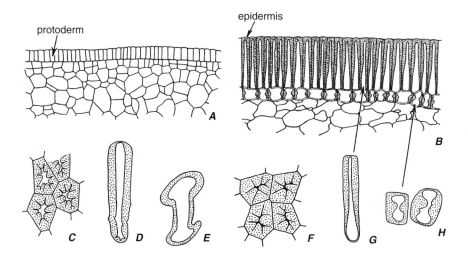

**FIGURE 8.14**

Sclereids of leguminous seed coats. **A**, **B**, outer parts of *Phaseolus* seed coat from transections of seeds in two stages of development. **B**, epidermis, a solid layer of macrosclereids. Subepidermal sclereids have most of the wall thickenings localized on anticlinal walls. **C–E**, sclereids of *Pisum*; **F–H**, of *Phaseolus*. **C**, **F**, groups of epidermal sclereids seen from the surface. **D**, **G**, epidermal sclereids; **E**, **H**, subepidermal sclereids. (A, B, ×240; C, F, ×595; D, E, G, H, ×300.)

attainment of great length by fibers of the primary plant body. Primary extraxylary fibers are initiated before the organ has elongated, and they can reach considerable length by elongating in unison with the other tissues in the growing organ. During this period of growth the walls of adjacent cells are so adjusted that no separation of the walls occurs. This method of growth is called **coordinated growth** (Chapter 5). The young fiber primordium increases in length without changing cellular contacts whether or not adjacent parenchymatous cells are dividing. The growth of primary extraxylary fibers in unison with the other tissues in the growing organ results in longer fibers commonly being found in longer organs (Aloni and Gad, 1982).

The great length attained by some primary extraxylary fibers is not the result of elongation by coordinated growth only. Somewhat later, the fiber primordium attains additional length by **intrusive growth** (Chapter 5). During intrusive growth the elongating cells grow at their apices (*apical intrusive growth*), usually at both

ends between the walls of other cells. During elongation, the fiber may become multinucleate as a result of repeated nuclear divisions not followed by formation of new walls. This is especially true of primary phloem fibers. While the fiber is still alive, its cytoplasm exhibits rotational streaming, a phenomenon apparently related to intercellular transport of materials (Worley, 1968).

Apical intrusive growth has been studied in detail in flax fibers (Schoch-Bodmer and Huber, 1951). By measuring young and old internodes and the fibers contained in these internodes, the authors calculated that by coordinated growth alone the fibers could become 1 to 1.8 cm long. Actually they found fibers ranging in length between 0.8 and 7.5 cm. Thus lengths over 1.8 cm must have been attained by apical intrusive growth. The growing tips of young fibers dissected out of living stems showed thin walls, contained dense cytoplasm (Fig. 8.15A–C) with chloroplasts, and were not plasmolyzable. When the tips ceased to grow, they became filled with secondary wall material (Fig. 8.15D–F).

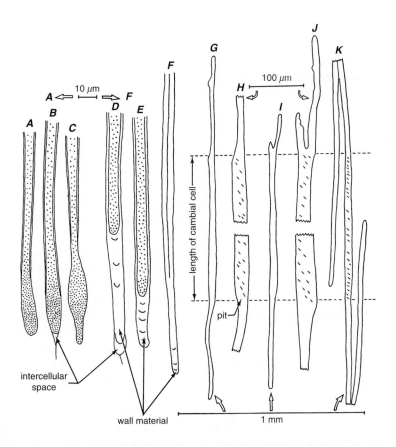

**FIGURE 8.15**

Apical intrusive growth in fibers from stems. **A–F**, from phloem of flax (*Linum perenne*), **G–J**, from xylem and, **K**, from phloem of *Sparmannia* (Tiliaceae). **H, J**, enlarged views of parts of **G, I**, respectively. **A–C**, the intrusively growing tips of fibers (below) have thin walls and dense cytoplasm. **D–F**, the tips of fibers have become filled with wall material after completion of growth. **G–K**, fibers have extended in both directions from the original position in the cambium (between broken lines). Pits occur only in the original cambial parts. The phloem fiber (**K**) is considerably longer than the xylem fibers (**G, I**). (From Esau, 1977; **A–F**, adapted from Schoch-Bodmer and Huber, 1951; **G–K**, adapted from Schoch-Bodmer, 1960.)

In contrast to primary fibers, which undergo both coordinated growth and intrusive growth, secondary fibers originate in the part of the organ that has ceased to elongate, and they can increase in length only by intrusive growth (Wenham and Cusick, 1975). The length of secondary phloem fibers and of secondary xylem fibers depends on the length of the cambial initials and on the amount of intrusive growth in the fiber primordia derived from those initials. If primary and secondary phloem fibers are present, the former are considerably longer. In *Cannabis* (hemp), for example, the length of the primary phloem fibers averaged about 13 mm, the secondary about 2 mm (Kundu, 1942).

Intrusive growth may be identified in transverse sections of stems and roots by the appearance of small cells—transections of growing tips—among the wider, not elongating parts of fiber primordia. The secondary vascular tissues of *Sparmannia* (Tiliaceae) offer a graphic illustration of this phenomenon (Fig. 8.16; Schoch-Bodmer and Huber, 1946). The orderly radial alignment of cells seen in the cambium is replaced by a mosaic pattern in the axial system of phloem. In a given transection, three to five growing fiber tips are added to each wider median portion of a fiber primordium (indicated by diagonal hatching in Fig. 8.16A) by intrusive elongation. The radial alignment in the axial system of xylem is less strongly affected because xylary fibers elongate less than do phloem fibers (Fig. 8.15G–K). As seen in radial longitudinal sections, the bipolar apical growth of fibers makes these cells extend above and below the horizontal levels of cambial cells among which they are initiated (Fig. 8.16B).

When during intrusive growth a fiber tip is obstructed by other cells, the tip curves or forks (Fig. 8.15I, J). Thus bent and forked ends in fibers (and sclereids) are additional evidence of intrusive growth. The intrusively growing parts usually fail to develop pits in their secondary walls and thus serve as a measure of the amount of apical elongation (Fig. 8.15G–K; Schoch-Bodmer, 1960).

Prolonged apical intrusive growth of fibers and some sclereids makes the secondary thickening of the walls in these cells a rather complex phenomenon. As mentioned previously, the secondary wall commonly develops over the primary after the latter ceases to expand (Chapter 4). In intrusively growing fibers and sclereids, the older part of the cell stops growing, whereas the apices continue to elongate. The older part of the cell (typically the median part) begins to form secondary wall layers before the growth of the tips is completed. From the median part of the cell, the secondary thickening progresses toward the tips and is completed after the tips cease to grow.

In rapidly growing stems of ramie (*Boehmeria nivea*), the longer primary phloem fibers (40–55 cm) extend, during their later stages of enlargement, through internodes that have ceased elongating (Aldaba, 1927). The increase in length of these fibers (initially about 20 μm long) is of the order of 2,500,000%, a gradual process that apparently requires months to complete. Secondary wall formation begins in the basal portions of the cells and continues upward toward the elongating tips in a series of concentric layers. When a fiber has completed its elongation, the inner tubular wall layers continue to grow upward, reaching the tip of the cell at successive intervals.

Sclereids arise either directly from cells that are early individualized as sclereids or through a belated sclerosis of apparently ordinary parenchyma cells. The primordia, or initials, of the terminal sclereids in the lamina of the *Mouriria huberi* leaf are clearly evident before the intercellular spaces appear in the mesophyll and while the small veins are entirely procambial (Foster, 1947). They arise from the same layer of cells as the procambial strands. The trichosclereids of the air roots of *Monstera* develop from cells early set aside by unequal, polarized divisions in files of cortical cells (Bloch, 1946). In contrast, the sclereids of the *Osmanthus* leaf are first evident in leaf blades 5 to 6 cm long, by which stage the blade is nearly one-half its full length (Fig. 8.17; Griffith, 1968). At this age a large part of the xylem and phloem of the major veins has matured, and the fibers associated with the veins are distinguishable but without conspicuous thickening. Sclerification of parenchyma cells in the secondary phloem commonly occurs in the nonconducting phloem, the part of the phloem no longer involved with long-distance transport (Chapter 14; Esau, 1969; Nanko, 1979). In the oaks (*Quercus*), for example, stone cells differentiate in several-years-old phloem, first in the rays and later in dilatation tissue (tissue involved with the increase in circumference of the bark) in clusters of variable size. In the nonconducting phloem of some woody angiosperms, fiber-sclereids develop from fusiform parenchyma cells or individual elements of parenchyma strands. The fiber-sclereids in the secondary phloem of *Pyrus communis* (Evert, 1961) and *Pyrus malus* (*Malus domestica*) (Evert, 1963) originate from parenchyma strands the second season after they are derived from the vascular cambium. At that time the individual elements of the strands undergo intensive intrusive growth and then form secondary walls. In the nonconducting secondary phloem of *Pereskia* (Cactaceae), some sclereids with multilayered secondary walls become subdivided by septa into compartments, each of which differentiates into a sclereid with a multilayered secondary wall (Fig. 8.6B; Bailey, 1961). Such sclereids are reminiscent of the septate fibers in bamboos (Parameswaran and Liese, 1977).

Sclereid primordia may not differ in appearance from neighboring parenchyma cells. Generally, primordia of idioblastic sclereids are distinguishable from neighboring cells by their large, conspicuous nuclei and often

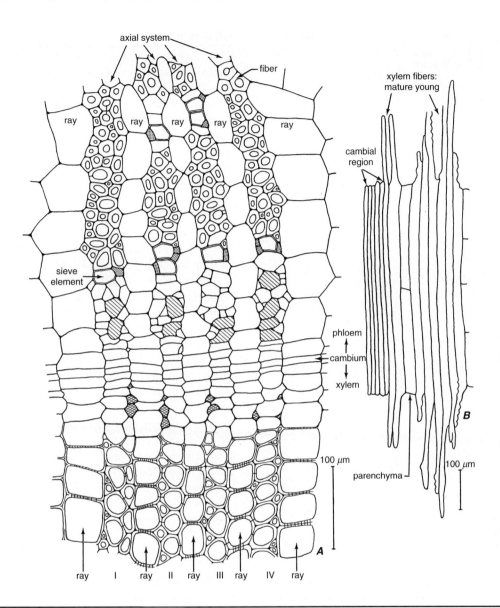

**FIGURE 8.16**

Development of fibers in secondary phloem and xylem of *Sparmannia* (Tiliaceae) as seen in transverse (**A**) and radial longitudinal (**B**) sections of stem. In **A**, I–IV are the files of cells in the axial (longitudinal) system. The files alternate with rays. Phloem and xylem are immature next to the cambium. Mature xylem has secondary walls. In mature phloem, dotted companion cells serve to identify the sieve elements; secondary walls mark the fibers. Cells with diagonal lines are median parts of young fiber cells. They are accompanied by small cells most of which are tips of intrusively growing fibers. The cross-hatched cells on the xylem side are intrusively growing tips of xylem fibers. **B**, xylem fibers extend beyond the cambial region in both directions. (From Esau, 1977; adapted from Schoch-Bodmer and Huber, 1946.)

dense cytoplasm (Boyd et al., 1982; Heide-Jørgensen, 1990).

## ■ FACTORS CONTROLLING DEVELOPMENT OF FIBERS AND SCLEREIDS

The factors controlling the development of fibers and sclereids have been the object of numerous experimen-

tal studies. Studies by Sachs (1972) and Aloni (1976, 1978) revealed that fiber development in strands is dependent on stimuli originating in young leaf primordia. Early removal of the primordia in *Pisum sativum* prevented fiber differentiation; changing the position of the leaves experimentally changed the position of the fiber strands as well (Sachs, 1972). The results of the *Pisum* study were confirmed in *Coleus*, in which it was also shown that primary phloem fiber induction is a

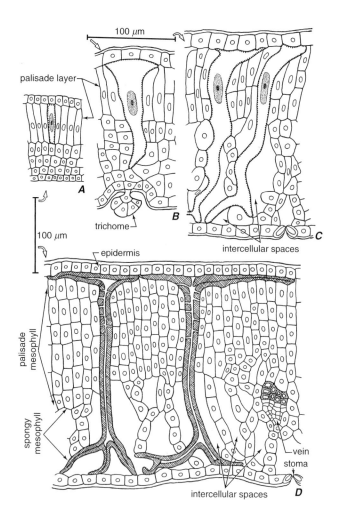

palisade layer

100 μm

100 μm

trichome

epidermis

intercellular spaces

palisade mesophyll

spongy mesophyll

vein

stoma

intercellular spaces

A

B

C

D

**FIGURE 8.17**

Development of sclereids in the leaf of *Osmanthus fragrans* (Oleaceae). **A–C**, differentiating sclereids, indicated by large nuclei and dots along the walls; **D**, mature sclereids, indicated by cross-hatched secondary walls. In all drawings, the mesophyll and epidermal cells are marked with circles or ovals. The narrow intercellular spaces characteristic of palisade parenchyma have been omitted. **A**, future sclereid is indicated symbolically; it was not yet differentiated from other palisade cells (drawing from primordium 23 mm long). **B**, young sclereid has extended beyond the palisade layer (blade approximately 5.5 cm long). **C**, two young sclereids have reached the lower epidermis by growing through the spongy mesophyll (blade 10–12 cm long). Enlargement of the sclereids involves both coordinated and apical intrusive growth. The thickness of the blade doubles after initiation of the sclereids; thus part of the growth of the sclereid occurs in unison with the palisade parenchyma. Growth of the branches and the portion of the wall in contact with the spongy mesophyll, however, involves apical intrusive growth. Deposition of the secondary wall in these sclereids is uniform and rapid, and does not occur until the leaf has reached full size. **D**, mature sclereids have some branches extended parallel with the epidermis and others projected into intercellular spaces. Pits in the secondary wall are located in the parts of sclereids that do not sever connections with adjacent cells during growth. (From Esau, 1977; adapted from Griffith, 1968.)

strictly polar one, in a downward direction from the leaves to the roots (Aloni, 1976, 1978). Moreover it was shown that the effect of the leaves on the differentiation of the primary phloem fibers in *Coleus* can be replaced by the exogenous application of auxin (IAA) in combination with gibberellin (GA₃) (Aloni, 1979). IAA alone induced the differentiation of only a few fibers; GA₃ alone had no effect on fiber differentiation. When various combinations of both hormones were applied, high concentrations of IAA stimulated rapid differentiation of thick-walled fibers, while high GA₃ levels resulted in long, thin-walled fibers. Both hormones also are required for the development of fibers in the secondary xylem of *Populus* (Digby and Wareing, 1966). Cytokinin originating in the roots also appears to play a regulatory role in the development of secondary xylem fibers (Aloni, 1982; Saks et al., 1984).

Several *Arabidopsis* mutants have been discovered that affect the development of the fibers in the interfascicular regions of the inflorescence stems (Turner and Somerville, 1997; Zhong et al., 1997; Turner and Hall, 2000; Burk et al., 2001). Of particular interest is the

*interfascicular fiberless1 (ifl1)* mutant, in which interfascicular (extraxylary) fibers fail to develop (Zhong et al., 1997), indicating that the *INTERFASCICULAR FIBERLESS1 (IFL1)* gene, which was found to be the same gene as *REVOLUTA (REV)* (Ratcliffe et al., 2000), is essential for normal differentiation of the interfascicular fibers. It is also required for normal development of the secondary xylem. The *IFL1/REV* gene is expressed in the interfascicular region in which the fibers differentiate as well as in the vascular regions (Zhong and Ye, 1999). An assay of auxin polar transport revealed that the flow of auxin along the inflorescence stems is drastically reduced in the *ifl1* mutants. Moreover an auxin transport inhibitor altered the normal differentiation of interfascicular fibers in inflorescence stems of wild-type plants (Zhong and Ye, 2001). The apparent correlation between the reduced auxin polar flow and the alteration in fiber differentiation in the *ifl1* mutants suggests that the *IFL1/REV* gene may be involved in controlling the flow of auxin along the interfascicular regions. Results of a separate experimental study (Little et al., 2002) in which the auxin supply was altered clearly indicate the need for IAA for wall thickening and lignification in the interfascicular fibers in the *Arabidopsis* inflorescence stem.

Cutting into leaves (*Camellia japonica*, Foard, 1959; *Magnolia thamnodes*, *Talauma villosa*, Tucker, 1975), which normally have marginal sclereids, induced differentiation of sclereids along the "new" margins. When sclerenchyma cylinders of eudicotyledonous stems were interrupted by cutting away on one side of an internode, the continuity of the cylinder was restored by the differentiation of sclereids within the wound callus (Warren Wilson et al., 1983). The results of these experiments were interpreted as evidence of the positional control of sclereid development. In the leaves, cells that normally would have become mesophyll cells specialized for photosynthesis were induced to develop into sclereids when brought into the proximity of a margin. In the stems, the arrangement of the regenerated sclereids tended to reflect the original sclerenchyma (mainly or largely fibers) cylinder in the unwounded stem. Investigations of hormonal factors indicated that auxin levels in the leaf influence sclereid development (Al-Talib and Torrey, 1961; Rao, A. N., and Singarayar, 1968). When auxin concentration was high, the development was suppressed, whereas at low concentrations of auxin the cell walls remained thin and did not become lignified. Interestingly differentiation of sclereids was induced in the pith of *Arabidopsis thaliana* by the removal of the developing inflorescences (Lev-Yadun, 1997). The pith of mature control plants had no sclereids.

# REFERENCES

ALDABA, V. C. 1927. The structure and development of the cell wall in plants. I. Bast fibers of *Boehmeria* and *Linum*. *Am. J. Bot.* 14, 16–24.

ALONI, R. 1976. Polarity of induction and pattern of primary phloem fiber differentiation in *Coleus*. *Am. J. Bot.* 63, 877–889.

ALONI, R. 1978. Source of induction and sites of primary phloem fibre differentiation in *Coleus blumei*. *Ann. Bot.* n.s. 42, 1261–1269.

ALONI, R. 1979. Role of auxin and gibberellin in differentiation of primary phloem fibers. *Plant Physiol.* 63, 609–614.

ALONI, R. 1982. Role of cytokinin in differentiation of secondary xylem fibers. *Plant Physiol.* 70, 1631–1633.

ALONI, R., and A. E. GAD. 1982. Anatomy of the primary phloem fiber system in *Pisum sativum*. *Am. J. Bot.* 69, 979–984.

AL-TALIB, K. H., and J. G. TORREY. 1961. Sclereid distribution in the leaves of *Pseudotsuga* under natural and experimental conditions. *Am. J. Bot.* 48, 71–79.

BAAS, P. 1986. Terminology of imperforate tracheary elements—In defense of libriform fibres with minutely bordered pits. *IAWA Bull.* n.s. 7, 82–86.

BAILEY, I. W. 1961. Comparative anatomy of the leaf-bearing Cactaceae. II. Structure and distribution of sclerenchyma in the phloem of *Pereskia*, *Pereskiopsis* and *Quiabentia*. *J. Arnold Arbor.* 42, 144–150.

BLOCH, R. 1946. Differentiation and pattern in *Monstera deliciosa*. The idioblastic development of the trichoscereids in the air root. *Am. J. Bot.* 33, 544–551.

BLYTH, A. 1958. Origin of primary extraxylary stem fibers in dicotyledons. *Univ. Calif. Publ. Bot.* 30, 145–232.

BOYD, D. W., W. M. HARRIS, and L. E. MURRY. 1982. Sclereid development in *Camellia* petioles. *Am. J. Bot.* 69, 339–347.

BURK, D. H., B. LIU, R. ZHONG, W. H. MORRISON, and Z.-H. YE. 2001. A katanin-like protein regulates normal cell wall biosynthesis and cell elongation. *Plant Cell* 13, 807–827.

BUTTERFIELD, B. G., and B. A. MEYLAN. 1976. The occurrence of septate fibres in some New Zealand woods. *N. Z. J. Bot.* 14, 123–130.

CARPENTER, C. H. 1963. *Papermaking fibers: A photomicrographic atlas of woody, non-woody, and man-made fibers used in papermaking.* Tech. Publ. 74. State University College of Forestry at Syracuse University, Syracuse, NY.

CHALK, L. 1983. Fibres. In: *Anatomy of the Dicotyledons*, 2nd ed., vol. II, *Wood Structure and Conclusion of the General Introduction*, pp. 28–38, C. R. Metcalfe and L. Chalk. Clarendon Press, Oxford.

DE BARY, A. 1884. Comparative anatomy of the vegetative organs of the phanerogams and ferns. Clarendon Press, Oxford.

DIGBY, J., and P. F. WAREING. 1966. The effect of applied growth hormones on cambial division and the differentiation of the cambial derivatives. *Ann. Bot.* n.s. 30, 539–548.

DUMBROFF, E. B., and H. W. ELMORE. 1977. Living fibres are a principal feature of the xylem in seedlings of *Acer saccharum* Marsh. *Ann. Bot.* n.s. 41, 471–472.

ESAU, K. 1969. *The Phloem. Handbuch der Pflanzenanatomie*, Band 5, Teil 2, *Histologie*. Gebrüder Borntraeger, Berlin, Stuttgart.

ESAU, K. 1977. *Anatomy of Seed Plants*, 2nd ed. Wiley, New York.

ESAU, K. 1979. Phloem. In: *Anatomy of the Dicotyledons*, 2nd ed., vol. I, *Systematic Anatomy of Leaf and Stem, with a Brief History of the Subject*, pp. 181–189, C. R. Metcalfe and L. Chalk. Clarendon Press, Oxford.

EVERT, R. F. 1961. Some aspects of cambial development in *Pyrus communis*. *Am. J. Bot.* 48, 479–488.

EVERT, R. F. 1963. Ontogeny and structure of the secondary phloem in *Pyrus malus*. *Am. J. Bot.* 50, 8–37.

FAHN, A., and B. LESHEM. 1963. Wood fibres with living protoplasts. *New Phytol.* 62, 91–98.

FISHER, J. B., and J. W. STEVENSON. 1981. Occurrence of reaction wood in branches of dicotyledons and its role in tree architecture. *Bot. Gaz.* 142, 82–95.

FOARD, D. E. 1959. Pattern and control of sclereid formation in the leaf of *Camellia japonica*. *Nature* 184, 1663–1664.

FOSTER, A. S. 1944. Structure and development of sclereids in the petiole of *Camellia japonica* L. *Bull. Torrey Bot. Club* 71, 302–326.

FOSTER, A. S. 1947. Structure and ontogeny of the terminal sclereids in the leaf of *Mouriria Huberi* Cogn. *Am. J. Bot.* 34, 501–514.

FOSTER, A. S. 1955. Structure and ontogeny of terminal sclereids in *Boronia serrulata*. *Am. J. Bot.* 42, 551–560.

FOSTER, A. S. 1956. Plant idioblasts: Remarkable examples of cell specialization. *Protoplasma* 46, 184–193.

GRIFFITH, M. M. 1968. Development of sclereids in *Osmanthus fragrans* Lour. *Phytomorphology* 18, 75–79.

GRITSCH, C. S., and R. J. MURPHY. 2005. Ultrastructure of fibre and parenchyma cell walls during early stages of culm development in *Dendrocalamus asper*. *Ann. Bot.* 95, 619–629.

HABERLANDT, G. 1914. *Physiological Plant Anatomy*. Macmillan, London.

HARRIS, M., ed. 1954. *Handbook of Textile Fibers*. Harris Research Laboratories, Washington, DC.

HARRIS, W. M. 1983. On the development of macrosclereids in seed coats of *Pisum sativum* L. *Am. J. Bot.* 70, 1528–1535.

HARRIS, W. M. 1984. On the development of osteosclereids in seed coats of *Pisum sativum* L. *New Phytol.* 98, 135–141.

HEIDE-JØRGENSEN, H. S. 1990. Xeromorphic leaves of *Hakea suaveolens* R. Br. IV. Ontogeny, structure and function of the sclereids. *Aust. J. Bot.* 38, 25–43.

HOLDHEIDE, W. 1951. Anatomie mitteleuropäischer Gehölzrinden (mit mikrophotographischem Atlas). In: *Handbuch der Mikroskopie in der Technik*, Band 5, Heft 1, pp. 193–367. Umschau Verlag, Frankfurt am Main.

IAWA Committee on Nomenclature. 1964. International glossary of terms used in wood anatomy. *Trop. Woods* 107, 1–36.

KARABOURNIOTIS, G. 1998. Light-guiding function of foliar sclereids in the evergreen sclerophyll *Phillyrea latifolia*: A quantitative approach. *J. Exp. Bot.* 49, 739–746.

KARABOURNIOTIS, G., N. PAPASTERGIOU, E. KABANOPOULOU, and C. FASSEAS. 1994. Foliar sclereids of *Olea europaea* may function as optical fibres. *Can. J. Bot.* 72, 330–336.

KUNDU, B. C. 1942. The anatomy of two Indian fibre plants, *Cannabis* and *Corchorus* with special reference to the fibre distribution and development. *J. Indian Bot. Soc.* 21, 93–128.

KUO-HUANG, L.-L. 1990. Calcium oxalate crystals in the leaves of *Nelumbo nucifera* and *Nymphaea tetragona*. *Taiwania* 35, 178–190.

KUO-HUANG, L.-L. 1992. Ultrastructural study on the development of crystal-forming sclereids in *Nymphaea tetragona*. *Taiwania* 37, 104–114.

LEV-YADUN, S. 1997. Fibres and fibre-sclereids in wild-type *Arabidopsis thaliana*. *Ann. Bot.* 80, 125–129.

LITTLE, C. H. A., J. E. MACDONALD, and O. OLSSON. 2002. Involvement of indole-3-acetic acid in fascicular and interfascicular cambial growth and interfascicular extraxylary fiber differentiation in *Arabidopsis thaliana* inflorescence stems. *Int. J. Plant Sci.* 163, 519–529.

MAGEE, J. A. 1948. Histological structure of the stem of *Zea mays* in relation to stiffness of stalk. *Iowa State Coll. J. Sci.* 22, 257–268.

MANN, L. K. 1952. Anatomy of the garlic bulb and factors affecting bud development. *Hilgardia* 21, 195–251.

MURDY, W. H. 1960. The strengthening system in the stem of maize. *Ann. Mo. Bot. Gard.* 67, 205–226.

MURPHY, R. J., and K. L. ALVIN. 1992. Variation in fibre wall structure in bamboo. *IAWA Bull.* n.s. 13, 403–410.

MURPHY, R. J., and K. L. ALVIN. 1997. Fibre maturation in the bamboo *Gigantochloa scortechinii*. *IAWA J.* 18, 147–156.

NANKO, H. 1979. Studies on the development and cell wall structure of sclerenchymatous elements in the secondary phloem of woody dicotyledons and conifers. Ph. D. Thesis. Department of Wood Science and Technology, Kyoto University, Kyoto, Japan.

NANKO, H., H. SAIKI, and H. HARADA. 1977. Development and structure of the phloem fiber in the secondary phloem of *Populus euramericana*. *Mokuzai Gakkaishi (J. Jpn. Wood Res. Soc.)* 23, 267–272.

NEEDLES, H. L. 1981. *Handbook of Textile Fibers, Dyes, and Finishes*. Garland STPM Press, New York.

OHTANI, J. 1987. Vestures in septate wood fibres. *IAWA Bull.* n.s. 8, 59–67.

PARAMESWARAN, N., and W. LIESE. 1969. On the formation and fine structure of septate wood fibres of *Ribes sanguineum*. *Wood Sci. Technol.* 3, 272–286.

PARAMESWARAN, N., and W. LIESE. 1977. Structure of septate fibres in bamboo. *Holzforschung* 31, 55–57.

PATEL, R. N. 1964. On the occurrence of gelatinous fibres with special reference to root wood. *J. Inst. Wood Sci.* 12, 67–80.

PURKAYASTHA, S. K. 1958. Growth and development of septate and crystalliferous fibres in some Indian trees. *Proc. Natl. Inst. Sci. India* 24B, 239–244.

RAO, A. N., and M. SINGARAYAR. 1968. Controlled differentiation of foliar sclereids in *Fagraea fragrans*. *Experientia* 24, 298–299.

RAO, T. A. 1991. *Compendium of Foliar Sclereids in Angiosperms: Morphology and Taxonomy*. Wiley Eastern Limited, New Delhi.

RAO, T. A., and S. DAS. 1979. Leaf sclereids—Occurrence and distribution in the angiosperms. *Bot. Not.* 132, 319–324.

RATCLIFFE, O. J., J. L. RIECHMANN, and J. Z. ZHANG. 2000. *INTERFASCICULAR FIBERLESS1* is the same gene as *REVOLUTA*. *Plant Cell* 12, 315–317.

ROLAND, J.-C., D. REIS, B. VIAN, B. SATIAT-JEUNEMAITRE, and M. MOSINIAK. 1987. Morphogenesis of plant cell walls at the supramolecular level: Internal geometry and versatility of helicoidal expression. *Protoplasma* 140, 75–91.

ROLAND, J.-C., D. REIS, B. VIAN, and S. ROY. 1989. The helicoidal plant cell wall as a performing cellulose-based composite. *Biol. Cell* 67, 209–220.

SACHS, T. 1972. The induction of fibre differentiation in peas. *Ann. Bot.* n.s. 36, 189–197.

SAKS, Y., P. FEIGENBAUM, and R. ALONI. 1984. Regulatory effect of cytokinin on secondary xylem fiber formation in an *in vivo* system. *Plant Physiol.* 76, 638–642.

SCHOCH-BODMER, H. 1960. Spitzenwachstum und Tüpfelverteilung bei sekundären Fasern von *Sparmannia*. *Beih. Z. Schweiz. Forstver.* 30, 107–113.

SCHOCH-BODMER, H., and P. HUBER. 1946. Wachstumstypen plastischer Pflanzenmembranen. *Mitt. Naturforsch. Ges. Schaffhausen* 21, 29–43.

SCHOCH-BODMER, H., and P. HUBER. 1951. Das Spitzenwachstum der Bastfasern bei *Linum usitatissimum* und *Linum perenne*. *Ber. Schweiz. Bot. Ges.* 61, 377–404.

SCHWENDENER, S. 1874. *Das mechanische Princip in anatomischen Bau der Monocotylen mit vergleichenden Ausblicken auf die übrigen Pflanzenklassen*. Wilhelm Engelmann, Leipzig.

SPERRY, J. S. 1982. Observations of reaction fibers in leaves of dicotyledons. *J. Arnold Arbor.* 63, 173–185.

STAFF, I. A. 1974. The occurrence of reaction fibres in *Xanthorrhoea australis* R. Br. *Protoplasma* 82, 61–75.

STARITSKY, G. 1970. The morphogenesis of the inflorescence, flower and fruit of *Pyrus nivalis* Jacquin var. *orientalis* Terpó. *Meded. Landbouwhogesch. Wageningen* 70, 1–91.

TOBLER, F. 1957. *Die mechanischen Elemente und das mechanische System*. Handbuch der Pflanzenanatomie, 2nd ed., Band 4, Teil 6, *Histologie*. Gebrüder Borntraeger, Berlin-Nikolassee.

TOMLINSON, P. B. 1961. *Anatomy of the Monocotyledons. 2. Palmae*. Clarendon Press, Oxford.

TUCKER, S. C. 1975. Wound regeneration in the lamina of magnoliaceous leaves. *Can. J. Bot.* 53, 1352–1364.

TURNER, S. R., and M. HALL. 2000. The *gapped xylem* mutant identifies a common regulatory step in secondary cell wall deposition. *Plant J.* 24, 477–488.

TURNER, S. R., and C. R. SOMERVILLE. 1997. Collapsed xylem phenotype of *Arabidopsis* identifies mutants deficient in cellulose deposition in the secondary cell wall. *Plant Cell* 9, 689–701.

VON TEICHMAN, I., and A. E. VAN WYK. 1993. Ontogeny and structure of the drupe of *Ozoroa paniculosa* (Anacardiaceae). *Bot. J. Linn. Soc.* 111, 253–263.

WARDROP, A. B. 1964. The reaction anatomy of arborescent angiosperms. In: *The Formation of Wood in Forest Trees*, pp. 405–456, M. H. Zimmermann, ed. Academic Press, New York.

WARREN WILSON, J., S. J. DIRCKS, and R. I. GRANGE. 1983. Regeneration of sclerenchyma in wounded dicotyledon stems. *Ann. Bot.* n.s. 52, 295–303.

WENHAM, M. W., and F. CUSICK. The growth of secondary wood fibres. *New Phytol.* 74, 247–261.

WOLKINGER, F. 1971. Morphologie und systematische Verbreitung der lebenden Holzfasern bei Sträuchern und Bäumen. III. Systematische Verbreitung. *Holzforschung* 25, 29–30.

WORLEY, J. F. 1968. Rotational streaming in fiber cells and its role in translocation. *Plant Physiol.* 43, 1648–1655.

ZHONG, R., and Z.-H. YE. 1999. *IFL1*, a gene regulating interfascicular fiber differentiation in *Arabidopsis*, encodes a homeodomain-leucine zipper protein. *Plant Cell* 11, 2139–2152.

ZHONG, R., and Z.-H. YE. 2001. Alteration of polar transport in the *Arabidopsis ifl1* mutants. *Plant Physiol.* 126, 549–563.

ZHONG, R., J. J. TAYLOR, and Z.-H. YE. 1997. Disruption of interfascicular fiber differentiation in an *Arabidopsis* mutant. *Plant Cell* 9, 2159–2170.

# CHAPTER NINE

# Epidermis

The term **epidermis** designates the outermost layer of cells on the primary plant body. It is derived from the Greek *epi*, for upon, and *derma*, for skin. In this book the term epidermis refers to the outermost layer of cells on all parts of the primary plant body, including roots, stems, leaves, flowers, fruits, and seeds. An epidermis is considered to be absent, however, on the rootcap and not differentiated as such on the apical meristems.

The epidermis of the shoot arises from the outermost cell layer of the apical meristem. In roots the epidermis may have a common origin with cells of the rootcap or differentiate from the outermost cell layer of the cortex (Chapter 6; Clowes, 1994). The difference in origin of the epidermis in shoots and roots has convinced some investigators that the surface layer of the root should have its own name, *rhizodermis*, or *epiblem* (Linsbauer, 1930; Guttenberg, 1940). Despite the differences in origin, continuity exists between the epidermis of the root and that of the shoot. If the term epidermis and that of protoderm, for the undifferentiated epidermis, are used in a solely morphologic-topographic sense, and the problem of origin is ignored, both terms may be used broadly to refer to the primary surface tissue of the entire plant.

Organs having little or no secondary growth usually retain their epidermis as long as they exist. A notable exception is found in long-lived monocots that have no secondary addition to the vascular system but replace the epidermis with a special kind of periderm (Chapter 15). In woody roots and stems the epidermis varies in longevity, depending on the time of formation of the periderm. Ordinarily the periderm arises in the first year of growth of woody stems and roots, but numerous tree species produce no periderm until their axes are many times thicker than they were at the completion of primary growth. In such plants the epidermis, as well as the underlying cortex, continues to grow and thus keeps pace with the increasing circumference of the vascular cylinder. The individual cells enlarge tangentially and divide radially. An example of such prolonged growth is found in stems of the striped maple (*Acer pensylvanicum*; syn. *A. striatum*) in which trunks about 20 years old may attain a thickness of about 20 cm and still remain clothed with the original epidermis (de Bary, 1884). The cells of such an old epidermis are not

---

*Esau's Plant Anatomy, Third Edition*, By Ray F. Evert.
Copyright © 2006 John Wiley & Sons, Inc.

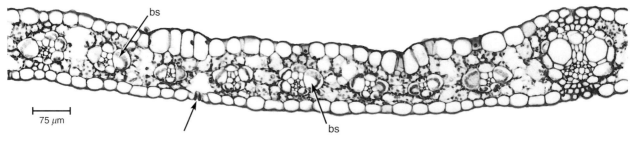

**FIGURE 9.1**

Transverse section of a maize (*Zea mays*) leaf showing a single-layered epidermis on both sides of the blade. A single stoma (arrow) can be seen here. The vascular bundles of various sizes are delimited from the mesophyll by prominent bundle sheaths (bs). (From Russell and Evert, 1985, Fig. 1. © 1985, Springer-Verlag.)

more than twice as wide tangentially as the epidermal cells in an axis 5 mm in thickness. This size relation clearly shows that the epidermal cells are dividing continuously while the stem increases in thickness. Another example is *Cercidium torreyanum*, a tree leafless most of the time but having a green bark and a persistent epidermis (Roth, 1963).

The epidermis is usually one layer of cells in thickness (Fig. 9.1). In some leaves the protodermal cells and their derivatives divide periclinally (parallel with the surface), resulting in a tissue consisting of several layers of ontogenetically related cells. (Sometimes only individual cells of the epidermis undergo periclinal divisions.) Such a tissue is referred to as a **multiple**, or **multiseriate**, **epidermis** (Figs. 9.2 and 9.3). The *velamen* (from the Latin word for cover) of the aerial and terrestrial roots of orchids is also an example of a multiple epidermis (Fig. 9.2). In leaves the outermost layer of a multiple epidermis resembles an ordinary uniseriate epidermis in having a cuticle; the inner layers commonly contain few or no chloroplasts. One of the functions ascribed to the inner layers is storage of water (Kaul, 1977). Representatives with a multiple epidermis may be found among the Moraceae (most species of *Ficus*), Pittosporaceae, Piperaceae (*Peperomia*), Begoniaceae, Malvaceae, Monocotyledoneae (palms, orchids), and others (Linsbauer, 1930). In some plants subepidermal layers resemble those of a multiple epidermis but are derived from the ground meristem. These layers are called **hypodermis** (from the Greek *hypo*, below, and *derma*, skin). A study of mature structures rarely permits the identification of the tissue either as multiple epidermis or as a combination of epidermis and a hypodermis. The origin of the subsurface layers can be properly revealed only by developmental studies.

The periclinal divisions initiating the multiple epidermis in leaves occur relatively late in development. In *Ficus*, for example, the leaf has a uniseriate epidermis until the stipules are shed. Then periclinal divisions

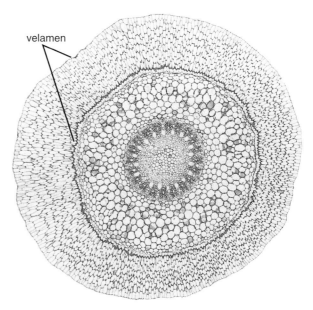

**FIGURE 9.2**

Transverse section of an orchid root showing the multiple epidermis, or velamen. (×25.)

occur in the epidermis (Fig. 9.3A). Similar divisions are repeated in the outer row of daughter cells, sometimes once, sometimes twice (Fig. 9.3B). During expansion of the leaf, anticlinal divisions also occur and, since these divisions are not synchronized in the different layers, the ontogenetic relationship between the layers becomes obscured (Fig. 9.3B, C). The inner layers expand more than the outer and the largest cells, called *lithocysts*, produce a calcified body, the **cystolith**, composed largely of calcium carbonate attached to a silicified stalk (Setoguchi et al., 1989; Taylor, M. G., et al., 1993). The stalk originates as a cylindrical ingrowth of the cell wall. The lithocysts do not divide but keep pace with the increasing depth of the epidermis and even overtake it

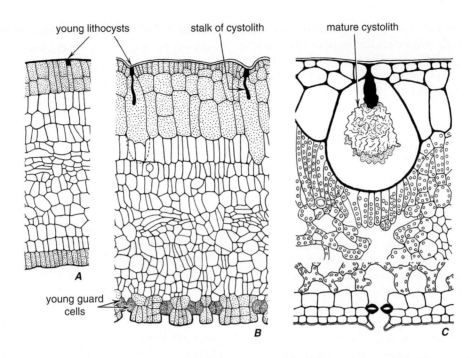

young lithocysts    stalk of cystolith    mature cystolith

*A*

young guard
cells

*B*

*C*

**FIGURE 9.3**

Multiple epidermis (on both leaf surfaces) in transverse sections of *Ficus elastica* leaves in three stages of development. Epidermis stippled in **A**, **B**, with thick walls in **C**. Part of the leaf is omitted in **C**. Cystolith development: **A**, wall thickens in the lithocyst; **B**, cellulose stalk appears; **C**, calcium carbonate is deposited on stalk. Unlike other epidermal cells, the lithocyst undergoes no periclinal divisions. (A, ×207; B, ×163; C, ×234.)

by expansion into the mesophyll (Fig. 9.3). In some plants (*Peperomia*, Fig. 7.4) the cells of the multiple epidermis remain arranged in radial rows and clearly reveal their common origin (Linsbauer, 1930).

The common functions of the epidermis of the aerial plant parts are considered to be reduction of water loss by transpiration, mechanical protection, and gaseous exchange through stomata. Because of the compact arrangement of the cells and the presence of the relatively tough cuticle, the epidermis also offers mechanical support and adds stiffness to the stems (Niklas and Paolillo, 1997). In stems and coleoptiles the epidermis, which is under tension, has been regarded as the tissue that controls elongation of the entire organ (Kutschera, 1992; see however, Peters and Tomos, 1996). The epidermis is also a dynamic storage compartment of various metabolic products (Dietz et al., 1994), and the site of light perception involved in circadian leaf movements and photoperiodic induction (Mayer et al., 1973; Levy and Dean, 1998; Hempel et al., 2000). In the seagrasses (Iyer and Barnabas, 1993) and other submerged aquatic angiosperms, the epidermis is the principal site of photosynthesis (Sculthorpe, 1967). The epidermis is an important protective layer against UV-B radiation-induced injuries in the mesophyll region of the leaf (Robberecht and Caldwell, 1978; Day et al., 1993; Bilger et al., 2001), and in some leaves the upper epidermal

cells act as lenses, focusing light upon the chloroplasts of the underlying palisade parenchyma cells (Bone et al., 1985; Martin, G., et al., 1989). Epidermal cells of both the shoot and root are involved with the absorption of water and solutes.

Although the mature epidermis is generally passive with regard to meristematic activity (Bruck et al., 1989), it often retains the potentiality for growth for a long time. As mentioned previously, in perennial stems in which the periderm arises late in life, or not at all, the epidermis continues to divide in response to the circumferential expansion of the axis. If a periderm is formed, the source of its meristem, the phellogen, may be the epidermis (Chapter 15). Adventitious buds can arise in the epidermis (Ramesh and Padhya, 1990; Redway, 1991; Hattori, 1992; Malik et al., 1993), and the regeneration of entire plants has been achieved from epidermal cells, including guard cells, in tissue culture (Korn, 1972; Sahgal et al., 1994; Hall et al., 1996; Hall, 1998). Thus even the protoplasts of highly differentiated guard cells can re-express their full genetic potential (totipotency).

The epidermis is a complex tissue composed of a wide variety of cell types, which reflect its multiplicity of functions. The groundmass of this tissue is composed of relatively unspecialized cells, the ordinary epidermal cells (also called ground cells, epidermal cells proper, unspecialized epidermal cells, pavement cells), and of more specialized cells dispersed throughout the mass. Among the more specialized cells are the guard cells of the stomata and a variety of appendages, the trichomes, including the root hairs, which develop from epidermal cells of the roots.

# ORDINARY EPIDERMAL CELLS

Mature **ordinary epidermal cells** (often simply referred to hereafter as epidermal cells) are variable in shape, but typically they are tabular, having little depth (Fig. 9.4). Some, such as the palisade-like epidermal cells of many seeds, are much deeper than they are wide. In elongated plant parts, such as stems, petioles, vein ribs of leaves, and leaves of most monocots, the epidermal cells are elongated parallel with the long axis of the plant part. In many leaves, petals, ovaries, and ovules, the epidermal cells have wavy vertical (anticlinal) walls. The pattern of waviness is controlled by local wall differentiation, which determines the pattern of wall expansion (Panteris et al., 1994).

Epidermal cells have living protoplasts and may store various products of metabolism. They contain plastids that usually develop only few grana and are, therefore, deficient in chlorophyll. Photosynthetically active chloroplasts, however, occur in the epidermis of plants living in deep shade, as well as in the epidermis of submerged water plants. Starch and protein crystals may be present in epidermal plastids, anthocyanins in vacuoles.

## Epidermal Cell Walls Vary in Thickness

Epidermal cell walls vary in thickness in different plants and in different parts of the same plant. In the thinner walled epidermis, the outer periclinal walls are frequently thicker than the inner periclinal and anticlinal walls. The periclinal walls in the leaves, hypocotyls, and epicotyls of some species have a crossed-polylamellate structure, in which lamellae with transversely oriented cellulose microfibrils alternate with lamellae in which the microfibrils are vertically oriented (Sargent, 1978; Takeda and Shibaoka, 1978; Satiat-Jeunemaitre et al., 1992; Gouret et al., 1993). An epidermis with exceedingly thick walls is found in the leaves of conifers (Fig. 9.5); the wall thickening, which is lignified and probably secondary, is so massive in some species that it almost occludes the lumina of the cells. Epidermal cell walls commonly are silicified as in grasses and sedges (Kaufmann et al., 1985; Piperno, 1988). Wall ingrowths typical of transfer cells commonly develop from the outer epidermal walls of the submerged leaves of seagrasses and freshwater plants (Gunning, 1977; Iyer and Barnabas, 1993).

Primary pit-fields and plasmodesmata generally occur in the anticlinal and inner periclinal walls of the epidermis, although the frequency of plasmodesmata between the epidermis and mesophyll of leaves is relatively low. For a time plasmodesmata were thought to occur in the outer epidermal walls and were called ***ectodesmata***. Subsequent research revealed that cytoplasmic strands do not occur in the outer walls but that bundles of interfibrillar spaces may extend from the plasma membrane to the cuticle within the cellulosic walls. These bundles need special treatment to make them visible. Microchannels, believed to contain pectin, have been

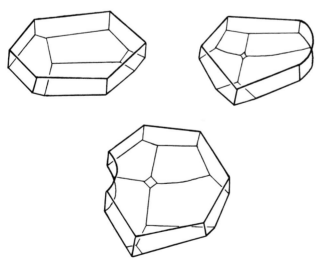

**FIGURE 9.4**

Three-dimensional aspect of epidermal cells of *Aloe aristata* (Liliaceae) leaf. The upper face in each drawing is the outer face of the cell. On the opposite side are the faces of contact with the subjacent mesophyll cells. (From Esau, 1977; redrawn from Matzke, 1947.)

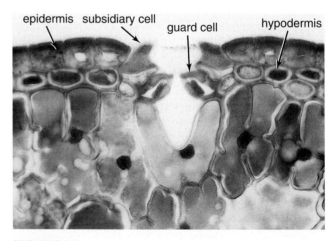

**FIGURE 9.5**

Conifer leaf, *Pinus resinosa*. Transverse section through outer part of a needle showing thick-walled epidermal cell and a stoma. (×450.)

reported in the outer epidermal wall of xerophytes (Lyshede, 1982). The term **teichode** (from the Greek words *teichos*, for wall, and *hodos*, for path) has been proposed as a substitution for both ectodesmata (Franke, 1971) and microchannels (Lyshede, 1982), neither of which are cytoplasmic structures. Teichodes have been implicated as pathways in foliar absorption and excretion (Lyshede, 1982).

### The Most Distinctive Feature of the Outer Epidermal Wall Is the Presence of a Cuticle

The **cuticle**, or **cuticular membrane**, consists predominately of two lipid components: insoluble cutin, which constitutes the matrix of the cuticle, and soluble waxes, some of which are deposited on the surface of the cuticle, the **epicuticular wax**, and others embedded in the matrix, the **cuticular**, or **intracuticular**, **wax**. The cuticle is characteristic of all plant surfaces exposed to air, even extending through the stomatal pores and lining the inner epidermal cell walls of the **substomatal chambers**, large substomatal intercellular spaces opposite the stomata (Fig. 9.5; Pesacreta and Hasenstein, 1999). It is the first protective barrier between the aerial surface of the plant and its environment and the principal barrier to the movement of water, including that of the transpiration stream, and solutes (Riederer and Schreiber, 2001). In exceptional cases, cuticles are also formed in cortical cells and give rise to a protective tissue called ***cuticular epithelium*** (Calvin, 1970; Wilson and Calvin, 2003).

The matrix of the cuticle may consist not only of one but of two lipid polymers, cutin and cutan (Jeffree, 1996; Villena et al., 1999). Unlike cutin, **cutan** is highly resistant to alkaline hydrolysis. Although the cuticles of some species appear to lack cutan (those of the tomato fruit, *Citrus* and *Erica* leaves), cutan may be the principal or only matrix polymer in some others, notably in *Beta vulgaris*. Cutan has been reported to be a constituent in fossilized plant cuticles, and mixed cutin/cutan cuticles have been reported in a number of extant species, including *Picea abies*, *Gossypium* sp., *Malus primula*, *Acer platanoides*, *Quercus robur*, *Agave americana*, and *Clivia miniata*.

Most cuticles consist of two more or less distinct regions, the cuticle proper and one or more cuticular layers (Fig. 9.6). The **cuticle proper** is the outermost region, containing cutin and embedded birefringent (cuticular) **waxes** but no cellulose. The process by which it is formed is called **cuticularization**. Epicuticular wax occurs on the surface of the cuticle proper, either in an amorphous form or as crystalline structures of various shapes (Fig. 9.7). Among the more common shapes are tubules, solid rodlets, filaments, plates, ribbons, and granules (Wilkinson, 1979; Barthlott et al., 1998; Meusel et al., 2000). Epicuticular wax imparts the "bloom" to many leaves and fruits. The bloom results from the reflection and scattering of light by the wax

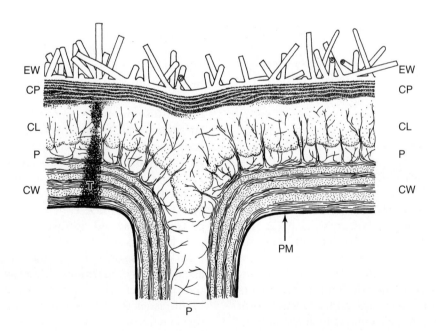

**FIGURE 9.6**

Generalized structure of a plant cuticle. Details: CL, cuticular layer or reticulate region, traversed by cellulose microfibrils; CP, cuticle proper, showing lamellate structure; CW, cell wall; EW, epicuticular wax; P, pectinaceous layer and middle lamella; PM, plasma membrane; T, teichode. (From Jeffree, 1986. Reprinted with permission of Cambridge University Press.)

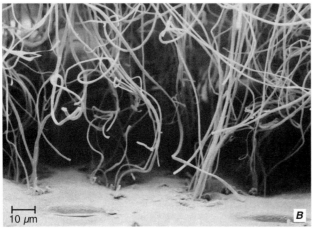

**FIGURE 9.7**

Surface views of epidermis showing epicuticular wax. **A**, plate-like wax projections on the adaxial surface of *Pisum* leaf. **B**, wax filaments on abaxial surface of sorghum (*Sorghum bicolor*) leaf sheath. (**A**, from Juniper, 1959. © 1959, with permission from Elsevier; **B**, from Jenks et al., 1994. © 1994 by The University of Chicago. All rights reserved.)

crystals. The epicuticular wax plays an important role in the reduction of water loss by the cuticle. The commercial practice of dipping grapes in chemicals that accelerate the drying of fruit causes a close adpression of wax platelets and their parallel orientation. This change probably facilitates the movement of water from the fruit to the atmosphere (Possingham, 1972). The epicuticular wax is also responsible for enhancing the ability of the epidermal surface to shed water (Eglinton and Hamilton, 1967; Rentschler, 1971; Barthlott and Neinhuis, 1997) and, consequently, limits the accumulation of contaminating particles and of water-borne spores of pathogens. An exceptionally thick layer of wax (up to 5 mm) occurs on the leaves of *Klopstockia cerifera*, the wax palm of the Andes (Kreger, 1958) and those of *Copernicia cerifera*, the Brazilian wax palm, from which carnauba wax is derived (Martin and Juniper, 1970).

The **cuticular layers** are found beneath the cuticle proper and are considered to be the outer portions of the cell wall encrusted to varying degrees with cutin. Cuticular wax, pectin, and hemicellulose may also occur in the cuticular layers. The process by which the cuticular layers are formed is called **cutinization**. Beneath the cuticular layers there commonly is a layer rich in pectin, the **pectin layer**, which bonds the cuticle to the outer walls. The pectin layer is continuous with the middle lamella between the anticlinal walls where the cuticle extends deeply, forming *cuticular pegs*.

The ultrastructure of the cuticle shows considerable variability. Two distinctive ultrastructural components may be found within the matrix: *lamellae* and *fibrillae* (Fig. 9.6). The fibrillae probably are mainly cellulosic. Plant species differ from one another in the presence or absence of one or the other of these components. On that basis, Holloway (1982) recognized six structural types of cuticle. When both components are present, the lamellate region corresponds to the cuticle proper and the fibrillae-containing, reticulate region to the cuticular layer(s). The ultrastructure of the cuticle appears to affect significantly cuticular permeability: cuticles with entirely reticulate structure are more permeable to certain substances than those with an outer lamellate region (Gouret et al., 1993; Santier and Chamel, 1998). Regardless, it is the cuticular waxes that form the main barrier to the diffusion of water and solutes across the cuticle, in large part by creating a tortuous pathway and hence an increased path length for diffusing molecules (Schreiber et al., 1996; Buchholz et al., 1998; Buchholz and Schönherr, 2000). Based on experimental evidence (Schönherr, 2000; Schreiber et al., 2001), Riederer and Schreiber (2001) have concluded that the bulk of the water crossing the cuticle diffuses as single molecules in a so-called lipophilic pathway composed of amorphous waxes. A minor fraction of the water may diffuse through water-filled polar pores of molecular dimensions, the pathway presumably followed by water-soluble organic compounds and by inorganic ions. Cuticular transpiration is not inversely related to thickness of cuticle as one might conclude intuitively (Schreiber and Riederer, 1996; Jordaan and Kruger, 1998). In fact, thick cuticles may show higher water permeabilities and diffusion coefficients than thin cuticles (Becker et al., 1986).

In at least some species the cuticle appears initially as an entirely amorphous, electron dense layer, called the **procuticle** (Fig. 9.8). Later the procuticle changes its ultrastructural appearance and is transformed into the cuticle proper typical of the species. The appearance of the cuticle proper is followed by that of cuticular layer(s), indicating that the cuticle proper is not a newly adcrusted layer (Heide-Jørgensen, 1991). By the time the cuticle is fully formed, it is several times thicker than the original procuticle. The cuticle varies in

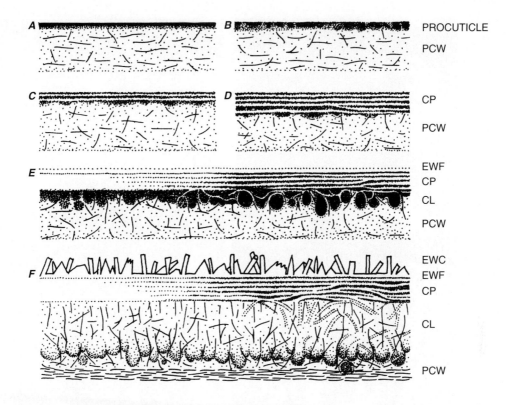

**FIGURE 9.8**

Development of the cuticle proper (CP) of a plant cuticle and early stages in the development of the cuticular layer (CL) within the primary cell wall (PCW). **A–D**, conversion of the procuticle to a lamellate cuticle proper. Globular lipids may be involved in further construction of the lamellate cuticle proper as in **E**. **E**, globular lipids coated with electron-lucent shells construct the cuticle proper/cuticular layer transition zone. Lamellae may become less regular. An amorphous film of epicuticular wax (EWF) apparently is present on the surface of the cuticle proper. **F**, incorporation of the primary cell wall (PCW) in the cuticular layer. Predominantly radial reticulations reach as far as the cuticle proper. Epicuticular wax crystals (EWC) begin to form before cessation of cell expansion. (From Jeffree, 1996, Fig. 2.12a–f. © Taylor and Francis.)

thickness, and its development is affected by environmental conditions (Juniper and Jeffree, 1983; Osborn and Taylor, 1990; Riederer and Schneider, 1990).

Cutin and waxes (or their precursors) are synthesized in the epidermal cells and must migrate to the surface through the cell walls. Neither the routes followed by these substances nor the mechanisms involved have been agreed upon. Some investigators presume that teichodes (ectodesmata, microchannels) function as pathways for cutin and waxes across the walls (Baker, 1982; Lyshede, 1982; Anton et al., 1994). Most attention has been given to the epicuticular waxes, whose precursors apparently are produced in endoplasmic reticulum and modified in the Golgi apparatus before being discharged from the cytoplasm by exocytosis (Lessire et al., 1982; Jenks et al., 1994; Kunst and Samuels, 2003). Although pores and channels have been detected in the cuticles of leaves and fruits of a fair number of taxa (Lyshede, 1982; Miller, 1985, 1986), such structures apparently are not ubiquitous. Neither pores nor channels could be found in either the wall or the cuticle of

the tubule- (epicuticular wax) forming cork cells of the *Sorghum bicolor* leaf (Fig. 9.9; Jenks et al., 1994). Some investigators believe that wax precursors follow no special pathway but rather diffuse through the wall and cuticle in a volatile solvent and then crystallize on the surface (Baker, 1982; Hallam, 1982). Neinhuis and co-workers (Neinhuis et al., 2001) have hypothesized that the wax molecules move together with the water vapor, permeating the cuticle in a process similar to steam distillation. At least one gene, the *Arabidopsis* gene *CUT1*, has been clearly established to function in wax production. It encodes a very-long-chain fatty acid-condensing enzyme required for cuticular wax production (Millar et al., 1999).

Cutin/cutan is highly inert and resistant to oxidizing maceration methods. The cuticle does not decay, since apparently no microorganisms possess cutin/cutan-degrading enzymes (Frey-Wyssling and Mühlethaler, 1959). Because of its chemical stability the cuticle is preserved as such in fossil material and is very useful in identification of fossil species (Edwards et al., 1982).

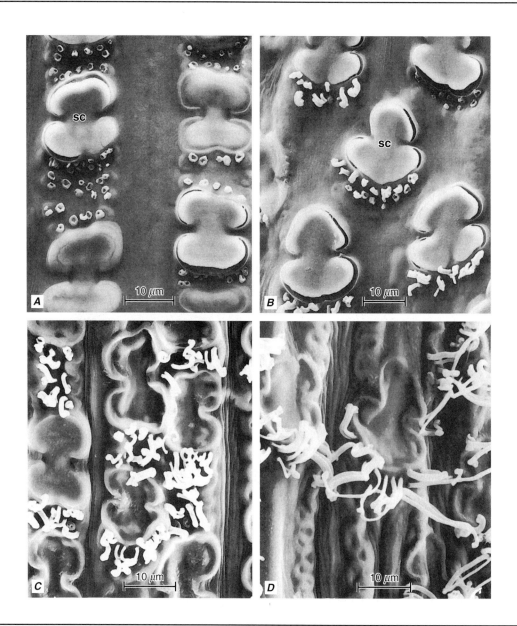

**FIGURE 9.9**

Development of epicuticular wax filaments on the abaxial surface of a sorghum (*Sorghum bicolor*) leaf sheath. **A**, wax filaments emerging from cork cells adjacent to silica cells (sc). Initially the filaments appear as circular secretions. **B**, with further development, the secretions appear as short cylinders. **C**, **D**, with continued development, the secretions form clusters of epicuticular wax filaments. (From Jenks et al., 1994. © 1994 by The University of Chicago. All rights reserved.)

Cuticular characters have also been shown to be useful in conifer taxonomy (Stockey et al., 1998; Kim et al., 1999; Ickert-Bond, 2000).

## ▌STOMATA

### Stomata Occur on All Aerial Parts of the Primary Plant Body

**Stomata** (singular: **stoma**) are openings (the stomatal pores, or apertures) in the epidermis, each bounded by two guard cells (Fig. 9.10), which by changes in shape bring about the opening and closing of the pore. The term *stoma* is Greek for mouth and, conventionally, it is used to designate both the pore and the two guard cells. In some species, the stomata are surrounded by cells that do not differ from other ground cells of the epidermis. These cells are called **neighboring cells**. In others, the guard cells are bordered by one or more cells that differ in size, shape, arrangement, and sometimes in content from the ordinary epidermal cells. These distinct cells are called **subsidiary cells** (Figs. 9.5, 9.13,

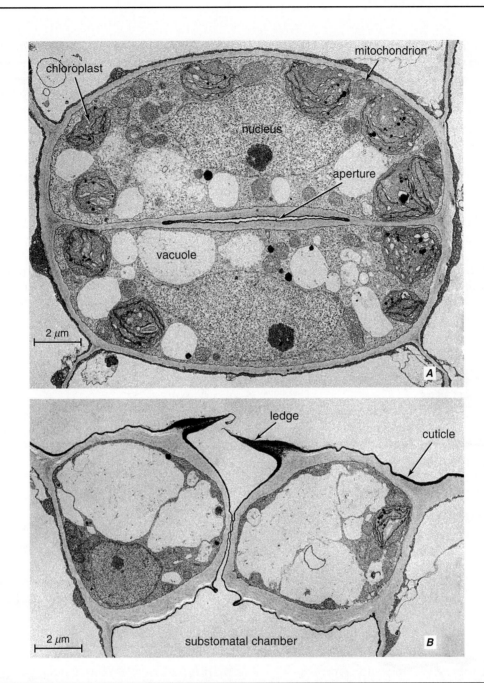

**FIGURE 9.10**

Electron micrographs of stomata from a sugar beet (*Beta vulgaris*) leaf seen from the surface (**A**) and in transverse section (**B**). (From Esau, 1977.)

9.14, 9.15, 9.17A, 9.20, and 9.21). The primary role of stomata is to regulate the exchange of water vapor and of $CO_2$ between the internal tissues of the plant and the atmosphere (Hetherington and Woodward, 2003).

Stomata occur on all aerial parts of the primary plant body but are most abundant on leaves. The aerial parts of some chlorophyll-free land plants (*Monotropa, Neottia*) and the leaves of the holoparasite family Balanophoraceae (Kuijt and Dong, 1990) lack stomata. Roots usually lack stomata. Stomata have been found on seedling roots of several species, including *Helianthus*

*annuus* (Tietz and Urbasch, 1977; Tarkowska and Wacowska, 1988) *Pisum arvense, Ornithopus sativus* (Tarkowska and Wacowska, 1988), *Pisum sativum* (Lefebvre, 1985), and *Ceratonia siliqua* (Christodoulakis et al., 2002). Stomatal density varies greatly in photosynthesizing leaves. It varies on different parts of the same leaf and on different leaves of the same plant, and it is influenced by environmental factors such as light and $CO_2$ levels. It has been suggested that environmental effects on both stomatal and trichome numbers may be mediated through cuticular wax composition (Bird and Gray, 2003). Studies have shown that the development of stomata in young leaves is regulated by a mechanism that senses light and $CO_2$ levels around mature leaves of the same plant rather than by the young leaves themselves (Brownlee, 2001; Lake et al., 2001; Woodward et al., 2002). The information sensed by the mature leaf must be relayed to the developing leaves via long-distance systemic signals. In leaves, stomata may occur on both surfaces (***amphistomatic leaf***) or on only one, either the upper (***epistomatic leaf***) or more commonly on the lower (***hypostomatic leaf***). Some examples of stomatal densities (per square millimeter, lower epidermis/upper epidermis), found in Willmer and Fricker (1996) are *Allium cepa* 175/175, *Arabidopsis thaliana* 194/103, *Avena sativa* 45/50, *Zea mays* 108/98, *Helianthus annuus* 175/120, *Nicotiana tabacum* 190/50, *Cornus florida* 83/0, *Quercus velu-*

*tina* 405/0, *Tilia americana* 891/0, *Larix decidua* 16/14, and *Pinus strobus* 120/120. In general, stomatal density is higher in xeromorphic leaves than in leaves of mesomorphic and hygromorphic (hydromorphic) plants (Roth, 1990). Among aquatic plants, stomata typically are distributed on all surfaces of emergent leaves and on only the upper surface of floating leaves. Submerged leaves generally lack stomata entirely (Sculthorpe, 1967). In the leaves of some species the stomata occur in distinct clusters rather than being more or less uniformly distributed as, for example, in *Begonia semperflorens* (2 to 4 per cluster) and *Saxifraga sarmentosa* (about 50 per cluster) (Weyers and Meidner, 1990).

Stomata vary in the level of their position in the epidermis (Fig. 9.11). They may occur at the same level as the adjacent epidermal cells, or they may be raised

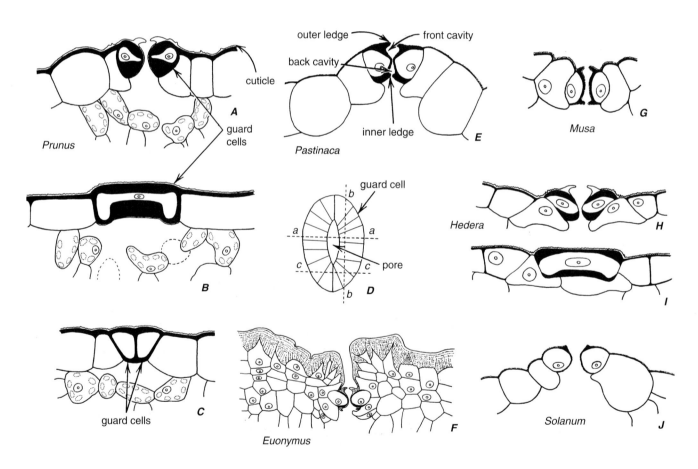

**FIGURE 9.11**

Stomata in the abaxial epidermis of foliage leaves. **A–C**, stomata and some associated cells from peach leaf sectioned along planes indicated in **D** by the broken lines aa, bb, and cc. **E–H, J**, stomata from various leaves cut along the plane aa. **I**, one guard cell of ivy cut along the plane bb. The stomata are raised in **A, E, J**. They are slightly raised in **H**, slightly sunken in **G**, and deeply sunken in **F**. The horn-like protrusions in the various guard cells are sectional views of ledges. Some stomata have two ledges (**E, F, G**); others only one (**A, H, J**). Ledges are cuticular in **A, F, H**. The euonymus leaf (**F**) has a thick cuticle; epidermal cells are partly occluded with cutin. (**A–D, F–J**, ×712; **E**, ×285.)

above or sunken below the surface of the epidermis. In some plants, stomata are restricted to depressions called **stomatal crypts**, which often contain prominently developed epidermal hairs (Fig. 9.12).

## Guard Cells Are Generally Kidney-shaped

The guard cells of eudicots are generally crescent-shaped with blunt ends (kidney-shaped) in surface view (Figs. 9.10A and 9.11D), and have **ledges** of wall material on the outer or both the outer and inner sides. In sectional views such ledges appear like horns. If two ledges are present, the outer ledge delimits a front chamber and the inner ledge delimits a rear chamber. Stomata with two ledges actually have three apertures, an outer and inner aperture formed by the ledges and a central aperture about midway between the other two formed by the opposing guard cell walls. The inner aperture rarely closes completely, and depending on the stage of pore formation, the outer or central aperture may be narrowest (Saxe, 1979). The guard cells are covered with a cuticle. As mentioned previously, the cuticle extends through the stomatal aperture(s) and into the substoma-

tal chamber. Apparently the guard cell cuticle differs in chemical composition from that of ordinary epidermal cells and is more permeable to water than the latter (Schönherr and Riederer, 1989). Each guard cell has a prominent nucleus, numerous mitochondria, and poorly developed chloroplasts, in which starch typically accumulates at night and decreases in amount during the day with increasing stomatal opening. The *vacuolar system* is dissected to variable degrees. The extent of vacuolar volume differs greatly between closed and open stomata, ranging from a very small fraction of cell volume in closed stomata to over 90% in open stomata. Kidney-shaped guard cells similar to those of eudicots also occur in some monocots and in gymnosperms.

In the Poaceae and a few other families of monocots, the guard cells are dumbbell-shaped; that is, they are narrow in the middle and enlarged at both ends (Fig. 9.13). The guard cell nucleus in the Poaceae is also dumbbell-shaped, being almost thread-like in the middle and ovoid at either end. Whether the dumbbell-shaped guard cells of other families of monocots have dumbbell-shaped nuclei remains to be determined (Sack, 1994). In the Poaceae most of the organelles, including the

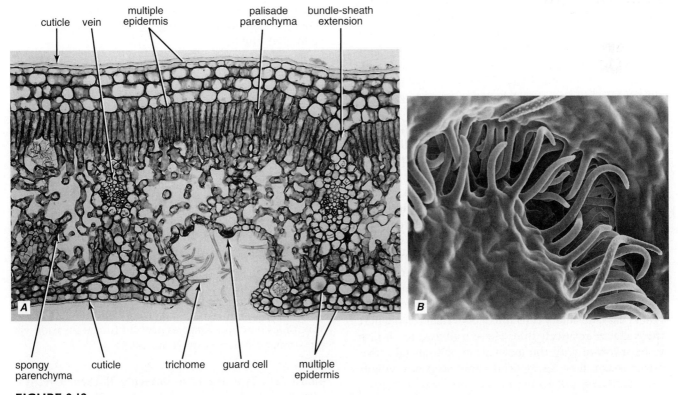

**FIGURE 9.12**

Oleander (*Nerium oleander*) leaf. **A**, transverse section showing a stomatal crypt on the lower side of the leaf. In oleander the stomata and trichomes are restricted to the crypts. The oleander leaf has a multiple epidermis. **B**, scanning electron micrograph of a stomatal crypt showing numerous trichomes lining the crypt. (**A**, ×177; **B**, ×725.)

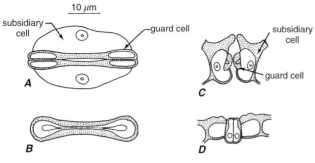

**FIGURE 9.13**

Dumbbell-shaped guard cells of rice (*Oryza;* Poaceae) shown (**A**) from the surface and (**B–D**) in sections made along planes indicated in Figure 9.11D by the broken lines aa, bb, and cc. **A**, guard cells are shown in a high focal plane so that the lumen is not visible in the narrow part of the cell. **B**, one guard cell cut through the plane bb, showing the dumbbell-shaped nucleus. **C**, cut through plane aa. **D**, cut through plane cc. (From Esau, 1977.)

vacuoles, are located in the bulbous ends of the cells. In addition the protoplasts of the two guard cells are interconnected through pores in the common wall between the enlarged ends. Because of this protoplasmic continuity, the guard cells must be considered as a single functional unit, in which changes in turgor are immediately realized. The pores appear to result from incomplete development of the wall (Kaufman et al., 1970a; Srivastava and Singh, 1972). There are two subsidiary cells, one on each side of the stoma (Figs. 9.13A and 9.14).

The stomata of most conifers are deeply sunken and appear as though suspended from the subsidiary cells that overarch them forming a funnel-like cavity called the **epistomatal chamber** (Figs. 9.5 and 9.15; Johnson and Riding, 1981; Riederer, 1989; Zellnig et al., 2002). In their median regions the guard cells are elliptical in cross section and have narrow lumina. At their ends the guard cells are triangular and the lumina are wider. A characteristic feature of these stomatal complexes is that the guard cell and subsidiary cell walls are partly lignified. The nonlignified regions of the guard cell walls occur at contact areas, so-called hinge areas, with other cells (subsidiary cells and hypodermal cells) where the walls are relatively thin. These wall features appear to be involved with the mechanism of stomatal movements in conifers. An especially thin strip of nonlignified guard cell wall also faces the pore. Lignified guard cells are rare in angiosperms (Kaufmann, 1927; Palevitz, 1981).

In the Pinaceae the epistomatal chamber typically is filled with epicuticular wax tubules, which form a

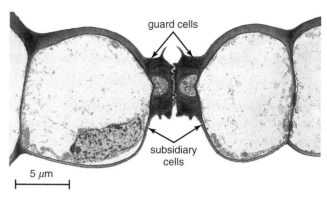

**FIGURE 9.14**

Transverse section through a closed stoma of maize (*Zea mays*) leaf. Each thick-walled guard cell is attached to a subsidiary cell.

porous "plug" over the stomata (Johnson and Riding, 1981; Riederer, 1989). The tubules are both guard cell and subsidiary cell in origin. Stomatal plugs occur also in other conifers (Podocarpaceae, Araucariaceae, and Cupressaceae; Carlquist, 1975; Brodribb and Hill, 1997) and in two families of vesselless angiosperms (Winteraceae and Trochodendraceae). In the vesselless angiosperms the stomata are plugged with alveolar material, which is wax-like in appearance but cutinaceous in composition (Bongers, 1973; Carlquist, 1975; Feild et al., 1998).

The function of stomatal plugs is not well understood (Brodribb and Hill, 1997). The most common suggestion is that the plugs serve primarily to restrict transpirational water loss. Although wax plugs clearly fulfill this role, Brodribb and Hill (1997) have suggested that the wax plugs of conifers may have evolved as an adaptation to wet conditions and serve to keep the pore free of water. This would facilitate gas exchange and enhance photosynthesis. Feild et al. (1998) similarly have concluded that the cutinaceous stomatal plugs in *Drimys winteri* (Winteraceae) are more important for promoting photosynthetic activity than for preventing water loss. Earlier Jeffree et al. (1971) calculated the restriction of gas exchange by wax plugs in the stomata of *Picea sitchensis.* They concluded that whereas the rate of transpiration was reduced about two-thirds, the rate of photosynthesis was reduced by only about one-third. Wax plugs may also serve to prevent fungal invasion via the stomatal pore (Meng et al., 1995).

## Guard Cells Typically Have Unevenly Thickened Walls with Radially Arranged Cellulose Microfibrils

Although the guard cells of the major taxa have their distinguishing characteristics, all share an outstanding feature—the presence of unevenly thickened walls.

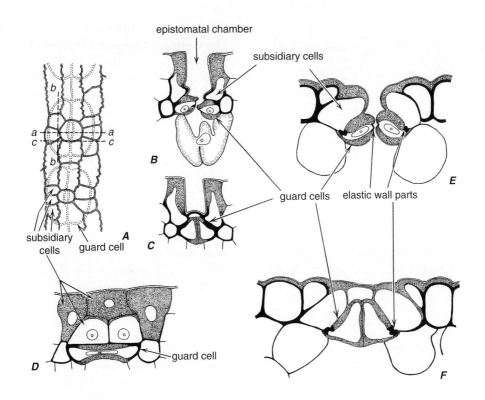

**FIGURE 9.15**

Stomata of conifer leaves. **A**, surface view of epidermis with two deeply sunken stomata from *Pinus merkusii*. Guard cells are overarched by subsidiary and other epidermal cells. Stomata and some associated cells of *Pinus* (**B–D**), and *Sequoia* (**E**, **F**). The broken lines in **A** indicate the planes along which the sections of stomata were made in **B–F**: aa, **B**, **E**; bb, **D**; cc, **C**, **F**. (A, ×182; B–D, ×308; E, F, ×588. A, adapted from Abagon, 1938.)

This feature appears to be related to the changes in shape and volume (and the concomitant changes in size of the stomatal aperture) brought about by changes in turgor within the guard cells. In kidney-shaped guard cells the wall away from the pore (***dorsal wall***) is generally thinner and therefore more flexible than the wall bordering the pore (***ventral wall***). The kidney-shaped guard cells are constrained at their ends where they are attached to one another; moreover these common guard cell walls remain almost constant in length during changes in turgor. Consequently increase in turgor causes the thin dorsal wall to bulge away from the aperture and the ventral wall facing the aperture to become straight or concave. The whole cell appears to bend away from the aperture and the aperture increases in size. Reversed changes occur under decreased turgor.

In dumbbell-shaped guard cells of the Poaceae the middle part has strongly unevenly thickened walls (their inner and outer walls are much thicker than the dorsal and ventral walls), whereas the bulbous ends have thin walls. In these guard cells increase in turgor causes a swelling of the bulbous ends and the consequent separation of the straight median portions from each other. Again, reversed changes occur under decreased turgor.

According to a different hypothesis, the radial arrangement of cellulose microfibrils (**radial micellation**) in the guard cell walls (indicated by radially arranged lines in Fig. 9.11D) plays a more important role in stomatal movement than the differential wall thickening (Aylor et al., 1973; Raschke, 1975). As the dorsal walls of kidney-shaped guard cells move outward with increase in turgor, the radial micellation transmits this movement to the wall bordering the pore (the ventral wall), and the pore opens. In dumbbell-shaped guard cells the microfibrils are predominantly axially arranged in the median portions. From the median portions microfibrils radiate out into the bulbous ends. The radial orientation of microfibrils in guard cell walls was recognized by polarization optics and electron microscopy (Raschke, 1975). Figure 9.16 depicts the results of some experiments with balloons that have been used in support of the role of radial micellation in stomatal movement. It is likely that wall thickenings and microfibril arrangement both contribute to stomatal movement (Franks et al., 1998).

Microtubule dynamics have been implicated in stomatal movement in *Vicia faba* (Yu et al., 2001). In fully open stomata the guard cell microtubules were found to be transversely oriented from the ventral wall to the

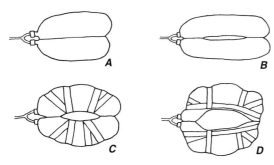

**FIGURE 9.16**

Models for studying the effect of radial arrangement of microfibrils in guard-cell walls on the opening of stomata. **A**, two latex cylinders connected at their ends and partially inflated. **B**, the same model at higher pressure. A narrow slit is visible. **C**, bands of tape simulate radial micellation on the cylinders, which are inflated. Slit wider than in **B**. **D**, radial micellation extends farther to the ends of the cylinders, and some tape is present along the "ventral wall." Inflation has induced the formation of a slit wider than in **C**. (From Esau, 1977; drawings adapted from photographs in Aylor, Parlange, and Krikorian, 1973.)

dorsal wall. During stomatal closure, in response to darkness, the microtubules became twisted and patched; in closed stomata, they appeared broken down into diffuse fragments. With reopening of the stomata in response to light, the microtubules became transversely oriented again. Although cortical microtubules are known to change orientation in response to stresses in the cell wall (Hejnowicz et al., 2000), treatment of the *Vicia faba* stomata with microtubule-stabilizing and microtubule-depolymerizing drugs suppressed light-induced opening and dark-induced closure of the stomata, leading Yu et al. (2001) to conclude that microtubule dynamics may be functionally involved with stomatal movement. Further support for the involvement of microtubules in guard cell function in *Vicia faba* comes from studies by Marcus et al. (2001) who concluded that microtubules are necessary for stomatal opening, more specifically, that they are required somewhere upstream to the ionic events (H$^+$ efflux and K$^+$ influx) that lead to stomatal opening, possibly participating in the signal transduction events leading to the ionic fluxes.

Volume increases in the guard cells are compensated in part by volume decreases in adjacent epidermal cells (subsidiary cells or neighboring cells) (Weyers and Meidner, 1990). Therefore it is the turgor difference between the guard cells and their immediate neighbors that actually determines the opening of the pore (Mansfield, 1983). Hence the stomatal complex should be considered as a functional unit.

## Blue Light and Abscisic Acid Are Important Signals in the Control of Stomatal Movement

Transport of potassium ions (K$^+$) between guard cells and subsidiary cells or neighboring cells is widely considered a principal factor in guard cell movement, the stoma being open in the presence of increased amounts of K$^+$. Some studies indicate that both K$^+$ and sucrose are primary guard cell osmotica, K$^+$ being the dominant osmoticum in the early-opening stages during the morning, and sucrose becoming the dominant osmoticum in early afternoon (Talbott and Zeiger, 1998). Uptake of K$^+$ by the guard cells is driven by a proton (H$^+$) gradient mediated by a blue-light activated plasma membrane H$^+$-ATPase (Kinoshita and Shimazaki, 1999; Zeiger, 2000; Assmann and Wang, 2001; Dietrich et al., 2001), and is accompanied by the uptake of chloride ions (Cl$^-$) and the accumulation of malate$^{2-}$, which is synthesized from starch in the guard cell chloroplasts. The elevation in the solute concentration results in a more negative water potential, which causes osmotic movement of water into the guard cells, guard cell swelling, and separation of the guard cells at the pore site. The guard cells in species of the genus *Allium* lack starch at all times (Schnabl and Ziegler, 1977; Schnable and Raschke, 1980), and apparently rely on Cl$^-$ alone to serve as the counterion for K$^+$. Stomatal closure occurs when Cl$^-$, malate$^{2-}$, and K$^+$ are lost from the guard cells. Water then moves down its water potential from the guard cell protoplast to the cell wall, reducing the turgor of the guard cells and causing closure of the stomatal pore.

The plant hormone abscisic acid (ABA) plays a crucial role as an endogenous signal that inhibits stomatal opening and induces stomatal closure (Zhang and Outlaw, 2001; Comstock, 2002). The primary sites of action of ABA appear to be specific ion channels in the guard cell plasma membrane and tonoplast that lead to the loss of both the K$^+$ and associated anions (Cl$^-$ and malate$^{2-}$) from both the vacuole and the cytosol. Experimental evidence indicates that ABA induces an increase in cytosolic pH and cytosolic Ca$^{2+}$, which act as second messengers in this system (Grabov and Blatt, 1998; Leckie et al., 1998; Blatt, 2000a; Wood et al., 2000; Ng et al., 2001). Several protein phosphatases and protein kinases also have been implicated in the regulation of channel activities (MacRobbie, 1998, 2000). Guard cells respond to a range of environmental stimuli such as light, CO$_2$ concentration, and temperature, in addition to plant hormones. The complex mechanism of stomatal movements is the subject of intensive studies and discussion, and is providing invaluable information to our understanding of signal transduction in plants (Hartung et al., 1998; Allen et al., 1999; Assmann and Shimazaki, 1999; Blatt, 2000b; Eun and Lee, 2000; Hamilton et al., 2000; Li and Assmann, 2000; Schroeder et al., 2001).

Although it long was presumed that the degree of stomatal opening is fairly homogeneous over the surface of a leaf, it is now known that despite nearly identical environmental conditions, stomata may be open in some areas of the leaf and closed in adjacent ones, resulting in patchy stomatal conductance (Mott and Buckley, 2000). **Stomatal patchiness** has been observed in a large number of species and families (Eckstein, 1997), and it is especially common but not limited to leaves that are divided into separate compartments by **bundle-sheath extensions**—panels of ground tissue extending from the bundle sheaths to the epidermis—associated with the network of veins (Fig. 7.3A; Terashima, 1992; Beyschlag and Eckstein, 2001). Such leaves are called *heterobaric leaves*. Little or no gas exchange occurs between the system of intercellular spaces of the different compartments in these leaves such that the leaf is essentially a collection of independent photosynthesizing and transpiring units (Beyschlag et al., 1992). The pattern and extent of patchiness can differ between upper and lower surfaces in amphistomatous leaves (Mott et al., 1993). Stress factors, particularly those that impose water stress on plants seem to play a major role in patchiness formation (Beyschlag and Eckstein, 2001; Buckley et al., 1999).

## Development of Stomatal Complexes Involves One or More Asymmetric Cell Divisions

Stomata begin to develop in a leaf shortly before the main period of meristematic activity in the epidermis is completed and continue to arise through a considerable part of the later expansion of the leaf by cell enlargement. In leaves with parallel venation, as in most monocots, and with the stomata arranged in longitudinal rows (Fig. 9.17A), the developmental stages of the stomata are observable in sequence in the successively more differentiated portions of the leaf. This sequence is basipetal, that is, from the tip of the leaf downward. The first stomata to mature are found at the leaf tip and newly initiated ones near the leaf base. In the netted-veined leaves, as in most eudicots (Fig. 9.17B), different developmental stages are mixed in a diffuse, or mosaic, fashion. A striking feature of young eudicot leaves is the tendency for precocious maturation of some stomata on the teeth of the leaves (Payne, W. W., 1979). These stomata may function as water pores of hydathodes (Chapter 16).

Stomatal development begins with an asymmetric, or unequal, anticlinal division of a protodermal cell. This division results in two cells, one that is usually larger and resembles the other protodermal cells and a second that is usually markedly smaller and contains densely staining cytoplasm and a large nucleus. The smaller of these two cells is called the **stomatal meristemoid**. In some species the sister cell of the meristemoid may divide asymmetrically again and give rise to another meristemoid (Rasmussen, 1981). Depending on the species, the meristemoid may function directly as the **guard mother cell** (guard-cell mother cell, stoma mother cell) or give rise to the guard mother cell after further divisions. The formation of the stomatal complex requires migration of the nucleus to specific sites in the parent cells before cell division and precise placement of the division planes. Consequently the stomatal complex has been the object of numerous ultrastructural studies aimed at determining the role of microtubules in positioning of the cell plate and in cell shaping (Palevitz and Hepler, 1976; Galatis, 1980, 1982; Palevitz, 1982; Sack, 1987).

An equal division of the guard mother cell gives rise to the two guard cells (Figs. 9.18A and 9.19A–C), which through differential wall deposition and expansion acquire their characteristic shape. The middle lamella at the site of the future pore swells (Fig. 9.18A, d), and the connection between the cells is weakened there. The cells then separate at the site and thus the stomatal opening is formed (Fig. 9.18A, e). The exact cause(s) of separation of the ventral walls at the pore site has not been identified, but three possibilities have been considered: enzymic hydrolysis of the middle lamella, tension brought about by increase in guard cell turgor, and formation of the cuticle, which eventually lines the newly formed pores (Sack, 1987). In *Arabidopsis* formation of the pore appears to involve the stretching of electron-dense material in the lens-shaped thickening at the pore site (Zhao and Sack, 1999). The guard mother cells occur at the same level as the adjacent epidermal cells. Various spatial readjustments occur between the guard cells and the adjacent epidermal cells and between the epidermis and mesophyll (Fig. 9.19) so that the guard cells may be elevated above or lowered below the surface of the epidermis. Even in the leaves of conifers, with their deeply sunken guard cells, the guard mother cells are at the same level as the other epidermal cells (Johnson and Riding, 1981). The substomatal chamber forms during stomatal development, before formation of the stomatal pore (Fig. 9.19E).

Although plasmodesmata occur in all walls of immature guard cells, they become sealed (truncated) with wall material as the wall thickens (Willmer and Sexton, 1979; Wille and Lucas, 1984; Zhao and Sack, 1999). The symplastic isolation of mature guard cells is further illustrated by the inability of fluorescent dyes microinjected into either guard cells or their adjacent cells to move across the common wall between them (Erwee et al., 1985; Palevitz and Hepler, 1985).

As indicated previously, subsidiary cells or neighboring cells may arise from the same meristemoid as the stoma or from cells that are not directly related ontogenetically with the guard mother cell. On this basis three major categories of stomatal ontogeny have been

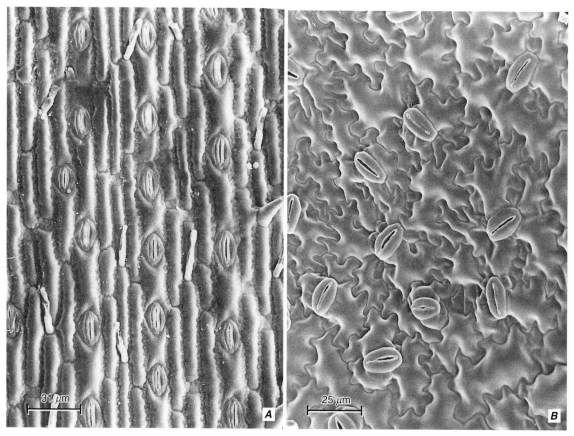

**FIGURE 9.17**

Surface views of stomata shown in scanning electron micrographs. **A**, maize (*Zea mays*) leaf showing the parallel arrangement of stomata typical of the leaves of monocots. In maize each pair of narrow guard cells is associated with two subsidiary cells, one on each side of the stoma. **B**, potato (*Solanum tuberosum*) leaf showing the random, or diffuse, arrangement of stomata typical of the leaves of eudicots. The kidney-shaped guard cells in potato are not associated with subsidiary cells. (**B**, courtesy of M. Michelle McCauley.)

recognized (Pant, 1965; Baranova, 1987, 1992): **mesogenous**, in which all of the subsidiary or neighboring cells have a common origin with the guard cells (Fig. 9.20); **perigenous**, in which none of the subsidiary or neighboring cells has a common origin with the guard cells (Fig. 9.21); **mesoperigenous**, in which at least one of the subsidiary or neighboring cells is directly related ontogenetically to the guard cells and the others are not.

In the development of a stoma with mesogenous subsidiary cells (Fig. 9.20), the precursor of the stomatal complex (the meristemoid) is formed by an asymmetric division of a protodermal cell, and two subsequent asymmetric divisions result in the partitioning of the precursor into a mother cell of the guard cells and two subsidiary cells. One more but equal division leads to the formation of two guard cells.

The origin of perigenous subsidiary cells is graphically illustrated in the differentiation of a grass stoma (Fig. 9.21). The meristemoid, which functions directly as the guard mother cell, is the short daughter cell formed through an asymmetric division of a protodermal cell. Before the guard mother cell divides, the subsidiary cells are formed along the sides of this short cell by asymmetric division of two contiguous cells (subsidiary mother cells). Division of the subsidiary mother cell is preceded by migration of its nucleus to an actin patch along the subsidiary mother cell wall flanking the guard mother cell. In the maize leaf subsidiary cell fate determinants apparently are localized to this actin patch and subsequently are transferred to the daughter nucleus in contact with the patch shortly after completion of mitosis. The daughter cell inheriting this nucleus is consequently determined to differentiate as the subsidiary cell (Gallagher and Smith, 2000). Growth adjustments after the formation of guard cells make the subsidiary cells appear as integral parts of the stomatal complex.

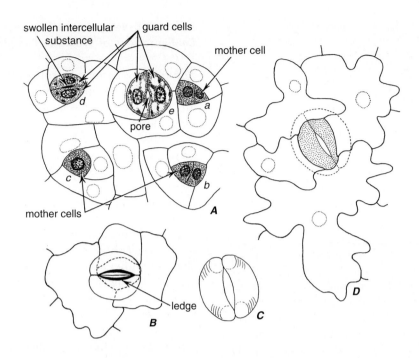

**FIGURE 9.18**

Stomata of *Nicotiana* (tobacco) in surface views. **A**, developmental stages: a, b, soon after division that resulted in formation of guard mother cell; c, guard mother cell has enlarged; d, guard mother cell has divided into two guard cells, still completely joined, but with swollen intercellular substance in position of future pore; e, young stoma with pore between guard cells. **B**, mature stoma seen from outer side of adaxial epidermis. **D**, similar stoma seen from inner side of abaxial epidermis. The guard cells are raised and thus appear above the epidermal cells in **B** and below them in **D**. **C**, guard cells as they appear from the inner side of the epidermis. (A, ×620; B–D, ×490.)

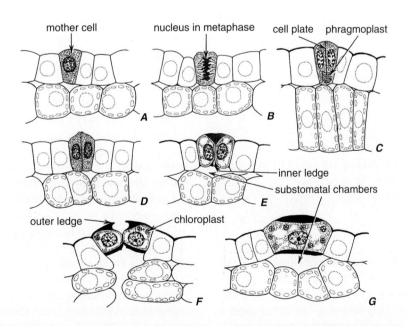

**FIGURE 9.19**

Development of stoma of *Nicotiana* (tobacco) leaf as seen in sections. **C**, from adaxial epidermis with some palisade cells; others from abaxial epidermis. **A–C**, guard mother cell before and during division into two guard cells. **D**, young guard cells with thin walls. **E**, guard cells have extended laterally and have begun to thicken their walls. Inner ledge and substomatal chamber have been formed. **F**, mature guard cells with upper and lower ledges and unevenly thickened walls. **G**, one mature guard cell cut parallel with its long axis and at right angles to the leaf surface. (All, ×490.)

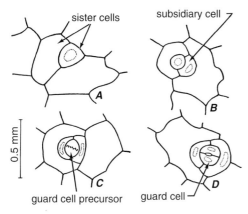

**FIGURE 9.20**

Development of stoma with mesogenous subsidiary cells in a leaf of *Thunbergia erecta*. **A**, epidermal cell has divided and given rise to a small precursor of the stomatal complex. **B**, the precursor has divided, setting apart one subsidiary cell. **C**, the second subsidiary cell and guard-cell precursor have been formed. **D**, the stomatal complex has been completed by division of the guard cell precursor. (From Esau, 1977; adapted from Paliwal, 1966.)

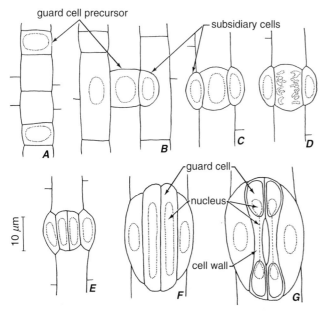

**FIGURE 9.21**

Development of stomatal complex in oat (*Avena sativa*) internode. The subsidiary cells are perigenous. **A**, the two short cells are guard-cell precursors. **B**, left, the nucleus of a long cell is in position to divide to form a subsidiary cell; right, subsidiary cell has been formed. **C**, guard-cell precursor before mitosis. **D**, guard-cell precursor in anaphase. **E**, stomatal complex of two guard cells and two subsidiary cells is still immature. **F**, cells of the stomatal complex have elongated. **G**, stomatal complex is mature. (From Esau, 1977; from photographs in Kaufman et al., 1970a.)

## Different Developmental Sequences Result in Different Configurations of Stomatal Complexes

The patterns formed by fully differentiated guard cells and the cells surrounding them, as seen from the surface, are used for taxonomic purposes. It is important to note, however, that mature stomatal complexes that look alike may have had different developmental pathways. Several classifications have been proposed for mature stomatal complexes in eudicots, with various degrees of complexity (Metcalfe and Chalk, 1950; Fryns-Claessens and Van Cotthem, 1973; Wilkinson, 1979; Baranova, 1987, 1992). Among the principal types of stomatal configurations are *anomocytic*, in which epidermal cells around the guard cells are not distinguishable from other epidermal cells, that is, subsidiary cells are lacking (Fig. 9.22A); *anisocytic*, in which the stoma is surrounded by three subsidiary cells, with one distinctly smaller than the other two (Fig. 9.22B; found in *Arabidopsis* and representative for the Brassicaceae); *paracytic*, in which the stoma is accompanied on either side by one or more subsidiary cells parallel to the long axis of the guard cells (Fig. 9.22C); *diacytic*, in which the stoma is enclosed by a pair of subsidiary cells whose common walls are at right angles to the guard cells (Fig. 9.22D); *actinocytic*, in which the stoma is surrounded by a circle of radiating cells whose long axes are perpendicular to the outline of the guard cells (Fig. 9.22E); *cyclocytic* (encyclocytic), in which the stoma is surrounded by one or two narrow rings of subsidiary cells, numbering four or more (Fig. 9.22F); *tetracytic*, in which the stoma is enclosed by four subsidiary cells, two lateral and two polar (terminal), also found in many monocots (Fig. 9.23). The same species may exhibit more than one type of stomatal complex, and the pattern may change during leaf development.

In most monocots the configuration of the stomatal complex is rather precisely related to the developmental sequence. Having examined about 100 species representing most families of the monocots, Tomlinson (1974) recognized the following main configurations of stomatal complexes resulting from specific developmental sequences (Fig. 9.23). The meristemoid arises through an **asymmetric division** of a protodermal cell (A). It is the smaller of the two cells and seems always to be the distal cell (toward the leaf apex). The meristemoid, which functions directly as the guard mother cell, normally is in contact with four **neighboring cells** (B). (Note that Tomlinson used the term neighboring cells to refer to the cells that lie next to the meristemoid when it is initiated.) These cells may not divide, whereupon they become *contact cells* directly, that is, cells that are in contact with the guard cells in the mature stomatal complex (F), as in Amaryllidaceae, Liliaceae, and Iridaceae. On the other hand, the neighboring cells may divide anticlinally and produce *derivatives*. The

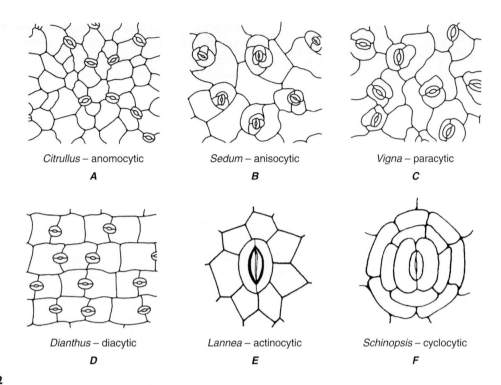

*Citrullus* – anomocytic
**A**

*Sedum* – anisocytic
**B**

*Vigna* – paracytic
**C**

*Dianthus* – diacytic
**D**

*Lannea* – actinocytic
**E**

*Schinopsis* – cyclocytic
**F**

**FIGURE 9.22**

Epidermis in surface views illustrating principal types of stomatal configurations. (**A–D**, from Esau, 1977; **E**, Fig. 10.3b and **F**, Fig. 10.3h redrawn from Wilkinson, 1979, *Anatomy of the Dicotyledons*, 2nd ed., vol. I, C. R. Metcalfe and L. Chalk, eds., by permission of Oxford University Press.)

orientation of these walls is of major importance in development of the stomatal complex: they may be exclusively oblique (C-E) or exclusively perpendicular and/or parallel to the files of protodermal cells (F-H). With division of the neighboring cells the stomatal complex now becomes definable as guard cells and a combination of neighboring cells and their derivatives (G) or as guard cells and derivatives of neighboring cells (E, H). Thus the contact cells of the stoma are either all derivatives (E-H) or a combination of derivatives and undivided neighboring cells (G). The type complex illustrated by G is the familiar grass type (Poaceae). It also occurs in a number of other families, including the Cyperaceae and Juncaceae; H is characteristic of many Commelinaceae, and E of the Palmae.

# TRICHOMES

**Trichomes** (from the Greek, meaning a growth of hair) are highly variable epidermal appendages (Figs. 9.24 and 9.25). They may occur on all parts of the plant and may persist throughout the life of the plant part or may fall off early. Some of the persisting trichomes remain alive; others die and become dry. Although trichomes vary widely in structure within families and smaller

groups of plants, they are sometimes remarkably uniform in a given taxon and have long been used for taxonomic purposes (Uphof and Hummel, 1962; Theobald et al., 1979).

Trichomes are usually distinguished from **emergences**, such as warts and prickles, which are formed from both epidermal and subepidermal tissue and typically are more massive than trichomes. The distinction between trichomes and emergences is not sharp, however, because some trichomes are elevated upon a base consisting of subepidermal cells. Thus a developmental study may be necessary to determine whether some outgrowths are solely epidermal in origin or both epidermal and subepidermal in origin.

## Trichomes Have a Variety of Functions

Plants growing in arid habitats tend to have hairier leaves than similar plants from more mesic habitats (Ehleringer, 1984; Fahn, 1986; Fahn and Cutler, 1992). Studies of arid-land plants indicate that increase in leaf pubescence (hairiness) reduces the transpiration rate by (1) increasing the reflection of solar radiation, which lowers leaf temperatures, and (2) increasing the boundary layer (the layer of still air through which water vapor must diffuse). Moreover the basal or stalk cells of the

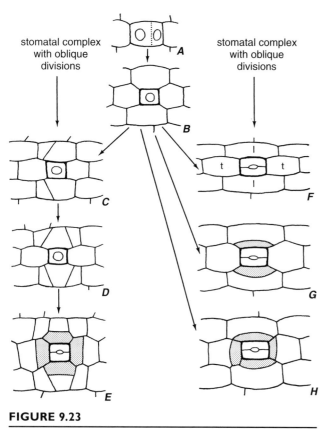

**FIGURE 9.23**

Examples of types of stomatal development in monocots. Diagrammatic. **A**, nonequational division results in the formation of **B**, a small guard-cell precursor surrounded by four neighboring cells in a cruciate arrangement. **C–E**, oblique and other divisions in neighboring cells result in the formation of four derivatives (stippled) in contact with the guard cells. **F–H**, no oblique divisions occur in the formation of stomatal complexes: **F**, original neighboring cells, two lateral (l) and two terminal (t), become contact cells; **G**, derivatives (stippled) of the two lateral neighboring cells and the two undivided terminal neighboring cells become contact cells; **H**, derivatives (stippled) of four neighboring cells become contact cells. **E**, palm type; **G**, grass type. (From Esau, 1977; adapted from Tomlinson, 1974.)

trichomes of at least some xeromorphic leaves are completely cutinized, precluding apoplastic water flow into the trichomes (Chapter 16; Fahn, 1986). Many "air plants" such as epiphytic bromeliads utilize foliar trichomes for the absorption of water and minerals (Owen and Thomson, 1991). In contrast, in the saltbush (*Atriplex*), salt-secreting trichomes remove salts from the leaf tissue, preventing an accumulation of toxic salts in the plant (Mozafar and Goodin, 1970; Thomson and Healey, 1984). During the early stages of leaf development polyphenol-containing trichomes may play a protective role against UV-B radiation damage (Karabourniotis and

Easseas, 1996). Trichomes may provide a defense against insects (Levin, 1973; Wagner, 1991). In numerous species, trichome density is negatively correlated with insect responses in feeding and oviposition and with nutrition of larvae. Hooked trichomes impale insects and their larvae (Eisner et al., 1998). Secretory (glandular) trichomes may provide a chemical defense (Chapter 16). Whereas some insect pests are poisoned by trichome secretions, others are rendered harmless by immobilization in the secretion (Levin, 1973).

### Trichomes May Be Classified into Different Morphological Categories

Some morphological trichome categories are (1) *papillae*, which are small epidermal outgrowths often considered distinct from trichomes; (2) *simple (unbranched) trichomes*, a large grouping of extremely common unicellular (Fig. 9.25C–F) and multicellular trichomes (Figs. 9.24I, J and 9.25A, B); (3) *two-* to *five-armed trichomes* of various shapes; (4) *stellate trichomes*, all of which are star-shaped although variable in structure (Fig. 9.24C, E, F); (5) *scales*, or *peltate*, *trichomes*, consisting of a discoid plate of cells often borne on a stalk or attached directly to the foot (Figs. 9.24A, B and 9.25G, H); (6) *dendritic (branched) trichomes*, which branch along an extended axis (Fig. 9.24D; Theobald et al., 1979); and (7) *root hairs*. In addition there are many specialized types of trichomes such as stinging hairs, pearl glands, cystolith-containing hairs (Fig. 9.25C, E, F), and water vesicles (Chapter 16). Anatomical features may also be used to facilitate the description of trichomes, features such as glandular (Fig. 9.25B, G, H) or nonglandular; unicellular or multicellular; uniseriate or multiseriate; surface features, if any; differences in wall thickness, if any; cuticle thickness; different cell types within the trichome, that is, base or foot (Fig. 9.25B, G), stalk, tip or head; and the presence of crystals, cystoliths, or other contents. An extensive glossary of plant trichome terminology was compiled by W. W. Payne (1978).

### A Trichome Is Initiated as a Protuberance from an Epidermal Cell

The development of trichomes varies in complexity in relation to their final form and structure. Multicellular trichomes show characteristic patterns of cell division and cell growth, some simple, others complex. Some aspects of the development of multicellular glandular trichomes are considered in Chapter 16. Here we consider developmental aspects of three unicellular trichomes: the cotton fiber, the root hair, and the branched trichome of *Arabidopsis*.

The Cotton Fiber   The unicellular cotton (*Gossypium*) trichome, commonly known as a **cotton fiber**, is

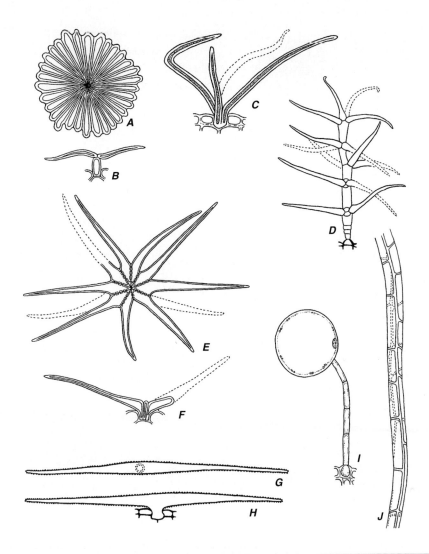

**FIGURE 9.24**

Trichomes. **A**, **B**, peltate scale of *Olea* in surface (**A**) and side (**B**) views. **C**, tufted, stellate hair of *Quercus*. **D**, dendritic hair of *Platanus*. **E**, **F**, stellate hair of *Sida* in surface (**E**) and side (**F**) views. **G**, **H**, two-armed, T-shaped unicellular hair of *Lobularia* in surface (**G**) and side (**H**) views. **I**, vesiculate hair of *Chenopodium*. **J**, part of multicellular shaggy hair of *Portulaca*. (A–C, I, ×210; D–H, J, ×105.)

initiated as a protuberance from a protodermal cell of the outer integument of the ovule (Ramsey and Berlin, 1976a, b; Stewart, 1975, 1986; Tiwari and Wilkins, 1995; Ryser, 1999). Development occurs synchronously for most of the trichomes, and their development can be divided into four somewhat overlapping phases. Phase 1, *fiber initiation* occurs at anthesis as the fiber initials appear as distinct protuberances on the surface of the ovule (Fig. 9.26A). Phase 2, *fiber elongation* begins soon afterward (Fig. 9.26B) and continues for 12 to 16 days after anthesis, depending on the cultivar. Whereas the cortical microtubules are randomly oriented in the fiber initials, they become oriented transversely to the long axis of the cell as the fiber begins to elongate. The fibers undergo dramatic elongation, reaching

lengths 1000 to 3000 times greater than their diameters (Peeters et al., 1987; Song and Allen, 1997). Elongation occurs via a diffuse mechanism; that is, it occurs throughout the length of the fiber (Fig. 9.27A), although it may be more rapid at the tip (Ryser, 1985). A large central vacuole usually resides in the basal part of the cell and the organelles appear to be dispersed more or less evenly throughout the cytosol (Tiwari and Wilkins, 1995). The primary walls of the cotton fibers are distinctly bilayered with a more electron-opaque outer layer consisting of pectins and extensin and a less electron-opaque inner layer of xyloglucans and cellulose (Vaughn and Turley, 1999). As is typical of cells with diffuse growth, new wall material is added throughout the cell surface. A cuticle extends over the wall of all

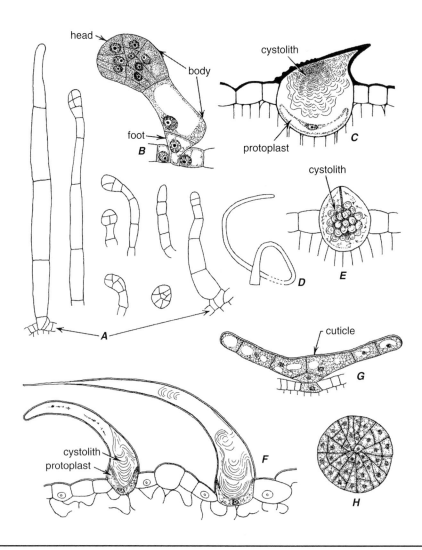

**FIGURE 9.25**

Trichomes. **A**, group of ordinary and glandular (with multicellular heads) hairs of *Nicotiana* (tobacco). **B**, enlarged view of glandular hair of tobacco, showing characteristic density of contents of glandular head. **C**, hooked hair with cystolith of *Humulus*. **D**, long coiled unicellular hair, and **E**, short bristle with cystolith of *Boehmeria*. **F**, hooked hairs with cystoliths of *Cannabis*. **G**, **H**, glandular peltate trichome of *Humulus* seen in sectional (**G**) and surface (**H**) views. (**H** from younger trichome than **G**.) (A, F, ×100; B, D, E, ×310; C–G, ×245; H, ×490.)

epidermal cells. Phase 3, **secondary wall formation** begins as the fiber approaches its final length and may continue for a further 20 to 30 days. The transition from primary wall formation during rapid cell elongation to the slowing of elongation and the onset of secondary wall formation is precisely correlated in the changing patterns of microtubules and wall microfibrils (Seagull, 1986, 1992; Dixon et al., 1994). With the beginning of secondary wall formation, the cortical microtubules begin changing their orientation from transverse to steeply pitched helices. In addition to cellulose the first layer of secondary wall contains some callose (Maltby et al., 1979). At maturity the secondary walls of cotton fibers consist of almost pure cellulose (Basra and Malik, 1984; Tokumoto et al., 2002). Those of the green-lint

mutant of cotton and some wild cotton species contain variable amounts of suberin and associated waxes, which typically are deposited in concentric layers that alternate with cellulosic layers (Ryser and Holloway, 1985; Schmutz et al., 1993). Hydrogen peroxide has been implicated as a signal in the differentiation of the secondary walls of the cotton fiber (Potikha et al., 1999). Phase 4, the **maturation** phase follows wall thickening. The fibers die, presumably by a process of programmed cell death, and become desiccated.

In an elegant study Ruan et al. (2001) found a correlation to exist between gating of the cotton fiber plasmodesmata and the expression of sucrose and K[+] transporter and expansin genes. The plasmodesmata that interconnect the cotton fibers with the underlying seed coat

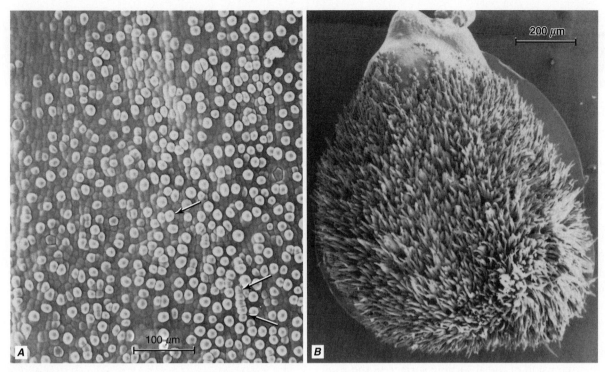

**FIGURE 9.26**

Scanning electron micrographs of developing cotton (*Gossypium hirsutum*) fibers. **A**, fiber initials on the chalazal half of an ovule on the evening of anthesis. The initials appear as tiny knobs. **B**, two days after anthesis, the ovule is covered by young fibers. (From Tiwari and Wilkins, 1995.)

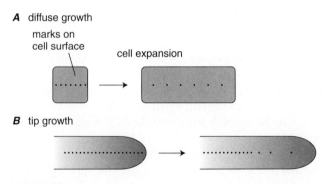

**FIGURE 9.27**

Cell elongation by diffuse and tip growth. Elongation of cotton fibers occurs uniformly throughout their length, that is, by diffuse growth (**A**). Elongation by root hairs and pollen tubes is confined to their tips; that is, root hairs and pollen tubes are tip-growing cells (**B**). If marks are placed on the surfaces of such cells, which then are allowed to grow, the relative distances between marks before and after further elongation has taken place will reflect the mechanism of elongation undertaken. (After Taiz and Zeiger, 2002. © Sinauer Associates.)

were dramatically down-regulated at the beginning of the elongation phase, completely blocking movement of the membrane-impermeant fluorescent solute carboxyfluorescein (CF) across that interface. As a result solute import into the developing fibers was shifted from an initially symplastic pathway to an apoplastic one. During the elongation phase the plasma membrane sucrose and $K^+$ transporter genes *GLSUT1* and *GhkT1* were expressed at maximal levels. Consequently fiber osmotic and turgor potentials were elevated, driving the phase of rapid elongation. The level of expansin mRNA was high only at the early period of elongation and decreased rapidly afterward. Overall, these results suggest that cotton fiber elongation is initially achieved by cell wall loosening and ultimately terminated by increased wall rigidity and loss of high turgor. The impermeability of the fiber plasmodesmata to CF was only temporary; symplastic continuity was reestablished at or near the end of the elongation phase. During the period of restricted CF import, most of the plasmodesmata were changed from unbranched to branched forms. The developing cotton fiber provides an excellent system for studies of cellulose biosynthesis and cell differentiation and growth (Tiwari and Wilkins, 1995; Pear et al., 1996; Song and Allen, 1997; Dixon et al., 2000; Kim, H. J., and Triplett, 2001).

Root Hairs The trichomes of roots, the **root hairs**, are tubular extensions of epidermal cells. In a study involving 37 species in 20 families, the root hairs varied between 5 and 17 micrometers in diameter and between 80 and 1500 micrometers in length (Dittmer, 1949). Root hairs typically are unicellular and unbranched (Linsbauer, 1930). The adventitious roots of *Kalanchoë fedtschenkoi* growing in air have multicellular root hairs, whereas the same kinds of roots growing in soil have unicellular ones (Popham and Henry, 1955). Root hairs are typical of roots but tubular outgrowths identical to root hairs may develop from epidermal cells on the lower portion of the hypocotyl of seedlings (Baranov, 1957; Haccius and Troll, 1961). Although root hairs typically are epidermal in origin, in the Commelinaceae (which includes *Rhoeo* and *Tradescantia*) "secondary root hairs" develop from cells of the exodermis several centimeters from the root tip in the region of older epidermal ("primary") root hairs (Pinkerton, 1936). In *schizoriza* (*scz*) mutants of *Arabidopsis* root hairs arise from the subepidermal layer of cells (Mylona et al., 2002). The principal function of root hairs is considered to be the extension of the absorbing surface of the root for the uptake of water and nutrients (Peterson and Farquhar, 1996). Root hairs have been identified as the sole producers of root exudate in *Sorghum* species (Czarnota et al., 2003).

Root hairs develop acropetally, that is, toward the apex of the root. Because of the acropetal sequence of initiation in most seedling taproots the root hairs show a uniform gradation in size, beginning with those nearest the apex and going back to those of mature length. Root hairs are initiated as small protuberances, or bulges (Fig. 9.28A), in the region of the root where cell division is subsiding. In *Arabidopsis*, root hairs always form at the end of the cell nearest the root apex (Schiefelbein and Sommerville, 1990; Shaw et al., 2000), and bulging at the initiation site is intimately linked to the acidification of the cell wall (Bibikova et al., 1997). Root hair initiation sites also show an accumulation of expansin (Baluška et al., 2000; Cho and Cosgrove, 2002) and an increase in xyloglucan and endotransglycosylase action (Vissenberg et al., 2001).

Unlike cotton fibers, which exhibit diffuse growth, root hairs are tip-growing cells (Fig. 9.27B; Galway et al., 1997). Like other tip-growing cells, most notably pollen tubes (Taylor, L. P., and Hepler, 1997; Hepler et al., 2001), elongating root hairs display a polarized organization of their contents with preferential localization of certain organelles to specific parts of the cell (Fig. 9.28). The apical part is enriched with secretory vesicles derived from Golgi bodies. The vesicles carry cell wall precursors that are released by exocytosis into the matrix of the developing wall. Calcium ($Ca^{2+}$) influx at the apex appears to be intimately linked with regulation of the secretory process through its effect on the actin

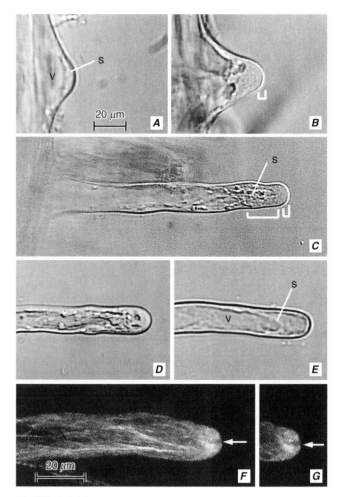

**FIGURE 9.28**

Differential interference contrast (**A–E**) and confocal (**F, G**) images of developing vetch (*Vicia sativa*) root hairs. **A,** emerging root hair, most of which is occupied by a large vacuole (v); s, cytoplasmic strands at the periphery. **B, C,** growing root hairs. Smooth region at tip contains Golgi vesicles (small bracket). The subapical region in **C** is traversed by cytoplasmic strands with many organelles (large bracket). **D,** root hair that is terminating growth with several small vacuoles close to tip. **E,** root hair full-grown with peripheral cytoplasm (s) and one large, central vacuole (v). **F, G,** immunolabeled bundles of actin filaments. The bundles are oriented parallel to the long axis of the cell. The very tip of the hair (cleft indicated by arrow) appears to be devoid of actin. (**A–E,** same magnification; **F, G,** same magnification. From Miller, D. D., et al., 1999. © Blackwell Publishing.)

component of the cytoskeleton (Gilroy and Jones, 2000). In growing root hairs, bundles of actin filaments extend the length of the root hairs in the cortical cytoplasm and loop back through a cytoplasmic strand traversing the vacuole (Figs. 9.28E, F and 9.29A; Ketelaar and Emons, 2001). The arrangement of actin filaments at the

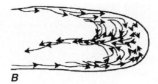

**FIGURE 9.29**

Schematic representations of the tip of a growing *Nicotiana tabacum* root hair. **A**, distribution of the actin filaments. **B**, reverse fountain streaming. (From Hepler et al., 2001. Reprinted, with permission, from the *Annual Review of Cell and Developmental Biology*, vol. 17. © 2001 by Annual Reviews. www.annualreviews.org)

tip is controversial. Some reports indicate that the actin filaments flare into a three-dimensional meshwork—an ***actin cap***—at the tip (Braun et al., 1999; Baluška et al., 2000), whereas others suggest that actin filaments are disorganized and few in number or absent at the tip (Figs. 9.28F, G and 9.29A; Cárdenas et al., 1998; Miller, D. D., et al., 1999). Cytoplasmic streaming in growing root hairs and pollen tubes is described as **reverse fountain streaming**, in which streaming moves acropetally along the sides of the cell and basipetally in the central strand (Fig. 9.29B; Geitmann and Emons, 2000; Hepler et al., 2001). The subapical part of the hair accumulates a large number of mitochondria and the basal region most of the other organelles. The nucleus migrates into the developing hair and, as long as the hair is growing, it is positioned some distance from the tip (Lloyd et al., 1987; Sato et al., 1995). Positioning of the nucleus is an actin-regulated process (Ketelaar et al., 2002). Upon completion of elongation the nucleus may assume a more or less random position (Meekes, 1985) or migrate to the base (Sato et al., 1995), and cytoplasmic polarity is lost. Now the bundles of actin filaments loop through the tip (Miller, D. D., et al., 1999), as evidenced by the circulation type of cytoplasmic streaming that occurs in fully grown hairs (Sieberer and Emons, 2000). The microtubules are longitudinally oriented in growing root hairs; as they approach the tip of the cell, they become randomly oriented (Lloyd, 1983; Traas et al., 1985). The microtubules apparently are responsible for the organization of actin filaments into bundles, which together with myosin are capable of transporting the secretory vesicles (Tominaga et al., 1997). The microtubules play a role in determining the growth direction of the cell (Ketelaar and Emons, 2001). The extension of the root hair wall proceeds rapidly (0.1 mm per hour in the radish root, Bonnett and Newcomb, 1966; $0.35 \pm 0.03\,\mu m$ per minute in *Medicago truncatula*, Shaw, S. L., et al., 2000). Root hairs typically are short-lived, their longevity commonly being measured

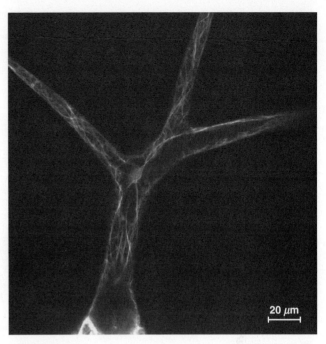

20 μm

**FIGURE 9.30**

Filamentous-actin cytoskeleton in *Arabidopsis* trichome. F-action is visualized in living trichomes using GFP fused to an actin-binding domain of the mouse Talin gene. (Courtesy of Jaideep Mathur.)

in days. Excellent reviews of root hair structure, development and function are provided by Ridge (1995), Peterson and Farquhar (1996), Gilroy and Jones (2000), and Ridge and Emons (2000).

The *Arabidopsis* Trichome Trichomes are the first epidermal cells to begin differentiating in the epidermis of developing leaf primordia, and those of *Arabidopsis* are no exception (Hülskamp et al., 1994; Larkin et al., 1996). Initiation and maturation of the trichomes proceed in an overall basipetal direction (tip to base) along the adaxial (upper) surface of the leaf primordium, although additional trichomes commonly are initiated between mature ones in portions of the leaf where the surrounding protodermal cells are still dividing as growth of the leaf continues. At maturity the leaf trichomes of *Arabidopsis* normally have three branches (Figs. 9.30 and 9.31B).

Trichome development in the *Arabidopsis* leaf may be divided into two growth phases (Hülskamp, 2000; Hülskamp and Kirik, 2000). The ***first phase*** is initiated when the trichome precursor stops dividing and begins to endoreduplicate (to undergo DNA replication in the absence of nuclear and cell divisions; Chapter 5). The incipient trichome first appears as a small protuberance on the surface of the leaf (Fig. 9.31A). After two or three

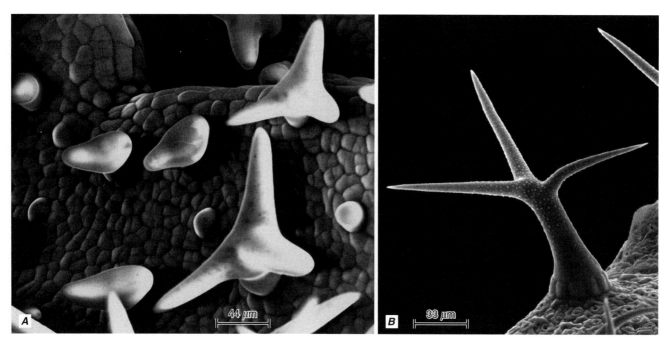

**FIGURE 9.31**

Scanning electron micrographs of the adaxial surface of *Arabidopsis* leaf showing (**A**) stages of trichome morphogenesis on a single leaf, and (**B**) a mature trichome with papillae. (Courtesy of Jaideep Mathur.)

endoreduplication cycles it grows out from the leaf surface and undergoes two successive branching events. The last, or fourth, round of endoreduplication occurs after the first branching event. The DNA content of the trichome now has increased 16 fold, from the 2C (C is the haploid DNA content) of normal protodermal cells to 32C (Hülskamp et al., 1994). The first two branches are aligned with the long axis (basal-distal) axis of the leaf (Fig. 9.31A). The distal branch then divides perpendicular to the first branching plane to produce the three-branched trichome (Fig. 9.31A, B). It is generally presumed that prior to branching—that is, during the tubular growth stage—the developing trichome increases in size largely by tip growth, and that afterward the trichome expands by diffuse growth. During the ***second phase***, which follows initiation of the three branches, the trichome undergoes rapid expansion and increases in size by a factor of 7 to 10 (Hülskamp and Kirik, 2000). As the trichome approaches maturity, the cell wall thickens and its surface becomes covered with papillae of unknown origin and function (Fig. 9.31B). The base of the mature trichome is surrounded by a ring of 8 to 12 rectangular cells that first become recognizable at about the time the trichome initiates branching (Hülskamp and Schnittger, 1998). The base of the trichome appears to have pushed under the surrounding cells to form a concavity, or socket; hence the surrounding cells are sometimes called ***socket cells***. Also termed

***accessory cells***, these cells are not closely related ontogenetically to the trichome (Larkin et al., 1996).

The cytoskeleton plays an essential role in trichome morphogenesis (Reddy and Day, I. S., 2000). During the first phase of trichome development, the microtubules play the predominant role; during the second phase the actin filaments do so. The microtubules are responsible for establishing the spatial patterning of trichome branches, the orientation of the microtubules playing a causative role in determining the direction of growth (Hülskamp, 2000; Mathur and Chua, 2000). The actin filaments (Fig. 9.30) play a dominant role during extension growth of the branches, targeting the delivery of cell wall components necessary for growth and serving to elaborate and maintain the already established branching pattern (Mathur et al., 1999; Szymanski et al., 1999; Bouyer et al., 2001; Mathur and Hülskamp, 2002).

Because of their simplicity and visibility, the leaf trichomes of *Arabidopsis* have provided an ideal genetic model system for the study of cell fate and morphogenesis in plants. An ever increasing number of genes that are required for trichome development are being identified. Based on the genetic analysis of the corresponding mutant phenotypes, a greater understanding of the sequence of regulatory and developmental steps for trichome morphogenesis is being achieved. Some excellent reviews of trichome morphogenesis in *Arabidopsis* are provided by Oppenheimer (1998), Glover (2000), and Hülskamp and colleagues (Hülskamp, 2000; Hülskamp and Kirik, 2000; Schwab et al., 2000).

# CELL PATTERNING IN THE EPIDERMIS

## The Spatial Distribution of Stomata and Trichomes in Leaves Is Nonrandom

It has long been known that the spatial distribution, or **patterning**, of the stomata and trichomes in the leaf epidermis is nonrandom and that a minimum spacing exists between them. The mechanisms that govern pattern formation, however, are just now being elucidated. Two proposed mechanisms have received the greatest attention: the cell lineage mechanism and the lateral inhibition mechanism. The **cell lineage mechanism** relies on a highly ordered series of cell divisions, usually asymmetric, that automatically result in different categories of cells. The ultimate fate of each of these cells can be predicted by its position in the lineage. The **lateral inhibition mechanism** does not rely on cell lineage but rather on interactions, or signaling, between developing epidermal cells to determine the fate of each cell. A third mechanism, the **cell cycle-dependent mechanism**, proposes that stomatal patterning is coupled to the cell cycle (Charlton, 1990; Croxdale, 2000).

There seems to be little doubt that a cell lineage-dependent mechanism is a major force driving stomatal patterning in the leaves of eudicots (Dolan and Okada, 1999; Glover, 2000; Serna et al., 2002). In *Arabidopsis*, for example, the ordered division pattern of the stomatal meristemoids results in a pair of guard cells surrounded by three clonally, or ontogenetically, related subsidiary cells, one distinctly smaller than the other two (anisocytic stomatal complex; Fig. 9.22B). Consequently each pair of guard cells is separated from another pair by at least one epidermal cell. Two *Arabidopsis* mutants, *two many mouths (tmm)* and *four lips (flp)*, have been identified that disrupt normal patterning and result in clustering of stomata (Yang and Sack, 1995; Geisler et al., 1998). It has been proposed that TMM is a component of a receptor complex, whose function is to sense positional clues during epidermal development (Nadeau and Sack, 2002). A third more recently discovered *Arabidopsis* stomatal mutant, *stomatal density and distribution1-1 (sdd1-1)*, exhibits a twofold to fourfold increase in stomatal density, a fraction of the additional stomata occurring in clusters (Berger and Altmann, 2000). Apparently the *SDD1* gene plays a role in the regulation of the number of cells entering the stomatal pathway and the number of asymmetric cell divisions that occur before stomatal development (Berger and Altmann, 2000; Serna and Fenoll, 2000). *SDD1* is expressed strongly in meristemoids/guard mother cells and weakly in cells bordering them. It has been proposed that *SDD1* generates a signal that moves from the meristemoids/guard mother cells to the bordering cells and either stimulates the development of bordering cells into ordinary epidermal cells or inhibits their conversion into additional (satellite) meristemoids (von Groll et al., 2002). The function of *SDD1* has been shown to be dependent on TMM activity (von Groll et al., 2002). (Incidentally, whereas stomatal patterning is nonrandom on the foliage leaves of wild type *Arabidopsis*, on the cotyledons of the same plant the stomatal pattern is random; Bean et al., 2002.)

In the leaves of the monocot *Tradescantia*, the activity of the epidermal cells can be separated into four major regions, or zones: a zone of proliferative divisions (the basal meristem), a zone without division where stomatal patterning takes place, a zone of stomatal development with divisions, and a zone in which only cell expansion takes place (Chin et al., 1995). As new cells are displaced from the basal meristem their position in the cell cycle apparently determines whether they will become stomatal or epidermal cells when they reach the patterning zone (Chin et al., 1995; Croxdale, 1998). Patterning of stomata in *Tradescantia* is also affected by late developmental events that may arrest up to 10% of the stomatal initials (guard mother cells) in their development (Boetsch et al., 1995). The stomatal initials that are arrested lie closer to their nearest neighboring initial than the average distance between stomata. Lateral inhibition may be involved here. The arrested initials switch developmental pathways and become ordinary epidermal cells.

Unlike stomatal patterning in the *Arabidopsis* leaf, the spacing of leaf trichomes does not rely on a cell lineage-based mechanism. As mentioned previously, the trichomes and surrounding accessory cells are not clonally related. There is no ordered cell division to provide intervening cells between trichomes. It is likely that interactions, or signaling, among developing epidermal cells determine which cells go on to develop into trichomes. Perhaps the developing trichomes recruit a set of accessory cells and inhibit other cells from trichome development (Glover, 2000).

Two genes, *GLABRA1 (GL1)* and *TRANSPARENT TESTA GLABRA1 (TTG1)*, have been identified as being required for the initiation of trichome development and proper trichome patterning in the *Arabidopsis* leaf. Both genes function as positive regulators of trichome development (Walker et al., 1999). Strong *gl1* and *ttg1* mutants produce no trichomes on the surfaces of their leaves (Larkin et al., 1994). A third gene, *GLABRA3 (GL3)*, also may play a role in the initiation of the leaf trichomes (Payne, C. T., et al., 2000). Two genes are known as negative regulators of trichome patterning in the *Arabidopsis* leaf, *TRIPTYCHON (TRY)* and *CAPRICE (CPC)* (Schellmann et al., 2002). Both genes are expressed in trichomes and act together during lateral inhibition of cells bordering the incipient trichomes.

Another gene involved early in trichome development is *GLABRA2 (GL2)*, which is expressed in trichomes throughout their development (Ohashi et al., 2002). Trichomes are produced in *gl2* mutants but their outgrowth is stunted and most do not branch (Hülskamp et al., 1994). Still another gene has been identified, *TRANSPARENT TESTA GLABRA2 (TTG2)*, that controls the early development of trichomes.

### There Are Three Main Types of Patterning in the Epidermis of Angiosperm Roots

**Type 1** In most angiosperms (almost all eudicots, some monocots) any protodermal cell of the root has the potential to form a root hair, and the root hairs are randomly arranged (Fig. 9.32A). Within the Poaceae, the subfamilies Arundinoideae, Bambusoideae, Chloridoideae, and Panicoideae exhibit this pattern (Row and Reeder, 1957; Clarke et al., 1979).

**Type 2** In the basal angiosperm family Nymphaeaceae and some monocots, the root hairs originate from the smaller product of an asymmetric division (Fig. 9.32B). These smaller and denser root-hair forming cells are called **trichoblasts** (Leavitt, 1904). In some families (Alismataceae, Araceae, Commelinaceae, Haemodora-

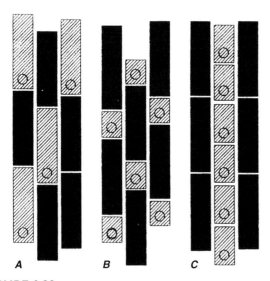

**FIGURE 9.32**

Three main types of root epidermal patterning in angiosperm roots. Hatched cells are root-hair cells and black cells are hairless cells. The circle indicates the location of the root-hair base. **A**, Type 1. Any protodermal cell can form a root hair. **B**, Type 2. The root hairs originate from the smaller product (trichoblast) of an asymmetric division. **C**, Type 3. There are discrete, vertical files composed entirely of shorter hair cells and longer hairless cells. (From Dolan, 1996, by permission of Oxford University Press.)

ceae, Hydrocharitaceae, Pontederiaceae, Typhaceae, and Zingiberaceae), the trichoblast is located at the proximal end (away from the root apex) of the initial protodermal cell. In others (Cyperaceae, Juncaceae, Poaceae, and Restianaceae), it is located at the distal end (toward the root apex) (Clowes, 2000). Prior to cytokinesis the nucleus migrates to either the proximal or distal end of the initial cell. The trichoblasts show considerable cytologic and biochemical differentiation. In *Hydrocharis*, for example, the trichoblasts differ from their long sister cells (**atrichoblasts**) in having larger nuclei and nucleoli, simpler plastids, more intense enzymic activity, and larger amounts of nucleohistone, total protein, RNA, and nuclear DNA (Cutter and Feldman, 1970a, b).

**Type 3** The third pattern, in which the cells are arranged in vertical files composed entirely of shorter **hair cells** or longer **nonhair**, or **hairless**, **cells** (Fig. 9.32C), is exemplified by *Arabidopsis* and other members of the Brassicaceae (Cormack, 1935; Bünning, 1951). Referred to as the *striped pattern* (Dolan and Costa, 2001), it also occurs in the Acanthaceae, Aizoaceae, Amaranthaceae, Basellaceae, Boraginaceae, Capparaceae, Caryophyllaceae, Euphorbiaceae, Hydrophyllaceae, Limnanthaceae, Plumbaginaceae, Polygonaceae, Portulacaceae, Resedaceae, and Salicaceae (Clowes, 2000; Pemberton et al., 2001). Both striped and nonstriped patterns are found among species of Onagraceae and Urticaceae (Clowes, 2000).

In the *Arabidopsis* root, hair and hairless cell types are specified in a distinct position-dependent pattern: root-hair cells are always positioned over the junction of the radial (anticlinal) walls between two cortical cells and hairless cells directly over cortical cells (Fig. 9.33; Dolan et al., 1994; Dolan, 1996; Schiefelbein et al., 1997). Several genes have been implicated in the establishment of the root epidermal pattern in *Arabidopsis*, including *TTG1*, *GL2*, *WEREWOLF (WER)*, and *CAPRICE (CPC)*. In *ttg1*, *gl2*, and *wer* mutants all epidermal cells produce root hairs, indicating that *TTG1*, *GL2*, and *WER* are negative regulators of root-hair development (Galway et al., 1994; Masucci et al., 1996; Lee and Schiefelbein, 1999). By contrast, *cpc* mutants do not form root hairs, whereas transgenic plants overexpressing *CPC* convert all of the root epidermal cells into hair-forming cells, indicating that *CPC*, which is predominately expressed in hairless cells, is a positive regulator of root-hair development (Wada et al., 1997, 2002). The expression of *CPC* is controlled by *TTG1* and *WER*, and *CPC* promotes differentiation of hair-forming cells by controlling *GL2*. It has been shown that CPC protein moves from hairless cells expressing *CPC* to hair-forming cells where it represses the *GL2* expression (Wada et al., 2002). As noted by Schiefelbein (2003), despite the very different distribution of hair cells in the root and shoot of *Arabidopsis* a

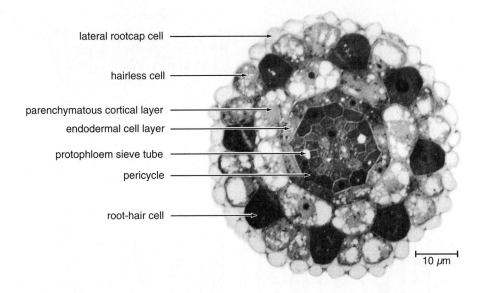

lateral rootcap cell

hairless cell

parenchymatous cortical layer

endodermal cell layer

protophloem sieve tube

pericycle

root-hair cell

10 μm

**FIGURE 9.33**

Transverse section of an *Arabidopsis* root. A single layer of lateral rootcap cells surrounds the epidermis. The densely staining epidermal cells positioned over the junction of the radial walls between adjacent cortical cells are root-hair cells. The markedly less dense epidermal cells are hairless cells. (Reprinted with permission from Schiefelbein et al., 1997. © American Society of Plant Biologists.)

similar molecular mechanism is responsible for patterning both cell types.

A clear relationship exists between symplastic communication and differentiation of the epidermis in the *Arabidopsis* root. Dye-coupling experiments indicate that initially, the epidermal cells of the root are symplastically coupled (Duckett et al., 1994). However, as they progress through the elongation zone and enter the region of differentiation, where they differentiate into hair cells or hairless cells, they become symplastically uncoupled. Mature root epidermal cells are symplastically isolated not only from each other but also from underlying cortical cells. The frequency of plasmodesmata within all tissues of the *Arabidopsis* root has been shown to decrease dramatically with root age (Zhu et al., 1998). The cells of the mature hypocotyl epidermis of *Arabidopsis* are symplastically coupled but isolated from the underlying cortex and from the root epidermis (Duckett et al., 1994).

# OTHER SPECIALIZED EPIDERMAL CELLS

In addition to guard cells and various kinds of trichomes, the epidermis may contain other kinds of specialized cells. The leaf epidermal system of Poaceae, for example, typically contains *long cells* and two kinds of *short cells*, silica cells and cork cells (Figs. 9.9 and 9.34). In some parts of the plant the short cells develop protrusions above the surface of the leaf in the form of papillae, bristles, spines, or hairs. The epidermal cells of Poaceae are arranged in parallel rows, and the composition of these rows varies in different parts of the plant (Prat, 1948, 1951). The inner face of the leaf sheath at its base, for example, has a homogeneous epidermis composed of long cells only. Elsewhere in the leaves combinations of the different types of cells are found. Rows containing long cells and stomata occur over the assimilatory tissue; only elongated cells or such cells combined with cork cells or bristles or with mixed pairs of short cells follow the veins. In the stem, too, the composition of the epidermis varies, depending on the level of the internode and on the position of the internode in the plant. Still another peculiar type of epidermal cell found in the Poaceae and other monocots is the bulliform cell.

## Silica and Cork Cells Frequently Occur Together in Pairs

Silica ($SiO_2 \cdot nH_2O$) is deposited in large quantities in the shoot system of grasses, and **silica cells** are so-called because, when they are fully developed, their lumina are filled with isotropic bodies of silica. The **cork cells** have suberized walls and often contain solid organic material. Apart from the frequency and distribution of the short cells, the shapes, or forms, of the **silica bodies** in the silica cells are very important for diagnostic and taxonomic purposes (Metcalfe, 1960; Ellis, 1979; Lanning and Eleuterius, 1989; Valdes-Reyna and Hatch,

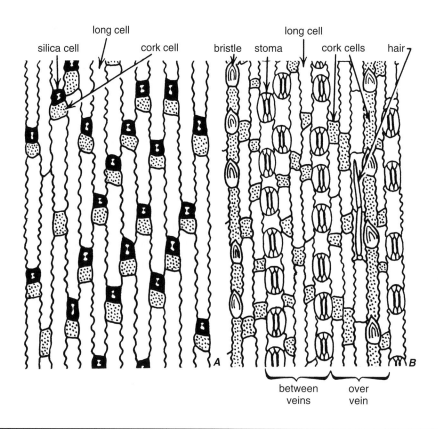

**FIGURE 9.34**

Epidermis of sugarcane (*Saccharum*) in surface view. **A**, epidermis of stem showing alternation of long cells with pairs of short cells: cork cells and silica cells. **B**, lower epidermis from a leaf blade, showing distribution of stomata in relation to various kinds of epidermal cells. (A, ×500; B, ×320. Adapted from Artschwager, 1940.)

1991; Ball et al., 1999). Also termed **phytoliths**, from the Greek meaning plant stones, silica bodies, or more exactly their various forms, have come to play an important role in archaeobotanical and geobotanical research (Piperno, 1988; Mulholland and Rapp, 1992; Bremond et al., 2004). According to Prychid et al. (2004), the most common type of silica body in monocots is the "druse-like" spherical body. Other forms include the "hat-shaped" type ("truncated conical"), "trough-shaped," and an amorphous, fragmentary type (silica sand). The shapes of the silica bodies do not necessarily conform with those of the silica cells containing them.

Silica-cork cell pairs arise from symmetrical, or equal, division of short cell initials in the basal (intercalary) meristem of the leaf and internode (Kaufman et al., 1970b, c; Lawton, 1980). Consequently the daughter cells initially are of equal size. The upper cell is the future silica cell, the lower one the future cork cell. The silica cell enlarges more rapidly than the cork cell and commonly bulges out from the surface of the epidermis and into the cork cell. Whereas the silica cell walls remain relatively thin, the cork cell walls become considerably thickened and suberized. As the silica cell approaches maturity, its nucleus breaks down and the

cell becomes filled with fibrillar material and contains an occasional lipid droplet, both substances presumably remnants of the protoplast. Finally, the lumen of the senescent silica cell becomes filled with silica, which polymerizes to form the silica body (Kaufman et al., 1985). The cork cell retains its nucleus and cytoplasm at maturity. In *Sorghum*, cork cells have been shown to secrete tubular filaments of epicuticular wax (Fig. 9.9; McWhorter et al., 1993; Jenks et al., 1994).

Silica bodies may occur in epidermal cells other than silica cells, including long epidermal cells and bulliform cells (Ellis, 1979; Kaufman et al., 1981, 1985; Whang et al., 1998). Silica deposits are found in abundance in the epidermal cell walls. In addition the intercellular spaces between subepidermal cells may become filled with silica. Several functions have been proposed for silica bodies and silica in cell walls. A proposed function for cell wall silica is that of providing support to the leaves. In Japan, silica in the form of slag is widely used as a siliceous fertilizer for rice plants. The leaves of rice plants so treated are more erect, allowing more light to reach the lower leaves and resulting in increased canopy photosynthesis. The presence of silica also increases resistance to various insects and pathogenic fungi and bacteria (Agarie et al., 1996). The hypothesis that silica bodies in silica cells might act as "windows" and silicified trichomes as "light pipes" to facilitate the transmission of light to photosynthetic mesophyll has been tested and found wanting (Kaufman et al., 1985; Agarie et al., 1996).

### Bulliform Cells Are Highly Vacuolated Cells

Bulliform cells, literally "cells shaped like bubbles," occur in all monocot orders except the Helobiae (Metcalfe, 1960). They either cover the entire upper surface of the blade or are restricted to grooves between the longitudinal veins (Fig. 9.35). In the latter location they form bands, usually several cells wide, between the veins. In transverse sections through such a band the cells often form a fan-like pattern because the median cells are usually the largest and are somewhat wedge-shaped. Bulliform cells may occur on both sides of the leaf. They are not necessarily restricted to the epidermis, but are sometimes accompanied by similar colorless cells in the subjacent mesophyll.

Bulliform cells are mainly water-containing cells and are colorless because they contain little or no chlorophyll. In addition tannins and crystals are rarely found in these cells, although, as mentioned previously, they may accumulate silica. Their radial walls are thin, but the outer wall may be as thick or thicker than those of the adjacent ordinary epidermal cells. The walls are composed of cellulose and pectic substances. The outer walls are cutinized and also bear a cuticle.

Controversy has surrounded the function of bulliform cells. Their sudden and rapid expansion during a certain stage of leaf development is presumed to bring about the unfolding of the blade, hence the term *expansion cells* at times applied to these cells. Another concept is that by changes in turgor, these cells play a role in the hygroscopic opening and closing movements of mature leaves, hence the alternative term *motor cells*. Still other workers doubt that the cells have any other function than that of water storage. Studies on the unfolding and the hygroscopic movements of leaves of certain grasses have shown that the bulliform cells are not actively or specifically concerned with these phenomena

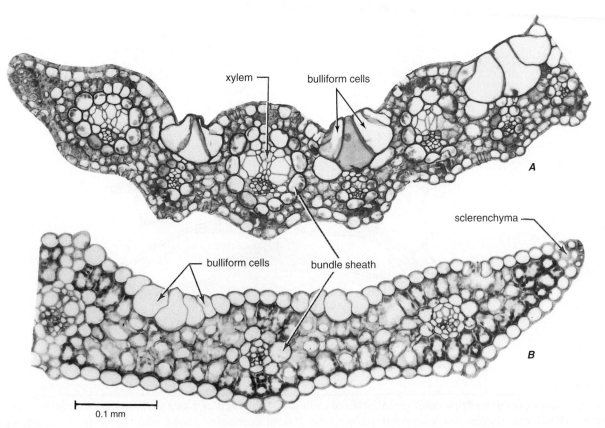

**FIGURE 9.35**

Transverse sections of grass leaf blades showing location of bulliform cells on upper side of the leaf. **A**, *Saccharum* (sugarcane), a C$_4$ grass, and **B**, *Avena* (oat), a C$_3$ grass. Note that the spatial association between mesophyll and vascular bundles is closer in sugarcane than in oat. (From Esau, 1977; slides courtesy of J. E. Sass.)

(Burström, 1942; Shields, 1951). Noting that the outer walls of bulliform cells are often quite thick and the lumina sometimes filled with silica, Metcalfe (1960) questioned how cells with such features could have an important motor function.

### Some Epidermal Hairs Contain Cystoliths

Undoubtedly the most familiar of cystoliths are the ellipsoidal cystoliths of *Ficus*, which, as mentioned previously, develop within lithocysts in the multiple epidermis of the leaf (Fig. 9.3). This type of cystolith formation was regarded by Solereder (1908) as the "true cystolith." Cystoliths also occur in the uniseriate epidermis of leaves, many in hairs. **Cystolith hairs** (Fig. 9.25C, E, F), or hair-like lithocysts, occur in several eudicot families, notably Moraceae (Wu and Kuo-Huang, 1997), Boraginaceae (Rao and Kumar, 1995; Rapisarda et al., 1997), Loasaceae, Ulmaceae, and Cannabaceae (Dayanandan and Kaufman, 1976; Mahlberg and Kim, 2004). Much of the information available on the distribution and composition of the cystoliths in hair-like lithocysts comes from studies dealing with the forensic identification of marijuana (*Cannabis sativa*) (Nakamura, 1969; Mitosinka et al., 1972; Nakamura and Thornton, 1973), for the presence of cystolith hairs is an important character in its identification.

Although the bodies of most cystoliths consist primarily of calcium carbonate, some contain abundant calcium carbonate and silica (Setoguchi et al., 1989; Piperno, 1988). Still others consist largely of silica (some species of Boraginaceae, Ulmaceae, Urticaceae, and Cecro-piaceae) (Nakamura, 1969; Piperno, 1988; Setoguchi et al., 1993). Inasmuch as the latter contain little or no calcium carbonate, they are not considered by all workers as cystoliths. Setoguchi et al. (1993), for example, refer to such cystolith-like containing cells as "silicified idioblasts."

Most detailed information on lithocyst-cystolith development comes from studies on the leaves and internodes of *Pilea cadierei* (Urticaceae) (Fig. 9.36;

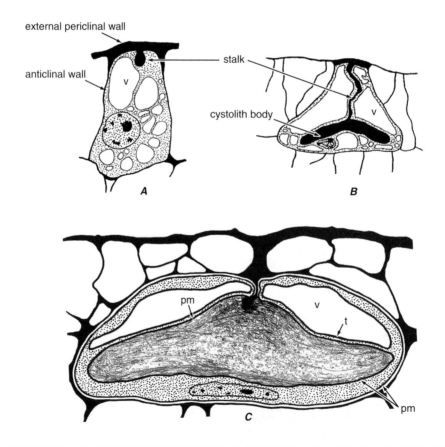

### FIGURE 9.36

Lithocyst development in *Pilea cadierei*. **A**, stalk of cystolith is initiated as a peg from thickened external periclinal wall. **B**, stalk of cystolith grows downward, pushing the plasma membrane ahead of it; the lithocyst enlarges greatly, and both it and the body of the cystolith become spindle-shaped. **C**, lithocyst near maturity. At maturity, the lithocyst cytoplasm occupies a thin boundary layer around the cell periphery and around the cystolith and its stalk. Details: pm, plasma membrane; t, tonoplast; v, vacuole. (**A, B**, adapted from Galatis et al., 1989; **C**, from photograph in Watt et al., 1987, by permission Oxford University Press.)

Watt et al., 1987; Galatis et al., 1989). The lithocysts in *P. cadierei* are initiated by asymmetric division of a protodermal cell. The smaller of the daughter cells may differentiate directly into a lithocyst, or it may undergo another division to produce a lithocyst. In either case, the incipient lithocyst becomes polarized, as the nucleus and most of the organelles come to lie close to the internal periclinal wall, while the external periclinal wall begins to thicken. When the external periclinal wall of the differentiating lithocyst becomes about twice as thick as that of the ordinary protodermal cells, the stalk of the cystolith is initiated as a peg, which grows downward, pushing the plasma membrane ahead of it. During stalk formation the lithocyst begins to vacuolate rapidly, and the vacuolar system comes to occupy the entire cell space except for the area in which the stalk and cystolith body are developing. While coordinating its growth with the dividing surrounding cells, the lithocyst elongates greatly, appearing to slide below the epidermis. Thus the once small, rectangular cell enlarges dramatically and becomes spindle-shaped. Development of the cystolith body is coordinated with that of the lithocyst as both elongate and increase in diameter together. The number and organization of the microtubules continually changes as differentiation of the lithocyst progresses, indicating that the microtubules play an important role in lithocyst morphogenesis (Galatis et al., 1989). When fully formed, the body of the spindle-shaped cystolith may measure up to 200 μm in length and 30 μm in diameter, attached at its midregion by the stalk to the external periclinal wall. At maturity the body of the cystolith is heavily impregnated with calcium carbonate. The bodies of some also contain silica and are covered in a sheath of siliceous material (Watt et al., 1987).

The physiological significance of cystoliths remains unclear. It has been suggested that cystolith formation may promote photosynthesis by enhancing the supply of carbon dioxide or that it may be the product of a detoxification mechanism similar to the formation of calcium granules in cells of molluscs (Setoguchi et al., 1989).

# ▍REFERENCES

ABAGON, M. A. 1938. A comparative anatomical study of the needles of *Pinus insularis* Endlicher and *Pinus merkusii* Junghun and De Vriese. *Nat. Appl. Sci. Bull.* 6, 29–58.

AGARIE, S., W. AGATA, H. UCHIDA, F. KUBOTA, and P. B. KAUFMAN. 1996. Function of silica bodies in the epidermal system of rice (*Oryza sativa* L.): Testing the window hypothesis. *J. Exp. Bot.* 47, 655–660.

ALLEN, G. J., K. KUCHITSU, S. P. CHU, Y. MURATA, and J. I. SCHROEDER. 1999. *Arabidopsis abi1-1* and *abi2-1* phosphatase mutations reduce abscisic acid-induced cytoplasmic calcium rises in guard cells. *Plant Cell* 11, 1785–1798.

ANTON, L. H., F. W. EWERS, R. HAMMERSCHMIDT, and K. L. KLOMPARENS. 1994. Mechanisms of deposition of epicuticular wax in leaves of broccoli, *Brassica oleracea* L. var. *capitata* L. *New Phytol.* 126, 505–510.

ARTSCHWAGER, E. 1940. Morphology of the vegetative organs of sugarcane. *J. Agric. Res.* 60, 503–549.

ASSMANN, S. M., and K.-I. SHIMAZAKI. 1999. The multisensory guard cell. Stomatal responses to blue light and abscisic acid. *Plant Physiol.* 119, 809–815.

ASSMANN, S. M., and X.-Q. WANG. 2001. From milliseconds to millions of years: Guard cells and environmental responses. *Curr. Opin. Plant Biol.* 4, 421–428.

AYLOR, D. E., J.-Y. PARLANGE, and A. D. KRIKORIAN. 1973. Stomatal mechanics. *Am. J. Bot.* 60, 163–171.

BAKER, E. A. 1982. Chemistry and morphology of plant epicuticular waxes. In: *The Plant Cuticle*, pp. 135–165, D. F. Cutler, K. L. Alvin, and C. E. Price, eds. Academic Press, London.

BALL, T. B., J. S. GARDNER, and N. ANDERSON. 1999. Identifying inflorescence phytoliths from selected species of wheat (*Triticum monococcum, T. dicoccon, T. dicoccoides,* and *T. aestivum*) and barley (*Hordeum vulgare* and *H. spontaneum*) (Gramineae). *Am. J. Bot.* 86, 1615–1623.

BALUŠKA, F., J. SALAJ, J. MATHUR, M. BRAUN, F. JASPER, J. ŠAMAJ, N.-H. CHUA, P. W. BARLOW, and D. VOLKMANN. 2000. Root hair formation: F-actin-dependent tip growth is initiated by local assembly of profilin-supported F-actin meshworks accumulated within expansin-enriched bulges. *Dev. Biol.* 227, 618–632.

BARANOV, P. A. 1957. Coleorhiza in Myrtaceae. *Phytomorphology* 7, 237–243.

BARANOVA, M. A. 1987. Historical development of the present classification of morphological types of stomates. *Bot. Rev.* 53, 53–79.

BARANOVA, M. 1992. Principles of comparative stomatographic studies of flowering plants. *Bot. Rev.* 58, 49–99.

BARTHLOTT, W., and C. NEINHUIS. 1997. Purity of the sacred lotus, or escape from contamination in biological surfaces. *Planta* 202, 1–8.

BARTHLOTT, W., C. NEINHUIS, D. CUTLER, F. DITSCH, I. MEUSEL, I. THEISEN, and H. WILHELMI. 1998. Classification and terminology of plant epicuticular waxes. *Bot. J. Linn. Soc.* 126, 237–260.

BASRA, A. S., and C. P. MALIK. 1984. Development of the cotton fiber. *Int. Rev. Cytol.* 89, 65–113.

BEAN, G. J., M. D. MARKS, M. HÜLSKAMP, M. CLAYTON, and J. L. CROXDALE. 2002. Tissue patterning of *Arabidopsis* cotyledons. *New Phytol.* 153, 461–467.

BECKER, M., G. KERSTIENS, and J. SCHÖNHERR. 1986. Water permeability of plant cuticles: Permeance, diffusion and partition coefficients. *Trees* 1, 54–60.

BERGER, D., and T. ALTMANN. 2000. A subtilisin-like serine protease involved in the regulation of stomatal density and distribution in *Arabidopsis thaliana. Genes Dev.* 14, 1119–1131.

BEYSCHLAG, W., and J. ECKSTEIN. 2001. Towards a causal analysis of stomatal patchiness: The role of stomatal size variability and hydrological heterogeneity. *Acta Oecol.* 22, 161–173.

BEYSCHLAG, W., H. PFANZ, and R. J. RYEL. 1992. Stomatal patchiness in Mediterranean evergreen sclerophylls. Phenomenology and consequences for the interpretation of the midday depression in photosynthesis and transpiration. *Planta* 187, 546–553.

BIBIKOVA, T. N., A. ZHIGILEI, and S. GILROY. 1997. Root hair growth in *Arabidopsis thaliana* is directed by calcium and an endogenous polarity. *Planta* 203, 495–505.

BILGER, W., T. JOHNSEN, and U. SCHREIBER. 2001. UV-excited chlorophyll fluorescence as a tool for the assessment of UV-protection by the epidermis of plants. *J. Exp. Bot.* 52, 2007–2014.

BIRD, S. M., and J. E. GRAY. 2003. Signals from the cuticle affect epidermal cell differentiation. *New Phytol.* 157, 9–23.

BLATT, M. R. 2000a. $Ca^{2+}$ signalling and control of guard-cell volume in stomatal movements. *Curr. Opin. Plant Biol.* 3, 196–204.

BLATT, M. R. 2000b. Cellular signaling and volume control in stomatal movements in plants. *Annu. Rev. Cell Dev. Biol.* 16, 221–241.

BOETSCH, J., J. CHIN, and J. CROXDALE. 1995. Arrest of stomatal initials in *Tradescantia* is linked to the proximity of neighboring stomata and results in the arrested initials acquiring properties of epidermal cells. *Dev. Biol.* 168, 28–38.

BONE, R. A., D. W. LEE, and J. M. NORMAN. 1985. Epidermal cells functioning as lenses in leaves of tropical rain-forest shade plants. *Appl. Opt.* 24, 1408–1412.

BONGERS, J. M. 1973. Epidermal leaf characters of the Winteraceae. *Blumea* 21, 381–411.

BONNETT, H. T., JR., and E. H. NEWCOMB. 1966. Coated vesicles and other cytoplasmic components of growing root hairs of radish. *Protoplasma* 62, 59–75.

BOUYER, D., V. KIRIK, and M. HÜLSKAMP. 2001. Cell polarity in *Arabidopsis* trichomes. *Semin. Cell Dev. Biol.* 12, 353–356.

BRAUN, M., F. BALUŠKA, M. VON WITSCH, and D. MENZEL. 1999. Redistribution of actin, profilin and phosphatidylinositol-4,5-bisphosphate in growing and maturing root hairs. *Planta* 209, 435–443.

BREMOND, L., A. ALEXANDRE, E. VÉLA, and J. GUIOT. 2004. Advantages and disadvantages of phytolith analysis for the reconstruction of Mediterranean vegetation: An assessment based on modern phytolith, pollen and botanical data (Luberon, France). *Rev. Palaeobot. Palynol.* 129, 213–228.

BRODRIBB, T., and R. S. HILL. 1997. Imbricacy and stomatal wax plugs reduce maximum leaf conductance in Southern Hemisphere conifers. *Aust. J. Bot.* 45, 657–668.

BROWNLEE, C. 2001. The long and short of stomatal density signals. *Trends Plant Sci.* 6, 441–442.

BRUCK, D. K., R. J. ALVAREZ, and D. B. WALKER. 1989. Leaf grafting and its prevention by the intact and abraded epidermis. *Can. J. Bot.* 67, 303–312.

BUCHHOLZ, A., and J. SCHÖNHERR. 2000. Thermodynamic analysis of diffusion of non-electrolytes across plant cuticles in the presence and absence of the plasticiser tributyl phosphate. *Planta* 212, 103–111.

BUCHHOLZ, A., P. BAUR, and J. SCHÖNHERR. 1998. Differences among plant species in cuticular permeabilities and solute mobilities are not caused by differential size selectivities. *Planta* 206, 322–328.

BUCKLEY, T. N., G. D. FARQUHAR, and K. A. MOTT. 1999. Carbon-water balance and patchy stomatal conductance. *Oecologia* 118, 132–143.

BÜNNING, E. 1951. Über die Differenziernugsvorgänge in der Cruciferenwurzel. *Planta* 39, 126–153.

BURSTRÖM, H. 1942. Über die Entfaltung und Einrollen eines mesophilen Grassblattes. *Bot. Not.* 1942, 351–362.

CALVIN, C. L. 1970. Anatomy of the aerial epidermis of the mistletoe, *Phoradendron flavescens. Bot. Gaz.* 131, 62–74.

CÁRDENAS, L., L. VIDALI, J. DOMÍNGUEZ, H. PÉREZ, F. SÁNCHEZ, P. K. HEPLER, and C. QUINTO. 1998. Rearrangement of actin microfilaments in plant root hairs responding to *Rhizobium etli* nodulation signals. *Plant Physiol.* 116, 871–877.

CARLQUIST, S. 1975. *Ecological Strategies of Xylem Evolution.* University of California Press, Berkeley.

CHARLTON, W. A. 1990. Differentiation in leaf epidermis of *Chlorophytum comosum* Baker. *Ann. Bot.* 66, 567–578.

CHIN, J., Y. WAN, J. SMITH, and J. CROXDALE. 1995. Linear aggregations of stomata and epidermal cells in *Tradescantia* leaves: Evidence for their group patterning as a function of the cell cycle. *Dev. Biol.* 168, 39–46.

CHO, H.-T., and D. J. COSGROVE. 2002. Regulation of root hair initiation and expansin gene expression in *Arabidopsis. Plant Cell* 14, 3237–3253.

CHRISTODOULAKIS, N. S., J. MENTI, and B. GALATIS. 2002. Structure and development of stomata on the primary root of *Ceratonia siliqua* L. *Ann. Bot.* 89, 23–29.

CLARKE, K. J., M. E. MCCULLY, and N. K. MIKI. 1979. A developmental study of the epidermis of young roots of *Zea mays* L. *Protoplasma* 98, 283–309.

CLOWES, F. A. L. 1994. Origin of the epidermis in root meristems. *New Phytol.* 127, 335–347.

CLOWES, F. A. L. 2000. Pattern in root meristem development in angiosperms. *New Phytol.* 146, 83–94.

COMSTOCK, J. P. 2002. Hydraulic and chemical signalling in the control of stomatal conductance and transpiration. *J. Exp. Bot.* 53, 195–200.

CORMACK, R. G. H. 1935. Investigations on the development of root hairs. *New Phytol.* 34, 30–54.

CROXDALE, J. 1998. Stomatal patterning in monocotyledons: *Tradescantia* as a model system. *J. Exp. Bot.* 49, 279–292.

CROXDALE, J. L. 2000. Stomatal patterning in angiosperms. *Am. J. Bot.* 87, 1069–1080.

CUTTER, E. G., and L. J. FELDMAN. 1970a. Trichoblasts in *Hydrocharis*. I. Origin, differentiation, dimensions and growth. *Am. J. Bot.* 57, 190–201.

CUTTER, E. G., and L. J. FELDMAN. 1970b. Trichoblasts in *Hydrocharis*. II. Nucleic acids, proteins and a consideration of cell growth in relation to endopolyploidy. *Am. J. Bot.* 57, 202–211.

CZARNOTA, M. A., R. N. PAUL, L. A. WESTON, and S. O. DUKE. 2003. Anatomy of sorgoleone-secreting root hairs of *Sorghum* species. *Int. J. Plant Sci.* 164, 861–866.

DAY, T. A., G. MARTIN, and T. C. VOGELMANN. 1993. Penetration of UV-B radiation in foliage: Evidence that the epidermis behaves as a non-uniform filter. *Plant Cell Environ.* 16, 735–741.

DAYANANDAN, P., and P. B. KAUFMAN. 1976. Trichomes of *Cannabis sativa* L. (Cannabaceae). *Am. J. Bot.* 63, 578–591.

DE BARY, A. 1884. *Comparative Anatomy of the Vegetative Organs of the Phanerogams and Ferns.* Clarendon Press, Oxford.

DIETRICH, P., D. SANDERS, and R. HEDRICH. 2001. The role of ion channels in light-dependent stomatal opening. *J. Exp. Bot.* 52, 1959–1967.

DIETZ, K.-J., B. HOLLENBACH, and E. HELLWEGE. 1994. The epidermis of barley leaves is a dynamic intermediary storage compartment of carbohydrates, amino acids and nitrate. *Physiol. Plant.* 92, 31–36.

DITTMER, H. J. 1949. Root hair variations in plant species. *Am. J. Bot.* 36, 152–155.

DIXON, D. C., R. W. SEAGULL, and B. A. TRIPLETT. 1994. Changes in the accumulation of α- and β-tubulin isotypes during cotton fiber development. *Plant Physiol.* 105, 1347–1353.

DIXON, D. C., W. R. MEREDITH JR., and B. A. TRIPLETT. 2000. An assessment of α-tubulin isotype modification in developing cotton fiber. *Int. J. Plant Sci.* 161, 63–67.

DOLAN, L. 1996. Pattern in the root epidermis: An interplay of diffusible signals and cellular geometry. *Ann. Bot.* 77, 547–553.

DOLAN, L., and S. COSTA. 2001. Evolution and genetics of root hair stripes in the root epidermis. *J. Exp. Bot.* 52, 413–417.

DOLAN, L., and K. OKADA. 1999. Signalling in cell type specification. *Semin. Cell Dev. Biol.* 10, 149–156.

DOLAN, L., C. M. DUCKETT, C. GRIERSON, P. LINSTEAD, K. SCHNEIDER, E. LAWSON, C. DEAN, S. POETHIG, and K. ROBERTS. 1994. Clonal relationships and cell patterning in the root epidermis of *Arabidopsis*. *Development* 120, 2465–2474.

DUCKETT, C. M., K. J. OPARKA, D. A. M. PRIOR, L. DOLAN, and K. ROBERTS. 1994. Dye-coupling in the root epidermis of *Arabidopsis* is progressively reduced during development. *Development* 120, 3247–3255.

ECKSTEIN, J. 1997. Heterogene Kohlenstoffassimilation in Blättern höherer Pflanzen als Folge der Variabilität stomatärer Öffnungsweiten. Charakterisierung und Kausalanalyse des Phänomens "stomatal patchiness." Ph.D. Thesis, Julius-Maximilians-Universität Würzburg.

EDWARDS, D., D. S. EDWARDS, and R. RAYNER. 1982. The cuticle of early vascular plants and its evolutionary significance. In: *The Plant Cuticle*, pp. 341–361, D. F. Cutler, K. L. Alvin, and C. E. Price, eds. Academic Press, London.

EGLINTON, G., and R. J. HAMILTON. 1967. Leaf epicuticular waxes. *Science* 156, 1322–1335.

EHLERINGER, J. 1984. Ecology and ecophysiology of leaf pubescence in North American desert plants. In: *Biology and Chemistry of Plant Trichomes*, pp. 113–132, E. Rodriguez, P. L. Healey, and I. Mehta, eds. Plenum Press, New York.

EISNER, T., M. EISNER, and E. R. HOEBEKE. 1998. When defense backfires: Detrimental effect of a plant's protective trichomes on an insect beneficial to the plant. *Proc. Natl. Acad. Sci. USA* 95, 4410–4414.

ELLIS, R. P. 1979. A procedure for standardizing comparative leaf anatomy in the Poaceae. II. The epidermis as seen in surface view. *Bothalia* 12, 641–671.

ERWEE, M. G., P. B. GOODWIN, and A. J. E. VAN BEL. 1985. Cell-cell communication in the leaves of *Commelina cyanea* and other plants. *Plant Cell Environ.* 8, 173–178.

ESAU, K. 1977. *Anatomy of Seed Plants*, 2nd ed. Wiley, New York.

EUN, S.-O., and Y. LEE. 2000. Stomatal opening by fusicoccin is accompanied by depolymerization of actin filaments in guard cells. *Planta* 210, 1014–1017.

FAHN, A. 1986. Structural and functional properties of trichomes of xeromorphic plants. *Ann. Bot.* 57, 631–637.

FAHN, A., and D. F. CUTLER. 1992. *Xerophytes. Encyclopedia of Plant Anatomy*, Band 13, Teil 3, Gebrüder Borntraeger, Berlin.

FEILD, T. S., M. A. ZWIENIECKI, M. J. DONOGHUE, and N. M. HOLBROOK. 1998. Stomatal plugs of *Drimys winteri* (Winteraceae) protect leaves from mist but not drought. *Proc. Natl. Acad. Sci. USA* 95, 14256–14259.

FRANKE, W. 1971. Über die Natur der Ektodesmen und einen Vorschlag zur Terminologie. *Ber. Dtsch. Bot. Ges.* 84, 533–537.

FRANKS, P. J., I. R. COWAN, and G. D. FARQUHAR. 1998. A study of stomatal mechanics using the cell pressure probe. *Plant Cell Environ.* 21, 94–100.

FREY-WYSSLING, A., and K. MÜHLETHALER. 1959. Über das submikroskopische Geschehen bei der Kutinisierung pflanzlicher Zellwände. *Vierteljahrsschr. Naturforsch. Ges. Zürich* 104, 294–299.

FRYNS-CLAESSENS, E., and W. VAN COTTHEM. 1973. A new classification of the ontogenetic types of stomata. *Bot. Rev.* 39, 71–138.

GALATIS, B. 1980. Microtubules and guard-cell morphogenesis in *Zea mays* L. *J. Cell Sci.* 45, 211–244.

GALATIS, B. 1982. The organization of microtubules in guard mother cells of *Zea mays*. *Can. J. Bot.* 60, 1148–1166.

GALATIS, B., P. APOSTOLAKOS, and E. PANTERIS. 1989. Microtubules and lithocyst morphogenesis in *Pilea cadierei*. *Can. J. Bot.* 67, 2788–2804.

GALLAGHER, K., and L. G. SMITH. 2000. Roles for polarity and nuclear determinants in specifying daughter cell fates after an asymmetric cell division in the maize leaf. *Curr. Biol.* 10, 1229–1232.

GALWAY, M. E., J. D. MASUCCI, A. M. LLOYD, V. WALBOT, R. W. DAVIS, and J. W. SCHIEFELBEIN. 1994. The *TTG* gene is required to specify epidermal cell fate and cell patterning in the *Arabidopsis* root. *Dev. Biol.* 166, 740–754.

GALWAY, M. E., J. W. HECKMAN JR., and J. W. SCHIEFELBEIN. 1997. Growth and ultrastructure of *Arabidopsis* root hairs: The *rhd3* mutation alters vacuole enlargement and tip growth. *Planta* 201, 209–218.

GEISLER, M., M. YANG, and F. D. SACK. 1998. Divergent regulation of stomatal initiation and patterning in organ and suborgan regions of *Arabidopsis* mutants *too many mouths* and *four lips*. *Planta* 205, 522–530.

GEITMANN, A., and A. M. C. EMONS. 2000. The cytoskeleton in plant and fungal tip growth. *J. Microsc.* 198, 218–245.

GILROY, S., and D. L. JONES. 2000. Through form to function: root hair development and nutrient uptake. *Trends Plant Sci.* 5, 56–60.

GLOVER, B. J. 2000. Differentiation in plant epidermal cells. *J. Exp. Bot.* 51, 497–505.

GOURET, E., R. ROHR, and A. CHAMEL. 1993. Ultrastructure and chemical composition of some isolated plant cuticles in relation to their permeability to the herbicide, diuron. *New Phytol.* 124, 423–431.

GRABOV, A., and M. R. BLATT. 1998. Co-ordination of signalling elements in guard cell ion channel control. *J. Exp. Bot.* 49, 351–360.

GUNNING, B. E. S. 1977. Transfer cells and their roles in transport of solutes in plants. *Sci. Prog. Oxf.* 64, 539–568.

GUTTENBERG, H. VON. 1940. *Der primäre Bau der Angiospermenwurzel. Handbuch der Pflanzenanatomie*, Band 8, Lief 39. Gebrüder Borntraeger, Berlin.

HACCIUS, B., and W. TROLL. 1961. Über die sogenannten Wurzelhaare an den Keimpflanzen von *Drosera-* und *Cuscuta-*Arten. *Beitr. Biol. Pflanz.* 36, 139–157.

HALL, R. D. 1998. Biotechnological applications for stomatal guard cells. *J. Exp. Bot.* 49, 369–375.

HALL, R. D., T. RIKSEN-BRUINSMA, G. WEYENS, M. LEFÈBVRE, J. M. DUNWELL, and F. A. KRENS. 1996. Stomatal guard cells are totipotent. *Plant Physiol.* 112, 889–892.

HALLAM, N. D. 1982. Fine structure of the leaf cuticle and the origin of leaf waxes. In: *The Plant Cuticle*, pp. 197–214, D. F. Cutler, K. L. Alvin, and C. E. Price, eds. Academic Press, London.

HAMILTON, D. W. A., A. HILLS, B. KÖHLER, and M. R. BLATT. 2000. $Ca^{2+}$ channels at the plasma membrane of stomatal guard cells are activated by hyperpolarization and abscisic acid. *Proc. Natl. Acad. Sci. USA* 97, 4967–4972.

HARTUNG, W., S. WILKINSON, and W. J. DAVIES. 1998. Factors that regulate abscisic acid concentrations at the primary site of action at the guard cell. *J. Exp. Bot.* 49, 361–367.

HATTORI, K. 1992. The process during shoot regeneration in the receptacle culture of chrysanthemum (*Chrysanthemum morifolium* Ramat.). Ikushu-gaku Zasshi (*Jpn. J. Breed.*) 42, 227–234.

HEIDE-JØRGENSEN, H. S. 1991. Cuticle development and ultrastructure: Evidence for a procuticle of high osmium affinity. *Planta* 183, 511–519.

HEJNOWICZ, Z., A. RUSIN, and T. RUSIN. 2000. Tensile tissue stress affects the orientation of cortical microtubules in the epidermis of sunflower hypocotyl. *J. Plant Growth Regul.* 19, 31–44.

HEMPEL, F. D., D. R. WELCH, and L. J. FELDMAN. 2000. Floral induction and determination: where in flowering controlled? *Trends Plant Sci.* 5, 17–21.

HEPLER, P. K., L. VIDALI, and A. Y. CHEUNG. 2001. Polarized cell growth in higher plants. *Annu. Rev. Cell Dev. Biol.* 17, 159–187.

HETHERINGTON, A. M., and F. I. WOODWARD. 2003. The role of stomata in sensing and driving environmental change. *Nature* 424, 901–908.

HOLLOWAY, P. J. 1982. Structure and histochemistry of plant cuticular membranes: An overview. In: *The Plant Cuticle*, pp. 1–32, D. F. Cutler, K. J. Alvin, and G. E. Price, eds. Academic Press, London.

HÜLSKAMP, M. 2000. Cell morphogenesis: how plants spit hairs. *Curr. Biol.* 10, R308–R310.

HÜLSKAMP, M., and V. KIRIK. 2000. Trichome differentiation and morphogenesis in *Arabidopsis*. *Adv. Bot. Res.* 31, 237–260.

HÜLSKAMP, M., and A. SCHNITTGER. 1998. Spatial regulation of trichome formation in *Arabidopsis thaliana*. *Semin. Cell Dev. Biol.* 9, 213–220.

HÜLSKAMP, M., S. MISÉRA, and G. JÜRGENS. 1994. Genetic dissection of trichome cell development in *Arabidopsis*. *Cell* 76, 555–566.

ICKERT-BOND, S. M. 2000. Cuticle micromorphology of *Pinus krempfii* Lecomte (Pinaceae) and additional species from Southeast Asia. *Int. J. Plant Sci.* 161, 301–317.

IYER, V., and A. D. BARNABAS. 1993. Effects of varying salinity on leaves of *Zostera capensis* Setchell. I. Ultrastructural changes. *Aquat. Bot.* 46, 141–153.

JEFFREE, C. E. 1986. The cuticle, epicuticular waxes and trichomes of plants, with reference to their structure, functions, and evolution. In: *Insects and the Plant Surface*, pp. 23–46, B. E. Juniper and R. Southwood, eds. Edward Arnold, London.

JEFFREE, C. E. 1996. Structure and ontogeny of plant cuticles. In: *Plant Cuticles: An Integrated* Functional Approach, pp. 33–82, G. Kerstiens, ed. BIOS Scientific Publishers, Oxford.

JEFFREE, C. E., R. P. C. JOHNSON, and P. G. JARVIS. 1971. Epicuticular wax in the stomatal antechamber of Sitka spruce and its effects on the diffusion of water vapour and carbon dioxide. *Planta* 98, 1–10.

JENKS, M. A., P. J. RICH, and E. N. ASHWORTH. 1994. Involvement of cork cells in the secretion of epicuticular wax filaments on *Sorghum bicolor* (L.) Moench. *Int. J. Plant Sci.* 155, 506–518.

JOHNSON, R. W., and R. T. RIDING. 1981. Structure and ontogeny of the stomatal complex in *Pinus strobus* L. and *Pinus banksiana* Lamb. *Am. J. Bot.* 68, 260–268.

JORDAAN, A., and H. KRUGER. 1998. Notes on the v ultrastructure of six xerophytes from southern Africa. *S. Afr. J. Bot.* 64, 82–85.

JUNIPER, B. E. 1959. The surfaces of plants. *Endeavour* 18, 20–25.

JUNIPER, B. E., and C. E. JEFFREE. 1983. *Plant Surfaces.* Edward Arnold, London.

KARABOURNIOTIS, G., and C. EASSEAS. 1996. The dense indumentum with its polyphenol content may replace the protective role of the epidermis in some young xeromorphic leaves. *Can. J. Bot.* 74, 347–351.

KAUFMAN, P. B., L. B. PETERING, C. S. YOCUM, and D. BAIC. 1970a. Ultrastructural studies on stomata development in internodes of *Avena sativa. Am. J. Bot.* 57, 33–49.

KAUFMAN, P. B., L. B. PETERING, and J. G. SMITH. 1970b. Ultrastructural development of cork-silica cell pairs in *Avena* internodal epidermis. *Bot. Gaz.* 131, 173–185.

KAUFMAN, P. B., L. B. PETERING, and S. L. SONI. 1970c. Ultrastructural studies on cellular differentiation in internodal epidermis of *Avena sativa. Phytomorphology* 20, 281–309.

KAUFMAN, P. B., P. DAYANANDAN, Y. TAKEOKA, W. C. BIGELOW, J. D. JONES, and R. ILER. 1981. Silica in shoots of higher plants. In: *Silicon and Siliceous Structures in Biological Systems*, pp. 409–449, T. L. Simpson and B. E. Volcani, eds. Springer-Verlag, New York.

KAUFMAN, P. B., P. DAYANANDAN, C. I. FRANKLIN, and Y. TAKEOKA. 1985. Structure and function of silica bodies in the epidermal system of grass shoots. *Ann. Bot.* 55, 487–507.

KAUFMANN, K. 1927. Anatomie und Physiologie der Spaltöffnungsapparate mit Verholzten Schliesszellmembranen. *Planta* 3, 27–59.

KAUL, R. B. 1977. The role of the multiple epidermis in foliar succulence of *Peperomia* (Piperaceae). *Bot. Gaz.* 138, 213–218.

KETELAAR, T., and A. M. C. EMONS. 2001. The cytoskeleton in plant cell growth: Lessons from root hairs. *New Phytol.* 152, 409–418.

KETELAAR, T., C. FAIVRE-MOSKALENKO, J. J. ESSELING, N. C. A. DE RUIJTER, C. S. GRIERSON, M. DOGTEROM, and A. M. C. EMONS.

2002. Positioning of nuclei in *Arabidopsis* root hairs: an actin-regulated process of tip growth. *Plant Cell* 14, 2941–2955.

KIM, H. J., and B. A. TRIPLETT. 2001. Cotton fiber growth in planta and in vitro. Models for plant cell elongation and cell wall biogenesis. *Plant Physiol.* 127, 1361–1366.

KIM, K., S. S. WHANG, and R. S. HILL. 1999. Cuticle micromorphology of leaves of *Pinus* (Pinaceae) in east and south-east Asia. *Bot. J. Linn. Soc.* 129, 55–74.

KINOSHITA, T., and K.-I. SHIMAZAKI. 1999. Blue light activates the plasma membrane $H^+$-ATPase by phosphorylation of the C-terminus in stomatal guard cells. *EMBO J.* 18, 5548–5558.

KORN, R. W. 1972. Arrangement of stomata on the leaves of *Pelargonium zonale* and *Sedum stahlii. Ann. Bot.* 36, 325–333.

KREGER, D. R. 1958. Wax. In: Der Stoffwechsel sekundärer Pflanzenstoffe. In: *Handbuch der Pflanzenphysiologie,* Band 10, pp. 249–269. Springer, Berlin.

KUIJT, J., and W.-X. DONG. 1990. Surface features of the leaves of *Balanophoraceae*—A family without stomata? *Plant Syst. Evol.* 170, 29–35.

KUNST, L., and A. L. SAMUELS. 2003. Biosynthesis and secretion of plant cuticular wax. *Prog. Lipid Res.* 42, 51–80.

KUTSCHERA, U. 1992. The role of the epidermis in the control of elongation growth in stems and coleoptiles. *Bot. Acta* 105, 246–252.

LAKE, J. A., W. P. QUICK, D. J. BEERLING, and F. I. WOODWARD. 2001. Signals from mature to new leaves. *Nature* 411, 154.

LANNING, F. C., and L. N. ELEUTERIUS. 1989. Silica deposition in some $C_3$ and $C_4$ species of grasses, sedges and composites in the USA. *Ann. Bot.* 63, 395–410.

LARKIN, J. C., D. G. OPPENHEIMER, A. M. LLOYD, E. T. PAPAROZZI, and M. D. MARKS. 1994. Roles of the *GLABROUS1* and *TRANSPARENT TESTA GLABRA* genes in *Arabidopsis* trichome development. *Plant Cell* 6, 1065–1076.

LARKIN, J. C., N. YOUNG, M. PRIGGE, and M. D. MARKS. 1996. The control of trichome spacing and number in *Arabidopsis. Development* 122, 997–1005.

LAWTON, J. R. 1980. Observations on the structure of epidermal cells, particularly the cork and silica cells, from the flowering stem internode of *Lolium temulentum* L. (Gramineae). *Bot. J. Linn. Soc.* 80, 161–177.

LEAVITT, R. G. 1904. Trichomes of the root in vascular cryptogams and angiosperms. *Proc. Boston Soc. Nat. Hist.* 31, 273–313.

LECKIE, C. P., M. R MCAINSH, L. MONTGOMERY, A. J. PRIESTLEY, I. STAXEN, A. A. R. WEBB, and A. M. HETHERINGTON. 1998. Second messengers in guard cells. *J. Exp. Bot.* 49, 339–349.

LEE, M. M., and J. SCHIEFELBEIN. 1999. WEREWOLF, a MYB-related protein in *Arabidopsis*, is a position-dependent regulator of epidermal cell patterning. *Cell* 99, 473–483.

LEFEBVRE, D. D. 1985. Stomata on the primary root of *Pisum sativum* L. *Ann. Bot.* 55, 337–341.

LESSIRE, R., T. ABDUL-KARIM, and C. CASSAGNE. 1982. Origin of the wax very long chain fatty acids in leek, *Allium porrum* L., leaves: A plausible model. In: *The Plant Cuticle*, pp. 167–179, D. F. Cutler, K. L. Alvin, and C. E. Price, eds. Academic Press, London.

LEVIN, D. A. 1973. The role of trichomes in plant defense. *Q. Rev. Biol.* 48, 3–15.

LEVY, Y. Y., and C. DEAN. 1998. Control of flowering time. *Curr. Opin. Plant Biol.* 1, 49–54.

LI, J., and S. M. ASSMANN. 2000. Protein phosphorylation and ion transport: A case study in guard cells. *Adv. Bot. Res.* 32, 459–479.

LINSBAUER, K. 1930. *Die Epidermis. Handbuch der Pflanzenanatomie,* Band 4, Lief 27. Borntraeger, Berlin.

LLOYD, C. W. 1983. Helical microtubular arrays in onion root hairs. *Nature* 305, 311–313.

LLOYD, C. W., K. J. PEARCE, D. J. RAWLINS, R. W. RIDGE, and P. J. SHAW. 1987. Endoplasmic microtubules connect the advancing nucleus to the tip of legume root hairs, but F-actin is involved in basipetal migration. *Cell Motil. Cytoskel.* 8, 27–36.

LYSHEDE, O. B. 1982. Structure of the outer epidermal wall in xerophytes. In: *The Plant Cuticle*, pp. 87–98, D. F. Cutler, K. L. Alvin, and C. E. Price, eds. Academic Press, London.

MACROBBIE, E. A. C. 1998. Signal transduction and ion channels in guard cells. *Philos. Trans. R. Soc. Lond. B* 353, 1475–1488.

MACROBBIE, E. A. C. 2000. ABA activates multiple $Ca^{2+}$ fluxes in stomatal guard cells, triggering vacuolar $K^+(Rb^+)$ release. *Proc. Natl. Acad. Sci. USA* 97, 12361–12368.

MAHLBERG, P. G., and E.-S. KIM. 2004. Accumulation of cannabinoids in glandular trichomes of *Cannabis* (Cannabaceae). *J. Indust. Hemp.* 9, 15–36.

MALIK, K. A., S. T. ALI-KHAN, and P. K. SAXENA. 1993. High-frequency organogenesis from direct seed culture in *Lathyrus*. *Ann. Bot.* 72, 629–637.

MALTBY, D., N. C. CARPITA, D. MONTEZINOS, C. KULOW, and D. P. DELMER. 1979. β-1,3-glucan in developing cotton fibers. *Plant Physiol.* 63, 1158–1164.

MANSFIELD, T. A. 1983. Movements of stomata. *Sci. Prog. Oxf.* 68, 519–542.

MARCUS, A. I., R. C. MOORE, and R. J. CYR. 2001. The role of microtubules in guard cell function. *Plant Physiol.* 125, 387–395.

MARTIN, G., S. A. JOSSERAND, J. F. BORNMAN, and T. C. VOGELMANN. 1989. Epidermal focussing and the light microenvironment within leaves of *Medicago sativa*. *Physiol. Plant.* 76, 485–492.

MARTIN, J. T., and B. E. JUNIPER. 1970. *The Cuticles of Plants*. St. Martin's, New York.

MASUCCI, J. D., W. G. RERIE, D. R. FOREMAN, M. ZHANG, M. E. GALWAY, M. D. MARKS, and J. W. SCHIEFELBEIN. 1996. The homeobox gene GLABRA 2 is required for position-dependent cell differentiation in the root epidermis of *Arabidopsis thaliana*. *Development* 122, 1253–1260.

MATHUR, J., and N.-H. CHUA. 2000. Microtubule stabilization leads to growth reorientation in *Arabidopsis* trichomes. *Plant Cell* 12, 465–477.

MATHUR, J., and M. HÜLSKAMP. 2002. Microtubules and microfilaments in cell morphogenesis in higher plants. *Curr. Biol.* 12, R669–R676.

MATHUR, J., P. SPIELHOFER, B. KOST, and N.-H. CHUA. 1999. The actin cytoskeleton is required to elaborate and maintain spatial patterning during trichome cell morphogenesis in *Arabidopsis thaliana*. *Development* 126, 5559–5568.

MATZKE, E. B. 1947. The three-dimensional shape of epidermal cells of *Aloe aristata*. *Am. J. Bot.* 34, 182–195.

MAYER, W., I. MOSER, and E. BÜNNING. 1973. Die Epidermis als Ort der Lichtperzeption für circadiane Laubblattbewegungen und photoperiodische Induktionen. *Z. Pflanzenphysiol.* 70, 66–73.

MCWHORTER, C. G., C. OUZTS, and R. N. PAUL. 1993. Micromorphology of Johnsongrass (*Sorghum halepense*) leaves. *Weed Sci.* 41, 583–589.

MEEKES, H. T. H. M. 1985. Ultrastructure, differentiation and cell wall texture of trichoblasts and root hairs of *Ceratopteris thalictroides* (L.) Brongn. (Parkeriaceae). *Aquat. Bot.* 21, 347–362.

MENG, F.-R., C. P. A. BOURQUE, R. F. BELCZEWSKI, N. J. WHITNEY, and P. A. ARP. 1995. Foliage responses of spruce trees to long-term low-grade sulphur dioxide deposition. *Environ. Pollut.* 90, 143–152.

METCALFE, C. R. 1960. *Anatomy of the Monocotyledons*, vol. I. Gramineae. Clarendon Press, Oxford.

METCALFE, C. R., and L. CHALK. 1950. *Anatomy of the Dicotyledons*, vol. II. Clarendon Press, Oxford.

MEUSEL, I., C. NEINHUIS, C. MARKSTÄDTER, and W. BARTHLOTT. 2000. Chemical composition and recrystallization of epicuticular waxes: Coiled rodlets and tubules. *Plant Biology* 2, 462–470.

MILLAR, A. A., S. CLEMENS, S. ZACHGO, E. M. GIBLIN, D. C. TAYLOR, and L. KUNST. 1999. *CUT1*, an *Arabidopsis* gene required for cuticular wax biosynthesis and pollen fertility, encodes a very-long-chain fatty acid condensing enzyme. *Plant Cell* 11, 825–838.

MILLER, D. D., N. C. A. DE RUIJTER, T. BISSELING, and A. M. C. EMONS. 1999. The role of actin in root hair morphogenesis: Studies with lipochito-oligosaccharide as a growth stimulator and cytochalasin as an actin perturbing drug. *Plant J.* 17, 141–154.

MILLER, R. H. 1985. The prevalence of pores and canals in leaf cuticular membranes. *Ann. Bot.* 55, 459–471.

MILLER, R. H. 1986. The prevalence of pores and canals in leaf cuticular membranes. II. Supplemental studies. *Ann. Bot.* 57, 419–434.

MITOSINKA, G. T., J. I. THORNTON, and T. L. HAYES. 1972. The examination of cystolithic hairs of *Cannabis* and other plants by means of the scanning electron microscope. *J. Forensic Sci. Soc.* 12, 521–529.

MOTT, K. A., and T. N. BUCKLEY. 2000. Patchy stomatal conductance: Emergent collective behaviour of stomata. *Trends Plant Sci.* 5, 258–262.

MOTT, K. A., Z. G. CARDON, and J. A. BERRY. 1993. Asymmetric patchy stomatal closure for the two surfaces of *Xanthium strumarium* L. leaves at low humidity. *Plant Cell Environ.* 16, 25–34.

MOZAFAR, A., and J. R. GOODIN. 1970. Vesiculated hairs: a mechanism for salt tolerance in *Atriplex halimus* L. *Plant Physiol.* 45, 62–65.

MULHOLLAND, S. C., and G. RAPP JR. 1992. A morphological classification of grass silica-bodies. In: *Phytolith Systematics: Emerging Issues*, pp. 65–89, G. Rapp Jr. and S. C. Mulholland, eds. Plenum Press, New York.

MYLONA, P., P. LINSTEAD, R. MARTIENSSEN, and L. DOLAN. 2002. *SCHIZORIZA* controls an asymmetric cell division and restricts epidermal identity in the *Arabidopsis* root. *Development* 129, 4327–4334.

NADEAU, J. A., and F. D. SACK. 2002. Control of stomatal distribution on the *Arabidopsis* leaf surface. *Science* 296, 1697–1700.

NAKAMURA, G. R. 1969. Forensic aspects of cystolith hairs of *Cannabis* and other plants. *J. Assoc. Off. Anal. Chem.* 52, 5–16.

NAKAMURA, G. R., and J. I. THORNTON. 1973. The forensic identification of marijuana: Some questions and answers. *J. Police Sci. Adm.* 1, 102–112.

NEINHUIS, C., K. KOCH, and W. BARTHLOTT. 2001. Movement and regeneration of epicuticular waxes through plant cuticles. *Planta* 213, 427–434.

NG, C. K.-Y., M. R. MCAINSH, J. E. GRAY, L. HUNT, C. P. LECKIE, L. MILLS, and A. M. HETHERINGTON. 2001. Calcium-based signalling systems in guard cells. *New Phytol.* 151, 109–120.

NIKLAS, K. J., and D. J. PAOLILLO JR. 1997. The role of the epidermis as a stiffening agent in *Tulipa* (Liliaceae) stems. *Am. J. Bot.* 84, 735–744.

OHASHI, Y., A. OKA, I. RUBERTI, G. MORELLI, and T. AOYAMA. 2002. Entopically additive expression of *GLABRA2* alters the frequency and spacing of trichome initiation. *Plant J.* 29, 359–369.

OPPENHEIMER, D. G. 1998. Genetics of plant cell shape. *Curr. Opin. Plant Biol.* 1, 520–524.

OSBORN, J. M., and T. N. TAYLOR. 1990. Morphological and ultrastructural studies of plant cuticular membranes. I. Sun and shade leaves of *Quercus velutina* (Fagaceae). *Bot. Gaz.* 151, 465–476.

OWEN, T. P., JR., and W. W. THOMSON. 1991. Structure and function of a specialized cell wall in the trichomes of the carnivo-

rous bromeliad *Brocchinia reducta. Can. J. Bot.* 69, 1700–1706.

PALEVITZ, B. A. 1981. The structure and development of stomatal cells. In: *Stomatal physiology.* pp. 1–23, P. G. Jarvis and T. A. Mansfield, eds. Cambridge University Press, Cambridge.

PALEVITZ, B. A. 1982. The stomatal complex as a model of cytoskeletal participation in cell differentiation. In: *The Cytoskeleton in Plant Growth and Development*, pp. 345–376, C. W. Lloyd, ed. Academic Press, London.

PALEVITZ, B. A., and P. K. HEPLER. 1976. Cellulose microfibril orientation and cell shaping in developing guard cells of *Allium*: The role of microtubules and ion accumulation. *Planta* 132, 71–93.

PALEVITZ, B. A., and P. K. HEPLER. 1985. Changes in dye coupling of stomatal cells of *Allium* and *Commelina* demonstrated by microinjection of Lucifer yellow. *Planta* 164, 473–479.

PALIWAL, G. S. 1966. Structure and ontogeny of stomata in some Acanthaceae. *Phytomorphology* 16, 527–539.

PANT, D. D. 1965. On the ontogeny of stomata and other homologous structures. *Plant Sci. Ser.* 1, 1–24.

PANTERIS, E., P. APOSTOLAKOS, and B. GALATIS. 1994. Sinuous ordinary epidermal cells: Behind several patterns of waviness, a common morphogenetic mechanism. *New Phytol.* 127, 771–780.

PAYNE, C. T., F. ZHANG, and A. M. LLOYD. 2000. *GL3* encodes a bHLH protein that regulates trichome development in *Arabidopsis* through interaction with GL1 and TTG1. *Genetics* 156, 1349–1362.

PAYNE, W. W. 1978. A glossary of plant hair terminology. *Brittonia* 30, 239–255.

PAYNE, W. W. 1979. Stomatal patterns in embryophytes: their evolution, ontogeny and interpretation. *Taxon* 28, 117–132.

PEAR, J. R., Y. KAWAGOE, W. E. SCHRECKENGOST, D. P. DELMER, and D. M. STALKER. 1996. Higher plants contain homologs of the bacterial *celA* genes encoding the catalytic subunit of cellulose synthase. *Proc. Natl. Acad. Sci. USA* 93, 12637–12642.

PEETERS, M.-C., S. VOETS, G. DAYATILAKE, and E. DE LANGHE. 1987. Nucleolar size at early stages of cotton fiber development in relation to final fiber dimension. *Physiol. Plant.* 71, 436–440.

PEMBERTON, L. M. S., S.-L. TSAI, P. H. LOVELL, and P. J. HARRIS. 2001. Epidermal patterning in seedling roots of eudicotyledons. *Ann. Bot.* 87, 649–654.

PESACRETA, T. C., and K. H. HASENSTEIN. 1999. The internal cuticle of *Cirsium horridulum* (Asteraceae) leaves. *Am. J. Bot.* 86, 923–928.

PETERS, W. S., and D. TOMOS. 1996. The epidermis still in control? *Bot. Acta* 109, 264–267.

PETERSON, R. L., and M. L. FARQUHAR. 1996. Root hairs: Specialized tubular cells extending root surfaces. *Bot. Rev.* 62, 1–40.

PINKERTON, M. E. 1936. Secondary root hairs. *Bot. Gaz.* 98, 147–158.

PIPERNO, D. R. 1988. *Phytolith Analysis: An Archeological and Geological Perspective.* Academic Press, San Diego.

POPHAM, R. A., and R. D. HENRY. 1955. Multicellular root hairs on adventitious roots of *Kalanachoe fedtschenkoi. Ohio J. Sci.* 55, 301–307.

POSSINGHAM, J. V. 1972. Surface wax structure in fresh and dried Sultana grapes. *Ann. Bot.* 36, 993–996.

POTIKHA, T. S., C. C. COLLINS, D. I. JOHNSON, D. P. DELMER, and A. LEVIN. 1999. The involvement of hydrogen peroxide in the differentiation of secondary walls in cotton fibers. *Plant Physiol* 119, 849–858.

PRAT, H. 1948. Histo-physiological gradients and plant organogenesis. Part I. General concept of a system of gradients in living organisms. *Bot. Rev.* 14, 603–643.

PRAT, H. 1951. Histo-physiological gradients and plant organogenesis. Part II. Histological gradients. *Bot. Rev.* 17, 693–746.

PRYCHID, C. J., P. J. RUDALL, and M. GREGORY. 2004. Systematics and biology of silica bodies in monocotyledons. *Bot. Rev.* 69, 377–440.

RAMESH, K., and M. A. PADHYA. 1990. In vitro propagation of neem, *Azadirachta indica* (A. Jus), from leaf discs. *Indian J. Exp. Biol.* 28, 932–935.

RAMSEY, J. C., and J. D. BERLIN. 1976a. Ultrastructural aspects of early stages in cotton fiber elongation. *Am. J. Bot.* 63, 868–876.

RAMSEY, J. C., and J. D. BERLIN. 1976b. Ultrastructure of early stages of cotton fiber differentiation. *Bot. Gaz.* 137, 11–19.

RAO, B. H., and K. V. KUMAR. 1995. Lithocysts as taxonomic markers of the species of *Cordia* L. (Boraginaceae). *Phytologia* 78, 260–263.

RAPISARDA, A., L. IAUK, and S. RAGUSA. 1997. Micromorphological study on leaves of some *Cordia* (Boraginaceae) species used in traditional medicine. *Econ. Bot.* 51, 385–391.

RASCHKE, K. 1975. Stomatal action. *Annu. Rev. Plant Physiol* 26, 309–340.

RASMUSSEN, H. 1981. Terminology and classification of stomata and stomatal development—A critical survey. *Bot. J. Linn. Soc.* 83, 199–212.

REDDY, A. S. N., and I. S. DAY. 2000. The role of the cytoskeleton and a molecular motor in trichome morphogenesis. *Trends Plant Sci.* 5, 503–505.

REDWAY, F. A. 1991. Histology and stereological analysis of shoot formation in leaf callus of *Saintpaulia ionantha* Wendl. (African violet). *Plant Sci.* 73, 243–251.

RENTSCHLER, E. 1971. Die Wasserbenetzbarkeit von Blattoberflächen und ihre submikroskopische Wachsstruktur. *Planta* 96, 119–135.

RIDGE, R. W. 1995. Recent developments in the cell and molecular biology of root hairs. *J. Plant Res.* 108, 399–405.

RIDGE, R. W., and A. M. C. EMONS, eds. 2000. *Root Hairs: Cell and Molecular Biology.* Springer, Tokyo.

RIEDERER, M. 1989. The cuticles of conifers: structure, composition and transport properties. In: *Forest Decline and Air Pollution: A Study of Spruce (Picea abies) on Acid Soils,* pp. 157–192, E.-D. Schulze, O. L. Lange, and R. Oren, eds. Springer-Verlag, Berlin.

RIEDERER, M., and G. SCHNEIDER. 1990. The effect of the environment on the permeability and composition of *Citrus* leaf cuticles. II. Composition of soluble cuticular lipids and correlation with transport properties. *Planta* 180, 154–165.

RIEDERER, M., and L. SCHREIBER. 2001. Protecting against water loss: Analysis of the barrier properties of plant cuticles. *J. Exp. Bot.* 52, 2023–2032.

ROBBERECHT, R., and M. M. CALDWELL. 1978. Leaf epidermal transmittance of ultraviolet radiation and its implications for plant sensitivity to ultraviolet-radiation induced injury. *Oecologia* 32, 277–287.

ROTH, I. 1963. Entwicklung der ausdauernden Epidermis sowie der primären Rinde des Stammes von *Cercidium torreyanum* in Laufe des sekunddären Dickenwachstums. *Österr. Bot. Z.* 110, 1–19.

ROTH, I. 1990. *Leaf Structure of a Venezuelan Cloud Forest in Relation to the Microclimate.* Encyclopedia of Plant Anatomy, Band 14, Teil 1. Gebrüder Borntraeger, Berlin.

ROW, H. C., and J. R. REEDER. 1957. Root-hair development as evidence of relationships among genera of Gramineae. *Am. J. Bot.* 44, 596–601.

RUAN, Y.-L., D. J. LLEWELLYN, and R. T. FURBANK. 2001. The control of single-celled cotton fiber elongation by developmentally reversible gating of plasmodesmata and coordinated expression of sucrose and K$^+$ transporters and expansin. *Plant Cell* 13, 47–60.

RUSSELL, S. H., and R. F. EVERT. 1985. Leaf vasculature in *Zea mays* L. *Planta* 164, 448–458.

RYSER, U. 1985. Cell wall biosynthesis in differentiating cotton fibres. *Eur. J. Cell Biol.* 39, 236–256.

RYSER, U. 1999. Cotton fiber initiation and histodifferentiation. In: *Cotton Fibers: Developmental Biology, Quality Improvement, and Textile Processing,* pp. 1–45, A. S. Basra, ed. Food Products Press, New York.

RYSER, U., and P. J. HOLLOWAY. 1985. Ultrastructure and chemistry of soluble and polymeric lipids in cell walls from seed coats and fibres of *Gossypium* species. *Planta* 163, 151–163.

SACK, F. D. 1987. The development and structure of stomata. In: *Stomatal Function,* pp. 59–89, E. Zeiger, G. D. Farquhar, and I. R. Cowan, eds. Stanford University Press, Stanford.

SACK, F. D. 1994. Structure of the stomatal complex of the monocot *Flagellaria indica. Am. J. Bot.* 81, 339–344.

SAHGAL, P., G. V. MARTINEZ, C. ROBERTS, and G. TALLMAN. 1994. Regeneration of plants from cultured guard cell protoplasts of *Nicotiana glauca* (Graham). *Plant Sci.* 97, 199–208.

SANTIER, S., and A. CHAMEL. 1998. Reassessment of the role of cuticular waxes in the transfer of organic molecules through plant cuticles. *Plant Physiol. Biochem* 36, 225–231.

SARGENT, C. 1978. Differentiation of the crossed-fibrillar outer epidermal wall during extension growth in *Hordeum vulgare* L. *Protoplasma* 95, 309–320.

SATIAT-JEUNEMAITRE, B., B. MARTIN, and C. HAWES. 1992. Plant cell wall architecture is revealed by rapid-freezing and deep-etching. *Protoplasma* 167, 33–42.

SATO, S., Y. OGASAWARA, and S. SAKURAGI. 1995. The relationship between growth, nucleus migration and cytoskeleton in root hairs of radish. In: *Structure and Function of Roots*, pp. 69–74, F. Baluška, M. Čiamporová, O. Gašparíková, and P. W. Barlow, eds. Kluwer Academic, Dordrecht.

SAXE, H. 1979. A structural and functional study of the coordinated reactions of individual *Commelina communis* L. stomata (Commelinaceae). *Am. J. Bot.* 66, 1044–1052.

SCHELLMANN, S., A. SCHNITTGER, V. KIRIK, T. WADA, K. OKADA, A. BEERMANN, J. THUMFAHRT, G. JÜRGENS, and M. HÜLSKAMP. 2002. *TRIPTYCHON* and *CAPRICE* mediate lateral inhibition during trichome and root hair patterning in *Arabidopsis*. *EMBO J.* 21, 5036–5046.

SCHIEFELBEIN, J. 2003. Cell-fate specification in the epidermis: A common patterning mechanism in the root and shoot. *Curr. Opin. Plant Biol.* 6, 74–78.

SCHIEFELBEIN, J. W., and C. SOMERVILLE. 1990. Genetic control of root hair development in *Arabidopsis thaliana*. *Plant Cell* 2, 235–243.

SCHIEFELBEIN, J. W., J. D. MASUCCI, and H. WANG. 1997. Building a root: The control of patterning and morphogenesis during root development. *Plant Cell* 9, 1089–1098.

SCHMUTZ, A., T. JENNY, N. AMRHEIN, and U. RYSER. 1993. Caffeic acid and glycerol are constituents of the suberin layers in green cotton fibres. *Planta* 189, 453–460.

SCHNABL, H., and K. RASCHKE. 1980. Potassium chloride as stomatal osmoticum in *Allium cepa* L., a species devoid of starch in guard cells. *Plant Physiol* 65, 88–93.

SCHNABL, H. AND H. ZIEGLER. 1977. The mechanism of stomatal movement in *Allium cepa* L. *Planta* 136, 37–43.

SCHÖNHERR, J. 2000. Calcium chloride penetrates plant cuticles via aqueous pores. *Planta* 212, 112–118.

SCHÖNHERR, J., and M. RIEDERER. 1989. Foliar penetration and accumulation of organic chemicals in plant cuticles. *Rev. Environ. Contam. Toxicol.* 108, 2–70.

SCHREIBER, L., and M. RIEDERER. 1996. Ecophysiology of cuticular transpiration: comparative investigation of cuticular water permeability of plant species from different habitats. *Oecologia* 107, 426–432.

SCHREIBER, L., T. KIRSCH, and M. RIEDERER. 1996. Transport properties of cuticular waxes of *Fagus sylvatica* L. and *Picea abies* (L.) Karst: Estimation of size selectivity and tortuosity from diffusion coefficients of aliphatic molecules. *Planta* 198, 104–109.

SCHREIBER, L., M. SKRABS, K. HARTMANN, P. DIAMANTOPOULOS, E. SIMANOVA, and J. SANTRUCEK. 2001. Effect of humidity on cuticular water permeability of isolated cuticular membranes and leaf disks. *Planta* 214, 274–282.

SCHROEDER, J. I., J. M. KWAK, and G. J. ALLEN. 2001. Guard cell abscisic acid signalling and engineeering drought hardiness in plants. *Nature* 410, 327–330.

SCHWAB, B., U. FOLKERS, H. ILGENFRITZ, and M. HÜLSKAMP. 2000. Trichome morphogenesis in *Arabidopsis*. *Philos. Trans. R. Soc. Lond. B.* 355, 879–883.

SCULTHORPE, C. D. 1967. *The Biology of Aquatic Vascular Plants.* Edward Arnold, London.

SEAGULL, R. W. 1986. Changes in microtubule organization and wall microfibril orientation during *in vitro* cotton fiber development: An immunofluorescent study. *Can. J. Bot.* 64, 1373–1381.

SEAGULL, R. W. 1992. A quantitative electron microscopic study of changes in microtubule arrays and wall microfibril orientation during *in vitro* cotton fiber development. *J. Cell Sci.* 101, 561–577.

SERNA, L., and C. FENOLL. 2000. Stomatal development in *Arabidopsis*: How to make a functional pattern. *Trends Plant Sci.* 5, 458–460.

SERNA, L., J. TORRES-CONTRERAS, and C. FENOLL. 2002. Clonal analysis of stomatal development and patterning in *Arabidopsis* leaves. *Dev. Biol.* 241, 24–33.

SETOGUCHI, H., M. OKAZAKI, and S. SUGA. 1989. Calcification in higher plants with special reference to cystoliths. In: *Origin, Evolution, and Modern Aspects of Biomineralization in Plants and Animals*, pp. 409–418, R. E. Crick, ed. Plenum Press, New York.

SETOGUCHI, H., H. TOBE, H. OHBA, and M. OKAZAKI. 1993. Silicon-accumulating idioblasts in leaves of Cecropiaceae (Urticales). *J. Plant Res.* 106, 327–335.

SHAW, S. L., J. DUMAIS, and S. R. LONG. 2000. Cell surface expansion in polarly growing root hairs of *Medicago truncatula*. *Plant Physiol.* 124, 959–969.

SHIELDS, L. M. 1951. The involution in leaves of certain xeric grasses. *Phytomorphology* 1, 225–241.

SIEBERER, B., and A. M. C. EMONS. 2000. Cytoarchitecture and pattern of cytoplasmic streaming in root hairs of *Medicago truncatula* during development and deformation by nodulation factors. *Protoplasma* 214, 118–127.

SOLEREDER, H. 1908. *Systematic Anatomy of the Dicotyledons: A Handbook for Laboratories of Pure and Applied Botany.* 2 vols. Clarendon Press, Oxford.

SONG, P., and R. D. ALLEN. 1997. Identification of a cotton fiber-specific acyl carrier protein cDNA by differential display. *Biochim. Biophy. Acta—Gene Struct. Express* 1351, 305–312.

SRIVASTAVA, L. M., and A. P. SINGH. 1972. Stomatal structure in corn leaves. *J. Ultrastruct. Res.* 39, 345–363.

STEWART, J. McD. 1975. Fiber initiation on the cotton ovule (*Gossypium hirsutum* L.). *Am. J. Bot.* 62, 723–730.

STEWART, J. McD. 1986. Integrated events in the flower and fruit. In: *Cotton Physiology*, pp. 261–300, J. R. Mauney and J. McD. Stewart, eds. Cotton Foundation, Memphis, TN.

STOCKEY, R. A., B. J. FREVEL, and P. WOLTZ. 1998. Cuticle micromorphology of *Podocarpus*, subgenus *Podocarpus*, section *Scytopodium* (Podocarpaceae) of Madagascar and South Africa. *Int. J. Plant Sci.* 159, 923–940.

SZYMANSKI, D. B., M. D. MARKS, and S. M. WICK. 1999. Organized F-actin is essential for normal trichome morphogenesis in *Arabidopsis*. *Plant Cell* 11, 2331–2347.

TAIZ, L., and E. ZEIGER. 2002. *Plant Physiology*, 3rd ed. Sinauer Associates, Sunderland, MA.

TAKEDA, K., and H. SHIBAOKA. 1978. The fine structure of the epidermal cell wall in Azuki bean epicotyl. *Bot. Mag. Tokyo* 91, 235–245.

TALBOTT, L. D., and E. ZEIGER. 1998. The role of sucrose in guard cell osmoregulation. *J. Exp. Bot.* 49, 329–337.

TARKOWSKA, J. A., and M. WACOWSKA. 1988. The significance of the presence of stomata on seedling roots. *Ann. Bot.* 61, 305–310.

TAYLOR, L. P., and P. K. HEPLER. 1997. Pollen germination and tube growth. *Annu. Rev. Plant Physiol. Plant Mol. Biol.* 48, 461–491.

TAYLOR, M. G., K. SIMKISS, G. N. GREAVES, M. OKAZAKI, and S. MANN. 1993. An X-ray absorption spectroscopy study of the structure and transformation of amorphous calcium carbonate from plant cystoliths. *Proc. R. Soc. Lond. B.* 252, 75–80.

TERASHIMA, I. 1992. Anatomy of non-uniform leaf photosynthesis. *Photosyn. Res.* 31, 195–212.

THEOBALD, W. L., J. L. KRAHULIK, and R. C. ROLLINS. 1979. Trichome description and classification. In: *Anatomy of the Dicotyledons*, 2nd ed., vol. I, *Systematic Anatomy of Leaf and Stem, with a Brief History of the Subject*, pp. 40–53, C. R. Metcalfe and L. Chalk. Clarendon Press, Oxford.

THOMSON, W. W., and P. L. HEALEY. 1984. Cellular basis of trichome secretion. In: *Biology and Chemistry of Plant Trichomes*, pp. 113–130, E. Rodriguez, P. L. Healey, and I. Mehta, eds. Plenum Press, New York.

TIETZ, A., and I. URBASCH. 1977. Spaltöffnungen an der keimwurzel von *Helianthus annuus* L. *Naturwissenschaften* 64, 533.

TIWARI, S. C., and T. A. WILKINS. 1995. Cotton (*Gossypium hirsutum*) seed trichomes expand via diffuse growing mechanism. *Can. J. Bot.* 73, 746–757.

TOKUMOTO, H., K. WAKABAYASHI, S. KAMISAKA, and T. HOSON. 2002. Changes in the sugar composition and molecular mass distribution of matrix polysaccharides during cotton fiber development. *Plant Cell Physiol* 43, 411–418.

TOMINAGA, M., K. MORITA, S. SONOBE, E. YOKOTA, and T. SHIMMEN. 1997. Microtubules regulate the organization of actin filaments at the cortical region in root hair cells of *Hydrocharis*. *Protoplasma* 199, 83–92.

TOMLINSON, P. B. 1974. Development of the stomatal complex as a taxonomic character in the monocotyledons. *Taxon* 23, 109–128.

TRAAS, J. A., P. BRAAT, A. M. EMONS, H. MEEKES, and J. DERKSEN. 1985. Microtubules in root hairs. *J. Cell Sci.* 76, 303–320.

UPHOF, J. C. TH., and K. HUMMEL. 1962. *Plant Hairs. Encyclopedia of Plant Anatomy*, Band 4, Teil 5. Gebrüder Borntraeger, Berlin.

VALDES-REYNA, J., and S. L. HATCH. 1991. Lemma micromorphology in the Eragrostideae (Poaceae). *Sida (Contrib. Bot.)* 14, 531–549.

VAUGHN, K. C., and R. B. TURLEY. 1999. The primary walls of cotton fibers contain an ensheathing pectin layer. *Protoplasma* 209, 226–237.

VILLENA, J. F., E. DOMÍNQUEZ, D. STEWART, and A. HEREDIA. 1999. Characterization and biosynthesis of non-degradable polymers in plant cuticles. *Planta* 208, 181–187.

VISSENBERG, K., S. C. FRY, and J.-P. VERBELEN. 2001. Root hair initiation is coupled to a highly localized increase of xyloglucan endotransglycosylase action in *Arabidopsis* roots. *Plant Physiol* 127, 1125–1135.

VON GROLL, U., D. BERGER, and T. ALTMANN. 2002. The subtilisin-like serine protease SDD1 mediates cell-to-cell signaling during *Arabidopsis* stomatal development. *Plant Cell* 14, 1527–1539.

WADA, T., T. TACHIBANA, Y. SHIMURA, and K. OKADA. 1997. Epidermal cell differentiation in *Arabidopsis* determined by a *Myb* homolog, *CPC*. *Science* 277, 1113–1116.

WADA, T., T. KURATA, R. TOMINAGA, Y. KOSHINO-KIMURA, T. TACHIBANA, K. GOTO, M. D. MARKS, Y. SHIMURA, and K. OKADA. 2002. Role of a positive regulator of roothair development, *CAPRICE*, in *Arabidopsis* root epidermal cell differentiation. *Development* 129, 5409–5419.

WAGNER, G. J. 1991. Secreting glandular trichomes: More than just hairs. *Plant Physiol.* 96, 675–679.

WALKER, A. R., P. A. DAVISON, A. C. BOLOGNESI-WINFIELD, C. M. JAMES, N. SRINIVASAN, T. L. BLUNDELL, J. J. ESCH, M. D. MARKS, and J. C. GRAY. 1999. The *TRANSPARENT TESTA GLABRA1* locus, which regulates trichome differentiation and anthocyanin biosynthesis in *Arabidopsis*, encodes a WD40 repeat protein. *Plant Cell* 11, 1337–1350.

WATT, W. M., C. K. MORRELL, D. L. SMITH, and M. W. STEER. 1987. Cystolith development and structure in *Pilea cadierei* (Urticaceae). *Ann. Bot.* 60, 71–84.

WEYERS, J. D. B., and H. MEIDNER. 1990. *Methods in Stomatal Research*. Longman Scientific & Technical, Harlow, Essex, England.

WHANG, S. S., K. KIM, and W. M. HESS. 1998. Variation of silica bodies in leaf epidermal long cells within and among seventeen species of *Oryza* (Poaceae). *Am. J. Bot.* 85, 461–466.

WILKINSON, H. P. 1979. The plant surface (mainly leaf). Part I: Stomata. In: *Anatomy of the Dicotyledons*, 2nd ed., vol. I, pp. 97–117, C. R. Metcalfe and L. Chalk. Clarendon Press, Oxford.

WILLE, A. C., and W. J. LUCAS. 1984. Ultrastructural and histochemical studies on guard cells. *Planta* 160, 129–142.

WILLMER, C., and M. FRICKER. 1996. *Stomata*, 2nd ed. Chapman and Hall, London.

WILLMER, C. M., and R. SEXTON. 1979. Stomata and plasmodesmata. *Protoplasma* 100, 113–124.

WILSON, C. A., and C. L. CALVIN. 2003. Development, taxonomlic significannce and ecological role of the cuticular epithelium in the Santalales. *IAWA J.* 24, 129–138.

WOOD, N. T., A. C. ALLAN, A. HALEY, M. VIRY-MOUSSAÏD, and A. J. TREWAVAS. 2000. The characterization of differential calcium signalling in tobacco guard cells. *Plant J.* 24, 335–344.

WOODWARD, F. I., J. A. LAKE, and W. P. QUICK. 2002. Stomatal development and $CO_2$: Ecological consequences. *New Phytol.* 153, 477–484.

WU, C.-C., and L.-L. KUO-HUANG. 1997. Calcium crystals in the leaves of some species of Moraceae. *Bot. Bull. Acad. Sin.* 38, 97–104.

YANG, M., and F. D. SACK. 1995. The *too many mouths* and *four lips* mutations affect stomatal production in *Arabidopsis*. *Plant Cell* 7, 2227–2239.

YU, R., R.-F. HUANG, X.-C. WANG, and M. YUAN. 2001. Microtubule dynamics are involved in stomatal movement of *Vicia faba* L. *Protoplasma* 216, 113–118.

ZEIGER, E. 2000. Sensory transduction of blue light in guard cells. *Trends Plant Sci.* 5, 183–185.

ZELLNIG, G., J. PETERS, M. S. JIMÉNEZ, D. MORALES, D. GRILL, and A. PERKTOLD. 2002. Three-dimensional reconstruction of the stomatal complex in *Pinus canariensis* needles using serial sections. *Plant Biol.* 4, 70–76.

ZHANG, S. Q., and W. H. OUTLAW JR. 2001. Abscisic acid introduced into the transpiration stream accumulates in the guard-cell apoplast and causes stomatal closure. *Plant Cell Environ.* 24, 1045–1054.

ZHAO, L., and F. D. SACK. 1999. Ultrastructure of stomatal development in *Arabidopsis* (Brassicaceae) leaves. *Am. J. Bot.* 86, 929–939.

ZHU, T., R. L. O'QUINN, W. J. LUCAS, and T. L. ROST. 1998. Directional cell-to-cell communication in the *Arabidopsis* root apical meristem. II. Dynamics of plasmodesmatal formation. *Protoplasma* 204, 84–93.

# CHAPTER TEN

# Xylem: Cell Types and Developmental Aspects

The **xylem** is the principal water-conducting tissue in a vascular plant. It is also involved in the transport of solutes, in support, and in food storage. Together with the phloem, the principal food-conducting tissue, the xylem forms a continuous vascular system extending throughout the plant body. As components of the vascular system, xylem and phloem are called **vascular tissues**. Sometimes the two together are spoken of as *the* vascular tissue. The term xylem was introduced by Nägeli (1858) and is derived from the Greek *xylon*, wood.

The vascular plants, also referred to as tracheophytes, form a monophyletic group consisting of two phyla of seedless vascular plants (Lycopodiophyta and Pteridophyta, which comprises the ferns, including the whisk ferns, and the horsetails), in addition to the gymnosperms and angiosperms, all with living representatives (Raven et al., 2005). In addition there are several entirely extinct phyla of vascular plants (Stewart and Rothwell, 1993; Taylor and Taylor, 1993). The terms vascular plants and tracheophytes refer to the characteristic conducting elements of the xylem, the **tracheary elements**. Because of their enduring rigid cell walls, the tracheary elements are more conspicuous than the sieve elements of the phloem, are better preserved in fossils, and may be studied with greater ease. It is the xylem therefore, rather than the phloem, that serves in the identification of vascular plants.

Developmentally the first xylem differentiates early in the ontogeny of the plant—in the embryo or young seedling (Gahan, 1988; Busse and Evert, 1999)—and as the plant grows, new xylem (together with the accompanying phloem) continuously develops from derivatives of the apical meristems. Thus the primary plant body, which is formed by the activity of the apical meristems, is permeated by a continuous system of vascular tissue. The vascular tissues that differentiate in the primary plant body are the **primary xylem** and the **primary phloem**. The meristematic tissue directly concerned with the formation of these tissues, and which is their immediate precursor, is the **procambium**. Ancient vascular plants, and many contemporary ones (small annuals of the eudicots and most monocots) as well, consist entirely of primary tissues.

In addition to primary growth, many plants undergo additional growth that thickens the stem and root after

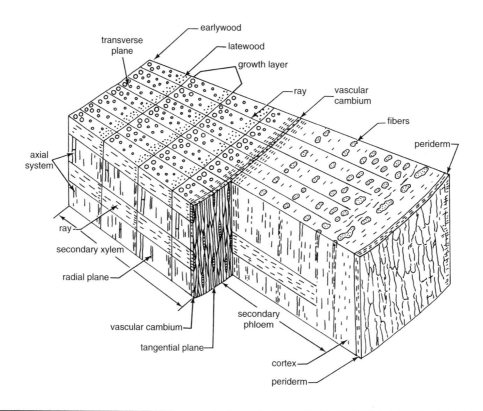

**FIGURE 10.1**

Block diagram illustrating the basic features of the secondary vascular tissues—secondary xylem and secondary phloem—and their spatial relation to one another and to vascular cambium, which gives rise to them. A periderm has replaced the epidermis as the dermal tissue system. (From Esau, 1977.)

primary growth (extension growth) is completed. Such growth is termed secondary growth. It results in part from the activity of the **vascular cambium**, the lateral meristem that produces the secondary vascular tissues, **secondary xylem** and **secondary phloem** (Fig. 10.1).

Structurally the xylem is a complex tissue containing at least tracheary elements and parenchyma cells and usually other types of cells, especially supporting cells. The principal cell types of the secondary xylem are listed in Table 10.1. The primary and the secondary xylem have histologic differences, but in many respects the two kinds of xylem intergrade with one another (Esau, 1943; Larson, 1974, 1976). Therefore, to be useful, the classification into primary xylem and secondary xylem must be considered broadly, relating these two components of the xylem tissue to development of the plant as a whole.

## ▌CELL TYPES OF THE XYLEM

### Tracheary Elements—Tracheids and Vessel Elements—Are the Conducting Cells of the Xylem

The term tracheary element is derived from "trachea," a name originally applied to certain primary xylem ele-

**TABLE 10.1 ■ Principal Cell Types in the Secondary Xylem**

| Cell Types | Principal Functions |
|---|---|
| Axial system | |
| Tracheary elements | |
| Tracheids } | Conduction of water; |
| Vessel elements } | transport of solutes |
| Fibers | |
| Fiber-tracheids } | Support; sometimes storage |
| Libriform fibers } | |
| Parenchyma cells } | |
| Radial (ray) system } | Food storage; translocation |
| Parenchyma cells } | of various substances |
| Tracheids in some conifers | |

ments resembling insect tracheae (Esau, 1961). Two fundamental types of tracheary elements occur in the xylem, the **tracheids** (Fig. 10.2A, B) and the **vessel elements**, or **vessel members** (Fig. 10.2C–F). Both are more or less elongated cells that have lignified secondary walls and are nonliving at maturity. They differ from one another in that tracheids are imperforate cells having only pit-pairs on their common walls, whereas

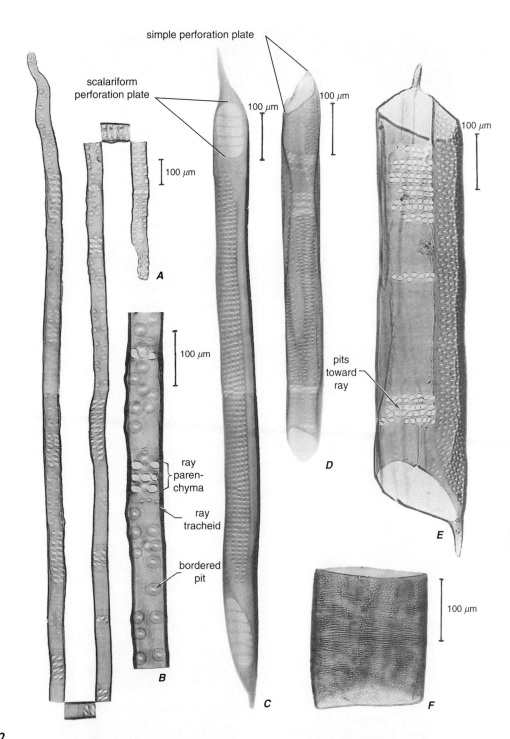

**FIGURE 10.2**

Tracheary elements. **A**, earlywood tracheid of sugar pine (*Pinus lambertiana*). **B**, enlarged part of **A**. **C–F**, vessel elements of tulip tree, *Liriodendron tulipifera* (**C**), beech, *Fagus grandifolia* (**D**), black cottonwood, *Populus trichocarpa* (**E**), tree-of-heaven, *Ailanthus altissima* (**F**). (From Carpenter, 1952; with permission from SUNY-ESF.)

vessel elements also have perforations, which are areas lacking both primary and secondary walls through which the vessel elements are interconnected.

The part of the vessel element wall bearing the perforation or perforations is called the **perforation plate** (IAWA Committee on Nomenclature, 1964; Wheeler et al., 1989). A perforation plate may have a single perforation (***simple perforation plate***; Figs. 10.2D–F and 10.3A) or several perforations (***multiple perforation plate***). The perforations in a multiple perforation plate may be elongated and arranged in a parallel series (***scalariform perforation plate***, from the Latin *scalaris*,

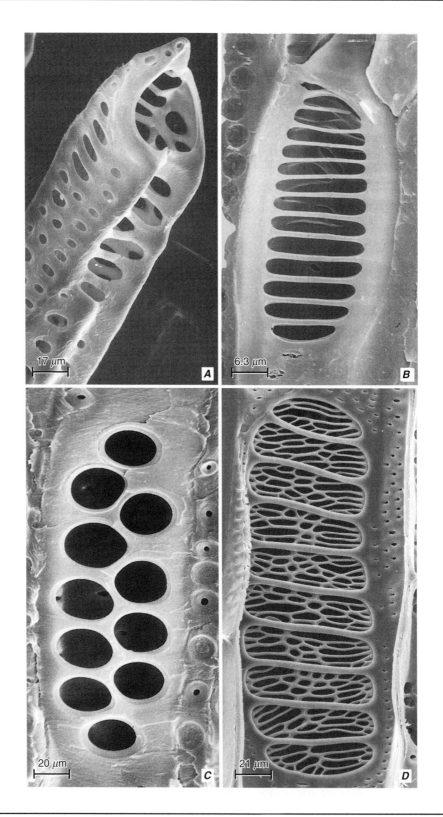

**FIGURE 10.3**

Perforation plates. Scanning electron micrographs of the perforated end walls of vessel elements from secondary xylem. **A**, a simple perforation plate, with its single large opening, in *Pelargonium* vessel element. **B**, the ladder-like bars of a scalariform perforation plate between vessel elements in *Rhododendron*. **C**, foraminate perforation plate, with its circular perforations, in *Ephedra*. **D**, contiguous scalariform and reticulate perforation plates in *Knema furfuracea*. (**A–C**, courtesy of P. Dayanandan; **D**, from Ohtani et al., 1992.)

ladder; Figs. 10.2C and 10.3B, D), or in a reticulate manner (***reticulate perforation plate***, from the Latin *rete*, net; Fig. 10.3D), or as a group of approximately circular holes (***foraminate perforation plate***; Fig. 10.3C; see Fig. 10.16). Multiple perforation plates are rarely found in woody species of low altitude tropical forests. They are more common in woody species of tropical high mountain floras and of temperate and mild-mesothermic climates characterized by low temperatures during winter, whereas species with scalariform perforation plates tend to be restricted to relatively nonseasonal mesic habitats, such as tropical cloud forests, summer-wet temperate forests, or boreal habitats where the soil never dries (Baas, 1986; Alves and Angyalossy-Alfonso, 2000; Carlquist, 2001).

Perforations generally occur on the end walls, with the vessel elements joined end-on-end (Fig. 10.4), forming long, continuous columns, or tubes, called **vessels**. Perforations may be present on the lateral walls too. Each vessel element of a vessel bears a perforation plate at each end, except for the uppermost vessel element and the lowermost one. The uppermost vessel element lacks a perforation plate at its upper end, and the lowermost vessel element lacks a perforation plate at its lower end. The movement of water and solutes from vessel to vessel occurs through the pit-pairs in their common walls. The length of a vessel has been defined as the maximum distance that water can travel without crossing from one vessel to an adjacent one through a pit membrane (Tyree, 1993).

A single vessel can consist of as few as two vessel elements (e.g., in the stem primary xylem of *Scleria*, Cyperaceae; Bierhorst and Zamora, 1965) or of hundreds or even thousands of vessel elements. In the latter case, vessel length cannot be determined by conventional microscopic methods. The approximate length of the longest vessels in a stem segment can be determined by forcing air through a piece of stem containing vessels that have been cut open at both ends (Zimmermann, 1982). The longest vessels of a species are slightly longer than the longest piece of stem through which air can be forced. Vessel-length distribution can be determined by forcing dilute latex paint through a piece of stem (Zimmermann and Jeje, 1981; Ewers and Fisher, 1989). The paint particles move from vessel element to vessel element via the perforations but are too large to penetrate the minute pores of the pit membranes. As water is lost laterally the paint particles accumulate in the vessels until the vessels are packed with them. The stem then is cut into segments of equal length, and the paint-containing vessels, which are easily identified with a stereo microscope, are counted at different distances from the point of injection. Presuming that the vessels are randomly distributed, the distribution of vessel lengths can be calculated. Air-flow-rate measurement at given pressure gradients can be used instead of paint to determine vessel-length distribution (Zimmermann, 1983).

The longest vessels occur in the earlywood of ring-porous species of woody eudicots. In ring-porous species, the vessels (pores) of the first-formed wood (earlywood) of a growth layer are especially wide (Fig. 10.1; Chapter 11). Some of these large-diameter vessels have been found to extend through almost the entire length of the tree's stem, although most were much shorter. A maximum length of 18 meters was measured in *Fraxinus americana* (Greenidge, 1952) and of 10.5 to 11.0 meters in *Quercus rubra* (Zimmermann and Jeje, 1981). In general, vessel lengths are correlated with vessel diameters: wide vessels are longer and narrow vessels are shorter (Greenidge, 1952; Zimmermann and Jeje, 1981). Analyses of vessel-length distribution have shown, however, that xylem contains many more short vessels than long ones.

A gradual increase in tracheary element size has been reported to occur from leaves to roots in trees and shrubs (Ewers et al., 1997). Both tracheid diameter and length increased from branches to trunk and down into the roots of *Sequoia sempervirens* (Bailey, 1958). In *Acer rubrum*, both vessel diameter and length gradually increased from twigs to branches, down the stem and into the roots (Zimmermann and Potter, 1982). Similarly, in *Betula occidentalis*, vessels were narrowest in twigs, intermediate in trunks, and widest in roots

**FIGURE 10.4**

Scanning electron micrograph showing parts of three vessel elements of a vessel in secondary xylem of red oak (*Quercus rubra*). Notice the rims (arrows) of the end walls between the vessel elements, which are arranged end on end. (Courtesy of Irvin B. Sachs.)

(Sperry and Saliendra, 1994). In general, roots have wider vessels than stems. Lianas are an exception, for their stem vessels are as wide as or wider than their root vessels (Ewers et al., 1997). The basipetal increase in vessel diameter is accompanied by a decrease in vessel density, that is, in the number of vessels per unit of transverse-sectional area.

### The Secondary Walls of Most Tracheary Elements Contain Pits

Simple and bordered pits are found in the secondary walls of tracheids and vessel elements of the latest-formed primary xylem and of the secondary xylem. The number and arrangement of these pits are highly variable, even on different wall facets, or surfaces, of the same cell, because they depend on the type of cell bordering the particular wall facet. Usually numerous bordered pit-pairs occur between contiguous tracheary elements (*intervascular pitting*; Fig. 10.5); few or no

pit-pairs may occur between tracheary elements and fibers; bordered, half-bordered, or simple pit-pairs are found between tracheary elements and parenchyma cells. In half-bordered pit-pairs the border is on the side of the tracheary element (Fig. 10.5K).

The bordered pits in tracheary elements show three main types of arrangement: scalariform, opposite, and alternate. If the pits are elongated transversely and arranged in vertical, ladder-like series, the pattern is called **scalariform pitting** (Fig. 10.5A–C). Circular or oval bordered pits arranged in horizontal pairs or short horizontal rows characterize **opposite pitting** (Fig. 10.5D, E). If such pits are crowded, their borders assume rectangular outlines in face view. When the pits are arranged in diagonal rows, the arrangement is **alternate pitting** (Figs. 10.5F, G and 10.8), and crowding results in borders that are polygonal (angular and with more than four sides) in outline in face view. Alternate pitting is clearly the most common type of pitting in eudicots.

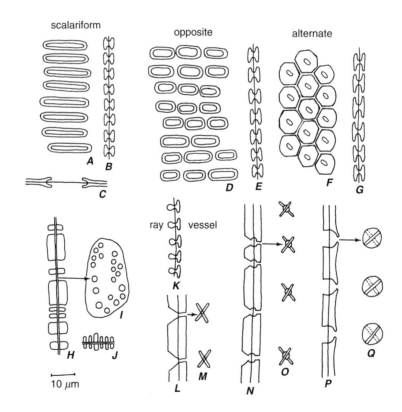

**FIGURE 10.5**

Pits and patterns of pitting. **A–C**, scalariform pitting in surface (**A**) and side (**B, C**) views (*Magnolia*). **D–E**, opposite pitting in surface (**D**) and side (**E**) views (*Liriodendron*). **F–G**, alternate pitting surface (**F**) and side (**G**) views (*Acer*). **A–G**, bordered pit-pairs in vessel members. **H–J**, simple pit-pairs in parenchyma cells in surface (**I**) and side (**H, J**) views; **H**, in side wall; **J**, in end wall (*Fraxinus*). **K**, half-bordered pit-pairs between a vessel and a ray cell in side view (*Liriodendron*). **L, M**, simple pit-pairs with slit-like apertures in side (**L**) and surface (**M**) views (libriform fiber). **N, O**, bordered pit-pairs with slit-like inner apertures extended beyond the outline of the pit border; **N**, side view, **O**, surface view (fiber-tracheid). **P, Q**, bordered pit-pairs with slit-like inner apertures included within the outline of the pit border; **P**, side view, **Q**, surface view (tracheid). **L–Q**, *Quercus*. (From Esau, 1977.)

The bordered pit-pairs of conifer tracheids have a particularly elaborate structure (Hacke et al., 2004). In the large, relatively thin-walled earlywood tracheids such pit-pairs commonly are circular in face view (Fig. 10.6A) and the borders enclose a conspicuous cavity (Fig. 10.6B). In the center of the pit membrane there is a thickening, the **torus** (plural: **tori**), which is somewhat larger in diameter than the pit apertures (Fig. 10.6A, B). It is surrounded by the thin part of the pit membrane, the **margo**, which consists of bundles of cellulose microfibrils, most of them radiating from the torus (Figs. 10.6A and 10.7). The open structure of the margo results from the removal of the noncellulosic matrix of the primary wall and middle lamella during cell maturation. Thickenings of middle lamella and primary wall, called **crassulae** (singular: **crassula**, from the Latin, little thickening), may occur between pit-pairs (not apparent in Fig. 10.6A). The margo is flexible and under certain conditions of stress it moves toward one or the other side of the border, closing the aperture with the torus (Fig. 10.6C). When the torus is in this position, the movement of water through the pit-pair is restricted. Such pit-pairs are said to be **aspirated**. The torus is characteristic of the bordered pits in Gnetophyta and Coniferophyta, but may be poorly developed. Tori or torus-like structures have been found in several species of eudicots (Parameswaran and Liese, 1981; Wheeler, 1983; Dute et al., 1990, 1996; Coleman

et al., 2004; Jansen et al., 2004). The margo of these pit membranes differs from that of conifers in that, instead of bundles of cellulose microfibrils radiating from the torus, the microfibrils form a dense meshwork containing many very small pores. No torus develops in the membrane of the half-bordered pit-pairs that occur in the walls between conifer tracheids and parenchyma cells.

In certain eudicots the pit cavities and/or apertures are wholly or partly lined with minute protuberances on the secondary wall (Jansen et al., 1998, 2001). Mostly branched or irregularly shaped, these protuberances are called **vestures**, and such pits are referred to as **vestured pits** (Fig. 10.8). Vestures may occur in all cell types of the secondary xylem. They are not only associated with pits, but can occur on the inner surface of the walls, at perforation plates, and on the helical thickenings (see below) of vessel walls (Bailey, 1933; Butterfield and Meylan, 1980; Metcalfe and Chalk, 1983; Carlquist, 2001). Vestures also occur on tracheid walls of gymnosperms and have been observed in two groups of monocots, namely some species of bamboo (Parameswaran and Liese, 1977) and of palms (Hong and Killmann, 1992). Minute unbranched protuberances, commonly called **warts**, also occur on tracheid walls in gymnosperms and vessel and fiber walls in angiosperms (Castro, 1988; Heady et al., 1994; Dute et al., 1996). Some workers consider there to be no difference

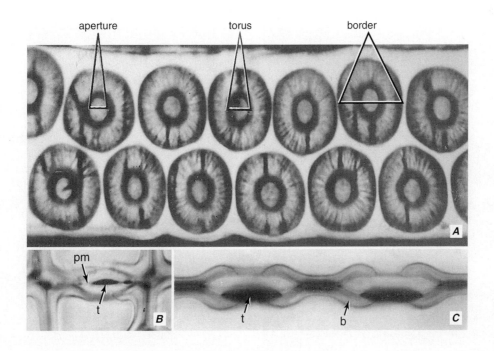

**FIGURE 10.6**

Bordered pits in conifer tracheids (**A**, *Tsuga*; **B**, *Abies*; **C**, *Pinus*). **A**, surface view of pits with thickening (torus) on pit membranes. **B**, **C**, pit-pairs in sectional views with torus (t) on pit membrane (pm) in median position (**B**) and appressed to the border (b in **C**; aspirated pit-pair). (A, ×1070; B, C, ×1425. A, from Bannan, 1941.)

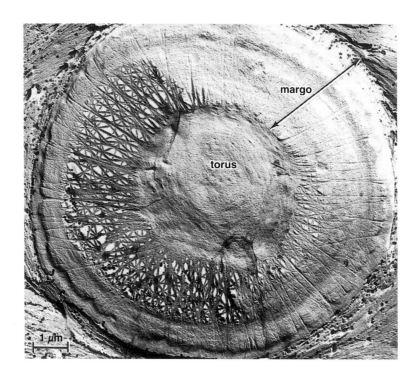

**FIGURE 10.7**

Scanning electron micrograph of bordered pit in earlywood tracheid of *Pinus pungens*. The border was cut away and the pit membrane is exposed. The pit membrane consists of an impermeable torus and a very porous margo. The microfibrils in the margo are predominantly in radial arrangement. (Courtesy of W. A. Côté Jr.)

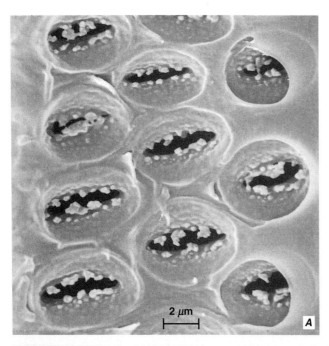

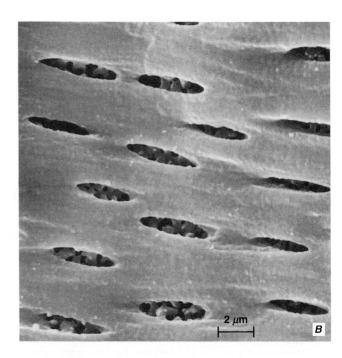

**FIGURE 10.8**

Vestured pits in vessel of *Gleditsia triacantha*. **A**, middle lamella view; **B**, view seen from vessel lumen. The arrangement of these pits is alternate. (Courtesy of P. Dayanandan.)

between vestures and warts and recommend that the terms warts and warty layer be replaced by the terms vestures and vestured layer (Ohtani et al., 1984).

Apparently most vestures consist largely of lignin (Mori et al., 1980; Ohtani et al., 1984; Harada and Côté, 1985). Lignin has been reported as lacking from the vestures in some members of the Fabaceae (Ranjani and Krishnamurthy, 1988; Castro, 1991). Other components of vestures are hemicellulose and small amounts of pectin; cellulose is lacking (Meylan and Butterfield, 1974; Mori et al., 1983; Ranjani and Krishnamurthy, 1988).

A striking correlation exists between the type of vessel perforation plate and vestured pits: virtually all taxa with vestured pits have simple perforation plates (Jansen et al., 2003). This correlation, among other factors, has led to the suggestion that vestured pits contribute to hydraulic safety. Results of one study support this suggestion. Evidence has been obtained that vestures limit the degree to which the pit membrane can be deflected from the center of the pit cavity, thus limiting the increase in porosity of the pit membrane that results from mechanical stress and reducing the probability of air seeding through the membrane (Choat et al., 2004).

Ridges, called **helical thickenings**, or **helical sculptures**, may form on the inner surface of the vessel elements in a roughly helical pattern without covering the pits (Fig. 10.9). Within the secondary xylem, helical thickenings are more common in the latewood (Carlquist and Hoekman, 1985). Helical thickenings appear to be more frequent in woody species of subtropical and temperate floras than in woody species of tropical floras (Van der Graaff and Baas, 1974; Baas, 1986; Alves and Angyalossy-Alfonso, 2000; Carlquist, 2001).

As noted by Sperry and Hacke (2004), tracheid and vessel walls—the xylem conduit walls—perform three important functions. They (1) permit waterflow between adjacent conduits, (2) prevent air entry from gas-filled (embolized) conduits to adjacent water-filled functional ones, and (3) prevent implosion (wall collapse; Cochard et al., 2004) under the significant negative pressures of the transpiration stream. These functions are fulfilled by the lignified secondary walls, which provide strength, and by the pits, which allow water flow between conduits.

## Vessels Are More Efficient Conduits of Water Than Are Tracheids

The greater efficiency of vessels as conduits of water (Wang et al., 1992; Becker et al., 1999) is due in part to the fact that water can flow relatively unimpeded from vessel element to vessel element through the perforations in their end walls. By contrast, water flowing from tracheid to tracheid must pass through the pit mem-

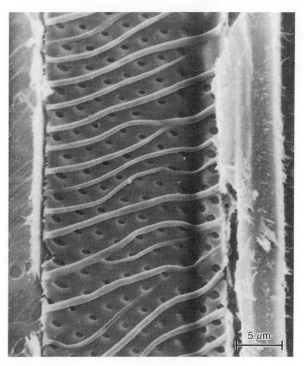

5 μm

**FIGURE 10.9**

Scanning electron micrograph of the secondary wall of a mature vessel of linden *(Tilia platyphyllos)* wood showing pits and helical thickenings. (From Vian et al., 1992.)

branes of the pit-pairs in their overlapping walls. The bordered pits in the tracheids of *Tsuga canadensis* have been estimated to account for about one-third of the total resistance to water flow though these conduits (Lancashire and Ennos, 2002). The torus-margo pit membrane of coniferous tracheids is more conductive, however, than the homogeneous vessel pit membrane (Hacke et al., 2004; Sperry and Hacke, 2004). The reason for the greater conductivity or efficiency of the torus-margo membrane is the presence of larger pores in the margo than in the pit membranes of vessels.

The wider and longer that vessels are, the higher is their hydraulic conductivity (or the lower their resistance to water flow). Of these two parameters, vessel width has by far the greater effect on conductivity (Zimmermann, 1982, 1983). The hydraulic conductivity of the vessel is roughly proportional to the fourth power of its radius (or diameter). Thus, if the relative diameters of three vessels are 1, 2, and 4, the relative volumes of water flowing through them under similar conditions would be 1, 16, and 256, respectively. Consequently wide vessels are very much more efficient water conductors than narrow vessels. However, whereas increased vessel diameter greatly increases efficiency of water conduction, at the same time it decreases safety.

With each 0.34 meter increase in height up the stems of chrysanthemum (*Dendranthema* × *grandiflorum*), the hydraulic conductivity was found to decrease by 50% (Nijsse et al., 2001). The decrease in conductivity was due to a decrease both in cross-sectional area and length of the vessels with stem height. With regard to the latter factor, higher in the stem the stream of water must traverse more interconduit connections—pit-pairs—per unit of stem length. The vessel lumina were calculated to account for about 70% of the hydraulic resistance, the pit-pairs at least part of the remaining 30% (Nijsse et al., 2001).

The columns of water in the conduits (vessels and/or tracheids) of the xylem are usually under tension and, consequently, are vulnerable to **cavitation**, that is, the formation of cavities within the conduits resulting in breakage of the water columns. Cavitation can precipitate an **embolism**, or blockage, of the conduit with air (Fig. 10.10). Beginning with a single vessel element, the entire vessel may soon become filled with water vapor and air. The vessel is now dysfunctional and no longer capable of conducting water. Inasmuch as wide vessels tend to be longer than narrow vessels, it would be safer for the plant to have fewer wide vessels than narrow ones (Comstock and Sperry, 2000). Because of the relatively large size of their xylem conduits, roots tend to be more vulnerable to water stress-induced cavitation than stems or twigs (Mencuccini and Comstock, 1997; Linton et al., 1998; Kolb and Sperry, 1999; Martínez-Vilalta et al., 2002).

Although the pit membranes provide significant resistance to the flow of water between conduits, they are very important to the safety of water transport. The surface tension of the air-water meniscus spanning the small pores in the pit membranes of the bordered pit-pairs between adjacent vessels usually prevents air bubbles from squeezing through the pores, helping to restrict them to a single vessel (Fig. 10.11; Sperry and Tyree, 1988). In conifer tracheids the passage of air is prevented by aspiration of the pit-pairs resulting in blockage of pit apertures by the tori. The margo pores are usually too large to contain an embolus.

Two phenomena—freezing and drought—are largely responsible for cavitation events (Hacke and Sperry, 2001). During winter and the growing season, most embolisms in temperate woody plants are associated with freeze-thaw events (Cochard et al., 1997). The xylem sap contains dissolved air. As the sap freezes, the dissolved gases freeze out as bubbles. Considerable evidence indicates that large diameter vessels are more vulnerable to freezing-induced embolism than narrow diameter vessels and conifer tracheids least of all (Sperry and Sullivan, 1992; Sperry et al., 1994; Tyree et al., 1994). As noted by Sperry and Sullivan (1992), this may explain the trend for decreasing conduit size with increasing latitude and altitude (Baas, 1986), the rarity

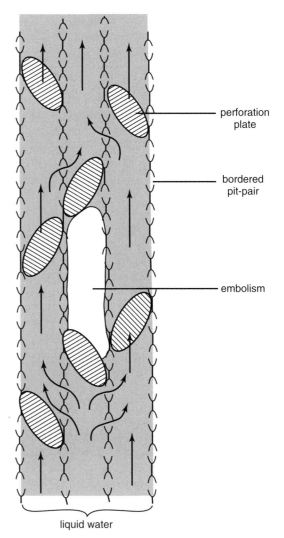

perforation plate

bordered pit-pair

embolism

liquid water

**FIGURE 10.10**

Embolized vessel element. An embolism consisting of water vapor has blocked the movement of water through a single vessel element. However, water is able to detour around the embolized element via the bordered pit-pairs between adjacent vessels. The vessel elements shown here are characterized by scalariform perforation plates. (From Raven et al., 2005.)

of woody vines, with their wide vessels, at high latitudes (Ewers, 1985; Ewers et al., 1990), and the dominance of conifers, with their narrow tracheids, in cold climates (see Maherali and DeLucia, 2000, and Stout and Sala, 2003, and literature cited therein, for discussions on xylem vulnerability in conifers).

Drought-induced water stress increases the tension of the xylem sap, that is, of the fluid contents of the xylem. When this tension exceeds the surface tension at the air-water meniscus spanning the pores in the pit

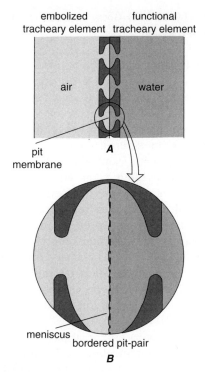

embolized functional
tracheary element tracheary element

air water

pit
membrane

*A*

meniscus
bordered pit-pair

*B*

**FIGURE 10.11**

Diagram showing bordered pit-pair between tracheary elements, one of which is embolized and thus nonfunctional (**A**). **B**, detail of a pit membrane. When a tracheary element is embolized, air is prevented from spreading to the adjacent functional tracheary element by the surface tension of the air-water meniscus spanning the pores in the pit membrane. (From Raven et al., 2005.)

membrane, air may be pulled into a functional conduit (Sperry and Tyree, 1988). This process is known as **air seeding** (Zimmermann, 1983; Sperry and Tyree, 1988). The largest pores are the most vulnerable to the penetration of air. A plant is susceptible to this mode of embolism any time even one of its vessels or tracheids becomes air-filled by physical damage (for example, by wind or herbivory). In conifers, air seeding probably occurs when the pressure difference between tracheids becomes large enough to rip the torus out of its position (Sperry and Tyree, 1990).

Considerable discourse has taken place over possible mechanisms involved with the recovery of hydraulic conductance following xylem embolism (Salleo et al., 1996; Holbrook and Zwieniecki, 1999; Tyree et al., 1999; Tibbetts and Ewers, 2000; Zwieniecki et al., 2001a; Hacke and Sperry, 2003). Two mechanisms have been attributed to the recovery of hydraulic conductivity by beech (*Fagus sylvatica*) trees that experience winter embolism (Cochard et al., 2001b). One mechanism is operative in early spring, before bud break, and is correlated with the occurrence of positive xylem pressures at the base of the trunk. The positive xylem pressures actively dissolve the embolisms. The second recovery mechanism is operative after bud break and is correlated with the renewal of cambial activity. At this time embolized vessels are replaced by new, functional vessels. As noted by Cochard et al. (2001b), the two mechanisms are complementary: the first occurs mostly in the root and the trunk, and the second mainly in young terminal shoots. In another study, winter embolism in branches of birch (*Betula* spp.) and alder (*Alnus* spp.) trees was reversed by refilling vessels with positive root pressures during spring, whereas branches of gambel oak (*Quercus gambelii*) trees relied on the production of new functional vessels to restore hydraulic conductance (Sperry et al., 1994). Like the beech tree, birch and alder trees are diffuse porous; gambel oak is ring porous.

Although positive root pressures long have been known to play a role in the refilling of embolized xylem conduits (Milburn, 1979), there have been reports that embolized vessels can refill in the absence of root pressure and when the xylem pressure is substantially negative (Salleo et al., 1996; Tyree et al., 1999; Hacke and Sperry, 2003). Embolisms have been reported to occur daily in many vessels in the shoots (Canny, 1997a, b) and roots (McCully et al., 1998; Buchard et al., 1999; McCully, 1999) of transpiring herbaceous plants. Whereas it is generally presumed that refilling with water of embolized vessels occurs after transpiration has ceased, refilling of the embolized vessels in the pertinent herbaceous plants reportedly takes place while the plants are still transpiring and the xylem sap is still under tension. The conclusions drawn from these studies have been criticized by several workers who contend that the observed embolisms are artifacts that result from the freezing procedure (cryo-microscopy) used in these studies (Cochard et al., 2001a; Richter, 2001; see, however, Canny et al., 2001).

Sculpturing of vessel walls and the nature of the perforation plates may influence vulnerability to embolism. It has been suggested, for example, that helical thickenings may reduce the occurrence of embolism events by virtue of increasing the surface area of vessels, and therefore increasing the bonding of water to vessel walls (Carlquist, 1983). Helical thickenings may also increase the conductive capacity of narrow vessels, which would provide a causal explanation for their prevalence in narrow latewood vessels (Roth, 1996). Scalariform perforation plates have been cited as a mechanism of trapping air bubbles in individual vessel elements and hence preventing blockage of entire vessels (Zimmermann, 1983; Sperry, 1985; Schulte et al., 1989; Ellerby and Ennos, 1998). Although the resistance of simple perforation plates to flow is lower than that of all but the simplest of scalariform perforation plates, scalariform perforation plates—even those with narrow perforations—are only slight obstructions to flow (Schulte et al., 1989). Regardless of the type of perforation plate, the vast majority of flow resistance in vessel elements appears to be due to the vessel wall (Ellerby and Ennos, 1998).

### Fibers Are Specialized as Supporting Elements in the Xylem

The fibers are long cells with secondary, commonly lignified, walls. The walls vary in thickness but are usually thicker than the walls of tracheids in the same wood. Two principal types of xylem fiber are recognized, the fiber-tracheids and the libriform fibers (Chapter 8). If both occur in the same wood, the libriform fiber is longer and commonly has thicker walls than the fiber-tracheid. The fiber-tracheids (Fig. 10.5N, O) have bordered pits with cavities smaller than the pit cavities of tracheids or vessels (Fig. 10.5P, Q) in the same wood. These pits have a pit canal with a circular outer aperture and an elongated or slit-like inner aperture (Chapter 4).

The pit in a libriform fiber has a slit-like aperture toward the cell lumen and a canal resembling a much flattened funnel, but no pit cavity (Fig. 10.5L, M). In other words, the pit has no border; it is simple. The reference to the pits of libriform fibers as simple implies a sharper distinction than actually exists. The fibrous xylem cells show a graduated series of pits between those with pronounced borders and those with vestigial borders or no borders. The intergrading forms with recognizable pit borders are placed, for convenience, in the fiber-tracheid category (Panshin and de Zeeuw, 1980).

Fibers of both categories may be septate (Chapter 8). Septate fibers (Fig. 8.6A; see Fig. 10.15), which are widely distributed in eudicots and are quite common in tropical hardwoods, usually retain their protoplasts in the mature active wood (Chapter 11), where they are concerned with the storage of reserve materials (Frison, 1948; Fahn and Leshem, 1963). Thus the living fibers approach xylem parenchyma cells in structure and function. The distinction between the two is particularly tenuous when the parenchyma cells develop secondary walls and septa. The retention of protoplasts by fibers is an indication of evolutionary advance (Bailey, 1953; Bailey and Srivastava, 1962), and where living fibers are present, the axial parenchyma is small in amount or absent (Money et al., 1950).

Another modification of fiber-tracheids and libriform fibers are the so-called gelatinous fibers (Chapter 8). Gelatinous fibers (Fig. 8.7; see Fig. 10.15) are common components of the reaction wood (Chapter 11) in eudicots.

### Living Parenchyma Cells Occur in Both the Primary and Secondary Xylem

In the secondary xylem the parenchyma cells are commonly present in two forms: **axial parenchyma** and **ray parenchyma** (see Fig. 10.16). The axial parenchyma cells are derived from the elongated fusiform initials of the vascular cambium, and consequently their long axes are oriented vertically in the stem or root. If the derivative of such a cambial cell differentiates into a parenchyma cell without transverse (or oblique) divisions, a **fusiform parenchyma cell** results. If such divisions occur, a **parenchyma strand** is formed. Parenchyma strands occur more commonly than fusiform parenchyma cells. Neither type undergoes intrusive growth. The ray parenchyma cells, which are derived from the relatively short ray initials of the vascular cambium, may have their long axes oriented either vertically or horizontally with regard to the axis of stem or root (Chapter 11).

The ray parenchyma and axial parenchyma cells of the secondary xylem typically have lignified secondary walls. The pit-pairs between parenchyma cells may be bordered, half-bordered, or simple (Fig. 10.5H–J; Carlquist, 2001). Some parenchyma cells become sclerified by deposition of thick secondary walls. These are sclerotic cells, or sclereids.

The parenchyma cells of the xylem have a variety of contents. They are particularly known for their storage of food reserves in the form of starch or fat. In many deciduous trees of the temperate zone, starch accumulates in late summer or early autumn and declines during dormancy as the starch is converted to sucrose at low winter temperatures (Zimmermann and Brown, 1971; Kozlowski and Pallardy, 1997a; Höll, 2000). The dissolution of starch during full dormancy may be primarily a protective action against frost injury (Essiamah and Eschrich, 1985). Starch is resynthesized and accumulates a second time at the end of dormancy in early spring. It subsequently decreases as reserves are utilized during the early season growth flush. The fat and storage protein contents of the parenchyma cells also vary seasonally (Fukazawa et al., 1980; Kozlowski and Pallardy, 1997b; Höll, 2000).

Tannins and crystals are common inclusions (Scurfield et al., 1973; Wheeler et al., 1989; Carlquist, 2001). The types of crystals and their arrangements may be sufficiently characteristic to serve in identification of woods. Prismatic (rhomboidal) crystals are the most common type of crystal in wood. Crystal-containing parenchyma cells frequently have lignified walls with secondary thickenings and may be chambered, or subdivided, by septa, each chamber containing a single crystal. Cells may secrete a layer of secondary wall material around the crystals. Generally, this layer of wall material is relatively thin, but in some instances, it may be so thick as to fill most of the cell lumen between the crystal and the primary wall. In herbaceous plants and young twigs of woody plants chloroplasts often occur in xylary parenchyma cells, particularly the ray parenchyma cells (Wiebe, 1975).

### In Some Species the Parenchyma Cells Develop Protrusions—Tyloses—That Enter the Vessels

In the secondary xylem both the axial and the ray parenchyma cells located next to the vessels may form outgrowths through the pit cavities and into the lumina of the vessels when the latter become inactive and lose their internal pressure (Fig. 10.12). These outgrowths are called **tyloses** (singular: **tylose**), and the parenchyma cells that give rise to them are referred to as **contact cells** (Braun, 1967, 1983) because they literally are in direct contact with the vessels (contact cells are considered further in Chapter 11). Contact cells are characterized by the presence of a loosely fibrillar cellulose-poor, pectin-rich wall layer that is deposited by the protoplast after completion of secondary wall formation (Czaninski, 1977; Gregory, 1978; Mueller and

Beckman, 1984). Called the **protective layer**, it commonly is deposited on all surfaces of the contact cell wall but is thickest on the side of the cell bordering the vessel, especially at the pit membrane.

During tylosis, or tylose formation, the protective layer balloons out as a tylose into the lumen of the vessel (Fig. 10.13). The nucleus and part of the cytoplasm of the parenchyma cell commonly migrate into the tylose. Growth of the tylose appears to be hormonally controlled (VanderMolen et al., 1987). Tyloses store a variety of substances and may develop secondary walls. Some even differentiate into sclereids. Tyloses are rarely found when the pit aperture on the vessel side is less than 10 μm in diameter (Chattaway, 1949), indicating that tylose formation may be physically limited by a minimal contact pit diameter (van der Schoot, 1989). In addition to secondary xylem, tyloses also occur in primary xylem (Czaninski, 1973; Catesson et al., 1982; Canny, 1997c; Keunecke et al., 1997).

Tyloses may be so numerous that they completely fill the lumen of the vessel element. In some woods, they are formed as the vessels cease to function (Fig. 10.12A, D). Tyloses are often induced to form prematurely by

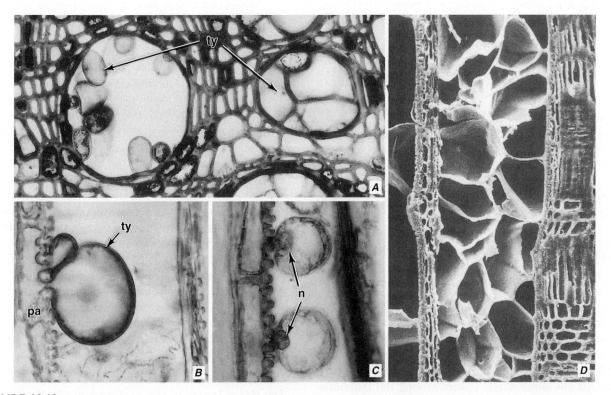

**FIGURE 10.12**

Tyloses (ty) in *Vitis* (grapevine, **A–C**) and *Carya ovata* (shagbark hickory, **D**) vessels as seen in transverse (**A**) and longitudinal (**B–D**) sections of xylem. **A**, left, young tyloses; right, vessel filled with tyloses. **B**, continuity between lumina of tyloses and parenchyma cell. **C**, nuclei (n) have migrated from parenchyma cells to tyloses. **D**, scanning electron micrograph of vessel filled with tyloses. (A, ×290; B, C, ×750; D, ×170. D, courtesy of Irvin B. Sachs.)

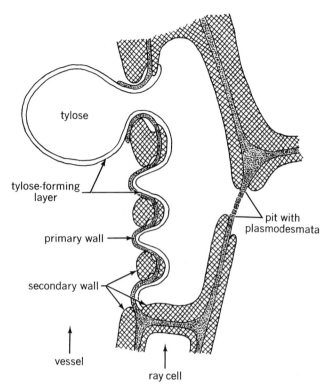

**FIGURE 10.13**

Diagram of ray cell that has formed a tylose protruding through a pit into the lumen of a vessel. The tylose-forming layer is also called protective layer. (From Esau, 1977.)

plant pathogens and may serve as a defensive mechanism by inhibiting the spread of the pathogen throughout the plant via the xylem (Beckman and Talboys, 1981; Mueller and Beckman, 1984; VanderMolen et al., 1987; Clérivet et al., 2000). In *Fusarium*-infected banana, a protective layer is not associated with tylose formation (VanderMolen et al., 1987).

# PHYLOGENETIC SPECIALIZATION OF TRACHEARY ELEMENTS AND FIBERS

The xylem occupies a unique position among plant tissues in that the study of its anatomy has come to play an important role with reference to taxonomy and phylogeny. The lines of specialization of the various structural features have been better established for the xylem than for any other single tissue. Among the individual lines, those pertaining to the evolution of the tracheary elements have been studied with particular thoroughness.

The tracheid is a more primitive element than the vessel element. It is the only kind of tracheary element found in the fossil seed plants (Stewart and Rothwell, 1993; Taylor and Taylor, 1993) and in most of the living seedless vascular plants and gymnosperms (Bailey and Tupper, 1918; Gifford and Foster, 1989).

The specialization of tracheary elements coincided with the separation of the functions of conduction and strengthening that occurred during the evolution of vascular plants (Bailey, 1953). In the less specialized state, conduction and support are combined in tracheids. With increased specialization, conducting elements—the vessel elements—evolved with greater efficiency in conduction than in support. In contrast, fibers evolved as primarily strengthening elements. Thus from primitive tracheids two lines of specialization diverged, one toward the vessels and the other toward the fibers (Fig. 10.14).

Vessel elements evolved independently in certain ferns, including the whisk ferns, *Psilotum nudum* and *Tmesipteris obliqua* (Schneider and Carlquist, 2000c; Carlquist and Schneider, 2001), *Equisetum* (Bierhorst, 1958), *Selaginella* (Schneider and Carlquist, 2000a, b), the Gnetophyta (Carlquist, 1996a), monocots, and "dicots" (Austrobaileyales, magnoliids, and eudicots). In the eudicots, vessel elements originated and underwent specialization first in the secondary xylem, then in the late primary xylem (metaxylem), and last in the early primary xylem (protoxylem). In the primary xylem of the monocots, origin and specialization of vessel elements also occurred first in the metaxylem, then in the protoxylem; furthermore, in the monocots, vessel elements appeared first in the root and later in stems, inflorescence axes, and leaves, in that order (Cheadle, 1953; Fahn, 1954). The relation between the first appearance of vessels and type of organ in eudicots has been explored less completely, but some data indicate an evolutionary lag in leaves, floral appendages, and seedlings (Bailey, 1954).

In the secondary xylem of eudicots, species with vessel elements arose from ones with tracheids bearing scalariform bordered pits (Bailey, 1944). Transition from a vesselless to a vessel-containing condition involved loss of pit membranes from a part of the wall bearing several bordered pits. Thus, a pitted wall part became a scalariform perforation plate (Fig. 10.14G, H). Remnants of membranes occur in the perforations of vessel elements of many primitive eudicots and are regarded as a primitive feature in eudicots (Carlquist 1992, 1996b, 2001). The tracheid-vessel transition is not a sharp one; all degrees of intermediacy may be found (Carlquist and Schneider, 2002).

## The Major Trends in the Evolution of the Vessel Element Are Correlated with Decrease in Vessel Element Length

1. ***Decrease in length.*** The most clearly established trend in evolution of the vessel elements is decrease

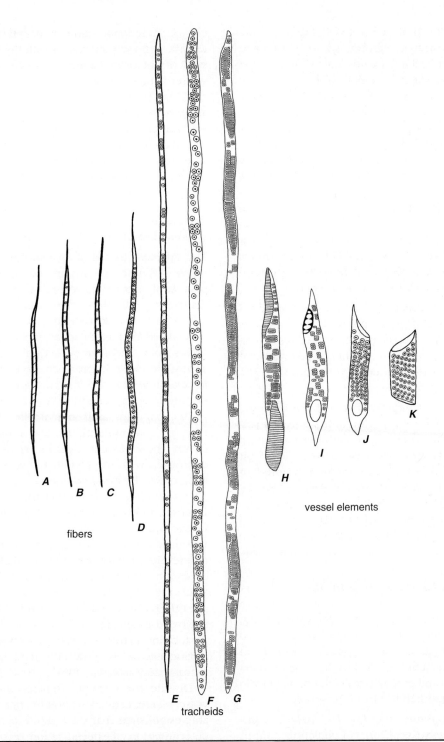

**FIGURE 10.14**

Main lines of specialization of tracheary elements and fibers. **E-G**, long tracheids from primitive woods. (**G**, reduced in scale.) **E**, **F**, circular bordered pits; **G**, elongated bordered pits in scalariform arrangement. **D-A**, evolution of fibers: decrease in length, reduction in size of pit borders, and change in shape and size of pit apertures. **H-K**, evolution of vessel elements: decrease in length, reduction in inclination of end walls, change from scalariform to simple perforation plates, and from opposite to alternate pit arrangement. (After Bailey and Tupper, 1918.)

in length (Fig. 10.14H–K). Longer vessel elements are found in more primitive groups (those with more numerous primitive floral features) and shorter vessel elements in more specialized ones (those with more numerous specialized floral features). The evolutionary sequence of vessel element types in the secondary xylem of eudicots began with long scalariformly pitted tracheids similar to those found in some primitive eudicots. These tracheids were succeeded by vessel elements of long narrow shape with tapering ends. The vessel elements then underwent a progressive decrease in length. The phylogenetic shortening of vessel elements is a particularly consistent characteristic and has occurred in all vascular plants that have developed vessels (Bailey, 1944). Other trends in vessel element evolution are defined by correlation with the decrease in vessel element length.

2. ***Inclined to transverse end walls.*** As the vessel elements shortened, their end walls became less inclined and finally transverse. Thus the vessel elements gradually acquired definite end walls of decreasing degree of inclination, in contrast to the tapering ends of tracheids.

3. ***Scalariform to simple perforation plates.*** In the more primitive state the perforation plate was scalariform, with numerous bars resembling a wall with scalariformly arranged bordered pits devoid of pit membranes. Increase in specialization resulted in removal of the borders and then a decrease in the number of bars until finally bars were totally eliminated. Thus a pitted wall part became a scalariform perforation plate, which later evolved into a simple perforation plate bearing a single opening (Fig. 10.14G–I).

4. ***Scalariform bordered pitting to alternate bordered pitting.*** The pitting of vessel walls also changed during the evolution. In intervessel pitting, bordered pit-pairs in scalariform series were replaced by bordered pit-pairs, first in opposite and later in alternate arrangement (Fig. 10.14H–K). The pit-pairs between vessels and parenchyma cells changed from bordered, through half-bordered, to simple.

5. ***Vessel outline from angular to rounded (as seen in transverse section).*** In eudicotyledonous vessels, angularity in outline is considered to be the primitive state and roundness a specialized condition. Interestingly a correlation exists between angularity and narrowness of vessels. Vessels that are rounded in outline tend to be wider.

Presumably phylogenetic specialization of the vessel element proceeded in the direction of increased conductive efficiency or safety, although the relationship between the trends and their adaptive value is not always obvious. For example, there is little agreement

on the functional value of decreased vessel element length, although shorter vessel elements are found in eudicots of drier habitats than in related eudicots of wetter habitats (Carlquist, 2001). The adaptive value of the trend from scalariform to opposite to alternate pits appears to be a gain in mechanical strength in the vessel wall rather than safety or conductivity (Carlquist, 1975). Although not as well defined a trend in vessel evolution as others, widening of vessel elements obviously resulted in greater conductive capacity.

### Deviations Exist in Trends of Vessel Element Evolution

The different trends of specialization of tracheary elements discussed in the preceding paragraphs are not necessarily closely correlated within specific groups of plants. Some of these trends may be accelerated and others retarded, so that the more and the less highly specialized characters occur in combinations. Moreover plants may secondarily acquire characteristics that appear primitive because of evolutionary loss. Vessels, for example, may be lost through nondevelopment of perforations in potential vessel elements. In aquatic plants, parasites, and succulents, vessels may fail to develop concomitantly with a reduction of vascular tissue. These vesselless plants are highly specialized as contrasted with the primitively vesselless angiosperms exemplified by *Trochodendron, Tetracentron, Drimys, Pseudowintera*, and others (Bailey, 1953; Cheadle, 1956; Lemesle, 1956). In some families, for example the Cactaceae and Asteraceae, evolutionary degeneration of vessel elements involved a decrease in width of cells and nondevelopment of perforations (Bailey, 1957; Carlquist, 1961). The resulting nonperforate cells, having the same kind of pitting as the vessel elements of the same wood, are referred to as **vascular tracheids**. Another deviating trend in specialization may be the development of perforation plates of a reticulate type in an otherwise phylogenetically highly advanced family such as the Asteraceae (Carlquist, 1961).

Despite these inconsistencies the major trends of vessel element specialization in angiosperms are so reliably established that they play a significant role in the determination of specialization of other structures in the xylem. Although the major trends in xylem evolution have generally been regarded as irreversible, the results of studies on ecological wood anatomy, which revealed that strong correlations exist between wood structure and macroclimatic environmental factors (e.g., temperature, seasonality, and water availability), cast doubt on the total irreversibility of the evolutionary trends (see discussion and references in Endress et al., 2000). The idea of irreversibility has also been challenged by cladistic analyses that indicate vessellessness is a derived state rather than being a primitive one (e.g., Young, 1981; Donoghue and Doyle, 1989; Loconte and Stevenson, 1991). It has been hypothesized that the vesselless condition in the Winteraceae has resulted from the loss of vessels as an adaptation to freezing-prone environments (Feild et al., 2002). Elegant and convincing defenses in support of the concept of irreversibility in general have been made by Baas and Wheeler (1996) and by Carlquist (1996b).

Whether or not the angiosperms were primitively vesselless remains a contentious issue (Herendeen et al., 1999; Endress et al., 2000). Thus far there is no evidence in the albeit sparse fossil record that angiosperms were originally vesselless. In fact vessel-bearing angiosperms with fairly advanced woods occur in the Middle and Upper Cretaceous (Wheeler and Baas, 1991), whereas the oldest vesselless angiosperm woods are from the Upper Cretaceous (Poole and Francis, 2000). Additional paleobotanical data may help to resolve this problem. The apparent lack of vessels in *Amborella*—considered by many as sister to all other angiosperms—suggests that the ancestral angiosperm condition was vesselless (Parkinson et al., 1999; Zanis et al., 2002; Angiosperm Phylogeny Group, 2003).

Although vessel elements evolved in angiosperms, tracheids were retained, and they too underwent phylogenetic changes. The tracheids became shorter, but not as short as the vessel elements, and the pitting of their walls became essentially similar to that of the associated vessel elements. The tracheids generally did not increase in width. Tracheids may be retained for reasons of conductive safety, although they are present in only a relatively small proportion of extant woods.

### Like Vessel Elements and Tracheids, Fibers Have Undergone a Phylogenetic Shortening

In the specialization of xylem fibers (Fig. 10.14D–A) the emphasis on mechanical function became apparent in the decrease in cell width and reduction in wall area occupied by the pit membrane. Concomitantly the pit borders became reduced and eventually disappeared. The inner apertures of the pit became elongated and then slit-like, paralleling the cellulose microfibrils that compose the wall. The evolutionary sequence was from tracheids, through fiber-tracheids, to libriform fibers. The two types of fiber intergrade with each other and also with the tracheids. Because of this lack of clear separation between fibers and tracheids the two kinds of elements have at times been grouped together under the term **imperforate tracheary elements** (Bailey and Tupper, 1918; Carlquist, 1986). Fibers are most highly specialized as supporting elements in those woods that have the most specialized vessel elements (Fig. 10.15), whereas such fibers are lacking in woods with tracheid-like vessel elements (Fig. 10.16). A further evolutionary advance results in the retention of protoplasts by septate fibers (Money et al., 1950).

The matter of evolutionary change in length of fibers is rather complex. The shortening of vessel elements is correlated with a shortening of the fusiform cambial initials (Chapter 12) from which the axial cells of the xylem are derived. Thus in woods with shorter vessel elements the fibers are derived ontogenetically from shorter initials than in more primitive woods with longer vessel elements. In other words, with increase in xylem specialization the fibers become shorter. Because, however, during ontogeny fibers undergo intrusive growth whereas vessel elements do so only slightly or not at all, the fibers are longer than the vessel elements in the mature wood, and of the two categories of fibers, the libriform fibers are the longer ones. Nevertheless, the fibers of specialized woods are shorter than their ultimate precursors, the primitive tracheids.

## ▌PRIMARY XYLEM

### Some Developmental and Structural Differences Exist between the Earlier and Later Formed Parts of the Primary Xylem

Developmentally the primary xylem usually consists of an earlier formed part, the **protoxylem** (from the Greek *proto*, first) and a later formed part, the **metaxylem** (from the Greek *meta*, after or beyond) (Figs. 10.17 and 10.18B). Although the two parts have some distinguishing characteristics, they merge with one another imperceptibly so that the delimitation of the two can be made only approximately.

The protoxylem differentiates in the parts of the primary plant body that have not completed their growth and differentiation. In fact in the stem and leaf the protoxylem usually matures before these organs undergo intensive elongation. Consequently the mature nonliving tracheary elements of the protoxylem are stretched and eventually destroyed. In the root the protoxylem elements frequently mature beyond the region of major elongation and hence persist longer than in the shoot.

The metaxylem commonly begins to differentiate in the still growing primary plant body, but matures largely after the elongation is completed. It is therefore less affected by the primary extension of the surrounding tissues than the protoxylem.

The protoxylem usually contains relatively few tracheary elements (tracheids or vessel elements) embedded in parenchyma that is considered to be part of the protoxylem. When the tracheary elements are destroyed they may become obliterated by surrounding parenchyma cells. The latter either remain thin walled or become lignified, with or without the ·deposition of secondary walls. In the shoot xylem of many monocots the stretched nonfunctioning elements are partly collapsed but not obliterated; instead, open canals, the so-

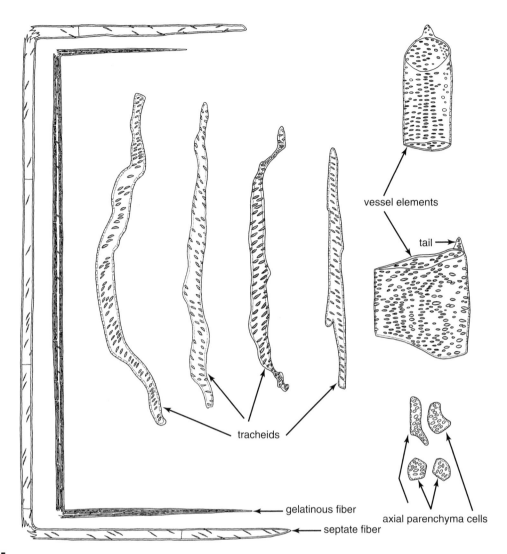

**FIGURE 10.15**

Isolated elements from secondary xylem of *Aristolochia brasiliensis,* a eudicotyledonous vine. Specialized wood with elements of axial system diverse in form. Fibers are libriform, with reduced pit borders. Some are thin-walled and septate; others have thick gelatinous walls. Tracheids are elongated and irregular in shape, with slightly bordered pits. Vessel elements are short and have simple perforations. Pits connecting vessel elements with other tracheary elements are slightly bordered; others are simple. Axial parenchyma cells are irregular in shape and have simple pits. Ray parenchyma cells are not shown. They are relatively large, with thin primary walls. (All, ×130.)

called **protoxylem lacunae**, surrounded by parenchyma cells appear in their place (see Fig. 13.33B). The secondary walls of the nonfunctioning tracheary elements may be seen along the margin of the lacuna.

The metaxylem is, as a rule, a more complex tissue than the protoxylem, and its tracheary elements are generally wider. In addition to tracheary elements and parenchyma cells, the metaxylem may contain fibers. The parenchyma cells may be dispersed among the tracheary elements or may occur in radial rows. In transverse sections the rows of parenchyma cells resemble

rays, but longitudinal sections reveal them as axial parenchyma. The radial seriation often encountered in the metaxylem, and also in the protoxylem, has at times led investigators to interpret the primary xylem of many plants as secondary, for radial seriation is characteristic of the secondary vascular tissue.

The tracheary elements of the metaxylem are retained after primary growth is completed but become nonfunctional after some secondary xylem is produced. In plants lacking secondary growth the metaxylem remains functional in mature plant organs.

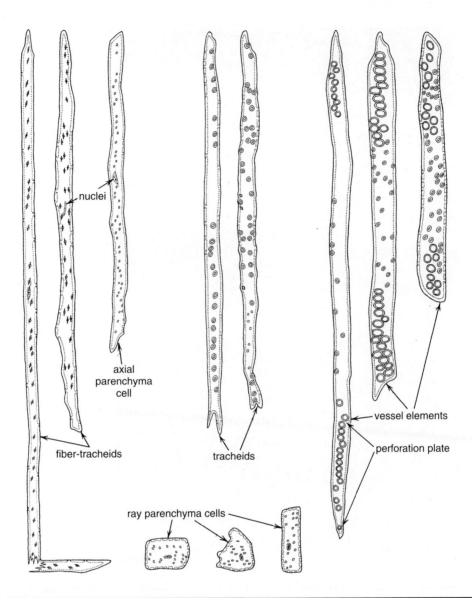

**FIGURE 10.16**

Isolated elements from secondary xylem of *Ephedra californica* (Gnetales). Primitive wood with relatively little morphologic differentiation among elements of axial system. Typical fibers are absent. Axial and ray parenchyma cells have secondary walls with simple pits. Fiber-tracheids have living contents and pits with reduced borders. Tracheids have pits with large borders. Vessel elements are slender, elongated, and have foraminate perforation plates. (All, ×155.)

### The Primary Tracheary Elements Have a Variety of Secondary Wall Thickenings

The different forms of wall appear in a specific ontogenetic series that indicates a progressive increase in the extent of the primary wall area covered by secondary wall material (Fig. 10.18). In the earliest tracheary elements the secondary walls may occur as rings (***annular*** thickenings) not connected with one another. The elements differentiating next have ***helical (spiral)*** thickenings. Then follow cells with thickenings that may be characterized as helices with coils interconnected (***scalariform*** thickenings). These are succeeded by cells with net-like, or ***reticulate***, thickenings, and finally by ***pitted*** elements.

Not all types of secondary thickenings are necessarily represented in the primary xylem of a given plant or plant part, and the different types of wall structure intergrade. The annular thickenings may be interconnected here and there, annular and helical or helical and scalariform thickenings may be combined in the same cell, and the difference between scalariform and reticulate is sometimes so tenuous that the thickening may best be called scalariform-reticulate. The pitted elements also intergrade with the earlier ontogenetic type.

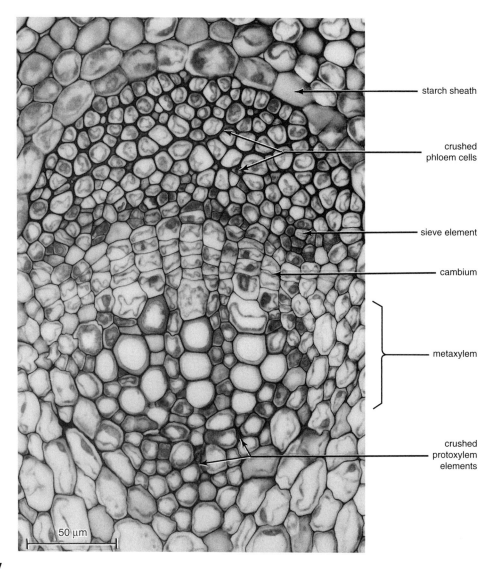

- starch sheath
- crushed phloem cells
- sieve element
- cambium
- metaxylem
- crushed protoxylem elements

50 µm

**FIGURE 10.17**

Vascular bundle from stem of *Medicago sativa* (alfalfa) in cross section. Illustrates primary xylem and phloem. The cambium has not yet produced secondary tissues. The earliest xylem (protoxylem) and phloem (protophloem) are no longer functioning in conduction. Their conducting cells have been obliterated. The functional tissues are metaxylem and metaphloem. (From Esau, 1977.)

The openings in a scalariform reticulum of the secondary wall may be comparable to pits, especially if a slight border is present. A border-like overarching of the secondary wall is common in the various types of secondary wall in the primary xylem. Rings, helices, and the bands of the scalariform-reticulate thickenings may be connected to the primary wall by narrow bases, so the secondary wall layers widen out toward the lumen of the cell and overarch the exposed primary wall parts (see Fig. 10.25A).

The intergrading nature of the secondary wall thickenings in the primary xylem makes it impossible to assign distinct types of wall thickenings to the protoxylem and the metaxylem with any degree of consistency. Most commonly the first tracheary elements to mature, that is, protoxylem elements, have the minimal amounts of secondary wall material. Annular and helical thickenings predominate. These types of thickening do not hinder materially the stretching of the mature protoxylem elements during the extension growth of the primary plant body. The evidence that such stretching occurs is easily perceived in the increase in distance between rings in older xylem elements, the tilting of rings, and the uncoiling of the helices (Fig. 10.19).

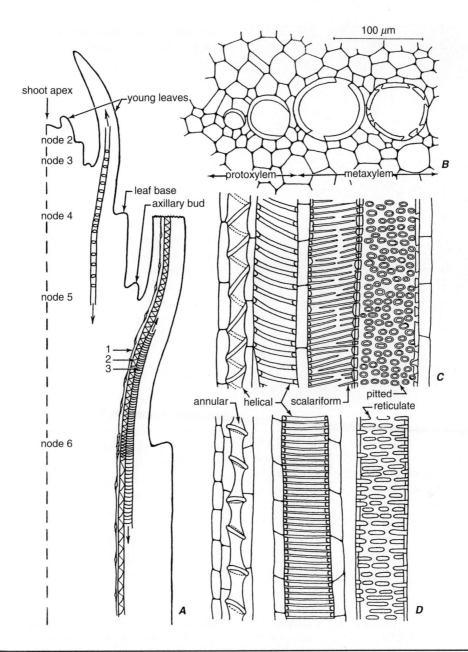

**FIGURE 10.18**

Details of structure and development of primary xylem. **A**, diagram of a shoot tip showing stages in xylem development at different levels. **B–D**, primary xylem of castor bean, *Ricinus,* in cross (**B**) and longitudinal (**C, D**) sections. (From Esau, 1977.)

The metaxylem, in the sense of xylem tissue maturing after the extension growth, may have helical, scalariform, reticulate, and pitted elements; one or more types of thickening may be omitted. If many elements with helical thickenings are present, the helices of the succeeding elements are less and less steep, a condition suggesting that some stretching occurs during the development of the earlier metaxylem elements.

Convincing evidence exists that the type of wall thickening in primary xylem is strongly influenced by the internal environment in which these cells differentiate. Annular thickenings develop when the xylem begins to mature before the maximum extension of the plant part occurs, as for example, in the shoots of normally elongating plants (Fig. 10.18A, nodes 3–5); they may be omitted if the first elements mature after this growth is largely completed, as is common in the roots. If the elongation of a plant part is suppressed before the first xylem elements mature, one or more of the early ontogenetic types of thickenings are omitted. On the

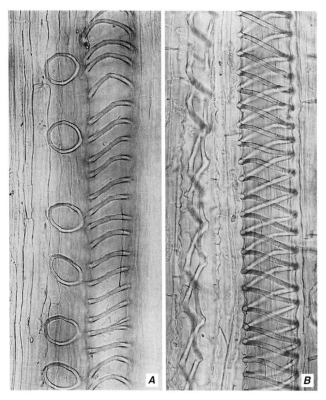

**FIGURE 10.19**

Parts of tracheary elements from the first-formed primary xylem (protoxylem) of the castor bean (*Ricinus communis*). **A**, tilted annular (the ring-like shapes at left) and helical wall thickenings in partly extended elements. **B**, double helical thickenings in elements that have been extended. The element on the left has been greatly extended, and the coils of the helices have been pulled far apart. (A, ×275; B, ×390.)

contrary, if elongation is stimulated, for example, by etiolating, more than the usual number of elements with annular and helical thickenings will be present.

According to a comprehensive study of mature and developing protoxylem and metaxylem of angiosperms (Bierhorst and Zamora, 1965), the elements with more extensive secondary thickenings than that represented by a helix deposit the secondary wall in two stages. First, a helical framework is built (first-order secondary wall). Then, additional secondary wall material is laid down as sheets or strands or both between the gyres of the helix (second-order secondary wall). This concept may be used to explain the effect of environment on the wall pattern in terms of inhibition or induction of second-order secondary wall deposition, depending on circumstances.

The intergrading of the different types of thickening of tracheary elements is not limited to the primary xylem. The delimitation between the primary and the secondary xylem may also be vague. To recognize the limits of the two tissues, it is necessary to consider many features, among these the length of tracheary elements—the last primary elements are often longer than the first secondary—and the organization of the tissue, particularly the appearance of the combination of ray and axial systems characteristic of secondary xylem. Sometimes the appearance of one or more identifying features of the secondary xylem is delayed, a phenomenon referred to as *paedomorphosis* (Carlquist, 1962, 2001).

In the primary xylem the protoxylem elements may be the narrowest, but not necessarily so. Successively differentiating metaxylem elements often are increasingly wider, whereas the first secondary xylem cells may be rather narrow and thus be distinct from those of the latest wide-celled metaxylem. On the whole, however, it is difficult to make precise distinctions between successive developmental categories of tissues.

## ▌TRACHEARY ELEMENT DIFFERENTIATION

Tracheary elements originate ontogenetically from either procambial cells (in the case of primary elements) or cambial derivatives (in the case of secondary elements). The primordial tracheary elements may or may not elongate before they develop secondary walls, but they usually expand laterally. Elongation of primordial tracheary elements is largely restricted to primary elements and is associated with the elongation, or extension, of the plant part in which they occur.

The differentiating tracheary element is a highly vacuolated cell with a nucleus and a full complement of organelles (Figs. 10.20 and 10.21). Early in differentiation of many tracheary elements, the nucleus undergoes dramatic changes in both size and ploidy level (Lai and Srivastava, 1976). Endoreduplication is common in somatic tissues of plants (Chapter 5; Gahan, 1988). Presumably it provides the differentiating tracheary element with additional gene copies to meet the heavy demand for the synthesis of cell wall and cytoplasmic components (O'Brien, 1981; Gahan, 1988).

After cell enlargement is completed, secondary wall layers are deposited in a pattern characteristic of the given type of tracheary element (Figs. 10.20B and 10.21). One of the earliest signs that the primordial tracheary element is about to embark upon differentiation is a change in distribution of cortical microtubules (Abe et al., 1995a, b; Chaffey et al., 1997a). At first, the microtubules are randomly arranged and spread evenly along the entire wall (Chaffey, 2000; Funada et al., 2000; Chaffey et al., 2002); during differentiation, their orientation changes dynamically. In expanding conifer tracheids, for instance, orientation of the cortical microtubules changes progressively from longitudinal to

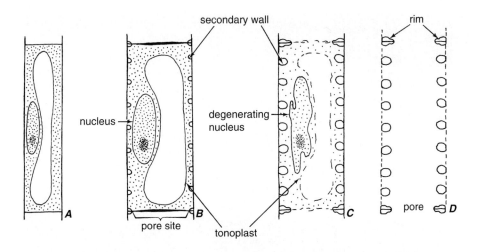

**FIGURE 10.20**

Diagrams illustrating development of a vessel element with a helical secondary thickening. **A**, cell without secondary wall. **B**, cell has attained full width, nucleus has enlarged, secondary wall has begun to be deposited, primary wall at the pore site has increased in thickness. **C**, cell at stage of lysis: secondary thickening completed, tonoplast ruptured, nucleus deformed, wall at pore site partly disintegrated. **D**, mature cell without protoplast, open pores at both ends, primary wall partly hydrolyzed between secondary thickenings. (From Esau, 1977.)

transverse, facilitating expansion of the radial wall (Funada et al., 2000; Funada, 2002). Further changes in orientation of the microtubules occur during secondary wall formation as the now helically arranged microtubules shift in orientation several times, ending in a flat S-helix (Funada et al., 2000; Funada, 2002). The changes in orientation of the microtubules is reflected in changes in the orientation of the cellulose microfibrils.

In differentiating vessel elements, cortical microtubules are concentrated in bands at the sites of secondary thickenings (Fig. 10.22). The endoplasmic reticulum is more conspicuous during deposition of the secondary wall thickenings than before, and its profiles are often seen between thickenings (Fig. 10.21). Golgi bodies and Golgi-derived vesicles are also conspicuous during secondary wall formation in both vessel elements and tracheids, as the Golgi apparatus plays an important role in the synthesis and delivery of matrix substances, notably hemicelluloses, to the developing wall (Awano et al., 2000, 2002; Samuels et al., 2002). The Golgi apparatus also delivers the rosettes, or cellulose synthase complexes, involved with the synthesis of cellulose microfibrils to the plasma membrane (Haigler and Brown, 1986).

Hosoo et al. (2002) reported a diurnal periodicity in the deposition of hemicellulose (glucomannans) in differentiating tracheids of *Cryptomaria japonica*. Whereas much amorphous material containing glucomannans was found on the innermost surface of developing secondary walls at night, the amorphous material was rarely observed during the day, when cellulose fibrils were clearly visible.

Secondary wall deposition is accompanied by lignification. At the beginning of secondary wall deposition, the primary wall of the primordial tracheary element is unlignified. In primary elements the primary wall typically remains unlignified (O'Brien, 1981; Wardrop, 1981). This stands in sharp contrast to the situation in tracheary elements of the secondary xylem in which all of their walls, except for the pit membranes between tracheary elements and the perforation sites between vessel elements, become lignified as differentiation continues (O'Brien, 1981; Czaninski, 1973; Chaffey et al., 1997b).

Development of the pit borders is initiated before secondary thickening of the wall begins (Liese, 1965; Leitch and Savidge, 1995). The "initial pit border" can be detected as concentrically oriented microfibrils at the periphery of the pit annulus (Liese, 1965; Murmanis and Sachs, 1969; Imamura and Harada, 1973). During early immunofluorescence studies circular bands of cortical microtubules were found around the inner margins of the developing pit borders in the tracheids of *Abies* and *Taxus* (Fig. 10.23; Uehara and Hogetsu, 1993; Abe et al., 1995a; Funada et al., 1997) and the vessel elements of *Aesculus* (Chaffey et al., 1997b). Subsequently, actin filaments and then both actin filaments and myosin were found to co-localize with the rings of microtubules at the bordered pits of *Aesculus* and *Populus* vessel elements and *Pinus* tracheids (Fig. 10.23; Chaffey et al., 1999, 2000, 2002; Chaffey, 2002; Chaffey and Barlow, 2002).

An early indication of pit development in conifer tracheids (Funada et al., 1997, 2000) and the vessel

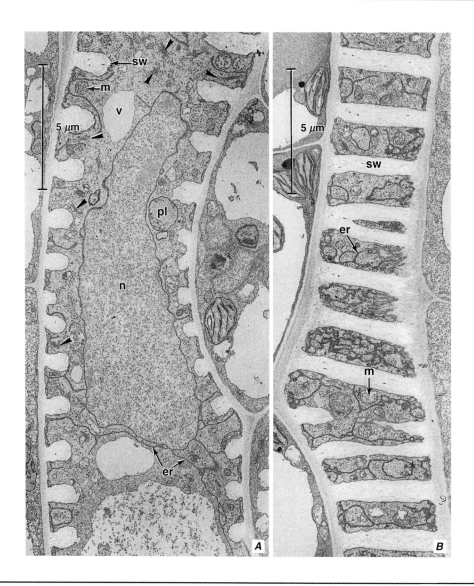

**FIGURE 10.21**

Differentiating tracheary elements in a leaf blade of sugar beet *(Beta vulgaris)*. The secondary thickening is helical (**A**) with transition to scalariform (**B**). **A**, section through cell lumen. **B**, section through the secondary thickening. Details: arrowheads, Golgi bodies; er, endoplasmic reticulum; m, mitochondrion; n, nucleus; pl, plastid; sw, secondary wall; v, vacuole. (From Esau, 1977.)

elements of *Aesculus* (Chaffey et al., 1997b, 1999; Chaffey, 2000) is the disappearance of microtubules at the sites where the bordered pits will eventually be formed. The alternate pattern of pit arrangement in the vessel elements of *Aesculus* is already detectable at this early time (Fig. 10.24). Each incipient pit border is subsequently delimited by the ring of microtubules, actin filaments, and myosin. As deposition of the secondary wall takes place around the opening and pre-existing pit membrane, the diameter of the ring and pit aperture decreases, possibly through the activity of the actin and myosin components which, it has been suggested, may constitute an acto-myosin contractile system (Chaffey and Barlow, 2002). In *Aesculus*, the sites of the vessel

contact pits can also be detected early by the presence of microtubule-free regions within an otherwise random array of microtubules (Chaffey et al., 1999). Unlike the ring of microtubules associated with the developing bordered pit, that associated with the developing contact pit, a nonbordered (simple) pit between the vessel element and adjacent (contact) ray cell does not decrease in diameter as secondary wall formation proceeds.

Portions of the primary wall that later are perforated in vessel elements are not covered by secondary wall material (Figs. 10.20B and 10.25C). The wall occupying the site of the future perforation is clearly set off from the secondary wall. It is thicker than the primary wall elsewhere and, in unstained thin sections, is much

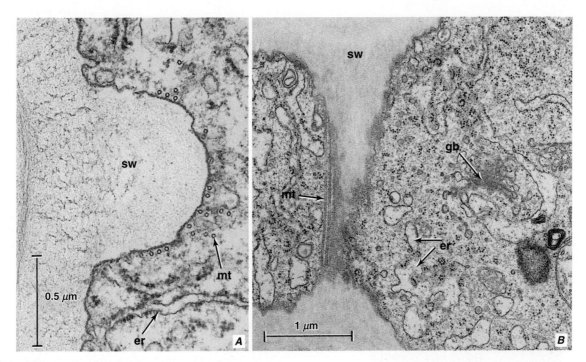

**FIGURE 10.22**

Parts of differentiating tracheary elements from leaves of **A**, bean *(Phaseolus vulgaris)* and, **B**, sugar beet *(Beta vulgaris)*. Microtubules associated with the secondary thickening are seen in cross section in **A**, in longitudinal section in **B**. Details: er, endoplasmic reticulum; gb, Golgi body; mt, microtubule; sw, secondary wall. (From Esau, 1977.)

lighter in appearance under the electron microscope than other wall parts of the same cell (Fig. 10.25C; Esau and Charvat, 1978). Thickening of the perforation sites in differentiating vessel elements of *Populus italica* and *Dianthus caryophyllus* has been shown to be mainly due to the addition of pectins and hemicelluloses (Benayoun et al., 1981). Rings of microtubules also are associated with development of the simple perforation plates in *Aesculus* and *Populus* (Chaffey, 2000; Chaffey et al., 2002). Actin filaments do not accompany these microtubules but the perforation sites in the *Populus* vessel elements are overlaid by a prominent meshwork of actin filaments (Chaffey et al., 2002).

Following deposition of the secondary wall, the cell undergoes autolysis affecting the protoplast and certain parts of the primary cell wall (Fig. 10.20C). The process of tracheary element death is an excellent example of programmed cell death (Chapter 5; Groover et al., 1997; Pennell and Lamb, 1997; Fukuda et al., 1998; Mittler, 1998; Groover and Jones, 1999). Structurally programmed cell death in tracheary elements involves the collapse and rupture of the large central vacuole, with the resultant release of hydrolytic enzymes (Fig. 10.20C). Degradation of both the cytoplasm and nucleus starts after the tonoplast ruptures. The hydrolases also reach

the cell walls and attack the primary wall parts not covered by lignified secondary wall layers, including pit membranes between tracheary elements and the primary wall at the perforation sites between vessel elements. Hydrolysis of the wall results in the removal of noncellulosic components (pectins and hemicelluloses), leaving a fine network of cellulose microfibrils (Figs. 10.20D and 10.25A). All lignified walls appear to be totally resistant to hydrolysis. Where tracheary elements border xylem parenchyma cells hydrolysis stops more or less at the region of the middle lamella. The hydrolytic removal of pectins at the pit membranes between vessels would seemingly preclude the presence there of "hydrogels," which have been proposed to play a role in the control of sap flow through the xylem (Zwieniecki et al., 2001b).

The hydrolysis of the nonlignified primary walls of the protoxylem elements of *Phaseolus vulgaris* and of *Glycine max* is followed by the secretion and incorporation of a glycine-rich protein (GRP1.8) into the hydrolyzed walls (Ryser et al., 1997). Thus the primary walls of the protoxylem elements are not merely the remnants of partial hydrolysis and passive elongation. Being unusually rich in protein, they have special chemical and physical properties. Glycine-rich protein has also

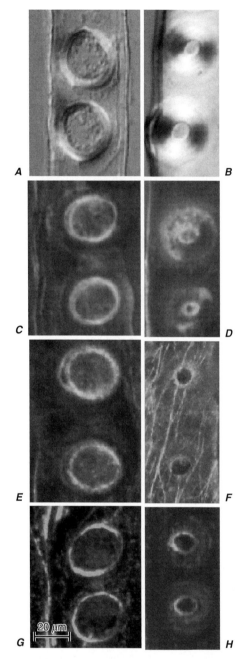

**FIGURE 10.23**

Immunofluorescent localizations of cytoskeletal proteins during development of bordered pits in radial walls of pine (*Pinus pinea*) tracheids. **A**, **B**, differential contrast images of tracheids showing early (**A**) and late (**B**) stages of bordered pit development. Initially wide in diameter (**A**), with development of the border, the opening to the pit becomes reduced to a narrow aperture in the mature tracheid. **C–H**, immunofluorescence localizations of tubulin (**C**, **D**), actin (**E**, **F**), and myosin (**G**, **H**) in early (**C**, **E**, **G**) and late (**D**, **F**, **H**) stages of bordered pit development. (All, same mag. From Chaffey, 2002; reproduced with permission of the New Phytologist Trust.)

been observed in the cell walls of isolated *Zinnia* leaf mesophyll cells regenerating into tracheary elements (see below) (Taylor and Haigler, 1993).

At the perforation sites the entire primary wall part disappears (Figs. 10.20D and 10.25B). The exact process by which the microfibrillar network at the perforation sites is removed remains unclear. In lysed scalariform perforation plates fine networks of fibrils can be seen stretching across the narrower perforations and at the lateral extremities of wider ones. Since the networks typically are not present in the conducting tissue, it is likely that they are removed by the transpiration stream (Meylan and Butterfield, 1981). This mechanism does not explain, however, perforation formation in isolated tracheary elements in culture (Nakashima et al., 2000).

### Plant Hormones Are Involved in the Differentiation of Tracheary Elements

It is well known that the polar flow of auxin from developing buds and young leaves toward the roots induces the differentiation of tracheary elements (Chapter 5; Aloni, 1987, 1995; Mattsson et al., 1999; Sachs, 2000). It has been suggested that a gradient of decreasing auxin concentration is responsible for the general increase in diameter of tracheary elements and decrease in their density from the leaves to the roots (Aloni and Zimmermann, 1983). As set forth in the six-point hypothesis (Aloni and Zimmermann, 1983), whereas high auxin levels near the young leaves induce narrow vessels because they differentiate rapidly, low auxin concentrations farther down lead to slower differentiation, more cell expansion before initiation of secondary wall deposition and, consequently, wider vessels. Studies on transgenic plants with altered levels of auxin confirm these general relations between auxin level and tracheary element differentiation (Klee and Estelle, 1991). Auxin-overproducing plants contain many more and smaller xylem elements than do control plants (Klee et al., 1987). Conversely, plants with lowered auxin levels contain fewer and generally larger tracheary elements (Romano et al., 1991).

Cytokinin from the roots may also be a limiting and controlling factor in vascular differentiation. It promotes tracheary element differentiation in a variety of plant species but acts only in combination with auxin (Aloni, 1995). In the presence of auxin, cytokinin stimulates early stages of vascular differentiation. Late stages of vascular differentiation may, however, occur in the absence of cytokinin. Studies of transgenic plants with overproduction of cytokinin confirm the involvement of cytokinin as a controlling factor in the differentiation of vessels (Aloni, 1995; Fukuda, 1996). In one study, cytokinin-overproducing plants contained many more and smaller vessels than did control plants (Li et al.,

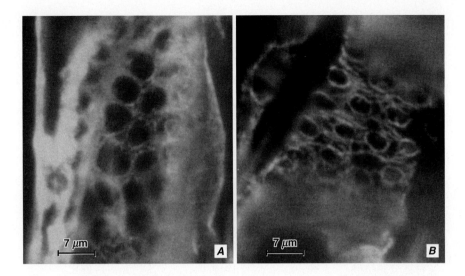

**FIGURE 10.24**

Immunolocalization of tubulin in developing vessel elements of *Aesculus hippocastanum* (horse chestnut). **A**, relatively early stage of bordered pit development. A ring of microtubules marks the site of the developing border, which surrounds a large microtubule-free zone. Note that the alternate pattern of pit arrangement is already apparent. **B**, at a later stage of development than that in **A**, the diameter of the ring of microtubules associated with the border is much reduced. (From Chaffey et al., 1997b.)

1992); in another, overproduction of cytokinin promoted a thicker vascular cylinder with more tracheary elements than did the controls (Medford et al., 1989). Collectively, Kuriyama and Fukuda (2001), Aloni (2001), and Dengler (2001) provide a comprehensive review of factors involved in the regulation of tracheary element and vascular development.

### Isolated Mesophyll Cells in Culture Can Transdifferentiate Directly into Tracheary Elements

Tracheary element differentiation has provided a useful model for the study of cell differentiation and programmed cell death in plants. Particularly useful has been the *Zinnia elegans* experimental system, in which single mesophyll cells—in the presence of auxin and cytokinin—can be made to transdifferentiate (i.e., dedifferentiate and then redifferentiate) into tracheary-like elements without intervening cell division (Fukuda 1996, 1997b; Groover et al., 1997; Groover and Jones, 1999; Milioni et al., 2001). Noting, however, that significant differences exist in the behavior of cortical microtubules and actin filaments during early stages of vascular differentiation of cambial derivatives in *Aesculus hippocastanum* and transdifferentiating *Zinnia* mesophyll cells, Chaffey and co-workers (Chaffey et al., 1997b) cautioned that it is questionable whether an in vitro system will be able to validate the findings from more natural systems.

A number of cytological, biochemical, and molecular markers for tracheary element differentiation have been identified in the *Zinnia* system and facilitate division of the transdifferentiation process into three stages (Fig. 10.26) (Fukuda, 1996, 1997b). *Stage I* immediately follows the induction of differentiation and corresponds to the dedifferentiation process. The latter involves wound-induced events and the activation of protein synthesis, both of which are regulated by hormones at a later stage in the transdifferentiation process. *Stage II* is defined by the accumulation of transcripts of the tracheary element differentiation-related genes *TED2*, *TED3*, and *TED4*. It also includes a marked increase in the transcription of other genes that encode components of the protein synthesis apparatus.

Dramatic changes occur to the cytoskeleton during stages I and II. The expression of tubulin genes begins in stage I and continues during stage II, bringing about an increase in the number of microtubules that are involved with secondary wall formation in stage III. Changes in actin organization during stage II result in the formation of thick actin cables that function in cytoplasmic streaming (Kobayashi et al., 1987).

*Stage III*, the maturation phase, involves secondary wall formation and autolysis. It is preceded by a rapid increase in brassinosteroids, which are necessary for the initiation of this final stage of tracheary element differentiation (Yamamoto et al., 2001). In addition, the calcium/calcium calmodulin (Ca/CaM) system may be

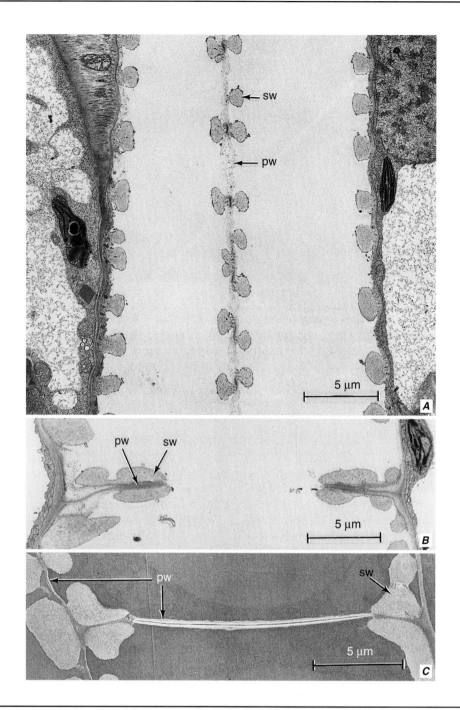

**FIGURE 10.25**

Parts of tracheary elements in longitudinal sections from leaves of **A**, **B**, tobacco (*Nicotiana tabacum*) and, **C**, bean (*Phaseolus vulgaris*) showing details of walls. In **A**, the wall between two tracheary elements (center) illustrates the effect of hydrolysis on the primary wall between the secondary thickenings: the primary wall is reduced to fibrils. In **B**, the end-wall perforation is delimited by a rim in which secondary thickenings are present. In **C**, the primary wall at the pore site has not yet disappeared. It is considerably thicker than the primary wall elsewhere and is supported by a secondarily thickened rim. Details: pw, primary wall; sw, secondary wall. (From Esau, 1977.)

involved in the entry into Stage III (Fig. 10.26). During stage III various enzymes associated with secondary wall formation and with cellular autolysis are activated (Fukuda, 1996; Endo et al., 2001). The hydrolytic enzymes accumulate in the vacuole where they are

sequestered from the cytosol. They are released from the vacuole upon its rupture. Among the hydrolytic enzymes is *Zinnia* endonuclease1, which has been shown to function directly in nuclear DNA degeneration (Ito and Fukuda, 2002). Two proteolytic enzymes

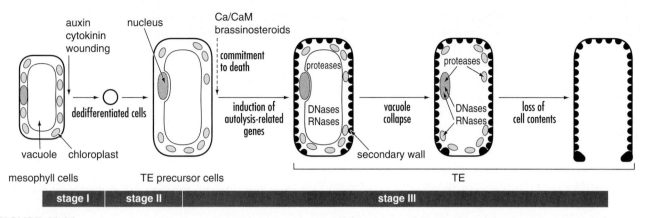

**FIGURE 10.26**

Model of tracheary element differentiation based on the *Zinnia* system. Mesophyll cells are induced to dedifferentiate and then to differentiate into tracheary elements (TE) by wounding and a combination of auxin and cytokinin. The transdifferentiation process is divided into the three stages shown here, and results in a mature tracheary element, with a perforation at one end. (Adapted from Fukuda, 1997a. Reprinted by permission from *Cell Death and Differentiation* 4, 684–688. © 1997 Macmillan Publishers Ltd.)

have been detected specifically in differentiating tracheary elements of *Zinnia*, namely cysteine protease and serine protease. These undoubtedly are only two of a complex set of proteases involved in the autolytic process. It has been suggested that a 40-kDa serine, which is secreted during secondary cell wall synthesis, may serve as a coordinating factor between secondary cell wall synthesis and programmed cell death (Groover and Jones, 1999). A subsequent study of the *Arabidopsis gapped xylem* mutant indicates, however, that the process of secondary cell wall formation and cell death are independently regulated in developing xylem elements (Turner and Hall, 2000). During the maturation phase many hydrolases are released from the tracheary elements into the extracellular space. Evidence indicates that TED4 protein released into the apoplast at that time serves to inhibit those hydrolases, protecting neighboring cells from undesirable injury (Endo et al., 2001). Perforation of the primary wall occurs at one end of single elements; in double elements, both of which have been derived from a single mesophyll cell, a perforation occurs in the common wall between them and at the end of one of the two elements, indicating that these elements formed in vitro have their own program to form perforations (Nakashima et al., 2000).

The time course of tracheary element differentiation has been determined for *Zinnia* cells cultured in inductive medium (Groover et al., 1997). Secondary cell wall formation takes an average of 6 hours to complete in a typical cell. Cytoplasmic streaming continues throughout the period of secondary wall formation but stops abruptly with its completion. Collapse of the large central vacuole begins with completion of secondary wall formation and takes only 3 minutes to be accomplished in a typical cell. After rupture of the tonoplast, the nucleus is rapidly degraded—within 10 to 20 minutes (Obara et al., 2001). Within several hours of tonoplast rupture, the dead cell is cleared of its contents. Remnants of chloroplasts may persist, however, for up to 24 hours.

# REFERENCES

ABE, H., R. FUNADA, H. IMAIZUMI, J. OHTANI, and K. FUKAZAWA. 1995a. Dynamic changes in the arrangement of cortical microtubules in conifer tracheids during differentiation. *Planta* 197, 418–421.

ABE, H., R. FUNADA, J. OHTANI, and K. FUKAZAWA. 1995b. Changes in the arrangement of microtubules and microfibrils in differentiating conifer tracheids during the expansion of cells. *Ann. Bot.* 75, 305–310.

ALONI, R. 1987. Differentiation of vascular tissues. *Annu. Rev. Plant Physiol.* 38, 179–204.

ALONI, R. 1995. The induction of vascular tissues by auxin and cytokinin. In: *Plant Hormones. Physiology, Biochemistry, and Molecular Biology*, pp. 531–546, P. J. Davies, ed. Kluwer Academic Dordrecht.

ALONI, R. 2001. Foliar and axial aspects of vascular differentiation: hypotheses and evidence. *J. Plant Growth Regul.* 20, 22–34.

ALONI, R., and M. H. ZIMMERMANN. 1983. The control of vessel size and density along the plant axis. A new hypothesis. *Differentiation* 24, 203–208.

ALVES, E. S., and V. ANGYALOSSY-ALFONSO. 2000. Ecological trends in the wood anatomy of some Brazilian species. I. Growth rings and vessels. *IAWA J.* 21, 3–30.

ANGIOSPERM PHYLOGENY GROUP. 2003. An update of the Angiosperm Phylogeny Group classification for the orders and families of flowering plants: APGII. *Bot. J. Linn. Soc.* 141, 399–436.

AWANO, T., K. TAKABE, M. FUJITA, and G. DANIEL. 2000. Deposition of glucuronoxylans on the secondary cell wall of Japanese beech as observed by immuno-scanning electron microscopy. *Protoplasma* 212, 72–79.

AWANO, T., K. TAKABE, and M. FUJITA. 2002. Xylan deposition on secondary wall of *Fagus crenata* fiber. *Protoplasma* 219, 106–115.

BAAS, P. 1986. Ecological patterns in xylem anatomy. In: *On the Economy of Plant Form and Function*, pp. 327–352, T. J. Givnish, ed. Cambridge University Press, Cambridge, New York.

BAAS, P., and E. A. WHEELER. 1996. Parallelism and reversibility in xylem evolution. A review. *IAWA J.* 17, 351–364.

BAILEY, I. W. 1933. The cambium and its derivative tissues. No. VIII. Structure, distribution, and diagnostic significance of vestured pits in dicotyledons. *J. Arnold Arbor.* 14, 259–273.

BAILEY, I. W. 1944. The development of vessels in angiosperms and its significance in morphological research. *Am. J. Bot.* 31, 421–428.

BAILEY, I. W. 1953. Evolution of the tracheary tissue of land plants. *Am. J. Bot.* 40, 4–8.

BAILEY, I. W. 1954. *Contributions to Plant Anatomy*. Chronica Botanica, Waltham, MA.

BAILEY, I. W. 1957. Additional notes on the vesselless dicotyledon, *Amborella trichopoda* Baill. *J. Arnold Arbor.* 38, 374–378.

BAILEY, I. W. 1958. The structure of tracheids in relation to the movement of liquids, suspensions and undissolved gases. In: *The Physiology of Forest Trees*, pp. 71–82, K. V. Thimann, ed. Ronald Press, New York.

BAILEY, I. W., and L. M. SRIVASTAVA. 1962. Comparative anatomy of the leaf-bearing Cactaceae. IV. The fusiform initials of the cambium and the form and structure of their derivatives. *J. Arnold Arbor.* 43, 187–202.

BAILEY, I. W., and W. W. TUPPER. 1918. Size variation in tracheary elements. I. A comparison between the secondary xylem of vascular cryptogams, gymnosperms and angiosperms. *Proc. Am. Acad. Arts Sci.* 54, 149–204.

BANNAN, M. W. 1941. Variability in wood structure in roots of native Ontario conifers. *Bull. Torrey Bot. Club* 68, 173–194.

BECKER, P., M. T. TYREE, and M. TSUDA. 1999. Hydraulic conductances of angiosperms versus conifers: Similar transport sufficiency at the whole-plant level. *Tree Physiol.* 19, 445–452.

BECKMAN, C. H., and P. W. TALBOYS. 1981. Anatomy of resistance. In: *Fungal Wilt Diseases of Plants*, pp. 487–521, M. E. Mace, A. A. Bell, and C. H. Beckman, eds. Academic Press, New York.

BENAYOUN, J., A. M. CATESSON, and Y. CZANINSKI. 1981. A cytochemical study of differentiation and breakdown of vessel end walls. *Ann. Bot.* 47, 687–698.

BIERHORST, D. W. 1958. Vessels in *Equisetum*. *Am. J. Bot.* 45, 534–537.

BIERHORST, D. W., and P. M. ZAMORA. 1965. Primary xylem elements and element associations of angiosperms. *Am. J. Bot.* 52, 657–710.

BRAUN, H. J. 1967. Entwicklung und Bau der Holzstrahlen unter dem Aspekt der Kontakt—Isolations—Differenzierung gegenüber dem Hydrosystem. I. Das Prinzip der Kontakt—Isolations—Differenzierung. *Holzforschung* 21, 33–37.

BRAUN, H. J. 1983. Zur Dynamik des Wassertransportes in Bäumen. *Ber. Dtsch. Bot. Ges.* 96, 29–47.

BUCHARD, C., M. McCULLY, and M. CANNY. 1999. Daily embolism and refilling of root xylem vessels in three dicotyledonous crop plants. *Agronomie* 19, 97–106.

BUSSE, J. S., and R. F. EVERT. 1999. Pattern of differentiation of the first vascular elements in the embryo and seedling of *Arabidopsis thaliana*. *Int. J. Plant Sci.* 160, 1–13.

BUTTERFIELD, B. G., and B. A. MEYLAN. 1980. *Three-dimensional Structure of Wood: An Ultrastructural Approach*, 2nd. ed. Chapman and Hall, London.

CANNY, M. J. 1997a. Vessel contents of leaves after excision—A test of Scholander's assumption. *Am. J. Bot.* 84, 1217–1222.

CANNY, M. J. 1997b. Vessel contents during transpiration—Embolisms and refilling. *Am. J. Bot.* 84, 1223–1230.

CANNY, M. J. 1997c. Tyloses and the maintenance of transpiration. *Ann. Bot.* 80, 565–570.

CANNY, M. J., C. X. HUANG, and M. E. McCULLY. 2001. The cohesion theory debate continues. *Trends Plant Sci.* 6, 454–455.

CARLQUIST, S. J. 1961. *Comparative Plant Anatomy: A Guide to Taxonomic and Evolutionary Application of Anatomical Data in Angiosperms*. Holt, Rinehart and Winston, New York.

CARLQUIST, S. 1962. A theory of paedomorphosis in dicotyledonous woods. *Phytomorphology* 12, 30–45.

CARLQUIST, S. J. 1975. *Ecological Strategies of Xylem Evolution*. University of California Press, Berkeley.

CARLQUIST, S. 1983. Wood anatomy of Onagraceae: Further species; root anatomy; significance of vestured pits and allied structures in dicotyledons. *Ann. Mo. Bot. Gard.* 69, 755–769.

CARLQUIST, S. 1986. Terminology of imperforate tracheary elements. *IAWA Bull.* n.s. 7, 75–81.

CARLQUIST, S. 1992. Pit membrane remnants in perforation plates of primitive dicotyledons and their significance. *Am. J. Bot.* 79, 660–672.

CARLQUIST, S. 1996a. Wood, bark, and stem anatomy of Gnetales: A summary. *Int. J. Plant Sci.* 157 (6; suppl.), S58–S76.

CARLQUIST, S. 1996b. Wood anatomy of primitive angiosperms: New perspectives and syntheses. In: *Flowering Plant Origin,*

*Evolution and Phylogeny,* pp. 68–90, D. W. Taylor and L. J. Hickey, eds. Chapman and Hall, New York.

CARLQUIST, S. J. 2001. *Comparative Wood Anatomy: Systematic, Ecological, and Evolutionary Aspects of Dicotyledon Wood,* rev. 2nd ed. Springer, Berlin.

CARLQUIST, S., and D. A. HOEKMAN. 1985. Ecological wood anatomy of the woody southern California flora. *IAWA Bull.* n.s. 6, 319–347.

CARLQUIST, S., and E. L. SCHNEIDER. 2001. Vessels in ferns: structural, ecological, and evolutionary significance. *Am. J. Bot.* 88, 1–13.

CARLQUIST, S., and E. L. SCHNEIDER. 2002. The tracheid-vessel element transition in angiosperms involves multiple independent features: Cladistic consequences. *Am. J. Bot.* 89, 185–195.

CARPENTER, C. H. 1952. *382 Photomicrographs of 91 Papermaking Fibers,* rev. ed. Tech. Publ. 74. State University of New York, College of Forestry, Syracuse.

CASTRO, M. A. 1988. Vestures and thickenings of the vessel wall in some species of *Prosopis* (Leguminosae). *IAWA Bull.* n.s. 9, 35–40.

CASTRO, M. A. 1991. Ultrastructure of vestures on the vessel wall in some species of *Prosopis* (Leguminosae-Mimosoideae). *IAWA Bull.* n.s. 12, 425–430.

CATESSON, A. M., M. MOREAU, and J. C. DUVAL. 1982. Distribution and ultrastructural characteristics of vessel contact cells in the stem xylem of carnation *Dianthus caryophyllus. IAWA Bull.* n.s. 3, 11–14.

CHAFFEY, N. J. 2000. Cytoskeleton, cell walls and cambium: New insights into secondary xylem differentiation. In: *Cell and Molecular Biology of Wood Formation,* pp. 31–42, R. A. Savidge, J. R. Barnett, and R. Napier, eds. BIOS Scientific, Oxford.

CHAFFEY, N. 2002. Why is there so little research into the cell biology of the secondary vascular system of trees? *New Phytol.* 153, 213–223.

CHAFFEY, N., and P. BARLOW. 2002. Myosin, microtubules, and microfilaments: Co-operation between cytoskeletal components during cambial cell division and secondary vascular differentiation in trees. *Planta* 214, 526–536.

CHAFFEY, N., P. BARLOW, and J. BARNETT. 1997a. Cortical microtubules rearrange during differentiation of vascular cambial derivatives, microfilaments do not. *Trees* 11, 333–341.

CHAFFEY, N. J., J. R. BARNETT, and P. W. BARLOW. 1997b. Cortical microtubule involvement in bordered pit formation in secondary xylem vessel elements of *Aesculus hippocastanum* L. (Hippocastanaceae): A correlative study using electron microscopy and indirect immunofluorescence microscopy. *Protoplasma* 197, 64–75.

CHAFFEY, N., J. BARNETT, and P. BARLOW. 1999. A cytoskeletal basis for wood formation in angiosperm trees: The involvement of cortical microtubules. *Planta* 208, 19–30.

CHAFFEY, N., P. BARLOW, and J. BARNETT. 2000. A cytoskeletal basis for wood formation in angiosperms trees: The involvement of microfilaments. *Planta* 210, 890–896.

CHAFFEY, N., P. BARLOW, and B. SUNDBERG. 2002. Understanding the role of the cytoskeleton in wood formation in angiosperm trees: Hybrid aspen *(Populus tremula* x *P. tremuloides)* as the model species. *Tree Physiol.* 22, 239–249.

CHATTAWAY, M. M. 1949. The development of tyloses and secretion of gum in heartwood formation. *Aust. J. Sci. Res. B, Biol. Sci.* 2, 227–240.

CHEADLE, V. I. 1953. Independent origin of vessels in the monocotyledons and dicotyledons. *Phytomorphology* 3, 23–44.

CHEADLE, V. I. 1956. Research on xylem and phloem—Progress in fifty years. *Am. J. Bot.* 43, 719–731.

CHOAT, B., S. JANSEN, M. A. ZWIENIECKI, E. SMETS, and N. M. HOLBROOK. 2004. Changes in pit membrane porosity due to deflection and stretching: The role of vestured pits. *J. Exp. Bot.* 55, 1569–1575.

CLÉRIVET, A., V. DÉON, I. ALAMI, F. LOPEZ, J.-P. GEIGER, and M. NICOLE. 2000. Tyloses and gels associated with cellulose accumulation in vessels are responses of plane tree seedlings *(Platanus* x *acerifolia)* to the vascular fungus *Ceratocystis fimbriata* f. sp *platani. Trees* 15, 25–31.

COCHARD H., M. PEIFFER, K. LE GALL, and A. GRANIER. 1997. Developmental control of xylem hydraulic resistances and vulnerability to embolism in *Fraxinus excelsior* L.: Impacts on water relations. *J. Exp. Bot.* 48, 655–663.

COCHARD, H., T. AMÉGLIO, and P. CRUIZIAT. 2001a. The cohesion theory debate continues. *Trends Plant Sci.* 6, 456.

COCHARD, H., D. LEMOINE, T. AMÉGLIO, and A. GRANIER. 2001b. Mechanisms of xylem recovery from winter embolism in *Fagus sylvatica. Tree Physiol.* 21, 27–33.

COCHARD, H., F. FROUX, S. MAYR, C. COUTAND. 2004. Xylem wall collapse in water-stressed pine needles. *Plant Physiol.* 134, 401–408.

COLEMAN, C. M., B. L. PRATHER, M. J. VALENTE, R. R. DUTE, and M. E. MILLER. 2004. Torus lignification in hardwoods. *IAWA J.* 25, 435–447.

COMSTOCK, J. P., and J. S. SPERRY. 2000. Theoretical considerations of optimal conduit length for water transport in vascular plants. *New Phytol.* 148, 195–218.

CZANINSKI, Y. 1973. Observations sur une nouvelle couche pariétale dans les cellules associées aux vaisseaux du Robinier et du Sycomore. *Protoplasma* 77, 211–219.

CZANINSKI, Y. 1977. Vessel-associated cells. *IAWA Bull.* 1977, 51–55.

DENGLER, N. G. 2001. Regulation of vascular development. *J. Plant Growth Regul.* 20, 1–13.

DONOGHUE, M. J., and J. A. DOYLE. 1989. Phylogenetic studies of seed plants and angiosperms based on morphological characters. In: *The Hierarchy of Life: Molecules and Morphology in Phylogenetic Analysis,* pp. 181–193, B. Fernholm, K. Bremer, and H. Jörnvall, eds. *Excerpta Medica,* Amsterdam.

DUTE, R. R., A. E. RUSHING, and J. W. PERRY. 1990. Torus structure and development in species of *Daphne*. *IAWA Bull.* n.s. 11, 401–412.

DUTE, R. R., J. D. FREEMAN, F. HENNING, and L. D. BARNARD. 1996. Intervascular pit membrane structure in *Daphne* and *Wikstroemia*—Systematic implications. *IAWA J.* 17, 161–181.

ELLERBY, D. J., and A. R. ENNOS. 1998. Resistances to fluid flow of model xylem vessels with simple and scalariform perforation plates. *J. Exp. Bot.* 49, 979–985.

ENDO, S., T. DEMURA, and H. FUKUDA. 2001. Inhibition of proteasome activity by the TED4 protein in extracellular space: A novel mechanism for protection of living cells from injury caused by dying cells. *Plant Cell Physiol.* 42, 9–19

ENDRESS, P. K., P. BAAS, and M. GREGORY. 2000. Systematic plant morphology and anatomy—50 years of progress. *Taxon* 49, 401–434.

ESAU, K. 1943. Origin and development of primary vascular tissues in seed plants. *Bot. Rev.* 9, 125–206.

ESAU, K. 1961. *Plants, Viruses, and Insects.* Harvard University Press, Cambridge, MA.

ESAU, K. 1977. *Anatomy of Seed Plants*, 2nd ed. Wiley, New York.

ESAU, K., and I. CHARVAT. 1978. On vessel member differentiation in the bean (*Phaseolus vulgaris* L.). *Ann. Bot.* 42, 665–677.

ESSIAMAH, S., and W. ESCHRICH. 1985. Changes of starch content in the storage tissues of deciduous trees during winter and spring. *IAWA Bull.* n.s. 6, 97–106.

EWERS, F. W. 1985. Xylem structure and water conduction in conifer trees, dicot trees, and lianas. *IAWA Bull.* n.s. 6, 309–317.

EWERS, F. W., and J. B. FISHER. 1989. Techniques for measuring vessel lengths and diameters in stems of woody plants. *Am. J. Bot.* 76, 645–656.

EWERS, F. W., J. B. FISHER, and S.-T. CHIU. 1990. A survey of vessel dimensions in stems of tropical lianas and other growth forms. *Oecologia* 84, 544–552.

EWERS, F. W., M. R. CARLTON, J. B. FISHER, K. J. KOLB, and M. T. TYREE. 1997. Vessel diameters in roots versus stems of tropical lianas and other growth forms. *IAWA J.* 18, 261–279.

FAHN, A. 1954. Metaxylem elements in some families of the Monocotyledoneae. *New Phytol.* 53, 530–540.

FAHN, A., and B. LESHEM. 1963. Wood fibres with living protoplasts. *New Phytol.* 62, 91–98.

FEILD, T. S., T. BRODRIBB, and N. M. HOLBROOK. 2002. Hardly a relict: Freezing and the evolution of vesselless wood in Winteraceae. *Evolution* 56, 464–478.

FOSTER, R. C. 1967. Fine structure of tyloses in three species of the Myrtaceae. *Aust. J. Bot.* 15, 25–34

FRISON, E. 1948. De la présence d'Amidon dans le Lumen des Fibres du Bois. *Bull. Agric. Congo Belge, Brussels*, 39, 869–874.

FUKAZAWA, K., K. YAMAMOTO, and S. ISHIDA. 1980. The season of heartwood formation in the genus *Pinus*. In: *Natural Variations of Wood Properties, Proceedings*, pp. 113–130. J. Bauch ed. Hamburg.

FUKUDA, H. 1996. Xylogenesis: Initiation, progression, and cell death. *Annu. Rev. Plant Physiol. Plant Mol. Biol.* 47, 299–325.

FUKUDA, H. 1997a. Programmed cell death during vascular system formation. *Cell Death Differ.* 4, 684–688.

FUKUDA, H. 1997b. Tracheary element differentiation. *Plant Cell* 9, 1147–1156.

FUKUDA, H., Y. WATANABE, H. KURIYAMA, S. AOYAGI, M. SUGIYAMA, R. YAMAMOTO, T. DEMURA, and A. MINAMI. 1998. Programming of cell death during xylogenesis. *J. Plant Res.* 111, 253–256.

FUNADA, R. 2002. Immunolocalisation and visualization of the cytoskeleton in gymnosperms using confocal laser scanning microscopy. In: *Wood Formation in Trees. Cell and Molecular Biology Techniques*, pp. 143–157, N. Chaffey, ed. Taylor and Francis, London.

FUNADA, R., H. ABE, O. FURUSAWA, H. IMAIZUMI, K. FUKAZAWA, and J. OHTANI. 1997. The orientation and localization of cortical microtubules in differentiating conifer tracheids during cell expansion. *Plant Cell Physiol.* 38, 210–212.

FUNADA, R., O. FURUSAWA, M. SHIBAGAKI, H. MIURA, T. MIURA, H. ABE, and J. OHTANI. 2000. The role of cytoskeleton in secondary xylem differentiation in conifers. In: *Cell and Molecular Biology of Wood Formation*, pp. 255–264, R. A. Savidge, J. R. Barnett, and R. Napier, eds. BIOS Scientific Oxford.

GAHAN, P. B. 1988. Xylem and phloem differentiation in perspective. In: *Vascular Differentiation and Plant Growth Regulators*, pp. 1–21, L. W. Roberts, P. B. Gahan, and R. Aloni, eds. Springer-Verlag, Berlin.

GIFFORD, E. M., and A. S. FOSTER. 1989. *Morphology and Evolution of Vascular Plants*, 3rd. ed. Freeman, New York.

GREENIDGE, K. N. H. 1952. An approach to the study of vessel length in hardwood species. *Am. J. Bot.* 39, 570–574.

GREGORY, R. A. 1978. Living elements of the conducting secondary xylem of sugar maple (*Acer saccharum* Marsh.). *IAWA Bull.* 1978, 65–69.

GROOVER, A., and A. M. JONES. 1999. Tracheary element differentiation uses a novel mechanism coordinating programmed cell death and secondary cell wall synthesis. *Plant Physiol.* 119, 375–384.

GROOVER, A., N. DEWITT, A. HEIDEL, and A. JONES. 1997. Programmed cell death of plant tracheary elements differentiating in vitro. *Protoplasma* 196, 197–211.

HACKE, U. G., and J. S. SPERRY. 2001. Functional and ecological xylem anatomy. *Perspect. Plant Ecol. Evol. Syst.* 4, 97–115.

HACKE, U. G., and J. S. SPERRY. 2003. Limits to xylem refilling under negative pressure in *Laurus nobilis* and *Acer negundo*. *Plant Cell Environ.* 26, 303–311.

HACKE, U. G., J. S. SPERRY, and J. PITTERMANN. 2004. Analysis of circular bordered pit function. II. Gymnosperm tracheids with torus-margo pit membranes. *Am. J. Bot.* 91, 386–400.

HAIGLER, C. H., and R. M. BROWN JR. 1986. Transport of rosettes from the Golgi apparatus to the plasma membrane in isolated mesophyll cells of *Zinnia elegans* during differentiation to tracheary elements in suspension culture. *Protoplasma* 134, 111–120.

HARADA, H., and W. A. CÔTÉ 1985. Structure of wood. In: *Biosynthesis and Biodegradation of Wood Components*, pp. 1–42, T. Higuchi, ed. Academic Press, Orlando, FL.

HEADY, R. D., R. B. CUNNINGHAM, C. F. DONNELLY, and P. D. EVANS. 1994. Morphology of warts in the tracheids of cypress pine (*Callitris* Vent.). *IAWA J.* 15, 265–281.

HERENDEEN, P. S., E. A. WHEELER, and P. BAAS. 1999. Angiosperm wood evolution and the potential contribution of paleontological data. *Bot. Rev.* 65, 278–300.

HOLBROOK, N. M., and M. A. ZWIENIECKI. 1999. Embolism repair and xylem tension: Do we need a miracle? *Plant Physiol.* 120, 7–10.

HÖLL, W. 2000. Distribution, fluctuation and metabolism of food reserves in the wood of trees. In: *Cell and Molecular Biology of Wood Formation*, pp. 347–362, R. A. Savidge, J. R. Barnett, and R. Napier, eds. BIOS Scientific, Oxford.

HONG, L. T., and W. KILLMANN. 1992. Some aspects of parenchymatous tissues in palm stems. In: *Proceedings, 2nd Pacific Regional Wood Anatomy Conference*, pp. 449–455, J. P. Rojo, J. U. Aday, E. R. Barile, R. K. Araral, and W. M. America, eds. The Institute, Laguna, Philippines.

HOSOO, Y., M. YOSHIDA, T. IMAI, and T. OKUYAMA. 2002. Diurnal difference in the amount of immunogold-labeled glucomannans detected with field emission scanning electron microscopy at the innermost surface of developing secondary walls of differentiating conifer tracheids. *Planta* 215, 1006–1012.

IAWA COMMITTEE ON NOMENCLATURE. 1964. International glossary of terms used in wood anatomy. *Trop. Woods* 107, 1–36.

IMAMURA, Y., and H. HARADA. 1973. Electron microscopic study on the development of the bordered pit in coniferous tracheids. *Wood Sci. Technol.* 7, 189–205.

ITO, J., and H. FUKUDA. 2002. ZEN1 is a key enzyme in the degradation of nuclear DNA during programmed cell death of tracheary elements. *Plant Cell* 14, 3201–3211.

JANSEN, S., E. SMETS, and P. BAAS. 1998. Vestures in woody plants: a review. *IAWA J.* 19, 347–382.

JANSEN, S., P. BAAS, and E. SMETS. 2001. Vestured pits: Their occurrence and systematic importance in eudicots. *Taxon* 50, 135–167.

JANSEN, S., P. BAAS, P. GASSON, and E. SMETS. 2003. Vestured pits: Do they promote safer water transport? *Int. J. Plant Sci.* 164, 405–413.

JANSEN, S., B. CHOAT, S. VINCKIER, F. LENS, P. SCHOLS, and E. SMETS. 2004. Intervascular pit membranes with a torus in the wood of *Ulmus* (Ulmaceae) and related genera. *New Phytol.* 163, 51–59.

KEUNECKE, M., J. U. SUTTER, B. SATTELMACHER, and U. P. HANSEN. 1997. Isolation and patch clamp measurements of xylem contact cells for the study of their role in the exchange between apoplast and symplast of leaves. *Plant Soil* 196, 239–244.

KLEE, H., and M. ESTELLE. 1991. Molecular genetic approaches to plant hormone biology. *Annu. Rev. Plant Physiol. Plant Mol. Biol.* 42, 529–551.

KLEE, H. J., R. B. HORSCH, M. A. HINCHEE, M. B. HEIN, and N. L. HOFFMANN. 1987. The effects of overproduction of two *Agrobacterium tumefaciens* T-DNA auxin biosynthetic gene products in transgenic petunia plants. *Genes Dev.* 1, 86–96.

KOBAYASHI, H., H. FUKUDA, and H. SHIBAOKA. 1987. Reorganization of actin filaments associated with the differentiation of tracheary elements in *Zinnia* mesophyll cells. *Protoplasma* 138, 69–71.

KOLB, K. J., and J. S. SPERRY. 1999. Transport constraints on water use by the Great Basin shrub, *Artemisia tridentata*. *Plant Cell Environ.* 22, 925–935.

KOZLOWSKI, T. T., and S. G. PALLARDY. 1997a. *Growth Control in Woody Plants*. Academic Press, San Diego.

KOZLOWSKI, T. T., and S. G. PALLARDY. 1997b. *Physiology of Woody Plants*, 2nd ed. Academic Press, San Diego.

KURIYAMA, H., and H. FUKUDA. 2001. Regulation of tracheary element differentiation. *J. Plant Growth Regul.* 20, 35–51.

LAI, V., and L. M. SRIVASTAVA. 1976. Nuclear changes during the differentiation of xylem vessel elements. *Cytobiologie* 12, 220–243.

LANCASHIRE, J. R., and A. R. ENNOS. 2002. Modelling the hydrodynamic resistance of bordered pits. *J. Exp. Bot.* 53, 1485–1493.

LARSON, P. R. 1974. Development and organization of the vascular system in cottonwood. In: *Proceedings, 3rd North American Forest Biology Workshop*, pp. 242–257, C. P. P. Reid and G. H. Fechner, eds. College of Forestry and Natural Resources, Colorado State University, Fort Collins.

LARSON, P. R. 1976. Development and organization of the secondary vessel system in *Populus grandidentata*. *Am. J. Bot.* 63, 369–381.

LEITCH, M. A., and R. A. SAVIDGE. 1995. Evidence for auxin regulation of bordered-pit positioning during tracheid differentiation in *Larix laricina*. *IAWA J.* 16, 289–297.

LEMESLE, R. 1956. Les éléments du xylème dans les Angiospermes à charactères primitifs. *Bull. Soc. Bot. Fr.* 103, 629–677.

LI, Y., G. HAGEN, and T. J. GUILFOYLE. 1992. Altered morphology in transgenic tobacco plants that overproduce cytokinins in specific tissues and organs. *Dev. Biol.* 153, 386–395.

LIESE, W. 1965. The fine structure of bordered pits in softwoods. In: *Cellular Ultrastructure of Woody Plants*, pp. 271–290, W. A. Côté Jr., ed. Syracuse University Press, Syracuse.

LINTON, M. J., J. S. SPERRY, and D. G. WILLIAMS. 1998. Limits to water transport in *Juniperus osteosperma* and *Pinus edulis*:

Implications for drought tolerance and regulation of transpiration. *Funct. Ecol.* 12, 906–911.

LOCONTE, H., and D. W. STEVENSON. 1991. Cladistics of the Magnoliidae. *Cladistics* 7, 267–296.

MAHERALI, H., and E. H. DELUCIA 2000. Xylem conductivity and vulnerability to cavitation of ponderosa pine growing in contrasting climates. *Tree Physiol.* 20, 859–867.

MARTÍNEZ-VILALTA, J., E. PRAT, I. OLIVERAS, and J. PIÑOL. 2002. Xylem hydraulic properties of roots and stems of nine Mediterranean woody species. *Oecologia* 133, 19–29.

MATTSSON, J., Z. R. SUNG, and T. BERLETH. 1999. Responses of plant vascular systems to auxin transport inhibition. *Development* 126, 2979–2991.

McCULLY, M. E. 1999. Root xylem embolisms and refilling. Relation to water potentials of soil, roots, and leaves, and osmotic potentials of root xylem sap. *Plant Physiol.* 119, 1001–1008.

McCULLY, M. E., C. X. HUANG, and L. E. C. LING. 1998. Daily embolism and refilling of xylem vessels in the roots of field-grown maize. *New Phytol.* 138, 327–342.

MEDFORD, J. I., R. HORGAN, Z. EL-SAWI, and H. J. KLEE. 1989. Alterations of endogenous cytokinins in transgenic plants using a chimeric isopentenyl transferase gene. *Plant Cell* 1, 403–413.

MENCUCCINI, M., and J. COMSTOCK. 1997. Vulnerability to cavitation in populations of two desert species, *Hymenoclea salsola* and *Ambrosia dumosa,* from different climatic regions. *J. Exp. Bot.* 48, 1323–1334.

METCALFE, C. R., and L. CHALK, eds. 1983. *Anatomy of the Dicotyledons,* 2nd. ed., vol. II. *Wood Structure and Conclusion of the General Introduction.* Clarendon Press, Oxford.

MEYER, R. W., and W. A. CÔTÉ JR. 1968. Formation of the protective layer and its role in tyloses development. *Wood Sci. Technol.* 2, 84–94.

MEYLAN, B. A., and B. G. BUTTERFIELD. 1974. Occurrence of vestured pits in the vessels and fibres of New Zealand woods. *N. Z. J. Bot.* 12, 3–18.

MEYLAN, B. A., and B. G. BUTTERFIELD. 1981. Perforation plate differentiation in the vessels of hardwoods. In: *Xylem Cell Development,* pp. 96–114, J. R. Barnett, ed. Castle House Publications, Tunbridge Wells, Kent.

MILBURN, J. A. 1979. *Water Flow in Plants.* Longman, London.

MILIONI, D., P.-E. SADO, N. J. STACEY, C. DOMINGO, K. ROBERTS, and M. C. McCANN. 2001. Differential expression of cell-wall-related genes during the formation of tracheary elements in the *Zinnia* mesophyll cell system. *Plant Mol. Biol.* 47, 221–238.

MITTLER, R. 1998. Cell death in plants. In: *When cells Die: A Comprehensive Evaluation of Apoptosis and Programmed Cell Death,* pp. 147–174, R. A. Lockshin, Z. Zakeri, and J. L. Tilly, eds. Wiley-Liss, New York.

MONEY, L. L., I. W. BAILEY, and B. G. L. SWAMY. 1950. The morphology and relationships of the Monimiaceae. *J. Arnold Arbor.* 31, 372–404.

MORI, N., M. FUJITA, H. HARADA, and H. SAIKI. 1983. Chemical composition of vestures and warts examined by selective extraction on ultrathin sections (in Japanese). *Kyoto Daigaku Nogaku bu Enshurin Hohoku (Bull. Kyoto Univ. For.)* 55, 299–306.

MUELLER, W. C., and C. H. BECKMAN. 1984. Ultrastructure of the cell wall of vessel contact cells in the xylem of xylem of tomato stems. *Ann. Bot.* 53, 107–114.

MURMANIS, L., and I. B. SACHS. 1969. Structure of pit border in *Pinus strobus* L. *Wood Fiber* 1, 7–17.

NÄGELI, C. W. 1858. Das Wachsthum des Stammes und der Wurzel bei den Gefässpflanzen und die Anordnung der Gefässsstränge im Stengel. *Beitr. Wiss. Bot.* 1, 1–56.

NAKASHIMA, J., K. TAKABE, M. FUJITA, and H. FUKUDA. 2000. Autolysis during in vitro tracheary element differentiation: Formation and location of the perforation. *Plant Cell Physiol.* 41, 1267–1271.

NIJSSE, J., G. W. A. M. VAN DER HEIJDEN, W. VAN IEPEREN, C. J. KEIJZER, and U. VAN MEETEREN. 2001. Xylem hydraulic conductivity related to conduit dimensions along chrysanthemum stems. *J. Exp. Bot.* 52, 319–327.

OBARA, K., H. KURIYAMA, and H. FUKUDA. 2001. Direct evidence of active and rapid nuclear degradation triggered by vacuole rupture during programmed cell death in zinnia. *Plant Physiol.* 125, 615–626.

O'BRIEN, T. P. 1981. The primary xylem. In: *Xylem Cell Development,* pp. 14–46, J. R. Barnett, ed. Castle House, Tunbridge Wells, Kent.

OHTANI, J., B. A. MEYLAN, and B. G. BUTTERFIELD. 1984. Vestures or warts—Proposed terminology. *IAWA Bull.* n.s. 5, 3–8.

OHTANI, J., Y. SAITOH, J. WU, K. FUKAZAWA, and S. Q. XIAO. 1992. Perforation plates in *Knema furfuracea* (Myristicaceae). *IAWA Bull.* n.s., 13, 301–306.

PANSHIN, A. J., and C. DE ZEEUW. 1980. *Textbook of Wood Technology: Structure, Identification, Properties, and Uses of the Commercial Woods of the United States and Canada,* 4th ed. McGraw-Hill, New York.

PARAMESWARAN, N., and W. LIESE. 1977. Occurrence of warts in bamboo species. *Wood Sci. Technol.* 11, 313–318.

PARAMESWARAN, N., and W. LIESE. 1981. Torus-like structures in interfibre pits of *Prunus* and *Pyrus*. *IAWA Bull.* n.s. 2, 89–93.

PARKINSON, C. L., K. L. ADAMS, and J. D. PALMER. 1999. Multigene analyses identify the three earliest lineages of extant flowering plants. *Curr. Biol.* 9, 1485–1488.

PENNELL, R. I., and C. LAMB. 1997. Programmed cell death in plants. *Plant Cell* 9, 1157–1168.

POOLE, I., and J. E. FRANCIS. 2000. The first record of fossil wood of Winteraceae from the Upper Cretaceous of Antarctica. *Ann. Bot.* 85, 307–315.

RANJANI, K., and K. V. KRISHNAMURTHY. 1988. Nature of vestures in the vestured pits of some Caesalpiniaceae. *IAWA Bull.* n.s. 9, 31–33.

RAVEN, P. H., R. F. EVERT, and S. E. EICHHORN. 2005. *Biology of Plants*, 7th ed. Freeman, New York.

RICHTER, H. 2001. The cohesion theory debate continues: The pitfalls of cryobiology. *Trends Plant Sci.* 6, 456–457.

ROMANO, C. P., M. B. HEIN, and H. J. KLEE. 1991. Inactivation of auxin in tobacco transformed with the indoleacetic acid-lysine synthetase gene of *Pseudomonas savastanoi*. *Genes Dev.* 5, 438–446.

ROTH, A. 1996. Water transport in xylem conduits with ring thickenings. *Plant Cell Environ.* 19, 622–629.

RYSER, U., M. SCHORDERET, G.-F. ZHAO, D. STUDER, K. RUEL, G. HAUF, and B. KELLER. 1997. Structural cell-wall proteins in protoxylem development: Evidence for a repair process mediated by a glycine-rich protein. *Plant J.* 12, 97–111.

SACHS, T. 2000. Integrating cellular and organismic aspects of vascular differentiation. *Plant Cell Physiol.* 41, 649–656.

SALLEO, S., M. A. LO GULLO, D. DE PAOLI, and M. ZIPPO. 1996. Xylem recovery from cavitation-induced embolism in young plants of *Laurus nobilis*: A possible mechanism. *New Phytol.* 132, 47–56.

SAMUELS, A. L., K. H. RENSING, C. J. DOUGLAS, S. D. MANSFIELD, D. P. DHARMAWARDHANA, and B. E. ELLIS. 2002. Cellular machinery of wood production: Differentiation of secondary xylem in *Pinus contorta* var. *latifolia*. *Planta* 216, 72–82.

SCHNEIDER, E. L., and S. CARLQUIST. 2000a. SEM studies on vessels of the homophyllous species of *Selaginella*. *Int. J. Plant Sci.* 161, 967–974.

SCHNEIDER, E. L., and S. CARLQUIST. 2000b. SEM studies on the vessels of heterophyllous species of *Selaginella*. *J. Torrey Bot. Soc.* 127, 263–270.

SCHNEIDER, E. L., and S. CARLQUIST. 2000c. SEM studies on vessels in ferns. 17. Psilotaceae. *Am. J. Bot.* 87, 176–181.

SCHULTE, P. J., A. C. GIBSON, and P. S. NOBEL. 1989. Water flow in vessels with simple or compound perforation plates. *Ann. Bot.* 64, 171–178.

SCURFIELD, G., A. J. MICHELL, and S. R. SILVA. 1973. Crystals in woody stems. *Bot. J. Linn. Soc.* 66, 277–289.

SPERRY, J. S. 1985. Xylem embolism in the palm *Rhapis excelsa*. *IAWA Bull.* n.s. 6, 283–292.

SPERRY, J. S., and U. G. HACKE. 2004. Analysis of circular bordered pit function. I. Angiosperm vessels with homogeneous pit membranes. *Am. J. Bot.* 91, 369–385.

SPERRY, J. S., and N. Z. SALIENDRA. 1994. Intra- and inter-plant variation in xylem cavitation in *Betula occidentalis*. *Plant Cell Environ.* 17, 1233–1241.

SPERRY, J. S., and J. E. M. SULLIVAN. 1992. Xylem embolism in response to freeze-thaw cycles and water stress in ring-porous, diffuse-porous, and conifer species. *Plant Physiol.* 100, 605–613.

SPERRY, J. S., and M. T. TYREE. 1988. Mechanism of water stress-induced xylem embolism. *Plant Physiol.* 88, 581–587.

SPERRY, J. S., and M. T. TYREE. 1990. Water-stress-induced xylem cavitation in three species of conifers. *Plant Cell Environ.* 13, 427–436.

SPERRY, J. S., K. L. NICHOLS, J. E. M. SULLIVAN, and S. E. EASTLACK. 1994. Xylem embolism in ring-porous, diffuse-porous, and coniferous trees of northern Utah and interior Alaska. *Ecology* 75, 1736–1752.

STEWART, W. N., and G. W. ROTHWELL. 1993. *Paleobotany and the Evolution of Plants*, 2nd ed. Cambridge University Press, New York.

STOUT, D. L., and A. SALA. 2003. Xylem vulnerability to cavitation in *Pseudotsuga menziesii* and *Pinus ponderosa* from contrasting habitats. *Tree Physiol.* 23, 43–50.

TAYLOR, J. G., and C. H. HAIGLER. 1993. Patterned secondary cell-wall assembly in tracheary elements occurs in a self-perpetuating cascade. *Acta Bot. Neerl.* 42, 153–163.

TAYLOR, T. N., and E. L. TAYLOR. 1993. *The Biology and Evolution of Fossil Plants*. Prentice Hall, Englewood Cliffs, NJ.

TIBBETTS, T. J., and F. W. EWERS. 2000. Root pressure and specific conductivity in temperate lianas: Exotic *Celastrus orbiculatus* (Celastraceae) vs. native *Vitis riparia* (Vitaceae). *Am. J. Bot.* 87, 1272–1278.

TURNER, S. R., and M. HALL. 2000. The *gapped xylem* mutant identifies a common regulatory step in secondary cell wall deposition. *Plant J.* 24, 477–488.

TYREE, M. T. 1993. Theory of vessel-length determination: The problem of nonrandom vessel ends. *Can. J. Bot.* 71, 297–302.

TYREE, M. T., S. D. DAVIS, and H. COCHARD. 1994. Biophysical perspectives of xylem evolution: Is there a tradeoff of hydraulic efficiency for vulnerability to dysfunction? *IAWA J.* 15, 335–360.

TYREE, M. T., S. SALLEO, A. NARDINI, M. A. LO GULLO, and R. MOSCA. 1999. Refilling of embolized vessels in young stems of laurel. Do we need a new paradigm? *Plant Physiol.* 120, 11–21.

UEHARA, K., and T. HOGETSU. 1993. Arrangement of cortical microtubules during formation of bordered pit in the tracheids of *Taxus*. *Protoplasma* 172, 145–153.

VAN DER GRAAFF, N. A., and P. BAAS. 1974. Wood anatomical variation in relation to latitude and altitude. *Blumea* 22, 101–121.

VANDERMOLEN, G. E., C. H. BECKMAN, and E. RODEHORST. 1987. The ultrastructure of tylose formation in resistant banana following inoculation with *Fusarium oxysporum* f. sp. *cubense*. *Physiol. Mol. Plant Pathol.* 31. 185–200.

VAN DER SCHOOT, C. 1989. Determinates of xylem-to-phloem transfer in tomato. Ph.D. Dissertation. Rijksuniversiteit te Utrecht, The Netherlands.

VIAN, B., J.-C. ROLAND, D. REIS, and M. MOSINIAK. 1992. Distribution and possible morphogenetic role of the xylans within the secondary vessel wall of linden wood. *IAWA Bull.* n.s. 13, 269–282.

WANG, J., N. E. IVES, and M. J. LECHOWICZ. 1992. The relation of foliar phenology to xylem embolism in trees. *Funct. Ecol.* 6, 469–475.

WARDROP, A. B. 1981. Lignification and xylogenesis. In: *Xylem Cell Development*, pp. 115–152. J. R. Barnett, ed. Castle House, Tunbridge Wells, Kent.

WHEELER, E. A. 1983. Intervascular pit membranes in *Ulmus* and *Celtis* native to the United States. *IAWA Bull.* n.s. 4, 79–88.

WHEELER, E. A., and P. BAAS. 1991. A survey of the fossil record for dicotyledonous wood and its significance for evolutionary and ecological wood anatomy. *IAWA Bull.* n.s. 12, 275–332.

WHEELER, E. A., P. BAAS, and P. E. GASSON, eds. 1989. IAWA list of microscopic features for hardwood identification. *IAWA Bull.* n.s. 10, 219–332.

WIEBE, H. H. 1975. Photosynthesis in wood. *Physiol. Plant.* 33, 245–246.

YAMAMOTO, R., S. FUJIOKA, T. DEMURA, S. TAKATSUTO, S. YOSHIDA, and H. FUKUDA. 2001. Brassinosteroid levels increase drastically prior to morphogenesis of tracheary elements. *Plant Physiol.* 125, 556–563.

YOUNG, D. A. 1981. Are the angiosperms primitively vesselless? *Syst. Bot.* 6, 313–330.

ZANIS, M. J., D. E. SOLTIS, P. S. SOLTIS, S. MATHEWS, and M. J. DONOGHUE. 2002. The root of the angiosperms revisited. *Proc. Natl. Acad. Sci. USA* 99, 6848–6853.

ZIMMERMANN, M. H. 1982. Functional xylem anatomy of angiosperm trees. In: *New Perspectives in Wood Anatomy*, pp. 59–70, P. Baas, ed. Martinus Nijhoff/W. Junk, The Hague.

ZIMMERMANN, M. H. 1983. *Xylem Structure and the Ascent of Sap.* Springer-Verlag, Berlin.

ZIMMERMANN, M. H., and C. L. BROWN. 1971. *Trees: Structure and Function.* Springer-Verlag, New York.

ZIMMERMANN, M. H., and A. JEJE. 1981. Vessel-length distribution in stems of some American woody plants. *Can. J. Bot.* 59, 1882–1892.

ZIMMERMANN, M. H., and D. POTTER. 1982. Vessel-length distributions in branches, stem, and roots of *Acer rubrum* L. *IAWA Bull.* n.s. 3, 103–109.

ZWIENIECKI, M. A., P. J. MELCHER, and N. M. HOLBROOK. 2001a. Hydraulic properties of individual xylem vessels of *Fraxinus americana*. *J. Exp. Bot.* 52, 257–264.

ZWIENIECKI, M. A., P. J. MELCHER, and N. M. HOLBROOK. 2001b. Hydrogel control of xylem hydraulic resistance in plants. *Science* 291, 1059–1062.

# CHAPTER ELEVEN

# Xylem: Secondary Xylem and Variations in Wood Structure

The **secondary xylem** is formed by a relatively complex meristem, the vascular cambium, consisting of vertically elongated fusiform initials and squarish or horizontally (radially) elongated ray initials (Chapter 12). The secondary xylem is therefore composed of two systems, the **axial** (vertical) and the **radial** (horizontal) systems (Fig. 11.1), an architecture not characteristic of the primary xylem. In the angiosperms the secondary xylem is commonly more complex than the primary in having a wider variety of component cells.

The sculpture of the secondary walls of the primary and secondary tracheary elements was considered in Chapter 10. There it was noted that elements of the late part of the metaxylem often intergrade with the secondary elements, since both may be similarly pitted. The type of pitting therefore may be of little or no use in distinguishing between the last-formed metaxylem and the first-formed secondary xylem.

Frequently the arrangement of cells, as seen in transverse sections, is stressed as a criterion for distinguishing the primary from the secondary xylem. The procambium and the primary xylem are said to have a haphazard cell arrangement, and the cambium and the secondary xylem, an orderly arrangement, with the cells aligned parallel with the radii of the secondary body. This distinction is highly unreliable, however, for in many plants the primary xylem shows just as definite radial seriation of cells as the secondary (Esau, 1943).

In many woody angiosperms the length of the tracheary elements reliably separates the primary from the secondary xylem, the length of the last-formed tracheary elements of the primary xylem being considerably longer than that of the first-formed tracheary elements of the secondary xylem (Bailey, 1944). Although the helically thickened tracheary elements are generally longer than the pitted elements of the same primary xylem, these pitted elements are still considerably longer than the first secondary tracheary elements. The difference in length between the last-formed primary elements and first-formed secondary elements may be caused both by the increase in length of the metaxylem cells during their differentiation and the lack of a comparable increase in length of the cambial derivatives and by possible transverse divisions of the procambial cells involved with their conversion to cambial cells just before the initiation of cambial activity. In gymno-

*Esau's Plant Anatomy, Third Edition,* By Ray F. Evert.

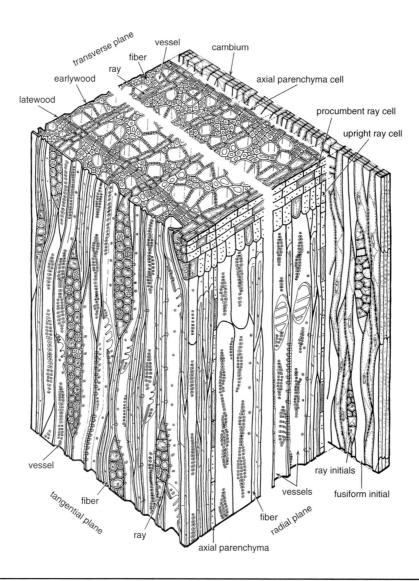

**FIGURE 11.1**

Block diagram of vascular cambium and secondary xylem of *Liriodendron tulipifera* L. (tulip tree), a woody angiosperm. The axial system consists of vessel elements with bordered pits in opposite arrangement and inclined end walls with scalariform perforation plates; fiber-tracheids with slightly bordered pits; and parenchyma strands in terminal position. The ray system contains heterocellular rays (marginal cells are upright, others procumbent), uniseriate and biseriate, of various heights. (Courtesy of I. W. Bailey; drawn by Mrs. J. P. Rogerson under the supervision of L. G. Livingston. Redrawn.)

sperms, too, the last primary elements typically are longer than the first secondary elements (Bailey, 1920).

The change from longer to shorter tracheary elements at the beginning of secondary growth is one of the steps in the establishment of mature characteristics of the secondary xylem. Various other changes accompany this step, for example, those involving the pitting, the ray structure, and the distribution of axial parenchyma. By these changes, the secondary xylem eventually reflects the evolutionary level characteristic of the species. Since the evolutionary specialization of the xylem progresses from the secondary to the primary xylem, in a given species the latter may be less advanced with regard to the evolutionary specialization. It appears that eudicots that are not truly woody—even if they possess secondary growth—show a protraction of primary xylem characteristics into their secondary xylem (**paedomorphosis**, Carlquist, 1962, 2001). One of the expressions of paedomorphosis is a gradual, instead of a sudden, change in length of tracheary elements.

# ▌BASIC STRUCTURE OF SECONDARY XYLEM

## The Secondary Xylem Consists of Two Distinct Systems of Cells, Axial and Radial

The arrangement of cells into the vertical, or axial, system, on the one hand, and the horizontal, or radial, system, on the other, constitutes one of the conspicuous characteristics of secondary xylem, or wood. The axial system and the rays are arranged as two interpenetrating systems closely integrated with each other in origin, structure, and function. In active xylem the rays most commonly consist of living cells. The axial system contains, depending on the plant species, one or more different kinds of tracheary elements, fibers, and parenchyma cells. The living cells of the rays and those of the axial system are interconnected with each other by numerous plasmodesmata, so that the wood is permeated by a continuous three-dimensional system—a symplastic continuum—of living cells (Chaffey and Barlow, 2001). Moreover this system often is connected, through the rays, with the living cells of the pith, the phloem, and the cortex (van Bel, 1990b; Sauter, 2000).

Each of the two systems has its characteristic appearance in the three kinds of sections employed in the study of wood. In the **transverse section (cross section),** that is, the section cut at right angles to the main axis of stem or root, the cells of the axial system are cut transversely and reveal their smallest dimensions (Figs. 11.2A and 11.3A). The rays—which are characterized as having length, width, and height—in contrast, are exposed in their longitudinal extent in a transverse section. When stems or roots are cut lengthwise, two kinds of longitudinal sections are obtained: **radial** (Figs. 11.2B and 11.3B; parallel to a radius) and **tangential** (Figs. 11.2C and 11.3C; perpendicular to a radius). Both show the longitudinal extent of cells of the axial system, but they give strikingly different views of the rays. Radial sections expose the rays as horizontal bands lying across the axial system. When a radial section cuts a ray through its median plane, it reveals the height of the ray. A tangential section cuts a ray approximately per-

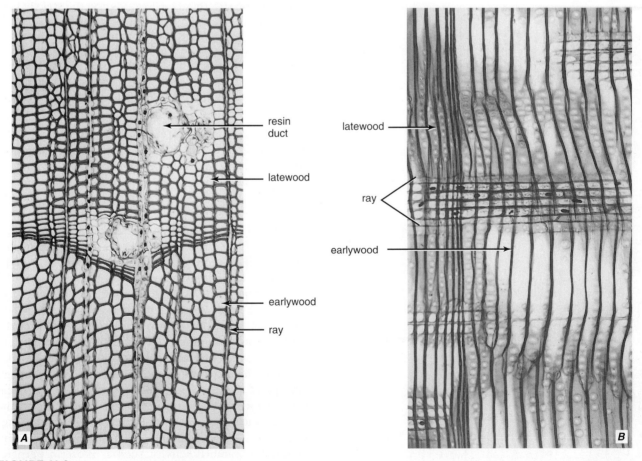

resin duct

latewood

earlywood

ray

latewood

ray

earlywood

**FIGURE 11.2**

Wood of white pine (*Pinus strobus*), a conifer, in (**A**) transverse, (**B**) radial, and (**C**) tangential sections. The wood of white pine is nonstoried. (All, ×110.)

pendicular to its horizontal extent and reveals its height and width. In tangential sections it is therefore easy to measure the height of the ray—this is usually done in terms of number of cells—and to determine whether the ray is one or more cells wide.

## Some Woods Are Storied and Others Are Nonstoried

The more or less orderly radial seriation of cells of the secondary xylem, as seen in transverse sections, is a result of the origin of these cells from periclinally, or tangentially, dividing cambial cells. In conifer wood this seriation is pronounced; in the wood of vessel-containing angiosperms it may be obscured by the ontogenetic enlargement of the vessel elements and the resultant lateral displacement of adjacent cells. Radial sections also reveal the radial seriation; in such sections the radial series of the axial system appear superimposed one upon the other in horizontal layers, or tiers. The tangential sections are varied in their appearance in different woods. In some, the horizontal layers are clearly displayed, and such wood is called **storied**, or **strati-**

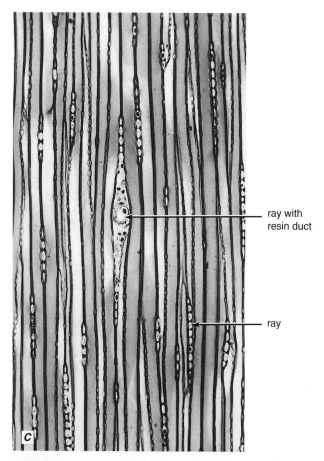

ray with
resin duct

ray

**FIGURE 11.2**

*(Continued)*

**fied**, wood (Fig. 11.4; *Aesculus, Cryptocarya, Diospyros, Ficus, Mansonia, Swietenia, Tabebuia, Tilia*, many *Asteraceae* and *Fabaceae*). In others, the cells of one tier unevenly overlap those of another. This type of wood is called **nonstoried**, or **nonstratified**, wood (Figs. 11.2C and 11.3C; *Acer, Fraxinus, Juglans, Mangifera, Manilkara, Ocotea, Populus, Pyrus, Quercus, Salix*, conifers). Tangential sections must be used to determine whether a wood is storied or nonstoried.

From an evolutionary aspect the storied woods are more highly specialized than the nonstoried. They are derived from vascular cambia with short fusiform initials and, hence, have short vessel elements. Because vessel elements and axial parenchyma cells elongate little, if at all, after they are derived from fusiform cambial initials, they show storying much more than do libriform fibers, fiber-tracheids, or tracheids. The apices of the nonperforate tracheary elements extend by intrusive growth beyond the limits of their own tier and thus partly efface its demarcation from other tiers. The storied condition is especially pronounced when the heights of the rays match that of the horizontal layer of the axial system, that is, when the rays also are storied (Fig. 11.4B). Many intermediate patterns are found between the strictly storied woods and the strictly nonstoried woods derived from cambia with long fusiform initials. Storied wood is found in eudicots only; it is unknown in conifers.

## Growth Rings Result from the Periodic Activity of the Vascular Cambium

The periodic activity of the vascular cambium (Chapter 12), which is a seasonal phenomenon in temperate regions related to changing day lengths and temperatures, produces **growth increments**, or **growth rings** (Fig. 11.5), in the secondary xylem. If such a growth layer represents one season's growth, it may be called an **annual ring**. Abrupt changes in available water and other environmental factors may be responsible for the production of more than one growth ring in a given year. Additional rings may also result from injuries by insects, fungi, or fire. Such an additional growth layer is called a **false annual ring** and the annual growth increment consisting of two or more rings is termed a **multiple annual ring**. In very suppressed or old trees the lower portions of the stem or of some branches may fail to produce xylem during a given year. Thus, although the age of a given portion of a woody branch or stem can be estimated by counting the growth rings, the estimates may be inaccurate if some rings are "missing" or if false annual rings are present. Trees that exhibit continuous cambial activity, such as those in perpetually wet tropical rainforests, may lack growth rings entirely (Alves and Angyalossy-Alfonso, 2000). It is therefore difficult to judge the age of such trees.

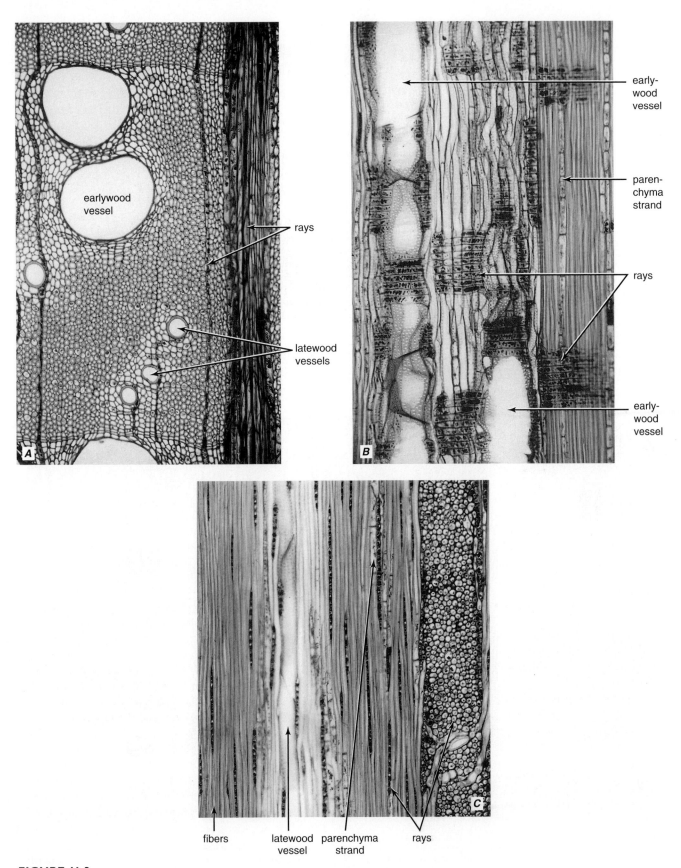

**FIGURE 11.3**

Wood of red oak (*Quercus rubra*) in (**A**) transverse, (**B**) radial, and (**C**) tangential sections. The wood of red oak is nonstoried. (All, ×100.)

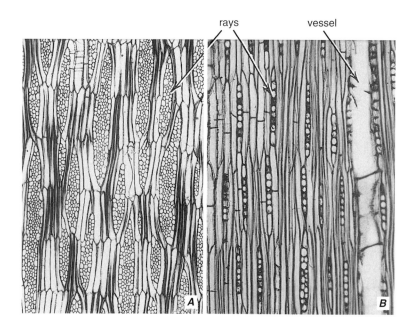

**FIGURE 11.4**

Storied wood, as revealed in tangential section. **A**, in *Triplochiton*, high multiseriate rays extend through more than one horizontal tier. **B**, in *Canavalia*, low uniseriate rays are each limited to one horizontal tier. (A, ×50; B, ×100. From Barghoorn, 1940, 1941.)

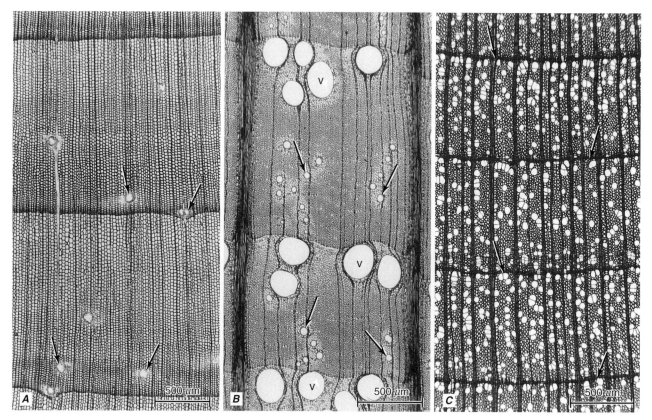

**FIGURE 11.5**

Growth rings of wood, in transverse sections. **A**, white pine (*Pinus strobus*). Lacking vessels, conifer wood is nonporous. Note the resin ducts (arrows), which occur largely in the latewood. **B**, red oak (*Quercus rubra*). As is characteristic of a ring-porous wood, the pores, or vessels (v), of the earlywood are distinctly larger than those of the latewood (arrows). **C**, tulip tree (*Liriodendron tulipifera*), a diffuse-porous wood. In tulip tree the ring boundaries are marked by bands of marginal parenchyma cells (arrows).

Growth rings occur in both deciduous and evergreen trees. Furthermore they are not confined to the temperate zone, with its striking contrast between the season of growth and season of dormancy. A distinct seasonality also occurs in many regions in the tropics that experience severe annual dry seasons as in much of Amazonia (Vetter and Botosso, 1989; Alves and Angyalossy-Alfonso, 2000) and Queensland, Australia (Ash, 1983) or annual flooding by great rivers such as the Amazon and the Rio Negro (Worbes, 1985, 1989). In the former regions most trees lose their leaves during the dry season and produce new ones shortly after the onset of the rainy season, the period during which growth takes place. Inundations result in anoxic soil conditions, which lead to reduced root activity and water uptake to the crown, and, consequently, to cambial dormancy and the formation of growth rings (Worbes, 1985, 1995).

The factors responsible for the periodicity of growth rings can differ among species growing side by side. Take, for example, the periodicity of growth rings in four species growing in a swamp forest remnant of the Atlantic rainforest in Rio de Janeiro, Brazil (Callado et al., 2001). Although all four species form annual growth rings, they exhibit different patterns of radial growth. In three of the species latewood formation was correlated with the period of leaf abscision, but it occurred at different times for each species. Flooding was a determinant of periodic growth in *Tabebuia cassinoides*, the only species that showed the growth rhythm expected for a wetland species; the photoperiod was indirectly responsible for the radial growth rhythm in *T. umbellata*, and endogenous rhythms accounted for the periodicity of radial growth in *Symphonia globulifera* and *Alchornea sidifolia*.

Growth rings are of varied degrees of distinctness, depending on the species of wood and also on the growing conditions (Schweingruber, 1988). The cause of visibility of the growth rings in a section of wood is the structural difference between the xylem produced in the early and the late parts of the growing season. In temperate woods the **earlywood** is less dense (with wider cells and proportionally thinner walls) than the **latewood** (with narrower cells and proportionally thicker walls) (Figs. 11.2A, 11.3A, and 11.5). In most species the earlywood of a given season merges more or less gradually with the latewood of the same season, but the boundary between the latewood of one season and the earlywood of the following season is ordinarily sharp. Such pronounced changes in cell wall thickness and dimensions are uncommon in tropical woods. Ring boundaries in many tropical woods are marked by bands of axial parenchyma cells produced at the beginning and/or at the end of a growing season (Boninsegna et al., 1989; Détienne, 1989; Gourlay, 1995; Mattos et al., 1999; Tomazello and da Silva Cardoso, 1999). Such bands are called **marginal parenchyma bands**. Their cells are often filled with various amorphous substances

or crystals. Marginal parenchyma bands also occur in many temperate trees (Fig. 11.5C).

The factors determining the change from the early-wood characteristics to those of the latewood are of continued interest to tree physiologists (Higuchi, 1997). Although several plant hormones have been implicated in earlywood and latewood formation, the case for auxin (IAA) involvement has been explored the most. The concentration of IAA in the cambial zone of a tree stem has been found to undergo seasonal changes, increasing from spring to summer and, then, decreasing to the spring level as autumn approaches. In winter, the IAA concentration in the dormant cambium is at a relatively low level. The transition from earlywood to latewood has been attributed to the decreasing levels of IAA (Larson, 1969). Accordingly, when changing growing conditions result in an earlier than usual decrease in the endogenous IAA concentration, the transition from earlywood to latewood formation occurs earlier. Latewood formation could not be attributed, however, to decreased IAA in the cambial region of *Picea abies* stems (Eklund et al., 1998), and in *Pinus sylvestris*, the auxin concentration was found to increase during the transition from earlywood to latewood (Uggla et al., 2001). Latewood formation in *Pinus radiata* and *P. sylvestris* has been attributed to by some workers to an increase in the level of endogenous abscisic acid in the cambial zone (Jenkins and Shepherd, 1974; Wodzicki and Wodzicki, 1980).

The width of individual growth rings may vary greatly from year to year as a function of such environmental factors as light, temperature, rainfall, available soil water, and length of the growing season (Kozlowski and Pallardy, 1997). The width of a growth ring can be a fairly accurate index of the rainfall of a particular year. Under favorable conditions—that is, during periods of adequate or abundant rainfall—the growth rings are wide; under unfavorable conditions, they are narrow. Recognition of these relations has led to the development of **dendrochronology**, that is, study of yearly growth patterns in trees and use of the information for evaluating past fluctuations in climate and dating past events in historical research (Schweingruber, 1988, 1993). The relative amounts of earlywood and latewood are affected by environmental conditions and specific differences.

## As Wood Becomes Older, It Gradually Becomes Nonfunctional in Conduction and Storage

The elements of the secondary xylem are variously specialized in relation to their function. The tracheary elements and the fibers, which are concerned, respectively, with the conduction of water and support, lose their protoplasts before they begin to perform their principal roles in the plant. The living cells, which store and transport food (parenchyma cells and certain fibers), are alive at the height of xylem activity. Eventually the living cells die. This stage is preceded by numerous changes in the wood that visibly differentiate the active sapwood from the inactive heartwood (Hillis, 1987; Higuchi, 1997).

**Sapwood**, by definition, is the part of the wood in a living tree that contains living cells and reserve materials. It may or may not be entirely functional in the conduction of water. For example, in a 45-year-old tree of *Quercus phellos*, 21 of the outermost growth rings contained living storage cells but only the two outermost rings were still involved with conduction (Ziegler, 1968). All 21 rings were part of the sapwood.

The most critical change during the conversion of sapwood into heartwood is the death of the parenchyma and other living cells of the wood. This is preceded by the removal of reserve substances or their conversion into heartwood substances. Thus **heartwood** is characterized by the absence of living cells and reserve substances. The innermost sapwood—the part of the wood in which heartwood formation takes place—is called the **transition zone**. Heartwood formation, a kind of programmed cell death, is a normal phenomenon in the life of the tree and results from physiological death due to internal factors. It occurs in the roots as well as the stems of many species but only in the region near the stem wood (Hillis, 1987). Once it is initiated, it continues throughout the life of the tree. With increasing age, the heartwood becomes infiltrated with various organic compounds such as phenolics, oils, gums, resins, and aromatic and coloring materials. These compounds are collectively referred to as *extractives* because they can be extracted from the wood in organic solvents (Hillis, 1987). Some of these substances impregnate the walls; others enter into the cell lumina as well.

At least two different types of heartwood formation can be distinguished (Magel, 2000; and literature cited therein). In Type 1, also called the *Robinia*-**Type**, the accumulation of phenolic extractives begins in the tissues of the transition zone. In Type 2, or the *Juglans*-**Type**, phenolic precursors of the heartwood extractives gradually accumulate in the aging sapwood tissues. Key enzymes involved with the biosynthesis of flavonoids (the largest group of plant phenolic compounds) and the genes encoding them are now being identified in space and time (Magel, 2000; Beritognolo et al., 2002; and literature cited therein). Two such enzymes are phenylalanine ammonia lyase (PAL) and chalcone synthase (CHS). PAL actually is involved with two separate events, one in relation to lignin formation in newly formed wood, and the other in relation to the formation of heartwood extractives. CHS, in contrast to PAL, is active exclusively in the transition zone. The activation of PAL and CHS is correlated with the accumulation of the flavonoids, which are synthesized de novo in the sapwood cells undergoing transformation to heartwood (Magel, 2000; Beritognolo et al., 2002). Although the

hydrolysis of storage starch provides some of the carbon for the formation of the phenolics, the bulk of phenolic synthesis is dependent on imported sucrose. In *Robinia*, enhanced enzymic degradation of sucrose coincided in time and location with increased activities of PAL and CHS and the accumulation of phenolic heartwood extractives, indicating a close involvement of sucrose metabolism with heartwood formation (Magel, 2000).

The conversion of sapwood to heartwood may also be accompanied by a change in moisture content. In most conifers, the moisture content of the heartwood is considerably lower than that of the sapwood. The situation in woody angiosperms varies among species and with the season. In many species, the moisture content of the heartwood differs little from that of the sapwood. In some species of certain genera (e.g., *Betula, Carya, Eucalyptus, Fraxinus, Juglans, Morus, Populus, Quercus, Ulmus*), the heartwoods contain more moisture than do the sapwoods.

In many woody angiosperms heartwood formation is accompanied by the development of tyloses in the vessels (Chapter 10; Chattaway, 1949). Examples of woods with abundant development of tyloses are those of *Astronium, Catalpa, Dipterocarpus, Juglans nigra, Maclura, Morus, Quercus* (white oak species), *Robinia*, and *Vitis*. Many genera never develop tyloses. In conifer wood the pit membranes having tori may become fixed so that the tori are appressed to the borders and close the apertures (aspirated pit-pairs, Chapter 10) and may be incrusted with lignin-like and other substances (Krahmer and Côté, 1963; Yamamoto, 1982; Fujii et al., 1997; Sano and Nakada, 1998). The aspiration of bordered pits appears to be related to processes causing the drying out of the central core of the wood (Harris, 1954). The various changes that occur during heartwood formation do not affect the strength of the wood but make it more durable than the sapwood, less easily attacked by various decay organisms, and less penetrable to various liquids (including artificial preservatives).

The proportion of sapwood and heartwood and the degree of visible and actual differences between the two are highly variable in different species and in different conditions of growth. In most trees the heartwood is usually darker in color than the surrounding sapwood. When freshly cut, the color of various heartwoods covers a broad spectrum, including the jet-black (ebony) in some species of *Diospyros* and in *Dalbergia melanoxylon*; purple in species of *Peltogyne*; red in *Simira* (*Sickingia*) and *Brosimum rubescens*; yellow in species of *Berberis* and *Cladrastis*; and orange in *Dalbergia retusa, Pterocarpus*, and *Soyauxia* (Hillis, 1987). Some trees have no clearly differentiated heartwood (*Abies, Ceiba, Ochroma, Picea, Populus, Salix*), others have thin sapwood (*Morus, Robinia, Taxus*), and still others have a thick sapwood (*Acer, Dalbergia, Fagus, Fraxinus*).

In some species the sapwood is early converted into heartwood; in others it shows greater longevity. Heartwood formation usually begins in *Robinia* species at 3 to 4 years, in some species of *Eucalyptus* at about 5 years, in several species of pine at 15 to 20 years, in European ash (*Fraxinus excelsior*) at 60 to 70 years, in beech at 80 to 100 years, and in *Alstonia scholaris* (Apocynaceae, West Africa) over 100 years (Dadswell and Hillis, 1962; Hillis, 1987).

Determining the depth of the sapwood and the pattern of sap velocity along the xylem radius are critical problems for those investigators interested in deriving estimates of canopy transpiration and forest water use (Wullschleger and King, 2000; Nadezhdina et al., 2002). As noted by Wullschleger and King (2000), "Failure to recognize that not all sapwood is functional in water transport will introduce systematic bias into estimates of both tree and stand water use."

## Reaction Wood Is a Type of Wood That Develops in Branches and Leaning or Crooked Stems

The formation of **reaction wood** is presumed to result from the tendency of the branch or stem to counteract the force inducing the inclined position (Boyd, 1977; Wilson and Archer, 1977; Timell, 1981; Hejnowicz, 1997;

Huang et al., 2001). In conifers, the reaction wood develops on the lower side of the branch or stem where compressive stresses are very high and is called **compression wood**. Compression wood also is formed by *Ginkgo* and the Taxales (Timell, 1983). In angiosperms and *Gnetum*, reaction wood develops on the upper side of branches and stems in zones where large tensile stresses exist and is called **tension wood.** A notable exception among the angiosperms is *Buxus microphylla*, which forms compression wood rather than tension wood on inclined stems (Yoshizawa et al., 1992).

Reaction wood differs from the normal wood in both anatomy and chemistry. It is not a common component of root wood. When found in roots, tension wood is evenly distributed around the circumference (Zimmermann et al., 1968; Höster and Liese, 1966). Compression wood forms in some gymnosperm roots only when they are exposed to light and then it is on the underside (Westing, 1965; Fayle, 1968).

Compression wood is produced by the increased activity of the vascular cambium on the lower side of the branch or leaning stem and typically results in the formation of eccentric growth rings. Portions of growth rings located on the lower side are generally much wider than those on the upper side (Fig. 11.6A). Hence

**FIGURE 11.6**

Reaction wood. **A**, transverse section of a pine (*Pinus* sp.) stem, showing compression wood with larger growth rings on the lower side. **B**, transverse section of a black walnut (*Juglans nigra*) stem, showing tension wood with larger growth rings on the upper side. The cracks in both stems are due to drying. (Courtesy of Regis B. Miller.)

compression wood causes straightening by expanding or pushing the stem or branch upright. The compression wood in conifers is typically denser and darker than the surrounding tissue, often appearing as a red-brown color on wood surfaces. Anatomically it is identified by its relatively short tracheids, which appear rounded in transverse sections (Fig. 11.7). The compression wood tracheids assume their rounded form during the final stage of primary wall formation, and at that time numerous schizogenous intercellular spaces arise in the tissue, except at the growth ring boundary (Lee and Eom, 1988; Takabe et al., 1992). Occasionally the tips of the tracheids are distorted. Compression wood tracheids typically lack an $S_3$ layer and the inner portion of their $S_2$ layer is deeply fissured with helical cavities (Figs. 11.7 and 11.8). Chemically, compression wood contains more lignin and less cellulose than normal wood. The compound middle lamella and outer part of the $S_2$ layer are highly lignified. The lengthwise shrinkage of compression wood upon drying is often 10 or more times as great as that of normal wood. Normal wood usually shrinks lengthwise not more than 0.1 to 0.3%. The difference in relative lengthwise shrinkage of normal and compression wood in a drying board often causes the board to twist and cup. Such wood is virtually useless except as fuel. The formation of compression wood has been shown to reduce the efficiency of xylem transport (Spicer and Gartner, 2002).

Tension wood is produced by the increased activity of the vascular cambium on the upper side of the branch or stem, and as in the case of compression wood, eccentric growth rings result. To straighten the stem, the tension wood must exert a pull. Tension wood is often difficult or impossible to identify without microscopic examination of wood sections. The most distinguishing feature of tension wood is the presence of **gelatinous fibers** (Fig. 11.9; Chapter 8), the inner secondary wall, or gelatinous layer (G layer), of which is nonlignified but rich in acidic polysaccharides, in addition to having large quantities of cellulose (Hariharan and Krishnamurthy, 1995; Jourez, 1997; Pilate et al., 2004). The gelatinous fibers may have two ($S_1$ + G) to four ($S_1$, $S_2$, $S_3$, G) secondary wall layers, the gelatinous layer usually being the innermost. The vessels of tension wood typically are reduced both in width and number. The ray and axial parenchyma may also be affected during tension wood production (Hariharan and Krishnamurthy, 1995). Lengthwise shrinkage of tension wood rarely exceeds 1%, but boards containing it twist out of shape in drying. When such logs are sawed green, tension wood tears loose in bundles of fibers, imparting a wooly appearance to the boards.

The secondary phloem adjacent, or attached, to tension wood may also contain gelatinous fibers (Nanko et al., 1982; Krishnamurthy et al., 1997). In the phloem

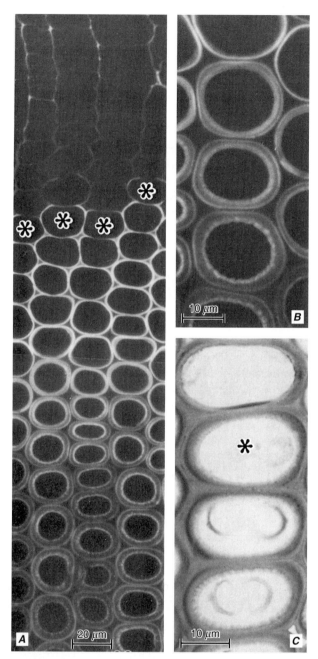

**FIGURE 11.7**

Compression wood tracheids in Todo fir (*Abies sachalinensis*, a conifer). **A**, fluorescence photomicrograph of differentiating compression wood. The fluorescence is intense only in the depositing secondary wall and is last seen at the inner surface of the cell wall. The asterisks mark tracheids at the start of $S_1$ deposition. **B**, fluorescence photomicrograph, showing tracheids undergoing $S_2$ deposition. **C**, light photomicrograph of differentiating tracheids stained for polysaccharides. The appearance of helical ridges and cavities in the inner portion of the $S_2$ layer coincides with the active lignification of the outer portion of the $S_2$ layer (asterisk). All transverse sections. (From Takabe et al., 1992.)

of *Populus euroamericana*, the walls of the gelatinous fibers consist of two lignified outer layers, the $S_1$ and $S_2$, and of as many as four alternately arranged unlignified (gelatinous) and lignified inner layers (Nanko et al., 1982).

There are some woody angiosperms—for example, *Lagunaria patersonii* (Scurfield, 1964), *Tilia cordata*, and *Liriodendron tulipifera* (Scurfield, 1965)—in which typical tension wood does not form. In these trees the leaning stems undergo asymmetric radial growth by increased production of both xylem and phloem on the upper sides of the stems. Gelatinous fibers are lacking, and the lignin content of the tension wood is similar to that of the normal wood. Clearly, gelatinous fibers are not necessary in these tree species for axis reorientation (Fisher and Stevenson, 1981; Wilson and Gartner, 1996).

Gelatinous fibers are not unique to branches and leaning stems. They also are found in the vertical stems of some species of *Fagus* (Fisher and Stevenson, 1981), *Populus* (Isebrands and Bensend, 1972), *Prosopis* (Robnett and Morey, 1973), *Salix* (Robards, 1966), and *Quercus* (Burkart and Cano-Capri, 1974). This reaction wood, with its gelatinous fibers, is probably associated with internal stresses that arise as new cells added by the cambium tend to shrink longitudinally during maturation of their walls (Hejnowicz, 1997). Indeed, as noted by Huang et al. (2001), in a normal vertically growing tree trunk, as newly formed xylem elements become lignified, they generate tension stress in the longitudinal direction and compressive stress in the tangential direction. This combination of stress is repeated with each new growth increment, resulting in a regular distribution of opposing stresses around the circumference of the trunk. As a result, tension stress arises in the outer part of a trunk and compressive stress in the inner part.

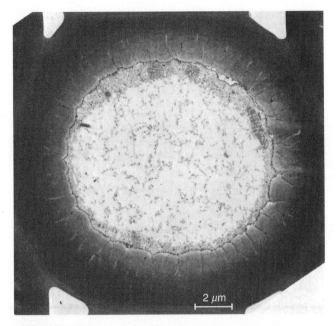

**FIGURE 11.8**

Transmission electron micrograph of compression wood tracheid in Todo fir (*Abies sacharinensis*) nearing the final stage of cell wall formation. Note helical ridges and cavities in the inner portion of the $S_2$ layer. (From Takabe et al., 1992.)

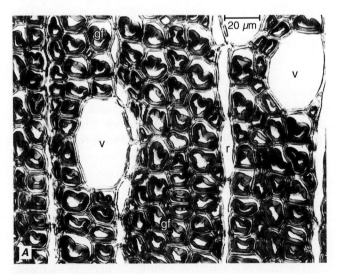
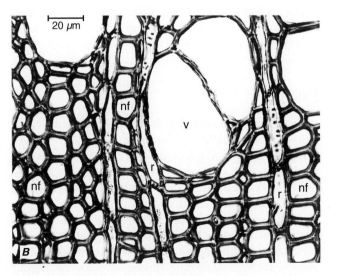

**FIGURE 11.9**

Transverse sections of tension wood (**A**) and normal wood (**B**) of poplar (*Populus euramericana*). The dark gelatinous layers have separated from the rest of the secondary wall in the gelatinous fibers (gf). Other details: nf, normal fiber; r, ray; v, vessel. (From Jourez, 1997.)

These stresses have been attributed to helping tree trunks withstand the force of wind strikes and resist cracking of the xylem by frost during severe winters (Mattheck and Kubler, 1995).

Research involving experimental modifications in position of plant axes has provided evidence that the stimulus of gravity and the distribution of endogenous growth substances are important factors in evoking the development of reaction wood (Casperson, 1965; Westing, 1968; Boyd, 1977). Early experiments with auxins and anti-auxins indicated that the tension wood in angiosperms is formed where auxin concentration is low (Morey and Cronshaw, 1968; Boyd, 1977). In contrast, the compression wood of conifers was found to form in regions of high auxin concentration (Westing, 1968; Sundberg et al., 1994). In a more recent study, employing high resolution analysis of endogenous IAA distribution across the cambial region tissues in both *Populus tremula* and *Pinus sylvestris* trees, it was demonstrated that reaction wood is formed without any obvious alterations in IAA balance (Hellgren et al., 2004). Gibberellic acid (GA$_3$) and ethylene also have been implicated in reaction wood formation (Baba et al., 1995; Dolan, 1997; Du et al., 2004). When reaction wood is induced for only a short time the cells formed at the beginning and end of the induction period may lack some of the anatomical characteristics typical of tension wood or compression wood, indicating that differentiation of reaction wood characteristics may be started or stopped during cell development (Boyd, 1977; Wilson and Archer, 1977). On the other hand, Casperson (1960) concluded that the response leading to the formation of tension wood in *Aesculus* hypocotyls occurred only in those fiber precursors stimulated at an early stage of their separation from the cambium. In *Acer saccharinum* some of the anatomical features of tension wood were already apparent in the primary xylem (Kang and Soh, 1992).

# WOODS

Woods are usually classified as either softwoods or hardwoods. The so-called **softwoods** are conifer woods, and the **hardwoods** are angiosperm wood. The two kinds of wood have basic structural differences, but the terms "softwood" and "hardwood" do not accurately express the relative density (weight per unit volume) or hardness of the wood. For example, one of the lightest and softest woods is balsa (*Ochroma lagopus*), a tropical hardwood. By contrast, the woods of some softwoods, such as slash pine (*Pinus elliotii*), are harder than some hardwoods. Conifer wood is homogeneous in structure—with long straight elements predominating. It is highly suitable for papermaking, where high toughness and strength are needed. Many commercially used hardwoods are especially strong, dense, and heavy because of a high proportion of fiber-tracheids and libriform fibers (*Astronium, Carya, Carpinus, Diospyros, Guaiacum, Manilkara, Ostrya, Quercus*). The main sources of commercial timbers are the conifers among the gymnosperms and the eudicots among the angiosperms. The arborescent, or tree-like, monocots do not produce a commercially important homogeneous body of secondary xylem (Tomlinson and Zimmermann, 1967; Butterfield and Meylan, 1980). Among the monocots, the bamboo culm, which has a high strength-weight ratio and is more resilient than conventional timbers, has long served as the most prominent "wood" of Asia. It is used for the construction of houses, furniture, utensils, in the making of paper, as a floor covering, and as fuel (Liese, 1996; Chapman, 1997; see Liese, 1998, for the anatomy of bamboo culms).

## The Wood of Conifers Is Relatively Simple in Structure

The wood of conifers is simpler and more homogeneous than that of most of the angiosperms (Figs. 11.2, 11.10, and 11.11). The chief distinction between the two kinds of wood is the absence of vessels in the conifers and their presence in most angiosperms. Another outstanding feature of conifer wood is the relatively small amount of parenchyma, particularly axial parenchyma.

## The Axial System of Conifer Woods Consists Mostly or Entirely of Tracheids

The tracheids are long cells averaging 2 to 5 mm in length (range: 0.5 to 11 mm; Bailey and Tupper, 1918), with their ends overlapping those of other tracheids (Fig. 11.2B; Chapter 10). The overlapping ends may be curved and branched because of intrusive growth. Basically the ends are wedge-shaped, with their pointed faces exposed in tangential sections and the blunt part of the wedges in radial sections. Fiber-tracheids may occur in the latewood, but libriform fibers are absent.

The tracheids are interconnected by circular or oval bordered pit-pairs in single, opposite (wide-lumened earlywood tracheids of Taxodiaceae and Pinaceae), or alternate (Araucariaceae) arrangement. The number of pits on each tracheid may vary from approximately 50 to 300 (Stamm, 1946). The pit-pairs are most abundant on the overlapping ends of the tracheids and are largely confined to the radial walls. The latewood tracheids may bear pits on their tangential walls. Helical thickenings (Chapter 10) on pitted walls have been encountered in tracheids of *Pseudotsuga, Taxus, Cephalotaxus,* and *Torreya* (Phillips, 1948).

The tracheids sometimes show thickenings—**crassulae**—of the middle lamella and primary wall along the upper and lower margins of the pit-pairs (Fig. 11.11A, B; Chapter 10). Other infrequently encountered wall sculptures are the **trabeculae**, small bars extending across

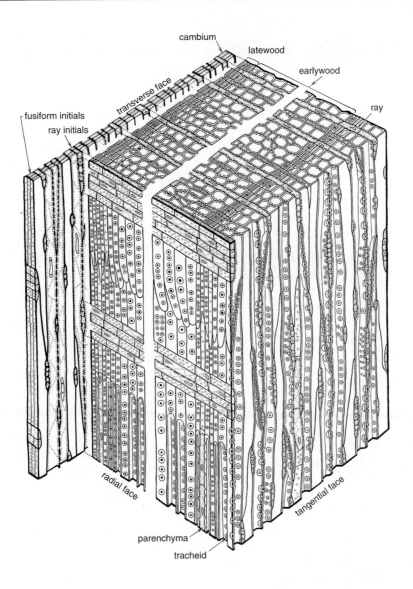

**FIGURE 11.10**

Block diagram of vascular cambium and secondary xylem of white cedar (*Thuja occidentalis* L.), a conifer. The axial system consists of tracheids and small amount of parenchyma. The ray system consists of low, uniseriate rays composed of parenchyma cells. (Courtesy of I. W. Bailey; drawn by Mrs. J. P. Rogerson under the supervision of L. G. Livingston. Redrawn.)

the lumina of the tracheids from one tangential wall to the other.

Axial parenchyma may or may not be present in conifer wood. In Podocarpaceae, Taxodiaceae, and Cupressaceae, parenchyma is occasionally present as single strands in the transition zone between the earlywood and latewood. As single strands, it is scanty or absent in the Pinaceae, Araucariaceae, and Taxaceae. In some genera, axial parenchyma or epithelial cells are restricted to that associated with resin ducts (*Cedrus, Keteleeria, Picea, Pinus, Larix, Pseudotsuga*). Secondary walls occur in epithelial cells in *Larix, Picea,* and *Pseudotsuga*.

## The Rays of Conifers May Consist of Both Parenchyma Cells and Tracheids

The rays of conifers are composed either of parenchyma cells alone, or of parenchyma cells and tracheids. Those composed of parenchyma cells alone are called **homocellular**, those containing both parenchyma cells and tracheids, **heterocellular** (Figs. 11.11D and 11.12). Ray tracheids resemble parenchyma cells in shape but lack protoplasts at maturity and have secondary walls with bordered pits. They are normally present in Pinaceae, except in *Abies, Keteleeria,* and *Pseudolarix,* and occasionally in *Sequoia* and most Cupressaceae (Phillips,

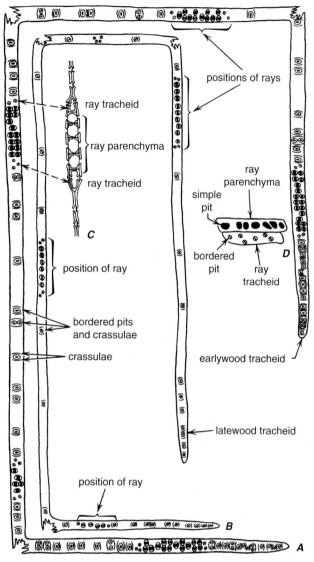

**FIGURE 11.11**

Elements from secondary xylem of *Pinus*. **A**, earlywood, and **B**, latewood tracheids. Radial walls in face views. **C**, ray in transverse section as seen in tangential section of wood. **D**, two ray cells as seen in a radial section of wood. Tracheids in **A**, **B**, show contact areas with rays. Small pits in these areas connect axial tracheids with ray tracheids. Large pits with partial borders connect ray parenchyma cells with axial tracheids. Elsewhere tracheids have pits with full borders. (All, ×100. **A**, **B**, **D**, adapted from Forsaith, 1926; with permission from SUNY-ESF.)

1948). The ray tracheids commonly occur along the margins (tops and/or bottoms) of the rays, one or more cells in depth, but may be interspersed among the layers of parenchyma cells.

Ray tracheids have lignified secondary walls. In some conifers these walls are thick and sculptured, with pro-

jections in the form of teeth or bands extending across the lumen of the cell. The ray parenchyma cells have living protoplasts in the sapwood and often darkly colored resinous deposits in the heartwood. They have only primary walls in Taxodiaceae, Araucariaceae, Taxaceae, Podocarpaceae, Cupressaceae, and Cephalotaxaceae (although the microfibrillar orientation of ray-cell walls of *Podocarpus amara* and *Tsuga canadensis* are interpreted as those typical of secondary walls; Wardrop and Dadswell, 1953) and have also secondary walls in Abietoideae (Bailey and Faull, 1934).

The rays of conifers are mostly one cell wide (Fig. 11.2C; **uniseriate**), occasionally two cells wide (**biseriate**) and from 1 to 20, or sometimes up to 50, cells high. The presence of a resin duct in a ray makes the normally uniseriate ray appear several cells wide except at the upper and lower limits (Fig. 11.2C). The rays containing resin ducts are called **fusiform rays.** The rays of conifers make up, on average, about 8% of the volume of the wood.

Each axial tracheid is in contact with one or more rays (Fig. 11.11A, B). The pit-pairs between the axial tracheids and ray parenchyma cells are half-bordered, with the border on the side of the tracheid; those between the axial tracheids and the ray tracheids are fully bordered. The pitting between ray parenchyma cells and axial tracheids form such characteristic patterns in radial sections that the **cross-field,** that is, the rectangle formed by the radial wall of a ray cell against an axial tracheid, is utilized in the classification and identification of conifer woods. The pit contacts between the ray parenchyma cells and axial tracheids are extensive, as are those between the axial parenchyma cells and axial tracheids when that cell combination is present. Thus both the axial and ray parenchyma cells are contact cells (Braun, 1970, 1984).

### The Wood of Many Conifers Contains Resin Ducts

Resin ducts appear as a constant feature in the axial and radial systems of the woods of such genera as *Pinus* (Figs. 11.2A, C and 11.5A), *Picea, Cathaya, Larix,* and *Pseudotsuga* (Wu and Hu, 1997). By contrast, resin ducts never occur in the woods of *Juniperus* and *Cupressus* (Fahn and Zamski, 1970). In still other genera such as *Abies, Cedrus, Pseudolarix,* and *Tsuga,* they arise only in response to injury. Normal ducts are elongated and occur singly (Figs. 11.2A and 11.5A); traumatic ducts generally are cyst-like and occur in tangential series (Fig. 11.13; Kuroda and Shimaji, 1983; Nagy et al., 2000). Some investigators consider all resin ducts in the wood traumatic (Thomson and Sifton, 1925; Bannan, 1936). The phenomena that induce the development of traumatic ducts are numerous. Some of these are formation of open and pressure wounds and injuries by frost and wind. Different groups of conifers are not alike in

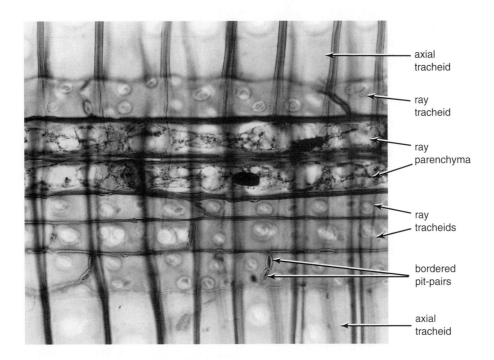

**FIGURE 11.12**

Radial section of white pine (*Pinus strobus*) wood, showing a portion of a ray consisting of parenchyma cells with protoplasts (the dark bodies are nuclei) and of ray tracheids with bordered pits in their walls. (×450.)

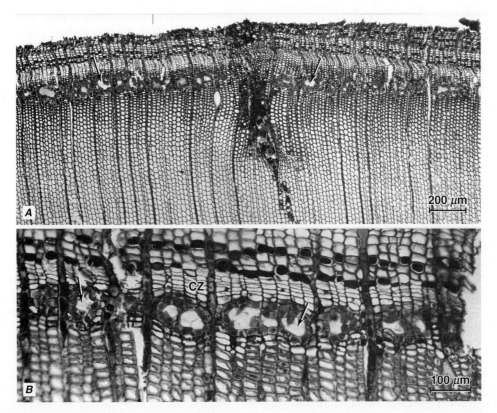

**FIGURE 11.13**

Traumatic resin ducts (arrows), bordering the cambial zone (cz), in the secondary xylem of Japanese hemlock (*Tsuga sieboldii*). The ducts were induced by the insertion of metal pins into the bark. **A**, 36 days after pinning, with abnormal tissue in the center; **B**, more detailed view 20 days after pinning. (Reprinted with permission from K. Kuroda and K. Shimaji. 1983. Traumatic resin canal formation as a marker of xylem growth. *Forest Science* 29, 653–659. © 1983 Society of American Foresters.)

their response to injuries. The genus *Pinus* appears to be least sensitive to external factors (Bannan, 1936).

Axial resin ducts commonly are located in the earlywood–latewood transition or latewood portions of growth rings (Figs. 11.2A and 11.5A; Wimmer et al., 1999; and literature cited therein). Their location and frequency may be influenced by both cambial age and climatic factors. In *Picea abies*, for instance, the majority of axial resin ducts in tree rings above 10 years are more likely to be found at the transition between earlywood and latewood, and those in tree rings at a young cambial age in the latewood (Wimmer et al., 1999). Summer temperature was found to affect the formation of the ducts most, with a direct relationship existing between high summer temperatures and high frequency of axial ducts.

Typically, resin ducts arise as schizogenous intercellular spaces by separation of parenchyma cells recently derived from the vascular cambium. Each radial duct originates at an axial duct and is continuous from the xylem into the phloem, although the ducts may not be open in the cambial region of species lined with thin-walled cells (Chattaway, 1951; Werker and Fahn, 1969; Wodzicki and Brown, 1973). The formation of radial ducts on the phloem side of the cambium may precede that of their counterparts on the xylem side. It has been suggested that the stimulus for duct formation first affects the ray initials and then is conducted inward by the rays to the xylem mother cells of the axial system. There the stimulus spreads vertically for a certain distance, causing the axial components to change into duct cells (Werker and Fahn, 1969). The radial ducts may continue to increase in length with cambial activity. Those of the axial system are variable in height. In the outermost growth ring of 10- to 23-year-old loblolly pine (*Pinus taeda*) trees, the axial resin ducts ranged in length from 20 to 510 mm (LaPasha and Wheeler, 1990).

During their development, the resin ducts form a lining, the **epithelium**, which generally is surrounded by a sheath of axial parenchyma cells, variously referred to as sheath cells, accompanying cells, or subsidiary cells (Wiedenhoeft and Miller, 2002). In *Pinus*, the epithelial cells are thin-walled (Fig. 11.2A), remain active for several years, and produce abundant resin. In *Pinus halepensis* and *Pinus taeda*, some of the axial cells bordering the epithelium are short-lived and deposit an inner suberized wall layer before collapsing (Werker and Fahn, 1969; LaPasha and Wheeler, 1990). In *Larix* and *Picea*, the epithelial cells have thick lignified walls and most of them die during the year of origin. These genera produce little resin. Thick, lignified walls have also been reported for the epithelial and bordering axial cells in *Pseudotsuga* (Fig. 11.14) and *Cathaya* (Wu and Hu, 1997). Eventually a resin duct may become closed by enlarging epithelial cells. These tylose-like intrusions

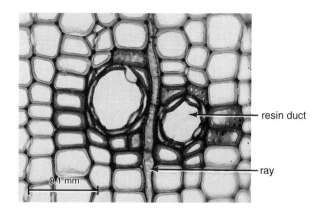

**FIGURE 11.14**

Transverse section of wood of *Pseudotsuga taxifolia*, showing two resin ducts with thick-walled epithelial cells. (From Esau, 1977.)

are called **tylosoids** (Record, 1934). They differ from tyloses in that they do not grow through pits.

Early studies of the connections between radial and axial resin ducts led to the concept of a three-dimensional anastomosing system of resin ducts within the wood. More recent studies indicate that such an extensive system may not exist, at least not in all conifer woods. In *Pinus halepensis*, for example, connections exist only between radial and axial ducts situated on the same radial plane, and not in every case where the two types of duct come close together (Werker and Fahn, 1969). Thus, in *Pinus halepensis*, there are many two-dimensional networks, each situated in a different radial plane. In *Pinus taeda*, axial and radial ducts often are in close proximity and even share epithelial cells, but direct openings between the two are rare (LaPasha and Wheeler, 1990).

### The Wood of Angiosperms Is More Complex and Varied Than That of Conifers

The complexity of the wood of angiosperms is due to the great variation in kind, size, form, and arrangement of its elements. The most complex angiosperm woods, such as that of oak, may contain vessel elements, tracheids, fiber-tracheids, libriform fibers, axial parenchyma, and rays of different sizes (Figs. 11.3 and 11.15). Some angiosperm woods are, however, less complicated in structure. Many Juglandaceae, for example, contain only fiber-tracheids among the imperforate nonliving cells (Heimsch and Wetmore, 1939). The wood of the vesselless angiosperms (Amborellaceae, Tetracentraceae, Trochodendraceae, Winteraceae) appears so similar to that of conifers that it has at times been erroneously interpreted as conifer wood. Vesselless angiosperm woods can, however, be distinguished from

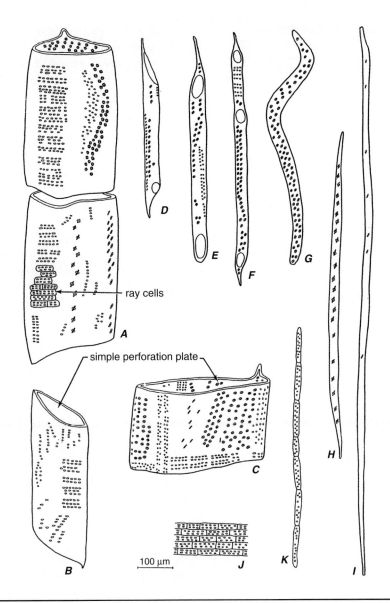

**FIGURE 11.15**

Cell types in secondary xylem as illustrated by dissociated wood elements of *Quercus*, oak. Various pits appear on cell walls. **A–C**, wide vessel elements. **D–F**, narrow vessel elements. **G**, tracheid. **H**, fiber-tracheid. **I**, libriform fiber. **J**, ray parenchyma cells. **K**, axial parenchyma strand. (From Esau, 1977; **A–I**, from photographs in Carpenter, 1952; with permission from SUNY-ESF.)

conifer wood by their tall broad rays (Wheeler et al., 1989).

Because of the complexity of structure of angiosperm woods many characters may be used in their identification (Wheeler et al., 1989; Wheeler and Baas, 1998). Some of the major features are size distribution of vessels in a growth layer (porosity); vessel arrangement and groupings; axial parenchyma arrangement and abundance; presence or absence of septate fibers; presence or absence of storied structure; size and types of rays; types of perforation plates in the vessels; and crystal size, arrangement, and abundance.

## On the Basis of Porosity, Two Main Types of Angiosperm Wood Are Recognized: Diffuse-porous and Ring-porous

The word **porous** is used by the wood anatomist to refer to the appearance of the vessels in transverse sections (Table 11.1). **Diffuse-porous** woods are woods in which the vessels, or **pores**, are rather uniform in size and distribution throughout a growth ring (Figs. 11.1 and 11.5C). In **ring-porous** woods the pores of the earlywood are distinctly larger than those of the latewood, resulting in a ring-like zone in the earlywood and

**TABLE 11.1 ■ Examples of Woods with Different Distributions of Vessels**

**Ring porous**

*Carya pecan* (pecan)
*Castanea dentata* (American chestnut)
*Catalpa speciosa*
*Celtis occidentalis* (hackberry)
*Fraxinus americana* (white ash)
*Gleditsia triacanthos* (honey locust)
*Gymnocladus dioicus* (Kentucky coffee tree)
*Maclura pomifera* (Osage orange)
*Morus rubra* (red mulberry)
*Paulownia tomentosa*
*Quercus* spp. (deciduous oaks)
*Robinia pseudoacacia* (black locust)
*Sassafras albidum*
*Ulmus americana* (American elm)

**Semi-ring porous or semi-diffuse porous**

*Diospyros virginiana* (persimmon)
*Juglans cinerea* (butternut)
*Juglans nigra* (black walnut)
*Lithocarpus densiflora* (tanbark oak)
*Populus deltoides* (cottonwood)
*Prunus serotina* (black cherry)
*Quercus virginiana* (live oak)
*Salix nigra* (black willow)

**Diffuse porous**

*Acer saccharinum* (silver maple)
*Acer saccharum* (sugar maple)
*Aesculus glabra* (buckeye)
*Aesculus hippocastanum* (horse chestnut)
*Alnus rubra* (red alder)
*Betula nigra* (red birch)
*Carpinus caroliniana* (blue beech)
*Cornus florida* (dogwood)
*Fagus grandifolia* (American beech)
*Ilex opaca* (holly)
*Liquidambar styraciflua* (American sweet gum)
*Liriodendron tulipifera* (tulip tree)
*Magnolia grandiflora* (evergreen magnolia)
*Nyssa sylvatica* (black gum)
*Platanus occidentalis* (American plane tree)
*Tilia americana* (basswood)
*Umbellularia californica* (California laurel)

Source: From Esau, 1977.

an abrupt transition between the earlywood and latewood of the same growth ring (Figs. 11.3A and 11.5B). Intergrading patterns occur between the types, and woods showing an intermediate condition between ring-porous and diffuse-porous may be called **semi-ring porous** or **semi-diffuse porous**. Moreover in a given species the distribution of vessels may vary in relation to environmental conditions and may change with increasing age of a tree. In *Populus euphratica*, the only *Populus* species native to Israel, vigorous shoot growth under conditions of ample water supply was associated with wide annual rings and diffuse-porous wood, whereas restricted shoot elongation of trees on dry sites was associated with narrow annual rings and ring-porous wood (Liphschitz and Waisel, 1970; Liphschitz, 1995). In the ring-porous oak, *Quercus ithaburensis,* intensive extension growth resulted in wide rings with diffuse-porous wood, whereas, under restricted extension growth, narrow rings and ring-porous wood were produced (Liphschitz, 1995). Carlquist (1980, 2001) has attempted to deal with such problems by taking into account all known types of cell variation seen within growth rings. He recognizes 15 different kinds of growth rings.

The ring-porous condition appears to be highly specialized and occurs in relatively few woods (Metcalfe and Chalk, 1983), most being species of the north temperate zone. Some wood anatomists consider the earlywood—the so-called pore zone—of ring-porous woods to be an additional tissue without an equivalent in the diffuse-porous woods (Studhalter, 1955), and the latewood to be comparable to the entire growth increment of diffuse-porous species (Chalk, 1936). It has been proposed that ring-porous species originated from diffuse-porous ones (Aloni, 1991; Wheeler and Baas, 1991). According to the limited-growth hypothesis of Aloni (1991), ring-porous species evolved from diffuse-porous species under selective pressures of limiting environments, which resulted in a decreased intensity of vegetative growth. The latter was accompanied by a reduction in auxin levels and an increased sensitivity of the cambium to relatively low auxin stimulation. Lev-Yadun (2000), noting that several species in the woody flora of Israel change porosity according to growth conditions, questioned the sensitivity aspect of the limited-growth hypothesis because it would require the cambium of such a tree to change its sensitivity to auxin as the porosity changes.

The physiological aspects also indicate the specialized nature of ring-porous wood. Ring-porous wood conducts water almost entirely in the outermost growth increment, with over 90% of the water being conducted in the wide, earlywood vessels (Zimmermann, 1983; Ellmore and Ewers, 1985) at peak velocities often 10 times greater than in diffuse-porous species (Huber, 1935). Because of their great widths, the earlywood vessels of ring-porous species are especially vulnerable to embolism formation (Chapter 10), and typically they become nonfunctional during the same year they are formed. Consequently, new earlywood vessels are produced rapidly each year before new leaves emerge (Ellmore and Ewers, 1985; Suzuki et al., 1996; Utsumi et al., 1996). In diffuse-porous species several growth increments are involved with water conduction at the same time, and new vessel formation is initiated after the onset of leaf expansion (Suzuki et al., 1996).

Ring porosity, with the formation of wide vessels early in the growing season, has long been regarded an adaptation to accommodate the high transpiration and flow rates prevalent at that time of year. The narrow, latewood vessels are more important later in the year when water stresses are greater and the wide, early-wood vessels are more likely to become embolized.

Within the main distributional patterns of vessels, minor variations occur in the spatial relation of the pores to each other. A pore is called **solitary** when the vessel is completely surrounded by other types of cells. A group of two or more pores appearing together form a **pore multiple**. This may be a **radial pore multiple**, with pores in a radial file, or a **pore cluster**, with an irregular grouping of pores. Although vessels or vessel groups may appear isolated in transverse sections of wood, in the three-dimensional space the vessels are interconnected in various planes (Fig. 11.16). In some species the vessels are interconnected only within individual growth increments, in others connections occur across the boundaries of growth increments (Braun, 1959; Kitin et al., 2004). According to Zimmermann (1983), vessel groups (vessel multiples) are safer than solitary vessels because they provide alternative paths for the xylem sap to bypass embolisms.

In a number of woody angiosperms the vessels are associated with **vasicentric tracheids,** generally irreg-

ularly shaped tracheids that occur around and adjacent to vessels (Fig. 11.3B; Carlquist, 1992, 2001). Although best known in the ring-porous woods of *Quercus* and *Castanea,* vasicentric tracheids also occur in diffuse-porous woods (e.g., many species of *Shorea* and *Eucalyptus*). They may be regarded as subsidiary conductive cells that take over the role of water transport when many of the vessels embolize at times of great water stress. Probably the safest conductive cells (the ones least likely to cavitate and embolize) found in vessel-containing wood are the vascular tracheids, which resemble narrow vessel elements and are formed at the end of a growth ring (Carlquist, 1992, 2001). Vascular tracheids would provide maximal safety for angiosperms found in regions with severe water stress conditions at the end of a growing season.

### The Distribution of Axial Parenchyma Shows Many Intergrading Patterns

Three general patterns, or distributions, of axial parenchyma can be recognized from transverse sections: apotracheal, paratracheal, and banded (Wheeler et al., 1989). Various combinations of these types may be present in a given wood. In the **apotracheal** type (*apo,* meaning from in Greek, in this instance, independence from) the axial parenchyma are not associated with the vessels, although some random contacts may exist. The apotracheal parenchyma is further divided into: **diffuse**, single parenchymatous strands or pairs of strands scattered among fibers (Fig. 11.17A) and **diffuse-in-aggregates,** parenchyma strands grouped into short discontinuous tangential or oblique lines (Fig. 11.17B). Diffuse apotracheal parenchyma may be **sparse**. In the **paratracheal** type (*para,* meaning beside in Greek), the axial parenchyma are associated with the vessels. The paratracheal parenchyma cells in direct contact with the vessels—the **contact cells**—have numerous prominent pit connections (contact pits) with the vessels. The physiological significance of the paratracheal contact cells will be considered along with that of the ray contact cells, below. The paratracheal parenchyma appears in the following forms: **scanty paratracheal**, occasional parenchyma cells associated with the vessels or an incomplete sheath of parenchyma around the vessels (Fig. 11.17C); **vasicentric**, parenchyma forming complete sheaths around the vessel (Fig. 11.17D); **aliform**, parenchyma surrounding or to one side of the vessel and with lateral extensions (Fig. 11.17E); and **confluent**, coalesced vasicentric or aliform parenchyma forming irregular tangential or diagonal bands (Fig. 11.17F). **Banded parenchyma** may be mainly independent of the vessels (Fig. 11.17G; apotracheal), associated with the vessels (Fig. 11.17H; paratracheal), or both. They may be straight, wavy, diagonal, continuous or discontinuous, and one to several cells wide. Bands over three cells wide generally are visible

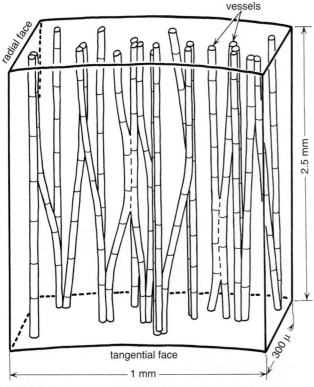

**FIGURE 11.16**

Network of vessels in *Populus* wood with lateral connections between vessels in both radial and tangential planes. The horizontal dimensions are represented on a larger scale than the vertical. The delimitations of vessel elements are approximate. (Adapted from Braun, 1959. © 1959, with permission from Elsevier.)

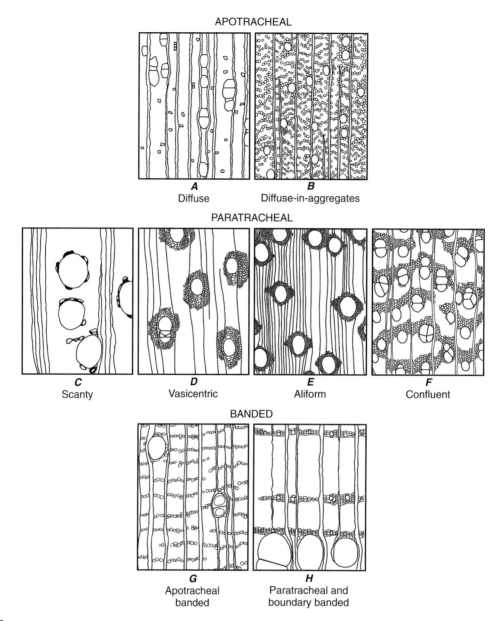

APOTRACHEAL

*A*
Diffuse

*B*
Diffuse-in-aggregates

PARATRACHEAL

*C*
Scanty

*D*
Vasicentric

*E*
Aliform

*F*
Confluent

BANDED

*G*
Apotracheal
banded

*H*
Paratracheal and
boundary banded

**FIGURE 11.17**

Distribution of axial parenchyma in wood of **A**, *Alnus glutinosa*; **B**, *Agonandra brasiliensis*; **C**, *Dillenia pulcherrima*; **D**, *Piptadeniastrum africanum*; **E**, *Microberlinia brazzavillensis*; **F**, *Peltogyne confertifolora*; **G**, *Carya pecan*; **H**, *Fraxinus* sp. All transverse sections. (**A–F**, from photographs in Wheeler et al., 1989; **G**, **H**, from Figure 9.8C, D, in Esau, 1977.)

to the unaided eye. Parenchyma bands at the ends of growth rings are called **marginal bands** (Fig. 11.5C) and may be restricted either to the end of a ring (**terminal parenchyma**) or to the beginning of one (**initial parenchyma**). According to Carlquist (2001), terminal parenchyma is the predominant form. Axial parenchyma may be absent or rare in a given wood. From the evolutionary aspect the apotracheal and diffuse patterns are primitive.

## The Rays of Angiosperms Typically Contain Only Parenchyma Cells

The ray parenchyma cells of the angiosperms vary in shape, but two fundamental forms may be distinguished: procumbent and upright (Fig. 11.18). **Procumbent ray cells** have their longest axes oriented radially and **upright ray cells** have their longest axes oriented vertically, or upright. Ray cells that appear square in radial sections of

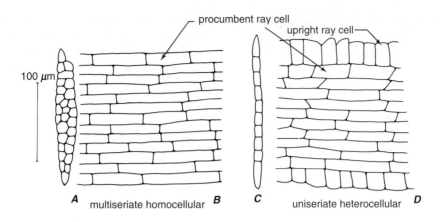

**FIGURE 11.18**

Two types of rays as seen in tangential (**A**, **C**) and radial (**B**, **D**) sections. **A**, **B**, *Acer saccharum*; **C**, **D**, *Fagus grandifolia*. (From Esau, 1977.)

the wood are called **square ray cells**, a modification of the upright. The two main types of ray parenchyma cells are often combined in the same ray, the upright cells appearing at the upper and lower margins of the ray. In the angiosperms, rays composed of one kind of cell are called **homocellular** (Fig. 11.18A, B), and those containing procumbent and upright cells, **heterocellular** (Fig. 11.18C, D).

In contrast to the predominantly uniseriate rays of conifers, those of the angiosperms may be one to many cells wide (Fig. 11.3C); that is, they may be **uniseriate** or **multiseriate** (multiseriate rays two cells wide commonly are called **biseriate rays**; Fig. 11.1), and range in height from one to many cells (from a few mm to 3 cm or more). The multiseriate rays frequently have uniseriate margins. Several individual rays may be so closely associated with one another that they appear to be one large ray. Such groups are called **aggregate rays** (e.g., many species of *Alnus*, *Carpinus*, *Corylus*, *Casuarina*, and some evergreen species of *Quercus*). Overall, the rays of angiosperms average about 17% of the volume of the wood, compared with the about 8% for conifer wood. Constituting such a large portion of the wood, the rays of angiosperms contribute substantially to the radial strength of the wood (Burgert and Eckstein, 2001).

The appearance of rays in radial and tangential sections can be used as the basis for their classification. Radial sections should be used to determine the cellular composition of rays and tangential sections to determine the width and height of rays. Individual rays may be homocellular or heterocellular. The entire ray system of a wood may consist of either homocellular or heterocellular rays or of combinations of the two types of rays.

The different ray combinations have a phylogenetic significance. The primitive ray tissue may be exemplified by that of the Winteraceae (*Drimys*). The rays are of two kinds: one homocellular—uniseriate composed of upright cells; the other heterocellular—multiseriate composed of radially elongated or nearly isodiametric cells in the multiseriate part and upright cells in the uniseriate marginal parts. Both kinds of ray are many cells in height. From such primitive ray structure other ray systems, more specialized, have been derived. For example, multiseriate rays may be eliminated (*Aesculus hippocastanum*) or increased in size (*Quercus*), or both multiseriate and uniseriate rays may be decreased in size (*Fraxinus*).

The evolution of rays strikingly illustrates the maxim that phylogenetic changes depend on successively modified ontogenies. In a given wood the specialized ray structure may appear gradually. The earlier growth layers may have more primitive ray structure than the latter because the vascular cambium commonly undergoes successive changes before it begins to produce a ray pattern of a more specialized type. In some specialized species with short fusiform initials the wood may be either entirely rayless or may develop rays only belatedly (Carlquist, 2001). Raylessness is an indicator of paedomorphosis. It results from a delay of horizontal subdivision of cambial initials that would bring about a distinction between fusiform cambial initials and ray initials. In totally rayless species virtually no such divisions occur for the duration of cambial activity, and most, perhaps all, are small shrubs or herbs.

The ray cells share some functions with the axial parenchyma cells and are also concerned with radial transport of substances between the xylem and the phloem (van der Schoot, 1989; van Bel, 1990a, b;

Lev-Yadun, 1995; Keunecke et al., 1997; Sauter, 2000; Chaffey and Barlow, 2001). As mentioned previously, ray and axial parenchyma cells form an extensive, three-dimensional symplastic continuum that permeates the vascular tissues and is continuous via the rays from xylem to phloem. The cytoskeleton (microtubules and actin filaments) has been implicated in the transport of substances within these cells and, in association with the acto-myosin of the plasmodesmata in their common walls, with intercellular transport (Chaffey and Barlow, 2001). Ray cells—both procumbent and upright—that are connected through pits with tracheary elements, like their counterparts among the paratracheal parenchyma cells, function as contact cells controlling the exchange of solutes (minerals, carbohydrates, and organic nitrogenous substances) between the storage parenchyma and the vessels. Typically, contact cells do not function as storage cells, although small quantities of starch may be found in some contact cells at certain times of the year (Czaninski, 1968; Braun, 1970; Sauter, 1972; Sauter et al., 1973; Catesson and Moreau, 1985). It is the paratracheal parenchyma cells and ray cells that have no contact with the vessels (**isolation cells**) that function as storage cells. During the spring mobilization of starch in deciduous trees of the temperate zone, the contact cells secrete sugars into the vessels for rapid transport to the buds. This process may also play a role in the refilling with water those vessels that had accumulated gases during winter (Améglio et al., 2004).

During the times of sugar secretion—most notably shortly before and during the period of bud swell—the contact cells exhibit high levels of respiratory activity and at the contact pits high levels of phosphatase activity. The secretion into and uptake from the vessels of solutes by the contact cells apparently is performed through substrate/proton cotransport mechanisms (van Bel and van Erven, 1979; Bonnemain and Fromard, 1987; Fromard et al., 1995). The contact cells therefore are analogous to the companion cells that serve in the sugar exchange with the sieve elements in the phloem (Chapter 13; Czaninski, 1987). They differ from companion cells, however, in their presence of lignified cell walls and of a pecto-cellulosic protective layer, which is involved with tylose formation (Chapter 10). Several functions have been suggested for the protective layer other than tylose formation (Schaffer and Wisniewski, 1989; van Bel and van der Schoot, 1988; Wisniewski and Davis, 1989). The one most relevant to the present discussion is that the protective layer is a means of maintaining apoplastic continuity along the entire surface of the protoplast, bringing the entire plasma membrane surface, not just the part of it in contact with the porous pit membrane, into contact with the apoplast (Barnett et al., 1993). The contact cells also differ from companion cells in their lack of plasmodesmata at the contact pits; companion cells have numerous pore-plasmodesmata connections in their common walls with the sieve elements (Chapter 13). The tangential walls of the ray cells contain numerous plasmodesmata, indicating that radial transport of sucrose and other metabolites in the rays is symplastic (Sauter and Kloth, 1986; Krabel, 2000; Chaffey and Barlow, 2001).

### Intercellular Spaces Similar to the Resin Ducts of Gymnosperms Occur in Angiosperm Woods

The intercellular spaces or ducts in angiosperm woods contain secondary plant products such as gums and resins (Chapter 17). They occur in both the axial and the radial systems (Wheeler et al., 1989) and vary in extent; some are more appropriately called intercellular cavities. The ducts and cavities may be schizogenous, but those formed in response to injury—**traumatic ducts** and **cavities**—commonly are lysigenous.

## ■ SOME ASPECTS OF SECONDARY XYLEM DEVELOPMENT

The derivatives that arise on the inner face of the cambium through tangential divisions of the cambial initials undergo complex changes during their development into the various elements of the xylem. The basic pattern of the secondary xylem, with its axial and radial systems, is determined by the structure of the cambium itself, since the cambium is composed of fusiform and ray initials. Also all of the changes in the relative proportions between these two systems—for example, the addition or the elimination of rays (Chapter 12)—originate in the cambium.

The derivatives of the ray initials generally undergo relatively little change during their differentiation. Ray cells enlarge radially as they emerge from the cambium, but the distinction between the upright and the procumbent cells is apparent in the cambium. Most ray cells remain parenchymatous, and although some develop secondary walls, their contents do not change much. Apparent exceptions among the angiosperms are **perforated ray cells**, cells within the rays that differentiate as vessel elements and connect axial vessels across the rays (Fig. 11.19; Carlquist, 1988; Nagai et al., 1994; Otegui, 1994; Machado and Angyalossy-Alfonso, 1995; Eom and Chung, 1996), and **radial fibers** such as those found in aggregate rays of *Quercus calliprinos* (Lev-Yadun, 1994b). A profound change also occurs in the ray tracheids of conifers, for they develop secondary walls with bordered pits and lose their protoplasts during maturation.

The ontogenetic changes that occur in the axial system vary with the type of cell, and each cell type has its own characteristic rate and duration for the differentiation processes. Typically the vessel elements and the cells in contact with them mature more rapidly than other cells in the developing xylem (Ridoutt and Sands,

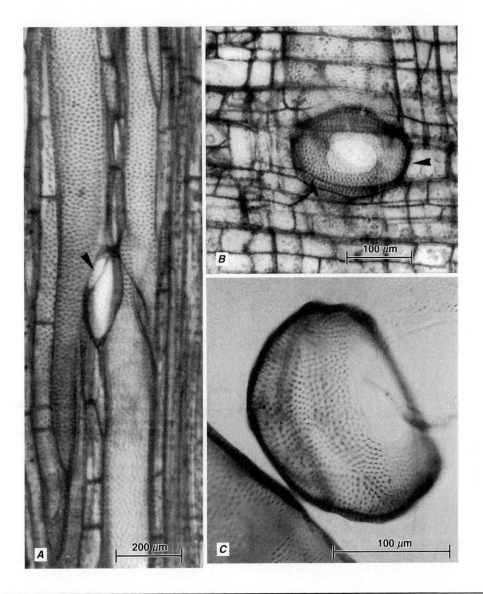

## FIGURE 11.19

Perforated ray cells, with simple perforations, in root wood of *Styrax camporium*. **A**, tangential section showing perforated ray cell (dart) interconnecting two vertical vessels. **B**, radial section, showing ray cell (dart) with perforation on radial wall. **C**, perforated ray cell from macerated wood. (From Machado et al., 1997.)

1994; Murakami et al., 1999; Kitin et al., 2003). The fibers take longer than other cell types, particularly the vessel elements, to mature (Doley and Leyton, 1968; Ridoutt and Sands, 1994; Murakami et al., 1999; Chaffey et al., 2002). Developing vessel elements elongate slightly, if at all, but they expand laterally, often so strongly that their ultimate width exceeds their height. Short, wide vessel elements are characteristic of highly specialized xylem. In many species of angiosperms the vessel elements expand in their median parts but not at the ends, which overlap those of the vertically adjacent elements. These ends are ultimately not occupied by the perforation and appear like elongate wall processes, *tails*, with or without pits.

Expansion of the vessel elements affects the arrangement and the shape of adjacent cells. These cells become crowded out of their original position and cease to reflect the radial seriation present in the cambial zone. The rays, too, may be deflected from their original positions. The cells in the immediate vicinity of an expanding vessel enlarge parallel with the surface of the vessel and assume a flattened appearance. But often these cells do not keep pace with the increase of the circumference of the vessel and become partly or completely separated from each other. As a result the expanding vessel element comes in contact with new cells. The expansion of a vessel element can be pictured as a phenomenon involving both coordinated and intrusive growth. As long as the cells next to the vessel element expand in unison with it, the common walls of the various cells undergo coordinated growth. During separation of adjacent cells, the vessel-element wall intrudes between the walls of other cells. When the future vessel element begins to expand in the xylem mother cell zone, production of cells ceases in one or more rows adjacent to the row containing the expanding cell. Divisions are resumed in these rows after the vessel has expanded and the cambium has been displaced outward.

The separation of the cells located next to an expanding vessel causes the development of cells having odd, irregular shapes. Some remain partially attached to each other and, as the vessel element continues to enlarge, these connections extend into long tubular structures. The parenchyma cells and the tracheids that are thus affected by developmental adjustments have received the names **disjunctive parenchyma cells** (Fig. 11.20) and **disjunctive tracheids**, respectively. These are modified growth forms of the xylem parenchyma cells and the tracheids of the axial system.

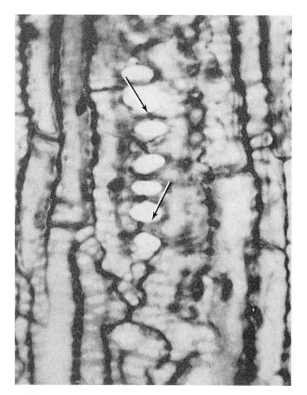

**FIGURE 11.20**

Longitudinal section of *Cucurbita* xylem showing the result of tearing apart of parenchyma cells that occurred near an expanding vessel. Arrows point to tubular structures connecting the disjunctive parenchyma cells. (×600. From Esau and Hewitt, 1940. *Hilgardia* 13 (5), 229–244. © 1940 Regents, University of California.)

In contrast to the vessel elements, the tracheids and fibers undergo relatively little width increase but often elongate much during differentiation. The degree of elongation of these elements in the different groups of plants varies widely. In the conifers, for example, the fusiform initials themselves are very long, and their derivatives elongate only slightly. In the angiosperms, on the contrary, the tracheids and the fibers become considerably longer than the meristematic cells. If the xylem contains tracheids, fiber-tracheids, and libriform fibers, the libriform fibers elongate the most, although the tracheids attain the largest volume because of their greater width. The elongation occurs through apical intrusive growth. In the extreme storied woods there may be little or no elongation of any kind of element (Record, 1934).

Woods containing no vessels retain a rather symmetric arrangement of cells, because in the absence of strongly expanding cells the original radial seriation characteristic of the cambial region is not much disturbed. There is some change in alignment resulting from apical intrusive growth of the axial tracheids.

Vessel elements, tracheids, and fiber-tracheids develop secondary walls and the end walls of the vessel elements become perforated. Ultimately the protoplasts disintegrate in those cells that are nonliving at maturity.

The fusiform meristematic cells that differentiate into the axial parenchyma typically do not elongate. If a parenchyma strand is formed, the fusiform cell divides transversely. No such divisions occur during the development of a fusiform parenchyma cell. In some plants the parenchyma cells develop secondary walls but do not die until the heartwood is formed. The parenchyma cells associated with resin and gum ducts in the axial system arise like axial parenchyma cells by transverse divisions of fusiform cells.

During development, each cell of the xylem must receive information about its position within the tissue and express the appropriate genes. The principal hormonal signal involved with the control of cambial activity and vascular development is auxin (IAA) (Little and Pharis, 1995). The apparent role of auxin in tracheary element differentiation, the transition from earlywood to latewood, and reaction wood formation has already been considered. In the intact plant the polar flow of auxin from expanding buds and young, growing leaves is essential for maintaining the vascular cambium and initiating the spatially organized patterns of vascular tissue (Aloni, 1987). Apparently not all of the auxin involved in secondary growth is derived from the growing shoots. The differentiating vascular tissues, and specifically xylem, appear to be important sources of auxin that maintain cambial activity after its initial reactivation under the influence of expanding buds (Sheldrake, 1971). Whereas auxin itself induces vessel elements, gibberellin, in the presence of auxin, may be the signal for fiber differentiation (Chapter 8; Aloni, 1979; Roberts et al., 1988).

It has been proposed that radial diffusion of the polarly transported auxin creates an auxin gradient across the cambial zone and its derivatives, and that this gradient establishes a positional signaling system from which the cambial derivatives interpret their radial position and, hence, express their genes (Sundberg et al., 2000; Mellerowicz et al., 2001; and literature cited therein). A steep concentration gradient of IAA has in fact been demonstrated across developing xylem and phloem in *Pinus sylvestris* (Uggla et al., 1996) and hybrid aspen (*Populus tremula* × *P. tremuloides*) (Tuominen et al., 1997). However, it is quite clear that the auxin gradient alone does not provide enough information to position either xylem or phloem mother cells or the cambial initials. Steep concentration gradients of soluble carbohydrates also occur across the cambium (Uggla et al., 2001). As noted by Mellerowicz et al. (2001), the presence of such gradients, together with accumulating evidence for the presence of sugar sensing in plants (Sheen et al., 1999), provide substantial support for the concept that auxin/sucrose ratios are determining factors in xylem and phloem differentiation (Warren Wilson and Warren Wilson, 1984).

A radial signal flow, which is independent of the axial flow, has also been invoked in the regulation of ray development (Lev-Yadun, 1994a; Lev-Yadun and Aloni, 1995). This signal flow is envisaged as occurring bidirectionally, with ethylene originating in the xylem flowing outward and controlling both the initiation of new rays and the enlargement of existing ones, and auxin flowing inward from the phloem being involved in the induction of vascular elements (ray tracheids, perforated ray cells) and fibers. The radial flow of ethylene would "disturb" radial auxin transport, however, and limit the formation of vascular elements and fibers in the generally parenchymatous rays (Lev-Yadun, 2000).

A great deal of information is needed before we will understand the complexity of the phenomenon of annual growth and the determination of the different cell types in the vascular tissues. Undoubtedly other growth regulators are involved and the activity of these substances is modified by nutritional conditions and the availability of water.

# IDENTIFICATION OF WOOD

The use of wood for purposes of identification requires a very sound knowledge of wood structure and of factors modifying that structure. The search for diagnostic features is best based on an examination of collections from more than one tree of the same species made with proper attention to the location of the sample on the tree. The wood acquires its mature character not at the beginning of cambial activity but in the later growth increments. That is because the wood produced during the early life of a part of the tree undergoes a progressive increase in dimensions and corresponding changes in form, structure, and disposition of cells in successive growth layers (Rendle, 1960). This **juvenile wood** is produced in the active crown region of the tree and is associated with the prolonged influence of the apical meristems on the vascular cambium. As the crown moves upward with continued growth, the cambium near the base of the tree becomes less influenced by the elongating crown region and begins to produce **mature wood**. With continued upward movement of the juvenile wood-producing crown, the production of mature wood progresses upward. Thus the wood of a twig would be of a different ontogenetic age than that of a trunk of the same tree. Furthermore, in certain sites, the wood has reaction wood properties that deviate more or less strongly from features considered to be typical of the taxon in question. Adverse or unusual environmental conditions and improper methods of preparation of samples for microscopy also may obscure the diagnostic features.

A further complicating aspect of wood identification is that the anatomical characteristics of woods are often less differentiated than the external features of the taxa involved. Although woods of large taxa differ considerably from one another, within groups of closely related taxa, such as species, or even genera, the wood may be so uniform that no consistent differences are detectable. Under such circumstances it is imperative to use a combination of gross, or macroscopic, and microscopic characters of woods, as well as odor and taste.

Some of the gross features of wood are color, grain, texture, and figure. **Color** in wood is variable both between different kinds of wood and within a species. The color of heartwood can be important in identifying a particular wood.

**Grain** in wood refers to the direction of alignment of the axial components—fibers, tracheids, vessel elements, and parenchyma cells—when considered en masse. For example, when all the axial components are oriented more or less parallel to the longitudinal axis of the trunk, the grain is said to be ***straight***. The term ***spiral grain*** is applied to a spiral arrangement of elements in a log or trunk, which has a twisted appearance after the bark has been removed (Fig. 11.21). (It has been suggested that spiral grain is an adaptation of trees to withstand stem breakage caused by wind-induced torsion; Skatter and Kucera, 1997.) If the orientation of the spiral is reversed at more or less regular intervals along a single radius, the grain is said to be ***interlocked***. The alignment of the axial components reflects the alignment of the cambial (fusiform) initials that gave rise to them (Chapter 12).

**FIGURE 11.21**

Trunk of a dead white oak (*Quercus alba*) tree from which the bark has fallen, revealing the spiral grain of the wood.

**Texture** of wood refers to the relative size and degree of size variation of elements within the growth rings. The texture of woods with wide bands of large vessels and broad rays, as in some ring-porous woods, can be described as ***coarse***, and that of woods with small vessels and narrow rays as ***fine***. Woods in which there is no perceptible difference between the earlywood and latewood can be described as having ***even*** texture, whereas those with distinct differences between the earlywood and latewood of a growth ring can be described as ***uneven***.

**Figure** refers to the patterns found on the longitudinal surfaces of wood. It depends on grain and texture and on the orientation of the surface that results from

sawing. In a restricted sense the term "figure" is used to refer to the more decorative woods, such as bird's-eye maple, prized in the furniture and cabinet-making industries.

For references on wood anatomical identification guides see Schweingruber and Bosshard (1978) and Schweingruber (1990), for Europe; Meylan and Butterfield (1978), for New Zealand; Panshin and de Zeeuw (1980) for North America; and Fahn et al. (1986) for Israel and adjacent regions. In addition, see Wheeler and Baas (1998), the IAWA List of Microscopic Features for Hardwood Identification (Wheeler et al., 1989), and the IAWA List of Microscopic Features for Softwood Identification (Richter et al., 2004).

## ❚ REFERENCES ❚

ALONI, R. 1979. Role of auxin and gibberellin in differentiation of primary phloem fibers. *Plant Physiol.* 63, 609–614.

ALONI, R. 1987. The induction of vascular tissues by auxin. In: *Plant Hormones and Their Role in Plant Growth and Development*, pp. 363–374, P. J. Davies, ed. Martinus Nijhoff, Dordrecht.

ALONI, R. 1991. Wood formation in deciduous hardwood trees. In: *Physiology of Trees*, pp. 175–197, A. S. Raghavendra, ed. Wiley, New York.

ALVES, E. S., and V. ANGYALOSSY-ALFONSO. 2000. Ecological trends in the wood anatomy of some Brazilian species. I. Growth rings and vessels. *IAWA J.* 21, 3–30.

AMÉGLIO, T., M. DECOURTEIX, G. ALVES, V. VALENTIN, S. SAKR, J.-L. JULIEN, G. PETEL, A. GUILLIOT, and A. LACOINTE. 2004. Temperature effects on xylem sap osmolarity in walnut trees: Evidence for a vitalistic model of winter embolism repair. *Tree Physiol.* 24, 785–793.

ASH, J. 1983. Tree rings in tropical *Callitris macleayana*. F. Muell. *Aust. J. Bot.* 31, 277–281.

BABA, K.-I., K. ADACHI, T. TAKE, T. YOKOYAMA, T. ITOH, and T. NAKAMURA. 1995. Induction of tension wood in GA$_3$-treated branches of the weeping type of Japanese cherry, *Prunus spachiana*. *Plant Cell Physiol.* 36, 983–988.

BAILEY, I. W. 1920. The cambium and its derivative tissues. II. Size variations of cambial initials in gymnosperms and angiosperms. *Am. J. Bot.* 7, 355–367.

BAILEY, I. W. 1944. The development of vessels in angiosperms and its significance in morphological research. *Am. J. Bot.* 31, 421–428.

BAILEY, I. W., and A. F. FAULL. 1934. The cambium and its derivative tissues. IX. Structural variability in the redwood *Sequoia sempervirens*, and its significance in the identification of the fossil woods. *J. Arnold Arbor.* 15, 233–254.

BAILEY, I. W., and W. W. TUPPER. 1918. Size variation in tracheary cells. I. A comparison between the secondary xylems of vascular cryptogams, gymnosperms and angiosperms. *Am. Acad. Arts Sci. Proc.* 54, 149–204.

BANNAN, M. W. 1936. Vertical resin ducts in the secondary wood of the Abietineae. *New Phytol.* 35, 11–46.

BARGHOORN, E. S., JR. 1940. The ontogenetic development and phylogenetic specialization of rays in the xylem of dicotyledons. I. The primitive ray structure. *Am. J. Bot.* 27, 918–928.

BARGHOORN, E. S., JR. 1941. The ontogenetic development and phylogenetic specialization of rays in the xylem of dicotyledons. II. Modification of the multiseriate and uniseriate rays. *Am. J. Bot.* 28, 273–282.

BARNETT, J. R., P. COOPER, and L. J. BONNER. 1993. The protective layer as an extension of the apoplast. *IAWA J.* 14, 163–171.

BERITOGNOLO, I., E. MAGEL, A. ABDEL-LATIF, J.-P. CHARPENTIER, C. JAY-ALLEMAND, and C. BRETON. 2002. Expression of genes encoding chalcone synthase, flavanone 3-hydroxylase and dihydroflavonal 4-reductase correlates with flavanol accumulation during heartwood formation in *Juglans nigra*. *Tree Physiol.* 22, 291–300.

BONINSEGNA, J. A., R. VILLALBA, L. AMARILLA, and J. OCAMPO. 1989. Studies on tree rings, growth rates and age-size relationships of tropical tree species in Misiones, Argentina. *IAWA Bull.* n.s. 10, 161–169.

BONNEMAIN, J.-L., and L. FROMARD. 1987. Physiologie comparée des cellules compagnes du phloème et des cellules associées aux vaisseaux. *Bull. Soc. Bot. Fr. Actual. Bot.* 134 (3/4), 27–37.

BOYD, J. D. 1977. Basic cause of differentiation of tension wood and compression wood. *Aust. For. Res.* 7, 121–143.

BRAUN, H. J. 1959. Die Vernetzung der Gefässe bei *Populus*. *Z. Bot.* 47, 421–434.

BRAUN, H. J. 1970. *Funktionelle Histologie der sekundären Sprossachse. I. Das Holz. Handbuch der Pflanzenanatomie*, Band 9, Teil 1. Gebrüder Borntraeger, Berlin.

BRAUN, H. J. 1984. The significance of the accessory tissues of the hydrosystem for osmotic water shifting as the second principle of water ascent, with some thoughts concerning the evolution of trees. *IAWA Bull.* n.s. 5, 275–294.

BURGERT, I., and D. ECKSTEIN. 2001. The tensile strength of isolated wood rays of beech (*Fagus sylvatica* L.) and its significance for the biomechanics of living trees. *Trees* 15, 168–170.

BURKART, L. F., and J. CANO-CAPRI. 1974. Tension wood in southern red oak *Quercus falcata* Michx. *Univ. Tex. For. Papers* 25, 1–4.

BUTTERFIELD, B. G., and B. A. MEYLAN. 1980. *Three-dimensional Structure of Wood: An Ultrastructural Approach*, 2nd ed. Chapman and Hall, London.

CALLADO, C. H., S. J. DA SILVA NETO, F. R. SCARANO, and C. G. COSTA. 2001. Periodicity of growth rings in some flood-prone trees of the Atlantic rain forest in Rio de Janeiro, Brazil. *Trees* 15, 492–497.

CARLQUIST, S. 1962. A theory of paedomorphosis in dicotyledonous woods. *Phytomorphology* 12, 30–45.

CARLQUIST, S. 1980. Further concepts in ecological wood anatomy, with comments on recent work in wood anatomy and evolution. *Aliso* 9, 499–553.

CARLQUIST, S. J. 1988. Comparative wood anatomy: systematic, ecological, and evolutionary aspects of dicotyledon wood. Springer-Verlag, Berlin, Heidelberg, New York.

CARLQUIST, S. 1992. Wood anatomy of Lamiaceae. A survey, with comments on vascular and vasicentric tracheids. *Aliso* 13, 309–338.

CARLQUIST, S. 2001. *Comparative Wood Anatomy: Systematic, Ecological, and Evolutionary Aspects of Dicotyledon Wood*, rev. 2nd ed. Springer, Berlin.

CARPENTER, C. H. 1952. *382 Photomicrographs of 91 Papermaking Fibers*, rev. ed. Tech. Publ. 74, State University of New York, College of Forestry, Syracuse.

CASPERSON, G. 1960. Über die Bildung von Zellwänden bei Laubhölzern I. Mitt. Festellung der Kambiumaktivität durch Erzeugen von Reaktionsholz. *Ber. Dtsch. Bot. Ges.* 73, 349–357.

CASPERSON, G. 1965. Zur Kambiumphysiologie von *Aesculus hippocastanum* L. *Flora* 155, 515–543.

CATESSON, A. M., and M. MOREAU. 1985. Secretory activities in vessel contact cells. *Isr. J. Bot.* 34, 157–165.

CHAFFEY, N., and P. BARLOW. 2001. The cytoskeleton facilitates a three-dimensional symplasmic continuum in the long-lived ray and axial parenchyma cells of angiosperm trees. *Planta* 213, 811–823.

CHAFFEY, N., E. CHOLEWA, S. REGAN, and B. SUNDBERG. 2002. Secondary xylem development in *Arabidopsis*: A model for wood formation. *Physiol. Plant.* 114, 594–600.

CHALK, L. 1936. A note on the meaning of the terms early wood and late wood. *Proc. Leeds Philos. Lit. Soc., Sci. Sect.*, 3, 325–326.

CHAPMAN, G. P., ed. 1997. *The Bamboos*. Academic Press, San Diego.

CHATTAWAY, M. M. 1949. The development of tyloses and secretion of gum in heartwood formation. *Aust. J. Sci. Res.; Ser. B, Biol. Sci.* 2, 227–240.

CHATTAWAY, M. M. 1951. The development of horizontal canals in rays. *Aust. J. Sci. Res., Ser. B, Biol. Sci.* 4, 1–11.

CZANINSKI, Y. 1968. Étude du parenchyme ligneux du Robinier (parenchyme à réserves et cellules associées aux vaisseau) au cours du cycle annuel. *J. Microscopie* 7, 145–164.

CZANINSKI, Y. 1987. Généralité et diversité des cellules associées aux éléments conducteurs. *Bull. Soc. Bot. Fr. Actual. Bot.* 134 (3/4), 19–26.

DADSWELL, H. E., and W. E. HILLIS. 1962. Wood. In: *Wood Extractives and Their Significance to the Pulp and Paper Industries*, pp. 3–55, W. E. Hillis, ed. Academic Press, New York.

DÉTIENNE, P. 1989. Appearance and periodicity of growth rings in some tropical woods. *IAWA Bull.* n.s. 10, 123–132.

DOLAN, L. 1997. The role of ethylene in the development of plant form. *J. Exp. Bot.* 48, 201–210.

DOLEY, D., and L. LEYTON. 1968. Effects of growth regulating substances and water potential on the development of secondary xylem in *Fraxinus*. *New Phytol.* 67, 579–594.

DU, S., H. UNO, and F. YAMAMOTO. 2004. Roles of auxin and gibberellin in gravity-induced tension wood formation in *Aesculus turbinata* seedlings. *IAWA J.* 25, 337–347.

EKLUND, L., C. H. A. LITTLE, and R. T. RIDING. 1998. Concentrations of oxygen and indole-3-acetic acid in the cambial region during latewood formation and dormancy development in *Picea abies* stems. *J. Exp. Bot.* 49, 205–211.

ELLMORE, G. S., and F. W. EWERS. 1985. Hydraulic conductivity in trunk xylem of elm, *Ulmus americana. IAWA Bull.* n.s. 6, 303–307.

EOM, Y. G., and Y. J. CHUNG. 1996. Perforated ray cells in Korean Caprifoliaceae. *IAWA J.* 17, 37–43.

ESAU, K. 1943. Origin and development of primary vascular tissues in seed plants. *Bot. Rev.* 9, 125–206.

ESAU, K. 1977. *Anatomy of Seed Plants*, 2nd ed. Wiley, New York.

ESAU, K., and WM. B. HEWITT. 1940. Structure of end walls in differentiating vessels. *Hilgardia* 13, 229–244.

FAHN, A., and E. ZAMSKI. 1970. The influence of pressure, wind, wounding and growth substances on the rate of resin duct formation in *Pinus halepensis* wood. *Isr. J. Bot.* 19, 429–446.

FAHN, A., E. WERKER, and P. BAAS. 1986. *Wood Anatomy and Identification of Trees and Shrubs from Israel and Adjacent Regions.* Israel Academy of Sciences and Humanities, Jerusalem.

FAYLE, D. C. F. 1968. *Radial Growth in Tree Roots.* University of Toronto Faculty of Forestry. Tech. Rep. 9.

FISHER, J. B., and J. W. STEVENSON. 1981. Occurrence of reaction wood in branches of dicotyledons and its role in tree architecture. *Bot. Gaz.* 142, 82–95.

FORSAITH, C. C. 1926. *The Technology of New York State Timbers.* New York State College of Forestry, Syracuse University Tech. Publ. 18.

FROMARD, L., V. BABIN, P. FLEURAT-LESSARD, J. C. FROMONT, R. SERRANO, and J. L. BONNEMAIN. 1995. Control of vascular sap pH by the vessel-associated cells in woody species (physiological and immunological studies). *Plant Physiol.* 108, 913–918.

FUJII, T., Y. SUZUKI, and N. KURODA. 1997. Bordered pit aspiration in the wood of *Cryptomeria japonica* in relation to air permeability. *IAWA J.* 18, 69–76.

GOURLAY, I. D. 1995. Growth ring characteristics of some African *Acacia* species. *J. Trop. Ecol.* 11, 121–140.

HARIHARAN, Y., and K. V. KRISHNAMURTHY. 1995. A cytochemical study of cambium and its xylary derivatives on the normal and tension wood sides of the stems of *Prosopis juliflora* (S. W.) DC. *Beitr. Biol. Pflanz.* 69, 459–472.

HARRIS, J. M. 1954. Heartwood formation in *Pinus radiata* (D. Don.). *New Phytol.* 53, 517–524.

HEIMSCH, C., JR., and R. H. WETMORE. 1939. The significance of wood anatomy in the taxonomy of the Juglandaceae. *Am. J. Bot.* 26, 651–660.

HEJNOWICZ, Z. 1997. Graviresponses in herbs and trees: A major role for the redistribution of tissue and growth stresses. *Planta* 203 (suppl. 1), S136–S146.

HELLGREN, J. M., K. OLOFSSON, and B. SUNDBERG. 2004. Patterns of auxin distribution during gravitational induction of reaction wood in poplar and pine. *Plant Physiol.* 135, 212–220.

HIGUCHI, T. 1997. *Biochemistry and Molecular Biology of Wood.* Springer-Verlag, Berlin.

HILLIS, W. E. 1987. *Heartwood and Tree Exudates.* Springer-Verlag, Berlin.

HÖSTER, H.-R., and W. LIESE. 1966. Über das Vorkommen von Reaktionsgewebe in Wurzeln und Ästen der Dikotyledonen. *Holzforschung* 20, 80–90.

HUANG, Y. S., S. S. CHEN, T. P. LIN, and Y. S. CHEN. 2001. Growth stress distribution in leaning trunks of *Cryptomeria japonica. Tree Physiol.* 21, 261–266.

HÜBER, B. 1935. Die physiologische Bedeutung der Ring- und Zerstreutporigkeit. *Ber. Dtsch. Bot. Ges.* 53, 711–719.

ISEBRANDS, J. G., and D. W. BENSEND. 1972. Incidence and structure of gelatinous fibers within rapid-growing eastern cottonwood. *Wood Fiber* 4, 61–71.

JENKINS, P. A., and K. R. SHEPHERD. 1974. Seasonal changes in levels of indoleacetic acid and abscisic acid in stem tissues of *Pinus radiata. N. Z. J. For. Sci.* 4, 511–519.

JOUREZ, B. 1997. Le bois de tension. 1. Définition et distribution das l'arbre. *Biotechnol. Agron. Soc. Environ.* 1, 100–112.

KANG, K. D., and W. Y. SOH. 1992. Differentiation of reaction tissues in the first internode of *Acer saccharinum* L. seedling positioned horizontally. *Korean J. Bot. (Singmul Hakhoe chi)* 35, 211–217.

KEUNECKE, M., J. U. SUTTER, B. SATTELMACHER, and U. P. HANSEN. 1997. Isolation and patch clamp measurements of xylem contact cells for the study of their role in the exchange between apoplast and symplast of leaves. *Plant Soil* 196, 239–244.

KITIN, P., Y. SANO, and R. FUNADA. 2003. Three-dimensional imaging and analysis of differentiating secondary xylem by confocal microscopy. *IAWA J.* 24, 211–222.

KITIN, P. B., T. FUJII, H. ABE, and R. FUNADA. 2004. Anatomy of the vessel network within and between tree rings of *Fraxinus lanuginosa* (Oleaceae). *Am. J. Bot.* 91, 779–788.

KOZLOWSKI, T. T., and S. G. PALLARDY. 1997. *Growth Control in Woody Plants.* Academic Press, San Diego.

KRABEL, D. 2000. Influence of sucrose on cambial activity. In: *Cell and Molecular Biology of Wood Formation*, pp. 113–125, R. A. Savidge, J. R. Barnett, and R. Napier, eds. BIOS Scientific, Oxford.

KRAHMER, R. L., and W. A. CÔTÉ JR. 1963. Changes in coniferous wood cells associated with heartwood formation. *TAPPI* 46, 42–49.

KRISHNAMURTHY, K. V., N. VENUGOPAL, V. NANDAGOPALAN, U. HARIHARAN, and A. SIVAKUMARI. 1997. Tension phloem in some legumes. *J. Plant Anat. Morphol.* 7, 20–23.

KURODA, K., and K. SHIMAJI. 1983. Traumatic resin canal formation as a marker of xylem growth. *For. Sci.* 29, 653–659.

LAPASHA, C. A., and E. A. WHEELER. 1990. Resin canals in *Pinus taeda*: Longitudinal canal lengths and interconnections between longitudinal and radial canals. *IAWA Bull.* n.s. 11, 227–238.

LARSON, P. R. 1969. Wood formation and the concept of wood quality. *Bull. Yale Univ. School For.* 74, 1–54.

LEE, P. W., and Y. G. EOM. 1988. Anatomical comparison between compression wood and opposite wood in a branch of Korean pine (*Pinus koraiensis*). *IAWA Bull.* n.s. 9, 275–284.

LEV-YADUN, S. 1994a. Experimental evidence for the autonomy of ray differentiation in *Ficus sycomorus* L. *New Phytol.* 126, 499–504.

LEV-YADUN, S. 1994b. Radial fibres in aggregate rays of *Quercus calliprinos* Webb.—Evidence for radial signal flow. *New Phytol.* 128, 45–48.

LEV-YADUN, S. 1995. Short secondary vessel members in branching regions in roots of *Arabidopsis thaliana. Aust. J. Bot.* 43, 435–438.

LEV-YADUN, S. 2000. Cellular patterns in dicotyledonous woods: their regulation. In: *Cell and Molecular Biology of Wood Formation*, pp. 315–324, R. A. Savidge, J. R. Barnett, and R. Napier, eds. BIOS Scientific, Oxford.

LEV-YADUN, S., and R. ALONI. 1995. Differentiation of the ray system in woody plants. *Bot. Rev.* 61, 45–84.

LIESE, W. 1996. Structural research on bamboo and rattan for their wider utilization. *J. Bamboo Res. (Zhu zi yan jiu hui kan)* 15, 1–14.

LIESE, W. 1998. *The Anatomy of Bamboo Culms*. Tech. Rep. 18. International Network for Bamboo and Rattan (INBAR), Beijing.

LIPHSCHITZ, N. 1995. Ecological wood anatomy: Changes in xylem structure in Israeli trees. In: *Wood Anatomy Research 1995*. Proceedings of the International Symposium on Tree Anatomy and Wood Formation, pp. 12–15, S. Wu, ed. International Academic Publishers, Beijing.

LIPHSCHITZ, N., and Y. WAISEL. 1970. Effects of environment on relations between extension and cambial growth of *Populus euphratica* Oliv. *New Phytol.* 69, 1059–1064.

LITTLE, C. H. A., and R. P. PHARIS. 1995. Hormonal control of radial and longitudinal growth in the tree stem. In: *Plant Stems: Physiology and Functional Morphology*, pp. 281–319, B. L. Gartner, ed. Academic Press, San Diego.

MACHADO, S. R., and V. ANGYALOSSY-ALFONSO. 1995. Occurrence of perforated ray cells in wood of *Styrax camporum* Pohl. (Styracaceae). *Rev. Brasil. Bot.* 18, 221–225.

MACHADO, S. R., V. ANGYALOSSY-ALFONSO, and B. L. DE MORRETES. 1997. Comparative wood anatomy of root and stem in *Styrax camporum* (Styracaceae). *IAWA J.* 18, 13–25.

MAGEL, E. A. 2000. Biochemistry and physiology of heartwood formation. In: *Cell and Molecular Biology of Wood Formation*, pp. 363–376, R. A. Savidge, J. R. Barnett, and N. Napier, eds. BIOS Scientific, Oxford.

MATTHECK, C., and H. KUBLER. 1995. *Wood: The Internal Optimization of Trees*. Springer-Verlag, Berlin.

MATTOS, P. PÓVOA DE, R. A. SEITZ, and G. I. BOLZON DE MUNIZ. 1999. Identification of annual growth rings based on periodical shoot growth. In: *Tree-Ring Analysis. Biological, Methodological, and Environmental Aspects*, pp. 139–145, R. Wimmer and R. E. Vetter, eds. CABI Publishing, Wallingford, Oxon.

MELLEROWICZ, E. J., M. BAUCHER, B. SUNDBERG, and W. BOERJAN. 2001. Unraveling cell wall formation in the woody dicot stem. *Plant Mol. Biol.* 47, 239–274.

METCALFE, C. R., and L. CHALK, eds. 1983. *Anatomy of the Dicotyledons*, 2nd ed., vol. II. *Wood Structure and Conclusion of the General Introduction*. Clarendon Press, Oxford.

MEYLAN, B. A., and B. G. BUTTERFIELD. 1978. *The Structure of New Zealand Woods*. Bull. 222, NZDSIR, Wellington.

MOREY, P. R., and J. CRONSHAW. 1968. Developmental changes in the secondary xylem of *Acer rubrum* induced by gibberellic acid, various auxins and 2,3,5-tri-iodobenzoic acid. *Protoplasma* 65, 315–326.

MURAKAMI, Y., R. FUNADA, Y. SANO, and J. OHTANI. 1999. The differentiation of contact cells and isolation cells in the xylem ray parenchyma of *Populus maximowiczii. Ann. Bot.* 84, 429–435.

NADEZHDINA, N., J. CERMÁK, and R. CEULEMANS. 2002. Radial patterns of sap flow in woody stems of dominant and understory species: Scaling errors associated with positioning of sensors. *Tree Physiol.* 22, 907–918.

NAGAI, S., J. OHTANI, K. FUKAZAWA, and J. WU. 1994. SEM observations on perforated ray cells. *IAWA J.* 15, 293–300.

NAGY, N. E., V. R. FRANCESCHI, H. SOLHEIM, T. KREKLING, and E. CHRISTIANSEN. 2000. Wound-induced traumatic resin duct development in stems of Norway spruce (Pinaceae): Anatomy and cytochemical traits. *Am. J. Bot.* 87, 302–313.

NANKO, H., H. SAIKI, and H. HARADA. 1982. Structural modification of secondary phloem fibers in the reaction phloem of *Populus euramericana. Mokuzai Gakkaishi (J. Jpn. Wood Res. Soc.)* 28, 202–207.

OTEGUI, M. S. 1994. Occurrence of perforated ray cells and ray splitting in *Rapanea laetevirens* and *R. lorentziana* (Myrsinaceae). *IAWA J.* 15, 257–263.

PANSHIN, A. J., and C. DE ZEEUW. 1980. *Textbook of Wood Technology: Structure, Identification, Properties, and Uses of the Commercial Woods of the United States and Canada*, 4th ed. McGraw-Hill, New York.

PHILLIPS, E. W. J. 1948. Identification of softwoods by their microscopic structure. *Dept. Sci. Ind. Res. For. Prod. Res. Bull.* No. 22. London.

PILATE, G., B. CHABBERT, B. CATHALA, A. YOSHINAGA, J.-C. LEPLÉ, F. LAURANS, C. LAPIERRE, and K. RUEL. 2004. Lignification and tension wood. *C.R. Biologies* 327, 889–901.

RECORD, S. J. 1934. *Identification of the timbers of temperate North America, including anatomy and certain physical properties of wood.* Wiley, New York.

RENDLE, B. J. 1960. Juvenile and adult wood. *J. Inst. Wood Sci.* 5, 58–61.

RICHTER, H. G., D. GROSSER, I. HEINZ, and P. E. GASSON, eds. 2004. IAWA list of microscopic features for softwood identification. *IAWA J.* 25, 1–70.

RIDOUTT, B. G., and R. SANDS. 1994. Quantification of the processes of secondary xylem fibre development in *Eucalyptus globulus* at two height levels. *IAWA J.* 15, 417–424.

ROBARDS, A. W. 1966. The application of the modified sine rule to tension wood production and eccentric growth in the stem of crack willow (*Salix fragilis* L.). *Ann. Bot.* 30, 513–523.

ROBERTS, L. W., P. B. GAHAN, and R. ALONI. 1988. *Vascular Differentiation and Plant Growth Regulators.* Springer-Verlag, Berlin.

ROBNETT, W. E., and P. R. MOREY. 1973. Wood formation in *Prosopis*: Effect of 2,4-D, 2,4,5-T, and TIBA. *Am. J. Bot.* 60. 745–754.

SANO, Y., and R. NAKADA. 1998. Time course of the secondary deposition of incrusting materials on bordered pit membranes in *Cryptomeria japonica*. *IAWA J.* 19, 285–299.

SAUTER, J. J. 1972. Respiratory and phosphatase activities in contact cells of wood rays and their possible role in sugar secretion. *Z. Pflanzenphysiol.* 67, 135–145.

SAUTER, J. J. 2000. Photosynthate allocation to the vascular cambium: facts and problems. In: *Cell and Molecular Biology of Wood Formation*, pp. 71–83, R. A. Savidge, J. R. Barnett, and R. Napier, eds. BIOS Scientific, Oxford.

SAUTER, J. J., and S. KLOTH. 1986. Plasmodesmatal frequency and radial translocation rates in ray cells of poplar (*Populus x canadensis* Moench "robusta"). *Planta* 168, 377–380.

SAUTER, J. J., W. ITEN, and M. H. ZIMMERMANN. 1973. Studies on the release of sugar into the vessels of sugar maple (*Acer saccharum*) *Can. J. Bot.* 51, 1–8.

SCHAFFER, K., and M. WISNIEWSKI. 1989. Development of the amorphous layer (protective layer) in xylem parenchyma of cv. Golden Delicious apple, cv. Loring Peach, and willow. *Am. J. Bot.* 76, 1569–1582.

SCHWEINGRUBER, F. H. 1988. *Tree Rings. Basics and Applications of Dendrochronology.* Reidel, Dordrecht.

SCHWEINGRUBER, F. H. 1990. *Anatomie europäischer Hölzer: Ein Atlas zur Bestimmung europäischer Baum-, Strauch-, und Zwergstrauchhölzer* (Anatomy of European woods: An atlas for the identification of European trees, shrubs, and dwarf shrubs). Verlag P. Haupt, Bern.

SCHWEINGRUBER, F. H. 1993. *Trees and Wood in Dendrochronology: Morphological, Anatomical, and Tree-ring Analytical Characteristics of Trees Frequently Used in Dendrochronology.* Springer-Verlag, Berlin.

SCHWEINGRUBER, F. H., and W. BOSSHARD. 1978. *Mikroskopische Holzanatomie: Formenspektren mitteleuropäischer Stamm-, und Zweighölzer zur Bestimmung von rezentem und subfossilem Material* (Microscopic wood anatomy: Structural variability of stems and twigs in recent and subfossil woods from Central Europe). Eidgenössische Anstalt für das Forstliche Versuchswesen, Birmensdorf.

SCURFIELD, G. 1964. The nature of reaction wood. IX. Anomalous cases of reaction anatomy. *Aust. J. Bot.* 12, 173–184.

SCURFIELD, G. 1965. The cankers of *Exocarpos cupressiformis* Labill. *Aust. J. Bot.* 13, 235–243.

SHEEN, J., L. ZHOU, and J.-C. JANG. 1999. Sugars as signaling molecules. *Curr. Opin. Plant Biol.* 2, 410–418.

SHELDRAKE, A. R. 1971. Auxin in the cambium and its differentiating derivatives. *J. Exp. Bot.* 22, 735–740.

SKATTER, S., and B. KUCERA. 1997. Spiral grain—an adaptation of trees to withstand stem breakage caused by wind-induced torsion. *Holz Roh- Werks.* 55, 207–213.

SPICER, R., and B. L. GARTNER. 2002. Compression wood has little impact on the water relations of Douglas-fir (*Pseudotsuga menziesii*) seedlings despite a large effect on shoot hydraulic properties. *New Phytol.* 154, 633–640.

STAMM, A. J. 1946. Passage of liquids, vapors, and dissolved materials through softwoods. *USDA Tech. Bull.* 929.

STUDHALTER, R. A. 1955. Tree growth. I. Some historical chapters. *Bot. Rev.* 21, 1–72.

SUNDBERG, B., H. TUOMINEN, and C. H. A. LITTLE. 1994. Effects of indole-3-acetic acid (IAA) transport inhibitors N-1-naphthylphthalamic acid and morphactin on endogenous IAA dynamics in relation to compression wood formation in 1-year-old *Pinus sylvestris* (L.) shoots. *Plant Physiol.* 106, 469–476.

SUNDBERG, B., C. UGGLA, and H. TUOMINEN. 2000. Cambial growth and auxin gradients. In: *Cell and Molecular Biology of Wood Formation*, pp. 169–188, R. A. Savidge, J. R. Barnett, and R. Napier, eds. BIOS Scientific, Oxford.

SUZUKI, M., K. YODA, and H. SUZUKI. 1996. Phenological comparison of the onset of vessel formation between ring-porous and diffuse-porous deciduous trees in a Japanese temperate forest. *IAWA J.* 17, 431–444.

TAKABE, K., T. MIYAUCHI, and K. FUKAZAWA. 1992. Cell wall formation of compression wood in Todo fir (*Abies sachalinensis*)—I. Deposition of polysaccharides. *IAWA Bull.* n.s. 13, 283–296.

THOMSON, R. G., and H. B. SIFTON. 1925. Resin canals in the Canadian spruce (*Picea canadensis* (Mill.) B. S. P.)—An anatomical study, especially in relation to traumatic effects and their bearing on phylogeny. *Philos. Trans. R. Soc. Lond. Ser. B* 214, 63–111.

TIMELL, T. E. 1981. Recent progress in the chemistry, ultrastructure, and formation of compression wood. *The Ekman-Days*

1981, Chemistry and Morphology of Wood and Wood Components, SPCI, Stockholm Rep. 38, vol. I, 99–147.

TIMELL, T. E. 1983. Origin and evolution of compression wood. Holzforschung 37, 1–10.

TOMAZELLO, M., and N. DA SILVA CARDOSO. 1999. Seasonal variations of the vascular cambium of teak (Tectona grandis L.) in Brazil. In: Tree-ring Analysis: Biological, Methodological, and Environmental Aspects, pp. 147–154, R. Wimmer and R. E. Vetter, eds. CABI Publishing, Wallingford, Oxon.

TOMLINSON, P. B., and M. H. ZIMMERMANN. 1967. The "wood" of monocotyledons. Bulletin [IAWA] 1967/2, 4–24.

TUOMINEN, H., L. PUECH, S. FINK, and B. SUNDBERG. 1997. A radial concentration gradient of indole-3-acetic acid is related to secondary xylem development in hybrid aspen. Plant Physiol. 115, 577–585.

UGGLA, C., T. MORITZ, G. SANDBERG, and B. SUNDBERG. 1996. Auxin as a positional signal in pattern formation in plants. Proc. Natl. Acad. Sci. USA 93, 9282–9286.

UGGLA, C., E. MAGEL, T. MORITZ, and B. SUNDBERG. 2001. Function and dynamics of auxin and carbohydrates during earlywood/latewood transition in Scots pine. Plant Physiol. 125, 2029–2039.

UTSUMI, Y., Y. SANO, J. OHTANI, and S. FUJIKAWA. 1996. Seasonal changes in the distribution of water in the outer growth rings of Fraxinus mandshurica var. japonica: A study by cryo-scanning electron microscopy. IAWA J. 17, 113–124.

VAN BEL, A. J. E. 1990a. Vessel-to-ray transport: Vital step in nitrogen cycling and deposition. In: Fast Growing Trees and Nitrogen Fixing Trees, pp. 222–231, D. Werner and P. Müller, eds. Gustav Fischer Verlag, Stuttgart.

VAN BEL, A. J. E. 1990b. Xylem-phloem exchange via the rays: the undervalued route of transport. J. Exp. Bot. 41, 631–644.

VAN BEL, A. J. E., and C. VAN DER SCHOOT. 1988. Primary function of the protective layer in contact cells: Buffer against oscillations in hydrostatic pressure in the vessels? IAWA Bull. n.s. 9, 285–288.

VAN BEL, A. J. E., and A. J. VAN ERVEN. 1979. A model for proton and potassium co-transport during the uptake of glutamine and sucrose by tomato internode disks. Planta 145, 77–82.

VAN DER SCHOOT, C. 1989. Determinates of xylem-to-phloem transfer in tomato. Ph.D. Dissertation. Rijksuniversiteit te Utrecht, The Netherlands.

VETTER, R. E., and P. C. BOTOSSO. 1989. Remarks on age and growth rate determination of Amazonian trees. IAWA Bull. n.s. 10, 133–145.

WARDROP, A. B., and H. E. DADSWELL. 1953. The development of the conifer tracheid. Holzforschung 7, 33–39.

WARREN WILSON, J., and P. M. WARREN WILSON. 1984. Control of tissue patterns in normal development and in regeneration. In: Positional Controls in Plant Development, pp. 225–280, P. Barlow and D. J. Carr, eds. Cambridge University Press, Cambridge.

WERKER, E., and A. FAHN. 1969. Resin ducts of Pinus halepensis Mill.: Their structure, development and pattern of arrangement. Bot. J. Linn. Soc. 62, 379–411.

WESTING, A. H. 1965. Formation and function of compression wood in gymnosperms. Bot. Rev. 31, 381–480.

WESTING, A. H. 1968. Formation and function of compression wood in gymnosperms. II. Bot. Rev. 34, 51–78.

WHEELER, E. A., and P. BAAS. 1991. A survey of the fossil record for dicotyledonous wood and its significance for evolutionary and ecological wood anatomy. IAWA Bull. n.s. 12, 275–332.

WHEELER, E. A., and P. BAAS. 1998. Wood identification—A review. IAWA J. 19, 241–264.

WHEELER, E. A., P. BAAS, and P. E. GASSON, eds. 1989. IAWA list of microscopic features for hardwood identification. IAWA Bull. n.s. 10, 219–332.

WIEDENHOEFT, A. C., and R. B. MILLER. 2002. Brief comments on the nomenclature of softwood axial resin canals and their associated cells. IAWA J. 23, 299–303.

WILSON, B. F., and R. R. ARCHER. 1977. Reaction wood: induction and mechanical action. Annu. Rev. Plant Physiol. 28, 23–43

WILSON, B. F., and B. L. GARTNER. 1996. Lean in red alder (Alnus rubra): Growth stress, tension wood, and righting response. Can. J. For. Res. 26, 1951–1956.

WIMMER, R., M. GRABNER, G. STRUMIA, and P. R. SHEPPARD. 1999. Significance of vertical resin ducts in the tree rings of spruce. In: Tree-ring Analysis: Biological, Methodological and Environmental Aspects, pp. 107–118, R. Wimmer and R. E. Vetter, eds. CABI Publishing, Wallingford, Oxon.

WISNIEWSKI, M., and G. DAVIS. 1989. Evidence for the involvement of a specific cell wall layer in regulation of deep supercooling of xylem parenchyma. Plant Physiol. 91, 151–156.

WODZICKI, T. J., and C. L. BROWN. 1973. Cellular differentiation of the cambium in the Pinaceae. Bot. Gaz. 134, 139–146.

WODZICKI, T. J., and A. B. WODZICKI. 1980. Seasonal abscisic acid accumulation in stem and cambial region of Pinus sylvestris, and its contribution to the hypothesis of a late-wood control system in conifers. Physiol. Plant. 48, 443–447.

WORBES, M. 1985. Structural and other adaptations to long-term flooding by trees in Central Amazonia. Amazoniana 9, 459–484.

WORBES, M. 1989. Growth rings, increment and age of trees in inundation forests, savannas and a mountain forest in the Neotropics. IAWA Bull. n.s. 10, 109–122.

WORBES, M. 1995. How to measure growth dynamics in tropical trees—A review. IAWA J. 16, 337–351.

WU, H., and Z.-H. HU. 1997. Comparative anatomy of resin ducts of the Pinaceae. Trees 11, 135–143.

WULLSCHLEGER, S. D., and A. W. KING. 2000. Radial variation in sap velocity as a function of stem diameter and sapwood thickness in yellow-poplar trees. *Tree Physiol.* 20, 511–518.

YAMAMOTO, K. 1982. Yearly and seasonal process of maturation of ray parenchyma cells in *Pinus* species. *Res. Bull. Coll. Exp. For. Hokkaido Univ.* 39, 245–296.

YOSHIZAWA, N., M. SATOH, S. YOKOTA, and T. IDEI. 1992. Formation and structure of reaction wood in *Buxus microphylla* var. *insularis* Nakai. *Wood Sci. Technol.* 27, 1–10.

ZIEGLER, H. 1968. Biologische Aspekte der Kernholzbildung (Biological aspects of heartwood formation). *Holz Roh- Werks.* 26, 61–68.

ZIMMERMANN, M. H. 1983. *Xylem Structure and the Ascent of Sap.* Springer-Verlag, Berlin.

ZIMMERMANN, M. H., A. B. WARDROP, and P. B. TOMLINSON. 1968. Tension wood in aerial roots of *Ficus benjamina* L. *Wood Sci. Technol.* 2, 95–104.

# CHAPTER TWELVE

# Vascular Cambium

The vascular cambium is the meristem that produces the secondary vascular tissues. It is a lateral meristem, for in contrast to the apical meristems, which are located at the tips of stems and roots, it occupies a lateral position in these organs. The vascular cambium, like the apical meristems (Chapter 6), consists of initial cells and their recent derivatives. In the three-dimensional aspect, the vascular cambium commonly forms a continuous cylindrical sheath about the xylem of stems and roots and their branches (Fig. 12.1). When the secondary vascular tissues of an axis are in discrete strands, the cambium may remain restricted to the strands in the form of strips. It also appears in strips in most petioles and leaf veins that undergo secondary growth. In the leaves (needles) of conifers, the vascular bundles increase somewhat in thickness after the first year through the activity of a vascular cambium (Strasburger, 1891; Ewers, 1982). In angiosperms, the larger veins may have primary and secondary vascular tissues; the smaller are usually entirely primary. Cambial activity is more pronounced in leaves of evergreen species than in those of the deciduous (Shtromberg, 1959).

## ORGANIZATION OF THE CAMBIUM

The cells of the vascular cambium do not fit the usual description of meristematic cells, as those that have dense cytoplasm, large nuclei, and an approximately isodiametric shape. Although the resting cambial cells are densely cytoplasmic, they contain many small vacuoles. Active cambial cells are highly vacuolated, consisting essentially of a single large central vacuole surrounded by a thin, parietal layer of dense cytoplasm.

### The Vascular Cambium Contains Two Types of Initials: Fusiform Initials and Ray Initials

Morphologically, cambial initials occur in two forms. One type of initial, the **fusiform initial** (Fig. 12.2A), is several times longer than wide; the other, the **ray initial** (Fig. 12.2B), is slightly elongated to nearly isodiametric. The term fusiform implies that the cell is shaped like a spindle. A fusiform cell, however, is an approximately prismatic cell in its middle part and wedge-shaped at the ends. The pointed end of the wedge is seen in

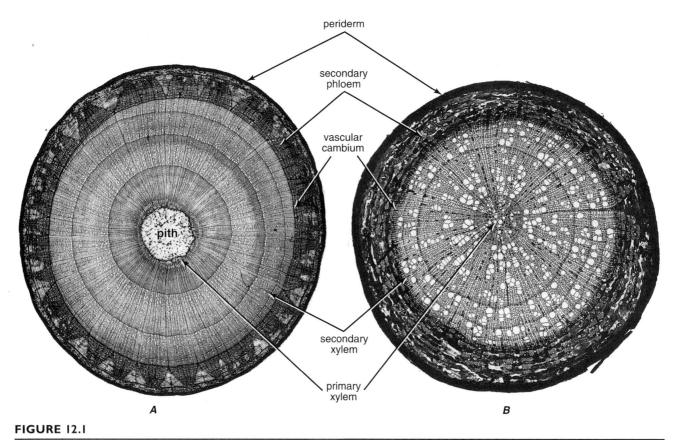

**FIGURE 12.1**

Transverse sections of *Tilia* stem (**A**) and root (**B**), each with periderm and several increments of secondary vascular tissues. The vascular cambium forms a continuous cylindrical sheath about the secondary xylem. (A, ×9.7; B, ×27.)

tangential sections, the truncated end in radial sections (Fig. 12.2A). The tangential sides of the cell are wider than the radial. The exact shape of the fusiform initials of *Pinus sylvestris* has been determined as that of long, pointed, tangentially flattened cells with an average of 18 faces (Dodd, 1948).

The fusiform initials give rise to all cells of the xylem and phloem that are arranged with their long axes parallel to the long axis of the organ in which they occur; in other words, they give rise to the longitudinal or axial systems of xylem and phloem (Fig. 12.2D). Examples of elements in these systems are tracheary elements, fibers, and axial parenchyma cells in the xylem; sieve elements, fibers, and axial parenchyma cells in the phloem. The ray initials give rise to the ray cells, that is, the elements of the radial system (the system of rays) of the xylem and the phloem (Fig. 12.2E; Chapters 11, 14).

The fusiform initials show a wide range of variation in their dimensions and volume. Some of these variations depend on the plant species. The following figures, expressed in millimeters, exemplify differ-

ences in the lengths of fusiform initials in several plants: *Sequoia sempervirens*, 8.70 (Bailey, 1923); *Pinus strobus*, 3.20; *Ginkgo*, 2.20; *Myristica*, 1.31; *Pyrus*, 0.53; *Populus*, 0.49; *Fraxinus*, 0.29; *Robinia*, 0.17 (Bailey, 1920a). Fusiform initials vary in length within species, partly in relation to growth conditions (Pomparat, 1974). They also show length modifications associated with developmental phenomena in a single plant. Generally, the length of fusiform initials increases with the age of the axis, but after reaching a certain maximum, it remains relatively stable (Bailey, 1920a; Boßhard, 1951; Bannan, 1960b; Ghouse and Yunus, 1973; Ghouse and Hashmi, 1980a; Khan, K. K., et al., 1981; Iqbal and Ghouse, 1987; Ajmal and Iqbal, 1992). After reaching their maximum length, the fusiform initials in some species (e.g., *Citrus sinensis*, Khan, M. I. H., et al., 1983) may undergo a gradual but slow decrease in length with increasing girth of the axis. In at least some species, fusiform initial length tends to increase from the top toward the base of the stem, reaching a maximum and then declining slightly at the base (Iqbal and Ghouse, 1979; Ridoutt and Sands,

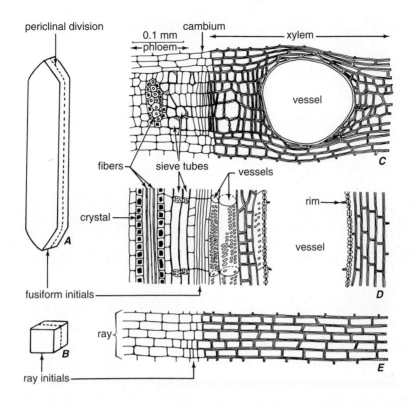

**FIGURE 12.2**

Vascular cambium in relation to derivative tissues. **A**, diagram of fusiform initial; **B**, of ray initial. In both, orientation of division concerned with formation of phloem and xylem cells (periclinal division) is indicated by broken lines. **C**, **D**, **E**, *Robinia pseudoacacia*; sections of stem include phloem, cambium, and xylem. **C**, transverse; **D**, radial (axial system only); **E**, radial (ray only). (From Esau, 1977.)

1993). The size of fusiform initials also may vary during the growing season (Paliwal et al., 1974; Sharma et al., 1979). The changes in the size of fusiform initials brings about similar changes in the xylem and phloem cells derived from these initials. The ultimate size of their derivatives, however, depends only partly on that of the cambial initials, because changes in size also occur during differentiation of cells.

### The Cambium May Be Storied or Nonstoried

The cambium may be storied (stratified), or nonstoried (nonstratified), depending on whether or not, *as seen in tangential sections*, the cells are arranged in horizontal tiers. In a **storied cambium** the fusiform initials are arranged in horizontal tiers, with the ends of the cells of one tier appearing at approximately the same level (Fig. 12.3). It is characteristic of plants with short fusiform initials. **Nonstoried cambia** are common in plants with long fusiform initials, which have strongly overlapping ends (Fig. 12.4). Intergrading types of arrangement occur in different plants. The cambium of *Fraxinus excelsior* is a mosaic of storied and nonstoried local areas (Krawczyszyn, 1977). The

storied cambium, which is more common in tropical species than in temperate ones, is considered to be phylogenetically more advanced than the nonstoried, the evolution from nonstoried to storied being accompanied by a shortening of fusiform initials (Bailey, 1923). Like the fusiform initials, the rays may be storied or nonstoried.

The fusiform cells of the vascular cambium are compactly arranged. Whether intercellular spaces continue radially between the xylem and phloem via the rays, however, has been the subject of long-lived debate (Larson, 1994). Intercellular spaces were found among ray initials in *Tectona grandis*, *Azadirachta indica*, and *Tamarindus indica*, but only when the cambium was inactive (Rajput and Rao, 1998a). In the active cambium the cells appeared compactly arranged. In an effort to resolve the question, both active and dormant cambia were examined in 15 temperate-zone species, including both eudicots (*Acer negundo*, *Acer saccharum*, *Cornus rasmosa*, *Cornus stolonifera*, *Malus domestica*, *Pyrus communis*, *Quercus alba*, *Rhus glabra*, *Robinia pseudoacacia*, *Salix nigra*, *Tilia americana*, *Ulmus americana*) and conifers (*Metasequoia glyptostroboides*, *Picea abies*, *Pinus pinea*). In all 15 species narrow,

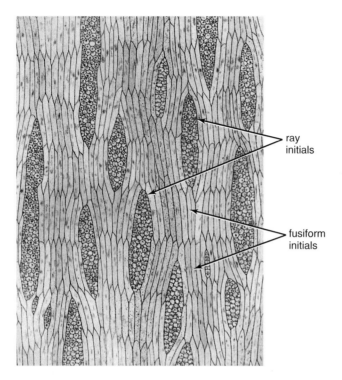

**FIGURE 12.3**

Storied cambium of black locust (*Robinia pseudoaca-cia*), as seen in tangential section. In a cambium such as this, the fusiform initials are arranged in horizontal tiers on tangential surfaces. (×125.)

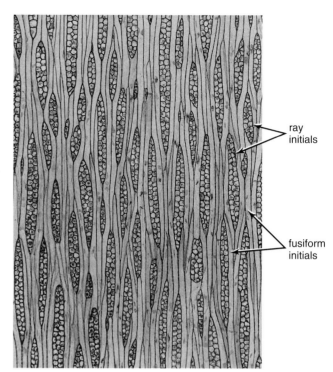

**FIGURE 12.4**

Nonstoried cambium of apple (*Malus domestica*), as seen in tangential section. In a cambium such as this, the fusiform initials are not arranged in horizontal tiers, as seen on tangential surfaces. (×125.)

radially oriented intercellular spaces were found within the rays and/or at the interfaces between vertically contiguous ray cells and fusiform cells of both the active and dormant cambium (Evert, unpublished data). A system of intercellular spaces was continuous between the secondary xylem and secondary phloem via the rays.

## FORMATION OF SECONDARY XYLEM AND SECONDARY PHLOEM

When the cambial initials produce xylem and phloem cells they divide periclinally (tangentially; Fig. 12.2A, B). At one time a derivative cell is produced inwardly toward the xylem, at another time outwardly toward the phloem, although not necessarily in alternation. Thus each cambial initial (Figs. 12.2C and 12.5) produces radial files of cells, one toward the inside, the other toward the outside, and the two files meet at the cambial initial. Such radial seriation may persist in the developing xylem and phloem, or it may be disturbed through various kinds of growth readjustments during differentiation of these tissues (Fig. 12.2C). These cambial divisions, which add cells to

the secondary vascular tissues, are also called **additive divisions**.

Additive divisions are not limited to the initials but are encountered also in varied numbers of derivatives. During the period of rest, xylem and phloem cells mature more or less close to the initials; sometimes only one cambial layer is left between the mature xylem and phloem elements (Fig. 12.6A). But some vascular tissue—frequently only phloem—may overwinter in an immature state (Fig. 12.6B).

During the height of cambial activity, cell addition occurs so rapidly that older cells are still meristematic when new cells are produced by the initials. Thus a wide zone of more or less undifferentiated cells accumulates. Within this zone, the **cambial zone**, only one cell in a given radial file is considered to be an initial in the sense that after it divides periclinally, one of the two resulting cells remains as an initial and the other is given off toward the differentiating xylem or phloem. The initials are difficult to distinguish from their recent derivatives in part because these derivatives divide periclinally one or more times before they begin to differentiate into xylem or phloem. The initial is, however, the only cell able to produce derivatives toward both the xylem and the phloem.

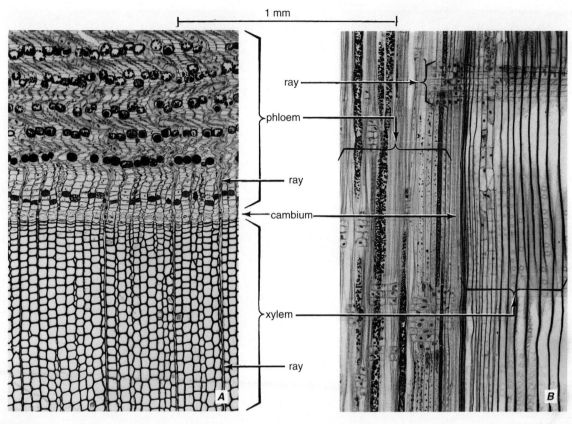

**FIGURE 12.5**

Vascular tissues and cambium in stem of pine (*Pinus* sp., a conifer) in cross (**A**) and radial (**B**) sections. (From Esau, 1977.)

The active cambial zone thus constitutes a more or less wide stratum of periclinally dividing cells organized into axial and radial systems. Within this stratum some workers visualize a single layer of cambial initials flanked along their two tangential walls by **phloem mother cells (phloem initials)** toward the outside and **xylem mother cells (xylem initials)** toward the inside, and they restrict use of the word cambium to this putative uniseriate layer of initials. Others, including the author of this book, use the terms cambium and cambial zone interchangeably. It is quite clear that the initial of a given radial file of cells in the cambial zone may not have an accurate tangential alignment with the initials in neighboring radial files (Evert, 1963a; Bannan, 1968; Mahmood, 1968; Catesson, 1987); quite likely there is never an uninterrupted, even layer of cambial initials around the axis (Timell, 1980; Włoch, 1981). Moreover a given initial may cease to participate in additive divisions and be displaced by its derivative, which then assumes the role of a cambial initial.

Cambial initials are not permanent entities in the cambium, but temporary, relatively short-lived tran-sients, each of which performs an "initial function" (Newman, 1956; Mahmood, 1990), a function that is perpetuated and inherited by one "heir" or cambial initial after the other (Newman, 1956). The cambium thus has many characters in common with the apical meristems (Chapter 6). In both, it is extremely difficult to delimit the initials from their recent derivatives, the derivatives in both being more or less meristematic, and in both, the initials are continually shifting positions and being displaced. It has been suggested that passing of the initial function from one cambial cell to another may help avoid the accumulation of harmful mutations that potentially could occur after hundreds or thousands of mitotic cycles in permanent initials of long-lived species (Gahan, 1988, 1989).

# ■ INITIALS VERSUS THEIR IMMEDIATE DERIVATIVES

The initials cannot be distinguished from their immediate derivatives by cytological features. This is true both of actively dividing cambia and of dormant cambia in which more than one layer of undifferentiated cells

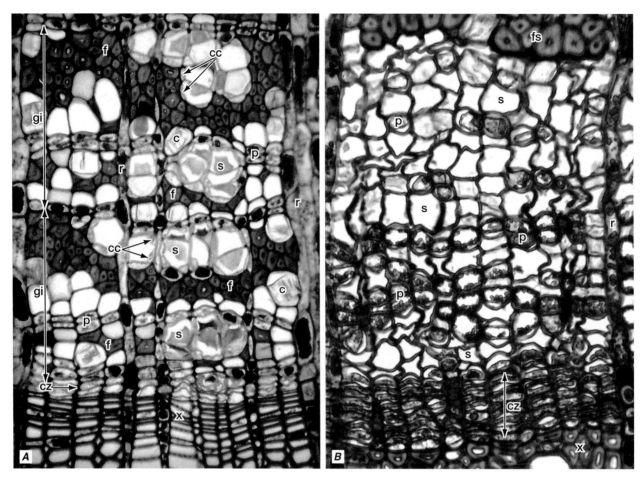

**FIGURE 12.6**

Transverse sections of vascular tissues and dormant cambia in (**A**) basswood (*Tilia americana*) and (**B**) apple (*Malus domestica*) stems. The dormant cambial zone in basswood consists of only one or two layers of cells, in apple it consists of several layers (5 to 11). Two growth increments (gi) of secondary phloem, with overwintering living sieve elements and companion cells (conducting phloem), can be seen in the basswood section (**A**). A single phloem increment—delimited above by a band of fiber-sclereids (fs)—is present in the apple section (**B**). The increment in the apple section consists entirely of nonconducting phloem: its sieve elements are dead and their companion cells (not discernible) have collapsed. Other details: c, crystal-containing cell; cc, companion cells; cz, cambial zone; f, fibers; p, parenchyma cell; r, ray; s, sieve element; x, xylem. (A, ×300; B, ×394.)

occur between completely differentiated elements of xylem and phloem. Most attempts to identify cambial initials have been made on conifers. The earliest such attempt was based on differences in the thickness of the tangential walls in the cambial cells of *Pinus sylvestris* (Sanio, 1873). Sanio noted that after cell-plate formation each of the new daughter cells enclosed ("emboxed") its protoplast with a new primary wall, explaining why the radial walls in the cambial zone are always much thicker than the tangential walls, and why the tangential walls vary in thickness. The initial cell in each radial file had an extra thick tangential wall. Sanio also noted that tangential walls meeting radial walls at rounded angles are older than those joining

radial walls at sharp angles. Using these criteria, Sanio recognized distinct groups of four cells in the cambial zone. Now called ***Sanio's four***, each group of four cells consists of the initial, its most immediate derivative, and two daughter cells. When xylem is being formed, the daughter cells divide once more, producing four xylem cells, referred to as the ***expanding***, or ***enlarging***, ***four*** (Fig. 12.7; Mahmood, 1968). The presence of Sanio's four in the cambial zone and of groups of expanding four in the differentiating xylem of conifers has since been confirmed (Murmanis and Sachs, 1969; Murmanis, 1970; Timell, 1980). Groups of four have not been recognized on the phloem side of the cambium; there, the cells appear to occur in pairs.

**FIGURE 12.7**

Theoretical sequence of events during the production of secondary xylem if each of the protoplasts of the new daughter cells is enclosed ("emboxed") by a new primary wall. The successive initials in xylem production are designated as $i$, $i_1$, $i_2$, and $i_3$; the xylem mother cells as $d$ and $d_1$; and the tissue cells derived from a pair of daughter cells as $t$. The original predivision initial $i$ is found in column **A**. Its division gives rise to the succeeding initial $i_1$ and mother cell $m$ (column **B**), each of which then enlarges to predivision size (column **C**). In column **D**, $i_1$ has divided into $i_2$ and $m_1$, and $m$ has divided to produce a pair of daughter cells $d$. The group of four cells in columns **D** and **E** correspond to Sanio's four. In columns **F** and **G** both Sanio's four and enlarging four can be recognized. Redrawn from A. Mahmood. 1968. *Australian Journal of Botany* 16, 177-195, with permission of CSIRO Publishing, Melbourne, Australia. © CSIRO.)

Except for *Quercus rubra* (Murmanis, 1977) and *Tilia cordata* (Włoch, 1989, as reported by Larson, 1994), Sanio's four have been identified only in conifers (Timell, 1980; Larson, 1994). Groups of expanding four have been found on the xylem side of the cambium in *Populus deltoides* (Isebrands and Larson, 1973) and *Tilia cordata* (Włoch and Zagórska-Marek, 1982; Włoch and Polap, 1994) and pairs of cells on the phloem side in *Tilia cordata* (Włoch and Zagórska-Marek, 1982; Włoch and Polap, 1994). Evidence for the presence of pairs of cells has also been found in the secondary phloem of *Pyrus malus (Malus domestica)* (Evert, 1963a). The failure or difficulty in identifying Sanio's four or groups of expanding four in other woody angiosperms (hardwoods) may be attributed to a combination of factors, including the relatively few layers of cells found in the dormant cambia of some hardwood species, the less orderly manner of cell division that occurs in

the active cambia of hardwoods (compared with the regular succession of cell division found in the cambia of conifers), and the distortion to the radial rows of cells that occurs just outside the actively dividing cambial cells of hardwoods from the extensive intrusive growth and lateral expansion of xylem derivatives.

The "emboxing" of daughter protoplasts, which has been used effectively with conifers to identify cambial initials, was questioned by Catesson and Roland (1981) in their study of several deciduous hardwoods. They could find no evidence for the deposition of a complete primary wall around each daughter protoplast following periclinal division (i.e., following formation of a new tangential wall). Instead, they found a heterogeneous distribution of polysaccharides around each of the daughter protoplasts, with polysaccharide lysis and deposition occurring simultaneously. Utilizing mild extraction and cytochemical techniques at the ultrastructural level, Catesson and Roland (1981; see also Roland, 1978) found young tangential cambial cell walls to be made up of a loose microfibrillar skeleton and a matrix rich in highly methylated pectins and the bulk of the radial walls to be hemicellulosic. The young tangential walls had no recognizable middle lamella, while the radial walls presented a classical tripartite structure (primary wall–middle lamella–primary wall), the middle lamella containing a high amount of acidic pectins (Catesson and Roland, 1981; Catesson, 1990). Large portions of the actively expanding radial walls—portions probably especially plastic and extensible—appeared completely devoid of cellulose. Immunolocalization studies on the vascular cambium of *Aesculus hippocastanum* taproots (Chaffey et al., 1997a) broadly support the views of Catesson and her co-workers (Catesson et al., 1994), with regard to the composition of cambial cell walls.

Other criteria besides differential wall thickness have been used in attempts to identify cambial initials. Bannan (1955) reported that the functioning initial could be identified in radial sections of *Thuja occidentalis* because it is slightly shorter than the adjoining derived xylem mother cells. Newman (1956) used the smallest cell in a ray, which he regarded as the ray initial, to identify the initials in neighboring rows of fusiform cells in *Pinus radiata*. Cambial cells that have recently undergone anticlinal division have been used to identify initials (Newman, 1956; Philipson et al., 1971), but anticlinal divisions of cambial initials are never frequent, and the cambial derivatives may also divide anticlinally (Cumbie, 1963; Bannan, 1968; Murmanis, 1970; Catesson, 1964, 1974).

As noted by Catesson (1994), the difficulty in recognizing the cambial initials is a consequence of a nearly total ignorance of the molecular events linked to derivative production and of the early steps of derivative differentiation. The first recognizable markers at light and electron microscope levels are cell enlargement and cell wall thickening. By that time the biochemical processes leading to cell determination and differentiation are already well under way. Preliminary studies of cell wall structure, composition, and development have provided some idea of the earliest cell wall changes occurring in cambial derivatives, including differences in the early biosynthesis of the microfibrillar skeletons in cell walls of derivatives on the phloem and xylem sides of the cambium (Catesson, 1989; Catesson et al., 1994; Baïer et al., 1994), and changes in the arrangement of the cortical microtubules from thick-walled, dormant cambial cells to thin-walled actively dividing cells to differentiating cambial derivatives (Chaffey et al., 1997b, 1998).

Biochemical and pectin immunolocalization studies on the vascular cambium of *Populus* spp. indicate that differences in pectin distribution and composition can be used as early markers of cell differentiation in both the xylem and the phloem (Guglielmino et al., 1997b; Ermel et al., 2000; Follet-Gueye et al., 2000). These studies confirm the results of an earlier one indicating that pectin distribution and calcium localization in cells on the xylem side of the cambium differ from those in cells on the phloem side at a very early stage of commitment (Baïer et al., 1994). Immunolocalization of pectin methylesterase, which controls the degree of methylation and hence the plasticity of cell walls, also revealed a different distribution of enzymes in actively dividing cambial cells and their immediate derivatives (Guglielmino et al., 1997a). Initially the enzymes occurred exclusively in the Golgi bodies, later in both Golgi bodies and in wall junctions, indicating that the activity of neutral pectin methylesterases might also be considered an early marker of differentiation in cambial derivatives (Micheli et al., 2000).

# ▌DEVELOPMENTAL CHANGES

As the core of secondary xylem increases in thickness, the cambium is displaced outward and its circumference increases. This increase is accomplished by division of cells, but in arborescent species it also involves complex phenomena of intrusive growth, loss of initials, and formation of ray initials from fusiform initials. The changes in the cambium are reflected by changes in the radial files of cells in the xylem or phloem as seen in serial tangential sections. By following these changes, it is possible to reconstruct the past events in the cambium.

Events in the cambium of conifers can be safely inferred from changes in tracheid numbers and orientation because the tracheids undergo relatively little elongation (apical intrusive growth) and lateral expansion during differentiation. By contrast, the cambial pattern, in general, is not well preserved in the secondary xylem of hardwoods. Fiber elongation in hardwoods typically is much greater than elongation of tracheids in conifers. That, in addition to the often considerable lateral expansion of differentiating vessel elements, precludes complete continuity in observation of cambial changes in such secondary xylem. In some hardwoods, however, the terminal layer of xylem in each annual ring preserves the cell pattern that existed in the cambium when that layer was formed (Hejnowicz and Krawczyszyn, 1969; Krawczyszyn, 1977; Włoch et al., 1993). Thus the terminal layers of xylem from successive annual rings may be used to determine periodic structural changes that have occurred in the cambium. In other hardwoods, changes may be followed through the orientation and relative positions (splitting and uniting) of the xylem rays (Krawczyszyn, 1977; Włoch and Szendera, 1992). In still others, developmental changes in the cambium may be determined from a study of serial tangential sections of the phloem, providing large quantities of relatively undistorted phloem with easily distinguished growth increments accumulate in the bark (Evert, 1961).

The divisions increasing the number of initials are called **multiplicative divisions** (Bannan, 1955). In species having storied cambia (cambia that have short fusiform initials), the multiplicative divisions are mostly radial anticlinal (Fig. 12.8A; Zagórska-Marek, 1984). Thus two cells appear side by side where one was present formerly, and each enlarges tangentially. Slight apical intrusive growth restores the pointed ends to the daughter cells. In herbaceous and shrubby eudicots the anticlinal divisions are frequently lateral; that is, they intersect twice the same mother cell wall (Fig. 12.8B; Cumbie, 1969). In species having nonstoried cambia (cambia with long initials), the initials divide by formation of more or less inclined, or oblique, anticlinal walls (Fig. 12.8C–E; *pseudotransverse divisions*), and each new cell elongates by apical intrusive growth. As a result of this growth the new sister cells come to lie side by side in the tangential plane (Fig. 12.8F, G), and they thus increase the circumference of the cambium. During the intrusive growth the ends of the cells may fork (Fig. 12.8H, I). The ray initials also divide radially anticlinally in species that have multiseriate rays. Although both xylem and phloem mother cells may sometimes divide anticlinally, anticlinal divisions creating new cambial initials are restricted to cambial initials: only cambial initials can beget cambial initials.

A wide range of variation exists in the ratio of ray initials to fusiform initials; for example, the fusiform initials constitute 25% of the cambial area in *Dillenia*

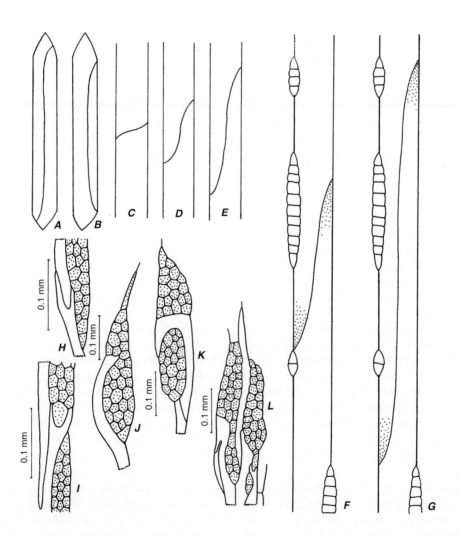

**FIGURE 12.8**

Division and growth of fusiform initials. Initial divided: **A**, by radial anticlinal wall; **B**, by lateral anticlinal wall; **C–E**, by various oblique anticlinal walls. **F**, **G**, oblique anticlinal division is followed by apical intrusive growth (growing apices are stippled). **H**, **I**, forking of fusiform initials during intrusive growth (*Juglans*). **J–L**, intrusion of fusiform initials into rays (*Liriodendron*). (All tangential views.) (From Esau, 1977.)

*indica* (Ghouse and Yunus, 1974) and 100% in rayless *Alseuosmia macrophylla* and *A. pusilla* (Paliwal and Srivastava, 1969). The ratio of ray to fusiform initials tends to increase with age of the stem but reaches a limit beyond which it does not change, resulting in a proportion of ray cells characteristic of the species (Ghouse and Yunus, 1976; Gregory, 1977).

### Formation of New Ray Initials from Fusiform Initials or Their Segments Is a Common Phenomenon

The addition of new ray initials maintains a relative constancy in the ratio between the rays and the axial components of the vascular cylinder (Braun, 1955). This constancy results from the addition of new rays as the column of xylem increases in girth; that is, new ray initials appear in the cambium. These new ray initials are derived from fusiform initials.

The initials of new uniseriate rays may arise as single cells, which are cut from the ends or sides of fusiform initials (conifers, Braun, 1955) or by transverse divisions of such initials (herbaceous and shrubby eudicots, Cumbie, 1967a, b, 1969). The origin of rays, however, may be a highly complicated process involving transverse subdivision of fusiform initials into several cells, loss of some of the products of these divisions, and the transformation of others into ray initials (Braun, 1955; Evert, 1961; Rao, 1988). In conifers and eudicots, new uniseriate rays begin as rays one or two cells high and only gradually attain the height typical for the species (Braun, 1955; Evert, 1961).

The increase in height of rays occurs through the union of newly formed ray initials with existing ones, through transverse divisions of the established ray initials, and through fusion of rays located one above the other (Fig. 12.9). In the formation of multiseriate rays, radial anticlinal divisions and fusions of laterally approximated rays are involved. Indications are that in the process of fusion some fusiform initials intervening between rays are converted into ray initials by transverse divisions; others are displaced toward the xylem or the phloem and are thus lost from the initial zone. The reverse process, a splitting of rays, also occurs. A common method of such splitting involves a breaking up of a panel of ray initials by a fusiform initial that intrudes among the ray initials (Fig. 12.8I–L). In some species, rays are dissected through the expansion of ray initials to fusiform size.

The phenomenon of loss of initials has been studied extensively in the conifers (Bannan, 1951-1962; Forward and Nolan, 1962; Hejnowicz, 1961) and less so in the angiosperms (Evert, 1961; Cumbie, 1963, 1984; Cheadle and Esau, 1964). The loss of fusiform initials is usually gradual. Before an initial is eliminated from the cambium, its precursors fail to enlarge normally—possibly even diminishing in size through loss of turgor—and become

abnormal in shape. Unequal periclinal divisions separate such cells into smaller and larger derivatives, the smaller of which remains the initial (Fig. 12.10C, G). Thus, gradually, the declining initial is reduced in size, particularly in length (Fig. 10.10D–F). Some of the short initials lapse into maturity; that is, they are lost outright from the cambium by maturing into xylem or phloem elements. Others become ray initials with or without further divisions. In transverse sections the loss of initials is revealed by discontinuities in the radial files of cells (Fig. 12.10A). The space released by a declining initial is filled by lateral expansion and/or by the intrusive growth of surviving initials. In *Hibiscus lasiocarpus* (Cumbie, 1963), *Aeschynomene hispida* (Butterfield, 1972), and *Aeschynomene virginica* (Cumbie, 1984)—all three herbaceous eudicots—there is no outright loss of fusiform initials, only the conversion of fusiform initials to ray initials.

The loss of fusiform initials is associated with the anticlinal divisions giving rise to new initials. The production of new initials typically results in numbers of cells far in excess of those necessary for adequate circumferential expansion. This excess production is accompanied by heavy loss, however, so that the net gain represents only a small part of the number produced. The loss appears to be related to vigor of growth. In *Thuja occidentalis*, the survival rate was found to be 20% when the annual xylem increment was 3 mm wide, whereas at the lowest rates the rate of loss and that of new production were almost equal (Bannan, 1960a). The accommodation to the increase in girth probably occurred through elongation of cells. In *Pyrus communis*, the outright loss of new fusiform initials was calculated to be 50%; roughly another 15% were transformed to ray initials (Fig. 12.11; Evert, 1961). Consequently only about 35% of the new initials that arose through anticlinal division survived and repeated the cycle of elongation and division. In *Liriodendron*, the loss of initials by maturation and by conversion into ray initials nearly equaled the addition of new fusiform initials to the cambium (Cheadle and Esau, 1964). Considerable evidence indicates that following anticlinal division, in both conifers and woody angiosperms, the longer sister cells and those with the most extensive ray contacts tend to survive (Bannan, 1956, 1968; Bannan and Bayly, 1956; Evert, 1961; Cheadle and Esau, 1964). It has been suggested that the fusiform initials with the greatest ray contact survive because they are in better position to compete for water, food materials, and other substances necessary for growth (Bannan, 1951), and that the selection of the longest sister fusiform initials contributes to the maintenance of an efficient cell length in the secondary vascular tissues (Bannan and Bayly, 1956).

As mentioned previously, anticlinal divisions are followed by intrusive elongation of the resulting cells. The

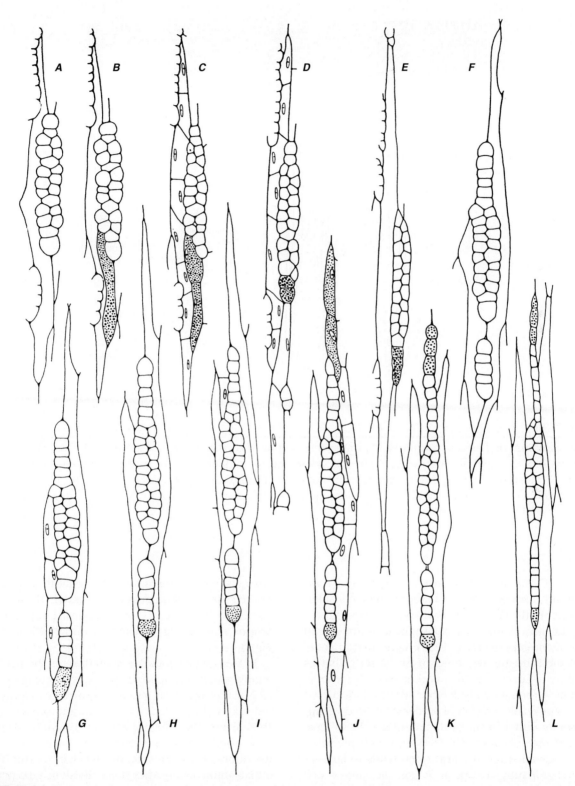

**FIGURE 12.9**

Drawings from serial tangential sections of the phloem of pear (*Pyrus communis*) to illustrate developmental changes in the cambium. In both series (**A-E**; **F-L**), each successive section is nearer the cambium. **A-D** and **F-K** represent derivatives of cambial initials; **E** and **L** are in the cambium. Stippled cells mark origin of ray initial. Parenchyma cells are with nuclei; sieve-tube elements, ray cells, and cambial cells are without nuclei. In the series **A-E**, new ray initials arose from a segment cut off the side of a fusiform initial (**B**). In series **F-L**, new ray initials arose in two ways: from a segment cut off the end of a fusiform initial (**G**) and through a reduction in length of a relatively short fusiform initial followed by its conversion to ray initials (**J**, **K**). Note the manner by which the ray in **L** attained its height. (All, ×260. From Evert, 1961.)

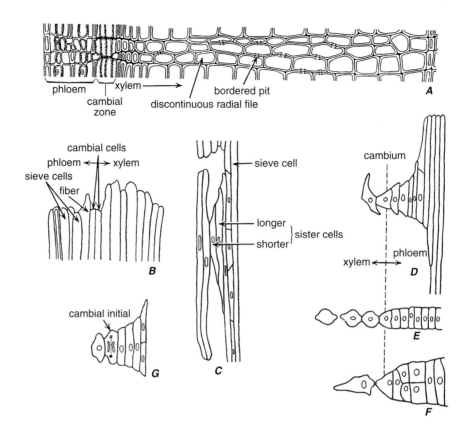

**FIGURE 12.10**

Vascular cambium of *Thuja occidentalis*. **A**, transverse section showing relation of xylem and phloem to cambium. The discontinuous radial file is represented in the xylem and the phloem but not in the cambium—loss of fusiform initial. **B–G**, radial sections. **B**, differences in length of cells in cambial zone. **C**, early stage in shortening of cambial cells by asymmetric periclinal division. **D**, earlier, and **E–G**, later stages in shortening of fusiform initials to dimensions of ray initials. (After Bannan, 1953, 1955.)

direction of this elongation may be polar. In *Thuja occidentalis*, for example, it was found to be considerably greater in the downward than in the upward directions (Bannan, 1956). In a subsequent study on 20 species of conifers, Bannan (1968) found that in some areas of the cambium the lower of two sister cells was more likely to survive, while in other areas the reverse was true. Although considerable variation occurred within a single tree, an overall tendency existed within a species for either lower or upper sister cells to have a better chance of survival. Cell elongation is predominantly basipetal when the lower cell tends to survive and predominantly acropetal when the upper cell survives.

Intrusive growth of fusiform cambial initials is generally thought of as occurring between radial walls, with little or no change in cell inclination. Under such circumstances the packets of cells originating from a given initial are located in the same radial file. Intrusive growth of initial cell ends may occur between periclinal walls of neighboring cell files, bring about

changes in cell inclination, and result in dislocation of packets in tangential planes. Under these circumstances a single file of cells can consist of packets with origins from different cambial initials (Włoch et al., 2001).

In trees with moderate growth rates the majority of multiplicative (anticlinal) divisions occur toward the end of the period of maximal growth concerned with the seasonal production of xylem and phloem (Braun, 1955; Evert, 1961, 1963b; Bannan, 1968). In plants with nonstoried cambia this timing in divisions means that the cambium contains, on the average, shorter fusiform initials immediately after these divisions take place and longer ones immediately before. Subsequently the new surviving cells elongate so that the average length of the initials increases until a new period of divisions ensues near the end of the growing season. This fluctuation in the average length of fusiform initials is reflected in the variation in length of their derivatives (Table 12.1). In young and vigorously growing trees, anticlinal divisions are less definitely restricted to the latter part of the

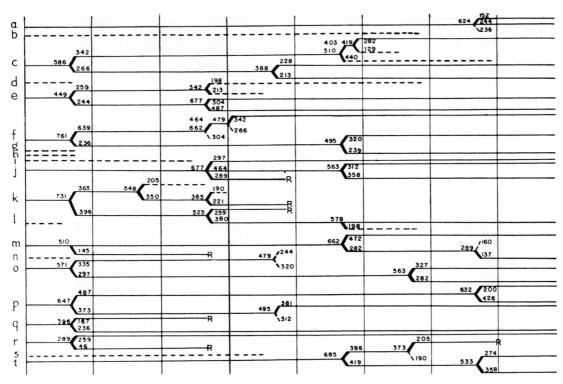

**FIGURE 12.11**

Diagram illustrating the developmental changes that took place over a seven-year period in one area of the vascular cambium of pear (*Pyrus communis*) as determined from serial tangential sections of the secondary phloem. Each lineal series of horizontal lines depicts the changes that took place within one group of related initials during the seven-year period. The forking of a horizontal line represents the division of an initial; a side branch indicates a division that produced a segment off the side of an initial. The broken lines mark failing initials, and the termination of these lines denotes the disappearance of the initials from the cambium. The letter R signifies the transformation of a fusiform initial to one or more ray initials. The vertical lines identify yearly growth increments. No attempt was made to indicate differences in widths of growth increments. The oldest growth increment (farthest from the cambium) is on the left. (From Evert, 1961.)

**TABLE 12.1** ■ **Combined Average Lengths of the First- and Last-formed Elements (Sieve-Tube Elements and Parenchyma Strands) of 7 Successive Growth Increments in a Defined Area of the Stem Secondary Phloem of *Pyrus communis***

| Average Lengths (μm) | |
|---|---|
| First-formed Elements | Last-formed Elements |
| 299 | 461 |
| 409 | 462 |
| 367 | 479 |
| 420 | 476 |
| 369 | 475 |
| 362 | 467 |
| 384 | 462 |

Source: From Evert, 1961.

growing season and may be frequent throughout the growing season.

### Domains Can Be Recognized within the Cambium

As mentioned previously, in nonstoried cambia, increase in girth of the cambium involves pseudotransverse, or oblique anticlinal, divisions followed by apical intrusive growth of the two daughter cells. The orientation of these two events may be either to the right (Z) or to the left (S) (Zagórska-Marek, 1995). The distribution of Z and S configurations on the cambial surface tends not to be random, so that areas exist where one or the other configuration prevails. Such areas are called **cambial domains** (Fig. 12.12). Often the inclination of the initials in the same domains cycles, or changes, with time from Z to S, and vice versa. The scale and temporal aspects of these changes determine whether the wood

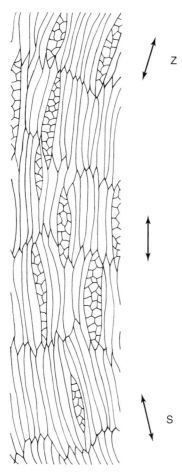

**FIGURE 12.12**

Diagram of a vascular cambium, as seen in tangential section, showing alternating domains. Below, the cell axes are inclined toward the left (S), in the middle of the diagram parallel to the stem axis, and above, toward the right (Z). (After Catesson, 1984. © Masson, Paris.)

grain is straight, spiral, wavy, or interlocked, or even of a more complex pattern (Krawczyszyn and Romberger, 1979; Harris, 1989; Włoch et al., 1993). In cambia producing straight-grained wood, the effect of nonrandomly oriented events is minimized by their low frequencies (Hejnowicz, 1971).

In species with storied cambia, the mechanism of reorientation, to the right (Z) or to the left (S), depends mainly on intrusive growth and the elimination of parts of initials as a result of unequal periclinal divisions (Hejnowicz and Zagórska-Marek, 1974; Włoch, 1976, 1981). Intrusive growth produces a new tip beside the original one, resulting in formation of a forked end. Eventually unequal periclinal division divides the initial into two cells unequal in size. The cell with the old tip loses the initial function, becoming either a xylem or a phloem mother cell (Włoch and Polap, 1994).

# SEASONAL CHANGES IN CAMBIAL CELL ULTRASTRUCTURE

Virtually all of the information available on the changes accompanying the seasonal cycle of meristematic activity in the vascular cambium at the ultrastructural level comes from studies of temperate tree species (Barnett, 1981, 1992; Rao, 1985; Sennerby-Forsse, 1986; Fahn and Werker, 1990; Catesson, 1994; Larson, 1994; Farrar and Evert, 1997a; Lachaud et al., 1999; Rensing and Samuels, 2004). In general, the changes are basically similar for hardwood and softwood species. Some of the changes—such as changes in degree of vacuolation and in storage products—are associated with cold acclimation (hardening) or deacclimation (dehardening) and have been described for other tissues (Wisniewski and Ashworth, 1986; Sagisaka et al., 1990; Kuroda and Sagisaka, 1993).

Cells of the dormant cambium are characterized by the density of their protoplasts and the thickness of their walls, most notably of their radial walls, which have a beaded appearance as viewed in tangential sections (Fig. 12.13A). The beaded appearance is due to the presence of deeply depressed primary pit-fields, which alternate with the thickened wall areas.

Both fusiform and ray cells of the dormant cambium contain numerous small vacuoles (Fig. 12.14). The vacuoles commonly contain proteinaceous material, others may contain polyphenols (tannins). Lipids in the form of droplets are common storage products of dormant cambial cells. Typically their cycle is opposite that of starch. For instance, whereas lipid droplets are numerous in dormant cambial cells of *Robinia pseudoacacia*, starch grains are absent from such cells (Farrar and Evert, 1997a). The reverse is true of cells in the active cambium. Hydrolysis of starch during the transition to dormancy may be a component of the freezing tolerance mechanism in the temperate zone trees, the resultant sugars serving as cryoprotectants.

During the transition to dormancy and thickening of the cambial cell walls, Golgi activity is high and the plasma membrane contains numerous invaginations. Gradually the Golgi bodies become inactive, and the plasma membrane assumes a smooth outline. Cyclosis stops. The dormant cambial cells contain numerous free ribosomes not aggregated as polysomes and mostly smooth tubular endoplasmic reticulum. Cambial cells contain all of the cytoplasmic components typical of parenchymatous cells.

Reactivation of the cambium is preceded by a resumption of cyclosis followed by the hydrolysis of storage products and coalescence of the numerous small vacuoles to form fewer larger vacuoles. The formation of fewer and larger vacuoles in cambial cells of *Populus trichocarpa* during reactivation has been shown to be

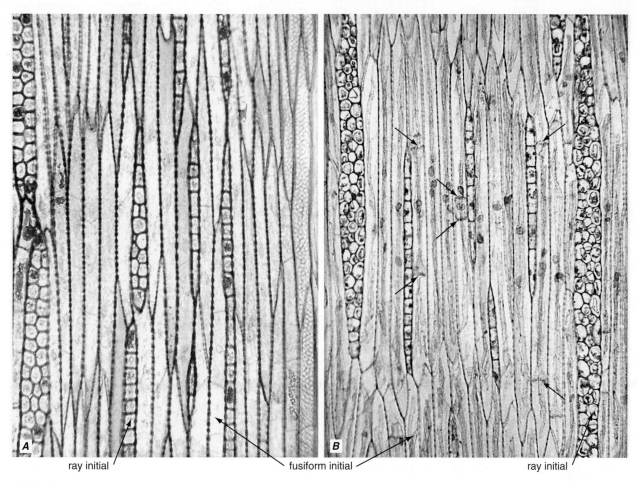

**FIGURE 12.13**

Dormant (**A**) and active (**B**) cambia of basswood (*Tilia americana*) as seen in tangential sections. Note the beaded appearance of the radial walls of the dormant fusiform cells in **A**, and the phragmoplasts (arrows) in dividing fusiform cells in **B**. (Both, ×400.)

associated with an increase uptake of K⁺, probably mediated by the activity of a plasma membrane H⁺-ATPase (Arend and Fromm, 2000). Concomitantly the plasma membrane becomes irregular in outline and begins to form numerous small invaginations. Some invaginations increase in size, protrude into the vacuole, and push the tonoplast inward. These invaginations, with their contents, eventually pinch off into the vacuole. This is a period of much membrane trafficking. Cambial reactivation is also preceded by a partial loosening of the radial walls, especially in cell junctions (Rao, 1985; Funada and Catesson, 1991). With the renewal of cambial activity, the radial walls of the cambial cells thin down (Fig. 12.13B).

The cortical microtubules of fusiform cambial cells are randomly arranged (Chaffey, 2000; Chaffey et al.,

2000; Funada et al., 2000; Chaffey et al., 2002). Bundles of actin filaments have been observed in fusiform cambial cells (Chaffey, 2000; Chaffey and Barlow, 2000; Funada et al., 2000). They are more or less longitudinally oriented or arranged as a series of parallel helices of low pitch. The actin filament bundles apparently extend the length of the cell. The cortical microtubules in ray cells of the cambial zone are also randomly arranged. By contrast, bundles of actin filaments are less frequent in ray cambial cells than in fusiform cambial cells, and they are randomly arranged (Chaffey and Barlow, 2000). These arrangements of microtubules and actin filaments persist throughout the seasonal cycle in both cambial cell types.

The most conspicuous feature of the fusiform cells of active cambia is the presence of a large central vacuole

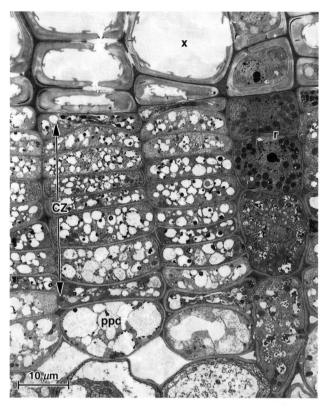

**FIGURE 12.14**

Electron micrograph of transverse section of dormant cambium of black locust (*Robinia pseudoacacia*). The cambial zone (cz) is bordered by mature xylem (x) above and by phloem parenchyma cells (ppc) immediately below. The two rows of fusiform cells are bordered by a ray (r) on the right. Note numerous small vacuoles (clear areas) in the fusiform cells. (From Farrar and Evert, 1997a, Fig. 2. © 1997, Springer-Verlag.)

(Fig. 12.15A). These cells also are characterized by the presence of mostly rough endoplasmic reticulum, ribosomes mostly aggregated as polysomes, and considerable Golgi activity. Table 12.2 summarizes, in general, some of the cytological changes that occur in the cambium during the seasonal cycle.

The nuclei of the fusiform cambial cells also exhibit seasonal variations. In conifers, the nuclei tend to be much longer and narrower during the fall and winter than during the spring and summer (Bailey, 1920b). In the hardwoods *Acer pseudoplatanus* (Catesson, 1980) and *Tectona grandis* (Dave and Rao, 1981), cessation of cambial activity is followed by a decrease in diameter of the nucleoli, which assume a resting appearance indicative of a state of low RNA synthesis. Similar changes in nuclear behavior have been observed in *Abies balsamea* (Mellerowicz et al., 1993). Fluctuations in DNA content also have been demonstrated in the fusiform cambial cells of balsam fir (Mellerowicz et al., 1989, 1990, 1992;

Lloyd et al., 1996). At the end of the growing season (September in Central New Brunswick, Canada) the interphase nuclei in balsam fir remained in the $G_1$ phase and at the 2C DNA level until after December, when DNA synthesis (S phase) was resumed. DNA levels were maximal at the beginning of cambial activity in April. They decreased during the cambial growing season and reached minimum levels in September.

The uninucleate condition of fusiform cambial cells was first recognized by Bailey (1919, 1920c) and since then has generally been accepted as such by other investigators. Occasional reports of multinucleate fusiform cells have appeared in the literature (Patel, 1975; Ghouse and Khan, 1977; Hashmi and Ghouse, 1978; Dave and Rao, 1981; Iqbal and Ghouse, 1987; Venugopal and Krishnamurthy, 1989). In all such cases the putative multinucleate condition was detected in tangential sections of cambia viewed with the light microscope. It is likely that the multinucleate appearance results from the narrow radial diameters of the exactly superimposed fusiform cells whose nuclei lie close to the same focal plane (Farrar and Evert, 1997b). Utilizing confocal laser scanning microscopy, which clearly allowed adjacent layers of cells in the cambium to be distinguished and the number of nuclei per cell to be determined, Kitin and co-workers (Kitin et al., 2002) were able to show that the fusiform cells in the cambium of *Kalopanax pictus* are exclusively uninucleate. The putative multinucleate condition of the fusiform cells in the pertinent tree species needs to be critically reexamined.

Little information is available on the distribution and frequency of plasmodesmata in the walls of cambial cells. In *Fraxinus excelsior*, the plasmodesmatal frequency has been reported to be highest in the tangential walls between ray cells and lowest in the tangential walls between fusiform cells (Goosen-de Roo, 1981). In *Robinia pseudoacacia* (Farrar, 1995), plasmodesmata are scattered throughout the tangential walls between fusiform cells; that is, they are not aggregated in primary pit-fields. By contrast, plasmodesmata in tangential walls between ray cells are aggregated in primary pit-fields. Moreover, plasmodesmata are aggregated in primary pit-fields in the radial walls between all cell combinations in the cambial region: between fusiform cells, between ray cells, and between fusiform cells and ray cells. The plasmodesmatal frequency (plasmodesmata per micrometer of cell wall interface) is highest in the tangential walls between ray cells. The lowest plasmodesmatal frequencies occur in the tangential walls between fusiform cells.

## ▌CYTOKINESIS OF FUSIFORM CELLS

As discussed previously (Chapter 4), long before the initiation of cytokinesis in relatively small vacuolated

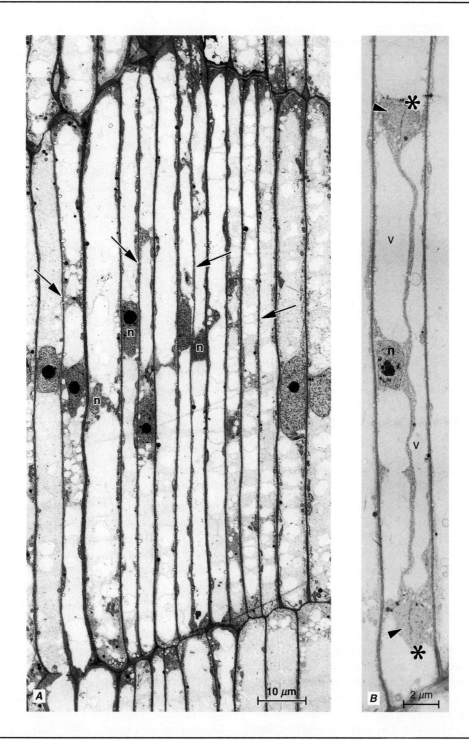

## FIGURE 12.15

Radial views of fusiform cells in the active cambium of black locust (*Robinia pseudoacacia*). **A**, view of cambial zone showing highly vacuolate, uninucleate fusiform cells. Arrows point to recently formed tangential walls. **B**, view of phragmoplast (arrowheads) and developing cell plate in dividing fusiform cell. The phragmosome is represented by the region of cytoplasm just in advance of the phragmoplast (asterisk). Other details: n, nucleus; v, vacuole. (From Farrar and Evert, 1997b, Figs. 2 and 17. © 1997, Springer-Verlag.)

plant cells the nucleus migrates to the center of the cell. The strands of cytoplasm supporting the nucleus then aggregate into a cytoplasmic plate, the phragmosome, that bisects the cell in the plane to be assumed later by the cell plate. In addition to nuclear position-ing and phragmosome formation, a preprophase band of microtubules typically is formed, marking the plane of the future cell plate. Thus both the phragmosome and preprophase band define the same plane.

**TABLE 12.2 ■ Cytological Changes in Cambium during the Seasonal Cycle**

| Physiological Stage | Activity | Transition to Dormancy | Dormancy | Reactivation |
|---|---|---|---|---|
| Nucleus | Dividing | $G_1$ stage | S stage | S or $G_2$ stage |
| Nucleolus diameter | Rather large | Decreasing | Rather small | Increasing |
| Vacuoles | Few, large | Several, fragmenting | Small, numerous | Numerous, coalescing |
| Cyclosis | Yes | Yes | No | Yes |
| Golgi bodies | Numerous, active | Numerous, active | Few, mostly inactive | Resumption of activity |
| ER | Rough | Rough | Mostly smooth | Rough |
| Ribosomes | Polysomes | Polysomes | Free | Polysomes |
| Actin filaments | Bundles in some species | NR | Bundles in some species | NR |
| Cortical microtubules | Random | NR | Helically arranged | NR |
| Plasma membrane | Irregular, some invaginations | Often large invaginations | Smooth | Irregular, some invaginations |
| Mitochondria | Round to oval | Round to elongated | Round to oval | Round to elongated |
| Plastids | Small with tubules or a few thylakoids | Small with tubules or a few thylakoids | Small with tubules or a few thylakoids | Small with tubules or a few thylakoids |

Source: Adapted from Lachaud et al., 1999.
Note: The presence of phytoferritin or of dense inclusions and the presence and seasonal distribution of starch in the plastids depend on plant species. NR = not recorded.

Cytokinesis of the fusiform cells in the vascular cambium is of special interest because of the great lengths of these highly vacuolated cells, compared with the relatively small dimensions of most vacuolated plant cells. (Fusiform cells may be several hundred times as long as they are wide radially.) Yet, when a fusiform cambial cell divides longitudinally, it must form a new cell wall along its entire length. In such a division the diameter of the phragmoplast initially is very much shorter than the long diameter, or length, of the cell (Fig. 12.16). Consequently the phragmoplast and the cell plate reach the longitudinal walls of the fusiform cell soon after mitosis, but the progress of the phragmoplast and the cell plate toward the ends of the cell is an extended process. Before the side walls are reached, the phragmoplast appears as a halo about the daughter nuclei in tangential sections of the cambium (Fig. 12.16A). After the side walls are intersected by the cell plate—but before the ends of the cell are reached—the phragmoplast appears as two bars intersecting the side walls (Fig. 12.16A). In radial sections, the phragmoplasts are seen in sectional view. There they have a roughly wedge-shaped outline, being bluntly convex in front and tapering at the rear along the cell plate (Figs. 12.15B and 12.16B–D).

Both callose and myosin have been immunolocalized in the cell plate of fusiform cambial cells, but not in the portion of the plate forming within the phragmoplast, in roots and shoots of *Populus tremula* × *P. tremuloides*, *Aesculus hippocastanum*, and *Pinus pinea* (Chaffey and Barlow, 2002). Tubulin and actin, by contrast, were largely confined to the phragmoplast, whereas actin filaments were localized alongside the growing cell plate, except for the portion of the plate forming within the phragmoplast. It has been suggested that an acto-myosin contractile system may play a role in pushing the phragmoplast toward the parental cell walls (Chaffey and Barlow, 2002).

Few ultrastructural studies have been published on cell division in large, highly vacuolated fusiform cells of the vascular cambium (Evert and Deshpande, 1970; Goosen-de Roo et al., 1980; Farrar and Evert, 1997b; Rensing et al., 2002). Those studies have revealed that the ultrastructure and sequence of events of mitosis and cytokinesis in dividing fusiform cells are essentially similar to those observed during the division of shorter cells, with two notable exceptions. In five of the species examined—*Tilia americana*, *Ulmus americana* (Evert and Deshpande, 1970), *Robinia pseudoacacia* (Farrar and Evert, 1997b), *Pinus ponderosa* and *P. contorta* (Rensing et al., 2002)—preprophase bands appear not to exist in the fusiform cells, although such bands were found in dividing ray cells of the three hardwood species. In addition, in the same five species, phragmo-

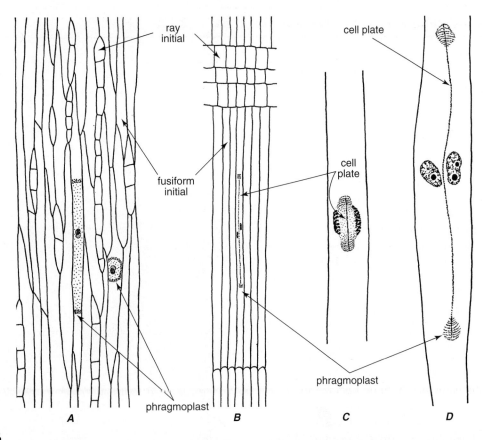

**FIGURE 12.16**

Cytokinesis in vascular cambium of *Nicotiana tabacum* as seen in tangential (**A**) and radial (**B–D**) sections of stem. Partly formed cell plates in surface (**A**) and side (**B**) views. **C**, early stage of division; **D**, later stage. (A, B, ×120; C, D, ×600.)

somes do not extend the length of the dividing cells. Rather in the fusiform cells of these species the phragmosome is represented by a broad cytoplasmic plate that migrates to the ends of the cell just in advance of the phragmoplast. It is pertinent to note that Oribe et al. (2001), utilizing immunofluorescence and confocal laser scanning microscopy, also found no evidence of the presence of preprophase bands in fusiform cambial cells of *Abies sachalinensis*; by contrast, preprophase bands were observed in ray cambial cells.

Preprophase bands, consisting of a relatively small number of microtubules, have been reported to occur in the fusiform cells of *Fraxinus excelsior* (Goosen-de Roo et al., 1980). Similar, relatively small, groups of microtubules were found along the radial walls of fusiform cells in *Robinia pseudoacacia*, but they were not interpreted as preprophase bands (Farrar and Evert, 1997b). On the other hand, apparently extended phragmosomes have been illustrated in radial sections of fusiform cells in the vascular cambium of *Fraxinus excelsior* (Goosen-de Roo et al., 1984). The tonoplast-bound phragmosomes consisted of a thin, perforated cytoplas-

mic layer located in the plane of the future cell plate and contained both microtubules and bundles of actin filaments. Although Arend and Fromm (2003) state that in fusiform cells of *Populus trichocarpa* "the phragmosome forms a long, dilated cytoplasmic strand throughout the cell," adequate documentation in support of this statement is wanting.

# ▌SEASONAL ACTIVITY

In woody perennials of temperate regions, periods of growth and reproduction alternate with periods of relative inactivity during winter. The seasonal periodicity also finds its expression in the cambial activity, and occurs in both deciduous and evergreen species. Production of new cells by the vascular cambium slows down or ceases entirely during dormancy, and the vascular tissues mature more or less closely to the cambial initials.

In the spring the dormant period is succeeded by reactivation of the vascular cambium. Workers long have recognized two phases in the resumption of cambial activity: (1) a phase of radial enlargement of the cambial cells ("swelling" of the cambium), during which the fusiform cells become highly vacuolated, followed by (2) the initiation of cell division (Larson, 1994). Although the resumption of cambial activity may be preceded by a decrease in density of their protoplasts, the cambial cells do not enlarge radially prior to cell division in all temperate zone species (Evert, 1961, 1963b; Derr and Evert, 1967; Deshpande, 1967; Davis and Evert, 1968; Tucker and Evert, 1969). In *Robinia pseudoacacia*, cell division begins prior to cell expansion, when many of the cambial cells are still densely cytoplasmic, contain numerous small vacuoles, and have abundant lipid droplets—in other words, when the cambial cells still have many of the characteristics of a dormant cambium (Fig. 12.17; Farrar and Evert, 1997b).

When the cambial cells expand, their radial walls become thinner and weaker. As a result the bark (all tissues outside the vascular cambium) may be easily separated from, or peeled off, the stem. Such separation of the bark from the wood is commonly called "slipping of the bark." The slippage of bark occurs not only through the cambial zone but also—and perhaps most often—through the differentiating xylem where the tracheary elements have attained their maximum diameters but are still without secondary walls. Slippage rarely occurs through the differentiating phloem. Swelling of the cambium and slippage of the bark are often used as an indication of radial growth, or of cambial activity. Slippage may occur, however, before cambial activity begins (Wilcox et al., 1956).

Some workers rely on the number of layers of undifferentiated cells in the cambial zone for recognizing cambial reactivation or the degree of cambial activity (e.g., Paliwal and Prasad, 1970; Paliwal et al., 1975; Villalba and Boninsegna, 1989; Rajput and Rao, 2000). It is difficult, however, to distinguish between cells that are still meristematic and those that are in early stages of differentiation.

The presence of differentiating xylem has also been used as an indication of cambial activity because of a long-held concept that xylem and phloem production begin simultaneously or that xylem production precedes that of phloem. In many species there is no regularity in the location of the first periclinal divisions, and two or more fusiform cells in a given radial file may begin to divide simultaneously. The first additive divisions may be made toward either the xylem or the phloem, depending on the plant species. Any comprehensive study of cambial activity therefore requires the consideration of both xylem and phloem production.

Recognition of the relative times of initiation and cessation of xylem and phloem production is often confounded by the presence in the cambial zone of over-

**FIGURE 12.17**

Radial view of fusiform cells in cambial zone of black locust (*Robinia pseudoacacia*). Cell division has just recently begun in this cambium, whose cells still contain numerous small vacuoles and abundant lipid droplets, both of which are characteristic of the dormant cambium. Arrow points to recently formed cell wall and arrowheads to a phragmoplast in a fusiform cell in the process of dividing. (From Farrar and Evert, 1997b, Fig. 1. © 1997, Springer-Verlag.)

**FIGURE 12.18**

Transverse sections showing xylem and phloem increments in stems of (**A**) trembling aspen (*Populus tremuloides*) and (**B**) white oak (*Quercus alba*) trees. Note that the size of the xylem increments greatly exceeds that of the phloem increments. Details: pi, phloem increment; xi, xylem increment. (Both, ×32. **A**, from Evert and Kozlowski, 1967.)

⟶

wintering phloem (see below) and/or xylem mother cells (Tepper and Hollis, 1967; Zasada and Zahner, 1969; Imagawa and Ishida, 1972; Suzuki et al., 1996), which are indistinguishable from the initials and which complete their differentiation in the spring. It is often uncertain whether authors have distinguished between the maturation of such overwintering elements and the production and subsequent differentiation of cells formed by new cambial activity. In addition, with regard to the cessation of production of vascular tissue, the terms production and differentiation are often used interchangeably, so a clear distinction is not always made between the production of new cells by cell division and the subsequent differentiation of these cells. Xylem and phloem differentiation may continue for some time after cell division has been completed; hence the presence of differentiating cells cannot be reliably used as an indication of cambial activity. Only the presence of mitotic figures and/or phragmoplasts can reliably be used as signs of cambial activity.

### The Size of the Xylem Increment Produced during One Year Generally Exceeds That of the Phloem

In *Eucalyptus camaldulensis* cell production toward the xylem was about four times that toward the phloem (Waisel et al., 1966), and in *Carya pecan*, about five times (Artschwager, 1950). Xylem to phloem ratios observed for some conifers were 6:1 in *Cupressus sempervirens* (Liphschitz et al., 1981), 10:1 in *Pseudotsuga menziesii* (Grillos and Smith, 1959), 14:1 in *Abies concolor* (Wilson, 1963), and 15:1 in a vigorously growing *Thuja occidentalis* (Bannan, 1955). On the other hand, in the tropical hardwood *Mimusops elengi* (Ghouse and Hashmi, 1983), almost equal amounts of xylem and phloem were produced, and in *Polyalthia longifolia*, phloem production exceeded that of xylem by at least 500 micrometers per year (Ghouse and Hashmi, 1978). Figure 12.18 shows the relative size of the last-formed xylem and phloem increments in *Populus tremuloides* and *Quercus alba*. In both species the xylem to phloem ratio was 10:1.

The size of a growth increment may vary greatly around the circumference of a stem from one part of a transverse section to another. In *Pyrus communis*, the size of the phloem increment produced during one year varied little from one part of a transverse section to another, whereas the seasonal amount of xylem varied greatly (Evert, 1961). Similarly, in *Thuja occidentalis*, the annual increment of phloem was about the same regardless of the size of the corresponding xylem increment (Bannan, 1955).

Although the relative size of the xylem and phloem increments were not recorded, the number of tracheids produced annually by *Picea glauca* trees growing in Alaska and New England were found to be the same, even though the period of cambial activity was much shorter in Alaska (65° N) than in New England (43° N) (Gregory and Wilson, 1968). The Alaska white spruce had adapted to the shorter growing season by increasing the rate of cell division in the cambial zone.

The rate of periclinal division of ray initials typically is low compared with that of fusiform initials. In *Pyrus malus (Malus domestica)* periclinal division of the ray initials did not begin until about a month and a half after periclinal division was initiated in the fusiform cells (Evert, 1963b). This coincided with the beginning of xylem production. Before then the ray cells of the cambial zone merely elongated radially, keeping up with the increase in radial growth that occurred primarily toward the phloem. Maximal division of ray initials took place in June, and cessation of division occurred by early July. After that time the newly formed ray cells elongated radially until radial growth was completed.

In most diffuse-porous species and conifers of temperate regions, the initiation of new phloem production and differentiation precedes that of new xylem production and differentiation. This is reflected in Tables 12.3 and 12.4, which, with the exception of *Pyrus communis* (from Davis, California) and *Malus domestica* (from Bozeman, Montana), include trees that grew largely within a 5 kilometer radius of the University of Wisconsin-Madison campus. Note that in the hardwood species lacking mature sieve elements during winter, the first sieve elements to differentiate in the spring arise from overwintering phloem mother cells (Table 12.3). Although some mature sieve elements are present year-round in the conifers, the first sieve elements to differentiate in the spring also are represented by overwintering phloem mother cells (Table 12.4).

Initiation of phloem production also precedes that of xylem production in the temperate diffuse-porous hardwoods *Acer pseudoplatanus* (Cockerham, 1930; Elliott, 1935; Catesson, 1964), *Salix fragilis* (Lawton, 1976), *Salix viminalis* (Sennerby-Forsse, 1986), and *Salix dasyclados* (Sennerby-Forsse and von Fircks, 1987). In contrast to the conifers listed in Table 12.4, the initiation of xylem production precedes that of phloem production in *Thuja occidentalis* (Bannan, 1955), *Pseudotsuga menziesii* (Grillos and Smith, 1959; Sisson, 1968), and *Juniperus californica* (Alfieri and Kemp, 1983).

In the ring-porous hardwood species, new phloem and xylem production and differentiation begin almost simultaneously (Table 12.3). This also is true for the diffuse-porous species *Tilia americana* and *Vitis riparia*, which differ from the other diffuse-porous species listed in Table 12.3 in their possession of large numbers of mature sieve elements that overwinter and function for one or more additional years (Chapter 14). Unlike its ring-porous hardwood counterparts, in the ring-porous gymnosperm *Ephedra californica*, xylem production and differentiation precede phloem production and differentiation (Alfieri and Mottola, 1983).

## A Distinct Seasonality in Cambial Activity Also Occurs in Many Tropical Regions

As mentioned previously (Chapter 11), a distinct seasonality in cambial activity also occurs in many tropical regions that experience severe annual dry seasons. Most of the detailed studies from these regions have been conducted on trees of India. Table 12.5 shows the results of some such studies. It is instructive to compare the results of these studies with those of their counterpart temperate hardwoods (Table 12.2). Note: (1) The relatively longer periods of cambial activity in the tropical species, compared with the temperate species; (2) only *Liquidambar formosana*, a subtropical species from Taiwan, does not have mature sieve elements present in the phloem year-round; (3) in the tropical diffuse-porous species, considerable variability exists in the relative times of initiation of new xylem and phloem production. *Polyalthia* exhibits two periods of phloem production, one before and the other after a period of xylem production. In *Mimusops* and *Delonix*, new xylem production is initiated before that of phloem, by little more than a month in *Mimusops* but by five months in *Delonix*; and (4) in the tropical ring-porous species, new xylem and phloem production and differentiation begin almost simultaneously, as in their temperate ring-porous counterparts.

In some tropical plant species the cambial cells divide more or less continuously, and the xylem and phloem elements undergo gradual differentiation. Based on the absence of discernible growth rings in the xylem, it has been estimated that about 75% of the trees growing in the rainforest of India exhibit continuous cambial activity (Chowdhury, 1961). The percentage of such trees drops to 43% in the rainforest of the Amazon Basin (Mainiere et al., 1983) and to only 15% in that of Malaysia (Koriba, 1958). In a study of the woody flora of South Florida, with a predominant West Indian element, 59% of the tropical species lacked growth rings, apparently the result of continuous cambial activity, even though the climate was markedly seasonal (Tomlinson and Craighead, 1972). In some tropical species (e.g., *Shorea*

**TABLE 12.3 ■ Cambial Activity and Times of Initiation of New Phloem (P) and Xylem (X) Production in Temperate-Zone Woody Angiosperms**

| | Jan | Feb | Mar | Apr | May | Jun | Jly | Aug | Sep | Oct | Nov | Dec |
|---|---|---|---|---|---|---|---|---|---|---|---|---|
| *Pyrus communis* | | | * | P | X | | | | | | | |
| *Malus sylvestris* | | | | *P | X | | | | | | | |
| *Populus tremuloides* | | | | * P | X | | | | | | | |
| *Parthenocissus inserta* | | | | *P | X | | | | | | | |
| *Rhus glabra*[a] | | | | *PX | | | | | | | | |
| *Robinia pseudoacacia*[a] | | | | *PX | | | | | | | | |
| *Celastrus scandens*[a] | | | | | P X | | | | | | | |
| *Acer negundo* | | | | P | X | | | | | | | |
| *Tilia americana* | | | | R | P X | | | | | | | |
| *Vitis riparia* | | | | R | | PX | | | | | | |
| *Quercus* spp[a] | | | | P X | | | | | | | | |
| *Ulmus americana*[a] | | | | P X | | | | | | | | |

a  Ring-porous species

*  First functional sieve elements arise from phloem mother cells that overwinter on outer margin of cambial zone

R  Reactivation

▢ No overwintering mature sieve elements

▣ Some mature sieve elements present year-round

Sources: *Pyrus communis*—Evert, 1960; *Pyrus malus*—Evert, 1963b; *Populus tremuloides*—Davis and Evert, 1968; *Parthenocissus inserta*—Davis and Evert, 1970; *Rhus glabra*—Evert, 1978; *Robinia pseudoacacia*—Derr and Evert, 1967; *Celastrus scandens*—Davis and Evert, 1970; *Acer negundo*—Tucker and Evert, 1969; *Tilia americana*—Evert, 1962; Deshpande, 1967; *Vitis riparia*—Davis and Evert, 1970; *Quercus* spp.—Anderson and Evert, 1965; *Ulmus Americana*—Tucker, 1968.

Note: In species with no overwintering mature sieve elements, the first functional sieve elements in spring originate from phloem mother cells that overwinter on the outer margin of the cambial zone. In two of the species (*Tilia americana* and *Vitis riparia*), with some mature sieve elements present year-round, the sieve elements that overwinter develop dormancy callose at their sieve plates and lateral sieve areas in late fall; the dormant sieve elements are reactivated in spring before the renewal of cambial activity. The times of cessation of phloem and xylem production and differentiation are not indicated.

spp.; Fujii et al., 1999) cell division in the cambium, although continuous year-round, slows down sufficiently seasonally, so indistinct growth boundaries can be discerned in the xylem. Virtually no information is available on the relative times of xylem and phloem production in tropical species exhibiting continuous cambial activity.

The annual course of cambial activity may serve as an indicator of the geographic origin of a species (Fahn, 1962, 1995; Liphschitz and Lev-Yadun, 1986). This is exemplified by the annual course of cambial activity in various woody plants growing in the Mediterranean and desert (Negev) regions of Israel. The range of temperature in these regions is such that cambial activity may occur year-round, provided that such activity is a genetic characteristic of the plant. In desert regions, however, the amount of available soil water becomes a major factor in the control of cambial activity. Plants of

**TABLE 12.4 ■ Cambial Activity and Times of Initiation of New Phloem (P) and Xylem (X) Production and Differentiation in Several Temperate-Zone Conifers**

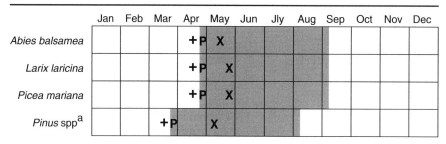

| | Jan | Feb | Mar | Apr | May | Jun | Jly | Aug | Sep | Oct | Nov | Dec |
|---|---|---|---|---|---|---|---|---|---|---|---|---|
| *Abies balsamea* | | | | +P | X | | | | | | | |
| *Larix laricina* | | | | +P | X | | | | | | | |
| *Picea mariana* | | | | +P | X | | | | | | | |
| *Pinus* spp[a] | | | +P | | X | | | | | | | |

a  *Pinus banksiana, P. resinosa, P. strobus*

+  Phloem mother cells on outer margin of
cambial zone begin to differentiate

▨  Some mature sieve elements present year-round

Sources: *Abies balsamea*—Alfieri and Evert, 1973; *Larix laricina*—Alfieri and Evert, 1973; *Picea mariana*—Alfieri and Evert, 1973; *Pinus* spp.—Alfieri and Evert, 1968.
Note: In these species some of the last-formed sieve cells remain functional through winter until new sieve elements differentiate in spring. The first new sieve elements in spring arise from phloem mother cells that overwinter on the outer margin of the cambial zone. The times of cessation of phloem and xylem production and differentiation are not indicated.

temperate Mediterranean origin (*Cedrus libani, Crataegus azarolus, Quercus calliprinos, Q. ithaburensis, Q. boissieri, Pistacia lentiscus,* and *P. palaestina*) growing in the Mediterranean region of Israel exhibit an annual cycle of cambial activity, with a dormant period, similar to that of their counterparts growing in the cool northern temperate zone (Fahn, 1995). Two plants of Australian origin (*Acacia saligna* and *Eucalyptus camaldulensis*), also growing in the Mediterranean region, exhibit cambial activity throughout most or all of the year, as do their Southern Hemisphere counterparts. Plants of Sudanian and Saharo-Arabian origin growing in the Negev also exhibit more or less continuous cambial activity. They survive in the desert either because they have deep roots and grow in wadies (stream beds that are dry except in the rainy season) or grow in sand dunes or salt marshes.

The annual rhythm of cambial activity was compared in *Proustia cuneifolia* and *Acacia caven*, two typical shrubs of the matorral in the semiarid region of central Chile (Aljaro et al., 1972). *Proustia*, a drought-deciduous shrub, shows a typical desert cambial rhythm, highly sensitive to rainfall, and with activity limited to periods of adequate rainfall (Fahn, 1964). It loses its leaves at the beginning of the dry season in early summer and remains dormant until the wet season begins in winter. *Acacia*, an evergreen, exhibits cambial activity almost year-round. Adaptation in *Acacia* is believed to consist in developing long roots capable of tapping under-

ground water. Although both shrubs grow together, they have different strategies for the same xeric conditions.

## ■ CAUSAL RELATIONS IN CAMBIAL ACTIVITY

Several aspects of hormonal involvement in cambial activity and the differentiation of cambial derivatives have been considered in the previous chapter and will not be reconsidered here. All five of the major groups of plant hormones (auxins, gibberellins, cytokinins, abscisic acid, ethylene) have been shown to be present in the cambial region and each, at one time or another, has been implicated in the control of cambial activity (Savidge, 1993; Little and Pharis, 1995; Ridoutt et al., 1995; Savidge, 2000; Sundberg et al., 2000; Mellerowicz et al., 2001; Helariutta and Bhaleroo, 2003). Considerable experimental evidence indicates, however, that auxin exerts the predominant role (Kozlowski and Pallardy, 1997).

Seasonal variation in IAA levels frequently has been described as the primary physiological factor regulating the cambium's seasonal activity—IAA biosynthesis in the spring by elongating shoots and expanding leaves being responsible for the renewal of cell division, and declining IAA levels in late summer and autumn resulting in the cessation of cambial activity (Savidge and Wareing, 1984; Little and Pharis, 1995). Experiments

**TABLE 12.5 ■ Cambial Activity and Times of Initiation of New Phloem (P) and Xylem (X) Production in Several Tropical Hardwoods**

| | Jan | Feb | Mar | Apr | May | Jun | Jly | Aug | Sep | Oct | Nov | Dec |
|---|---|---|---|---|---|---|---|---|---|---|---|---|
| *Liquidambar formosana* | | | * P | X | | | | | | | | |
| *Polyalthia longifolia* | | | | + | P | | X | | | P | | |
| *Mimusops elengi* | | | | + | | | X | P | | | | |
| *Delonix regia* | | | | + | | | | X | | P | | |
| *Grewia tiliaefolia*[a] | | | | | R+ | P X | | | | | | |
| *Pterocarya stenoptera*[a] | | | | + | P X | | | | | | | |
| *Tectona grandis*[a] | | | | | | P R X | | | | | | |

a   Ring-porous species

*   First functional sieve elements arise from phloem mother cells on outer margin of cambial zone

+   Phloem mother cells (in *Pterocarya,* partially differentiated sieve elements) on outer margin of cambial zone begin to differentiate

R   Reactivation

   Mature sieve elements not present year-round

   Some mature sieve elements present year-round

Sources: *Liquidambar formosana*—Lu and Chang, 1975; *Polyalthia longifolia*—Ghouse and Hashmi, 1978; *Mimusops elengi*—Ghouse and Hashmi, 1980b, 1983; *Delonix regia*—Ghouse and Hashmi, 1980c; *Grewia tiliaefolia*—Deshpande and Rajendrababu, 1985; *Pterocarya stenoptera*—Zhang et al., 1992; *Tectona grandis*—Rao and Dave, 1981; Rajput and Rao, 1998b; Rao and Rajput, 1999.
Note: Of the species represented here, only *Liquidambar formosana* does not have mature sieve elements present year-round. *Polyalthia longifolia* exhibits two periods of phloem production. The times of cessation of phloem and xylem production and differentiation are not indicated.

with several species indicate that the transition from activity to dormancy in the cambium is not regulated by changes in the concentration of IAA or ABA, which are known to stimulate and inhibit cambial activity, respectively, but rather by changes in the sensitivity of cambial cells to IAA (Lachaud, 1989; Lloyd et al., 1996).

The cessation of cambial activity and the onset of dormancy in woody species of temperate regions are induced by short days and cold temperatures (Kozlowski and Pallardy, 1997). In late summer–early autumn, short days induce the initial stage of dormancy called rest, during which the cambium is incapable of responding to IAA even though environmental conditions may be favorable for growth. Then, in early winter, the cambium enters the quiescent stage of dormancy, which is induced by chilling. Given favorable environmental conditions (suitable temperatures, adequate water), the quiescent cambium is capable of responding to IAA.

Sucrose has been shown to play a major role in cambial metabolism, the demand for it being greatest during the period of rapid cell division and cell growth in spring and summer (Sung et al., 1993a, b; Krabel, 2000). Plasma membrane $H^+$-ATPase has been localized in the cambial zone, in differentiating xylem elements, and in ray cells surrounding the vessels in the mature xylem of *Populus* spp. (Arend et al., 2002). It has been suggested that the plasma membrane $H^+$-ATPase, which is up-regulated and activated by auxin, plays a role in the uptake of sucrose via symport into the rapidly growing cambial cells (Arend et al., 2002). Throughout the growing season, sucrose synthase is the dominant enzyme for sucrose metabolism (Sung et al., 1993a, b).

Resumption of cambial activity has long been related to new primary growth from buds. In many diffuse-porous hardwoods cambial activity generally is depicted as beginning beneath the expanding buds and from there slowly spreading basipetally toward the main branches, the trunk, and the roots. By contrast, in ring-porous hardwoods and conifers, reactivation events are depicted as occurring well before bud break and spreading rapidly throughout the trunk.

The difference in growth patterns between diffuse-porous hardwoods, on the one hand, and conifers and ring-porous hardwoods, on the other, is not so clear-cut. No fundamental difference in the pattern of cambial reactivation was found between ring-porous *Quercus robur* and diffuse-porous *Fagus sylvatica* (Lachaud and Bonnemain, 1981). In both species, cambial reactivation was found to proceed downward from swelling buds in the branches and to occur simultaneously throughout the trunk. Similar patterns were found for ring-porous *Castanea sativa* and the diffuse-porous *Betula verrucosa* and *Acer campestre* (Boutin, 1985). In diffuse-porous *Salix viminalis*, cambial activity preceded bud break by almost two months (Sennerby-Forsse, 1986). Occasional reports of exceptions to a basipetal spread of cambial reactivation are found throughout the earlier literature. In several instances radial growth was reported to begin simultaneously in many parts of the tree and, in others, in older parts in advance of younger parts (Hartig, 1892, 1894; Mer, 1892; Chalk, 1927; Lodewick, 1928; Fahn, 1962). Much of the research on cambial growth has been undertaken with a primary interest in wood formation (Atkinson and Denne, 1988; Suzuki et al., 1996). Since wood formation is a consequence of cambial activity, it is not unlikely that many of the reports describing the beginning of radial growth actually describe the beginning of xylem production.

A detailed study of the initiation of cambial activity in diffuse-porous *Tilia americana* revealed that the beginning of cell division and the beginning of vascular differentiation are not restricted to regions in the neighborhood of buds (Deshpande, 1967). The initiation of cell division occurred in many different areas of the cambium at all levels of the tree. The first mitoses were few, scattered and discontinuous, and difficult to detect in transverse sections, requiring examination of a great many longitudinal sections. The first cell divisions occurred in the cambium at the same time as mitotic activity began in the buds. Beginning of differentiation of newly produced cambial derivatives into xylem and phloem elements was also widespread and occurred throughout the shoot system in areas previously "awakened." Further cambial activity apparently was influenced by the expanding shoots. A marked acceleration in cambial activity took place in one-year-old shoots beneath the foliating (leaf-forming) buds, most notably beneath the bud traces. Soon a gradient of cambial activity was established along the axis, with greater activity taking place in the one-year-old stem and lesser activity occurring in successively older stems. Gradually acceleration of cambial activity spread to lower levels of the tree. What in the past has been considered a basipetal initiation of cambial activity may rather be a basipetal acceleration of cambial activity.

That cambial activity may be initiated without auxin or a stimulus emanating from the buds finds support from results of girdling and bark isolation studies. In a nine-year-old *Pinus sylvestris* stem girdled during the winter, cambial activity occurred below the girdle in the next spring (Egierszdorff, 1981). It was concluded that auxin stored in the trunk over the winter permitted the initiation of divisions independently of the supply of auxin from the top. Studies involving the isolation of circular patches of bark on the sides of *Populus tremuloides* (Evert and Kozlowski, 1967) and *Acer saccharum* (Evert et al., 1972) trees at breast height at various times during the dormant and growing seasons also indicate that a stimulus moving downward from expanding buds is not required to initiate cambial activity. In all of the trembling aspen trees and in half of the sugar maples, isolation of the bark during the dormant season (in November, February, or March) did not prevent initiation of cambial activity in the isolated areas. Normal cambial activity and phloem and xylem development were prevented in the isolated areas, however, indicating that normal activity and development require a supply of currently translocated regulatory substances from the shoots.

# REFERENCES

AJMAL, S., and M. IQBAL. 1992. Structure of the vascular cambium of varying age and its derivative tissues in the stem of *Ficus rumphii* Blume. *Bot. J. Linn. Soc.* 109, 211–222.

ALFIERI, F. J., and R. F. EVERT. 1968. Seasonal development of the secondary phloem in *Pinus*. *Am. J. Bot.* 55, 518–528.

ALFIERI, F. J., and R. F. EVERT. 1973. Structure and seasonal development of the secondary phloem in the Pinaceae. *Bot. Gaz.* 134, 17–25.

ALFIERI, F. J., and R. I. KEMP. 1983. The seasonal cycle of phloem development in *Juniperus californica*. *Am. J. Bot.* 70, 891–896.

ALFIERI, F. J., and P. M. MOTTOLA. 1983. Seasonal changes in the phloem of *Ephedra californica* Wats. *Bot. Gaz.* 144, 240–246.

ALJARO, M. E., G. AVILA, A. HOFFMANN, and J. KUMMEROW. 1972. The annual rhythm of cambial activity in two woody species of the Chilean "matorral." *Am. J. Bot.* 59, 879–885.

ANDERSON, B. J., and R. F. EVERT. 1965. Some aspects of phloem development in *Quercus alba*. *Am. J. Bot.* 52 (Abstr.), 627.

AREND, M., and J. FROMM. 2000. Seasonal variation in the K, Ca and P content and distribution of plasma membrane H+-ATPase in the cambium of *Populus trichocarpa*. In: *Cell and Molecular Biology of Wood Formation*, pp. 67–70, R. A. Savidge, J. R. Barnett, and R. Napier, eds. BIOS Scientific, Oxford.

AREND, M., and J. FROMM. 2003. Ultrastructural changes in cambial cell derivatives during xylem differentiation in poplar. *Plant Biol.* 5, 255–264.

AREND, M., M. H. WEISENSEEL, M. BRUMMER, W. OSSWALD, and J. H. FROMM. 2002. Seasonal changes of plasma membrane H$^+$-ATPase and endogenous ion current during cambial growth in poplar plants. *Plant Physiol.* 129, 1651–1663.

ARTSCHWAGER, E. 1950. The time factor in the differentiation of secondary xylem and secondary phloem in pecan. *Am. J. Bot.* 37, 15–24.

ATKINSON, C. J., and M. P. DENNE. 1988. Reactivation of vessel production in ash (*Fraxinus excelsior* L.) trees. *Ann. Bot.* 61, 679–688.

BAÏER, M., R. GOLDBERG, A.-M. CATESSON, M. LIBERMAN, N. BOUCHEMAL, V. MICHON, and C. HERVÉ DU PENHOAT. 1994. Pectin changes in samples containing poplar cambium and inner bark in relation to the seasonal cycle. *Planta* 193, 446–454.

BAILEY, I. W. 1919. Phenomena of cell division in the cambium of arborescent gymnosperms and their cytological significance. *Proc. Natl. Acad. Sci. USA* 5, 283–285.

BAILEY, I. W. 1920A. The cambium and its derivative tissues. II. Size variations of cambial initials in gymnosperms and angiosperms. *Am. J. Bot.* 7, 355–367.

BAILEY, I. W. 1920B. The cambium and its derivative tissues. III. A reconnaissance of cytological phenomena in the cambium. *Am. J. Bot.* 7, 417–434.

BAILEY, I. W. 1920C. The formation of the cell plate in the cambium of higher plants. *Proc. Natl. Acad. Sci. USA* 6, 197–200.

BAILEY, I. W. 1923. The cambium and its derivatives. IV. The increase in girth of the cambium. *Am. J. Bot.* 10, 499–509.

BANNAN, M. W. 1951. The annual cycle of size changes in the fusiform cambial cells of *Chamaecyparis* and *Thuja*. *Can. J. Bot.* 29, 421–437.

BANNAN, M. W. 1953. Further observations on the reduction of fusiform cambial cells in *Thuja occidentalis* L. *Can. J. Bot.* 31, 63–74.

BANNAN, M. W. 1955. The vascular cambium and radial growth in *Thuja occidentalis* L. *Can. J. Bot.* 33, 113–138.

BANNAN, M. W. 1956. Some aspects of the elongation of fusiform cambial cells in *Thuja occidentalis* L. *Can. J. Bot.* 34, 175–196.

BANNAN, M. W. 1960A. Cambial behavior with reference to cell length and ring width in *Thuja occidentalis* L. *Can. J. Bot.* 38, 177–183.

BANNAN, M. W. 1960B. Ontogenetic trends in conifer cambium with respect to frequency of anticlinal division and cell length. *Can. J. Bot.* 38, 795–802.

BANNAN, M. W. 1962. Cambial behavior with reference to cell length and ring width in *Pinus strobus* L. *Can. J. Bot.* 40, 1057–1062.

BANNAN, M. W. 1968. Anticlinal divisions and the organization of conifer cambium. *Bot. Gaz.* 129, 107–113.

BANNAN, M. W., and I. L. BAYLY. 1956. Cell size and survival in conifer cambium. *Can. J. Bot.* 34, 769–776.

BARNETT, J. R. 1981. Secondary xylem cell development. In: *Xylem Cell Development*, pp. 47–95, J. R. Barnett, ed. Castle House, Tunbridge Wells, Kent.

BARNETT, J. R. 1992. Reactivation of the cambium in *Aesculus hippocastanum* L.: A transmission electron microscope study. *Ann. Bot.* 70, 169–177.

BOßHARD, H. H. 1951. Variabilität der Elemente des Eschenholzes in Funktion von der Kambiumtätigkeit. *Schweiz. Z. Forstwes.* 102, 648–665.

BOUTIN, B. 1985. Étude de la réactivation cambiale chez un arbre ayant un bois à zones poreuses (*Castanea sativa*) et deux autres au bois à pores diffus (*Betula verrucosa, Acer campestre*). *Can. J. Bot.* 63, 1335–1343.

BRAUN, H. J. 1955. Beiträge zur Entwicklungsgeschichte der Markstrahlen. *Bot. Stud.* 4, 73–131.

BUTTERFIELD, B. G. 1972. Developmental changes in the vascular cambium of *Aeschynomene hispida* Willd. *N. Z. J. Bot.* 10, 373–386.

CATESSON, A.-M. 1964. Origine, fonctionnement et variations cytologiques saisonnières du cambium de *l'Acer pseudoplatanus* L. (Acéracées). *Ann. Sci. Nat. Bot. Biol. Vég. Sér.* 12, 5, 229–498.

CATESSON, A. M. 1974. Cambial cells. In: *Dynamic Aspects of Plant Ultrastructure*, pp. 358–390, A. W. Robards, ed. McGraw-Hill, New York.

CATESSON, A.-M. 1980. The vascular cambium. In: *Control of Shoot Growth in Trees*, pp. 12–40, C. H. A. Little, ed. Maritimes Forest Research Centre, Fredericton, N. B.

CATESSON, A.-M. 1984. La dynamique cambiale. *Ann. Sci. Nat. Bot. Biol. Vég. Sér.* 13, 6, 23–43.

CATESSON, A. M. 1987. Characteristics of radial cell walls in the cambial zone. A means to locate the so-called initials? *IAWA Bull.* n.s. 8 (Abstr.), 309.

CATESSON, A.-M. 1989. Specific characters of vessel primary walls during the early stages of wood differentiation. *Biol. Cell.* 67, 221–226.

CATESSON, A. M. 1990. Cambial cytology and biochemistry. In: *The Vascular Cambium*, pp. 63–112, M. Iqbal, ed. Research Studies Press, Taunton, Somerset, England.

CATESSON, A.-M. 1994. Cambial ultrastructure and biochemistry: Changes in relation to vascular tissue differentiation and the seasonal cycle. *Int. J. Plant Sci*, 155, 251–261.

CATESSON, A. M., and J. C. ROLAND. 1981. Sequential changes associated with cell wall formation and fusion in the vascular cambium. *IAWA Bull.* n.s. 2, 151–162.

CATESSON, A. M., R. FUNADA, D. ROBERT-BABY, M. QUINET-SZÉLY, J. CHU-BÂ, and R. GOLDBERG. 1994. Biochemical and cytochemical cell wall changes across the cambial zone. *IAWA J.* 15, 91–101.

CHAFFEY, N. J. 2000. Cytoskeleton, cells walls and cambium: New insights into secondary xylem differentiation. In: *Cell and Molecular Biology of Wood Formation*, pp. 31–42, R. A. Savidge, J. R. Barnett, and R. Napier, eds. BIOS Scientific, Oxford.

CHAFFEY, N., and P. W. BARLOW. 2000. Actin in the secondary vascular system of woody plants. In: *Actin: A Dynamic Framework for Multiple Plant Cell Functions*, pp. 587–600, C. J. Staiger, F. Baluška, D. Volkman, and P. W. Barlow. eds. Kluwer Academic, Dordrecht.

CHAFFEY, N., and P. BARLOW. 2002. Myosin, microtubules, and microfilaments: Co-operation between cytoskeletal components during cambial cell division and secondary vascular differentiation in trees. *Planta* 214, 526–536.

CHAFFEY, N., J. BARNETT, and P. BARLOW. 1997A. Endomembranes, cytoskeleton, and cell walls: Aspects of the ultrastructure of the vascular cambium of taproots of *Aesculus hippocastanum* L. (Hippocastanaceae). *Int. J. Plant Sci.* 158, 97–109.

CHAFFEY, N., J. BARNETT, and P. BARLOW. 1997B. Arrangement of microtubules, but not microfilaments, indicates determination of cambial derivatives. In: *Biology of Root Formation and Development*, pp. 52–54, A. Altman and Y. Waisel, eds. Plenum Press, New York.

CHAFFEY, N. J., P. W. BARLOW, and J. R. BARNETT. 1998. A seasonal cycle of cell wall structure is accompanied by a cyclical rearrangement of cortical microtubules in fusiform cambial cells within taproots of *Aesculus hippocastanum* (Hippocastanaceae). *New Phytol.* 139, 623–635.

CHAFFEY, N., P. BARLOW, and J. BARNETT. 2000. Structure-function relationships during secondary phloem development in an angiosperm tree, *Aesculus hippocastanum*: Microtubules and cell walls. *Tree Physiol.* 20, 777–786.

CHAFFEY, N., P. BARLOW, and B. SUNDBERG. 2002. Understanding the role of the cytoskeleton in wood formation in angiosperm trees: Hybrid aspen (*Populus tremula* x *P. tremuloides*) as the model species *Tree Physiol.* 22, 239–249.

CHALK, L. 1927. The growth of the wood of ash (*Fraxinus excelsior* L. and *F. oxycarpa* Willd.) and Douglas fir (*Pseudotsuga Douglasii* Carr.). *Q. J. For.* 21, 102–122.

CHEADLE, V. I., and K. ESAU. 1964. Secondary phloem of *Liriodendron tulipifera*. *Calif. Univ. Publ. Bot.* 36, 143–252.

CHOWDHURY, K. A. 1961. Growth rings in tropical trees and taxonomy. *10th Pacific Science Congress Abstr.*, 280.

COCKERHAM, G. 1930. Some observations on cambial activity and seasonal starch content in sycamore (*Acer pseudo platanus*). *Proc. Leeds Philos. Lit. Soc., Sci. Sect.*, 2, 64–80.

CUMBIE, B. G. 1963. The vascular cambium and xylem development in *Hibiscus lasiocarpus*. *Am. J. Bot.* 50, 944–951.

CUMBIE, B. G. 1967a. Development and structure of the xylem in *Canavalia* (Leguminosae). *Bull. Torrey Bot. Club* 94, 162–175.

CUMBIE, B. G. 1967B. Developmental changes in the vascular cambium in *Leitneria floridana*. *Am. J. Bot.* 54, 414–424.

CUMBIE, B. G. 1969. Developmental changes in the vascular cambium of *Polygonum lapathifolium*. *Am. J. Bot.* 56, 139–146.

CUMBIE, B. G. 1984. Origin and development of the vascular cambium in *Aeschynomene virginica*. *Bull. Torrey Bot. Club* 111, 42–50.

DAVE, Y. S., and K. S. RAO. 1981. Seasonal nuclear behavior in fusiform cambial initials of *Tectona grandis* L. f. *Flora* 171, 299–305.

DAVIS, J. D., and R. F. EVERT. 1968. Seasonal development of the secondary phloem in *Populus tremuloides*. *Bot. Gaz.* 129, 1–8.

DAVIS, J. D., and R. F. EVERT. 1970. Seasonal cycle of phloem development in woody vines. *Bot. Gaz.* 131, 128–138.

DERR, W. F., and R. F. EVERT. 1967. The cambium and seasonal development of the phloem in *Robinia pseudoacacia*. *Am J. Bot.* 54, 147–153.

DESHPANDE, B. P. 1967. Initiation of cambial activity and its relation to primary growth in *Tilia americana* L. Ph.D. Dissertation. University of Wisconsin, Madison.

DESHPANDE, B. P., and T. RAJENDRABABU. 1985. Seasonal changes in the structure of the secondary phloem of *Grewia tiliaefolia*, a deciduous tree from India. *Ann. Bot.* 56, 61–71.

DODD, J. D. 1948. On the shapes of cells in the cambial zone of *Pinus silvestris* L. *Am. J. Bot.* 35, 666–682.

EGIERSZDORFF, S. 1981. The role of auxin stored in scotch pine trunk during spring activation of cambial activity. *Biol. Plant.* 23, 110–115.

ELLIOTT, J. H. 1935. Seasonal changes in the development of the phloem of the sycamore, *Acer Pseudo Platanus* L. *Proc. Leeds Philos. Lit. Soc., Sci. Sect.*, 3, 55–67.

ERMEL, F. F., M.-L. FOLLET-GUEYE, C. CIBERT, B. VIAN, C. MORVAN, A.-M. CATESSON, and R. GOLDBERG. 2000. Differential localization of arabinan and galactan side chains of rhamnogalacturonan I in cambial derivatives. *Planta* 210, 732–740.

ESAU, K. 1977. *Anatomy of Seed Plants*, 2nd ed. Wiley, New York.

EVERT, R. F. 1960. Phloem structure in *Pyrus communis* L. and its seasonal changes. *Univ. Calif. Publ. Bot.* 32, 127–194.

EVERT, R. F. 1961. Some aspects of cambial development in *Pyrus communis*. *Am. J. Bot.* 48, 479–488.

EVERT, R. F. 1962. Some aspects of phloem development in *Tilia americana*. *Am. J. Bot.* 49 (Abstr.), 659.

EVERT, R. F. 1963a. Ontogeny and structure of the secondary phloem in *Pyrus malus*. *Am. J. Bot.* 50, 8–37.

EVERT, R. F. 1963b. The cambium and seasonal development of the phloem in *Pyrus malus*. *Am. J. Bot.* 50, 149–159.

EVERT, R. F. 1978. Seasonal development of the secondary phloem in *Rhus glabra* L. *Botanical Society of America, Miscellaneous Series*, Publ. 156 (Abstr.), 25.

EVERT, R. F., and B. P. DESHPANDE. 1970. An ultrastructural study of cell division in the cambium. *Am. J. Bot.* 57, 942–961.

EVERT, R. F., and T. T. KOZLOWSKI. 1967. Effect of isolation of bark on cambial activity and development of xylem and phloem in trembling aspen. *Am. J. Bot.* 54, 1045–1055.

EVERT, R. F., T. T. KOZLOWSKI, and J. D. DAVIS. 1972. Influence of phloem blockage on cambial growth of sugar maple. *Am. J. Bot.* 59, 632–641.

EWERS, F. W. 1982. Secondary growth in needle leaves of *Pinus longaeva* (bristlecone pine) and other conifers: Quantitative data. *Am. J. Bot.* 69, 1552–1559.

FAHN, A. 1962. Xylem structure and the annual rhythm of cambial activity in woody species of the East Mediterranean regions. *News Bull.* [IAWA] 1962/1, 2–6.

FAHN, A. 1964. Some anatomical adaptations of desert plants. *Phytomorphology* 14, 93–102.

FAHN, A. 1995. Seasonal cambial activity and phytogeographic origin of woody plants: A hypothesis. *Isr. J. Plant Sci.* 43, 69–75.

FAHN, A., and E. WERKER. 1990. Seasonal cambial activity. In: *The Vascular Cambium*, pp. 139–157, M. Iqbal, ed. Research Studies Press, Taunton, Somerset, England.

FARRAR, J. J. 1995. Ultrastructure of the vascular cambium of *Robinia pseudoacacia*. Ph.D. Dissertation. University of Wisconsin, Madison.

FARRAR, J. J., and R. F. EVERT. 1997a. Seasonal changes in the ultrastructure of the vascular cambium of *Robinia pseudoacacia*. *Trees* 11, 191–202.

FARRAR, J. J., and R. F. EVERT. 1997b. Ultrastructure of cell division in the fusiform cells of the vascular cambium of *Robinia pseudoacacia*. *Trees* 11, 203–215.

FOLLET-GUEYE, M. L., F. F. ERMEL, B. VIAN, A. M. CATESSON, and R. GOLDBERG. 2000. Pectin remodelling during cambial derivative differentiation. In: *Cell and Molecular Biology of Wood Formation*, pp. 289–294, R. A. Savidge, J. R. Barnett, and R. Napier, eds. BIOS Scientific, Oxford.

FORWARD, D. F., and N. J. NOLAN. 1962. Growth and morphogenesis in Canadian forest species. VI. The significance of specific increment of cambial area in *Pinus resinosa* Ait. *Can. J. Bot.* 40, 95–111.

FUJII, T., A. T. SALANG, and T. FUJIWARA. 1999. Growth periodicity in relation to the xylem development in three *Shorea* spp. (Dipterocarpaceae) growing in Sarawak. In: *Tree-ring Analysis: Biological, Methodological and Environmental Aspects*, pp. 169–183, R. Wimmer and R. E. Vetter, eds. CABI Publishing, Wallingford, Oxon.

FUNADA, R., and A. M. CATESSON. 1991. Partial cell wall lysis and the resumption of meristematic activity in *Fraxinus excelsior* cambium. *IAWA Bull.* n.s. 12, 439–444.

FUNADA, R., O. FURUSAWA, M. SHIBAGAKI, H. MIURA, T. MIURA, H. ABE, and J. OHTANI. 2000. The role of cytoskeleton in secondary xylem differentiation in conifers. In: *Cell and Molecular Biology of Wood Formation*, pp. 255–264, R. A. Savidge, J. R. Barnett and R. Napier, eds. BIOS Scientific, Oxford.

GAHAN, P. B. 1988. Xylem and phloem differentiation in perspective. In: *Vascular Differentiation and Plant Growth Regulators*, pp. 1–21, L. W. Roberts, P. B. Gahan, and R. Aloni, eds. Springer-Verlag, Berlin.

GAHAN, P. B. 1989. How stable are cambial initials? *Bot. J. Linn. Soc.* 100, 319–321.

GHOUSE, A. K. M., and S. HASHMI. 1978. Seasonal cycle of vascular differentiation in *Polyalthia longifolia* (Annonaceae). *Beitr. Biol. Pflanz.* 54, 375–380.

GHOUSE, A. K. M., and S. HASHMI. 1980A. Changes in the vascular cambium of *Polyalthia longifolia* Benth. et Hook. (Annonaceae) in relation to the girth of the tree. *Flora* 170, 135–143.

GHOUSE, A. K. M., and S. HASHMI. 1980B. Seasonal production of secondary phloem and its longevity in *Mimusops elengi* L. *Flora* 170, 175–179.

GHOUSE, A. K. M., and S. HASHMI. 1980C. Longevity of secondary phloem in *Delonix regia* Rafin. *Proc. Indian Acad. Sci. B, Plant Sci.* 89, 67–72.

GHOUSE, A. K. M., and S. HASHMI. 1983. Periodicity of cambium and of the formation of xylem and phloem in *Mimusops elengi* L., an evergreen member of tropical India. *Flora* 173, 479–487.

GHOUSE, A. K. M., and M. I. H. KHAN. 1977. Seasonal variation in the nuclear number of fusiform cambial initials in *Psidium guajava* L. *Caryologia* 30, 441–444.

GHOUSE, A. K. M., and M. YUNUS. 1973. Some aspects of cambial development in the shoots of *Dalbergia sissoo* Roxb. *Flora* 162, 549–558.

GHOUSE, A. K. M., and M. YUNUS. 1974. The ratio of ray and fusiform initials in some woody species of the Ranalian complex. *Bull. Torrey Bot. Club* 101, 363–366.

GHOUSE, A. K. M., and M. YUNUS. 1976. Ratio of ray and fusiform initials in the vascular cambium of certain leguminous trees. *Flora* 165, 23–28.

GOOSEN-DE ROO, L. 1981. Plasmodesmata in the cambial zone of *Fraxinus excelsior* L. *Acta Bot. Neerl.* 30, 156.

GOOSEN-DE ROO, L., C. J. VENVERLOO, and P. D. BURGGRAAF. 1980. Cell division in highly vacuolated plant cells. In: *Electron Microscopy 1980*, vol. 2, *Biology: Proc. 7th Eur. Congr. Electron Microsc.*, pp. 232–233. Hague, The Netherlands.

GOOSEN-DE ROO, L., R. BAKHUIZEN, P. C. VAN SPRONSEN, and K. R. LIBBENGA. 1984. The presence of extended phragmosomes containing cytoskeletal elements in fusiform cambial cells of *Fraxinus excelsior* L. *Protoplasma* 122, 145–152.

GREGORY, R. A. 1977. Cambial activity and ray cell abundance in *Acer saccharum*. *Can J. Bot.* 55, 2559–2564.

GREGORY, R. A., and B. F. WILSON. 1968. A comparison of cambial activity of white spruce in Alaska and New England. *Can. J. Bot.* 46, 733–734.

GRILLOS, S. J., and F. H. SMITH. 1959. The secondary phloem of Douglas fir. *For. Sci.* 5, 377–388.

GUGLIELMINO, N., M. LIBERMAN, A. M. CATESSON, A. MARECK, R. PRAT, S. MUTAFTSCHIEV, and R. GOLDBERG. 1997A. Pectin methylesterases from poplar cambium and inner bark: Localization, properties and seasonal changes. *Planta* 202, 70–75.

GUGLIELMINO, N., M. LIBERMAN, A. JAUNEAU, B. VIAN, A. M. CATESSON, and R. GOLDBERG. 1997b. Pectin immunolocalization and calcium visualization in differentiating derivatives from poplar cambium. *Protoplasma* 199, 151–160.

HARRIS, J. M. 1989. *Spiral Grain and Wave Phenomena in Wood Formation.* Springer-Verlag, Berlin.

HARTIG, R. 1892. Ueber Dickenwachsthum und Jahrringbildung. *Bot. Z.* 50, 176–180, 193–196.

HARTIG, R. 1894. Untersuchungen über die Entstehung und die Eigenschaften des Eichenholzes. *Forstlich-Naturwiss. Z.* 3, 1–13, 49–68, 172–191, 193–203.

HASHMI, S., and A. K. M. GHOUSE. 1978. On the nuclear number of the fusiform initials of *Polyalthia longifolia* Benth. and Hook. *J. Indian Bot. Soc.* 57 (suppl.; Abstr.), 24.

HEJNOWICZ, Z. 1961. Anticlinal divisions, intrusive growth, and loss of fusiform initials in nonstoried cambium. *Acta Soc. Bot. Pol.* 30, 729–748.

HEJNOWICZ, Z. 1971. Upward movement of the domain pattern in the cambium producing wavy grain in *Picea excelsa*. *Acta Soc. Bot. Pol.* 40, 499–512.

HEJNOWICZ, Z., and J. KRAWCZYSZYN. 1969. Oriented morphogenetic phenomena in cambium of broadleaved trees. *Acta Soc. Bot. Pol.* 38, 547–560.

HEJNOWICZ, Z., and B. ZAGÓRSKA-MAREK. 1974. Mechanism of changes in grain inclination in wood produced by storeyed cambium. *Acta Soc. Bot. Pol.* 43, 381–398.

HELARIUTTA, Y., and R. BHALERAO. 2003. Between xylem and phloem: The genetic control of cambial activity in plants. *Plant Biol.* 5, 465–472.

IMAGAWA, H., and S. ISHIDA. 1972. Study on the wood formation in trees. Report III. Occurrence of the overwintering cells in cambial zone in several ring-porous trees. *Res. Bull. Col. Exp. For. Hokkaido Univ. (Enshurin Kenkyu hohoku)* 29, 207–221.

IQBAL, M., and A. K. M. GHOUSE. 1979. Anatomical changes in *Prosopis spicigera* with growing girth of stem. *Phytomorphology* 29, 204–211.

IQBAL, M., and A. K. M. GHOUSE. 1987. Anatomy of the vascular cambium of *Acacia nilotica* (L.) Del. var. *telia* Troup (Mimosaceae) in relation to age and season. *Bot. J. Linn. Soc.* 94, 385–397.

ISEBRANDS, J. G., and P. R. LARSON. 1973. Some observations on the cambial zone in cottonwood. *IAWA Bull.* 1973/3, 3–11.

KHAN, K. K., Z. AHMAD, and M. IQBAL. 1981. Trends of ontogenetic size variation of cambial initials and their derivatives in the stem of *Bauhinia parviflora* Vahl. *Bull. Soc. Bot. Fr. Lett. Bot.* 128, 165–175.

KHAN, M. I. H., T. O. SIDDIQI, and A. H. KHAN. 1983. Ontogenetic changes in the cambial structure of *Citrus sinensis* L. *Flora* 173, 151–158.

KITIN, P., Y. SANO, and R. FUNADA. 2002. Fusiform cells in the cambium of *Kalopanax pictus* are exclusively mononucleate. *J. Exp. Bot.* 53, 483–488.

KORIBA, K. 1958. On the periodicity of tree-growth in the tropics, with reference to the mode of branching, the leaf-fall, and the formation of the resting bud. *Gardens' Bull. Straits Settlements* 17, 11–81.

KOZLOWSKI, T. T., and S. G. PALLARDY. 1997. *Growth Control in Woody Plants.* Academic Press, San Diego.

KRABEL, D. 2000. Influence of sucrose on cambial activity. In: *Cell and Molecular Biology of Wood Formation*, pp. 113–125, R. A. Savidge, J. R. Barnett, and R. Napier, eds. BIOS Scientific, Oxford.

KRAWCZYSZYN, J. 1977. The transition from nonstoried to storied cambium in *Fraxinus excelsior*. I. The occurrence of radial anticlinal divisions. *Can J. Bot.* 55, 3034–3041.

KRAWCZYSZYN, J., and J. A. ROMBERGER. 1979. Cyclical cell length changes in wood in relation to storied structure and interlocked grain. *Can. J. Bot.* 57, 787–794.

KURODA, H., and S. SAGISAKA. 1993. Ultrastructural changes in cortical cells of apple (*Malus pumila* Mill.) associated with cold hardiness. *Plant Cell Physiol.* 34, 357–365.

LACHAUD, S. 1989. Participation of auxin and abscisic acid in the regulation of seasonal variations in cambial activity and xylogenesis. *Trees* 3, 125–137.

LACHAUD, S., and J.-L. BONNEMAIN. 1981. Xylogenèse chez les Dicotylédones arborescentes. I. Modalités de la remise en activité du cambium et de la xylogenèse chez les Hêtres et les Chênes âgés. *Can. J. Bot.* 59, 1222–1230.

LACHAUD, S., A.-M. CATESSON, and J.-L. BONNEMAIN. 1999. Structure and functions of the vascular cambium. *C.R. Acad. Sci. Paris, Sci. de la Vie* 322, 633–650.

LARSON, P. R. 1994. *The Vascular Cambium: Development and Structure.* Springer-Verlag. Berlin.

LAWTON, J. R. 1976. Seasonal variation in the secondary phloem from the main trunks of willow and sycamore trees. *New Phytol.* 77, 761–771.

LIPHSCHITZ, N., and S. LEV-YADUN. 1986. Cambial activity of evergreen and seasonal dimorphics around the Mediterranean. *IAWA Bull.* n.s. 7, 145–153.

LIPHSCHITZ, N., S. LEV-YADUN, and Y. WAISEL. 1981. The annual rhythm of activity of the lateral meristems (cambium and phellogen) in *Cupressus sempervirens* L. *Ann. Bot.* 47, 485–496.

LITTLE, C. H. A., and R. P. PHARIS. 1995. Hormonal control of radial and longitudinal growth in the tree stem. In: *Plant Stems: Physiology and Functional Morphology*, pp. 281–319, B. L. Gartner, ed. Academic Press, San Diego.

LLOYD, A. D., E. J. MELLEROWICZ, R. T. RIDING, and C. H. A. LITTLE. 1996. Changes in nuclear genome size and relative ribosomal RNA gene content in cambial region cells of *Abies balsamea* shoots during the development of dormancy. *Can. J. Bot.* 74, 290–298.

LODEWICK, J. E. 1928. *Seasonal activity of the cambium in some northeastern trees*. Bull. N.Y. State Col. For. Syracuse Univ. Tech. Publ. No. 23.

LU, C.-Y., and S.-H. T. CHANG. 1975. Seasonal activity of the cambium in the young branch of *Liquidambar formosana* Hance. *Taiwania* 20, 32–47.

MAHMOOD, A. 1968. Cell grouping and primary wall generations in the cambial zone, xylem, and phloem in *Pinus*. *Aust. J. Bot.* 16, 177–195.

MAHMOOD, A. 1990. The parental cell walls. In: *The Vascular Cambium*, pp. 113–126, M. Iqbal, ed. Research Studies Press, Taunton, Somerset, England.

MAINIERE, C., J. P. CHIMELO, and V. A. ALFONSO. 1983. *Manual de identificação das principais madeiras comerciais brasileiras*. Ed. Promocet., Publicação IPT 1226, São Paulo.

MELLEROWICZ, E. J., R. T. RIDING, and C. H. A. LITTLE. 1989. Genomic variability in the vascular cambium of *Abies balsamea*. *Can. J. Bot.* 67, 990–996.

MELLEROWICZ, E. J., R. T. RIDING, and C. H. A. LITTLE. 1990. Nuclear size and shape changes in fusiform cambial cells of *Abies balsamea* during the annual cycle of activity and dormancy. *Can J. Bot.* 68, 1857–1863.

MELLEROWICZ, E. J., R. T. RIDING, and C. H. A. LITTLE. 1992. Periodicity of cambial activity in *Abies balsamea*. II. Effects of temperature and photoperiod on the size of the nuclear genome in fusiform cambial cells. *Physiol. Plant.* 85, 526–530.

MELLEROWICZ, E. J., R. T. RIDING, and C. H. A. LITTLE. 1993. Nucleolar activity in the fusiform cambial cells of *Abies balsamea* (Pinaceae): Effect of season and age. *Am. J. Bot.* 80, 1168–1174.

MELLEROWICZ, E. J., M. BAUCHER, B. SUNDBERG, and W. BOERJAN. 2001. Unravelling cell wall formation in the woody dicot stem. *Plant Mol. Biol.* 47, 239–274.

MER, E. 1892. Reveil et extinction de l'activité cambiale dans les arbres. *C.R. Séances Acad. Sci.* 114, 242–245.

MICHELI, F., M. BORDENAVE, and L. RICHARD. 2000. Pectin methylesterases: Possible markers for cambial derivative differentiation? In: *Cell and Molecular Biology of Wood Formation*, pp. 295–304, R. A. Savidge, J. R. Barnett, and R. Napier, eds. BIOS Scientific, Oxford.

MURMANIS, L. 1970. Locating the initial in the vascular cambium of *Pinus strobus* L. by electron microscopy. *Wood Sci. Technol.* 4, 1–14.

MURMANIS, L. 1977. Development of vascular cambium into secondary tissue of *Quercus rubra* L. *Ann. Bot.* 41, 617–620.

MURMANIS, L., and I. B. SACHS. 1969. Seasonal development of secondary xylem in *Pinus strobus* L. *Wood Sci. Technol.* 3, 177–193.

NEWMAN, I. V. 1956. Pattern in meristems of vascular plants—I. Cell partition in the living apices and in the cambial zone in relation to the concepts of initial cells and apical cells. *Phytomorphology* 6, 1–19.

ORIBE, Y., R. FUNADA, M. SHIBAGAKI, and T. KUBO. 2001. Cambial reactivation in locally heated stems of the evergreen conifer *Abies sachalinensis* (Schmidt) Masters. *Planta* 212, 684–691.

PALIWAL, G. S., and N. V. S. R. K. PRASAD. 1970. Seasonal activity of cambium in some tropical trees. I. *Dalbergia sissoo*. *Phytomorphology* 20, 333–339.

PALIWAL, G. S., and L. M. SRIVASTAVA. 1969. The cambium of *Aleuosmia*. *Phytomorphology* 19, 5–8.

PALIWAL, G. S., V. S. SAJWAN, and N. V. S. R. K. PRASAD. 1974. Seasonal variations in the size of the cambial initials in *Polyalthia longifolia*. *Curr. Sci.* 43, 620–621.

PALIWAL, G. S., N. V. S. R. K. PRASAD, V. S. SAJWAN, and S. K. AGGARWAL. 1975. Seasonal activity of cambium in some tropical trees. II. *Polyalthia longifolia*. *Phytomorphology* 25, 478–484.

PATEL, J. D. 1975. Occurrence of multinucleate fusiform initials in *Solanum melongea* L. *Curr. Sci.* 44, 516–517.

PHILIPSON, W. R., J. M. WARD, and B. G. BUTTERFIELD. 1971. *The Vascular Cambium: Its Development and Activity*. Chapman and Hall, London.

POMPARAT, M. 1974. Étude des variations de la longueur des trachéides de la tige et de la racine du Pin maritime au cours de l'année: Influence des facteurs édaphiques sur l'activité cambiale. Ph.D. Thèse. Université de Bordeaux.

RAJPUT, K. S., and K. S. RAO. 1998A. Occurrence of intercellular spaces in cambial rays. *Isr. J. Plant Sci.* 46, 299–302.

RAJPUT, K. S., and K. S. RAO. 1998B. Seasonal anatomy of secondary phloem of teak (*Tectona grandis* L. Verbenaceae) growing in dry and moist deciduous forests. *Phyton (Horn)* 38, 251–258.

RAJPUT, K. S., and K. S. RAO. 2000. Cambial activity and development of wood in *Acacia nilotica* (L.) Del. growing in different forests of Gujarat State. *Flora* 195, 165–171.

RAO, K. S. 1985. Seasonal ultrastructural changes in the cambium of *Aesculus hippocastanum* L. *Ann. Sci. Nat. Bot. Biol. Vég., Sér 13*, 7, 213–228.

RAO, K. S. 1988. Cambial activity and developmental changes in ray initials of some tropical trees. *Flora* 181, 425–434.

RAO, K. S., and Y. S. DAVE. 1981. Seasonal variations in the cambial anatomy of *Tectona grandis* (Verbenaceae). *Nord. J. Bot.* 1, 535–542.

RAO, K. S., and K. S. RAJPUT. 1999. Seasonal behaviour of vascular cambium in teak (*Tectona grandis*) growing in moist deciduous and dry deciduous forests. *IAWA J.* 20, 85–93.

RENSING, K. H., and A. L. SAMUELS. 2004. Cellular changes associated with rest and quiescence in winter-dormant vascular cambium of *Pinus contorta*. *Trees* 18, 373–380.

RENSING, K. H., A. L. SAMUELS, and R. A. SAVIDGE. 2002. Ultrastructure of vascular cambial cell cytokinesis in pine seedlings preserved by cryofixation and substitution. *Protoplasma* 220, 39–49.

RIDOUTT, B. G., and R. SANDS. 1993. Within-tree variation in cambial anatomy and xylem cell differentiation in *Eucalyptus globulus*. Trees 8, 18–22.

RIDOUTT, B. G., R. P. PHARIS, and R. SANDS. 1995. Identification and quantification of cambial region hormones of *Eucalyptus globulus*. Plant Cell Physiol. 36, 1143–1147.

ROLAND, J.-C. 1978. Early differences between radial walls and tangential walls of actively growing cambial zone. IAWA Bull. 1978/1, 7–10.

SAGISAKA, S., M. ASADA, and Y. H. AHN. 1990. Ultrastructure of poplar cortical cells during the transition from growing to wintering stages and vice versa. Trees 4, 120–127.

SANIO, K. 1873. Anatomie der gemeinen Kiefer (*Pinus silvestris* L.). II. 2. Entwickelungsgeschichte der Holzzellen. Jahrb. Wiss. Bot. 9, 50–126.

SAVIDGE, R. A. 1993. Formation of annual rings in trees. In: *Oscillations and Morphogenesis*, pp. 343–363, L. Rensing, ed. Dekker, New York.

SAVIDGE, R. A. 2000. Biochemistry of seasonal cambial growth and wood formation—An overview of the challenges. In: *Cell and Molecular Biology of Wood Formation*, pp. 1–30, R. A. Savidge, J. R. Barnett, and R. Napier, eds. BIOS Scientific, Oxford.

SAVIDGE, R. A., and P. F. WAREING. 1984. Seasonal cambial activity and xylem development in *Pinus contorta* in relation to endogenous indol-3-yl-acetic and (S)-abscisic acid levels. Can. J. For. Res. 14, 676–682.

SENNERBY-FORSSE, L. 1986. Seasonal variation in the ultrastructure of the cambium in young stems of willow (*Salix viminalis*) in relation to phenology. Physiol. Plant. 67, 529–537.

SENNERBY-FORSSE, L., and H. A. VON FIRCKS. 1987. Ultrastructure of cells in the cambial region during winter hardening and spring dehardening in *Salix dasyclados* Wim. grown at two nutrient levels. Trees 1, 151–163.

SHARMA, H. K., D. D. SHARMA, and G. S. PALIWAL. 1979. Annual rhythm of size variations in cambial initials of *Azadirachta indica* A. Juss. Geobios 6, 127–129.

SHTROMBERG, A. YA. 1959. Cambium activity in leaves of several woody dicotyledenous plants. Dokl. Akad. Nauk SSSR 124, 699–702.

SISSON, W. E., JR. 1968. Cambial divisions in *Pseudotsuga menziesii*. Am. J. Bot. 55, 923–926.

STRASBURGER, E. 1891. *Ueber den Bau und die Verrichtungen der Leitungsbahnen in den Pflanzen. Histologische Beiträge*, Heft 3. Gustav Fischer, Jena.

SUNDBERG, B., C. UGGLA, and H. TUOMINEN. 2000. *Cambial growth and auxin gradients*. In: *Cell and Molecular Biology of Wood Formation*, pp. 169–188, R. A. Savidge, J. R. Barnett, and R. Napier, eds. BIOS Scientific, Oxford.

SUNG, S.-J. S., P. P. KORMANIK, and C. C. BLACK. 1993A. Vascular cambial sucrose metabolism and growth in loblolly pine (*Pinus taeda* L.) in relation to transplanting trees. Tree Physiol. 12, 243–258.

SUNG, S.-J. S., P. P. KORMANIK, and C. C. BLACK. 1993B. Understanding sucrose metabolism and growth in a developing sweetgum plantation. In: *Proc. 22nd Southern Forest Tree Improvement Conference*, June 14–17, 1993, pp. 114–123. Sponsored Publ. No. 44 of the Southern Forest Tree Improvement Committee.

SUZUKI, M., K. YODA, and H. SUZUKI. 1996. Phenological comparison of the onset of vessel formation between ring-porous and diffuse-porous deciduous trees in a Japanese temperate forest. IAWA J. 17, 431–444.

TEPPER, H. B., and C. A. HOLLIS. 1967. Mitotic reactivation of the terminal bud and cambium of white ash. Science 156, 1635–1636.

TIMELL, T. E. 1980. Organization and ultrastructure of the dormant cambial zone in compression wood of *Picea abies*. Wood Sci. Technol. 14, 161–179.

TOMLINSON, P. B., and F. C. CRAIGHEAD SR. 1972. Growth-ring studies on the native trees of sub-tropical Florida. In: *Research Trends in Plant Anatomy—K.A. Chowdhury Commemoration Volume*, pp. 39–51, A. K. M. Ghouse and Mohd Yunus, eds. Tata McGraw-Hill, Bombay.

TUCKER, C. M. 1968. Seasonal phloem development in *Ulmus americana*. Am. J. Bot. 55 (Abstr.), 716.

TUCKER, C. M., and R. F. EVERT. 1969. Seasonal development of the secondary phloem in *Acer negundo* Am. J. Bot. 56, 275–284.

VENUGOPAL, N., and K. V. KRISHNAMURTHY. 1989. Organisation of vascular cambium during different seasons in some tropical timber trees. Nord. J. Bot. 8, 631–638.

VILLALBA, R., and J. A. BONINSEGNA. 1989. Dendrochronological studies on *Prosopis flexuosa* DC. IAWA Bull. n.s. 10, 155–160.

WAISEL, Y., I. NOAH, and A. FAHN. 1966. Cambial activity in *Eucalyptus camaldulensis* Dehn.: II. The production of phloem and xylem elements. New Phytol. 65, 319–324.

WILCOX, H., F. J. CZABATOR, G. GIROLAMI, D. E. MORELAND, and R. F. SMITH. 1956. *Chemical debarking of some pulpwood species*. State Univ. N.Y. Col. For. Syracuse. Tech. Publ. 77.

WILSON, B. F. 1963. Increase in cell wall surface area during enlargement of cambial derivatives in *Abies concolor*. Am. J. Bot. 50, 95–102.

WISNIEWSKI, M., and E. N. ASHWORTH. 1986. A comparison of seasonal ultrastructural changes in stem tissues of peach (*Prunus persica*) that exhibit contrasting mechanisms of cold hardiness. Bot. Gaz. 147, 407–417.

WŁOCH, W. 1976. Cell events in cambium, connected with the formation and existence of a whirled cell arrangement. Acta Soc. Bot. Pol. 45, 313–326.

WŁOCH, W. 1981. Nonparallelism of cambium cells in neighboring rows. Acta Soc. Bot. Pol. 50, 625–636.

WŁOCH, W. T. 1989. Chiralne zderzenia komórkowe i wzór domenowy w kambium lipy (Chiral cell events and domain pattern in the cambium of lime). Dr. Hab. Univ. Śląski w Katowicach.

WŁOCH, W., and E. POLAP. 1994. The intrusive growth of initial cells in re-arrangement of cells in cambium of *Tilia cordata* Mill. *Acta Soc. Bot. Pol.* 63, 109–116.

WŁOCH, W., and W. SZENDERA. 1992. Observation of changes of cambial domain patterns on the basis of primary ray development in *Fagus silvatica* L. *Acta Soc. Bot. Pol.* 61, 319–330.

WŁOCH, W., and B. ZAGÓRSKA-MAREK. 1982. Reconstruction of storeyed cambium in the linden. *Acta Soc. Bot. Pol.* 51, 215–228.

WŁOCH, W., J. KARCZEWSKI, and B. OGRODNIK. 1993. Relationship between the grain pattern in the wood, domain pattern and pattern of growth activity in the storeyed cambium of trees. *Trees* 7, 137–143.

WŁOCH, W., E. MAZUR, and P. KOJS. 2001. Intensive change of inclination of cambial initials in *Picea abies* (L.) Karst. tumours. *Trees* 15, 498–502.

ZAGÓRSKA-MAREK, B. 1984. Pseudotransverse divisions and intrusive elongation of fusiform initials in the storeyed cambium of *Tilia. Can. J. Bot.* 62, 20–27.

ZAGÓRSKA-MAREK, B. 1995. Morphogenetic waves in cambium and figured wood formation. In: *Encyclopedia of Plant Anatomy* Band 9, Teil 4, *The Cambial Derivatives*, pp. 69–92, M. Iqbal, ed. Gebrüder Borntraeger, Berlin.

ZASADA, J. C., and R. ZAHNER. 1969. Vessel element development in the earlywood of red oak (*Quercus rubra*). *Can. J. Bot.* 47, 1965–1971.

ZHANG, Z.-J., Z.-R. CHEN, J.-Y. LIN, and Y.-T. ZHANG. 1992. Seasonal variations of secondary phloem development in *Pterocarya stenoptera* and its relation to feeding of *Kerria yunnanensis. Acta Bot. Sin. (Chih wu hsüeh pao)* 34, 682–687.

# CHAPTER THIRTEEN

# Phloem: Cell Types and Developmental Aspects

The phloem, although correctly called the principal food-conducting tissue of vascular plants, plays a much greater role than that in the life of the plant. A wide range of substances are transported in the phloem. Among those substances are sugars, amino acids, micronutrients, lipids (primarily in the form of free fatty acids; Madey et al., 2002), hormones (Baker, 2000), the floral stimulus (florigen; Hoffmann-Benning et al., 2002), and numerous proteins and RNAs (Schobert et al., 1998), some of which, in addition to the hormones, floral stimulus, and sucrose (Chiou and Bush, 1998; Lalonde et al., 1999), serve as informational or signaling molecules (Ruiz-Medrano et al., 2001). Dubbed the "information superhighway" (Jorgensen et al., 1998), the phloem plays a major role in inter-organ communication and in the coordination of growth processes within the plant. Long-distance signaling in plants occurs predominantly through the phloem (Crawford and Zambryski, 1999; Thompson and Schulz, 1999; Ruiz-Medrano et al., 2001; van Bel and Gaupels, 2004). The phloem also transports a large volume of water and may serve as the principal source of water for fruits, young leaves, and storage organs such as tubers (Ziegler, 1963; Pate, 1975; Lee,

1989, 1990; Araki et al., 2004; Nerd and Neumann, 2004).

As a rule, the phloem is spatially associated with the xylem in the vascular system (Fig. 13.1) and, like the xylem, may be classified as primary or secondary on the basis of its time of appearance in relation to the development of the plant or organ as a whole. The **primary phloem** is initiated in the embryo or young seedling (Gahan, 1988; Busse and Evert, 1999), is constantly added to during the development of the primary plant body, and completes its differentiation when the primary plant body is fully formed. The primary phloem is derived from the procambium. The **secondary phloem** (Chapter 14) originates from the vascular cambium and reflects the organization of this meristem in its possession of axial and radial systems. The phloem rays are continuous through the cambium with those of the xylem, providing a pathway for radial transport of substances between the two vascular tissues.

Although the phloem commonly occupies a position external to the xylem in stem and root or abaxial (on the lower side) in leaves and leaf-like organs, in many eudicot families (e.g., Apocynaceae, Asclepiadaceae,

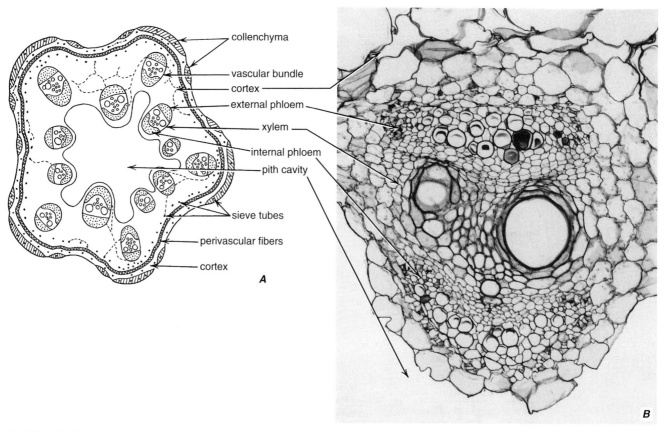

**FIGURE 13.1**

**A,** transverse section of a *Cucurbita* stem. Herbaceous vine with discrete vascular bundles, each having phloem on opposite sides of the xylem (bicollateral bundles). The vascular region is delimited on the outside by sclerenchyma (perivascular fibers). The cortex is composed of parenchyma and collenchyma. There is an epidermis. A cavity has replaced the pith. Small strands of extrafascicular sieve tubes and companion cells traverse parenchyma of the vascular region and cortex. **B,** transverse section of *Cucurbita* vascular bundle showing external and internal phloem. Typically a vascular cambium develops between the external phloem and the xylem but not between the internal phloem and the xylem. (A, ×8; B, ×130.)

Convolvulaceae, Cucurbitaceae, Myrtaceae, Solanaceae, Asteraceae) part of the phloem is located on the opposite side as well (Fig. 13.1). The two types of phloem are called **external phloem** and **internal phloem,** or **intraxylary phloem,** respectively. The internal phloem is largely primary in development (in some perennial species the addition of internal phloem is prolonged into the secondary stage of growth of the axis) and begins to differentiate later than the external phloem and usually also later than the protoxylem (Esau, 1969). A notable exception is found in the minor veins of *Cucurbita pepo* leaves, in which the adaxial (on the upper side) phloem differentiates in advance of the abaxial phloem (Turgeon and Webb, 1976). In certain families (e.g., Amaranthaceae, Chenopodiaceae, Nyctaginaceae, Salvadoraceae) the cambium, in addition to producing phloem outward and xylem inward, periodically forms some strands or layers of phloem toward the

interior of the stem so that the phloem strands become embedded in the xylem. Such phloem strands are referred to as **included phloem,** or **interxylary phloem.**

Sieve tubes that form lateral connections between their counterparts in the primary phloem of longitudinal vascular bundles in internodes and petioles are common in many species of seed plants (Figs. 13.1A and 13.2; Aloni and Sachs, 1973; Oross and Lucas, 1985; McCauley and Evert, 1988; Aloni and Barnett, 1996). Referred to as **phloem anastomoses,** they also connect the internal and external phloem in stems (Esau, 1938; Fukuda, 1967; Bonnemain, 1969) and the adaxial with the abaxial phloem in leaves (Artschwager, 1918; Hayward, 1938; McCauley and Evert, 1988). In a study on the functional significance of the phloem anastomoses in stems of *Dahlia pinnata* (Aloni and Peterson, 1990), it was found that the phloem anastomoses do not

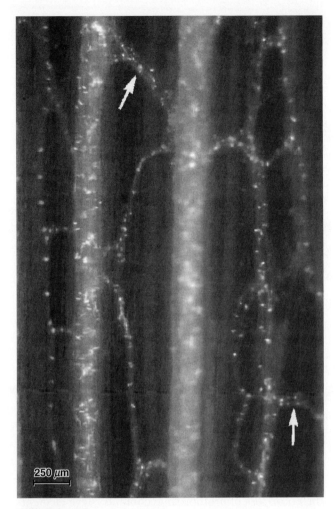

**FIGURE 13.2**

Phloem anastomoses (two marked by arrows) as seen in a thick section of a *Dahlia pinnata* internode after clearing and staining with aniline blue. The photograph was taken with an epifluorescent microscope. The numerous dots indicate the sites of callose, which occur at the lateral sieve areas and sieve plates of the sieve tubes. Two longitudinal vascular bundles, interconnected by phloem anastomoses, can be seen here. In *Dahlia* there are about 3000 phloem anastomoses per internode. (Courtesy of Roni Aloni.)

function under normal conditions. When the longitudinal strands were severed, however, the anastomoses began to function in transport. It was concluded that although the phloem anastomoses of the *Dahlia* internodes are capable of functioning, they serve mainly as an emergency system that provides alternative pathways for assimilates around the stem (Aloni and Peterson, 1990).

The overall development and structure of the phloem tissue parallels those of the xylem, but the distinct function of the phloem is associated with structural charac-

teristics peculiar to this tissue. The phloem tissue is less sclerified and less persisting than the xylem tissue. Because of its usual position near the periphery of stem and root, the phloem becomes much modified in relation to the increase in circumference of the axis during secondary growth, and portions of it no longer involved with conduction eventually may be cut off by a periderm (Chapter 15). The old xylem, in contrast, remains relatively unchanged in its basic structure.

# CELL TYPES OF THE PHLOEM

Primary and secondary phloem tissues contain the same categories of cells. The primary phloem, however, is not organized into two systems, the axial and the radial; it has no rays. The basic components of the phloem are the sieve elements and various kinds of parenchyma cells. Fibers and sclereids are common phloem components. Laticifers, resin ducts, and various idioblasts, specialized morphologically and physiologically, may also be present in the phloem. In this chapter only the principal cell types are considered in detail. The summary illustration (Fig. 13.3) and the list of phloem cells in Table 13.1 are based on the characteristic composition of the secondary phloem.

The principal conducting cells of the phloem are the **sieve elements**, so called because of the presence in their walls of areas (**sieve areas**) penetrated by pores. Among seed plants the sieve elements may be segregated into the less specialized **sieve cells** (Fig. 13.4A) and the more specialized **sieve-tube elements**, or **sieve-tube members** (Fig. 13.4B–H). This classification parallels that of tracheary elements into the less specialized tracheids and the more specialized vessel elements. The term **sieve tube** designates a longitudinal series of sieve-tube elements, just as the term vessel denotes a longitudinal series of vessel elements. In both classifications the characteristics of wall structure—pits and perforation plates in the tracheary elements, sieve areas and sieve plates (see below) in the sieve elements—may serve to distinguish the elements of the two kinds of categories. However, whereas vessel elements are found in angiosperms, the Gnetophyta, and certain seedless vascular plants, sieve-tube elements occur only in angiosperms. Moreover, use of the term sieve cell is restricted to gymnospermous sieve elements, which—considered later in this chapter—are remarkably uniform in structure and development. The sieve elements of the seedless vascular plants, or vascular cryptogams, show much variation in structure and development and are simply referred to by the general term sieve element (Evert, 1990a).

Young sieve elements contain all of the cellular components characteristic of young plant cells. As they differentiate, the sieve elements undergo profound changes,

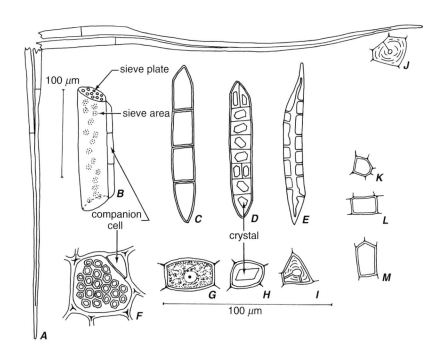

**FIGURE 13.3**

Cell types in the secondary phloem of a eudicot, *Robinia pseudoacacia*. **A–E**, longitudinal views; **F–J**, transverse sections. **A, J**, fiber. **B**, sieve-tube element and companion cells. **F**, sieve-tube element in plane of sieve plate and companion cell. **C, G**, phloem parenchyma cells (parenchyma strand in **C**). **D, H**, crystal-containing parenchyma cells. **E, I**, sclereids. **K–M**, ray cells in tangential (**K**), radial (**L**), and transverse (**M**) sections of phloem. (From Esau, 1977.)

**TABLE 13.1 ■ Cell Types of the Secondary Phloem**

| Cell Types | Principal Functions |
|---|---|
| Axial system | |
| Sieve elements | |
| Sieve cells (in gymnosperms) | Long-distance conduction of food materials; long-distance signaling |
| Sieve-tube elements, with companion cells (in angiosperms) | |
| Sclerenchyma cells | |
| Fibers | Support; sometimes storage of food materials |
| Sclereids | |
| Parenchyma cells | Storage and radial translocation of food substances |
| Radial (ray) system | |
| Parenchyma | |

Source: Esau, 1977.

the major ones being breakdown of the nucleus and tonoplast and formation of wall areas, the sieve areas, with pores that increase the degree of protoplasmic continuity between vertically or laterally adjoining sieve elements. Whereas tracheary elements undergo programmed cell death—a total autophagy—resulting in the entire loss of protoplasmic contents, sieve elements experience a **selective autophagy** (Fig. 13.5). At maturity the sieve element protoplast retains a plasma membrane, endoplasmic reticulum, plastids, and mitochondria, all of which occupy a parietal position (along the wall) within the cell.

# ▌ THE ANGIOSPERMOUS SIEVE-TUBE ELEMENT

The angiospermous sieve-tube element is characterized by the presence of **sieve plates**, wall parts bearing sieve areas with pores that are larger than those of sieve areas on other wall parts of the same cell. With relatively few exceptions (e.g., protophloem elements in roots of *Nicotiana tabacum*, Esau and Gill, 1972; metaphloem elements in the aerial stem of the holoparasite *Epifagus virginiana*, Walsh and Popovich, 1977; sieve elements of many palms, Parthasarathy, 1974a, b; *Lemna minor*, Melaragno and Walsh, 1976; and all members of the Poaceae, Evert et al., 1971b; Kuo et al., 1972; Eleftheriou, 1990), the protoplasts of sieve-tube elements contain **P-protein** (phloem protein, formerly called slime). In addition to sieve plates and P-protein, sieve-tube elements typically are associated with **companion cells**, specialized parenchyma cells closely related to the

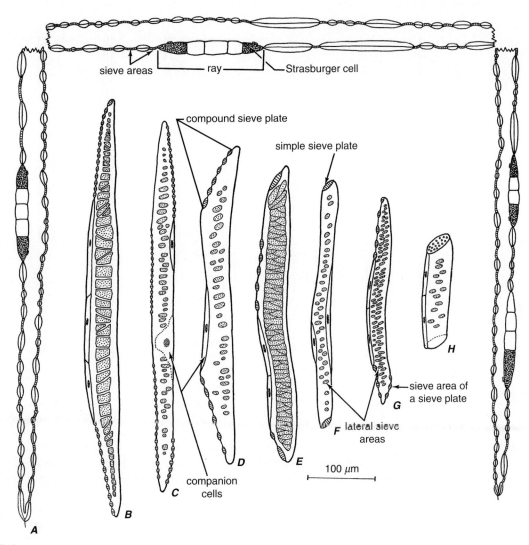

**FIGURE 13.4**

Variations in structure of sieve elements. **A**, sieve cell of *Pinus pinea*, with associated rays, as seen in tangential section. Others are sieve-tube elements with companion cells from tangential sections of phloem of the following species: **B**, *Juglans hindsii*; **C**, *Malus domestica*; **D**, *Liriodendron tulipifera*; **E**, *Acer pseudoplatanus*; **F**, *Crypto-carya rubra*; **G**, *Fraxinus americana*; **H**, *Wisteria* sp. In **B–G**, the sieve plates appear in side views and their sieve areas are thicker than the intervening wall regions because of deposition of callose. (From Esau, 1977.)

sieve-tube elements both ontogenetically and function-ally. The term **sieve tube–companion cell complex**, or **sieve element–companion cell complex**, com-monly is used to refer to a sieve-tube element and its associated companion cell(s).

### In Some Taxa the Sieve-Tube Element Walls Are Remarkably Thick

The walls of sieve-tube elements commonly are described as primary and with standard microchemical tests usually give positive reactions for only cellulose and pectin (Esau, 1969). In the leaves of grasses, the last-formed sieve tubes of the longitudinal bundles typi-cally have relatively thick cell walls (Fig. 13.6). In some species—*Triticum aestivum* (Kuo and O'Brien, 1974), *Aegilops comosa* (Eleftheriou, 1981), *Saccharum offici-narum* (Colbert and Evert, 1982), *Hordeum vulgare* (Dannenhoffer et al., 1990)—these walls are lignified. Although variable in thickness, the walls of sieve-tube elements usually are distinctly thicker than those of the surrounding parenchymatous cells, a character that may facilitate recognition of the sieve-tube element.

In many species the walls of the sieve-tube elements consist of two morphologically distinct layers, a rela-tively thin outer layer and a more or less thick inner layer. In fresh sections the distinct inner layer exhibits a shiny or glistening appearance and, hence, received the name

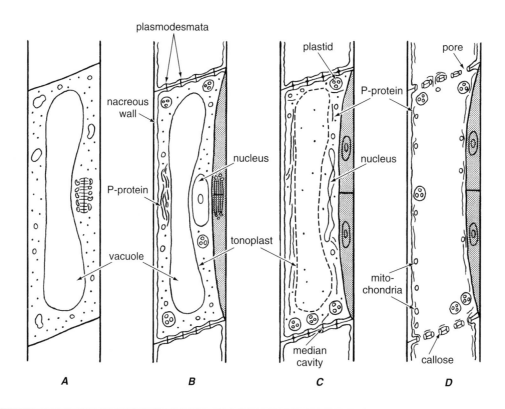

**FIGURE 13.5**

Diagrams illustrating differentiation of a sieve-tube element. **A**, precursor of sieve-tube element in division. **B**, after division: sieve-tube element with nacreous wall and P-protein body; dividing companion cell precursor (stippled). **C**, nucleus degenerating, tonoplast partly broken down, P-protein dispersed; median cavities in future sieve plates; two companion cells (stippled). **D**, mature sieve-tube element; pores in sieve plates open; they are lined with callose and some P-protein. In addition to plastids, mitochondria are present. No endoplasmic reticulum is shown. (From Esau, 1977.)

**nacreous** (having a pearly luster) wall. The nacreous layer contains less cellulose than the outer wall layer and is pectin-poor (Esau and Cheadle, 1958; Botha and Evert, 1981). Sometimes the nacreous layer is so thick as almost to occlude the lumen of the cell. Although some workers have classified this wall layer as secondary, its behavior is quite variable. In primary sieve-tube elements, it commonly is transitory in nature and becomes reduced in thickness as the cell approaches maturity and then disappears at about the time the cell reaches maturity. In secondary phloem sieve-tube elements, the nacreous layer may or may not be reduced in thickness with age (Fig. 13.7; Esau and Cheadle, 1958; Gilliland et al., 1984). The nacreous layer does not extend into the region of the sieve areas and sieve plates.

Through the use of mild extraction procedures for the removal of noncellulosic wall components and of electron microscopy, the nacreous thickenings of certain eudicots have been shown to have a polylamellate structure, the concentrically arranged lamellae consisting of densely packed microfibrils (Deshpande, 1976; Catesson, 1982). The nacreous walls in sieve-tube elements of the seagrasses also exhibit a polylamellate appearance without extraction (Kuo, 1983).

After fixation with glutaraldehyde-osmium tetroxide and staining with uranyl acetate and lead citrate, the inner surface of the wall of sieve-tube elements often appears considerably more electron dense than the rest of the wall (Fig. 13.8; Evert and Mierzwa, 1989). This region, which often shows netted and/or striate patterns, apparently is a pectin-rich layer of nonmicrofibrillar material and, unlike the nacreous thickening, extends into the sieve areas and sieve plates (Lucas and Franceschi, 1982; Evert and Mierzwa, 1989). In sieve tubes of the leaf-blade veins of *Hordeum vulgare* this electron-dense inner wall region is thickest at the lateral sieve areas and the sieve plates where it is permeated by a labyrinth of tubules formed by the plasma membrane (Evert and Mierzwa, 1989). Along the lateral walls between sieve areas this inner wall region is permeated by numerous microvilli-like evaginations of the plasma membrane, greatly increasing the cell wall–plasma membrane interface and giving it the appearance of a brush border.

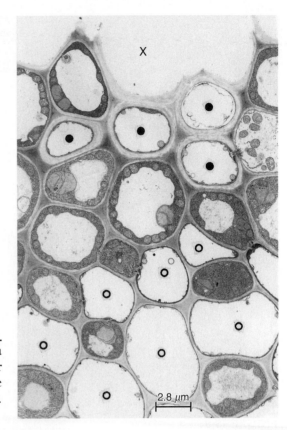

## FIGURE 13.6

Electron micrograph of portion of a large vascular bundle from a barley (*Hordeum vulgare*) leaf. Note the thick walls of the four last-formed sieve tubes (solid dots) bordering the xylem, and the relatively thin walls of the sieve tubes (open dots) formed earlier. Other detail: x, xylem. (From Dannenhoffer et al., 1990.)

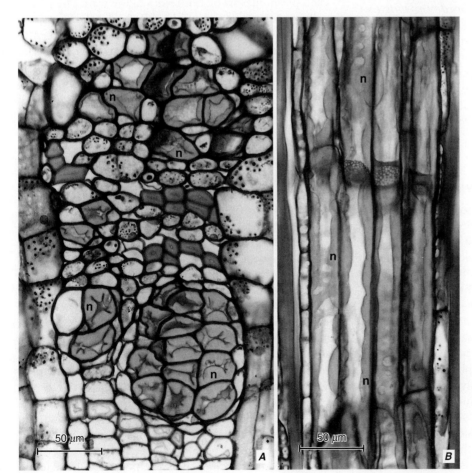

## FIGURE 13.7

Transverse (**A**) and radial longitudinal (**B**) sections of the secondary phloem of *Magnolia kobus*. Note the thick inner wall layer (n, nacreous layer) of the sieve tubes. (From Evert, 1990b, Figs. 16.19 and 16.20. © 1990, Springer-Verlag.)

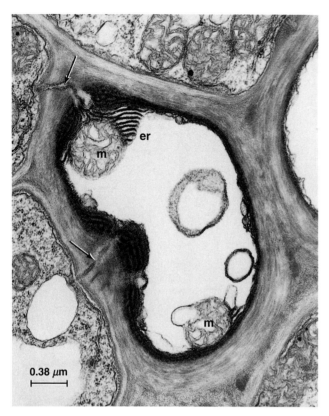

**FIGURE 13.8**

Electron micrograph of a barley (*Hordeum vulgare*) sieve tube, as seen in transverse section through a longitudinal bundle of leaf. The inner surface of the wall, which is markedly more electron dense than the rest of the wall, is thickest at the sites of pore-plasmodesmata connections (arrows) with parenchymatous elements. Other details: er, endoplasmic reticulum; m, mitochondrion. (From Evert and Mierzwa, 1989, Fig. 2. © 1989, Springer-Verlag.)

### Sieve Plates Usually Occur on End Walls

As mentioned previously, in the angiosperms the size of the sieve-area pores varies considerably on the walls of the same cell (Fig. 13.9A–C). The diameter of pores in the sieve areas ranges from a fraction of a micrometer (little wider than a plasmodesma) to 15 μm and probably more in some eudicots (Esau and Cheadle, 1959). Sieve areas with the larger pores usually occur on the end walls, those with the smaller pores on the side, or lateral, walls. Hence the sieve plates usually occur on the end walls, the sieve-tube elements being arranged end on end, forming a sieve tube. Sieve plates may occur on side walls. Some sieve plates bear only a single sieve area (Fig. 13.9A; **simple sieve plate**), while others bear two or more (Fig. 13.9D, E; **compound sieve plate**).

In routine preparations of conducting phloem, the sieve pores typically are lined with the wall constituent **callose** (Chapter 4). Most, if not all, of the callose asso-

ciated with conducting sieve-tube elements is deposited there in response to mechanical injury or some other kind of stimulation (Evert and Derr, 1964; Esau, 1969; Eschrich, 1975). Not all of the callose associated with sieve pores is such **wound callose**. Callose normally accumulates at the sieve plates and lateral sieve areas of senescing sieve elements (Fig. 13.10). This **definitive callose** disappears some time after the sieve element dies. Callose usually accumulates at the sieve plates and lateral sieve areas of secondary phloem sieve-tube elements that function for more than one growing season (Davis and Evert, 1970). In temperate regions this **dormancy callose** is deposited in the fall and then is removed in early spring during reactivation of the dormant, overwintering sieve elements.

### Callose Apparently Plays a Role in Sieve-Pore Development

In young sieve-tube elements, the sieve area (or areas) of the incipient sieve plate is penetrated by variable numbers of plasmodesmata, each of which is associated with a cisterna of endoplasmic reticulum on both sides of the wall (Fig. 13.11A). The pore sites first become distinguishable from the rest of the wall with the appearance of callose beneath the plasma membrane around each plasmodesma on both sides of the wall. The paired callose deposits, commonly called **platelets**, assume the form of collars or cones interrupted in the center where the plasmodesma is located (Fig. 13.11B, C). The platelets undergo rapid enlargement and initially may exceed the rest of the wall in their rate of thickening. Thickening of the cellulosic-pectin portion of the wall may overtake the callose platelets; then the pore sites appear as depressions in the plate. The presence of the callose platelets at the pore sites apparently precludes further deposition of cellulose there so that the cellulosic wall parts sandwiched between platelets remain thin. Localization of callose platelets at the pore sites and thickening of the wall are among the earliest indicators of sieve-element development.

Perforation of the pore sites begins at about the time of nuclear degeneration. The removal of wall material begins in the region of the middle lamella surrounding the plasmodesma (Fig. 13.11D, E). In some instances, a median cavity is formed initially and then further simultaneous removal of the callose platelets and of the wall substance sandwiched between them results in formation of the pore (Deshpande, 1974, 1975). In others, lysis in the region of the middle lamella results in a merging of the opposing callose platelets so that initially the young pore is lined with callose (Esau and Thorsch, 1984, 1985). The cisternae of endoplasmic reticulum remain closely appressed to the plasma membrane bordering the callose platelets throughout pore development, only to be removed as the pores attain their full

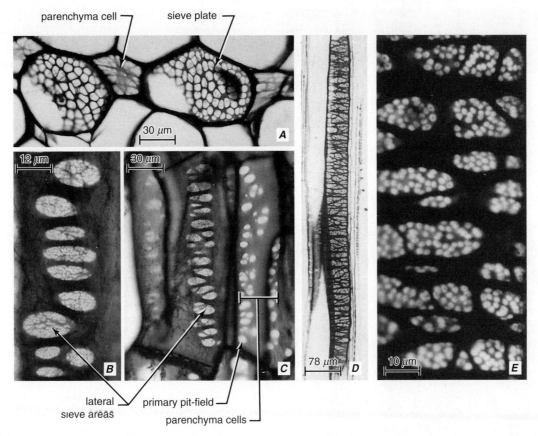

parenchyma cell    sieve plate

30 μm

A

12 μm

30 μm

B

C

78 μm  D

10 μm  E

lateral
sieve areas    primary pit-field
parenchyma cells

**FIGURE 13.9**

**A**, simple sieve plates of *Cucurbita* in surface view. **B**, **C**, lateral sieve areas in sieve-tube elements and primary pit-fields in parenchyma cells of *Cucurbita* in surface view. **D**, surface view of compound sieve plate of *Cocos*, a monocot with sieve areas in reticulate arrangement. **E**, part of similar sieve plate. Light spots are callose cylinders. (**A–C**, from Esau et al., 1953; **D**, **E**, from Cheadle and Whitford, 1941.)

size (Fig. 13.11F). Development of the lateral sieve-area pores is essentially similar to that of the sieve-plate pores (Evert et al., 1971a).

Whether callose is universally involved with sieve-pore formation is problematic. Callose was not found at any stage of sieve-pore development in the root protophloem of the small aquatic monocotyledon *Lemna minor* (Walsh and Melaragno, 1976). Callose could be induced to form, however, in response to injury.

### Changes in the Appearance of the Plastids and the Appearance of P-protein Are Early Indicators of Sieve-Tube Element Development

Initially the young sieve-tube element protoplast (Fig. 13.12) resembles the protoplast of other procambial cells or recent cambial derivatives. Both young, nucleate sieve-tube elements and their neighboring nucleate cells contain Golgi bodies, plastids, and mitochondria. Variable numbers of vacuoles are delimited from the cytosol

by tonoplasts. The cytoplasm is rich in free ribosomes and contains a network of rough endoplasmic reticulum. Microtubules, mostly oriented at right angles to the long axis of the cell, occur next to the plasma membrane, bordering a thin cell wall. Longitudinally oriented bundles of actin filaments are fairly numerous. With the exception of the microtubules, the various cellular components are more or less randomly distributed throughout the cell.

Changes in the appearance of the plastids, which initially are similar in appearance to those of neighboring cells, are early indicators of sieve-tube development. As a sieve-tube plastid matures, its stroma becomes less dense, and inclusions characteristic of the plastid type may appear (Fig. 13.13A–D). Until then, it is often difficult to distinguish the plastids from the mitochondria. In mature sieve-tube elements the stroma is electron-transparent, and often the internal membranes (thylakoids) are sparse. Sieve-tube plastids occur in two basic types, S-type (S, starch) and P-type (P, protein) (Behnke,

**FIGURE 13.10**

Longitudinal view (tangential section) of nonfunctional sieve-tube elements with massive deposits of definitive callose (arrows) at the sieve plates and lateral sieve areas in the secondary phloem of elm (*Ulmus americana*). Other details: c, crystalliferous cells; f, fiber; r, ray. (×400. From Evert et al., 1969.)

1991a). The **S-type** occurs in two forms, one of which contains only starch (Fig. 13.13A, C); the other is devoid of any inclusion. The **P-type** exists in six forms and contains one or two kinds of proteinaceous inclusions (crystals, Fig. 13.13B, D, and/or filaments, Fig. 13.13E). Two of the six also contain starch. All of the monocots have P-type plastids, and those containing only cuneate protein crystals (Fig. 13.13B, D) are dominant (Eleftheriou, 1990). Unlike ordinary starch, sieve-tube starch stains brownish red rather than blue-black with iodine (I₂KI). The sieve-tube starch in *Phaseolus vulgaris* is a highly branched molecule of the amylopectin type with numerous $\alpha(1{\rightarrow}6)$ linkages (Palevitz and Newcomb, 1970). Plastid differences in sieve elements are taxonomically useful (Behnke, 1991a, 2003).

Another early indicator of sieve-tube element development is the appearance of P-protein, which first becomes discernible with the light microscope as dis-

crete bodies, one or more per cell (Fig. 13.14A, B). The P-protein bodies appear after the precursor of the sieve-tube element has divided to give rise to one or more companion cells. Most species have **dispersive P-protein bodies**. Small at first, these P-protein bodies increase in size (Fig. 13.14A, B) and eventually begin to disperse, forming strands or networks in the parietal layer of cytoplasm. By this time the nucleus has begun to degenerate. After the tonoplast disappears, the dispersed P-protein is found in a parietal position in the cell lumen and sieve-plate pores (Figs. 13.15 and 13.16D; Evert et al., 1973c; Fellows and Geiger, 1974; Fisher, D. B., 1975; Turgeon et al., 1975; Lawton, D. M., and Newman, 1979; Deshpande, 1984; Deshpande and Rajendrababu, 1985; Russin and Evert, 1985; Knoblauch and van Bel, 1998; Ehlers et al., 2000), provided that care has been taken to disturb the phloem as little as possible during sampling. Otherwise, with the release of the high hydrostatic pressures of the sieve-tube contents at the time the sieve tubes are severed, the P-protein may become dispersed throughout the lumen or accumulate, upon surging, as **slime plugs** on the sides of the sieve plates away from the sites of pressure release.

At the electron microscope level the P-protein often appears in filaments of tubular form, with subunits arranged helically (Fig. 13.16A–C). The P-protein filaments in *Cucurbita maxima* are composed of two very abundant proteins: phloem protein 1 (PP1), a 96 kDa protein filament, and phloem protein 2 (PP2), a 25 kDa dimeric lectin that binds covalently to PP1. Protein and mRNA localization patterns indicate that PP1 and PP2 are synthesized in companion cells of differentiating and mature sieve element–companion cell complexes and that polymerized forms of P-protein accumulate in the sieve-tube elements during differentiation (Bostwick et al., 1992; Clark et al., 1997; Dannenhoffer et al., 1997; Golecki et al., 1999). Apparently the PP1 and PP2 subunits synthesized in companion cells are transported into the sieve-tube elements via the pore-plasmodesmata connections in their common walls. Thus far the role of the filamentous P-protein remains uncertain. It has been suggested that the PP1 serves to seal the sieve-plate pores of injured elements, representing the sieve tubes first line of defense against the loss of assimilates, with wound callose shoring up the defenses at variable rates (Evert, 1990b). The role of the lectin (PP2) is no less uncertain. PP2 subunits have been found to move in the assimilate stream from source to sink (see below) and to cycle between sieve elements and companion cells (Golecki et al., 1999; Dinant et al., 2003). PP2-like genes have been identified in 16 genera of seed plants, including a gymnosperm (*Picea taeda*) and four genera of Poaceae, none of which contain PP1. A PP2-like gene was also found in a nonvascular plant, the moss *Physcomitrella patens*. It appears that PP2-like

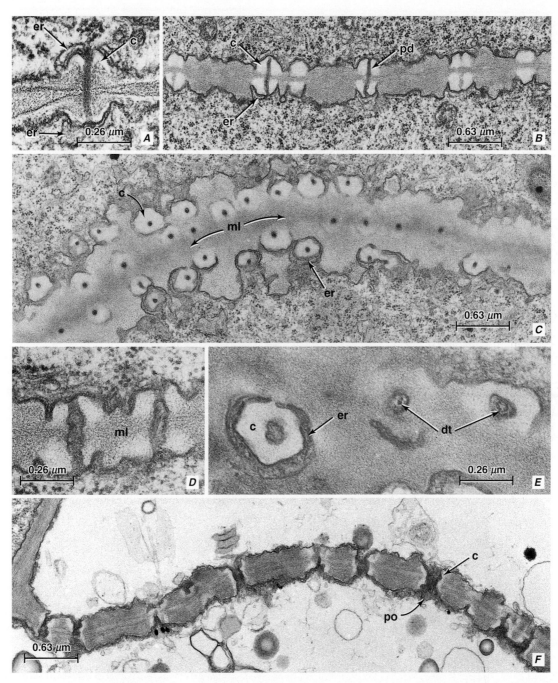

**FIGURE 13.11**

Developing sieve plates in sieve-tube elements from internodes of cotton (*Gossypium hirsutum*), as seen in sectional (**A, B, D, F**) and surface views (**C, E**). **A**, a plasmodesma, which marks the site of a future pore. Some callose (c) has been deposited beneath cisternae of endoplasmic reticulum (er). **B, C**, callose platelets (c) enclose the plasmodesmata (pd) at the pore sites. **D, E**, pores have begun to develop with widening of the plasmodesmatal canal. **F**, mature sieve plate with open pores (po) lined with small amounts of callose and filled with P-protein. Other details: dt, desmotubule; ml, middle lamella. (From Esau and Thorsch, 1985.)

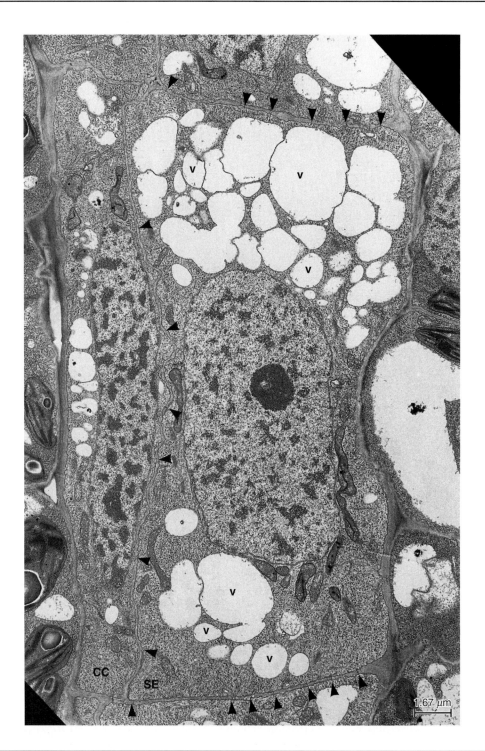

**FIGURE 13.12**

Longitudinal view of young sieve-tube element (SE) and companion cell (CC) from leaf of tobacco (*Nicotiana tabacum*). Arrowheads mark discernible plasmodesmata in the two ends (future sieve plates) of the sieve-tube element and in the common wall between sieve-tube element and companion cell (sites of future pore–plasmodesmata connections). Numerous small vacuoles (v) occur above and below the nucleus of the sieve-tube element. (From Esau and Thorsch, 1985.)

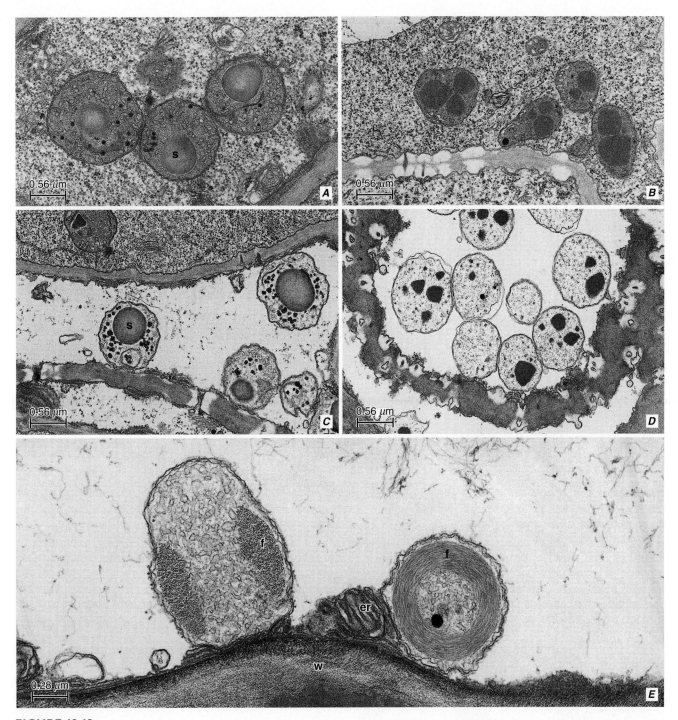

**FIGURE 13.13**

Sieve-tube plastids. Immature (**A**) and mature (**C**) S-type plastids in bean (*Phaseolus*) root tip; immature (**B**) and mature (**D**) P-type plastids with cuneate protein crystals (dense inclusions), in onion (*Allium*) root tip. **E**, P-type plastids, with filamentous protein inclusions (f), in sieve tube of spinach (*Spinacia*) leaf. Other details: er, endoplasmic reticulum; s, starch; w, wall.

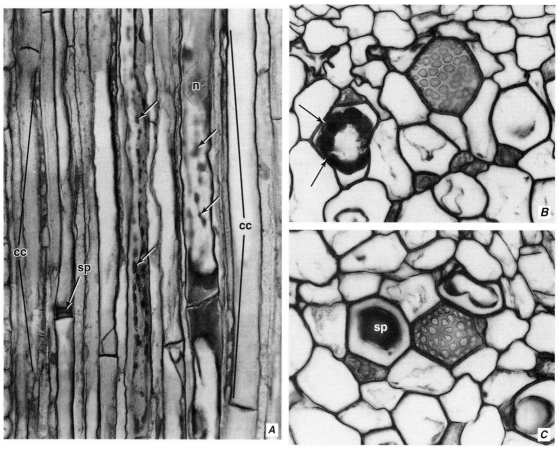

**FIGURE 13.14**

Immature and mature sieve-tube elements in the stem phloem of squash (*Cucurbita maxima*), as seen in longitudinal (**A**) and transverse (**B, C**) sections. **A**, two immature sieve-tube elements (on right and in center) contain numerous P-protein bodies (arrows). The P-protein bodies in the sieve-tube element on the right have begun to disperse in the parietal layer of cytoplasm. The nucleus (n) in this element has begun to degenerate and is barely discernible. A strand of companion cells (cc) accompanies the mature sieve-tube elements, far right and left. A slime plug (sp) can be seen in the sieve-tube element on lower left. **B**, two immature sieve tubes. Large P-protein bodies (arrows) can be seen in sieve tube on the left, an immature (simple) sieve plate in face view in the one on the right, above. The small, dense cells are companion cells. **C**, two mature sieve-tube elements. A slime plug (sp) can be seen in the sieve-tube element on the left, a mature sieve plate in the one on the right. The small dense cells are companion cells. (A, ×300; B, C, ×750.)

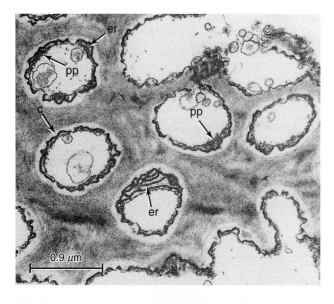

**FIGURE 13.15**

Electron micrograph of portion of mature *Cucurbita maxima* sieve plate in surface view. The pores are lined by a narrow callose cylinder (c) and plasma membrane (not labeled). Elements of endoplasmic reticulum (er) and P-protein (pp) are also found along margins of pores. (From Evert et al., 1973c, Fig. 2. © 1973, Springer-Verlag.)

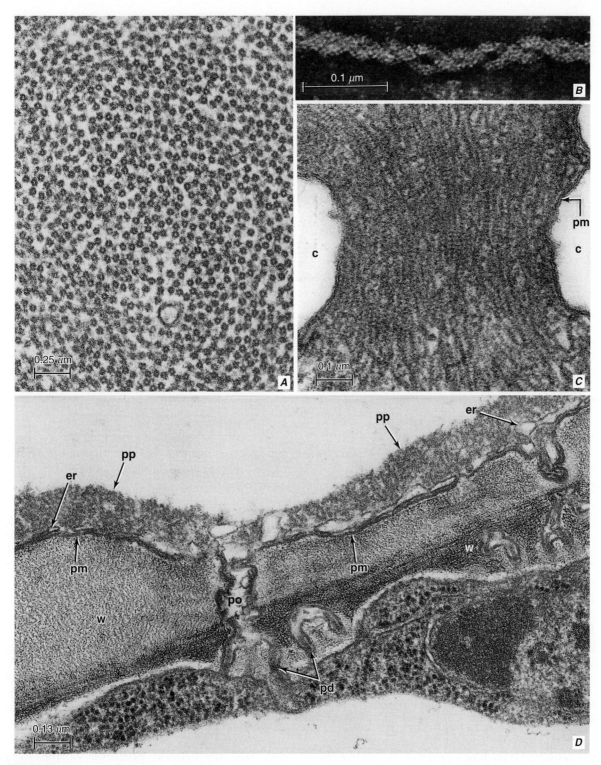

**FIGURE 13.16**

P-protein. Sieve-tube elements of *Poinsettia* (**A**), *Nicotiana tabacum* (**B**), *Nelumbo nucifera* (**C**), and *Cucurbita pepo* (**D**). **A**, portion of a P-protein body showing tubular filaments. **B**, high magnification of negatively stained phloem exudate reveals the double-stranded structure of the P-protein filament. **C**, P-protein accumulated in a sieve-plate pore shows horizontal striations in the extended filaments; callose (c) lining the pore beneath the plasma membrane (pm). **D**, transverse section showing portions of wall (w) and parietal layer of cytoplasm of mature sieve-tube element (above). The parietal layer in this view consists of the plasma membrane (pm), discontinuous profiles of endoplasmic reticulum (er), and P-protein (pp). Pore–plasmodesmata connections can be seen in the sieve-tube element (pore-side)–companion cell (plasmodesmata-side) wall. Other details: po, pore; pd, plasmodesmata. (**B**, reprinted from Cronshaw et al., 1973. © 1973, with permission from Elsevier; **C**, from Esau, 1977; **D**, from Evert et al., 1973c, Fig. 6. © 1973, Springer-Verlag.)

proteins may have properties that are not exclusively related to PP1 or vascular-specific functions (Dinant et al., 2003). It has been suggested that PP2 may serve to immobilize bacteria and fungi at wound sites or as an anchor for the organelles that persist along the walls in mature, conducting sieve-tube elements. Minute, clamp-like structures, which have been proposed to be responsible for the peripheral positioning of the components in mature sieve elements, have been found in *Vicia faba* and *Lycopersicon esculentum* sieve-tube elements (Ehlers et al., 2000). The chemical nature of these "clamps" is unknown.

In some taxa (basically woody families), the P-protein bodies disperse only partially or not at all (**nondispersive P-protein bodies**, Fig. 13.17; see also Fig. 13.36A; Behnke, 1991b). Cytoplasmic inclusions once regarded as extruded nucleoli are examples of these (Deshpande and Evert, 1970; Esau, 1978a; Behnke and Kiristis, 1983). Often cited as examples of nondispersive P-protein bodies are the tailed or tailless spindle-shaped crystalline P-protein bodies of the Fabaceae, previously called persistent slime bodies by light microscopists (Esau, 1969). It has been shown, however, that these P-protein bodies are able to undergo rapid and reversible calcium-controlled conversions from the condensed "resting state" into a dispersed state, in which they occlude the sieve tubes (Knoblauch et al., 2001). Dispersal of the crystalloids is triggered by plasma membrane leakage and abrupt turgor changes. It has been suggested that the ability of the P-protein to cycle between dispersal and condensation may provide an efficient mechanism to control sieve tube conductivity (Knoblauch et al., 2001). Four major forms of nondispersive P-protein bodies can be recognized in eudicotyledonous sieve-tube elements: spindle-shaped, compound-spherical, rod-shaped, and rosette-like (Behnke, 1991b). The great majority of nondispersive protein bodies are of cytoplasmic origin. Nuclear nondispersive protein bodies have been found in two eudicot families, the Boraginaceae and Myristicaceae (Behnke, 1991b), and in the monocot family Zingiberaceae (Behnke, 1994).

### Nuclear Degeneration May Be Chromatolytic or Pycnotic

One of the major events in the final stages of sieve-element ontogeny is the degeneration of the nucleus. In most angiosperms—both eudicots (Evert, 1990b) and monocots (Eleftheriou, 1990)—nuclear degeneration is by **chromatolysis**, a process involving the loss of stainable contents (chromatin and nucleoli) and eventual rupture of the nuclear envelope (Fig. 13.18B). **Pycnotic degeneration**, during which the chromatin forms a very dense mass prior to rupture of the nuclear envelope, has been reported to occur primarily in differentiating protophloem sieve-tube elements.

At about the time the nuclei begin to degenerate, the cisternae of endoplasmic reticulum begin to form stacks (Figs. 13.18A and 13.19A). During the stacking process the endoplasmic reticulum begins to migrate toward the wall, and the ribosomes disappear from the surfaces that face one another in a stack, although electron-dense material, possibly enzymes, accumulate between the cisternae (Fig. 13.19A, B). Ribosomes on the outer surfaces of the membrane stacks disappear concomitantly with the free ribosomes of the cytoplasm. With increasing maturation of the sieve-tube element, the now entirely smooth endoplasmic reticulum may undergo further modification into convoluted, lattice-like, and tubular forms. In most fully mature sieve-tube elements the endoplasmic reticulum is represented largely by a complex network—a parietal, anastomosing system—that lies next to the plasma membrane, along with the surviving organelles and the P-protein. Only two kinds of organelles are retained, the plastids and mitochondria (Fig. 13.19C). Neither microtubules nor actin filaments have been discerned in electron micrographs of mature sieve-tube elements, although both actin and profilin, which has been implicated in the regulation of actin filament polymerization (Staiger et al., 1997), have been found at high levels in sieve-tube exudate (Guo et al., 1998; Schobert et al., 1998).

The two delimiting membranes, plasma membrane and tonoplast, show contrasting behavior. Whereas the plasma membrane persists as a selectively permeable membrane, the tonoplast breaks down and the delimitation between vacuole and parietal cytoplasm disappears. With clearing of the lumina of the superimposed sieve-tube elements and development of unoccluded sieve-plate pores between them, the sieve tube becomes an ideal conduit for the flow of solution of the assimilate stream (Figs. 13.15 and 13.20).

## ▌COMPANION CELLS

Sieve-tube elements are characteristically associated with specialized parenchyma cells called **companion cells**. Typically, companion cells are derived from the same mother cell as their associated sieve-tube elements, so that the two kinds of cells are closely related ontogenetically (Fig. 13.5). In the formation of the companion cells the meristematic precursor of the sieve-tube element divides longitudinally one or more times. One of the resulting cells, usually distinguished by being larger, differentiates into the sieve-tube element. One or more companion cells may be associated with a single sieve-tube element, and the companion cells may occur on one or more sides of the sieve-tube element wall. In some taxa the companion cells occur in vertical series (**companion cell strands;** Fig. 13.21B, C), the result of divisions of their immediate precursor. Companion cells

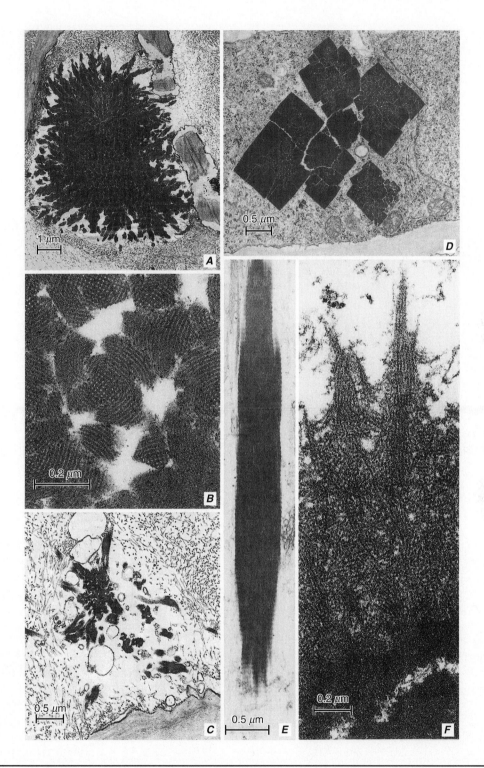

**FIGURE 13.17**

Nondispersive P-protein bodies. **A**, *Quercus alba*. Compound-spherical body near sieve plate in mature sieve-tube element. **B**, *Quercus alba*. Detail of spherical body. **C**, *Rhus glabra*. Compound-spherical body in mature sieve-tube element. Described as "stellate," by Deshpande and Evert (1970). **D**, *Robinia pseudoacacia*. Transverse view of spindle-shaped body in immature sieve-tube element. **E**, *R. pseudoacacia*. Longitudinal view of spindle-shaped body in mature sieve-tube element. **F**, *Tilia americana*. Portion of compound-spherical body. The peripheral region (above) is composed of rod-like components; the more dense, central region (below) shows little or no substructure. The spherical bodies of *Quercus* and *Tilia* were once regarded as extruded nucleoli. (**A–C**, and **F**, reprinted from Deshpande and Evert, 1970. © 1970, with permission from Elsevier; **D**, **E**, from Evert, 1990b, Figs. 6.16 and 6.17. © 1990, Springer-Verlag.)

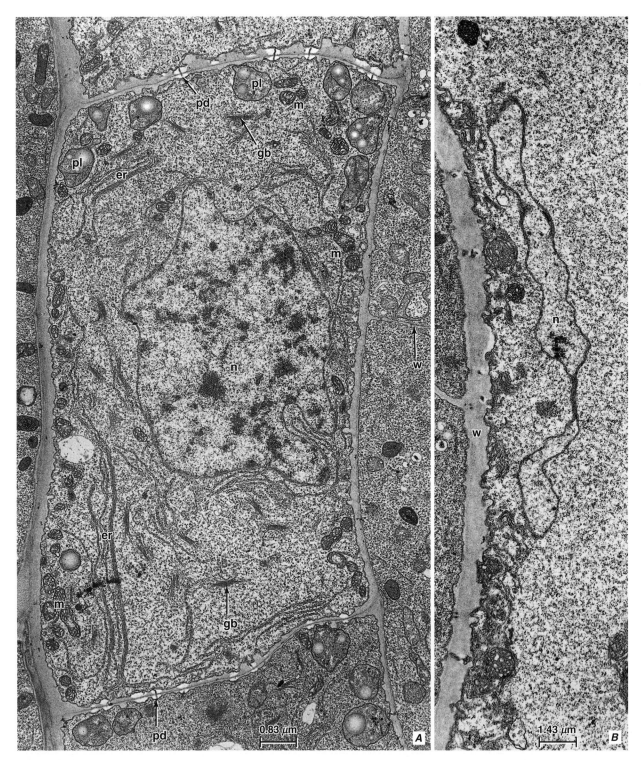

**FIGURE 13.18**

**A**, immature protophloem sieve-tube element in root of tobacco (*Nicotiana tabacum*). Stacking of the endoplasmic reticulum (er) has begun, and most of the plastids (pl) and mitochondria (m) have become distributed along the wall. The nucleus (n) has begun to lose stainable contents, and the pore sites of the developing sieve plates in both end walls are marked by the presence of pairs of callose platelets. A single plasmodesma (pd) traverses the platelets, one platelet on either side of the wall. Other details: gb, Golgi body; w, wall between parenchyma cells. **B**, partly collapsed nucleus (n) in immature sieve-tube element at later stage than in **A**. The organelles now are located along the wall (w). (Reprinted from Esau and Gill, 1972. © 1972, with permission from Elsevier.)

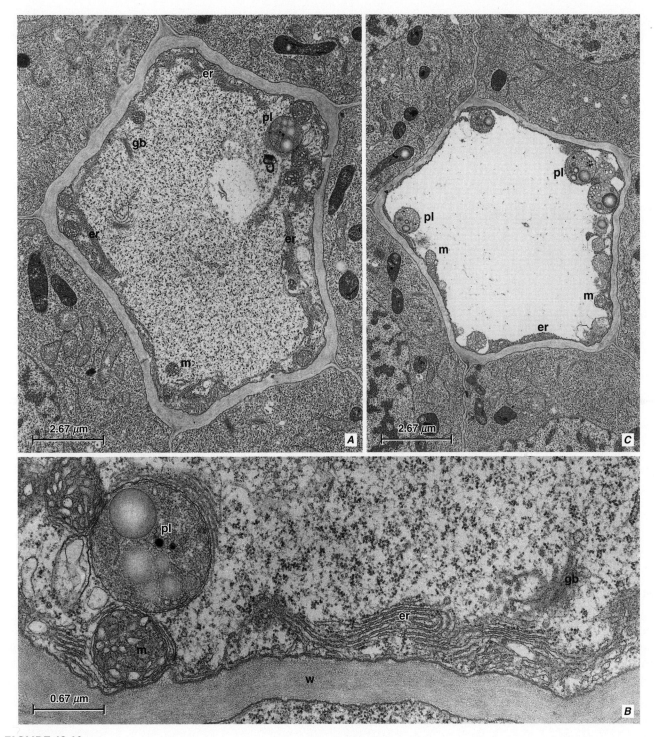

**FIGURE 13.19**

Transverse sections of immature (**A, B**) and mature (**C**) protophloem sieve-tube elements in root of tobacco (*Nicotiana tabacum*). **A**, the stacked endoplasmic reticulum (er) and organelles (mitochondria, m, and plastids, pl) already are in a peripheral position. Golgi bodies (gb) and abundant ribosomes are still present. **B**, detail of stacked endoplasmic reticulum. **C**, the mature sieve element has a clear appearance. Other detail: w, wall. (**A, B**, reprinted from Esau and Gill, 1972. © 1972, with permission from Elsevier.)

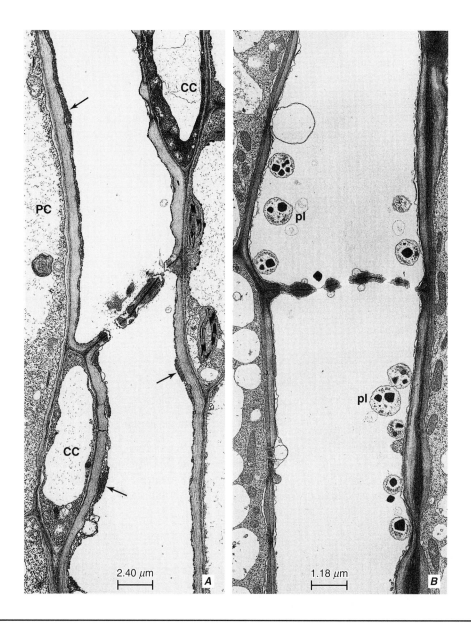

**FIGURE 13.20**

Longitudinal sections of portions of mature sieve-tube elements, showing parietal distribution of cytoplasmic components and sieve plates with unoccluded pores. **A**, *Cucurbita maxima*. Unlabeled arrows point to P-protein. Other details: CC, companion cell; PC, parenchyma cell. **B**, *Zea mays*. Typical of monocotyledonous sieve-tube elements, those of maize contain P-type plastids (pl), with cuneate protein crystals. Maize, a member of the Poaceae, lacks P-protein. (**A**, from Evert et al., 1973c, Fig. 11. © 1973, Springer-Verlag; **B**, courtesy of Michael A. Walsh.)

also vary in size. Some—both individual cells and strands—are as long as the sieve-tube element with which they are related (Fig. 13.21A); others are shorter than the sieve-tube element (Fig. 13.21D–I; Esau, 1969). The ontogenetic relation of companion cells to sieve-tube elements is usually regarded as a specific characteristic of these cells, although some parenchymatous elements commonly regarded as companion cells may not be derived from the same mother cell as their associated sieve-tube element (e.g., in longitudinal veins of the maize leaf blade; Evert et al., 1978). The relation is,

however, typical in angiosperms, and the presence of companion cells is included in the definition of the sieve-tube element as contrasted with the sieve cell.

Whereas the sieve-tube element protoplast undergoes a selective autophagy and assumes a clear appearance during its ontogeny, the companion cell protoplast commonly increases in density as it approaches maturity. This increase in density is due in part to an increase in density in the ribosome (polysome) population and partly to an increase in density of the cytosol itself (Behnke, 1975; Esau, 1978b). The mature companion

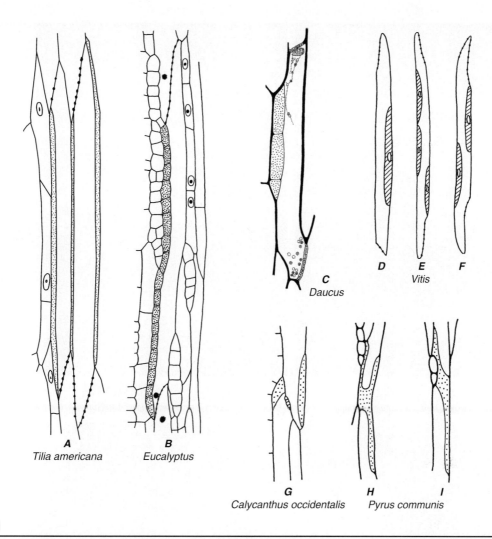

**FIGURE 13.21**

Companion cells (longitudinal views). **A**, sieve-tube elements of *Tilia americana* with companion cells (stippled) that extend the length of the element, from sieve plate to sieve plate. **B**, sieve-tube element of *Eucalyptus*, with long strand of companion cells. The dense bodies near the sieve plates are nondispersive P-protein bodies once believed to be extruded nucleoli. **C**, sieve-tube element of *Daucus* (carrot), with strand of three companion cells. The small bodies near sieve plates are plastids with starch; the large body is P-protein. **D–F**, portions of sieve-tube elements of *Vitis*; companion cells hatched. **G**, sieve-tube elements with companion cells of *Calycanthus occidentalis*; **H**, **I**, portions of sieve-tube elements of *Pyrus communis*, with companion cells. (A, ×255; B, ×230; C, ×390; D–F, ×95; G–I, ×175. **A**, from Evert, 1963. © 1963 by The University of Chicago. All rights reserved; **B**, from Esau, 1947; **C**, adapted from Esau, 1940. *Hilgardia* 13(5), 175–226. © 1940 Regents, University of California; **D–F**, from Esau, 1948. *Hilgardia* 18(5), 217–296. © 1948 Regents, University of California; **G**, reproduced by permission University of California Press: Cheadle and Esau, 1958. *Univ. Calif. Publ. Bot.* © 1958, The Regents of the University of California; **H**, **I**, reproduced by permission University of California Press: Evert, 1960. *Univ. Calif. Publ. Bot.* © 1960, The Regents of the University of California.)

cell also contains numerous mitochondria, rough endoplasmic reticulum, plastids, and a prominent nucleus. Companion cell plastids typically lack starch, although some exceptions exist (e.g., in *Cucurbita*, Esau and Cronshaw, 1968; *Amaranthus*, Fisher, D. G., and Evert, 1982; *Solanum*, McCauley and Evert, 1989). The cells are vacuolated to various degrees.

Companion cells are intimately connected with their associated sieve-tube elements by numerous cytoplasmic connections, consisting of a pore on the sieve-tube element side of the wall and much-branched plasmodesmata on the companion cell side (Fig. 13.22). During development of these connections callose appears at the site of the future pore on the sieve-tube element side of

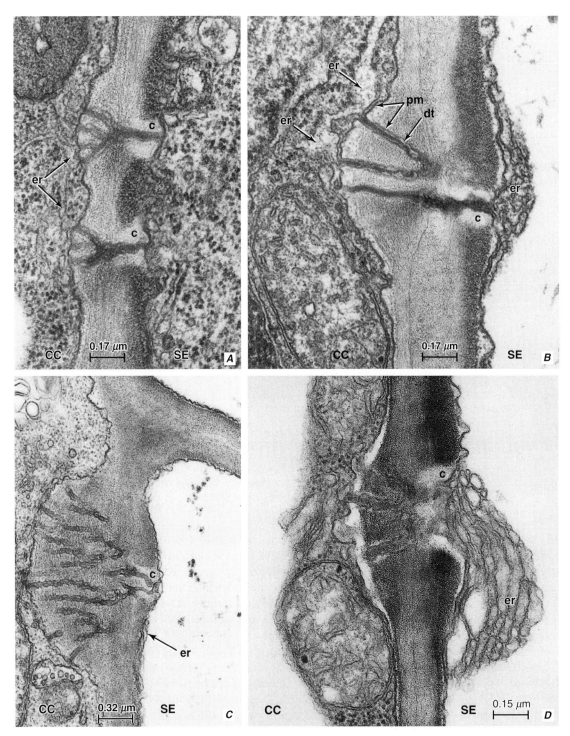

**FIGURE 13.22**

Longitudinal views of pore–plasmodesmata connections. **A**, immature and **B**, mature connections in sieve-tube element–companion cell walls of cotton (*Gossypium hirsutum*) internodes. Note branched plasmodesmata in companion-cell side of wall. **A**, during development, callose deposition is limited to the plasmodesmata (future pores) on the sieve-tube element side of the wall. **B**, in this mature sieve-tube element the pore is partially constricted by (presumably) wound callose. **C**, pore–plasmodesmata connections in the thickened portion of sieve-tube element–companion cell wall in minor vein of cottonwood (*Populus deltoides*) leaf. The plasmodesmata are highly branched in the companion cell wall. **D**, pore–plasmodesmata connections in the common wall between sieve-tube element and companion cell in a leaf vein of barley (*Hordeum vulgare*). An aggregate of endoplasmic reticulum is associated with the pore on the sieve-tube element side. Details: c, callose; CC, companion cell; dt, desmotubule; er, endoplasmic reticulum; pm, plasma membrane; SE, sieve-tube element. (**A**, **B**, from Esau and Thorsch, 1985; **C**, from Russin and Evert, 1985; **D**, from Evert et al., 1971b, Fig. 4. © 1971, Springer-Verlag.)

the wall (Fig. 13.22A). Pore formation is initiated with development of a median cavity in the region of the middle lamella, and formation of branched plasmodesmata is associated with a buildup of cell wall on the companion cell side (Deshpande, 1975; Esau and Thorsch, 1985). Hence these branched plasmodesmata are not secondary plasmodesmata but rather modified primary plasmodesmata (Chapter 4). It is generally presumed that the parietal network of endoplasmic reticulum of the mature sieve-tube element is connected with the endoplasmic reticulum of the companion cell via the desmotubules in the companion cell wall.

Typically the walls of companion cells are neither sclerified nor lignified, and commonly the companion cells collapse when their associated sieve-tube elements die. Sclerification of companion cells has been reported in nonconducting phloem of *Carpodetus serratus* (Brook, 1951) and *Tilia americana* (Evert, 1963). In the minor veins of mature leaves of many herbaceous eudicots, the companion cells possess irregular ingrowths of wall material, which is typical of transfer cells (see below; Pate and Gunning, 1969).

Inasmuch as the mature sieve-tube element lacks a nucleus and ribosomes at maturity, it long has been presumed that these elements depend on the companion cells for their livelihood, the informational molecules, proteins, and ATP necessary for their maintenance being delivered to them via the pore–plasmodesmata connections (referred to as "pore–plasmodesma units" by some workers; van Bel et al., 2002) in the sieve tube-companion cell walls. The interdependence of these two cells is further supported by the fact that both cease to function and die at the same time. Clearly the companion cell is the life-support system of the sieve-tube element.

Microinjection of either companion cells or sieve tubes with fluorescently labeled probes have revealed that the size exclusion limit of plasmodesmata in mature sieve tube–companion cell complexes is relatively large—between 10 and 40 kDa—and that movement between companion cell and sieve-tube element occurs in both directions (Kempers and van Bel, 1997). Substantial evidence indicates that proteins synthesized in the companion cells cycle between the companion cells and sieve-tube elements (Thompson, 1999), and transgenic plants expressing green fluorescent protein (GFP)—presumably synthesized in companion cells—have demonstrated that movement of the GFP throughout the plant in the assimilate stream (Imlau et al., 1999). Only a few of the estimated 200 endogenous soluble proteins present in phloem exudate or sieve-tube sap have been identified. Among them are ubiquitin and chaperones, which have been implicated in protein turnover in mature sieve-tube elements (Schobert et al., 1995). Whereas some of the phloem proteins might serve as long-distance signaling molecules, many likely play a role in the maintenance of the sieve-tube elements.

# THE MECHANISM OF PHLOEM TRANSPORT IN ANGIOSPERMS

Originally proposed by Ernst Münch (1930) and modified by others (see below; Crafts and Crisp, 1971; Eschrich et al., 1972; Young et al., 1973; van Bel, 1993), the **osmotically generated pressure-flow mechanism** is widely accepted to explain the flow of assimilates through angiospermous sieve tubes between the **sources** of assimilates and the sites of utilization, or **sinks**, of those assimilates. Assimilates are said to follow a source-to-sink pattern. The principal sources (net exporters) of assimilates are photosynthesizing leaves, although storage tissues may also serve as important sources. All plant parts unable to meet their own nutritional needs may act as sinks (net importers of assimilates), including meristematic tissues, below ground parts (e.g., roots, tubers, rhizomes), fruits, seeds, and most parenchymatous cells of cortex, pith, xylem, and phloem.

Simply explained, the osmotically generated pressure-flow mechanism operates as follows (Fig. 13.23). Sugars entering the sieve tubes at the source bring about an increase in solute concentration there. With the increase in solute concentration, the water potential is decreased, and water from the xylem enters the sieve tube by osmosis. The removal of sugar at the sink has the opposite effect. There the solute concentration falls, the water potential is increased, and water leaves the sieve tube. With the movement of water into the sieve tube at the source and out of it at the sink, the sugar molecules are carried passively by the water along the concentration gradient by a volume, or mass, flow between source and sink (Eschrich et al., 1972).

In the original Münch model of pressure flow, the sieve tubes were regarded as impermeable conduits. In fact the sieve tube between source and sink is bounded by a selectively permeable membrane, the plasma membrane, not only at source and sink but all along the pathway (Eschrich et al., 1972; Phillips and Dungan, 1993). The presence of a selectively permeable membrane is essential for osmosis, the generating force, for the mechanism to operate; hence the need for a living conduit. With regard to the osmotically generated pressure-flow mechanism, the plasma membrane is the most important component of the cell. Water enters and leaves the sieve tube by osmosis along its entire length. Few, if any, of the original water molecules entering the sieve tube at the source find their way to the sink, because they are exchanged with other water molecules that enter the sieve tube from the phloem apoplast along the transport pathway (Eschrich et al., 1972;

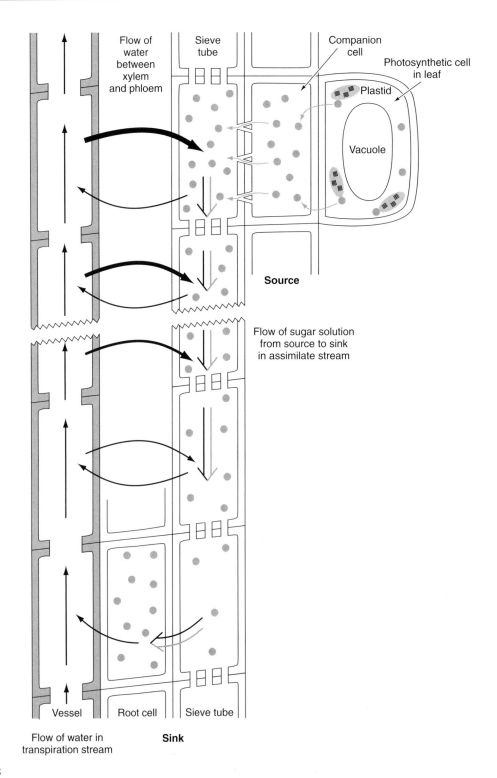

**FIGURE 13.23**

Diagram of osmotically generated pressure-flow mechanism. Dots represent sugar molecules that have their origin in photosynthesizing cells in the leaf (the source). Sugar is loaded into the sieve tube via the companion cells at the source. With the increased concentration of sugars, the water potential is decreased and water enters the sieve tube by osmosis. Sugar is removed (unloaded) at the sink, and the sugar concentration falls; as a result the water potential increases, and water leaves the sieve tube. With the movement of water into the sieve tube at the source and out of it at the sink, the sugar molecules are carried passively by the water along the concentration gradient between source and sink. Water enters and leaves the sieve tube all along the pathway from source to sink. Evidence indicates that few, if any, of the original water molecules entering the sieve tube at the source make it to the sink, because they are exchanged with other water molecules that enter the sieve tube from the phloem apoplast along the pathway. (After Raven et al., 2005.)

Phillips and Dungan, 1993). Water exiting the sieve tube at the sink is recirculated in the xylem (Köckenberger et al., 1997). All along the pathway photoassimilates originating in the leaves are removed to maintain mature tissues and to supply the needs of growing tissues (e.g., the vascular cambium and its immediate derivatives). In addition, substantial amounts of photoassimilates commonly escape, or leak, from the sieve tubes along the pathway (Hayes et al., 1987; Minchin and Thorpe, 1987).

The functioning of the phloem in the distribution of photoassimilates throughout the plant depends on the cooperation between the sieve-tube elements and their companion cells (van Bel, 1996; Schulz, 1998; Oparka and Turgeon, 1999). The nature of the cooperation is reflected in part by the relative size of the sieve tubes and companion cells along the pathway. In the **collection phloem** of minor veins of source leaves (the small veins embedded in the mesophyll tissue, or photosynthetic ground tissue) the companion cells are typically larger than their often diminutive sieve-tube elements (Figs. 13.24–13.26; Evert, 1977, 1990b). This size difference is considered a reflection of the active role played by the companion cells in the collection, or uptake (against a concentration gradient) of photoassimilates, which then are transferred to the sieve-tube elements via the pore–plasmodesmata connections in the sieve tube–companion cell walls. This active process is called **phloem loading** (see below).

In the **release phloem** of terminal sinks the companion cells are greatly reduced in size or absent altogether (Offler and Patrick, 1984; Warmbrodt, 1985a, b; Hayes et al., 1985). In most sink tissues (e.g., of developing roots and leaves) **unloading** takes place symplastically. The actual unloading process is probably passive, not requiring an expenditure of energy by the companion cells. Transport into sink tissues, called **post-phloem** or **post-sieve-tube transport** (Fisher, D. B., and Oparka, 1996; Patrick, 1997), however, depends on metabolic energy. In symplastic unloaders, energy is needed to maintain the concentration gradient between the sieve-tube–companion cell complexes and sink cells. In apoplastic unloaders, energy is needed to accumulate sugars to high concentrations in sink cells, such as those of sugar beet roots and sugarcane stems, although apoplastic unloading in mature sugarcane internodes has been questioned (Jacobsen et al., 1992). In potato, apoplastic unloading of sucrose was found to predominate in stolons undergoing extensive growth; however, during the first visible signs of tuberization, a transition occurred from apoplastic to symplastic unloading (Viola et al., 2001; see also Kühn et al., 2003).

In **transport phloem**, the cross-sectional area of the sieve tubes is greater than that of their counterparts in collection and release phloem, and the companion cells

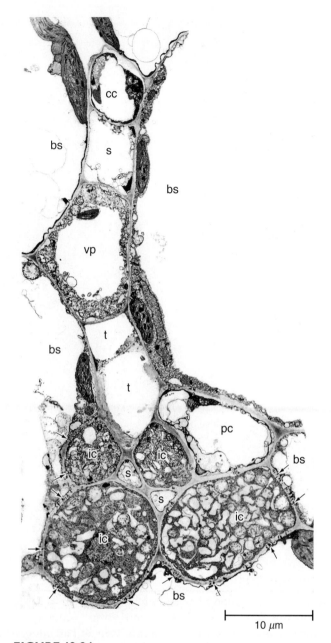

**FIGURE 13.24**

Transverse section through a minor vein of *Cucumis melo* leaf. In this plane of section the abaxial (lower) phloem contains two diminutive sieve tubes (s) bordered by four intermediary cells (ic) in addition to a parenchyma cell (pc). The adaxial phloem consists of a single sieve tube (s) and companion cell (cc). Note the numerous plasmodesmata (arrows) in the common wall between the intermediary cells and bundle-sheath cells (bs). This is a type 1 minor vein, and a symplastic loader. Other details: t, tracheary element; vp, vascular parenchyma cell. (From Schmitz et al., 1987, Fig. 1. © 1987, Springer-Verlag.)

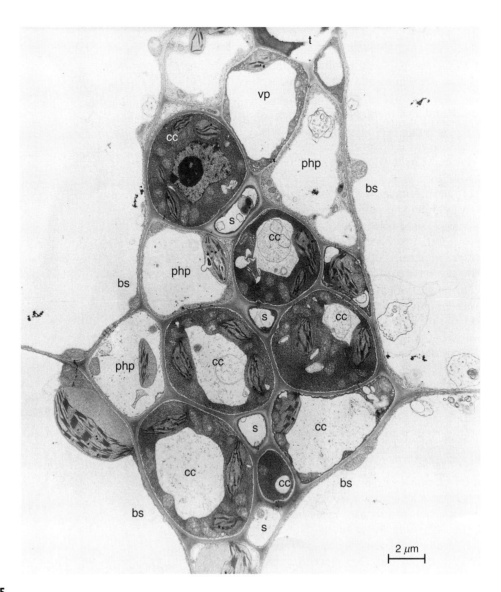

**FIGURE 13.25**

Transverse section of portion of a minor vein of a sugar beet (*Beta vulgaris*) leaf. In this plane of section the vein contains four sieve tubes (s) and seven "ordinary" companion cells (cc), that is, companion cells without wall ingrowths. This is a type 2a minor vein and an apoplastic phloem loader. Other details: bs, bundle-sheath cell; php, phloem parenchyma cell; t, tracheary element; vp, vascular parenchyma cell. (From Evert and Mierzwa, 1986.)

are intermediate in size between those in the collection and release phloem or are entirely absent (Figs. 13.1B and 13.14B, C). The transport phloem has a dual task. One is to deliver photoassimilates to the terminal sinks. This necessitates retention of sufficient photoassimilate to maintain a pressure flow. As mentioned previously, leakage of photoassimilate from the sieve tubes is a common phenomenon along the transport phloem, or pathway, between source and sink. It is believed that the companion cells are involved with the retrieval of leaked photoassimilates. Retention of photoassimilates in transport phloem is enhanced by the near symplastic isolation of the sieve tube-companion cell complexes

there (van Bel and van Rijen, 1994; van Bel, 1996). The second task of the transport phloem is to provide nourishment to heterotrophic tissues along the pathway, including axial sinks such as cambial tissues.

# THE SOURCE LEAF AND MINOR VEIN PHLOEM

As mentioned previously, mature photosynthesizing leaves are the principal sources of the plant. In most angiosperms other than monocots, the vascular bundles, or veins, of the leaf are arranged in a branching pattern, with successively smaller veins branching from

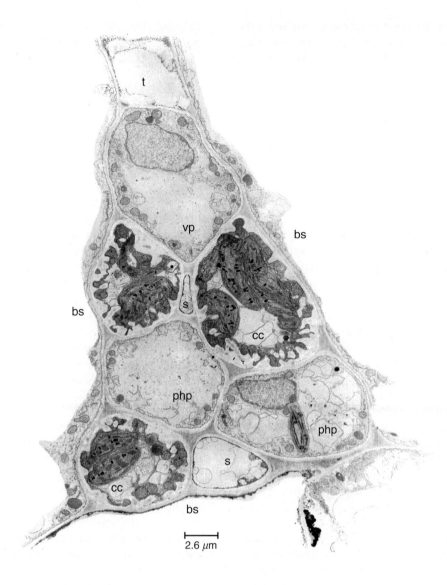

**FIGURE 13.26**

Transverse section of portion of a minor vein of a marigold (*Tagetes patula*) leaf. In this plane of section the vein contains two sieve tubes (s) and three companion cells (cc) with wall ingrowths, that is, the companion cells are transfer cells, or A-type cells (Pate and Gunning, 1969). This is a type 2b minor vein and an apoplastic phloem loader. Other details: bs, bundle-sheath cell; php, phloem parenchyma cell; t, tracheary element; vp, vascular parenchyma cell.

somewhat larger ones. This type of vein arrangement is known as **reticulate**, or **netted**, **venation**. Often the largest vein extends along the long axis of the leaf as a midvein. The midvein, along with its spatially associated ground tissue, makes up the so-called midrib of such leaves. Other, somewhat smaller veins branching from the midvein typically are associated with rib tissue too. All veins associated with ribs (protrusions most commonly on the underside of the leaf) are called **major veins**. The small veins of the leaf that are more or less embedded in mesophyll tissue and not associated with ribs are called **minor veins**. The minor veins are com-

pletely enclosed by a bundle sheath consisting of compactly arranged cells. In eudicot leaves the bundle-sheath cells commonly are parenchymatous and may or may not have chloroplasts. Xylem occurs commonly on the upper side of a vein and phloem on the lower side (Figs. 13.25 and 13.26).

The minor veins play the principal role in the collection of photoassimilate. Before being taken up by the sieve tube–companion cell complexes of the minor veins, photoassimilate produced by photosynthesis in the mesophyll cells and destined to be exported from the leaf must first traverse the bundle sheaths enclosing the veins. From the sieve tubes of the minor veins the photoassimilate, which is in solution in the sieve-tube sap, flows into successively larger veins and eventually into the major veins—transport veins—for export from the leaf. Thus the assimilate stream of the leaf is analogous to a watershed, with small streams feeding into successively larger streams.

## Several Types of Minor Veins Occur in Dicotyledonous Leaves

The minor veins of "dicotyledonous" (magnoliids and eudicots) leaves vary in their structure and in the degree of symplastic continuity of their sieve tube–companion cell complexes with other cell types of the leaf. In some plants, the frequency of plasmodesmata between the bundle-sheath cells and companion cells is abundant or moderate, whereas in others, plasmodesmata are infrequent at that interface (Gamalei, 1989, 1991). On that basis, two general types of minor veins have been recognized (Gamalei, 1991). Those with abundant plasmodesmata at the bundle sheath–companion cell interface (>10 plasmodesmata per μm² of interface) are termed **type 1** and those with few plasmodesmata at that interface are termed **type 2**. The type 1 minor veins are also termed **open**, and the type 2, **closed**. The veins with moderate plasmodesmatal contacts between the bundle-sheath cells and companion cells (<10 plasmodesmata per μm² of interface) are intermediate between types 1 and 2 and are termed **type 1-2a**. (*Arabidopsis thaliana* is a type 1-2a species; Haritatos et al., 2000.) Two subcategories of type 2 are distinguished: **type 2a**, with sporadic plasmodesmatal contacts (<1 per μm² of interface), and **type 2b**, with virtually no plasmodesmatal contacts (<0.1 per μm² of interface). Thus the span in plasmodesmatal frequency at the bundle sheath–companion cell interface between type 1 and type 2b is about three orders of magnitude.

Because of the great variation in frequency of plasmodesmata at the bundle sheath–companion cell interface of the minor veins among species, the concept arose that two mechanisms of phloem loading exist, symplastic and apoplastic (van Bel, 1993). The type 1 species, with an abundance of plasmodesmata linking their minor vein companion cells to the bundle sheaths became regarded as symplastic loaders, and the type 2 species, with a paucity of such connections, as apoplastic loaders (Gamalei, 1989, 1991, 2000; van Bel, 1993; Grusak et al., 1996; Turgeon, 1996).

Although the mechanism of apoplastic loading has long been understood (see below), an explanation for symplastic loading involving active transport through plasmodesmata is wanting. As noted by Turgeon and Medville (2004), "Active transport of small molecules through plasmodesmata is unknown, and diffusion against a concentration gradient is impossible."

## Type 1 Species with Specialized Companion Cells, Termed Intermediary Cells, Are Symplastic Loaders

The minor veins of some type 1 species are characterized by the presence of specialized companion cells called **intermediary cells** (Fig. 13.24). Typically these are especially large cells having dense cytoplasm with an extensive labyrinth of endoplasmic reticulum, numerous small vacuoles, rudimentary plastids, and fields of highly branched plasmodesmata leading into them from the bundle-sheath cells (Turgeon et al., 1993). Only eight families with "true" intermediary cells have thus far been identified: Acanthaceae, Celastraceae, Cucurbitaceae, Hydrangaceae, Lamiaceae, Oleaceae, Scrophulariaceae, and Verbenaceae (see references in Turgeon and Medville, 1998, and Turgeon et al., 2001).

The presence of intermediary cells is always correlated with the transport of large quantities of raffinose and stachyose, in addition to some sucrose (Turgeon et al., 1993). Species with intermediary cells are regarded as symplastic loaders (Turgeon, 1996; Beebe and Russin, 1999), and a **polymer trap** mechanism has been proposed to explain phloem loading involving the intermediary cells (Turgeon, 1991; Haritatos et al., 1996). Briefly, sucrose synthesized in the mesophyll diffuses via plasmodesmata from the mesophyll cells to the bundle-sheath cells and into the intermediary cells. In the intermediary cells, raffinose and stachyose are synthesized from the sucrose, thus maintaining a diffusion gradient between the mesophyll cells and the intermediary cells. Raffinose and stachyose molecules are too large to diffuse back through the plasmodesmata into the bundle-sheath cells and, hence, accumulate to high concentrations in the intermediary cells. From the intermediary cells the raffinose and stachyose diffuse into the sieve tubes via the pore-plasmodesmata connections in their common walls and are carried away in the assimilate stream by mass flow.

Data on type 1 species lacking intermediary cells are limited. The few such species that have been investigated are apoplastic loaders. These include *Liriodendron tulipifera* (Magnoliaceae) (Goggin et al., 2001), *Clethra barbinervis* and *Liquidambar struraciflua* (both Altingiaceae) (Turgeon and Medville, 2004). All three species transport sucrose almost exclusively. Obviously, plasmodesmatal frequency alone cannot be used as an indicator of phloem-loading strategy. The results of such studies have led Turgeon and Medville (2004) to propose that symplastic loading may be restricted to species that transport polymers, such as raffinose-family oligosaccharides in quantity, and that other species, no matter how numerous their minor vein plasmodesmata, may load via the apoplast.

## Species with Type 2 Minor Veins Are Apoplastic Loaders

As indicated previously, the mechanism of apoplastic loading is well established. Sucrose is the principal transport sugar of apoplastic loaders. Apoplastic loading of sucrose molecules involves sucrose-proton co-transport, which is energized by a plasma membrane ATPase and mediated by a sucrose transporter located on the plasma membrane (Lalonde et al., 2003). In potato, tomato, and tobacco, the leaf sucrose transporter (SUT1) localizes to the plasma membrane of the sieve element, not to that of the companion cell (Kühn et al., 1999), whereas in *Arabidopsis* (Stadler and Sauer, 1996; Gottwald et al., 2000) and *Plantago major* (Stadler et al., 1995), the sucrose transporter (SUC2) is specifically expressed in the companion cells. A plasma membrane $H^+$-ATPase has also been localized to *Arabidopsis* companion cells (DeWitt and Sussman, 1995). The difference in location of the sucrose transporter may indicate that in some apoplastic loaders the uptake of sucrose occurs via the sieve element plasma membrane and in others via the plasma membrane of the companion cells. Thus, unlike those type 1 species, in which sugar is concentrated by energy used to synthesize raffinose and stachyose in intermediary cells, the type 2 species use energy to concentrate sugar via a sucrose-proton co-transport at the plasma membrane.

The companion cells of type 2a minor veins have smooth walls, and commonly they are referred to as ordinary companion cells (Fig. 13.25). Those of type 2b minor veins have wall ingrowths, and therefore they are transfer cells (Fig. 13.26).

In some species, two types of transfer cell occur in the minor vein phloem. Designated A-type and B-type (Pate and Gunning, 1969), the former are companion cells and the latter phloem parenchyma cells. Either A-cells or B-cells or both may be present in a minor vein (Fig. 7.5), depending on the species.

The minor veins of *Arabidopsis thaliana*, a type 1-2a species, have B-type cells and ordinary companion cells (Haritatos et al., 2000). Noting that the B-type cells make numerous contacts with both the bundle-sheath cells and companion cells, Haritatos et al. (2000) suggested that the most likely route of sucrose transport in *Arabidopsis* minor veins is from the bundle sheath to the phloem parenchyma cells (the B-type cells) through plasmodesmata, followed by efflux into the apoplast across the plasma membrane bordering the wall ingrowths and carrier-mediated uptake into the sieve element–companion cell complexes.

## The Collection of Photoassimilate by the Minor Veins in Some Leaves May Not Involve an Active Step

In some plants the mechanism by which sucrose enters the sieve tube–companion cell complexes of the minor veins seems not to involve an active step, that is, not to involve phloem loading per se. These are plants with open minor veins that transport large amounts of sucrose and only small amounts of raffinose and stachyose. Willow (*Salix babylonica*; Turgeon and Medville, 1998) and cottonwood (*Populus deltoides*; Russin and Evert, 1985) are representative of such plants. Both are type 1 plants, according to Gamalei (1989). In willow and cottonwood, no evidence could be found for the accumulation of sucrose against a concentration gradient in the minor vein phloem. Apparently the sucrose diffuses symplastically along a concentration gradient from the mesophyll to the sieve tube–companion cell complexes of the minor veins (Turgeon and Medville, 1998). The lack of a loading step corresponds to the Münch (1930) model of phloem transport. Münch considered the chloroplasts of the mesophyll cells as the "source" of the concentration gradient, and presumed that once the sugar entered the sieve tubes in the minor veins, it would be carried away by a mass flow of solution, presumably because the hydrostatic pressure in the sieve tubes of the stem would be lower than that of the sieve tubes in the leaf.

## Some Minor Veins Contain More Than One Kind of Companion Cell

Thus far our discussion of minor veins has placed them in clear-cut categories, each characterized by the presence of a specific type of companion cell. In some species, however, the minor veins contain more than one type of companion cell. For example, both intermediary cells and ordinary companion cells are found in minor veins of the Cucurbitaceae (Fig. 13.24; *Cucurbita pepo*, Turgeon et al., 1975; *Cucumis melo*, Schmitz et al., 1987), *Coleus blumei* (Fisher, D. G., 1986), and *Euonymus fortunei* (Turgeon et al., 2001). This combination of companion cell types is also found in several Scrophulariaceae (*Alonsoc meridonalis*, Knop et al., 2001; and *Alonsoc warscewiczii*, *Mimulus cardinalis*, *Verbascum chaixi*, Turgeon et al., 1993). The minor veins of some Scrophulariaceae contain intermediary cells and transfer cells (*Nemesia strumosa*, *Rhodochiton atrosanguineum*, Turgeon et al., 1993), and those of others modified intermediary cells and transfer cells (*Ascarina* spp., Turgeon et al., 1993; Knop et al., 2001). Some modified intermediary cells in *Ascarina scandens* even had a few wall ingrowths (Turgeon et al., 1993). The presence of minor veins with more than one type of companion cell suggests that more than one mechanism of phloem loading may operate in single veins of some plants (Knop et al., 2004).

### The Minor Veins in Leaf Blades of the Poaceae Contain Two Types of Metaphloem Sieve Tubes

The vascular system of grass leaves, unlike that of "dicotyledonous" leaves, with its netted arrangement, consists of longitudinal strands interconnected by transverse bundles. This vein arrangement is called **striate**, or **parallel**, **venation**. In any given transverse section of the leaf blade, three types of longitudinal vascular bundle can be recognized—large, intermediate, and small—on the basis of their size, the composition of their xylem and phloem, and the nature of their contiguous tissues (Colbert and Evert, 1982; Russell and Evert, 1985; Dannenhoffer et al., 1990). Although all longitudinal bundles are able to transport photoassimilate for some distance down the blade, it is the large bundles that are primarily involved with longitudinal transport and export of photoassimilate from the leaf. The small bundles, on the other hand, are the bundles primarily involved with phloem loading and the collection of photoassimilate. Photoassimilate collected in the small bundles, which do not extend into the leaf sheath, is transferred laterally to larger bundles via the transverse veins for export from the blade (Fritz et al., 1983, 1989). In the blade the intermediate bundles also are involved with the collection of photoassimilates; hence both intermediate and small bundles may be considered minor veins.

The metaphloem of the minor veins contains two kinds of sieve tube, thin-walled and thick-walled (Fig. 13.27; Kuo and O'Brien, 1974; Miyake and Maeda, 1976; Evert et al., 1978; Colbert and Evert, 1982; Eleftheriou, 1990; Botha, 1992; Evert et al., 1996b). More important than the relative thickness of their walls is the fact that **thin-walled sieve tubes**, which are first-formed, are associated with companion cells. The **thick-walled sieve tubes**, which are the last vascular elements to mature, lack companion cells.

In mature maize (Evert et al., 1978) and sugarcane (Robinson-Beers and Evert, 1991) leaves the thin-walled sieve tubes and their companion cells—the sieve tube–companion cell complexes—are virtually isolated symplastically from the rest of the leaf. The thick-walled sieve tubes lack companion cells but are connected symplastically to vascular parenchyma cells that also abut the vessel elements of the xylem. Microautoradiographic studies on the maize leaf indicate that the thin-walled sieve tubes are capable themselves of accumulating photoassimilates from the apoplast, whereas the thick-walled sieve tubes are involved with the retrieval of sucrose transferred to them by the vascular parenchyma cells bordering the vessels (Fritz et al., 1983).

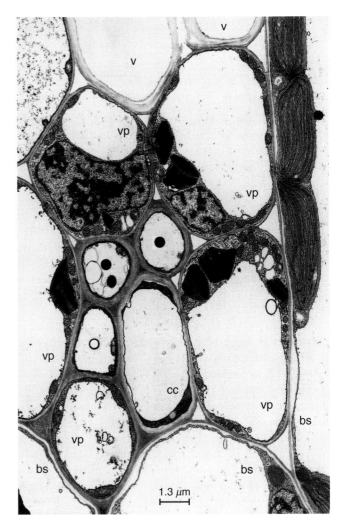

**FIGURE 13.27**

Transverse section of a small vascular bundle in a maize (*Zea mays*) leaf. This small bundle contains a single thin-walled sieve tube (open dot) and its associated companion cell (cc) and two thick-walled sieve tubes (solid dots), which are separated from the vessels (v) by vascular parenchyma cells (vp). The bundle is surrounded by a bundle sheath (bs). (From Evert et al., 1996a. © 1996 by The University of Chicago. All right reserved.)

### ■ THE GYMNOSPERMOUS SIEVE CELL

Sieve cells typically are considerably elongated cells (in the secondary phloem of conifers, 1.5 to 5 mm long) with tapering end walls not easily distinguished from the lateral walls (Fig. 13.4A). The sieve areas are more numerous on the overlapping ends of the sieve cells but have essentially the same degree of differentiation as those of the lateral walls. In other words, unlike sieve-tube elements, sieve cells lack sieve plates. Moreover, unlike the open sieve pores of sieve-tube elements, the sieve pores of sieve cells are traversed by numerous elements of tubular endoplasmic reticulum. In addition, whereas sieve-tube elements typically contain P-protein, sieve cells lack P-protein at all stages of their development. Sieve cells also lack companion cells but are associated functionally with **Strasburger cells**, or **albuminous cells** (Fig. 13.28), which are analogous to the companion cell. Strasburger cells rarely are ontogenetically related to their associated sieve cells.

Most of the information on the sieve cell comes from studies on the secondary phloem of conifers. However, in most respects, the development and structure of other gymnospermous sieve cells parallel those of coniferous sieve cells (Behnke, 1990; Schulz, 1990).

### The Walls of Sieve Cells Are Characterized as Primary

Considerable variation exists in the thickness of sieve cell walls. With the exception of the Pinaceae, the walls of gymnospermous sieve cells are characterized as primary. In the secondary phloem sieve cells of Pina-

ceae the walls are thick and are interpreted as having secondary thickening (Abbe and Crafts, 1939). This thickening, which has a lamellate appearance (Fig. 13.29), does not cover the sieve areas but forms a border around them. Distinct secondary walls have not been recorded in the sieve cells of any other gymnospermous taxa.

### Callose Does Not Play a Role in Sieve-Pore Development in Gymnosperms

The sieve areas of gymnosperms develop from portions of the wall traversed by numerous plasmodesmata. In contrast with sieve-pore development in angiosperms, neither small endoplasmic reticulum cisternae nor callose platelets are involved with sieve-pore development in gymnosperms (Evert et al., 1973b; Neuberger and Evert, 1975, 1976; Cresson and Evert, 1994).

Prior to pore formation, selective deposition of wall material, similar in appearance to that of the primary

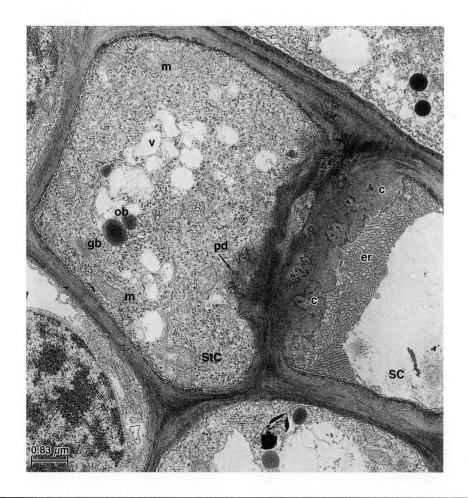

**FIGURE 13.28**

Transverse section showing connections between Strasburger cell (StC) and mature sieve cell (SC) in the hypocotyl of *Pinus resinosa*. Plasmodesmata (pd) occur on the Strasburger-cell side of the wall and pores on the sieve-cell side. The pores are occluded with callose (c) and bordered by a massive aggregate of endoplasmic reticulum (er). Other details: gb, Golgi body; m, mitochondrion; ob, oil body. (From Neuberger and Evert, 1975.)

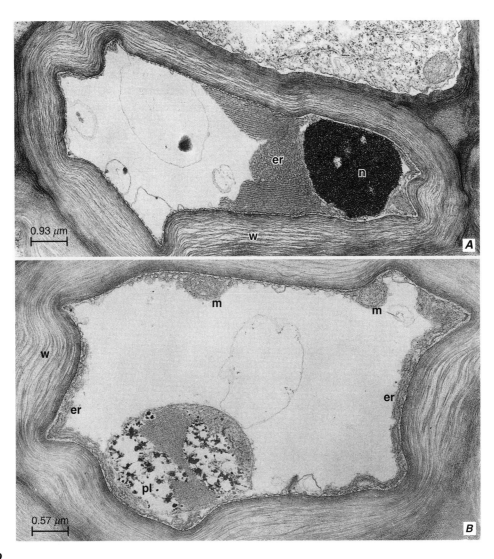

**FIGURE 13.29**

Transverse sections of mature sieve cells in the hypocotyl of *Pinus resinosa*. **A**, shows necrotic nucleus (n) bordered by large aggregate of endoplasmic reticulum (er). **B**, illustrates the typical distribution of cellular components of mature sieve cells, including endoplasmic reticulum (er), mitochondria (m), and plastids (pl). Note lamellate appearance of the walls (w). (From Neuberger and Evert, 1974.)

wall, results in thickening of the sieve-area wall. Very early in sieve-cell differentiation, median cavities appear in the region of the middle lamella in association with the plasmodesmata. As sieve-area differentiation proceeds, the median cavities gradually increase in size and merge to form a single large (compound) median cavity. Concomitantly, the plasmodesmata widen more or less uniformly throughout their length and aggregates of smooth tubular endoplasmic reticulum appear opposite the developing pores. These membrane aggregates persist at the sieve areas throughout the life of the sieve cell (Neuberger and Evert, 1975, 1976; Schulz, 1992). Numerous elements of tubular endoplasmic reticulum traverse the plasma-membrane lined pores and median cavities, unifying the aggregates on either side of the wall (Fig. 13.30). Note that unlike the angiospermous sieve pores, which are continuous across the common wall, the sieve pores of gymnosperms extend only half-way to the median cavity. Callose may or may not line the pores of conducting sieve cells, and when present, it is probably wound callose. Definitive callose typically accumulates at the sieve areas of senescing sieve cells. It eventually disappears after the sieve cell dies.

### Little Variation Exists in Sieve-Cell Differentiation among Gymnosperms

As with young sieve-tube elements, young sieve cells contain all of the components characteristic of young

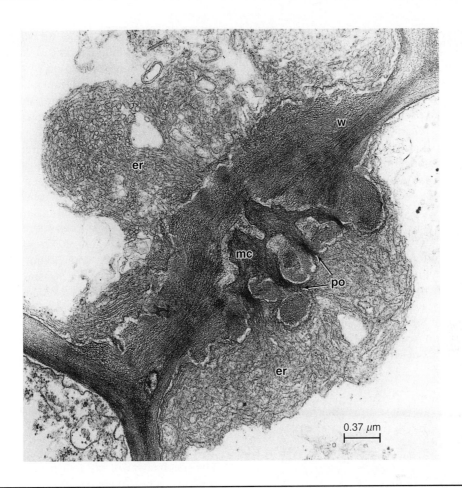

**FIGURE 13.30**

Oblique section of sieve area in wall (w) between mature sieve cells in hypocotyl of *Pinus resinosa*. Massive aggregates of tubular endoplasmic reticulum (er) border the sieve area on both sides. Endoplasmic reticulum (er) can be seen traversing the pores (po) and entering the median cavity (mc), which contains much endoplasmic reticulum. (From Neuberger and Evert, 1975.)

nucleate plant cells. Similarly the sieve cell undergoes a selective autophagy during maturation, resulting in disorganization and/or disappearance of most cellular components, including nucleus, ribosomes, Golgi bodies, actin filaments, microtubules, and tonoplast.

The first visible indication of sieve-cell differentiation is an increase in thickness of the wall (Evert et al., 1973a; Neuberger and Evert, 1976; Cresson and Evert, 1994). Among the protoplasmic components the plastids early show the most marked changes. These two features enable one to distinguish between very young sieve cells and their neighboring parenchymatous elements. Both S- and P-type plastids occur in sieve cells. S-type plastids are found in all taxa but the Pinaceae, which contain only the P-type (Fig. 13.29B; Behnke, 1974, 1990; Schulz, 1990).

Nuclear degeneration in sieve cells is pycnotic, with the degenerate nucleus commonly persisting as an electron-dense mass (Fig. 13.29A), sometimes with portions of the nuclear envelope intact (Behnke and Paliwal,

1973; Evert et al., 1973a; Neuberger and Evert, 1974, 1976). Besides nuclear pycnosis and modification of the plastids the most impressive changes to occur to the cytoplasmic components during sieve-cell differentiation involves the endoplasmic reticulum. The original rough endoplasmic reticulum of the young sieve cell loses its ribosomes and is incorporated into an extensive, newly formed system of smooth tubular endoplasmic reticulum. As indicated earlier, aggregates of tubular endoplasmic reticulum appear early opposite developing sieve areas and persist there throughout the life of the mature cell. In *Pinus* (Neuberger and Evert, 1975) and *Ephedra* (Cresson and Evert, 1994) the aggregates associated with the sieve areas are interconnected with one another longitudinally by a network of parietal endoplasmic reticulum. Thus the endoplasmic reticulum of the mature sieve cell constitutes an extensive system that is also continuous with that of neighboring sieve cells via the sieve-area pores and median cavities.

# STRASBURGER CELLS

The counterpart of the companion cell in gymnosperm phloem is the **Strasburger cell**, named after Eduard Strasburger who gave it the name "Eiweisszellen," or **albuminous cell**. The principal feature distinguishing the Strasburger cell from other parenchymatous elements of the phloem is its symplastic connections with the sieve cells. These connections are reminiscent of the connections between sieve-tube elements and companion cells: pores on the sieve-cell side and branched plasmodesmata on the Strasburger-cell side (Figs. 13.28 and 13.31). The sieve cell–Strasburger cell connections have fairly large median cavities containing numerous elements of smooth tubular endoplasmic reticulum, which is continuous with the parietal network of endoplasmic reticulum on the sieve-cell side via the large aggregates of tubular endoplasmic reticulum bordering the pores. On the Strasburger-cell side, tubules of endoplasmic reticulum are continuous with the desmotubules of the plasmodesmata in the Strasburger-cell wall.

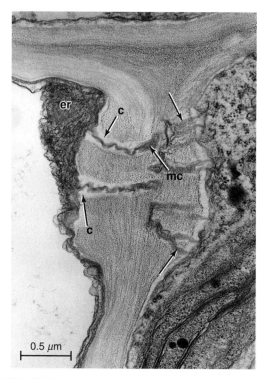

**FIGURE 13.31**

Pore–plasmodesmata connections between a sieve cell (left) and a Strasburger cell (right) in young stem of *Ephedra viridis*. Arrows point to branched plasmodesmata in Strasburger-cell wall. Callose (c) constricts the pores, obscuring their contents, but does not extend into the median cavity (mc). An aggregate of endoplasmic reticulum (er) is associated with the pores. (From Cresson and Evert, 1994.)

Like companion cells, Strasburger cells contain numerous mitochondria and a large ribosome (polysome) population, in addition to other cellular components characteristic of nucleate plant cells. As mentioned previously, unlike the companion cells, the Strasburger cell rarely is ontogenetically related to its associated sieve cell.

Presumably the role of the Strasburger cell is similar to that of companion cells: maintenance of its associated sieve element. Histochemical data strongly implicate the Strasburger cell with a role in long-distance transport of substances in the sieve cells (Sauter and Braun, 1968, 1972; Sauter, 1974). Marked increases in both respiratory (Sauter and Braun, 1972; Sauter, 1974) and acid phosphatase (Sauter and Braun, 1968, 1972; Sauter, 1974) activity occurred in Strasburger cells at the time their associated sieve cells reached maturity. No increase in activity could be detected in Strasburger cells that bordered immature sieve cells or in other parenchymatous elements of the phloem that lack connections with the sieve cells. The Strasburger cells die when their associated sieve cells die.

# THE MECHANISM OF PHLOEM TRANSPORT IN GYMNOSPERMS

The mechanism of phloem transport in gymnosperms remains to be elucidated. With the sieve areas covered with aggregates of endoplasmic reticulum and most of the pore space occupied by tubular elements of endoplasmic reticulum, the resultant resistance to a volume flow would seem to be incompatible with a pressure-flow mechanism. Yet the transport velocity of assimilates in the phloem of the conifer *Metasequoia glyptostroboides,* 48 to 60 cm per hour (Willenbrink and Kollmann, 1966), essentially falls within the range of velocities, 50 to 100 cm per hour, commonly cited for angiosperms (Crafts and Crisp, 1971; Kursanov, 1984). The answer to the puzzle will undoubtedly be solved when the role of the sieve-cell endoplasmic reticulum is known. Being such a prominent component of the sieve-cell protoplast, it is inconceivable that the endoplasmic reticulum, which forms a continuous system that extends from one sieve cell to the next, does not play a prominent role in long-distance transport.

An active role for the endoplasmic reticulum is supported by the localization of the enzymes nucleoside triphosphatase and glycerophosphatase at the endoplasmic reticulum aggregates bordering the sieve areas (Sauter, 1976, 1977). Such a role for the endoplasmic reticulum is further supported by its staining with the cationic dye DiOC (Schulz, 1992). DiOC presumably marks membranes that have a significant membrane potential with a negative charge inside (Matzke and Matzke, 1986). It has been suggested that the endoplasmic reticulum of the sieve cells is able to regulate the long-distance gradient of assimilates by reestablishing the gradient in each sink (Schulz, 1992). Schulz (1992) noted that (1) the activity of nucleoside triphosphatases at the endoplasmic reticulum complexes, or aggregates, (2) the presence of a proton gradient across these membranes, and (3) the high membrane surface suggest that phloem transport in gymnosperms does not only depend on loading in source leaves and unloading in sinks but also requires energy-consumptive steps within the transport path.

# ▌PARENCHYMA CELLS

The phloem contains variable numbers of parenchyma cells other than companion cells and Strasburger cells. Parenchyma cells containing various substances, such as starch, tannins, and crystals, are regular components of the phloem. Crystal-forming parenchyma cells may be subdivided into small cells, each containing a single crystal (Fig. 13.3D). Such chambered **crystalliferous cells** are commonly associated with fibers or sclereids and have lignified walls with secondary thickenings (Nanko et al., 1976).

The parenchyma cells of the primary phloem are elongated and are oriented, like the sieve elements, with their long axes parallel with the longitudinal extent of the vascular tissue. In the secondary phloem (Chapter 14), parenchyma cells occur in two systems, the axial and the radial systems, and are classified as **axial parenchyma cells**, or **phloem parenchyma cells**, and **ray parenchyma cells**, respectively. The axial parenchyma may occur in parenchyma strands or as single fusiform parenchyma cells. A strand results from division of a precursor cell into two or more cells. The ray parenchyma cells constitute the phloem rays.

In many eudicots some parenchyma cells may arise from the same mother cells as the sieve-tube elements (but before the companion cells are formed). Parenchyma cells, especially those ontogenetically related to the sieve-tube elements, may die at the end of the functioning period of the associated sieve-tube elements. Parenchyma cells thus may intergrade with companion cells in their relation to the sieve-tube elements, and the two are not always unequivocally distinguishable from one another even at the electron microscope level (Esau, 1969). The more closely related ontogenetically parenchyma cells are to the sieve-tube elements, the more closely they resemble companion cells both in appearance and in frequency of cytoplasmic connections with the sieve-tube elements. The symplastic connection of parenchyma cells with sieve-tube elements is brought about largely, however, through the companion cells.

In conducting phloem, the phloem parenchyma and ray parenchyma cells apparently have primary unligni-

fied walls. In some instances, where the parenchyma cells are in contact with fibers, they may develop lignified secondary walls. After the tissue ceases to conduct, the parenchyma cells may remain relatively unchanged, or they may become sclerified. In many plants a phellogen eventually arises in the phloem (Chapter 15). It is formed by the phloem parenchyma and the ray parenchyma.

# ▌SCLERENCHYMA CELLS

The fundamental structure of fibers and sclereids, their origin, and their development were considered in Chapter 8. Fibers are common components of both primary and secondary phloem. In the primary phloem, fibers occur in the outermost part of the tissue; in the secondary phloem, in various distributional patterns among the other phloem cells of the axial system. In some plants, the fibers are typically lignified; in others, they are not. The pits in their walls are usually simple, but may be slightly bordered. The fibers may be septate or nonseptate and may be living or nonliving at maturity. Living fibers serve as storage cells as they do in the xylem. Gelatinous fibers also occur in the phloem. In many species, primary and secondary fibers are long and are used as a commercial source of fiber (*Linum, Cannabis, Hibiscus*).

Sclereids are also frequently found in phloem. They may occur in combination with fibers or alone, and they may be present in both axial and radial systems of the secondary phloem. Sclereids typically differentiate in older parts of the phloem as a result of sclerification of parenchyma cells. This sclerification may or may not be preceded by intrusive growth of the cells. During such growth the sclereids often become branched or may be elongated. The distinction between fibers and sclereids is not always sharp, especially if the sclereids are long and slender. The intermediate cell types are called fiber-sclereids.

# ▌LONGEVITY OF SIEVE ELEMENTS

The behavior of sieve elements and neighboring cells during the transition of the phloem from a conducting to a nonconducting condition has long been recognized (Esau, 1969). Commonly the earliest sign of the initiation of cessation of function of sieve elements is the appearance of definitive callose at the sieve areas. The callose, which may accumulate in massive amounts, usually disappears entirely some time after the protoplasmic contents of the element have degenerated. As mentioned previously, death of the sieve elements is accompanied by death of their companion cells or Strasburger cells, and sometimes of other parenchymatous cells as well. With the loss of turgor pressure by the degenerating sieve elements and growth adjustments within the tissue, the sieve elements and closely associated parenchymatous cells may collapse and become obliterated. The sieve elements may remain open, however, and become filled with air. **Tylosoids** (tylose-like protrusions from contiguous parenchymatous cells) may invade the lumina of the dead sieve elements or simply push the sieve element wall to one side, causing collapse of the sieve element (e.g., in *Vitis*, Esau, 1948; in the metaphloem of palms, Parthasarathy and Tomlinson, 1967). In *Smilax rotundifolia* it is the companion cells that form tylosoids, which then may become sclerified (Ervin and Evert, 1967). In a study of the secondary phloem of six forest trees from Nigeria, tylosoids were found in the secondary phloem of all trees that normally formed tyloses (Lawton and Lawton, 1971).

The shortest-lived sieve elements are those of the protophloem, which soon are replaced by metaphloem sieve elements. In plant parts with little or no secondary growth, most of the metaphloem sieve elements remain functional for the life of the part in which they occur, a matter of months. In the rhizomes of *Polygonatum canaliculatum* and *Typha latifolia* and the aerial stems of *Smilax hispida* and *Smilax latifolia* (all four species are perennial monocots), many metaphloem sieve-tube elements remain functional for two or more years (Ervin and Evert, 1967, 1970). In *Smilax hispida* some mature 5-year-old sieve-tube elements were still alive (Ervin and Evert, 1970). This pales when compared with the decades-old living metaphloem sieve-tube elements of some palms (Parthasarathy, 1974b).

In many temperate species of woody angiosperms the secondary phloem sieve-tube elements function only during the season in which they are formed, becoming nonfunctional in the fall, so the phloem lacks mature living sieve tubes during winter (Chapter 14). Similar growth patterns have been reported for some subtropical and tropical species. On the other hand, in some woody angiosperms large numbers of secondary sieve-tube elements may function for two or more seasons (e.g., for five years in *Tilia americana*, Evert, 1962; 10 years in *Tilia cordata*, Holdheide, 1951), becoming dormant in late fall and becoming reactivated in spring. In the secondary phloem of needle leaves of *Pinus longaeva*, individual sieve cells live 3.8 to 6.5 years (Ewers, 1982).

## ▌TRENDS IN SPECIALIZATION OF SIEVE-TUBE ELEMENTS

The evolutionary changes of sieve-tube elements have been studied comprehensively in the metaphloem of the monocotyledons (Cheadle and Whitford, 1941; Cheadle, 1948; Cheadle and Uhl, 1948). In 219 species of 158 genera of 33 families of monocots, only sieve-tube ele-

ments were found. These sieve-tube elements showed the following trends in evolutionary specialization: (1) a gradual change in the orientation of the end walls from very oblique, or inclined, to transverse, (2) a progressive localization of highly specialized sieve areas (areas with larger pores) on the end walls, (3) a stepwise change from compound to simple sieve plates, and (4) a progressive decrease in conspicuousness of the sieve areas on the side walls. The specialization has progressed from the leaf toward the root; that is, the most highly specialized sieve-tube elements occur in leaves, inflorescence axes, corms, and rhizomes, and the less highly evolved ones in the roots. Essentially similar results were obtained in an extensive study of the metaphloem of palms (Parthasarathy, 1966). Thus the direction of specialization of sieve-tube elements in monocots is opposite to that in which the evolution of the tracheary elements has occurred (Chapter 10).

Discussions of the phylogenetic trends in specialization of sieve-tube elements in the woody angiosperms (largely eudicots) pertains to those of the secondary phloem (Zahur, 1959; Roth, 1981; den Outer, 1983, 1986). Included among the trends in specialization of these elements is a decrease in length, the most reliable and consistent measure of the degree of specialization for vessel elements in the secondary xylem of woody angiosperms (Chapter 10; Bailey, 1944; Cheadle, 1956). The phylogenetic decrease in length of vessel elements is correlated with the shortening of the fusiform initials leading to storied cambia. Indeed, an analysis of the characters associated with storied structure in the secondary phloem of 49 species of woody eudicots revealed highly specialized sieve-tube elements. These elements were generally short with simple sieve plates on slightly oblique to transverse end walls and had poorly developed and relatively few sieve areas in the side walls (den Outer, 1986). However, the phylogenetic decrease in length, so well established for vessel elements, is less direct and less consistent in the evolution of sieve-tube elements. That is because in many species of woody angiosperms an ontogenetic decrease in length of the phloem mother cells by transverse divisions (secondary partitions) obscures the length relations between the sieve-tube elements and the fusiform initials (Esau and Cheadle, 1955; Zahur, 1959).

One might expect the less specialized sieve-tube elements to resemble the sieve cells of gymnosperms. To the extent that the primitive sieve-tube elements are relatively long and with a poor distinction between the sieve areas on end and side walls, that is true. However, no angiospermous sieve elements have sieve areas similar to the endoplasmic-reticulum containing sieve areas of gymnospermous sieve cells. The structure of sieve cells among gymnosperms is remarkably uniform and stands in sharp contrast to that of the P-protein containing sieve-tube elements, which typically are associated with ontogenetically related companion cells. Only the secondary phloem sieve elements of *Austrobaileya scandens* among angiosperms have been reported as lacking sieve plates (Srivastava, 1970). The sieve pores in the greatly inclined end walls of these cells are similar in size to those in the lateral walls (see Fig. 14.7B). Noting, however, that the sieve-area pores in the *Austrobaileya* sieve elements are not interconnected by a common median cavity in the region of the middle lamella, as are those in gymnospermous sieve cells (see below), Behnke (1986) chose to designate the end walls, with their multiple sieve areas, compound sieve plates. Regardless, these sieve elements contain P-protein and have companion cells. Having at least two of the three characters shared by sieve-tube elements—the other being sieve plates—the sieve elements of *Austrobaileya* may be regarded as primitive sieve-tube elements.

# SIEVE ELEMENTS OF SEEDLESS VASCULAR PLANTS

Before the use of the electron microscope in the study of phloem tissue, the distinction between sieve-tube elements and sieve cells was made largely on the differences in the size and distribution of the sieve pores, that is, on the presence (sieve-tube elements) or absence (sieve cells) of sieve plates. With few exceptions the sieve elements of seedless vascular plants, or vascular cryptogams, clearly lacked sieve plates and, accordingly, were classified as sieve cells. With the electron microscope, it was found that considerable variation exists in the distribution, size, contents, and development of the sieve-element pores in the vascular cryptogams, and that none of the sieve areas in this diverse group of plants is similar to those, with median cavities and pores that extend only half-way, of gymnospermous sieve cells. Although the sieve pores in the sieve-element walls of some vascular cryptogams (the ferns, including the whisk ferns, and the horsetails) are traversed by many endoplasmic reticulum membranes (Fig. 13.32A, B, D), these membranes are not similar in appearance to the masses of tubular elements associated with the sieve areas in gymnospermous sieve cells. In some vascular cryptogams (the lycopods) the sieve pores are virtually free of any endoplasmic reticulum membranes (Fig. 13.32C). In addition, the vascular cryptogams lack parenchyma cells analogous to Strasburger cells. All things considered, the sieve elements of the vascular cryptogams are not similar to those of the gymnosperms, hence the reason for restricting use of the term sieve cell to the sieve elements of gymnosperms. Nor are the sieve elements of the vascular cryptogams similar to the angio-

spermous sieve-tube element, for all lack P-protein at all stages of their development and all lack parenchyma cells analogous to companion cells.

The sieve elements of the vascular cryptogams undergo a selective autophagy, resulting in degeneration of the nucleus and in the loss of the same cytoplasmic components, as do their gymnosperm and angiosperm counterparts. With the exception of the lycopods, the most distinctive feature of the sieve elements in all other groups of seedless vascular plants is the presence of **refractive spherules**, single membrane-bound, electron-dense proteinaceous bodies (Fig. 13.32A, B, D) that appear highly refractive when viewed in unstained sections with the light microscope. Both the endoplasmic reticulum and the Golgi apparatus have been implicated with the formation of refractive spherules. (See Evert, 1990a and the literature cited therein for detailed considerations of the sieve elements in vascular cryptogams.)

# PRIMARY PHLOEM

The primary phloem is classified into protophloem and metaphloem on the same basis that the primary xylem is classified into protoxylem and metaxylem. The **protophloem** matures in plant parts that are still undergoing extension growth, and its sieve elements are stretched and soon become nonfunctional. Eventually they are completely obliterated (Figs. 13.33 and 13.34). The **metaphloem** differentiates later and, in plants without secondary growth, constitutes the only conducting phloem in adult plant parts.

The protophloem sieve elements of angiosperms are usually narrow and inconspicuous, but they are enucleate and have sieve areas with callose. Companion cells may be absent from the protophloem in both roots and shoots. Often the nucleate cells associated with the first sieve elements are not sufficiently distinct to be recognized as companion cells, even though they may have been derived from the same precursor as the sieve element (Esau, 1969). Many grass roots develop **protophloem poles** consisting of a protophloem sieve element and two distinct companion cells that flank the sieve element internally, that is, on the surface opposite the pericycle (Fig. 13.35; Eleftheriou, 1990). The absence of companion cells in a given protophloem strand is not necessarily consistent. Some protophloem sieve elements may have companion cells and others not, as in the protophloem of *Lepidium sativum*, *Sinapsis alba*, and *Cucurbita pepo* roots (Resch, 1961) and that of the large vascular bundles in maize (Evert and Russin, 1993) and barley (Evert et al., 1996b) leaves.

In many eudicots the protophloem sieve elements occur among conspicuously elongated living cells. In

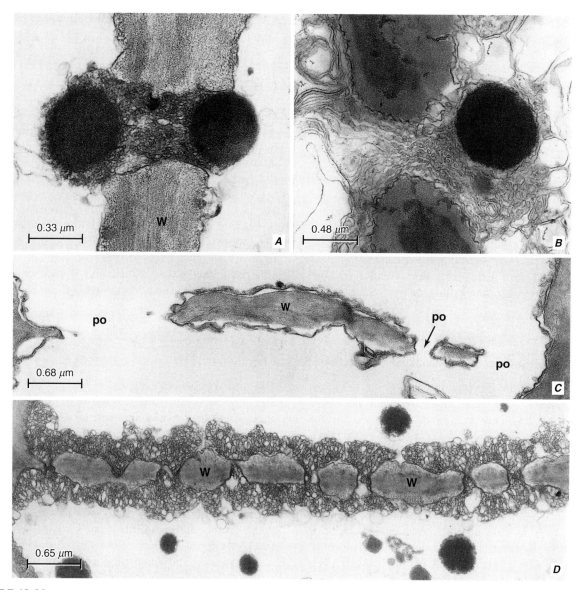

**FIGURE 13.32**

Electron micrographs of sieve-area pores in some seedless vascular plants. **A**, in the eusporangiate fern *Botrychium virginianum* the pores are filled with numerous membranes, apparently tubular endoplasmic reticulum. **B**, endoplasmic reticulum-filled pore in the wall between mature sieve elements in aerial stem of the horsetail *Equisetum hyemale*. **C**, unoccluded pores (po) in the wall between mature sieve elements in the corm of the quillwort *Isoetes muricata*. **D**, endoplasmic reticulum-filled pores in the wall between mature sieve elements in an aerial shoot of the whisk fern *Psilotum nudum*. The electron-dense bodies in **A**, **B**, and **D** are refractive spherules. (**A**, from Evert, 1976. Reproduced with the permission of the publisher; **B**, from Dute and Evert, 1978. By permission of Oxford University Press; **C**, from Kruatrachue and Evert, 1977; **D**, from Perry and Evert, 1975.)

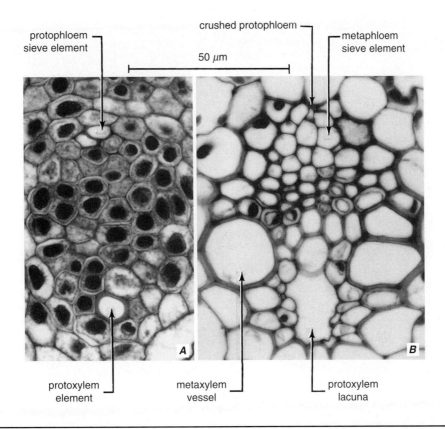

protophloem
sieve element

crushed protophloem

metaphloem
sieve element

50 μm

protoxylem
element

metaxylem
vessel

protoxylem
lacuna

A

B

**FIGURE 13.33**

Transverse sections of vascular bundles of oat (*Avena sativa*) in two stages of differentiation. **A**, first elements of protophloem and protoxylem have matured. **B**, metaphloem and metaxylem are mature; protophloem is crushed; protoxylem is replaced by a lacuna. (**A**, from Esau, 1957a; **B**, from Esau, 1957b. *Hilgardia* 27 (1), 15–69. © 1957 Regents, University of California.)

numerous species these elongated cells are fiber primordia. While the sieve elements cease to function and are obliterated, the fiber primordia increase in length, develop secondary walls, and mature as fibers called **primary phloem fibers**, or **protophloem fibers** (Fig. 13.36). Such fibers are found on the periphery of the phloem region in numerous eudicot stems and are often erroneously called pericyclic fibers (Chapter 8). In the leaf blades and the petioles of eudicots the protophloem cells remaining after the destruction of the sieve tubes often differentiate into long cells with collenchymatically thickened unlignified walls. The strands of these cells appear, in transverse sections, like bundle caps delimiting the vascular bundles on their abaxial sides. This type of transformation of the protophloem in leaves is widely distributed and occurs also in those species that have primary phloem fibers in the stem (Esau, 1939). Primary phloem fibers occur in roots also. With its transformation the protophloem loses all similarity to phloem tissue.

The first-formed phloem elements in gymnosperms have caused difficulties in their recognition as sieve elements. Referred to as **precursory phloem cells** (liber précurseur, Chauveaud, 1902a, b), these cells intergrade with both parenchyma cells and sieve elements. Chauveaud (1910) recognized the precursory phloem only in roots, hypocotyls, and cotyledons. Subsequently it was found that the first phloem elements may have no readily distinguishable sieve areas in the shoots of gymnosperms as well, and they were compared with the precursory phloem cells of Chauveaud (Esau, 1969).

Since the metaphloem matures after the growth in length of the surrounding tissues is completed, it is retained as a conducting tissue longer than the protophloem. Some herbaceous eudicots and most monocots produce no secondary tissues and depend entirely on the metaphloem for assimilate transport after their bodies are fully developed. In woody and herbaceous species having cambial secondary growth the metaphloem sieve elements become inactive after the secondary conducting elements differentiate. In such plants the metaphloem sieve elements may be partly crushed or completely obliterated.

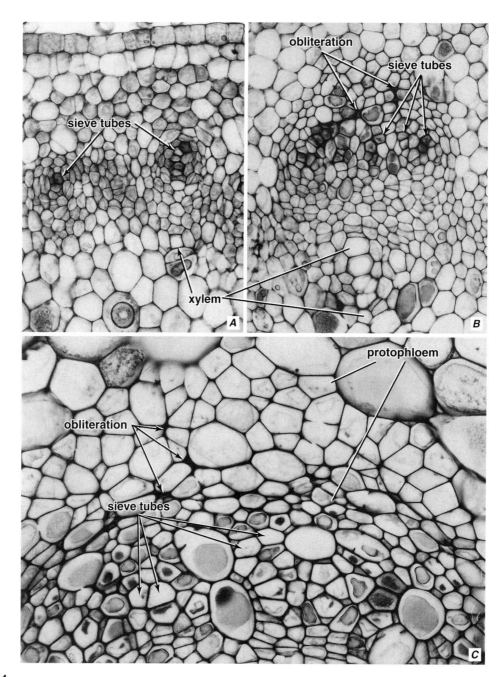

**FIGURE 13.34**

Differentiation of primary phloem as seen in transverse sections of shoots of *Vitis vinifera*. **A**, two procambial bundles, one with one sieve tube and the other with several. **B**, vascular bundle with many protophloem sieve tubes. Some of these are obliterated. Protoxylem present in **A**, **B**. **C**, protophloem sieve tubes obliterated and metaphloem is differentiated (lower half of figure). Protophloem represented by primordia of fibers. Metaphloem consists of sieve tubes, companion cells, phloem parenchyma, and much enlarged tannin-containing parenchyma cells. (All, ×600. From Esau, 1948. *Hilgardia* 18 (5), 217–296. © 1948 Regents, University of California.)

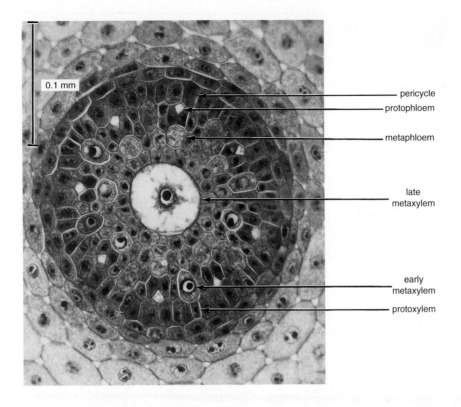

**FIGURE 13.35**

Transverse section of vascular cylinder in differentiating root of barley (*Hordeum vulgare*). The vascular cylinder has eight mature protophloem sieve tubes, each accompanied by two companion cells. The metaphloem sieve tubes are immature. In the xylem, which is all immature, the metaxylem elements are more highly vacuolated than the protoxylem elements. (From Esau, 1957b. *Hilgardia* 27 (1), 15–69. © 1957 Regents, University of California.)

The sieve elements of the metaphloem are commonly longer and wider than those of the protophloem, and their sieve areas are more distinct. Companion cells and phloem parenchyma are regularly present in the metaphloem of the eudicots (Fig. 13.36A). In the monocots the sieve tubes and companion cells often form strands containing no phloem parenchyma among them, although such cells may be present on the periphery of the strands (Cheadle and Uhl, 1948). In such phloem the sieve elements and companion cells form a regular pattern, a feature that is considered to be phylogenetically advanced (Carlquist, 1961). A monocotyledonous type of metaphloem, without phloem parenchyma cells among the sieve tubes, may be found in herbaceous eudicots (Ranunculaceae).

The metaphloem of eudicots usually lacks fibers (Esau, 1950). If primary phloem fibers occur in eudicots, they arise in the protophloem (Figs. 13.34C and 13.36B), but not in the metaphloem, even if such elements are later formed in the secondary phloem. In herbaceous species the old metaphloem may become strongly sclerified. Whether the cells undergoing such sclerification should be classified as fibers or as sclerotic phloem parenchyma is problematical. In monocots, sclerenchyma encloses the vascular bundles as bundle sheaths and may also be present in the metaphloem (Cheadle and Uhl, 1948).

The delimitation between protophloem and metaphloem is sometimes rather clear, as, for example, in the aerial roots of monocots having only sieve tubes in the protophloem and distinct companion cells associated with the sieve tubes in the metaphloem. In eudicots the two tissues usually merge gradually, and their delimitation must be based on a developmental study.

In plants having secondary phloem the distinction between this tissue and the metaphloem may be quite uncertain. The delimitation of the two tissues is particularly difficult if radial seriation of cells occurs in both tissues. An exception has been found in *Prunus* species and *Citrus limonia,* in which the last cells initiated on the phloem side by the procambium mature as large parenchyma cells and sharply delimit the primary from the secondary phloem (Schneider, 1945, 1955).

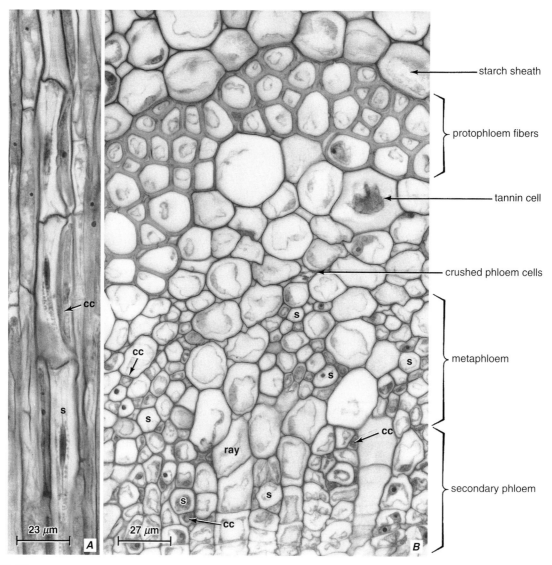

**FIGURE 13.36**

Phloem from bean (*Phaseolus vulgaris*) leaf. **A**, longitudinal section showing part of a sieve tube with two complete and two incomplete sieve-tube elements (s). Companion cell at cc. Spindle-shaped nondispersive P-protein bodies in three of the elements. **B**, transverse section including no longer conducting protophloem with fibers, metaphloem, and some secondary phloem. Sieve element at s, companion cell at cc. (From Esau, 1977.)

# REFERENCES

ABBE, L. B., and A. S. CRAFTS. 1939. Phloem of white pine and other coniferous species. *Bot. Gaz.* 100, 695–722.

ALONI, R., and J. R. BARNETT. 1996. The development of phloem anastomoses between vascular bundles and their role in xylem regeneration after wounding in *Cucurbita* and *Dahlia*. *Planta* 198, 595–603.

ALONI, R., and C. A. PETERSON. 1990. The functional significance of phloem anastomoses in stems of *Dahlia pinnata* Cav. *Planta* 182, 583–590.

ALONI, R., and T. SACHS. 1973. The three-dimensional structure of primary phloem systems. *Planta* 113, 345–353.

ARAKI, T., T. EGUCHI, T. WAJIMA, S. YOSHIDA, and M. KITANO. 2004. Dynamic analysis of growth, water balance and sap fluxes through phloem and xylem in a tomato fruit: Short-term effect of water stress. *Environ. Control Biol.* 42, 225–240.

ARTSCHWAGER, E. F. 1918. Anatomy of the potato plant, with special reference to the ontogeny of the vascular system. *J. Agric. Res.* 14, 221–252.

BAILEY, I. W. 1944. The development of vessels in angiosperms and its significance in morphological research. *Am. J. Bot.* 31, 421–428.

BAKER, D. A. 2000. Long-distance vascular transport of endogenous hormones in plants and their role in source-sink regulation. *Isr. J. Plant Sci.* 48, 199–203.

BEEBE, D. U., and W. A. RUSSIN. 1999. Plasmodesmata in the phloem-loading pathway. In: *Plasmodesmata: Structure, Function, Role in Cell Communication*, pp. 261–293, A. J. E. van Bel and W. J. P. van Kesteren, eds. Springer, Berlin.

BEHNKE, H.-D. 1974. Sieve-element plastids of Gymnospermae: their ultrastructure in relation to systematics. *Plant Syst. Evol.* 123, 1–12.

BEHNKE, H.-D. 1975. Companion cells and transfer cells. In: *Phloem Transport*, pp. 187–210, S. Aronoff, J. Dainty, P. R. Gorham, L. M. Srivastava, and C. A. Swanson, eds. Plenum Press, New York.

BEHNKE, H.-D. 1986. Sieve element characters and the systematic position of *Austrobaileya*, Austrobaileyaceae—With comments to the distinction and definition of sieve cells and sieve-tube members. *Plant Syst. Evol.* 152, 101–121.

BEHNKE, H.-D. 1990. Cycads and gnetophytes. In: *Sieve Elements. Comparative Structure, Induction and Development*, pp. 89–101, H.-D. Behnke, and R. D. Sjolund, eds. Springer-Verlag, Berlin.

BEHNKE, H.-D. 1991a. Distribution and evolution of forms and types of sieve-element plastids in dicotyledons. *Aliso* 13, 167–182.

BEHNKE, H.-D. 1991b. Nondispersive protein bodies in sieve elements: A survey and review of their origin, distribution and taxonomic significance. *IAWA Bull.* n.s. 12, 143–175.

BEHNKE, H.-D. 1994. Sieve-element plastids, nuclear crystals and phloem proteins in the Zingiberales. *Bot. Acta* 107, 3–11.

BEHNKE, H.-D. 2003. Sieve-element plastids and evolution of monocotyledons, with emphasis on Melanthiaceae sensu lato and Aristolochiaceae-Asaroideae, a putative dicotyledon sister group. *Bot. Rev.* 68, 524–544.

BEHNKE, H.-D., and U. KIRITSIS. 1983. Ultrastructure and differentiation of sieve elements in primitive angiosperms. I. Winteraceae. *Protoplasma* 118, 148–156.

BEHNKE, H.-D., and G. S. PALIWAL. 1973. Ultrastructure of phloem and its development in *Gnetum gnemon*, with some observations on *Ephedra campylopoda*. *Protoplasma* 78, 305–319.

BONNEMAIN, J.-L. 1969. Le phloème interne et le phloème inclus des dicotylédones: Leur histogénèse leur physiologie. *Rev. Gén. Bot.* 76, 5–36.

BOSTWICK, D. E., J. M. DANNENHOFFER, M. I. SKAGGS, R. M. LISTER, B. A. LARKINS, and G. A. THOMPSON. 1992. Pumpkin phloem lectin genes are specifically expressed in companion cells. *Plant Cell* 4, 1539–1548.

BOTHA, C. E. J. 1992. Plasmodesmatal distribution, structure and frequency in relation to assimilation in $C_3$ and $C_4$ grasses in southern Africa. *Planta* 187, 348–358.

BOTHA, C. E. J., and R. F. EVERT. 1981. Studies on *Artemisia afra* Jacq.: The phloem in stem and leaf. *Protoplasma* 109, 217–231.

BROOK, P. J. 1951. Vegetative anatomy of *Carpodetus serratus* Forst. *Trans. R. Soc. N. Z.* 79, 276–285.

BUSSE, J. S., and R. F. EVERT. 1999. Pattern of differentiation of the first vascular elements in the embryo and seedling of *Arabidopsis thaliana*. *Int. J. Plant Sci.* 160, 1–13.

CARLQUIST, S. 1961. *Comparative Plant Anatomy: A Guide to Taxonomic and Evolutionary Application of Anatomical Data in Angiosperms*. Holt, Rinehart and Winston, New York.

CATESSON, A.-M. 1982. Cell wall architecture in the secondary sieve tubes of *Acer* and *Populus*. *Ann. Bot.* 49, 131–134.

CHAUVEAUD, G. 1902a. De l'existénce d'éléments précurseurs des tubes criblés chez les Gymnospermes. *C.R. Acad. Sci.* 134, 1605–1606.

CHAUVEAUD, G. 1902b. Développement des éléments précurseurs des tubes criblés dans le *Thuia orientalis*. *Bull. Mus. Hist. Nat.* 8, 447–454.

CHAUVEAUD, G. 1910. Recherches sur les tissus transitoires du corps végétatif des plantes vasculaires. *Ann. Sci. Nat. Bot. Sér.* 9, 12, 1–70.

CHEADLE, V. I. 1948. Observations on the phloem in the Monocotyledoneae. II. Additional data on the occurrence and phylogenetic specialization in structure of the sieve tubes in the metaphloem. *Am. J. Bot.* 35, 129–131.

CHEADLE, V. I. 1956. Research on xylem and phloem—Progress in fifty years. *Am. J. Bot.* 43, 719–731.

CHEADLE, V. I., and K. ESAU. 1958. Secondary phloem of the Calycanthaceae. *Univ. Calif. Publ. Bot.* 24, 397–510.

CHEADLE, V. I., and N. W. UHL. 1948. The relation of metaphloem to the types of vascular bundles in the Monocotyledoneae. *Am. J. Bot.* 35, 578–583.

CHEADLE, V. I., and N. B. WHITFORD. 1941. Observations on the phloem in the Monocotyledoneae. I. The occurrence and phylogenetic specialization in structure of the sieve tubes in the metaphloem. *Am. J. Bot.* 28, 623–627.

CHIOU, T.-J., and D. R. BUSH. 1998. Sucrose is a signal molecule in assimilate partitioning. *Proc. Natl. Acad. Sci. USA* 95, 4784–4788.

CLARK, A. M., K. R. JACOBSEN, D. E. BOSTWICK, J. M. DANNENHOFFER, M. I. SKAGGS, and G. A. THOMPSON. 1997. Molecular characterization of a phloem-specific gene encoding the filament protein, phloem protein I (PPI), from *Cucurbita maxima*. *Plant J.* 12, 49–61.

COLBERT, J. T., and R. F. EVERT. 1982. Leaf vasculature in sugarcane (*Saccharum officinarum* L.). *Planta* 156, 136–151.

CRAFTS, A. S., and C. E. CRISP. 1971. *Phloem transport in plants*. Freeman, San Francisco.

CRAWFORD, K. M., and P. C. ZAMBRYSKI. 1999. Plasmodesmata signaling: Many roles, sophisticated statutes. *Curr. Opin. Plant Biol.* 2, 382–387.

CRESSON, R. A., and R. F. EVERT. 1994. Development and ultrastructure of the primary phloem in the shoot of *Ephedra viridis* (Ephedraceae). *Am. J. Bot.* 81, 868–877.

CRONSHAW, J., J. GILDER, and D. STONE. 1973. Fine structural studies of P-proteins in *Cucurbita, Cucumis,* and *Nicotiana. J. Ultrastruct. Res.* 45, 192–205.

DANNENHOFFER, J. M., W. EBERT Jr., and R. F. EVERT. 1990. Leaf vasculature in barley, *Hordeum vulgare* (Poaceae). *Am. J. Bot.* 77, 636–652.

DANNENHOFFER, J. M., A. SCHULZ, M. I. SKAGGS, D. E. BOSTWICK, and G. A. THOMPSON. 1997. Expression of the phloem lectin is developmentally linked to vascular differentiation in cucurbits. *Planta* 201, 405–414.

DAVIS, J. D., and R. F. EVERT. 1970. Seasonal cycle of phloem development in woody vines. *Bot. Gaz.* 131, 128–138.

DEN OUTER, R. W. 1983. Comparative study of the secondary phloem of some woody dicotyledons. *Acta Bot. Neerl.* 32, 29–38.

DEN OUTER, R. W. 1986. Storied structure of the secondary phloem. *IAWA Bull.* n.s. 7, 47–51.

DESHPANDE, B. P. 1974. Development of the sieve plate in *Saxifraga sarmentosa* L. *Ann. Bot.* 38, 151–158.

DESHPANDE, B. P. 1975. Differentiation of the sieve plate of *Cucurbita:* A further view. *Ann. Bot.* 39, 1015–1022.

DESHPANDE, B. P. 1976. Observations on the fine structure of plant cell walls. III. The sieve tube wall in *Cucurbita. Ann. Bot.* 40, 443–446.

DESHPANDE, B. P. 1984. Distribution of P-protein in mature sieve elements of *Cucurbita maxima* seedlings subjected to prolonged darkness. *Ann. Bot.* 53, 237–247.

DESHPANDE, B. P., and R. F. EVERT. 1970. A reevaluation of extruded nucleoli in sieve elements. *J. Ultrastruct. Res.* 33, 483–494.

DESHPANDE, B. P., and T. RAJENDRABABU. 1985. Seasonal changes in the structure of the secondary phloem of *Grewia tiliaefolia,* a deciduous tree from India. *Ann. Bot.* 56, 61–71.

DEWITT, N. D., and M. R. SUSSMAN. 1995. Immunocytological localization of an epitope-tagged plasma membrane proton pump ($H^+$-ATPase) in phloem companion cells. *Plant Cell* 7, 2053–2067.

DINANT, S., A. M. CLARK, Y. ZHU, F. VILAINE, J.-C. PALAUQUI, C. KUSIAK, and G. A. THOMPSON. 2003. Diversity of the superfamily of phloem lectins (phloem protein 2) in angiosperms. *Plant Physiol.* 131, 114–128.

DUTE, R. R., and R. F. EVERT. 1978. Sieve-element ontogeny in the aerial shoot of *Equisetum hyemale* L. *Ann. Bot.* 42, 23–32.

EHLERS, K., M. KNOBLAUCH, and A. J. E. van BEL. 2000. Ultrastructural features of well-preserved and injured sieve elements: Minute clamps keep the phloem transport conduits free for mass flow. *Protoplasma* 214, 80–92.

ELEFTHERIOU, E. P. 1981. A light and electron microscopy study on phloem differentiation of the grass *Aegilops comosa* var. *thessalica.* Ph.D. Thesis. University of Thessaloniki. Thessaloniki, Greece.

ELEFTHERIOU, E. P. 1990. Monocotyledons. In: *Sieve Elements. Comparative Structure, Induction and Development,* pp. 139–159, H.-D. Behnke and R. D. Sjolund, eds. Springer-Verlag, Berlin.

ERVIN, E. L., and R. F. EVERT. 1967. Aspects of sieve element ontogeny and structure in *Smilax rotundifolia. Bot. Gaz.* 128, 138–144.

ERVIN, E. L., and R. F. EVERT. 1970. Observations on sieve elements in three perennial monocotyledons. *Am. J. Bot.* 57, 218–224.

ESAU, K. 1938. Ontogeny and structure of the phloem of tobacco. *Hilgardia* 11, 343–424.

ESAU, K. 1939. Development and structure of the phloem tissue. *Bot. Rev.* 5, 373–432.

ESAU, K. 1940. Developmental anatomy of the fleshy storage organ of *Daucus carota. Hilgardia* 13, 175–226.

ESAU, K. 1947. A study of some sieve-tube inclusions. *Am. J. Bot.* 34, 224–233.

ESAU, K. 1948. Phloem structure in the grapevine, and its seasonal changes. *Hilgardia* 18, 217–296.

ESAU, K. 1950. Development and structure of the phloem tissue. II. *Bot. Rev.* 16, 67–114.

ESAU, K. 1957a. Phloem degeneration in Gramineae affected by the barley yellow-dwarf virus. *Am. J. Bot.* 44, 245–251.

ESAU, K. 1957b. Anatomic effects of barley yellow dwarf virus and maleic hydrazide on certain Gramineae. *Hilgardia* 27, 15–69.

ESAU, K. 1969. *The Phloem. Encyclopedia of Plant Anatomy.* Histologie, Band 5, Teil 2. Gebrüder Borntraeger, Berlin.

ESAU, K. 1977. *Anatomy of Seed Plants,* 2nd ed. Wiley, New York.

ESAU, K. 1978a. The protein inclusions in sieve elements of cotton (*Gossypium hirsutum* L.). *J. Ultrastruct. Res.* 63, 224–235.

ESAU, K. 1978b. Developmental features of the primary phloem in *Phaseolus vulgaris* L. *Ann. Bot.* 42, 1–13.

ESAU, K., and V. I. CHEADLE. 1955. Significance of cell divisions in differentiating secondary phloem. *Acta Bot. Neerl.* 4, 348–357.

ESAU, K., and V. I. CHEADLE. 1958. Wall thickening in sieve elements. *Proc. Natl. Acad. Sci. USA* 44, 546–553.

ESAU, K., and V. I. CHEADLE. 1959. Size of pores and their contents in sieve elements of dicotyledons. *Proc. Natl. Acad. Sci. USA* 45, 156–162.

ESAU, K., and J. CRONSHAW. 1968. Plastids and mitochondria in the phloem of *Cucurbita. Can. J. Bot.* 46, 877–880.

ESAU, K., and R. H. GILL. 1972. Nucleus and endoplasmic reticulum in differentiating root protophloem of *Nicotiana tabacum. J. Ultrastruct. Res.* 41, 160–175.

ESAU, K., and J. THORSCH. 1984. The sieve plate of *Echium* (Boraginaceae): Developmental aspects and response of P-protein to protein digestion. *J. Ultrastruct. Res.* 86, 31–45.

ESAU, K., and J. THORSCH. 1985. Sieve plate pores and plasmodesmata, the communication channels of the symplast: Ultrastructural aspects and developmental relations. *Am. J. Bot.* 72, 1641–1653.

ESAU, K., V. I. CHEADLE, and E. M. GIFFORD JR. 1953. Comparative structure and possible trends of specialization of the phloem. *Am. J. Bot.* 40, 9–19.

ESCHRICH, W. 1975. Sealing systems in phloem. In: *Encyclopedia of Plant Physiology*, n.s. vol. 1. *Transport in plants*. 1. *Phloem Transport*, pp. 39–56, Springer-Verlag, Berlin.

ESCHRICH, W., R. F. EVERT, and J. H. YOUNG. 1972. Solution flow in tubular semipermeable membranes. *Planta* 107, 279–300.

EVERT, R. F. 1960. Phloem structure in *Pyrus communis* L. and its seasonal changes. *Univ. Calif. Publ. Bot.* 32, 127–196.

EVERT, R. F. 1962. Some aspects of phloem development in *Tilia americana*. *Am. J. Bot.* 49 (Abstr.), 659.

EVERT, R. F. 1963. Sclerified companion cells in *Tilia americana*. *Bot. Gaz.* 124, 262–264.

EVERT, R. F. 1976. Some aspects of sieve-element structure and development in *Botrychium virginianum*. *Isr. J. Bot.* 25, 101–126.

EVERT, R. F. 1977. Phloem structure and histochemistry. *Annu. Rev. Plant Physiol.* 28, 199–222.

EVERT, R. F. 1980. Vascular anatomy of angiospermous leaves, with special consideration of the maize leaf. *Ber. Dtsch. Bot. Ges.* 93, 43–55.

EVERT, R. F. 1990a. Seedless vascular plants. In: *Sieve Elements. Comparative Structure, Induction and Development*, pp. 35–62, H.-D. Behnke, and R. D. Sjolund, eds. Springer-Verlag, Berlin.

EVERT, R. F. 1990b. Dicotyledons. In: *Sieve Elements. Comparative Structure, Induction and Development*, pp. 103–137, H.-D. Behnke, and R. D. Sjolund, eds. Springer-Verlag, Berlin.

EVERT, R. F., and W. F. DERR. 1964. Callose substance in sieve elements. *Am. J. Bot.* 51, 552–559.

EVERT, R. F., and R. J. MIERZWA. 1986. Pathway(s) of assimilate movement from mesophyll cells to sieve tubes in the *Beta vulgaris* leaf. In: *Plant Biology*, vol. 1, *Phloem Transport*, pp. 419–432, J. Cronshaw, W. J. Lucas, and R. T. Giaquinta, eds. Alan R. Liss, New York.

EVERT, R. F., and R. J. MIERZWA. 1989. The cell wall-plasmalemma interface in sieve tubes of barley. *Planta* 177, 24–34.

EVERT, R. F., and W. A. RUSSIN. 1993. Structurally, phloem unloading in the maize leaf cannot be symplastic. *Am. J. Bot.* 80, 1310–1317.

EVERT, R. F., C. M. TUCKER, J. D. DAVIS, and B. P. DESHPANDE. 1969. Light microscope investigation of sieve-element ontogeny and structure in *Ulmus americana*. *Am. J. Bot.* 56, 999–1017.

EVERT, R. F., B. P. DESHPANDE, and S. E. EICHHORN. 1971a. Lateral sieve-area pores in woody dicotyledons. *Can. J. Bot.* 49, 1509–1515.

EVERT, R. F., W. ESCHRICH, and S. E. EICHHORN. 1971b. Sieveplate pores in leaf veins of *Hordeum vulgare*. *Planta* 100, 262–267.

EVERT, R. F., C. H. BORNMAN, V. BUTLER, and M. G. GILLILAND. 1973a. Structure and development of the sieve-cell protoplast in leaf veins of *Welwitschia*. *Protoplasma* 76, 1–21.

EVERT, R. F., C. H. BORNMAN, V. BUTLER, and M. G. GILLILAND. 1973b. Structure and development of sieve areas in leaf veins of *Welwitschia*. *Protoplasma* 76, 23–34.

EVERT, R. F., W. ESCHRICH, and S. E. EICHHORN. 1973c. P-protein distribution in mature sieve elements of *Cucurbita maxima*. *Planta* 109, 193–210.

EVERT, R. F., W. ESCHRICH, and W. HEYSER. 1978. Leaf structure in relation to solute transport and phloem loading in *Zea mays* L. *Planta* 138, 279–294.

EVERT, R. F., W. A. RUSSIN, and A. M. BOSABALIDIS. 1996a. Anatomical and ultrastructural changes associated with sink-to-source transition in developing maize leaves. *Int. J. Plant Sci.* 157, 247–261.

EVERT, R. F., W. A. RUSSIN, and C. E. J. BOTHA. 1996b. Distribution and frequency of plasmodesmata in relation to photoassimilate pathways and phloem loading in the barley leaf. *Planta* 198, 572–579.

EWERS, F. W. 1982. Developmental and cytological evidence for mode of origin of secondary phloem in needle leaves of *Pinus longaeva* (bristlecone pine) and *P. flexilis*. *Bot. Jahrb. Syst. Pflanzengesch. Pflanzengeogr.* 103, 59–88.

FELLOWS, R. J., and D. R. GEIGER. 1974. Structural and physiological changes in sugar beet leaves during sink to source conversion. *Plant Physiol.* 54, 877–885.

FISHER, D. B. 1975. Structure of functional soybean sieve elements. *Plant Physiol.* 56, 555–569.

FISHER, D. B., and K. J. OPARKA. 1996. Post-phloem transport: Principles and problems. *J. Exp. Bot.* 47, 1141–1154.

FISHER, D. G. 1986. Ultrastructure, plasmodesmatal frequency, and solute concentration in green areas of variegated *Coleus blumei* Benth. leaves. *Planta* 169, 141–152.

FISHER, D. G., and R. F. EVERT. 1982. Studies on the leaf of *Amaranthus retroflexus* (Amaranthaceae): ultrastructure, plasmodesmatal frequency and solute concentration in relation to phloem loading. *Planta* 155, 377–387.

FRITZ, E., R. F. EVERT, and W. HEYSER. 1983. Microautoradiographic studies of phloem loading and transport in the leaf of *Zea mays* L. *Planta* 159, 193–206.

FRITZ, E., R. F. EVERT, and H. NASSE. 1989. Loading and transport of assimilates in different maize leaf bundles. Digital image analysis of $^{14}$C-microautoradiographs. *Planta* 178, 1–9.

FUKUDA, Y. 1967. Anatomical study of the internal phloem in the stems of dicotyledons, with special reference to its histogenesis. *J. Fac. Sci. Univ. Tokyo, Sect. III. Bot.* 9, 313–375.

GAHAN, P. B. 1988. Xylem and phloem differentiation in perspective. In: *Vascular Differentiation and Plant Growth Regulators*, pp.

1–21, L. W. Roberts, P. B. Gahan, and R. Aloni, eds. Springer-Verlag, Berlin.

GAMALEI, Y. 1989. Structure and function of leaf minor veins in trees and shrubs. A taxonomic review. *Trees* 3, 96–110.

GAMALEI, Y. 1991. Phloem loading and its development related to plant evolution from trees to herbs. *Trees* 5, 50–64.

GAMALEI, Y. 2000. Comparative anatomy and physiology of minor veins and paraveinal parenchyma in the leaves of dicots. *Bot. Zh.* 85, 34–49.

GILLILAND, M. G., J. VAN STADEN, and A. G. BRUTON. 1984. Studies on the translocation system of guayule (*Parthenium argentatum* Gray). *Protoplasma* 122, 169–177.

GOGGIN, F. L., R. MEDVILLE, and R. TURGEON. 2001. Phloem loading in the tulip tree: Mechanisms and evolutionary implications. *Plant Physiol.* 125, 891–899.

GOLECKI, B., A. SCHULZ, and G. A. THOMPSON. 1999. Translocation of structural P proteins in the phloem. *Plant Cell* 11, 127–140.

GOTTWALD, J. R., P. J. KRYSAN, J. C. YOUNG, R. F. EVERT, and M. R. SUSSMAN. 2000. Genetic evidence for the *in planta* role of phloem-specific plasma membrane sucrose transporters. *Proc. Natl. Acad. Sci. USA* 97, 13979–13984.

GRUSAK, M. A., D. U. BEEBE, and R. TURGEON. 1996. Phloem loading. In: *Photoassimilate Distribution in Plants and Crops: Source-Sink Relationships*, pp. 209–227, E. ZAMSKI and A. A. SCHAFFER, eds. Dekker, New York.

GUO, Y. H., B. G. HUA, F. Y. YU, Q. LENG, and C. H. LOU. 1998. The effects of microfilament and microtubule inhibitors and periodic electrical impulses on phloem transport in pea seedling. *Chinese Sci. Bull.* (*Kexue tongbao*) 43, 312–315.

HARITATOS, E., F. KELLER, and R. TURGEON. 1996. Raffinose oligosaccharide concentrations measured in individual cell and tissue types in *Cucumis melo* L. leaves: Implications for phloem loading. *Planta* 198, 614–622.

HARITATOS, E., R. MEDVILLE, and R. TURGEON. 2000. Minor vein structure and sugar transport in *Arabidopsis thaliana*. *Planta* 211, 105–111.

HAYES, P. M., C. E. OFFLER, and J. W. PATRICK. 1985. Cellular structures, plasma membrane surface areas and plasmodesmatal frequencies of the stem of *Phaseolus vulgaris* L. in relation to radial photosynthate transfer. *Ann. Bot.* 56, 125–138.

HAYES, P. M., J. W. PATRICK, and C. E. OFFLER. 1987. The cellular pathway of radial transfer of photosynthates in stems of *Phaseolus vulgaris* L.: Effects of cellular plasmolysis and *p*-chloromercuribenzene sulphonic acid. *Ann. Bot.* 59, 635–642.

HAYWARD, H. E. 1938. Solanaceae. *Solanum tuberosum*. In: *The Structure of Economic Plants*, pp. 514–549. Macmillan, New York.

HOFFMANN-BENNING, S., D. A. GAGE, L. MCINTOSH, H. KENDE, and J. A. D. ZEEVAART. 2002. Comparison of peptides in the phloem sap of flowering and non-flowering *Perilla* and lupine plants using microbore HPLC followed by matrix-assisted laser desorption/ionization time-of-flight spectrometry. *Planta* 216, 140–147.

HOLDHEIDE, W. 1951. Anatomie mitteleuropäischer Gehölzrinden (mit mikrophotographischem Atlas). In: *Handbuch der Mikroskopie in der Technik*, Band 5, Heft 1, pp. 193–367, H. Freund, ed. Umschau Verlag, Frankfurt am Main.

IMLAU, A., E. TRUERNIT, and N. SAUER. 1999. Cell-to-cell and long-distance trafficking of the green fluorescent protein in the phloem and symplasmic unloading of the protein into sink tissues. *Plant Cell* 11, 309–322.

JACOBSEN, K. R., D. G. FISHER, A. MARETZKI, and P. H. MOORE. 1992. Developmental changes in the anatomy of the sugarcane stem in relation to phloem unloading and sucrose storage. *Bot. Acta* 105, 70–80.

JORGENSEN, R. A., R. G. ATKINSON, R. L. S. FORSTER, and W. J. LUCAS. 1998. An RNA-based information superhighway in plants. *Science* 279, 1486–1487.

KEMPERS, R., and A. J. E. VAN BEL. 1997. Symplasmic connections between sieve element and companion cell in the stem phloem of *Vicia faba* have a molecular exclusion limit of at least 10 kDa. *Planta* 201, 195–201.

KNOBLAUCH, M., and A. J. E. VAN BEL. 1998. Sieve tubes in action. *Plant Cell* 10, 35–50.

KNOBLAUCH, M., W. S. PETERS, K. EHLERS, and A. J. E. VAN BEL. 2001. Reversible calcium-regulated stopcocks in legume sieve tubes. *Plant Cell* 13, 1221–1230.

KNOP, C., O. VOITSEKHOVSKAJA, and G. LOHAUS. 2001. Sucrose transporters in two members of the Scrophulariaceae with different types of transport sugar. *Planta* 213, 80–91.

KNOP, C., R. STADLER, N. SAUER, and G. LOHAUS. 2004. AmSUT1, a sucrose transporter in collection and transport phloem of the putative symplastic phloem loader *Alonsoa meridionalis*. *Plant Physiol.* 134, 204–214.

KÖCHENKERGER, W., J. M. POPE, Y. XIA, K. R. JEFFREY, E. KOMOR, and P. T. CALLAGHAN. 1997. A non-invasive measurement of phloem and xylem water flow in castor bean seedlings by nuclear magnetic resonance microimaging. *Planta* 201, 53–63.

KRUATRACHUE, M., and R. F. EVERT. 1977. The lateral meristem and its derivatives in the corm of *Isoetes muricata*. *Am. J. Bot.* 64, 310–325.

KÜHN, C., L. BARKER, L. BURKLE, and W. B. FROMMER. 1999. Update on sucrose transport in higher plants. *J. Exp. Bot.* 50 (spec. iss.), 935–953.

KÜHN, C., M.-R. HAJIREZAEI, A. R. FERNIE, U. ROESSNER-TUNALI, T. CZECHOWSKI, B. HIRNER, and W. B. FROMMER. 2003. The sucrose transporter *StSUT1* localizes to sieve elements in potato tuber phloem and influences tuber physiology and development. *Plant Physiol.* 131, 102–113.

KUO, J. 1983. The nacreous walls of sieve elements in sea grasses. *Am. J. Bot.* 70, 159–164.

KUO, J., and T. P. O'BRIEN. 1974. Lignified sieve elements in the wheat leaf. *Planta* 117, 349–353.

KUO, J., T. P. O'BRIEN, and S.-Y. ZEE. 1972. The transverse veins of the wheat leaf. *Aust. J. Biol. Sci.* 25, 721–737.

KURSANOV, A. L. 1984. *Assimilate Transport in Plants*, 2nd rev. ed. Elsevier, Amsterdam.

LALONDE, S., E. BOLES, H. HELLMANN, L. BARKER, J. W. PATRICK, W. B. FROMMER, and J. M. WARD. 1999. The dual function of sugar carriers: transport and sugar sensing. *Plant Cell* 11, 707–726.

LALONDE, S., M. TEGEDER, M. THRONE-HOLST, W. B. FROMMER, and J. W. PATRICK. 2003. Phloem loading and unloading of sugars and amino acids. *Plant Cell Environ.* 26, 37–56.

LAWTON, D. M., and Y. M. NEWMAN. 1979. Ultrastructure of phloem in young runner-bean stem: Discovery, in old sieve elements on the brink of collapse, of parietal bundles of P-protein tubules linked to the plasmalemma. *New Phytol.* 82, 213–222.

LAWTON, J. R., and J. R. S. LAWTON. 1971. Seasonal variations in the secondary phloem of some forest trees from Nigeria. *New Phytol.* 70, 187–196.

LEE, D. R. 1989. Vasculature of the abscission zone of tomato fruit: implications for transport. *Can. J. Bot.* 67, 1898–1902.

LEE, D. R. 1990. A unidirectional water flux model of fruit growth. *Can. J. Bot.* 68, 1286–1290.

LUCAS, W. J., and V. R. FRANCESCHI, 1982. Organization of the sieve-element walls of leaf minor veins. *J. Ultrastruct. Res.* 81, 209–221.

MADEY, E., L. M. NOWACK, and J. E. THOMPSON. 2002. Isolation and characterization of lipid in phloem sap of canola. *Planta* 214, 625–634.

MATZKE, M. A., and A. J. M. MATZKE. 1986. Visualization of mitochondria and nuclei in living plant cells by the use of a potential-sensitive fluorescent dye. *Plant Cell Environ.* 9, 73–77.

MCCAULEY, M. M., and R. F. EVERT. 1988. The anatomy of the leaf of potato, *Solanum tuberosum* L. "Russet Burbank." *Bot. Gaz.* 149, 179–195.

MCCAULEY, M. M., and R. F. EVERT. 1989. Minor veins of the potato (*Solanum tuberosum* L.) leaf: Ultrastructure and plasmodesmatal frequency. *Bot. Gaz.* 150, 351–368.

MELARAGNO, J. E., and M. A. WALSH. 1976. Ultrastructural features of developing sieve elements in *Lemna minor* L.—The protoplast. *Am. J. Bot.* 63, 1145–1157.

MINCHIN, P. E. H., and M. R. THORPE. 1987. Measurement of unloading and reloading of photo-assimilate within the stem of bean. *J. Exp. Bot.* 38, 211–220.

MIYAKE, H., and E. MAEDA. 1976. The fine structure of plastids in various tissues in the leaf blade of rice. *Ann. Bot.* 40, 1131–1138.

MÜNCH, E. 1930. *Die Stoffbewegungen in der Pflanze*. Gustav Fischer, Jena.

NANKO, H., H. SAIKI, and H. HARADA. 1976. Cell wall development of chambered crystalliferous cells in the secondary phloem of *Populus euroamericana*. *Bull. Kyoto Univ. For. (Kyoto Daigaku Nogaku bu Enshurin hokoku)* 48, 167–177.

NERD, A., and P. M. NEUMANN. 2004. Phloem water transport maintains stem growth in a drought-stressed crop cactus (*Hylocereus undatus*). *J. Am. Soc. Hortic. Sci.* 129, 486–490.

NEUBERGER, D. S., and R. F. EVERT. 1974. Structure and development of the sieve-element protoplast in the hypocotyl of *Pinus resinosa*. *Am. J. Bot.* 61, 360–374.

NEUBERGER, D. S., and R. F. EVERT. 1975. Structure and development of sieve areas in the hypocotyl of *Pinus resinosa*. *Protoplasma* 84, 109–125.

NEUBERGER, D. S., and R. F. EVERT. 1976. Structure and development of sieve cells in the primary phloem of *Pinus resinosa*. *Protoplasma* 87, 27–37.

OFFLER, C. E., and J. W. PATRICK. 1984. Cellular structures, plasma membrane surface areas and plasmodesmatal frequencies of seed coats of *Phaseolus vulgaris* L. in relation to photosynthate transfer. *Aust. J. Plant Physiol.* 11, 79–99.

OPARKA, K. J., and R. TURGEON. 1999. Sieve elements and companion cells—Traffic control centers of the phloem. *Plant Cell* 11, 739–750.

OROSS, J. W., and W. J. LUCAS. 1985. Sugar beet petiole structure: Vascular anastomoses and phloem ultrastructure. *Can. J. Bot.* 63, 2295–2304.

PALEVITZ, B. A., and E. H. NEWCOMB. 1970. A study of sieve element starch using sequential enzymatic digestion and electron microscopy. *J. Cell Biol.* 45, 383–398.

PARTHASARATHY, M. V. 1966. Studies on metaphloem in petioles and roots of Palmae. Ph.D. Thesis. Cornell University, Ithaca.

PARTHASARATHY, M. V. 1974a. Ultrastructure of phloem in palms. I. Immature sieve elements and parenchymatic elements. *Protoplasma* 79, 59–91.

PARTHASARATHY, M. V. 1974b. Ultrastructure of phloem in palms. II. Structural changes, and fate of the organelles in differentiating sieve elements. *Protoplasma* 79, 93–125.

PARTHASARATHY, M. V., and P. B. Tomlinson. 1967. Anatomical features of metaphloem in stems of *Sabal*, *Cocos* and two other palms. *Am. J. Bot.* 54, 1143–1151.

PATE, J. S. 1975. Exchange of solutes between phloem and xylem and circulation in the whole plant. In: *Encyclopedia of Plant Physiology*, n.s. vol. 1. *Transport in Plants*. I. *Phloem Transport*, pp. 451–473, Springer-Verlag, Berlin.

PATE, J. S., and B. E. S. GUNNING. 1969. Vascular transfer cells in angiosperm leaves. A taxonomic and morphological survey. *Protoplasma* 68, 135–156.

PATRICK, J. W. 1997. Phloem unloading: Sieve element unloading and post-sieve element transport. *Annu. Rev. Plant Physiol. Plant Mol. Biol.* 48, 191–222.

PERRY, J. W., and R. F. EVERT. 1975. Structure and development of the sieve elements in *Psilotum nudum. Am. J. Bot.* 62, 1038–1052.

PHILLIPS, R. J., and S. R. DUNGAN. 1993. Asymptotic analysis of flow in sieve tubes with semi-permeable walls. *J. Theor. Biol.* 162, 465–485.

RAVEN, P. H., R. F. EVERT, and S. E. EICHHORN. 2005. *Biology of Plants*, 7th ed. Freeman, New York.

RESCH, A. 1961. Zur Frage nach den Geleitzellen im Protophloem der Wurzel. *Z. Bot.* 49, 82–95.

ROBINSON-BEERS, K., and R. F. EVERT. 1991. Ultrastructure of and plasmodesmatal frequency in mature leaves of sugarcane. *Planta* 184, 291–306.

ROTH, I. 1981. Structural patterns of tropical barks. In: *Encyclopedia of Plant Anatomy*, Band 9, Teil 3. Gebrüder Borntraeger, Berlin.

RUIZ-MEDRANO, R., B. XOCONOSTLE-CÁZARES, and W. J. LUCAS. 2001. The phloem as a conduit for inter-organ communication. *Curr. Opin. Plant Biol.* 4, 202–209.

RUSSELL, S. H., and R. F. EVERT. 1985. Leaf vasculature in *Zea mays* L. *Planta* 164, 448–458.

RUSSIN, W. A., and R. F. EVERT. 1985. Studies on the leaf of *Populus deltoides* (Salicaceae): Ultrastructure, plasmodesmatal frequency, and solute concentrations. *Am. J. Bot.* 72, 1232–1247.

SAUTER, J. J. 1974. Structure and physiology of Strasburger cells. *Ber. Dtsch. Bot. Ges.* 87, 327–336.

SAUTER, J. J. 1976. Untersuchungen zur Lokalisation von Glycerophosphatase- und Nucleosidtriphosphatase-Aktivität in Siebzellen von *Larix. Z. Pflanzenphysiol.* 79, 254–271.

SAUTER, J. J., 1977. Electron microscopical localization of adenosine triphosphatase and β-glycerophosphatase in sieve cells of *Pinus nigra* var. *austriaca* (Hoess) Battoux. *Z. Pflanzenphysiol.* 81, 438–458.

SAUTER, J. J. and H. J. BRAUN. 1968. Histologische und cytochemische Untersuchungen zur Funktion der Baststrahlen von *Larix decidua* Mill., unter besonderer Berücksichtigung der Strasburger-Zellen. *Z. Pflanzenphysiol.* 59, 420–438.

SAUTER, J. J., and H. J. BRAUN. 1972. Cytochemische Untersuchung der Atmungsaktivität in den Strasburger-Zellen von *Larix* und ihre Bedeutung für den Assimilattransport. *Z. Pflanzenphysiol.* 66, 440–458.

SCHMITZ, K., B. CUYPERS, and M. MOLL. 1987. Pathway of assimilate transfer between mesophyll cells and minor veins in leaves of *Cucumis melo* L. *Planta* 171, 19–29.

SCHNEIDER, H. 1945. The anatomy of peach and cherry phloem. *Bull. Torrey Bot. Club* 72, 137–156.

SCHNEIDER, H. 1955. Ontogeny of lemon tree bark. *Am. J. Bot.* 42, 893–905.

SCHOBERT, C., P. GROßMANN, M. GOTTSCHALK, E. KOMOR, A. PECSVARADI, and U. ZUR MIEDEN. 1995. Sieve-tube exudate from *Ricinus communis* L. seedlings contains ubiquitin and chaperones. *Planta* 196, 205–210.

SCHOBERT, C., L. BAKER, J. SZEDERKÉNYI, P. GROßMANN, E. KOMOR, H. HAYASHI, M. CHINO, and W. J. LUCAS. 1998. Identification of immunologically related proteins in sieve-tube exudate collected from monocotyledonous and dicotyledonous plants. *Planta* 206, 245–252.

SCHULZ, A. 1990. Conifers. In: *Sieve Elements. Comparative Structure, Induction and Development*, pp. 63–88, H.-D. Behnke and R. D. Sjolund, eds. Springer-Verlag, Berlin.

SCHULZ, A. 1992. Living sieve cells of conifers as visualized by confocal, laser-scanning fluorescence microscopy. *Protoplasma* 166, 153–164.

SCHULZ, A. 1998. Phloem. Structure related to function. *Prog. Bot.* 59, 429–475.

SRIVASTAVA, L. M. 1970. The secondary phloem of *Austrobaileya scandens. Can. J. Bot.* 48, 341–359.

STADLER, R., and N. SAUER. 1996. The *Arabidopsis thaliana AtSUC2* gene is specifically expressed in companion cells. *Bot. Acta* 109, 299–306.

STADLER, R., J. BRANDNER, A. SCHULZ, M. GAHRTZ, and N. SAUER. 1995. Phloem loading by the PmSUC2 sucrose carrier from *Plantago major* occurs into companion cells. *Plant Cell* 7, 1545–1554.

STAIGER, C. J., B. C. GIBBON, D. R. KOVAR, and L. E. ZONIA. 1997. Profilin and actin-depolymerizing factor: Modulators of actin organization in plants. *Trends Plant Sci.* 2, 275–281.

THOMPSON, G. A. 1999. P-protein trafficking through plasmodesmata. In: *Plasmodesmata: Structure, Function, Role in Cell Communication*, pp. 295–313, A. J. E. van Bel and W. J. P. van Kesteren, eds. Springer, Berlin.

THOMPSON, G. A., and A. SCHULZ. 1999. Macromolecular trafficking in the phloem. *Trends Plant Sci.* 4, 354–360.

TURGEON, R. 1991. Symplastic phloem loading and the sink-source transition in leaves: A model. In: *Recent Advances in Phloem Transport and Assimilate Compartmentation*, pp. 18–22, J. L. Bonnemain, S. Delrot, W. J. Lucas, and J. Dainty, eds. Ouest Editions, Nantes, France.

TURGEON, R. 1996. Phloem loading and plasmodesmata. *Trends Plant Sci.* 1, 413–423.

TURGEON, R., and R. MEDVILLE. 1998. The absence of phloem loading in willow leaves. *Proc. Natl. Acad. Sci. USA* 95, 12055–12060.

TURGEON, R., and R. MEDVILLE. 2004. Phloem loading. A reevaluation of the relationship between plasmodesmatal frequencies and loading strategies. *Plant Physiol.* 136, 3795–3803.

TURGEON, R., and J. A. WEBB. 1976. Leaf development and phloem transport in *Cucurbita pepo*: Maturation of the minor veins. *Planta* 129, 265–269.

TURGEON, R., J. A. WEBB, and R. F. EVERT. 1975. Ultrastructure of minor veins of *Cucurbita pepo* leaves. *Protoplasma* 83, 217–232.

TURGEON, R., D. U. BEEBE, and E. GOWAN. 1993. The intermediary cell: Minor-vein anatomy and raffinose oligosaccharide synthesis in Scrophulariaceae. *Planta* 191, 446–456.

TURGEON, R., R. MEDVILLE, and K. C. NIXON. 2001. The evolution of minor vein phloem and phloem goading. *Am. J. Bot.* 88, 1331–1339.

VAN BEL, A. J. E. 1993. The transport phloem. Specifics of its functioning. *Prog. Bot.* 54, 134–150.

VAN BEL, A. J. E. 1996. Interaction between sieve element and companion cell and the consequences for photoassimilate distribution. Two structural hardware frames with associated physiological software packages in dicotyledons? *J. Exp. Bot.* 47 (spec. iss.), 1129–1140.

VAN BEL, A. J. E., and F. GAUPELS. 2004. Pathogen-induced resistance and alarm signals in the phloem. *Mol. Plant Pathol.* 5, 495–504.

VAN BEL, A. J. E., and H. V. M. VAN RIJEN. 1994. Microelectrode-recorded development of the symplasmic autonomy of the sieve element/companion cell complex in the stem phloem of *Lupinus luteus* L. *Planta* 192, 165–175.

VAN BEL, A. J. E., K. EHLERS, and M. KNOBLAUCH. 2002. Sieve elements caught in the act. *Trends Plant Sci.* 7, 126–132.

VIOLA, R., A. G. ROBERTS, S. HAUPT, S. GAZZANI, R. D. HANCOCK, N. MARMIROLI, G. C. MACHRAY, and K. J. OPARKA. 2001. Tuberization in potato involves a switch from apoplastic to symplastic phloem unloading. *Plant Cell* 13, 385–398.

WALSH, M. A., and J. E. MELARAGNO. 1976. Ultrastructural features of developing sieve elements in *Lemna minor* L.—Sieve plate and lateral sieve areas. *Am. J. Bot.* 63, 1174–1183.

WALSH, M. A., and T. M. POPOVICH. 1977. Some ultrastructural aspects of metaphloem sieve elements in the aerial stem of the holoparasitic angiosperm *Epifagus virginiana* (Orobanchaceae). *Am. J. Bot.* 64, 326–336.

WARMBRODT, R. D. 1985a. Studies on the root of *Hordeum vulgare* L.—Ultrastructure of the seminal root with special reference to the phloem. *Am. J. Bot.* 72, 414–432.

WARMBRODT, R. D. 1985b. Studies on the root of *Zea mays* L.—Structure of the adventitious roots with respect to phloem unloading. *Bot. Gaz.* 146, 169–180.

WILLENBRINK, J., and R. KOLLMANN. 1966. Über den Assimilat-transport im Phloem von *Metasequoia*. *Z. Pflanzenphysiol.* 55, 42–53.

YOUNG, J. H., R. F. EVERT, and W. ESCHRICH. 1973. On the volume-flow mechanism of phloem transport. *Planta* 113, 355–366.

ZAHUR, M. S. 1959. Comparative study of secondary phloem of 423 species of woody dicotyledons belonging to 85 families. Cornell Univ. Agric. Exp. Stan. Mem. 358. New York State College of Agriculture, Ithaca.

ZIEGLER, H. 1963. Die Ferntransport organischer Stoffe in den Pflanzen. *Naturwissenschaften* 50, 177–186.

# CHAPTER FOURTEEN

# Phloem: Secondary Phloem and Variations in Its Structure

Inasmuch as the secondary phloem is derived from the vascular cambium, the arrangement of cells in the secondary phloem parallels that in the secondary xylem. A vertical or axial system of cells, derived from the fusiform initials of the cambium, is interpenetrated by the horizontal or radial system derived from the ray initials (Fig. 14.1). The principal components of the axial system are sieve elements (either sieve cells or sieve-tube elements, the latter with companion cells), parenchyma cells, and fibers. Those of the radial system are ray parenchyma cells.

Storied, nonstoried, and intermediate types of arrangement of phloem cells may be found in different species of plants. As in the xylem, the type of arrangement is determined, first, by the morphology of the cambium (i.e., whether or not it is storied) and, second, by the degree of elongation of the various elements of the axial system during tissue differentiation.

Many conifers and woody angiosperms show a division of secondary phloem into annual growth increments (Huber, 1939; Holdheide, 1951; Srivastava, 1963), although this division is less clear than in the secondary xylem and may be obscured by growth conditions. The growth increments in the phloem are distinguishable if the cells of the early phloem expand more strongly than those of the late phloem. In the Pinaceae, several layers of relatively wide early-phloem sieve cells are formed, followed by a more or less continuous tangential band of parenchyma, and finally by several layers of somewhat narrower late-phloem sieve cells (Holdheide, 1951; Alfieri and Evert, 1968, 1973). In *Ulmus americana*, a temperate-zone hardwood, the sieve-tube elements, which occur in more or less distinct tangential bands, intergrade in width, from relatively wide early-phloem elements to narrower late-phloem elements. In *Citharexylum myrianthum* (Verbenaceae) and *Cedrela fissilis* (Meliaceae), both tropical hardwoods of Brazil, the late-formed sieve elements, which occur in scattered groups, are markedly narrower radially than earlier-formed elements and can be used reliably to delimit growth increments in the phloem (Veronica Angyallossy, personal communication). In *Pyrus communis* and *Malus domestica*, a band of future fiber-sclereids and crystalliferous cells overwinter in a meristematic state near the cambium and when mature can serve as a marker for delimiting the successive growth layers

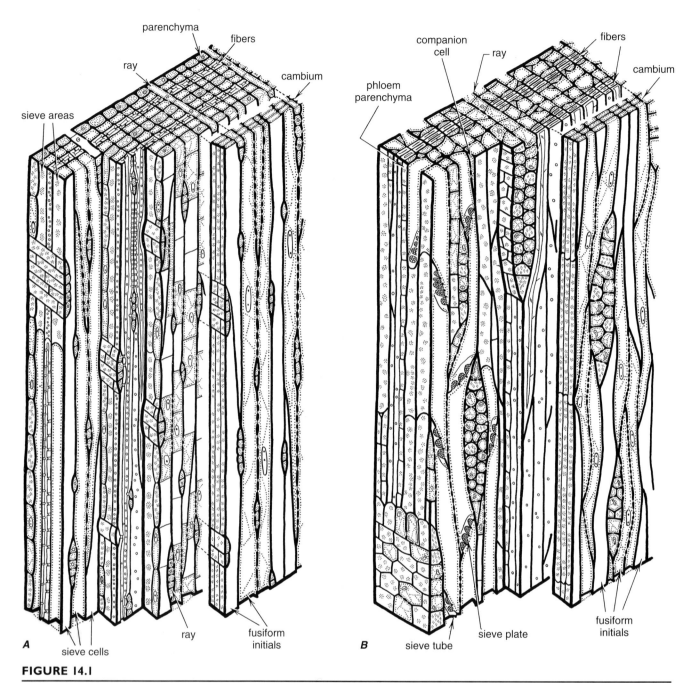

**FIGURE 14.1**

Block diagrams of secondary phloem and vascular cambium of **A**, *Thuja occidentalis* (white cedar), a conifer, and of **B**, *Liriodendron tulipifera* (tulip tree), a hardwood. (Courtesy of I. W. Bailey; drawn by Mrs. J. P. Rogerson under the supervision of L. G. Livingston. Redrawn.)

(Evert, 1960, 1963). Many gymnosperms and angiosperms form tangential bands of fibers in the secondary phloem. The number of these bands is not necessarily constant from season to season, and therefore they cannot be safely used to determine the age of the phloem tissue. In conifers with phloem fibers, however, the fibers may be wider and have thicker walls in the early phloem, narrower and with thinner walls in the late phloem. Such a pattern has been observed in *Chamaecyparis lawsoniana* (Huber, 1949), *Juniperus communis*

(Holdheide, 1951), and *Thuja occidentalis* (Bannan, 1955). In angiosperms, the first-formed band of phloem fibers in a given increment is often wider than the last-formed band (e.g., in *Robinia pseudoacacia,* Derr and Evert, 1967; and *Tilia americana,* in which the first-formed band of sieve-tube elements is also often wider than subsequently formed bands, Evert, 1962). The collapse of the sieve elements in the nonconducting phloem and the concomitant modifications in some other cells—notably the enlargement of the parenchyma cells—

contribute toward obscuring the structural differences that might exist in the different parts of a growth layer at its inception.

The phloem rays are continuous with the xylem rays, since both arise from a common group of ray initials in the cambium. Together they constitute the vascular rays. Near the cambium the phloem and xylem rays of common origin are usually identical in height and width. However, the older part of the phloem ray, which is displaced outward by the expansion of the secondary plant body, may increase in width, sometimes very considerably. Before the phloem rays become dilated, their variations in form and size are similar to those of xylem rays in the same species. Phloem rays may be uniseriate, biseriate, or multiseriate. They vary in height, and small and large rays may be present in the same species.

The rays may be composed of one kind of cell, or they may contain both kinds of cell, procumbent or upright. Phloem rays do not attain the same lengths as xylem rays because the vascular cambium produces less phloem than xylem. They are also not as long because the outer portions of the phloem commonly are sloughed through the activity of the phellogen, or cork cambium, the lateral meristem that produces the periderm (Chapter 15).

Sometimes, with reference to stems and roots, it is convenient to treat as a unit the phloem and all the tissues located outside it. The nontechnical term **bark** is employed for this purpose (Fig. 14.2). In stems and roots possessing only primary tissues, bark most commonly refers to the primary phloem and the cortex. In axes in a secondary state of growth, it may include the primary and the secondary phloem, various amounts of cortex, and the periderm. In old stems and roots, the bark may consist entirely of secondary tissues, of layers of dead secondary phloem sandwiched between layers of periderm, and of the living tissues inside the innermost periderm (see Fig. 15.1). The innermost periderm and the tissues of the axis isolated by it may be combined under the designation of **outer bark**. The subjacent living phloem may be designated the **inner bark**. Outer and inner bark cannot be distinguished in barks that have only a superficial periderm. Only the inner bark is considered in this chapter.

# ∎ CONIFER PHLOEM

In conifers the secondary phloem parallels the secondary xylem in the relative simplicity of its structure (Fig. 14.1A; Srivastava, 1963; den Outer, 1967). The axial system contains sieve cells and parenchyma cells, some of which may be differentiated as Strasburger cells. In the Pinaceae, the Strasburger cells of the axial system commonly occur as radial plates of cells, which are derivatives of declining fusiform initials (Srivastava,

1963). Fibers and sclereids may also be present. The rays are uniseriate and contain parenchyma cells and Strasburger cells, if these cells are present in the species. Most commonly the Strasburger cells are located at the margins of the rays, although occasionally they may be found in the middle of the ray. Where ray-Strasburger cells occur on the phloem side of the cambium, ray tracheids occur on the xylem side, except for *Abies*, which rarely has ray tracheids. The cell arrangement is nonstoried. Resin ducts normally are present in both axial and radial systems. Resin ducts also may be induced to form in reaction to mechanical or chemical wounding or to injury by insects or pathogens. These traumatic phloem resin ducts form tangentially anastomosing networks (Yamanaka, 1984, 1989; Kuroda, 1998).

The sieve cells of conifers are slender, elongated elements comparable to the fusiform initials from which they are derived (Figs. 13.4A and 14.1A). They overlap each other at their ends, and each is in contact with several rays. The sieve areas occur almost exclusively on the radial walls. They are particularly abundant on the ends, which overlap those of other sieve cells. In contrast to the axial system of the secondary phloem of angiosperms, that of the conifers consists mainly of sieve elements, comprising up to 90% of the axial system in some Pinaceae.

The axial parenchyma cells occur chiefly in longitudinal strands. They store starch at certain times of the year but are particularly conspicuous when they contain resinous and polyphenolic inclusions. The polyphenolic substances in the axil parenchyma cells of *Picea abies* have been shown to play a major role in defense against invasive organisms such as the blue-stain fungus, *Ceratocystis polonica* (Franceschi et al., 1998). Calcium oxalate crystals are common in the secondary phloem of conifers (Hudgins et al., 2003). In Pinaceae the crystals accumulate intracellulary, in crystalliferous parenchyma; in the non-Pinaceae crystal accumulation is extracellular, in the cell walls. In conifer stems the crystals, in combination with fiber rows (where present), provide an effective barrier against small bark-boring insects (Hudgins et al., 2003).

The distribution of cells in conifer phloem shows two major patterns. In the Pinaceae, which consistently lack fibers, the pattern is determined by the relative arrangement of the sieve cells and axial parenchyma cells. The axial parenchyma cells may form uniseriate tangential bands with bands of sieve cells so wide that the sieve cells become the basic elements of the axial system (Fig. 14.3A). The sieve cells may form much narrower bands, only one to three cells in width. The parenchyma bands are sometimes irregular, or the parenchyma cells are so sparsely distributed that a radial alternation with bands of sieve cells is barely discernible. In the Taxodiaceae, Cupressaceae, and parts of Podocarpaceae and Taxaceae, fibers are consistently

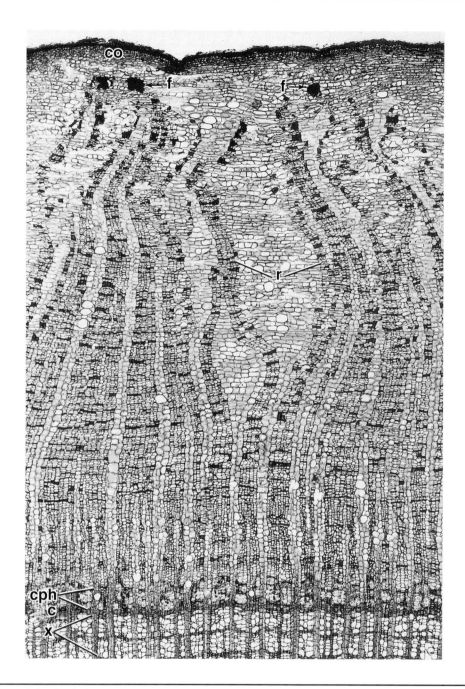

**FIGURE 14.2**

Transverse section of the bark from an 18-year-old stem of *Liriodendron tulipifera*. Except for the cortex (co) and remnants of primary phloem, as evidenced by the presence of primary phloem fibers (f), this bark—all tissues outside the vascular cambium—consists almost entirely of secondary phloem; the original periderm is still present outside the cortex. Only a very narrow portion (0.1 mm wide) of the phloem next to the cambium (c) contains mature, living sieve tubes, that is, is conducting phloem (cph), the rest of the phloem is nonconducting. Some rays (r) are dilated. Other detail: x, xylem. (×37. Reproduced by permission University of California Press: Cheadle and Esau, 1964. *Univ. Calif. Publ. Bot.* © 1964, The Regents of the University of California.)

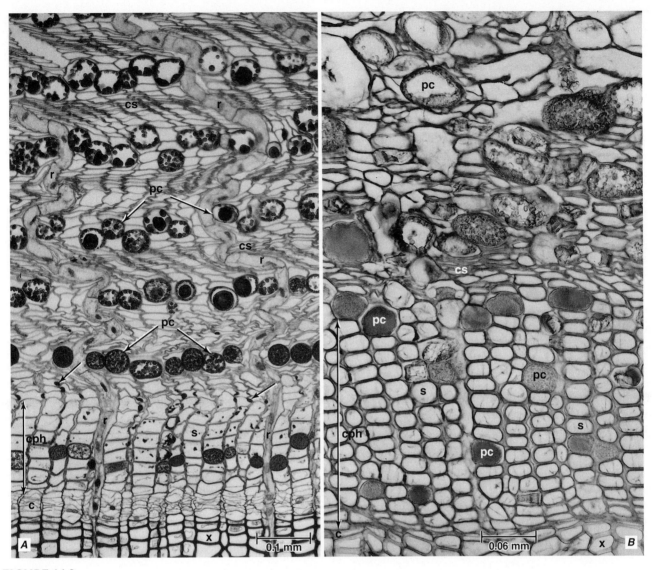

**FIGURE I4.3**

Transverse sections of secondary phloem of *Pinus*. Conducting phloem (cph) much smaller in amount than the non-conducting phloem (only partly shown in these figures), in which all sieve cells are crushed (cs) and only axial parenchyma cells (pc) and ray parenchyma cells (r) are intact. In **A** (*Pinus strobus*), the axial parenchyma cells are arranged in tangential bands and separate early phloem from late phloem in each increment. Portions of 7 growth increments can be seen here. In **B** (*Pinus* sp.), the axial parenchyma cells are scattered. Other details: c, cambium; s, sieve cell; x, xylem; in **A**, unlabeled arrows point to definitive callose. (**A**, from Esau, 1977; **B**, from Esau, 1969. www.schweizerbart.de)

present, and the characteristic pattern is based on a regular sequence of alternating tangential uniseriate bands of fibers, sieve cells, parenchyma cells, sieve cells, fibers, and so forth, in that order (see Fig. 14.5B). Disturbances in the pattern occur, and the sequence is less regular in some families than in others. A specific pattern may not be discernible (some Araucariaceae, e.g., *Agathis australis*, Chan, 1986) or one may develop as the stem ages.

Considering the gymnosperms as a whole, den Outer (1967) recognized an evolutionary trend in the secondary phloem toward increasing organization, regularity, and repetition. He grouped the tissues into three types:

**1. *Pseudotsuga taxifolia* type** (some Pinaceae belong to this type), with *Tsuga canadensis* subtype (the other Pinaceae are grouped in this subtype). The axial system consists mainly of sieve cells, among which the few parenchyma cells occur either in discontinuous tangential bands (Fig. 14.3A) or scattered, forming a parenchyma-cell net (Fig. 14.3B). In the *Pseudotsuga taxifolia* type the ray parenchyma cells are connected symplastically to the sieve cells only indirectly via

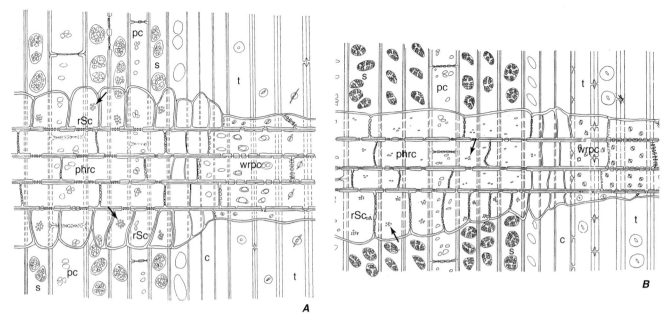

**FIGURE 14.4**

Radial views showing distribution of the symplastic connections between the sieve cells (s) and cells of the rays near the cambium. **A**, in *Pseudotsuga taxifolia* such connections (arrows) occur between the sieve cells and ray-Strasburger cells (rSc), but not between sieve cells and phloem-ray parenchyma cells (phrc). **B**, in *Tsuga canadensis* symplastic connections (arrows) occur between sieve cells and all cells of the rays, both ray-Strasburger cells and phloem-ray parenchyma cells. Other details: c, cambium; pc, axial parenchyma cell; t, tracheid; wrpc, wood-ray parenchyma cell. (From den Outer, 1967.)

the Strasburger cells (Fig. 14.4A), whereas in the *Tsuga canadensis* subtype both the ray parenchyma cells and Strasburger cells are connected directly to the sieve cells (Fig. 14.4B). With the exception of radial plates of cells, the Strasburger cells occur almost exclusively in the rays.

**2.  *Ginkgo biloba* type** (includes, in addition to *Ginkgo biloba*, Cycadaceae, Araucariaceae, and parts of Podocarpaceae and Taxaceae). The axial system consists of bands of sieve cells and bands of parenchyma cells in nearly equal proportions (Fig. 14.5A). Only phloem-Strasburger cells occur; they form long longitudinal strands lying within the phloem-parenchyma cell bands.

**3.  *Chamaecyparis pisifera* type** (includes Cupressaceae, Taxodiaceae, and parts of Podocarpaceae and Taxaceae). The axial system consists of a regular sequence of alternating cell types (Fig. 14.5B). The Strasburger cells lie usually scattered among the parenchyma cells, singly, never in long longitudinal strands.

Thus, in this putative evolutionary series, the level of organization of the secondary phloem increases. It begins with an axial system consisting mostly of sieve cells and a scattering of parenchyma cells and sclereids

and culminates in one composed of a regular, repeating sequence of tangential uniseriate layers of fibers, sieve cells, parenchyma cells, fibers, and so on.

# ANGIOSPERM PHLOEM

The secondary phloem of angiosperms shows a wider diversity of patterns of cell arrangement and more variations in the components than the phloem of conifers. Storied, intermediate, and nonstoried arrangements of cells are encountered, and the rays may be uniseriate, biseriate, or multiseriate. Sieve-tube elements, companion cells, and parenchyma cells are constant elements of the axial system, but fibers may be absent (Fig. 14.6A; *Aristolochia, Austrobaileya, Calycanthus* spp., *Drimys* spp., *Rhus typhina*). Both systems may contain sclereids, secretory elements of schizogenous and lysigenous origins, laticifers, and various idioblasts, individualized cells with specialized contents, such as oil, mucilage, tannin, and crystals. Crystal formation is common and may occur in strand parenchyma cells, in ray parenchyma cells, or in sclerenchyma cells. Chambered crystalliferous parenchyma strands typically surround bundles of fibers in the secondary phloem. Crystals may be so abundant as to contribute considerable mechanical support to the bark, even resulting in a hard bark in the absence of sclerenchyma (Roth, 1981). The bark of

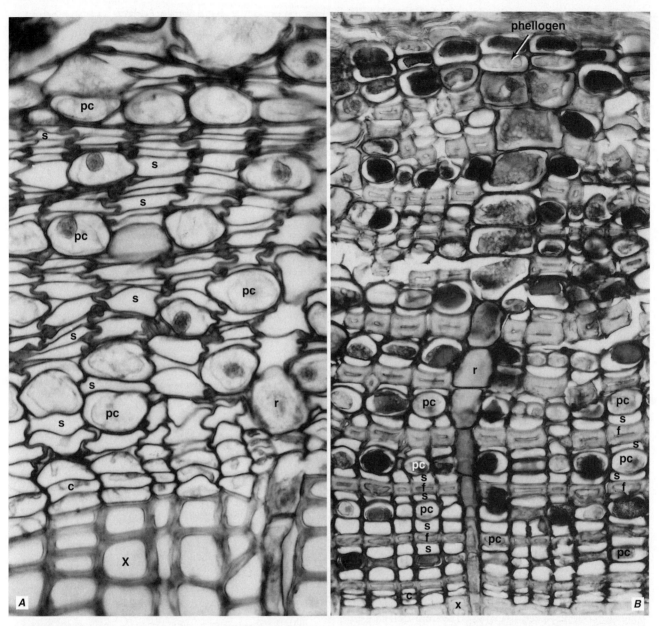

**FIGURE 14.5**

Transverse sections of secondary phloem. **A**, from a twig of *Ginkgo biloba*. Sieve cells (s) and axial parenchyma cells (pc) in layers. Sieve cells partly collapsed, particularly in older phloem. Parenchyma cells turgid. In older stems the layers, or bands, of sieve cells and axial parenchyma are more obvious than here. **B**, from stem of *Taxodium distichum*. Cells alternate radially in sequence of fiber (f), sieve cell (s), parenchyma cell (pc), sieve cell (s), fiber (f), and so on. In nonconducting phloem (above) sieve cells are crushed by enlarged parenchyma cells. Other details: **c**, cambium; r, ray; x, xylem. (A, ×600; B, ×400. From Esau, 1969. www.schweizerbart.de)

trees of tropical humid forests, in particular, have well-developed secretory systems (Roth, 1981).

### The Patterns Formed by the Fibers Can Be of Taxonomic Significance

One of the most conspicuous differences in the appearance of the bark and secondary phloem of different species results from the distribution of the fibers, and the patterns formed by them are useful for identification (Holdheide, 1951; Chattaway, 1953; Chang, 1954; Zahur, 1959; Roth, 1981; Archer and van Wyk, 1993; den Outer, 1993). In certain angiosperms, the fibers are scattered, or irregularly dispersed, among other cells of the axial system (Fig. 14.6B; *Campsis, Tecoma, Nicotiana, Cephalanthus, Laurus*). In others, the fibers occur in

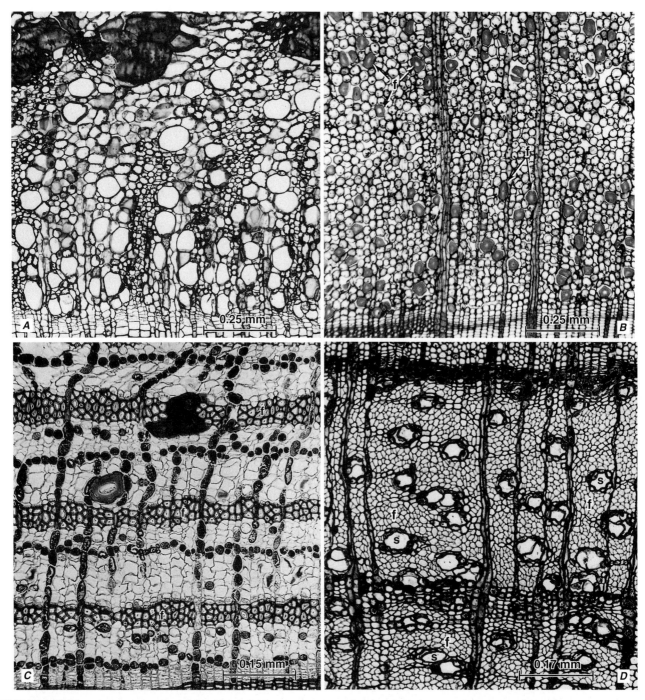

**FIGURE 14.6**

Secondary phloem of woody angiosperms in transverse sections. **A**, *Drimys winteri*, which lacks fibers. The large cells are secretory cells. **B**, *Campsis*, fibers (f) scattered single. **C**, *Castanea*, fibers (f) in parallel tangential bands. **D**, *Carya*, fibers (f) surround groups of sieve elements (s) and parenchymatous cells.

tangential bands, more or less regularly alternating with bands containing the sieve tubes and parenchymatous components of the axial system (Fig. 14.6C; *Castanea, Corchorus, Liriodendron, Magnolia, Robinia, Tilia, Vitis*), or they may be somewhat scattered. Fibers may be so abundant that the sieve tubes and parenchymatous cells occur as small groups surrounded by fibers (Fig. 14.6D; *Carya, Eucalyptus, Ursiniopsis*). In some species, sclerenchyma cells, usually sclereids or fiber-sclereids, differentiate only in the nonconducting part of the phloem (*Prunus, Pyrus, Sorbus, Laburnum, Aesculus*). The septate fibers of *Vitis* are living cells concerned with storage of starch.

## Secondary Sieve-Tube Elements Show Considerable Variation in Form and Distribution

The sieve tubes and axial parenchyma cells show varied spatial interrelationships. Sometimes the sieve tubes occur in long, continuous radial files, or they may form tangential bands with similar bands of parenchyma. In phloem having tangential bands of fibers alternating with bands of sieve-tube elements and associated parenchyma, the sieve tubes are commonly separated by parenchyma from the fibers and the rays. The parenchyma bands form a continuous network together with the rays (Ziegler, 1964; Roth, 1981). Considerable variation exists in the proportion of parenchyma cells to other cell types comprising the axial system. In tropical lianas, such as *Datura*, the parenchyma cells are so abundant that they constitute the ground mass of the conducting phloem (den Outer, 1993).

Many woody angiosperms have nonstoried phloem, with elongated sieve-tube elements bearing mostly compound sieve plates on the inclined end walls (Fig. 14.7A; *Betula, Quercus, Populus, Aesculus, Tilia, Juglans,*

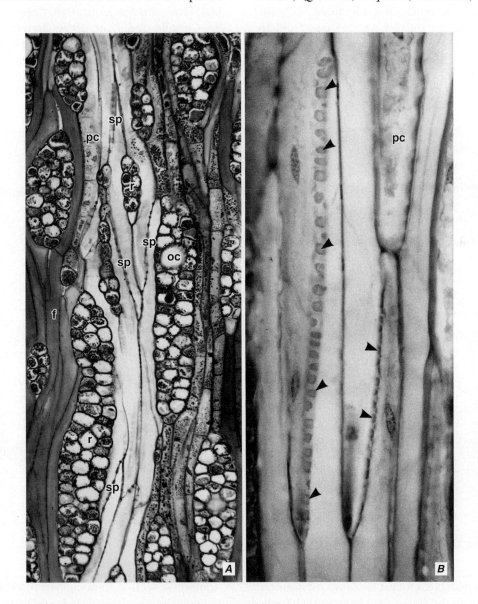

## FIGURE 14.7

Tangential sections of nonstoried secondary phloem of *Liriodendron tulipifera* (**A**) and *Austrobaileya scandens* (**B**), both of which have sieve-tube elements with compound sieve plates (sp) on inclined end walls. Several sieve plates can be seen in **A**, a single sieve plate in **B**. The greatly inclined sieve plates of *Austrobaileya* have numerous sieve areas (darts). Other details: f, fiber; pc, axial parenchyma cell; oc, oil cell in ray; r, ray. (A, ×100; B, ×413. **B**, Plate 3D from Esau, 1979, *Anatomy of the Dicotyledons*, 2nd ed., vol. I, C. R. Metcalfe and L. Chalk, eds., by permission of Oxford University Press.)

*Liriodendron*). In some genera, the sieve areas of the sieve plates are distinctly more differentiated than the lateral sieve areas. In others, with greatly inclined end walls, as in those of *Austrobaileya* (Fig. 14.7B; Behnke, 1986), the Winteraceae (Behnke and Kiritsis, 1983), and the Pomoideae (Evert, 1960, 1963), there is less distinction between the two kinds of sieve areas. Slightly inclined (*Fagus, Acer*) and transverse (Fig. 14.8; *Fraxinus, Ulmus, Robinia*) end walls usually bear simple sieve plates. The individual sieve-tube elements in such plants are relatively short, and if the phloem is derived from a cambium with short initials, the phloem may be more or less distinctly storied (Fig. 14.8B; den Outer, 1986).

If the sieve-tube elements possess inclined end walls, the ends of the cells are roughly wedge-shaped and are so oriented that the wide side of the wedge is exposed in the radial section, the narrow in the tangential. The compound sieve plates are borne on the wide sides of the wedge-like ends and are therefore seen in face views

in the radial sections and in sectional views in the tangential sections.

As mentioned previously, the secondary phloem rays are comparable to the xylem rays of the same species but may become dilated in the older parts of the tissue. The degree of this dilatation is highly variable. The extreme dilatation of certain rays is one of the most conspicuous characteristics of the phloem of *Tilia* (see Fig. 14.16). The wide rays separate the axial system together with the undilated rays into blocks narrowed down toward the periphery of the stem.

Sieve-tube elements, either solitary or in groups, are found in the phloem rays of some eudicots (e.g., in Fig. 14.9, *Vitis vinifera*, Esau, 1948; *Calycanthus occidentalis*, Cheadle and Esau, 1958; *Strychnos nux-vomica, Leucosceptrum cannum, Dahlia imperialis, Gynura angulosa*, Chavan et al., 1983; *Erythrina indica, Acacia nilotica, Tectona grandis*, Rajput and Rao, 1997). These "ray sieve elements" are analogous to the so-called

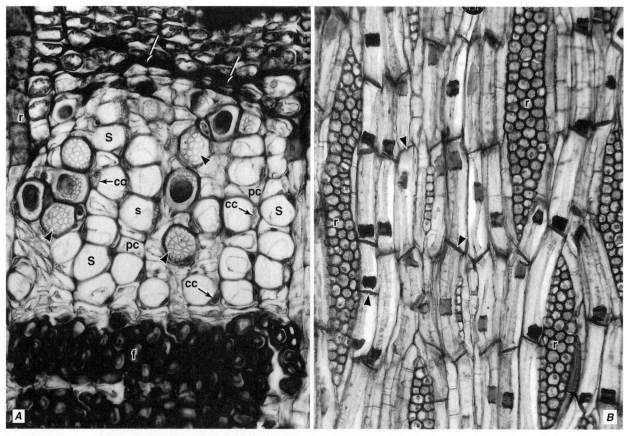

**FIGURE 14.8**

Secondary phloem of black locust (*Robinia pseudoacacia*). **A**, transverse section showing part of a band of functional sieve tubes. Many of the sieve-tube elements (s) have been sectioned near the plane of their end walls, revealing their simple sieve plates (darts). The sieve-tube elements of the previous year's phloem increment (above) have been crushed and obliterated (arrows); below is a band of fibers (f). **B**, tangential section, revealing the storied nature of the phloem. Darts point to the sieve plates. The dark bodies in the sieve-tube elements are nondispersive P-protein bodies. Other details: cc, companion cell; pc, parenchyma cell; r, ray. (A, ×370; B, ×180. A, from Evert, 1990b, Fig. 6.1. © 1990, Springer-Verlag; **B**, courtesy William F. Derr.)

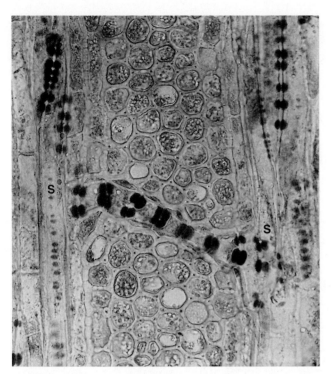

**FIGURE 14.9**

Tangential section of the secondary phloem of *Vitis vinifera*, showing a strand of short sieve-tube elements in a ray. Massive accumulations of callose (appears dark here) mark the sieve plates. Detail: s, sieve-tube elements of axial system. (×290. From Esau, 1948. *Hilgardia* 18 (5), 217–296. © 1948 Regents, University of California.)

perforated ray cells (vessel elements) encountered in the xylem rays of some species (Chapter 11). Both kinds of element may serve to interconnect their respective conducting elements on either side of a ray. Radially arranged sieve-tube elements also occur within some phloem rays (Rajput and Rao, 1997; Rajput, 2004). Whether the sieve-tube elements have their origin from ray cells per se or from potential ray cells (Cheadle and Esau, 1958) is problematic. Similar cells were encountered in the phloem rays of *Malus domestica* and, through the examination of serial tangential sections, were found to be derived from declining fusiform initials, some of which eventually were converted to ray initials that then gave rise only to ray parenchyma cells (Evert, 1963). Perhaps the same is true for perforated ray cells.

Herbaceous eudicots possessing secondary growth (*Nicotiana*, *Gossypium*) may have secondary phloem resembling that of woody species. Some herbaceous species, like the vine *Cucurbita*, have a secondary phloem scarcely distinguishable from the primary except in having larger cells. *Cucurbita* has external and internal phloem (Fig. 13.1), and commonly only the

external phloem is augmented by secondary growth. The secondary phloem consists of wide sieve tubes, narrow companion cells, and phloem parenchyma cells of intermediate size. There are neither fibers nor rays. The sieve plates are simple with large pores. The lateral walls bear sieve areas that resemble primary pit-fields (Fig. 13.9C). They are much less specialized than the single sieve areas of the simple sieve plates. In transverse sections the small companion cells often appear as though cut out of the sides of the sieve tubes. Longitudinally, the companion cells usually extend from sieve plate to sieve plate, sometimes as a single cell, other times as a strand of two or more cells.

Secondary phloem of relatively simple structure is found in eudicot storage organs, such as those of the carrot, the dandelion, and the beet. Storage parenchyma predominates in this kind of phloem, and the sieve tubes and companion cells appear as strands anastomosing within the parenchyma.

# ▌DIFFERENTIATION IN THE SECONDARY PHLOEM

The derivatives of the vascular cambium commonly undergo some divisions before the various phloem elements begin to differentiate. As mentioned in Chapter 12, evidence indicates that most cambial derivatives divide periclinally at least once, giving rise to pairs of cells on the phloem side. In *Thuja occidentalis*, which produces a regular sequence of alternating cell types in the axial system, the derivative of a cambial initial commonly divides periclinally once to produce two cells of which the outer generally becomes a sieve cell and the inner either a parenchyma strand or a fiber, according to the sequence in the pattern (Fig. 14.10; Bannan, 1955). Occasionally a cambial derivative differentiates into a phloem cell without dividing periclinally. In woody angiosperms the divisions preceding sieve-tube element differentiation are varied in number and orientation. At least the divisions (one or more) forming the companion cell occur. In *Malus domestica* the cambial derivatives divide periclinally at least once. These divisions may yield pairs of sieve-tube elements (with their companion cells), pairs of parenchyma strands, or a pair consisting of a sieve-tube element (with companion cell) and a parenchyma strand (Fig. 14.11; Evert, 1963). In some species the divisions of the cambial derivatives are mainly anticlinal, including transverse, oblique, and/or longitudinal, resulting in assemblages containing various combinations of sieve-tube elements (with companion cells) and parenchyma cells, or of only sieve-tube elements and companion cells (Esau and Cheadle, 1955; Cheadle and Esau, 1958, 1964; Esau, 1969). Among the anticlinal divisions are some that result in an ontogenetic reduction of the potential length of the sieve-tube elements.

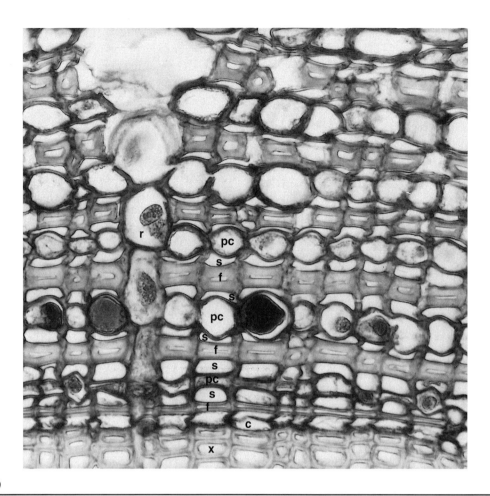

**FIGURE 14.10**

Transverse section of secondary phloem from stem of *Thuja occidentalis*. The axial system consists of a regular alternating sequence of cell types: fiber (f), sieve cell (s), parenchyma cell (pc), sieve cell (s), fiber (f), and so on. The final periclinal division of a cambial derivative commonly gives rise to either a sieve cell and a fiber or a sieve cell and the precursor of a parenchyma strand. The sieve cell is the outer cell in both instances. Other details: c, camlium; r, ray; x, xylem. (×600. From Esau, 1969. www.schweizerbart.de)

The differentiation of a phloem cell as a specific element of the phloem tissue begins after the various cell divisions are completed and the cell has been programmed for differentiation. The fusiform cells that give rise to axial parenchyma cells commonly subdivide by transverse or oblique divisions (parenchyma strand formation), or they differentiate directly into long, fusiform parenchyma cells. Typically the various phloem cells increase in size but the parenchyma cells and sieve elements undergo mainly lateral expansion, whereas the fibers may elongate by apical intrusive growth. In both gymnosperms and angiosperms the sieve elements undergo little or no increase in length, so at maturity the sieve elements are approximately as long as the cambial initials. Thus in angiosperms the sieve-tube elements also correspond in length to the vessel elements in the secondary xylem. Ray cells commonly change little during differentiation, except that they expand somewhat.

### Sclerenchyma Cells in the Secondary Phloem Commonly Are Classified as Fibers, Sclereids, and Fiber-Sclereids

There has been considerable deliberation over the classification of sclerenchyma cells in the secondary phloem because there are no clear-cut criteria to categorize them either on the basis of their timing of maturation or their morphological characteristics (Esau, 1969). Although there are many intergrading types, the sclerenchyma cells of the secondary phloem commonly are classified as fibers, sclereids, and fiber-sclereids as follows:

**Fibers** are narrow, elongated sclerenchyma cells that develop close to the cambium and reach maturity in the conducting phloem (Fig. 14.12). They may or may not undergo intrusive growth, may or may not have lignified walls, and may or may not retain their protoplasts at maturity. Although

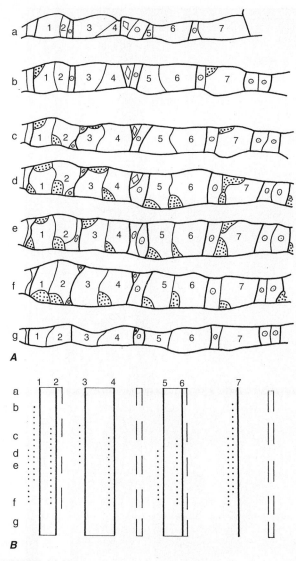

**FIGURE 14.11**

Secondary phloem of *Malus domestica*. Analysis of a radial file of cambial derivatives. **A**, drawings a–g illustrate the radial file in transverse sections taken at levels indicated by the positions of a–g in **B**. In **A**, parenchyma cells are with nuclei, sieve elements numbered, and companion cells stippled. In **B**, the numbered solid lines represent sieve elements, the dotted lines companion cells, and the frequently broken lines parenchyma cells. Assemblages are indicated, in part, by horizontal lines that connect related cells. (×393. From Evert, 1963.)

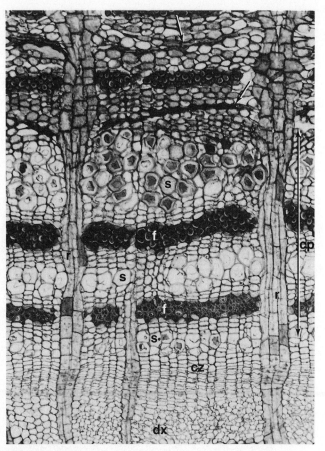

**FIGURE 14.12**

Transverse section of secondary phloem of black locust (*Robinia pseudoacacia*) showing mostly conducting phloem (cp). Cambial activity has resulted thus far in the production of three new bands of sieve tubes (s) and two of fibers (f). Sieve tubes of the nonconducting phloem (indicated by arrows) have collapsed. Other details: cz, cambial zone; dx, differentiating xylem; r, ray. (×150.)

crystal-containing cells may accompany sclereids or fiber-sclereids, chambered crystalliferous cells typically are found along the margins of fiber bands.

**Sclereids** develop mainly in the nonconducting phloem by modification of already differentiated axial or ray parenchyma cells. Some, however, are early individualized as sclereid primordia near the cambium and mature in the conducting phloem. The typical sclereid is shorter than a fiber, has a larger cell lumen, and a thick, often multilayered cell wall traversed by conspicuous simple pits that branch (ramiform pits). They range from unbranched brachysclereids (stone cells) to irregular twisted forms such as those found in *Abies* (Fig. 14.13) and *Eucalyptus* bark (Chattaway, 1953, 1955a).

**Fiber-sclereids** originate from axial parenchyma cells in the nonconducting phloem. These cells undergo intrusive growth, so at maturity they may be indistinguishable from true fibers.

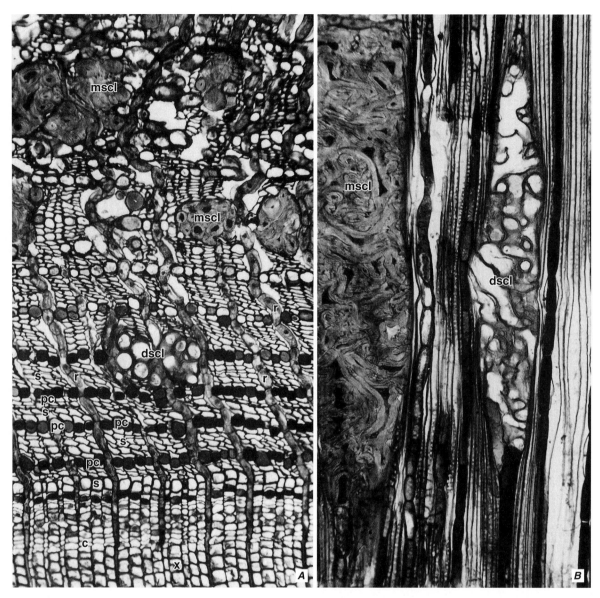

**FIGURE 14.13**

Secondary phloem of *Abies sachalinensis* var. *mayriana* in transverse (**A**) and radial longitudinal (**B**) sections of stem. One-cell-wide tangential layers of axial parenchyma cells (pc) alternate with layers of sieve cells (s). In non-conducting phloem, parenchyma cells give rise to irregular twisted sclereids. Other details: c, cambium; dscl, differentiating sclereids; mscl, mature sclereids; r, rays; x, xylem. (Both, ×92. From Esau, 1969. www.schweizerbart.de)

## The Conducting Phloem Constitutes Only a Small Part of the Inner Bark

The phloem is considered to be differentiated into a conducting tissue involved with long-distance transport when the sieve elements become enucleate and develop the other specialized characteristics, including open sieve-area pores, of mature sieve elements. The width of the yearly increment of the **conducting phloem** produced in one season varies with the species and growth conditions and, as discussed in Chapter 12, is generally considerably narrower than the corresponding increment of xylem. Moreover, in many temperate-zone deciduous species of woody angiosperms a given increment of phloem functions for only a single season. During winter in such species there are no living mature sieve elements and, hence, no conducting phloem (see Table 12.3). In these species the first functional sieve elements in the spring arise from phloem mother cells that overwintered on the outer margin of the cambial zone (Fig. 14.14).

In other species of woody angiosperms—both deciduous and evergreen of temperate and tropical zones—and in conifers (see Tables 12.3, 12.4, and 12.5) at least some functional sieve elements are present in the phloem

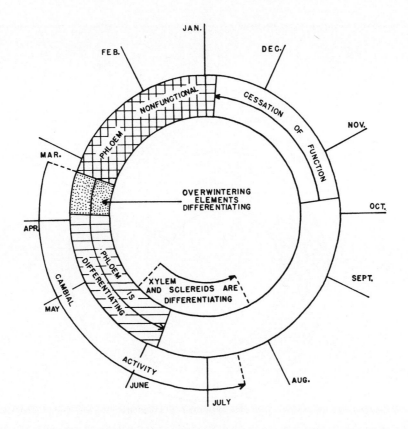

**FIGURE 14.14**

Diagram interpreting seasonal growth of secondary phloem in *Pyrus communis* (pear) stem. (Reproduced by permission University of California Press: Evert, 1960. *Univ. Calif. Publ. Bot.* © 1960, The Regents of the University of California.)

year-round. The details vary. In most conifers all but the last-formed sieve cells cease to function during the same season in which they are derived from the vascular cambium. However, the former overwinter and remain functional until new sieve cells differentiate in the spring (Alfieri and Evert, 1968, 1973). In *Juniperus californica*, all sieve cells of the previous year's phloem increment overwinter in a mature, functional state (Alfieri and Kemp, 1983). A pattern similar to that found in most conifers is exhibited by the temperate-zone ring-porous hardwoods *Quercus alba* (Anderson and Evert, 1965) and *Ulmus americana* (Tucker, 1968). Relatively large numbers of sieve elements may remain functional for two or more seasons in certain species. In some temperate species dormancy callose develops in the fall on the sieve areas of sieve elements that will overwinter in a dormant state. When the sieve elements are reactivated in the spring, the dormancy callose is removed. This pattern occurs, for example, in *Tilia* phloem, whose sieve elements may function for as many as 5 years in *T. americana* (Evert, 1962) and 10 years in *T. cordata* (Holdheide, 1951), in *Carya ovata*, with sieve elements that function for 2 to 6 years (Davis, 1993b), and in *Vitis* (Esau, 1948; Davis and Evert, 1970) and in the biennial

canes of *Rubus allegheniensis* (Davis, 1993a), with sieve elements that function for 2 years (Fig. 14.15). A similar pattern of dormancy and reactivation apparently occurs in the phloem of *Grewia tiliaefolia*, a tropical deciduous tree from India (Deshpande and Rajendrababu, 1985). Auxin has been implicated in the removal of dormancy callose from the secondary sieve tubes of *Magnolia kobus* (Aloni and Peterson, 1997).

Because of the relatively narrow width of the yearly increment of phloem and its usually short functioning life, the layer of conducting phloem occupies only a small proportion of the bark. Huber (1939) found the width in millimeters of the conducting phloem to be 0.23 to 0.325 for *Larix* and 0.14 to 0.27 for *Picea*. Some examples of the width in millimeters of the conducting phloem in deciduous species are 0.2 for *Fraxinus americana* and 0.35 for *Tectona grandis* (Zimmermann, 1961); 0.2 to 0.3 for *Quercus*, *Fagus*, *Acer*, and *Betula*; 0.4 to 0.7 for *Ulmus* and *Juglans*; and 0.8 to 1.0 for *Salix* and *Populus* (Holdheide, 1951). All these examples are of temperate-zone trees except for that of *Tectona grandis* (teak), whose bark contains a modest amount of conducting phloem. Teak seemingly stands in sharp contrast with the Dipterocarpaceae, which were

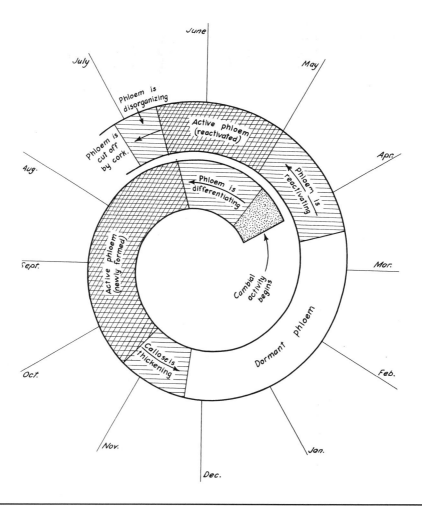

**FIGURE 14.15**

Diagram interpreting seasonal growth of secondary phloem in *Vitis vinifera* (grapevine) stem. (From Esau, 1948. *Hilgardia* 18 (5), 217–296. © 1948 Regents, University of California.)

reported to have 5 to 6 mm-wide bands of conducting phloem (Whitmore, 1962). The latter report has served to fuel the belief that the bark of tropical trees typically contains substantially wider bands of conducting phloem than their temperate-zone counterparts. The accuracy of Whitmore's (1962) measurements has been questioned, however (Esau, 1969). Roth (1981) has noted that the conducting phloem occupies only a very small part of the inner bark width of trees of Venezuelan Guayana. The same is true of the inner bark of *Citharexylum myrianthum* and *Cedrela fissilis* of Brazil (Veronica Angyallossy, personal communication). Sieve tubes occupy from 25% to 50% of the conducting phloem in woody angiosperms.

## NONCONDUCTING PHLOEM

The part of the phloem in which the sieve elements have ceased to function may be referred to as **nonconducting phloem**. This term is preferable to the ambigu-

ous term nonfunctioning phloem because the part of the inner bark in which the sieve elements are dead and no longer conducting commonly retains living axial and ray parenchyma cells. These cells continue to store starch, tannins, and other substances until the tissue is separated from the living part of the bark by phellogen activity.

There may be a lag between the time the sieve elements stop conducting and their actual death, but the various signs of the inactive state of the sieve elements are readily detected. The sieve areas are either covered with a mass of callose (definitive callose) or entirely free of this substance; callose eventually disappears in old inactive sieve elements. The contents of the sieve elements may be completely disorganized, or they may be absent and the cells filled with gas. The companion cells and some parenchyma cells of angiosperms and the Strasburger cells of conifers cease to function when their associated sieve elements die. The determination of the nonconducting state of the phloem is particularly certain if the sieve elements are more or less collapsed or obliterated. In some species, there is no obvious boundary between collapsed and noncollapsed phloem (e.g., in *Euonymus bungeamus*, Lin and Gao, 1993; arborescent Leguminosae, Costa et al., 1997; *Eucalyptus globulus*, Quilhó et al., 1999). In others, the sieve elements remain intact for several years after they die and may not collapse until after they have been separated from the inner bark by phellogen activity. Therefore the suggestion that the terms collapsed and noncollapsed phloem be used in place of conducting and nonconducting phloem (Trockenbrodt, 1990) has not been adopted in this book.

### The Nonconducting Phloem Differs Structurally from the Conducting Phloem

The phenomena that bring about the structural differences between the conducting and nonconducting phloem may be placed into four categories: (1) the breakdown of sieve elements and some of the cells associated with them; (2) the dilatation growth resulting from enlargement and division of parenchyma cells, axial or ray cells or both (cell enlargement alone may occur); (3) the sclerification, that is, development of secondary walls in parenchyma cells; and (4) the accumulation of crystals (Esau, 1969). The characteristics of the nonconducting phloem as a whole are varied in different plants, reflecting the manner and degree to which each of the four phenomena is expressed. In certain angiosperms, like *Liriodendron, Tilia, Populus*, and *Juglans*, the shape of the functionless sieve tubes changes little, although their companion cells collapse. In others, like *Aristolochia* and *Robinia* the sieve-tube elements and associated cells collapse completely, and since they occur in tangential bands, the crushed cells alternate more or less regularly with tangential bands of turgid parenchyma cells (Fig. 14.12). In still others the collapse of the sieve elements is accompanied by a conspicuous shrinkage of the tissue and bending of rays. In conifers the collapse of the old sieve cells is very marked. The nonconducting phloem of *Pinus* shows dense masses of collapsed sieve cells interspersed with intact phloem parenchyma cells (Fig. 14.3), and the rays are bent and folded. In conifers having fibers in the phloem, the sieve cells are crushed between the fibers and the enlarging phloem parenchyma cells (Fig. 14.5B). In *Vitis vinifera* the inactive sieve tubes become filled with tyloses-like proliferations (tylosoids) from the axial parenchyma cells (Esau, 1948).

### Dilatation Is the Means by Which the Phloem Is Adjusted to the Increase in Circumference of the Axis Resulting from Secondary Growth

Both axial and ray parenchyma may participate in dilatation growth but uniform participation of both is rare (Holdheide, 1951). Sometimes the ray cells merely

extend tangentially, but more commonly the number of cells is increased in the tangential direction by radial divisions. These divisions may be restricted to the median part of a ray, giving the impression of a localized (dilatation) meristem (Fig. 14.16; Holdheide, 1951; Schneider, 1955). Usually only some rays become dilated; others remain as wide as they were at the time of origin in the cambium.

The dilatation growth of axial parenchyma is frequently a less conspicuous phenomenon than that of rays, but it is very common (Esau, 1969). In the conifers, the enlargement and proliferation of axial parenchyma cells is the main form of dilatation (Liese and Matte,

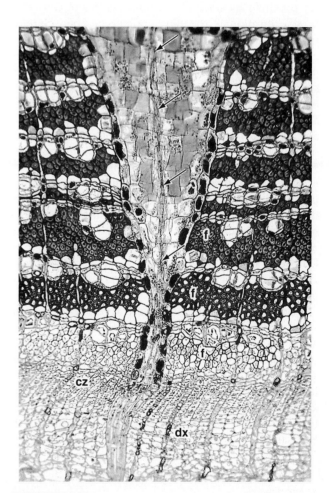

**FIGURE 14.16**

Transverse section of *Tilia americana* (basswood) stem undergoing secondary growth. In the current year's phloem three new tangential bands of phloem fibers (f) have been produced; the two nearest the wide cambial zone (cz) are still differentiating. A band of differentiating sieve-tube elements, with P-protein bodies, can be seen bordering (to the outside) the newest band of fibers. A dilatation meristem (arrows) extends the length of the dilated ray. Other detail: dx, differentiating xylem. (×125.)

1962) and may continue for many years. Some enlargement of axial parenchyma cells commonly occurs in connection with the collapse of inactive sieve elements. Axial parenchyma cells may also proliferate to the extent of forming wedges of tissue resembling dilated rays as in *Eucalyptus* (Chattaway, 1955b) and the Dipterocarpaceae (Whitmore, 1962). Enlargement of axial parenchyma may continue for some time after the phloem has been cut off by a periderm (Chattaway, 1955b; Esau, 1964). The dilatation of phloem generally is interrupted when a phellogen arises in the phloem and separates it from the inner bark.

Sclerification of the nonconducting phloem is almost always associated with dilatation growth. Both axial and ray parenchyma cells may become sclerified. One kind of sclerification is the development of fiber-sclereids from axial parenchyma cells that undergo intrusive growth and deposit secondary walls in the innermost part of the nonconducting phloem. Some examples of species with fiber-sclereids, as given by Holdheide (1951), are *Ulmus scabra, Pyrus communis, Malus domestica, Sorbus aucuparia, Prunus padus, Fraxinus excelsior,* and *Fagus sylvatica*. Sclereid development is common in both the rays and the dilatation tissue in the axial system. In most cases sclerification is preceded by cell enlargement. Intrusive growth also may occur, resulting in bent or undulated walls. Sclerification of the nonconducting phloem may continue indefinitely, producing masses of sclereids. Sclerenchyma in *Fagus* may constitute 60% of the tissue (Holdheide, 1951). In certain tropical barks, such as the *Licania* type, sclerenchyma may cover over 90% of the cross-sectional area in the nonconducting phloem (Roth, 1973).

The nonconducting phloem also accumulates various substances such as crystals and phenolic substances. Although crystals occur in the conducting phloem, they usually accumulate in cells bordering those undergoing sclerification, and therefore the crystals may be particularly conspicuous in nonconducting phloem. The types and distribution of crystals are sufficiently characteristic to be useful in comparative studies (Holdheide, 1951; Patel and Shand, 1985; Archer and van Wyk, 1993).

# ▮ REFERENCES

ALFIERI, F. J., and R. F. EVERT. 1968. Seasonal development of the secondary phloem in *Pinus*. Am. J. Bot. 55, 518–528.

ALFIERI, F. J., and R. F. EVERT. 1973. Structure and seasonal development of the secondary phloem in the Pinaceae. Bot. Gaz. 134, 17–25.

ALFIERI, F. J., and R. I. KEMP. 1983. The seasonal cycle of phloem development in *Juniperus californica*. Am. J. Bot. 70, 891–896.

ALONI, R., and C. A. PETERSON. 1997. Auxin promotes dormancy callose removal from the phloem of *Magnolia kobus* and callose

accumulation and earlywood vessel differentiation in *Quercus robur*. J. Plant Res. 110, 37–44.

ANDERSON, B. J., and R. F. EVERT. 1965. Some aspects of phloem development in *Quercus alba*. Am. J. Bot. 52 (Abstr.), 627.

ARCHER, R. H., and A. E. VAN WYK. 1993. Bark structure and intergeneric relationships of some southern African Cassinoideae (Celastraceae). IAWA J. 14, 35–53.

BANNAN, M. W. 1955. The vascular cambium and radial growth in *Thuja occidentalis* L. Can. J. Bot. 33, 113–138.

BEHNKE, H.-D. 1986. Sieve element characters and the systematic position of *Austrobaileya, Austrobaileyaceae*—With comments to the distribution and definition of sieve cells and sieve-tube members. Plant Syst. Evol. 152, 101–121.

BEHNKE, H.-D., and U. KIRITSIS. 1983. Ultrastructure and differentiation of sieve elements in primitive angiosperms. I. Winteraceae. Protoplasma 118, 148–156.

CHAN, L.-L. 1986. The anatomy of the bark of *Agathis* in New Zealand. IAWA Bull. n.s. 7, 229–241.

CHANG, Y.-P. 1954. Anatomy of common North American pulpwood barks. TAPPI Monograph Ser. No. 14. Technical Association of the Pulp and Paper Industry, New York.

CHATTAWAY, M. M. 1953. The anatomy of bark. I. The genus *Eucalyptus*. Aust. J. Bot. 1, 402–433.

CHATTAWAY, M. M. 1955a. The anatomy of bark. III. Enlarged fibres in the bloodwoods (*Eucalyptus* spp.) Aust. J. Bot. 3, 28–38.

CHATTAWAY, M. M. 1955b. The anatomy of bark. VI. Peppermints, boxes, ironbarks, and other eucalypts with cracked and furrowed barks. Aust. J. Bot. 3, 170–176.

CHAVAN, R. R., J. J. SHAH, and K. R. PATEL. 1983. Isolated sieve tube(s)/elements in the barks of some angiosperms. IAWA Bull. n.s. 4, 255–263.

CHEADLE, V. I., and K. ESAU. 1958. Secondary phloem of the Calycanthaceae. Univ. Calif. Publ. Bot. 24, 397–510.

CHEADLE, V. I., and K. ESAU. 1964. Secondary phloem of *Liriodendron tulipifera*. Univ. Calif. Publ. Bot. 36, 143–252.

COSTA, C. G., V. T. RAUBER CORADIN, C. M. CZARNESKI, and B. A. DA S. PEREIRA. 1997. Bark anatomy of arborescent Leguminosae of cerrado and gallery forest of Central Brazil. IAWA J. 18, 385–399.

DAVIS, J. D. 1993a. Secondary phloem development cycle in biennial canes of *Rubus allegheniensis*. Am. J. Bot. 80 (Abstr.), 22.

DAVIS, J. D. 1993b. Seasonal secondary phloem development in *Carya ovata*. Am. J. Bot. 80 (Abstr.), 23.

DAVIS, J. D., and R. F. EVERT. 1970. Seasonal cycle of phloem development in woody vines. Bot. Gaz. 131, 128–138.

DEN OUTER, R. W. 1967. Histological investigations of the secondary phloem of gymnosperms. Meded. Landbouwhogesch. Wageningen 67-7, 1–119.

DEN OUTER, R. W. 1986. Storied structure of the secondary phloem. IAWA Bull. n.s. 7, 47–51.

DEN OUTER, R. W. 1993. Evolutionary trends in secondary phloem anatomy of trees, shrubs and climbers from Africa (mainly Ivory Coast). *Acta Bot. Neerl.* 42, 269–287.

DERR, W. F., and R. F. EVERT. 1967. The cambium and seasonal development of the phloem in *Robinia pseudoacacia. Am. J. Bot.* 54, 147–153.

DESHPANDE, B. P., and T. RAJENDRABABU. 1985. Seasonal changes in the structure of the secondary phloem of *Grewia tiliaefolia,* a deciduous tree from India. *Ann. Bot.* 56, 61–71.

ESAU, K. 1948. Phloem structure in the grapevine, and its seasonal changes. *Hilgardia* 18, 217–296.

ESAU, K. 1964. Structure and development of the bark in dicotyledons. In: *The Formation of Wood in Torest Trees,* pp. 37–50, M. H. Zimmermann, ed. Academic Press, New York.

ESAU, K. 1969. *The Phloem. Encyclopedia of Plant Anatomy. Histology,* Band. 5, Teil 2. Gebrüder Borntraeger, Berlin.

ESAU, K. 1977. *Anatomy of Seed Plants,* 2nd ed. Wiley, New York.

ESAU, K. 1979. Phloem. In: *Anatomy of the Dicotyledons,* 2nd ed., vol. I. *Systematic Anatomy of Leaf and Stem, with a Brief History of the Subject,* pp. 181–189, C. R. Metcalfe and L. Chalk, eds. Clarendon Press, Oxford.

ESAU, K., and V. I. CHEADLE. 1955. Significance of cell divisions in differentiating secondary phloem. *Acta Bot. Neerl.* 4, 348–357.

EVERT, R. F. 1960. Phloem structure in *Pyrus communis* L. and its seasonal changes. *Univ. Calif. Publ. Bot.* 32, 127–196.

EVERT, R. F. 1962. Some aspects of phloem development in *Tilia americana. Am. J. Bot.* 49 (Abstr.), 659.

EVERT, R. F. 1963. Ontogeny and structure of the secondary phloem in *Pyrus malus. Am. J. Bot.* 50, 8–37.

EVERT, R. F. 1990. Dicotyledons. In: *Sieve Elements. Comparative Structure, Induction and Development,* pp. 103–137, H.-D. Behnke and R. D. Sjolund, eds. Springer-Verlag, Berlin.

FRANCESCHI, V. R., T. KREKLING, A. A. BERRYMAN, and E. CHRISTIANSEN. 1998. Specialized phloem parenchyma cells in Norway spruce (Pinaceae) bark are an important site of defense reactions. *Am. J. Bot.* 85, 601–615.

HOLDHEIDE, W. 1951. Anatomie mitteleuropäischer Gehölzrinden (mit mikrophotographischem Atlas). In: *Handbuch der Mikroskopie in der Technik,* Band 5, Heft 1, pp. 193–367, H. Freund, ed. Umschau Verlag, Frankfurt am Main.

HUBER, B. 1939. Das Siebröhrensystem unserer Bäume und seine jahrezeitlichen Veränderungen. *Jahrb. Wiss. Bot.* 88, 176–242.

HUBER, B. 1949. Zur Phylogenie des Jahrringbaues der Rinde. *Svensk Bot. Tidskr.* 43, 376–382.

HUDGINS, J. W., T. KREKLING, and V. R. FRANCESCHI. 2003. Distribution of calcium oxalate crystals in the secondary phloem of conifers: a constitutive defense mechanism? *New Phytol.* 159, 677–690.

KURODA, K. 1998. Seasonal variation in traumatic resin canal formation in *Chamaecyparis obtusa* phloem. *IAWA J.* 19, 181–189.

LIESE, W., and V. MATTE. 1962. Beitrag zur Rindenanatomie der Gattung *Dacrydium. Forstwiss. Centralbl.* 81, 268–280.

LIN, J.-A., and X.-Z. GAO. 1993. Anatomical studies on secondary phloem of *Euonymus bungeanus. Acta Bot. Sin. (Chih wu hsüeh pao)* 35, 506–512.

PATEL, R. N., and J. E. SHAND. 1985. Bark anatomy of *Nothofagus* species indigenous to New Zealand. *N. Z. J. Bot.* 23, 511–532.

QUILHÓ, T., H. PEREIRA, and H. G. RICHTER. 1999. Variability of bark structure in plantation-grown *Eucalyptus globulus. IAWA J.* 20, 171–180.

RAJPUT, K. S. 2004. Occurrence of radial sieve elements in the secondary phloem rays of some tropical species. *Isr. J. Plant Sci.* 52, 109–114.

RAJPUT, K. S., and K. S. RAO. 1997. Occurrence of sieve elements in phloem rays. *IAWA J.* 18, 197–201.

ROTH, I. 1973. Estructura anatómica de la corteza de algunas especies arbóreas Venezolanas de *Rosaceae. Acta Bot. Venez.* 8, 121–161.

ROTH, I. 1981. Structural patterns of tropical barks. In: *Encyclopedia of Plant Anatomy,* Band. 9, Teil 3. Gebrüder Borntraeger, Berlin.

SCHNEIDER, H. 1955. Ontogeny of lemon tree bark. *Am. J. Bot.* 42, 893–905.

SRIVASTAVA, L. M. 1963. Secondary phloem in the Pinaceae. *Univ. Calif. Publ. Bot.* 36, 1–142.

TROCKENBRODT, M. 1990. Survey and discussion of the terminology used in bark anatomy. *IAWA Bull.* n.s. 11, 141–166.

TUCKER, C. M. 1968. Seasonal phloem development in *Ulmus americana. Am. J. Bot.* 55 (Abstr.), 716.

WHITMORE, T. C. 1962. Studies in systematic bark morphology. I. Bark morphology in Dipterocarpaceae. *New Phytol.* 61, 191–207.

YAMANAKA, K. 1984. Normal and traumatic resin-canals in the secondary phloem of conifers. *Mokuzai gakkai shi (J. Jpn. Wood Res. Soc.)* 30, 347–353.

YAMANAKA, K. 1989. Formation of traumatic phloem resin canals in *Chamaecyparis obtusa. IAWA Bull.* n.s. 10, 384–394.

ZAHUR, M. S. 1959. Comparative study of secondary phloem of 423 species of woody dicotyledons belonging to 85 families. Cornell Univ. Agric. Expt. Stan. Mem. 358. New York State College of Agriculture, Ithaca.

ZIEGLER, H. 1964. Storage, mobilization and distribution of reserve material in trees. In: *The Formation of Wood in Forest Trees,* pp. 303–320, M. H. Zimmermann, ed. Academic Press, New York.

ZIMMERMANN, M. H. 1961. Movement of organic substances in trees. *Science* 133, 73–79.

# CHAPTER FIFTEEN

# Periderm

**Periderm** is a protective tissue of secondary origin. It replaces the epidermis in stems and roots that increase in thickness by secondary growth. Structurally the periderm consists of three parts: the **phellogen**, or **cork cambium**, the meristem that produces the periderm; the **phellem**, commonly called **cork**, produced by the phellogen toward the outside; and the **phelloderm**, a tissue often resembling cortical or phloem parenchyma and consisting of the inner derivatives of the phellogen.

The term periderm should be distinguished from the nontechnical term bark (Chapter 14). Although the word bark is employed loosely, and often inconsistently, it is a useful term if properly defined. **Bark** may be used most appropriately to designate all tissues outside the vascular cambium. In the secondary state, bark includes the secondary phloem, the primary tissues that may still be present outside the secondary phloem, the periderm, and the dead tissues outside the periderm. As the periderm develops, it separates, by means of a nonliving layer of cork cells, variable amounts of primary and secondary tissues of the axis from the subjacent living tissues. The tissue layers thus separated die, bringing

about a distinction between the nonliving **outer bark** and living **inner bark** (Fig. 15.1). The technical term for outer bark is **rhytidome**. The conducting phloem is the innermost part of the living bark. As mentioned in Chapter 14, the term bark is sometimes used for stems and roots in the primary state of growth. It then includes the primary phloem, the cortex, and the epidermis. However, because of the radially alternate arrangement of xylem and phloem in roots in the primary state, the primary phloem of a root cannot be conveniently included with the cortex under the term bark.

The structure and the development of periderm are better known in stems than in roots. Therefore most of the information on periderm presented in this chapter pertains to stems, unless roots are mentioned specifically.

## OCCURRENCE

Periderm formation is a common phenomenon in roots and stems of woody angiosperms and gymnosperms. Periderm occurs also in herbaceous eudicots, especially

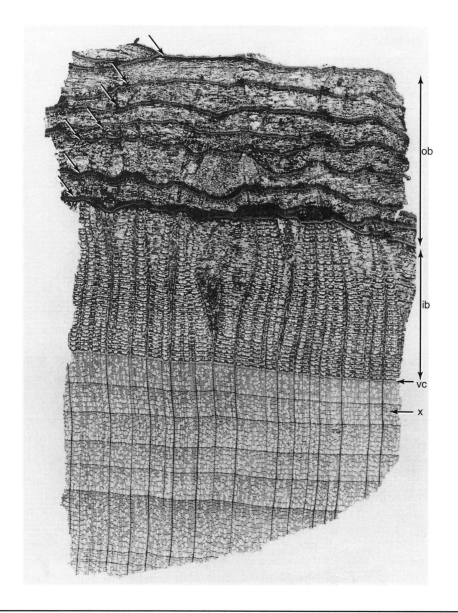

**FIGURE 15.1**

Transverse section of bark and some secondary xylem from an old stem of basswood (*Tilia americana*). Several periderms (arrows) can be seen traversing the outer bark (ob) in the upper third of the section. The periderms in basswood form overlapping layers as is characteristic of a scale bark. To the inside of the outer bark is the inner bark (ib), consisting largely of nonconducting phloem. The conducting phloem comprises a small layer of cells contiguous to the vascular cambium (vc). The inner bark is quite distinct from the more lightly stained xylem (x) in the lower third of the section. (×11.)

in the oldest parts of stem and root. Some monocots have periderm, and others a different kind of secondary protective tissue. Leaves normally produce no periderm, although scales of winter buds in some gymnosperms and woody angiosperms are an exception.

The formation of periderm in stems of woody plants may be considerably delayed, as compared with that of the secondary vascular tissues. It may never occur, even though the stem continues to increase in thickness. In such instances the tissues outside the vascular cambium, including the epidermis, keep pace with the increase in

axis circumference (species of *Acacia, Acer, Citrus, Eucalyptus, Ilex, Laurus, Menispermum, Viscum*). The individual cells divide radially and enlarge tangentially.

Periderm develops along surfaces that are exposed after abscission of plant parts, such as leaves and branches. Periderm formation is also an important stage in the development of protective layers near injured or dead (necrosed) tissues (wound periderm or wound cork), whether resulting from mechanical wounding (Tucker, 1975; Thomson et al., 1995; Oven et al., 1999) or invasion of parasites (Achor et al., 1997; Dzerefos and

Witkowski, 1997; Geibel, 1998). In several families of eudicots, periderm is formed in the xylem—interxylary cork—in relation to a normal dying back of annual shoots or a splitting of perennating roots and stems (Moss and Gorham, 1953; Ginsburg, 1963). The longitudinal splitting of the strap-shaped leaves of *Welwitschia mirabilis* occurs in areas of mesophyll breakdown and periderm formation (Salema, 1967). Periderms, in the shape of tubes, may form inside the bark under natural conditions or in response to wounding, isolating strands of phloem fibers (Evert, 1963; Aloni and Peterson, 1991; Lev-Yadun and Aloni, 1991). In apples and pears, russeting results from the replacement by periderm of the outer layers of the fruit over part or all of its surface (Gil et al., 1994).

## ▌CHARACTERISTICS OF THE COMPONENTS

### The Phellogen Is Relatively Simple in Structure

In contrast to the vascular cambium, the phellogen has only one form of cell. In transverse section the phellogen commonly appears as a continuous tangential layer (lateral meristem) of rectangular cells (Fig. 15.2A), each with its derivatives in a radial file extending outward through the cork cells and inward through the phelloderm cells. In longitudinal sections the phellogen cells are rectangular or polygonal in outline (Fig. 15.2B), sometimes rather irregular. It is sometimes difficult to distinguish phellogen cells from newly formed phelloderm cells (Wacowska, 1985).

### Several Kinds of Phellem Cells May Arise from the Phellogen

The phellem cells are often approximately prismatic in shape (Fig. 15.3A, B), although they may be rather irregular in the tangential plane (Fig. 15.3F). They may be elongated vertically (Fig. 15.3E, F), radially (Fig. 15.3B–E), or tangentially (Fig. 15.3A, narrow cells). They are usually compactly arranged; that is, the tissue lacks intercellular spaces. Notable exceptions are found in some trees of tropical humid forests (e.g., *Alseis labatioides* and *Coutarea hexandra*, Rubiaceae; *Parkia pendula*, Mimosaceae), in which intercellular spaces arise between radial files of cork cells, forming cork aerenchyma (Roth, 1981). Flooding may result in increased phellogen activity and the production of loosely arranged radial files of cork cells (Fig. 15.4; Angeles et al., 1986; Angeles, 1992). In flooded *Ulmus americana* stems the system of intercellular spaces of the cork was continuous with that of the cortex via intercellular spaces in the phellogen (Angeles et al., 1986). Cork cells are nonliving at maturity. They are then filled with air, fluid, or solid contents; some are colorless and others pigmented.

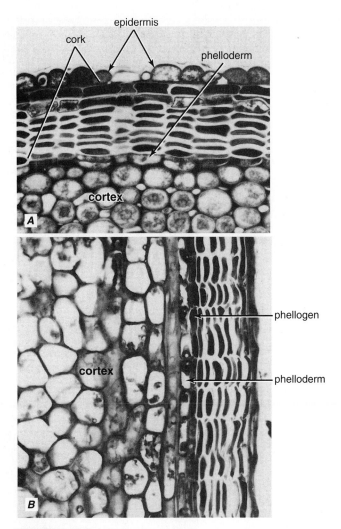

**FIGURE 15.2**

Periderm consisting largely of cork in transverse (**A**) and longitudinal (**B**) sections of dormant *Betula* twig. (Both, ×430.)

Cork cells typically have suberized cell walls. The suberin usually occurs as a distinct lamella that covers the inner surface of the original primary cellulose wall. The suberin lamella has a layered appearance under the electron microscope because it consists of alternating electron-dense layers and electron-lucent layers (Fig. 4.5; Thomson et al., 1995). Cork cells may have either thick or thin walls. In thick-walled cells a lignified cellulose layer occurs on the inner surface of the suberin lamella, which thus is embedded between two cellulose layers. Cork cells may have evenly or unevenly thickened walls. Some have U-shaped wall thickenings, with either the inner or outer tangential wall together with adjoining parts of the radial walls being thickened. In many *Pinus* species the thick-walled cells develop into heavily lignified stone cells (Fig. 15.5). The distinctly lamellate walls contain numerous ramiform (branched

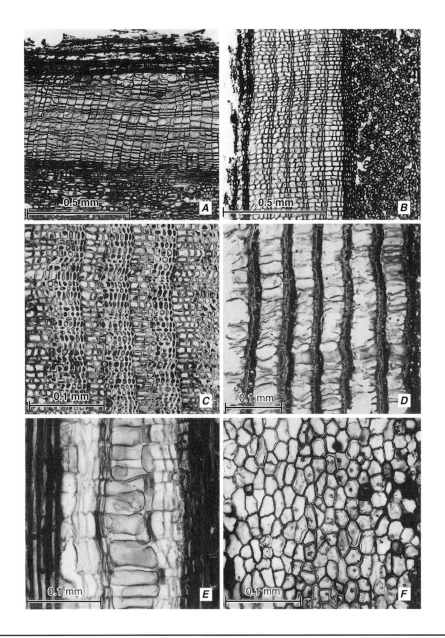

**FIGURE 15.3**

Variation in structure of phellem in stems. **A**, **B**, *Rhus typhina*. Phellem in transverse (**A**) and radial (**B**) sections of stem shows growth layers revealed in the alternation of narrower and wider cells. **C**, birch (*Betula populifolia*). Phellem with thick cell walls and conspicuous growth layers; radial section. **D**, *Rhododendron maximum*. Heterogeneous phellem consisting of cells of different sizes; sclereids compose some of the layers of small cells; radial section. **E**, **F**, *Vaccinium corymbosum*. Phellem in radial (**E**, light-colored cells in the middle) and tangential (**F**) sections. Phellem cells vary in form in **E**. (From Esau, 1977.)

simple) pits and possess many irregular projections along their margins. In tangential sections these sclereids resemble irregularly rounded, interlocked cogwheels (Howard, 1971; Patel, 1975). Cork cell walls may be brown or yellow, or they may remain colorless.

In many species the phellem consists of cork cells and of nonsuberized cells called **phelloids**, that is, phellem-like cells. Like the cork cells, these nonsuberized cells may have either thick or thin walls, and they may differentiate as sclereids (Fig. 15.3D). In the bark of

*Melaleuca* the phellogen gives rise to alternating layers of suberized and nonsuberized cells (Chiang and Wang, 1984). The suberized cells remain radially flattened but the nonsuberized cells elongate radially soon after they are produced by the phellogen. The suberized cells are characterized by the presence of Casparian strips in their anticlinal walls.

In some plants the phellem consists of thin-walled cells and thick-walled cells, often arranged in alternating tangential bands of one or more cell layers (species

**FIGURE 15.4**

Transverse section of *Ulmus americana* stem that had been flooded for 15 days. The activity of the phellogen increased in response to flooding, forming loosely arranged, filamentous-like strands of cork cells (arrows). Other details: ph, phloem; x, xylem. (×80. From Angeles et al., 1986.)

of *Eucalyptus* and *Eugenia*, Chattaway, 1953, 1959; species of *Pinus, Picea, Larix*, Srivastava, 1963; *Betula populifolia; Robinia pseudoacacia*, Waisel et al., 1967; some Cassinoideae of southern Africa, Archer and Van Wyk, 1993). There are many examples of layered cork among tropical trees (Roth, 1981). In some, the layers may simply be distinguished by their cell content. Phellem may consist entirely of thick-walled cells (*Ceratonia siliqua; Torrubia cuspidata, Diplotropis purpurea*, Roth, 1981) or only of thin-walled cells (species of *Abies, Cedrus*, Srivastava, 1963; *Pseudotsuga*, Srivastava, 1963; Krahmer and Wellons, 1973).

The layered appearance of phellem often makes it possible to distinguish growth increments in this tissue. In some species such as *Betula populifolia*, the growth increments are discernible because each consists of two distinct layers or bands of cells, one thick-walled, the other thin-walled (Fig. 15.3C). In other species consisting of only one type of cell, the increments may be discernible because of differences in the radial dimensions of the cells, as exhibited by the phellem of *Betula papy-*

*rifera* (Chang, 1954) and *Rhus typhina* (Fig. 15.3A, B). In *Pseudotsuga menziesii*, growth increments are discernible because of the presence of denser and darker appearing zones of cork cells at the end of the increments resulting from severe folding and crushing of their radial walls (Krahmer and Wellons, 1973; Patel, 1975). In *Picea glauca*, each growth increment, which consists of bands of thick- and thin-walled cells, ends with one or more layers of crystal-containing cells (Grozdits et al., 1982). It is questionable whether all such growth increments represent annual increments.

Cork used commercially as bottle cork comes from the cork oak, *Quercus suber*, which is native to the Mediterranean region. Consisting of thin-walled cells with air-filled lumina, it is highly impervious to water and gasses and resistant to oil. It is light in weight and has thermal insulating qualities. The first phellogen arises in the first year of growth in the cell layer immediately under the epidermis (Graça and Pereira, 2004). The first cork produced by the cork oak tree is of little commercial value. When the tree is about 20 years old, the original periderm is removed, and a new phellogen is formed in the cortex just a few millimeters below the site of the first one. The cork produced by the new phellogen accumulates very rapidly, and after about nine years is thick enough to be stripped from the tree (Costa et al., 2001). Once again a new phellogen arises beneath the previous one, and after about another nine years the cork can be stripped again. This procedure may be repeated at about nine-year intervals until the tree is 150 or more years old. After several strippings the new phellogens arise in the nonconducting phloem. Mature cork is a compressible, resilient tissue. The commercially valuable properties—imperviousness to water and insulating qualities—also make the cork effective as a protective layer on the plant surface. The dead tissue that becomes isolated by the periderm adds to the insulating effect of the cork.

### Considerable Variation Exists in the Width and Composition of Phelloderm

The phelloderm is commonly depicted as consisting of cells resembling cortical or phloem parenchyma cells and distinguishable from the latter only by their position in the same radial files as the phellem cells. In fact, cells similar in appearance to those of the phellem may be found in the phelloderm, although those of the phelloderm do not have suberized walls. Many conifers have phelloderms consisting of both parenchymatous and sclerenchymatous elements (Fig. 15.5). Sclerification of all or part of phelloderm is common in barks of tropical trees. The sclereids may have evenly thickened walls or U-shaped wall thickenings, and layers of thin-walled unlignified cells may alternate with layers of lignified sclerenchyma cells.

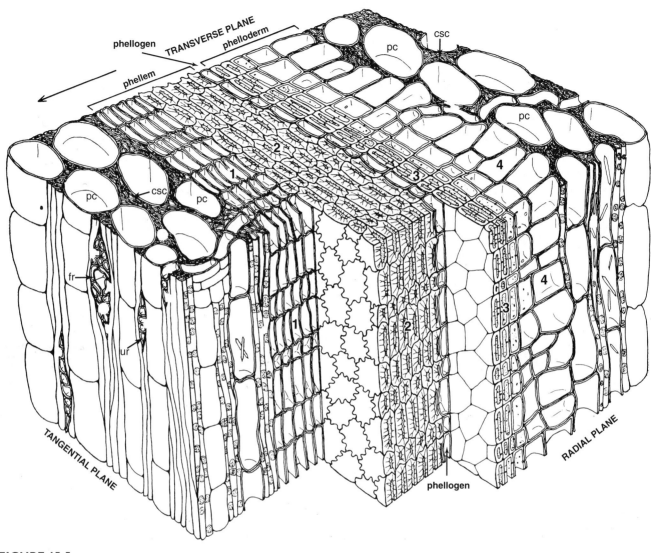

**FIGURE 15.5**

Block diagram of portion of the outer bark (rhytidome) tissues from the stem of a southern pine. The arrow points to the stem exterior. A single periderm, consisting of phellem, phellogen, and phelloderm, is shown here, on either side of which is nonconducting phloem consisting of crushed sieve cells (csc) and enlarged axial parenchyma cells (pc). The phellem consists of thin-walled cells (1) and thick-walled stone cells (2), which in tangential view resemble interlocked cogwheels. The phelloderm consists of unexpanded thick-walled cells (3) and of expanded thin-walled cells (4). Other details: fr, fusiform ray; ur, uniseriate ray. (From Howard, 1971.)

Some plants have no phelloderm at all. In others this tissue is one to three or more cells in depth (Fig. 15.6). The number of phelloderm cells in the same layer of periderm may change somewhat as the stem ages. In *Tilia*, for example, the phelloderm may be one cell deep in the first year, two in the second, and three or four later. The subsequent periderms formed beneath the first, in later years, contain as much phelloderm as the first or less. A relatively wide phelloderm was observed in stems and roots of certain Cucurbitaceae (Dittmer and Roser, 1963). In certain gymnosperms the phelloderm is very broad; in *Ginkgo* as many as 40 cell layers

could be counted. In the barks of many tropical trees, the periderms have very thin phellems and the phelloderm is the principal protective layer. In *Myrcia amazonia*, for example, the phellem is only one cell layer thick. Enormously thick phelloderms have been observed in some tropical trees. For example, in *Ficus* sp. the phelloderm occupied more than a third of the entire bark width, and in *Brosimum* sp. two-thirds of the entire width (Roth, 1981).

Unlike the typical compact arrangement of the phellem cells, the phelloderm cells have numerous intercellular spaces among them. In addition the phel-

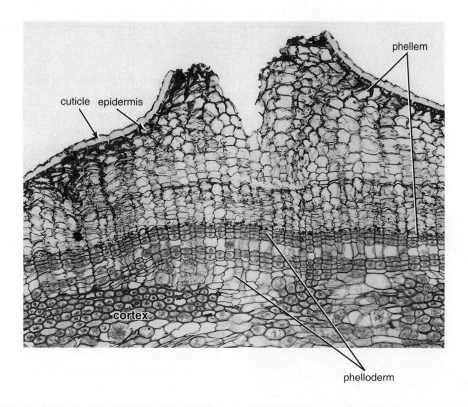

**FIGURE 15.6**

Transverse section through the stem of *Aristolochia* (Dutchman's pipe), showing the periderm, the phelloderm of which consists of several layers of cells. (×140.)

loderm cells—especially those of the first periderms—may contain numerous chloroplasts and be photosynthetically active. This apparently is a common feature in conifers (Godkin et al., 1983). Chloroplasts also have been found in the phelloderm of *Alstonia scholaris* (Santos, 1926), *Citrus limon* (Schneider, 1955), and *Populus tremuloides* (Pearson and Lawrence, 1958) bark. The parenchymatous elements of the phelloderm may perform a storage function, mainly of starch. The phelloderm may also give rise to new phellogen layers, as reported for that of the lemon tree (Schneider, 1955).

## DEVELOPMENT OF PERIDERM

### The Sites of Origin of the Phellogen Are Varied

With regard to the origin of the phellogen, it is necessary to distinguish between the first periderm and the subsequent periderms, which arise beneath the first and replace it as the axis continues to increase in circumference. In the stem the phellogen of the first periderm may be initiated at different depths outside the vascular cambium. In most stems the first phellogen arises in the subepidermal layer (Fig. 15.7A). In a few plants epidermal cells give rise to the phellogen (*Malus, Pyrus,*

*Nerium oleander, Myrsine australis, Viburnum lantana*). Sometimes the phellogen is formed partly from epidermis, partly from subepidermal cells. In some stems, the second or third cortical layer initiates the development of periderm (*Quercus suber, Robinia pseudoacacia, Gleditschia triancanthos* and other Fabaceae; species *of Aristolochia, Pinus,* and *Larix*). In still others, the phellogen arises near the vascular region or directly within the phloem (Fig 15.8; Caryophyllaceae, Cupressaceae, Ericaceae, Chenopodiaceae, *Berberis, Camellia, Puncia, Vitis*). If the first periderm is followed by others these are formed repeatedly—but rarely each season—in successively deeper layers of the cortex or phloem. As mentioned previously, cork development may occur within the xylem (interxylary cork, Moss and Gorham, 1953; Ginsburg, 1963).

The first phellogen is initiated either uniformly around the circumference of the axis or in localized areas and becomes continuous by a lateral spread of meristematic activity. When the initial activity is localized the first divisions are frequently those concerned with formation of lenticels (see below). From the margins of these structures the divisions spread around the stem circumference. In *Acer negundo* four to six years may be required before the phellogen forms a continuous ring around the stem (Wacowska, 1985). In some species

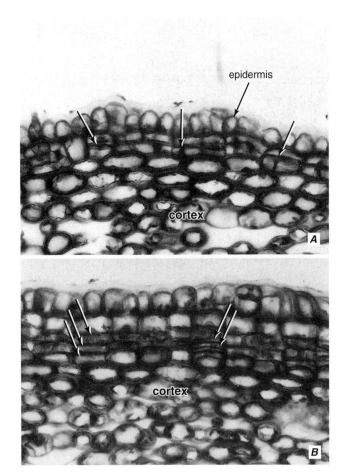

**FIGURE I5.7**

Transverse sections of *Prunus* stem showing earlier (**A**) and later (**B**) stages of development of periderm by periclinal divisions (arrows) in subepidermal layer. (Both, ×430.)

a positive correlation exists between the first phellogen initiation sites and the location of trichomes, with the first divisions involved with initiation of the phellogen occurring immediately below the trichomes (Arzee et al., 1978). The sequent periderms appear most commonly as discontinuous but overlapping layers (Figs. 15.1 and 15.9B). These approximately shell-shaped layers originate beneath cracks of overlying periderms (Fig. 15.9A). The sequent periderms may also be continuous around the circumference or at least for considerable parts of the circumference (Fig. 15.9C).

Secondary growth of vascular tissues and the formation of periderm are common in roots of woody angiosperms and conifers. In most roots the first periderm originates deep in the axis, usually in the pericycle, but it may appear near the surface as, for example, in some trees and perennial herbaceous plants in which the root cortex serves for food storage. Like stems, the roots also

may produce periderm layers of successively greater depths in the axis.

### The Phellogen Is Initiated by Divisions of Various Kinds of Cells

Depending on the position of the phellogen, the cells involved with its initiation may be cells of the epidermis, subepidermal parenchyma or collenchyma, parenchyma of the pericycle or phloem, including that of the phloem rays. Usually these cells are indistinguishable from other cells of the same categories. All are living cells and therefore potentially meristematic. The initiating divisions may begin in the presence of chloroplasts and various storage substances such as starch and tannins, and while the cells still have thick primary walls, as in collenchyma. Eventually the chloroplasts change into leucoplasts and the starch, tannins, and wall thickenings disappear. Sometimes the subepidermal cells, in which phellogen is to arise, have no collenchymatous thickenings and show an orderly and compact arrangement.

The phellogen is initiated by periclinal divisions (Fig. 15.7). The first periclinal division in a given cell forms two apparently similar cells. Frequently the inner of these two cells divides no further and is then regarded as a phelloderm cell, while the outer functions as the phellogen cell and divides. The outer of the two products of the second division matures into the first cork cell, while the inner remains meristematic and divides again. Sometimes the first division results in the formation of a cork cell and a phellogen cell. Although most of the successive divisions are periclinal, the phellogen keeps pace with the increase in circumference of the axis by periodic divisions of its cells in the radial anticlinal plane.

### The Time of Appearance of the First and Subsequent Periderms Varies

The first periderm commonly appears during the first year of growth of the stem and root. The subsequent, deeper periderms may be initiated later the same year or many years later or may never appear. In addition to specific differences, environmental conditions influence the appearance of both the initial and the sequent periderms. Availability of water, temperature, and intensity of light all affect the timing of the development of periderms (De Zeeuw, 1941; Borger and Kozlowski, 1972a, b, c; Morgensen, 1968; Morgensen and David, 1968; Waisel, 1995).

The first superficial periderm may be retained for life or for many years in species of *Betula, Fagus, Abies, Carpinus, Anabasis, Haloxylon,* and *Quercus* and in many species of tropical trees (Roth, 1981). In the carob tree (*Ceratonia siliqua*) the initiation of sequent

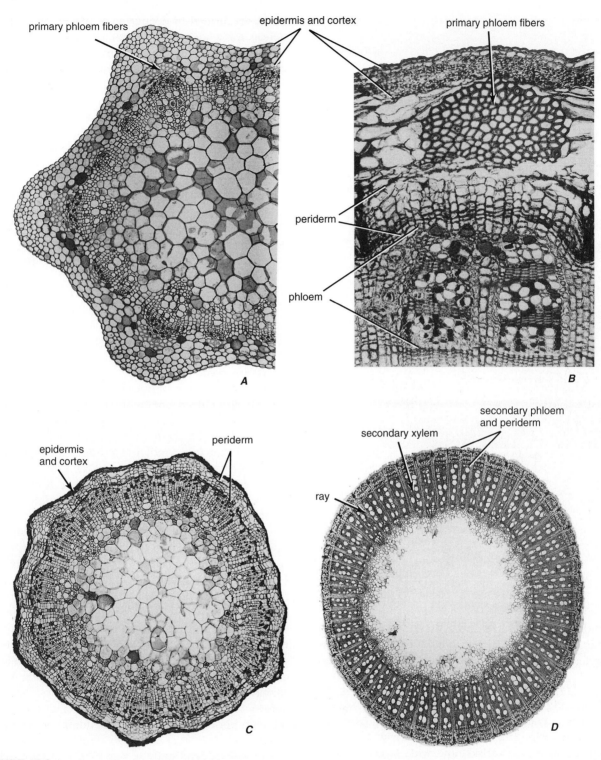

**FIGURE 15.8**

Origin of the first periderm in the grapevine (*Vitis vinifera*) as seen in transverse sections. **A**, seedling stem without periderm. **B**, older seedling stem with periderm that originated in the primary phloem. The strands of thick-walled cells are primary phloem fibers. The nonsclerified cells outside the periderm have died and collapsed. **C**, older seedling stem with periderm, which forms a complete cylinder around the stem. The epidermis and cortex have been obliterated. **D**, one-year-old cane with periderm outside the secondary phloem. (A, ×90; B, ×115; C, ×50; D, ×10. **A**, Plate 4B, and **C**, **D**, Plate 5A, B from Esau, 1948. *Hilgardia* 18 (5), 217–296. © 1948 Regents, University of California.)

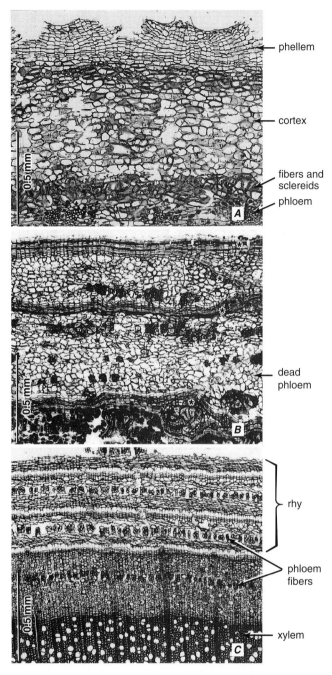

**FIGURE 15.9**

Periderm and rhytidome in transverse sections of stems. **A**, *Talauma*. Phellem with deep cracks. **B**, *Quercus alba* (white oak). Rhytidome with narrow layers of sequent periderms (asterisks) and wide layers of dead phloem tissue. **C**, *Lonicera tatarica*. Rhytidome (rhy) in which periderm layers alternate with layers derived from secondary phloem containing phloem fibers. (From Esau, 1977.)

periderms is restricted to tree parts estimated to be more than 40 years old (Arzee et al., 1977). In *Fagus sylvatica*, the original periderm and old phloem may remain on the bark for as long as 200 years. An initial

periderm formed in deeper parts of the axis may also persist for a long time (*Ribes, Berberis, Punica*). In *Melaleuca*, three to four sequent periderms are formed during the first year, each originating in the secondary phloem (Chiang and Wang, 1984). In most trees the first periderms are replaced by sequent periderms within a few years. In apple and pear trees the first periderm is generally replaced in the sixth to eighth year of growth, and in *Pinus sylvestris* in the eighth to tenth year. Temperate-zone trees tend to produce more sequent periderms than do tropical trees.

The period(s) of activity of the phellogen and the vascular cambium may or may not correspond. The activity of the two lateral meristems occur independently in *Robinia pseudoacacia* (Waisel et al., 1967), *Acacia raddiana* (Arzee et al., 1970), *Abies alba* (Golinowski, 1971), *Cupressus sempervirens* (Liphschitz et al., 1981), *Pinus pinea, Pinus halepensis* (Liphschitz et al., 1984), and *Pistacia lentiscus* (Liphschitz et al., 1985). In *Ceratonia siliqua* (Arzee et al., 1977), *Quercus boissieri*, and *Quercus ithaburensis* (Arzee et al., 1978), by contrast, the activity of both meristems coincides.

First and sequent periderms have long been considered to differ from one another only in their time of origin. A series of studies using cryofixation and chemical techniques on the bark of several conifers showed, however, that two kinds of periderm are involved in rhytidome formation (Mullick, 1971). The initial periderm and some sequent periderms are brown, other sequent periderms are reddish purple. In addition to the color the two kinds of periderm have other specific physical and chemical characters, and they also differ in their position in the rhytidome. The reddish purple sequent periderms occur next to the dead phloem embedded in the rhytidome and appear to serve in protecting living tissues from effects associated with cell death. The brown sequent periderms appear sporadically and are separated from the dead phloem by the reddish-purple periderms. The brown first periderm and the brown sequent periderms are similar in all their characteristics. Both act in protecting living tissues against the external environment, the first periderm before the formation of rhytidome and the brown sequent periderms after the shedding of rhytidome layers.

In a subsequent study it was found, in four species of conifers, that wound and pathological periderms and periderms formed at abscission zones and on old resin blisters were of the reddish purple-sequent type. Inasmuch as all reddish purple-pigmented periderms, including the usual sequent periderm, abut necrotic tissues, it was suggested that they constitute a single category of periderm, the **necrophylactic**. The brown periderms, the first and the sequent, which protect living tissues against the external environment, constitute a second category, the **exophylactic** (Mullick and Jensen, 1973).

# MORPHOLOGY OF PERIDERM AND RHYTIDOME

The external appearance of axes bearing periderm or rhytidome is highly variable (Fig. 15.10). This variation depends partly on the characteristics and manner of growth of the periderm itself and partly on the amount and kind of tissue separated by the periderm from the axis. The characteristic external appearance of bark can provide valuable taxonomic information, especially in the identification of tropical trees (Whitmore, 1962a, b; Roth, 1981; Yunus et al., 1990; Khan, 1996).

Rhytidome formation occurs by the successive development of periderms. Consequently, barks that have only a superficial periderm do not form a rhytidome. In such barks only a small amount of primary tissue is cut off, involving either a part of or the entire epidermis or possibly one or two cortical layers. This tissue is

**FIGURE 15.10**

Bark from four species of deciduous hardwoods. **A**, shaggy bark of the shagbark hickory (*Carya ovata*). **B**, deeply furrowed bark of the black oak (*Quercus velutina*). **C**, thin, peeling "bark" of the paper birch (*Betula papyrifera*). The peeling in the paper birch actually takes place in the boundary between narrow phellem cells and broad ones. The horizontal streaks on the surface of the bark are lenticels. **D**, scaly bark of the sycamore, or buttonwood (*Platanus occidentalis*).

eventually sloughed away, and the phellem is exposed. If the exposed cork tissue is thin, it commonly has a smooth surface. In the paper birch (*Betula papyrifera*), for example, the periderm peels off into thin, papery layers (Fig. 15.10C) through the boundary between narrow phellem cells and broad ones (Chang, 1954). If the exposed cork tissue is thick, the surface is cracked and fissured. Massive cork usually shows layers that seem to represent annual increments.

The stems of some eudicots (*Ulmus* sp.) produce a so-called winged cork, a form that results from a symmetrical longitudinal splitting of the cork in relation to the uneven expansion of different sectors of the stem (Smithson, 1954). Winged cork may also result from intensive localized activity of phellogen considerably in advance of the formation of the periderm elsewhere (*Euonymus alatus*, Bowen, 1963). Cork prickles, which form on stems of some tropical trees (certain Rutaceae, Bombacaceae, Euphorbiaceae, Fabaceae) result from the phellogen producing phellem in excess at certain spots. They are pure cork formations and consist of thick-walled lignified cells. Because of their scarce occurrence, they are an excellent characteristic in bark identification (Roth, 1981).

On the basis of manner of origin of the successive layers of periderm, two forms of rhytidome, or outer bark, are distinguished, scale bark and ring bark. **Scale bark** occurs when the sequent periderms are formed as overlapping layers, each cutting out a "scale" of tissue (*Pinus, Pyrus, Quercus, Tilia*). **Ring bark** is less common and results from the formation of successive periderms approximately concentrically around the axis (Cupressaceae, *Lonicera, Clematis, Vitis*). This type of outer bark is associated with plants in which the first periderm originates in deep layers of the axis. A scale bark with very large individual scales (*Platanus*) may be regarded as being intermediate between the scale bark and the ring bark.

In some rhytidomes parenchyma and soft cork cells predominate. Others contain large amounts of fibers usually derived from the phloem. The presence of fibrous tissue lends a characteristic aspect to the bark (Holdheide, 1951; Roth, 1981). If fibers are absent, the bark breaks into individual scales or shells (*Pinus, Acer pseudoplatanus*). In fibrous bark a netlike pattern of splitting occurs (*Tilia, Fraxinus*).

The scaling off of the bark may have different structural bases. If thin-walled cells of cork or of phelloids are present in the periderms of the rhytidome, the scales may exfoliate along these lines. Breaks in the rhytidome can occur also through cells of nonperidermal tissues. In *Eucalyptus*, breaks occur through phloem parenchyma cells (Chattaway, 1953), and in *Lonicera trataria*, between fibers and parenchyma of the phloem. Cork is frequently a strong tissue and renders the bark persistent, even if deep cracks develop (species of *Betula, Pinus, Quercus, Robinia, Salix, Sequoia*). Such barks wear off without forming scales.

## POLYDERM

A special type of protective tissue called **polyderm** occurs in roots and underground stems of Hypericaceae, Myrtaceae, Onagraceae, and Rosaceae (Nelson and Wilhelm, 1957; Tippett and O'Brien, 1976; Rühl and Stösser, 1988; McKenzie and Peterson, 1995). It arises from a meristem originating in the pericycle and consists of alternating layers of suberized cells, one cell deep, and of nonsuberized cells, several cells deep (Fig. 15.11). The polyderm may accumulate 20 or more such alternating layers, but only the outermost layers are dead. In the living part, the nonsuberized cells function as storage cells. In submerged parts of aquatic plants, the polyderm may develop intercellular spaces and then function as aerenchyma.

## PROTECTIVE TISSUE IN MONOCOTYLEDONS

In herbaceous monocots, the epidermis is permanent and serves as the only protective tissue on the plant axis. Should the epidermis be ruptured, the underlying cortical cells become secondarily suberized as suberin lamellae typical of cork cells are deposited on the cellulose walls. Such behavior is common in the Poaceae, Juncaceae, Typhaceae, and other families.

Monocots rarely produce a periderm similar to that in other angiosperms (Solereder and Meyer, 1928). The palm *Roystonea* produces such a periderm, whose phellem consists of compactly arranged cells with thick lignified walls. Some phelloderm cells also become sclerified and lignified (Chiang and Lu, 1979).

In most woody monocots, including the palms, a special type of protective tissue is formed by repeated division of cortical parenchyma cells and subsequent suberization of the products of division (Tomlinson, 1961, 1969). Parenchyma cells in successively deeper positions undergo similar divisions and suberization. Thus the cork arises without formation of an initial layer, or phellogen. Because the linear files of cells form tangential bands as seen in transverse sections, the tissue is referred to as **storied cork** (Fig. 15.12). As the formation of cork progresses inward, nonsuberized cells may become embedded among the cork cells. Thus a tissue analogous to the rhytidome of the woody angiosperms is formed (*Dracaena, Cordyline, Yucca*).

## WOUND PERIDERM

Wounding induces a series of metabolic events and related cytological responses that lead, under favorable conditions, to a complete closure of the wound (Bostock

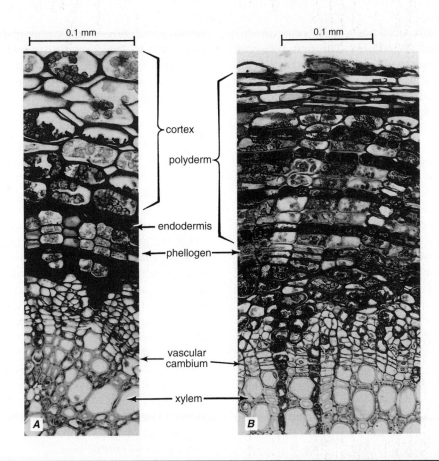

**FIGURE 15.11**

Polyderm of root of strawberry (*Fragaria*) in transverse sections. **A**, root in early stage of secondary growth. Phellogen has been initiated, but the cortex is still intact. **B**, older root. Wide layer of polyderm has been formed by the phellogen. The cells composing the darkly stained bands in the polyderm are suberized. These cells alternate with nonsuberized cells. Both kinds of cells are living. Nonliving suberized cells form the outer covering. No cortex is present. (From Nelson and Wilhelm, 1957. *Hilgardia* 26 (15), 631–642. © 1957 Regents, University of California.)

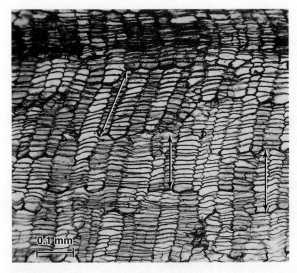

**FIGURE 15.12**

Storied cork of *Cordyline terminalis* in cross section. (From Esau, 1977; slide courtesy of Vernon I. Cheadle.)

and Sterner, 1989). Wound healing is a developmental process requiring synthesis of DNA and proteins (Borchert and McChesney, 1973). Dramatic ultrastructural changes evident in cells bordering the injured cell layer (Barckhausen, 1978) collectively indicate heightened transcriptional, translational, and secretory activities. The sequence of events that occur during wound healing in the stems of gymnosperms (Mullick and Jensen, 1976; Oven et al., 1999) and woody angiosperms (Biggs and Stobbs, 1986; Trockenbrodt, 1994; Hawkins and Boudet, 1996; Woodward and Pocock, 1996; Oven et al., 1999) is similar to that exhibited by a wounded potato tuber, probably the most extensively studied object in this regard (Thomson et al., 1995; Schreiber et al., 2005).

Formation of the wound, or necrophylactic, periderm is preceded by a sealing of the newly exposed surface by an impervious layer of cells commonly referred to as the **boundary layer**. The boundary layer is derived from cells present at the time of wounding. It develops immediately below the dead (necrosed) cells on the surface of the wound (Fig. 15.13). The first apparent response (within 15 minutes) to wounding in the potato tuber is the deposition of callose at the sites of plasmodesmata on the boundary-cell side of the walls bordering necrosed cells (Thomson et al., 1995). This wound callose thus seals the symplastic connections at this interface.

Lignification precedes suberization of the boundary-cell walls. The middle lamellas and primary walls of the boundary-layer cells first become lignified. Lignification is followed by suberization of the walls, that is, by the formation of a suberin lamella along all inner surfaces of the previously lignified walls. The ligno-suberized boundary layer provides an impervious barrier to moisture loss and microbial invasion of the living tissue below and helps to maintain conditions favorable to the formation of the wound periderm. Why does lignification precede suberization in wound healing? Evidence on the role of phenolic compounds in disease resistance indicates a close correlation between the deposition of

lignin and lignin-related compounds in cell walls and resistance to infection by fungal and bacterial plant pathogens (Nicholson and Hammerschmidt, 1992). This resistance was attributed mainly to the presumed toxicity of lignin-related precursors to pathogens as well as the barrier effect resulting from lignification of cell walls.

Periclinal divisions in cells beneath the boundary layer mark the initiation of a wound phellogen. Newly formed cork cells can be distinguished from boundary cells by their radial alignment within the developing wound periderm. Lignification-suberization of the wound cork cells follow the same sequence as that exhibited by the boundary-layer cells.

Successful development of wound periderm is important in horticultural practice when plant parts used in propagation must be cut (e.g., potato tubers, sweet potato roots). Experiments in which wound-healing phenomena in sliced potato tubers were retarded by chemical treatment showed the importance of wound periderm in protection from infection by decay organisms (Audia et al., 1962). Environmental conditions markedly influence the development of wound periderm (Doster and Bostock, 1988; Bostock and Stermer, 1989). The ability to develop wound periderm in response to invasion by parasites may distinguish resistant from susceptible plants.

Plant taxa vary with regard to the anatomical aspects of wound healing, as they do in details of natural development of the protective tissue (El Hadidi, 1969; Swamy and Sivaramakrishna, 1972; Barckhausen, 1978). In general, monocots are less responsive to wounding than the eudicots. In eudicots and certain monocots (Liliales, Araceae, Pandanaceae) healing includes formation of both boundary layer and wound periderm. In other monocots no wound periderm is detectable. Among these, the Zingiberales produce a slightly suberized boundary layer, whereas the Arecaceae and the Poaceae form a lignified boundary layer.

# LENTICELS

A **lenticel** may be defined as a limited part of the periderm in which the phellogen is more active than elsewhere and produces a tissue that, in contrast to the phellem, has numerous intercellular spaces. The lenticel phellogen itself also has intercellular spaces. Because of this relatively open arrangement of cells, the lenticels are regarded as structures permitting the entry of air though the periderm (Groh et al., 2002).

Lenticels are usual components of periderm of stems and roots. Exceptions are found among stems with a regular formation of periderms around the entire circumference and that shed their outer layers of bark annually (species of *Vitis, Lonicera, Tecoma, Clematis,*

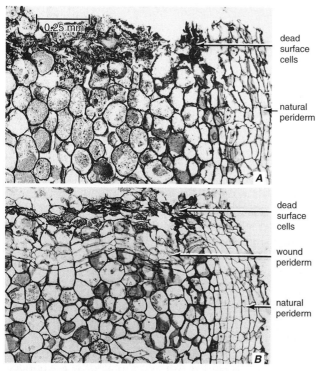

**FIGURE 15.13**

Wound periderm formation in root of sweet potato (*Ipomoea batatas*). **A**, broken end of wound covered with dead cells. **B**, a wound periderm has developed beneath the dead surface and has become connected (right) with the natural periderm. (Both, same magnification. From Morris and Mann, 1955. *Hilgardia* 24 (7), 143–183. © 1955 Regents, University of California.)

*Rubus*, and some others, mostly vines). Lenticels ("cork warts") occur on the surfaces of leaves of certain taxa (Roth, 1992, 1995). The small dots on the surface of apples, pears, and plums are examples of lenticels on fruits.

Outwardly a lenticel often appears as a vertically or horizontally elongated mass of loose cells that protrudes above the surface through a fissure in the periderm (Fig. 15.10C). Lenticels vary in size from structures barely visible without magnification to those 1 cm and more in length. They occur singly or in rows. Vertical rows of lenticels frequently occur opposite the wide vascular rays, but in general, there is no constant positional relation between lenticels and rays.

The phellogen of a lenticel is continuous with that of the corky periderm but usually bends inward so that it appears more deeply situated (Fig. 15.14). The loose tissue formed by the lenticel phellogen toward the outside is the **complementary**, or **filling**, **tissue** (Wutz, 1955); the tissue formed toward the inside is the phelloderm.

The degree of difference between the filling tissue and the neighboring phellem varies in different species. In the gymnosperms the filling tissue is composed of the same types of cells as the phellem. The main difference between the two is that the tissue of the lenticel has intercellular spaces. Lenticel cells may also have thinner walls and be radially elongated instead of radially flattened like the phellem cells of so many species. In lenticels of the potato tuber, scanning electron microscopy has revealed waxy outgrowths on the cell walls facing the intercellular spaces (Hayward, 1974). This wax may serve in the regulation of water loss from the tuber and in deterring the entry of water, and possible pathogens, through the lenticels.

## Three Structural Types of Lenticels Are Recognized in Woody Angiosperms

The first and simplest type of lenticel in woody angiosperms is exemplified by species of *Liriodendron*, *Magnolia*, *Malus*, *Persea* (Fig. 15.14A, B), *Populus*, *Pyrus*, and *Salix*, and has a filling tissue composed of suberized cells. This tissue, though having intercellular spaces, may be more or less compact and may show annual growth layers, with thinner-walled looser tissue appearing earlier, and thicker-walled more compact tissue later.

Lenticels of the second type, as found in species of *Fraxinus*, *Quercus*, *Sambucus* (Fig. 15.14C), and *Tilia*, consist mainly of a mass of more or less loosely structured nonsuberized filling tissue. At the end of the

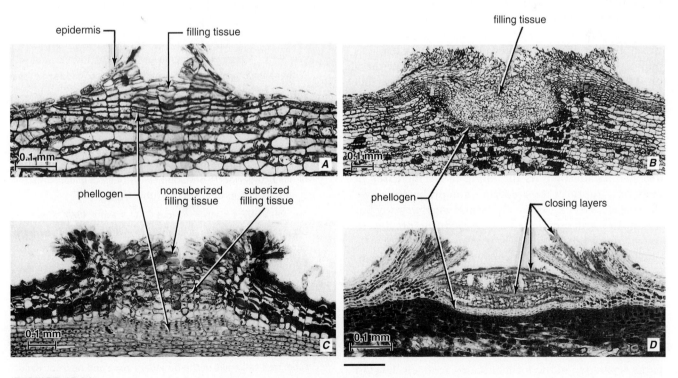

**FIGURE 15.14**

Lenticels in transverse sections of stems. **A, B,** avocado (*Persea americana*). Young lenticel in **A,** older in **B.** No closing layers present. **C,** elderberry (*Sambucus canadensis*). Lenticel with compact layer of suberized cells interior to loosely structured nonsuberized filling tissue. **D,** beech (*Fagus grandifolia*). Lenticel with closing layers. (**A, B, D,** from Esau, 1977.)

season the filling tissue is succeeded by a more compact layer of suberized cells.

The third type, illustrated by lenticels of species of *Betula, Fagus* (Fig. 15.14D), *Prunus*, and *Robinia*, shows the highest degree of specialization. The filling tissue is layered because loose nonsuberized tissue regularly alternates with compact suberized tissue. The compact tissue forms the **closing layers**, each one to several cells in depth, that hold together the loose tissue, usually in layers several cells deep. Several strata of each kind of tissue may be produced yearly. The closing layers are successively broken by the new growth.

In Norway spruce (*Picea abies*), a conifer, the lenticel phellogen generally produces a single new closing layer each year (Rosner and Kartush, 2003). The production of new filling tissue, which begins in spring, eventually disrupts the closing layer formed during the previous growing season. Differentiation of a new closing layer occurs in late summer. Thus the lenticels are most permeable between the time of rupture of the previously formed closing layer and that of differentiation of the new one. That period corresponds with the most active period of wood production in Norway spruce (Rosner and Kartush, 2003).

### The First Lenticels Frequently Appear under Stomata

In periderms initiated in the subepidermal layer, the first lenticels usually arise beneath stomata. They may appear before the stem ceases its primary growth and before the periderm is initiated (Fig. 15.14A), or lenticels may arise simultaneously at the termination of primary growth. The parenchyma cells about the substomatal chamber divide in various planes, chlorophyll disappears, and a colorless loose tissue is formed. The divisions successively occur deeper and deeper in the cortical parenchyma and become oriented periclinally. Thus a periclinally dividing meristem, the lenticel phellogen, is established. As the filling tissue increases in amount, it ruptures the epidermis and protrudes above the surface. The exposed cells die and weather away but are replaced by others developing from the phellogen. By divisions producing cells toward the interior, the phellogen beneath the lenticel forms some phelloderm, usually more than under the cork.

Lenticels are maintained in the periderm as long as the periderm continues to grow, and new ones arise from time to time by change in the activity of the phellogen from formation of phellem to that of lenticel tissue. The deeper periderms also have lenticels. In barks separating in the form of scales, lenticels develop in the newly exposed periderm. If the bark is adherent and fissured, the lenticels occur at the bottom of the furrows. Lenticels on rough bark surfaces are not readily seen. The lenticels of the rhytidome are basically similar to those of the initial periderm, but their phellogen is less active, and therefore they are not as well differentiated. If cork tissue is massive, the lenticels are continued through the whole thickness of the tissue, a feature well illustrated by the commercial cork (*Quercus suber*), in which the lenticels are visible as brown powdery streaks in transverse and radial sections. Because these lenticels are porous, bottle corks are cut vertically from the cork sheet so that the cylindrical lenticels extend transversely through them.

## REFERENCES

ACHOR, D. S., H. BROWNING, and L. G. ALBRIGO. 1997. Anatomical and histochemical effects of feeding by *Citrus* leafminer larvae (*Phyllocnistis citrella* Stainton) in *Citrus* leaves. *J. Am. Soc. Hortic. Sci.* 122, 829–836.

ALONI, R., and C. A. PETERSON. 1991. Naturally occurring periderm tubes around secondary phloem fibres in the bark of *Vitis vinifera* L. *IAWA Bull. n.s.* 12, 57–61.

ANGELES, G. 1992. The periderm of flooded and non-flooded *Ludwigia octovalvis* (Onagraceae). *IAWA Bull. n.s.* 13, 195–200.

ANGELES, G., R. F. EVERT, and T. T. KOZLOWSKI. 1986. Development of lenticels and adventitious roots in flooded *Ulmus americana* seedlings. *Can. J. For. Res.* 16, 585–590.

ARCHER, R. H., and A. E. VAN WYK. 1993. Bark structure and intergeneric relationships of some southern African Cassinoideae (Celastraceae). *IAWA J.* 14, 35–53.

ARZEE, T., Y. WAISEL, and N. LIPHSCHITZ. 1970. Periderm development and phellogen activity in the shoots of *Acacia raddiana* Savi. *New Phytol.* 69, 395–398.

ARZEE, T., E. ARBEL, and L. COHEN. 1977. Ontogeny of periderm and phellogen activity in *Ceratonia siliqua* L. *Bot. Gaz.* 138, 329–333.

ARZEE, T., D. KAMIR, and L. COHEN. 1978. On the relationship of hairs to periderm development in *Quercus ithaburensis* and *Q. infectoria*. *Bot. Gaz.* 139, 95–101.

AUDIA, W. V., W. L. SMITH JR., and C. C. CRAFT. 1962. Effects of isopropyl *N*-(3-chlorophenyl) carbamate on suberin, periderm, and decay development by Katahdin potato slices. *Bot. Gaz.* 123, 255–258

BARCKHAUSEN, R. 1978. Ultrastructural changes in wounded plant storage tissue cells. In: *Biochemistry of Wounded Plant Tissues*, pp. 1–42, G. Kahl, ed. Walter de Gruyter, Berlin.

BIGGS, A. R., and L. W. STOBBS. 1986. Fine structure of the suberized cell walls in the boundary zone and necrophylactic periderm in wounded peach bark. *Can. J. Bot.* 64, 1606–1610.

BORCHERT, R., and J. D. MCCHESNEY. 1973. Time course and localization of DNA synthesis during wound healing of potato tuber tissue. *Dev. Biol.* 35, 293–301.

BORGER, G. A., and T. T. KOZLOWSKI. 1972a. Effects of water deficits on first periderm and xylem development in *Fraxinus pennsylvanica*. *Can. J. For. Res.* 2, 144–151.

BORGER, G. A., and T. T. KOZLOWSKI. 1972b. Effects of light intensity on early periderm and xylem development in *Pinus resinosa, Fraxinus pennsylvanica*. and *Robinia pseudoacacia. Can. J. For. Res.* 2, 190–197.

BORGER, G. A., and T. T. KOZLOWSKI. 1972c. Effects of temperature on first periderm and xylem development in *Fraxinus pennsylvanica, Robinia pseudoacacia*, and *Ailanthus altissima. Can. J. For. Res.* 2, 198–205.

BOSTOCK, R. M., and B. A. STERMER. 1989. Perspectives on wound healing in resistance to pathogens. *Annu. Rev. Phytopathol.* 27, 343–371.

BOWEN, W. R. 1963. Origin and development of winged cork in *Euonymus alatus. Bot. Gaz.* 124, 256–261.

CHANG, Y.-p. 1954. *Anatomy of common North American pulpwood barks*. TAPPI Monograph Ser. No. 14. Technical Association of the Pulp and Paper Industry, New York.

CHATTAWAY, M. M. 1953. The anatomy of bark. I. The genus *Eucalyptus. Aust. J. Bot.* 1, 402–433.

CHATTAWAY, M. M. 1959. The anatomy of bark. VII. Species of *Eugenia (sens. lat.). Trop. Woods* 111, 1–14.

CHIANG, S. H. T. [TSAI-CHIANG, S. H.] and C. Y. LU. 1979. Lateral thickening of the stem of *Roystonea regia. Proc. Natl. Sci. Council Rep. China* 3, 404–413.

CHIANG, S. H. T., and S. C. WANG. 1984. The structure and formation of *Melaleuca* bark. *Wood Fiber Sci.* 16, 357–373.

COSTA, A., H. PEREIRA, and A. OLIVEIRA. 2001. A dendroclimatological approach to diameter growth in adult cork-oak trees under production. *Trees* 15, 438–443.

DE ZEEUW, C. 1941. *Influence of exposure on the time of deep cork formation in three northeastern trees*. Bull. N.Y. State Col. For. Syracuse Univ. Tech. Publ. No. 56.

DITTMER, H. J., and M. L. ROSER. 1963. The periderm of certain members of the Cucurbitaceae. *Southwest. Nat.* 8, 1–9.

DOSTER, M. A., and R. M. BOSTOCK. 1988. Effects of low temperature on resistance of almond trees to *Phytophthora* pruning wound cankers in relation to lignin and suberin formation in wounded bark tissue. *Phytopathology* 78, 478–483.

DZEREFOS, C. M., and E. T. F. WITKOWSKI. 1997. Development and anatomy of the attachment structure of woodrose-producing mistletoes. *S. Afr. J. Bot.* 63, 416–420.

EL HADIDI, M. N. 1969. Observations on the wound-healing process in some flowering plants. *Mikroskopie* 25, 54–69.

ESAU, K. 1948. Phloem structure in the grapevine, and its seasonal changes. *Hilgardia* 18, 217–296.

ESAU, K. 1977. *Anatomy of Seed Plants*, 2nd ed. Wiley, New York.

EVERT, R. F. 1963. Ontogeny and structure of the secondary phloem in *Pyrus malus. Am. J. Bot.* 50, 8–37.

GEIBEL, M. 1998. Die Valsa—Krankheit beim Steinobst—biologische grundlagen und Resistenzforschung. *Erwerbsobstbau* 40, 74–79.

GIL, G. F., D. A. URQUIZA, J. A. BOFARULL, G. MONTENEGRO, and J. P. ZOFFOLI. 1994. Russet development in the "Beurre Bosc" pear. *Acta Hortic.* 367, 239–247.

GINSBURG, C. 1963. Some anatomic features of splitting of desert shrubs. *Phytomorphology* 13, 92–97.

GODKIN, S. E., G. A. GROZDITS, and C. T. KEITH. 1983. The periderms of three North American conifers. Part 2. Fine structure. *Wood Sci. Technol.* 17, 13–30.

GOLINOWSKI, W. O. 1971. The anatomical structure of the common fir (*Abies alba* Mill.). I. Development of bark tissues. *Acta Soc. Bot. Pol.* 40, 149–181.

GRAÇA, J., and H. PEREIRA. 2004. The periderm development in *Quercus suber. IAWA J.* 25, 325–335.

GROH, B., C. HÜBNER, and K. J. LENDZIAN. 2002. Water and oxygen permeance of phellems isolated from trees: The role of waxes and lenticels. *Planta* 215, 794–801.

GROZDITS, G. A., S. E. GODKIN, and C. T. KEITH. 1982. The periderms of three North American conifers. Part I. Anatomy. *Wood Sci. Technol.* 16, 305–316.

HAWKINS, S., and A. BOUDET. 1996. Wound-induced lignin and suberin deposition in a woody angiosperm (*Eucalyptus gunnii* Hook.): Histochemistry of early changes in young plants. *Protoplasma* 191, 96–104.

HAYWARD, P. 1974. Waxy structures in the lenticels of potato tubers and their possible effects on gas exchange. *Planta* 120, 273–277.

HOLDHEIDE, W. 1951. Anatomie mitteleuropäischer Gehölzrinden (mit mikrophotographischem Atlas). In: *Handbuch der Mikroskopie in der Technik*, Band 5, Heft 1, pp. 195–367, H. Freund, ed. Umschau Verlag, Frankfurt am Main.

HOWARD, E. T. 1971. Bark structure of the southern pines. *Wood Sci.* 3, 134–148.

KHAN, M. A. 1996. Bark: A pointer for tree identification in field conditions. *Acta Bot. Indica* 24, 41–44.

KRAHMER, R. L., and J. D. WELLONS. 1973. Some anatomical and chemical characteristics of Douglas-fir cork. *Wood Sci.* 6, 97–105.

LEV-YADUN, S., and R. ALONI. 1991. Wound-induced periderm tubes in the bark of *Melia azedarach, Ficus sycomorus*, and *Platanus acerifolia. IAWA Bull.* n.s. 12, 62–66.

LIPHSCHITZ, N., S. LEV-YADUN, and Y. WAISEL. 1981. The annual rhythm of activity of the lateral meristems (cambium and phellogen) in *Cupressus sempervirens* L. *Ann. Bot.* 47, 485–496.

LIPHSCHITZ, N., S. LEV-YADUN, E. ROSEN, and Y. WAISEL. 1984. The annual rhythm of activity of the lateral meristems (cambium and phellogen) in *Pinus halepensis* Mill. and *Pinus pinea* L. *IAWA Bull.* n.s. 5, 263–274.

LIPHSCHITZ, N., S. LEV-YADUN, and Y. WAISEL. 1985. The annual rhythm of activity of the lateral meristems (cambium and phellogen) in *Pistacia lentiscus* L. *IAWA Bull.* n.s. 6, 239–244.

MCKENZIE, B. E., and C. A. PETERSON. 1995. Root browning in *Pinus banksiana* Lamb. and *Eucalyptus pilularis* Sm. 2. Anatomy and permeability of the cork zone. *Bot. Acta* 108, 138–143.

MORGENSEN, H. L. 1968. Studies on the bark of the cork bark fir: *Abies lasiocarpa* var. *arizonica* (Merriam) Lemmon. I. Periderm ontogeny. *J. Ariz. Acad. Sci.* 5, 36–40.

MORGENSEN, H. L., and J. R. DAVID. 1968. Studies on the bark of the cork fir: *Abies lasiocarpa* var. *arizonica* (Merriam) Lemmon. II. The effect of exposure on the time of initial rhytidome formation. *J. Ariz. Acad. Sci.* 5, 108–109.

MORRIS, L. L., and L. K. MANN. 1955. Wound healing, keeping quality, and compositional changes during curing and storage of sweet potatoes. *Hilgardia* 24, 143–183.

MOSS, E. H., and A. L. GORHAM. 1953. Interxylary cork and fission of stems and roots. *Phytomorphology* 3, 285–294.

MULLICK, D. B. 1971. Natural pigment differences distinguish first and sequent periderms of conifers through a cryofixation and chemical techniques. *Can. J. Bot.* 49, 1703–1711.

MULLICK, D. B., and G. D. JENSEN. 1973. New concepts and terminology of coniferous periderms: Necrophylactic and exophylactic periderms. *Can. J. Bot.* 51, 1459–1470.

MULLICK, D. B., and G. D. JENSEN. 1976. Rates of non-suberized impervious tissue development after wounding at different times of the year in three conifer species. *Can J. Bot.* 54, 881–892.

NELSON, P. E., and S. WILHELM. 1957. Some anatomic aspects of the strawberry root. *Hilgardia* 26, 631–642.

NICHOLSON, R. L., and R. HAMMERSCHMIDT. 1992. Phenolic compounds and their role in disease resistance. *Annu. Rev. Phytopathol.* 30, 369–389.

OVEN, P., N. TORELLI, W. C. SHORTLE, and M. ZUPANČIČ. 1999. The formation of a ligno-suberized layer and necrophylactic periderm in beech bark (*Fagus sylvatica* L.). *Flora* 194, 137–144.

PATEL, R. N. 1975. Bark anatomy of radiata pine, Corsican pine, and Douglas fir grown in New Zealand. *N. Z. J. Bot.* 13, 149–167.

PEARSON, L. C., and D. B. LAWRENCE. 1958. Photosynthesis in aspen bark. *Am. J. Bot.* 45, 383–387.

ROSNER, S., and B. KARTUSH. 2003. Structural changes in primary lenticels of Norway spruce over the seasons. *IAWA J.* 24, 105–116.

ROTH, I. 1981. *Structural Patterns of Tropical Barks. Encyclopedia of Plant Anatomy*, Band 9, Teil 3. Gebrüder Borntraeger, Berlin.

ROTH, I. 1992. *Leaf Structure: Coastal Vegetation and Mangroves of Venezuela. Encyclopedia of Plant Anatomy*, Band 14, Teil 2. Gebrüder Borntraeger, Berlin.

ROTH, I. 1995. *Leaf Structure: Montane Regions of Venezuela with an Excursion into Argentina. Encyclopedia of Plant Anatomy*, Band 14, Teil 3. Gebrüder Borntraeger, Berlin.

RÜHL, K., and R. STÖSSER. 1988. Peridermausbildung und Wundreaktion an Ruten verschiedener Himbeersorten (*Rubus idaeus* L.). *Mitt. Klosterneuburg* 38, 21–29.

SALEMA, R. 1967. On the occurrence of periderm in the leaves of *Welwitschia mirabilis*. *Can. J. Bot.* 45, 1469–1471.

SANTOS, J. K. 1926. Histological study of the bark of *Alstonia scholaris* R. Brown from the Philippines. *Philipp. J. Sci.* 31, 415–425.

SCHNEIDER, H. 1955. Ontogeny of lemon tree bark. *Am. J. Bot.* 42, 893–905.

SCHREIBER, L., R. FRANKE, and K. HARTMANN. 2005. Wax and suberin development of native and wound periderm of potato (*Solanum tuberosum* L.) and its relation to peridermal transpiration. *Planta* 220, 520–530.

SMITHSON, E. 1954. Development of winged cork in *Ulmus* x *hollandica* Mill. *Proc. Leeds Philos. Lit. Soc., Sci. Sect.,* 6, 211–220.

SOLEREDER, H., and F. J. MEYER. 1928. *Systematische Anatomie der Monokotyledonen*. Heft III. Gebrüder Borntraeger, Berlin.

SRIVASTAVA, L. M. 1963. Secondary phloem in the Pinaceae. *Univ. Calif. Publ. Bot.* 36, 1–142.

SWAMY, B. G. L., and D. SIVARAMAKRISHNA. 1972. Wound healing responses in monocotyledons. I. Responses in vivo. *Phytomorphology* 22, 305–314.

THOMSON, N., R. F. EVERT, and A. KELMAN. 1995. Wound healing in whole potato tubers: A cytochemical, fluorescence, and ultrastructural analysis of cut and bruise wounds. *Can. J. Bot.* 73, 1436–1450.

TIPPETT, J. T., and T. P. O'BRIEN. 1976. The structure of eucalypt roots. *Aust. J. Bot.* 24, 619–632.

TOMLINSON, P. B. 1961. *Anatomy of the Monocotyledons. II. Palmae.* Clarendon Press, Oxford.

TOMLINSON, P. B. 1969. *Anatomy of the Monocotyledons. III. Commelinales-Zingiberales.* Clarendon Press, Oxford.

TROCKENBRODT, M. 1994. Light and electron microscopic investigations on wound reactions in the bark of *Salix caprea* L. and *Tilia tomentosa* Moench. *Flora* 189, 131–140.

TUCKER, S. C. 1975. Wound regeneration in the lamina of magnoliaceous leaves. *Can. J. Bot.* 53, 1352–1364.

WACOWSKA, M. 1985. Ontogenesis and structure of periderm in *Acer negundo* L. and x *Fatshedera lizei* Guillaum. *Acta Soc. Bot. Pol.* 54, 17–27.

WAISEL, Y. 1995. Developmental and functional aspects of the periderm. In: *Encyclopedia of Plant Anatomy*, Band 9, Teil 4, *The Cambial Derivatives*, pp. 293–315. Gebrüder Borntraeger, Berlin.

WAISEL, Y., N. LIPHSCHITZ, and T. ARZEE. 1967. Phellogen activity in *Robinia pseudoacacia* L. *New Phytol.* 66, 331–335.

WHITMORE, T. C. 1962a. Studies in systematic bark morphology. I. Bark morphology in Dipterocarpaceae. *New Phytol.* 61, 191–207.

WHITMORE, T. C. 1962b. Studies in systematic bark morphology. III. Bark taxonomy in Dipterocarpaceae. *Gardens' Bull. Singapore* 19, 321–371.

WOODWARD, S., and S. POCOCK. 1996. Formation of the ligno-suberized barrier zone and wound periderm in four species of European broad-leaved trees. *Eur. J. For. Pathol.* 26, 97–105.

WUTZ, A. 1955. Anatomische Untersuchungen über System und periodische Veränderungen der Lenticellen. *Bot. Stud.* 4, 43–72.

YUNUS, M., D. YUNUS, and M. IQBAL. 1990. Systematic bark morphology of some tropical trees. *Bot. J. Linn. Soc.* 103, 367–377.

# CHAPTER SIXTEEN

# External Secretory Structures

**Secretion** refers to the complex phenomena of separation of substances from the protoplast or their isolation in parts of the protoplast. The secreted substances may be surplus ions that are removed in the form of salts, surplus assimilates that are eliminated as sugars or as cell wall substances, secondary products of metabolism that are not utilizable or only partially utilizable physiologically (alkaloids, tannins, essential oils, resins, various crystals), or substances that have a special physiological function after they are secreted (enzymes, hormones). The removal of substances that no longer participate in the metabolism of a cell is sometimes referred to as **excretion**. In the plant, however, no sharp distinction can be made between excretion and secretion (Schnepf, 1974). The same cell may accumulate both nonutilizable secondary metabolites and primary metabolites that are utilized again. Furthermore the exact role of many of the secondary metabolites, perhaps of most, is not known. In this book the term secretion is used to include both secretion in the strict sense and excretion. Secretion covers both removal of material from the cell (either to the surface of the plant or into internal spaces) and accumulation of secreted materials in some compartment of the cell.

Discussions of secretion phenomena in plants usually emphasize activities of specialized secretory structures such as glandular hairs, nectaries, resin ducts, laticifers, and others. In reality secretory activities occur in all living cells as part of the normal metabolism. Secretion characterizes various steps in the accumulation of temporary deposits in vacuoles and other organelles, in mobilization of enzymes involved in synthesis and breakdown of cellular components, in interchange of materials between organelles, and in phenomena of transport between cells. The ubiquity of secretory processes in the living plant must not be lost sight of when the specialized secretory structures are studied.

The visibly differentiated secretory structures occur in many forms. Highly differentiated secretory structures consisting of many cells are referred to as **glands** (Fig. 16.1F); the simpler ones are qualified as glandular, such as glandular hairs, glandular epidermis, or glandular cells (Fig. 16.1A–E). The distinction is vague, however,

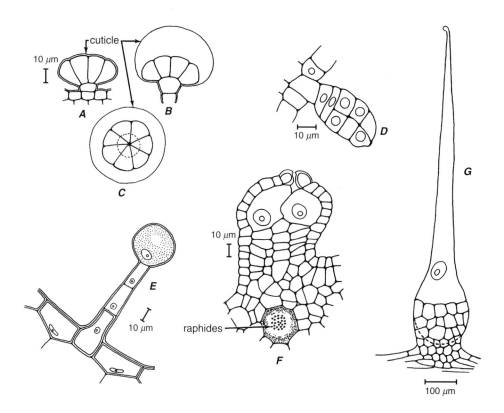

**FIGURE 16.1**

Secretory structures. **A–C,** glandular trichomes from the leaf of lavender (*Lavandula vera*) with cuticle undistended (**A**) and distended (**B, C**) by accumulation of secretion. **D,** glandular trichome from leaf of cotton (*Gossypium*). **E,** glandular trichome with unicellular head from stem of *Pelargonium*. **F,** pearl gland from leaf of grapevine (*Vitis vinifera*). **G,** stinging hair of nettle (*Urtica urens*). (From Esau, 1977.)

and a variety of secretory structures, large and small, the hair-like and the more elaborate ones, are often called glands.

Glands vary greatly with regard to the kind of substances they secrete. The substances that are secreted may be supplied directly or indirectly to the glands by the vascular tissues, as in the case of salt glands, hydathodes, and nectaries. Such substances are unmodified or only slightly modified by the secretory structures themselves. Conversely, the secreted substances may be synthesized by the constituent cells of the secretory structures, as with mucilage cells, oil glands, and the epithelial cells of resin ducts. Glands may be highly specific in their activities as is indicated by the predominance of one compound or one group of compounds in the material exported by a given gland (Fahn, 1979a, 1988; Kronestedt-Robards and Robards, 1991). Some glands secrete mainly **hydrophilic** (water-loving) **substances**, others release mainly **lipophilic** (water-hating) **substances**. Still other glands secrete fair amounts of both hydrophilic and lipophilic substances; hence it is not always possible to classify a given gland as strictly hydrophilic or strictly lipophilic (Corsi and Bottega, 1999; Werker, 2000).

The cells involved with the process of secretion typically contain dense protoplasts with abundant mitochondria. The frequency of other cellular components varies according to the particular substance secreted (Fahn, 1988). For example, mucilage-secreting cells are characterized by the presence of abundant Golgi bodies, which are involved in mucilage production and the elimination of mucilage from the protoplast by exocytosis. The most common ultrastructural feature of cells secreting lipophilic substances is the presence of abundant endoplasmic reticulum, much of which is spatially associated with plastids containing osmiophilic material. Both the plastids and the endoplasmic reticulum (and possibly other cellular components) participate in the synthesis of lipophilic substances. The endoplasmic reticulum may also be involved with intracellular transport of the lipophilic substances from their sites of synthesis to the plasma membrane.

Our understanding of the processes involved with the elimination of secretions from cells comes largely through the study of ultrastructural changes associated with the development of secretory cells. The methods of elimination of secretions from the protoplast can take place in various ways. One method, called

**granulocrine secretion**, is through the fusion of secretory vesicles with the plasma membrane or, in other words, by exocytosis. A second method, termed **eccrine secretion**, involves a direct passage of small molecules or ions through the plasma membrane. This process is passive if it is controlled by concentration gradients, active if it requires metabolic energy. Cells secreting hydrophilic substances, for example, those in glands secreting salts or carbohydrates, may be differentiated as transfer cells characterized by wall ingrowths that increase the surface of the plasma membrane (Pate and Gunning, 1972). Both granulocrine secretion and eccrine secretion are said to be of the **merocrine** (to separate) type. The secretory substances of some glands are discharged completely only upon degeneration or lysis of the secretory cells. This so-called **holocrine** type of secretion may be preceded by merocrine secretion.

In many plants the flow of secreted substances back into the plant via the apoplast, or cell wall, is prevented by cutinization of the walls in an endodermis-like layer of cells located beneath the secretory cells. In secretory trichomes the side walls of the stalk cells typically are cutinized. The presence of this gasket-like apoplastic barrier indicates that the flow of secretory substances or their precursors into the secretory cells must follow a symplastic pathway. These cutinized cells have been called "barrier cells," a descriptive name.

The remainder of this chapter will be devoted to specific examples of secretory structures found on the surface of the plant. In the following chapter (Chapter 17), examples of internal secretory structures (i.e., secretory structures embedded in various tissues) will be presented.

# ▌ SALT GLANDS

Plants growing in saline habitats have developed numerous adaptations to salt stress (Lüttge, 1983; Batanouny, 1993). The secretion of ions by salt glands is the best known mechanism for regulating the salt content of plant shoots. The composition of the secreted salt solution depends on the composition of the root environment. In addition to $Na^+$ and $Cl^-$, other ions found in the secreted solutions of salt glands are $Mg^{2+}$, $K^+$, $SO_4^{2-}$, $NO_3^-$, $PO_4^{3-}$, $Br^-$, and $HCO_3^-$ (Thomson et al., 1988). A sharp distinction cannot be made between salt glands and hydathodes, as the fluid secreted from hydathodes often also includes salts. Unlike hydathodes, there is no direct connection between the salt glands and the vascular bundles. Salt glands typically are found in halophytes (plants that grow in saline environments), and occur in at least 11 families of eudicotyledons and in one family of monocotyledons, the Poaceae (Gramineae) (Fahn, 1988, 2000; Batanouny, 1993). They vary in structure and in methods of salt release.

## Salt Bladders Secrete Ions into a Large Central Vacuole

The salt glands of the Chenopodiaceae, including essentially all *Atriplex* (saltbush) species, are trichomes consisting of one or more **stalk cells** and a large terminal **bladder cell**, which at maturity contains a large central vacuole (Fig. 16.2). The bladder and stalk cells are covered externally by a cuticle and the side walls of the stalk cell (or the lowest stalk cell, if the stalk consists of more than one cell) become completely cutinized (Thomson and Platt-Aloia, 1979). A symplastic continuum exists between the bladder cells and the mesophyll cells of the leaves. Part of the ions carried in the transpiration stream are eventually delivered through the protoplast and plasmodesmata to the bladder cells. There the ions are secreted into the central vacuole. Eventually the bladder cell collapses, and the salt is deposited on the surface of the leaf (holocrine secretion). A considerable positive gradient of salt concentration exists from the mesophyll cells to the bladder cells, indicating that the delivery of ions into the bladder-cell vacuoles is an energy consuming process (Lüttge, 1971; Schirmer and Breckle, 1982; Batanouny, 1993).

## Other Glands Secrete Salt Directly to the Outside

The Two-Celled Glands of the Poaceae The simplest anatomically of glands eliminating salt directly to the outside are those of the Poaceae. The most extensively

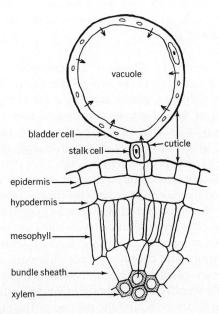

**FIGURE 16.2**

Diagram of salt-secreting trichome shown attached to part of leaf of *Atriplex* (saltbush). The long arrow indicates the path of movement of ions from the xylem to the bladder cell of the trichome. Short arrows indicate release of ions into the vacuole. (From Esau, 1977.)

studied ultrastructurally are those of *Spartina*, *Cynodon*, *Distichlis* (Thomson et al., 1988), and *Sporobolus* (Naidoo and Naidoo, 1998). Also called microhairs, the glands consist of only two cells, a **basal cell** and a **cap cell** (Fig. 16.3). A continuous cuticle covers the outer protruding portion of the glands and their adjoining epidermal cells. In contrast to other salt glands and secretory trichomes in general, the side walls of the basal cell are not cutinized. Neither cap cell nor basal cell contains a large central vacuole. The two most distinguishing features of the cap cell are a large nucleus and an expanded cuticle, which separates under fluid pressure from the outer cap cell wall, forming a **collecting chamber**. Fine openings or pores, through which the salty water is eliminated, penetrate the cuticle in this region. The most distinctive feature of the basal cell is the presence of numerous and extensive invaginations of the plasma membrane, termed **partitioning membranes**, which extend into the basal cell from the wall between basal and cap cells. The partitioning membranes are closely associated with mitochondria. It is presumed that these membranes play a role in the overall process of secretion, and that secretion from the cap cell protoplast to the permeable cap cell wall and the collecting chamber is eccrine. Cytochemical localization of ATPase activity in the salt glands of *Sporobolus* indicates that uptake of ions into the basal cell and secretion from the cap cell are both active processes (Naidoo and Naidoo, 1999). Symplastic continuity, as indicated by the presence of plasmodesmata, occurs between basal cell and neighboring mesophyll and epidermal cells as well as between basal cell and cap cell (Oross et al., 1985).

The Multicellular Glands of Eudicotyledons The salt glands of many eudicots are multicellular. Those of *Tamarix aphylla* (Tamaricaceae) consist of eight cells each, six of which are **secretory** and two are basal **collecting cells** (Fig. 16.4; Thomson and Liu, 1967; Shimony and Fahn, 1968; Bosabalidis and Thomson, 1984). The group of secretory cells is enclosed by a cuticle except where the lowermost secretory cells are connected by plasmodesmata with the collecting cells. Because the uncutinized wall portions of the lowest secretory cells are continuous with those of the underlying collecting cells, they are termed **transfusion zones**. Symplastic continuity exists between the subtending mesophyll cells and all cells of the gland. Wall ingrowths amplify the plasma membrane surface area of the secretory cells, and the portion of the cuticle on top of the gland contains pores through which salty water is exuded.

Many similarities exist between the salt glands of the mangrove *Avicennia* and those of *Tamarix*. The salt glands of *Avicennia* consist of 2 to 4 collecting cells, 1 stalk cell, and 8 to 12 secretory cells (Drennan et al.,

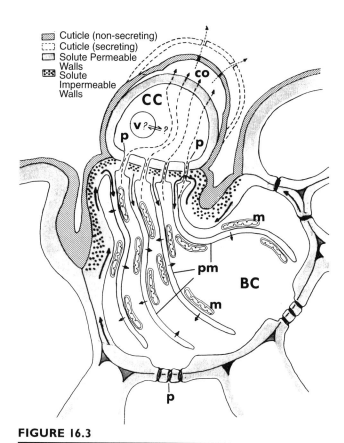

**FIGURE 16.3**

Model of structure-function relations in the two-celled salt gland of Bermuda grass (*Cynodon*). Plasmodesmata (p) occur between the basal cell (BC) and all adjoining cells, including the cap cell (CC). The only impermeable part of the gland wall occurs in the neck region of the basal cell, where the wall is lignified. The protoplast of the basal cell is characterized by the presence of numerous, long invaginations of the plasma membrane (pm) that originate near the juncture of the two gland cells. These partitioning membranes are closely associated with many mitochondria (m) and microtubules (not shown here). The cap cell, which is relatively unspecialized compared with the basal cell, contains a normal complement of organelles, including vacuoles (v) of varying size. Short arrows indicate proposed energy-requiring transmembrane flux of solutes from the lumina of the partitioning membranes to the basal cell cytoplasm; long arrows, the pathway of passive transport to the partitioning membranes; long dashes, the pathway of diffusive flow through the gland symplast; short dashes, pressurized flow of salt solution from the collecting chamber (co) through pores in the distended cuticle. (From Oross et al., 1985; reproduced with the permission of the publisher.)

1987). The cuticle overarches the top of the gland, forming a collecting chamber between the secretory cells and the inner surface of the cuticle, which contains numerous narrow pores. The side walls of the stalk cells

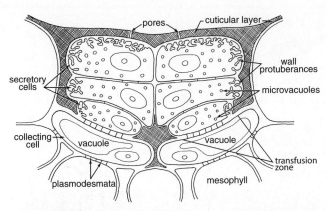

**FIGURE 16.4**

Diagram of a salt-secreting gland of *Tamarix aphylla* (tamarisk). The complex of eight cells—six of which are secretory and two the so-called collecting cells—is embedded in the epidermis and is in contact with the mesophyll (below). Cuticle and cutinized wall are indicated jointly (cuticular layer) by cross hatching. (From Esau, 1977; constructed from data in Thomson et al., 1969.)

are completely cutinized and the protoplasts adhere tenaciously to them. Collective evidence indicates that a symplastic pathway is the predominant course followed by salt to and through the gland complex, and that salty water secretion is an active process, involving membrane H⁺-ATPase where ATP hydrolysis drives ion transport across the plasma membrane of the secretory cells (Drennan et al., 1992; Dschida et al., 1992; Balsamo et al., 1995).

# HYDATHODES

**Hydathodes** are structures that discharge liquid water with various dissolved substances from the interior of the leaf to its surface, a process called **guttation**. The water of guttation is forced out of the leaves by root pressure. Structurally, hydathodes are modified parts of leaves, usually located at leaf tips or margins, especially at the teeth. In the usual form the hydathode consists of (1) the terminal tracheids of one to three vein endings, (2) the **epithem**, composed of thin-walled, chloroplast-deficient parenchyma cells located above or distal to the vein endings, (3) a sheath—a continuation of the bundle sheath—that extends to the epidermis, and (4) openings, called **water pores**, in the epidermis (Fig. 16.5). The epithem may have prominent intercellular spaces or it may be compactly arranged, with small intercellular spaces (Brouillet et al., 1987). Some epithem cells are known to differentiate as transfer cells provided with wall ingrowths (Perrin, 1971). The sheath cells often contain tannin-like substances; in some plants

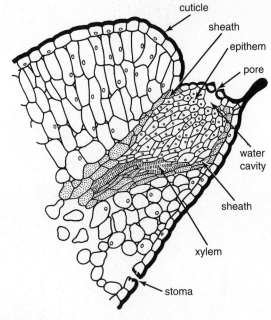

**FIGURE 16.5**

Hydathode of leaf of *Saxifraga lingulata* in longitudinal section. Tannin-containing sheath cells are stippled. (After Häusermann and Frey-Wyssling, 1963.)

their walls are suberized or have Casparian strips (Sperlich, 1939). The water pores are usually diminutive stomata that are permanently open and incapable of opening and closing.

Considerable variation exists in hydathode structure (Perrin, 1972). In some leaves the epithem is sparse (e.g., in *Sparganium emersum*, Pedersen et al., 1977; and *Solanum tuberosum*, McCauley and Evert, 1988; Fig. 16.6) or lacking, as in *Triticum aestivum* and *Oryza sativa* (Maeda and Maeda, 1987, 1988) and in other Poaceae. The vein endings in the hydathodes of wheat and rice also lack sheaths. In some submerged freshwater plants, both eudicots and monocots, the openings of the hydathodes are not represented by nonfunctional stomata (water pores) but by **apical openings** that result from the degradation of one to several water pores or of many ordinary epidermal cells (Pedersen et al., 1977).

Typical hydathodes occur along the leaf margins, usually singly at the leaf tip or the tip of a tooth. In some species of Crassulaceae and Moraceae, and almost all species of Urticaceae, hydathodes are distributed over the entire leaf surface, not only at the margin. In some species of Crassulaceae a minor vein anywhere in the lamina may turn (mostly upward) toward the leaf surface and terminate in a **laminar hydathode**. There are about 300 hydathodes in a typical leaf of *Crassula argentea* (Rost, 1969). In the Urticaceae, laminar hydathodes are associated with minor vein junctions (Lersten

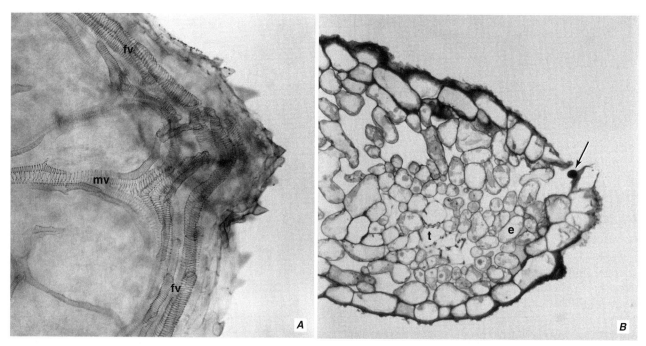

**FIGURE 16.6**

Hydathode in the leaf of potato (*Solanum tuberosum*). **A**, clearing of the tip of a terminal leaflet showing midvein (mv) and fimbrial veins (fv) converging to form a hydathode. **B**, transverse section through hydathode at tip of terminal leaflet. Giant stoma associated with hydathode is open (arrow); one guard cell is collapsed. Other details: e, epithem; t, tracheary element. (A, ×161; B, ×276. From McCauley and Evert, 1988. *Bot. Gaz.* © 1988 by The University of Chicago. All rights reserved.)

and Curtis, 1991). According to Tucker and Hoefert (1968), the only known hydathodes that develop from a shoot apex are those found at the tips of *Vitis vinifera* tendrils.

The water emanating from hydathodes may contain various salts, sugars, and other organic substances. Hydathodes of *Populus deltoides* leaves guttate water containing varying concentrations of sugar, which Curtis and Lersten (1974) called nectar. They considered that *Populus* has a rather unspecialized type of hydathode that can also act as a nectary under certain conditions. Guttation products may cause injury to plants through accumulation and concentrations or through interaction with pesticides (Ivanoff, 1963).

Haberlandt (1918) distinguished between passive hydathodes, such as those described above, and active hydathodes. The so-called active hydathodes, termed **trichome-hydathodes**, are glandular trichomes that secrete solutions of salts and other substances (Fig. 16.7; Heinrich, 1973; Ponzi and Pizzolongo, 1992).

Although hydathodes generally are associated with the discharge of water from the plant, those in many xerophytic *Crassula* species have been shown to function in absorption of condensed fog or dew water (Martin and von Willert, 2000). Moreover it has been proposed that the hydathodes in the leaf teeth of

*Populus balsamifera* function in the retrieval of solutes from the transpiration stream (Wilson et al., 1991). After a guttation event, as in the early morning, pathogenic bacteria suspended in the guttation fluid may be drawn back into the hydathode interior where they multiply and then invade the xylem, causing disease (Guo and Leach, 1989; Carlton et al., 1998; Hugouvieux et al., 1998). Guttation from hydathodes of submerged aquatic plants has been implicated in the mechanism driving an acropetal (upward) transport of water through the plant (Pedersen et al., 1977).

# NECTARIES

**Nectaries** are secretory structures that release an aqueous fluid (**nectar**) with a high sugar content. Two main categories of nectaries can be distinguished: floral nectaries and extrafloral nectaries. **Floral nectaries** are directly associated with pollination. Through their secretion of nectar, they provide a reward to insects and other animals that serve as pollinators (Baker and Baker, 1983a, b; Cruden et al., 1983; Galetto and Bernardello, 2004; Raven et al., 2005). The floral nectaries occupy various locations on the flower (Fig. 16.8; Fahn, 1979a, 1998). They are found on sepals, petals, stamens, ovaries,

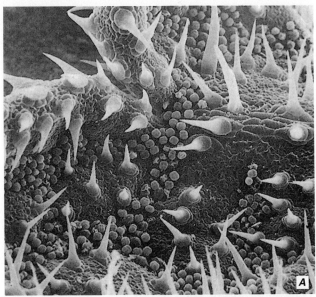

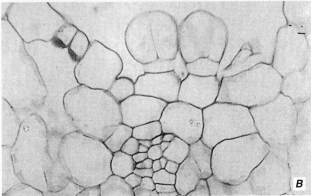

**FIGURE 16.7**

Trichome-hydathodes of the *Rhinanthus minor* leaf. **A,** scanning electron micrograph showing numerous trichome-hydathodes on lower surface of leaf. The larger, pointed structures are unicellular hairs. **B,** longitudinal view showing mature trichome-hydathodes, which are six-celled structures. Each trichome consists of four cap cells, a foot cell, and a basal epidermal cell. (A, ×125; B, ×635. From Ponzi and Pizzolongo, 1992.)

or the receptacle. A general trend of migration of the nectary during evolution from the perianth towards the ovary, style and, in some cases, to the stigma has been deduced from comparative studies (Fahn, 1953, 1979a). **Extrafloral nectaries** are not usually associated with pollination. They attract insects, particularly ants, that prey on or exclude the plant's herbivores (Pemberton, 1998; Pemberton and Lee, 1996; Keeler and Kaul, 1984; Heil et al., 2004). In *Stryphnodendron microstachyum,* a neotropical tree, the ants also collect the spores of the rust fungus *Pestalotia,* thereby reducing the incidence of the pathogen's attack on the leaves (de la Fuente and Marquis, 1999). Extrafloral nectaries occur on the

vegetative plant parts, the pedicels of flowers, and the outer surfaces of the outer floral parts (Zimmermann, 1932; Elias, 1983). In Australian members of the genus *Acacia,* which lack floral nectaries, the extrafloral nectaries attract both ants and pollinators (Marginson et al., 1985).

In eudicotyledonous flowers the nectar may be secreted by the basal parts of stamens (Fig. 16.8C) or by a ring-like nectary below the stamens (Fig. 16.8E; Caryophyllales, Polygonales, Chenopodiales). The nectary may be a ring or a disc at the base of the ovary (Fig. 16.8D, F; Theales, Ericales, Polemoniales, Solanales, Lamiales) or a disc between the stamens and the ovary (Fig. 16.8G). Several discrete glands may occur at the base of the stamens (Fig. 16.8L). In the Tiliales, the nectaries consist of multicellular glandular hairs, usually packed close together to form a cushion-like growth (Fig. 16.8I). Such nectaries occur on various floral parts, frequently on sepals. In the perigynous Rosaceae, the nectary is located between the ovary and the stamens, lining the interior of the floral cup (Fig. 16.8J). In the epigynous flower of the Umbellales, the nectary occurs on the top of the ovary (Fig. 16.8H). In the Asteraceae, it is a tubular structure at the top of the ovary, encircling the base of the style. In most of the insect-pollinated genera of the Lamiales, Berberidales, and Ranunculales, the nectaries are modified stamens, or **staminodes** (Fig. 16.8K). The nectary on the petals of *Frasera* consists of a cup with a glandular floor and a wall provided with numerous hair-like sclerified processes that plug the apical opening and force the bumble bee to work along the sides of the gland (Davies, 1952). In *Euphorbia* species (Euphorbiaceae), the lobed extrafloral nectary (**cyathial nectary**) is attached to the involucre investing inflorescence (Fig. 16.9; Arumugasamy et al., 1990). In some *Ipomoea* species (Convolvulaceae), the extrafloral nectaries consist of recessed chambers lined with secretory trichomes and connected to the surface only by a duct (Keeler and Kaul, 1979, 1984). These **crypt nectaries** are reminiscent of the **septal nectaries** (Fig. 16.8A, B; Rudall, 2002; Sajo et al., 2004) found only in the monocots. The septal nectaries have the structure of pockets with a glandular lining consisting of trichomes. They arise in parts of the ovary where the carpel walls are incompletely fused. If they are deeply embedded in the ovary, they have outlets in the form of canals leading to the surface of the ovary.

The secretory tissue of a nectary may be restricted to the epidermis, or it may be several layers of cells deep. Secretory epidermal cells may resemble ordinary epidermal cells morphologically, be represented by trichomes, or be elongated like palisade cells. Most nectaries consist of an epidermis and specialized parenchyma. The tissue composing the nectary is called **nectariferous tissue.** The epidermis of many floral nectaries contain stomata that are permanently open and, in that

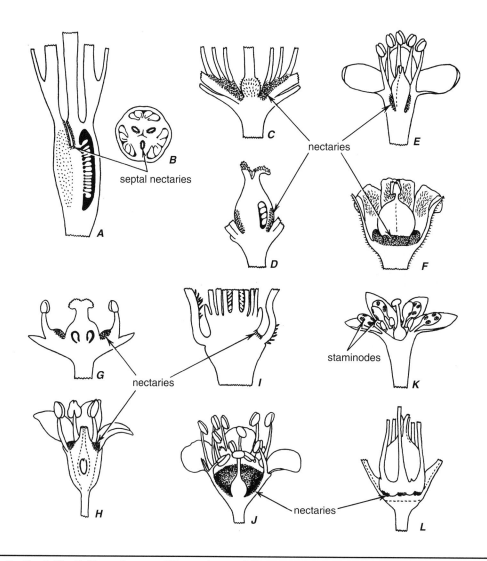

**FIGURE 16.8**

Nectaries. Longitudinal (**A, C–L**) and cross (**B**) sections of flowers. Septal, in Liliales, *Narcissus* (**A**) and *Gladiolus* (**B**); **C**, external, at base of stamens (*Thea*, Theales); **D**, ring at base of ovary (*Euyra*, Theales); **E**, ring below stamens (*Coccoloba*, Polygonales); **F**, disc below ovary (*Jatropha*, Euphorbiales); **G**, disc between ovary and stamens (*Perrottetia*, Celastrales; **H**, disc above inferior ovary (*Mastixia*, Umbellales); **I**, cushion of hairs at base of sepal (*Corchorus*, Tiliales); **J**, lining floral cup (*Prunus*, Rosales); **K**, modified stamens, staminodes (*Cinnamomum*, Laurales); **L**, glands at bases of stamens (*Linum*, Geraniales). (Adapted from Brown, 1938.)

regard, are similar to the water pores of hydathodes. In nectaries such stomata are termed **modified stomata** (Davis and Gunning, 1992, 1993). The modified stomata provide the pathway for the release of the nectar secreted by the underlying nectariferous tissue to the surface. In the extrafloral nectaries of *Sambucus nigra* (Caprifoliaceae), the nectar is secreted into a lysigenous cavity and released through a rupture in the epidermis (Fahn, 1987). Vascular tissue occurs more or less close to the secretory tissue. Sometimes this is merely a vascular bundle of the organ on which the nectary occurs, but many nectaries have their own vascular bundles, often consisting of phloem only. Laticifers may be present in nectaries (Tóth-Soma et al., 1995/96).

Active secretory cells in nectaries have dense cytoplasm and small vacuoles, which often contain tannins. Numerous mitochondria with well-developed cristae indicate that these cells respire intensively. In most nectaries the endoplasmic reticulum is highly developed and may be stacked or convoluted. This endoplasmic reticulum reaches its maximum volume and is associated with vesicles at the stage of nectar secretion. Numerous active Golgi bodies may also be present. In some nectaries (*Lonicera japonica*, Caprifoliaceae; Fahn and Rachmilevitz, 1970) vesicles derived from endoplasmic reticulum, rather than those released by Golgi bodies, are thought to be concerned with the secretion of sugar. The walls of the secretory cells often

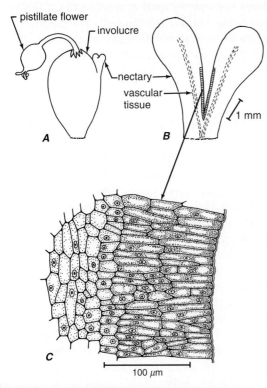

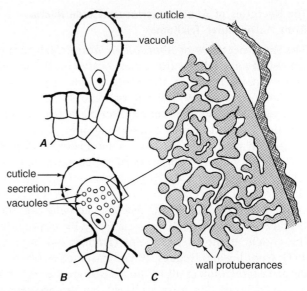

**FIGURE 16.9**

Extrafloral nectary of poinsettia (*Euphorbia pulcherrima*). **A**, view of pistillate flower and involucre with nectary. **B**, lobed nectary is attached to involucre investing inflorescence. **C**, detail of secretory issue. The extrafloral nectaries of poinsettia attract pollinators. (From Esau, 1977.)

**FIGURE 16.10**

Details of nectary of *Lonicera japonica*. **A**, **B**, nectar-secreting hairs from inner epidermis of corolla tube shown before secretion (**A**) and during secretion (**B**). **C**, part of cell wall with protuberances that characterize an actively secreting hair. The black line outlining the protuberances indicates the plasma membrane. The apparently detached protuberances are to be thought of as connected to the wall at levels other than that of the drawing. (From Esau, 1977; constructed from data in Fahn and Rachmilevitz, 1970.)

have wall ingrowths indicative of transfer cells (Fahn, 1979a, b, 2000). The secretory cells in *Maxillaria coccinea* (Orchidaceae) are collenchymatous (Stpiczyńska et al., 2004).

The two main mechanisms cited in the literature for the transport of nectar from the protoplast of the secretory cells are granulocrine and eccrine. Holocrine secretion rarely is cited for nectaries but has been well documented for the floral nectaries of two *Helleborus* species (Vesprini et al., 1999). In these nectaries the nectar is released by rupture of the wall and cuticle of each epidermal cell. This nectar has a high sugar content, mainly sucrose, and also contains lipids and proteins.

The outer surface of the nectaries is covered by a cuticle. In nectaries that secrete through trichomes, the side walls of the lower part of unicellular trichomes or those of the lowest cells (stalk cells) of multicellular trichomes are completely cutinized (Fahn, 1979a, b), as is characteristic of almost all secretory trichomes. The three floral nectaries described here demonstrate some of the variation in nectary structure and mode of secretion.

### The Nectaries of *Lonicera japonica* Exude Nectar from Unicellular Trichomes

In the floral nectaries of *Lonicera japonica* the nectar-secreting cells are short unicellular hairs, or trichomes, located in a limited area of the inner epidermis of the corolla tube (Fahn and Rachmilevitz, 1970). Each of these hairs consists of a narrow stalk-like part and an upper spherical head (Fig. 16.10). The stalk protrudes above the neighboring ordinary epidermal cells. In addition to the secretory epidermis the nectariferous tissue consists of subepidermal parenchyma, which abuts the vascular bundles of the corolla tube. Young hairs have a single large vacuole and a tightly fitting cuticle composed of thick and thin areas. In actively secreting hairs, the vacuole volume is reduced, small vacuoles replace the single large one, and the cuticle of the head is detached. It is presumed that expansion of the cuticle occurs in the thin areas and that the nectar diffuses through these areas. At this stage the upper part of the secretory cell wall bears numerous ingrowths, which form an extensive plasma membrane-lined labyrinth. Extensive stacks of endoplasmic reticulum-lined cisternae apparently give rise to vesicles that fuse with the plasma membrane and release the nectar (granulocrine secretion).

## The Nectaries of *Abutilon striatum* Exude Nectar from Multicellular Trichomes

The floral nectaries of *Abutilon striatum* (Malvaceae) consist of multicellular trichomes located on the lower inner (adaxial) side of the fused sepals (Fig. 16.11; Findlay and Mercer, 1971). Each trichome is single-celled at its base, stalk, and tip but multiseriate between stalk and tip (Fig. 16.12A). An extensive system of vascular strands, in which the phloem predominates, underlies each nectary. Only two layers of subglandular parenchyma cells separate the trichomes from the nearest sieve tubes. The nectary trichomes of *Abutilon* secrete copious amounts of sucrose, fructose, and glucose, the nectar emerging through transient (short-lived) pores in the cuticle at the trichome tips (Findlay and Mercer, 1971; Gunning and Hughes, 1976). Inasmuch as plasmodesmata interconnect all of the cells from the phloem to the trichome tips, it seems likely that the pre-nectar moves symplastically all the way, that is, from the phloem to the trichome tips, where it is secreted as nectar. During the secretory phase the trichome cells contain an extensive system of endoplasmic reticulum, dubbed "secretory reticulum" by Robards and Stark (1988). Robards and Stark (1988) proposed that secretion from the *Abutilon* nectaries is neither eccrine nor granulocrine. The presence of the secretory reticulum and physiological data has led them to conclude that pre-nectar is actively loaded into the secretory reticulum of all trichome cells (Fig. 16.12B). The resultant increase in hydrostatic pressure within the reticulum then effects the opening of "sphincters," which connect the lumen of the reticulum to the outside of the plasma membrane. The forcibly expelled nectar then moves apoplastically

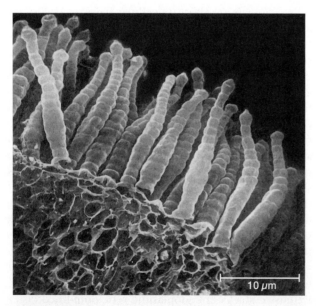

**FIGURE 16.11**

Nectar-secreting trichomes of *Abutilon pictum*. (From Fahn, 2000. © 2000, with permission from Elsevier.)

under the cuticle until it reaches the transient cuticle pores that overlay the tip cell. The floral nectaries of *Hibiscus rosa-sinensis* exhibit close morphological, structural, and presumably physiological similarity to those of *Abutilon* (Sawidis et al., 1987a, b, 1989; Sawidis, 1991).

## The Nectaries of *Vicia faba* Exude Nectar via Stomata

The floral nectary of *Vicia faba* consists of a disc that encircles the base of the gynoecium and bears a prominent projection on the free-stamen side of the flower (Fig. 16.13A; Davis et al., 1988). (In many legumes, including *Vicia*, 9 of the 10 stamens are united in one bundle and the tenth stamen is solitary, or free. In papilinoid legumes, floral nectaries occur most frequently as a disc, although considerable variation exists in their morphology; Waddle and Lersten, 1973.) Whereas several large, modified stomata occur at the tip of the projection (Fig. 16.13B), stomata are absent elsewhere on the projection and on the disc. Epidermal cells of the projection possess wall ingrowths along their outer walls. The disc consists of 9 or 10 subepidermal layers of relatively small and closely packed nectariferous (parenchyma) cells. Wall ingrowths develop next to intercellular spaces at the base of the projection and in cells of the projection itself, where the intercellular spaces are larger and the wall ingrowths more abundant than in the disc. The nectary is vascularized exclusively by phloem, which originates from the vascular bundles destined for the stamens. Some sieve tubes end in the disc, but most of the phloem enters the projection, forming a central strand (Fig. 16.14) that extends to within 12 cells below the projection tip. The sieve-tube elements are accompanied by large, densely staining companion cells with wall ingrowths. Although the Golgi apparatus is not particularly well developed in the epidermal and nectariferous cells of the projection, the endoplasmic reticulum cisternae are quite prominent and often in close association with the plasma membrane of these cells. These ultrastructural features favor the existence of a granulocrine secretion mechanism. This stands in contrast with the condition in the foliar nectary of *Trifolium pratense*, another legume, in which proliferation of the endoplasmic reticulum apparently does not occur. Accordingly, Eriksson (1977) concluded that an eccrine secretory mechanism is operative in the *Trifolium* nectary. Razem and Davis (1999) arrived at a similar conclusion for the floral nectary of *Pisum sativum*. Only sucrose was found in the floral nectar of *Vicia faba* (Davis et al., 1988).

## The Most Common Sugars in Nectar Are Sucrose, Glucose, and Fructose

Based on the quantitative relationships between sucrose and glucose/fructose, three nectar types can be

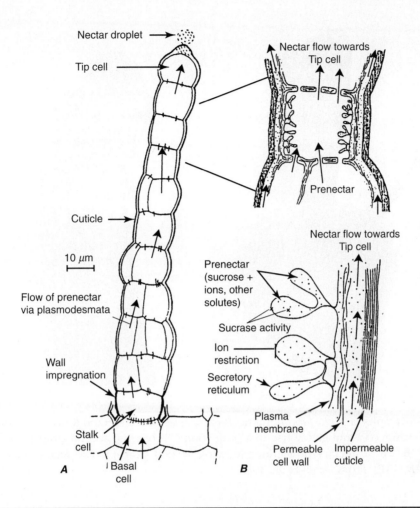

**FIGURE 16.12**

Model of structure-function relations in the nectar-secreting trichome of *Abutilon striatum*. **A**, it has been proposed that pre-nectar moves into the symplast of the trichome via the numerous plasmodesmata in the transverse wall of the stalk cell, whose lateral walls are cutinized and hence impermeable. **B**, in each of the trichome cells some pre-nectar is loaded from the cytoplasm into the secretory reticulum. At this stage a filtration effect takes place, defining the chemical composition of the secreted product. The sucrose is partially hydrolyzed to glucose and fructose. Whether this takes place at the membrane or within the cisternal cavity has not been determined. As loading into the secretory reticulum continues, a hydrostatic pressure builds up until a minute pulse of nectar is forced into the freely permeable cell wall (apoplast) between the plasma membrane and cuticle. The continuing buildup of pressure within this compartment ultimately reaches the level where the pores open in the cuticle over the tip cell and a pulse of nectar is released to the exterior. (From Robards and Stark, 1988.)

distinguished: (1) sucrose-dominant, (2) glucose-dominant, and (3) that with an equal ratio of sucrose to glucose/fructose (Fahn, 1979a, 2000). Small quantities of other substances such as amino acids, organic acids, proteins (mainly enzymes), lipids, mineral ions, phosphates, alkaloids, phenolics, and antioxidants also may be present (Baker and Baker, 1983a, b; Fahn, 1979a; Bahadur et al., 1998).

Nectar has its origin in the phloem as sieve-tube sap, which moves symplastically from the sieve tubes to the secretory cells. Along the way the pre-nectar may be modified in the nectariferous tissue by enzymic activity, or even after its secretion by reabsorption of nectar

(Nicolson, 1995; Nepi et al., 1996; Koopowitz and Marchant, 1998; Vesprini et al., 1999). The reabsorption of nectar helps to minimize nectar theft by opportunists that do not effect pollination and to recover some of the energy stored in the sugars.

A unique manner of nectar release is found in some neotropical, mainly Andean, genera of the Melastomataceae (Vogel, 1997). Most members of this family have no floral nectaries, yet many of them exude copious nectar from the stamen filaments. The nectar has its origin as sieve-tube sap that leaks from the central vascular bundle of the filament. It finds its way out of the filament through slit-like ruptures in the filament tissue. Being more or

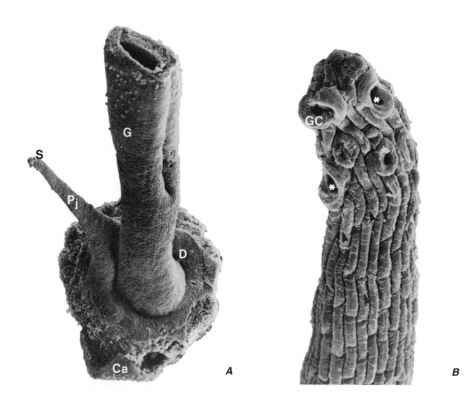

**FIGURE 16.13**

Scanning electron micrographs of *Vicia faba* floral nectary. **A**, base of a dissected flower, showing a nectary disk (D) surrounding the gynoecium (G), and giving rise to a projection (Pj) bearing several stomata (S) at its tip; part of the calyx (Ca) is evident. **B**, tip of nectary projection, showing several stomata. The guard cells (GC) surround large pores (asterisks). (Both, ×310. From Davis et al., 1988.)

less pure sieve-tube sap, the nectar consists predominantly of sucrose. The composition of this nectar contrasts sharply with that of the nectar of *Capsicum annuum* (Solanaceae; Rabinowitch et al., 1993) and *Thryptomene calycina* (Myrtaceae; Beardsell et al., 1989), which contains only fructose and glucose. Relatively few plants produce nectar containing no sucrose.

A correlation exists between the type of vascular tissue supplying the nectaries and the sugar concentration. Nectaries supplied by phloem alone secrete higher concentrations than those vascularized by both xylem and phloem or primarily by xylem. In flowers of several species of Brassicaceae, including *Arabidopsis thaliana*, with two pairs of nectaries (Fig. 16.15; lateral and median), the lateral nectaries were found to produce on average 95% of the total nectar carbohydrate (Davis et al., 1998). The lateral nectaries are supplied with an abundance of phloem (Fig. 16.16), whereas the median nectaries receive a comparatively small number of sieve tubes. The gene *CRABS CLAW (CRC)* has been shown to be required for the initiation of nectary development in *Arabidopsis thaliana; crc* flowers lack nectaries (Bowman and Smyth, 1999). However, whereas *CRABS CLAW* is essential for nectary formation, its ectopic expression is not sufficient to induce ectopic nectary

formation (Baum et al., 2001). Multiple factors have been shown to act in restricting the nectaries to the flower in *Arabidopsis*, surprisingly, among them the *LEAFY* and *UNUSUAL FLORAL ORGANS* genes (Baum et al., 2001).

The composition of the nectar secreted may differ between male and female flowers of the same plant and even between nectaries of the same flower. Moreover the sugar composition can change considerably during the period of nectar secretion, as demonstrated for the floral nectaries of *Strelitzia reginae* (Kronestedt-Robards et al., 1989). In the study of individual flowers of Brassicaceae species, nectar from the lateral nectaries consistently contained higher quantities of glucose than fructose, whereas that from the median nectaries possessed higher quantities of fructose than glucose (Davis et al., 1998). In *Cucurbita pepo* the female flower produces sweeter nectar with a lower protein content than the male flower (Nepi et al., 1996).

In many plants the pre-nectar originating in the phloem accumulates as starch grains in plastids of the nectariferous cells. Upon hydrolysis the starch serves as the main source of sugar at anthesis (Durkee et al., 1981; Zer and Fahn, 1992; Belmonte et al., 1994; Nepi et al., 1996; Gaffal et al., 1998).

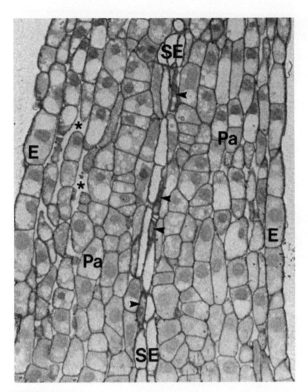

**FIGURE 16.14**

Longitudinal section of *Vicia faba* nectary projection showing central strand of sieve elements (SE) with associated companion cells (arrowheads). Parenchyma cells (Pa) and intercellular spaces (asterisks) surround the phloem strand. Other detail: E, epidermis. (×315. From Davis et al., 1988.)

### Structures Intermediate between Nectaries and Hydathodes Also Exist

As mentioned previously, the hydathodes of *Populus deltoides* leaves, which guttate water containing varying concentrations of sugar, were considered by Curtis and Lersten (1974) to act as nectaries under certain conditions. Hydathodes intergrade both structurally and functionally with extrafloral nectaries in many plants (Janda, 1937; Frey-Wyssling and Häusermann, 1960; Pate and Gunning, 1972; Elias and Gelband, 1977; Belin-Depoux, 1989). In the leaves of *Impatiens balfourii* a gradation occurs in the structures on the leaf teeth from nectary to hydathode (Elias and Gelband, 1977). The intermediate structures have the elongate form and stomata of hydathodes and a rounded tip with raphides and cells characteristic of the nectaries, which led Elias and Gelband (1977) to postulate that the foliar nectaries of *Impatiens* evolved from hydathodes. (The reader is referred to Vogel's 1998 discussion on hydathodes as the likely evolutionary precursors of floral nectaries.) Regardless of any possible evolutionary relationship between hydathodes and nectaries, the principal source

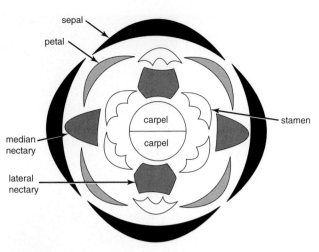

**FIGURE 16.15**

Stylized floral diagram of *Brassica rapa* and *B. napus*, showing location of lateral and median nectaries. (After Davis et al., 1996, by permission of Oxford University Press.)

of secreted solutes in hydathodes is the transpiration stream, and the ultimate vascular tissue in them is xylem. In nectaries the principal source of the sugars is the assimilate stream, and in many nectaries the ultimate vascular tissue is phloem only.

## ▌COLLETERS

**Colleters**—the term is derived from the Greek *colla*, glue, referring to the sticky secretions from these structures—are common on bud scales and young leaves (Thomas, 1991). The fluid they produce is mucilaginous or resinous in nature and insoluble in water. Colleters develop on young foliar organs, and their sticky secretion permeates and covers the entire bud. When the bud opens and the leaves expand, the colleters commonly dry up and fall off. The probable function of colleters is to provide a protective coating for the dormant buds and to protect the developing meristem and young, differentiating leaves or their stipules.

Colleters are not trichomes. They are emergences formed from both epidermal and subepidermal tissues. Based on their morphology, several types of colleters can be recognized. The most common type, designated the **standard type** by Lersten (1974a, b) in his study of the colleters in the Rubiaceae, consists of a multiseriate axis of elongate cells ensheathed by a palisade-like epidermis (the secretory epithelial layer) whose cells are closely appressed to each other and covered with a thin cuticle (Fig. 16.17A). Standard type colleters of the Apocynaceae are differentiated into a long head and a short stalk, which is devoid of secretory cells. In *Allamanda* the stalk is green and photosynthetic, whereas

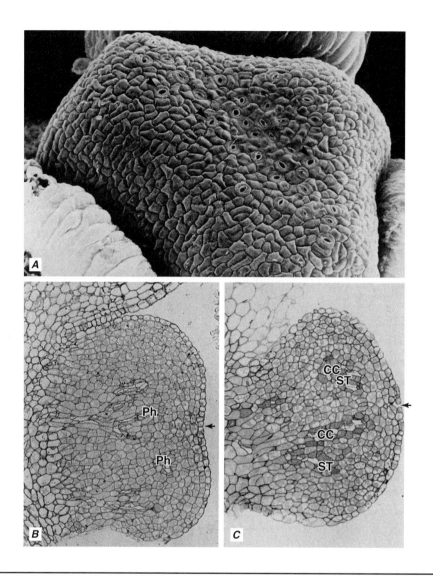

**FIGURE 16.16**

Lateral floral nectary of *Brassica napus*. **A**, scanning electron micrograph of lateral nectary showing numerous open stomata on surface of the nectary. Note the slight depression in the middle of the nectary. **B**, longitudinal section showing that the phloem strands (Ph) penetrate the gland interior. The arrowhead points to the depression at the surface of the nectary. **C**, oblique section through one of the nectary lobes showing sieve-tube elements (ST) and densely staining companion cells (CC). Note open stoma (arrowhead) on the nectary surface. (A, ×200; B, ×110; C, ×113. From Davis et al., 1986.)

the head is pale, yellowish and glandular in nature (Ramayya and Bahadur, 1968). Additional colleter types recognized by Lersten (1974a, b) in the Rubiaceae on the basis of their morphology are the **reduced standard colleter**, with epidermal cells that are quite short (Fig. 16.17B); the **dendroid colleter**, consisting of a filamentous stalk from which radiate many elongated epidermal cells (Fig. 16.17C); and the **brushlike colleter**, lacking an elongate axis but with elongate epidermal cells (Fig. 16.17D). Other morphological types of colleter have been recognized in the Rubiaceae (Robbrecht, 1987) and in other taxa. In some species of *Piriqueta* (Turneraceae) a morphological transition is apparent

between the colleters and extrafloral nectaries of the leaves (González, 1998). However, none of these colleters are vascularized, nor do they secrete a conspicuous sugary secretion. Different types of crystals commonly occur in colleters, and they are of taxonomic significance.

The secretory cells of colleters contain abundant mitochondria and Golgi bodies and an extensive system of endoplasmic reticulum (both rough and smooth) (Klein et al., 2004). Undoubtedly, the mechanism involved in the release of the secretory material is granulocrine. Typically the secretory material accumulates beneath the cuticle, which eventually ruptures, although

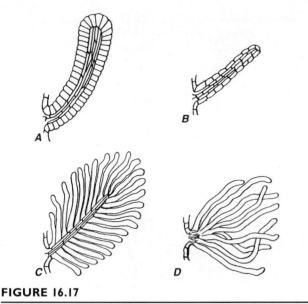

**FIGURE 16.17**

Different types of colleters. **A**, the most common standard type. **B**, the reduced standard. **C**, the dendroid. **D**, the brushlike. (From Lersten, 1974a. © Blackwell Publishing.)

there are other modes of exudation from colleters (Thomas, 1991).

An interesting symbiotic relationship exists between bacteria and the leaves of certain species of the Rubiaceae and Myrsinaceae (Lersten and Horner, 1976; Lersten, 1977). This relationship manifests itself in the form of **bacterial leaf nodules**. The bacteria live in the mucilaginous fluid secreted by colleters. Some of the bacteria-laden mucilage enters the substomatal chambers of young developing leaves and nodule development is initiated. Additional bacteria-laden mucilage becomes trapped in the ovules of developing flowers, and the bacteria become incorporated into the seed. The bacteria thus are carried internally from one generation to the next.

# ▌OSMOPHORES

The fragrance of flowers is commonly produced by volatile substances—mainly terpenoids and aromatic compounds—distributed throughout the epidermis of perianth parts (Weichel, 1956; Vainstein et al., 2001). In some plants, however, the fragrance originates in special glands known as **osmophores**, a term derived from the Greek words *osmo*, odor, and *pherein*, to bear. The term was first used by Arcangeli in 1883 (as cited in Vogel, 1990) for the fragrant spadix of certain members of the Araceae. The fragrances are attractive to the flowers' pollinators. Some bees (male euglossine bees) presumably use the fragrance produced by osmophores of the subtribe Stanhopeinae (Orchidaceae) as a precursor for a sex pheromone (Dressler, 1982).

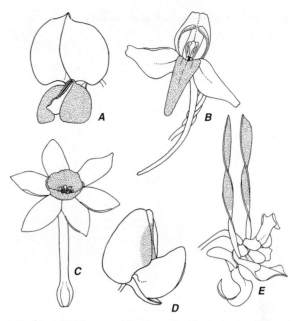

**FIGURE 16.18**

Flowers tested by neutral-red staining for location of osmophores (stippled)—flower parts containing secretory tissue responsible for emission of fragrance. **A**, *Spartium junceum*; **B**, *Platanthera bifolia*; **C**, *Narcissus jonquilla*; **D**, *Lupinus Cruckshanksii;* **E**, *Dendrobium minax.* (After Vogel, 1962.)

Examples of osmophores are found in Asclepiadaceae, Aristolochiaceae, Calycanthaceae, Saxifragaceae, Solanaceae, Araceae, Burmanniaceae, Iridaceae, and Orchidaceae. Various floral parts may be differentiated as osmophores, and they may assume the form of flaps, cilia, or brushes. The extension of the spadix, called the appendix, of the Araceae (Weryszko-Chmielewska and Stpiczyńska, 1995; Skubatz et al., 1996) and the insect-attracting tissue in the flowers of Orchidaceae (Pridgeon and Stern, 1983, 1985; Curry et al., 1991; Stpiczyńska, 1993) are osmophores. Osmophores may be identified by staining them with neutral red in whole flowers submerged in a solution of the dye (Fig. 16.18; Stern et al., 1986).

Osmophores consist of glandular tissue usually several cell layers in depth (Fig. 16.19). The outer layer is formed by the epidermis, which is covered by a very thin cuticle. Two to five subepidermal layers may differ considerably in density from the subjacent ground tissue or intergrade imperceptibly with it (Curry et al., 1991). The glandular tissue may be compact or it may be permeated by intercellular spaces. In *Ceropegia elegans* (Asclepiadaceae) the lower glandular layers are vascularized with vein endings consisting of phloem only, a condition found in other osmophores, according to Vogel (1990). Vogel (1990) has suggested that in *Ceropegia* the epidermal layer of the osmophore has the task

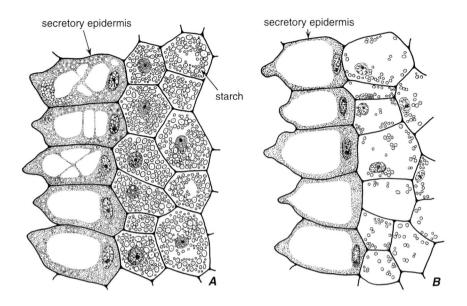

**FIGURE 16.19**

Sections through secretory tissue of osmophores of a *Ceropegia stapeliaeformis* flower. **A**, at beginning of secretory activity, with much starch. **B**, after emission of fragrance: secretory cells with reduced cytoplasmic density, and starch depleted in tissue below epidermis. (After photographs by Vogel, 1962.)

of accumulation and release of the fragrance substance and that a major part of fragrance synthesis takes place in the other glandular layers. Stomata occur in the surface of the osmophore.

The cells of the osmophore contain numerous amyloplasts and mitochondria. Endoplasmic reticulum, particularly smooth-surfaced, is abundant, but Golgi bodies are scarce. Starch grains and lipid droplets are abundant in the glandular cells at the beginning of secretory activity. The presence of lipid droplets in the cytosol and of plastoglobuli in the amyloplasts commonly are associated with fragrance production (Curry et al., 1991). The emission of the volatile secretions is of short duration and is associated with a utilization of large amounts of storage products. At post-anthesis, the cells of the osmophore are extremely vacuolate (Fig. 16.19B), and few amyloplasts, mitochondria, and endoplasmic reticulum cisternae remain (Pridgeon and Stern, 1983; Stern et al., 1987).

Several cellular components—rough and smooth endoplasmic reticulum, plastids, and mitochondria—have been implicated in synthesis of the terpenoid components of the fragrances (Pridgeon and Stern, 1983, 1985; Curry, 1987). Both granulocrine and eccrine secretion have been reported (Kronestedt-Robards and Robards, 1991). In addition, ultrastructural evidence indicates that in the appendix osmophore of *Sauromatum guttatum* (the voodoo lily) the endoplasmic reticulum may fuse with the plasma membrane, forming a channel for the release of volatiles to the exterior of the cell (Skubatz et al., 1996). It is believed that heat released

by the inflorescence of members of the Araceae during anthesis serves to volatize the lipids (Meeuse and Raskin, 1988; Skubatz et al., 1993; Skubatz and Kunkel, 1999). At least nine different chemical classes are liberated during the thermogenic activity in the voodoo lily (Skubatz et al., 1996).

## GLANDULAR TRICHOMES SECRETING LIPOPHILIC SUBSTANCES

Glandular trichomes secreting lipophilic substances are found in many eudicot families (e.g., Asteraceae, Cannabaceae, Fagaceae, Geraniaceae, Lamiaceae, Plumbaginaceae, Scrophulariaceae, Solanaceae, and Zygophyllaceae). Among the lipophilic substances secreted by the trichomes are terpenoids (such as essential oils and resins), fats, waxes, and flavonoid aglycones. Terpenoids are the most commonly occurring lipophilic substances in such trichomes. They serve a variety of functions in plants, including deterring herbivores, attracting pollinators (Duke, 1991; Lerdau et al., 1994; Paré and Tumlinson, 1999; Singsaas, 2000), and, because of their sticky condition, aiding in the dispersal of certain fruits (Heinrich et al., 2002).

Probably the best-studied of glandular trichomes secreting lipophilic substances are those of the mint family, Lamiaceae. Two main types of glandular trichomes are found in the Lamiaceae, peltate and capitate (Fig. 16.20). **Peltate trichomes** consist of a basal cell, a short stalk cell, the lateral walls of which are com-

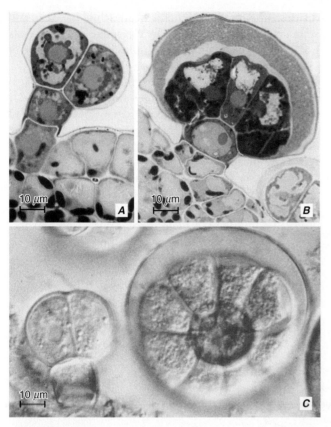

**FIGURE 16.20**

Glandular trichomes of *Leonotis leonurus* (Lamiaceae). **A**, sectional view of a fully developed capitate trichome. **B**, sectional view of a mature peltate trichome, showing secretory material within subcuticular space. **C**, fresh leaf section, showing lateral view of capitate trichome (left) and top view of peltate trichome (right). (From Ascensão et al., 1995, by permission of Oxford University Press.)

pletely cutinized, and a broad head of 4 to 18 secretory cells arranged in a single layer in one or two concentric rings. The secretory product of the peltate trichome accumulates in a large subcuticular space formed by the detachment of the cuticle together with the outer layer of the cell wall (Danilova and Kashina, 1988; Werker et al., 1993; Ascensão et al., 1995; Ascensão et al., 1999; Gersbach, 2002). Typically it remains in this space until some external force ruptures the cuticle. The **capitate trichomes** consist of a basal cell, a stalk one to several cells long, and an ovoid or spherical head of 1 to 4 cells. Most species of Lamiaceae have two types of capitate trichome, short-stalked and long-stalked (Mattern and Vogel, 1994; Bosabalidis, 2002). At best, only a slight cuticle elevation has been observed above the head cells of capitate trichomes. Pores are known to occur in the cuticle of some (Amelunxen, 1964; Ascensão and Pais, 1998). The head cells in at least two *Nepeta* species

(*N. racemosa*, Bourett et al., 1994; *N. cataria*, Kolalite, 1998) exhibit wall ingrowths typical of transfer cells. Some Lamiaceae have additional types of glandular trichomes. *Plectranthus ornatus*, for instance, has five morphological types (Ascensão et al., 1999). In *Plectranthus ornatus* (Ascensão et al., 1999) and *Leonotis leonurus* (Ascensão et al., 1997) the peltate trichomes secrete an oleoresin, consisting of essential oils and resiniferous acids, and flavonoid aglycones. In *Leonotis* the capitate trichomes secrete polysaccharides and proteins and small amounts of essential oils and flavonoids.

Monoterpenes are frequent constituents of essential oils and resins. They comprise the major components of the essential oils of the Lamiaceae (Lawrence, 1981). Monoterpene biosynthesis has been specifically localized to the secretory cells of the glandular trichomes in spearmint (*Mentha spicata*) and peppermint (*Mentha piperita*) leaves (Gershenzon et al., 1989; McCaskill et al., 1992; Lange et al., 2000). In peppermint, monoterpene accumulation is restricted to leaves 12 to 20 days of age, the period of maximal leaf expansion (Gershenzon et al., 2000). During this active secretory state numerous leucoplasts sheathed by abundant smooth endoplasmic reticulum populate the head cells, an ultrastructural syndrome shared with other essential oil- and resin-secreting glands. In the glandular trichomes of peppermint, it has been demonstrated that the pathway of monoterpene biosynthesis originates in the leucoplast (Turner, G., et al., 1999, 2000). Apparently most of the monoterpene biosynthetic enzymes in peppermint are developmentally regulated at the level of gene expression (Lange and Croteau, 1999; McConkey et al., 2000).

# ▌GLANDULAR TRICHOME DEVELOPMENT

Glandular trichomes begin to develop at the initial stages of leaf development. In *Ocimum basilicum* (Lamiaceae), morphologically well-developed glandular trichomes, interspersed among younger trichomes, were clearly present on leaf primordia just 0.5 mm long (Werker et al., 1993). Because of their asynchronous initiation, glandular trichomes at different stages of development occur side by side (Danilova and Kashina, 1988). Development of new trichomes continues as long as a portion of the protoderm remains meristematic. New gland production generally ceases in a given portion of the leaf as soon as it begins to expand. Inasmuch as this occurs last at the base of the leaf, the leaf base is the last part to bear immature glands. The genus *Fagonia*, which belongs to the desert family Zygophyllaceae, may be an exception. Glandular trichomes at various stages of development have been reported to occur side by side in young and fully expanded *Fagonia* leaves (Fahn and Shimony, 1998). In some leaves leaf expansion is accompanied by a decrease in density of the glandular trichomes (Werker et al., 1993; Ascensão et al., 1997; Fahn and Shimony, 1996). Some workers contend that the number of gland initials is fixed at the time of leaf emergence (Werker and Fahn, 1981; Figueiredo and Pais, 1994; Ascensão et al., 1997), whereas others report that an increase occurs in trichome numbers throughout leaf development (Turner, J. C., et al., 1980; Croteau et al., 1981; Maffei et al., 1989).

Development of peltate glandular trichomes is fairly uniform among Lamiaceae, and is exemplified here by the trichomes of *Origanum* (Fig. 16.21; Bosabalidis and Exarchou, 1995; Bosabalidis, 2002). It begins with a single protodermal cell. After elongating somewhat, the protodermal cell divides periclinally and asymmetrically twice to give rise to a basal cell, a stalk cell, and the initial cell of the trichome head. The latter cell then divides anticlinally to form the head, while the basal and stalk cells increase in size. Once formed, the head cells become involved with the synthesis and secretion of essential oil, which accumulates beneath the cuticle. During intense secretion the cuticle over the apical walls of the head cells becomes fully detached and a large dome-shaped subcuticular cavity becomes filled with the essential oil. When secretion is completed, the head cells and stalk cell degenerate. The basal cell retains its protoplast.

Ascensão and Pais (1998) and Figueiredo and Pais (1994) distinguish three stages in the development of the glandular trichomes in *Leonotis leonurus* (Lamiaceae) and *Achillea millefolium* (Asteraceae), respectively: presecretory, secretory, and postsecretory. The presecretory stage begins with the protodermal cell and ends when the trichome is fully formed. The events associated with the secretory and postsecretory stages are self-obvious.

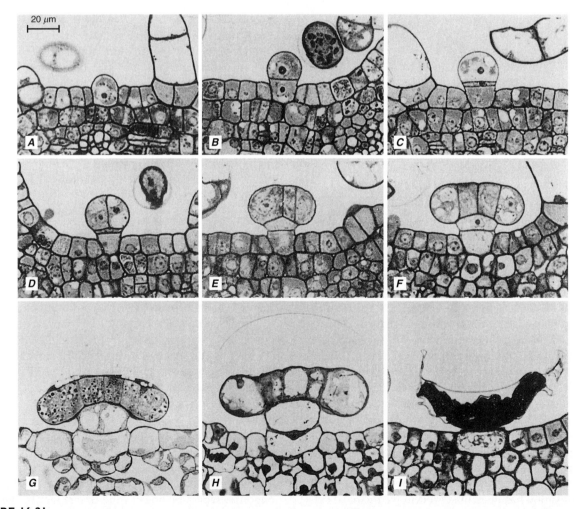

**FIGURE 16.21**

Successive stages of glandular trichome development in the *Origanum × intercedens* leaf, as seen in transverse sections of leaf. (From Bosabalidis and Exarchou, 1995. © 1995 by The University of Chicago. All rights reserved.)

# THE GLANDULAR STRUCTURES OF CARNIVOROUS PLANTS

**Carnivorous plants** are plants that can attract and trap insects and then digest them and absorb the products of digestion (Fahn, 1979a; Joel, 1986; Juniper et al., 1989). Several trap morphologies or mechanisms have evolved for the capture of prey, including pitchers (*Nepenthes*, *Darlingtonia*, *Sarracenia*), suction traps (*Utricularia*, *Biovularia*, *Polypompholyx*), adhesive traps (*Pinguicula*, *Drosera*), and snap traps (*Dionaea*, *Aldrovanda*). In addition several types of glands occur in carnivorous plants, most commonly alluring glands, mucilage glands, and digestive glands. The alluring glands generally are nectaries. In some plants, mucilage and digestive enzymes are secreted by the same type of gland; in *Drosera*, for instance, the stalked glands, or tentacles, secrete both mucilage and enzymes and function also in absorption of the digestive products. Several different glandular trichomes occur on the trap of the bladderwort *Utricularia* (Fineran, 1985), including four-armed (quadrifids) and two-armed (bifids) trichomes on the inside of the trap, squat dome-shaped external glands on the outside, and closely arranged epithelial cells that line the threshold of the doorway (Fig. 16.22). The internal glands function in the removal of excess water from the trap lumen after firing of the trap (i.e., after the sudden sucking in of prey and a related increase in volume of the trap) and in solute transport and digestive activities. The external glands of the trap excrete water and those of the doorway secrete mucilage, which seals the door after firing. The mucilage may also attract prey.

The leaves of the butterwort *Pinguicula* possess two kinds of glands, stalked glands and sessile glands. The stalk glands produce mucilage, which is used to capture prey. They consist of a large basal cell located in the epidermis, a long stalk cell, and a columellar cell that supports a head typically composed of 16 radiating secretory cells. The head cells are characterized by the presence of large mitochondria with well-developed cristae, a conspicuous population of Golgi bodies with many associated vesicles, and wall ingrowths on their radial walls. Whereas the outer surfaces of the secretory cells have at best a poorly developed cuticle, the free walls of the basal cell are completely cutinized.

Most attention has been given to the sessile, digestive glands of *Pinguicula* (Heslop-Harrison and Heslop-Harrison, 1980, 1981; Vassilyev and Muravnik, 1988a, b). These glands are composed of three functional compartments: (1) a basal reservoir cell, (2) an

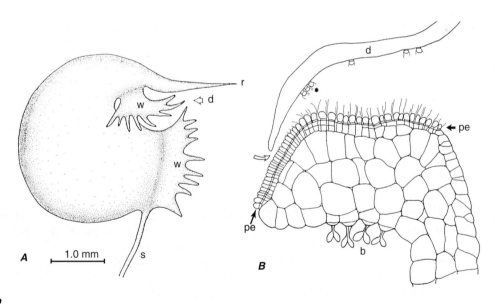

**FIGURE 16.22**

The trap of bladderwort (*Utricularia*). **A**, diagram of external morphology of trap. The doorway (d) is on the dorsal side distal to the stalk (s). Several wings (w) and a rostrum (r) occur near the doorway and are believed to aid in funneling prey toward the entrance. **B**, structure of doorway in longitudinal section (exterior of trap to the right, interior to the left). The door (d) consists of two layers of cells (not shown) with groups of dome-shaped glands on its outer surface. The door is shown ajar (clear arrow) as if trap were beginning to fire. Pavement epithelium (pe) consists of orderly array of closely packed glandular hairs, which form a pseudo-epidermal layer that extends from entrance of doorway to inner lip of threshold. The door glands and terminal cells of the pavement epithelium have ruptured cuticles extending out from them after the release of mucilage. The threshold is several cells thick and bears bifid trichomes (b) on its undersurface. (From Fineran, 1985. Reproduced with the permission of the publisher.)

intervening cell of endodermal character, and (3) a group of secretory head cells. The lateral walls of the intervening (endodermoid) cell are completely cutinized, forming "Casparian strips" to which the plasma membrane is tightly attached. The intervening cell contains abundant storage lipid and numerous mitochondria. At maturity the bounding cuticle on the secretory head cells is discontinuous. During maturation, the secretory head cells form a special layer, called the slime layer, between the plasma membrane and the cell wall. This layer serves for the storage of digestive enzymes apparently synthesized on the rough endoplasmic reticulum of the head cells and transferred into the slime layer and vacuoles. During maturation the rough endoplasmic reticulum of the secretory cells undergoes a striking fourfold increase in volume. It has been suggested that digestive enzymes may be transferred directly from the endoplasmic reticulum into the vacuoles and slime layer through continuity of the reticulum membranes with the tonoplast and plasma membrane (Vassilyev and Muravnik, 1988a). The Golgi apparatus also is believed to be involved with the secretion of enzymes in *Pinguicula*. The secretory cells remain highly active throughout the period of prey digestion and absorption of nutrients. After the digestion and absorption cycle, destructive processes characteristic of senescent cells are initiated in the glands (Vassilyev and Muravnik, 1988b).

# STINGING HAIRS

**Stinging hairs** are known to occur in four families of eudicots: Urticaceae, Euphorbiaceae, Loasaceae, and Hydrophyllaceae. Although commonly interpreted as trichomes, they might more correctly be considered emergences because they are outgrowths that involve both epidermal and subepidermal cell layers (Thurston, 1974, 1976).

The stinging hairs of *Urtica* (nettle) consist of four morphologically distinct regions: (1) a spherical tip, (2) a neck, (3) a shaft that resembles a fine capillary tube, and (4) a bulbous base supported by a pedestal derived from epidermal and subepidermal cells (Fig. 16.1G; Thurston, 1974; Corsi and Garbari, 1990). The apical wall of the stinging hair is silicized and extremely fragile so that, when the hair is touched, the spherical tip breaks off at the neck, leaving a sharp edge. This edge readily penetrates the skin, and pressure on the unsilicified bulbous base forces liquid into the wound.

The stinging hairs of *Urtica* contain histamine, acetylcholine, and serotonin. It is questionable, however, whether these substances are those responsible for irritation (Thurston and Lersten, 1969; Pollard and Briggs, 1984). It has long been presumed that stinging hairs serve as a protection against herbivores.

# REFERENCES

AMELUNXEN, F. 1964. Elektronenmikroskopische Untersuchungen an den DRÜSENHAAREN VON *Mentha piperita* L. *Planta Med.* 12, 121–139.

ARUMUGASAMY, K., R. B. SUBRAMANIAN, and J. A. INAMDAR. 1990. Cyathial nectaries of *Euphorbia neriifolia* L.: ultrastructure and secretion. *Phytomorphology* 40, 281–288.

ASCENSÃO, L., and M. S. PAIS. 1998. The leaf capitate trichomes of *Leonotis leonurus*: Histochemistry, ultrastructure, and secretion. *Ann. Bot.* 81, 263–271.

ASCENSÃO, L., N. MARQUES, and M. S. PAIS. 1995. Glandular trichomes on vegetative and reproductive organs of *Leonotis leonurus* (Lamiaceae). *Ann. Bot.* 75, 619–626.

ASCENSÃO, L., N. MARQUES, and M. S. PAIS. 1997. Peltate glandular trichomes of *Leonotis leonurus* leaves: Ultrastructure and histochemical characterization of secretions. *Int. J. Plant Sci.* 158, 249–258.

ASCENSÃO, L., L. MOTA, and M. DE M. CASTRO. 1999. Glandular trichomes on the leaves and flowers of *Plectranthus ornatus*: Morphology, distribution and histochemistry. *Ann. Bot.* 84, 437–447.

BAHADUR, B., C. S. REDDI, J. S. A. RAJU, H. K. JAIN, and N. R. SWAMY. 1998. Nectar chemistry. In: *Nectary Biology: Structure, Function and Utilization*, pp. 21–39, B. Bahadur, ed. Dattsons, Nagpur, India.

BAKER, H. G., and I. BAKER. 1983a. A brief historical review of the chemistry of floral nectar. In: *The Biology of Nectaries*, pp. 126–152, B. Bentley and T. Elias, eds. Columbia University Press, New York.

BAKER, H. G., and I. BAKER. 1983b. Floral nectar sugar constituents in relation to pollinator type. In: *Handbook of Experimental Pollination Biology*, pp. 117–141, C. E. Jones and R. J. Little, eds. Scientific and Academic Editions, New York.

BALSAMO, R. A., M. E. ADAMS, and W. W. THOMSON. 1995. Electrophysiology of the salt glands of *Avicennia germinans*. *Int. J. Plant Sci.* 156, 658–667.

BATANOUNY, K. H. 1993. Adaptation of plants to saline conditions in arid regions. In: *Towards the Rational Use of High Salinity Tolerant Plants*, vol. I, *Deliberations about high salinity tolerant plants and ecosystems*, pp. 387–401, H. Lieth and A. A. Al Masoom, eds. Kluwer Academic, Dordrecht.

BAUM, S. F., Y. ESHED, and J. L. BOWMAN. 2001. The *Arabidopsis* nectary is an ABC-independent floral structure. *Development* 128, 4657–4667.

BEARDSELL, D. V., E. G. WILLIAMS, and R. B. KNOX. 1989. The structure and histochemistry of the nectary and anther secretory tissue of the flowers of *Thryptomene calycina* (Lindl.) Stapf (Myrtaceae). *Aust. J. Bot.* 37, 65–80.

BELIN-DEPOUX, M. 1989. Des hydathodes aux nectaires chez les plantes tropicales. *Bull. Soc. Bot. Fr. Actual. Bot.* 136, 151–168.

BELMONTE, E., L. CARDEMIL, and M. T. K. ARROYO. 1994. Floral nectary structure and nectar composition in *Eccremocarpus scaber* (Bignoniaceae), a hummingbird-pollinated plant of central Chile. *Am. J. Bot.* 81, 493–503.

BOSABALIDIS, A. M. 2002. Structural features of *Origanum* sp. In: *Oregano: The Genera Origanum and Lippia*, pp. 11–64, S. E. Kintzios, ed. Taylor and Francis, London.

BOSABALIDIS, A. M., and F. EXARCHOU. 1995. Effect of NAA and GA$_3$ on leaves and glandular trichomes of *Origanum × intercedens* Rech.: Morphological and anatomical features. *Int. J. Plant Sci.* 156, 488–495.

BOSABALIDIS, A. M., and W. W. THOMSON. 1984. Light microscopical studies on salt gland development in *Tamarix aphylla* L. *Ann. Bot.* 54, 169–174.

BOURETT, T. M., R. J. HOWARD, D. P. O'KEEFE, and D. L. HALLAHAN. 1994. Gland development on leaf surfaces of *Nepeta racemosa. Int. J. Plant Sci.* 155, 623–632.

BOWMAN, J. L., and D. R. SMYTH. 1999. *CRABS CLAW,* a gene that regulates carpel and nectary development in *Arabidopsis,* encodes a novel protein with zinc finger and helix-loop-helix domains. *Development* 126, 2387–2396.

BROUILLET, L., C. BERTRAND, A. CUERRIER, and D. BARABÉ. 1987. Les hydathodes des genres *Begonia* et *Hillebrandia* (Begoniaceae). *Can. J. Bot.* 65, 34–52.

BROWN, W. H. 1938. The bearing of nectaries on the phylogeny of flowering plants. *Proc. Am. Philos. Soc.* 79, 549–595.

CARLTON, W. M., E. J. BRAUN, and M. L. GLEASON. 1998. Ingress of *Clavibacter michiganensis* subsp. *michiganensis* into tomato leaves through hydathodes. *Phytopathology* 88, 525–529.

CORSI, G., and S. BOTTEGA. 1999. Glandular hairs of *Salvia officinalis*: New data on morphology, localization and histochemistry in relation to function. *Ann. Bot.* 84, 657–664.

CORSI, G., and F. GARBARI. 1990. The stinging hair of *Urtica membranacea* Poiret (Urticaceae). I. Morphology and ontogeny. *Atti Soc. Tosc. Sci. Nat., Mem., Ser. B,* 97, 193–199.

CROTEAU, R., M. FELTON, F. KARP, and R. KJONAAS. Relationship of camphor biosynthesis to leaf development in sage (*Salvia officinalis*). 1981. *Plant Physiol.* 67, 820–824.

CRUDEN, R. W., S. M. HERMANN, and S. PETERSON. 1983. Patterns of nectar production and plant animal coevolution. In: *The Biology of Nectaries*, pp. 80–125, B. Bentley and T. Elias, eds. Columbia University Press, New York.

CURRY, K. J. 1987. Initiation of terpenoid synthesis in osmophores of *Stanhopea anfracta* (Orchidaceae): A cytochemical study. *Am. J. Bot.* 74, 1332–1338.

CURRY, K. J., L. M. McDOWELL, W. S. JUDD, and W. L. STERN. 1991. Osmophores, floral features, and systematics of *Stanhopea* (Orchidaceae). *Am. J. Bot.* 78, 610–623.

CURTIS, J. D., and N. R. LERSTEN. 1974. Morphology, seasonal variation, and function of resin glands on buds and leaves of *Populus deltoides* (Salicaceae). *Am. J. Bot.* 61, 835–845.

DANILOVA, M. F., and T. K. KASHINA, 1988. Ultrastructure of peltate glands in *Perilla ocymoides* and their possible role in the synthesis of steroid hormones and gibberellins. *Phytomorphology* 38, 309–320.

DAVIES, P. A. 1952. Structure and function of the mature glands on the petals of *Frasera carolinensis. Kentucky Acad. Sci. Trans.* 13, 228–234.

DAVIS, A. R., and B. E. S. GUNNING. 1992. The modified stomata of the floral nectary of *Vicia faba* L. 1. Development, anatomy and ultrastructure. *Protoplasma* 166, 134–152.

DAVIS, A. R., and B. E. S. GUNNING. 1993. The modified stomata of the floral nectary of *Vicia faba* L. 3. Physiological aspects, including comparisons with foliar stomata. *Bot. Acta* 106, 241–253.

DAVIS, A. R., R. L. PETERSON, and R. W. SHUEL. 1986. Anatomy and vasculature of the floral nectaries of *Brassica napus* (Brassicaceae). *Can. J. Bot.* 64, 2508–2516.

DAVIS, A. R., R. L. PETERSON, and R. W. SHUEL. 1988. Vasculature and ultrastructure of the floral and stipular nectaries of *Vicia faba* (Leguminosae). *Can. J. Bot.* 66, 1435–1448.

DAVIS, A. R., L. C. FOWKE, V. K. SAWHNEY, and N. H. LOW. 1996. Floral nectar secretion and ploidy in *Brassica rapa* and *B. napus* (Brassicaceae). II. Quantified variability of nectary structure and function in rapid-cycling lines. *Ann. Bot.* 77, 223–234.

DAVIS, A. R., J. D. PYLATUIK, J. C. PARADIS, and N. H. LOW. 1998. Nectar-carbohydrate production and composition vary in relation to nectary anatomy and location within individual flowers of several species of Brassicaceae. *Planta* 205, 305–318.

DE LA FUENTE, M. A. S., and R. J. MARQUIS. 1999. The role of ant-tended extrafloral nectaries in the protection and benefit of a Neotropical rainforest tree. *Oecologia* 118, 192–202.

DRENNAN, P. M., P. BERJAK, J. R. LAWTON, and N. W. PAMMENTER. 1987. Ultrastructure of the salt glands of the mangrove, *Avicennia marina* (Forssk.) Vierh., as indicated by the use of selective membrane staining. *Planta* 172, 176–183.

DRENNAN, P. M., P. BERJAK, and N. W. PAMMENTER. 1992. Ion gradients and adenosine triphosphatase localization in the salt glands of *Avicennia marina* (Forsskål) Vierh. *S. Afr. J. Bot.* 58, 486–490.

DRESSLER, R. L. 1982. Biology of the orchid bees (Euglossini). *Annu. Rev. Ecol. Syst.* 13, 373–394.

DSCHIDA, W. J., K. A. PLATT-ALOIA, and W. W. THOMSON. 1992. Epidermal peels of *Avicennia germinans* (L.) Stearn: A useful system to study the function of salt glands. *Ann. Bot.* 70, 501–509.

DUKE, S. O. 1991. Plant terpenoids as pesticides. In: *Handbook of Natural Toxins*, vol. 6, *Toxicology of Plant and Fungal Compounds*, pp. 269–296, R. F. Keeler and A. T. Tu, eds. Dekker, New York.

DURKEE, L. T., D. J. GAAL, and W. H. REISNER. 1981. The floral and extra-floral nectaries of *Passiflora*. I. The floral nectary. *Am. J. Bot.* 68, 453–462.

ELIAS, T. S. 1983. Extrafloral nectaries: their structure and distribution. In: *The Biology of Nectaries*, pp. 174–203, B. Bentley and T. Elias, eds. Columbia University Press, New York.

ELIAS, T. S., and H. GELBAND. 1977. Morphology, anatomy, and relationship of extrafloral nectaries and hydathodes in two species of *Impatiens* (Balsaminaceae). *Bot. Gaz.* 138, 206–212.

ERIKSSON, M. 1977. The ultrastructure of the nectary of red clover (*Trifolium pratense*). *J. Apic. Res.* 16, 184–193.

ESAU, K. 1977. *Anatomy of Seed Plants*, 2nd ed. Wiley, New York.

FAHN, A. 1953. The topography of the nectary in the flower and its phylogenetic trend. *Phytomorphology* 3, 424–426.

FAHN, A. 1979a. *Secretory Tissues in Plants*. Academic Press, London.

FAHN, A. 1979b. Ultrastructure of nectaries in relation to nectar secretion. *Am. J. Bot.* 66, 977–985.

FAHN, A. 1987. Extrafloral nectaries of *Sambucus niger* L. *Ann. Bot.* 60, 299–308.

FAHN, A. 1988. Secretory tissues in vascular plants. *New Phytol.* 108, 229–257.

FAHN, A. 1998. Nectaries structure and nectar secretion. In: *Nectary Biology: Structure, Function and Utilization*, pp. 1–20, B. Bahadur, ed. Dattsons, Nagpur, India.

FAHN, A. 2000. Structure and function of secretory cells. *Adv. Bot. Res.* 31, 37–75.

FAHN, A., and T. RACHMILEVITZ. 1970. Ultrastructure and nectar secretion in *Lonicera japonica*. In: *New Research in Plant Anatomy*, pp. 51–56, N. K. B. Robson, D. F. Cutler, and M. Gregory, eds. Academic Press, London.

FAHN, A., and C. SHIMONY. 1996. Glandular trichomes of *Fagonia* L. (Zygophyllaceae) species: Structure, development and secreted materials. *Ann. Bot.* 77, 25–34.

FAHN, A., and C. SHIMONY. 1998. Ultrastructure and secretion of the secretory cells of two species of *Fagonia* L. (Zygophyllaceae). *Ann. Bot.* 81, 557–565.

FIGUEIREDO, A. C., and M. S. S. PAIS. 1994. Ultrastructural aspects of the glandular cells from the secretory trichomes and from the cell suspension cultures of *Achillea millefolium* L. ssp. *millefolium*. *Ann. Bot.* 74, 179–190.

FINDLAY, N., and F. V. MERCER. 1971. Nectar production in *Abutilon*. I. Movement of nectar through the cuticle. *Aust. J. Biol. Sci.* 24, 647–656.

FINERAN, B. A. 1985. Glandular trichomes in *Utricularia*: a review of their structure and function. *Isr. J. Bot.* 34, 295–330.

FREY-WYSSLING, A., and E. HÄUSERMANN. 1960. Deutung der gestaltlosen Nektarien. *Ber. Schweiz. Bot. Ges.* 70, 150–162.

GAFFAL, K. P., W. HEIMLER, and S. EL-GAMMAL. 1998. The floral nectary of *Digitalis purpurea* L., structure and nectar secretion. *Ann. Bot.* 81, 251–262.

GALETTO, L., and G. BERNARDELLO. 2004. Floral nectaries, nectar production dynamics and chemical composition in six *Ipomoea* species (Convolvulaceae) in relation to pollinators. *Ann. Bot.* 94, 269–280.

GERSBACH, P. V. 2002. The essential oil secretory structures of *Prostanthera ovalifolia* (Lamiaceae). *Ann. Bot.* 89, 255–260.

GERSHENZON, J., M. MAFFEI, and R. CROTEAU. 1989. Biochemical and histochemical localization of monoterpene biosynthesis in the glandular trichomes of spearmint (*Mentha spicata*). *Plant Physiol.* 89, 1351–1357.

GERSHENZON, J., M. E. McCONKEY, and R. B. CROTEAU. 2000. Regulation of monoterpene accumulation in leaves of peppermint. *Plant Physiol.* 122, 205–213.

GONZÁLEZ, A. M. 1998. Colleters in *Turnera* and *Piriqueta* (Turneraceae). *Bot. J. Linn. Soc.* 128, 215–228.

GUNNING, B. E. S., and J. E. HUGHES. 1976. Quantitative assessment of symplastic transport of pre-nectar into the trichomes of *Abutilon* nectaries. *Aust. J. Plant Physiol.* 3, 619–637.

GUO, A., and J. E. LEACH. 1989. Examination of rice hydathode water pores exposed to *Xanthomonas campestris* pv. *oryzae*. *Phytopathology* 79, 433–436.

HABERLANDT, G. 1918. Physiologische Pflanzenanatomie, 5th ed. W. Engelman, Leipzig.

HÄUSERMANN, E., and A. FREY-WYSSLING. 1963. Phosphatase-Aktivität in Hydathoden. *Protoplasma* 57, 371–380.

HEIL, M., A. HILPERT, R. KRÜGER, and K. E. LINSENMAIR. 2004. Competition among visitors to extrafloral nectaries as a source of ecological costs of an indirect defence. *J. Trop. Ecol.* 20, 201–208.

HEINRICH, G. 1973. Die Feinstruktur der Trichom-Hydathoden von *Monarda fistulosa*. *Protoplasma* 77, 271–278.

HEINRICH, G., H. W. PFEIFHOFER, E. STABENTHEINER, and T. SAWIDIS. 2002. Glandular hairs of *Sigesbeckia jorullensis* Kunth (Asteraceae): Morphology, histochemistry and composition of essential oil. *Ann. Bot.* 89, 459–469.

HESLOP-HARRISON, Y., and J. HESLOP-HARRISON. 1980. Chloride ion movement and enzyme secretion from the digestive glands of *Pinguicula*. *Ann. Bot.* 45, 729–731.

HESLOP-HARRISON, Y., and J. HESLOP-HARRISON. 1981. The digestive glands of *Pinguicula*: Structure and cytochemistry. *Ann. Bot.* 47, 293–319.

HUGOUVIEUX, V., C. E. BARBER, and M. J. DANIELS. 1998. Entry of *Xanthomonas campestris* pv. *campestris* into hydathodes of *Arabidopsis thaliana* leaves: A system for studying early infection events in bacterial pathogenesis. *Mol. Plant-Microbe Interact.* 11, 537–543.

IVANOFF, S. S. 1963. Guttation injuries in plants. *Bot. Rev.* 29, 202–229.

JANDA, C. 1937. Die extranuptialen Nektarien der Malvaceen. *Österr. Bot. Z.* 86, 81–130.

JOEL, D. M. 1986. Glandular structures in carnivorous plants: Their role in mutual and unilateral exploitation of insects. In: *Insects and the Plant Surface*, pp. 219–234, B. Juniper and R. Southwood, eds. Edward Arnold, London.

JUNIPER, B. E., R. J. ROBINS, and D. M. JOEL. 1989. *The Carnivorous Plants.* Academic Press, London.

KEELER, K. H., and R. B. KAUL. 1979. Morphology and distribution of petiolar nectaries in *Ipomoea* (Convolvulaceae). *Am. J. Bot.* 66, 946–952.

KEELER, K. H., and R. B. KAUL. 1984. Distribution of defense nectaries in *Ipomoea* (Convolvulaceae). *Am. J. Bot.* 71, 1364–1372.

KLEIN, D. E., V. M. GOMES, S. J. DA SILVA-NETO, and M. DA CUNHA. 2004. The structure of colleters in several species of *Simira* (Rubiaceae). *Ann. Bot.* 94, 733–740.

KOLALITE, M. R. 1998. Comparative analysis of ultrastructure of glandular trichomes in two *Nepeta cataria* chemotypes (*N. cataria* and *N. cataria* var. *citriodora*). *Nord. J. Bot.* 18, 589–598.

KOOPOWITZ, H., and T. A. MARCHANT. 1998. Postpollination nectar reabsorption in the African epiphyte *Aerangis verdickii. Am. J. Bot.* 85, 508–512.

KRONESTEDT-ROBARDS, E., and A. W. ROBARDS. 1991. Exocytosis in gland cells. In: *Endocytosis, Exocytosis and Vesicle Traffic in Plants*, pp. 199–232, C. R. Hawes, J. O. D. Coleman, and D. E. Evans, eds. Cambridge University Press, Cambridge.

KRONESTEDT-ROBARDS, E. C., M. GREGER, and A. W. ROBARDS. 1989. The nectar of the *Strelitzia reginae* flower. *Physiol. Plant.* 77, 341–346.

LANGE, B. M., and R. CROTEAU. 1999. Isopentenyl diphosphate biosynthesis via a mevalonate-independent pathway: Isopentenyl monophosphate kinase catalyzes the terminal enzymatic step. *Proc. Natl. Acad. Sci. USA* 96, 13714–13719.

LANGE, B. M., M. R. WILDUNG, E. J. STAUBER, C. SANCHEZ, D. POUCHNIK, and R. CROTEAU. 2000. Probing essential oil biosynthesis and secretion by functional evaluation of expressed sequence tags from mint glandular trichomes. *Proc. Natl. Acad. Sci. USA* 97, 2934–2939.

LAWRENCE, B. M. 1981. Monoterpene interrelationships in the *Mentha* genus: A biosynthetic discussion. In: *Essential Oils*, pp. 1–81, B. D. Mookherjee and C. J. Mussinan, eds. Allured Publishing, Wheaton, IL.

LERDAU, M., M. LITVAK, and R. MONSON. 1994. Plant chemical defense: Monoterpenes and the growth-differentiation balance hypothesis. *Trends Ecol. Evol.* 9, 58–61.

LERSTEN, N. R. 1974a. Colleter morphology in *Pavetta, Neorosea* and *Tricalysia* (Rubiaceae) and its relationship to the bacterial leaf nodule symbiosis. *Bot. J. Linn. Soc.* 69, 125–136.

LERSTEN, N. R. 1974b. Morphology and distribution of colleters and crystals in relation to the taxonomy and bacterial leaf nodule symbiosis of *Psychotria* (Rubiaceae). *Am. J. Bot.* 61, 973–981.

LERSTEN, N. R. 1977. Trichome forms in *Ardisia* (Myrsinaceae) in relation to the bacterial leaf nodule symbiosis. *Bot. J. Linn. Soc.* 75, 229–244.

LERSTEN, N. R., and J. D. CURTIS. 1991. Laminar hydathodes in Urticaceae: Survey of tribes and anatomical observations on *Pilea pumila* and *Urtica dioica. Plant Syst. Evol.* 176, 179–203.

LERSTEN, N. R., and H. T. HORNER JR. 1976. Bacterial leaf nodule symbiosis in angiosperms with emphasis on Rubiaceae and Myrsinaceae. *Bot. Rev.* 42, 145–214.

LÜTTGE, U. 1971. Structure and function of plant glands. *Annu. Rev. Plant Physiol.* 22, 23–44.

LÜTTGE, U. 1983. Mineral nutrition: salinity. *Prog. Bot.* 45, 76–88.

MAEDA, E., and K. MAEDA. 1987. Ultrastructural studies of leaf hydathodes: I. Wheat (*Triticum aestivum*) leaf tips. *Nihon Sakumotsu Gakkai kiji* (*Jpn. J. Crop Sci.*) 56, 641–651.

MAEDA, E., and K. MAEDA. 1988. Ultrastructural studies of leaf hydathodes: II. Rice (*Oryza sativa*) leaf tips. *Nihon Sakumotsu Gakkai kiji* (*Jpn. J. Crop Sci.*) 57, 733–742.

MAFFEI, M., F. CHIALVA, and T. SACCO. 1989. Glandular trichomes and essential oils in developing peppermint leaves. I. Variation of peltate trichome number and terpene distribution within leaves. *New Phytol.* 111, 707–716.

MARGINSON, R., M. SEDGLEY, T. J. DOUGLAS, and R. B. KNOX. 1985. Structure and secretion of the extrafloral nectaries of Australian acacias. *Isr. J. Bot.* 34, 91–102.

MARTIN, C. E., and D. J. VON WILLERT. 2000. Leaf epidermal hydathodes and the ecophysiological consequences of foliar water uptake in species of *Crassula* from the Namib Desert in Southern Africa. *Plant Biol.* 2, 229–242.

MATTERN, V. G., and S. VOGEL. 1994. Lamiaceen-Blüten duften mit dem Kelch—Prüfung einer Hypothese. I. Anatomische Untersuchungen: Vergleich der Laub-und Kelchdrüsen. *Beitr. Biol. Pflanz.* 68, 125–156.

MCCASKILL, D., J. GERSHENZON, and R. CROTEAU. 1992. Morphology and monoterpene biosynthetic capabilities of secretory cell clusters isolated from glandular trichomes of peppermint (*Mentha piperita* L.). *Planta* 187, 445–454.

MCCAULEY, M. M., and R. F. EVERT. 1988. The anatomy of the leaf of potato, *Solanum tuberosum* L. 'Russet Burbank.' *Bot. Gaz.* 149, 179–195.

MCCONKEY, M. E., J. GERSHENZON, and R. B. CROTEAU. 2000. Developmental regulation of monoterpene biosynthesis in the glandular trichomes of peppermint. *Plant Physiol.* 122, 215–223.

MEEUSE, B. J. D., and I. RASKIN. 1988. Sexual reproduction in the arum lily family, with emphasis on thermogenicity. *Sex. Plant Reprod.* 1, 3–15.

NAIDOO, Y., and G. NAIDOO. 1998. *Sporobolus virginicus* leaf salt glands: Morphology and ultrastructure. *S. Afr. J. Bot.* 64, 198–204.

NAIDOO, Y., and G. NAIDOO. 1999. Cytochemical localisation of adenosine triphosphatase activity in salt glands of *Sporobolus virginicus* (L.) Kunth. *S. Afr. J. Bot.* 65, 370–373.

NEPI, M., E. PACINI, and M. T. M. WILLEMSE. 1996. Nectary biology of *Cucurbita pepo:* Ecophysiological aspects. *Acta Bot. Neerl.* 45, 41–54.

NICOLSON, S. W. 1995. Direct demonstration of nectar reabsorption in the flowers of *Grevillea robusta* (Proteaceae). *Funct. Ecol.* 9, 584–588.

OROSS, J. W., R. T. LEONARD, and W. W. THOMSON. 1985. Flux rate and a secretion model for salt glands of grasses. *Isr. J. Bot.* 34, 69–77.

PARÉ, P. W., and J. H. TUMLINSON. 1999. Plant volatiles as a defense against insect herbivores. *Plant Physiol.* 121, 325–331.

PATE, J. S., and B. E. S. GUNNING. 1972. Transfer cells. *Annu. Rev. Plant Physiol.* 23, 173–196.

PEDERSEN, O., L. B. JØRGENSEN, and K. SAND-JENSEN. 1977. Through-flow of water in leaves of a submerged plant is influenced by the apical opening. *Planta* 202, 43–50.

PEMBERTON, R. W. 1998. The occurrence and abundance of plants with extrafloral nectaries, the basis for antiherbivore defensive mutualisms, along a latitudinal gradient in east Asia. *J. Biogeogr.* 25, 661–668.

PEMBERTON, R. W., and J.-H. LEE. 1996. The influence of extrafloral nectaries on parasitism of an insect herbivore. *Am. J. Bot.* 83, 1187–1194.

PERRIN, A. 1971. Présence de "cellules de transfert" au sein de l'épithème de quelques hydathodes. *Z. Pflanzenphysiol.* 65, 39–51.

PERRIN, A. 1972. Contribution à l'étude de l'organization et du fonctionnement des hydathodes; recherches anatomiques, ultrastructurales et physiologiques Thesis. Univ. Claude Bernard, Lyon, France.

POLLARD, A. J., and D. BRIGGS. 1984. Genecological studies of *Urtica dioica* L. III. Stinging hairs and plant-herbivore interactions. *New Phytol.* 97, 507–522.

PONZI, R., and P. PIZZOLONGO. 1992. Structure and function of *Rhinanthus minor* L. trichome hydathode. *Phytomorphology* 42, 1–6.

PRIDGEON, A. M., and W. L. STERN. 1983. Ultrastructure of osmophores in *Restrepia* (Orchidaceae). *Am. J. Bot.* 70, 1233–1243.

PRIDGEON, A. M., and W. L. STERN. 1985. Osmophores of *Scaphosepalum* (Orchidaceae). *Bot. Gaz.* 146, 115–123.

RABINOWITCH, H. D., A. FAHN, T. MEIR, and Y. LENSKY. 1993. Flower and nectar attributes of pepper (*Capsicum annuum* L.) plants in relation to their attractiveness to honeybees (*Apis mellifera* L.) *Ann. Appl. Biol.* 123, 221–232.

RAMAYYA, N., and B. BAHADUR. 1968. Morphology of the "Squamellae" in the light of their ontogeny. *Curr. Sci.* 37, 520–522.

RAVEN, P. H., R. F. EVERT, and S. E. EICHHORN. 2005. *Biology of Plants*, 7th ed. Freeman, New York.

RAZEM, F. A., and A. R. DAVIS. 1999. Anatomical and ultrastructural changes of the floral nectary of *Pisum sativum* L. during flower development. *Protoplasma* 206, 57–72.

ROBARDS, A. W., and M. STARK. 1988. Nectar secretion in *Abutilon:* a new model. *Protoplasma* 142, 79–91.

ROBBRECHT, E. 1987. The African genus *Tricalysia* A. Rich. (Rubiaceae). 4. A revision of the species of sectio *Tricalysia* and sectio *Rosea. Bull. Jard. Bot. Natl. Belg.* 57, 39–208.

ROST, T. L. 1969. Vascular pattern and hydathodes in leaves of *Crassula argentea* (Crassulaceae). *Bot. Gaz.* 130, 267–270.

RUDALL, P. 2002. Homologies of inferior ovaries and septal nectaries in monocotyledons. *Int. J. Plant Sci.* 163, 261–276.

SAJO, M. G., P. J RUDALL, and C. J. PRYCHID. 2004. Floral anatomy of Bromeliaceae, with particular reference to the evolution of epigyny and septal nectaries in commelinid monocots. *Plant Syst. Evol.* 247, 215–231.

SAWIDIS, TH. 1991. A histochemical study of nectaries of *Hibiscus rosa-sinensis. J. Exp. Bot.* 42, 1477–1487.

SAWIDIS, TH., E. P. ELEFTHERIOU, and I. TSEKOS. 1987a. The floral nectaries of *Hibiscus rosa-sinensis.* I. Development of the secretory hairs. *Ann. Bot.* 59, 643–652.

SAWIDIS, TH., E. P. ELEFTHERIOU, and I. TSEKOS. 1987b. The floral nectaries of *Hibiscus rosa-sinensis.* II. Plasmodesmatal frequencies. *Phyton (Horn)* 27, 155–164.

SAWIDIS, TH., E. P. ELEFTHERIOU, and I. TSEKOS. 1989. The floral nectaries of *Hibiscus rosa-sinensis.* III. A morphometric and ultrastructural approach. *Nord. J. Bot.* 9, 63–71.

SCHIRMER, U. and S.-W. BRECKLE. 1982. The role of bladders for salt removal in some Chenopodiaceae (mainly *Atriplex* species). In: *Contributions to the Ecology of Halophytes*, pp. 215–231, D. N. Sen and K. S. Rajpurohit, eds. W. Junk, The Hague.

SCHNEPF, E. 1974. Gland cells. In: *Dynamic Aspects of Plant Ultrastructure*, pp. 331–357, A. W. Robards, ed. McGraw-Hill, London.

SHIMONY, C., and A. FAHN. 1968. Light- and electron-microscopical studies on the structure of salt glands of *Tamarix aphylla* L. *Bot. J. Linn. Soc.* 60, 283–288.

SINGSAAS, E. L. 2000. Terpenes and the thermotolerance of photosynthesis. *New Phytol.* 146, 1–3.

SKUBATZ, H., and D. D. KUNKEL. 1999. Further studies of the glandular tissue of the *Sauromatum guttatum* (Araceae) appendix. *Am. J. Bot.* 86, 841–854.

SKUBATZ, H., D. D. KUNKEL, and B. J. D. MEEUSE. 1993. Ultrastructural changes in the appendix of the *Sauromatum guttatum* inflorescence during anthesis. *Sex. Plant Reprod.* 6, 153–170.

SKUBATZ, H., D. D. KUNKEL, W. N. HOWALD, R. TRENKLE, and B. MOOKHERJEE. 1996. The *Sauromatum guttatum* appendix as an osmophore: Excretory pathways, composition of volatiles and attractiveness to insects. *New Phytol.* 134, 631–640.

SPERLICH, A. 1939. *Das trophische Parenchym. B. Exkretionsgewebe. Handbuch der Pflanzenanatomie*, Heft 3, Band 4, *Histologie*. Gebrüder Borntraeger, Berlin.

STERN, W. L., K. J. CURRY, and W. M. WHITTEN. 1986. Staining fragrance glands in orchid flowers. *Bull. Torrey Bot. Club* 113, 288–297.

STERN, W. L., K. J. CURRY, and A. M. PRIDGEON. 1987. Osmophores of *Stanhopea* (Orchidaceae). *Am. J. Bot.* 74, 1323–1331.

STPICZYŃSKA, M. 1993. Anatomy and ultrastructure of osmophores of *Cymbidium tracyanum* Rolfe (Orchidaceae). *Acta Soc. Bot. Pol.* 62, 5–9.

STPICZYŃSKA, M., K. L. DAVIES, and A. GREGG. 2004. Nectary structure and nectar secretion in *Maxillaria coccinea* (Jacq.) L. O. Williams ex Hodge (Orchidaceae). *Ann. Bot.* 93, 87–95.

THOMAS, V. 1991. Structural, functional and phylogenetic aspects of the colleter. *Ann. Bot.* 68, 287–305.

THOMSON, W. W., and L. L. LIU. 1967. Ultrastructural features of the salt gland of *Tamarix aphylla* L. *Planta* 73, 201–220.

THOMSON, W. W., and K. PLATT-ALOIA. 1979. Ultrastructural transitions associated with the development of the bladder cells of the trichomes of *Atriplex. Cytobios* 25, 105–114.

THOMSON, W. W., W. L. BERRY, and L. L. LIU. 1969. Localization and secretion of salt by the salt glands of *Tamarix aphylla. Proc. Natl. Acad. Sci. USA* 63, 310–317.

THOMSON, W. W., C. D. FARADAY, and J. W. OROSS. 1988. Salt glands. In: *Solute Transport in Plant Cells and Tissues*, pp. 498–537, D. A. Baker and J. L. Hall, eds. Longman Scientific and Technical, Harlow, Essex.

THURSTON, E. L. 1974. Morphology, fine structure, and ontogeny of the stinging emergence of *Urtica dioica. Am. J. Bot.* 61, 809–817.

THURSTON, E. L. 1976. Morphology, fine structure and ontogeny of the stinging emergence of *Tragia ramosa* and *T. saxicola* (Euphorbiaceae). *Am. J. Bot.* 63, 710–718.

THURSTON, E. L., and N. R. LERSTEN. 1969. The morphology and toxicology of plant stinging hairs. *Bot. Rev.* 35, 393–412.

TÓTH-SOMA, L. T., N. M. DATTA, and Z. SZEGLETES. 1995/1996. General connections between latex and nectar secretional systems of *Asclepias syriaca* L. *Acta Biol. Szeged.* 41, 37–44.

TUCKER, S. C., and L. L. HOEFERT. 1968. Ontogeny of the tendril in *Vitis vinifera. Am. J. Bot.* 55, 1110–1119.

TURNER, G., J. GERSHENZON, E. E. NIELSON, J. E. FROELICH, and R. CROTEAU. 1999. Limonene synthase, the enzyme responsible for monoterpene biosynthesis in peppermint, is localized to leucoplasts of oil gland secretory cells. *Plant Physiol.* 120, 879–886.

TURNER, G. W., J. GERSHENZON, and R. B. CROTEAU. 2000. Development of peltate glandular trichomes of peppermint. *Plant Physiol.* 124, 665–679.

TURNER, J. C., J. K. HEMPHILL, and P. G. MAHLBERG. 1980. Trichomes and cannabinoid content of developing leaves and bracts of *Cannabis sativa* L. (Cannabaceae). *Am. J. Bot.* 67, 1397–1406.

VAINSTEIN, A., E. LEWINSOHN, E. PICHERSKY, and D. WEISS. 2001. Floral fragrance: New inroads into an old commodity. *Plant Physiol.* 127, 1383–1389.

VASSILYEV, A. E., and L. E. MURAVNIK. 1988a. The ultrastructure of the digestive glands in *Pinguicula vulgaris* L. (Lentibulariaceae) relative to their function. I. The changes during maturation. *Ann. Bot.* 62, 329–341.

VASSILYEV, A. E., and L. E. MURAVNIK. 1988b. The ultrastructure of the digestive glands in *Pinguicula vulgaris* L. (Lentibulariaceae) relative to their function. II. The changes on stimulation. *Ann. Bot.* 62, 343–351.

VESPRINI, J. L., M. NEPI, and E. PACINI. 1999. Nectary structure, nectar secretion patterns and nectar composition in two *Helleborus* species. *Plant Biol.* 1, 560–568.

VOGEL, S. 1962. Duftdrüsen im Dienste der Bestäubung. Über Bau und Funktion der Osmophoren. Mainz: Abh. Mathematisch-Naturwiss. Klasse 10, 1–165.

VOGEL, S. 1990. The role of scent glands in pollination. On the structure and function of osmophores. S. S. Renner, sci. ed. Smithsonian Institution Libraries and National Science Foundation, Washington, DC.

VOGEL, S. 1997. Remarkable nectaries: structure, ecology, organophyletic perspectives. I. Substitutive nectaries. *Flora* 192, 305–333.

VOGEL, S. 1998. Remarkable nectaries: Structure, ecology, organophyletic perspectives. IV. Miscellaneous cases. *Flora* 193, 225–248.

WADDLE, R. M., and N. R. LERSTEN, 1973. Morphology of discoid floral nectaries in Leguminosae, especially tribe Phaseoleae (Papilionoideae). *Phytomorphology* 23, 152–161.

WEICHEL, G. 1956. Natürliche Lagerstätten ätherischer Öle. In: *Die ätherischen Öle*, 4th ed., Band 1, pp. 233–254, E. Gildemeister and Fr. Hoffmann, eds. Akademie-Verlag, Berlin.

WERKER, E. 2000. Trichome diversity and development. *Adv. Bot. Res.* 31, 1–35.

WERKER, E., and A. FAHN. 1981. Secretory hairs of *Inula viscosa* (L.) Ait.—Development, ultrastructure, and secretion. *Bot. Gaz.* 142, 461–476.

WERKER, E., E. PUTIEVSKY, U. RAVID, N. DUDAI, and I. KATZIR. 1993. Glandular hairs and essential oil in developing leaves of *Ocimum basilicum* L. (Lamiaceae). *Ann. Bot.* 71, 43–50.

WERYSZKO-CHMIELEWSKA, E., and M. STPICZYŃSKA. 1995. Osmophores of *Amorphophallus rivieri* Durieu (*Araceae*). *Acta Soc. Bot. Pol.* 64, 121–129.

WILSON, T. P., M. J. CANNY, and M. E. McCULLY. 1991. Leaf teeth, transpiration and the retrieval of apoplastic solutes in balsam poplar. *Physiol. Plant.* 83, 225–232.

ZER, H., and A. FAHN. 1992. Floral nectaries of *Rosmarinus officinalis* L. Structure, ultrastructure and nectar secretion. *Ann. Bot.* 70, 391–397.

ZIMMERMANN, J. G. 1932. Über die extrafloralen Nektarien der Angiospermen. *Beih. Bot. Centralbl.* 49, 99–196.

# CHAPTER SEVENTEEN

# Internal Secretory Structures

In the previous chapter, specific examples of secretory structures found on the surface of the plant were described. This chapter is devoted to consideration of internal secretory structures (Fig. 17.1), beginning with specific types of internal secretory cells.

## INTERNAL SECRETORY CELLS

On the basis of the variability and location of the secretory tissues in the vascular plant body, Fahn (2002) has suggested that during the course of evolution the protective secretory tissues proceeded from the leaf mesophyll—as occurring in some pteridophytes—in two directions. One direction was from the mesophyll to the outside, to the epidermis and its trichomes, as in many angiosperms; the other was to the inside, to the primary and secondary phloem, and in a few conifers, to the secondary xylem too.

Internal secretory cells have a wide variety of contents: oils, resins, mucilages, gums, tannins, and crystals. The secretory cells often appear as specialized cells. They are then called idioblasts, more specifically secretory idioblasts. The secretory cells may be much enlarged, especially in length, and are then called sacs or tubes. The secretory cells are usually classified on the basis of their contents, but many secretory cells contain mixtures of substances, and in many the contents have not been identified. Nevertheless, the secretory cells, as well as the secretory cavities and ducts, are useful for diagnostic purposes in taxonomic work (Metcalfe and Chalk, 1950, 1979).

Crystal-containing cells (Chapter 3) are often treated as secretory idioblasts (Foster, 1956; Metcalfe and Chalk, 1979). Some do not differ from other parenchyma cells in the tissue, but others are considerably modified, for example, as the lithocysts (cystolith-containing cells) of *Ficus elastica* leaves (Chapter 9) and mucilage-containing raphide cells (Fig. 17.1B). Crystal-forming cells may die after the deposition of the crystal, or crystals, is completed. In the secondary vascular tissues a cell that forms crystals may become divided into smaller cells. In another modification the crystal is walled off by cellulose from the living part of the protoplast.

Cells containing the enzyme myrosinase have been identified in such families as Capparidaceae,

*Esau's Plant Anatomy, Third Edition,* By Ray F. Evert.

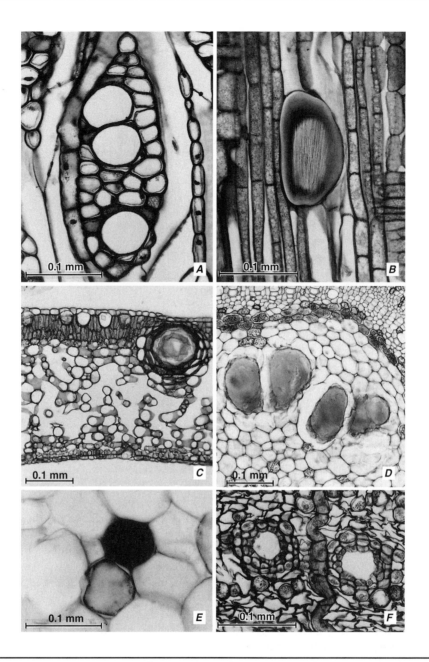

**FIGURE 17.1**

Various internal secretory structures. **A**, oil cells in tangential section of tulip tree (*Liriodendron*) of phloem ray. **B**, idioblast containing mucilage and raphides in radial section of *Hydrangea paniculata* phloem. **C**, secretory cavity (oil gland) in lemon (*Citrus*) leaf (upper part of leaf to the right). **D**, mucilage ducts in pith of basswood (*Tilia*) stem in transverse section. **E**, tannin cells in pith of elderberry (*Sambucus*) stem in transverse section. **F**, schizogenous secretory ducts in transverse section of nonconducting phloem of *Rhus typhina*. (**A–C**, **E**, **F**, from Esau, 1977.)

Resedaceae, and Brassicaceae, but they are mainly found in the Brassicaceae (Fahn, 1979; Bones and Iversen, 1985; Rask et al., 2000). The myrosinase is located in the large central vacuole of idioblasts called **myrosin cells**. Myrosinase hydrolyses glucosinolates to aglucons that decompose to form toxic products such as isothiocyanates (mustard gas), nitrites, and epithionitrites, which may play important roles in the defense of the plant against insects and microorganisms (Rask et al., 2000). Only damage to the tissues can bring about this

reaction because the enzyme and substrate occur in different cells, the myrosinase in the myrosin cell and the thioglucosides in what appear as ordinary parenchyma cells. In *Arabidopsis* the myrosin cells occur in the phloem parenchyma, and the glucosinolate-containing cells, called **S-cells** because they are highly enriched in sulfur, are located in the ground tissue external to the phloem (Fig. 17.2; Koroleva et al., 2000; Andréasson et al., 2001). The two cells, however, are sometimes in direct contact with one another (Andréasson et al.,

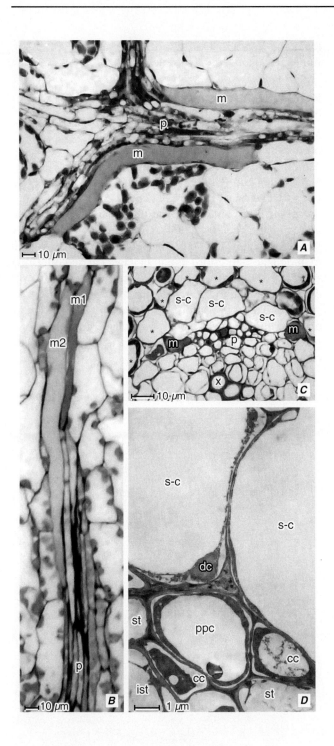

**FIGURE 17.2**

Myrosin cells (m) in young rosette leaf (**A**, **B**) and pedicel (**C**, **D**) of *Arabidopsis* leaf. S-cells (s-c) are shown in **C** and **D** (**A–C**, light micrographs; **D**, electron micrograph). **A**, paradermal section of leaf blade showing phloem (p) and two long, relatively broad myrosin cells. **B**, paradermal section of leaf blade showing obliquely sectioned vascular bundle with portion of phloem (p) and two adjacent myrosin cells (m1 and m2). **C**, transverse section showing myrosin cells associated with the phloem (p) of a vascular bundle. S-cells are located between the phloem (p) and cells (asterisks) of the starch sheath; x, xylem. **D**, transverse section showing portions of two highly vacuolated S-cells bordering the outer surface of the phloem. Three sieve-tube elements (st), one of which is immature (ist), two companion cells (cc), and a phloem parenchyma cell (ppc) can be seen here. (Reprinted with permission from Andréasson et al., 2001. © American Society of Plant Biologists.)

The cell wall of mature **oil cells** has three distinct layers: an external (primary wall) layer, a suberized layer (suberin lamella), and an inner (tertiary wall) layer (Maron and Fahn, 1979; Baas and Gregory, 1985; Mariani et al., 1989; Bakker and Gerritsen, 1990; Bakker et al., 1991; Platt and Thomson, 1992). After the inner wall layer has been deposited, an **oil cavity** is formed. This cavity is surrounded by plasma membrane and attached to a bell-like protrusion, called the **cupule**, of the inner wall layer (Fig. 17.3; Maron and Fahn, 1979; Bakker and Gerritsen, 1990; Platt and Thomson, 1992). Oil, which most likely is synthesized in plastids and released into the cytosol, is secreted into the oil cavity via the plasma membrane. As the oil cavity increases in size, the protoplast gradually becomes appressed against the inner wall layer. At maturity the enlarged oil cell shows complete degeneration of the protoplast and the oil, mixed with remnants of cytoplasmic components, occupies the entire volume of the cell (Fig. 17.4A). The suberized wall layer (Fig. 17.4B) seals off the oil cell, preventing leakage of potentially toxic substances into the surrounding cells.

Avocado (*Persea americana*) contains oil cells in the leaves, seeds, roots, and fruits (Platt and Thomson, 1992). During fruit ripening, hydrolytic enzymes (cellulase and polygalacturonase) degrade the primary walls of the parenchyma cells and fruit softening occurs. The suberized wall of the idioblast oil cells is immune to the activity of these enzymes, however, and remains intact during ripening (Platt and Thomson, 1992). Several insecticidal compounds have been identified in the oil

2001). Myrosinase activity has also been detected in the guard cells of *Arabidopsis* (Husebye et al., 2002).

## Oil Cells Secrete Their Oils into an Oil Cavity

Some plant families, for example, Calycanthaceae, Lauraceae, Magnoliaceae, Winteraceae, and Simaroubaceae, have secretory cells with oily contents (Metcalfe and Chalk, 1979; Baas and Gregory, 1985). (The first four families are magnoliids.) Superficially these cells appear like large parenchyma cells (Fig. 17.1A), and are known to occur in vascular and ground tissues of stem and leaf.

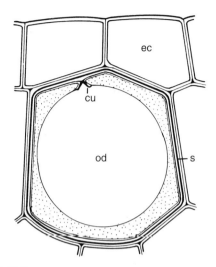

**FIGURE 17.3**

Schematic drawing of an oil cell in the *Laurus nobilis* (Lauraceae) leaf. Details: cu, cupule; ec, epidermal cell; od, oil droplet; s, suberin wall layer. (From Maron and Fahn, 1979. © Blackwell Publishing.)

from the avocado fruit (Rodriguez-Saona et al., 1998; Rodriguez-Saona and Trumble, 1999).

### Mucilage Cells Deposit Their Mucilage between the Protoplast and the Cellulosic Cell Wall

**Mucilage cells** occur in a large number of "dicotyledonous" families (magnoliids and eudicots), including the Annonaceae, Cactaceae, Lauraceae, Magnoliaceae, Malvaceae, and Tiliaceae (Metcalfe and Chalk, 1979; Gregory and Baas, 1989). They may occur in all parts of the plant body and usually differentiate very close to meristematic regions. Their cellulosic walls are usually thin and unlignified. Only the Golgi bodies are involved in mucilage secretion, the mucilage carried in Golgi vesicles passing through the plasma membrane by exocytosis (Trachtenberg and Fahn, 1981). With progressive mucilage deposition, the lumen of the cell may become almost occluded with mucilage and the protoplast confined to thread-like regions (Fig. 17.5). The protoplast eventually degenerates.

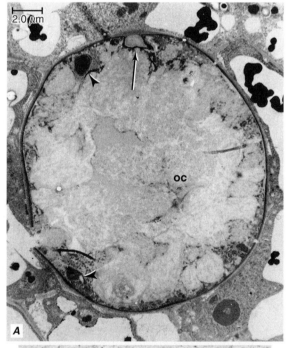

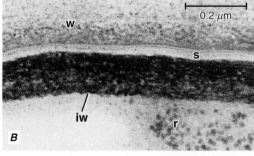

**FIGURE 17.4**

Electron micrographs of mature oil cell in leaf of *Cinnamomum burmanni* (Lauraceae) leaf. **A**, this oil cell, with an enlarged oil cavity (oc) and cytoplasmic remnants, is no longer enclosed by a plasma membrane. A cupule (arrow) and some degenerating organelles (darts) are evident. **B**, detail of cell wall showing suberized layer (s). Other details: iw, electron-dense inner wall layer; w, wall. (From Bakker et al., 1991.)

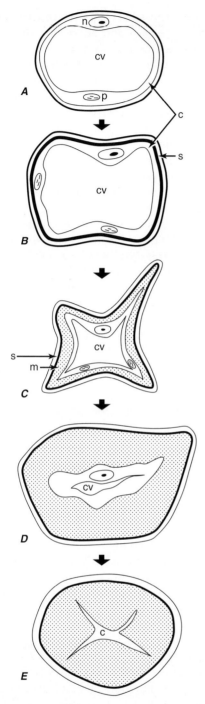

**FIGURE 17.5**

Schematic representation of developmental stages of mucilage cell in *Cinnamomum burmanni*. **A**, stage 1: young cell with typical cytoplasmic components and central vacuole (cv) devoid of deposits. **B**, stage 2: idioblast with suberized wall layer (s). **C**, stage 3a: collapsed cell with mucilage (m) deposited against suberized layer. **D**, stage 3b: protoplast moves inward due to prolonged mucilage deposition. The central vacuole disappears. **E**, stage 3c: nearly mature cell filled with mucilage surrounding degenerating cytoplasm. Other details: c, cytoplasm; n, nucleus; p, plastid. (Redrawn from Bakker et al., 1991.)

Oil and mucilage cells in the magnoliids (Magnoliales and Laurales) share a number of identical features, one of the most important being the presence of a suberized layer in the cell wall (Bakker and Gerritsen, 1989, 1990; Bakker et al., 1991). Of these two types of secretory cell, the presence of a suberized layer had long been generally accepted as being typical only for oil cells (Baas and Gregory, 1985). The eudicots, such as *Hibiscus* (Malvales), lack a suberized wall layer in their mucilage cells. It has been hypothesized that magnoliids bearing mucilage cells with suberized walls inherited the ability to deposit a suberized wall layer from magnoliids with the originally occurring oil cells. The presence of a suberized layer in the mucilage cells today is regarded an ancestral remnant without a function. The more advanced eudicots apparently lost the ability to deposit a suberized layer in their mucilage cells (Bakker and Baas, 1993).

Mucilage-containing cells often include raphide crystals. In these **mucilage-crystal idioblasts** both the crystals and the mucilage occur in the large central vacuole of the cell (Fig. 17.1B). The crystals are formed first, then the mucilage accumulates around them (Kausch and Horner, 1983, 1984; Wang et al., 1994).

A number of functions have been attributed to mucilage cells, but virtually no experimental data are available to support them (for review, see Gregory and Baas, 1989).

## Tannin Is the Most Conspicuous Inclusion in Numerous Secretory Cells

Tannin is a common secondary metabolite in parenchyma cells (Chapter 3), but some cells contain this substance in great abundance. Such cells may be conspicuously enlarged. The tannin cells often form connected systems and may be associated with vascular bundles. **Tannin idioblasts** occur in many families (Crassulaceae, Ericaceae, Fabaceae, Myrtaceae, Rosaceae, Vitaceae). Examples of tannin cells can be found in leaves of *Sempervivum tectorum* and species of *Echeveria* and of tube-like tannin cells in the pith (Fig. 17.1E) and phloem of *Sambucus* stems. The tube-like tannin cells in *Sambucus* are coenocytic (multinucleate). They originate from mononucleate cells (mother tannin cells) in the first internode by synchronous mitotic divisions without cytokinesis (Fig. 17.6; Zobel, 1985a, b). Mature tannin tubes as long as the internodes were found in *Sambucus racemosa*, the longest being 32.8 cm (Zobel, 1986b).

The tannin compounds in the tanniniferous cells are oxidized to brown and reddish brown phlobaphenes, which are readily perceived under the microscope. The tannins are sequestered in the vacuoles of the cells. Rough endoplasmic reticulum is the most likely site for tannin synthesis (Parham and Kaustinen, 1977; Zobel,

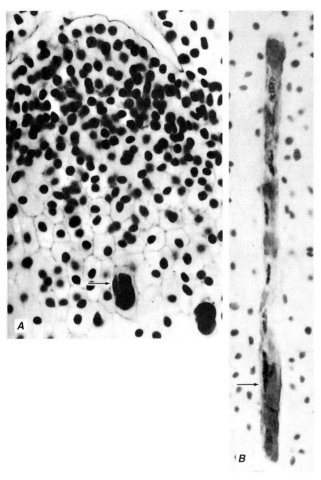

**FIGURE 17.6**

Tube-like tannin cells in *Sambucus racemosa*. **A**, the tube-like tannin cells originate from mono-nucleate tannin cells such as these shown in the first internode of the shoot. The nucleus is discernible in one of the two tannin cells shown here (arrow). **B**, a multinucleate tannin cell with prophase nuclei. The larger nucleus (arrow) in this cell probably resulted from fusion of two smaller nuclei. (A, ×185; B, ×170. From Zobel, 1985a, by permission of Oxford University press.)

1986a; Rao, K. S., 1988). Small tannin-containing vesicles apparently derived from the endoplasmic reticulum fuse with the tonoplast and their contents, the tannins, are deposited in the vacuole. Cells in the ground tissue of the fruit of *Ceratonia siliqua* contain solid tannoids, inclusions of tannins combined with other substances. Tannins are probably the most important deterrents to herbivores feeding on angiosperms.

# ▌ SECRETORY CAVITIES AND DUCTS

Secretory cavities and ducts (canals) differ from secretory cells in that they secrete substances into intercellular spaces. They are glands consisting of relatively large intercellular spaces commonly lined by special-ized secretory (epithelial) cells. **Secretory cavities** are short secretory spaces and **secretory ducts** are long secretory spaces. In some plants (*Lysimachia, Myrsine, Ardisia*) resinous material is secreted by parenchyma cells into ordinary intercellular spaces and forms a granular layer along the wall. The contents of the cavities and ducts may consist of terpenoids or carbohydrates or of terpenoids together with carbohydrates and other substances.

Three developmental types of secretory cavities and ducts have been recognized: schizogenous, lysigenous, and schizolysigenous. **Schizogenous** cavities and ducts form by a separation of cells, resulting in a space lined with secretory cells composing the epithelium. **Lysigenous** cavities and ducts result from a dissolution (autolysis) of cells. In these cavities and ducts the secretory product is formed in the cells that eventually break down and release the product into the resultant space (holocrine secretion). Partly disintegrated cells occur along the periphery of the space. The development of **schizolysigenous** cavities and ducts initially is schizogenous, but lysigeny occurs in later stages as the epithelial cells lining the space undergo autolysis, further enlarging the space. There is some inconsistency in use of the schizolysigenous category of duct development in those cases in which only occasional epithelial cells undergo autolysis after the secretory phase. Some investigators consider such ducts as schizolysigenous, but others regard them as schizogenous.

## The Best-Known Secretory Ducts Are the Resin Ducts of Conifers

The resin ducts of conifers occur in vascular tissues (Chapter 10) and ground tissues of all plant organs and are, structurally, long intercellular spaces lined with resin-producing epithelial cells (Werker and Fahn, 1969; Fahn and Benayoun, 1976; Fahn, 1979; Wu and Hu, 1994). With one possible exception, their development is quite uniform: throughout the plant, in both primary and secondary tissues, the ducts form schizogenously. Only the resin ducts in the bud scales of *Pinus pinaster* have been reported to follow a schizolysigenous pattern of development (Charon et al., 1986).

Ducts similar to those of conifers occur in Anacardiaceae (Fig. 17.1F and 17.7), Asteraceae, Brassicaceae, Fabaceae, Hypericaceae, and Simaroubaceae (Metcalfe and Chalk, 1979). Disagreement has existed among investigators over the manner of duct formation in some taxa. Take, for example, the resin ducts in *Parthenium argentatum*, guayule (Asteraceae). Most accounts of duct formation in guayule—as for members of the Asteraceae, in general—indicate that it is exclusively schizogenous (Lloyd, 1911; Artschwager, 1945; Gilliland et al., 1988; Łotocka and Geszprych, 2004). Joseph et al. (1988) report, however, that whereas ducts originat-

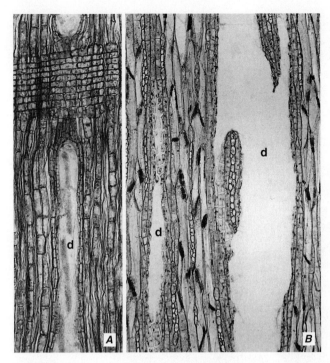

**FIGURE 17.7**

Secretory ducts (d) of *Rhus glabra* (Anacardiaceae) in radial (**A**) and tangential (**B**) sections. (Both, ×120. From Fahn and Evert, 1974.)

ing in the vascular cambium of guayule are initiated and develop schizogenously, those in primary tissue develop schizolysigenously.

Development of the secretory ducts in Anacardiaceae apparently vary among plant parts and from species to species. In *Lannea coromandelica*, for instance, development of the ducts in the primary phloem of the stem reportedly is schizogenous, whereas that of their counterparts in the secondary phloem and phelloderm is lysigenous (Venkaiah and Shah, 1984; Venkaiah, 1992). Schizogenous development has been reported for the gum-resin ducts in the secondary phloem of *Rhus glabra* (Fahn and Evert, 1974) and schizolysigenous for those in the primary phloem of *Anacardium occidentale* (Nair et al., 1983) and *Semecarpus anacardium* (Bhatt and Ram, 1992). According to Joel and Fahn (1980), the resin ducts in the primary phloem and pith of *Mangifera indica* develop lysigenously. In their report, Joel and Fahn (1980) listed three main characteristics of lysigenous ducts that can be used to distinguish clearly between schizogenous and lysigenous ducts: (1) the presence of disorganized cytoplasm in the duct lumen, (2) the presence of wall remnants in the duct lumen attached to the living epithelial cells, and (3) the presence of specific intercellular spaces at the cell corners facing the duct lumen.

Certain tendencies apparently exist in the distribution pattern of gum and gum-resin producing ducts and

cavities in different tissues of the plant body (Babu and Menon, 1990). Typically, ducts formed in the pith are unbranched and nonanastomosing, whereas ducts and cavities arising in the secondary xylem and secondary phloem tend to branch and anastomose tangentially (Fig. 17.7B).

Several organelles appear to be involved in resin synthesis. Those most commonly implicated are the plastids, which are sheathed by endoplasmic reticulum. Osmiophilic droplets have been observed in the stroma of plastids, in the plastid envelope, within the endoplasmic reticulum near the plastids, and on both sides of the plasma membrane. Osmiophilic droplets also have been observed in the mitochondria, and in some cases even the nuclear envelope (Fahn and Evert, 1974; Fahn, 1979, 1988b; Wu, H., and Hu, 1994; Castro and DeMagistris, 1999). Most investigators favor a granulocrine method of resin secretion, either by exocytosis or by plasma membrane invaginations, which surround the resin droplets and detach them from the protoplast (Fahn, 1988a; Babu et al., 1990; Arumugasamy et al., 1993; Wu, H., and Hu, 1994). Some eccrine elimination may also occur (Bhatt and Ram, 1992; Nair and Subrahmanyam, 1998). The Golgi complex clearly is involved with the synthesis and secretion of the polysaccharide gum, which is deposited by exocytosis as new wall layers. The gum in the duct lumen originates directly from the outer wall layers (Fig. 17.8), while at the same time new wall material is added to the inner surface of the wall (Fahn and Evert, 1974; Bhatt and Shah, 1985; Bhatt, 1987; Venkaiah, 1990, 1992). In a quantitative ultrastructural study of the epithelial cells of secretory ducts in the primary phloem of *Rhus toxicodendron*, Vassilyev (2000) concluded that the rough endoplasmic reticulum and Golgi apparatus are involved in glycoprotein secretion by means of exocytosis of large granular vesicles, and the smooth, tubular endoplasmic reticulum to be primarily responsible for terpene synthesis and intracellular transport. Peroxisomes are proposed to be involved in the regulation of terpene synthesis. The plastids apparently do not participate actively in the secretory process.

### Development of Secretory Cavities Appears to Be Schizogenous

Secretory cavities occur in such families as Apocynaceae, Asclepiadaceae, Asteraceae, Euphorbiaceae, Fabaceae, Malvaceae, Myrtaceae, Rutaceae, and Tiliaceae (Metcalfe and Chalk, 1979). As with duct development, different opinions exist over the manner of cavity development in some taxa. Prime examples are the secretory cavities (oil glands) in *Citrus*, which have been reported to develop schizogenously by some workers, schizolysigenously by others, and lysigenously by still others (Thomson et al., 1976; Fahn, 1979; Bosabalidis and Tsekos, 1982; Turner et al., 1998).

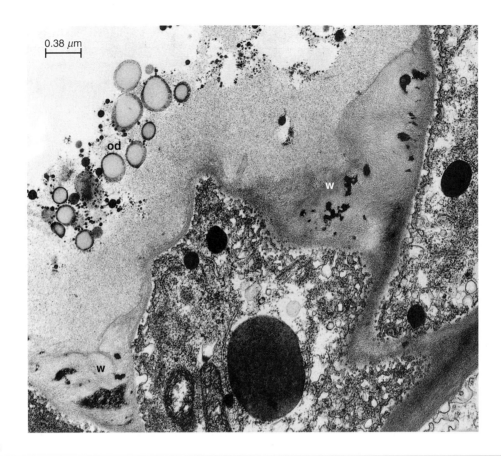

**FIGURE 17.8**

Recently divided epithelial cell of mature *Rhus glabra* secretory duct in transverse section. Osmiophilic droplets (od) such as those seen in these cells are secreted into the duct lumen. Concomitantly the wall layers (w) facing the lumen disintegrate and form, together with the secreted osmiophilic droplets, the gum-resin. (From Fahn and Evert, 1974.)

The concept of lysigenous cavity development has been questioned by Turner and colleagues (Turner et al., 1998; Turner, 1999). During a study of secretory cavity development in *Citrus limon*, they found that the lysigenous appearance of these cavities is the result of fixation artifacts (Turner et al., 1998). In standard aqueous fixatives the epithelial cells undergo rapid destructive swelling, causing the schizogenous secretory cavities of lemon to appear lysigenous. In an earlier study, Turner (1994) compared dry-mounted and wet-mounted cavities or ducts from 10 species of seed plants (one species each of *Cycas*, *Ginkgo*, *Sequoia*, *Hibiscus*, *Hypericum*, *Myoporum*, *Philodendron*, *Prunus*, and two species of *Eucalyptus*). Whereas all dry-mounted specimens resembled schizogenous cavities or ducts, significant swelling of epithelial cells was observed in water mounts of seven of the species. Moreover, rapid destructive swelling of epithelial cells was evident in secretory cavities of *Myoporum*. The case for lysigenous gland development, in general, merits reexamination.

During their study of oil gland (cavity) development in the embryo of *Eucalyptus* (Myrtaceae), Carr and Carr (1970) found that contrary to recent literature at that time, development of the oil glands in *Eucalyptus* is entirely schizogenous, not schizolysigenous (Fig. 17.9). The gland arises by division of a single epidermal cell and becomes differentiated into epithelial and casing (sheath) cells. Some of the latter may be contributed by a subepidermal cell. The formation of the oil cavity as an intercellular space results from separation of the epithelial cells. There is no resorption of cells. Carr and Carr (1970) noted that the mature epithelial cells have very delicate cell walls and are difficult to fix, embed, and section without extensive damage. In well-preserved preparations, all of the cells, including the epithelial cells, of the oil gland are still present in aged and senescing cotyledons. During senescence all of the cells of the oil gland, like those of the rest of the cotyledons, undergo degenerative changes.

The schizogenous oil cavities in *Psoralea bituminosa* and *P. macrostachya* (Fabaceae), which appear as dots on the leaves, are traversed by many elongated cells. Cavity development begins with anticlinal divisions in localized groups of protodermal cells (Turner, 1986). These cells then elongate, those in the center elongating most, and form a hemispherical protuber-

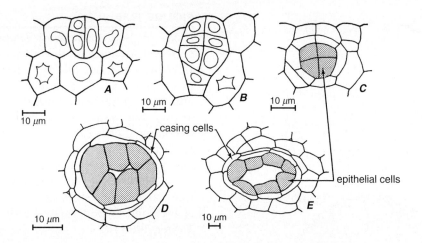

**FIGURE 17.9**

Development of epidermal oil glands in the embryo of *Eucalyptus* shown in longitudinal (**A–C**) and transverse (**D, E**) sections. **A, B**, two stages in division of a gland initial and its derivatives. **C**, after completion of divisions: secretory cells (stippled) are surrounded by casing cells. **D**, schizogenous formation of cavity between secretory cells. **E**, mature gland with secretory cells forming the epithelium around the oil cavity. (From photomicrographs in D. J. Carr and G. M. Carr. 1970. *Australian Journal of Botany* 18, 191-212, with permission of CSIRO Publishing, Melbourne, Australia. © CSIRO.)

ance on the surface of the leaf (Fig. 17.10). As development proceeds, these elongated cells, or trabeculae, separate from each other (schizogeny) in the center of the protuberance but remain attached to each other top and bottom. During leaf expansion the protuberance sinks until its outer surface is flush with the surface of the leaf. The result is a secretory cavity lined with an epithelium of modified epidermal cells (Fig. 17.11). Another example of internal oil cavities of epidermal origin and composition is found in some species of *Polygonum* (Polygonaceae) (Curtis and Lersten, 1994).

As mentioned previously, schizogenous secretory spaces are common in the Asteraceae, and most have been described as ducts or canals. Lersten and Curtis (1987) have cautioned that such descriptions usually have been based on examination of only transverse sections, which are generally not adequate to describe ducts. In *Solidago canadensis* the foliar secretory spaces begin as discrete cavities separated from one another only by epithelial cells. Elongation of these cavities, accompanied by stretching and separation of septa, gives a false impression at maturity of an indefinitely long duct instead of a series of tubular cavities (Lersten and Curtis, 1989).

## Secretory Ducts and Cavities May Arise under the Stimulus of Injury

Secretory ducts and cavities resulting from normal development may be difficult to distinguish from ducts and cavities arising under the stimulus of injury (wound-

ing, mechanical pressure, attack by insects or microorganisms, physiological disturbances such as water stress, environmental factors). Resin, gum-resin, and gum ducts and cavities are frequently **traumatic formations** in the secondary phloem (Fig. 17.12) and secondary xylem (Fig. 11.13) of conifers and woody angiosperms. Their development and contents may parallel those constituting normal features of these tissues (e.g., the normal and traumatic vertical resin ducts in *Pinus halepensis*; Fahn and Zamski, 1970). In *Picea abies* the traumatic resin ducts formed in reaction to wounding and fungal infection appear to play an important role in the development and maintenance of enhanced resistance to the pathogenic bluestain fungus *Ceratocystis polonica* (Christiansen et al., 1999).

The traumatic gum-resin ducts in the secondary xylem of *Ailanthus excelsa* (Simaroubaceae) and the traumatic gum ducts in the secondary phloem of *Moringa oleifera* (Moringaceae) are initiated by autolysis of axial parenchyma cells (Babu et al., 1987; Subrahmanyam and Shah, 1988). The lumina of both ducts are lined by epithelial cells that eventually undergo autolysis and release their contents into the duct. The traumatic gum ducts formed in the secondary xylem of *Sterculia urens* (Sterculiaceae) also result from breakdown of xylem cells but their irregular lumina are without any distinct epithelial cells (Setia, 1984). In young stems of *Citrus* plants infected with the brown-rot fungus *Phytophthora citrophthora* gum ducts are initiated schizogenously in xylem mother cells, and the cells lining the duct lumen differentiate into epithelial

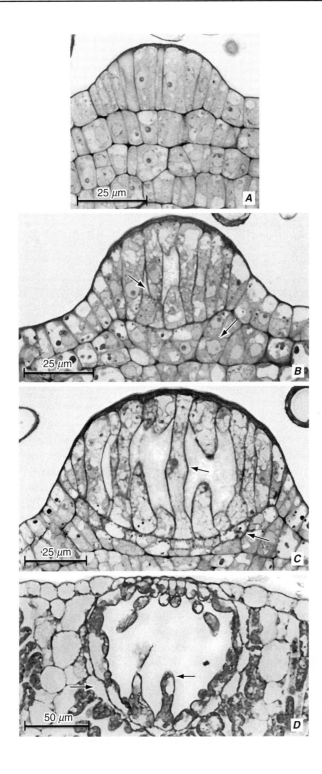

◄

**FIGURE 17.10**

Stages in schizogenous development of trabeculate secretory cavities in the leaves of *Psoralea macrostachya*. **A**, palisade-like protodermal cells at early stage of development. **B**, protodermal cells have elongated and are beginning to separate schizogenously (upper arrow). Below the developing trabeculae subtending hypodermal cells (lower arrow) have recently divided. **C**, further separation of the trabeculae (upper arrow) has occurred and cells of the hypodermal layer (lower arrow) have expanded laterally. **D**, mature cavity in transverse section showing portions of trabeculae (right arrow) and hypodermal sheath (left arrow). (From Turner, 1986.)

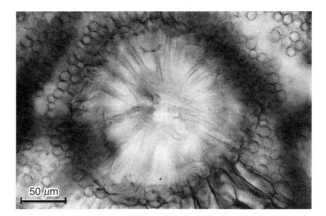

**FIG. 17.11**

A mature trabeculate cavity of *Psoralea macrostachya* in a leaf clearing. (From Turner, 1986.)

cells (Gedalovich and Fahn, 1985a). At the end of the secretory phase the walls of many of the epithelial cells break, and the gum still present in the cells is released into the lumen. The production of ethylene by the infected tissue is believed to be the direct cause of the formation of the gum ducts (Gedalovich and Fahn, 1985b).

The secretion of gum as a result of injury is known as **gummosis**, and different views have been expressed on the way the gum is produced (Gedalovich and Fahn, 1985a; Hillis, 1987; Fahn, 1988b). Some investigators have attributed formation of the gum to cell-wall decomposition; others have concluded that the gum is the product of the secretory cells lining the duct lumen.

## Kino Veins Are a Special Type of Traumatic Duct

Any discussion of traumatic ducts would be incomplete without mention of **kino veins**, which form frequently in the wood of the genus *Eucalyptus* in response to wounding or fungal infection (Fig. 17.13; Hillis, 1987). Kino veins also occur in the secondary phloem of some members of the subgenus *Symphyomyrtus* of *Eucalyptus* (Tippett, 1986). In both xylem and phloem the kino veins are initiated by lysigenous breakdown of parenchyma bands produced by the vascular cambium. Although referred to as gum in times past, **kino** contains polyphenols, some of which are tannins. The

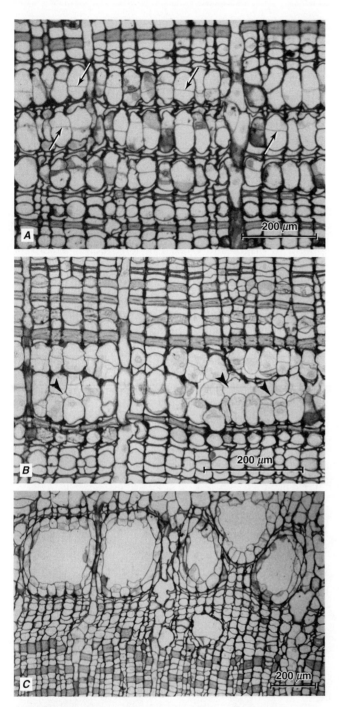

**FIGURE 17.12**

Traumatic resin canals in secondary phloem of *Chamaecyparis obtusa* in transverse section. Formation of the canals was induced by mechanical wounding. **A**, between day 7–9 after wounding, expanding axial parenchyma cells began to divide periclinally; arrows point to newly formed periclinal walls. **B**, 15 days after wounding, the centrally situated cells began to separate schizogenously from each other (arrowheads), forming canals. **C**, traumatic phloem resin canals 45 days after wounding. (From Yamanaka, 1989.)

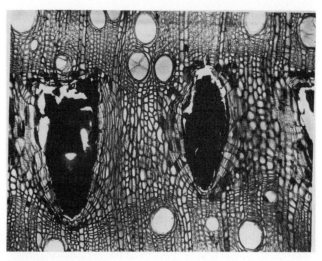

**FIGURE 17.13**

Mature kino veins as seen in transverse section of *Eucalyptus maculata* wood. (×43. Courtesy of Jugo Ilic, CSIRO, Australia.)

polyphenols accumulate in the traumatic parenchyma and are released into the future duct lumen when the parenchyma cells break down (holocrine secretion). The kino veins commonly form a dense, tangentially anastomosing network. At about the same time as the first kino is being released, the parenchyma cells surrounding the duct lumen divide repeatedly and form a peripheral "cambium." Derivatives of this "cambium" accumulate polyphenols. They eventually break down also, adding to the kino already present in the duct lumen. In the final stage the peripheral "cambium" produces several layers of suberized cells in the form of a typical periderm (Skene, 1965). Ethylene, either of microbial or host origin, may play a role in stimulating kino vein formation after injury (Wilkes et al., 1989).

# ▋ LATICIFERS

**Laticifers** are cells or series of connected cells containing a fluid called **latex** (plural, **latices**) and forming systems that permeate various tissues of the plant body. The word laticifer and its adjectival form laticiferous are derived from the word latex, meaning juice in Latin. Although the structures bearing latex may be single cells or series of fused cells, both kinds often produce complex systems of tube-like growth form in which recognition of the limits of individual cells is highly problematical. The term laticifer therefore appears most useful if applied to either a single cell or a structure resulting from fusion of cells. A single-celled laticifer can be qualified, on the basis of origin, as a **simple laticifer**, and the structure derived from union of cells as a **compound laticifer**.

The laticifers vary widely in their structure, as does latex in its composition. Latex may be present in ordinary parenchyma cells—in the pericarp of *Decaisnea insignis* (Hu and Tien, 1973), in the leaf of *Solidago* (Bonner and Galston, 1947)—or it may be formed in branching (*Euphorbia*) or anastomosing (*Hevea*) systems of tubes. The ordinary parenchyma cells with latex and the elaborate laticiferous systems intergrade with each other through intermediate types of structures of various degrees of morphologic specialization. Idioblastic laticifers (*Jatropha*, Dehgan and Craig, 1978) also intergrade with certain idioblasts that contain tannins, mucilages, proteinaceous and other compounds. The situation is further complicated by the occurrence of tanniniferous tubes (Myristicaceae, Fujii, 1988) and of schizogenous (Kisser, 1958) and lysigenous (*Mammillaria*, Wittler and Mauseth, 1984a, b) canals containing latex. Thus laticifers cannot be delimited precisely.

Latex-containing plants have been estimated to include some 12,500 species in 900 genera. The plants concerned include more than 22 families, mostly of eudicots but a few of monocots (Metcalfe, 1983). Latici-fers also occur in the gymnosperm *Gnetum* (Behnke and Herrmann, 1978; Carlquist, 1996; Tomlinson, 2003; Tomlinson and Fisher, 2005) and the fern *Regnellidium* (Labouriau, 1952). The plants containing latex range from small herbaceous annuals such as the spurges (*Euphorbia*) to large trees like the rubber-yielding *Hevea*. They occur in all parts of the World, but arborescent types are most common in the tropical floras.

## On the Basis of Their Structure, Laticifers Are Grouped in Two Major Classes: Articulated and Nonarticulated

**Articulated laticifers** (i.e., jointed laticifers) are compound in origin and consist of longitudinal chains of cells in which the walls separating the individual cells either remain intact, become perforated, or are completely removed (Fig. 17.14). The perforation or resorption of the end walls gives rise to laticifers that are tube-like in form and resemble xylem vessels in origin. Articulated laticifers may arise in both primary and secondary bodies of the plant. The **nonarticulated latici-**

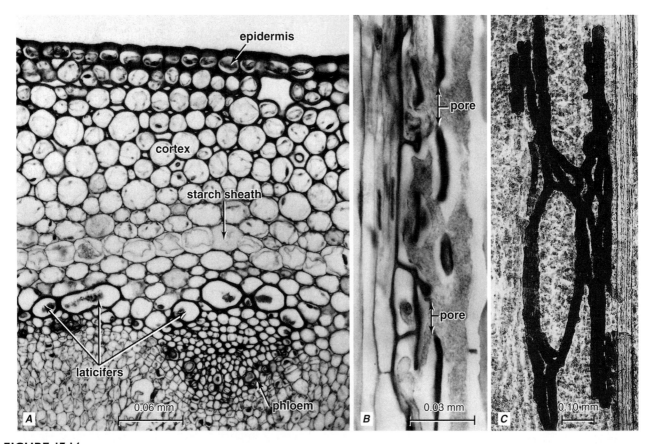

**FIGURE 17.14**

Articulated anastomosing laticifers in *Lactuca scariola*. **A**, transverse section of stem. Laticifers are outside the phloem. **B, C**, longitudinal views of laticifers in partly macerated tissue (**B**) and section (**C**) of stem. Perforations can be seen in the walls of the laticifer in **B**. (**C**, from Esau, 1977.)

**fers** originate from single cells that through continued growth develop into tube-like structures, often much branched, but typically they undergo no fusions with other similar cells (Fig. 17.15). They are simple in origin, and typically arise in the primary plant body.

Both articulated and nonarticulated laticifers vary in degree of complexity of their structure. Some of the articulated laticifers consist of long, cell chains or compound tubes not connected with each other laterally; others form lateral anastomoses with similar cell chains or tubes, all united into a netlike structure or reticulum.

The two forms of laticifers are called **articulated non-anastomosing laticifers** and **articulated anastomosing laticifers**, respectively.

The nonarticulated laticifers also vary in degree of complexity in their structure. Some develop into long, more or less straight tubes, whereas others branch repeatedly, each cell thus forming an immense system of tubes. The appropriate names for these two forms of laticifers are **nonarticulated unbranched laticifers** and **nonarticulated branched laticifers**, respectively. The latter include the longest of plant cells.

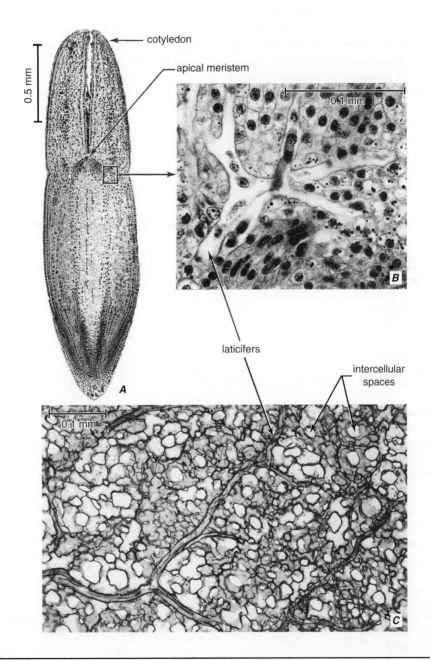

#### FIGURE 17.15

Nonarticulated branched laticifers of *Euphorbia* sp. **A**, embryo. Rectangle indicates a locus of origin of laticifers. **B**, section through laticifers showing multinucleate condition. **C**, laticifers branching within spongy parenchyma as seen in a paradermal section of a leaf. (From Esau, 1977; slide for **A**, **B**, courtesy of K. C. Baker.)

The list of various types of laticifers given in Table 17.1 shows that the type of laticiferous element is not constant in a given family. In the Euphorbiaceae, for example, *Euphorbia* has nonarticulated laticifers, whereas *Hevea* has articulated laticifers. The leaves of most *Jatropha* species (also Euphorbiaceae) contain both nonarticulated and articulated laticifers, in addition to idioblastic laticiferous cells that intergrade with the laticifers (Dehgan and Craig, 1978). In *Hevea*, *Manihot*, and *Cnidoscolus*, the articulated laticifers in the leaves form branches that undergo intrusive growth and ramify throughout the mesophyll (Rudall, 1987). These branches lack septa and are virtually impossible to distinguish from the nonarticulated laticifers in *Euphorbia*. Similar ramifying branches also occur in the pith and cortex of the stems.

Systematic comparative studies of laticifers are scarce, and the possible phylogenetic significance of the variations in the degree of their specialization has not yet been revealed. A systematic survey of the occurrence of articulated, anastomosing laticifers in leaves and inflorescences of 75 genera of Araceae, the largest family of laticiferous monocots, affirmed the significance of laticifer morphology in the systematics of that family (French,

## TABLE 17.1 ■ Examples of the Various Types of Laticifers by Family and Genus

### Articulated anastomosing

Asteraceae, tribe Cichorieae (*Cichorium, Lactuca, Scorzonera, Sonchus, Taraxacum, Tragopogon*)
Campanulaceae, the Lobelioideae
Caricaceae (*Carica papaya*)
Papaveraceae (*Argemone, Papaver*)
Euphorbiaceae (*Hevea, Manihot*)
Araceae, the Colocasioideae

### Articulated nonanastomosing

Convolvulaceae (*Convolvulus, Dichondra, Ipomoea*)
Papaveraceae (*Chelidonium*)
Sapotaceae (*Achras sapota, Manilkara zapota*)
Araceae, the Calloideae, Aroideae, Lasioideae, Philodendroideae
Liliaceae (*Allium*)
Musaceae (*Musa*)

### Nonarticulated branched

Euphorbiaceae (*Euphorbia*)
Asclepiadaceae (*Asclepias, Cryptostegia*)
Apocynaceae (*Nerium oleander, Allamanda violaceae*)
Moraceae (*Ficus, Broussonetia, Maclura, Morus*)
Cyclanthaceae (*Cyclanthus bipartitus*)

### Nonarticulated unbranched

Apocynaceae (*Vinca*)
Urticaceae (*Urtica*)
Cannabaceae (*Humulus, Cannabis*)

1988). Anastomosing laticifers are limited to the subfamily Colocasioideae and to *Zomicarpa* (Aroideae). Chemical, in addition to morphological, features of laticifers are of potential application as an aid in delimiting taxa and interpreting evolutionary trends (Mahlberg et al., 1987; Fox and French, 1988). It is generally presumed that articulated and nonarticulated laticifers evolved independently of one another and represent polyphyletic origins within vascular plants. However, inasmuch as both types of laticifer may undergo intrusive growth, as noted above, they may not be as divergent as commonly believed (Rudall, 1987). Although laticifers are regarded as recently evolved cell types, the fossil record indicates that nonarticulated forms were already present in an arborescent plant of the Eocene (Mahlberg et al., 1984).

## Latex Varies in Appearance and in Composition

The term latex refers to the fluid that can be extracted from a laticifer. Latex may be clear in appearance (*Morus, Humulus, Nerium oleander*, most Araceae) or milky (*Asclepias, Euphorbia, Ficus, Lactuca*). It is yellow-brown in *Cannabis* and yellow or orange in the Papaveraceae. Latex contains various substances in solution and colloidal suspension: carbohydrates, organic acids, salts, sterols, fats, and mucilages. Among the common components of latex are the terpenoids, rubber (*cis*-1,4-polyisoprene) being one of their well-known representatives. Rubber originates as particles in the cytosol (Coyvaerts et al., 1991; Bouteau et al., 1999), whereas other terpenoid-containing particles originate in small vesicles. As the laticifers approach maturity, the various particles are released into a large central vacuole (d'Auzac et al., 1982). Many other substances are found in latices, such as cardiac glycosides (in representatives of the Apocynales), alkaloids (morphine, codeine, and papaverine in the opium poppy, *Papaver somniferum*), cannabinoids (*Cannabis sativa*), sugar (in representatives of Asteraceae), large amounts of protein (*Ficus callosa*), and tannins (*Musa*, Aroideae). Crystals of oxalates and malate may be abundant in latex. Starch grains occur in laticifers of some genera of Euphorbiaceae (Biesboer and Mahlberg, 1981a; Mahlberg, 1982; Rudall, 1987; Mahlberg and Assi, 2002) and in those of *Thevetia peruviana* (Apocynaceae, Kumar and Tandon, 1990). The starch grains in *Thevetia* are osteoid (bone-shaped), whereas those in the Euphorbiaceae assume various forms—rod, spindle, osteoid, discoid, and intermediate forms—and may become very large (Fig. 17.16). Small starch grains have been observed in the plastids of differentiating *Allamanda violacea* (Apocynaceae) laticifers (Inamdar et al., 1988). Thus, latices contain a wide array of secondary metabolites, none of which is mobilizable, or able to participate further in the metabolism of the cell. Moreover, the mobilization of latex

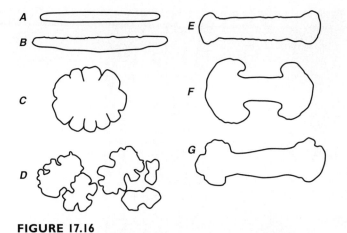

**FIGURE 17.16**

Shapes of starch grains in nonarticulated laticifers of Euphorbiaceae. **A**, rod-shaped grain of *Euphorbia lathuris*. **B**, spindle-shaped grain from *E. myrsinites*. **C**, **D**, discoid grains from *E. lactea*. **E**, slightly osteoid grain from *E. heterophylla*. **F**, **G**, osteoid grains from *E. pseudocactus* and *Pedilanthus tithymaloides*, respectively. (From photographs in Biesboer and Mahlberg, 1981b.)

starch has not been observed (Nissen and Foley, 1986; Spilatro and Mahlberg, 1986).

A variety of enzymes occur in latices, including the proteolytic enzyme papain in *Carica papaya* and lysosomal hydrolases such as acid phosphatase, acid RNAase, and acid protease in *Asclepias curassavica* (Giordani et al., 1982). Both cellulase and pectinase activity have been detected in the latex of the articulated laticifers of *Lactuca sativa* (Giordani et al., 1987). In other studies, however, cellulase activity was encountered exclusively in the latices of articulated laticifers (*Carica papaya, Musa textilis, Achras sapota*, and various *Hevea* species; Sheldrake, 1969; Sheldrake and Moir, 1970) and pectinase activity in those of nonarticulated laticifers (*Asclepias syriaca*, Wilson et al., 1976; *Nerium oleander*, Allen and Nessler, 1984). These results led to the suggestion that cellulase is involved with the removal of end walls during differentiation of articulated laticifers, and pectinase with the intrusive growth of the nonarticulated laticifers (Sheldrake, 1969; Sheldrake and Moir, 1970; Wilson et al., 1976; Allen and Nessler, 1984).

Laticifers often harbor bacteria and trypanosomatid flagellates of the genus *Phytomonas*. Those of apparently healthy *Chamaesyce thymifolia* plants have been found harboring both kinds of organisms (Da Cunha et al., 1998, 2000). An obligate laticifer-inhabiting bacterium (a rickettsial relative) has been found in association with papaya bunchy top disease (PBT), a major disease of *Carica papaya* in the American tropics, and long thought to be caused by a phytoplasma (Davis et al., 1998a, b). Should this disease be proven to be the cause of PBT, it would represent the first example of a leaf-hopper transmitted, laticifer-inhabiting plant pathogen. Trypanosome parasites from the laticifers of *Euphorbia pinea* are transmitted by the squash bug *Stenocephalus agilis*, and have been successfully cultured in vitro in liquid medium (Dollet et al., 1982). Not all such attempts at culturing laticifer-inhabiting trypanosomatid flagellates have been successful (Kastelein and Parsadi, 1984). The association of the laticifer-inhabiting *Phytomonas staheli* with hartrot disease of coconut and oil palm has been clearly established (Parthasarathy et al., 1976; Dollet et al., 1977) but the pathogenicity of this organism remains to be proved.

The contents of laticifers are under considerable turgor (Tibbitts et al., 1985; Milburn and Ranasinghe, 1996). Thus, whenever a laticifer is severed, a turgor gradient is established and the latex flows toward the open end (Bonner and Galston, 1947). This flow eventually ceases, and subsequently, the turgor is restored. The flow of latex from the cut laticifer is reminiscent of the surging of sieve-tube sap that takes place when a sieve tube is severed (Chapter 13). In both instances this phenomenon contributes to the difficulty of obtaining critical preservation of the mature protoplast (Condon and Fineran, 1989a).

## Articulated and Nonarticulated Laticifers Apparently Differ from One Another Cytologically

Initially both articulated and nonarticulated laticifers exhibit prominent nuclei and dense cytoplasm rich in ribosomes, rough endoplasmic reticulum, Golgi bodies, and plastids. Differentiation of nonarticulated laticifers is accompanied by nuclear divisions resulting in a coenocytic condition (Stockstill and Nessler, 1986; Murugan and Inamdar, 1987; Roy and De, 1992; Balaji et al., 1993). Articulated laticifers, in which the series of cells have fused by dissolution of their common walls, commonly are characterized as multinucleate, but the so-called multinucleate condition in these laticifers is not the result of multiplication of nuclei, rather from the fusion of their protoplasts.

As development proceeds, numerous vesicles, often called small vacuoles, appear in both kinds of laticifer (Fig. 17.17). Apparently endoplasmic reticulum in origin, they contain a variety of substances—some contain latex particles, others alkaloids, papain—depending on the species. Many are lysosomal vesicles involved with gradual degeneration of many, most, or all of the cytoplasmic organelles by autophagic processes. The lysosomal vesicles, or microvacuoles, in *Hevea* latex are usually referred to as **lutoids** (Wu, J.-L., and Hao, 1990; d'Auzac et al., 1995). As autophagy continues, the remaining cytoplasmic components become peripheral in distribution, resulting in formation of a large central vacuole filled with a variety of substances.

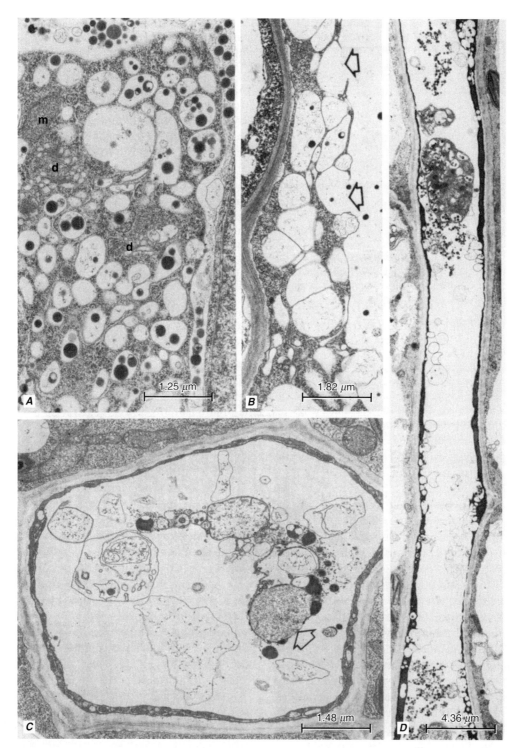

**FIGURE 17.17**

Late stages in differentiation of nonarticulated laticifers in poinsettia (*Euphorbia pulcherrima*). **A,** a mass of the cytoplasm isolated within the central vacuole. Numerous latex particles occur in the small vacuoles and in the surrounding central vacuole (top and right). Mitochondrion (m), dictyosomes (d), or Golgi bodies, and ribosomes are also present in the cytoplasm. **B,** portion of the peripheral cytoplasm abutting the large central vacuole in a laticifer approaching maturity. Some small peripheral vacuoles have just fused with the central vacuole (open arrows) and released their latex particles. **C,** transverse section of a laticifer nearing the end of differentiation, with cytoplasmic debris at various stages of degeneration, including a nucleus (open arrow) in the large central vacuole. **D,** longitudinal section of maturing region of a laticifer. The continuous central vacuole contains clusters of latex particles and remnants of degenerating cytoplasm. The peripheral cytoplasm of the laticifer is more electron dense than that of the adjacent parenchyma cells. (From Fineran, 1983, by permission of Oxford University Press.)

The extent of degeneration of cytoplasmic components apparently differs between articulated and nonarticulated laticifers, although in that regard variation apparently exists among nonarticulated laticifers. Most nonarticulated laticifers have a large central vacuole at maturity, and the presence of both plasma membrane and tonoplast has been noted for those of *Nelumbo nucifera* (Esau and Kosakai, 1975), *Asclepias syriaca* (Wilson and Mahlberg, 1980), *Euphorbia pulcherrima* (Fig. 17.17D; Fineran, 1982, 1983), and *Nerium oleander* (Stockstill and Nessler, 1986). In these laticifers the liquid matrix of the latex may be regarded as the cell sap of the laticifer. Only in *Chamaesyce thymifolia* are the nonarticulated laticifers reported to lack a tonoplast when "completely differentiated" (Da Cunha et al., 1998). Although reduced in number, some organelles and nuclei apparently persist in the mature portions of nonarticulated laticifers. The persistent nuclei in *Euphorbia pulcherrima* have been described as degenerate (Fineran, 1983). In *Nerium oleander* the mature protoplast is described as containing nuclei of normal appearance and the "usual complement of organelles" (Stockstill and Nessler, 1986). By contrast, in *Chamaesyce thymifolia* total degeneration of nuclei and organelles is reported to occur (Da Cunha et al., 1998). It is entirely possible that some of the cytological variation reported for nonarticulated laticifers may be a reflection of the degree to which their protoplasts have been successfully preserved.

Accounts of differentiation of articulated laticifers are rather uniform. Autophagy results in total elimination of nuclei and cellular organelles. As articulated laticifers approach maturity the tonoplast disappears and the lumen of the cell becomes filled with vesicles and latex particles (Fig. 17.18; Condon and Fineran, 1989a, b; Griffing and Nessler, 1989). Only the plasma membrane remains intact and functional (Zhang et al., 1983; Alva et al., 1990; Zeng et al., 1994).

Most laticifers have nonlignified primary walls variable in thickness. Lignification of laticifer walls has been recorded by some workers (Dressler, 1957; Carlquist, 1996). In a few species the laticifers develop very thick walls (in stems of *Euphorbia abdelkuri*, Rudall, 1987) considered by some workers as secondary (Solereder, 1908). A. R. Rao and Tewari (1960) reported that in *Codiaeum variegatum* the laticifers in young leaves become sclereids in older ones, and A. R. Rao and Malaviya (1964) suggested that sclereids in leaves of many Euphorbiaceae may be sclerified laticifers. Rudall (1994), noting that the leaves of many euphorbiaceous genera lack laticifers but contain highly branched sclereids resembling laticifers in structure and distribution, suggested that in some instances such sclereids may be homologous with laticifers. The walls of the articulated laticifers in the Convolvulaceae are suberized; that is, they contain suberin lamellae (Fineran et al., 1988).

Plasmodesmata have been observed only rarely in the common walls between laticifers and other cell types.

## Laticifers Are Widely Distributed in the Plant Body, Reflecting Their Mode of Development

Nonarticulated Laticifers The branched nonarticulated laticifers of the Euphorbiaceae, Asclepiadaceae, Apocynaceae, and Moraceae arise during the development of the embryo in the form of relatively few primordia, or initials, then grow concomitantly with the plant into branched systems permeating the whole plant body (Fig. 17.15) (Mahlberg, 1961, 1963; Cass, 1985; Murugan and Inamdar, 1987; Rudall, 1987, 1994; van Veenendaal and den Outer, 1990; Roy and De, 1992; Da Cunha et al., 1998). The laticifer initials appear in the embryo as the cotyledons are being initiated, and are located in the plane of the embryo that later represents the cotyledonary node. In some species, the initials arise in the outer region of the future vascular cylinder (i.e., from the procambium that will develop into protophloem); in others, they arise just outside the future vascular cylinder. In either case, the laticifer initials are closely associated spatially with the phloem. The number of initials varies both between and within species. In some *Euphorbia* species, only 4 initials have been recognized; in others, 8 and 12; and in still others, many initials distributed in arcs or in a complete circle. Five to 7 initials occur in *Jatropha dioica* (Cass, 1985). Eight initials have been recognized in *Morus nigra* (Moraceae, van Veenendaal and den Outer, 1990). In *Nerium oleander* (Apocynaceae) usually 28 laticifer initials are present (Fig. 17.19; Mahlberg, 1961). The initials form protrusions in various directions, and the apices of these protrusions push their way intercellularly among the surrounding cells by intrusive growth, in a manner resembling the growth of a fungal hypha. Typically the laticifer initials penetrate downward into the root and upward into the cotyledons and toward the shoot apex. Additional branches rapidly penetrate the cortex, extending as far as the subepidermal layer; others penetrate the pith.

When the seed is mature, the embryo has a system of tubes arranged in a characteristic manner. In *Euphorbia*, for example, one set of tubes extends from the cotyledonary node downward, following the periphery of the vascular cylinder of the hypocotyl. Another set passes downward within the cortex, usually near its periphery. The two sets of tubes end near the root meristem at the base of the hypocotylary axis. A third set is prolonged into the cotyledons where the tubes branch, sometimes profusely. A fourth set of tubes extends inward and upward from the nodal initials toward the shoot apex of the epicotyl where the tubes form a ring-like network. The terminations of this network reach into the third or fourth layers beneath the surface of the

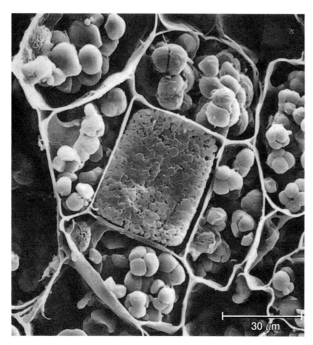

**FIGURE 17.18**

Scanning electron micrograph of a mature articulated laticifer in rhizome cortex of *Calystegia silvatica* (Convolvulaceae). The turgid laticifer is packed full with spherical latex particles. Amyloplasts are plentiful in neighboring parenchyma cells. (From Condon and Fineran, 1989b. © 1989 by The University of Chicago. All rights reserved.)

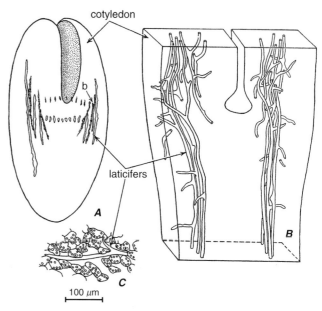

**FIGURE 17.19**

Nonarticulated laticifers of *Nerium oleander*. **A**, immature embryo 550 μm long. Young laticifers at cotyledonary node. They occur along periphery of vascular region. Beginning of laticifer branching at b. **B**, 75 μm wide section of mature embryo 5 mm long. Laticifers extend from node into cotyledons and hypocotyl. Short branches extend into mesophyll of cotyledons and cortex of hypocotyl. **C**, branch of laticifer in proliferated mesophyll of cultured embryo. It extends through intercellular spaces. (**A, B**, after Mahlberg, 1961; **C**, from photograph in Mahlberg, 1959.)

apical meristem. Thus there are laticifer terminations in the immediate vicinity of both apical meristems, that of the shoot and that of the root. Few, if any, branches penetrate toward the center of the future pith of the *Euphorbia* embryo.

When a seed germinates and the embryo develops into a plant, the laticifers keep pace with this growth by continuously penetrating the meristematic tissues formed by the active apical meristems. When axillary buds or lateral roots arise, they also are penetrated by the intrusively growing tips of the laticifers. At the nodes the laticifers enter the leaves and pith via the leaf trace gap. Since the laticifer tips penetrate the tissues close to the apical meristems, the tube portions below the tips occur for a time in growing tissues and extend in unison with them. Thus the laticifers elongate at their apices by intrusive growth and subsequently extend with the surrounding tissues by coordinated growth (Lee and Mahlberg, 1999).

In leaves, the nonarticulated laticifers ramify throughout the mesophyll but tend to follow the course of the veins. In some species the laticifers extend directly to the epidermis, frequently intruding between epidermal cells (*Baloghia lucida, Codiaeum variegatum*; Fig.

17.20A), or into the bases of trichomes (*Croton* spp.; Fig. 17.20B) (Rudall, 1987, 1994).

If the plant produces secondary tissues, the nonarticulated laticifers grow into these also. In *Cryptostegia* (Asclepiadaceae), for example, the secondary phloem is penetrated by prolongations from the cortical and primary phloem laticifers (Artschwager, 1946). Moreover, the continuity between the laticifer branches in the pith and cortex, established through the interfascicular regions during primary growth, is seemingly not ruptured by the activity of the vascular cambium during secondary growth. The parts of the laticifer located in the cambium appear to extend by localized growth (intercalary growth) and eventually become embedded in secondary phloem and xylem (Blaser, 1945). In *Croton* spp., laticifers have been recorded penetrating from primary tissues into the vascular cambium and the secondary xylem (Rudall, 1989). In *Croton conduplicatus*, ray initials in the vascular cambium occasionally are converted to laticifer initials and intruded between cells of the phloem rays in the manner of a nonarticulated laticifer (Rudall, 1989). A secondary laticiferous system,

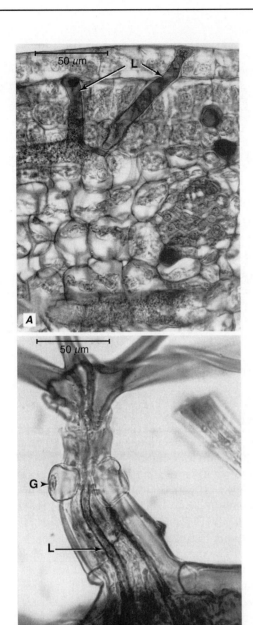

**FIGURE 17.20**

Branched nonarticulated laticifers (L) intruding (**A**) between epidermal cells in *Codiaceum variegatum* leaf and (**B**) into base of stellate hair, with glandular region (G), in *Croton* sp. (From Rudall, 1994.)

produced by the vascular cambium, has also been recorded in *Morus nigra* (van Veenendaal and den Outer, 1990). Prior to the latter two reports, it was generally accepted that nonarticulated laticifers originate only from primary tissues, as distinct from articulated

laticifers, which may be both primary and secondary in origin.

Nonarticulated laticifers are not uncommon in the rays of the secondary xylem, occurring in genera of Apocynaceae, Asclepiadaceae, Euphorbiaceae, and Moraceae (Wheeler et al., 1989). It generally is presumed that the laticifers penetrate the rays from the pith. Axial nonarticulated laticifers (interspersed among the fibers) are known only from the secondary xylem of Moraceae. In some lianoid species of *Gnetum*, nonarticulated laticifers have been observed in the conjunctive tissue—parenchyma between successive vascular cylinders—where they follow a vertical course (Carlquist, 1996).

Nonarticulated unbranched laticifers show a simpler pattern of growth than the branched type (Zander, 1928; Schaffstein, 1932; Sperlich, 1939). The primordia of these laticifers have been recognized, not in the embryo, but in the developing shoot (*Vinca, Cannabis*) or in the shoot and root (*Eucommia*). New primordia arise repeatedly beneath the apical meristems, and each elongates into an unbranched tube, apparently by a combination of intrusive and coordinated growth. In the shoot the tubes may extend for some distance in the stem and also diverge into the leaves (*Vinca*). Laticifers may arise in the leaves also, independently of those formed in the stem (*Cannabis, Eucommia*).

**Articulated Laticifers** The initials of articulated laticifers may or may not be apparent in the mature embryo, but they become clearly visible soon after the seed begins to germinate. In the Cichorieae (Scott, 1882; Baranova, 1935), Euphorbiaceae (Scott, 1886; Rudall, 1994), and Papaveraceae (Thureson-Klein, 1970), the initials appear in the protophloem region of the procambial tissue or peripheral to it in both the cotyledons and the hypocotyledonary axis. Those of the cotyledons are most well developed at this stage. The initials are arranged in more or less discrete longitudinal rows but the formation of lateral protuberances results in an anastomosing system. In *Hevea brasiliensis* the walls between lateral protuberances become perforated before the transverse walls between initials (Scott, 1886). Where the rows of initials lie side by side, parts of the common wall become resorbed. With breakdown of the lateral and transverse walls, the cells are united into a system of compound tubes. As the plant develops from the embryo, the tubes are extended by differentiation of further meristematic cells into laticiferous elements. Thus the laticifers differentiate acropetally in the newly formed plant parts, and they are prolonged not only within the axis but also in the leaves and, later, in the flowers and fruits. The direction of differentiation is similar to that of the nonarticulated branched laticifers, but it occurs by successive conversion of cells into laticiferous elements instead of by apical intrusive growth. As mentioned previously, the articulated laticifers of *Hevea*, *Manihot*, and *Cnidoscolus* form elongated intrusive branches in the same manner as nonarticulated laticifers. These branches, which ramify throughout the leaf mesophyll, also penetrate the cortex and pith of all three genera (Rudall, 1987). The development of articulated laticifers of the nonanastomosing kind is similar to that of the anastomosing laticifers, except that no lateral connections are established among the various tubes (Fig. 17.21; Karling, 1929). In some genera (*Allium*, Fig. 17.22, Hayward, 1938; *Ipomoea*, Hayward, 1938; Alva et al., 1990) the cells of the nonanastomosing kind retain their end walls.

Articulated laticifers, like their nonarticulated counterparts, show various arrangements within the plant body and a frequent association with the phloem. In the primary body of the Cichorieae (Asteraceae subfamily) laticifers appear on the outer periphery of the phloem and within the phloem itself. In species with internal phloem, laticifers are associated with this tissue also. The external and the internal laticifers are interconnected across the interfascicular areas.

The Cichorieae produce laticifers also during secondary growth, mainly in the secondary phloem. This development has been followed in some detail in the fleshy roots of *Tragopogon* (Scott, 1882), *Scorzonera* (Baranova, 1935), and *Taraxacum* (Fig. 17.23; Artschwager and McGuire, 1943; Krotkov, 1945). Longitudinal rows of derivatives from fusiform cambial initials fuse into tubes through resorption of the end walls. Lateral connections are established—directly or by means of protuberances—among the tubes differentiating in the same tangential plane. The tissue formed by the cambium consists of a series of concentric layers of laticifers, parenchyma cells, and sieve tubes (with accompanying companion cells). The laticifers of one concentric layer are rarely joined with those of another concentric layer. Rays of parenchyma traverse the whole tissue in a radial direction. The laticiferous system that makes *Hevea brasiliensis* (Euphorbiaceae) such an outstanding rubber producer is the secondary system that develops in the secondary phloem (Fig. 17.24), which also consists of alternating layers of articulated laticifers, parenchyma cells, and sieve tubes (Hébant and de Faÿ, 1980; Hébant et al., 1981). In the secondary phloem of *Manilkara zapota* (Sapotaceae), the source of commercial latex from which chicle is obtained, some rays are composed entirely of laticifers (Mustard, 1982). With dilatation of these rays in older branches, their ends anastomose, lose their identity, and form a secondary laticiferous mass internal to the periderm. The laticifers of the rays and axial systems are interconnected.

The laticifers of *Papaver somniferum* occur in the phloem and become particularly well developed in the mesocarp about two weeks after the petals fall

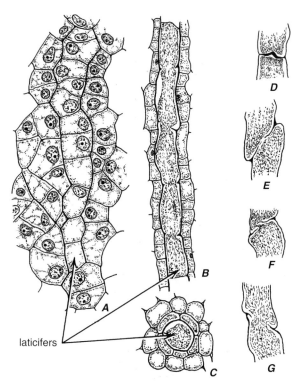

**FIGURE 17.21**

Development of articulated laticifer in *Achras sapota* (Sapotaceae) in longitudinal (**A**, **B**, **D**–**G**) and transverse (**C**) sections. **A**, vertical file of young laticifer cells (from arrow upward) with end walls intact. **B**, file of cells has been converted into laticiferous vessel by partial dissolution of end walls. Remnants of end walls indicate articulations between members of laticifer. In **C**, flattened cells ensheath laticifer. **D**–**G**, stages in perforation of end wall. Wall to be perforated first becomes swollen (**D**), and then breaks down (**E**–**G**). (Adapted from Karling, 1929.)

(Fairbairn and Kapoor, 1960). At this time, the capsules are harvested for the extraction of opium. In the leaves of the Cichorieae the laticifers accompany the vascular bundles, ramify more or less profusely in the mesophyll, and reach the epidermis. The epidermal hairs of the floral involucres of the Cichorieae become directly connected with laticifers by a breakdown of the separating walls, and as a result the latex readily issues from these hairs when they are broken (Sperlich, 1939). In effect the hairs actually represent terminations of the laticiferous system.

Among the monocots, the laticifers of *Musa* are associated with the vascular tissues and also occur in the cortex (Skutch, 1932). In *Allium* the laticifers are entirely separated from the vascular tissue (Fig. 17.21; Hayward, 1938). They lie near the abaxial, or lower, surface of the leaves or scales, between the second and third layers of parenchyma. The *Allium* laticifers have the form of

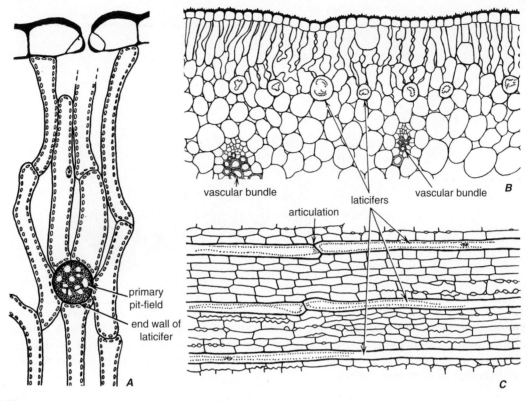

**FIGURE 17.22**

Articulated laticifers in *Allium*. **A**, transverse section through fleshy scale of *A. cepa*, showing epidermis with stoma, a few mesophyll cells, and a laticifer with end wall in surface view. **B**, **C**, laticifers in *A. sativum* leaves in transverse (**B**) and tangential (**C**) sections. **B**, palisade parenchyma beneath epidermis. Laticifers occur in third layer of mesophyll not in contact with vascular bundles. **C**, laticifers appear like continuous tubes except in places where end wall (articulation) between superposed cells is visible. The end wall is not perforated. (A, ×300; B, C, ×79. **B**, **C**, drawn from photomicrographs courtesy of L. K. Mann.)

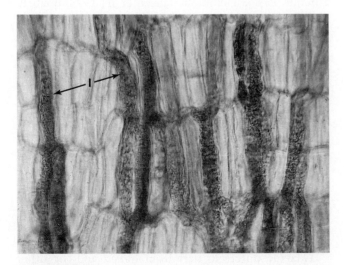

**FIGURE 17.23**

Articulated anastomosing laticifers (l) of *Taraxacum kok-saghyz* in longitudinal section of secondary phloem of root. (×280. From Artschwager and McGuire, 1943.)

longitudinal chains of cells arranged parallel in the upper parts of foliar organs and converging toward their bases. The individual cells are considerably elongated. The end walls, which are not perforated, have conspicuous primary pit-fields. Although *Allium* laticifers are classified as nonanastomosing, they form some interconnections at the bases of the leaves or scales.

### The Principal Source of Commercial Rubber Is the Bark of the Para Rubber Tree, *Hevea brasiliensis*

Natural rubber is produced in over 2500 plant species (Bonner, 1991), but few contain enough to make extraction worthwhile. Among plants of secondary importance as rubber producers are the Russian dandelion (*Taraxacum kok-saghyz*, Asteraceae), Ceara rubber tree (*Manihot glaziovii*, Euphorbiaceae), African rubber tree (*Funtumia elastica*, Apocynaceae), the African woody vines *Landolphia*, *Clitandra*, and *Carpodinus* (all three Apocynaceae), the woody climber *Cryptostegia grandiflora* (Asclepiadaceae), which is native to Madagascar but fairly widespread as a weed, and guayule

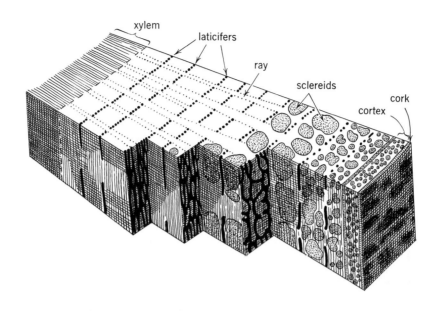

**FIGURE 17.24**

Block diagram of bark of *Hevea brasiliensis*, depicting arrangement of articulated laticifers in secondary phloem. Layers containing sieve tubes and associated parenchyma cells alternate with those in which laticifers (shown in solid black) differentiate. Parenchymatous secondary phloem rays traverse tissue radially. In tangential sections, laticifers of a given growth zone form a reticulum. Sclereids occur in old phloem where sieve tubes and laticifers are nonfunctional. (Adapted from Vischer, 1923.)

(*Parthenium argentatum*, Asteraceae), a woody desert shrub that has been developed as a crop for arid regions of the world. *Hevea brasiliensis* remains the only important source of natural rubber.

Most of the rubber-laden latex obtained from the bark of *H. brasiliensis* during tapping comes from the layers of laticifers located in the nonconducting phloem. The conducting phloem is restricted to a narrow band about 0.2 to 1.0 mm wide, next to the vascular cambium (Hébant and de Faÿ, 1980; Hébant et al., 1981). Care is taken during tapping not to penetrate the conducting phloem so as to avoid injury to the cambium.

High turgor pressure in the laticifers is essential for the flow of latex during tapping, and it requires a ready transfer of water from the phloem apoplast to the laticifer (Jacob et al., 1998). The lutoids, the enzyme-containing microvacuoles, which represent a component of the latex, play an important role in stopping latex flow after tapping (Siswanto, 1994; Jacob et al., 1998). Most of the lutoids are destroyed during tapping by physical stress, releasing coagulation factors that eventually halt flow. Latex regeneration between tappings depends on the influx of carbohydrates—primarily sucrose, the initial molecule for polyisoprene synthesis—from the sieve tubes of the conducting phloem.

A study on the distribution of plasmodesmata in the secondary phloem of *Hevea* indicates that numerous plasmodesmata occur between ray and axial parenchyma cells but are rare or lacking between the latici-

fers and the parenchyma cells ensheathing them (de Faÿ et al., 1989). Hence, although sucrose may follow a symplastic pathway from the conducting phloem to the vicinity of the laticifers of the nonconducting phloem, at the parenchyma cell-laticifer interface the sucrose must enter the apoplast before it can be taken up by the laticifers. Considerable evidence indicates that active uptake of sugar by the laticifers involves sucrose-$H^+$ and glucose-$H^+$ cotransports mediated by $H^+$-ATPase at the laticifer plasma membrane (Jacob et al., 1998; Bouteau et al., 1999). Interestingly the lutoid membrane has been found also to contain proton pumps (Cretin, 1982; d'Auzac et al., 1995). Whereas lutoids originate from the endoplasmic reticulum, rubber first appears as particles in the cytosol of the young laticifer. Molecular studies have begun to provide information on gene expression in the laticifers of *Hevea* (Coyvaerts et al., 1991; Chye et al., 1992; Adiwilaga and Kush, 1996) and other species (Song et al., 1991; Pancoro and Hughes, 1992; Nessler, 1994; Facchini and De Luca, 1995; Han et al., 2000). Jasmonic acid has been implicated in development of the articulated anastomosing laticifers in *Hevea* (Hao and Wu, 2000), and cytokinins and auxins in development of the nonarticulated laticifers in *Calotropis* (Datta and De, 1986; Suri and Ramawat, 1995, 1996).

Guayule differs in part from *Hevea* and the other rubber-producing plants listed above in that it lacks laticifers. Rubber formation in guayule occurs in the cytosol of parenchyma cells of stem and root, and apparently all of the parenchyma cells in guayule have the potential to synthesize rubber (Gilliland and van Staden, 1983; Backhaus, 1985). Rubber particles have been found in meristematic tissue of shoot apices and in stem parenchyma of cortex, pith, and phloem rays. In cortex and pith, rubber particles first appear in the epithelial cells surrounding the resin ducts. At maturity, most of the rubber particles of cells associated with the resin ducts occur in the vacuole (Fig. 17.25; Backhaus and Walsh, 1983).

Although the rubber content of tapped *Hevea* latex amounts to about 25% dry weight by volume, it accounts for only about 2% of the total dry weight of the plant (Leong et al., 1982). By contrast, guayule can accumulate up to 22% of its dry weight as rubber in its many parenchyma cells (Anonymous, 1977). Nevertheless, the yields from *Hevea* are much higher than those from guayule (Leong et al., 1982) because the replenishment of rubber that occurs following tapping is more efficient than regrowth and filling of new tissue by guayule.

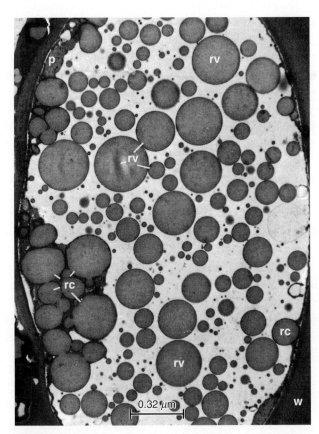

**FIGURE 17.25**

Rubber particles in epithelial cell of guayule (*Parthenium argentatum*). Most of the rubber particles occur in the vacuole (rv), compared with the number of those found in the peripheral cytoplasm (rc). Other details: p, plastid; w, cell wall. (From Backhaus and Walsh, 1983. © 1983 by The University of Chicago. All rights reserved.)

The mean diameter of rubber particles in *Hevea* is 0.96 μm and in guayule, 1.41 μm (Cornish et al., 1993). The particles are composed of a spherical homogeneous core of rubber enclosed by a monolayer biomembrane that serves as an interface between the hydrophobic rubber interior and the aqueous cytosol (Cornish et al., 1999). The biomembrane also prevents aggregation of the particles.

### The Function of Laticifers Is Not Clear

Laticifers have been an object of intensive study since the early days of plant anatomy (de Bary, 1884; Sperlich, 1939). Because of their distribution in the plant body and their liquid, often milky contents that flow out readily when the plant is cut, the laticiferous system was compared by early botanists with the circulatory system of animals. One of the common views advanced was that laticifers were concerned with food conduction. However, no actual movement of substances has been observed in laticifers, only a local and spasmodic one. Laticifers have been described also as food-storage elements, but it is quite clear that the food substances found in some latices, most notably starch, are not readily mobilized (Spilatro and Mahlberg, 1986; Nissen and Foley, 1986). Because the starch grains commonly accumulate at wounds, it has been suggested that the laticifer amyloplast may have evolved a secondary function as a component in a wound healing mechanism (Spilatro and Mahlberg, 1990). Rubber, which costs the plant large amounts of energy to synthesize, represents another enigma. Since it cannot be metabolized, of what value is it to the plant? It has been suggested that rubber occurs in plants as a response to their production of excess photosynthate and, therefore, is in all probability a metabolic "overspill" (Paterson-Jones et al., 1990). The potential of utilizing latex- and rubber-producing plants as sinks for atmospheric carbon dioxide, a greenhouse gas, has been noted by Hunter (1994). Laticifers undoubtedly serve as systems to sequester toxic secondary metabolites, which may function as protection against herbivores (Da Cunha et al., 1998; Raven et al., 2005). Inasmuch as laticifers accumulate many substances that are commonly recognized as excretory in general, laticifers appear to fit best the class of excretory structures.

## REFERENCES

ADIWILAGA, K., and A. KUSH. 1996. Cloning and characterization of cDNA encoding farnesyl diphosphate synthase from rubber tree (*Hevea brasiliensis*). *Plant Mol. Biol.* 30, 935–946.

ALLEN, R. D., and C. L. NESSLER. 1984. Cytochemical localization of pectinase activity in laticifers of *Nerium oleander* L. *Protoplasma* 119, 74–78.

ALVA, R., J. MÁRQUEZ-GUZMÁN, A. MARTÍNEZ-MENA, and E. M. ENGLEMAN. 1990. Laticifers in the embryo of *Ipomoea purpurea* (Convolvulaceae). *Phytomorphology* 40, 125–129.

ANDRÉASSON, E., L. B. JØRGENSEN, A.-S. HÖGLUND, L. RASK, and J. MEIJER. 2001. Different myrosinase and idioblast distribution in *Arabidopsis* and *Brassica napus*. *Plant Physiol.* 127, 1750–1763.

ANONYMOUS. 1977. Guayule: *An Alternative Source of Natural Rubber.* [National Research Council (U. S.). Panel on Guayule.] National Academy of Sciences, Washington, DC.

ARTSCHWAGER, E. 1945. Growth studies on guayule (*Parthenium argentatum*). USDA, Washington, DC. Tech. Bull. No. 885.

ARTSCHWAGER, E. 1946. Contribution to the morphology and anatomy of *Cryptostegia* (*Cryptostegia grandiflora*). USDA, Washington, DC. Tech. Bull. No. 915.

ARTSCHWAGER, E., and R. C. MCGUIRE. 1943. Contribution to the morphology and anatomy of the Russian dandelion (*Taraxacum kok-saghyz*). USDA, Washington, DC. Tech. Bull. No. 843.

ARUMUGASAMY, K., K. UDAIYAN, S. MANIAN, and V. SUGAVANAM. 1993. Ultrastructure and oil secretion in *Hiptage sericea* Hook. *Acta Soc. Bot. Pol.* 62, 17–20.

D'AUZAC, J., H. CRÉTIN, B. MARIN, and C. LIORET. 1982. A plant vacuolar system: The lutoids from *Hevea brasiliensis* latex. *Physiol. Vég.* 20, 311–331.

D'AUZAC, J., J.-C. PRÉVÔT, and J.-L. JACOB. 1995. What's new about lutoids? A vacuolar system model from *Hevea* latex. *Plant Physiol. Biochem.* 33, 765–777.

BAAS, P., and M. GREGORY. 1985. A survey of oil cells in the dicotyledons with comments on their replacement by and joint occurrence with mucilage cells. *Isr. J. Bot.* 34, 167–186.

BABU, A. M., and A. R. S. MENON. 1990. Distribution of gum and gum-resin ducts in plant body: Certain familiar features and their significance. *Flora* 184, 257–261.

BABU, A. M., G. M. NAIR, and J. J. SHAH. 1987. Traumatic gum-resin cavities in the stem of *Ailanthus excelsa* Roxb. *IAWA Bull.* n.s. 8, 167–174.

BABU, A. M., P. JOHN, and G. M. NAIR. 1990. Ultrastructure of gum-resin secreting cells in the pith of *Ailanthus excelsa* Roxb. *Acta Bot. Neerl.* 39, 389–398.

BACKHAUS, R. A. 1985. Rubber formation in plants—A mini-review. *Isr. J. Bot.* 34, 283–293.

BACKHAUS, R. A., and S. WALSH. 1983. The ontogeny of rubber formation in guayule, *Parthenium argentatum* Gray. *Bot. Gaz.* 144, 391–400.

BAKKER, M. E., and P. BAAS. 1993. Cell walls in oil and mucilage cells. *Acta Bot. Neerl.* 42, 133–139.

BAKKER, M. E., and A. F. GERRITSEN. 1989. A suberized layer in the cell wall of mucilage cells of *Cinnamomum*. *Ann. Bot.* 63, 441–448.

BAKKER, M. E., and A. F. GERRITSEN. 1990. Ultrastructure and development of oil idioblasts in *Annona muricata* L. *Ann. Bot.* 66, 673–686.

BAKKER, M. E., A. F. GERRITSEN, and P. J. VAN DER SCHAAF. 1991. Development of oil and mucilage cells in *Cinnamomum burmanni*. An ultrastructural study. *Acta Bot. Neerl.* 40, 339–356.

BALAJI, K., R. B. SUBRAMANIAN, and J. A. INAMDAR. 1993. Occurrence of non-articulated laticifers in *Streblus asper* Lour. (Moraceae). *Phytomorphology* 43, 235–238.

BARANOVA, E. A. 1935. Ontogenez mlechnoc systemy tau-sagyza (*Scorzonera tau-saghyz* Lipsch. et Bosse.) (Ontogenese des Milchsaftsystems bei (*Scorzonera tau-saghyz* Lipsch. et Bosse.) *Bot. Zh.* SSSR 20, 600–616.

BEHNKE, H.-D., and S. HERRMANN. 1978. Fine structure and development of laticifers in *Gnetum gnemon* L. *Protoplasma* 95, 371–384.

BHATT, J. R. 1987. Development and structure of primary secretory ducts in the stem of *Commiphora wightii* (Burseraceae). *Ann. Bot.* 60, 405–416.

BHATT, J. R., and H. Y. M. RAM. 1992. Development and ultrastructure of primary secretory ducts in the stem of *Semecarpus anacardium* (Anacardiaceae). *IAWA Bull.* n.s. 13, 173–185.

BHATT, J. R., and J. J. SHAH. 1985. Ethephon (2-chloroethylphosphonic acid) enhanced gum-resinosis in mango, *Mangifera indica* L. *Indian J. Exp. Biol.* 23, 330–339.

BIESBOER, D. D., and P. G. MAHLBERG. 1981a. A comparison of alpha-amylases from the latex of three selected species of *Euphorbia* (Euphorbiaceae). *Am. J. Bot.* 68, 498–506.

BIESBOER, D. D., and P. G. MAHLBERG. 1981b. Laticifer starch grain morphology and laticifer evolution in *Euphorbia* (Euphorbiaceae). *Nord. J. Bot.* 1, 447–457.

BLASER, H. W. 1945. Anatomy of *Cryptostegia grandiflora* with special reference to the latex system. *Am. J. Bot.* 32, 135–141.

BONES, A., and T.-H. IVERSEN. 1985. Myrosin cells and myrosinase. *Isr. J. Bot.* 34, 351–376.

BONNER, J. 1991. The history of rubber. In: *Guayule: Natural Rubber. A Technical Publication with Emphasis on Recent Findings*, pp. 1–16, J. W. Whitworth and E. E. Whitehead, eds. Office of Arid Lands Studies, University of Arizona, Tucson, and USDA, Washington, DC.

BONNER, J., and A. W. GALSTON. 1947. The physiology and biochemistry of rubber formation in plants. *Bot. Rev.* 13, 543–596.

BOSABALIDIS, A., and I. TSEKOS. 1982. Ultrastructural studies on the secretory cavities of *Citrus deliciosa* Ten. II. Development of the essential oil-accumulating central space of the gland and process of active secretion. *Protoplasma* 112, 63–70.

BOUTEAU, F., O. DELLIS, U. BOUSQUET, and J. P. RONA. 1999. Evidence of multiple sugar uptake across the plasma membrane of laticifer protoplasts from *Hevea*. *Bioelectrochem. Bioenerg.* 48. 135–139.

CARLQUIST, S. 1996. Wood, bark and stem anatomy of New World species of *Gnetum*. *Bot. J. Linn. Soc.* 120, 1–19.

CARR, D. J., and S. G. M. CARR. 1970. Oil glands and ducts in *Eucalyptus* l'Hérit. II. Development and structure of oil glands in the embryo. *Aust. J. Bot.* 18, 191–212.

CASS, D. D. 1985. Origin and development of the non-articulated laticifers of *Jatropha dioica*. *Phytomorphology* 35, 133–140.

CASTRO, M. A., and A. A. DE MAGISTRIS. 1999. Ultrastructure of foliar secretory cavity in *Cupressus arizonica* var. *glabra* (Sudw.) Little (Cupressaceae). *Biocell* 23, 19–28.

CHARON, J., J. LAUNAY, and E. VINDT-BALGUERIE. 1986. Ontogenèse des canaux sécréteurs d'origine primaire dans le bourgeon de *Pin maritime*. *Can. J. Bot.* 64, 2955–2964.

CHRISTIANSEN, E., P. KROKENE, A. A. BERRYMAN, V. R. FRANCESCHI, T. KREKLING, F. LIEUTIER, A. LÖNNEBORG, and H. SOLHEIM. 1999. Mechanical injury and fungal infection induce acquired resistance in Norway spruce. *Tree Physiol.* 19, 399–403.

CHYE, M.-L., C.-T. TAN, and N.-H. CHUA. 1992. Three genes encode 3-hydroxy-3-methylglutaryl-coenzyme A reductase in *Hevea brasiliensis*: *hmg1* and *hmg3* are differentially expressed. *Plant Mol. Biol.* 19, 473–484.

CONDON, J. M., and B. A. FINERAN. 1989a. The effect of chemical fixation and dehydration on the preservation of latex in *Calystegia silvatica* (Convolvulaceae). Examination of exudate and latex *in situ* by light and scanning electron microscopy. *J. Exp. Bot.* 40, 925–939.

CONDON, J. M., and B. A. FINERAN. 1989b. Distribution and organization of articulated laticifers in *Calystegia silvatica* (Convolvulaceae). *Bot. Gaz.* 150, 289–302.

CORNISH, K., D. J. SILER, O.-K. GROSJEAN, and N. GOODMAN. 1993. Fundamental similarities in rubber particle architecture and function in three evolutionarily divergent plant species. *J. Nat. Rubb. Res.* 8, 275–285.

CORNISH, K., D. F. WOOD, and J. J. WINDLE. 1999. Rubber particles from four different species, examined by transmission electron microscopy and electron-paramagnetic-resonance spin labeling, are found to consist of a homogeneous rubber core enclosed by a contiguous, monolayer biomembrane. *Planta* 210, 85–96.

COYVAERTS, E., M. DENNIS, D. LIGHT, and N.-H. CHUA. 1991. Cloning and sequencing of the cDNA encoding the rubber elongation factor of *Hevea brasiliensis*. *Plant Physiol.* 97, 317–321.

CRETIN, H. 1982. The proton gradient across the vacuo-lysosomal membrane of lutoids from the latex of *Hevea brasiliensis*. I. Further evidence for a proton-translocating ATPase on the vacuo-lysosomal membrane of intact lutoids. *J. Membrane Biol.* 65, 175–184.

CURTIS, J. D., and N. R. LERSTEN. 1994. Developmental anatomy of internal cavities of epidermal origin in leaves of *Polygonum* (Polygonaceae). *New Phytol.* 127, 761–770.

DA CUNHA, M., C. G. COSTA, R. D. MACHADO, and F. C. MIGUENS. 1998. Distribution and differentiation of the laticifer system in *Chamaesyce thymifolia* (L.) Millsp. (Euphorbiaceae). *Acta Bot. Neerl.* 47, 209–218.

DA CUNHA, M., V. M. GOMES, J. XAVIER-FILHO, M. ATTIAS, W. DE SOUZA, and F. C. MIGUENS. 2000. The laticifer system of *Chamaesyce thymifolia*: a closed environment for plant trypanosomatids. *Biocell* 24, 123–132.

DATTA, S. K., and S. DE. 1986. Laticifer differentiation of *Calotropis gigantea*. R. Br. Ex Ait. in cultures. *Ann. Bot.* 57, 403–406.

DAVIS, M. J., J. B. KRAMER, F. H. FERWERDA, and B. R. BRUNNER. 1998a. Association of a bacterium and not a phytoplasma with papaya bunchy top disease. *Phytopathology* 86, 102–109.

DAVIS, M. J., Z. YING, B. R. BRUNNER, A. PANTOJA, and F. H. FERWERDA. 1998b. Rickettsial relative associated with papaya bunchy top disease. *Curr. Microbiol.* 36, 80–84.

DE BARY, A. 1884. *Comparative Anatomy of the Vegetative Organs of the Phanerogams and Ferns.* Clarendon Press, Oxford.

DE FAŸ, E., C. SANIER, and C. HÉBANT. 1989. The distribution of plasmodesmata in the phloem of *Hevea brasiliensis* in relation to laticifer loading. *Protoplasma* 149, 155–162.

DEHGAN, B., and M. E. CRAIG. 1978. Types of laticifers and crystals in *Jatropha* and their taxonomic implications. *Am. J. Bot.* 65, 345–352.

DOLLET, M., J. GIANNOTTI, and M. OLLAGNIER. 1977. Observation de protozaires flagellés dans les tubes cribles de Palmiers à huile malades. *C. R. Acad. Sci., Paris, Sér. D* 284, 643–645.

DOLLET, M., D. CAMBRONY, and D. GARGANI. 1982. Culture axénique *in vitro* de *Phytomonas* sp. (Trypanosomatidae) d'*Euphorbe*, transmis par *Stenocephalus agilis* Scop (Coreide). *C. R. Acad. Sci., Paris, Sér. III* 295, 547–550.

DRESSLER, R. 1957. The genus *Pedilanthus* (Euphorbiaceae). *Contributions from the Gray Herbarium of Harvard University* 182, 1–188.

ESAU, K. 1977. *Anatomy of Seed Plants*, 2nd ed. Wiley, New York.

ESAU, K., and H. KOSAKAI. 1975. Laticifers in *Nelumbo nucifera* Gaertn.: Distribution and structure. *Ann. Bot.* 39, 713–719.

FACCHINI, P. J., and V. DE LUCA. 1995. Phloem-specific expression of tyrosine/dopa decarboxylase genes and the biosynthesis of isoquinoline alkaloids in opium poppy. *Plant Cell* 7, 1811–1821.

FAHN, A. 1979. *Secretory Tissues in Plants.* Academic Press, London.

FAHN, A. 1988a. Secretory tissues in vascular plants. *New Phytol.* 108, 229–257.

FAHN, A. 1988b. Secretory tissues and factors influencing their development. *Phyton (Horn)* 28, 13–26.

FAHN, A. 2002. Functions and location of secretory tissues in plants and their possible evolutionary trends. *Isr. J. Plant Sci.* 50(suppl. 1), S59–S64.

FAHN, A., and J. BENAYOUN. 1976. Ultrastructure of resin ducts in *Pinus halepensis*. Development, possible sites of resin synthesis, and mode of its elimination from the protoplast. *Ann. Bot.* 40, 857–863.

FAHN, A., and R. F. EVERT. 1974. Ultrastructure of the secretory ducts of *Rhus glabra* L. *Am. J. Bot.* 61, 1–14.

FAHN, A., and E. ZAMSKI. 1970. The influence of pressure, wind, wounding and growth substances on the rate of resin duct formation in *Pinus halepensis* wood. *Isr. J. Bot.* 19, 429–446.

FAIRBAIRN, J. W., and L. D. KAPOOR. 1960. The laticiferous vessels of *Papaver somniferum* L. *Planta Med.* 8, 49–61.

FINERAN, B. A. 1982. Distribution and organization of non-articulated laticifers in mature tissues of poinsettia (*Euphorbia pulcherrima* Willd.). *Ann. Bot.* 50, 207–220.

FINERAN, B. A. 1983. Differentiation of non-articulated laticifers in poinsettia (*Euphorbia pulcherrima* Willd.). *Ann. Bot.* 52, 279–293.

FINERAN, B. A., J. M. CONDON, and M. INGERFELD. 1988. An impregnated suberized wall layer in laticifers of the Convolvulaceae, and its resemblance to that in walls of oil cells. *Protoplasma* 147, 42–54.

FOSTER, A. S. 1956. Plant idioblasts: Remarkable examples of cell specialization. *Protoplasma* 46, 184–193.

FOX, M. G., and J. C. FRENCH. 1988. Systematic occurrence of sterols in latex of Araceae: Subfamily Colocasioideae. *Am. J. Bot.* 75, 132–137.

FRENCH, J. C. 1988. Systematic occurrence of anastomosing laticifers in Araceae. *Bot. Gaz.* 149, 71–81.

FUJII, T. 1988. Structure of latex and tanniniferous tubes in tropical hardwoods. (Japanese with English summary.) *Bull. For. For. Prod. Res. Inst.* No. 352, 113–118

GEDALOVICH, E., and A. FAHN. 1985a. The development and ultrastructure of gum ducts in *Citrus* plants formed as a result of brown-rot gummosis. *Protoplasma* 127, 73–81.

GEDALOVICH, E., and A. FAHN. 1985b. Ethylene and gum duct formation in *Citrus*. *Ann. Bot.* 56, 571–577.

GILLILAND, M. G., and J. VAN STADEN. 1983. Detection of rubber in guayule (*Parthenium argentatum* Gray) at the ultrastructural level. *Z. Pflanzenphysiol.* 110, 285–291.

GILLILAND, M. G., M. R. APPLETON, and J. VAN STADEN. 1988. Gland cells in resin canal epithelia in guayule (*Parthenium argentatum*) in relation to resin and rubber production. *Ann. Bot.* 61, 55–64.

GIORDANI, R., F. BLASCO, and J.-C. BERTRAND. 1982. Confirmation biochimique de la nature vacuolaire et lysosomale du latex des laticifères non articulés d'*Asclepias curassavica*. *C. R. Acad. Sci., Paris, Sér. III* 295, 641–646.

GIORDANI, R., G. NOAT, and F. MARTY. 1987. Compartmentation of glycosidases in a light vacuole fraction from the latex of *Lactuca sativa* L. In: *Plant Vacuoles: Their Importance in Solute Compartmentation in Cells and Their Applications in Plant Biotechnology*. NATO ASI Series, vol. 134, pp. 383–391. B. Marin, ed. Plenum Press, New York.

GREGORY, M., and P. BAAS. 1989. A survey of mucilage cells in vegetative organs of the dicotyledons. *Isr. J. Bot.* 38, 125–174.

GRIFFING, L. R., and G. L. NESSLER. 1989. Immunolocalization of the major latex proteins in developing laticifers of opium poppy (*Papaver somniferum*). *J. Plant Physiol.* 134, 357–363.

HAN, K.-H., D. H. SHIN, J. YANG, I. J. KIM, S. K. OH, and K. S. CHOW. 2000. Genes expressed in the latex of *Hevea brasiliensis*. *Tree Physiol.* 20, 503–510.

HAO, B.-Z., and J.-L. WU. 2000. Laticifer differentiation in *Hevea brasiliensis*: Induction by exogenous jasmonic acid and linolenic acid. *Ann. Bot.* 85, 37–43.

HAYWARD, H. E. 1938. *The Structure of Economic Plants*. Macmillan, New York.

HÉBANT, C., and E. DE FAŸ. 1980. Functional organization of the bark of *Hevea brasiliensis* (rubber tree): A structural and histoenzymological study. *Z. Pflanzenphysiol.* 97, 391–398.

HÉBANT, C., C. DEVIC, and E. DE FAŸ. 1981. Organisation fonctionnelle du tissu producteur de l'*Hevea brasiliensis*. *Caoutchoucs et Plastiques* 614, 97–100.

HILLIS, W. E. 1987. *Heartwood and Tree Exudates*. Springer-Verlag, Berlin.

HU, C.-H., and L.-H. TIEN. 1973. The formation of rubber and differentiation of cellular structures in the secretory epidermis of fruits of *Decaisnea fargesii* Franch. *Acta Bot. Sin.* 15, 174–178.

HUNTER, J. R. 1994. Reconsidering the functions of latex. *Trees* 9, 1–5.

HUSEBYE, H., S. CHADCHAWAN, P. WINGE, O. P. THANGSTAD, and A. M. BONES. 2002. Guard cell- and phloem idioblast-specific expression of thioglucoside glucohydrolase 1 (myrosinase) in *Arabidopsis*. *Plant Physiol.* 128, 1180–1188.

INAMDAR, J. A., V. MURUGAN, and R. B. SUBRAMANIAN. 1988. Ultrastructure of non-articulated laticifers in *Allamanda violacea*. *Ann. Bot.* 62, 583–588.

JACOB, J. L., J. C. PRÉVÔT, R. LACOTE, E. GOHET, A. CLÉMENT, R. GALLOIS, T. JOET, V. PUJADE-RENAUD, and J. D'AUZAC. 1998. Les mécanismes biologiques de la production de caoutchouc par *Hevea brasiliensis*. *Plant. Rech. Dév.* 5, 5–13.

JOEL, D. M., and A. FAHN. 1980. Ultrastructure of the resin ducts of *Mangifera indica* L. (Anacardiaceae). 1. Differentiation and senescence of the shoot ducts. *Ann. Bot.* 46, 225–233.

JOSEPH, J. P., J. J. SHAH, and J. A. INAMDAR. 1988. Distribution, development and structure of resin ducts in guayule (*Parthenium argentatum* Gray). *Ann. Bot.* 61, 377–387.

KARLING, J. S. 1929. The laticiferous system of *Achras zapota* L. I. A preliminary account of the origin, structure, and distribution of the latex vessels in the apical meristem. *Am. J. Bot.* 16, 803–824.

KASTELEIN, P., and M. PARSADI. 1984. Observations on cultures of the protozoa *Phytomonas* sp. (Trypanosomatidae) associated with the laticifer *Allamanda cathartica* L. (Apocynaceae). *De Surinaamse Landbouw* 32, 85–89.

KAUSCH, A. P., and H. T. HORNER. 1983. The development of mucilaginous raphide crystal idioblasts in young leaves of *Typha angustifolia* L. (Typhaceae). *Am. J. Bot.* 70, 691–705.

KAUSCH, A. P., and H. T. HORNER. 1984. Differentiation of raphide crystal idioblasts in isolated root cultures of *Yucca torreyi* (Agavaceae). *Can. J. Bot.* 62, 1474–1484.

KISSER, J. G. 1958. Die Ausscheidung von ätherischen Ölen und Harzen. In: *Handbuch der Pflanzenphysiologie*, Band 10, *Der Stoffwechsel sekundärer Pflanzenstoffe*, pp. 91–131, Springer-Verlag, Berlin.

KOROLEVA, O. A., A. DAVIES, R. DEEKEN, M. R. THORPE, A. D. TOMOS, and R. HEDRICH. 2000. Identification of a new glucosinolate-rich cell type in *Arabidopsis* flower stalk. *Plant Physiol.* 124, 599–608.

KROTKOV, G. A. 1945. A review of literature on *Taraxacum kok-saghyz* Rod. *Bot. Rev.* 11, 417–461.

KUMAR, A., and P. TANDON. 1990. Investigation on the in vitro laticifer differentiation in *Thevetia peruviana* L. *Phytomorphology* 40, 113–117.

LABOURIAU, L. G. 1952. On the latex of *Regnellidium diphyllum* Lindm. *Phyton (Buenos Aires)* 2, 57–74.

LEE, K. B., and P. G. MAHLBERG. 1999. Ultrastructure and development of nonarticulated laticifers in seedlings of *Euphorbia maculata* L. *J. Plant Biol. (Singmul Hakhoe chi)* 42, 57–62.

LEONG, S. K., W. LEONG, and P. K. YOON. 1982. Harvesting of shoots for rubber extraction in *Hevea*. *J. Rubb. Res. Inst. Malaysia* 30, 117–122.

LERSTEN, N. R., and J. D. CURTIS. 1987. Internal secretory spaces in Asteraceae: A review and original observations on *Conyza canadensis* (Tribe Astereae). *La Cellule* 74, 179–196.

LERSTEN, N. R., and J. D. CURTIS. 1989. Foliar oil reservoir anatomy and distribution in *Solidago canadensis* (Asteraceae, tribe Astereae). *Nord. J. Bot.* 9, 281–287.

LLOYD, F. E. 1911. Guayule (*Parthenium argentatum* Gray): A rubber-plant of the Chihuahuan Desert. Carnegie Institution of Washington, Washington, DC., Publ. No. 139.

ŁOTOCKA, B., and A. GESZPRYCH. 2004. Anatomy of the vegetative organs and secretory structures of *Rhaponticum carthamoides* (Asteraeae). *Bot. J. Linn. Soc.* 144, 207–233.

MAHLBERG, P. G. 1959. Karyokinesis in the non-articulated laticifers of *Nerium oleander* L. *Phytomorphology* 9, 110–118.

MAHLBERG, P. G. 1961. Embryogeny and histogenesis in *Nerium oleander*. II. Origin and development of the non-articulated laticifer. *Am. J. Bot.* 48, 90–99.

MAHLBERG, P. G. 1963. Development of non-articulated laticifer in seedling axis of *Nerium oleander*. *Bot. Gaz.* 124, 224–231.

MAHLBERG, P. G. 1982. Comparative morphology of starch grains in latex from varieties of poinsettia, *Euphorbia pulcherrima* Willd. (Euphorbiaceae). *Bot. Gaz.* 143, 206–209.

MAHLBERG, P. G., and L. A. ASSI. 2002. A new shape of plastid starch grains from laticifers of *Anthostema* (Euphorbiaceae). *S. Afr. J. Bot.* 68, 231–233.

MAHLBERG, P. G., D. W. FIELD, and J. S. FRYE. 1984. Fossil laticifers from Eocene brown coal deposits of the Geiseltal. *Am. J. Bot.* 71, 1192–1200.

MAHLBERG, P. G., D. G. DAVIS, D. S. GALITZ, and G. D. MANNERS. 1987. Laticifers and the classification of *Euphorbia*: The chemotaxonomy of *Euphorbia esula* L. *Bot. J. Linn. Soc.* 94, 165–180.

MANN, L. K. 1952. Anatomy of the garlic bulb and factors affecting bulb development. *Hilgardia* 21, 195–251.

MARIANI, P., E. M. CAPPELLETTI, D. CAMPOCCIA, and B. BALDAN. 1989. Oil cell ultrastructure and development in *Liriodendron tulipifera* L. *Bot. Gaz.* 150, 391–396.

MARON, R., and A. FAHN. 1979. Ultrastructure and development of oil cells in *Laurus nobilis* L. leaves. *Bot. J. Linn. Soc.* 78, 31–40.

METCALFE, C. R. 1983. Laticifers and latex. In: *Anatomy of the Dicotyledons*, 2nd ed., vol. II, *Wood Structure and Conclusion of the General Introduction*, pp. 70–81, C. R. Metcalfe and L. Chalk, eds. Clarendon Press, Oxford.

METCALFE, C. R., and L. CHALK. 1950. *Anatomy of the Dicotyledons*, 2 vols. Clarendon Press, Oxford.

METCALFE, C. R., and L. CHALK, eds. 1979. *Anatomy of the Dicotyledons*. vol. I. *Systematic Anatomy of Leaf and Stem, with a Brief History of the Subject*. Clarendon Press, Oxford.

MILBURN, J. A., and M. S. RANASINGHE. 1996. A comparison of methods for studying pressure and solute potentials in xylem and also in phloem laticifers of *Hevea brasiliensis*. *J. Exp. Bot.* 47, 135–143.

MURUGAN, V., and J. A. INAMDAR. 1987. Studies in the laticifers of *Vallaris solanacea* (Roth) O. Ktze. *Phytomorphology* 37, 209–214.

MUSTARD, M. J. 1982. Origin and distribution of secondary articulated anastomosing laticifers in *Manilkcara zapota* van Royen (Sapotaceae). *J. Am. Soc. Hortic. Sci.* 107, 355–360.

NAIR, M. N. B., and S. V. SUBRAHMANYAM. 1998. Ultrastructure of the epithelial cells and oleo-gumresin secretion in *Boswellia serrata* (Burseraceae). *IAWA J.* 19, 415–427.

NAIR, G. M., K. VENKAIAH, and J. J. SHAH. 1983. Ultrastructure of gum-resin ducts in cashew *(Anacardium occidentale)*. *Ann. Bot.* 51, 297–305.

NESSLER, C. L. 1994. Sequence analysis of two new members of the major latex protein gene family supports the triploid-hybrid origin of the opium poppy. *Gene* 139, 207–209.

NISSEN, S. J., and M. E. FOLEY. 1986. No latex starch utilization in *Euphorbia esula* L. *Plant Physiol.* 81, 696–698.

PANCORO, A., and M. A. HUGHES. 1992. *In-situ* localization of cyanogenic β-glucosidase (linamarase) gene expression in leaves of cassava *(Manihot esculenta* Cranz) using non-isotopic riboprobes. *Plant J.* 2, 821–827.

PARHAM, R. A., and H. M. KAUSTINEN. 1977. On the site of tannin synthesis in plant cells. *Bot. Gaz.* 138, 465–467.

PARTHASARATHY, M. V., W. G. VAN SLOBBE, and C. SOUDANT. 1976. Trypanosomatid flagellate in the phloem of diseased coconut palms. *Science* 192, 1346–1348.

PATERSON-JONES, J. C., M. G. GILLILAND, and J. VAN STADEN. 1990. The biosynthesis of natural rubber. *J. Plant Physiol.* 136, 257–263.

PLATT, K. A., and W. W. THOMSON. 1992. Idioblast oil cells of avocado: Distribution, isolation, ultrastructure, histochemistry, and biochemistry. *Int. J. Plant Sci.* 153, 301–310.

RAO, A. R., and M. MALAVIYA. 1964. On the latex-cells and latex of *Jatropha*. *Proc. Indian Acad. Sci., Sect. B*, 60, 95–106.

RAO, A. R., and J. P. TEWARI. 1960. On the morphology and ontogeny of the foliar sclereids of *Codiaeum variegatum* Blume. *Proc. Natl. Inst. Sci. India, Part B, Biol. Sci.* 26, 1–6.

RAO, K. S. 1988. Fine structural details of tannin accumulations in non-dividing cambial cells. *Ann. Bot.* 62, 575–581.

RASK, L., E. ANDRÉASSON, B. EKBOM, S. ERIKSSON, B. PONTOPPI-DAN, and J. MEIJER. 2000. Myrosinase: Gene family evolution and herbivore defense in Brassicaceae. *Plant Mol. Biol.* 42, 93–114.

RAVEN, P. H., R. F. EVERT, and S. E. EICHHORN. 2005. *Biology of Plants*, 7th ed. Freeman, New York.

RODRIGUEZ-SAONA, C. R., and J. T. TRUMBLE. 1999. Effect of avocadofurans on larval survival, growth, and food preference of the generalist herbivore, *Spodoptera exigua*. *Entomol. Exp. Appl.* 90, 131–140.

RODRIGUEZ-SAONA, C., J. G. MILLAR, D. F. MAYNARD, and J. T. TRUMBLE. 1998. Novel antifeedant and insecticidal compounds from avocado idioblast cell oil. *J. Chem. Ecol.* 24, 867–889.

ROY, A. T., and D. N. DE. 1992. Studies on differentiation of laticifers through light and electron microscopy in *Calotropis gigantea* (Linn.) R. Br. *Ann. Bot.* 70, 443–449.

RUDALL, P. J. 1987. Laticifers in Euphorbiaceae—A conspectus. *Bot. J. Linn. Soc.* 94, 143–163.

RUDALL, P. 1989. Laticifers in vascular cambium and wood of *Croton* spp. (Euphorbiaceae). *IAWA Bull.* n.s. 10, 379–383.

RUDALL, P. 1994. Laticifers in Crotonoideae (Euphorbiaceae): Homology and evolution. *Ann. Mo. Bot. Gard.* 81, 270–282.

SCHAFFSTEIN, G. 1932. Untersuchungen an ungegliederten Mil-chröhren. *Beih. Bot. Zentralbl.* 49, 197–220.

SCOTT, D. H. 1882. The development of articulated laticiferous vessels. *Q. J. Microsc. Sci.* 22, 136–153.

SCOTT, D. H. 1886. On the occurrence of articulated laticiferous vessels in *Hevea*. *J. Linn. Soc. Lond., Bot.* 21, 566–573.

SETIA, R. C. 1984. Traumatic gum duct formation in *Sterculia urens* Roxb. in response to injury. *Phyton (Horn)* 24, 253–255.

SHELDRAKE, A. R. 1969. Cellulase in latex and its possible significance in cell differentiation. *Planta* 89, 82–84.

SHELDRAKE, A. R., and G. F. J. MOIR. 1970. A cellulase in *Hevea* latex. *Physiol. Plant.* 23, 267–277.

SISWANTO. 1994. Physiological mechanism related to latex production of *Hevea brasiliensis*. *Bul. Biotek. Perkebunan* 1, 23–29.

SKENE, D. S. 1965. The development of kino veins in *Eucalyptus obliqua* L'Hérit. *Aust. J. Bot.* 13, 367–378.

SKUTCH, A. F. 1932. Anatomy of the axis of the banana. *Bot. Gaz.* 93, 233–258.

SOLEREDER, H. 1908. *Systematic Anatomy of the Dicotyledons: A Handbook for Laboratories of Pure and Applied Botany*. 2 vols. Clarendon Press, Oxford.

SONG, Y.-H., P.-F. WONG, and N.-H. CHUA. 1991. Tissue culture and genetic transformation of dandelion. *Acta Horti.* 289, 261–262.

SPERLICH, A. 1939. *Das trophische Parenchym. B. Exkretionsgewebe. Handbuch der Pflanzenanatomie*, Band 4, Teil 2, *Histologie*. Gebrüder Borntraeger, Berlin.

SPILATRO, S. R., and P. G. MAHLBERG. 1986. Latex and laticifer starch content of developing leaves of *Euphorbia pulcherrima*. *Am. J. Bot.* 73, 1312–1318.

SPILATRO, S. R., and P. G. MAHLBERG. 1990. Characterization of starch grains in the nonarticulated laticifer of *Euphorbia pulcherrima* (Poinsettia). *Am. J. Bot.* 77, 153–158.

STOCKSTILL, B. L., and C. L. NESSLER. 1986. Ultrastructural observations on the nonarticulated, branched laticifers in *Nerium oleander* L. (Apocynaceae). *Phytomorphology* 36, 347–355.

SUBRAHMANYAM, S. V., and J. J. SHAH. 1988. The metabolic status of traumatic gum ducts in *Moringa oleifera* Lam. *IAWA Bull.* n. s. 9, 187–195.

SURI, S. S., and K. G. RAMAWAT. 1995. *In vitro* hormonal regulation of laticifer differentiation in *Calotropis procera*. *Ann. Bot.* 75, 477–480.

SURI, S. S., and K. G. RAMAWAT. 1996. Effect of *Calotropis* latex on laticifers differentiation in callus cultures of *Calotropis procera*. *Biol. Plant.* 38, 185–190.

THOMSON, W. W., K. A. PLATT-ALOIA, and A. G. ENDRESS. 1976. Ultrastructure of oil gland development in the leaf of *Citrus sinensis* L. *Bot. Gaz.* 137, 330–340.

THURESON-KLEIN, Å. 1970. Observations on the development and fine structure of the articulated laticifers of *Papaver somniferum*. *Ann. Bot.* 34, 751–759.

TIBBITTS, T. W., J. BENSINK, F. KUIPER, and J. HOBÉ. 1985. Association of latex pressure with tipburn injury of lettuce. *J. Am. Soc. Hortic. Sci.* 110, 362–365.

TIPPETT, J. T. 1986. Formation and fate of kino veins in *Eucalyptus* L'Hérit. *IAWA Bull.* n.s. 7, 137–143.

TOMLINSON, P. B. 2003. Development of gelatinous (reaction) fibers in stems of *Gnetum gnemon* (Gnetales). *Am. J. Bot.* 90, 965–972.

TOMLINSON, P. B., and J. B. FISHER. 2005. Development of non-lignified fibers in leaves of *Gnetum gnemon* (Gnetales). *Am. J. Bot.* 92, 383–389.

TRACHTENBERG, S., and A. FAHN. 1981. The mucilage cells of *Opuntia ficus-indica* (L.) Mill.—Development, ultrastructure, and mucilage secretion. *Bot. Gaz.* 142, 206–213.

TURNER, G. W. 1986. Comparative development of secretory cavities in tribes Amorpheae and Psoraleeae (Leguminosae: Papilionoideae). *Am. J. Bot.* 73, 1178–1192.

TURNER, G. W. 1994. Development of essential oil secreting glands from leaves of *Citrus limon* Burm. f., and a reexamination of the lysigenous gland hypothesis. Ph.D. Dissertation, University of California, Davis.

TURNER, G. W. 1999. A brief history of the lysigenous gland hypothesis. *Bot. Rev.* 65, 76–88.

TURNER, G. W., A. M. BERRY, and E. M. GIFFORD. 1998. Schizogenous secretory cavities of *Citrus limon* (L.) Burm. f. and a reevaluation of the lysigenous gland concept. *Int. J. Plant Sci.* 159, 75–88.

VAN VEENENDAAL, W. L. H., and R. W. DEN OUTER. 1990. Distribution and development of the non-articulated branched laticifers of *Morus nigra* L. (Moraceae). *Acta Bot. Neerl.* 39, 285–296.

VASSILYEV, A. E. 2000. Quantitative ultrastructural data of secretory duct epithelial cells in *Rhus toxicodendron*. *Int. J. Plant Sci.* 161, 615–630.

VENKAIAH, K. 1990. Ultrastructure of gum-resin ducts in *Ailanthus excelsa* Roxb. *Fedds. Repert.* 101, 63–68.

VENKAIAH, K. 1992. Development, ultrastructure and secretion of gum ducts in *Lannea coromandelica* (Houtt.) Merr. (Anacardiaceae). *Ann. Bot.* 69, 449–457.

VENKAIAH, K., and J. J. SHAH. 1984. Distribution, development and structure of gum ducts in *Lannea coromandelica* (Houtt.) Merr. *Ann. Bot.* 54, 175–186.

VISCHER, W. 1923. Über die konstanz anatomischer und physiologischer Eigenschaften von *Hevea brasiliensis* Müller Arg. (Euphorbiaceae). *Verh. Natforsch. Ges. Basel* 35 (I), 174–185.

WANG, Z.-Y., K. S. GOULD, and K. J. PATTERSON. 1994. Structure and development of mucilage-crystal idioblasts in the roots of five *Actinidia* species. *Int. J. Plant Sci.* 155, 342–349.

WERKER, E., and A. FAHN. 1969. Resin ducts of *Pinus halepensis* Mill.—Their structure, development and pattern of arrangement. *Bot. J. Linn. Soc.* 62, 379–411.

WHEELER, E. A., P. BAAS, and P. E. GASSON, eds. 1989. IAWA list of microscopic features for hardwood identification. *IAWA Bull.* n.s. 10, 219–332.

WILKES, J., G. T. DALE, and K. M. OLD. 1989. Production of ethylene by *Endothia gyrosa* and *Cytospora eucalypticola* and its possible relationship to kino vein formation in *Eucalyptus maculata*. *Physiol. Mol. Plant Pathol.* 34, 171–180.

WILSON, K. J., and P. G. MAHLBERG. 1980. Ultrastructure of developing and mature nonarticulated laticifers in the milkweed *Asclepias syriaca* L. (Asclepiadaceae). *Am. J. Bot.* 67, 1160–1170.

WILSON, K. J., C. L. NESSLER, and P. G. MAHLBERG. 1976. Pectinase in *Asclepias* latex and its possible role in laticifer growth and development. *Am. J. Bot.* 63, 1140–1144.

WITTLER, G. H., and J. D. MAUSETH. 1984a. The ultrastructure of developing latex ducts in *Mammillaria heyderi* (Cactaceae). *Am. J. Bot.* 71, 100–110.

WITTLER, G. H., and J. D. MAUSETH. 1984b. Schizogeny and ultrastructure of developing latex ducts in *Mammillaria guerreronis* (Cactaceae). *Am. J. Bot.* 71, 1128–1138.

WU, H., and Z.-H. HU. 1994. Ultrastructure of the resin duct initiation and formation in *Pinus tabulae formis*. *Chinese J. Bot.* 6, 123–128.

WU, J.-I., and B.-Z. HAO. 1990. Ultrastructural observation of differentiation laticifers in *Hevea brasiliensis*. *Acta Bot. Sin.* 32, 350–354.

YAMANAKA, K. 1989. Formation of traumatic phloem resin canals in *Chamaecyparis obtusa*. *IAWA Bull.* n.s. 10, 384–394.

ZANDER, A. 1928. Über Verlauf und Entstehung der Milchröhren des Hanfes *(Cannabis sativa)*. *Flora* 123, 191–218

ZENG, Y., B.-R. JI, and B. YU. 1994. Laticifer ultrastructural and immunocytochemical studies of papain in *Carica papaya*. *Acta Bot. Sin.* 36, 497–501.

ZHANG, W.-C., W.-M. YAN, and C.-H. LOU. 1983. Intracellular and intercellular changes in constitution during the development of laticiferous system in garlic scape. *Acta Bot. Sin.* 25, 8–12.

ZOBEL, A. M. 1985a. Ontogenesis of tannin coenocytes in *Sambucus racemosa* L. I. Development of the coenocytes from mononucleate tannin cells. *Ann. Bot.* 55, 765–773.

ZOBEL, A. M. 1985b. Ontogenesis of tannin coenocytes in *Sambucus racemosa* L. II. Mother tannin cells. *Ann. Bot.* 56, 91–104.

ZOBEL, A. M. 1986a. Localization of phenolic compounds in tannin-secreting cells from *Sambucus racemosa* L. shoots. *Ann. Bot.* 57, 801–810.

ZOBEL, A. M. 1986b. Ontogenesis of tannin-containing coenocytes in *Sambucus racemosa* L. III. The mature coenocyte. *Ann. Bot.* 58, 849–858.

# Addendum: Other Pertinent References Not Cited in the Text

## CHAPTERS 2 AND 3

ALDRIDGE, C., J. MAPLE, and S. G. MØLLER. 2005. The molecular biology of plastid division in higher plants. *J. Exp. Bot.* 56, 1061–1077. **(Review)**

ANIENTO, F., and D. G. ROBINSON. 2005. Testing for endocytosis in plants. *Protoplasma* 226, 3–11. **(Review)**

BAAS, P. W., A. KARABAY, and L. QIANG. 2005. Microtubules cut and run. *Trends Cell Biol.* 15, 518–524. **(Opinion)**

BALUŠKA, F., J. ŠAMAJ, A. HLAVACKA, J. KENDRICK-JONES, and D. VOLKMANN. 2004. Actin-dependent fluid-phase endocytosis in inner cortex cells of maize root apices. *J. Exp. Bot.* 55, 463–473. **(Specialized actin- and myosin VIII-enriched membrane domains carry out a tissue-specific form of fluid-phase endocytosis in maize root apices. The loss of microtubules did not inhibit this process.)**

BECK, C. F. 2005. Signaling pathways from the chloroplast to the nucleus. *Planta* 222, 743–756. **(Review)**

BISGROVE, S. R., W. E. HABLE, and D. L. KROPF. 2004. +TIPs and microtubule regulation. The beginning of the plus end in plants. *Plant Physiol.* 136, 3855–3863. **(Update)**

BOURSIAC, Y., S. CHEN, D.-T. LUU, M. SORIEUL, N. VAN DEN DRIES, and C. MAUREL. 2005. Early effects of salinity on water transport in *Arabidopsis* roots. Molecular and cellular features of aquaporin expression. *Plant Physiol.* 139, 790–805. **(Exposure of roots to salt induced changes in aquaporin expression at multiple levels, including a coordinated transcriptional down-regulation and subcellular localization of both plasma membrane intrinsic proteins [PIPS] and tonoplast intrinsic proteins [TIPS]. These mechanisms may act in concert to regulate water transport, mostly in the long term [≥6 h].)**

BRANDIZZI, F., S. L. IRONS, and D. E. EVANS. 2004. The plant nuclear envelope: new prospects for a poorly understood structure. *New Phytol.* 163, 227–246. **(Review)**

BROWN, R. C., and B. E. LEMMON. 2001. The cytoskeleton and spatial control of cytokinesis in the plant life cycle. *Protoplasma* 215, 35–49. **(Review)**

CROFTS, A. J., H. WASHIDA, T. W. OKITA, M. OGAWA, T. KUMAMARU, and H. SATOH. 2004. Targeting of proteins to endoplasmic reticulum-derived compartments in plants. The importance of RNA localization. *Plant Physiol.* 136, 3414–3419. **(Update)**

DIXIT, R., R. CYR, and S. GILROY. 2006. Using intrinsically fluorescent proteins for plant cell imaging. *Plant J.* 45, 599–615. **(Review)**

DRØBAK, B. K., V. E. FRANKLIN-TONG, and C. J. STAIGER. 2004. The role of the actin cytoskeleton in plant cell signaling. *New Phytol.* 163, 13–30. **(Review)**

EHRHARDT, D., and S. L. SHAW. 2006. Microtubule dynamics and organization in the plant cortical array. *Annu. Rev. Plant Biol.* 57. **In Press. (Review)**

EPIMASHKO, S., T. MECKEL, E. FISCHER-SCHLIEBS, U. LÜTTGE, and G. THIEL. 2004. Two functionally different vacuoles for static and dynamic purposes in one plant mesophyll leaf cell. *Plant J.* 37, 294–300. **(Two large independent types of vacuoles occur in the mesophyll cells of the common ice plant, *Mesembryanthemum crystallinum*, in which photosynthesis proceeds via crassulacean acid metabolism. One sequesters permanently large amounts of NaCl for osmotic purpose and for protecting the protoplast from NaCl toxicity; the other stores the nocturnally acquired $CO_2$ as malate and re-mobilizes the malate in the daytime.)**

FRANCESCHI, V. R., and P. A. NAKATA. 2005. Calcium oxalate in plants: formation and function. *Annu. Rev. Plant Biol.* 56, 41–71. **(Review)**

GALILI, G. 2004. ER-derived compartments are formed by highly regulated processes and have special functions in plants. *Plant Physiol.* 136, 3411–3413. **(State of the field)**

GELDNER, N. 2004. The plant endosomal system—its structure and role in signal transduction and plant development. *Planta* 219, 547–560. **(Review)**

GUNNING, B. E. S. 2005. Plastid stromules: video microscopy of their outgrowth, retraction, tensioning, anchoring, branching, bridging, and tip-shedding. *Protoplasma* 225, 33–42.

GUTIERREZ, C. 2005. Coupling cell proliferation and development in plants. *Nature Cell Biol.* 7, 535–541. **(Review)**

HARA-NISHIMURA, I., R. MATSUSHIMA, T. SHIMADA, and M. NISHIMURA. 2004. Diversity and formation of endoplasmic reticulum-derived compartments in plants. Are these compartments specific to plant cells? *Plant Physiol.* 136, 3435–3439. **(Update)**

HASHIMOTO, T., and T. KATO. 2006. Cortical control of plant microtubules. *Curr. Opin. Plant Biol.* 9, 5–11. **(Review)**

HAWES, C. 2005. Cell biology of the plant Golgi apparatus. *New Phytol.* 165, 29–44. **(Review)**

HERMAN, E., and M. SCHMIDT. 2004. Endoplasmic reticulum to vacuole trafficking of endoplasmic reticulum bodies provides an alternate pathway for protein transfer to the vacuole. *Plant Physiol.* 136, 3440–3446. **(Update)**

HOWITT, C. A., and B. J. POGSON. 2006. Carotenoid accumulation and function in seeds and non-green tissues. *Plant Cell Environ.* 29, 435–445. **(Review)**

HSIEH, K., and A. H. C. HUANG. 2004. Endoplasmic reticulum, oleosins, and oils in seeds and tapetum cells. *Plant Physiol.* 136, 3427–3434. **(Update)**

HUGHES, N. M., H. S. NEUFELD, and K. O. BURKEY. 2005. Functional role of anthocyanins in high-light winter leaves of the evergreen herb *Galax urceolata. New Phytol.* 168, 575–587. **(Results suggest that anthocyanins function as light attenuators and may also contribute to the antioxidant pool in winter leaves.)**

HUSSEY, P. J., ed. 2004. *The Plant Cytoskeleton in Cell Differentiation and Development. Annual Plant Reviews,* vol. 10. Blackwell/CRC Press, Oxford/Boca Raton. **(Review)**

HUSSEY, P. J., T. KETELAAR, and M. DEEKS. 2006. Control of the actin cytoskeleton in plant cell growth. *Annu. Rev. Plant Biol.* 57. **On line. (Review)**

JOLIVET, P., E. ROUX, S. D'ANDREA, M. DAVANTURE, L. NEGRONI, M. ZIVY, and T. CHARDOT. 2004. Protein composition of oil bodies in *Arabidopsis thaliana* ecotype WS. *Plant Physiol. Biochem.* 42, 501–509. **(Oleosins represented up to 79% of oil body proteins; an 18.5 kDa oleosin was the most abundant among them.)**

JÜRGENS, G. 2004. Membrane trafficking in plants. *Annu. Rev. Cell Dev. Biol.* 20, 481–504. **(Review)**

KAWASAKI, M., M. TANIGUCHI, and H. MIYAKE. 2004. Structural changes and fate of crystalloplastids during growth of calcium oxalate crystal idioblasts in Japanese yam (*Dioscorea japonica* Thunb.) tubers. *Plant Prod. Sci.* 7, 283–291. **(Crystalloplastids similar to small vacuoles and/or vesicles incorporated into the central vacuoles of crystal idioblasts apparently are involved in calcium oxalate crystal formation.)**

KIM, H., M. PARK, S. J. KIM, and I. HWANG. 2005. Actin filaments play a critical role in vacuolar trafficking at the Golgi complex in plant cells. *Plant Cell* 17, 888–902. **(The roles played by actin filaments in intercellular trafficking were investigated with the use of latrunculin B, an inhibitor of actin filament assembly, or actin mutants that disrupt actin filaments when overexpressed.)**

KLYACHKO, N. L. 2004. Actin cytoskeleton and the shape of the plant cell. *Russ. J. Plant Physiol.* 51, 827–833. **(Review)**

KREBS, A., K. N. GOLDIE, and A. HOENGER. 2004. Complex formation with kinesin motor domains affects the structure of microtubules. *J. Mol. Biol.* 335, 139–153. **(The interaction between kinesin and tubulin indicates that microtubules play an active role in intracellular processes through modulations of their core structure.)**

LEE, M. C. S., E. A. MILLER, J. GOLDBERG, L. ORCI, and R. SCHEKMAN. 2004. Bi-directional protein transport between the ER and Golgi. *Annu. Rev. Cell Dev. Biol.* 20, 87–123. **(Review)**

LEE, Y.-R. J., and B. LIU. 2004. Cytoskeletal motors in *Arabidopsis.* Sixty-one kinesins and seventeen myosins. *Plant Physiol.* 136, 3877–3883. **(Update)**

LERSTEN, N. R., and H. T. HORNER. 2004. Calcium oxalate crystal macropattern development during *Prunus virginiana* (Rosaceae) leaf growth. *Can. J. Bot.* 82, 1800–1808. (**This study describes in detail the initiation and progressive development of all components of the foliar crystal macropattern of choke-cherry. Druses are confined to stem, petiole, and leaf veins, whereas prismatics are localized in stipules, bud scales, and leaf lamina.**)

MACKENZIE, S. A. 2005. Plant organellar protein targeting: a traffic plan still under construction. *Trends Cell Biol.* 548–554. (**Review**)

MALIGA, P. 2004. Plastid transformation in higher plants. *Annu. Rev. Plant Biol.* 55, 289–313. (**Review**)

MAPLE, J., and S. G. MØLLER. 2005. An emerging picture of plastid division in higher plants. *Planta* 223, 1–4. (**Review**)

MATHUR, J. 2006. Local interactions shape plant cells. *Curr. Opin. Cell Biol.* 18, 40–46. (**Review**)

MAZEN, A. M. A. 2004. Calcium oxalate crystals in leaves of *Corchorus olitorius* as related to accumulation of toxic metals. *Russ. J. Plant Physiol.* 51, 281–285. (**X-ray microanalysis of Ca oxalate crystals in leaves from plants exposed to 5 µg/ml Al incorporated Al into the crystals, suggesting a possible contribution for Ca oxalate-crystal formation in sequestering and tolerance of some toxic metals.**)

MECKEL, T., A. C. HURST, G. THIEL, and U. HOMANN. 2005. Guard cells undergo constitutive and pressure-driven membrane turnover. *Protoplasma* 226, 23–29. (**Review**)

MIYAGISHIMA, S.-Y. 2005. Origin and evolution of the chloroplast division machinery. *J. Plant Res.* 118, 295–306. (**Review**)

MØLLER, S. G., ed. 2004. *Plastids. Annual Plant Reviews*, vol. 13. Blackwell/CRC Press, Oxford/Boca Raton. (**Review**)

MOTOMURA, H., T. FUJII, and M. SUZUKI. 2006. Silica deposition in abaxial epidermis before the opening of leaf blades of *Pleioblastus chino* (Poaceae, Bambusoideae). *Ann. Bot.* 97, 513–519. (**The cell types in the leaf epidermis of bamboo are classified into three groups according to the pattern of silica deposition.**)

OVEČKA, M., I. LANG, F. BALUŠKA, A. ISMAIL, P. ILLÉS, and I. K. LICHTSCHEIDL. 2005. Endocytosis and vesicle trafficking during tip growth of root hairs. *Protoplasma* 226, 39–54. (**With the use of the fluorescent endocytosis marker dyes FM1–43 and FM4–64, endocytosis was localize in the tips of living root hairs of *Arabidopsis thaliana* and *Triticum aestivum*. Endoplasmic reticulum was not involved in trafficking pathways of endosomes. The actin cytoskeleton was involved with the endocytosis, as well as with further membrane trafficking.**)

PARK, M., S. J. KIM, A. VITALE, and I. HWANG. 2004. Identification of the protein storage vacuole and protein targeting to the vacuole in leaf cells of three plant species. *Plant Physiol.* 134, 625–639. (**Protein trafficking to protein storage vacuoles [PSV] was investigated in cells of *Nicotiana tabacum, Phaseolus vulgaris,* and *Arabidopsis* leaves. Proteins can be transported to the PSV by Golgi-dependent and Golgi-independent pathways, depending on the individual cargo proteins.**)

REISEN, D., F. MARTY, and N. LEBORGNE-CASTEL. 2005. New insights into the tonoplast architecture of plant vacuoles and vacuolar dynamics during osmotic stress. *BMC Plant Biol.* 5, 13 [13 pp.]. (**3-D processing of a GFP-labeled tonoplast provides visual constructions of the plant cell vacuole and elaborates on the nature of tonoplast folding and architecture. The unity of the vacuole is maintained during acclimation to osmotic stress.**)

ROSE, A., S. PATEL, and I. MEIER. 2004. The plant nuclear envelope. *Planta* 218, 327–336. (**Review**)

SAKAI, Y., and S. TAKAGI. 2005. Reorganized actin filaments anchor chloroplasts along the anticlinal walls of *Vallisneria* epidermal cells under high-intensity blue light. *Planta* 221, 823–830. (**High-intensity blue light [BL] induced dynamic reorganization of actin filaments in the cytoplasmic layers that face the outer periclinal wall and the anticlinal wall [A side]. The BL-induced avoidance response of chloroplasts apparently includes both photosynthesis-dependent and actin-dependent anchorage of chloroplasts on the A side of epidermal cells.**)

ŠAMAJ, J., N. D. READ, D. VOLKMANN, D. MENZEL, and F. BALUŠKA. 2005. The endocytic network in plants. *Trends in Cell Biol.* 15, 425–433. (**Review**)

SHAW, S. L. 2006. Imaging the live plant cell. *Plant J.* 45, 573–598. (**Review**)

SHEAHAN, M. B., D. W. MCCURDY, and R. J. ROSE. 2005. Mitochondria as a connected population: ensuring continuity of the mitochondrial genome during plant cell dedifferentiation through massive mitochondrial fusion. *Plant J.* 44, 744–755. (**This highly informative study indicates that developmentally regulated fusion ensures continuity of the mitochondrial genome.**)

SHEAHAN, M. B., R. J. ROSE, and D. W. MCCURDY. 2004. Organelle inheritance in plant cell division: the actin cytoskeleton is required for unbiased inheritance of chloroplasts, mitochondria and endoplasmic reticulum in dividing protoplasts. *Plant J.* 37, 379–390.

SMITH, L. G., and D. G. OPPENHEIMER. 2005. Spatial control of cell expansion by the plant cytoskeleton. *Annu. Rev. Cell Dev. Biol.* 21, 271–295. (**Review**)

STEPINSKI, D. 2004. Ultrastructural and autoradiographic studies of the role of nucleolar vacuoles in soybean root meristem. *Folia Histochem. Cytobiol.* 42, 57–61. (**It is hypothesized that nucleolar vacuoles may be involved in the intensification of pre-ribosome transport outside the nucleolus.**)

TAKEMOTO, D., and A. R. HARDHAM. 2004. The cytoskeleton as a regulator and target of biotic interactions in plants. *Plant Physiol.* 136, 3864–3876. **(Update)**

TIAN, W.-M., and Z.-H. HU. 2004. Distribution and ultrastructure of vegetative storage proteins in Leguminosae. *IAWA J.* 25, 459–469. **(See this article and articles cited therein for the presence of storage proteins in both temperate and tropical woody plants.)**

TREUTTER, D. 2005. Significance of flavonoids in plant resistance and enhancement of their biosynthesis. *Plant Biol.* 7, 581–591. **(Review)**

VITALE, A., and G. HINZ. 2005. Sorting of proteins to storage vacuoles: how many mechanisms? *Trends Plant Sci.* 10, 316–323. **(Review)**

WADA, M., and N. SUETSUGU. 2004. Plant organelle positioning. *Curr. Opin. Plant Biol.* 7, 626–631. **(Review)**

WASTENEYS, G. O. 2004. Progress in understanding the role of microtubules in plant cells. *Curr. Opin. Plant Biol.* 7, 651–660. **(Review)**

WASTENEYS, G. O., and M. FUJITA. 2006. Establishing and maintaining axial growth: wall mechanical properties and the cytoskeleton. *J. Plant Res.* 119, 5–10. **(Review)**

WASTENEYS, G. O., and M. E. GALWAY. 2003. Remodeling the cytoskeleton for growth and form: an overview with some new views. *Annu. Rev. Plant Biol.* 54, 691–722. **(Review)**

YAMADA, K., T. SHIMADA, M. NISHIMURA, and I. HARA-NISHIMURA. 2005. A VPE family supporting various vacuolar functions in plants. *Physiol. Plant.* 123, 369–375. **(Review)**

# ▌CHAPTER 4

ABE, H., and R. FUNADA. 2005. Review—The orientation of cellulose microfibrils in the cell walls of tracheids in conifers. A model based on observations by field emission-scanning electron microscopy. *IAWA J.* 26, 161–174. **(Review)**

BALUŠKA, F., J. ŠAMAJ, P. WOJTASZEK, D. VOLKMANN, and D. MENZEL. 2003. Cytoskeleton-plasma membrane-cell wall continuum in plants. Emerging links revisited. *Plant Physiol.* 133, 482–491. **(Review)**

BASKIN, T. I. 2005. Anisotropic expansion of the plant cell wall. 2005. *Annu. Rev. Cell Dev. Biol.* 21, 203–222. **(Review)**

BRUMMELL, D. A. 2006. Cell wall disassembly in ripening fruit. *Funct. Plant Biol.* 33, 103–119. **(Review)**

BURTON, R. A., N. FARROKHI, A. BACIC, and G. B. FINCHER. 2005. Plant cell wall polysaccharide biosynthesis: real progress in the identification of participating genes. *Planta* 221, 309–312. **(Progress report)**

CHANLIAUD, E., J. DE SILVA, B. STRONGITHARM, G. JERONIMIDIS, and M. J. GIDLEY. 2004. Mechanical effects of plant cell wall enzymes on cellulose/xyloglucan composites. *Plant J.* 38, 27–37. **(Direct in vitro evidence is provided for the involvement of cell wall xyloglucan-specific enzymes in**

**mechanical changes underlying plant cell wall remodeling and growth processes.)**

DIXIT, R., and R. J. CYR. 2002. Spatio-temporal relationships between nuclear-envelope breakdown and preprophase band disappearance in cultured tobacco cells. *Protoplasma* 219, 116–121. **(A causal relationship apparently exists between nuclear-envelope breakdown and disappearance of the preprophase band.)**

DONALDSON, L., and P. XU. 2005. Microfibril orientation across the secondary cell wall of radiata pine tracheids. *Trees* 19, 644–653.

FLEMING, A. J., ed. 2005. *Intercellular Communication in Plants. Annual Plant Reviews*, vol. 16. Blackwell/CRC Press, Oxford/Boca Raton.

FRY, S. C. 2004. Primary cell wall metabolism: tracking the careers of wall polymers in living plant cells. *New Phytol.* 161, 641–675. **(Review)**

JAMET, E., H. CANUT, G. BOUDART, and R. F. PONT-LEZICA. 2006. Cell wall proteins: a new insight through proteomics. *Trends Plant Sci.* 11, 33–39. **(Review)**

JÜRGENS, G. 2005. Cytokinesis in higher plants. *Annu. Rev. Plant Biol.* 56, 281–299. **(Review)**

JÜRGENS, G. 2005. Plant cytokinesis: fission by fusion. *Trends Cell Biol.* 15, 277–283. **(Review)**

KAWAMURA, E., R. HIMMELSPACH, M. C. RASHBROOKE, A. T. WHITTINGTON, K. R. GALE, D. A. COLLINGS, and G. O. WASTENEYS. 2006. MICROTUBULE ORGANIZATION 1 regulates structure and function of microtubule arrays during mitosis and cytokinesis in the *Arabidopsis* root. *Plant Physiol.* 140, 102–114. **(Quantitative analysis of *mor1-1*-generated defects in preprophase bands, spindles, and phragmoplasts of dividing vegetative cells suggest that microtubule length is a critical determinant of spindle and phragmoplast structure, orientation, and function.)**

KIM, I., K. KOBAYASHI, E. CHO, and P. C. ZAMBRYSKI. 2005. Subdomains for transport via plasmodesmata corresponding to the apical-basal axis are established during *Arabidopsis* embryogenesis. *Proc. Natl. Acad. Sci. USA* 102, 11945–11950. **(Evidence is presented indicating that cell-to-cell communication via plasmodesmata conveys positional information critical to establishment of the axial body pattern during embryogenesis in *Arabidopsis*.)**

KIM, I., and P. C. ZAMBRYSKI. 2005. Cell-to-cell communication via plasmodesmata during *Arabidopsis* embryogenesis. *Curr. Opin. Plant Biol.* 8, 593–599. **(Review)**

LLOYD, C., and J. CHAN. 2004. Microtubules and the shape of plants to come. *Nature Rev. Mol. Cell Biol.* 5, 13–22. **(Review)**

MARCUS, A. I., R. DIXIT, and R. J. CYR. 2005. Narrowing of the preprophase microtubule band is not required for cell division plane determination in cultured plant cells. *Protoplasma* 226,

169–174. **(Although the prophase band microtubules do not directly mark the division site in cultured tobacco BY-2 cells, they are required for accurate spindle positioning.)**

MARRY, M., K. ROBERTS, S. J. JOPSON, I. M. HUXHAM, M. C. JARVIS, J. CORSAR, E. ROBERTSON, and M. C. MCCANN. 2006. Cell–cell adhesion in fresh sugar-beet root parenchyma requires both pectin esters and calcium cross-links. *Physiol. Plant.* 126, 243–256. **(Cell–cell adhesion in sugar-beet root parenchyma depends on both ester and Ca²⁺ cross-linked polymers.)**

MULDER, B. M., and A. M. C. EMONS. 2001. A dynamic model for plant cell wall architecture formation. *J. Math. Biol.* 42, 261–289. **(A dynamic mathematical model is presented that explains plant cell wall architecture.)**

OPARKA, K. J. 2004. Getting the message across: how do plant cells exchange macromolecular complexes? *Trends Plant Sci.* 9, 33–41. **(Review)**

OTEGUI, M. S., K. J. VERBRUGGHE, and A. R. SKOP. 2005. Midbodies and phragmoplasts: analogous structures involved in cytokinesis. *Trends in Cell Biol.* 15, 404–413. **(Review)**

PANTERIS, E., P. APOSTOLAKOS, H. QUADER, and B. GALATIS. 2004. A cortical cytoplasmic ring predicts the division plane in vacuolated cells of *Coleus*: the role of actomyosin and microtubules in the establishment and function of the division site. *New Phytol.* 163, 271–286. **(The division plane is predicted by a cortical cytoplasmic ring [CCR], rich in actin filaments and endoplasmic reticulum, formed at interphase. The nucleus migrates to the CCR before entering the phragmosome. During preprophase, a preprophase microtubule band is organized in the CCR. Actomyosin and microtubules play crucial roles in the establishment and function of the division site.)**

PETER, G., and D. NEALE. 2004. Molecular basis for the evolution of xylem lignification. *Curr. Opin. Plant Biol.* 7, 737–742. **(Review)**

PETERMAN, T. K., Y. M. OHOL, L. J. MCREYNOLDS, and E. J. LUNA. 2004. Patellin1, a novel Sec14-like protein, localizes to the cell plate and binds phosphoinositides. *Plant Physiol.* 136, 3080–3094. **(The findings suggest a role for patellin1 in membrane-trafficking events associated with cell-plate expansion or maturation and point to the involvement of phosphoinositides in cell-plate biogenesis.)**

POPPER, Z. A., and S. C. FRY. 2004. Primary cell wall composition of pteridophytes and spermatophytes. *New Phytol.* 164, 165–174. **(Review)**

RALET, M.-C., G. ANDRÉ-LEROUX, B. QUÉMÉNER, and J.-F. THIBAULT. 2005. Sugar beet (*Beta vulgaris*) pectins are covalently cross-linked through diferulic bridges in the cell wall. *Phytochemistry* 66, 2800–2814. **(Direct evidence indicates that pectic arabinans and galactans are covalently cross-linked [intra- or intermolecularly] through dehydrodiferulates in sugar beet cell walls.)**

REFRÉGIER, G., S. PELLETIER, D. JAILLARD, and H. HÖFTE. 2004. Interaction between wall deposition and cell elongation in dark-grown hypocotyl cells in *Arabidopsis*. *Plant Physiol.* 135, 959–968. **(The rate of cell wall synthesis was not coupled to the elongation rate of epidermal cells. Polysaccharides were axially oriented in thin walls. Innermost cellulose microfibrils were transversely oriented in both slowly and rapidly growing cells, indicating that transversely deposited microfibrils reoriented in deeper layers of the expanding wall.)**

REIS, D., and B. VIAN. 2004. Helicoidal pattern in secondary cell walls and possible role of xylans in their construction. *C.R. Biol.* 327, 785–790. **(Review)**

ROBERTS, A. G., and K. J. OPARKA. 2003. Plasmodesmata and the control of symplastic transport. *Plant Cell Environ.* 26, 103–124. **(Review)**

ROS-BARCELÓ, A. 2005. Xylem parenchyma cells deliver the H₂O₂ necessary for lignification in differentiating xylem vessels. *Planta* 220, 747–756. **(In *Zinnia elegans* stems non-lignified xylem parenchyma cells are the source of H₂O₂ necessary for the polymerization of cinnamyl alcohols in the secondary cell wall of lignifying xylem vessels.)**

ROUDIER, F., A. G. FERNANDEZ, M. FUJITA, R. HIMMELSPACH, G. H. H. BORNER, G. SCHINDELMAN, S. SONG, T. I. BASKIN, P. DUPREE, G. O. WASTENEYS, and P. N. BENFEY. 2005. COBRA, an *Arabidopsis* extracellular glycosyl-phosphatidyl inositol-anchored protein, specifically controls highly anisotropic expansion through its involvement in cellulose microfibril orientation. *Plant Cell* 17, 1749–1763. **(COBRA has been implicated in cellulose microfibril deposition in rapidly elongating root cells. It is distributed mainly near the cell surface in transverse bands that parallel cortical microtubules.)**

RUIZ-MEDRANO, R., B. XOCONOSTLE-CAZARES, and F. KRAGLER. 2004. The plasmodesmal transport pathway for homeotic proteins, silencing signals and viruses. *Curr. Opin. Plant Biol.* 7, 641–650. **(Review)**

SAXENA, I. M., and R. M. BROWN JR. 2005. Cellulose biosynthesis: current views and evolving concepts. *Ann. Bot.* 96, 9–21. **(Review)**

SEDBROOK, J. C. 2004. MAPs in plant cells: delineating microtubule growth dynamics and organization. *Curr. Opin. Plant Biol.* 7, 632–640. **(Review)**

SEGUÍ-SIMARRO, J. M., J. R. AUSTIN II, E. A. WHITE, and L. A. STAEHELIN. 2004. Electron tomographic analysis of somatic cell plate formation in meristematic cells of *Arabidopsis* preserved by high-pressure freezing. *Plant Cell* 16, 836–856. **(Cell-plate assembly sites, consisting of a filamentous ribosome-excluding cell-plate assembly matrix [CPAM] and Golgi-derived vesicles, are formed at** the equatorial planes of phragmoplast initials, which arise from clusters of microtubules during late anaphase. It is suggested that **CPAM**, which is found only around growing cell-plate regions, is responsible for regulating growth of the cell plate.)

SEGUÍ-SIMARRO, J. M., and L. A. STAEHELIN. 2006. Cell cycle-dependent changes in Golgi stacks, vacuoles, clathrin-coated vesicles and multivesicular bodies in meristematic cells of *Arabidopsis thaliana*: A quantitative and spatial analysis. *Planta* 223, 223–236. **(Among the notable cell cycle-dependent changes reported in this article is that involving the vacuolar system. During early telophase cytokinesis the vacuoles form sausage like tubular compartments with a 50% reduced surface area and an 80% reduced volume compared to prometaphase cells. It is postulated that this transient reduction in vacuole volume during early telophase provides a means for increasing the volume of the cytosol to accommodate the forming phragmoplast microtubule array and associated cell plate-forming structures.)**

SOMERVILLE, C., S. BAUER, G. BRININSTOOL, M. FACETTE, T. HAMANN, J. MILNE, E. OSBORNE, A. PAREDEZ, S. PERSSON, T. RAAB, S. VORWERK, and H. YOUNGS. 2004. Toward a systems approach to understanding plant cell walls. *Science* 306, 2206–2211.

VISSENBERG, K., S. C. FRY, M. PAULY, H. HÖFTE, and J.-P. VERBELEN. 2005. XTH acts at the microfibril-matrix interface during cell elongation. *J. Exp. Bot.* 56, 673–683.

YASUDA, H., K. KANDA, H. KOIWA, K. SUENAGA, S.-I. KIDOU, and S.-I. EJIRI. 2005. Localization of actin filaments on mitotic apparatus in tobacco BY-2 cells. *Planta* 222, 118–129. **(Similar results obtained in both staining with rhodamine-phalloidin and immunostaining with actin antibody strongly indicate the participation of actin in the organization of the spindle body or in the process of chromosome segregation.)**

ZAMBRYSKI, P. 2004. Cell-to-cell transport of proteins and fluorescent tracers via plasmodesmata during plant development. *J. Cell Biol.* 162, 165–168. **(Mini-review)**

ZAMBRYSKI, P., and K. CRAWFORD. 2000. Plasmodesmata: gatekeepers for cell-to-cell transport of developmental signals in plants. *Annu. Rev. Cell Dev. Biol.* 16, 393–421. **(Review)**

# ▌CHAPTERS 5 AND 6

ABE, M., Y. KOBAYASHI, S. YAMAMOTO, Y. DAIMON, A. YAMAGUCHI, Y. IKEDA, H. ICHINOKI, M. NOTAGUCHI, K. GOTO, and T. ARAKI. 2005. FD, A bZIP protein mediating signals from the floral pathway integrator FT at the shoot apex. *Science* 309, 1052–1056. **(It is shown that, in *Arabidopsis*, FLOWERING LOCUS T [FT]**, a protein encoded by the *FLOW-ERING LOCUS T [FT]* gene in leaves, can interact with **FD**—a bZIP transcription factor present only in the shoot apex—to activate floral identity genes such as *APETALA1 [API]*. See also Huang et al., 2005, and Wigge et al., 2005.)**

ADE-ADEMILUA, O. E., and C. E. J. BOTHA. 2005. A re-evaluation of plastochron index determination in peas—a case for using leaflet length. *S. Afr. J. Bot.* 71, 76–80. **(It is proposed that leaflet growth should be used as a measure of the plastochron index in peas.)**

ANGENENT, G. C., J. STUURMAN, K. C. SNOWDEN, and R. KOES. 2005. Use of *Petunia* to unravel plant meristem functioning. *Trends Plant Sci.* 10, 243–250. **(Review)**

BEEMSTER, G. T. S., S. VERCRUYSSE, L. DEVEYLDER, M. KUIPER, and D. INZÉ. 2006. The *Arabidopsis* leaf as a model system for investigating the role of cell cycle regulation in organ growth. *J. Plant Res.* 1129, 43–50.

BERNHARDT, C., M. ZHAO, A. GONZALEZ, A. LLOYD, and J. SCHIEFELBEIN. 2005. The bHLH genes *GL3* and *EGL3* participate in an intercellular regulatory circuit that controls cell patterning in the *Arabidopsis* root epidermis. *Development* 132, 291–298. **(An analysis of the expression of *GL3* and *EGL3* during root epidermis development revealed that *GL3* and *EGL3* gene expression and RNA accumulation occur preferentially in the developing hair cells. GL3 protein was found to move from the hair cells to the non-hair cells. The results of this study suggest that GL3/EGL3 accumulation in cells that adopt the non-hair fate is dependent on specification of the hair cell fate.)**

BOZHKOV, P. V., M. F. SUAREZ, L. H. FILONOVA, G. DANIEL, A. A. ZAMYATNIN JR., S. RODRIGUEZ-NIETO, B. ZHIVOTOVSKY, and A. SMERTENKO. 2005. Cysteine protease mcII-Pa executes programmed cell death during plant embryogenesis. *Proc. Natl. Acad. Sci. USA* 102, 14463–14468. **(The results of this study establish metacaspase as an executioner of programmed cell death [PCD] during embryo patterning and provide a functional link between PCD and embryogenesis in plants.)**

CANALES, C., S. GRIGG, and M. TSIANTIS. 2005. The formation and patterning of leaves: recent advances. *Planta* 2221, 752–756. **(Review)**

CARLES, C. C., D. CHOFFNES-INADA, K. REVILLE, K. LERTPIRI-YAPONG, and J. C. FLETCHER. 2005. *ULTRAPETALA1* encodes a SAND domain putative transcriptional regulator that controls shoot and floral meristem activity in *Arabidopsis*. *Development* 132, 897–911.

CASTELLANO, M. M., and R. SABLOWSKI. 2005. Intercellular signalling in the transition from stem cells to organogenesis in meristems. *Curr. Opin. Plant Biol.* 8, 26–31.

CHANG, C., and A. B. BLEECKER. 2004. Ethylene biology. More than a gas. *Plant Physiol.* 136, 2895–2899. **(State of the field)**

CHENG, Y., and X. CHEN. 2004. Posttranscriptional control of plant development. *Curr. Opin. Plant Biol.* 7, 20–25. **(Review)**

DEL RÍO, L. A., F. J. CORPAS, and J. B. BARROSO. 2004. Nitric oxide and nitric oxide synthase activity in plants. *Phytochemistry* 65, 783–792. **(Review)**

DHONUKSHE, P., J. KLEINE-VEHN, and J. FRIML. 2005. Cell polarity, auxin transport, and cytoskeleton-mediated division planes: who comes first? *Protoplasma* 226, 67–73. **(Review)**

DOLAN, L. and J. DAVIES. 2004. Cell expansion in roots. *Curr. Opin. Plant Biol.* 7, 33–39. **(Review)**

EVANS, L. S., and R. K. PEREZ. 2004. Diversity of cell lengths in intercalary meristem regions of grasses: location of the proliferative cell population. *Can. J. Bot.* 82, 115–122. **(Not all parenchyma cells of the intercalary meristems are rapidly proliferating.)**

FLEMING, A. J. 2005. Formation of primordia and phyllotaxy. *Curr. Opin. Plant Biol.* 8, 53–58. **(Review)**

FLEMING, A. J. 2006. The co-ordination of cell division, differentiation and morphogenesis in the shoot apical meristem: a perspective. *J. Exp. Bot.* 57, 25–32. **(Data obtained from a series of experiments support an organismal view of plant morphogenesis and the idea that the cell wall plays a key role in the mechanism by which this is achieved.)**

FLEMING, A. J. 2006. The integration of cell proliferation and growth in leaf morphogenesis. *J. Plant Res.* 119, 31–36. **(Review)**

FLETCHER, J. C. 2002. Shoot and floral meristem maintenance in *Arabidopsis*. *Annu. Rev. Plant Biol.* 53, 45–66. **(Review)**

FRIML, J., P. BENFEY, E. BENKOVÁ, M. BENNETT, T. BERLETH, N. GELDNER, M. GREBE, M. HEISLER, J. HEJÁTKO, G. JÜRGENS, T. LAUX, K. LINDSEY, W. LUKOWITZ, C. LUSCHNIG, R. OFFRINGA, B. SCHERES, R. SWARUP, R. TORRES-RUIZ, D. WEIJERS, and E. ZAŽÍMALOVÁ. 2006. Apical-basal polarity: why plant cells don't stand on their heads. *Trends Plant Sci.* 11, 12–14. **(The authors critique the anatomical apical-basal terminology.)**

GRAFI, G. 2004. How cells dedifferentiate: a lesson from plants. *Dev. Biol.* 268, 1–6. **(Review)**

GRANDJEAN, O., T. VERNOUX, P. LAUFS, K. BELCRAM, Y. MIZUKAMI, and J. TRAAS. 2004. In vivo analysis of cell division, cell growth, and differentiation at the shoot apical meristem in *Arabidopsis*. *Plant Cell* 16, 74–87. **(Confocal microscopy combined with green fluorescence protein marker lines and vital dyes were used to visualize living shoot apical meristems. The effects of several mitotic drugs on meristem development indicate that DNA synthesis plays an important role in growth and patterning.)**

GRAY, J., ed. 2004. Programmed cell death in plants. Blackwell/CRC Press, Oxford/Boca Raton.

HAIGLER, C. H., D. ZHANG, and C. G. WILKERSON. 2005. Biotechnological improvement of cotton fibre maturity. *Physiol. Plant.* 124, 285–294. **(Review)**

HAKE, S., H. M. S. SMITH, H. HOLTAN, E. MAGNANI, G. MELE, and J. RAMIREZ. 2004. The role of *knox* genes in plant development. *Annu. Rev. Cell Dev. Biol.* 20, 125–151. **(Review)**

HARA-NISHIMURA, I., N. HATSUGAI, S. NAKAUNE, M. KUROYANAGI, and M. NISHIMURA. 2005. Vacuolar processing enzyme: an executor of plant cell death. *Curr. Opin. Plant Biol.* 8, 404–408. **(Review)**

HAUBRICK, L. L., and S. M. ASSMANN. 2006. Brassinosteroids and plant function: some clues, more puzzles. *Plant Cell Environ.* 29, 446–457. **(Review)**

HÖRTENSTEINER, S. 2006. Chlorophyll degradation during senescence. *Annu. Rev. Plant Biol.* 57. **On line. (Review)**

HUANG, T., H. BÖHLENIUS, S. ERIKSSON, F. PARCY, and O. NILSSON. 2005. The mRNA of the *Arabidopsis* gene *FT* moves from leaf to shoot apex and induces flowering. *Science* 309, 1694–1696. **(The data suggest that the *FT* mRNA is an important component of the elusive "florigen" signal that moves from leaf to shoot apex via the phloem sieve tubes. It is possible the FT protein is also moving and is responsible for the floral induction. See also Abe et al., 2005, and Huang et al., 2005.)**

INGRAM, G. C. 2004. Between the sheets: inter-cell-layer communication in plant development. *Philos. Trans. R. Soc. Lond. B* 359, 891–906. **(Review)**

IVANOV, V. B. 2004. Meristem as a self-renewing systems: maintenance and cessation of cell proliferation. *Russ. J. Plant Physiol.* 51, 834–847. **(Review)**

JAKOBY, M., and A. SCHNITTGER. 2004. Cell cycle and differentiation. *Curr. Opin. Plant Biol.* 7, 661–669. **(Review)**

JENIK, P. D., and M. K. BARTON. 2005. Surge and destroy: the role of auxin in plant embryogenesis. *Development* 132, 3577–3585. **(Review)**

JIANG, K., T. BALLINGER, D. LI, S. ZHANG, and L. FELDMAN. 2006. A role for mitochondria in the establishment and maintenance of the maize root quiescent center. *Plant Physiol.* 140, 1118–1125. **(Mitochondria in the quiescent center [QC] of the maize [*Zea mays*] root showed marked reductions in the activities of tricarboxylic acid cycle enzymes, and pyruvate dehydrogenase activity was not detected there. The authors postulate that modifications of mitochondrial function are central to the establishment and maintenance of the QC.)**

JIANG, K., and L. J. FELDMAN. 2005. Regulation of root apical meristem development. *Annu. Rev. Cell Dev. Biol.* 21, 485–509. **(Review)**

JIMÉNEZ, V. M. 2005. Involvement of plant hormones and plant growth regulators on *in vitro* somatic embryogenesis. *Plant Growth Regul.* 47, 91–110. **(Review)**

JING, H.-C., J. HILLE, and P. P. DIJKWEL. 2003. Ageing in plants: conserved strategies and novel pathways. *Plant Biol.* 5, 455–464. **(Review)**

JONGEBLOED, U., J. SZEDERKÉNYI, K. HARTIG, C. SCHOBERT, and E. KOMOR. 2004. Sequence of morphological and physiological events during natural ageing and senescence of a castor bean leaf: sieve tube occlusion and carbohydrate back-up precede chlorophyll degradation. *Physiol. Plant.* 120, 338–346. **(Phloem blockage precedes and may be causal for chlorophyll degradation in leaf senescence.)**

JÖNSSON, H., M. HEISLER, G. V. REDDY, V. AGRAWAL, V. GOR, B. E. SHAPIRO, E. MJOLSNESS, and E. M. MEYEROWITZ. 2005. Modeling the organization of the *WUSCHEL* expression domain in the shoot apical meristem. *Bioinformatics* 21 (suppl. 1): i232–i240. **(Two models are presented to account for the organization of the *WUSCHEL* expression domain in the shoot apical meristem of *Arabidopsis thaliana*.)**

JORDY, M.-N. 2004. Seasonal variation of organogenetic activity and reserves allocation in the shoot apex of *Pinus pinaster*. Ait. *Ann. Bot.* 93, 25–37. **(It is concluded that depending on the sites of accumulation within the shoot apical meristem and on the stage of the annual growth cycle, lipids, starch, and tannins may be involved in different processes, for example, energy and structural materials released by lipid synthesis in spring contributing to stem elongation and/or cell-to-cell communication.)**

KAWAKATSU, T., J.-I. ITOH, K. MIYOSHI, N. KURATA, N. ALVAREZ, B. VEIT, and Y. NAGATO. 2006. *PLASTOCHRON2* regulates leaf initiation and maturation in rice. *Plant Cell* 18, 612–625. **(The authors propose a model in which the plastochron is determined by signals from immature leaves that act non-cell-autonomously in the shoot apical meristem to inhibit the initiation of new leaves.)**

KEPINSKI, S. 2006. Integrating hormone signaling and patterning mechanisms in plant development. *Curr. Opin. Plant Biol.* 9, 28–34. **(Review)**

KESKITALO, J., G. BERGQUIST, P. GARDESTRÖM, and S. JANSSON. 2005. A cellular timetable of autumn senescence. *Plant Physiol.* 139, 1635–1648. **(Changes in pigment, metabolite and nutrient content, photosynthesis, and cell and organelle integrity were followed in senescing leaves of a free-growing aspen tree [*Populus tremula*] in autumn.)**

KIEFFER, M., Y. STERN, H. COOK, E. CLERICI, C. MAULBETSCH, T. LAUX, and B. DAVIES. 2006. Analysis of the transcription factor WUSCHEL and its functional homologue in *Antirrhinum* reveals a potential mechanism for their roles in meristem maintenance. *Plant Cell* 18, 560–573. **(The results of this study suggest that WUS functions by recruiting transcriptional corepressors to repress target genes that promote differentiation, thereby ensuring stem cell maintenance.)**

KONDOROSI, E., and A. KONDOROSI. 2004. Endoreduplication and activation of the anaphase-promoting complex during symbiotic cell development. *FEBS Lett.* 567, 152–157. **(Endoreduplication is an integral part of symbiotic cell differentiation during nitrogen-fixing nodule development.)**

KWAK, S.-H., R. SHEN, and J. SCHIEFELBEIN. 2005. Positional signaling mediated by a receptor-like kinase in *Arabidopsis. Science* 307, 1111–1113.

KWIATKOWSKA, D. 2004. Structural integration at the shoot apical meristem: models, measurements, and experiments. *Am. J. Bot.* 91, 1277–1293. **(Review of mechanical aspects of shoot apical meristem growth.)**

KWIATKOWSKA, D., and J. DUMAIS. 2003. Growth and morphogenesis at the vegetative shoot apex of *Anagallis arvensis* L. *J. Exp. Bot.* 54, 1585–1595. **(The geometry and expansion of the shoot apex surface are analyzed using a nondestructive replica method and a 3-D reconstruction algorithm.)**

LACROIX, C., B. JEUNE, and D. BARABÉ. 2005. Encasement in plant morphology: an integrative approach from genes to organisms. *Can. J. Bot.* 83, 1207–1221. **(Review)**

LARKIN, J. C., M. L. BROWN, and J. SCHIEFELBEIN. 2003. How do cells know what they want to be when they grow up? Lessons from epidermal patterning in *Arabidopsis. Annu. Rev. Plant Biol.* 54, 403–430. **(Review)**

LAZAR, G., and H. M. GOODMAN. 2006. *MAX1*, a regulator of the flavonoid pathway, controls vegetative axillary bud outgrowth in *Arabidopsis. Proc. Natl. Acad. Sci. USA* 103, 472–476. **(The results of this study lead the authors to speculate that *MAX1* could repress axillary bud outgrowth via regulating flavonoid-dependent auxin retention in the bud and underlying stem.)**

LEIVA-NETO, J. T., G. GRAFI, P. A. SABELLI, R. A. DANTE, Y. WOO, S. MADDOCK, W. J. GORDON-KAMM, and B. A. LARKINS. 2004. A dominant negative mutant of cyclin-dependent kinase A reduces endoreduplication but not cell size or gene expression in maize endosperm. *Plant Cell* 16, 1854–1869. **(A reduced level of endoreduplication did not affect cell size and had little effect on the level of endosperm gene-expression.)**

LJUNG, K., A. K. HULL, J. CELENZA, M. YAMADA, M. ESTELLE, J. NORMANLY, and G. SANDBERG. 2005. Sites and regulation of auxin biosynthesis in *Arabidopsis* roots. *Plant Cell* 17, 1090–1104. **(An important source of auxin has been identified in the meristematic region of the primary root tip and the tips of emerged lateral roots. A model is presented for how the primary root is supplied with auxin during early seedling development.)**

LUMBA, S., and P. MCCOURT. 2005. Preventing leaf identity theft with hormones. *Curr. Opin. Plant Biol.* 8, 501–505. **(Review)**

MATHUR, J. 2006. Local interactions shape plant cells. *Curr. Opin. Cell Biol.* 18, 40–46. **(Review)**

MᴄSᴛᴇᴇɴ, P., and O. Lᴇʏsᴇʀ. 2005. Shoot branching. *Annu. Rev. Plant Biol.* 56, 353–374. **(Review)**

Müssɪɢ, C. 2005. Brassinosteroid-promoted growth. *Plant Biol.* 7, 110–117. **(Review)**

Nᴇᴍᴏᴛᴏ, K., I. Nᴀɢᴀɴᴏ, T. Hᴏɢᴇᴛsᴜ, and N. Mɪʏᴀᴍᴏᴛᴏ. 2004. Dynamics of cortical microtubules in developing maize internodes. *New Phytol.* 162, 95–103. **(The orientation of cortical microtubules in cells of the intercalary meristem originated from cells with randomly orientated microtubules and remained unchanged throughout the proliferation of the internodal cells.)**

Pᴏɴᴄᴇ, G., P. W. Bᴀʀʟᴏᴡ, L. J. Fᴇʟᴅᴍᴀɴ, and G. I. Cᴀssᴀʙ. 2005. Auxin and ethylene interactions control mitotic activity of the quiescent centre, root cap size, and pattern of cap cell differentiation in maize. *Plant Cell Environ.* 28, 719–732. **(Control of root cap size, shape, and structure was found to involve interactions between the rootcap [RC] and the quiescent centre [QC]. Results of experiments with ethylene and the polar auxin inhibitor 1-N-naphthylphthalamic acid [NPA] suggests that the QC ensures an ordered internal distribution of auxin and thereby regulates not only the planes of growth and division in both the root apex proper and the RC meristem, but also regulates cell fate in the RC. Ethylene apparently regulates the auxin redistribution system that resides in the RC.)**

Rᴀɴɢᴀɴᴀᴛʜ, R. M. 2005. Asymmetric cell divisions in flowering plants—one mother, "two-many" daughters. *Plant Biol.* 7, 425–448. **(Review)**

Rᴇᴅᴅʏ, G. V., M. G. Hᴇɪsʟᴇʀ, D. W. Eʜʀʜᴀʀᴅᴛ, and E. M. Mᴇʏᴇʀᴏᴡɪᴛᴢ. 2004. Real-time lineage analysis reveals oriented cell divisions associated with morphogenesis at the shoot apex of *Arabidopsis thaliana*. *Development* 131, 4225–4237. **(A live-imaging technique based on confocal microscopy has been utilized to analyze growth in real time by monitoring individual cell divisions in the shoot apical meristem [SAM] of *Arabidopsis thaliana*. The analysis revealed that cell division activity in the SAM is subject to temporal activity and coordinated across clonally distinct layers of cells.)**

Rᴇᴅᴅʏ, G. V., and E. M. Mᴇʏᴇʀᴏᴡɪᴛᴢ. 2005. Stem-cell homeostasis and growth dynamics can be uncoupled in the *Arabidopsis* shoot apex. *Science* 310, 663–667. **(It is shown that the CLAVATA3 [CLV3] gene restricts its own domain of expression [the central zone, CZ] by preventing differentiation of peripheral zone [PZ], which surround the CZ, into CZ cells and restricts overall shoot apical meristem [SAM] size by a separate, long-range effect on cell division rate.)**

Rᴇɪɴʜᴀʀᴅᴛ, D. 2005. Phyllotaxis—a new chapter in an old tale about beauty and magic numbers. *Curr. Opin. Plant Biol.* 8, 487–493. **(Review)**

Rᴇɪɴʜᴀʀᴅᴛ, D., E.-R. Pᴇsᴄᴇ, P. Sᴛɪᴇɢᴇʀ, T. Mᴀɴᴅᴇʟ, K. Bᴀʟᴛᴇɴsᴘᴇʀɢᴇʀ, M. Bᴇɴɴᴇᴛᴛ, J. Tʀᴀᴀs, J. Fʀɪᴍʟ, and C. Kᴜʜʟᴇᴍᴇɪᴇʀ. 2003. Regulation of phyllotaxis by polar auxin transport. *Nature* 426, 255–260. **(The results of this study show that PIN1 and auxin play a central role in phyllotactic patterning in *Arabidopsis*. PIN1, on the other hand, responds to phyllotactic patterning information, indicating that phyllotaxis involves a feedback mechanism. Based on these results and other experimental data, the authors propose a model for the regulation of phyllotaxis in *Arabidopsis*.)**

Rᴏᴅʀíǫᴜᴇᴢ-Rᴏᴅʀíɢᴜᴇᴢ, J. F., S. Sʜɪsʜᴋᴏᴠᴀ, S. Nᴀᴘsᴜᴄɪᴀʟʏ-Mᴇɴᴅɪᴠɪʟ, and J. G. Dᴜʙʀᴏᴠsᴋʏ. 2003. Apical meristem organization and lack of establishment of the quiescent center in Cactaceae roots with determinate growth. *Planta* 217, 849–857. **(Establishment of a quiescent center is required for the maintenance of the apical meristem and indeterminate root growth.)**

Sᴀᴍᴘᴇᴅʀᴏ, J., R. D. Cᴀʀᴇʏ, and D. J. Cᴏsɢʀᴏᴠᴇ. 2006. Genome histories clarify evolution of the expansin superfamily: new insight from the poplar genome and pine ESTs. *J. Plant Res.* 119, 11–21. **(Review)**

Sᴄʜɪʟᴍɪʟʟᴇʀ, A. L., and G. A. Hᴏᴡᴇ. 2005. Systemic signaling in the wound response. *Curr. Opin. Plant Biol.* 8, 369–377. **(Brief review on the role of jasmonic acid and systemin in the response to wounding.)**

Sʜᴏsᴛᴀᴋ, S. 2006. (Re)defining stem cells. *BioEssays* 28, 301–308. **(The author discusses the confusion currently existing in use of the term stem cell.)**

Sᴛᴇғғᴇɴs, B., and M. Sᴀᴜᴛᴇʀ. 2005. Epidermal cell death in rice is regulated by ethylene, gibberellin, and abscisic acid. *Plant Physiol.* 139, 713–721. **(Induction of programmed cell death [PCD] of epidermal cells covering adventitious root primordia in deepwater rice [Oryza sativa] is induced by submergence. Induction of PCD is dependent on ethylene signaling and is further promoted by gibberellin [GA], the ethylene and GA acting in a synergistic manner. Abscisic acid was shown to delay ethylene-induced as well as GA-promoted cell death.)**

Sᴜɢɪʏᴀᴍᴀ, S.-I. 2005. Polyploidy and cellular mechanisms changing leaf size: Comparison of diploid and autotetraploid populations in two species of *Lolium*. *Ann. Bot.* 96, 931–938. **(Polyploidy increased leaf size by increasing cell size.)**

Tᴀɴᴀᴋᴀ, M., K. Tᴀᴋᴇɪ, M. Kᴏᴊɪᴍᴀ, H. Sᴀᴋᴀᴋɪʙᴀʀᴀ, and H. Mᴏʀɪ. 2006. Auxin controls local cytokinin biosynthesis in the nodal stem in apical dominance. *Plant J.* 45, 1028–1036. **(The authors demonstrate that auxin negatively regulates local cytokinin [CK] biosynthesis in the nodal stem by controlling the expression level of the pea [Pisum sativum L.] gene *adenosine phosphate-isopentenyltransferase* [PsIPT], which encodes a key enzyme in CK biosynthesis.)**

Tᴇᴀʟᴇ, W. D., I. A. Pᴀᴘᴏɴᴏᴠ, F. Dɪᴛᴇɴɢᴏᴜ, and K. Pᴀʟᴍᴇ. 2005. Auxin and the developing root of *Arabidopsis thaliana*. *Physiol. Plant.* 123, 130–138. **(Review)**

VALLADARES, F., and D. BRITES. 2004. Leaf phyllotaxis: does it really affect light capture? *Plant Ecol.* 174, 11–17.

VAN DOORN, W. G. 2005. Plant programmed cell death and the point of no return. *Trends Plant Sci.* 10, 478–483. **(Review)**

VAN DOORN, W. G., and E. J. WOLTERING. 2005. Many ways to exit? Cell death categories in plants. *Trends Plant Sci.* 10, 117–122. **(Review)**

VANISREE, M., C.-Y. LEE, S.-F. LO, S. M. NALAWADE, C. Y. LIN, and H.-S. TSAY. 2004. Studies on the production of some important secondary metabolites form medicinal plants by plant tissue cultures. *Bot. Bull. Acad. Sin.* 45, 1–22. **(Review)**

VANNESTE, S., L. MAES, I. DE SMET, K. HIMANEN, M. NAUDTS, D. INZÉ, and T. BEECKMAN. 2005. Auxin regulation of cell cycle and its role during lateral root initiation. *Physiol. Plant.* 123, 139–146. **(Review)**

VEIT, B. 2004. Determination of cell fate in apical meristems. *Curr. Opin. Plant Biol.* 7, 57–64. **(Review)**

WARD, S. P., and O. LEYSER. 2004. Shoot branching. *Curr. Opin. Plant Biol.* 7, 73–78. **(Review)**

WEIJERS, D., and G. JÜRGENS. 2005. Auxin and embryo axis formation: the ends in sight? *Curr. Opin. Plant Biol.* 8, 32–37. **(Review)**

WIGGE, P. A., M. C. KIM, K. E. JAEGER, W. BUSCH, M. SCHMID, J. U. LOHMANN, and D. WEIGEL. 2005. Integration of spatial and temporal information during floral induction in *Arabidopsis*. *Science* 309, 1056–1059. **(See summary with Abe et al., 2005.)**

WILLIAMS, L., and J. C. FLETCHER. 2005. Stem cell regulation in the *Arabidopsis* shoot apical meristem. *Curr. Opin. Plant Biol.* 8, 582–586. **(Review)**

WOODWARD, A. W., and B. BARTEL. 2005. Auxin: regulation, action, and interaction. *Ann. Bot.* 95, 707–735. **(Review)**

# CHAPTERS 7 AND 8

AGEEVA, M. V., B. PETROVSKÁ, H. KIEFT, V. V. SAL'NIKOV, A. V. SNEGIREVA, J. E. G. VAN DAM, W. L. H. VAN VEENENDAAL, A. M. C. EMONS, T. A. GORSHKOVA, and A. A. M. VAN LAMMEREN. 2005. Intrusive growth of flax phloem fibers is of intercalary type. *Planta* 222, 565–574. **(The primary phloem fibers of *Linum usitatissimum* initially undergo coordinated growth, followed by intrusive growth. Evidence indicates that the intrusive growth phase is accomplished by a diffuse mode of cell elongation, not by tip growth. The intrusively growing fiber is multinucleate and, lacking plasmodesmata, is symplastically isolated.)**

ANGELES, G., S. A. OWENS, and F. W. EWERS. 2004. Fluorescence shell: a novel view of sclereid morphology with the confocal laser scanning microscope. *Microsc. Res. Techniq.* 63, 282–288. **(CLSM was used to observe sclereids from stems of *Avicennia germinans* and from fruits of *Pyrus calleryana* and *P. communis*. The use of CLSM for extended focus fluorescence images made it easy to illustrate and quantify the degree of branching of the pits and the number of cell wall facets.)**

EVANS, D. E. 2003. Aerenchyma formation. *New Phytol.* 161, 35–49.

GORSHKOVA, T., and C. MORVAN. 2006. Secondary cell-wall assembly in flax phloem fibres: role of galactans. *Planta* 223, 149–158. **(Review)**

GOTTSCHLING, M., and H. H. HILGER. 2003. First fossil record of transfer cells in angiosperms. *Am. J. Bot.* 90, 957–959.

GRITSCH, C. S., G. KLEIST, and R. J. MURPHY. 2004. Developmental changes in cell wall structure of phloem fibres of the bamboo *Dendrocalamus asper*. *Ann. Bot.* 94, 497–505. **(The multilayered nature of cell wall structure varied considerably between individual cells and was not specifically related to the thickness of the cell wall.)**

MALIK, A. I., T. D. COLMER, H. LAMBERS, and M. SCHORTEMEYER. 2003. Aerenchyma formation and radial $O_2$ loss along adventitious roots of wheat with only the apical root portion exposed to $O_2$ deficiency. *Plant Cell Environ.* 26, 1713–1722. **(Aerenchyma formed when only part of the root system was exposed to $O_2$ deficiency was shown to be functional in conducting $O_2$.)**

PURNOBASUKI, H., and M. SUZUKI. 2005. Aerenchyma tissue development and gas-pathway structure in root of *Avicennia marina* (Forsk.) Vierh. *J. Plant Res.* 118, 285–294. **(Aerenchyma development was due to the formation of schizogenous intercellular spaces. Cell separation occurred between the longitudinal cell columns, resulting in the formation of long intercellular spaces along the root axis. These long intercellular spaces were interconnected by abundant small pores or canals of schizogenous origin.)**

SEAGO, J. L., JR., L. C. MARSH, K. J. STEVENS, A. SOUKUP, O. VOTRUBOVÁ, and D. E. ENSTONE. 2005. A re-examination of the root cortex in wetland flowering plants with respect to aerenchyma. *Ann. Bot.* 96, 565–579. **(Review)**

TOMLINSON, P. B., and J. B. FISHER. 2005. Development of non-lignified fibers in leaves of *Gnetum gnemon* (Gnetales). *Am. J. Bot.* 92, 383–389. **(The nonlignified thick-walled fibers in the leaves of *Gnetum gnemon* may have a hydraulic function, in addition to a mechanical one.)**

# CHAPTER 9

ASSMANN, S. M., and T. I. BASKIN. 1998. The function of guard cells does not require an intact array of cortical microtubules. *J. Exp. Bot.* 49, 163–170. **(Guard cells in epidermal peels mediated stomatal opening in response to light or fusicoccin, and mediated stomatal closure in response to darkness and calcium, regardless of the presence of 1mM colchicine, which depolymerized most microtubules.)**

BERGMANN, D. C. 2004. Integrating signals in stomatal development. *Curr. Opin. Plant Biol.* 7, 26–32. **(Review)**

BÜCHSENSCHÜTZ, K., I. MARTEN, D. BECKER, K. PHILIPPAR, P. ACHE, and R. HEDRICH. 2005. Differential expression of $K^+$ channels between guard cells and subsidiary cells within the maize stomatal complex. *Planta* 222, 968–976. **(Interaction between subsidiary cells and guard cells is based on overlapping as well as differential expression of $K^+$ channels in the two cell types of the maize stomatal complex.)**

DRISCOLL, S. P., A. PRINS, E. OLMOS, K. J. KUNERT, and C. H. FOYER. 2006. Specification of adaxial and abaxial stomata, epidermal structure and photosynthesis to $CO_2$ enrichment in maize leaves. *J. Exp. Bot.* 57, 381–390. **(The results of this study indicate that maize leaves adjust their stomatal densities through changes in epidermal cell numbers rather than stomatal numbers.)**

FAN, L.-M., Z.-X. ZHAO, and S. M. ASSMANN. 2004. Guard cells: a dynamic signaling model. *Curr. Opin. Plant Biol.* 7, 537–546. **(Review)**

GALATIS, B., and P. APOSTOLAKOS. 2004. The role of the cytoskeleton in the morphogenesis and function of stomatal complexes. *New Phytol.* 161, 613–639. **(Review)**

GAO, X.-Q., C.-G. LI, P.-C. WEI, X.-Y. ZHANG, J. CHEN, and X.-C. WANG. 2005. The dynamic changes of tonoplasts in guard cells are important for stomatal movement in *Vicia faba*. *Plant Physiol.* 139, 1207–1216.

HERNANDEZ, M. L., H. J. PASSAS, and L. G. SMITH. 1999. Clonal analysis of epidermal patterning during maize leaf development. *Dev. Biol.* 216, 646–658. **(Results of the clones analyzed clearly show that lineage does not account for the linear patterning of stomata and bulliform cells, implying that position information must direct the differentiation patterns of these cell types in maize.)**

HOLROYD, G. H., A. M. HETHERINGTON, and J. E. GRAY. 2002. A role for the cuticular waxes in the environmental control of stomatal development. *New Phytol.* 153, 433–439. **(Review)**

ICPN WORKING GROUP: M. MADELLA, A. ALEXANDRE, and T. BALL. 2005. International code for phytolith nomenclature 1.0. *Ann. Bot.* 96, 253–260. **(This paper proposes an easy to follow, internationally accepted protocol to describe and name phytoliths.)**

KOIWAI, H., K. NAKAMINAMI, M. SEO, W. MITSUHASHI, T. TOYOMASU, and T. KOSHIBA. 2004. Tissue-specific localization of an abscisic acid biosynthetic enzyme, AAO3, in *Arabidopsis*. *Plant Physiol.* 134, 1697–1707. **(Results indicate that ABA synthesized in the vascular system is transported to various target tissues and cells. Guard cells are able to synthesize ABA.)**

KOUWENBERG, L. L. R., W. M. KÜRSCHNER, and H. VISSCHER. 2004. Changes in stomatal frequency and size during elongation of *Tsuga heterophylla* needles. *Ann. Bot.* 94, 561–569. **(Stomata first appear in the apical region of the needle and then spread basipetally. Although the number of stomatal rows does not change during needle development, stomatal density decreases nonlinearly with increasing needle area, until about 50% of final needle area. Stomatal and epidermal cell formation continues until completion of needle maturation.)**

LAHAV, M., M. ABU-ABIED, E. BELAUSOV, A. SCHWARTZ, and E. SADOT. 2004. Microtubules of guard cells are light sensitive. *Plant Cell Physiol.* 45, 573–582. **(Microtubules [MTs] in guard cells of *Commelina communis* leaves incubated in the light were organized in parallel straight and dense bundles; in the dark they were less straight and oriented randomly near the stomatal pores. Similarly in *Arabidopsis* guard cells MTs were organized in parallel arrays in the light but disorganized in the dark.)**

LUCAS, J. R., J. A. NADEAU, and F. D. SACK. 2006. Microtubule arrays and *Arabidopsis* stomatal development. *J. Exp. Bot.* 57, 71–79. **(During stomatal development in *Arabidopsis* the preprophase bands of microtubules are correctly placed away from stomata and from two types of precursor cells, indicating that all three cell types participate in an intercellular signaling pathway that orients the division site.)**

MILLER, D. D., N. C. A. DE RUIJTER, and A. M. C. EMONS. 1997. From signal to form: aspects of the cytoskeleton-plasma membrane-cell wall continuum in root hair tips. *J. Exp. Bot.* 48, 1881–1896. **(Review)**

MIYAZAWA, S.-I., N. J. LIVINGSTON, and D. H. TURPIN. 2006. Stomatal development in new leaves is related to the stomatal conductance of mature leaves in poplar (*Populus trichocarpa* x *P. deltoides*). *J. Exp. Bot.* 57, 373–380. **(The results of this study suggest that epidermal cell development and stomatal development are regulated by different physiological mechanisms. The stomatal conductance of mature leaves apparently has a regulatory effect on the stomatal development of expanding leaves.)**

MOTOMURA, H., T. FUJII, and M. SUZUKI. 2004. Silica deposition in relation to ageing of leaf tissues in *Sasa veitchii* (Carrière) Rehder (Poaceae: Bambusoideae). *Ann. Bot.* 93, 235–248. **(Two hypotheses on silica deposition were tested: first, that silica deposition occurs passively as a result of water uptake by plants, and second, that silica deposition is controlled by plants. The results indicate that the deposition process differed, depending on the cell type.)**

NADEAU, J. A., and F. D. SACK. 2003. Stomatal development: cross talk puts mouths in place. *Trends Plant Sci.* 8, 294–299. **(Review)**

PEI, Z.-M., and K. KUCHITSU. 2005. Early ABA signaling events in guard cells. *J. Plant Growth Regul.* 24, 296–307. **(Review)**

PIGHIN, J. A., H. ZHENG, L. J. BALAKSHIN, I. P. GOODMAN, T. L. WESTERN, R. JETTER, L. KUNST, and A. L. SAMUELS. 2004. Plant cuticular lipid export requires an ABC transporter. *Science*

306, 702–704. (**The ABC transporter CER5 localized in the plasma membrane of *Arabidopsis* epidermal cells is involved with wax export to the cuticle.**)

RICHARDSON, A., R. FRANKE, G. KERSTIENS, M. JARVIS, L. SCHREIBER, and W. FRICKE. 2005. Cuticular wax deposition in growing barley (*Hordeum vulgare*) leaves commences in relation to the point of emergence of epidermal cells from the sheaths of older leaves. *Planta* 222, 472–483. (**The results indicate that cuticular layers are deposited along the growing barley leaf independent of cell age or developmental stage. Rather, the reference point for wax deposition appears to be the point of emergence of cells into the atmosphere.**)

SCHREIBER, L. 2005. Polar paths of diffusion across plant cuticles: new evidence for an old hypothesis. *Ann. Bot.* 95, 1069–1073. (**Botanical briefing**)

SERNA, L. 2005. Epidermal cell patterning and differentiation throughout the apical-basal axis of the seedling. *J. Exp. Bot.* 56, 1983–1989. (**Review**)

SERNA, L., J. TORRES-CONTRERAS, and C. FENOLL. 2002. Specification of stomatal fate in *Arabidopsis*: evidences for cellular interactions. *New Phytol.* 153, 399–404. (**Review**)

SHI, Y.-H., S.-W. ZHU, X.-Z. MAO, J.-X. FENG, Y.-M. QIN, L. ZHANG, J. CHENG, L.-P. WEI, Z.-Y. WANG, and Y.-X. ZHU. 2006. Transcriptome profiling, molecular biological, and physiological studies reveal a major role for ethylene in cotton fiber cell elongation. *Plant Cell* 18, 651–664. (**The results of this study indicate that ethylene plays a major role in promoting cotton fiber elongation, and that ethylene may promote cell elongation by increasing the expression of sucrose synthase, tubulin, and expansion genes.**)

SHPAK, E. D., J. M. MCABEE, L. J. PILLITTERI, and K. U. TORII. 2005. Stomatal patterning and differentiation by synergistic interactions of receptor kinases. *Science* 309, 290–293. (**The findings of this study suggest that the ERECTA [ER]-family leucine-rich repeat-like kinases [LRR-RLKs] together act as negative regulators of stomatal development in *Arabidopsis*.**).

TANAKA, Y., T. SANO, M. TAMAOKI, N. NAKAJIMA, N. KONDO, and S. HASEZAWA. 2005. Ethylene inhibits abscisic acid-induced stomatal closure in *Arabidopsis*. *Plant Physiol.* 138, 2337–2343. (**The results indicate that ethylene delays stomatal closure by inhibiting the ABA signaling pathway.**)

VALKAMA, E., J.-P. SALMINEN, J. KORICHEVA, and K. PIHLAJA. 2004. Changes in leaf trichomes and epicuticular flavonoids during leaf development in three birch taxa. *Ann. Bot.* 94, 233–242. (**The densities of both glandular and nonglandular trichomes decreased markedly with leaf expansion while the total number of trichomes per leaf remained constant. In addition concentrations of most of the individual leaf surface flavonoids correlated positively with glandular trichome density within species.**

Apparently the functional role of trichomes is probably most important at early stages of birch leaf development.**)

VAN BRUAENE, N., G. JOSS, and P. VAN OOSTVELDT. 2004. Reorganization and in vivo dynamics of microtubules during *Arabidopsis* root hair development. *Plant Physiol.* 136, 3905–3919. (**This study provides new insights into the mechanisms of microtubule [MT] [re]organization during root hair development in *Arabidopsis thaliana*. The data show how MTs reorient after apparent contact with other MTs and support a model for MT alignment based on repeated reorientation of dynamic MT growth.**)

WU, Y., A. C. MACHADO, R. G. WHITE, D. J. LLEWELLYN, and E. S. DENNIS. 2006. Expression profiling identifies genes expressed early during lint fibre initiation in cotton. *Plant Cell Physiol.* 47, 107–127. (**Both the GhMyb25 transcription factor and homeodomain gene were predominantly ovule specific and were up-regulated on the day of anthesis in fibre initials relative to adjacent non-fibre ovule epidermal cells. DNA content measurements indicate that the fibre initials undergo DNA endoreduplication.**)

YANG, H.-M., J.-H. ZHANG, and X.-Y. ZHANG. 2005. Regulation mechanisms of stomatal oscillation. *J. Integr. Plant Biol.* 47, 1159–1172. (**Review**)

# CHAPTERS 10 AND 11

BUCCI, S. J., F. G. SCHOLZ, G. GOLDSTEIN, F. C. MEINZER, and L. DA S. L. STERNBERG. 2003. Dynamic changes in hydraulic conductivity in petioles of two savanna tree species: factors and mechanisms contributing to the refilling of embolized vessels. *Plant Cell Environ.* 26, 1633–1645. (**This study presents evidence that embolism formation and repair are two distinct phenomena controlled by different variables, the degree of embolism being a function of tension, and the rate of refilling a function of internal pressure imbalances.**)

BURGESS, S. S. O., J. PITTERMANN, and T. E. DAWSON. 2006. Hydraulic efficiency and safety of branch xylem increases with height in *Sequoia sempervirens* (D. Don) crowns. *Plant Cell Environ.* 29, 229–239. (**Measurements of resistance of branch xylem to embolism show an increase in safety with height. An expected decrease in xylem efficiency, however, was not observed. The absence of a safety-efficiency tradeoff may be explained in part by opposing height trends in the pit aperture and conduit diameter of tracheids and the major and semi-independent roles these play in determining xylem safety and efficiency, respectively.**)

COCHARD, H., F. FROUX, S. MAYR, and C. COUTAND. 2004. Xylem wall collapse in water-stressed pine needles. *Plant Physiol.* 134, 401–408. (**When severely dehydrated, tracheid walls completely collapsed, but the lumina still appeared** to be filled with sap. Further dehydration resulted in embolized tracheids and relaxation of the walls. Wall collapse in dehydrated needles was rapidly reversed upon rehydration.)

CUTLER, D. F., P. J. RUDALL, P. E. GASSON, and R. M. O. GALE. 1987. *Root Identification Manual of Trees and Shrubs: A Guide to the Anatomy of Roots of Trees and Shrubs Hardy in Britain and Northern Europe*. Chapman and Hall, London.

FAYLE, D. C. F. 1968. *Radial Growth in Tree Roots: Distribution, Timing, Anatomy*. University of Toronto, Faculty of Forestry, Toronto.

GABALDÓN, C., L. V. GÓMEZ ROS, M. A. PEDREÑO, and A. ROS-BARCELÓ. 2005. Nitric oxide production by the differentiating xylem of *Zinnia elegans*. *New Phytol*. 165, 121–130. (**NO production was mainly located in both phloem and xylem regardless of the cell differentiation status. However, there was evidence that plant cells, which are just predetermined to irreversibly trans-differentiate into xylem elements, show a burst in NO production. This burst is sustained as long as secondary cell wall synthesis and cell autolysis are in progress.**)

GANSERT, D. 2003. Xylem sap flow as a major pathway for oxygen supply to the sapwood of birch (*Betula pubescens* Ehr.). *Plant Cell Environ*. 26, 1803–1814. (**Sap flow contributed about 60% to the total oxygen supply to the sapwood. It not only affected the oxygen status of the sapwood but also had an effect on radial $O_2$ transport between stem and atmosphere.**)

HACKE, U. G., J. S. SPERRY, J. K. WHEELER, and L. CASTRO. 2006. Scaling of angiosperm xylem structure with safety and efficiency. *Tree Physiol*. 26, 689–701. (**The authors tested the hypothesis that greater cavitation resistance correlates with less total inter-vessel pit area per vessel [the pit area hypothesis] and evaluated a trade-off between cavitation safety and efficiency. Fourteen species of diverse growth form and family affinity were added to published data [29 species total]. A safety versus efficiency trade-off was evident, and the pit area hypothesis was supported by a strong relationship [$r^2 = 0.77$] between increasing cavitation resistance and diminishing pit membranes per vessel.**)

HSU, L. C. Y., J. C. F. WALKER, B. G. BUTTERFIELD, and S. L. JACKSON. 2006. Compression wood does not form in the roots of *Pinus radiata*. *IAWA J*. 27, 45–54. (**Compression wood was not observed in either tap or lateral roots farther than 300 mm from the base of the stem of *Pinus radiata*. Buried roots apparently lack the ability to develop compression wood.**)

JACOBSEN, A. L., F. W. EWERS, R. B. PRATT, W. A. PADDOCK III, and S. D. DAVIS. 2005. Do xylem fibers affect vessel cavitation resistance? *Plant Physiol*. 139, 546–556. (**Possible mechanical and hydraulic costs to increased cavitation resistance were examined among six co-occurring chaparral shrub species in southern California. A cor-

relation was found between cavitation resistance and fiber wall area, suggesting a mechanical role for fibers in cavitation resistance.**)

KARAM, G. N. 2005. Biomechanical model of the xylem vessels in vascular plants. *Ann. Bot*. 95, 1179–1186. (**The morphology of the xylem vessel through the different phases of growth apparently follows optimal engineering design principles.**)

KOJS, P., W. WŁOCH, and A. RUSIN. 2004. Rearrangement of cells in storeyed cambium of *Lonchocarpus sericeus* (Poir.) DC connected with the formation of interlocked grain in the xylem. *Trees* 18, 136–144. (**The mechanism of formation of the regular interlocked grain was investigated. New contacts between cells are formed by means of intrusive growth of the ends of cells belonging to one storey between the tangential walls of cells of the neighboring storey and unequal periclinal divisions, which change the shape of the initials.**)

LI, A.-M., Y.-R. WANG, and H. WU. 2004. Cytochemical localization of pectinase: the cytochemical evidence for resin ducts formed by schizogeny in *Pinus massoniana*. *Acta Bot. Sin*. 46, 443–450. (**Cytochemical evidence indicates that pectinase is involved in the schizogenous development of resin ducts in *Pinus massoniana*.**)

LOPEZ, O. R., T. A. KURSAR, H. COCHARD, and M. T. TYREE. 2005. Interspecific variation in xylem vulnerability to cavitation among tropical tree and shrub species. *Tree Physiol*. 25, 1553–1562. (**Stem xylem vulnerability to cavitation was investigated in nine tropical species with different life histories and habitat associations. The results support the functional dependence of drought tolerance on xylem resistance to cavitation.**)

MAUSETH, J. D. 2004. Wide-band tracheids are present in almost all species of Cactaceae. *J. Plant Res*. 117, 69–76. (**Wide-band tracheids [WBTs]—short, broad barrel-shaped tracheids with annular or helical secondary-wall thickenings that project deeply into the tracheid lumen—are present in the wood of almost all species of Cactaceae. They probably originated only once in the Cactaceae or in the Cactaceae/Portulacaceae clade. WBTs can both expand and contract, and they are believed to reduce the risk of cavitation that would occur if they had rigid walls and fixed volumes.**)

MAUSETH, J. D., and J. F. STEVENSON. 2004. Theoretical considerations of vessel diameter and conductive safety in populations of vessels. *Int. J. Plant Sci*. 165, 359–368. (**A comparison is made of the conductive safety of various vessel populations.**)

McELRONE, A. J., W. T. POCKMAN, J. MARTÍNEZ-VILALTA, and R. B. JACKSON. 2004. Variation in xylem structure and function in stems and roots of trees to 20 m depth. *New Phytol*. 163, 507–517.

MILES, A. 1978. *Photomicrographs of World Woods*. HM Stationery Office, London.

MOTOSE, H., M. SUGIYAMA, and H. FUKUDA. 2004. A proteoglycan mediates inductive interaction during plant vascular development. *Nature* 429, 873–878. (**A glycoprotein signal molecule has been identified in cultured *Zinnia* cells. Named xylogen, it conveys information enabling the formation of vessels [continuous strands of vessel elements].**)

OLSON, M. E. 2005. Commentary: typology, homology, and homoplasy in comparative wood anatomy. *IAWA J.* 26, 507–522.

ORIBE, Y., R. FUNADA, and T. KUBO. 2003. Relationships between cambial activity, cell differentiation and the localization of starch in storage tissues around the cambium in locally heated stems of *Abies sachalinensis* (Schmidt) Masters. *Trees* 17, 185–192. (**Results suggest that the extent of both cell division and cell differentiation depends on the amount of starch in storage tissues around the vascular cambium in locally heated stems of this evergreen conifer growing in a cool-temperature zone.**)

PITTERMANN, J., J. S. SPERRY, U. G. HACKE, J. K. WHEELER, and E. H. SIKKEMA. 2005. Torus-margo pits help conifers compete with angiosperms. *Science* 310, 1924. (**The pit-area resistance of conifers was 59 times lower than that of the angiosperm average. This compensates for the short length and much lower pit area of tracheids and results in comparable resistivities of conifer tracheids and angiosperm vessels.**)

RYSER, U., M. SCHORDERET, R. GUYOT, and B. KELLER. 2004. A new structural element containing glycine-rich proteins and rhamnogalacturonan I in the protoxylem of seed plants. *J. Cell Sci.* 117, 1179–1190. (**The polysaccharide-rich primary cell wall of living and elongating protoxylem elements is progressively modified and finally replaced by a protein-rich wall in dead and passively stretched elements.**)

SALLEO, S., M. A. LO GULLO, P. TRIFILÒ, and A. NARDINI. 2004. New evidence for a role of vessel-associated cells and phloem in the rapid xylem refilling of cavitated stems of *Larus nobilis*. *Plant Cell Environ.* 27, 1065–1076. (**A mechanism for xylem refilling based upon starch to sugar conversion and transport into embolized conduits, assisted by pressure-driven radial mass flow, is proposed.**)

SANO, Y. 2004. Intervascular pitting across the annual ring boundary in *Betula platyphylla* var. *japonica* and *Fraxinus mandshurica* var. *japonica*. *IAWA J.* 25, 129–140. (**Unilaterally compound pits were present in the intervascular common wall at the annual ring boundary in both species.**)

SPERRY, J. S. 2003. Evolution of water transport and xylem structure. *Int. J. Plant Sci.* 164 (3, suppl.), S115–S127. (**Review**)

SPERRY, J. S., U. G. HACKE, and J. K. WHEELER. 2005. Comparative analysis of end wall resistivity in xylem conduits. *Plant Cell Environ.* 28, 456–465. (**The hydraulic resistivity—pressure gradient/flow rate—through the end walls of xylem conduits was estimated in a vessel-bearing fern, a tracheid-bearing gymnosperm, one vesselless angiosperm, and four vessel-bearing angiosperms. The results suggest that end wall and lumen resistivities are nearly co-limiting in vascular plants.**)

STILLER, V., J. S. SPERRY, and R. LAFITTE. 2005. Embolized conduits of rice (*Oryza sativa*, Poaceae) refill despite negative xylem pressure. *Am. J. Bot.* 92, 1970–1974.

VAZQUEZ-COOZ, J., and R. W. MEYER. 2004. Occurrence and lignification of libriform fibers in normal and tension wood of red and sugar maple. *Wood Fiber Sci.* 36, 56–70. (**In normal and tension wood of *Acer rubrum* and *A. saccharum* libriform fibers occur in interrupted wavy bands and have larger lumina than fiber-tracheids; intercellular spaces are common.**)

WATANABE, Y., Y. SANO, T. ASADA, and R. FUNADA. 2006. Histochemical study of the chemical composition of vestured pits in two species of *Eucalyptus*. *IAWA J.* 27, 33–43. (**It appears that the vestures of the vessel elements and fibers in the wood of *Eucalyptus camaldulensis* and *E. globus* consist mainly of alkali-soluble polyphenols and polysaccharides.**)

WHEELER, J. K., J. S. SPERRY, U. G. HACKE, and N. HOANG. 2005. Inter-vessel pitting and cavitation in woody Rosaceae and other vesselled plants: a basis for a safely versus efficiency trade-off in xylem transport. *Plant Cell Environ.* 28, 800–812. (**No relationship was found between pit resistance and cavitation pressure. An inverse relationship was found, however, between pit area per vessel and vulnerability to cavitation.**)

WIMMER, R. 2002. Wood anatomical features in tree-rings as indicators of environmental change. *Dendrochronologia* 20, 21–36. (**Review**)

YANG, J., D. P. KAMDEM, D. E. KEATHLEY, and K.-H. HAN. 2004. Seasonal changes in gene expression at the sapwood-heartwood transition zone of black locust (*Robinia pseudoacacia*) revealed by cDNA microarray analysis. *Tree Physiol.* 24, 461–474. (**When samples from the transition zone of mature trees collected in summer and fall were compared, 569 genes showed differential expression patterns: 293 genes were up-regulated in summer [July 5] and 276 genes were up-regulated in fall [November 27]. More than 50% of the secondary and hormone metabolism-related genes on the microarrays were up-regulated in summer. Twenty-nine out of 55 genes involved in signal reduction were differentially regulated, suggesting that the ray parenchyma cells located in the innermost part of the trunk wood react to seasonal changes.**)

YE, Z.-H. 2002. Vascular tissue differentiation and pattern formation in plants. *Annu. Rev. Plant Biol.* 53, 183–202. (**Review**)

YOSHIDA, S., H. KURIYAMA, and H. FUKUDA. 2005. Inhibition of transdifferentiation into tracheary elements by polar auxin transport inhibitors through intracellular auxin depletion. *Plant Cell Physiol.* 46, 2019–2028.

# CHAPTER 12

ESPINOSA-RUIZ, A., S. SAXENA, J. SCHMIDT, E. MELLEROWICZ, P. MISKOLCZI, L. BAKO, and R. P. BHALERAO. 2004. Differential stage-specific regulation of cyclin-dependent kinases during cambial dormancy in hybrid aspen. *Plant J.* 38, 603–615. (**Cambial dormancy in woody plants can be divided into two stages, ecodormant and endodormant. Whereas trees in the ecodormant state resume growth upon exposure to growth-promotive signals, those in the endodormant state fail to respond to such signals. Results of this study, in which the regulation of cyclin-dependent kinases were analyzed, indicate that the eco- and endodormant stages of cambial dormancy involve a stage-specific regulation of the cell cycle effectors at multiple levels.**)

KOJS, P., A. RUSIN, M. IQBAL, W. WŁOCH, and J. JURA. 2004. Readjustments of cambial initials in *Wisteria floribunda* (Willd.) DC. for development of storeyed structure. *New Phytol.* 163, 287–297. (**The mechanism of storeyed cambial structure formation in *W. floribunda* involves both anticlinal cell divisions and concomitant intrusive growth of the ends of cambial cells of one cell packet along the tangential walls of cells of the neighboring packet.**)

LEÓN-GÓMEZ, C., and A. MONROY-ATA. 2005. Seasonality in cambial activity of four lianas from a Mexican lowland tropical rainforest. *IAWA J.* 26, 111–120. (**In all four species—*Machaerium cobanense*, *M. floribundum*, *Gouania lupuloides*, and *Trichostigma octandrum*—the cambium is active throughout the year. In all but *T. octandrum*, cambial activity was higher in the rainy period than in the dry period. Cambial activity in *T. octandrum* was not significantly associated with the wet or dry season.**)

MWANGE, K.-N'K., H.-W. HOU, Y.-Q. WANG, X.-Q. HE, and K.-M. CUI. 2005. Opposite patterns in the annual distribution and time-course of endogenous abscisic acid and indole-3-acetic acid in relation to the periodicity of cambial activity in *Eucommia ulmoides* Oliv. *J. Exp. Bot.* 56, 1017–1028. (**In the active period [AP], ABA underwent an abrupt decease, reaching its lowest level in the summer. It peaked in the winter. IAA showed a reverse pattern to that of ABA: it sharply increased in the AP but noticeably decreased with the commencement of the first quiescence [in autumn]. Laterally most of the ABA was located in mature tissues, whereas the IAA essentially was located in the cambial region. Results of experimental studies suggest that in *E. ulmoides*, ABA and IAA probably interact in the cambial region.**)

MYSKOW, E., and B. ZAGORSKA-MAREK. 2004. Ontogenetic development of storied ray pattern in cambium of *Hippophae rhamnoides* L. *Acta Soc. Bot. Pol.* 73, 93–101. (**The development of the storied arrangement of both the fusiform initials and rays in *Hippophae rhamnoides* involves, first, anticlinal longitudinal divisions and restricted intrusive growth of fusiform initials, followed by the initiation of secondary rays largely from segmentation of fusiform initials. Highly controlled vertical migration of rays on the cambial surface also contributes to a storied pattern.**)

SCHRADER, J., R. MOYLE, R. BHALERAO, M. HERTZBERG, J. LUNDEBERG, P. NILSSON, and R. P. BHALERAO. 2004. Cambial meristem dormancy in trees involves extensive remodelling of the transcriptome. *Plant J.* 40, 173–187. (**Purified active and dormant cambial cells from *Populus tremula* were used to generate meristem-specific cDNA libraries and for microarray experiments to define global transcriptional changes underlying cambial dormancy. A significant reduction was found in the complexity of the cambial transcriptome in the dormant state. Among others, the findings indicate that the cell cycle machinery is maintained in a skeletal state in the dormant cambium, and that down-regulation of *PttPIN1* and *PttPIN2* transcripts explains the reduced basipetal polar auxin transport during dormancy.**)

# CHAPTERS 13 AND 14

AMIARD, V., K. E. MUEH, B. DEMMIG-ADAMS, V. EBBERT, R. TURGEON, and W. W. ADAMS, III. 2005. Anatomical and photosynthetic acclimation to the light environment in species with differing mechanisms of phloem loading. *Proc. Natl. Acad. Sci. USA* 102, 12968–12973. (**The ability to regulate photosynthesis in response to the light environment [growth under low light or high light or when transferred from low to high light] was compared between apoplastic loaders [pea and spinach] and symplastic loaders [pumpkin and *Verbascum phoeniceum*]).**)

AYRE, B. G., F. KELLER, and R. TURGEON. 2003. Symplastic continuity between companion cells and the translocation stream: long-distance transport is controlled by retention and retrieval mechanisms in the phloem. *Plant Physiol.* 131, 1518–1528. (**A model is proposed in which the transport of oligosaccharides is an adaptive strategy to improve retention of photoassimilate, and thus efficiency of translocation, in the phloem.**)

BARLOW, P. 2005. Patterned cell determination in a plant tissue: the secondary phloem of trees. *BioEssays* 27, 533–541. (**It is hypothesized that in conjunction with the positional values conferred by the graded radial distribution of auxin, cell divisions at particular positions within the cambium are sufficient to determine not only each of the phloem cell types but also their recurrent pattern of differentiation within each radial file.**)

Bové, J. M., and M. Garnier. 2003. Phloem- and xylem-restricted plant pathogenic bacteria. *Plant Sci.* 164, 423–438. **(Review)**

Carlsbecker, A., and Y. Helariutta. 2005. Phloem and xylem specification: pieces of the puzzle emerge. *Curr. Opin. Plant Biol.* 8, 512–517. **(Review)**

Dunisch, O., M. Schulte, and K. Kruse. 2003. Cambial growth of *Swietenia macrophylla* King studied under controlled conditions by high resolution laser measurements. *Holzforschung* 57, 196–206. **(Radial cell enlargement after cambial dormancy occurred first in sieve tubes with contact to ray parenchyma cells; on the xylem side, radial cell enlargement of vessels and paratracheal parenchyma was induced almost simultaneously along the shoot circumference. Radial cell enlargement of phloem and xylem derivatives formed first after cambial reactivation was induced almost simultaneously along the shoot axis.)**

Franceschi, V. R., P. Krokene, E. Christiansen, and T. Krekling. 2005. Anatomical and chemical defenses of conifer bark against bark beetles and other pests. *New Phytol.* 167, 353–376. **(Review)**

Franceschi, V. R., P. Krokene, T. Krekling, and E. Christiansen. 2000. Phloem parenchyma cells are involved in local and distant defense responses to fungal inoculation or bark-beetle attack in Norway spruce (Pinaceae). *Am. J. Bot.* 87, 314–316.

Garnier, M., S. Jagoueix-Eveillard, and X. Foissac. 2003. Walled bacteria inhabiting the phloem sieve tubes. *Recent Res. Dev. Microbiol.* 7, 209–223. **(Mallicutes [spiroplasmas and phytoplasmas], which lack a cell wall, are the most common bacteria inhabiting sieve tubes. Walled bacteria also inhabit sieve tubes. They belong to different subclasses of the subdivision Proteobacteria. This paper presents an overview of the phloem-restricted Proteobacteria.)**

Gould, N., M. R. Thorpe, O. Koroleva, and P. E. H. Minchin. 2005. Phloem hydrostatic pressure relates to solute loading rate: a direct test of the Münch hypothesis. *Funct. Plant Biol.* 32, 1019–1026. **(The role of solute uptake in creating the hydrostatic pressure associated with phloem flow was tested in mature leaves of barley and sow thistle.)**

Hancock, R. D., D. McRae, S. Haupt, and R. Viola. 2003. Synthesis of L-ascorbic acid in the phloem. *BMC Plant Biol.* 3 (7) [13 pp.]. **(Active L-ascorbic acid synthesis was detected in the phloem-rich vascular exudates of *Cucurbita pepo* fruits and demonstrated in isolated phloem strands from *Apium graveolens*.)**

Höltta, T., T. Vesala, S. Sevanto, M. Perämäki, and E. Nikinmaa. 2006. Modeling xylem and phloem water flows in trees according to cohesion theory and Münch hypothesis. *Trees* 20, 67–78. **(Water and solute flows in the coupled system of xylem and phloem were modeled together, with predictions for xylem and whole stem diameter changes. With the model, the authors were able to produce water circulation between xylem and phloem as presented by the Münch hypothesis.)**

Hsu, Y.-S., S.-J. Chen, C.-M. Lee, and L.-L. Kuo-Huang. 2005. Anatomical characteristics of the secondary phloem in branches of *Zelkova serrata* Makino. *Bot. Bull. Acad. Sin.* 46, 143–149. **(No obvious difference in thickness existed between the secondary phloem in the upper side [reaction phloem] and lower side [opposite phloem] of leaning branches. Gelatinous fibers, which occurred in both reaction phloem and opposite phloem, formed earlier and occupied a larger area ratio in the upper side. In addition, the sieve elements in the upper side were longer and wider than those in the lower side.)**

Langhans, M., R. Ratajczak, M. Lützelschwab, W. Michalke, R. Wächter, E. Fischer-Schliebs, and C. I. Ullrich. 2001. Immunolocalization of plasma-membrane $H^+$-ATPase and tonoplast-type pyrophosphatase in the plasma membrane of the sieve element-companion cell complex in the stem of *Ricinus communis* L. *Planta* 213, 11–19. **(The plasma-membrane [PM] $H^+$-ATPase and the tonoplast-type pyrophosphatase [PPase] were immunolocalized by epifluorescence and confocal laser scanning microscopy [CLSM] upon single or double labeling with specific monoclonal and polyclonal antibodies. Quantitative fluorescence evaluation by CLSM revealed both pumps simultaneously in the sieve-element PM.)**

Lough, T. J., and W. J. Lucas. 2006. Integrative plant biology: role of phloem long-distance macromolecular trafficking. *Annu. Rev. Plant Biol.* 57. **On line. (Review)**

Machado, S. R., C. R. Marcati, B. Lange de Morretes, and V. Angyalossy. 2005. Comparative bark anatomy of root and stem in *Styrax camporum* (Styracaceae) *IAWA J.* 26, 477–487.

Minchin, P. E. H., and A. Lacointe. 2005. New understanding on phloem physiology and possible consequences for modelling long-distance carbon transport. *New Phytol.* 166, 771–779. **(Review)**

Narváez-Vasquez, J., and C. A. Ryan. 2004. The cellular localization of prosystemin: a functional role for phloem parenchyma in systemic wound signaling. *Planta* 218, 360–369. **(The phloem parenchyma cells in the leaf blades, petioles, and stems of *Lycopersicon esculentum* [*Solanum lycopersicum*] are the sites for the synthesis and processing of prosystemin, the first line of defense signaling in response to herbivore and pathogen attacks.)**

Thompson, M. V. 2006. Phloem: the long and the short of it. *Trends Plant Sci.* 11, 26–32. **(The author presents three metaphors for phloem transport intended to help construct an accurate theoretical framework of the temporal, long-distance behavior of the phloem—a framework that does not depend on a turgor differential as an important control variable. Noting that** the molecular regulation of phloem solute exchange will make sense only in light of its anatomy-dependent, long-distance behavior, the author emphasizes the need of a significant recommitment to studying the quantitative anatomy of phloem.)

TURGEON, R. 2006. Phloem loading: how leaves gain their independence. *BioScience* 56, 15–24. **(Review)**

VAN BEL, A. J. E. 2003. The phloem, a miracle of ingenuity. *Plant Cell Environ.* 26, 125–149. **(Review)**

VOITSEKHOVSKAJA, O. V., O. A. KOROLEVA, D. R. BATASHEV, C. KNOP, A. D. TOMOS, Y. V. GAMALEI, H.-W. HELDT, and G. LOHAUS. 2006. Phloem loading in two Scrophulariaceae species. What can drive symplastic flow via plasmodesmata? *Plant Physiol.* 140, 383–395. **(It is concluded that in both *Alonsoa meridionalis* and *Asarina barclaiana* apoplastic phloem loading is an indispensable mechanism and that symplastic entrance of solutes into the phloem may occur by mass flow.)**

VON ARX, G., and H. DIETZ. 2006. Growth rings in the roots of temperate forbs are robust annual markers. *Plant Biol.* 8, 224–233. **(The growth rings in the secondary xylem of the roots of northern temperate forbs were found to represent robust annual growth increments. Hence, they can reliably be used in herb-chronological studies of age- and growth-related questions in plant ecology.)**

WALZ, C., P. GIAVALISCO, M. SCHAD, M. JUENGER, J. KLOSE, and J. KEHR. 2004. Proteomics of cucurbit phloem exudate reveals a network of defence proteins. *Phytochemistry* 65, 1795–1804. **(A total of 45 different proteins were identified from phloem exudates of *Cucumis sativus* and *Cucurbita maxima*; the majority of these are involved in stress and defense reactions.)**

WU, H., and X.-F. ZHENG. 2003. Ultrastructural studies on the sieve elements in root protophloem of *Arabidopsis thaliana*. *Acta Bot. Sin.* 45, 322–330.

ZHANG, L.-Y., Y.-B. PENG, S. PELLESCHI-TRAVIER, Y. FAN, Y.-F. LU, Y.-M. LU, X.-P. GAO, Y.-Y. SHEN, S. DELROT, and D.-P. ZHANG. 2004. Evidence for apoplasmic phloem unloading in developing apple fruit. *Plant Physiol.* 135, 574–586. **(Structural and experimental data clearly indicate that phloem unloading in the apple fruit is apoplastic and provide information on both the structural and molecular features involved in the process.)**

## CHAPTER 15

LANGENFELD-HEYSER, R. 1997. Physiological functions of lenticels. In: *Trees—Contributions to Modern Tree Physiology*, pp. 43–56. H. Rennenberg, W. Eschrich, and H. Ziegler, eds. Backhuys Publishers, Leiden, The Netherlands.

MANCUSO, S., and A. M. MARRAS. 2003. Different pathways of the oxygen supply in the sapwood of young *Olea europaea*

trees. *Planta* 216, 1028–1033. **(In the daylight hours, almost all the oxygen present in the sapwood was delivered by the transpiration stream, not by gaseous transport via lenticels.)**

WAISEL, Y. 1995. Developmental and functional aspects of the periderm. In: *The Cambial Derivatives*, pp. 293–315, M. Iqbal, ed. Gebrüder Borntraeger, Berlin.

## CHAPTERS 16 AND 17

BIRD, D. A., V. R. FRANCESCHI, and P. J. FACCHINI. 2003. A tale of three cell types: alkaloid biosynthesis is localized to sieve elements in opium poppy. *Plant Cell* 15, 2626–2635. **(Immunofluorescence labeling using affinity-purified antibodies showed that three key enzymes, one of which is codeinone reductase, involved in the biosynthesis of morphine and the related alkaloid sanguinarine are restricted to the parietal region of sieve elements adjacent or proximal to laticifers.)**

CARTER, C., S. SHAFIR, L. YEHONATAN, R. G. PALMER, and R. THORNBURG. 2006. A novel role for proline in plant floral nectars. *Naturwissenschaften* 93, 72–79. **(The nectar of ornamental tobacco and two insect pollinated wild perennial species of soybean were found to contain high levels of proline. Because insects such as honeybees prefer proline-rich nectars, the authors hypothesize that some plants offer such nectars as a mechanism to attract visiting pollinators.)**

CHEN, C.-C., and Y.-R. CHEN. 2005. Study on laminar hydathodes of *Ficus formosana* (Moraceae). I. Morphology and ultrastructure. *Bot. Bull. Acad. Sin.* 46, 205–215. **(The laminar hydathodes of *F. formosana* are distributed in two linear rows, one each between margin and midrib, on the adaxial surface. They are located on meshworks of veins with several vein ends, and consist of epithem, tracheids, a bounding sheath layer, and water pores [permanently opened guard cells]. Numerous invaginations of the plasma membrane were observed in the epithem cells, indicative of endocytosis.)**

DAVIES, K. L., M. STPICZYŃSKA, and A. GREGG. 2005. Nectar-secreting floral stomata in *Maxillaria anceps* Ames & C. Schweinf. (Orchidaceae). *Ann. Bot.* 96, 217–227. **(Nectar appears as droplets that are exuded by modified stomata whose apertures become almost completely covered by a cuticular layer.)**

DE LA BARRERA, E., and P. S. NOBEL. 2004. Nectar: properties, floral aspects, and speculations on origin. *Trends Plant Sci.* 9, 65–69.

EL MOUSSAOUI, A., M. NIJS, C. PAUL, R. WINTJENS, J. VINCENTELLI, M. AZARKAN, and Y. LOOZE. 2001. Revisiting the enzymes stored in the laticifers of *Carica papaya* in the context of their possible participation in the plant defence mechanism. *Cell. Mol. Life Sci.* 58, 556–570. **(Review. Golgi vesicles may contribute to a granulocrine process.)**

FEILD, T. S., T. L. SAGE, C. CZERNIAK, and W. J. D. ILES. 2005. Hydathodal leaf teeth of *Chloranthus japonicus* (Chloranthaceae) prevent guttation-induced flooding of the mesophyll. *Plant Cell Environ.* 28, 1179–1190.

HORNER, H. T., R. A. HEALY, T. CERVANTES-MARTINEZ, and R. G. PALMER. 2003. Floral nectary fine structure and development in *Glycine max* L. (Fabaceae). *Int. J. Plant Sci.* 164, 675–690. (**The nectaries exhibit holocrine secretion different from that reported for other legume taxa and most other nonlegume taxa.**)

KLEIN, D. E., V. M. GOMES, S. J. DA SILVA-NETO, and M. DA CUNHA. 2004. The structure of colleters in several species of *Simira* (Rubiaceae). *Ann. Bot.* 94, 733–740. (**The colleters in each of the species examined show a different pattern of distribution and have taxonomic importance below the genus level.**)

KOLB, D., and M. MÜLLER. 2004. Light, conventional and environmental scanning electron microscopy of the trichomes of *Cucurbita pepo* subsp. *pepo* var. *styriaca* and histochemistry of glandular secretory products. *Ann. Bot.* 94, 515–526. (**Four different types of trichome occur on the leaves of the Styrian oil pumpkin [*Cucurbita pepo* var. *styriaca*]; three are glandular and one nonglandular. The three glandular trichomes are capitate. The nonglandular trichome is described as "columnar-digit." Histochemical reactions revealed that the secretion material contained terpenes, flavones, and lipids.**)

LEITÃO, C. A. E., R. M. S. A. MEIRA, A. A. AZEVEDO, J. M. DE ARAÚJO, K. L. F. SILVA, and R. G. COLLEVATTI. 2005. Anatomy of the floral, bract, and foliar nectaries of *Triumfetta semitriloba* (Tiliaceae). *Can. J. Bot.* 83, 279–286. (**The nectaries of *T. semitriloba* are of a specialized type. A secretory epidermis, consisting of pluricellular and multiserial nectariferous trichomes, covers a nectariferous parenchyma vascularized by phloem and xylem.**)

MONACELLI, B., A. VALLETTA, N. RASCIO, I. MORO, and G. PASQUA. 2005. Laticifers in *Camptotheca acuminata* Decne: distribution and structure. *Protoplasma* 226, 155–161. (**Laticifers are reported for the first time in a member [*Camptotheca acuminata*] of the Nyssaceae. They are nonarticulated unbranched laticifers and occur in leaf and stem. None were found in the roots.**)

PILATZKE-WUNDERLICH, I., and C. L. NESSLER. 2001. Expression and activity of cell-wall-degrading enzymes in the latex of opium poppy, *Papaver somniferum* L. *Plant Mol. Biol.* 45, 567–576. (**An abundance of transcripts encoding latex-specific pectin-degrading enzymes was found in the articulated laticifers of *Papaver somniferum*. These enzymes apparently play an important role in development of the laticifers.**)

RUDGERS, J. A. 2004. Enemies of herbivores can shape plant traits: selection in a facultative ant-plant mutualism. *Ecology* 85, 192–205. (**Experimental evidence indicates that ant associates can influence the evolution of extrafloral nectary traits.**)

RUDGERS, J. A., and M. C. GARDENER. 2004. Extrafloral nectar as a resource mediating multispecies interactions. *Ecology* 85, 1495–1502.

SERPE, M. D., A. J. MUIR, C. ANDÈME-ONZIGHI, and A. DRIOUICH. 2004. Differential distribution of callose and a $(1{\rightarrow}4)\beta$-D-galactan epitope in the laticiferous plant *Euphorbia heterophylla* L. *Int. J. Plant Sci.* 165, 571–585. (**The nonarticulated laticifer walls differ from those of their surrounding cells. For example, the level of a $(1{\rightarrow}4)\beta$-D-galactan epitope was much lower in laticifers than in other cells, and an anti-$(1{\rightarrow}3)\beta$-D-galactan antibody that recognizes callose did not label laticifer walls and walls immediately adjacent to them. The antibody did, however, produce a punctuated labeling pattern in most other cells.**)

SERPE, M.D., A. J. MUIR, and A. DRIOUICH. 2002. Immunolocalization of $\beta$-D-glucans, pectins, and arabinogalactan-proteins during intrusive growth and elongation of nonarticulated laticifers in *Asclepias speciosa* Torr. *Planta* 215, 357–370. (**Laticifer elongation is associated with development of a homogalacturonan-rich middle lamella between laticifers and their neighboring cells. In addition the walls deposited by the laticifers differ from those of the cells surrounding them. These and other results indicate that laticifer penetration causes changes in the walls of meristematic cells and that there are differences in wall composition within laticifers and between laticifers and their surrounding cells.**)

SERPE, M. D., A. J. MUIR, and A. M. KEIDEL. 2001. Localization of cell-wall polysaccharides in nonarticulated laticifers of *Asclepias speciosa* Torr. *Protoplasma* 216, 215–226. (**The nonarticulated laticifers of *Asclepias speciosa* have distinctive cytochemical properties that change along their length.**)

WAGNER, G. J., E. WANG, and R. W. SHEPHERD. 2004. New approaches for studying and exploiting an old protuberance, the plant trichome. *Ann. Bot.* 93, 3–11. (**Botanical briefing**)

WEID, M., J. ZIEGLER, and T. M. KUTCHAN. 2004. The roles of latex and the vascular bundle in morphine biosynthesis in the opium poppy, *Papaver somniferum. Proc. Natl. Acad. Sci. USA* 101, 13957–13962. (**Immunolocalization of five enzymes involved in alkaloid formation is reported. Codeinone reductase localized to the laticifers, the site of morphinan alkaloid accumulation.**)

WIST, T. J., and A. R. DAVIS. 2006. Floral nectar production and nectary anatomy and ultrastructure of *Echinacea purpurea* (Asteraceae). *Ann. Bot.* 97, 177–193. (**The floral nectaries of *Echinacea purpurea* were supplied by phloem only. Both sieve elements and companion cells were found adjacent to the epidermis. The abundance of mitochondria in the nectaries suggests an eccrine mechanism of secretion, although Golgi vesicles may contribute to a granulocrine process.**)

# Glossary

**A**

**abaxial**  Directed away from the axis. Opposite of *adaxial*. With regard to a leaf, the lower, or "dorsal," surface.

**accessory bud**  A bud located above or on either side of the main axillary bud.

**accessory cell**  See *subsidiary cell*.

**acicular crystal**  Needle-shaped crystal.

**acropetal development** (or **differentiation**)  Produced or becoming differentiated in a succession toward the apex of an organ. The opposite of basipetal but means the same as *basifugal*.

**actin filament**  A helical protein filament, 5 to 7 nanometers (nm) thick, composed of globular actin molecules; a major constituent of all eukaryotic cells. Also called *microfilament*.

**actinocytic stoma**  Stoma surrounded by a circle of radiating cells.

**adaxial**  Directed toward the axis. Opposite of *abaxial*. With regard to a leaf, the upper, or "ventral," surface.

**adventitious**  Refers to structures arising not at their usual sites, as roots originating on stems or leaves instead of on other roots, buds developing on leaves or roots instead of in leaf axils on shoots.

**aerenchyma**  Parenchyma tissue containing particularly large intercellular spaces of *schizogenous, lysigenous,* or *rhexigenous* origin.

**aggregate ray**  In secondary vascular tissues; a group of small rays arranged so as to appear to be one large ray.

**albuminous cell**  See *Strasburger cell*.

**aleurone**  Granules of protein (aleurone grains) present in seeds, usually restricted to the outermost layer, the *aleurone layer* of the endosperm. (*Protein bodies* is the preferred term for aleurone grains.)

**aleurone layer**  Outermost layer of endosperm in cereals and many other taxa that contains protein bodies and enzymes concerned with endosperm digestion.

**aliform paratracheal parenchyma**  In secondary xylem; vasicentric groups of axial parenchyma cells having tangential wing-like extensions as seen in transverse section. See also *paratracheal parenchyma* and *vasicentric paratracheal parenchyma*.

**alternate pitting**  In tracheary elements; pits in diagonal rows.

**amyloplast**   A colorless plastid (*leucoplast*) that forms starch grains.

**anastomosis**   Refers to cells or strands of cells that are interconnected with one another as, for example, the veins in a leaf.

**analogy**   Means having the same function as but a different phylogenetic origin than another entity.

**anatomy**   The study of the internal structure of organisms; *morphology* is the study of their external structure.

**angiosperm**   A group of plants whose seeds are borne within a mature ovary (fruit).

**angstrom**   (originally *ångström*) A unit of length equal to one-tenth of a nanometer (nm). Symbol A or Å.

**angular collenchyma**   A form of collenchyma in which the primary wall thickening is most prominent in the angles where several cells are joined.

**anisocytic stoma**   A stomatal complex in which three subsidiary cells, one distinctly smaller than the other two, surround the stoma.

**anisotropic**   Having different properties along different axes; optical anisotropy causes polarization and double refraction of light.

**annual ring**   In secondary xylem; growth ring formed during one season. The term is deprecated because more than one growth ring may be formed during a single year.

**annular cell wall thickening**   In tracheary elements of the xylem; secondary wall deposited in the form of rings.

**anomalous secondary growth**   A term of convenience referring to types of secondary growth that differ from the more familiar ones.

**anomocytic stoma**   A stoma without subsidiary cells.

**anther**   Pollen-bearing part of the stamen.

**anthocyanin**   A water-soluble blue, purple, or red flavonoid pigment occurring in the vacuolar cell sap.

**Anthophyta**   The phylum of angiosperms, or flowering plants.

**anticlinal**   Commonly refers to orientation of cell wall or plane of cell division; perpendicular to the nearest surface. Opposite of *periclinal*.

**apex** (pl. **apices**), or **summit**   Tip, topmost part, pointed end of anything. In shoot or root the tip containing the apical meristem.

**apical cell**   Single cell that occupies the distal position in an apical meristem of root or shoot and is usually interpreted as the initial cell in the apical meristem; typical of seedless vascular plants.

**apical dominance**   Influence exerted by a terminal bud in suppressing the growth of lateral, or axillary, buds.

**apical meristem**   A group of meristematic cells at the apex of root or shoot that by cell division produces the precursors of the primary tissues of root or shoot; may be *vegetative*, initiating vegetative tissues and organs, or *reproductive*, initiating reproductive tissues and organs.

**apoplast**   Cell wall continuum and intercellular spaces of a plant or plant organ; the movement of substances via the cell walls is called *apoplastic movement* or *transport*.

**apoptosis**   Programmed cell death in animal cells mediated by a group of protein-degrading enzymes called caspases; involves a programmed series of events that leads to dismantling of the cell contents.

**apotracheal parenchyma**   In secondary xylem; axial parenchyma typically independent of the vessels (pores). Includes *diffuse* and *diffuse-in-aggregates*.

**apposition**   Growth of cell wall by successive deposition of wall material, layer upon layer. Opposite of *intussusception*.

**articulated laticifer**   Laticifer composed of more than one cell with common walls intact or partly or entirely removed; anastomosing or nonanastomosing; a *compound laticifer*.

**aspirated pit**   In gymnosperm wood; bordered pit in which the pit membrane is laterally displaced and the torus blocks the aperture.

**astrosclereid**   A branched, or ramified, type of sclereid.

**axial organ**   Root, stem, inflorescence, or flower axis without its appendages.

**axial parenchyma**   Parenchyma cells in the axial system of secondary vascular tissues; as contrasted with ray parenchyma cells.

**axial system**   All secondary vascular cells derived from the fusiform cambial initials and oriented with their longest diameter parallel with the main axis of stem or root. Other terms: *vertical system* and *longitudinal system*.

**axial tracheid**   Tracheid in the axial system of secondary xylem; as contrasted with ray tracheid.

**axil**   Upper angle between a stem and a twig or a leaf.

**axillary bud**   Bud in the axil of a leaf.

**axillary meristem**   Meristem located in the axil of a leaf and giving rise to an axillary bud.

**B**

**banded parenchyma**   In secondary xylem; axial parenchyma in concentric bands as seen in transverse section, mainly independent (apotracheal) of vessels (pores).

**bark**   A nontechnical term applied to all tissues outside the vascular cambium or the xylem; in older trees may be divided into dead outer bark and living inner bark, which consists of secondary phloem. See also *rhytidome*.

**bars of Sanio**   See *crassulae*.

**basifugal development** See *acropetal development*.

**basipetal development** (or **differentiation**) Produced or becoming differentiated in a succession toward the base of an organ. The opposite of *acropetal* and *basifugal*.

**bast fiber** Originally phloem fiber, now any extraxylary fiber.

**bicollateral vascular bundle** A bundle having phloem on two sides of the xylem.

**biseriate ray** A ray in secondary vascular tissue, two cells in width.

**blind pit** A pit without a complementary pit in an adjacent wall, which may face a lumen of a cell or an intercellular space.

**bordered pit** A pit in which the secondary wall overarches the pit membrane.

**bordered pit-pair** An intercellular pairing of bordered pits.

**boundary parenchyma** See *marginal bands*.

**brachysclereid** A short, roughly isodiametric sclereid, resembling a parenchyma cell in shape; a *stone cell*.

**branch gap** In the nodal region of a stem; a region of parenchyma in the vascular cylinder of the stem located where the branch traces are bent toward the branch. Usually confluent with the gap of the leaf subtending the branch.

**branch root** See *lateral root*.

**branch traces** Vascular bundles connecting the vascular tissue of the branch and that of the main stem. They are leaf traces of the first leaves (prophylls) on the branch.

**branched pit** See *ramiform pit*.

**bulliform cell** An enlarged epidermal cell present, with other similar cells, in longitudinal rows in leaves of grasses. Also called *motor cell* because of its presumed participation in the mechanism of rolling and unrolling of leaves.

**bundle cap** Sclerenchyma or collenchymatous parenchyma appearing like a cap on the xylem and/or phloem side of a vascular bundle as seen in transverse section.

**bundle sheath** Layer or layers of cells enclosing a vascular bundle in a leaf; may consist of parenchyma or sclerenchyma.

**bundle sheath extension** A plate of ground tissue extending from a bundle sheath to the epidermis in a leaf; may be present on one or on both sides of the bundle and may consist of parenchyma or sclerenchyma.

## C

**callose** A polysaccharide, β-1,3 glucan, yielding glucose on hydrolysis. Common wall constituent in the sieve areas of sieve elements; also develops rapidly in reaction to injury in sieve elements and parenchyma cells.

**callus** A tissue composed of large thin-walled cells developing as a result of injury, as in wound healing or grafting, and in tissue culture. (The use of callus for accumulations of callose on sieve areas is deprecated.)

**callus tissue** See *callus*.

**calyptrogen** In root apex; meristem giving rise to the rootcap independently of the initials of cortex and central cylinder.

**cambial initials** Cells so localized in the vascular cambium or phellogen that their periclinal divisions can contribute cells either to the outside or to the inside of the axis; in vascular cambium, classified into *fusiform initials* (source of axial cells of xylem and phloem) and *ray initials* (source of the ray cells).

**cambium** A meristem with products of periclinal divisions commonly contributed in two directions and arranged in radial files. Term preferably applied only to the two lateral meristems, the *vascular cambium* and the *cork cambium*, or *phellogen*.

**Casparian strip**, or **band** A band-like wall formation within primary walls that contains suberin and lignin; typical of endodermal cells in roots, in which it occurs in radial and transverse anticlinal walls.

**cell** Structural and physiological unit of a living organism. The plant cell consists of protoplast and cell wall; in nonliving state, of cell wall only, or cell wall and some nonliving inclusions.

**cell plate** A partition appearing at telophase between the two nuclei formed during mitosis (and some meioses) and indicating the early stage of the division of a cell (*cytokinesis*) by means of a new cell wall; is formed in the *phragmoplast*.

**cell wall** More or less rigid outermost layer of plant cells, which encloses the protoplast. In higher plants, composed of cellulose and other organic and inorganic substances.

**cellulose** A polysaccharide, β-1,4 glucan—the main component of cell walls in most plants; consists of long chain-like molecules whose basic units are anhydrous glucose residues of the formula $C_6H_{10}O_5$.

**central cylinder** A term of convenience applied to the vascular tissues and associated ground tissue in stem and root. Refers to the same part of stem and root that is designated *stele*.

**central mother cells** Rather large vacuolated cells in subsurface position in apical meristem of shoot in gymnosperms.

**centrifugal development** Produced or developing successively farther away from the center.

**centripetal development** Produced or developing successively closer to the center.

**chimera** A shoot apical meristem composed of cells of different genotypes. In *periclinal chimeras*, cells of different genetic composition are arranged in periclinal layers.

**chlorenchyma** Parenchyma tissue containing chloroplasts; leaf mesophyll and other green parenchyma.

**chloroplast** A chlorophyll-containing plastid with thylakoids organized into grana and intergrana (or stroma) thylakoids, and embedded in a stroma.

**chromatolysis** Nuclear degeneration involving the loss of stainable contents (chromatin and nucleoli) and eventual rupture of the nuclear envelope.

**chromoplast** A plastid containing pigments other than chlorophyll, usually yellow and orange carotenoid pigments.

**circular bordered pit** A bordered pit with circular aperture.

**cisterna** (pl. **cisternae**) A flattened, saclike membranous compartment as in endoplasmic reticulum, Golgi body, or thylakoid.

**collateral vascular bundle** A bundle having phloem only on one side of the xylem, usually the abaxial side.

**collenchyma** A supporting tissue composed of more or less elongated living cells with unevenly thickened, nonlignified primary walls. Common in regions of primary growth in stems and leaves.

**colleter** A multicellular appendage (*emergence*) formed from both epidermal and subepidermal tissues. They produce sticky secretions, and are common on bud scales and young leaves.

**columella** Central part of a rootcap in which the cells are arranged in longitudinal files.

**companion cell** A specialized parenchyma cell associated with a sieve-tube element in angiosperm phloem and arising from the same mother cell as the sieve-tube element.

**complementary tissue** See *filling tissue.*

**complex tissue** A tissue consisting of two or more cell types; epidermis, periderm, xylem, and phloem are complex tissues.

**compound laticifer** See *articulated laticifer.*

**compound middle lamella** A collective term applied to two primary walls and middle lamella; usually used when the true middle lamella is not distinguishable from the primary walls. May also include the earliest secondary wall layers.

**compound sieve plate** A sieve plate composed of several sieve areas in either scalariform or reticulate arrangement.

**compression wood** Reaction wood in conifers, which is formed on the lower sides of branches and leaning or crooked stems and characterized by dense structure, strong lignification and certain other features.

**conducting tissue** See *vascular tissue.*

**confluent paratracheal parenchyma** In secondary xylem; coalesced aliform groups of axial parenchyma cells forming irregular tangential or diagonal bands, as seen in transverse section. See also *para-*

*tracheal parenchyma* and *aliform paratracheal parenchyma.*

**contact cell** A paratracheal parenchyma cell or a ray parenchyma cell in direct contact with the vessels and physiologically associated with them. Analogous to companion cell in the phloem.

**coordinated growth** Growth of cells in a manner that involves no separation of walls, as opposed to *intrusive growth.*

**cork** See *phellem.*

**cork cambium** See *phellogen.*

**cork cell** A phellem cell derived from the phellogen, nonliving at maturity, and having suberized walls; protective in function because the walls are highly impervious to water.

**corpus** The core in an apical meristem covered by the tunica and showing volume growth by divisions of cells in various planes.

**cortex** Primary ground tissue region between the vascular system and the epidermis in stem and root. Term also used with reference to peripheral region of a cell protoplast.

**cotyledon** Seed leaf; generally absorbs food in monocotyledons and stores food in other angiosperms.

**crassulae** (sing. **crassula**) Thickenings of intercellular material and primary wall along the upper and lower margins of a pit-pair in the tracheids of gymnosperms. Also called *bars of Sanio.*

**cristae** (sing. **crista**) Crest-like infoldings of the inner membrane in a mitochondrion.

**cross-field** A term of convenience for the rectangle formed by the walls of a ray cell against an axial tracheid; as seen in radial section of the secondary xylem of conifers.

**crystal sand** A mass of very fine free crystals.

**crystalloid** Protein crystal that is less angular than a mineral crystal and swells in water.

**cuticle** Waxy or fatty layer on outer wall of epidermal cells, formed of cutin and wax.

**cuticularization** Process of formation of the cuticle.

**cutin** A complex fatty substance considerably impervious to water; present in plants as an impregnation of epidermal walls and as a separate layer, the *cuticle*, on the outer surface of the epidermis.

**cutinization** Process of impregnation with cutin.

**cyclocytic stoma** Stoma surrounded by one or two narrow rings of subsidiary cells, numbering four or more. Also called *encyclocytic.*

**cyclosis** Streaming of cytoplasm in a cell.

**cystolith** A concretion of calcium carbonate on an outgrowth of a cell wall. Occurs in a cell called *lithocyst.*

**cytochimera** A chimera having combinations of cell layers with diploid and polyploid nuclei. See also *chimera.*

**cytohistological zonation** Presence of regions in the apical meristem having distinctive cytological characteristics. The term is meant to imply that a cytological zonation results in a subdivision into distinguishable tissue regions.

**cytokinesis**   The process of division of a cell as distinguished from the division of the nucleus, or *karyokinesis (mitosis)*.

**cytological zonation**   See *cytohistological zonation.*

**cytology**   The science dealing with the cell.

**cytoplasm**   Living matter of a cell, exclusive of the nucleus.

**cytoplasmic ground substance**   See *cytosol.*

**cytoskeleton**   Flexible, three-dimensional network of microtubules and actin filaments (microfilaments) within cells.

**cytosol**   Cytoplasmic matrix of the cytoplasm in which the nucleus, various organelles, and membrane systems are embedded. Also referred to as *cytoplasmic ground substance* and *hyaloplasm*.

**D**

**decussate**   Arrangement of leaves in pairs that alternate with one another at right angles.

**dedifferentiation**   A reversal in differentiation of a cell or tissue that is presumed to occur when a more or less completely differentiated cell resumes meristematic activity.

**derivative**   A cell produced by division of a meristematic cell in such a way that it enters the path of differentiation into a body cell; its sister cell may remain in the meristem.

**dermal issue**   See *dermal tissue system.*

**dermal tissue system**   Outer covering tissue of a plant; epidermis or periderm.

**dermatogen**   Meristem forming the epidermis and arising from independent initials in the apical meristem. One of the three histogens, *plerome, periblem,* and *dermatogen,* according to Hanstein.

**desmotubule**   The tubule traversing a plasmodesmatal canal and uniting the endoplasmic reticulum of the two adjacent cells.

**detached meristem**   A meristem, with a potential to give rise to an axillary bud, appearing detached from the apical meristem because of the vacuolation of intervening cells.

**determinate growth**   Growth of limited duration, characteristic of floral meristems and leaves.

**development**   Change in form and complexity of an organism or part of an organism from its beginning to maturity; combined with growth.

**diacytic stoma**   A stomatal complex in which one pair of subsidiary cells, with their common walls at right angles to the long axis of the guard cells, surrounds the stoma.

**diaphragms in pith**   Transverse layers (diaphragms) of firm-walled cells alternating with regions of soft-walled cells that may collapse with age.

**dicotyledons**   Obsolete term used to refer to all angiosperms other than monocotyledons; characterized by having two cotyledons. See also *eudicotyledons* and *magnoliids.*

**dictyosome**   See *Golgi body.*

**differentiation**   A physiological and morphological change occurring in a cell, a tissue, an organ, or a plant during development from a meristematic, or juvenile, stage to a mature, or adult, stage. Usually associated with an increase in specialization.

**diffuse apotracheal parenchyma**   Axial parenchyma in secondary xylem occurring as single cells or as strands distributed irregularly among the fibers, as seen in transverse section. See also *apotracheal parenchyma.*

**diffuse-porous wood**   Secondary xylem in which the pores (vessels) are distributed fairly uniformly throughout a growth layer or change in size gradually from earlywood to latewood.

**dilatation**   Growth of parenchyma by cell division in pith, rays, or axial system in vascular tissues; causes the increase in circumference of bark in stem and root.

**distal**   Farthest from the point of origin or attachment. Opposite of *proximal.*

**distichous**   Arrangement of leaves in two vertical rows; two-ranked arrangement.

**dorsal**   Equivalent to *abaxial* in botanical usage.

**druse**   A globular, compound, calcium-oxalate crystal with numerous crystals projecting from its surface.

**duct**   An elongated space formed by separation of cells from one another (schizogenous origin), by dissolution of cells (lysigenous origin), or by a combination of the two processes (schizolysigenous origin); usually concerned with secretion.

**E**

**earlywood**   Wood formed in first part of a growth layer and characterized by a lower density and larger cells than the latewood. Term replaces *spring wood.*

**eccrine secretion**   Secretion leaves the cell as individual molecules passing through the plasma membrane and cell wall. Compare with *granulocrine secretion.*

**ectodesma**   See *teichode.*

**elaioplast**   A leucoplast type of plastid forming and storing oil.

**embryogenesis** (or **embryogeny**)   Formation of embryo.

**embryoid**   An embryo, often indistinguishable from a normal one, developing not from an egg but from a somatic cell, often in tissue culture.

**encyclocytic**   See *cyclocytic.*

**endocytosis**   Uptake of material into cells by means of invagination of the plasma membrane; if solid material is involved, the process is called phagocytosis; if dissolved material is involved it is called pinocytosis.

**endodermis**   Layer of ground tissue forming a sheath around the vascular region and having the Casparian strip in its anticlinal walls; may have secondary walls later. It is the innermost layer of the cortex in roots and stems of seed plants.

**endodermoid**   Resembling the endodermis.

**endogenous**   Arising from a deep-seated tissue, as a lateral root.

**endomembrane system**   Collectively, the cellular membranes that form a continuum (plasma membrane, tonoplast, endoplasmic reticulum, Golgi bodies, and nuclear envelope).

**endoplasmic reticulum**   (usually abbreviated to ER) A system of membranes forming cisternoid or tubular compartments that permeate the cytosol. The cisternae appear like paired membranes in sectional profiles. The membranes may be coated with ribosomes (rough ER) or be free of ribosomes (smooth ER).

**endoreduplication (endoreplication)**   A DNA replication cycle in which no mitosis-like structural changes take place; during endoreduplication, polytene chromosomes are formed.

**enucleate**   Lacking a nucleus.

**epiblem**   Term used sometimes for the epidermis of the root. See also *rhizodermis*.

**epicotyl**   Upper part of the axis of an embryo or seedling, above the cotyledons (seed leaves) and below the next leaf or leaves. See also *plumule*.

**epidermis**   The outer layer of cells in the plant body, primary in origin. If it is multiseriate (*multiple epidermis*), only the outermost layer differentiates as a typical epidermis.

**epithelium**   A compact layer of cells, often secretory in function, covering a free surface or lining a cavity.

**epithem**   Mesophyll of a hydathode concerned with secretion of water.

**ergastic substances**   Passive products of protoplast such as starch grains, fat globules, crystals, and fluids; occur in cytoplasm, organelles, vacuoles, and cell walls.

**eudicotyledons**   One of two major classes of angiosperms. Eudicotyledones, formerly grouped with the magnoliids, a diverse group of archaic flowering plants, as "dicots"; abbreviated as eudicot.

**eukaryotic**   (also *eucaryotic*) Refers to organisms having membrane-bound nuclei, genetic material organized into chromosomes, and membrane-bound cytoplasmic organelles. Opposite of *prokaryotic*.

**eumeristem**   Meristem composed of relatively small cells, approximately isodiametric in shape, compactly arranged, and having thin walls, a dense cytoplasm, and large nuclei; word means "true meristem."

**exocytosis**   A cellular process in which particulate matter or dissolved substances are enclosed in a vesicle

and transported to the cell surface; there, the membrane of the vesicle fuses with the plasma membrane, expelling the vesicle's contents to the outside.

**exodermis**   Outer layer, one or more cells in depth, of the cortex in some roots; a type of *hypodermis*, the walls of which may be suberized and/or lignified.

**exogenous**   Arising in superficial tissue, as an axillary bud.

**expansins**   A novel class of proteins involved with the loosening of cell wall structure.

**external phloem**   Primary phloem located externally to the primary xylem.

**extrafloral nectary**   Nectary occurring on a plant part other than a flower. See also *nectary*.

**extraxylary fibers**   Fibers in various tissue regions other than the xylem.

**F**

**false annual ring**   One of more than one growth layers formed in the secondary xylem during one growth season, as seen in transverse section.

**fascicle**   A bundle.

**fascicular cambium**   Vascular cambium originating from procambium within a vascular bundle, or fascicle.

**festucoid**   Pertaining to the Festucoideae a subfamily of grasses.

**fiber**   An elongated, usually tapering sclerenchyma cell with a lignified or nonlignified secondary wall; may or may not have a living protoplast at maturity.

**fiber-sclereid**   A sclerenchyma cell with characteristics intermediate between those of a fiber and a sclereid.

**fiber-tracheid**   A fiber-like tracheid in the secondary xylem; commonly thick walled, with pointed ends and bordered pits that have lenticular to slit-like apertures.

**fibril**   Submicroscopic threads composed of cellulose molecules that constitute the form in which cellulose occurs in the wall.

**file meristem**   See *rib meristem*.

**filiform**   Thread-like.

**filiform sclereid**   A much elongated, slender sclereid resembling a fiber.

**filling tissue**   Loose tissue formed by the lenticel phellogen toward the outside; may or may not be suberized. Also called *complementary tissue*.

**flank meristem**   A misnomer used with reference to the peripheral region of an apical meristem. The use of the word flank implies that the entity is two-sided. The term should be replaced with *peripheral meristem*.

**floral nectary**   See *nectary*.

**florigen**   A hypothetical hormone presumed to be concerned with the induction of flowering.

**founder cells**   Group of cells in the peripheral zone of the apical meristem involved with the initiation of a leaf primordium.

**fundamental tissue** See *ground tissue.*

**fundamental tissue system** See *ground tissue system.*

**fusiform cell** An elongated cell tapering at the ends.

**fusiform initial** In vascular cambium; an elongated cell with approximately wedge-shaped ends that gives rise to the elements of the axial system in the secondary vascular tissues.

**G**

**gelatinous fiber** A fiber with a so-called gelatinous layer (G-layer), an innermost secondary wall layer that can be distinguished from the outer secondary wall layer(s) by its high cellulose content and lack of lignin.

**genome** Totality of genetic information contained in the nucleus, plastid, or mitochondrion.

**genomics** Field of genetics that studies the content, organization, and function of genetic information in whole genomes.

**genotype** Genetic constitution of an organism; contrasted with *phenotype.*

**germination** Resumption of growth by the embryo in a seed; also beginning of growth of a spore, pollen grain, bud, or other structure.

**gland** A multicellular secretory structure.

**glandular hair** A trichome having a unicellular or multicellular head composed of secretory cells; usually borne on a stalk of nonglandular cells.

**glyoxysome** A peroxisome containing enzymes necessary for the conversion of fats into carbohydrates.

**Golgi apparatus** A term used to refer collectively to all the Golgi bodies of a given cell. Also called *Golgi complex.*

**Golgi body** A group of flat, disk-shaped sacs, or cisternae, that are often branched into tubules at their margins; serve as collecting and packaging centers for the cell and concerned with secretory activities. Also called *dictyosomes.*

**grana** (sing. **granum**) Subunits of chloroplasts seen as green granules with the light microscope and as stacks of disk-shaped cisternae, the *thylakoids*, with the electron microscope; the grana contain the chlorophylls and carotenoids and are the sites of the light reactions in photosynthesis.

**granulocrine secretion** Secretion passes an inner cytoplasmic membrane, usually that of a vesicle, and is extruded from the cell after the vesicle fuses with the plasma membrane and releases its contents to the outside. Compare with *eccrine secretion.*

**gravitropism** Growth in which the direction is determined by gravity.

**ground meristem** A primary meristem, or meristematic tissue, derived from the apical meristem and giving rise to the ground tissues.

**ground tissue** Tissues other than the vascular tissues, the epidermis, and the periderm. Also called *fundamental tissue.*

**ground tissue system** The total complex of ground tissues of the plant.

**growth** Irreversible increase in size by cell division and/or cell enlargement.

**growth layer** A layer of secondary xylem or secondary phloem produced during a single growth period, which may extend through one season (*annual ring*) or part of one season (*false annual ring*) if more than one layer is formed in one season. Also called *growth increment.*

**growth ring** A growth layer of secondary xylem or secondary phloem as seen in transverse section of stem or root; may be an *annual ring* or a *false annual ring.*

**guard cells** A pair of cells flanking the stomatal pore and causing the opening and closing of the pore by change in turgor.

**gum** A nontechnical term applied to material resulting from breakdown of plant cells, mainly of their carbohydrates.

**gum duct** A duct that contains gum.

**gummosis** A symptom of a disease characterized by the formation of gum, which may accumulate in cavities or ducts or appear on the surface of the plant.

**guttation** Exudation from leaves of water derived from the xylem; caused by root pressure.

**gymnosperm** A seed plant with seeds not enclosed in an ovary; the conifers are the most familiar group.

**H**

**hadrom** (or *hadrome*) The tracheary elements and the associated parenchymatous cells of the xylem tissue; the specifically supporting cells (fibers and sclereids) are excluded. See also *leptom.*

**half-bordered pit-pair** A pit-pair consisting of a bordered and a simple pit.

**haplocheilic stoma** Stomatal type in gymnosperms; subsidiary cells are not related to the guard cells ontogenetically.

**hardwood** A name commonly applied to the wood of a magnoliid or eudicot tree.

**heartwood** Inner layers of secondary xylem that have ceased to function in storage and conduction and in which reserve materials have been removed or converted into heartwood substances; generally darker colored than the functioning *sapwood.*

**helical cell wall thickening** In tracheary elements of the xylem; secondary wall deposited on the primary or secondary wall as a continuous helix. Also referred to as *spiral cell wall thickening.*

**hemicellulose** A general term for a heterogeneous group of noncrystalline glycans that are tightly bound in the cell wall.

**heterocellular ray**   A ray in secondary vascular tissues composed of cells of more than one form; in angiosperms, of procumbent and square or upright cells; in conifers, of parenchyma cells and ray tracheids.

**heterogeneous ray tissue system**   Rays in secondary vascular tissues all heterocellular or combinations of homocellular and heterocellular rays. Term not used for conifers.

**hilum**   (1) The central part of a starch grain around which the layers of starch are arranged concentrically; (2) the scar left by the detached funiculus on a seed.

**histogen**   Hanstein's term for a meristem in shoot or root tip that forms a definite tissue system in the plant body. Three histogens were recognized: *dermatogen*, *periblem*, and *plerome*. See definitions of these terms.

**histogen concept**   Hanstein's concept stating that the three primary tissue systems in the plant—the epidermis, the cortex, and the vascular system with the associated ground tissue—originate from distinct meristems, the histogens, in the apical meristems. See *histogen*.

**histogenesis**   The formation of tissues (hence, *histogenetic*) having to do with origin or formation of tissues.

**histogenetic**   See *histogenesis*.

**homocellular ray**   A ray in secondary vascular tissues composed of cells of one form only: in angiosperms, of procumbent, or square, or upright cells; in conifers, of parenchyma cells only.

**homogeneous ray tissue system**   Rays in secondary vascular tissues all homocellular, composed of procumbent cells only. Term not used for conifers.

**homology**   A condition indicative of the same phylogenetic, or evolutionary, origin, but not necessarily the same in present structure and/or function.

**horizontal parenchyma**   See *ray parenchyma*.

**horizontal system**   See *ray system*.

**hormone**   An organic substance produced usually in minute amounts in one part of an organism, from which it is transported to another part of that organism on which it has a specific effect; hormones function as highly specific chemical signals between cells.

**hyaloplasm**   See *cytosol*.

**hydathode**   A structural modification of vascular and ground tissues, usually in a leaf, that permits the release of water through a pore in the epidermis; may be secretory in function. See *epithem*.

**hydromorphic**   Refers to the structural features of *hydrophytes*.

**hydrophyte**   A plant that requires a large supply of water and may grow partly or entirely submerged in water.

**hygromorphic**   Synonym of *hydromorphic*.

**hyperplasia**   Refers most commonly to an excessive multiplication of cells.

**hypertrophy**   Refers most commonly to abnormal enlargement. Hypertrophy of a cell or its parts involves no cell division. Hypertrophy of an organ may involve both enlargement of cells and abnormal cell multiplication (*hyperplasia*).

**hypocotyl**   Axial part of embryo or seedling located between the cotyledon or cotyledons and the radicle.

**hypocotyl-root axis**   Axial part of embryo or seedling comprising the hypocotyl and the root meristem or the radicle, if one is present.

**hypodermis**   A layer or layers of cells beneath the epidermis distinct from the underlying ground tissue cells.

**hypophysis**   The uppermost cell of suspensor from which part of the root and rootcap in the embryo of angiosperms are derived.

## I

**idioblast**   A cell in a tissue that markedly differs in form, size, or contents from other cells in the same tissue.

**included phloem**   Secondary phloem included in the secondary xylem of certain eudicots. Term replaces *interxylary phloem*.

**increment**   In growth, an addition to the plant body by the activity of a meristem.

**indeterminate growth**   Unrestricted or unlimited growth, as with a vegetative apical meristem that produces an unrestricted number of lateral organs indefinitely.

**initial**   (1) Cell in a meristem that by division gives rise to two cells, one of which remains in the meristem, the other is added to the plant body; (2) sometimes used to designate a cell in its earliest stage of specialization. More appropriate term for (2), *primordium*.

**initial parenchyma**   See *marginal bands*.

**inner bark**   In older trees, the living part of the bark; the bark inside the innermost periderm. See also *bark*.

**intercalary growth**   Growth by cell division that occurs some distance from the meristem in which the cells originated.

**intercalary meristem**   Meristematic tissue derived from the apical meristem and continuing meristematic activity some distance from that meristem; may be intercalated between tissues that are no longer meristematic.

**intercellular space**   A space between two or more cells in a tissue; may have *schizogenous*, *lysigenous*, *schizolysigenous*, or *rhexigenous* origin.

**intercellular substance**   See *middle lamella*.

**interfascicular cambium**   Vascular cambium arising between vascular bundles (fascicles) in the interfascicular parenchyma.

**interfascicular region**   Tissue region located between the vascular bundles (fascicles) in a stem. Also called *medullary* or *pith ray.*

**intermediary cell**   Especially large companion cells with fields of highly branched plasmodesmata leading into them from the bundle-sheath cells; their presence is correlated with the transport of large quantities of raffinose and stachyose.

**internal phloem**   The primary phloem located internally from the primary xylem. Term replaces *intraxylary phloem.*

**internode**   Region between successive nodes of a stem.

**interpositional growth**   See *intrusive growth.*

**intervascular pitting**   Pitting between tracheary elements.

**interxylary cork**   Cork that develops within the xylem tissue.

**interxylary phloem**   See *included phloem.*

**intraxylary phloem**   See *internal phloem.*

**intrusive growth**   A type of growth in which a growing cell intrudes between other cells that separate from each other along the middle lamella in front of the tip of the growing cell. Also called *interpositional growth.*

**intussusception**   Growth of cell wall by interpolation of new wall material within previously formed wall. Opposite of *apposition.*

**isodiametric**   Regular in form, with all diameters equally long.

**isolation cells**   In secondary xylem, paratracheal parenchyma cells and ray cells that have no contact with the vessels; function as storage cells.

**isotropic**   Having the same properties along all axes. Optically isotropic material does not affect the light.

**K**

**karyokinesis**   Division of a nucleus as distinguished from the division of the cell, or *cytokinesis.* Also called *mitosis.*

**L**

**L1, L2, L3 layers**   The outer cell layers of angiosperm apical meristems with a tunica-corpus organization.

**lacuna** (pl. **lacunae**)   Space. Usually air space between cells, which may be *schizogenous, lysigenous, schizolysigenous,* or *rhexigenous* in origin. Also used with reference to the *leaf gap.*

**lacunar collenchyma**   A collenchyma characterized by intercellular spaces and cell wall thickenings facing the spaces.

**lamella**   A thin plate or layer.

**lamellar collenchyma**   A collenchyma in which cell wall thickenings are deposited mainly on tangential walls.

**lamina of leaf**   Expanded part of the leaf. Also called *blade* of the leaf.

**latewood**   The secondary xylem formed in the later part of a growth layer; denser and composed of smaller cells than the earlywood. Term replaces *summer wood.*

**lateral meristem**   A meristem located parallel with the sides of the axis; refers to the *vascular cambium* and *phellogen,* or *cork cambium.*

**lateral root**   A root arising from another, older root; also called *branch root,* or *secondary root,* if the older root is the primary root, or taproot.

**latex** (pl. **latices**)   A fluid, often milky, contained in laticifers; consists of a variety of organic and inorganic substances, often including rubber.

**laticifer**   A cell or a cell series containing a characteristic fluid called latex.

**laticiferous cell**   A nonarticulated, or simple, laticifer.

**laticiferous vessel**   An articulated, or compound, laticifer in which the cell walls between contiguous cells are partly or completely removed.

**leaf buttress**   A lateral protrusion below the apical meristem constituting the initial stage in the development of a leaf primordium.

**leaf fibers**   Technical designation of fibers derived from monocotyledons, chiefly from their leaves.

**leaf primordium**   A lateral outgrowth from the apical meristem that eventually will become a leaf.

**leaf sheath**   The lower part of a leaf that invests the stem more or less completely.

**leaf trace**   A vascular bundle in the stem extending between its connection with a leaf and that with another vascular unit in the stem; a leaf may have one or more leaf traces.

**leaf trace gap**   A region of parenchyma in the vascular cylinder of a stem located above the level where a leaf trace diverges toward the leaf. Also called *lacuna,* an interfascicular region; it involves no interruption of vascular connections.

**lenticel**   An isolated region in the periderm distinguished from the phellem in having intercellular spaces; the tissue may or may not be suberized.

**leptom** (or *leptome*)   The sieve elements and the associated parenchymatous cells of the phloem tissue; the supporting cells (fibers and sclereids) are excluded. See also *hadrom.*

**leucoplast**   A colorless plastid.

**libriform fiber**   A xylem fiber commonly with thick walls and simple pits; usually the longest cell in the tissue.

**lignification**   Impregnation with lignin.

**lignins**   Phenolic polymers deposited mainly in cell walls of supporting and conducting tissues; formed from the polymerization of three main monomeric units, the monolignols *p*-coumaryl, coniferyl, and sinapyl alcohols.

**lithocyst**  A cell containing a *cystolith*.

**longitudinal parenchyma**  See *axial parenchyma*.

**longitudinal system**  In secondary vascular tissues. See *axial system*.

**lumen**  Space bounded by (1) the plant cell wall; (2) the thylakoid space in chloroplasts; (3) the narrow, transparent space of endoplasmic reticulum.

**lutoids**  Vesicles, also called vacuoles, in laticifers bounded by a single membrane and containing a spectrum of hydrolytic enzymes capable of degrading most of the organic compounds in the cell.

**lysigenous**  As applied to an intercellular space, originating by a dissolution of cells.

**lysis**  A process of disintegration or degradation.

**lysosomal compartment**  A region in the cell protoplast or cell wall where acid hydrolases, capable of digesting cytoplasmic constituents and metabolites, are localized. Bounded by a single membrane in the protoplast and usually constituting the vacuolar system. Another term, *lytic compartment*.

**lysosome**  An organelle bounded by a single membrane and containing acid hydrolytic enzymes capable of breaking down proteins and other organic macromolecules; in plants, represented by vacuoles. See also *lysosomal compartment*.

**lytic compartment**  See *lysosomal compartment*.

**M**

**maceration**  Artificial separation of cells of a tissue by causing a disintegration of the middle lamella.

**macrofibril**  An aggregation of *microfibrils* in a cell wall visible with the light microscope.

**macrosclereid**  Elongated sclereid with unevenly distributed secondary wall thickening; common in seed epidermis of Fabaceae.

**magnoliids**  A clade, or evolutionary line, of angiosperms leading to the eudicots. The leaves of most magnoliids possess ester-containing oil cells.

**major veins**  Larger leaf vascular bundles, which are associated with ribs; they are largely involved with the transport of substances into and out of the leaf.

**mantle**  Outer layers of the kind of apical meristem that shows a layered arrangement of cells.

**marginal bands**  Parenchyma bands at the ends of growth rings in secondary xylem; may be restricted to the end of a ring (*terminal parenchyma*) or to the beginning of one (*initial parenchyma*).

**mass meristem**  A meristematic tissue in which the cells divide in various planes so that the tissue increases in volume.

**matrix**  Generally refers to a medium in which something is embedded.

**mechanical tissue**  See *supporting tissue*.

**medulla**  Synonym for *pith*.

**medullary ray**  See *interfascicular region*.

**meiosis**  Two successive nuclear divisions in which the chromosome number is reduced from diploid to haploid and segregation of the genes occurs.

**meristem**  Embryonic tissue region, primarily concerned with formation of new cells.

**meristematic cell**  A cel synthesizing protoplasm and producing new cells by division; varies in form, size, wall thickness, and degree of vacuolation, but has only a primary cell wall.

**meristemoid**  A cell or a group of cells constituting an active locus of meristematic activity in a tissue composed of somewhat older, differentiating cells.

**merophyte**  Immediate unicellular derivative of an apical cell and the multicellular structural units derived from them.

**mesomorphic**  Refers to structural features of *mesophytes*.

**mesophyll**  Photosynthetic parenchyma of a leaf blade located between the two epidermal layers.

**mesophyte**  A plant requiring an environment that is neither too wet nor too dry.

**mestome sheath**  An endodermoid sheath of a vascular bundle; the inner of two sheaths of leaves of Poaceae, mainly those of the festucoid subfamily.

**metacutisation**  Deposition of suberin lamellae in outer cells of root tips that cease to be active in growth and absorption at the end of seasonal growth. *Late suberization*.

**metaphloem**  Part of the primary phloem that differentiates after the protophloem and before the secondary phloem, if any of the latter is formed in a given taxon.

**metaxylem**  Part of the primary xylem that differentiates after the protoxylem and before the secondary xylem, if any of the latter is formed in a given taxon.

**micelles**  Regions in cellulose microfibrils in which the cellulose molecules are arranged parallel to each other so that a crystalline lattice structure is present.

**microbody**  See *peroxisome*.

**microfibril**  A thread-like component of the cell wall consisting of cellulose molecules and visible only with the electron microscope.

**microfilament**  See *actin filament*.

**micrometer**  One-thousandth millimeter; also called *micron*. Symbol μm.

**micron**  See *micrometer*.

**microtubules**  Nonmembranous tubules about 25 nanometers (nm) in diameter and of indefinite length. Located in the cytoplasm in a nondividing eukaryotic cell, usually near the cell wall, and form the meiotic or mitotic spindle and the phragmoplast in a dividing cell.

**middle lamella**  Layer of intercellular material, chiefly pectic substances, cementing together the primary walls of contiguous cells.

**minor veins** Small leaf vascular bundles, which are located in the mesophyll and enclosed by a bundle sheath; they are involved with the distribution of the transpiration stream and the uptake of the products of photosynthesis.

**mitochondrion (pl. mitochondria)** Double membrane-bound cell organelle concerned with respiration; carries enzymes and is the major source of ATP in non-photosynthetic cells.

**mitosis** See *karyokinesis*.

**monocotyledon** A plant whose embryo has one cotyledon; one of the two great classes of angiosperms, the Monocotyledones; often abbreviated as monocot; the other great class, the Eudicotyledones.

**morphogenesis** Development of form; the sum of phenomena of development and differentiation of tissues and organs.

**morphology** Study of form and its development.

**mother cell** See *precursory cell*.

**motor cell** See *bulliform cell*.

**mucilage cell** Cell containing mucilages or gums or similar carbohydrate material characterized by the property of swelling in water.

**mucilage duct** A duct containing mucilage or gum or similar carbohydrate material. See also *duct*.

**multiperforate perforation plate** In vessel element of the xylem; a perforation plate that has more than one perforation.

**multiple epidermis** A tissue two or more cell layers deep derived from the protoderm; only the outermost layer differentiates as a typical epidermis.

**multiseriate ray** A ray in secondary vascular tissues that is few to many cells wide.

**myrosin cell** Cell containing myrosinases, enzymes that hydrolyze glucosinolates. Occur mainly in the Brassicaceae.

**N**

**nacré wall** See *nacreous wall*.

**nacreous wall** A nonlignified wall thickening that is often found in sieve elements and resembles a secondary wall when it attains considerable thickness; designation based on glistening appearance of the wall in fresh tissue.

**nanometer** One millionth of a millimeter; symbol nm. Equal to 10 angstroms.

**nectary** A multicellular glandular structure secreting a liquid containing organic substances including sugar. Occurs in flowers (*floral nectary*) and vegetative plant parts (*extrafloral nectary*).

**netted venation** Veins in a leaf blade form an anastomosing system, the whole resembling a net; also called *reticulate venation*.

**node** That part of the stem at which one or more leaves are attached; not sharply delimited anatomically.

**nonarticulated laticifer** A simple laticifer consisting of a single, commonly multinucleate, cell; may be branched or unbranched.

**nonporous wood** Secondary xylem having no vessels.

**nonstoried cambium** Vascular cambium in which the fusiform initials and rays are not arranged in horizontal tiers on tangential surfaces. Also called *nonstratified cambium*.

**nonstoried wood** Secondary xylem in which the axial cells and rays are not arranged in horizontal tiers on tangential surfaces. Also called *nonstratified wood*.

**nonstratified cambium** See *nonstoried cambium*.

**nonstratified wood** See *nonstoried wood*.

**nuclear envelope** Double membrane enclosing the nucleus of a cell.

**nucleoid** A region of DNA in prokaryotic cells, mitochondria, and chloroplasts.

**nucleolar organizer region** A special area on a certain chromosome associated with the formation of the nucleolus.

**nucleolus (pl. nucleoli)** A small, spherical body found in the nucleus of eukaryotic cells, which is composed of rRNA in the process of being transcribed from copies of rRNA genes; the site of production of ribosomal subunits.

**nucleoplasm** Ground substance of the nucleus.

**nucleus** In biology, organelle in a eukaryotic cell bounded by a double membrane and containing the chromosomes, nucleoli, and nucleoplasm.

**O**

**ontogeny** Development of an organism, organ, tissue, or cell from inception to maturity.

**opposite pitting** Pits in tracheary elements disposed in horizontal pairs or in short horizontal rows.

**organ** A distinct and visibly differentiated part of a plant, such as root, stem, leaf, or part of a flower.

**organelle** A distinct body within the cytoplasm of a cell, specialized in function; specifically, membrane-bound.

**organism** Any individual living thing, either unicellular or multicellular.

**orthostichy** A vertical line along which is attached a series of leaves or scales on an axis of a shoot or shoot-like organ. Often incorrectly applied to a steep helix, or *parastichy*.

**osteosclereid** Bone-shaped sclereid having a columnar middle part and enlargements at both ends.

**outer bark** In older trees, the dead part of the bark; the innermost periderm and all tissues outside it; also called *rhytidome*. See also *bark*.

**P**

**paedomorphosis** Delay in evolutionary advance in some characteristics as compared with others resulting in a combination of juvenile and advanced characteristics in the same cell, tissue, or organ.

**palisade parenchyma** Leaf mesophyll parenchyma characterized by elongated form of cells and their arrangement with their long axes perpendicular to the surface of the leaf.

**panicoid** Pertaining to the Panicoideae, a subfamily of grasses.

**papilla** (pl. **papillae**) A soft protuberance on an epidermal cell; a type of trichome.

**paracytic stoma** A stomatal complex in which one or more subsidiary cells flank the stoma parallel with the long axes of the guard cells.

**paradermal** Parallel with the epidermis. Refers specifically to a section made parallel with the surface of a flat organ such as a leaf; it is also a *tangential section.*

**parallel venation** Main veins in a leaf blade arranged approximately parallel to one another, although converging at base and apex of leaf.

**parastichy** A helix along which is attached a series of leaves or scales on an axis of a shoot or shoot-like organ. See also *orthostichy.*

**paratracheal parenchyma** Axial parenchyma in secondary xylem associated with vessels and other tracheary elements. Includes *aliform, confluent,* and *vasicentric.*

**parenchyma** Tissue composed of parenchyma cells.

**parenchyma cell** Typically a not distinctly specialized cell with a nucleate protoplast concerned with one or more of the various physiological and biochemical activities in plants. Varies in size, form, and wall structure.

**parietal cytoplasm** Cytoplasm located next to the cell wall.

**pectic substances** A group of complex carbohydrates, derivatives of polygalacturonic acid, occurring in plant cell walls; particularly abundant as a constituent of the middle lamella.

**peltate hair** A trichome consisting of a discoid plate of cells borne on a stalk or attached directly to the basal foot cell.

**perforation plate** Part of a wall of a vessel element that is perforated.

**periblem** The meristem forming the cortex. One of the three histogens, *plerome, periblem,* and *dermatogen,* according to Hanstein.

**periclinal** Commonly refers to orientation of cell wall or plane of cell division; parallel with the circumference or the nearest surface of an organ. Opposite of *anticlinal.* See also *tangential.*

**periclinal chimera** See *chimera.*

**pericycle** Part of ground tissue of the stele located between the phloem and the endodermis. In seed plants, regularly present in roots, absent in most stems.

**pericyclic fiber** See *perivascular fiber.*

**pericyclic sclerenchyma** See *perivascular sclerenchyma.*

**periderm** Secondary protective tissue that replaces the epidermis in stems and roots, rarely in other organs. Consists of *phellem* (cork), *phellogen* (cork cambium), and *phelloderm.*

**perimedullary region** or **zone** Peripheral region of the pith (medulla). Also called *medullary sheath.*

**perinuclear space** Space between the two membranes forming the nuclear envelope.

**perivascular fiber** A fiber located along the outer periphery of the vascular cylinder in the axis of a seed plant and not originating in the phloem. Alternate term, *pericyclic fiber.*

**perivascular sclerenchyma** Sclerenchyma located along the outer periphery of the vascular cylinder and not originating in the phloem. Alternate term, *pericyclic sclerenchyma.*

**peroxisome** A spherical, single membrane-bound organelle; some are involved in photorespiration and others (called *glyoxysomes*) with the conversion of fats to sugars during seed germination. Also called *microbody.*

**phellem (cork)** Protective tissue composed of nonliving cells with suberized walls and formed centrifugally by the phellogen (cork cambium) as part of the periderm. Replaces the epidermis in older stems and roots of many seed plants.

**phelloderm** A tissue resembling cortical parenchyma produced centripetally by the phellogen (cork cambium) as part of the periderm of stems and roots in seed plants.

**phellogen (cork cambium)** A lateral meristem forming the periderm, a secondary protective tissue common in stems and roots of seed plants. Produces phellem (cork) centrifugally, phelloderm centripetally by periclinal divisions.

**phelloid cell** A cell within the phellem (cork) but distinct from the cork cell in having no suberin in its walls. May be a sclereid.

**phenotype** Physical appearance of an organism resulting from interaction between its *genotype* (genetic constitution) and the environment.

**phlobaphenes** Anhydrous derivatives of tannins. Amorphous yellow, red, or brown substances very conspicuous when present in cells.

**phloem** Principal food-conducting tissue of the vascular plant composed mainly of sieve elements, various kinds of parenchyma cells, fibers, and sclereids.

**phloem elements** Cells of the phloem tissue.

**phloem initial** A cambial cell on the phloem side of the cambial zone that is the source of one or more cells arising by periclinal divisions and differentiating into phloem elements with or without additional divisions in various planes. Sometimes called *phloem mother cell.*

**phloem mother cell**   A cambial derivative that is the source of certain elements of the phloem tissue, such as, a sieve-tube element and its companion cells or a strand of phloem parenchyma cells. Used also in a wider sense synonymously with *phloem initial.*

**phloem parenchyma**   Parenchyma cells located in the phloem. In secondary phloem refers to axial parenchyma.

**phloem ray**   That part of a vascular ray that is located in the secondary phloem.

**phloic procambium**   That part of procambium that differentiates into primary phloem.

**photoperiodism**   Response to duration and timing of day and night expressed in the character of growth, development, and flowering in plants.

**photorespiration**   Oxygenase activity of Rubisco combined with the salvage pathway, consuming $O_2$ and releasing $CO_2$; occurs when Rubisco binds $O_2$ instead of $CO_2$.

**photosynthetic cell**   A chloroplast-containing cell engaged in photosynthesis.

**phragmoplast**   Fibrous structure (light microscope view) that arises between the daughter nuclei at telophase and within which the initial partition (*cell plate*), dividing the mother cell in two (*cytokinesis*), is formed. Appears at first as a spindle connected to the two nuclei, but later spreads laterally in the form of a ring. Consists of microtubules.

**phragmosome**   Layer of cytoplasm formed across the cell where the nucleus becomes located and divides. The equatorial plane of the subsequently appearing phragmoplast coincides with the plane of the cytoplasmic layer.

**phyllochron**   Interval between the visible appearance or emergence of successive leaves in the intact plant.

**phyllotaxy** (or **phyllotaxis**)   Mode in which the leaves are arranged on the axis of a shoot.

**phylogeny**   Evolutionary relationships among organisms; the developmental history of a group of organisms.

**phytomeres**   Units, or modules, repetitively produced by the vegetative shoot apex. Each phytomere consists of a node, with its attached leaf, a subjacent internode, and a bud at the base of the internode.

**pinocytosis**   See *endocytosis.*

**pit**   A recess or cavity in the cell wall where the primary wall is not covered by secondary wall. Pit-like structures in the primary wall are designated *primordial pits, primary pits,* or *primary pit-fields.* A pit is usually a member of a pit-pair.

**pit aperture**   Opening into the pit from the interior of the cell. If a pit canal is present in a bordered pit, two apertures are recognized, the *inner,* from the cell lumen into the canal, and the *outer,* from the canal into the pit cavity.

**pit canal**   Passage from the cell lumen to the chamber of a bordered pit. (Simple pits in thick walls usually have canal-like cavities.)

**pit cavity**   Entire space within a pit from pit membrane to the cell lumen or to the outer pit aperture if a pit canal is present.

**pit-field**   See *primary pit-field.*

**pit membrane**   Part of the intercellular layer and primary cell wall that limits a pit cavity externally.

**pit-pair**   Two complementary pits of two adjacent cells. Essential components are two *pit cavities* and the *pit membrane.*

**pith**   Ground tissue in the center of a stem or root. Homology of pith in root and stem is uncertain.

**pith ray**   See *interfascicular region.*

**plasma membrane**   Single membrane delimiting the cytoplasm next to the cell wall. A type of unit membrane. Also called *plasmalemma.*

**plasmalemma**   See *plasma membrane.*

**plasmodesma** (pl. **plasmodesmata**)   A connection of protoplasts of two contiguous cells through a channel in the cell wall. This plasma membrane-lined channel typically is traversed by a tubular strand of tightly constricted endoplasmic reticulum called a *desmotubule,* which is continuous with the endoplasmic reticulum in the contiguous cells. The region between the plasma membrane and the desmotubule is called the *cytoplasmic sleeve.*

**plastid**   Organelle with a double membrane in the cytoplasm of many eukaryotes. May be concerned with photosynthesis (*chloroplast*) or starch storage (*amyloplast*), or contain yellow or orange pigments (*chromoplast*). See also *leucoplast.*

**plastochron** (or *plastochrone*)   The time interval between the inception of two successive repetitive events, as origin of leaf primordia, attainment of certain stage of development of a leaf, etc. Variable in length as measured in time units.

**plastoglobule**   Globule in a plastid with lipid as the basic component.

**plate collenchyma**   See *lamellar collenchyma.*

**plate meristem**   A meristematic tissue consisting of parallel layers of cells dividing only anticlinally with reference to the wide surface of the tissue. Characteristic of ground meristem of plant parts that assume a flat form as a leaf.

**plerome**   The meristem forming the core of the axis composed of the primary vascular tissues and associated ground tissue such as pith and interfascicular regions. One of the three histogens, *plerome, periblem,* and *dermatogen,* according to Hanstein.

**plumule**   Portion of the young shoot above the cotyledon(s); the first bud of an embryo. See also *epicotyl.*

**polyderm**   A type of protective tissue in which suberized cells alternate with nonsuberized parenchyma cells and both kinds of cell have living protoplasts.

**polymerization** Chemical union of monomers, such as glucose or nucleotides, resulting in the formation of polymers, such as starch, cellulose, or nucleic acid.

**polysaccharide** A carbohydrate composed of many monosaccharide units joined in a chain, for example, starch, cellulose.

**polysome** (or **polyribosome**) Aggregation of ribosomes apparently concerned with protein synthesis as a group.

**pore** A term of convenience for the transverse section of a vessel in the secondary xylem.

**pore cluster** See *pore multiple*.

**pore multiple** In secondary xylem; a group of two or more pores (transverse sections of vessels) crowded together and flattened along the surfaces of contact. *Radial pore multiple*, pores in radial file; *pore cluster*, irregular grouping.

**porous wood** Secondary xylem having vessels.

**P-protein** Phloem protein; a proteinaceous substance found in cells of angiosperm phloem, especially in sieve-tube elements; formerly called slime.

**precursory cell** A cell giving rise to others by division.

**preprophase band** A ring-like band of microtubules, found just beneath the plasma membrane, that delimits the equatorial plane of the future mitotic spindle of a cell preparing to divide.

**primary body (of plant)** Part of the plant, or entire plant if no secondary growth occurs, that arises from the embryo and the apical meristems and their derivative meristematic tissues and is composed of primary tissues.

**primary cell wall** Version based on studies with the light microscope: cell wall formed chiefly while the cell is increasing in size. Version based on studies with the electron microscope: cell wall in which the cellulose microfibrils show various orientations—from random to more or less parallel—that may change considerably during the increase in size of the cell. The two versions do not necessarily coincide in delimiting primary from secondary wall.

**primary growth** Growth of successively formed roots and vegetative and reproductive shoots from the time of their initiation by the apical meristems and until the completion of their expansion. Has its inception in the apical meristems and continues in their derivative meristems, protoderm, ground meristem, and procambium, as well as in the partly differentiated primary tissues.

**primary meristem** Often used for each of the three meristematic tissues derived from the apical meristem: protoderm, ground meristem, and procambium.

**primary metabolites** Molecules that are found in all plant cells and that are necessary for the life of the plant; examples are simple sugars, amino acids, proteins, and nucleic acids.

**primary phloem** Phloem tissue differentiating from procambium during primary growth and differentiation of a vascular plant. Commonly divided into the earlier *protophloem* and the later *metaphloem*. Not differentiated into axial and ray systems.

**primary phloem fibers** Fibers located on the outer periphery of the vascular region and originating in the primary phloem, usually the protophloem. Often called *pericyclic fibers*.

**primary pit** See *primary pit-field*.

**primary pit-field** A thin area of the primary cell wall and middle lamella within the limits of which one or more pit-pairs develop if a secondary wall is formed. Also called *primordial pit* and *primary pit*.

**primary plant body** See *primary body*.

**primary root** Taproot. Root developing in continuation of the radicle of the embryo.

**primary thickening meristem** A meristem derived from the apical meristem and responsible for the primary increase in thickness of the shoot axis. May appear as a distinct mantle-like zone. Often found in monocotyledons.

**primary tissues** Tissues derived from the embryo and the apical meristems.

**primary vascular tissues** Xylem and phloem differentiating from procambium during primary growth and differentiation of a vascular plant.

**primary wall** See *primary cell wall*.

**primary xylem** Xylem tissue differentiating from procambium during primary growth and differentiation of a vascular plant. Commonly divided into the earlier *protoxylem* and the later *metaxylem*. Not differentiated into axial and ray systems.

**primordial pit** See *primary pit-field*.

**primordium** (pl. **primordia**) An organ, a cell, or an organized series of cells in their earliest stage of differentiation, for example, leaf primordium, sclereid primordium, vessel element primordium.

**procambium** Primary meristem or meristematic tissue that differentiates into the primary vascular tissue. Also called *provascular tissue*.

**procumbent ray cell** In secondary vascular tissues; a ray cell having its longest axis in radial direction.

**prodesmogen** A meristem precursory to desmogen (*procambium*). The term has the same connotation as *residual meristem*.

**programmed cell death** The genetically controlled, or programmed, series of changes in a living cell or organism that leads to its death.

**prokaryotic** (also *procaryotic*) Refers to an organism, the cells of which lack a membrane-bound nucleus and membrane-bound organelles; Bacteria and Archaea.

**prolamellar body** Semicrystalline body found in plastids arrested in development by the absence of light.

**promeristem** Initiating cells and their most recent derivatives in an apical meristem. Also called *protomeristem*.

**prophyll** First or one of two first leaves on a lateral shoot.

**proplastid** A plastid in its earliest stages of development.

**protoderm** Primary meristem or meristematic tissue giving rise to the epidermis; also epidermis in meristematic state. May or may not arise from independent initials in the apical meristem.

**protomeristem** See *promeristem*.

**protophloem** First-formed elements of the phloem in a plant organ. First part of the primary phloem.

**protophloem poles** Term of convenience for loci of phloem elements that are the first to mature in the vascular system of a plant organ. Applied to views in transverse sections.

**protoplasm** Living substance. Inclusive term for all living contents of a cell or an entire organism.

**protoplast** Organized living unit of a single cell including protoplasmic and nonprotoplasmic contents of a cell but excluding the cell wall.

**protoxylem** First-formed elements of the xylem in a plant organ. First part of the primary xylem.

**protoxylem lacuna** Space surrounded by parenchyma cells in the protoxylem of a vascular bundle. Appears in some plants after the tracheary elements of protoxylem are stretched and torn.

**protoxylem poles** Term of convenience for loci of xylem elements that are the first to mature in the vascular system of a plant organ. Applied to views in transverse sections.

**provascular tissue** See *procambium*.

**proximal** Situated near the point of origin or attachment. Opposite of *distal*.

**pycnotic degeneration** Nuclear degeneration during which the chromatin forms a very dense mass prior to rupture of the nuclear envelope.

**Q**

**quarter-sawed oak** Oak wood sawed along a radial plane so that the radial surface showing the wide rays characteristic of this wood is exposed.

**quiescent center** Initial region in the apical meristem that has reached a state of relative inactivity; common in roots.

**R**

**radial parenchyma** See *ray parenchyma*.

**radial pore multiple** See *pore multiple*.

**radial section** A longitudinal section coinciding with a radius of a cylindrical body, such as stem.

**radial seriation** Arrangement of units, such as cells, in an orderly sequence in a radial direction. Characteristic of cambial derivatives.

**radial system** See *ray system*.

**radicle** Embryonic root. Forms the basal continuation of the hypocotyl in an embryo.

**ramified** Branched.

**ramiform pit** Pit that appears to be branched because it is formed by a coalescence of two or more simple pits during the increase in thickness of the secondary wall.

**raphides** Needle-shaped crystals commonly occurring in bundles.

**ray** A panel of tissue variable in height and width, formed by the ray initials in the vascular cambium and extending radially in the secondary xylem and secondary phloem.

**ray initial** A meristematic ray cell in the vascular cambium that gives rise to ray cells of the secondary xylem and secondary phloem.

**ray parenchyma** Parenchyma cells of a ray in secondary vascular tissues. Contrasted with *axial parenchyma*.

**ray system** Total of all rays in the secondary vascular tissues. Also called *horizontal system* and *radial system*.

**ray tracheid** Tracheid in a ray. Found in the secondary xylem of certain conifers.

**reaction wood** Wood with more or less distinctive anatomical characteristics formed in parts of leaning or crooked stems and on lower (conifers) or upper (magnoliids and eudicots) sides of branches. See *compression wood* and *tension wood*.

**redifferentiation** A reversal in differentiation in a cell or tissue and subsequent differentiation into another type of cell or tissue.

**residual meristem** Used in the sense of residuum of the least differentiated part of the apical meristem. A tissue that is relatively more highly meristematic than the associated differentiating tissues beneath the apical meristem. Gives rise to procambium and to interfascicular ground tissue.

**resin canal** See *resin duct*.

**resin duct** A duct of schizogenous origin lined with resin-secreting cells (*epithelial cells*) and containing resin.

**reticulate cell wall thickening** In tracheary elements of the xylem; secondary cell wall deposited on the primary so as to form a net-like pattern.

**reticulate perforation plate** In vessel element of the xylem; a type of multiperforate plate in which the bars delimiting the perforations form a net-like pattern.

**reticulate sieve plate** A compound sieve plate with sieve areas arranged so as to form a net-like pattern.

**reticulate venation** See *netted venation*.

**reticulum** A net.

**retting** Freeing fiber bundles from other tissues by utilizing the action of microorganisms causing, in a suitable moist environment, the disintegration of the thin-walled cells surrounding the fibers.

**rhexigenous**   As applied to an intercellular space, originating by rupture of cells.

**rhizodermis**   Primary surface layer of the root. Use of the term implies that this layer is not homologous with the epidermis of the shoot. See also *epiblem*.

**rhytidome**   A technical term for the outer bark, which consists of periderm and tissues isolated by it, namely cortical and phloem tissues.

**rib**   An elongate protrusion, as those along the large veins on the underside of a leaf.

**rib meristem**   A meristematic tissue in which the cells divide perpendicular to the longitudinal axis of an organ and produce a complex of parallel, vertical files ("ribs") of cells. Particularly common in ground meristem of organs assuming a cylindrical form. Also called *file meristem*.

**ribosome**   A cell component composed of protein and RNA and concerned with protein synthesis. Occurs in the cytosol, nucleus, plastids, and mitochondria.

**ring bark**   A type of rhytidome resulting from the formation of successive periderms approximately concentrically around the axis.

**ring-porous wood**   Secondary xylem in which the pores (vessels) of the earlywood are distinctly larger than those of the latewood and form a well-defined zone or ring in a transverse section of wood.

**rootcap**   A thimble-like mass of cells covering the apical meristem of the root.

**root hair**   A type of trichome on root epidermis that is a simple extension of an epidermal cell and is concerned with absorption of soil solution.

## S

**sapwood**   Outer part of the wood of stem or root containing living cells and reserves; it may or may not function in the conduction of water. Generally lighter colored that the *heartwood*.

**scalariform cell wall thickening**   In tracheary elements of the xylem; secondary wall deposited on the primary so as to form a ladder-like pattern. Similar to a helix of low pitch with the coils interconnected at intervals.

**scalariform perforation plate**   In vessel element of the xylem; a type of multiperforate plate in which elongated perforations are arranged parallel to one another so that the cell wall bars between them form a ladder-like pattern.

**scalariform pitting**   In tracheary elements of the xylem; elongated pits arranged parallel to one another so as to form a ladder-like pattern.

**scalariform-reticulate cell wall thickening**   In tracheary elements of the xylem; secondary thickening intermediate between scalariform and reticulate.

**scalariform sieve plate**   A compound sieve plate with elongated sieve areas arranged parallel to one another in a ladder-like pattern.

**scale bark**   A type of rhytidome in which the sequent periderms develop as restricted overlapping strata, each cutting out a scale-like mass of tissue.

**schizogenous**   As applied to an intercellular space, originating by separation of cell walls along the middle lamella.

**schizolysigenous**   As applied to an intercellular space, originating by a combination of two processes, separation and degradation of cell walls.

**sclereid**   A sclerenchyma cell, varied in form, but typically not much elongated, and having thick, lignified secondary walls with many pits.

**sclerenchyma**   A tissue composed of sclerenchyma cells. Also a collective term for sclerenchyma cells in the plant body or plant organ. Includes *fibers*, *fiber-sclereids*, and *sclereids*.

**sclerenchyma cell**   Cell variable in form and size and having more or less thick, often lignified, secondary walls. Belongs to the category of supporting cells and may or may not be devoid of protoplast at maturity.

**sclerification**   Act of becoming changed into sclerenchyma, that is, developing secondary walls, with or without subsequent lignification.

**sclerotic parenchyma cell**   A parenchyma cell that through deposition of a thick secondary wall becomes changed into a sclereid.

**scutellum** (pl. **scutella**)   Cotyledon in Poaceae embryo specialized for absorption of endosperm.

**secondary body**   Part of the plant body that is added to the primary body by the activity of the lateral meristems, vascular cambium and phellogen. Consists of secondary vascular tissues and periderm.

**secondary cell wall**   Version based on studies with the light microscope: cell wall deposited in some cells over the primary wall after the primary wall ceases to increase in surface. Version based on studies with the electron microscope: cell wall in which the cellulose microfibrils show a definite parallel orientation. The two versions do not necessarily coincide in delimiting secondary from primary wall.

**secondary growth**   In gymnosperms, most magnoliids and eudicots, and some monocots. A type of growth characterized by an increase in thickness of stem and root and resulting from formation of secondary vascular tissues by the vascular cambium. Commonly supplemented by activity of the cork cambium (phellogen) forming periderm.

**secondary metabolites**   Molecules that are restricted in their distribution, both within the plant and among different plants; important for the survival and propagation of the plants that produce them; there are three major classes—alkaloids, terpenoids, and phenolics. Also called *secondary products*.

**secondary phloem** Phloem tissue formed by the vascular cambium during secondary growth in a vascular plant. Differentiated into axial and ray systems.

**secondary phloem fiber** A fiber located in the axial system of secondary phloem.

**secondary plant body** See *secondary body*.

**secondary root** See *branch root*.

**secondary thickening** Used for both deposition of secondary cell wall material and secondary increase in thickness of stems and roots.

**secondary tissues** Tissues produced by vascular cambium and phellogen during secondary growth.

**secondary vascular tissues** Vascular tissues (both xylem and phloem) formed by the vascular cambium during secondary growth in a vascular plant. Differentiated into axial and ray systems.

**secondary wall** See *secondary cell wall*.

**secondary xylem** Xylem tissue formed by the vascular cambium during secondary growth in a vascular plant. Differentiated into axial and ray systems.

**secretory cavity** Commonly refers to a space lysigenous in origin and containing secretion derived from the cells that broke down in the formation of the cavity.

**secretory cell** A living cell specialized with regard to secretion or excretion of one or more, often organic, substances.

**secretory duct** Commonly refers to a duct schizogenous in origin and containing a secretion derived from the cells (epithelial cells) lining the duct. See *epithelium*.

**secretory hair** See *glandular hair*.

**secretory structure** Any of a great variety of structures, simple or complex, external or internal, that produces a secretion.

**seed coat** Outer coat of the seed derived from the integument or integuments. Also called *testa*.

**septate fiber** A fiber with thin transverse walls (septa), which are formed after the cell develops a secondary wall thickening.

**septum** (pl. **septa**) A partition.

**sheath** A sheet-like structure enclosing or encircling another. Applied to tubular or enrolled part of an organ, such as a leaf sheath, and to a tissue layer surrounding a complex of another tissue, as a bundle sheath enclosing a vascular bundle.

**shell zone** In axillary bud primordia; a zone of parallel curving layers of cells, the entire complex shell-like in form. A result of orderly cell division along the proximal limits of the primordium.

**shoot** Above-ground portions, such as the stem and leaves, of a vascular plant.

**sieve area** A portion of the sieve-element wall containing clusters of pores through which the protoplasts of adjacent sieve elements are interconnected.

**sieve cell** A type of sieve element that has relatively undifferentiated sieve areas (with narrow pores), rather uniform in structure on all walls; that is, there are no sieve plates; found in phloem of gymnosperms.

**sieve element** Cell in the phloem tissue concerned with mainly longitudinal conduction of food materials. Classified into gymnospermous *sieve cell* and angiospermous *sieve-tube element*.

**sieve field** Old term for a relatively undifferentiated sieve area found on wall parts other than the sieve plates.

**sieve pitting** An arrangement of small pits in sieve-like clusters.

**sieve plate** Part of the wall of a sieve-tube element bearing one (*simple sieve plate*) or more (*compound sieve plate*) highly differentiated sieve areas.

**sieve tube** A series of sieve-tube elements arranged end to end and interconnected through sieve plates.

**sieve-tube element** One of the series of cellular components of a sieve tube. It shows a more or less pronounced differentiation between sieve plates (wide pores) and lateral sieve areas (narrow pores). Also *sieve-tube member* and the obsolete sieve-tube segment.

**sieve-tube member** See *sieve-tube element*.

**silica cell** Cell filled with silica, as in epidermis of grasses.

**simple laticifer** Laticifer that is a single cell. *A non-articulated laticifer*.

**simple perforation plate** In vessel element of the xylem; a perforation plate with a single perforation.

**simple pit** A pit in which the cavity becomes wider, remains of constant width, or only gradually becomes narrower during the growth in thickness of the secondary wall, that is, toward the lumen of the cell.

**simple pit-pair** An intercellular pairing of two simple pits.

**simple sieve plate** Sieve plate composed of one sieve area.

**simple tissue** A tissue composed of a single cell type; parenchyma, collenchyma, and sclerenchyma are simple tissues.

**slime** See *P-protein*.

**slime body** An aggregation of P-protein.

**slime plug** An accumulation of P-protein on a sieve area, usually with extensions into the sieve-area pores.

**softwood** A name commonly applied to the wood of a conifer.

**solitary pore** A pore (transverse section of a vessel in secondary xylem) surrounded by cells other than vessel elements.

**specialization** Change in structure of a cell, a tissue, plant organ, or entire plant associated with a restriction of functions, potentialities, or adaptability to varying conditions. May result in greater efficiency with regard to certain specific functions. Some specializations are irreversible, others reversible.

**specialized** Refers to (1) organisms having special adaptations to a particular habitat or mode of life; (2) cells or tissues having a characteristic function distinguishing them from other cells or tissues, more generalized in their function.

**spindle fibers** Bundles of microtubules, some of which extend from the kinetochores of the chromosomes to the poles of the spindle.

**spiral cell wall thickening** See *helical cell wall thickening*.

**spongy parenchyma** Leaf mesophyll parenchyma with conspicuous intercellular spaces.

**spring wood** See *earlywood*.

**square ray cell** In secondary vascular tissues, a ray cell approximately square as seen in radial section. (Considered to be of the same morphological type as the *upright ray cell*.)

**starch** An insoluble carbohydrate, the chief food storage substance of plants, composed of anhydrous glucose residues of the formula $C_6H_{10}O_5$ into which it easily breaks down.

**starch sheath** Applied to the innermost region (one or more cell layers) of the cortex when this region is characterized by conspicuous and rather stable accumulation of starch.

**stele (column)** Conceived by P. Van Tieghem as a morphologic unit of the plant body comprising the vascular system and the associated ground tissue (pericycle, interfascicular regions, and pith). The *central cylinder* of the axis (stem and root).

**stellate** Star shaped.

**stereom** (or *stereome*) Collective term for supporting tissue as contrasted with the conducting tissues *hadrom* and *leptom*.

**stoma** (pl. **stomata**) An opening in the epidermis of leaves and stems bordered by two guard cells and serving in gas exchange; also used to refer to the entire stomatal apparatus—the guard cells plus their included pore.

**stomatal complex** Stoma and associated epidermal cells that may be ontogenetically and/or physiologically related to the guard cells. Also called *stomatal apparatus*.

**stomatal crypt** A depression in the leaf, the epidermis of which bears stomata.

**stone cell** See *brachysclereid*.

**storied cambium** Vascular cambium in which the fusiform initials are arranged in horizontal tiers on tangential surfaces; the rays may also be so arranged. Also called *stratified cambium*.

**storied cork** Protective tissue found in the monocotyledons. The suberized cells occur in radial files, each consisting of several cells—all of which are derived from one cell.

**storied wood** Wood in which the axial cells are arranged in horizontal tiers on tangential surfaces; the rays may also be so arranged. (Rays alone may be storied.) Also called *stratified wood*.

**Strasburger cell** In gymnosperm phloem; certain ray and axial parenchyma cells spatially and functionally associated with the sieve cells, thus resembling the companion cells of angiosperms but not originating from the same precursory cells as the sieve cells. Also called *albuminous cells*.

**stratified cambium** See *storied cambium*.

**stratified wood** See *storied wood*.

**striate venation** See *parallel venation*.

**stroma** The ground substance of plastids.

**styloid** An elongated crystal with pointed or square ends.

**subapical initial** A cell beneath the protoderm at the apex of a leaf primordium that appears to function as an initial of the interior tissue of the leaf. Questionable concept.

**suberin** Fatty substance in the cell wall of cork tissue and in the Casparian strip of the endodermis.

**suberization** Impregnation of the cell wall with suberin or deposition of suberin lamellae on the wall.

**subsidiary cell** An epidermal cell associated with a stoma and at least morphologically distinguishable from the epidermal cells composing the groundmass of the tissue. Also called *accessory cell*.

**summer wood** See *latewood*.

**supernumerary cambium layer** Vascular cambium originating in phloem or pericycle outside the regularly formed vascular cambium. Characteristic of some plants with anomalous type of secondary growth.

**supporting cell** See *supporting tissue*.

**supporting tissue** Refers to tissue composed of cells with more or less thickened walls, primary (collenchyma) or secondary (sclerenchyma) that adds strength to the plant body. Also called *mechanical tissue*.

**suspensor** An extension at the base of the embryo that anchors the embryo in the embryo sac.

**symplast** Interconnected protoplasts and their plasmodesmata; the movement of substances in the symplast is called symplastic movement, or symplastic transport.

**symplastic growth** See *coordinated growth*.

**syndetocheilic** Stomatal type in gymnosperms; subsidiary cells (or their precursors) are derived from the same protodermal cell as the guard-cell mother cell.

**T**

**tabular** Having the form of a tablet or slab.

**tangential** In the direction of the tangent; at right angles to the radius. May coincide with *periclinal*.

**tangential section** A longitudinal section cut at right angles to a radius. Applicable to cylindrical structures such as stem or root, but used also for leaf blades when the section is made parallel with the expanded surface. Substitute term for leaf, *paradermal*.

**tannin** General term for a heterogeneous group of phenol derivatives. Amorphous, strongly astringent substance widely distributed in plants, and used in tanning, dyeing, and preparation of ink.

**taproot** First, or primary, root of a plant forming a direct continuation of the radicle of the embryo.

**taxon** (pl. **taxa**) Any one of the categories (species, genus, family, etc.) into which living organisms are classified.

**teichode** A linear space in the outer epidermal wall in which the fibrillar structure is more loose and open than elsewhere in the wall. Replaces the term *ectodesma*.

**tension wood** Reaction wood in angiosperms, formed on the upper sides of branches and leaning or crooked stems and characterized by lack of lignification and often by high content of gelatinous fibers.

**terminal parenchyma** See *marginal bands*.

**thylakoids** Sac-like membranous structures (cisternae) in a chloroplast combined into stacks (grana) and present singly in the stroma (stroma thylakoids) as interconnections between grana.

**tissue** Group of cells organized into a structural and functional unit. Component cells may be alike (simple tissue) or varied (complex tissue).

**tissue system** A tissue or tissues in a plant or plant organ structurally and functionally organized into a unit. Commonly three tissue systems are recognized, *dermal, vascular,* and *fundamental* (ground tissue system).

**tonoplast** A single cytoplasmic membrane bordering the vacuole. A kind of *unit membrane.*

**torus** (pl. **tori**) Central thickened part of the pit membrane in a bordered pit consisting mainly of middle lamella and two primary walls. Typical of bordered pits in conifers and some other gymnosperms; also found in several species of eudicots.

**totipotent** Potential of a plant cell to develop into an entire plant.

**trabecula** (pl. **trabeculae**) A rod-like or spool-shaped part of a cell wall extending radially across the lumen of a cell. In initials and derivatives of vascular cambium in seed plants.

**trachea** Old term for xylem vessel, implying a resemblance to an animal trachea.

**tracheary element** General term for a water-conducting cell, tracheid or vessel element.

**tracheid** A tracheary element of the xylem that has no perforations, as contrasted with a vessel element. May occur in primary and in secondary xylem. May have any kind of secondary wall thickening found in tracheary elements.

**transection** See *transverse section.*

**transfer cell** Parenchyma cell with wall ingrowths (or invaginations) that increase the surface of the plasma membrane. Appears to be specialized for short-distance transfer of solutes. Cells without wall ingrowths may also function as transfer cells.

**transition zone** With reference to an apical meristem, a zone of orderly dividing cells disposed about the inner limit of the promeristem or, more specifically, of the group of central mother cells. Is transitional between the apical meristem and the subapical primary meristematic tissues.

**transverse division (of cell)** With reference to cell, division perpendicular to the longitudinal axis of the cell. With reference to plant part, division of the cell perpendicular to the long axis of the plant part.

**transverse section** A cross section. Section taken perpendicular to the longitudinal axis of an entity. Also called *transection.*

**traumatic resin duct** A resin duct developing in response to injury.

**trichoblast** Commonly used for a cell in root epidermis that gives rise to a root hair.

**trichome** An outgrowth from the epidermis. Trichomes vary in size and complexity and include hairs, scales, and other structures and may be glandular.

**trichosclereid** A type of branched sclereid, usually with hair-like branches extending into intercellular spaces.

**tropism** Refers to movement or growth in response to an external stimulus, the site of which determines the direction of the movement or growth.

**tunica** Peripheral layer or layers in an apical meristem of a shoot with cells that divide in the anticlinal plane and thus contribute to the growth in surface of the meristem. Forms a mantle over the corpus.

**tunica-corpus concept** A concept of the organization of apical meristem of shoot according to which this meristem is differentiated into two regions distinguished by their method of growth: the peripheral, tunica, one or more layers of cells showing surface growth (anticlinal divisions); the interior, corpus, a mass of cells showing volume growth (divisions in various planes).

**tylose** (pl. **tyloses**) In xylem, an outgrowth from a parenchyma cell (axial or one in a ray) through a pit cavity into a tracheary cell, partially or completely blocking the lumen of the latter. Growth typically is preceded by a deposition of a special wall layer on the side of the parenchyma cell that forms the wall of the tylose.

**tylosoid** An outgrowth resembling a tylose. Examples are outgrowths of parenchyma cells into sieve elements in phloem and of epithelial cells into intercellular resin ducts.

## U

**undifferentiated** In ontogeny, still in a meristematic state or resembling meristematic structures. In a mature state, relatively unspecialized.

**uniseriate ray**   In secondary vascular tissues, ray one cell wide.

**unit membrane**   A historical concept of basic membrane structure visualizing two layers of protein enclosing an inner layer of lipid, the three layers forming a unit. The term continues to be useful for describing sectioned membranes (profiles), exhibiting two dark lines separated by a clear space, as seen with the electron microscope.

**upright ray cell**   In secondary vascular tissues, ray cell oriented axially (vertically in the axis) with its longest dimension.

## V

**vacuolar membrane**   See *tonoplast*.

**vacuolation**   Ontogenetically, the development of vacuoles in a cell; in mature state, the presence of vacuoles in a cell.

**vacuole**   Multifunctional organelles bounded by a single membrane, the *tonoplast*, or *vacuolar membrane*. Some vacuoles function primarily as storage organelles, others as lytic compartments. Involved in uptake of water during germination and growth and maintenance of water in the cell.

**vacuome**   Collective term for the total of all vacuoles in a cell, tissue, or plant.

**vascular**   Refers to plant tissue or region consisting of or giving rise to conducting tissue, xylem and/or phloem.

**vascular bundle**   A strand-like part of the vascular system composed of xylem and phloem.

**vascular cambium**   Lateral meristem that forms the secondary vascular tissues, secondary phloem and secondary xylem, in stem and root. Is located between those two tissues and, by periclinal divisions, gives off cells toward both tissues.

**vascular cylinder**   Vascular region of the axis. Term used synonymously with *stele* or *central cylinder* or in a more restricted sense excluding the pith.

**vascular meristem**   General term applicable to *procambium* and *vascular cambium*.

**vascular ray**   A ray in secondary xylem or secondary phloem.

**vascular system**   Total of the vascular tissues in their specific arrangement in a plant or plant organ.

**vascular tissue**   A general term referring to either or both vascular tissues, xylem and phloem.

**vasicentric paratracheal parenchyma**   Axial parenchyma in secondary xylem forming complete sheaths around vessels. See *paratracheal parenchyma*.

**vein**   A strand of vascular tissue in a flat organ, as a leaf. Hence, leaf venation.

**vein rib**   In a leaf, ridge of ground tissue occurring along a larger (major) vein, usually on the lower side of the leaf.

**velamen**   A multiple epidermis covering the aerial roots of some tropical epiphytic orchids and aroids. Occurs in some terrestrial roots also.

**venation**   Arrangement of veins in the leaf blade.

**vertical parenchyma**   See *axial parenchyma*.

**vertical system**   In secondary vascular tissues. See *axial system*.

**vessel**   A tube-like series of vessel elements, the common walls of which have perforations.

**vessel element**   One of the cellular components of a vessel. Also *vessel member* and the obsolete vessel segment.

**vessel member**   See *vessel element*.

**vestured pit**   Bordered pit with projections from the overhanging secondary wall on the side facing the cavity.

## W

**wall**   See *cell wall*.

**water vesicle**   A type of trichome. An enlarged, highly vacuolated epidermal cell.

**wood**   Usually secondary xylem of gymnosperms, magnoliids, and eudicots, but also applied to any other xylem.

**wound cork**   See *wound periderm*.

**wound gum**   Gum formed as a result of some injury. See *gum*.

**wound periderm**   Periderm formed in response to wounding or other injury.

## X

**xeromorphic**   Refers to structural features typical of xerophytes.

**xerophyte**   A plant adapted to a dry habitat.

**xylem**   Principal water-conducting tissue in vascular plants characterized by the presence of tracheary elements. The xylem may also serve as a supporting tissue, especially the secondary xylem (wood).

**xylem elements**   Cells composing the xylem tissue.

**xylem fiber**   A fiber of the xylem tissue. Two types are recognized in the secondary xylem, *fiber-tracheid* and *libriform fiber*.

**xylem initial**   A cambial cell on the xylem side of the cambial zone that is the source of one or more cells arising by periclinal divisions and differentiating into xylem elements either with or without additional divisions in various planes. Sometimes called *xylem mother cell*.

**xylem mother cell**   A cambial derivative that is the source of certain elements of the xylem, such as axial parenchyma cells forming a parenchyma strand. Used also in a wider sense synonymously with *xylem initial*.

**xylem ray**   That part of a vascular ray that is located in the secondary xylem.

**xylotomy**   Anatomy of xylem.

# Author Index

(Bold-face type indicates bibliographic references.)

Aaziz, R. 52, **58**, 90, **91**
Abagon, M. A., 223, **243**
Abbe, E. C., 146, **162**
Abbe, L. B., 387, **398**
Abdel-Latif, A., **317**
Abdul-Karim, T., **248**
Abe, H., 72, 81–83, **91**, **92**, 276, 277, 283, **286**, **318**, **351**
Abel, S., 83, **91**
Abeles, F. B., 123, **123**
Achor, D. S., **428**, **442**
Adachi, K., **316**
Adam, J. S., 160, **165**
Adams, K. L., 32, **37**, **288**
Adams, M. E., **466**
Adiwilaga, K., 494, **495**
Adler, H. T., **173**
Agarie, S., 241, **243**
Agata, W., **243**
Aggarwal, S. K., **353**
Aguirre, M., **39**
Ahmad, Z., **352**

Ahn, S. M., **126**
Ahn, Y. H., **354**
Aida, G. P., **37**
Aida, M., 11, **12**, **170**
Ait-ali, T., **129**
Aitken, J., **94**
Ajmal, S., 324, **348**
Alabadi, D., 12, **12**
Alabouvette, J., **173**
Alami, I., **285**
Albersheim, P., 65, **93**, **95**, **97–99**
Albrigo, L. G., **442**
Aldaba, V. C., 204, **207**
Aldington, S., 68, **91**, **94**
Alexandre, A., **244**
Alfieri, F. J., **59**, 344, 346, **348**, 407, 421, **424**
Alfonso, V. A., **353**
Ali-Khan, S. T., **248**
Aljaro, M. E., 346, **348**
Allan, A. C., **253**
Allen, G. J., 224, **243**, **251**

Allen, G. S., 156, **165**
Allen, N. S., **59**
Allen, R. D., 231, 233, **251**, 487, **495**
Allen, S., **62**
Aloni, E., **124**
Aloni, R., 121, 122, **124**, 203, 205, 206, **207**, **209**, 280, 281, **283**, 308, 314, 315, **316**, **319**, **320**, 358, 359, **398**, 421, **424**, **429**, **442**, **443**
Alonso, J., **39**
Al-Talib, K. H., 207, **207**
Altmann, T., 237, 244, **252**
Altschuler, Y., **61**
Alva, R., 489, 492, **496**
Alvarez, R. J., **244**
Alvarez-Buylla, E. R., **166**
Alvernaz, J., **62**
Alves, E. S., 259, 263, **284**, 294, 297, **316**
Alves, G., **316**
Alvin, K. L., 195, 196, **208**
Amakawa, T., **128**
Amarilla, L., **317**

# Subject Index

(Numbers in bold face indicate figures located apart from the description of the subject in the figures and those involving taxa.)

*Esau's Plant Anatomy, Third Edition,* By Ray F. Evert.
Copyright © 2006 John Wiley & Sons, Inc.